Drug Design
From Structure and Mode-of-Action to Rational Design Concepts

Gerhard Klebe

Drug Design

From Structure and Mode-of-Action to Rational Design Concepts

Springer

Gerhard Klebe
Institute of Pharmaceutical Chemistry
Philipps-University Marburg
Marburg, Germany

ISBN 978-3-662-68997-4 ISBN 978-3-662-68998-1 (eBook)
https://doi.org/10.1007/978-3-662-68998-1

Translation from the German language edition: "Wirkstoffdesign: Entwurf und Wirkung von Arzneistoffen" by Gerhard Klebe, © Der/die Herausgeber bzw. der/die Autor(en), exklusiv lizenziert an Springer-Verlag GmbH, DE, ein Teil von Springer Nature 2024. Published by Springer Berlin Heidelberg. All Rights Reserved.

Cover illustration: Crystal structures of human protein kinase A with ligands derived from fragment discovery and optimization (PDB codes: 3OOG, 3OVV, 3OXT, 3P0M) summarized in the Ph.D. thesis of Dr. Helene Köster, Univ. Marburg (2012), (https://d-nb.info/102718376X/34).

This Springer imprint is published by the registered company Springer-Verlag GmbH, DE, part of Springer Nature.
The registered company address is: Heidelberger Platz 3, 14197 Berlin, Germany

Scientists have searched for a perpetuum mobile.
They have found it: It is science itself.

Victor Hugo

Preface and Acknowledgement

The present *Drug Design* textbook is based on the German version first written by Hans-Joachim Böhm, Hugo Kubinyi, and me in 1996. After several years of success on the market, the German version was completely rewritten and significantly expanded. I thank my two former coauthors for allowing me to reuse some of their passages from the original version in this edition.

Meanwhile, a third German edition is available. Several attempts have been made to translate this book into English to make it available to a wider audience. The reason for this was that the author was repeatedly asked why such a successful book was not available in English. An analysis of the textbook market revealed that no similar compendium covering the same area of interest was available. Springer agreed to a translation project, and in 2013, Dr. Leila Telan, a gifted bilingual medicinal chemist and physician, produced a first translation of the German second edition. It was available in the Springer Reference series, but not yet with the accessibility the author desired as an affordable textbook for students. Even a translation into Chinese as a textbook was achieved in 2018. Thus, to realize the idea of an English textbook version, the author took the opportunity to translate the third German edition with the help of the earlier handbook copy. This book is intended for students of chemistry, pharmacy, biochemistry, biology, chemical biology, and medicine who are interested in the design of new drugs and the structural basis of drug action. It is also tailored for practitioners in the pharmaceutical industry who want a more comprehensive overview of various aspects of the drug discovery process.

In the 15 years since the second German edition was published, many areas of drug design have matured and are now an integral part of drug discovery worldwide. The field has continued to benefit enormously from the steady increase in computing power. Many techniques, such as virtual screening or mining large databases, have become significantly faster and more comprehensive. However, for a quantum leap in performance and reliability, a better understanding of the biophysical principles of molecular recognition still needs to be incorporated into the methods. Machine learning and artificial intelligence have brought many algorithms to higher levels of performance. One example is protein structure prediction from a sequence. It has now reached a level of reliability that qualifies it for predicting new spatial structures. Real innovations have been observed in the field of experimental structure determination. Cryo-electron microscopy allows us to gain insight into large biomolecules such as ion channels or G-protein-coupled receptors. Although these biomolecules have been drug targets for decades, there has been a lack of structural information to truly understand their mechanisms of action. Today, it is even more obvious that models cannot replace this. The search for lead structures is increasingly based on fragment-based lead discovery methods. If you compare drug design today with the early days 40 years ago, we are tackling much more difficult targets, such as modulating protein communication with small molecules. So the real advances are in addressing increasingly complex mechanisms of action. We are also thinking more about the temporal and energetic sequence of binding processes and how the dynamics of the molecules involved determine the mode of action and the therapeutic profile of putative drug molecules. All the more reason for a textbook author to ask the question: How can one provide an overview of this ever-growing field? An exemplary selection of the examples to be considered is essential. They must be representative of many other works that remain unnamed. Of course, it is particularly appealing to take up and present the new aspects. However, the origins of drug design must not be lost. This is the only way to understand how the field has evolved. Older methods, which still play an important role in daily work, should not be completely ignored. This book is, therefore, a compromise between the old and the new. It tries to pass on my own enthusiasm and fascination for this field of research to young scientists. What better way to do this than with a textbook? It can present content from a different perspective and show cross-references between individual aspects in a way that is hardly possible in an original research paper or a review article, especially since such reports are often only really comprehensible to experts.

Numerous colleagues have contributed to the success of this book by critically reviewing various passages and chapters and by providing many comments. In particular, I would like to thank my colleagues in Marburg, Prof. Dr. Andreas Heine and Prof. Dr. Klaus Reuter, for their fruitful collaboration and many discussions over more than 20 years, and some of the concepts we developed together have found their way into this book. In Marburg, I would also like to thank Dr. Daniel Hilger, Prof. Dr. Jens Kockskämper, Dr. Dzung Nguyen, and Dr. Doro Vornicescu for many suggestions and for proofreading individual chapters. Outside Marburg, I would like to thank Prof. Dr. Paul Czodrowski (Mainz, Germany), Dr. Stefan Duhr (Munich, Germany), Prof. Dr. Richard Engh (Tromsø, Norway), Prof. Dr. Oliver Ernst (Toronto, Canada), Dr. Michael Hennig (Villigen, Switzerland), Dr. Wolfgang Jahnke (Basel, Switzerland), Dr. Gerhard Müller (Munich, Germany), Prof. Dr. John Ladbury (Leeds, United Kingdom), Prof. Dr. Milton Stubbs (Halle, Germany), Prof. Dr. Matthias Rarey (Hamburg, Germany), Dr. Ulrich Rant (Munich, Germany), Prof. Dr. Peter Tonge (Stony Brook, New York, USA), and Prof. Dr. Daniel Wilson (Hamburg, Germany) for their comments and careful editing.

Dr. Christoph Sager (Basel, Switzerland) kindly programmed the platform and made the videos for the illustrations available on YouTube. I would also like to thank Ms. Susanne Dathe (Springer-Verlag Heidelberg) for her exemplary support during the production of this book.

Gerhard Klebe
Frankfurt, Germany

Chemical Structures of Amino Acids, Molecular Graphics and Introduction

Chemical Structures, Three-Letter and One-Letter Codes of Amino Acids

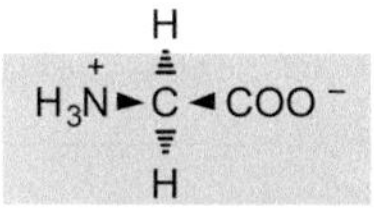
Glycine (Gly) G

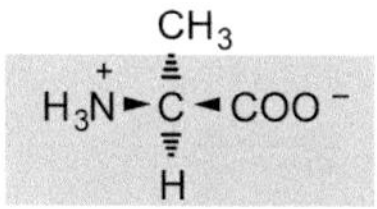
Alanine (Ala) A

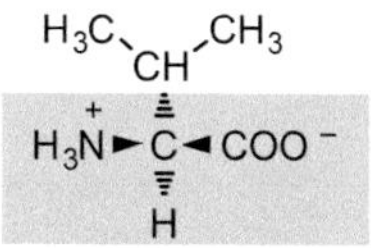
Valine (Val) V

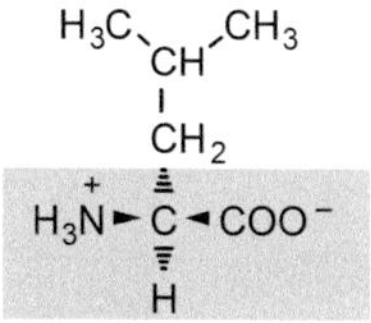
Leucine (Leu) L

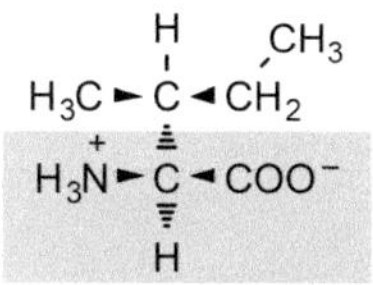
Isoleucine (Ile) I

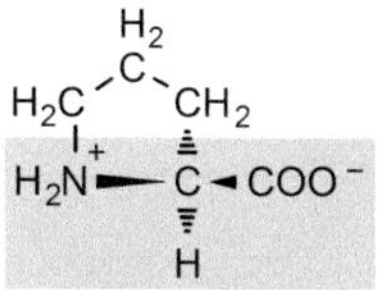
Proline (Pro) P

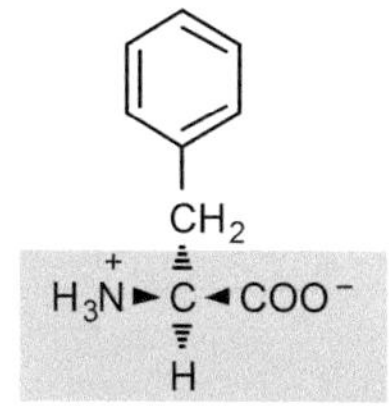
Phenylalanine (Phe) F

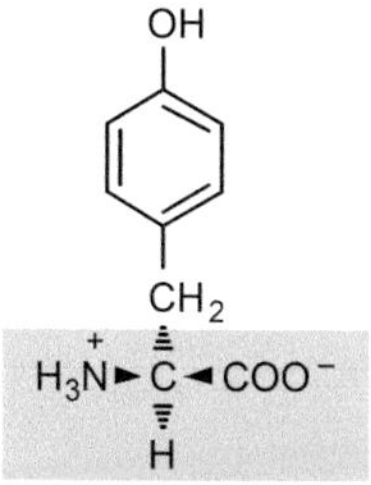
Tyrosine (Tyr) Y

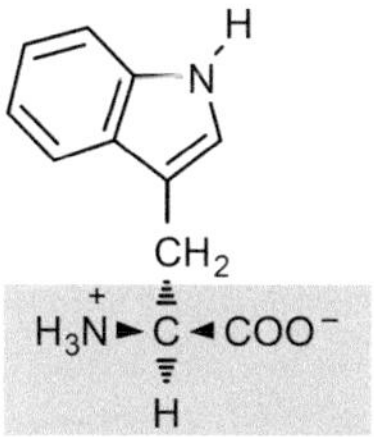
Tryptophan (Trp) W

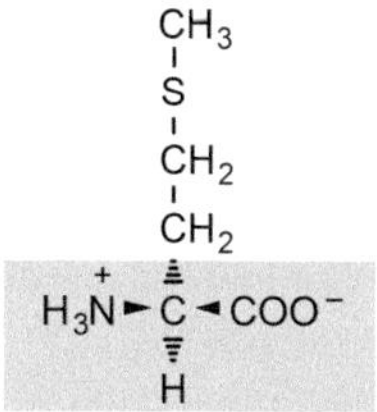
Methionine (Met) M

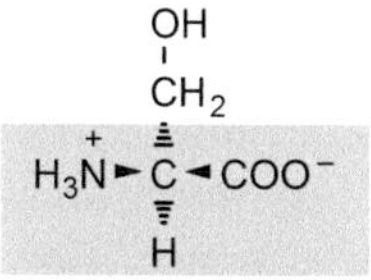
Serine (Ser) S

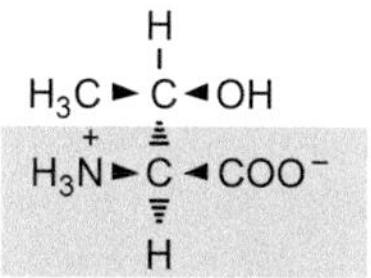
Threonine (Thr) T

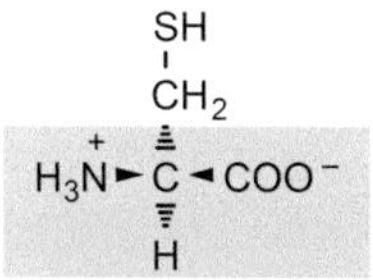
Cysteine (Cys) C

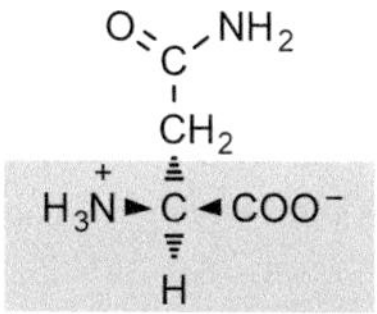
Asparagine (Asn) N

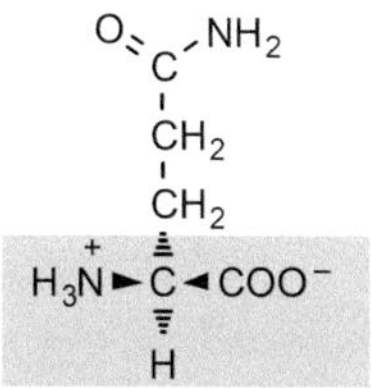
Glutamine (Gln) Q

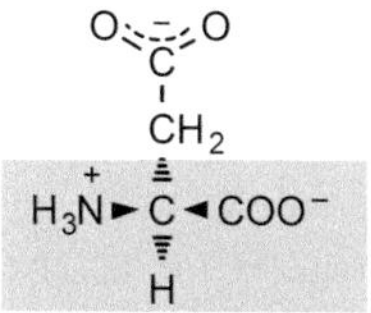
Aspartate (Asp) D

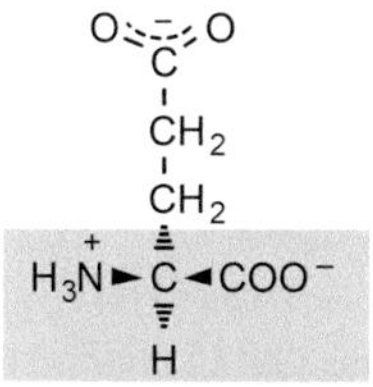
Glutamate (Glu) E

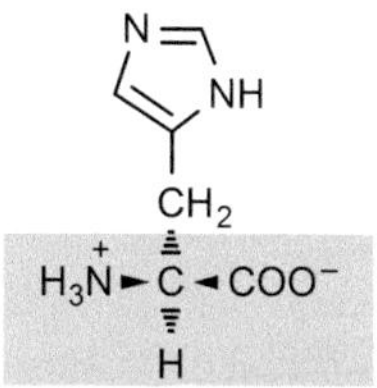
Histidine (His) H

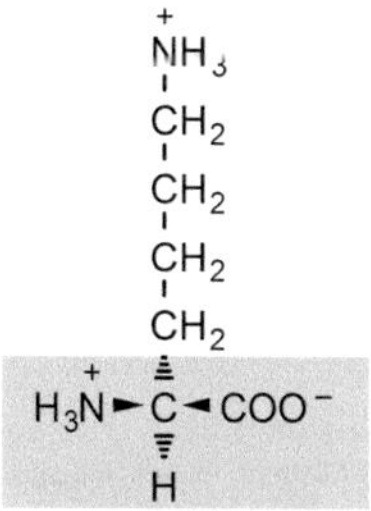
Lysine (Lys) K

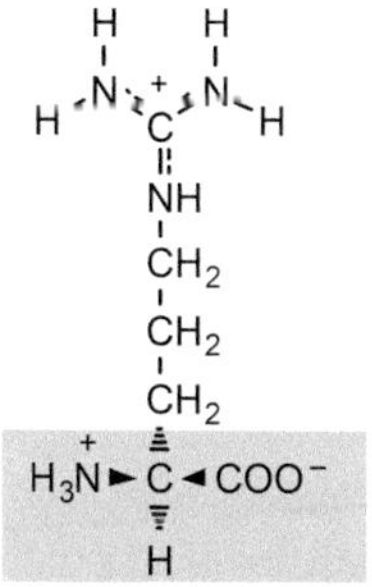
Arginine (Arg) R

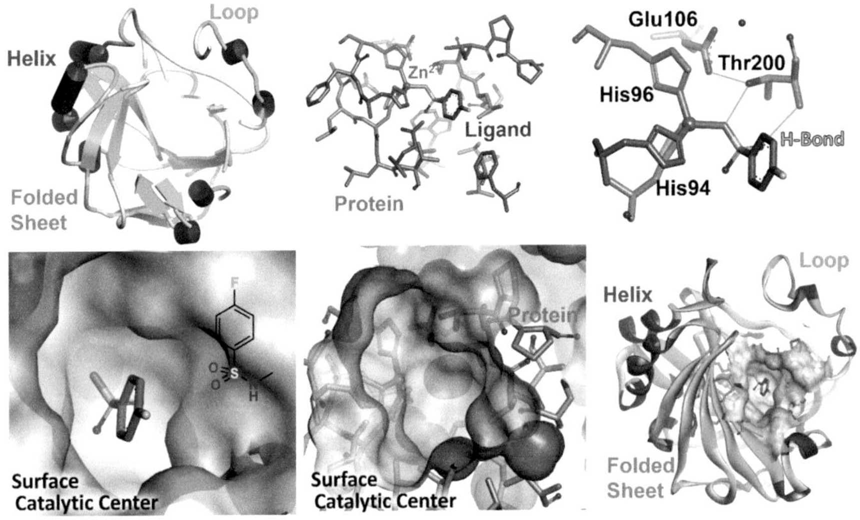

▸ https://sn.pub/dfOc6R

This figure explains how the protein structures and bound ligands are represented in many of the figures in this book. In the *upper left*, the protein is schematically represented by the course of its main chain. Segments of the polymer chain with a folded sheet structure (as *arrows*) are highlighted in *light blue*, helical segments (as *cylinders*) in *red*, and loop regions in *green*. At the *top center*, amino acid residues in the active site are shown in a stick representation. Unless otherwise noted, protein carbon atoms are shown in *orange*, ligand carbon atoms in *gray*, oxygen atoms in *red*, nitrogen atoms in *blue*, sulfur atoms in *yellow*, phosphorus atoms in *orange*, fluorine atoms in *turquoise*, chlorine atoms in *green*, bromine atoms in *brown*, iodine atoms in *purple*, and metal ions in *grayish blue*. Hydrogen atoms are *white*, but are usually omitted for clarity. At the *top right*, amino acids are identified by a three-letter code and their position in the sequence (e.g., His 94). Hydrogen bonds between the ligand (here *para*-fluorophenylsulfonamide) and the amino acids of the protein are indicated by thin *light-green lines*. In the *lower left* corner, around the binding pocket (a section of the structure is shown), the solvent-accessible surface has been calculated (see Sect. 15.6) and is indicated as a *pale gray-white area*. In the *center below* is an analogous representation of the surface, now transparent, along with the amino acid residues of the binding pocket. At the *bottom right* is an overall view of the protein (in this case carbonic anhydrase II, see Sect. 25.7 for details), with the indicated binding pocket of the catalytic center blocked by an inhibitor. It coordinates to the zinc ion and forms three hydrogen bonds to the protein. The polymer chain is shown as a continuous ribbon model. The color coding, with *light blue*, *red* and *green* sections, corresponds to that in the figure above left.

The illustrations were created using Discovery Studio Visualizer V20.1.0.19295 from Dassault Systemes Biovia Corp., Copyright 2019. Together with this book, the reader can access these figures as videos in the form of moving images using the provided QR codes or Springer's tiny URLs linking to ▸ https://drugdesign.ch/en. In the electronic pdf version, the URLs are linked directly and can be accessed by clicking on them. The videos have been uploaded to YouTube and can be viewed on standard mobile phones, tablets, or desktop computers using the browsers installed on these devices. The settings in these browsers will need to be adjusted to display YouTube videos. The book contains many internal links to Chapters, Sections, Figures or Tables, highlighted in blue. In the pdf version they can be activated by clicking on them. To return, please use the [ALT + <] key combination.

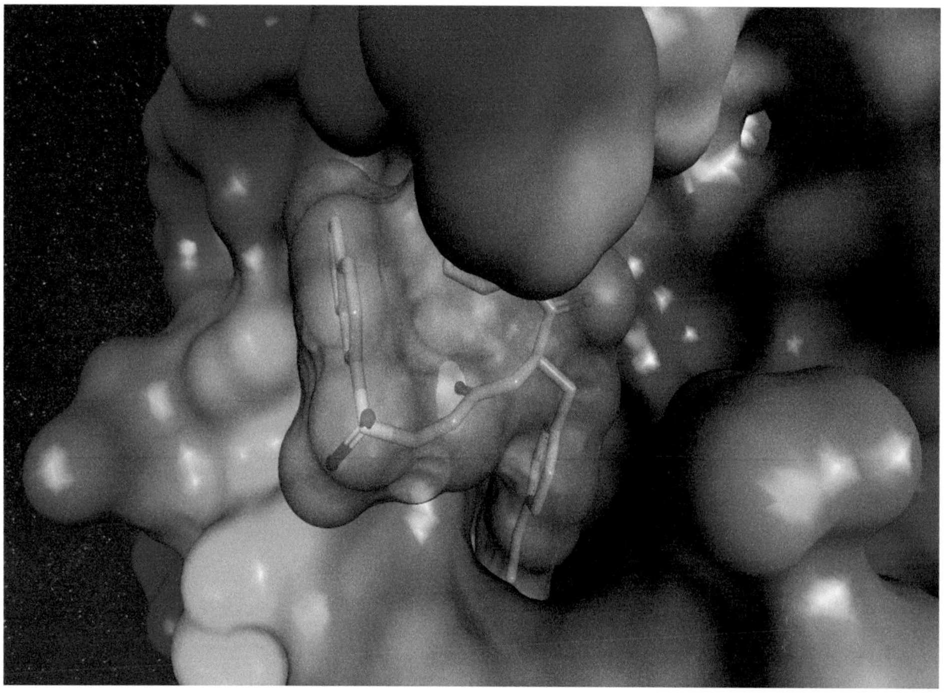

When considering the diversity of all possible chemical compounds, a comparison with the infinite vastness of the universe is often made. If one does not limit oneself to the compounds that have actually been synthesized to date, but also generates molecules on the computer that can in principle be synthesized but have not yet been produced, gigantically large numbers of molecules can be generated. If one also takes into account the many stereoisomers that can be generated for complex organic molecules, the number becomes even larger. The art of rational drug design is now to fish out of the vast universe of all conceivable chemical compounds the right one that inhibits the target protein under consideration in a highly potent and selective manner. This figure shows NAPAP, one of the first highly potent thrombin inhibitors. For many years, it has served as a lead structure for the development of inhibitors of trypsin-like serine proteases. It binds deeply into the so-called S_1 pocket of the enzyme thrombin, which plays a central role in blood clotting. The inhibitor NAPAP is shown here atom color-coded, with a translucent white surface. Its binding site in the protein is represented by a plastic surface that changes its color from blue (outside), to green, and to red (deeply buried), depending on the depth of the depression of the binding site. The structures of protein kinase A with four ligands are shown on the cover of this book. Starting with a weakly binding millimolar fragment (red surface), this initial hit was optimized in several iterative design cycles to a nanomolar ligand (blue surface). The spatial structures of proteins and ligands are sketched in the background, symbolizing the universe of all possible molecular structures. This book explains how the structures of ligands and proteins can be used to develop potent drug candidates using computer modeling, chemical synthesis, screening, and binding assays. (▶ https://sn.pub/4NppHn)

Introduction

Drug design is science, technology, and art all in one. An invention is the result of a creative act; a discovery is the recognition of a pre-existing reality. Design involves both processes, but emphasizes a focused approach based on existing knowledge and technologies. In addition, the creativity and intuition of the researcher play a crucial role.

Drugs are all substances that exert an effect on a system by inducing a particular effect. In the context of this book, they are substances that exhibit a biochemical or pharmacological effect, in most cases as drug medications, that achieve a therapeutic effect in humans.

The idea of rational drug design is not new. Already more than a century ago, organic compounds were synthesized in a targeted manner in order to produce new drugs. Scientists have always worked with models and design hypotheses. In the early days, these models were often flawed or even wrong. As our understanding of the molecular basis of diseases has steadily improved, the models and hypotheses have become better and more reliable. However, we are still far from the dream of designing a finished drug using rational concepts on the drawing board.

In the case of artistic design, such as a poster or a commodity, or in the case of engineering, such as the design of a car, a computer, or a machine, the result is usually directly predictable. In contrast, the design of active substances is even today not sufficiently predictable. The consequences of even the smallest structural changes of the biological properties and the mode of action of a biologically active molecule are too multifaceted and yet too poorly understood.

Until now, scientists have relied on the principle of trial and error in order to find new medicines. The rules empirically derived have resulted in a knowledge base for the rational design of active substances, which individual researchers have put into practice with greater or lesser success. Today, many new methods from genetic engineering, structural biology, biophysics, and informatics are available for the search for novel drugs. The work of recent years has contributed above all to our understanding of the molecular mechanisms of action of many known drugs, although these compounds were often previously found by simpler methods and sometimes even discovered by chance. This book attempts to present many of these modes of action or mechanisms.

Methodological advances in experimental structure determination now allow the routine determination of the three-dimensional structure of proteins, RNAs and DNAs, as well as their ligand complexes. As shown in many of the illustrations in this book (see the general explanation of how to "read" these illustrations on page X), in combination with the commented videos provided via the Internet and accessible via QR codes, they make a very important contribution to the reader's understanding of the targeted design of active agents. Three-dimensional (3D) structures down to atomic resolution are known of more than a million small molecules and more than 200,000 proteins, nucleic acids, and protein–ligand complexes. Their number continues to grow exponentially. Methods for predicting the spatial structures of small molecules are mature. Semiempirical and quantum chemical calculations on active agents are routinely used. The sequencing of the human genome is complete, and new genome data from other organisms are being added weekly.

Recently, significant progress has been made in predicting the 3D structures of proteins from their amino acid sequence. With the help of artificial intelligence, computer programs evaluate the enormous treasure of experimentally determined structural data. What is still lacking is a comparably powerful and detailed database that reveals the structure of water molecules in the interior and on the surface of biomolecular complexes. Yet, their properties are essential for understanding how drugs bind and act. Predictions of the protonation states of amino acids in protein-binding sites and of ligand functional groups, particularly their changes during ligand binding, are still in their infancy. The potential of RNA molecules for drug therapy is slowly being recognized. Concepts for structural–biological classification and *de novo* prediction of the spatial structures of RNA molecules, including their solvate structures and protonation states, still need to be developed. Only then will it be possible to use machine learning and artificial intelligence to provide solutions for drug design through computer algorithms, similar to the current protein structure prediction from sequence.

The structure-based and computer-aided design of new drugs is now an integral part of practical drug research. Thanks to a dramatic increase in computing power, computer programs now support the search, modeling, and targeted design of new active agents. The mere increase in throughput can possibly enable these techniques to exhaustively design and filter for potential drug candidates. However, to take the methodology to a higher level of reliability and relevance, a deeper understanding of drug–target interactions is essential. How do the local water structure and the mutual polarization of the binding partners affect the biological properties? Restricting ourselves solely to the interactions of the pharmaceuticals with the residues in the binding pocket of a single biological target structure also does not adequately capture the complexity of drug action. The binding kinetics and residence time of a drug at its target structure are just as important as pharmacokinetic, toxicological, and metabolic properties. How important are modulations of the dynamics of the formed complexes for drug action and therapeutic success? What other target structures are influenced in addition to the actual target? How troublesome, or perhaps even how important, are adverse side effects? When can a patient be offered personalized therapy and individualized dosing? Today, the gene sequencing of each and every one of us is possible at a financially affordable cost and in a reasonable amount of time. But how can this knowledge be used for personalized medicine? Future drug research will have to answer these questions for patients without losing sight of the cost–benefit ratio.

So why is the development of a new drug still so complicated and why has the time it takes hardly been reduced in the last 40 years? The cost of developing and launching a drug to market has risen steadily and continues to do so. The current cost range is between US$ 200 million and 4000 million. Only large pharmaceutical companies, especially those in the top sales and profit segment, can still afford such an investment. There is always the risk of failure in the late stages of clinical trials or of misjudging of the therapeutic potential of a new compound. There are often enough setbacks. On a more optimistic note is the observation that drug discovery is now tackling much more challenging targets. Ten or twenty years ago, these were still considered inaccessible or simply "undruggable." Examples include the successful disruption or targeted enhancement of protein–protein contact surfaces by small molecules or the targeting of overregulated or pathogenic proteins for selective proteolytic degradation. The body's own immune system is enhanced to selectively eliminate virus-infected or degenerated cancer cells.

There is, however, a crucial hurdle to overcome when using drug molecules to interfere with the control mechanisms of biological processes: In order to compete with the natural ligands of enzymes and receptors, or to mimic their properties, they must be sufficiently precise and efficient in terms of both the mechanism of action and the site of action. In the case of endogenous active compounds, our organism uses different principles. Substances such as the body's own hormones act predominantly systemically, meaning they are released at one site in the body and transported via the bloodstream to one or more completely different sites of action. It is only there that they develop their effect. This requires a high degree of specificity and selectivity of action. Substances such as neurotransmitters are also used by our body in many places for very different tasks. This can only be achieved in a controlled manner if they are used strictly locally. Our organism uses a highly developed spatial compartmentalization for this purpose. Substances such as neurotransmitters are formed close to where they are needed, where they exert their task, and then they are immediately removed again. Drug research is also aimed at interfering with these mechanisms and possibly correcting them, but we want to use substances that can be taken orally if possible. As with hormones, this is only possible if a very high specificity and selectivity can be achieved. In addition to this, sufficient stability and bioavailability from the gastrointestinal tract must also be achieved. Under these conditions, it is extremely difficult to achieve a tailored therapeutic influence on target structures for which the body itself uses only relatively small and unselective ligands. Inevitably, our active agents must become larger in order to achieve the required selectivity of binding. It is obvious that efficacy profiles of such substances will be somewhat different, and usually they will increase in size for selectivity reasons. All these aspects open up enormous perspectives for drug development, but they do not make it any easier today.

The purpose of this book is to describe the development of new drugs under these difficult and ever-changing conditions. Design methods are presented, and drug development is illustrated with known mechanisms of action and selected case studies. Drug discovery is a multidisciplinary field in which chemists, pharmacists, physicians, technologists, molecular biologists, biochemists, pharmacologists, toxicologists, and clinicians work together to pave the way for a compound to become a new therapeutic drug. For these reasons, the majority of drug developments still takes place in industry. Only industry has the financial resources and, above all, the organizational structures needed for all the disciplines involved to successfully work together. Only in this way can research be channeled in a targeted manner. However, the basic principles and forward-looking innovations in drug research are increasingly being developed in academia, not least for reasons of cost and critical risk assessment. Interestingly, more and more university research initiatives are focusing on the development of drugs for infectious diseases, rare diseases, or diseases of the Third World. The inevitably commercially oriented pharmaceutical industry in the industrialized countries has increasingly withdrawn from these areas. This is all the more alarming when one considers that our improved quality of life and increased life expectancy are largely due to the victory over devastating infectious diseases. The coronavirus pandemic that we have just lived through has brought this back to our minds in a very vivid way.

Rising research and development costs, an already high standard of therapy in many indications, much greater safety awareness and, as a result, more stringent regulatory requirements have led to a steady decline in the number of new chemical entities (NCEs) introduced annually into therapy over the past few decades: from 70–100 NCEs in 1960–1969, to 60–70 in 1970–1979, to an average of 50 in 1980–1989, to 40–45 in the 1990s and the two decades of the new millennium. However, in addition to the expansion of indications for drugs that have been known for a long time, it is the new developments that have brought significant progress in therapy.

A paradigm shift is often talked about in pharmaceutical research. This refers to the application of new technologies and knowledge. As far as the structure of the market is concerned, the process of concentration through company takeovers, acquisitions, and mergers into giant Big Pharma companies has slowed considerably. Fortunately, this has given way to a very dynamic, almost unmanageable scene of small, highly flexible biotech and start-up companies that are stimulating pharmaceutical research with new, innovative strategies. Big pharmaceutical companies are outsourcing high-risk research concepts to these small biotechs and using their services up to and including the development of clinical candidates. In addition, prescribing practices are changing across the healthcare sector. In the past, the physician alone was responsible for therapy, sometimes in consultation with the pharmacist. Today, cost pressures, negative lists, health insurance companies, hospital or drugstore purchasing organizations, the ubiquitous Internet, and even public opinion are increasingly influencing therapy.

This book is a textbook on drug discovery, the principles of action, and the way new drugs are discovered and developed. It differs from the classic textbooks on medicinal and pharmaceutical chemistry both in its structure and in its aims. It covers the basic principles, methods, successes, and obstacles in the search for new drugs. Rather than discussing classes of drugs in terms of their indications, it focuses on the route to the active compound and the structural requirements for its action on a particular target protein or family of target proteins. As the title suggests, the book is aimed at students and scientists in the fields of chemistry, pharmacy, biochemistry, biology, and medicine who are interested in the art of designing new drugs using knowledge of the structural basis of their activity at the site of action.

The first part begins with an introduction to the history of drug research. This is followed by a description of serendipity as a concept in drug discovery that is difficult to plan but always highly valued. Selected examples from classical drug research are presented as examples. A discussion of the fundamentals of drug action, the thermodynamics of ligand–receptor interactions, and the influence of the three-dimensional spatial structure of a drug on its efficacy round off this part.

In the second part, the search for new lead structures and their optimization and the use of prodrug strategies are introduced. New screening technologies are discussed, as well as the systematic modification of structures using the concept of bioisosterism and a peptidomimetic approach.

The third part describes experimental and theoretical methods used in drug discovery. Combinatorial chemistry has provided access to a large number of test compounds. Gene technology can produce the target proteins in their pure form and has helped to characterize the properties and function of these proteins from the molecular level to the cellular assembly to the organismal level. It has bridged the gap between understanding the effects of drug therapy on the complex microstructure of a cell and the systems biology of an organism. The spatial structure of proteins and protein–ligand complexes is accessible through X-ray crystallography, cryo-electron microscopy, and NMR spectroscopy. The structural principles of proteins and DNA are becoming better understood and are increasingly providing access to the binding geometry of drugs. Computational methods and molecular dynamics simulations, including complex conformational analysis, have also increased our understanding and modeling perspectives of targeted drug design.

The fourth part introduces design techniques such as pharmacophore and receptor modeling, and discusses the methods and uses of quantitative structure–activity relationships (QSAR). Insights into drug transport and distribution in biological systems are provided, and various structure-based design techniques are presented. A drug design case study from the author's research concludes the first part of this textbook.

The fifth part focuses on the core question of drug design: How do drugs actually work and how does this translate into the design of new drugs? Enzymes, receptors, channels, transporters, and surface proteins are divided into separate chapters and discussed as a group of target structures. The spatial architecture of the protein and the modes of action are used to explain in detail why a drug works and why it must have a particular geometry and structure in order to work. These chapters illustrate the contributions of structure-based design and medicinal chemistry optimization to the discovery of new drugs, and highlight different aspects of the drug discovery process.

The concept of this book means that many important drugs are not covered, or are only mentioned in passing. The same applies to receptor theory, pharmacokinetics and metabolism, the basics of genetic engineering and statistical methods. The biochemical, molecular biological, and pharmacological fundamentals of the mode of action of drugs, which are important for understanding drug design, are only briefly discussed. Other disciplines, such as pharmaceutical formulation, toxicology, and clinical trials, which are critical to the development of a compound into a medicine and its use in patients, are not covered in this book.

The selection of examples from therapeutic areas has been made subjectively and for didactic reasons based on case studies, and to highlight different aspects of drug discovery. An attempt has been made to provide a balanced presentation of drug design methods and their practical application.

The interested reader need not read the book chronologically. If the reader is interested only in drugs and their mode of action, they may start with Chap. 22. The reader who is more interested in methodology, or the medicinal chemist who wants to learn the basics of drug design, can concentrate on Chaps. 4–21.

There are many cross-references throughout the text to help the reader find the passages needed for a more detailed understanding at a particular point in other sections.

The following bibliography lists particularly recommended monographs and, in alphabetical order, journals and series on the subject, which are not mentioned individually in the later chapters. The reader will also find a summary of the most important aspects after each chapter. Suggestions for further reading and many original articles are given at the end of each chapter.

Literature

Monographs

L. Brunton, J. Lazo, K. Parker, Goodman & Gilman's the pharmacological basis of therapeutics, 11th edn, McGraw-Hill, Europe (2005)

C. R. Ganellin, S. M. Roberts (Eds.), Medicinal chemistry. The role of organic chemistry in drug research, 2nd edn, Academic Press, London (1993)

F. D. King (ed), Medicinal chemistry: principles and practice, 2nd edn, The Royal Society of Chemistry, Cambridge (2003)

K. Stromgaard, P. Krogsgaard-Larsen, U. Madsen (Eds.), Textbook of Drug Design and Discovery, 5th edn, CRC Press, Taylor & Francis Group, Boca Raton (2017)

D. Lednicer (Ed.), Chronicles of drug discovery, vol 3. American Chemical Society, Washington, DC and earlier volumes from this series (1993)

T. L. Lemke, D. A. Williams, Foye's principles of medicinal chemistry, 6th edn., Williams & Wilkins, Baltimore (2008)

S. Hongmao, A Practical Guide to Rational Drug Design, 1st edn, Woodhead Publishing (2015)

B. E. Blass, Basic Principles of Drug Discovery and Development, 2nd edn, Academic Press (2021)

R. Mannhold, H. Kubinyi, G. Folkers (Eds.), Methods and principles in medicinal chemistry. Wiley-VCH, Weinheim, Series with Guest Editors

R. A. Maxwell, S. B. Eckhardt, Drug discovery. A casebook and analysis. Humana Press, Clifton (1990)

E. Mutschler, H. Derendorf, Drug action, basic principles and therapeutic aspects. CRC Press: Boca Raton/Ann Arbor/London/Tokyo (1995)

R. B. Silverman, M. W. Holladay, The Organic Chemistry of Drug Design and Drug Action, 3rd edn., Academic Press, (2014)

C. G. Wermuth, N. Koga, H. König, B. W. Metcalf (Eds.), Medicinal chemistry for the 21st century. Blackwell Scientific, Oxford (1992)

Journals and Series

ACS Chemical Biology
ACS Medicinal Chemistry Letters
Annual Reports in Medicinal Chemistry
Chemistry & Biology
ChemMedChem
Drug Discovery Today
Drug News and Perspectives
European Journal of Medicinal Chemistry
Journal of Enzyme Inhibition and Medicinal Chemistry
Journal of Computer-Aided Molecular Design
Journal of Medicinal Chemistry
Methods and Principles in Medicinal Chemistry
Nature Communications
Nature Reviews Drug Discovery
Perspectives in Drug Discovery and Design

Pharmacochemistry Library
Progress in Drug Research
Quantitative Structure–Activity Relationships
Reviews in Computational Chemistry
Science
Scientific American
Trends in Pharmacological Sciences

Nowadays the Internet, discussion platforms, and the tremendously valuable tool of Wikipedia are available to everyone and provide access to an enormous source of information.

Chapter Abstract Videos

For each chapter, a short video provides an overview of what is covered in that chapter. They can be accessed via the following QR codes or short URLs.

1 https://sn.pub/vrehez	2 https://sn.pub/ti9pj7	3 https://sn.pub/vqibjb	4 https://sn.pub/8ink1c
5 https://sn.pub/815bp7	6 https://sn.pub/keziu0	7 https://sn.pub/k19jip	8 https://sn.pub/tts2en
9 https://sn.pub/wippd2	10 https://sn.pub/b19kz8	11 https://sn.pub/qhucur	12 https://sn.pub/54f7da
13 https://sn.pub/uhz9e7	14 https://sn.pub/8oi55k	15 https://sn.pub/bwhyti	16 https://sn.pub/mkznv9
17 https://sn.pub/0bzx5y	18 https://sn.pub/j7pt6b	19 https://sn.pub/ssi0og	20 https://sn.pub/0alfsy
21 https://sn.pub/ya1k72	22 https://sn.pub/01mfsx	23 https://sn.pub/gqj47j	24 https://sn.pub/wrrga8
25 https://sn.pub/cy407x	26 https://sn.pub/e3r70u	27 https://sn.pub/x4cyhg	28 https://sn.pub/l9lvml
29 https://sn.pub/fc8htp	30 https://sn.pub/ekembs	31 https://sn.pub/ksv7om	32 https://sn.pub/cizkno

Contents

About the author

Gerhard Klebe
is Professor of Medicinal Chemistry at Philipps University in Marburg, Germany. He retired in April 2020. He has been teaching pharmaceutical and medicinal chemistry, drug action and drug design for almost 30 years. His research has focused on structure-activity relationships, 3D-QSAR methods, conformational and pharmacophore analysis, docking methods, database analysis, protein crystallography, structure-based drug design, and biophysical characterization of protein-ligand interactions (https://agklebe.pharmazie.uni-marburg.de/?id=&lang=en). The thermodynamic characterization and the involvement of water molecules in the binding process have been a particular focus. A number of well known and worldwide applied computer tools have been developed in his laboratory (CoMSIA, AFMoC, ReLiBase, Cavbase, Drugscore, Mobile). With his research group, he has worked on a significant number of the drug targets also covered in the book. In total, his group has contributed more than 1600 crystal structures to the publicly available PDB. Together with colleagues at the Bessy synchrotron in Berlin, his group established a dedicated beamline for fragment-based lead discovery, including many tools for further hit-to-lead optimization. Out of his research group came the start-up company CrystalsFirst (www.crystalsfirst.com), which focuses on crystallographic fragment screening using sophisticated protein crystal stabilization techniques and combining crystal screening with computational design. Prior to his appointment in Marburg, he taught drug design and crystallography at the University of Heidelberg and worked for more than a decade in an industrial setting in drug discovery at BASF AG in Ludwigshafen, Germany.

Foundations in Drug Research

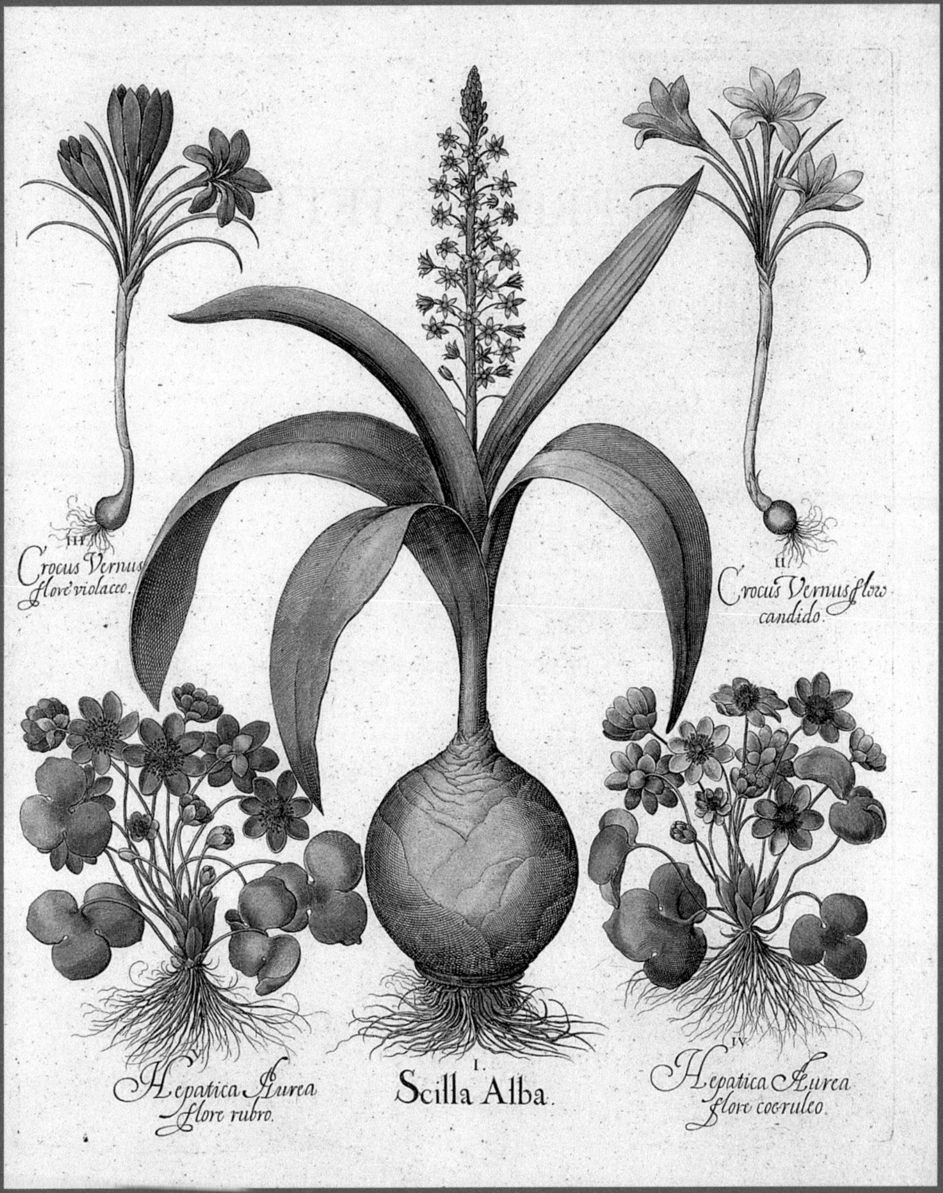

This colored copper engraving from the probably most beautiful plant book, the *Hortus Eystettensis* von Basilius Besler, Eichstätt, 1613, shows the sea onion, *Scilla alba* (today: *Urginea maritima L.*). It was already known to the ancient Egyptians, Greeks, and Romans as a remedy for many illnesses, e.g., dropsy (today: congestive heart failure). It was also worshipped as a general repellent against misfortune. It was not until the 20th century that the cardioactive glycosides scillaren and proscillavidin contained in the sea onion were isolated in pure form and introduced into therapy as such or in the form of the more bioavailable derivative meproscillarin (Clift®).

Contents

Drug Research: Yesterday, Today, and Tomorrow

Contents

G. Klebe, *Drug Design*, https://doi.org/10.1007/978-3-662-68998-1_1

The direct path to medicines is an old dream of mankind. Even the alchemists searched for the elixir, the *Arcanum*, that would cure all diseases. It has still not been found. On the contrary, drug therapy has become even more complicated as our knowledge of the various causes of disease has become more complex.

Nevertheless, the **success of drug research** has been impressive. For hundreds of years, alcohol, opium, and solanaceous alkaloids (from thorn apples) were the only drugs used to prepare for surgery. Today, general anesthesia, neuroleptanalgesia, and local anesthetics make surgical and dental procedures completely painless. Until this century, plagues and infectious diseases killed more people than all the wars. Today, thanks to hygiene, vaccines, chemotherapeutics, and antibiotics, these diseases have been suppressed, at least in industrialized countries. The dangerously increasing number of therapy-resistant bacterial and viral pathogens (e.g., tuberculosis) has created new problems and makes the development of new drugs urgently necessary. H_2-receptor antagonists and proton pump inhibitors have dramatically reduced the number of surgical procedures for the treatment of gastric and duodenal ulcers. Combinations of these inhibitors with antibiotics have brought even more progress by allowing causal therapy (Sect. 3.5). Cardiovascular diseases, diabetes, and psychiatric diseases (diseases of the central nervous system, CNS) are mostly treated symptomatically, that is, the cause of the disease is not addressed, but rather the negative effects of the disease on the organism. Treatment is often limited to slowing the progression of the disease or improving quality of life. Synthetic corticosteroids have led to significant pain reduction and retardation of pathologic bone degeneration associated with chronic inflammatory diseases (e.g., rheumatoid and chronic polyarthritis). The spectrum of cancer therapy ranges from cure, especially in combination with surgery and radiotherapy, to complete failure of all therapeutic measures.

Highlights such as **inhibitors of BCR-ABL kinase** for a hereditary form of blood cancer are unfortunately still the exception (Sect. 26.4). Recent developments that stimulate the body's own **immune system to fight cancer** are promising (Sect. 31.7).

In the case of bacterial infectious diseases such as tuberculosis or viral diseases, the threatening increase in the number of therapy-resistant pathogens is constantly creating new problems. Strongly modified viral variants can suddenly become a massive threat. At the turn of the year 2019/2020, the world had to learn this the hard way. We witnessed how the new highly contagious **SARS-CoV-2 virus** spread across the globe in the blink of an eye. Most likely, the virus was transmitted to humans from a pet market in Wuhan, China, with bats and other animals as intermediate hosts. But rumors that the outbreak was caused by an accident or carelessness in a high-security virus research laboratory added to the already tense, often politically motivated speculations. From continent to continent, health systems were stretched to the limit and sometimes hopelessly overwhelmed. Mankind experienced in all its severity what the exponential spread of a virus really means. Rapidly increasing numbers of victims were terrified country after country. Social and economic life came to a standstill for weeks. Both acute medicine and any kind of antiviral therapy seemed overwhelmed. Nothing but the robustness of one's own immune system had the power to counteract the virus. What sounds like a horror scenario in an apocalyptic novel suddenly became reality. The World Health Organization (WHO) reports that by 2024, 777 million people worldwide had been infected with the virus and 7.1 million had died from the infection. This bitter experience alone is a painful reminder of the urgent need to develop new medicines and therapies. This is especially true if we want to maintain the quality of life that we have already achieved. The mutability of some viruses and the fact that they can easily jump between animals and humans should be warning enough. It seems almost miraculous that a pandemic like the one in 2020 has not happened before. Perhaps the Spanish flu of a hundred years ago was similar, with a highly pathogenic influenza virus. This makes it all the more likely that the pandemic we have now experienced will not remain a singular event. As if by miracle, two companies managed to develop highly effective **mRNA vaccines** within just one year (Sect. 32.4), which are now offering protection to mankind. In the field of protease inhibitors, too, a therapeutic agent was made available in record time (Sect. 23.9). In the face of adversity, mankind seems to be able to lower bureaucratic hurdles, at least for a short time. We are capable of extraordinary innovation and results-oriented decision-making. In retrospect, this was only possible because the scientific community had years of experience in drug discovery. It would be desirable if this spirit from the pandemic era could be carried over into future research to combat cancer and neglected infectious diseases.

The **history of drug discovery** can be divided into several sequential phases:

- The beginning, when empirical methods were the only source of new medicines,
- Targeted isolation of active compounds from plants,
- The beginning of a systematic search for new synthetic materials with biological effects and the introduction of animal models as surrogates for patients,
- The use of molecular and other *in vitro* test systems as precise models and as a replacement for animal experiments,
- The introduction of experimental and theoretical methods such as protein crystallography, molecular modeling, and quantitative structure–activity relationships for the targeted structure-based and computer-aided design of drugs, and
- The discoveries of new targets and the validation of their therapeutic value through genomic, transcrip-

tomic, and proteomic analysis, knock-in and knockout animal models, and gene silencing with siRNA.

Each preceding phase loses its importance as the next phase arrives. Interestingly, in modern drug discovery, the phases run in the opposite direction. First, a target structure is discovered in the sequenced genome of an organism and its function is modulated to validate this target as a candidate for drug therapy. This is followed by structure-based and computational design of a compound in close collaboration with multiple *in vitro* assays to elucidate the activity and spectrum of activity. The next step is animal testing to establish clinical relevance, and finally clinical trials to confirm the suitability of a test substance as a medicine for patients.

1.1 It All Began with Traditional Medicines

The beginnings of drug therapy can be found in traditional medicine. The narcotic effect of poppy milk, the use of autumn crocus (*Colchicum autumnale*) for gout, and the diuretic effect of squill (*Urginia maritime*) for dropsy (now called congestive heart failure) have been known since ancient times. The dried herbs and extracts of these and other plants have been a major source of medicine for more than 5000 years. The oldest written records of their use date back to 3000 BCE.

Around 1550 BCE, the ancient Egyptian *Papyrus Ebers* listed about 800 prescriptions, many of which included additional rituals to invoke the help of the gods. The five-volume book *De Materia Medica* by Dioskurides (Greek physician, first century CE) is the most scientifically rigorous work of antiquity. It contains descriptions of 800 medicinal plants, 100 animal products, and 90 minerals. Its influence extended into late Arabic medicine and early modern times.

The most famous medicine of antiquity was undoubtedly **theriac**. Its predecessor, mithridatum, was used by the king of Pontus, Mithridates VI (120–63 BCE), as an antidote for poisonings of all kinds. Theriac can be traced back to Andromachus, the private physician of the Emperor Nero, and originally contained 64 ingredients. This preparation was widely used until the eighteenth century. It was prepared in many variations with up to 100 ingredients. In some cities, it was even prepared under state control to ensure that no ingredient was left out! Its use became a panacea for all diseases. In addition, every imaginable wonder drug was used; some examples are rain worm oil, unicorn powder, stomach stones, human cranium powder, mummy dust, and many more.

Traditional Chinese medicine was already very advanced in ancient times. A peculiarity of its formulation was and is the circumstances responsible for the effect of four different qualities. The chief (*jun*) is the carrier of the effect, the adjutant (*chen*) supports the effect or induces another effect. The assistant (*zuo*) may also support the main effect or serve to ameliorate side effects, and one or

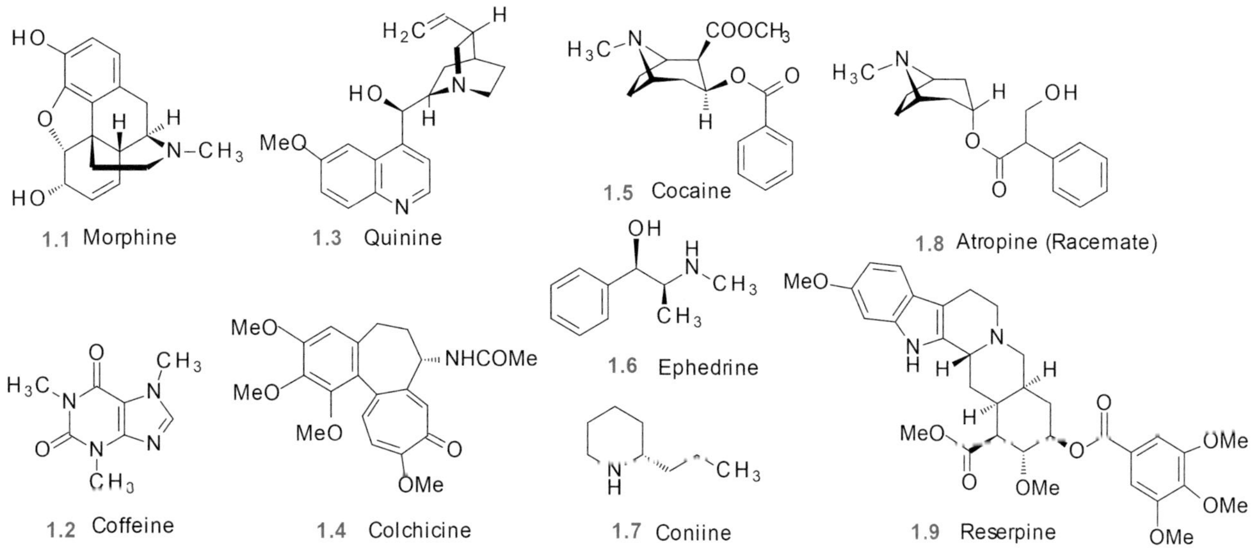

Fig. 1.1 Many important natural products were isolated in the nineteenth century, and a few were synthesized. Morphine **1.1** was isolated from opium by Friedrich Wilhelm Adam Sertürner in 1806, caffeine **1.2** was isolated from coffee, and quinine **1.3** was isolated from cinchona bark by Friedlieb Runge in 1819. Quinine was discovered independently by Pierre Joseph Pelletier and Joseph Bienaimé Caventou, who one year later isolated colchicine **1.4** from autumn crocus. Cocaine **1.5** was extracted from coca leaves by Albert Niemann in 1860, and ephedrine **1.6** was extracted from the Chinese plant Ma Huang (*Ephedra vulgaris*) by Nagayoshi Nagai. In 1886, the first alkaloid, coniine **1.7**, which is found in hemlock, was synthesized by Albert Ladenburg; in 1901 atropine **1.8** was synthesized from deadly nightshade by Richard Willstätter. Reserpine **1.9** from *Rauwolfia serpentina* was first prepared in the middle of the twentieth century, and its structure was elucidated

more messengers (*shi*) moderate the desired effect. The Chinese *Pen-Ts'ao* school (first and second century CE), whose goal was to live as long as possible without aging (!), recommended the following dosage regimen:

> "When treating a disease with a medicine, if a strong effect is desired, one should begin with a dose that is not larger than a grain of millet. If the disease is healed, no more medicine should be given. If the disease is not healed, the dose should be doubled. If that does not heal the disease, the dose should be increased tenfold. When the disease is healed, the therapy should always be discontinued."

The Chinese *Materia Medica*, published by Li Shizhen in 1590, consists of 52 volumes. It contains nearly 1900 medical principles, plants, insects, animals, and minerals, with 10,000 detailed recipes for their preparation. The *Chinese Pharmacopeia* of 1990 contains only two volumes. One of these volumes contains 784 traditional medicines, while the other contains 967 medicines from "Western" medicine.

Paracelsus (born Theophrastus Bombastus von Hohenheim; 1493/1494–1541) made a great breakthrough for scientific medical research. He understood the human body as a "chemical laboratory" and held the ingredients of medicines themselves, the *Quinta Essentia*, responsible for their healing effects. Nevertheless, until the beginning of the nineteenth century, all therapeutic principles were based on either plant, animal, or mineral extracts; only in the rarest cases were pure organic compounds used. This changed radically with the advent of organic chemistry. The great age of **natural products** from plants (for examples see **1.1**–**1.9**, ◘ Fig. 1.1) and the active substances derived from them had begun. Premature hopes placed in some of these substances at the turn of the last century, for example in heroin (Sect. 3.3) or cocaine (Sect. 3.4), were very quickly dashed, but natural products from plants laid the foundation for our modern pharmacy and form an exceedingly large part of it. Natural products and their analogues and derivatives are also well represented among today's best-selling drugs.

1.2 Animal Experiments as a Starting Point for Drug Research

The wealth of experience of **traditional medicine** is based on many thousands of years of sometimes accidental, sometimes intentional observations of their therapeutic effects on humans. Planned studies on animals were relatively rare. The biophysical experiment of Luigi Galvani, a professor of anatomy in Bologna, first described in his book *De viribus electricitatis in motu musculari* in 1791, has become famous. As early as 1780, his students had observed frog legs twitching when the nerve was dissected while a static electricity generator was used. Such devices were standard equipment in many laboratories at the time. He wanted to show in standardized experiments whether the twitching was also caused by thunderstorms. He hung the legs on an iron window grill with a copper hook—they twitched already when they touched the grill. The voltage difference between the two metals was enough to stimulate the nerve, even without an electrical discharge.

The systematic study of the biological effects of plant extracts, animal venoms, and synthetic substances on animals began in the mid-1800s. In 1847, the first department of pharmacology was established at the Imperial University in Dorpat (today Tartu, Estonia). The famous pharmacologist Sir James W. Black, who developed the first β-blocker (an antihypertensive drug, Sect. 29.3) at ICI and later helped develop the first H_2 antagonists (see gastrointestinal ulcer drugs, Sect. 3.5) at Smith, Kline & French, compared pharmacological testing to a prism: what pharmacologists see in the properties of their substances depends directly on the model used to test the substances.

Like a prism, the models distort our vision in different ways. There is no such thing as a depressed rabbit or a schizophrenic rat. Even if such animals existed, they would not be able to share their subjective perceptions and emotions with us. Genetically modified animals (Sect. 12.5), such as the Alzheimer mouse, are also approximations of reality that have been distorted through a different prism, to use Black's analogy. This reality is often underestimated in industrial practice. Scientists tend to optimize their experiments on a particular, isolated model. In doing so, many factors and characteristics that are essential for a drug, such as selectivity or bioavailability, are not sufficiently taken into account.

There is no way out of this dilemma. We need simple *in vitro* models (Sect. 1.5) to test large series of potentially active compounds, and we need the animal models to correlate the data and make predictions about therapeutic effects in humans. Historically, therapeutic advances have been made when a new *in vivo* or *in vitro* pharmacological model was available for a new effect (see H_2 receptor antagonists, Sect. 3.5).

Typical errors in the selection of models and in the interpretation and comparison of experimental results arise from different modes of application and the correlation of results obtained in different animal species. It does not make sense to optimize the therapeutic range of a compound in one species and the toxicology in another. Furthermore, comparing effects after a fixed dose without determining an effective dose also distorts the results because very potent and weak substances fall outside the measurement range. Measuring the effect strictly according to a schedule is also questionable because it does not capture the latency period, the time before an effect is seen, nor the time of maximum biological effect. Whole-animal models usually involve the use of auxiliary medications, which can also influence experimental results. Anesthetized animals often give completely different results than conscious animals.

1.3 The Battle Against Infectious Disease

Plague and infectious diseases, most notably malaria and tuberculosis, have killed more people over time than all the wars in human history. Twenty-two million people died in the first wave of the 1918 influenza ("Spanish flu"). Until the middle of the twentieth century, millions of people died each year from malaria, and unfortunately these numbers are rising again today (Sect. 3.2). Until the turn of the century, ipecac (*Psychotria ipecacuanha*) and cinchona (*Cinchona officinalis L.*) were the only therapeutic approaches to this disease. The impressive successes in the fight against the plague are largely due to the last 80 years of drug research. We owe this to the sulfonamides (Sect. 2.3) and their combinations with dihydrofolate reductase inhibitors (Sect. 27.2), the antibiotics (Sects. 2.4, 6.4, and 32.7), and the synthetic tuberculostatic drugs (Sect. 6.5). When Selman A. Waksman (1888–1973) received the Nobel Prize for the discovery of streptomycin (Sect. 6.4), a little girl congratulated him with a bouquet of flowers. She was the first patient with meningeal tuberculosis to be healed with streptomycin. Today we cannot imagine the atmosphere of a tuberculosis hospital from our own experience, but only from Thomas Mann's *The Magic Mountain* (German: *Zauberberg*).

But infectious diseases, including tuberculosis, are on the rise. In the past, many antibiotics were overused. This, combined with the spread of resistant pathogens in hospitals, means that many cases can only be treated with very specific antibiotics. If resistance develops to these antibiotics, all our weapons will be blunt. New viral infections are on the horizon. Before the advent of acquired immune deficiency syndrome (AIDS), there were very few cases of pneumonia caused by the fungus *Pneumocystis jirovecii* (formerly *Pneumocystis carinii*). This type of pneumonia is the main cause of death in AIDS patients and immunosuppressed patients after organ transplantation. Great efforts have been made to find drugs to treat AIDS and its complications (Sect. 24.3). On the other hand, many widespread tropical diseases, such as malaria and Chagas disease, are still inadequately researched, and the spread of resistance to currently available drugs is a growing global problem. Because these diseases are rampant in parts of the world where people lack the economic resources to afford chemotherapy, more and more pharmaceutical companies have withdrawn from these areas of research for economic reasons. The chances of recovering the development costs from the population of the Third World are poor. This is where global policy must provide some structure so that these people can benefit from the technological advances of modern drug research. One example of this is the Bill and Melinda Gates Foundation, which is dedicated to the treatment and eradication of diseases throughout the world, with a particular focus on developing countries. Improved hygiene has also helped reduce the risk of infections such as traumatic fever and *Shigella* dysentery (discussed in Chap. 21). Above all, vaccines have contributed to the eradication of many infectious diseases. Hopes are still pinned on new and combined vaccines for the prevention of AIDS, malaria, and gastrointestinal ulcers, the latter now known to be caused by the bacterium *Helicobacter pylori* (Sect. 3.5). The Coronavirus pandemic in 2020 has made us once again aware of the importance of developing a vaccine in time to immunize the population.

1.4 Biological Concepts in Drug Research

Acetylcholine **1.10** (◘ Fig. 1.2), synthesized by Adolf v. Bayer in 1869, is a **neurotransmitter**, that is, a transmitter of nerve impulses. In 1921, Otto Loewi, a pharmacologist, demonstrated its biological action in an elegant experiment. Two isolated frog hearts were perfused with the same solution. The vagal nerve of one of the hearts was stimulated, resulting in a slowing of the heart rate, known as bradycardia. Shortly thereafter, the second heart also began to beat more slowly, a clear indication of humoral signaling. Soon after, acetylcholine was identified as the responsible "vagus substance." Acetylcholine itself cannot be used therapeutically because it is metabolized too quickly by acetylcholine esterases (Sect. 23.7).

In 1901, Thomas Bell Aldrich (1861–1938) and Jokichi Takamine isolated the first **human hormone**, adrenaline (also called epinephrine) **1.11** (◘ Fig. 1.2). This hormone and its *N*-desmethyl derivative, noradrenaline (also called norepinephrine) **1.12**, are produced in a central location, the adrenal glands, and are released under stress conditions to the entire system except the CNS and placenta, which have their own barriers to most polar compounds. These substances cause different reactions in different parts of the organism, where they react with the corresponding receptors. The specificity is poor, and a plethora of pharmacodynamic effects result: pulse and blood pressure rise, and the organism is prepared for "flight"—which has been an extremely important function throughout evolution.

Noradrenaline and adrenaline are **neurotransmitters** (Sect. 29.3), as are acetylcholine, biogenic amines **1.13**–**1.15**, amino acids **1.16**–**1.19**, and peptides such as **1.20** and **1.21** (◘ Fig. 1.2). Neurotransmitters are produced locally in nerve cells, stored, and released upon nerve stimulation. After interacting with receptors on the

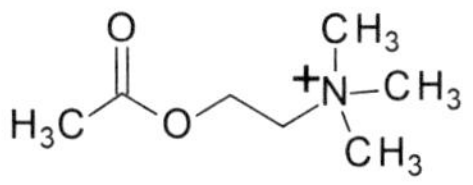

1.10 Acetylcholine

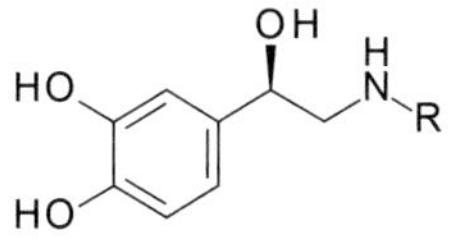

1.11 Adrenaline, R = CH_3

1.12 Noradrenaline, R = H

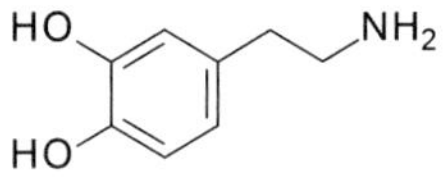

1.13 Dopamine

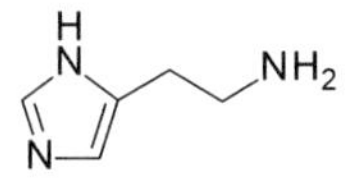

1.14 Histamine

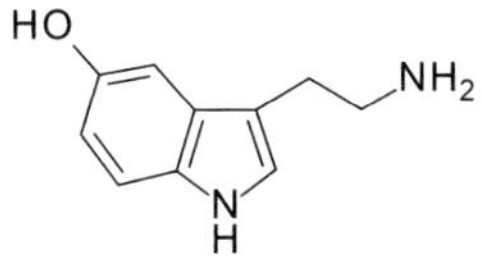

1.15 Serotonin

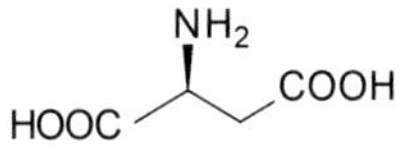

1.16 Aspartic acid

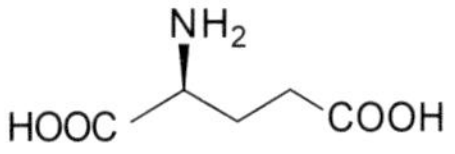

1.17 Glutamic acid

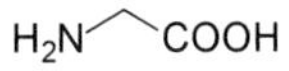

1.18 Glycine

1.19 γ-Aminobutyric acid

Tyr-Gly-Gly-Phe-Met **1.20** Met-Enkephalin

Tyr-Gly-Gly-Phe-Leu **1.21** Leu-Enkephalin

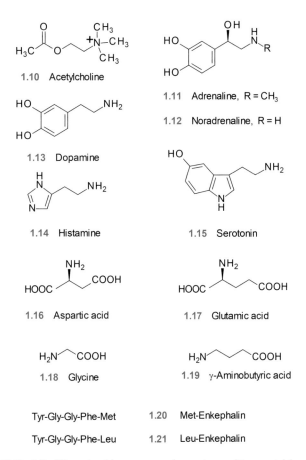

Fig. 1.2 The natural hormones und neurotransmitters acetylcholine **1.10**, adrenaline **1.11**, noradrenaline **1.12**, dopamine **1.13**, histamine **1.14**, and serotonin **1.15**, the excitatory amino acids glutamic acid **1.16** and aspartic acid **1.17**, the inhibitory amino acid glycine **1.18** and γ-aminobutyric acid (GABA) **1.19**, and several peptides, such as the enkephalins **1.20** and **1.21**, substance P and others serve as lead structures for drugs for a variety of cardiovascular and CNS diseases (see Chaps. 3, 29, and 30)

neighboring nerve cell, they are rapidly metabolized or taken up again by the same neuron that released them. Depending on the name of the neurotransmitter, we speak of the adrenergic, cholinergic, and dopaminergic (etc.) systems. The effect produced by adrenaline is called adrenergic, and an antagonist of this system is called antiadrenergic. However, this nomenclature is not always strictly observed. It is common to see combinations of the name of the neurotransmitter with the term **agonist** or **antagonist**, or sometimes **blocker** instead of antagonist, for example, a dopamine agonist, a histamine antagonist, or a β-blocker for antagonists of β-adrenergic receptors (Sect. 29.3). A plethora of drugs have been derived from the structural variations of neurotransmitters.

At the end of the 1920s, the **steroid hormones** were isolated and their structures determined (Sect. 28.5). Overall, the discoveries of the mid-twentieth century heralded the "golden age" of drug discovery. The systematic variation of the principles responsible for biological activity and our increasing knowledge of the **mode of action** led to the synthesis of enzyme inhibitors, receptor agonists and antagonists, which, together with natural product derivatives from plants, make up the majority of our modern pharmacopeia.

1.5 *In Vitro* Models and Molecular Test Systems

About 50 years ago, we started thinking about testing compounds in simple ***in vitro* models**. These models involve biological testing in test tubes rather than on animals. There are many compelling reasons to avoid **animal testing**. They are increasingly criticized by the public for many good reasons. They are also time consuming, costly, and difficult to standardize. Initially, cell culture models were preferred, such as tumor cell cultures for testing cytostatic therapies or embryonic chicken heart cells for testing cardiac agents. Later, receptor binding studies were added. The first molecular test models were enzyme inhibitor assays, in which the inhibitory activity of a molecule on a specific target protein could be evaluated in the absence of disturbing side effects (Chap. 7). Advances in gene technology (Chap. 12) have not only simplified the preparation of the enzyme, but also allowed receptor binding studies to be performed on standardized materials. It is now possible to accurately evaluate the full spectrum of activity of any compound on any enzyme, receptor of any type or subtype, ion channel, or transporter. This has become routine in academic and industrial drug discovery. Before **biological screening** begins, the following questions must be answered: What is the therapeutic target to be addressed and how can it be achieved? Therapeutic concepts are based on the pathophysiology and the causes of its alteration. Regulatory interventions with drugs are intended to restore normal physiological conditions as closely as possible. But there is a problem. Nature works on two orthogonal principles: the **specificity of the mode of action** and an accentuated spatial separation of effects; the **compartmentalization**. Adrenaline, produced in the adrenal glands, affects the entire body except the brain. When it is released there, it acts only at the synapse between two nerve cells. When it comes to specificity, chemists can beat Nature most of the time, but when it comes to spatial separation, they fail by a wide margin.

Advances in gene technology (Chap. 12) allow us to study compounds much more precisely than in the past, but the use of isolated enzymes and binding studies is far from the reality of animal models and even further from humans. In analogy to the difference between an animal experiment and an isolated organ experiment, a well-established correlation between the results obtained in cell culture and an *in vitro* assay and the desired therapeutic effect is a prerequisite for the successful use of the *in vitro* model. Quantitative activity-activity relationships between different biological effects (Chap. 19) provide the link between animal models and humans.

One modern researcher stands out in the field of CNS drugs, but also in the fields of cardiovascular drugs and antihistamines. Paul Janssen (1926–2003) was the director of Janssen Pharmaceuticals in Beerse, Belgium. In the years following World War II, his company discovered more than 70 new compounds, took them through preclinical and clinical development, and established them as therapeutics. In the process, his company established itself as the most successful in pharmaceutical history. His recipe for success was no secret. Paul Janssen was a master of structural variation, a "Beethoven of drug discovery." The systematic combination of pharmacologically interesting structural building blocks and the elegant evaluation of receptor binding studies, *in vitro* models, and animal experiments were the basis of his success.

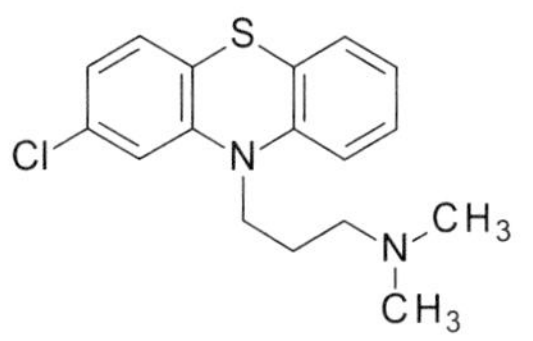

1.22 Chlorpromazine

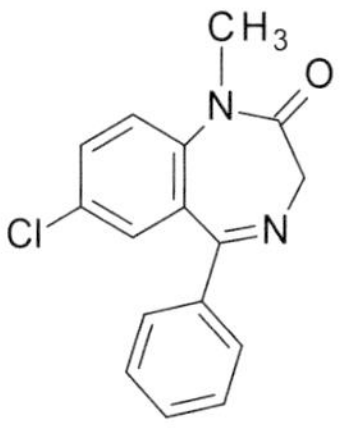

1.23 Diazepam

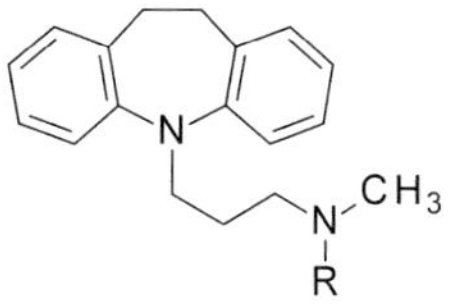

1.24 Imipramine, R = CH_3
1.25 Desipramine, R = H

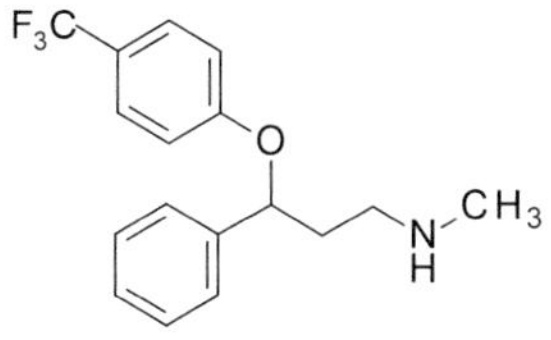

1.26 Fluoxetine

Fig. 1.3 A revolution in the therapy of psychiatric illness was brought about by the discovery of potent neuroleptics such as chlorpromazine **1.22**, tranquilizers such as diazepam **1.23**, and antidepressants such as imipramine **1.24**. For the first time, these compounds allowed targeted treatment of schizophrenia, chronic anxiety, and depression. Examples of newer antidepressants with specific modes of action on transport systems (Sect. 22.7) for noradrenaline and serotonin are desipramine **1.25** and fluoxetine **1.26**, respectively

1.6 The Successful Therapy of Psychiatric Illness

Until the middle of the last century, psychiatric hospitals were purely custodial institutions; they were almost indistinguishable from prisons in terms of restricting an individual's personal freedom. The discovery of neuroleptics, antidepressants, anticonvulsants, and sedatives revolutionized psychiatry. Typical examples of this class of drugs are shown in Fig. 1.3. With the repertoire of drugs available today, schizophrenia, chronic anxiety, and depression preponderate open-ward psychiatry. Many patients can be treated in an ambulatory setting.

In 1933, Manfred Sakel (1901–1957), who worked at the Psychiatric University Hospital in Vienna, noticed that individuals with schizophrenia who were given insulin to stimulate their appetite became calmer. Encouraged by this result, he increased the dose to the point of hypoglycemic coma, a form of deep unconsciousness induced by low blood sugar. Insulin shock, pentetrazole, and electroshock became standard treatment for psychotic illness for the next two decades, an impressive and frightening demonstration of the lack of therapeutic alternatives.

This situation changed in the 1950s with the discovery of reserpine **1.9** (Fig. 1.1), an herbal natural product. This substance acts by depleting the reserves of the neurotransmitters noradrenaline, serotonin, and dopamine in nerve cells. Reserpine was the first substance to show a pronounced neuroleptic effect, that is, it is sedating and calming, and it was the first compound used in psychotic disorders for which the biological effect could be explained by a mode of action. Reserpine has also been used as an antihypertensive drug. Because of its very broad and unspecific action, it is rarely used today for psychiatric disorders or arterial hypertension.

The role of dopamine **1.13** (Fig. 1.2) in the etiology of schizophrenia became clear with the discovery of chlorpromazine **1.22** (Fig. 1.3), a substance that showed a favorable clinical effect. Unlike reserpine, which is nonspecific, chlorpromazine is a pure dopamine antagonist. The use of chlorpromazine and analogous tricyclic neuroleptics induced symptoms similar to those seen in Parkinson's disease. This was the first indication that an endogenous dopamine deficiency was the cause of this disease.

Chlordiazepoxide (Librium®, Sect. 2.7), the first **benzodiazepine** tranquilizer, was discovered by chance. Only a year after its introduction and for many years after that, the chemically closely related drug diazepam **1.23** (Valium®, Fig. 1.3) was the world's best-selling drug. The Rolling Stones commemorated it in their multifaceted song "Mother's Little Helper." Many companies embarked on elaborate synthetic programs, and chemists and pharmacologists applied their entire arsenal of methods. Success justified their efforts. Substances with different modes of action were created: more tranquilizers, sedatives, hypnotics, and even antagonists. Even today, benzodiazepines (Sect. 30.7) are among the most popular and widely used drugs.

The first **antidepressant**, iproniazid (Sect. 27.8), was also an accidental discovery. It works by inhibiting the metabolism of the biogenic amines dopamine, serotonin, noradrenaline, and adrenaline by inhibiting the enzyme monoamine oxidase (Sect. 27.8). In addition to other

1

severe side effects, the first nonspecific representatives caused hypertensive crises and, when taken with certain foods, a few fatalities occurred. Tyramine, a substance found in cheese, wine, and beer (hence the term "cheese effect"), was not properly metabolized. This caused a life-threatening increase in noradrenaline, a hormone that raises blood pressure.

The antidepressant imipramine **1.24** (◘ Fig. 1.3) resulted from the synthesis of analogues of chlorpromazine. Interestingly, and despite its close structural relationship, it is not a neuroleptic, but rather acts in the opposite way. It blocks the transporter for noradrenaline and serotonin (Sect. 30.10), thus, preventing the re-uptake of these neurotransmitters from the synaptic gap. Desipramine **1.25** and fluoxetine **1.26** are even more selective in that they inhibit only the noradrenaline or serotonin transporter of nerve cells.

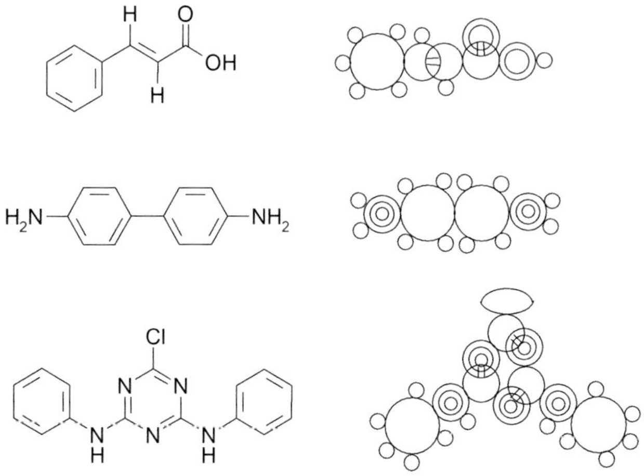

◘ **Fig. 1.4** Loschmidt's book *Constitutionsformeln der Organischen Chemie* (1861) contains structures that anticipate both the formulation of the benzene ring as well as the modern modeling structure. Kekulé must have known about this book because he disparaged it in a letter to Emil Erlenmeyer in January 1862 in that he referred to it as *Confusionsformeln*. Loschmidt did not become famous for his book, but rather because he carried out an experiment in 1865 that determined the number of molecules in a mole to be 6.02×10^{23}, a constant that was later named after him

1.7 Modeling and Computer-Aided Design

To model the properties and reactions of molecules, and especially their intermolecular interactions, a powerful tool is available: the **computer**. In addition to solving complex numerical problems, it is the translation of the results into **color graphics** that perfectly matches the human ability to comprehend images more quickly and easily than text or columns of numbers. This is not surprising. Our brains process text sequentially, but images are grasped in parallel. X-ray crystallography, electron microscopy, and multidimensional NMR spectroscopy (Chap. 13) contribute to our understanding of molecules, as do quantum mechanical and force field calculations (Chap. 15).

Is **molecular modeling** an invention of modern times? Yes and no. Friedrich August Kekulé (1829–1896) is said to have derived his cyclic structure for benzene from a vision of a snake circling around itself and biting its own tail (the snake, *Uroborus*, is an ancient alchemical symbol). The dream that became famous, however, may have come from the memory of the book *Constitutionsformeln der Organischen Chemie* by the Austrian schoolteacher Joseph Loschmidt (1821–1895; ◘ Fig. 1.4). The notation for the constitution of organic compounds introduced by Kekulé has become widely accepted and has greatly stimulated organic chemistry. However, Loschmidt would have been pleased to see images from molecular modeling that closely resemble his formulas. Today, we are increasingly focusing on the three-dimensional character, steric requirements, and electronic properties of molecules. The first **structure-based design** was performed on hemoglobin, the red blood pigment, in the research group of Peter Goodford (Sect. 20.1). Hemoglobin's affinity for oxygen is modulated by so-called allosteric effector molecules that bind to the core of the tetrameric protein. From the three-dimensional structure, Goodford deduced simple dialdehydes and their bisulfite addition products. These substances bind to hemoglobin in the predicted manner and shift the oxygen-binding curve in the expected direction.

The first drug developed by using a structure-based approach is the antihypertensive agent **captopril**, an angiotensin-converting enzyme (ACE) inhibitor (Sect. 25.4). Although the lead structure was a snake venom, the decisive breakthrough was made after modeling the binding site. For this, the binding site of carboxypeptidase, another zinc protease, was used because its three-dimensional structure was known at the time.

The road to a new drug is difficult and tedious. A nested overview of the interplay between the different methods and disciplines from a modern point of view is illustrated in the scheme in ◘ Fig. 1.5. In the last few years, molecular modeling (Chap. 15) and particularly the modeling of ligand–receptor interactions (Chap. 4) have gained importance. Although modeling is employed predominantly for the targeted structure modification of lead compounds, it is also suitable for the structure-based and computer-aided design of drugs (Chap. 20) and lead structure discovery (Sect. 7.6). Examples of these approaches are given in Chaps. 21, 23–32.

In addition to modeling and computer-aided design, **structure–activity relationship analysis** (Chap. 18) has contributed to the understanding of the correlation between the chemical structure of compounds and their biological effects. By using these methods, the influence of lipophilic, electronic, and steric factors on the variation

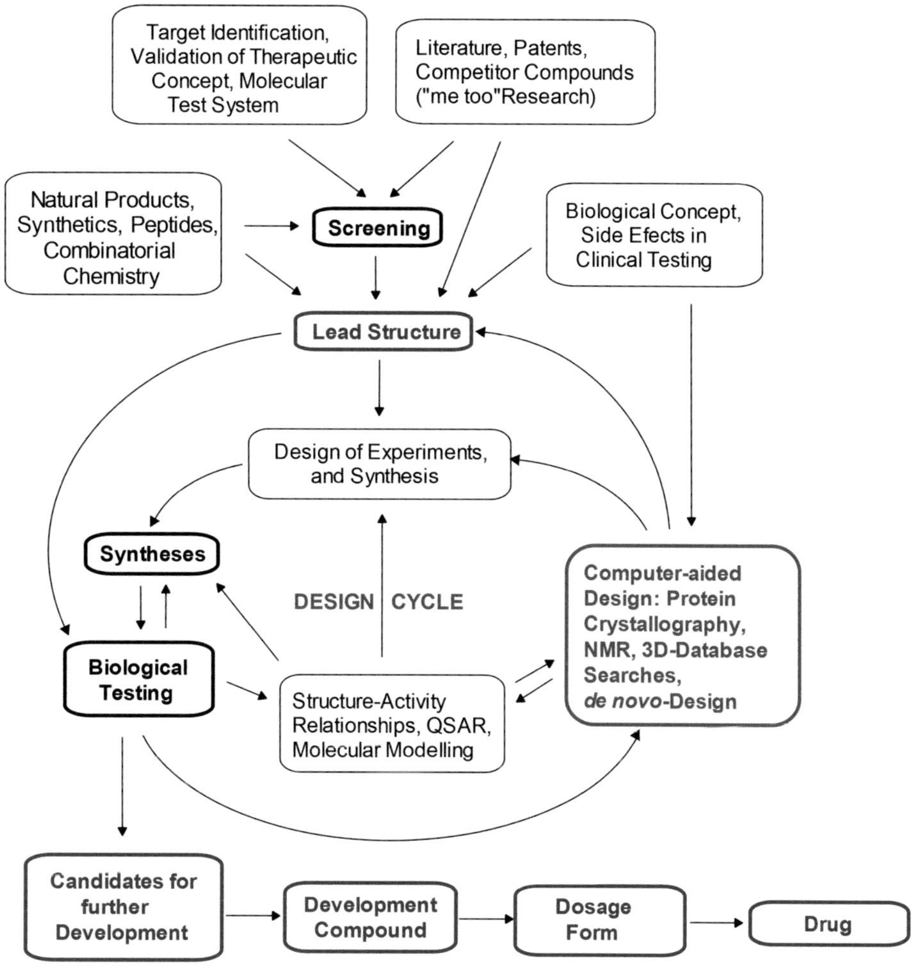

Fig. 1.5 The way to a drug is long. The upper part of the figure shows routes to lead structures. The middle part describes the design cycle, which in practically all cases must be repeatedly re-iterated. Each of these phases is described in detail in the following chapters. Iterative optimization results in candidates for further development such as preclinical and toxicological studies. It is from these studies that the actual candidates are selected. Formulation, clinical trials, and registration then lead to a new medicine. The last phases are not presented in this book

of the biological activity, transport, and distribution of drugs in biological systems could be systematized for the first time on statistically significant foundations.

1.8 The Results of Drug Research and the Drug Market

The development of different methods in drug research has already been described in the last section. Table 1.1 gives a short historical overview of the most prominent results.

The assessment of a **drug's efficacy and safety** has reached an extraordinarily high level today. In some ways, this has been a benefit to our goal of finding new medicines, but it has also been an obstacle. **Acetylsalicylic acid** (Aspirin®) is undoubtedly a valuable drug. Today, this compound would have great difficulty passing clinical trials. Acetylsalicylic acid is an irreversible enzyme inhibitor, it is relatively weakly effective, it causes gastric bleeding at high doses, and it has a very short biological half-life. Each of these problems would be a strong argument against its further development today. It would probably have failed in screening. In a risk–benefit analysis, however, it is better than most alternatives. What is the problem? It probably lies in the analytical–deterministic mindset that dominates science and, therefore, also drug research. What is often overlooked is that such an approach cannot always do justice to the complexity of the human system that we are dealing with in drug therapy.

Despite public healthcare systems that act as a barrier between the supplier and the consumer, the pharmaceutical market, with **global sales** of more than $ 1450 billion, is highly competitive. Two forces influence this market: the state of science and technology and the needs of patients. A small number of drugs account for the majority of sales. Constantly changing "hit lists" of the best-selling drugs can be found on the internet. Due to mergers of established pharmaceutical companies over

Table 1.1 Important milestones in drug research

Year	Substance	Indication/Mode of action	Year	Substance	Indicaton/Mode of action
1806	Morphine	Hypnotic	1996	Saquinavir	HIV protease inhibitor
1875	Salicylic acid	Antiinflammatory	1996	Ritonavir	HIV protease inhibitor
1884	Cocaine	Stimulant, local anesthetic	1996	Indinavir	HIV Protease inhibitor
1888	Phenacetin	Analgetic and Antipyretic	1996	Nevirapine	HIV reverse transcriptase inhibitor
1889	Acetylsalicylic acid	Analgetic and Antipyretic	1997	Sibutramine	Obesity (uptake inhibitor)
1903	Barbiturate	Sedative	1997	Orlistat	Obesity (lipase inhibitor)
1909	Arsphenamine	Antisyphilic	1997	Tolcapon	Parkinson's disease (COMT inhibitor)
1921	Procaine	Local anesthetic	1998	Sildenafil	Erectile dysfunction (PDE5 inhibitor)
1922	Insulin	Antidiabetic	1998	Montelukast	Broncholytic (leucotriene receptor antagonist)
1928	Estrone	Female sex hormone	1999	Infliximab	Antirheumatic (TNFα antagonist)
1928	Penicillin	Antibiotic	2000	Celecoxib	Analgesic (COX-2 inhibitor)
1935	Sulfachrysoidine	Bacteriostatic	2000	Verteporfin	Macular degeneration (photodynamic therapy)
1944	Streptomycin	Antibiotic	2001	Imatinab	Acute myeloid leukemia (kinase inhibitor)
1945	Chloroquine	Antimalarial	2002	Boscutan	Aterial hypertension (endothelin-1 receptor antagonist)
1952	Chlorpromazine	Neuroleptic	2002	Aprepitant	Antiemetic (neurokinin receptor antagonist)
1956	Tolbutamide	Oral antidiabetic	2003	Enfuvirtide	HIV fusion inhibitor (oligopeptide)
1960	Chlordiazepoxide	Tranquilizer	2004	Ximelagatran	Coagulation inhibitor (thrombin inhibitor)
1962	Verapamil	Calcium channel blocker	2004	Bortezomib	Multiple myeloma (proteasome inhibitor)
1963	Propranolol	Antihypertensive (beta blocker)	2005	Bevacizumab	Cytostatic (angiogenesis inhibitor)
1964	Furosemide	Diuretic	2006	Natalizumab	Multiple sclerosis (monoclonal antibody, integrin inhibitor)
1971	L-Dopa	Parkinson's disease	2006	Aliskiren	Antihypertensive (renin inhibitor)
1973	Tamoxifen	Breast cancer (estrogen receptor antagonist)	2007	Maraviroc	HIV fusion inhibitor (CCR5 antagonist)
1975	Nifedipine	Calcium channel blocker	2007	Sitagliptin	Type 2 diabetes (DPPVI inhibitor)
1976	Cimetidine	Gastrointestinal ulcer (H_2 blocker)	2008	Raltegravir	HIV integrase inhibitor
1981	Captopril	Antihypertensive (ACE inhibitor)	2009	Rivaroxaban	Oral anticoagulant (FXa inhibitor)
1981	Ranitidine	Gastrointestinal ulcer (H_2 blocker)	2010	Mifamurtide	Drug against osteosarcoma (bone cancer)
1983	Cyclosporin A	Immunsuppressant	2011	Fingolimod	Immunomodulating drug (multiple sclerosis therapy)
1984	Enalapril	Antihypertensive (ACE inhibitor)	2012	Ivacaftor	Mucoviscidosis (or Cystic fibrosis)
1985	Mefloquine	Antimalarial	2013	Boceprevir	Antiviral hepatitis C-virus (NS3/4A protease inhibitor)
1986	Fluoxetine	Antidepressant (5-HT transport inhibitor)	2014	Bedaquilin	ATP synthase inhibitor (lung tuberculosis)
1987	Artemisinin	Antimalarial	2015	Evolocumab	Monoclonal antibody against proprotein convertase PCSK9
1987	Lovastatine	Cholesterol biosynthesis inhibitor	2016	Dalbavancin	Antibiotic (lipoglycopeptide), skin infection
1988	Omeprazole	Gastrointestinal ulcer (H/K-ATPase inhibitor)	2017	Nusinersen	Antisense oligonucleotide, spinal muscle atrophy
1990	Ondansetron	Antiemetic (5-HT_3 blocker)	2018	Kymriah	CAR-T-cell therapy, blood cancer, Non-Hodgkin lymphoma
1991	Sumatriptan	Migraine (5-HT_1 agonist)	2019	Volanesorsen	Antisense oligonucleotide against Chylomicronemia syndrome
1993	Risperidone	Antipsychotic (D_2/5-HT_2 blocker)	2020	mRNA vaccines	Covid-19 vaccines to produce viral *spike* protein
1994	Famciclovir	Antiviral/herpes (DNA-Polymerase inhibitor)	2021	Asciminib	Allosteric inhibitor of BCR-ABL kinase
1995	Losartan	Arterial hypertension (ATII antagonist)	2022	Nirmatrelvir	First cysteine protease inhibitor (Covid-19 main protease)
1995	Dorzolamide	Glaucoma (carbonic anhydase inhibitor)	2023	Casgevy	First CRISPR-Cas9 gene-editing therapy

the last 30–40 years, the market has shrunk to a small number of "**BigPharma**" companies. It is often the case that a single drug can make or break a company. Often just two or three drugs account for more than 50% of a large company's sales. A historical example is Glaxo. It went from midfield to the top with ranitidine. Astra experienced a similar boom with omeprazole. Today, after the merger with Zeneca, it is one of the biggest players in the field. Sankyo also had a single drug, lovastatin, that boosted sales. With its drugs sildenafil (Viagra®) and atorvastatin (Sortis®/Lipitor®), Pfizer's profits soared to unimaginable heights.

Over the past three decades, we have seen an increasing concentration of pharmaceutical companies, with the market becoming an oligopoly dominated by multinational corporations. Bear in mind that sales giants such as GlaxoSmithKline (GSK), Novartis, Sanofi, Bayer, Bristol-Myers-Squibb or AstraZeneca were created by mergers only about 20 years ago. Companies such as Pfizer and Roche have grown significantly through acquisitions. The importance of research for pharmaceutical companies is evident when one considers that typically 15–20% of turnover is invested in this area. It is unclear whether an equibilirium in the pharmaceutical market has been reached or whether the landscape will continue to change. In the meantime, many small start-up biotech companies are shaking up the scene. Globalization is also playing its part. In addition, countries that have not yet reached the necessary level of development in their research are now entering the market. Political circumstances and risky dependencies, which are now viewed as problematic, will also contribute to a rethinking of these developments.

1.9 A Subject of Conflict: Pharmaceuticals

Drugs are at the center of public interest. While for decades only the doctor, sometimes in consultation with the pharmacist, prescribed the use of a drug, today's patients, alarmed by publications in the lay press, by package inserts, or by more or less serious and well-researched information on the internet, want to take control of the administration of a drug or at least participate in the decision-making process. Influencers are increasingly shaping opinion, and the use of artificial intelligence may make it more and more difficult to distinguish between serious information and fake news.

An example illustrates the issues. Psychotropic drugs have a powerful effect on personality and behavior. At least since the introduction of Valium® (diazepam), these drugs have been in the media spotlight. They are invaluable in the treatment of psychiatric disorders. On the other hand, the risk of abuse and addiction is particularly high. Some of these drugs are even used for self-medication without strict adherence to indication guidelines. Fluoxetine **1.26** (Prozac®, ◘ Fig. 1.3) was introduced by Eli Lilly in 1988 and brought about a clear advance in the treatment of depression. There are now over ten popular science books with controversial content on this one drug alone. Peter Kramer's book *Listening to Prozac* takes a generally sympathetic tone, claiming that depressed patients feel better and more "in harmony" with their personalities after treatment with fluoxetine. This book was on the *New York Times* bestseller list for over 21 weeks. Peter Breggin's book *Talking Back to Prozac* polemically criticized fluoxetine, the company Eli Lilly, and the U.S. Food and Drug Administration (FDA). Side effects, risks, and especially the addictive potential were emphasized. Both books contain correct statements, but both may lead to the wrong conclusions. Prozac® is a valuable drug for the treatment of clinically manifest depression; however, for the treatment of mundane unhappiness or as a general stimulant, it is a drug with many risks.

To conduct a **risk–benefit analysis** of a drug, it is important to consider not only the desired effect, but also the severity of the disease and the objective and subjective side effects. In oncology, even severe side effects are often tolerated for the possibility of improving the patient's condition. If a terminal cancer patient is denied effective pain management because of the risk of addiction, this must be considered malpractice. On the other hand, many people are reckless with highly potent medications. Misuse of antibiotics, faith in the almighty power of tranquilizers and antidepressants, or chronic use of analgesics and laxatives can do more harm than good.

1.10 Synopsis

- Drug research can be divided into several sequential phases starting with empirical observations of the uptake of natural products from food, the development of *in vitro* test systems, increasing understanding of structures and modes of action, to *in vivo* models and gene technology.
- It all started with traditional medicines. The first prescriptions date back to the ancient Egyptians and to traditional Chinese medicine.
- Paracelsus founded scientific medical research and understood humans to be a "chemical laboratory." The ingredients of drugs were first held responsible for healing effects.
- With the advent of organic chemistry, the first therapeutic principles based on pure organic compounds became available. The great age of natural products from plants and their active ingredients began.
- Systematic studies on animals began in the 19th century can be seen as a starting point for drug research. *In vitro* models are needed to test large series of potentially active compounds, but animal models are required to correlate the data and make predictions about the therapeutic effects in humans.
- Our present life expectancy would not be possible without the successful fight against infectious diseases. The broad application of antibiotics and the spread of resistant pathogens, however, have led to situations in which the best weapons against infectious diseases are becoming increasingly ineffective. Research against widespread tropical diseases has been neglected, and the currently increasing resistance to available medications represents a worldwide problem.
- The elucidation of biological concepts, pathways, and regulatory cycles by endogenous compounds has strongly stimulated drug research. Many developed drugs have arisen from structural variations of neurotransmitters, hormones, steroids, or natural substrates.
- Systematic substance testing began with the establishment of *in vitro* models that replaced biological testing on animals by assays in test tubes. Gene technology has made it possible to prepare sufficient amounts of pure proteins for testing.
- The discovery of neuroleptics, antidepressants, anticonvulsives, and sedatives has revolutionized the treatment of psychiatric diseases.
- Molecular modeling and computer-aided design along with structural biology give access to rational considerations on drug action. The first structure-based design project was carried out on hemoglobin, and the first drug developed by using a structure-based approach was the antihypertensive captopril.
- The assessment of drug efficacy and safety has reached an extraordinarily high standard today. The worldwide drug market, with more 1450 billion US dollars in sales per year, is large and highly competitive. Only a few drugs command a large portion of sales and determine the particular dynamics in the market; the current tendency is corporate contraction to fewer and bigger companies. Often a single drug can make or break a company.
- Drugs remain in the focal point of public interest. It is no longer the physician alone who influences the prescription of medication; multiple sources of information have an impact and inform the patient. A proper risk–benefit analysis of a medication, taking into consideration not only the desired therapeutic effect but also the severity of an illness, is needed.

1

Bibliography and Further Reading

General Literature

R. Schmitz, Geschichte der Pharmazie, Bd. 1, GOVI-Verlag, Eschborn (1998)

C. Friedrich and W.-D. Müller-Jahncke, Von der Frühen Neuzeit bis zur Gegenwart, Vol. 2, GOVI-Verlag, Eschborn (2005)

W.-D. Müller-Jahnke and C. Friedrich, Arzneimittelgeschichte, Wissenschaftliche Verlagsgesellschaft, Stuttgart (2005)

S. H. Barondes, Molecules and Mental Illness, Scientific American Library, W. H. Freeman and Company, New York (1993)

R. M. Restak, Receptors, Bantam Books, New York (1994)

T. J. Perun and C. L. Propst, Eds., Computer-Aided Drug Design. Methods and Applications, Marcel Dekker, New York (1989)

C. R. Beddell, Eds., The Design of Drugs to Macromolecular Targets, John Wiley & Sons, Chichester (1992)

E. C. Herrmann and R. Franke, Eds., Computer Aided Drug Design in Industrial Research, Ernst Schering Research Foundation Workshop 15, Springer-Verlag, Berlin (1995)

K. Müller, Eds., De Novo Design, Persp. Drug Discov. Design, Vol. 3, Escom, Leiden (1995)

B. Werth, The Billion-Dollar Molecule. One Company's Quest for the Perfect Drug, Touchstone, New York (1994)

D. Fischer and J. Breitenbach, Eds., Die Pharmaindustrie, Spektrum Akademischer Verlag, Heidelberg, Berlin (2003)

A. Giannis, Naturstoffe im Dienst der Medizin - Von der Tragödie zur Therapie, Springer-Verlag GmbH, (2023)

Special Literature

E. Mutschler, Arzneimittel – Erfolge, Misserfolge, Hoffnungen, Deutsche Apoth.-Ztg. **127**, 2025–2033 (1987)

D. J. Newman and G. M. Cragg, Natural Products as Sources of New Drugs over the Last 25 Years, J. Nat. Prod. **70**, 461–477 (2007)

C. R. Noe and A. Bader, Facts Are Better Than Dreams, Chem. Britain **29**, 126–128 (1993), (formulas of Kekulé and Loschmidt)

C. R. Beddell, P. J. Goodford, F. E. Norrington *et al.*, Compounds Designed to Fit a Site of Known Structure in Human Hemoglobin, Br. J. Pharmac., **57**, 201–209 (1976)

P. Kramer, Listening to Prozac, Viking, New York (1993)

P. R. Breggin and G. R. Breggin, Talking Back to Prozac, St. Martin's Press, New York (1994)

https://data.who.int/dashboards/covid19/cases (Last accessed Nov. 29, 2024)

In the Beginning, There Was Serendipity

Contents

G. Klebe, *Drug Design*, https://doi.org/10.1007/978-3-662-68998-1_2

"A lucky accident dropped the medicine into our hands;" this is how a publication on August 14, 1886, from Arnold Cahn and Paul Hepp in the *Centralblatt für Klinische Medizin* began. The history of drug research is punctuated by lucky accidents. As a general rule, detailed knowledge of biological systems was absent. Therefore, it is not surprising that the working hypotheses were often wrong, and the obtained results differed from expectations. The case of accidental success fell into the background over time. Today, happenstance as a strategy has been replaced by the arduous and ambitious goal of preparing drugs by using a straightforward approach. The only exception to this is the kind of shotgun-style testing of large and diverse chemical compound libraries, including microbial and plant extracts that is done with the goal of finding new lead structures. In this case, serendipity is desired to find as large and diverse a palette of lead structures (Chaps. 6, 7, 8, 9).

2.1 Acetanilide Instead of Naphthalene: A New, Valuable Antipyretic

Back to Cahn and Hepp. What happened? There are several legends about this lucky accident. The most plausible version is that the antipyretic effect of naphthalene, a compound widely available from coal tar, was tested. The substance indeed showed fever-lowering qualities. The responsible substance however, was not naphthalene but rather something entirely different: acetanilide **2.1** (◘ Fig. 2.1). Further experiments confirmed the efficacy. Shortly thereafter, the company Kalle & Co. introduced it to the market with the name "Antifebrin."

Phenacetin **2.2** (◘ Fig. 2.1) was subsequently developed based upon a targeted approach. At the time, Bayer in Elberfeld had 30 t of *p*-nitrophenol, a side product from dye production, on their waste heap. The then 25-year-old Carl Duisberg, who later became the chairman of Bayer Farbenfabriken AG and who also took a leading role in the foundation of I.G. Farbenindustrie in 1924, wanted to use *p*-nitrophenol for the preparation of acetanilide as it could easily be reduced to *p*-aminophenol. The known toxicity of phenol groups led to the design of *p*-ethoxyacetanilide **2.2** (phenacetin), which actually did have the desired qualities and served as an analgesic for headaches and as an antipyretic for a century. Unfortunately its metabolite **2.4**, which still contains the ethoxy group, leads to the production of methemoglobin, an oxidized form of the red blood pigment that is incapable of carrying oxygen. Furthermore, chronic misuse by, for instance, taking kilogram quantities of phenacetin over a lifetime, leads to kidney damage. Paradoxically, the main metabolite of phenacetin, *p*-hydroxyacetanilide **2.5** (◘ Fig. 2.1, acetaminophen in American English, or paracetamol in UK English) is actually responsible for the effect, and it is less toxic and better tolerated. In the USA alone, paracetamol achieved over US$ 1.3 billion in annual sales. This is even more than for acetylsalicylic acid.

2.2 Anesthetics and Sedatives: Pure Accidental Discovery

In 1799, Humphry Davy (1778–1829) discovered the euphoric effect of nitrous oxide (N_2O), which was appropriately named "laughing gas." The dentist Horace Wells (1815–1848) saw a traveling theater production of a "sniffing party" with N_2O in 1844 in which a participant suffered from a flesh wound, apparently without pain. To test this effect, Wells had one of his own teeth extracted, also without pain. He then repeated the procedure on many people, with success. However, one public demonstration went awry, and this drove him to suicide four years later. It's frightening how things keep repeating themselves. Recently, nitrous oxide has been increasingly used as a modern party drug. Its strong euphoric effect, sometimes combined with drowsiness and even

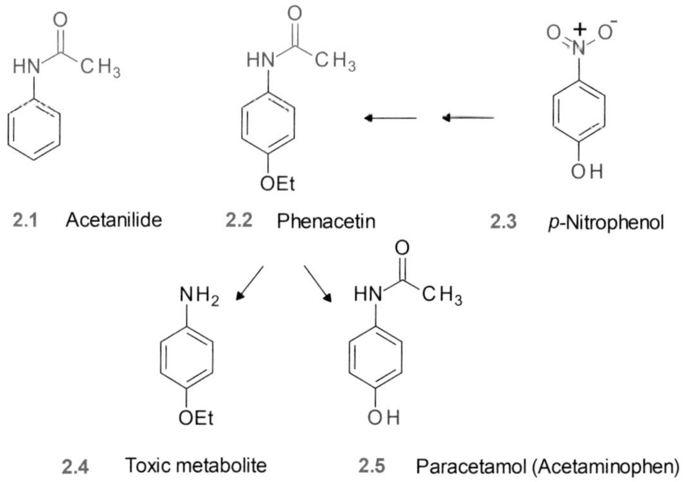

◘ **Fig. 2.1** By starting with the accidently discovered acetanilide **2.1**, Carl Duisberg planned the synthesis of phenacetin **2.2** from nitrophenol **2.3**. In contrast to the toxic metabolite **2.4**, the main metabolite, paracetamol (Amer. acetaminophen) **2.5** is well tolerated

hallucinations, is obviously appealing. Unfortunately, the risks are underestimated as frequent application can severely damage the nervous system.

The same effect was observed in 1842 by Crawford W. Long (1815–1878) with ether, but he did not report it immediately. After administering ether, he was able to remove an ulcer from the neck of a volunteer. William T. Morton (1819–1868) successfully carried out the first ether anesthesia in the same hospital as Wells. Starting in 1847, chloroform was used as an anesthetic. A few years later, anesthesia became standard for surgical procedures, a real blessing for the suffering of humanity.

Oskar Liebreich (1839–1908) wanted to develop a depot form of chloroform **2.6** in 1868. Because chloral hydrate can be cleaved with base in an aqueous milieu, he hoped that this could also happen in the body. Chloral hydrate is in fact a sedative, but this is because of its active metabolite, trichloroethanol **2.8** (◘ Fig. 2.2), and not because it releases chloroform.

In 1885, Oswald Schmiedeberg (1838–1921) tested urethane **2.9** (ethylcarbamate, ◘ Fig. 2.3) because he thought that it would release ethanol in the organism. Urethane itself is the active agent. Its optimization later led to isoamylcarbamate **2.10** (Hedonal®, 1899). Based on this, open and cyclic carbamates and ureas were investigated. In 1903, the first barbiturate sedative, barbital (Veronal®) resulted. In the decades that followed, a wealth of better-tolerated barbiturates with a broader pharmacokinetic spectrum were introduced.

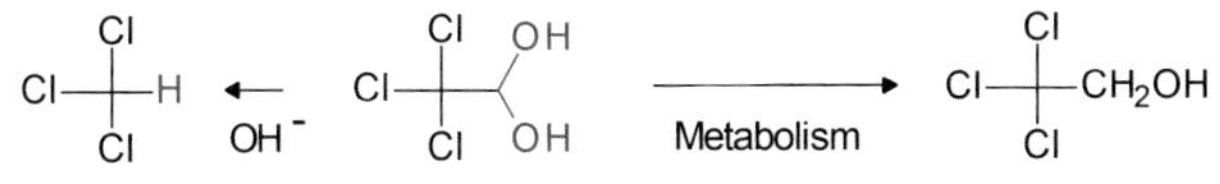

2.6 Chloroform 2.7 Chloral hydrate 2.8 Trichlorethanol

◘ **Fig. 2.2** The anesthetic chloroform **2.6** is formed upon treatment of chloral hydrate **2.7** with base. However, this reaction does not work *in vivo*. The active metabolite of **2.7** is trichloroethanol **2.8**

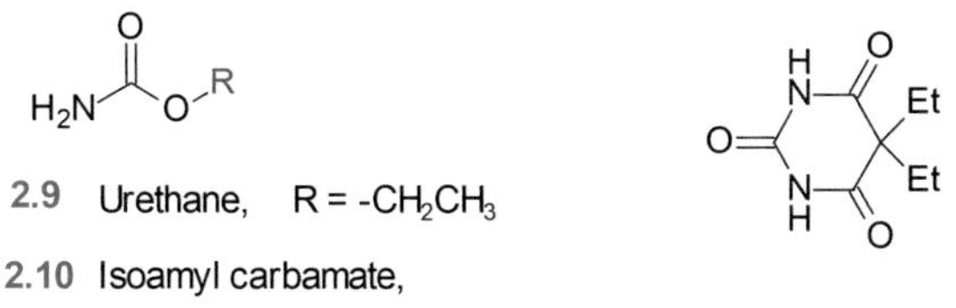

2.9 Urethane, R = $-CH_2CH_3$

2.10 Isoamyl carbamate, R = $-CH_2CH_2CH(CH_3)_2$

2.11 Barbital

◘ **Fig. 2.3** The hypothetical "prodrug" of ethanol, urethane **2.9**, led to the development of isoamylcarbamate **2.10**, which in turn led to the first barbiturate, barbital **2.11**

2.3 Fruitful Synergies: Dyes and Pharmaceuticals

Dyes and pharmaceuticals have stimulated one another. The first synthetic dye was the result of a failed drug synthesis. In 1856, August Wilhelm v. Hoffman assigned the task of synthesizing quinine, an alkaloid used for treating malaria (Sect. 3.2), to the then 17-year-old William Henry Perkins (1838–1907). By starting with only the molecular formula, it was anticipated that the oxidation of an allyl-substituted toluidine would deliver the desired product. However, now that the structural formula is known, we understand that this could not possibly have worked! Upon oxidation of aniline that was contaminated with *o*- and *p*-toluidine, Perkins isolated a dark precipitate. It contained a dye, mauveine **2.12** (◘ Fig. 2.4) that colored silks a brilliant mauve. Other dyes were prepared in rapid succession. The development and later proliferation of the dye industry in England and Germany in the second half of the nineteenth century can be traced back to this accidental discovery. Towards the end of the nineteenth century increasing competition and a difficult economic situation in the dye market inspired the reactionary expansion into industrial pharmaceutical research. In 1896, a pharmaceutical research laboratory was founded in the 33-year-old Bayer Farbenfabrik. At that time innumerable synthetic dyes were known; therefore, it is not surprising that these substances were tested for pharmacological effects.

Of all people, wine adulterators played an important role in the discovery of the first synthetic laxative. To stop people from selling Trester wine (so-called *Nachwein*) as a natural wine (*Naturwein*), in 1900 the dye phenolphthalein was added as an easily detectable indicator. The Hungarian pharmacologist Zoltán von Vámossy (1868–1953) investigated the effects of this compound. Back then, the conventions of the pharmacologists were

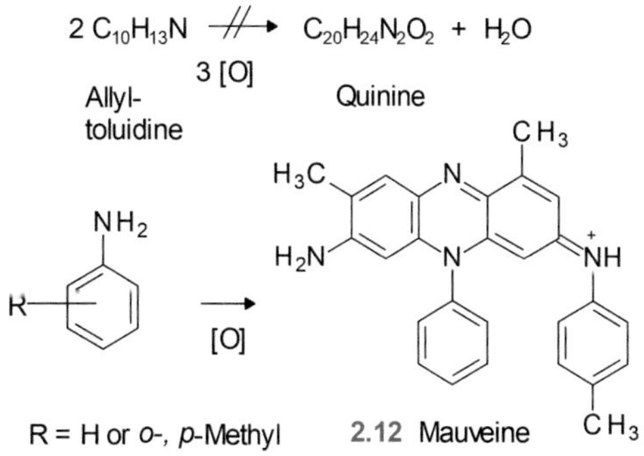

◘ **Fig. 2.4** An unsuccessful quinine synthesis founded the dye industry. The chemical structures of many organic compounds were still entirely unknown in the middle of the nineteenth century. The attempt to prepare quinine via a simple route (*upper* reaction) could not have worked. The oxidation of an impure aniline (*below*) gave mauveine **2.12** in 1856, which was used to dye silk a brilliant mauve color. It was the first synthetic dye!

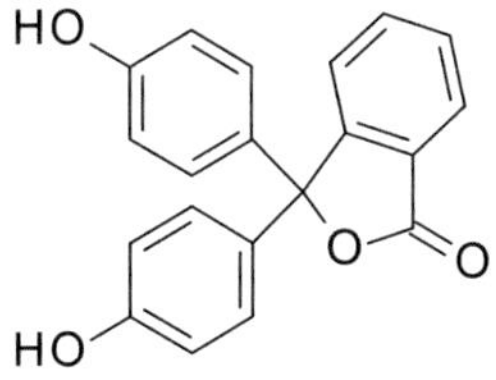

2.13 Phenolphthalein

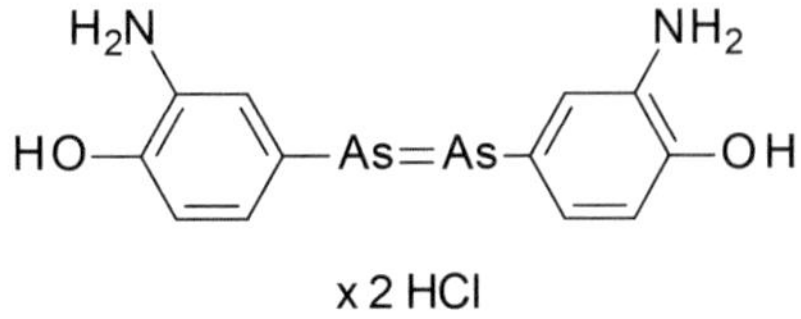

2.14 Arsphenamine

Fig. 2.5 The laxative effect of phenolphthalein became apparent while testing it as an additive for cheap wines. The antisyphilis compound arsphenamine **2.14** (Salvarsan®, here shown as monomer) is simply an azo dye in which the –N=N– group was exchanged for an –As=As– group

still rather primitive. The intravenous application of 0.01–0.03 g to rabbits caused death "with loud shrieking, convulsions, and paralysis." Vámossy then decided to feed 1–2 g to a rabbit and 5 g to a 4 kg lap dog. Because these oral doses were all well tolerated, Vámossy took 1.5 g of phenolphthalein himself, and a friend took 1.0 g. The effects were explosive: rumbling in the bowels, diarrhea, and for two additional days loose stools. It was later established that 150–200 mg would have been a therapeutic dose.

An entire range of antibacterial and antiparasitic dyes are based on the work of Robert Koch (1843–1910). He showed that bacteria and parasites accumulate dyes specifically. Based on this, Paul Ehrlich (1854–1915) hoped to kill pathogens selectively with suitably chosen dyes. In 1891, he cured two mild cases of malaria by treating the patients with methylene blue. In the following years he tested hundreds of different pigments, and thousands more analogues were later synthesized in the laboratories of Bayer and Hoechst. In 1909, Paul Ehrlich pursued a rational design when he exchanged both of the nitrogen atoms of an –N=N– group of an azo dye for arsenic atoms. Arsphenamine **2.14** (Salvarsan®, Fig. 2.5) was the first effective compound to treat syphilis: the first chemotherapeutic. It became an extraordinary economic success for the company Hoechst.

The breakthrough with chemotherapeutics was made by the physician Gerhard Domagk (1895–1964). At the age of 31, he took over the newly formed department of experimental pathology at Bayer in Elberfeld. Azo dyes bearing sulfonamide groups had already been designed by the chemists Fritz Mietzsch and Josef Klarer, but they showed no *in vitro* activity; Domagk tested these substances in *streptococci*-infected mice. By using this model, he found the first active substances in 1932. Sulfamidochrysoidine **2.15** (Protonsil®, Fig. 2.6), a dark-red dye that could treat even severe *streptococci* infections, was first made in 1935. The sulfonamides became world famous a year later when the son of the US president Theodore D. Roosevelt, Jr. was treated with such a compound to cure a severe sinus infection. But even here a false hypothesis led to success. It was not the azo dye itself, but rather its metabolite sulfanilamide **2.16** that was effective. Sulfanilamide replaces *p*-aminobenzoic acid **2.17** (Fig. 2.6), which is needed for the bacterial synthesis of the enzymatic cofactor, dihydrofolic acid (Sect. 27.2).

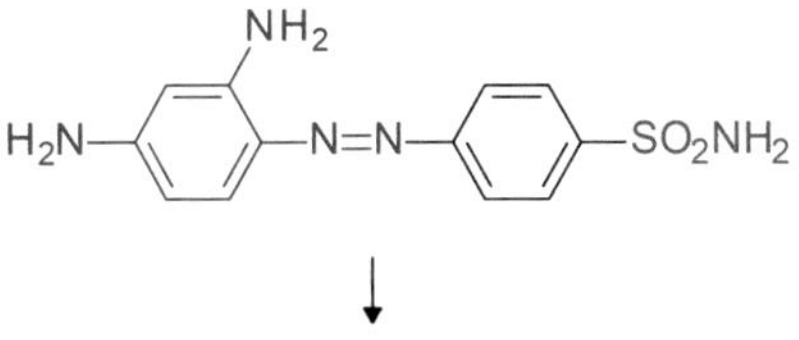

2.15 Sulfachrysoidine

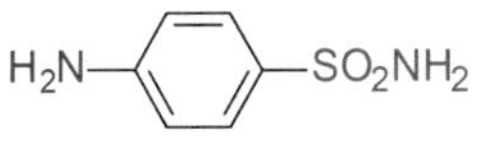

2.16 Sulfanilamide

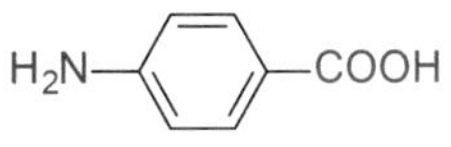

2.17 *p*-Aminobenzoic acid

Fig. 2.6 The red azo dye sulfamidochrysoidine **2.15** is effective only after cleavage to the colorless sulfanilamide **2.16**, which is a bacterial antimetabolite of *p*-aminobenzoic acid **2.17**

2.4 Fungi Kill Bacteria and Help with Syntheses

The discovery of the antibiotic effect of *Penicillium notatum* by Alexander Fleming (1881–1955) in 1928 is the most famous example of a serendipitous discovery. Fleming noticed that a spoiled *staphylococcus* culture had been contaminated with a fungal infection. In the area around the fungus, no bacteria could grow. Further investigations showed that this fungus could also curb other bacteria. Fleming called the still-unknown agent **penicillin**. It was not until 1940 that it was isolated and characterized by Ernst Boris Chain (1906–1979) and Howard Florey (1910–1985). In 1941, an English policeman was the first patient to be treated with penicillin. Despite a temporary improvement, and even though penicillin could be isolated from his urine, he died after a few days as no more penicillin was available for his continued therapy. The fungus *Penicillium chrysogenum*, which produces more penicillin than *Penicillium notatum* and is easier to cultivate, was isolated from a moldy melon in Illinois. The tedious route to the structural elucidation of penicillin and the successful work to systematically vary its structure are scientific masterworks of the first order. There were even more difficult problems to conquer to optimize its production and its biotechnological mass production. Today, the modified penicillins **2.18** and cephalosporins **2.19**

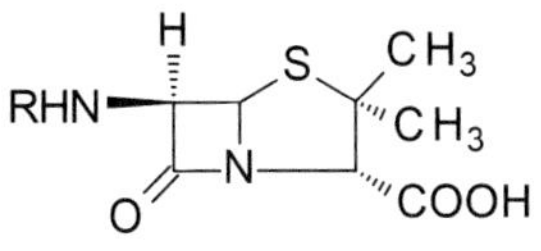

2.18 Penicillins

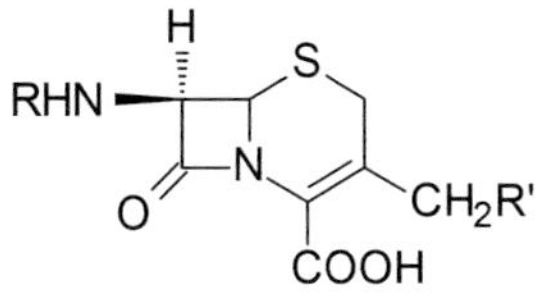

2.19 Cephalosporins

Fig. 2.7 Fleming's accidental discovery of the antibiotic effects of a fungus has delivered a wide palette of penicillins **2.18** and cephalosporins **2.19**, each with different R groups

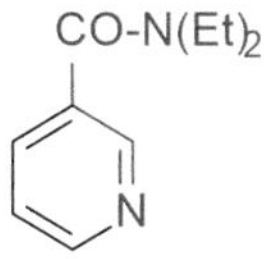

2.20 *N,N*-diethyl nicotinamide

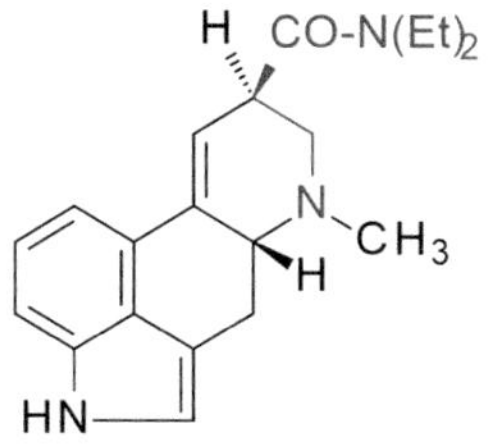

2.21 LSD

Fig. 2.8 *N,N*-Diethyl nicotinamide **2.20** is a centrally active derivative of nicotinic acid. Hofmann wanted to synthesize a general stimulant analogously by preparing the *N,N*-diethyl amide of lysergic acid. The result was the hallucinogen lysergic acid diethyl amide **2.21** (LSD)

(Fig. 2.7), which make up a broad range of antibiotics with outstanding bioavailability, are available. The newer analogues have a broader spectrum of activity against many pathogens and are distinguished by a generally improved stability to the penicillin-degrading enzyme β-lactamase (Sect. 23.7). Fleming was a researcher to whom Pasteur's thesis "chance favors the prepared mind" fully applies. One day in 1921 while working in his laboratory with a cold, he tried a rather headstrong experiment. He added a drop from his own nasal mucus to a bacterial culture and found a few days later that the bacteria had been killed. This "experiment" led to the discovery of lysozyme, an enzyme that hydrolyzes the bacterial cell wall. As a therapy, it is unfortunately unsuitable because it does not attack most human pathogens.

Chance and a fungus played an important role in the industrial synthesis of corticosteroids. An important step in the synthesis is the introduction of an oxygen atom at a particular position in the steroid scaffold, position 11. In 1952, chemists at the Upjohn company sought after a soil bacteria that could hydroxylate a steroid in this position. Just when they finally decided to set an agar plate on the window bank of the laboratory, *Rhizopus arrhizus* landed exactly there. This fungus transforms progesterone (Sect. 28.5) to 11α-hydroxyprogesterone. With its help, the yield could be increased to 50%. The closely related fungus *Rhizopus nigricans* even afforded 90% of the desired product.

2.5 The Discovery of the Hallucinogenic Effect of LSD

In the 1930s, Albert Hoffmann (1906–2008) was working on the partial synthesis of ergoline alkaloids at Sandoz. In 1938, he wanted to find a way to transfer the respiratory and cardiovascular stimulatory effect of *N,N*-diethyl nicotinamide **2.20** into this class of compounds. Analogous to **2.20**, he prepared *N,N*-diethyl lysergamide **2.21** (Fig. 2.8) with the hope of maintaining the stimulatory circulatory and respiratory effects. The substances showed no particular effect other than the experimental animals being agitated under anesthesia. Therefore, they were not pursued at first. Hoffman prepared the substances for a second time five years later because he wanted to investigate them more thoroughly. Upon the purification procedure and recrystallization, he reported feeling "*a strange agitation combined with a slight dizziness.*" At home he fell into "*a not-unpleasant inebriated condition that was characterized by extremely animated fantasies ... after about 2 hours, the condition went away.*" Hoffman suspected a connection to the compounds he prepared and conducted a self-experiment with 0.25 mg a few days later. That was the smallest dose with which he expected to see an effect. The outcome was dramatic, the experience was the same as the first time, but much more intense. He had a technician accompany him home on his bicycle. During the ride, his condition took on a threatening form, and he fell into a severe crisis dominated by dizziness and anxiety. The world took on a grotesque form. Later it was determined that 0.02–0.1 mg is enough to cause hallucinations. The substance was temporarily marketed as Delyside® for use in psychotherapy and to treat anxiety and compulsive disorders.

2.6 The Synthetic Route Determines the Structure

The structure of the first calcium channel blocker, verapamil **2.22** was determined by its synthesis (Fig. 2.9). Verapamil counteracts the effects of β-adrenergic agonists, but it is not a β-blocker. It was only after its introduction to the market that Albrecht Fleckenstein clarified its mode of action: it blocks the inwards membrane voltage-dependent flow of calcium ions through the calcium channels (Sect. 30.4) in cardiac and endothelial cells. The hypotonic effect was initially seen as a side effect, but in the following years it became the most important reason for use. The second group of therapeutically important calcium channel blockers, nifedipine **2.23** was inspired by a synthetic principle, i.e., the Hantzsch synthesis of dihydropyridines (Fig. 2.9) from 1882. Remarkably, the pharmacological experiments on nifedipine had to be carried out in a darkened room because of its photosensitivity. All the more reason to applaud its development into a medicine despite this characteristic.

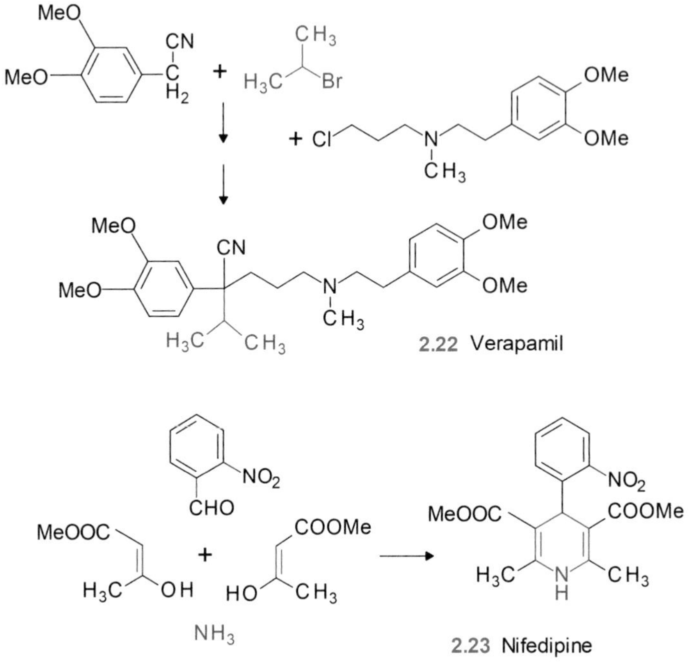

Fig. 2.9 Ferdinand Dengel, a chemist at the former Knoll AG, wanted to prepare a cardiovascular drug by alkylating a nitrile. To avoid a double substitution, he started with the sterically demanding isopropyl group. The result was the first calcium channel blocker, verapamil **2.22**. The isopropyl group is the optimal alkyl group because it stabilizes the biologically active conformation. The synthetic route played an important role in the development of the second calcium channel blocker, nifedipine **2.23**. In 1948, Friedrich Bosser at Bayer was given the task of finding new substances that dilate the coronary arteries. After years of work, he turned in 1964 to the easily prepared dihydropyridines, which surprisingly displayed the desired effects. In this case, the space-filling nitro group promotes the biologically active conformation (Sects. 17.9 and 30.4)

2.7 Surprising Rearrangements Lead to Medicines

Leo Sternbach (1908–2005), a chemist at Hoffman La Roche, was involved in a program in the mid-1950s to find structurally novel tranquilizers. Sternbach remembered a synthetic program on pigments from a decade before in which *N*-oxide **2.24** (Fig. 2.10) was also prepared. Its reaction with secondary amines delivered the expected products, which were pharmacologically absolutely uninteresting. The work was practically ended in 1957, and the laboratory was being cleaned up when it was noticed that a crystalline base and its hydrochloride salt had precipitated from a solution. The substance was the product of a reaction between *N*-oxide **2.24** and methylamine, but it was never tested due to other priorities. The subsequent pharmacological testing convincingly showed outstanding qualities. It was only later established that an unexpected ring rearrangement reaction had occurred to afford chlordiazepoxide **2.25** (Librium®, Fig. 2.10).

There are other examples of this sort. In 1974, W. Berney was working on spirodihydronaphthalenes **2.26** (Fig. 2.11) with the goal of preparing CNS-active substances. Upon acid treatment, he obtained a compound that was highly potent *in vitro* and *in vivo* against a series of human pathogenic fungi in a routine broad screening at Sandoz Research Institute in Vienna. In 1985, the substance was introduced as naftifine **2.27**, and later a more potent analogue, terbinafine **2.28** (Fig. 2.11) followed. Both substances showed a previously unknown mode of action. They damage the membrane of fungi in that they block the ergosterol biosynthesis. This happens in a very early step due to the inhibition of the enzyme squalene epoxidase.

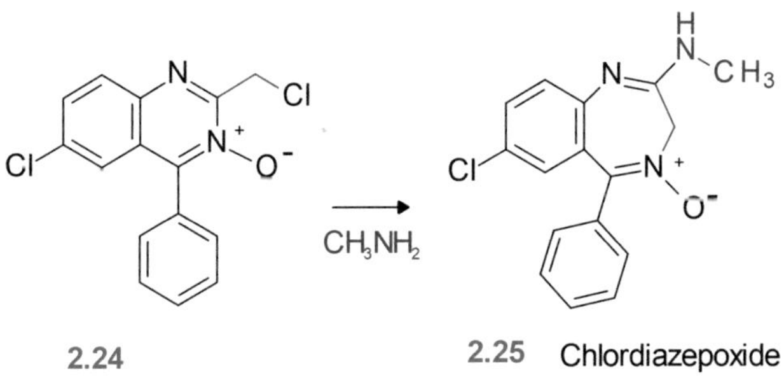

Fig. 2.10 Treatment of **2.25** with methylamine delivers the rearrangement product chlordiazepoxide **2.25** (Librium®) instead of the expected one. This first test compound became the first of the benzodiazepine class to be marketed

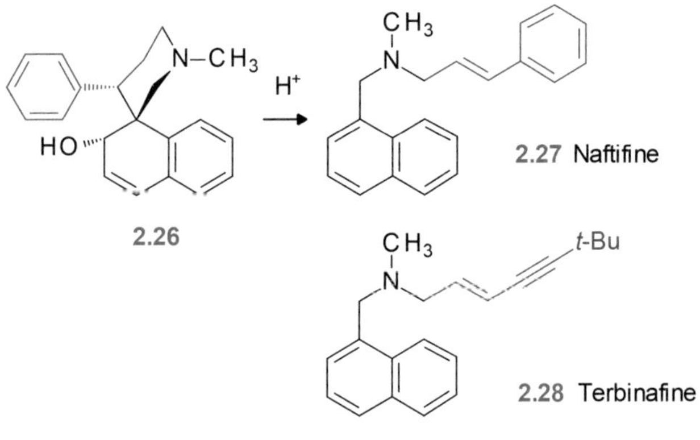

Fig. 2.11 Instead of CNS activity, naftifine **2.27**, prepared from spiro compound **2.26**, is an antimycotic. A comparison with the more potent terbinafine **2.28** shows that the phenyl group can advantageously be replaced with a *tert*-butylethinyl group

2.8 A Long List of Accidents

The list of accidental discoveries, from which a few are described here, can be extended *ad infinitum*. A few more examples are briefly mentioned without chemical formulas.

- Pethidine (Sect. 3.3), the first fully synthetic opiate analgesic, was synthesized in the 1930s as part of an anticonvulsives research program, by starting from atropine.
- The suitability of antihistamines for the prevention of motion sickness was discovered in Boston because of a treatment for a skin rash. A patient reported that her motion sickness, which always occurred when riding a Boston street car, went away. The "clinical trial" was carried out in 1947 on hundreds of sailors on the transatlantic voyage of the USNS *General Ballou*.
- Haloperidol (Sect. 3.3) was meant to be an analgesic; it turned out to be a neuroleptic.
- Imipramine is structurally very similar to the neuroleptic chlorpromazine (Sects. 1.6 and 8.5). Nonetheless, it has the opposite effect and is an antidepressant.
- Phenylbutazone was meant to be an additive used to dissolve the anti-inflammatory aminophenazone. The substance turned out to be an anti-inflammatory agent itself as did its metabolite oxyphenbutazone.
- An attempt to isolate the causative agent of bipolar disorder from the urine of patients afforded only uric acid. Because uric acid is poorly soluble, lithium urate was tested. This led to the discovery of the antidepressant effect of lithium salts.
- Clonidine was meant to be a local treatment for the runny nose that accompanies the common cold. Instead of the expected effect, a profound hypotonic effect was surprisingly found. Despite intensive structural variations, none of clonidine's analogues have surpassed its potency.
- Levamisole was developed as a broad-spectrum anthelmintic (antiworm agent). Instead, an immunomodulatory effect was accidently found that now stands in the therapeutic foreground.
- Praziquantel was originally meant to be an antidepressant. Because of its high polarity, it cannot cross the blood–brain barrier. An outstanding suitability for the treatment of the tropical disease schistosomiasis (or bilharziosis) was found through broad biological testing.
- A chemist at Searle who was working on dipeptides licked his fingers while flipping through the pages of a book. The sweet taste that he noticed turned out to be caused by the artificial sweetener aspartame. Saccharine was also found in a very similar way. In the case of cyclamate, a smoker noticed a sweet taste to his cigarettes.
- Even today when one would think that rational concepts dominate drug research, the lucky accident still helps to make "blockbusters." In the pursuit of a phosphodiesterase inhibitor to hinder the degradation of cyclic guanosine monophosphate (cGMP), an improved treatment for angina pectoris was not found (Sect. 25.8). Instead it became conspicuous that the male subjects in the clinical trial did not want to give up the substance. After the side effect of a stronger penile erection was recognized, the side effect became the main effect. The compound sildenafil was marketed for the treatment of erectile dysfunction as Viagra® and developed into a billion dollar product.

2.9 Where Would We Be Without Serendipity?

In the English-speaking world, a word is in use that is difficult to translate into other languages: serendipity. This term, as an expression of a lucky accident, was coined by Sir Horace Walpole in 1754. It is derived from a Persian fairytale in which three princes of Serendip (earlier Ceylon, today Sri Lanka) have accidental and unexpected luck and make interesting discoveries entirely analogously to the many examples in this chapter. Serendipity has played an exceedingly important role in general in science, and especially in drug research. How would our modern medicine supply look without all of these lucky accidents? By no means should an arbitrary approach be taken, and an accidental discovery be counted upon. To the contrary, chemists and pharmacologists have always developed concrete ideas as to how and why particular structural variations on a lead compound should be pursued. Some of these hypotheses were correct, and others were false. One thing that they always had in common that helped the researchers was that when a hypothesis failed, or an unexpected result was found, they recognized the potential consequences of the result, drew the correct conclusions, and did the right things. The following chapters will show numerous examples of successful targeted drug design in cases in which the correct working hypothesis was realized. The search for a new active substance is, however, not a process that can be pushed through by a purely technically oriented management. As a general rule, short-term planning and bureaucratic control have only negative consequences. On the other hand, the search for new medicines requires a concerted effort from many different groups of specialists who must work together in a suitable organizational structure. The subsequent preclinical and clinical development of a newly found active substance is an extremely expensive and time-consuming process that must be carefully planned, carried out, and controlled. For this, other instruments are necessary than are used for drug discovery.

2

2.10 Synopsis

- The history of early drug research is full of lucky accidents. Many active principles of substances were discovered by serendipity, but mostly success can be attributed to an outstanding researcher with a "prepared mind" who observed important effects.
- Dyes and pharmaceuticals, both developed in the early stages of the up-coming chemical industry, especially stimulated each other in very fruitful synergies.
- The discovery by Alexander Fleming of the first antibiotic principle, the penicillins, as a defense mechanism of a fungus against bacteria is one of the most famous examples of a serendipitous discovery.
- The partial synthesis of ergoline alkaloids led to the discovery of the hallucinogenic effects of LSD. In those days, researchers frequently conducted self-experiments to first test active principle in humans.
- Unexpected synthetic products, surprising structural rearrangements, and initially false working hypotheses produced new, pharmacologically interesting substances with surprising or outstanding qualities.
- Even today, where rational concepts and the understanding of mode of action dominates drug research, the lucky accident can still help to make "blockbusters" as proven recently by the example of sildenafil (Viagra®).

Bibliography and Further Reading

General Literature

A. Burger, A Guide to the Chemical Basis of Drug Design, John Wiley & Sons, New York (1983)

G. de Stevens, Serendipity and Structured Research in Drug Discovery, Fortschr. Arzneimittelforsch., **30**, 189–203 (1986)

E. Verg, Meilensteine. 125 Jahre Bayer, 1863–1988, Bayer AG (1988)

R. M. Roberts, Serendipity. Accidental Discoveries in Science, John Wiley & Sons, New York (1989)

W. Sneader, Chronology of Drug Introductions, in: Comprehensive Medicinal Chemistry, C. Hansch, P. G. Sammes and J. B. Taylor, Eds., Vol. 1, P. D. Kennewell, Ed., Pergamon Press, Oxford, pp. 7–80 (1990)

R. M. Restak, Receptors, Bantam Books, New York (1994)

H. Kubinyi, Chance Favors the Prepared Mind. From Serendipity to Rational Drug Design, J. Receptor & Signal Transduction Research, **19**, 15–39 (1999)

T. A. Ban, The Role of Serendipity in Drug Discovery, Dialogues in Clinical Neuroscience, **8**, 335–344 (2006)

E. Hargrave-Thomas, B. Yu, J. Reynisson, Serendipity in anticancer drug discovery, World J. Clin. Oncol., **3**, 1–6 (2012)

Special Literature

A. Cahn and P. Hepp, Das Antifebrin, ein neues Fiebermittel, Centralblatt für Klinische Medizin, **7**, 561–564 (1886)

Z. von Vámossy, Ist Phenolphthalein ein unschädliches Mittel zum Kenntlichmachen von Tresterweinen? Chemiker-Zeitung, **24**, 679–680 (1900)

L. H. Sternbach, The Benzodiazepine Story, Fortschr. Arzneimittelforsch., **22**, 229–266 (1978)

A. Hofmann, LSD – mein Sorgenkind, dtv / Klett-Cotta (1993)

A. Stütz, Allylamine derivates—a new class of active substances in antifungal chemotherapy, Angew. Chem. Int. Ed. Engl., **26**, 320–328 (1987)

Classical Drug Research

Contents

G. Klebe, *Drug Design*, https://doi.org/10.1007/978-3-662-68998-1_3

The 100 years of pharmaceutical research from 1880 to 1980 were punctuated by trial and error, but also by elegant ideas and their translation into therapeutically valuable principles. Many lead structures were found by accident (see Chap. 2), while others came from traditional medicines or from biochemical concepts. In contrast to modern drug research, classical design was the result of rather limited knowledge of the pathophysiology and cellular and molecular etiology of disease, and it was restricted to animal testing. Nevertheless, this phase was extremely successful, especially in its last 50 years until about 1980. The targeted fight against infectious diseases and the successful treatment of many psychiatric and other important diseases can be attributed to this period of drug development. With this came a significant increase in quality of life and life expectancy. In the following sections, some selected examples are used to demonstrate different aspects of classical pharmaceutical research. They show how for known drugs, the field of application can expand over the years or how the therapy in an indication area can change through the discovery of new targets. The treatment of malaria illustrates that the development of resistance forces the constant search for new active substances with novel modes of action.

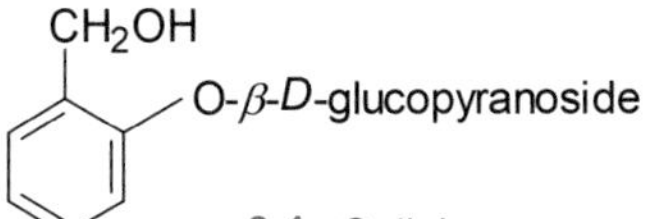

3.1 Salicin

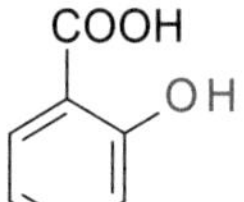

3.2 Salicylic acid

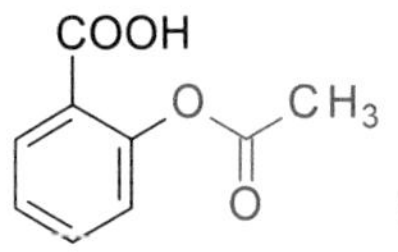

3.3 Acetylsalicylic acid

Fig. 3.1 Salicylic acid **3.2** is the oxidation and cleavage product of salicin **3.1**, which is isolated from willow bark. Acetylsalicylic acid (ASA) **3.3** is not simply a prodrug of salicylic acid, but rather a drug with its own mode of action

3.1 Aspirin: A Never-Ending Story

The history of acetylsalicylic acid (ASA, Aspirin®) reflects the progress of pharmaceutical research like no other example. This is especially true for the elucidation of the mode of action, and the newly found targeted therapies that resulted. Willow bark extracts have been used since antiquity for the treatment of inflammation. When Napoleon marched across Europe between 1806 and 1813, the bark was even used as a substitute for cinchona bark (Sect. 3.2). Salicin **3.1**, a glucoside of the *o*-hydroxybenzylalcohol saligenin, is responsible for the effect. Upon hydrolysis and oxidation, the actual active compound, salicylic acid **3.2** (Fig. 3.1), is formed.

In 1897, the then 29-year-old Bayer chemist Felix Hoffmann began a systematic search for derivatives of salicylic acid after a request from his father, who suffered from severe rheumatoid arthritis. High doses of salicylic acid caused unpleasant gastric irritation and vomiting. Hoffmann prepared simple derivatives of salicylic acid and was successful within the year. On October 10, 1897, he synthesized acetylsalicylic acid **3.3** (ASA, Fig. 3.1) for the first time in a pure form.

It was a lucky strike. Although ASA has a very short half-life in plasma, it is a highly analgesic, antipyretic, and anti-inflammatory agent. The clinical trial was carried out at the Diakonissenkrankenhaus in Halle an der Saale on 50 patients. On February 1, 1899, Bayer registered ASA as Aspirin® (A for acetyl and *Spiraea*, another plant that contains salicylic acid) as a trademark under the number 36,433. From then on, it was sold as 1 g of powder in envelopes, and shortly thereafter as tablets. Detractors alleged that it was only developed in tablet form so that Bayer could emboss their famous Bayer cross onto the tablets. Aspirin quickly gained a leading place in drug therapy. One-hundred years after its market introduction, 40,000 tonnes of ASA are produced and pressed into tablets every year, worldwide. At the end of 1994, the Bayer plant in Bitterfeld produced 400,000 Aspirin® tablets per hour, 3.5 billion per year. The importance that the trademark Aspirin had for Bayer became clear in 1994 when the company paid US$ 1 billion to take over the self-medication business from Sterling-Winthrop, which included the trademark rights for Aspirin, which had been lost in 1918.

The Spanish philosopher José Ortega y Gasset called the previous century the "Age of Aspirin." In his book *The Rising of the Masses*, he wrote:

> "The ordinary person lives today more easily, comfortably and safely than the most powerful of the past. Why should he care that he is not richer than others when the world is [richer] and roads, trains, hotels, telegraphs, personal safety, and Aspirin® are at his disposal."

Jaroslaw Hasek, Kurt Tucholsky, Giovanni Guareschi, Graham Greene, John Steinbeck, Agatha Christie, Truman Capote, Hans Helmut Kirst, and Edgar Wallace also wrote about Aspirin. The singer Enrico Caruso treated his headaches with only "German Aspirin," out of principle. Even Franz Kafka and Thomas Mann raved about its outstanding effects in their letters. In 1986 on an official visit to Germany, Queen Elizabeth II said that:

> "German successes span the entire breadth of human life. From philosophy, music and literature, to the discovery of X-rays and the mass production of Aspirin®."

The compliment was wonderful, but one must also consider that all of these scientific discoveries are slightly more than 100 years old! ASA was considered to be a prodrug of salicylic acid and a drug of unknown mode of action until John Robert Vane (Nobel Prize 1982) and Sergio H. Ferreira discovered in 1971 that salicylic acid and other nonsteroidal anti-inflammatory drugs inhibit prostaglandin G/H synthase (cyclooxygenase, COX, Sect. 27.9). COX, a ubiquitously present, membrane-bound enzyme transforms arachidonic acid **3.4** over a cyclic endoperoxide into PGH_2 **3.5**, which in turn is transformed into prostacyclin **3.6**, thromboxane A_2 **3.7**, and other prostaglandins. Large quantities of prostaglandins are produced in inflamed tissue; thus, inhibition of cyclooxygenase mitigates the cause of the process itself (◘ Fig. 3.2).

ASA is in fact a metabolic precursor of salicylic acid. In contrast to other anti-inflammatory drugs, including salicylic acid, however, it has an astonishing mode of action (Sect. 27.9). It has been known for some time that ASA selectively acetylates the hydroxyl group of the amino acid serine 530 of cyclooxygenase. In 1995, the three-dimensional complex structure of a bromine analogue was elucidated for the first time (◘ Fig. 27.41). This emphasizes that ASA, analogously to other COX inhibitors, docks near the arachidonic acid binding site. Therefore, despite its relatively weak binding, ASA is in an outstanding position to acetylate this serine. Serine 530 is not involved in the catalytic mechanism, but the additional volume of the acetyl group impedes arachidonic acid's entrance to the binding site and therefore the synthesis of the prostaglandin precursors. A COX mutant that carries an alanine instead of a serine at position 530 is enzymatically fully active and is inhibited by every other anti-inflammatory compound. However, as expected, ASA only weakly inhibits this mutant.

Stimulation for the continued research on nonsteroidal anti-inflammatory drugs was generated by the discovery in 1991 of a second cyclooxygenase, COX-2. All anti-inflammatory drugs until then were unselective, or they exerted their effect overwhelmingly over COX-1 and only slightly over COX-2. The most important side effect of ASA and other anti-inflammatory drugs is the gastrointestinal damage that can occur at high doses. This results from the inhibition of the COX-1-dependent synthesis of prostacyclin **3.6**, which protects the gastric mucosa. In contrast to the ubiquitously occurring COX-1, COX-2 is responsible for the rapid synthesis of prostaglandins in inflamed tissue. It has been possible to bring many drugs to the market that are more than 1000-fold more selective for COX-2 than COX-1 (Sect. 27.9).

But do not worry, Aspirin® will live forever. Its success is growing in another market. Even at low doses ASA inhibits the synthesis of thromboxane A_2 **3.7**, which initiates the coagulation of platelets (thrombocytes). Because of the irreversible inhibition of cyclooxygenase by ASA and the inability of platelets, which lack a nucleus, to resynthesize their enzymes, even a single contact with the substance is enough to suppress synthesis for about a week, the lifetime of a thrombocyte. In tissues other than thrombocytes, the enzyme is continuously resynthesized. Therefore, the physiological adversary to thromboxane, the aggregation-inhibiting prostacyclin that is produced in the walls of the vasculature, can be replenished (◘ Fig. 3.2).

In terms of the condition of increased clotting tendency, ASA thus shifts the biosynthesis from the "bad" thromboxane to the "good" prostacyclin. This effect suggests a therapeutic application of ASA in cases of increased tendency to clot, e.g., before and after myocardial infarction and stroke. Understanding the mechanism of the antithrombotic effect, it has been suggested that the doses used for therapy should be reduced by a factor of 10. This simultaneously reduces the risk of gastric and intestinal bleeding as possible side effects. These considerations led to the recommendation made in the early 2000s, which has in the meantime become highly con-

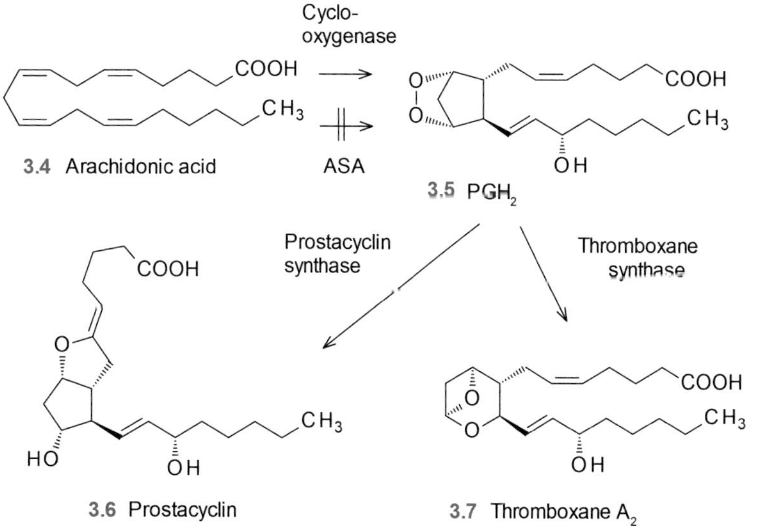

◘ **Fig. 3.2** Arachidonic acid **3.4** undergoes an oxidative cyclization and a peroxidase reaction in the prostaglandin biosynthesis to give the primary product PGH_2 **3.5**. Finally, prostacyclin synthase transforms PGH_2 into prostacyclin **3.6**, which protects the gastric mucosa, dilates blood vessels, and inhibits platelet (thrombocyte) aggregation. The platelet thromboxane synthase transforms PGH_2 into thromboxane A_2, which promotes aggregation. ASA irreversibly inhibits cyclooxygenase. By using low ASA doses, thromboxane A_2 synthesis in the platelets is more strongly inhibited than the production of prostacyclin in the vascular walls troversial, to take ASA as a preventive treatment during long-distance flights. Lack of mobility, cramped seating, coupled with dry air and reduced pressure in the cabin, leads to dehydration. This causes the blood to "thicken" and results in changes in its flow velocity. This "economy class syndrome" could lead to so-called "jet legs" and could increase the risk of thromboembolism and venous thrombosis. ASA should actually only have a preventive effect on arterial thromboses; therefore, this therapy recommendation seems rather controversial. Nevertheless, studies keep coming up that try to establish a benefit in venous thrombosis. A comprehensive ASPREE study involving 19,114 healthy seniors without cardiovascular risk factors were randomized to receive either in the first half 100 mg ASA daily or in the second half a placebo. The study found no benefit for primary prevention. This makes the success of such prevention highly questionable. By contrast, the use of ASA before surgical interventions is not recommended. Surgeons do not want their patients to have an increased bleeding tendency due to reduced clotting ability during surgery.

But other observations do give ASA the potential of a "preventive drug." A 6-year observation of 600,000 volunteers is worth an entry in the *Guinness Book of Records*. According to their evaluation, ASA appears to reduce the risk of fatal colon cancer by 40%. However, it must be taken into account in such studies that because of possible stomach and intestinal bleeding as side effects of ASA, colonoscopy was probably carried out more intensively in the treated group than in the untreated reference group. It is quite conceivable that colon cancer was, thus, detected more frequently at an early stage that could still be subjected to surgical treatment. The mode of action of the tumor-protective effect of ASA has still not really been clarified. Both apoptotic effects and the influence of ASA on the metastasis of tumors have been described. The risk reduction of tumors in different tissues when taking ASA appears to be different. In the meantime, studies have even become known which, for example, relate the cancer-preventive profile of ASA to the methylation status of promoters of certain breast cancer genes. This should be seen as an indication that even the epigenetic status (Sect. 12.14) of individual patients can be decisive for the effect.

A chewable aspirin tablet has been available since 1992. Here ASA is buffered with calcium carbonate, absorption is faster, and side effects are reduced. But far be it from anyone who thinks this is as far as it goes. In 2014, Bayer launched a new formulation with sodium carbonate and highly dispersive silicon dioxide as excipients. In the acidic environment of the stomach, the active ingredient quickly dissolves into very small particles. This so-called MicroActive technology allows even faster absorption with accelerated onset of action. ASA has had an unbelievable career, particularly if one considers that it would never have had a chance to be approved under modern criteria. Its short plasma half-life, the irreversible protein inhibition, and the high dosage would have met today's exclusion criteria. A definitive endpoint in this hypothetical development using the contemporary criteria would be the observation of teratogenicity in rats. A pathological result in toxicity studies with this animal model would definitely lead to discontinuation, because who would dare to wager that a teratogenic effect occurs in rodents, but not in humans. Aspirin®—really a never-ending story!

3.2 Malaria: Success and Failure

The treatment of malaria begins with the discovery of cinchona, around which there are numerous legends. The nicest and most frequently cited version is that of the fever-stricken Countess Cinchon, the wife of the Spanish viceroy in Lima, Peru, who was healed by the physician Juan de Vega in 1638. On the advice of the town magistrate of Loja, Quinquina the "bark of the barks" (therefore, the confusing name "cinchona bark") was brought in from 800 km away. The Countess was allegedly healed and from then on distributed the powder herself. In the older works, the cinchona bark was also called "Countess powder" or "Jesuit powder." Perhaps it was also true that the Indians, who were forced into compulsory service in the silver mines by their Christian conquerors, chewed the bark to fight off shivering in the cold. The clever Jesuits took note of these observations and thought that chewing the bark would also help with the shivering that comes from a malarial fever episode. Cinchona then came back to Europe with the Jesuits.

Malaria, the remittent fever, is a widespread tropical and subtropical disease. Because it is transmitted by the anopheles mosquito, it occurs particularly in wetlands. Even the city Buenos Aires (Span. "good airs") was badly hit by malaria (Ital. *mala aria* = "bad airs"). Alexander the Great, the Gothic King Alarich, and the German Emperors Otto II and Heinrich IV died of it. Even Albrecht Dürer (1471–1528) apparently suffered from malaria. He sent his private physician a drawing of himself in which he was wearing only a loincloth. His right hand is over his spleen with the additional text that *do der gelb Fleck ist vnd mit dem Finger drawff dewt, do ist mir we* (there where the yellow spot is and where the finger points, is where it hurts). In Europe malaria was still widespread until the middle of the twentieth century. In northern Germany, the last epidemics were in the years 1896, 1918, and 1926.

The miasma, emissions from the ground, swamps, and corpses, were long seen as the source of malaria and other epidemics. The Roman author Marcus Terentius Varrus (116–27 BCE) suspected back then that small invisible organisms might be responsible. Towards the end of the nineteenth century, the anopheles mosquito was

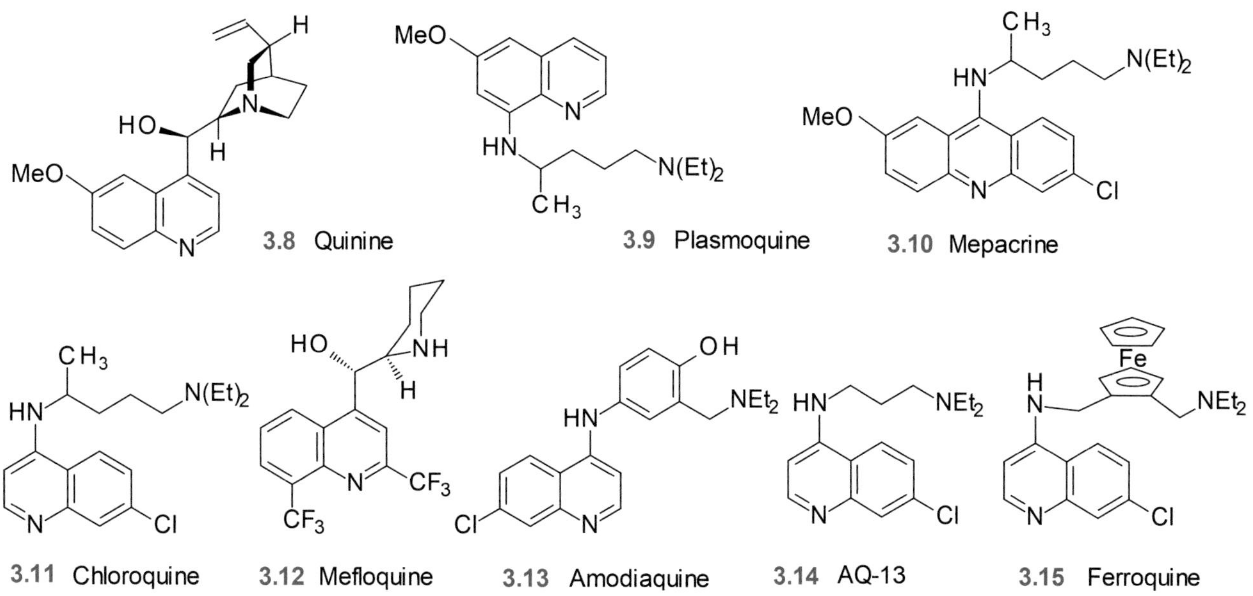

Fig. 3.3 Simple synthetic analogues with antimalarial effects were derived from quinine **3.8**. Plasmoquine **3.9** still contains the methoxyquinoline ring of quinine, but it is in a different position. The later-developed analogues mepacrine **3.10** and chloroquine **3.11** show strong similarity to quinine. The newer derivatives mefloquine **3.12** and amodiaquine **3.13** are also structurally closely related to quinine. With AQ-13 **3.14** and ferroquine **3.15**, a ferrocene sandwich complex, two new substances from the chloroquine family are available

identified as the vector, and a plasmodium was recognized as the cause of malaria.

Around 1930, there were about 700 million people infected, and in 2003 the number was estimated to be 300–500 million. Up to 1.2 million people die every year, mostly children under the age of 5, and many others suffer permanent injury. Psychiatric changes also result. The term "spleen" for eccentricity originally came from the enlarged spleen that malaria causes.

The active substance in the cinchona bark, the alkaloid quinine **3.8** (Fig. 3.3), was isolated in 1820. Aside from the positive therapeutic effects, it also had considerable side effects. Nonetheless, up until a few years ago it was the most important antimalarial, particularly for the parenteral treatment of severe malaria. The first synthetic alternative, plasmoquine **3.9**, became available in 1927, but it is seldom used due to its side effects. The later-developed, more potent analogues **3.10**–**3.12** show a clear structural relationship to the lead structure quinine (Fig. 3.3). It was only through the protection from malaria that the exploitation of the colonies was possible.

The World Health Organization (WHO) initiated a global malaria eradication program in 1955 mainly through the use of the insecticide dichlorodiphenyltrichloroethane **3.16** (DDT, Fig. 3.4). The success was overwhelming; the number of cases and fatalities was reduced to practically zero (Table 3.1). In 1953, it was estimated that five million lives had been saved since 1942. In India alone, the number of cases went from 75 million to 750,000, and the number of annual fatalities was reduced to 1500. DDT has saved more lives than all antimalarial drugs put together! The acute toxicity of DDT is actually not a problem for mammals and humans. Unfortunately, it turned out that DDT decomposes extremely slowly in the environment, and it enriches as it moves its way up the food chain, especially in birds and fish. It also accumulates in human fat and in breast milk. The chronic toxicity comes from long-term retention of one year or more, which is a serious problem.

The moving book *Silent Spring* by Rachel Carson was published in 1962. Despite warnings from experts, DDT spraying for mosquitoes was stopped in Sri Lanka in 1963, and the number of malaria cases raced to 2.4 million by 1968/1969. By then it was too late to use DDT again because the mosquitoes had become resistant, and this was certainly also partially due to the residual DDT that remained in the environment in the intervening years.

Further investigations showed that a DDT metabolite, dichlorodiphenyldichloroethylene **3.17** (DDE, Fig. 3.4) has surprisingly strong antiandrogenous effects, that is, it blocks the effects of male hormones. Therefore, DDE is responsible for the DDT-dependent reproductive and developmental disorders that are seen in some species, perhaps also in humans. It is remarkable that the effect of this metabolite was only discovered 50 years after DDT was introduced.

Not only the mosquitoes became resistant to DDT, the parasite also became resistant to the drugs. For this reason, the history of the chemotherapeutic developments for malaria has been a rollercoaster ride of new promising compounds, and the more or less quick development and distribution of resistant parasites.

3

Chloroquine **3.11** was prepared in 1934 in the Bayer laboratories, but was judged to be "too toxic." It was "rediscovered" by the Americans and deployed as a malaria therapeutic *par excellence*. Efficacious, well tolerated, and above all else inexpensive to produce, it, along with the above-described mosquito extermination with DDT and landscaping measures, brought us within reach of a victory over malaria. It is regarded as an inhibitor of hemoglobin utilization. The malaria pathogen degrades hemoglobin in infected red blood cells to obtain proteins for its metabolism. Chloroquine inhibits the crystallization of hemozoin, a degradation product of hemoglobin. If hemozoin can no longer be crystallized, the parasite dies. However, as early as in the 1960s, in various places in Southeast Asia, Oceania and South America, independently and almost simultaneously resistant parasites emerged. They possessed a mutated transport protein in the membrane of their gastriole that recognizes chloroquine as a substrate. By using this protein, they were able to expel chloroquine from its target. In the meantime, resistant parasites have spread throughout almost the entire geographic range of malaria. Chloroquine lost its once phenomenal status for the treatment of malaria tropica.

Since then, a malaria therapeutic with similar qualities as chloroquine has been sought by researchers, until now, however, without success. The structurally related amodiaquine **3.13** (◘ Fig. 3.3) is in fact effective against weakly chloroquine-resistant strains, but it is largely ineffective against highly resistant strains (especially in Southeast Asia). Moreover, upon long-term use as a prophylaxis, it carries the risk of irreversible liver damage or life-threatening agranulocytosis. Research has produced two new substances from the chloroquine family, AQ-13 **3.14** and ferroquine **3.15** (◘ Fig. 3.3), an exotic-looking iron-sandwich complex.

In the short term, it appeared that the antifolate combination of sulfadoxine/pyrimethamine **3.18/3.19** (Fansidar®) could replace chloroquine (◘ Fig. 3.5), but the first resistance occurred much faster than with chloroquine. Starting from the point of origin in Southeast Asia, the resistance has spread throughout the world.

The wars of the last century have also promoted the search for new antimalarial drugs. Tremendous effort was made at the Walter Reed Army Institute of Research in the USA. Over the course of 40 years, and particularly during World War II and the Vietnam War, more than 250,000 substances were tested for an antimalarial effect. Considering the exerted effort, success was modest: the two aryl amino alcohols halofantrine **3.20** and mefloquine **3.12**, and the 8-aminoquinoline tafenoquine **3.21**, which was approved in the US in 2018. After its introduction, halofantrine was withdrawn from the market because it caused lethal arrhythmias (Sect. 30.3). In Southeast Asia, the resistance to mefloquine developed so quickly that it can only be used in combination with artesunate **3.22**. Because mefloquine has been used sparingly due to its price, most of the parasite strains are still sensitive to it. For this reason, today mefloquine is one of the most important malaria prophylactics for Western tourists. Artesunate is a partial-synthetic derivative of dihydroartemisinin **3.24**, which is produced by reducing artemisinin. It is isolated from annual mugwort (*Artemisia annua*). In the early 1970s, the Chinese scientist Tu Youyou extracted the compound from the plant and tested its effectiveness against malaria. In 2015, Youyou was awarded the Nobel Prize for Medicine for this achievement. The mechanism of action of arte-

◘ **Table 3.1** Number of malaria cases in different countries before and after the introduction of DDT **3.16** (◘ Fig. 3.5). The numbers in parentheses are the years. (From T. H. Jukes (1974) Naturwiss. **61**, 6–16)

Country	Cases of malaria (year)	
	Before DDT	After DDT
Italy	411,602 (1946)	37 (1969)
Spain	19,644 (1950)	28 (1969)[a]
Yugoslavia	169,545 (1937)	15 (1969)[a]
Bulgaria	144,631 (1946)	10 (1969)[a]
Romania	338,198 (1948)	4 (1969)[a]
Turkey	1,188,969 (1950)	2173 (1969)
India	~ 75 million per year	~ 750,000 (1969)
Sri Lanka	2.8 million (1946)	110 (1961)
		31 (1962)
		17 (1963)
		2.5 million (1968/1969)[b]
Taiwan	> 1 million (1945)	9 (1969)
Venezuela	817,115 (1943)	800 (1958)
Mauritius	46,395 (1948)	17 (1969)

[a] Imported cases
[b] After DDT spraying was discontinued in 1963

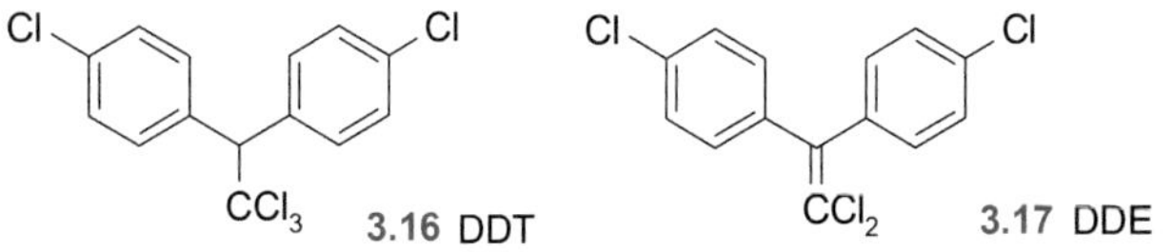

3.16 DDT 3.17 DDE

◘ **Fig. 3.4** The insecticide *p,p'*-dichlorodiphenyltrichloroethane **3.16** (DDT) saved more human life than all of antimalarials put together. The latest investigations show though that the antiandrogenic effects of the main metabolite *p,p'*-dichlorodiphenyldichloroethylene **3.17** (DDE) is possibly the main culprit responsible for reproductive disorders found in animals, including perhaps humans

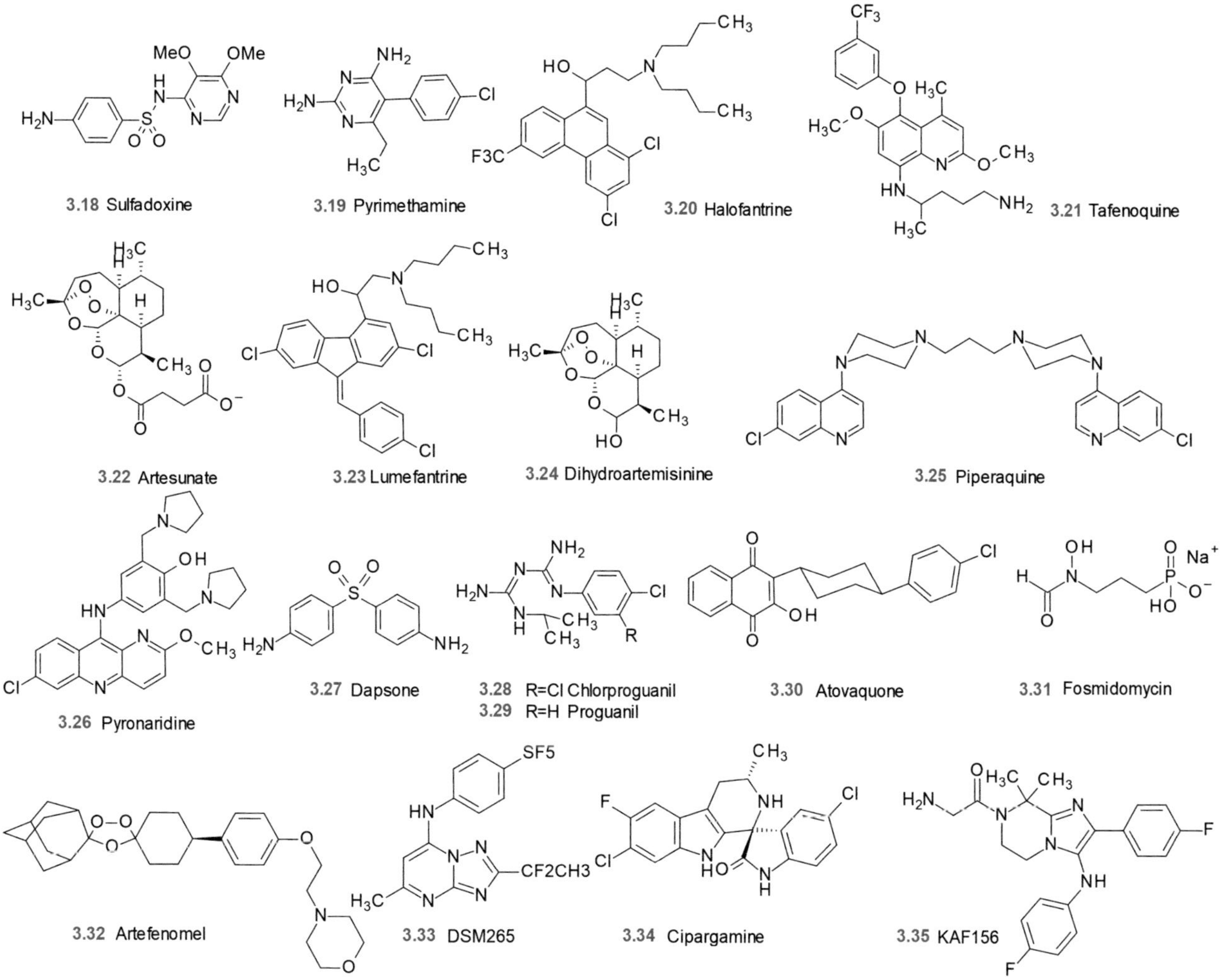

Fig. 3.5 Recent research on antimalarial drugs has described numerous products, often used in combination. Fansidar, a combination of sulfadoxine **3.18** and pyrimethamine **3.19**, was initially considered the drug of choice. However, the development of resistance has rendered even this promising drug useless. At present, hopes are pinned on the artemisinin derivatives **3.22** and **3.24**, and another member of this family, artefenomel **3.32**, is available. Fosmidomycin **3.31** inhibits the mevalonate-independent biosynthetic pathway for isoprenoid synthesis, and DSM265 **3.33** inhibits dihydroorotate dehydrogenase, an enzyme essential for pyrimidine neosynthesis in malaria pathogens. Cipargamine **3.34** blocks a parasite-specific membrane pump, and the mechanism of action of KAF156 **3.35** remains to be elucidated

misinins is still not fully understood. What is unusual is their endoperoxide structure, which is essential for their action. Cleavage of the endoperoxide bridge generates reactive oxygen species (ROS) that can induce apoptosis. Oxidative stress increases in infected red blood cells. As a result, the level of unfolded proteins increases and hemoglobin degradation is impaired. Activation of caspase-like enzymes and DNA fragmentation indicate the initiation of apoptotic cell death. Thus, artemisinins inhibit the nutritional process of the parasites.

The artemisinins are currently the most effective antimalarial drugs for rapid control of the parasites. Unfortunately, resistance to these compounds has begun to emerge. Today, artemisinin-based combination therapy is the WHO recommendation. Combinations are made with whatever is available, including substances to which massive resistance has already been observed. One combination is made with the arylaminoalcohol lumefantrine **3.23**, which was developed in China and is still very effective. Dihydroartemisinin/piperaquine (Eurartesim®) **3.24/3.25** and artesunate/pyronaridine (Pyramax®) **3.22/3.26** have also been approved as combination preparations (Fig. 3.5). Both artemisinin combination partners were developed in China in the 1960s and 1980s, respectively. Although pyronaridine has an aza-acridine rather than a quinoline parent scaffold, they belong to the same class of drugs as chloroquine. Resistance to these two drugs is already widespread in Southeast Asia. The combination of dapsone/chloroproguanil (Lap-Dap®) **3.27/3.28** belongs to a long-established class of antifolates. Again, most Southeast Asian strains are already resistant. It was withdrawn from the market in 2008 due to toxicity concerns. Real novelties in the mechanism of action are rare. In 1997, the very expensive combination atovaquone/proguanil **3.30**/**3.29** (Malarone®) was introduced, which synergistically inhibits the mitochondrial respiratory chain. A promising candidate was the antibiotic fosmidomycin **3.31** which inhibits a parasite-specific, mevalonate-independent pathway for isoprenoid synthesis. Increased efforts are needed to find new compounds. Artefenomel **3.32** is a new analogue of the artemisinin family in clinical development. It is characterized by a significantly longer half-life and has the potential to achieve therapeutic success with a single dose. Inhibitors of dihydroorotate dehydrogenase have also been sought for some time. The enzyme is essential for the malaria pathogen in the pyrimidine resynthesis of DNA building blocks. With DSM265 **3.33**, a hopeful candidate may be in clinical trials. Cipargamine **3.34**, a spiroindolone, can inhibit a parasite's own membrane pump, thereby permanently disrupting the osmotic balance of the pathogens. Furthermore, KAF156 **3.35**, a new imidazole piperazine, is in clinical trials for which the mechanism of action still needs to be clarified. Thus, several substances are currently being developed as promising candidates with alternative action profiles. Only this way can we be prepared for the time when resistance to artemisinins increasingly manifests itself.

As an alternative to drug therapy, vaccine developments have been advanced for many years. Although plasmodia are single-cell organisms, they prove to be extremely adaptable, also due to the change of host. Equipped with more than 5400 genes, they repeatedly manage to change their surface structures so strongly through variation in protein expression that our immune system must constantly re-adjust to seemingly changed parasites. Nevertheless, the first field studies with the vaccine Mosquirix®, which contains an artificially produced fusion protein of certain surface areas of the parasite as an antigen, started in 2019. In humans, this is intended to trigger antibody production (Sect. 32.3) in order to eliminate the parasite before it spreads in the red blood cells in the event of disease. A further development of Mosquirix® with optimized antigen and better adjuvant is the R21/Matrix-M malaria vaccine. It provides about 80% protection. Approval in 2023 for children aged 5 to 36 months in Ghana is based on a Phase III study involving 4800 children. Work is also underway on living vaccines that activate $CD8^+$ T cells to eliminate infected cells in the human body (Sect. 31.7). In addition, a number of mRNA-based malaria vaccine candidates are currently under development at BioNTech. It remains to be seen whether this will achieve the grand goal of one day eradicating malaria. However, it should not be forgotten that exposure prophylaxis with mosquito nets or repellents still plays an important role in the control of malaria.

3.3 Morphine Analogues: A Molecule Cut to Pieces

Research on opiates has taught us how complex natural products can be systematically simplified, and structurally abbreviated analogues can be prepared that have the identical effect, but sometimes with even better specificity. It has also shown that there is sometimes no obvious solution for a specific problem. Separation of the analgesic and addictive qualities could not, or only inadequately be achieved.

The narcotic, analgesic, and euphoric effects of opium, which is isolated from poppies, have been known for at least 5000 years. Opium was used for operations, but is also a commonly abused drug. The importance of its abuse in the cultural history of humanity is illustrated, among other places, in the "Opium Wars" of the nineteenth century. In 1840, the Chinese wanted to stop the English from importing opium and burned 20,000 cases of it; this led to a 2-year-long war between the two countries.

In 1804/1805, the pharmacy assistant Friedrich Wilhelm Adam Sertürner of the Hof-Apotheke in Paderborn, Germany, isolated the compound with the sleep-inducing principle. He named it morpheum (later morphine) after *Morpheus*, the Greek god of dreams and son of *Hypnos*. Morphine addiction took on a whole new dimension after 1853 with the invention of the hypodermic needle and syringe by Charles G. Pravaz and Alexander Wood. As a result, morphine and heroin addiction spread widely, and in the history of humanity it is one of many examples of the misuse of a beneficial discovery.

Morphine **3.36** (◘ Fig. 3.6) is one of the few examples of a natural product that is still used today in its original form. It belongs to the most potent known analgesics. If it is administered according to the correct dose and schedule, the danger of addiction is low. The addictive potential is often overestimated by physicians such that patients with severe pain are often inadequately treated with opiates. Morphine is also a prime example of the success of systematic structural variation in the direction of more-easily manufactured, simpler analogues as well as more selective activity. The first modified products were simple derivatives such as the methyl ether codeine **3.37**, which is also found in poppies. Codeine is weaker than morphine, but it is bioavailable after oral administration. It has a pronounced antitussive effect and a low addictive potential. Unfortunately, the opposite is true for the potent, fast-acting diacetyl derivative heroin **3.38**. It has enormous addictive potential. Today it seems ironic that at the end of the nineteenth century Heinrich Dreser, a senior pharmacologist at Bayer, wanted to discontinue the development of Aspirin® because of a suspected cardiotoxicity in favor of developing heroin as a well-tolerated and potent cough medicine (sic!), at least until he realized the mistake. Arthur Eichengrün at Bayer tested ASA on himself without any side effects. He was finally able to convince Carl Duisberg to continue with ASA, which ultimately led to the drug's approval for therapy. Eichengrün must therefore be regarded as another father of ASA. Of all the morphine derivatives, codeine and heroin are the most widespread: codeine is in numerous combination preparations, and heroin is in the drug scene. Some *n*-alkyl derivatives of morphine and close analogues, for instance, naloxone **3.39**, are opiate antagonists, that is, they inhibit the effect of morphine (◘ Fig. 3.6).

The structural elucidation of morphine took more than 120 years, and its total synthesis, and ultimate structural proof, was completed in 1952 by Marshall Gates and Gilg Tschudi. Morphine contains five rings: an aromatic benzene ring, two unsaturated six-membered rings, the nitrogen-containing piperidine ring, and an oxygen-containing five-membered ring. Systematic structural modifications had the goal of simplifying the structure, for example, by opening one or more rings, or removing them altogether.

In 1939, the potent analogue pethidine **3.40** (◘ Fig. 3.7) was the first fully synthetic analgesic, though it was originally based on the spasmolytic atropine **3.41**. Despite this, it is recognized to be a morphine analogue. In levomethadone **3.42**, the piperidine ring of pethidine is opened, an oxygen atom from the ester group is removed, and another aromatic ring is added. There are thousands of other analogues, some of which have been introduced to therapy. Aside from the deconstruction of morphine, the construction of additional rings has surprisingly led to more potent analogues, for example, etorphine **3.43** (◘ Fig. 3.7).

For a long time, it was a complete mystery why our bodies would have extra receptors for the contents of poppy plants, so-called opiate receptors. The solution came with the discovery of the endogenous morphine-like peptides Met- and Leu-enkephalin (Sect. 10.2), which are the natural ligands for these receptors. The discovery stimulated an intensive search for orally active peptides or peptidomimetics devoid of addictive potential. The result of the work was more than sobering. Although orally active analogues were found, their addictive potential was identical to that of morphine and most morphine-derived analogues.

A few synthetic analogues have, in addition to agonistic activity, a weak antagonistic effect as well. The potential for these substances to be abused by addicts is less than with the classical morphine analogues. Combination preparations of agonists and antagonists are also available. With appropriate use, the analgesic effect of the agonist dominates because it is present in excess. If the medicine is injected intravenously, the more-strongly binding antagonist displaces the agonist, and the desired euphoric effect never sets in.

The work with regard to improved selectivity was also successful. Today cough medicines and antidiarrhea medicines, for example, loperamide **3.44** (◘ Fig. 3.8), are available that have no central morphine-like effects. This substance is able to pass through the blood–brain barrier but is immediately expelled by an active transporter. Upon inhibition of these transporters, for instance, when coupled with quinidine, loperamide also has classical opiate effects. Its structure unites elements of pethidine **3.40** and levomethadone **3.42**.

In this section, only a few representatives of the many thousand structural modifications of morphine can be discussed. The approach of Paul Janssen should not remain unmentioned though. He started with pethidine **3.40** with the goal of preparing a strong analgesic, but instead experienced unexpected success in another area. The result was the neuroleptic haloperidol **3.45** (◘ Fig. 3.8), a drug for the treatment of schizophrenia, the mode of action of which is mediated by an antagonistic effect at the dopamine D_2 receptor (Sect. 29.1).

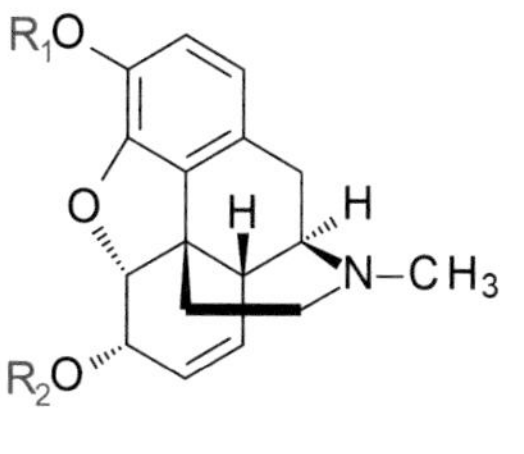

3.36 Morphine, $R_1 = R_2 = H$
3.37 Codeine, $R_1 = Me$, $R_2 = H$
3.38 Heroin, $R_1 = R_2 = Acetyl$

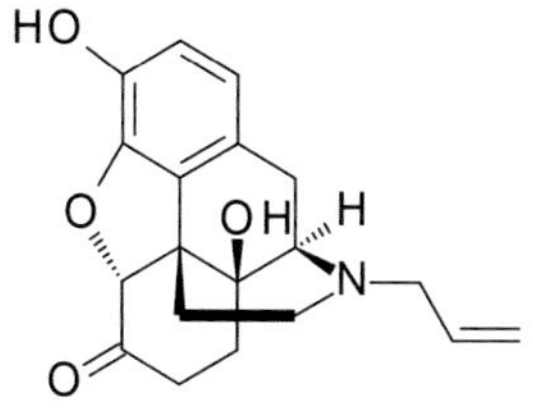

3.39 Naloxone

◘ **Fig. 3.6** Morphine **3.36** and codeine **3.37** served as lead structures for heroin **3.38**, which has better CNS bioavailability, and naloxone **3.39**, a morphine antagonist

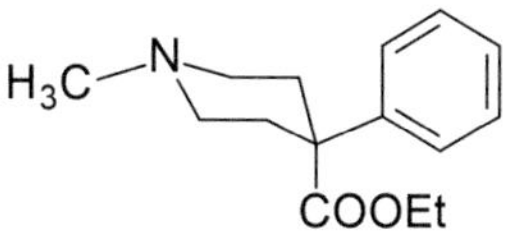

3.40 Pethidine

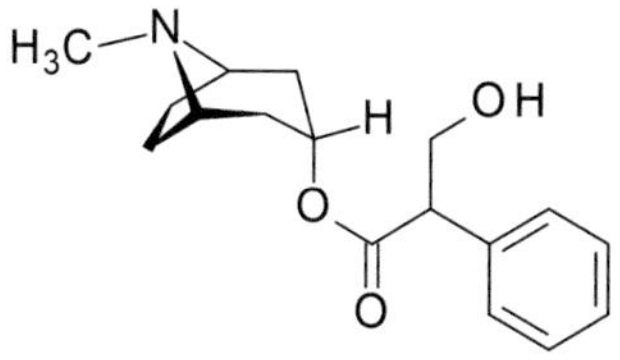

3.41 Atropine

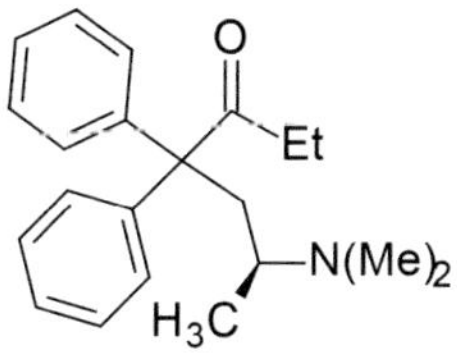

3.42 Levomethadone

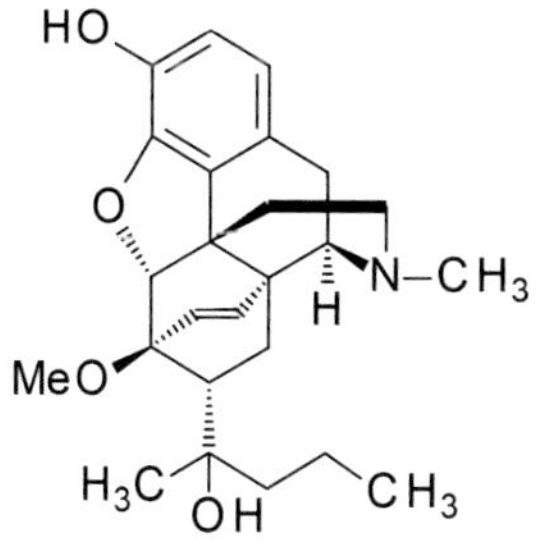

3.43 Etorphine

◘ **Fig. 3.7** The architecture of morphine was dissected in many ways. The strongly potent pethidine **3.40**, the first fully synthetic opiate analgesic, was discovered in the 1930s in a search for anticonvulsives by varying the structure of atropine **3.41**. It is recognizable, however, that pethidine retains the benzene ring of morphine as well as its piperidine ring. Levomethadone **3.42** is derived from pethidine. The addition of another ring led to substances with a potency that surpasses morphine by orders of magnitude. Etorphine **3.43** is 2000- to 10,000-times more potent than morphine in animals. Since 1963, it is used in African wildlife preserves to immobilize large animals such as elephants and rhinoceroses

3

Fig. 3.8 Structural derivatives of morphine and its analogues have led to selective antidiarrhea agents, loperamide **3.44**, for instance, as well as neuroleptics such as haloperidol **3.45**

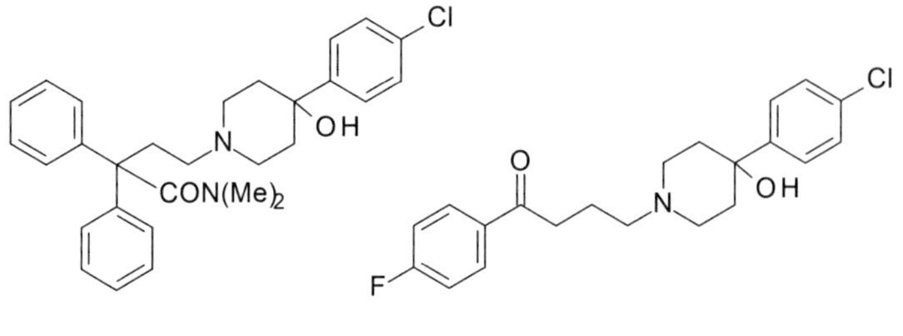

3.44 Loperamide **3.45** Haloperidol

3.4 Cocaine: Drug and Valuable Lead Structure

No other substance sparkles in so many ways as cocaine. It is at the pinnacle of all illegal drugs. In 2017 alone, investigative authorities seized 552 tonnes of cocaine worldwide. The smuggling methods are becoming rougher and rougher, for example, people are sent on the journey between continents as living transporters with swallowed cocaine-filled condoms. The amount that ultimately ends up for consumption in Europe is estimated at around 150 tonnes per year. **Cocaine** was also the chemical starting material for a wide palette of valuable local anesthetics and antiarrhythmics. We can thank the lead structure cocaine for local anesthesia, pain-free dentistry, and nerve block anesthesia for smaller surgical procedures. The translation of the quite positive central effects of cocaine onto analogues devoid of addictive potential is still in progress. The example of morphine leads one to fear that this goal might not be possible.

Coca leaves and cocaine **3.46** (■ Fig. 3.9) belong to the oldest known drugs. Chewing dried coca leaves has a long tradition in Peru and Bolivia. In 1744, Garcilaso de la Vega wrote that coca "*satisfies hunger, gives new energy to the tired and exhausted, and lets the unhappy forget their troubles.*" The Scottish author Robert Louis Stevenson (*Treasure Island*) wrote in his novella *The Strange Case of Dr. Jekyll and Mr. Hyde* about a personality split that a doctor undergoes under the influence of drugs. He wrote the first draft of this novella in only three days and nights, while under the influence of cocaine.

In 1863 the American chemist Angelo Mariani (1838–1914) patented a mixture of coca extract and wine as *Vin Mariani*. It made him a rich man. In 1886, the pharmacist John S. Pemberton developed a coca-containing stimulant and headache remedy that he named Coca-Cola. He sold the rights in 1891 to a colleague, A. G. Chandler, who founded the Coca-Cola Company one year later. Up until 1906, Coca-Cola indeed contained a small amount of cocaine, but today it only contains the harmless stimulant caffeine. Back at the turn of the twentieth century, cocaine was already fashionable, particularly in artistic circles. The Viennese psychiatrist Sigmund Freud (1856–1939) experimented with cocaine intensively and rather uncritically. He considered it to be a wonder drug, took it himself regularly, and recommended it generously for use in therapy, for the treatment of stomach aches, and for a depressed mood. Later, after massive criticism from his colleagues he turned away from it. In the Andes of South America, one of the important cultivation areas of the coca bush, coca sweets and chewing gum are sold in supermarkets alongside coca tea. They are used by the general population as a treatment for high altitude sickness.

Cocaine causes the release of dopamine from its transporter (see Sect. 22.7). Usually it is sniffed, occasionally it is intravenously injected, or it is mixed in drinks or taken orally. Sniffing delivers it quickly to the brain where it displaces dopamine from the binding site of the transporter and this causes increased dopamine release into the synaptic gap. The free base, which is made by mixing it with sodium bicarbonate (*crack*) is absorbed very quickly through the lungs by smoking it, and causes euphoria that is even distinctly stronger than when the salt (*coke*, *powder*, *snow*) is sniffed. Because cocaine does not bind for long, the transporter is quickly reloaded with dopamine. The same effect can be induced again after a little while. Other cocaine analogues that bind for longer do not allow the effect to be repeated for hours. Psychological dependence occurs very quickly, even after the first use in the case of crack cocaine. Physical withdrawal symptoms, as seen with heroin addicts, usually do not occur.

The credit for discovering the local anesthetic effect of cocaine does not go to Freud but rather a friend of his, the ophthalmologist Carl Koller (1857–1944). Freud had planned to investigate this effect but in 1884 he wanted to first visit a friend of his, Martha Bernays, in New York. Koller picked up on Freud's suggestion and carried out the decisive experiment on the eye in his absence. The synthetic benzoic acid esters and anilides that were initially used as **local anesthetics** were not derived from cocaine **3.46**, but rather from *p*-aminobenzoic acid esters; benzocaine **3.47** was already used in therapy in 1902. A structural relationship to cocaine is, however, easily seen in modern local anesthetics such as lidocaine **3.48** and mepivacaine **3.49** (■ Fig. 3.9 and Sect. 30.4).

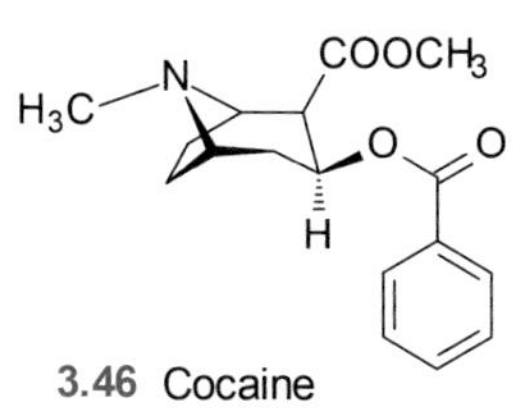

3.46 Cocaine

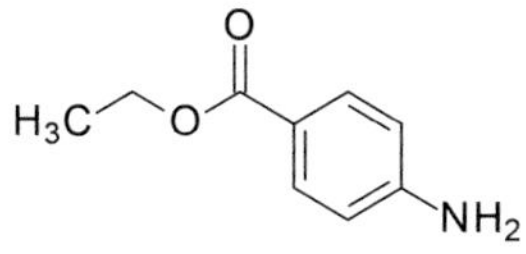

3.47 Benzocaine

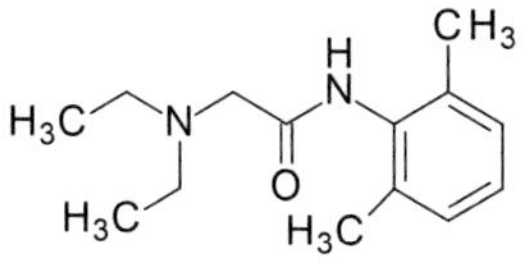

3.48 Lidocaine

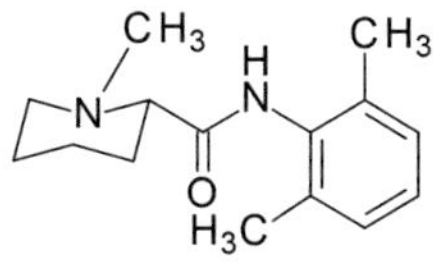

3.49 Mepivacaine

Fig. 3.9 The local anesthetic effect of cocaine **3.46** was recognized early on. The independently found lead structure benzocaine **3.47** and the basic moiety of cocaine were models for synthetic local anesthetics. The structural relationship is clearly recognizable in lidocaine **3.48**, which also acts as an antiarrhythmic, and in mepivacaine **3.49**

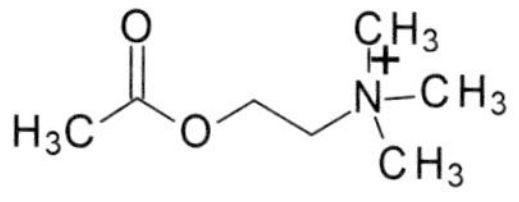

3.50 Acetylcholine

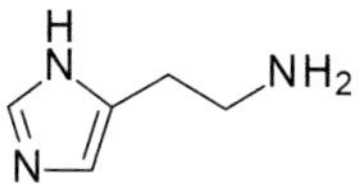

3.51 Histamine

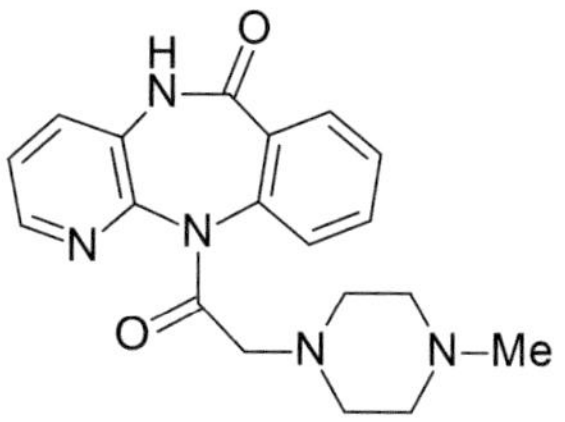

3.52 Pirenzepine

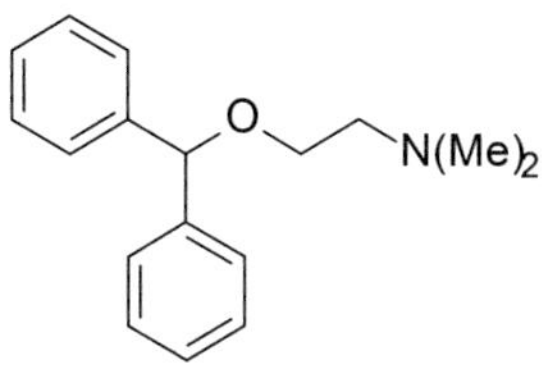

3.53 Diphenhydramine

Fig. 3.10 Acetylcholine **3.50** and histamine **3.51** stimulate acid production in the stomach. The acetylcholine receptor antagonist pirenzepine **3.52** was the first drug specifically for ulcer therapy. Classical H_1 antihistamines such as diphenhydramine **3.53** cannot antagonize histamine in the stomach

3.5 H_2 Antagonists: Ulcer Therapy Without Surgery

The history of the treatment of gastroduodenal ulcers is long and educational. Basic research clarified the important mechanisms without providing a new drug. The development of the therapy occurred in several phases. Again and again, new wasn't necessarily better. In the beginning, the treatment consisted of **antacids**, and later anticholinergics. In severe cases, only surgery helped. The H_2 antagonists made the breakthrough to purely pharmaceutical treatment. Now we are experiencing the victory lap of the proton pump inhibitors, which are used in different combinations with antibiotics. Perhaps in the future this will be augmented or even replaced by a vaccine.

Gastric and duodenal ulcers are usually chronic illnesses and are widespread in the general population. Any damage to the mucosal membrane of the stomach leads to damage to the underlying cells through proteolytic enzymes and gastric acid. Acetylcholine **3.50**, histamine **3.51**, and gastrin, a mixture of peptides with 17 (*little* gastrin) and 34 (*big* gastrin) amino acids, stimulate the production of acid (Fig. 3.10).

For decades the treatment of gastroduodenal ulcers was based on reducing the amount of acid, for instance, with sodium bicarbonate, calcium carbonate, basic magnesium salts, and aluminum oxide hydrate. Advanced ulcers had to be treated surgically. **Anticholinergics**, antagonists of the acetylcholine receptor should, in principle, have been suitable for ulcer treatment; however, unspecific antagonists are out of the question because of their severe side effects. It was not until pirenzepine **3.52** (Fig. 3.10), a selective so-called M_1 antagonist, was developed that this class could be used in therapy. Here the undesirable side effects of unspecific anticholinergics are only apparent at relative high doses.

The role of histamine in acid secretion was initially called into question because the classical antihistamines, later defined as H_1 antihistamines, did not reduce acid secretion. These substances, for instance, diphenhydramine **3.53** (Fig. 3.10), antagonize histamine in the intestines, lungs, and in allergic reactions. Today a wide palette of different histamine antagonists is available for the treatment of allergic rhinitis (hay fever). The most important side effect, particularly with older substances, is a more or less pronounced sedation.

Histamine-induced gastric acid secretion, the effect on the heart, and uterus contractions are not inhibited by diphenhydramine and other analogues. It was first suspected in 1948 that there might be two different **histamine receptors, H_1 and H_2**. The H_1-type is inhibited by diphenhydramine, but the H_2-type, which is responsible for the above-mentioned effects is not. Both belong to the family of G-protein-coupled receptors (Sect. 29.1). In the meantime, two additional members of the family, the H_3 and H_4 receptors, have been discovered. In 1964, James W. Black (1924–2010) at Smith Kline & French in England began to develop three models to test the inhibition of these other effects of the H_2-mediated effect of histamine. One was an *in vivo* model measuring gastric perfusion on anesthetized rats, and two were *in vitro* models evaluating the histamine-induced stimulation of a guinea pig heart and a rat uterus. James Black later received not only the Nobel Prize, but was also knighted by Queen Elizabeth II, two rather unusual honors for an industrial pharmaceutical researcher.

Despite all strategies that were available for the development of receptor antagonists, the search for an H_2

antagonist was to no avail for years. The American management in Philadelphia became impatient and wanted to stop the program. The first promising result came just in the nick of time. Because all lipophilic analogues were ineffective, the earlier more polar compounds that had already been investigated were reinvestigated. A compound that had already been synthesized in 1928 and determined to be ineffective, N_α-guanylhistamine **3.54** (◘ Fig. 3.11), now appeared to be a weak antagonist. The effect had been overlooked because **3.54** is actually a partial agonist and, therefore, shows a weak histamine-like effect. Within a few days the first lead structure, *S*-(2-imidazoyl-4-yl-ethyl)isothiourea **3.55**, with interesting activity was identified.

The extension of the side chains of both of these compounds delivered partial agonists, the antagonistic effects of which were too weak. It was only in 1972 after they abandoned the hypothesis that the basic nitrogen in the side chain was necessary for activity that they, after chain elongation and an *N*-methyl substitution of the thiourea, arrived at the first clinically useful H_2 antagonist burimamide **3.56**. Human trials confirmed the efficacy, but the bioavailability was poor. The next milestone was achieved with the development of metiamide **3.57** (◘ Fig. 3.11), which is 5- to 10-times more potent than burimamide and clinically demonstrated the desired ulcer-healing effect. In some patients, however, granulocytopenia occurred, which is a dangerous suppression of white blood cells and cannot be tolerated.

The medical need was great. It was not foreseeable whether the observed effect was a result of H_2 antagonism. We have the company to thank for taking on the risk of further research. The sulfur atom of the thiourea was suspect. An isosteric exchange for an oxygen atom delivered a less-potent urea analogue. Exchange for an =NH group led back to guanidine, which was strongly basic, but a potent antagonist nonetheless. Substitution of the imino group for an NO_2 or a CN group led to less-basic analogues, the antagonistic potency of which was comparable to metiamide. The somewhat more active of the two analogues, **cimetidine 3.58** (◘ Fig. 3.11), was clinically tested. In November 1976 and in August 1977, it was introduced in England and the USA, respectively. By 1979, it was available in over 100 countries. Shortly thereafter in 1983, cimetidine (Tagamet®) became the most-prescribed drug in many countries, and its sales reached about US$1 billion.

Such a successful drug makes other companies restless. There are many cases in the history of pharmaceutical research in which a major new concept was adapted by developments in other companies. Other examples of this are the structurally entirely different calcium channel blockers verapamil and nifedipine (Sects. 2.6 and 30.4) and the angiotensin-converting enzyme inhibitors captopril and enalapril (Sect. 25.4).

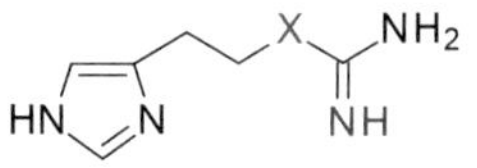

3.54 X = -NH-

3.55 X = -S-

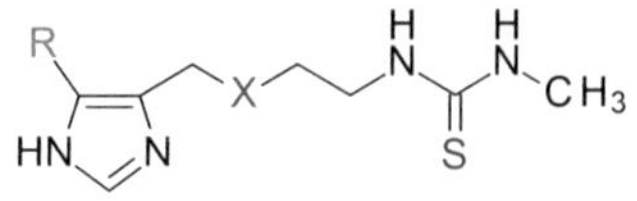

3.56 Burimamide, R = H, X = -CH$_2$-

3.57 Metiamide, R = CH$_3$, X = -S-

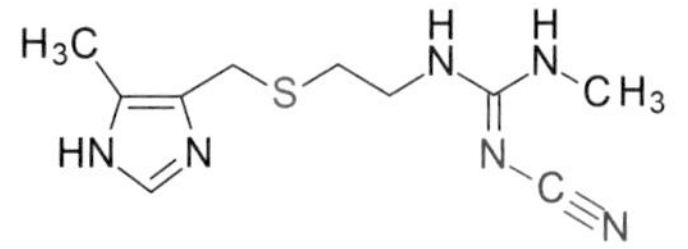

3.58 Cimetidine

◘ **Fig. 3.11** N_α-Guanylhistamine **3.54** and *S*-(2-imidazolyl-4-yl-ethyl)isothiourea **3.55** served as lead structures for H_2-type antihistamines. The first clinically tested H_2 antagonists, burimamide **3.56** and metiamide **3.57**, were unsuitable for treatment. Only the development of cimetidine **3.58** led to a breakthrough and an exceedingly successful therapy

The same happened in the development of the H_2 antagonists. Ulcer therapy had been researched since 1960 at Allen and Hansburys, a subsidiary of Glaxo. One of the first lead structures **3.59** (◘ Fig. 3.12), an aminotetrazole with about the same potency as burimamide, was systematically varied without success. Their research management also wanted to stop the project to concentrate on the anticholinergics. The breakthrough came upon replacement of the tetrazole ring with a furan. It was not exactly an obvious idea because the previously synthesized compounds always had at least one nitrogen atom in the ring. The $-CH_2SCH_2CH_2-$ chain was taken over from metiamide **3.57**, and a dimethylaminomethylene group was added to improve water solubility; the result was AH 18665 **3.60** (◘ Fig. 3.12).

The chemists also synthesized a cyanoguanidine AH 18801 **3.61** that was comparable to cimetidine **3.58** in terms of potency. The substance's characteristics were, however, unsatisfactory: the melting point was too low. The nitrovinyl analogue **3.62** brought success in this respect. It was synthesized and was an oil! That was not seen as a prohibitive problem because it was redeemingly 10-times more potent than cyanoguanidine **3.61** in the rat. **Ranitidine 3.62** (◘ Fig. 3.12) was developed as a drug and introduced in 1981 as Zantac® and Sostril®. Compared to cimetidine, ranitidine was 4- to 5-times more efficacious in humans and had the advantage that it was more selective. In 1987, ranitidine overtook cimetidine. In 1994 with US$ 4 billion in sales, it became the most economically successful drug in annual sales at that time. Within a few years, Glaxo was catapulted to the pinnacle of the world rankings of pharmaceutical corporations. Glaxo used this opportunity. The research of this company and its strategy in drug development belong to "the finest" in the branch today. Through mergers and acquisitions with competitors, Glaxo, "GSK" as it is known today, has become one of the largest pharmaceutical corporations on the market.

In the meantime, an antitumor effect in colon, gastric, and renal cancer has been reported for cimetidine. Apparently it suppresses tumor-mediated interleukin-1-induced selectin activation (Sect. 31.3).

It is understandable from the chemical structure that cimetidine has a high affinity for cytochrome P450 enzymes, particularly CYP 3A4 (Sect. 27.6). As a consequence, interactions with other drugs that depend on CYP 3A4 for metabolism are common. What was first seen as an indispensable imidazole moiety in **3.58** blocks the catalytic iron center in the P450 enzymes. Ranitidine **3.62** carries a furan ring in the same position and lacks P450 inhibition. After cimetidine and ranitidine, very few other drugs have made their way to the market. Nizatidine **3.63** and famotidine **3.64** contain a thiazole ring as a heterocycle (◘ Fig. 3.12). In **3.64**, the electron-withdrawing group of the guanidine moiety is replaced by a sulfonamide group.

It is true even for the H_2 blockers that good drugs are replaced by better ones. After being prompted to acid stimulation, the cells use an H^+/K^+-ATPase active enzyme to pump protons out of the cell in exchange for potassium at the cost of energy. If "the faucet is turned off" at this step, not only the histamine-induced acid production, but also the acetylcholine- and gastrin-mediated acid production is stopped. Omeprazole **3.65** is a prodrug that has been developed, which, upon rearrangement, acts as an irreversible inhibitor of this proton pump (Sect. 30.9). The effect of omeprazole therefore lasts longer, and the reduction in acid secretion is stronger than with H_2 antagonists. Gastric and duodenal ulcers heal more quickly and reliably. These substances also hit it big. At the end of the twentieth century, Losec®, Antra® (both from Astra), and Prilosec® (Merck & Co., USA) had combined global sales of over US$ 6 billion despite the fact that they were introduced to the market much later than ranitidine. The enantiomerically pure form esomeprazole (Nexium®) even reached US$ 7 billion in sales in 2007.

At first, the use of an enantiomerically pure substance seemed questionable, as the stereogenic center on the sulfur atom is lost during chemical activation (Sect. 9.5, ◘ Fig. 9.13). However, the manufacturing company was able to demonstrate better bioavailability for the *S*-stereoisomer, which justifies the use of the enantiomerically pure form. Besides omeprazole, pantoprazole **3.66** in particular was able to achieve high market shares. Together with the follow-up products lansoprazole **3.67** and rabeprazole **3.68**, several alternatives are available that differ in their pK_a value and have different rates of activation due to their acid lability. Since

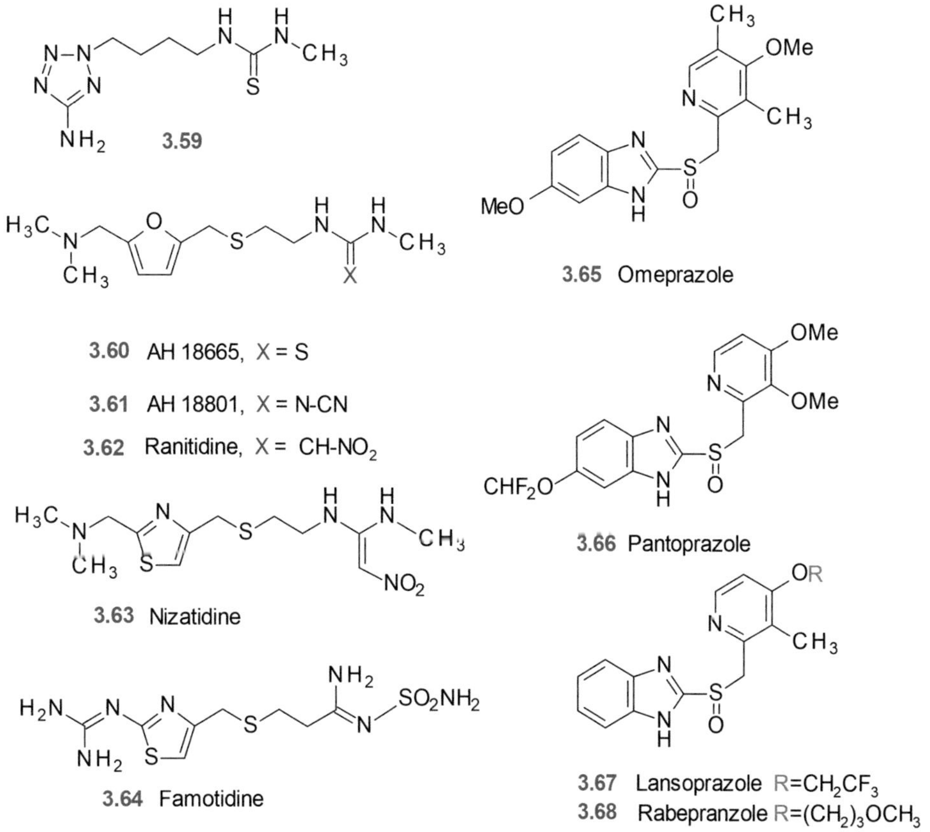

◘ **Fig. 3.12** The lead structures **3.59–3.61** were steps on the way to ratinidine **3.62**, which in the 1980s was the economically most important drug. Nizatidine **3.63** and famotidine **3.64** represent newer developments. Omeprazole **3.65** is a proton pump inhibitor. Further compounds (**3.66–3.68**) with a slightly different substitution pattern follow the same mode of action (Sect. 30.9)

activation occurs in an acidic environment, the active ingredient must first pass through the stomach, protected in a firm tablet, to then be absorbed via the intestine. The activation required for the effect finally takes place after transportation into the gastric cells of the stomach at the low pH prevailing there. In recent years, proton pump inhibitors that no longer irreversibly bind to the protein and also do not require acid-catalyzed activation (Sect. 30.9) have been developed.

Initially, there was concern that prolonged use of proton pump inhibitors would reduce the protective function of the stomach in killing harmful germs by raising the pH of the stomach. However, a comprehensive study of 17,600 patients with pantoprazole showed that there was no statistically significant difference in inflammation other than enteric infections compared to a placebo group.

That is not even the end of the story. Although in principle it had been known since 1983, the relevance of the **bacteria *Helicobacter pylori*** for the etiology of ulcers was first discussed in 1994 at a conference of the US National Institutes of Health (NIH). This bacterium infects a large portion of the population in childhood. Frequently, it is spread within a family; a kiss can be enough to infect someone. It causes gastrointestinal damage in a portion of those infected, which can lead to an ulcer. In the meantime, it is held responsible not only for ulcers but also for at least two different forms of gastric cancer. It survives assault by many antibacterial agents as well as the acidic milieu of the stomach. It has a urease that releases ammonia in its immediate vicinity, which in turn neutralizes the gastric acid.

The drugs of choice to treat such infections are combinations of H_2 blockers, proton pump inhibitors, and antibiotics. *H. pylori* seems to quickly develop antibiotic resistance though. At the beginning of 1995, the first animal model, a mouse with a sustained *H. pylori* infection, became available. This promoted further research in this important area. A vaccine is currently under development. A portion of the vaccinated patients exerted enough of an immune response to defend against the bacteria. For practical use, however, its reliability must be improved. Perhaps in the foreseeable future, we will have an ulcer therapy that is completely different, for instance, an oral vaccine that delivers life-long protection. The revolution is in sight: a one-time treatment without repeated gastroscopy. Patients will be delighted. Others will see this dramatic change in therapy with mixed emotions.

3.6 Synopsis

- Even though the period of classical drug research was strongly governed by trial and error, it has been exceptionally successful. Many leads were found by accident or from traditional medicine, though limited knowledge of pathophysiology or molecular disease etiology was available. In many cases, their mechanism of action and interaction with a target protein were later elucidated.
- Acetylsalicylic acid or Aspirin® is one of our oldest but also most prototypical drugs. Originating from bark extracts and chemically modified to improve taste and tolerance, it achieves its actual potency and mode of action by irreversibly inhibiting cyclooxygenase.
- Since then two isoforms of cyclooxygenase have been characterized, one is constitutionally present, and the other is induced in inflamed tissue. Acetylsalicylic acid inhibits both unselectively, giving rise to some undesirable side effects.
- Due to irreversible inhibition of COX in platelets, Aspirin has an influence on the ratio of synthesized thromboxane and prostacyclin, which decreases the coagulation tendency of blood. As a consequence, Aspirin is recommended as "preventive medicine" to protect against thrombosis or to reduce mortality of heart attack.
- Malaria is a widespread tropical/subtropical disease transmitted by the anopheles mosquito and caused by the plasmodium parasite accessing erythrocytes in humans. The disease had been nearly eradicated by fighting the mosquito with the insecticide DDT. One of the oldest active substances to use against the parasite is quinine, which is isolated from cinchona bark.
- After stopping DDT spraying for the mosquitos, malaria raged again. Increasing resistance of the parasite to known drugs occurred, and the development of new chemotherapeutics for malaria has been a rollercoaster ride of promising compounds and the development of resistant parasites.
- Morphine, isolated from poppies and used as the unchanged natural product, is a potent analgesic. When administered correctly, the risk of addiction is low. Its complex structure of five fused rings has been simplified and cut into pieces to give more-easily accessible analogues with higher selectivity.
- Cocaine, which is the active ingredient in coca leaves, is one of our oldest drugs. Upon replacement of dopamine from its transporter in the synaptic gap, its euphoric effect is achieved. The cocaine structure served as a lead structure for the development of anesthetics.
- Ulcer therapy went through several phases of drug development, leading to active substances with increasingly efficient modes of action to reduce production of gastric acid.
- Starting with antacids and rather unspecific anticholinergics, selective H_2 antagonists were real breakthroughs in the pure pharmaceutical treatment of ulcers. They act upon the H_2 receptor, a member of G-protein-coupled receptors (GPCRs). A protein that pumps protons for acid release is stimulated through

these receptors. Proton pump inhibitors such as omeprazole directly block the function of the proton-secreting H^+/K^+-ATPase that builds up the acidic milieu.

- The bacterium *Helicobacter pylori* causes gastrointestinal damage leading to ulcers. It can be eradicated by combining a proton pump inhibitor with an antibiotic. A vaccine against the bacterium could deliver life-long protection.

Bibliography and Further Reading

General Literature

G. Ehrhart and H. Ruschig, Arzneimittel. Entwicklung, Wirkung, Darstellung, 2nd ed., Verlag Chemie GmbH, Weinheim, 1972
A. Burger, A Guide to the Chemical Basis of Drug Design, John Wiley & Sons, New York, 1983
E. Verg, Meilensteine. 125 Jahre Bayer, 1863–1988, Bayer AG, 1988
W. Sneader, Drug Prototypes and their Exploitation, John Wiley & Sons, Chichester, 1996
J. Ryan, A. Newman and M. Jacobs, Editors, The Pharmaceutical Century. Ten Decades of Drug Discovery, Supplement to ACS Publications, American Chemical Society, Washington, 2000
W. Sneader, Drug Discovery. A History. John Wiley & Sons, Chichester, 2005
A. Giannis, Naturstoffe im Dienst der Medizin - Von der Tragödie zur Therapie, Springer-Verlag GmbH, (2023)

Special Literature

Aspirin – eine unendliche Geschichte, Research. Das Bayer-Forschungsmagazin, Heft 6, pp 4–21 (1992)
W.-D. Müller-Jahnke and C. Friedrich, Ein Siegeszug mit Hindernissen – 110 Jahre „Megastar“ Aspirin, Gesch. d. Pharm. **62**, 1–11 (2010)
C. Patrono, Aspirin and Human Platelets: From Clinical Trials to Acetylation of Cyclooxygenase and Back, Trends Pharm. Sci. **10**, 453–458 (1989)
B. Battistini, R. Botting and Y. S. Bakhle, COX-1 and COX-2: Toward the Development of More Selective NSAIDs, Drug News & Perspectives **7**, 501–512 (1994)
A. Jull et al., Low dose aspirin as adjuvant treatment for venous leg ulceration: pragmatic, randomised, double blind, placebo controlled trial (Aspirin4VLU), BMJ **359**, j5157 (2017)
T.W. Wakefield et al. An Aspirin a Day to Keep the Clots Away. Can Aspirin Prevent Recurrent Thrombosis in Extended Treatment for Venous Thromboembolism? Circulation **130**, 1031–1033 (2014)
J.J. McNeil et al. Effect of Aspirin on Cardiovascular Events and Bleeding in the Healthy Elderly, N. Engl. J. Med. **379**,1509–1518 (2018)
P.M. Rothwell, M. Wilson, C.E. Elwin, B. Norrving, A. Algra, C.P. Warlow, T.W. Meade. Long-term effect of aspirin on colorectal cancer incidence and mortality: 20-year follow-up of five randomised trials, Lancet **376**, 1741–1750 (2010)
R. H. Schirmer and K. Becker, Malaria – Geschichte und Geschichten, Futura **4**, 15–21 (1993)
J. Wiesner, R. Ortmann, H. Jomaa and M. Schlitzer, New Antimalarial Drugs. Angew. Chem., Int. Ed. Engl. **42**, 5274–5293 (2003)
M. Schlitzer, Malaria Chemotherapeutics Part I: History of Antimalarial Drug Development, Currently Used Therapeutics, and Drugs in Clinical Development, ChemMedChem **2**, 944–986 (2007)
H. M. Ismaila et al., Artemisinin activity-based probes identify multiple molecular targets within the asexual stage of the malaria parasites Plasmodium falciparum 3D7, Proc. Nat. Acad. Sci, USA, **113**, 2080–2085 (2016)
W. R. Kelce *et al.*, Persistent DDT Metabolite p,p′-DDE is a Potent Androgen Receptor Antagonist, Nature **375**, 581–585 (1995)
K.-L. Täschner and W. Richtberg, Koka und Kokain. Konsum und Wirkung, 2. Auflage, Deutscher Ärzte-Verlag, Köln, 1988
H. Kubas and H. Stark, Medizinische Chemie von Histamin-H_2-Rezeptorantagonisten, Pharm. u. Z. **36**, 24–32 (2007)
University of Oxford: R21/Matrix-M™ malaria vaccine developed by University of Oxford receives regulatory clearance for use in Ghana. https://www.ox.ac.uk/news/2023-04-13-r21matrix-m-malaria-vaccine-developed-university-oxford-receives-regulatory (Last accessed Nov. 29, 2024)

Protein–Ligand Interactions as the Basis for Drug Action

Contents

G. Klebe, *Drug Design*, https://doi.org/10.1007/978-3-662-68998-1_4

The following question must be answered before a drug can be specifically designed: How do drugs act in general? For example: How does Aspirin® relieve a headache? Why do β-blockers lower blood pressure? Where does a calcium channel blocker act? How does cocaine work? How do sulfonamides prevent the growth of bacterial pathogens? A drug must bind to a specific target molecule in the body to exert its pharmacological action. This is usually a protein, but nucleic acids in the form of RNA and DNA can also be targets for active compounds. An important requirement for binding is that the drug has the correct size and shape to optimally fit into a cavity on the surface of the protein, a binding pocket. In addition, the surface properties of the ligand and the protein must be complementary so that specific interactions can be formed. In 1894, Emil Fischer compared the exact fit of a substrate to the catalytic center of an enzyme to the picture of a **lock and key**. In 1913, Paul Ehrlich formulated the *Corpora non agunt nisi fixata*, which literally means "bodies do not act unless they are bound." His point was that drugs designed to kill bacteria or parasites must be "fixed," that is, bound by certain structures. Both concepts are the starting point for rational drug discovery. In a broad sense, they are still valid today. After ingestion, a **drug** must reach its target tissue where it interacts with a **biological target macromolecule**. Specific drugs must have sufficient affinity for the binding site of this macromolecule. Only in this way can they achieve the required selectivity and develop the desired biological effect, ideally without significant side effects.

The most important terms related to the modes of action of drugs are briefly defined in ◘ Table 4.1. These terms are described in detail in Chaps. 23–32 using multiple examples of target structures. Drugs often act as inhibitors of enzymes or as agonists or antagonists on receptors. Enzyme inhibitors and receptor antagonists occupy a binding site (often named **orthosteric site**) and prevent the substrate or endogenous ligand from docking there. Agonists have an additional property called **intrinsic activity**. As a result, the receptor takes on a three-dimensional structure that elicits a response from a downstream process (for details see Chaps. 22, 29 and 30).

Although ion channels, pores, and transport systems are also receptors in the broadest sense, they are considered a separate group. The term "receptor" is often used

◘ **Table 4.1** Brief definitions of the most important terms

Term	Definition
Ligand	A (usually small) molecule that binds to a biological macromolecule
Enzyme	An endogenous biocatalyst that can transform one or more substrates into one or more products
Substrate	A ligand that is the starting material for an enzymatic reaction
Inhibitor	A ligand that prevents the binding of a substrate either directly (competitive) or indirectly (allosteric), reversibly or irreversibly
Receptor	A membrane-bound or soluble protein (or a protein complex) that initiates an effect after binding an agonist
Agonist	A receptor ligand that exhibits an intrinsic activity, that is, it causes a receptor response
Antagonist	A receptor ligand that either directly (competitive) or indirectly (allosteric) prevents the binding of an agonist
Partial agonist	A weak agonist that has a high affinity to the binding site, and in this way also acts partly as an antagonist
Inverse agonist	A ligand that stabilizes the inactive conformation of a receptor or ion channel
Functional antagonist	A substance that prevents a receptor response by another mode of action
Allosteric effector	A ligand that influences the function of a protein by causing a change in the 3D structure of the protein
Interface antagonist	A low molecular weight ligand that disturbs or blocks the formation of a protein–protein interface
Molecular glue	A low molecular weight ligand that stabilizes and "glues" a protein–protein interface together
Ion channel	A pore in a protein that allows specific ions to flow in and out across the cell membrane along a concentration gradient. Opening and closing is affected by binding a ligand or by a membrane potential change
Transporter	A protein that transports molecules or ions across the cell membrane against the concentration gradient by consuming energy
Antimetabolite	A substance that interferes with the biosynthesis of a central metabolic product either as a false substrate or as an inhibitor

loosely as a general term for any biological macromolecule that interacts with a drug.

Beyond the binding of a small-molecule ligand, biomolecules often communicate with each other by recognizing and forming large shared surface contacts. It is through these contacts that the primary attack and entry of viruses, bacteria, and parasites into the host cell takes place. Many cells receive a signal via surface receptors when they bind a macromolecule (Chaps. 29–32). Even the rolling behavior of leukocytes in the vasculature is controlled by such surface receptors. These systems are increasingly being exploited for drug therapy by using active macromolecular substances known as biologicals or biopharmaceuticals (Chap. 32). They are increasingly being introduced as therapeutics in our pharmaceutical arsenal.

4.1 The Lock-and-Key Principle

In the early 1880s, Emil Fischer studied the cleavage of glucosides with various enzymes that differed only in the stereochemistry of the glycosidic carbon atom. He noticed that certain glucosides could only be cleaved by one group of enzymes. Other glucosides could only be cleaved by another group of enzymes. He drew the correct conclusions from his observations and published them in an article in *Berichte der Deutschen Chemischen Gesellschaft* (Reports of the German Chemical Society) in 1894:

> » "The limited effect of enzymes on the glucosides can also be explained by the assumption that a chemical process can be initiated only by those [enzymes] that have a similar geometric construction that approximates that of the molecule [substrates]. To use a picture, I want to say that enzymes and glucosides must fit together like a lock and key to be able to exert a chemical effect upon each other. This idea has gained plausibility and value for stereochemistry research after the phenomena was transferred from the biological to the chemical field."

In the same year he refined this picture:

> » "Apparently here the geometrical construction exerts such a large influence on the play of chemical affinities that the comparison of the two molecules undergoing an interaction seems to me to be comparable to a lock and key. If the fact that some yeasts can ferment a larger number of hexoses than others is to be explained, the picture can be completed by differentiating between master and special keys."

Emil Fischer did not pursue this picture any further and later even complained that it was often quoted out of context. He was interested in the configuration of sugars, but not in the configuration of isomeric glucosides. He expressed a rather distanced attitude to purely theoretical considerations. In 1912, he wrote in a letter, "I myself do not take much pleasure in theoretical things." This is remarkably modest for a man who exerted such great influence with his image of a **lock and key**! Emil Fischer would certainly have been pleased and proud to see the results of X-ray structure determinations of protein–ligand complexes, such as retinol (vitamin A) bound to the retinol-binding protein, responsible for the transport of this rather hydrophobic molecule (◘ Fig. 4.1).

The binding of ligands to proteins spans a wide range of specificity and this can occur with different selectivities with respect to the bound ligands. Originally, the terms **specificity** and **selectivity** originate from the study of enzymes. Specificity refers to the ability of an enzyme to preferentially convert one or a few substances to any other analogue. Specific ligands therefore require a particular chemical substructure that enables the enzyme to catalytically convert them. Selectivity refers to the selective and potent binding of a substrate that ideally fits into the binding site on a given target protein, filtering out that substrate (or ligand) relative to a large number of other candidates. The extent to which high specificity or high selectivity is required depends strongly on the biological and cellular context in which a protein performs its function.

The following examples will illustrate these aspects. Enzyme binding sites can be highly specific, especially for chemically related substrates. For example, they discriminate exceedingly between closely related analogues. In protein biosynthesis, exquisite recognition of distinct chemical moieties is required, not even the slightest mishap may occur. Friedrich Cramer investigated the mechanism for the recognition of the amino acids valine and leucine. These amino acids differ in their side chains only in that a methyl group is replaced by an ethyl group. The smaller valine residue should fit easily into the "lock" for a leucine, although it may not bind as strongly. A clear distinction, which is absolutely necessary for error-free protein synthesis, can only occur through repeated recognition. This is indeed the case. An energy-intensive, iterative, and carefully scrutinized verification process reduces the error quotient to less than 1:200,000. Because of this strict feedback and control process, even the correct binding partner is sometimes unsuccessful. Over 80% are rejected as "questionable." The result is a process with an accuracy of about 1:40,000.

Due to its function as a transport protein, retinol-binding protein does not discriminate between substrates with high selectivity. The extreme precision as in protein synthesis is apparently not necessary for its proper function. In addition to the "stretched" retinol isomer, the "folded" retinol isomer and chemically related substances also bind to this protein. Also, other proteins discriminate very little. Examples of less selective proteins include digestive enzymes (Sect. 23.3), metabolic enzymes (e.g., cytochromes; Sect. 27.6), or the glycoprotein

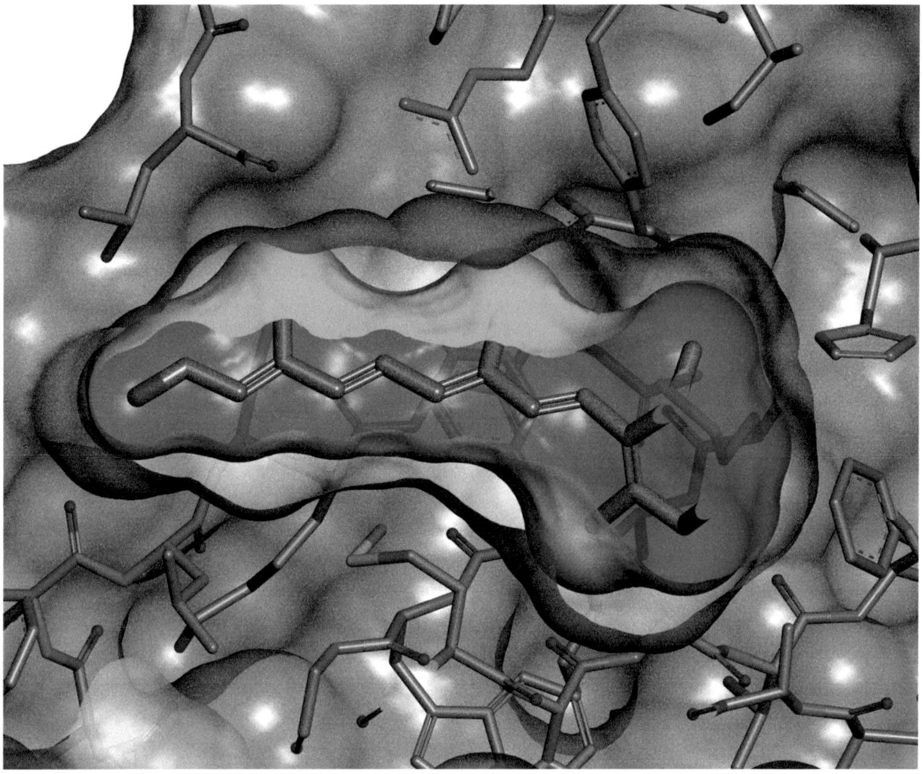

Fig. 4.1 Like a key in a lock, vitamin A (retinol, *gray* carbon atoms) fits into the binding pocket of its transport protein. The surface of the ligand is shown in *green, inside blue*. The amino acids of the protein in the direct vicinity of the binding pocket are shown (carbon atoms in *orange*, oxygen in *red*, nitrogen in *blue*). To improve the clarity, the back of the binding site and the residues in front of the binding site have been omitted. (► https://sn.pub/ty7wbA)

GP 170, which is responsible for drug resistance of tumor cells (Sect. 30.10). One bacterial transport protein, oligopeptide-binding protein A, can bind any peptide with two to five amino acids with approximately the same affinity; this is an extreme case of "chemical promiscuity."

Linus Pauling applied the lock-and-key principle to the **transition states of enzymatically catalyzed reactions**. Flexible adaptation often occurs during substrate binding. The transition state of the reaction binds more strongly to the enzyme than either the substrate or the product (Sect. 22.3) and it is stabilized by the functional groups of the binding site. The "lock-and-key" principle has been repeatedly challenged because of the mobility of the ligand in the binding site; however, even with a modern high-security lock, the pins are still mobile and play an essential role in the mechanism.

In the 1950s, Daniel E. Koshland proposed the **theory of "induced fit,"** which states that the ligand induces a conformational change in the protein by binding to it. The theory assumes a specific effect, such as enzymatic cleavage of the substrate. This mechanism does not contradict the lock-and-key principle because, as mentioned earlier, even a high-security lock has moving parts. Small, induced adaptations play an essential role in the ligand–receptor complex. Even the relocation of entire protein domains has been observed. The adaptability of a protein is usually related to its function. Proteins often need to be flexible enough to perform their biological functions and to recognize the right ligands. In the case of G-protein-coupled receptors, we are learning more and more about the importance of intrinsic dynamic behavior for proper functioning (Sect. 29.4).

There are two fundamentally different starting points for the rational design of ligands, which differ in the information content of the system. Either the exact three-dimensional structure of the binding site is known or it is unknown. In the first case, the lock is known and the key "only" needs to be cut and filed to the correct size (Chap. 20). In the other case, the active and inactive analogues represent the matching and mismatching keys. By comparing the keys and systematically varying them, better fitting keys can be designed (Chap. 17). In the following section, the binding of a small-molecule drug ("ligand") to a macromolecular receptor is examined in more detail. These drug targets can be located outside the cell, inside the cell, or embedded in the cell membrane. Therefore, we will briefly discuss the structure and function of the cell membrane before focusing on the protein–ligand interaction.

4.2 The Essential Role of the Membrane

Most biological processes in our body take place inside cells. These cells are surrounded by a membrane that protects the cellular contents from "leaking." The membrane also prevents unwanted xenobiotics from entering the cell and mediates contacts between cells. Membranes are also found within the cell, where they form substructures (called compartments) and separate individual cellular components. In mammalian cells, the outer **membrane consists of a lipid bilayer** in which proteins and cholesterol molecules are embedded (◻ Fig. 4.2). All molecules can move relatively freely, which is why it is called a fluid mosaic membrane.

Such lipid membranes act as barriers for polar substances and as permeable layers for nonpolar molecules. The importance of membranes for drug transport and distribution is discussed in detail in Chap. 19. Here, only the important function of the lipid membrane for the activity of drug molecules is discussed. Membrane-embedded proteins belong to completely different classes. These include membrane-anchored and membrane-resident enzymes, the large class of G-protein-coupled receptors (Chap. 29), ion channels, pores, and transporters (Chap. 30), and surface receptors (Chap. 31).

Because of the phosphate and ethanolamine head groups, the two outer layers of the lipid bilayer are highly polar. The alkyl chains are on the inside, where the membrane is nonpolar. Many drugs are also nonpolar and accumulate here in higher concentrations than in solution. Amphiphilic (soap-like) molecules, i.e., substances that are both nonpolar and polar, arrange themselves in the membrane so that the nonpolar part is on the inside (◻ Fig. 4.2). This orientation within the membrane is particularly important when the polar group is a positively charged nitrogen atom, which can form additional electrostatic interactions with the phosphate group of the lipids.

Meanwhile, this concept has been experimentally proven by many independent methods. For many receptors, it is accepted that the ligand binds to a site inside the protein that is accessible only from the inner layer of the membrane (e.g., lipases, Sect. 23.7; or cyclooxygenases, Sect. 27.9). Therefore, the enrichment and arrangement of an active molecule in the membrane plays an important role to optimally access the binding site. If, on the other hand, the molecule assumes an incorrect orientation, its docking to the binding site can be hindered.

4.3 The Binding Constant K_i Describes the Strength of Protein–Ligand Interactions

We would like to quantify the affinity or the strength of the binding of a ligand to its target protein. For this purpose, the binding of the ligand to its protein is regarded as a chemical reaction for which a dynamic chemical equilibrium is established after a certain interval of time. Once this equilibrium is reached, as many ligand molecules bind to the protein as dissociate from it. Macroscopically, the system no longer appears to change from the outside; equilibrium is established. For such a situation, we can formulate a law of mass action:

$$\text{protein} + \text{ligand} \rightleftharpoons \text{protein–ligand complex}$$
$$K_\mathrm{d} = \frac{[\text{ligand}]\times[\text{protein}]}{[\text{ligand–protein complex}]} = 1/K_\mathrm{a} \tag{4.1}$$

This constancy of the concentrations of ligand and protein molecules in the numerator divided by the concentration of the formed complex in the denominator is expressed as a characteristic quantity, the so-called equilibrium or binding constant. It is defined as the **dissociation constant K_d** (Eq. 4.1). Its reciprocal value is called the association constant K_a. In the case of enzymes, the so-called **inhibition constant K_i** is very often determined in an assay (Sect. 7.2) according to Michaelis–Menten kinetics known from biochemistry. Although not defined in exactly the same way, the quantities are usually considered and discussed as equivalent in drug design.

Instead of a K_i value for the inhibition of an enzyme, an **IC_{50} value** is often given in case of enzyme inhibition. The IC_{50} value measures the concentration of an inhibitor required to reduce the turn-over rate of the considered enzyme reaction by half. Unlike the K_i value, the IC_{50} value depends on the concentration of the enzyme and the substrate. Therefore, the IC_{50} value is also influenced by the affinity of the substrate for the enzyme, since substrate

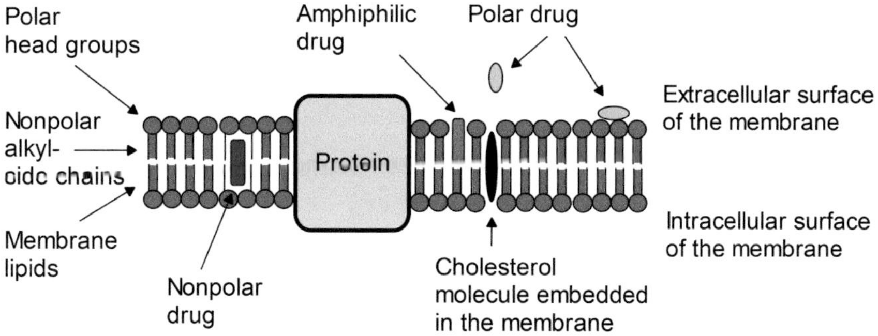

◻ **Fig. 4.2** Membranes from mammalian cells are constructed from a lipid double layer ("bilayer"), in which proteins (*yellow*) and individual cholesterol molecules (*black*) are embedded. The individual lipid molecules (*orange*) orient their polar groups to the exterior of the membrane, and their alkyl chains to the interior. Therefore, polar drugs (*light blue*) accumulate on the outside of the membrane. Nonpolar drugs (*red*) are enriched in the interior of the membrane. Amphiphilic drugs (*violet*) are oriented into the membrane according to their structure. Despite this, all of the molecules can move relatively freely. Therefore, this is called a "fluid mosaic membrane"

and inhibitor compete for the same binding site. The IC_{50} value can be converted to a K_i value using the Cheng–Prusoff equation. Experience has shown that IC_{50} and K_i values are parallel to a first approximation, so that the more easily determined IC_{50} value is well suited to characterize a ligand in comparison to other compounds. However, they should only be used if the data were determined in the same laboratory and under identical assay conditions.

The affinity thus describes the strength of the interaction between protein and ligand. It is defined by the ratio of the amount of free to protein-bound ligand. It has the dimension of a concentration with the unit mol/L (M). The smaller its value, the more strongly the ligand binds to the protein, because then it is mainly the formed complex that is present. If the concentration of the ligand is significantly lower than the binding constant, only a very small proportion of the protein molecules will be occupied by bound ligand molecules. A biological effect, such as the inhibition of an enzyme, can then hardly be observed. If the free ligand concentration at equilibrium corresponds exactly to the value of the binding constant, half of all protein molecules present will be occupied by ligand molecules.

From chemical thermodynamics, we know an important relationship which connects the binding constant (or precisely the dissociation constant K_d) with the so-called **Gibbs free energy (ΔG)**. With this expression, the binding constant and the change in Gibbs free energy of the chemical reaction under consideration, that is the formation of our complex, can be converted into each other (Eq. 4.2). Just as altitude data on a geographic map only make sense if they are related to a reference point such as sea level, energy data refer to the Gibbs free energy $\Delta G°$ under so-called standard conditions (all binding partners are present in molar amounts, at T = 298 K and a pressure of 1 atm).

$$\Delta G = \Delta G° - RT \ln K_d \quad (4.2)$$

In Eq. 4.2, R is the gas constant and T is the absolute temperature in Kelvin. If chemical equilibrium has been reached, $\Delta G = 0$, resulting in Eq. 4.3:

$$\Delta G° = RT \ln K_d \quad (4.3)$$

The superscript zero to indicate standard conditions is generally omitted in the specification of ΔG, which is also the case in the following text. A binding constant of $K_d = 10^{-9}$ M = 1 nM, a respectable value for an active compound, corresponds to a Gibbs free binding energy of −53.4 kJ/mol at body temperature (to transform to kcal, divide by 4.184). At 1 µM it is −35.61 kJ/mol, at 1 mM − 17.80 kJ/mol. A change of K_d by one order of magnitude means a change of the free energy of binding by 5.94 kJ/mol (these values have to be compared with the molecular association energies; ◘ Table 4.2).

It should be noted at this point that in chemistry a "**system-egoistic approach**" is applied. This means that a negative energy value means that the system is releasing energy, e.g., in an exothermic reaction. A positive value means that the system absorbs energy from outside. For the change in Gibbs free energy, a negative value means that the reaction runs spontaneously into an equilibrium, a positive value means that the reaction does not run at all.

But back to thermodynamics. As described above, the binding constant is related to changes in the Gibbs free energy. But why do we need such a "free energy" instead of simply considering a set of energy terms to describe the energetics leading to the formation of a protein–ligand complex?

Simply changing a few energy contributions does not provide a complete answer to why a particular process, such as the formation of a protein–ligand complex, occurs **spontaneously**. The following example illustrates this point. If we take a hot and a cold piece of metal and bring them into contact, everyone knows that heat will flow from the hot metal to the cold one. The opposite is not observed, even though the energy content of the entire system would remain unchanged for this process. Why does energy spontaneously flow from a hot object to a cold one and not vice versa? This has to do with the tendency of all-natural processes to distribute energy evenly. In a hot piece of metal, the metal atoms vibrate very strongly around their resting positions, which is why the piece of metal is hot. Some vibrational degrees of freedom are strongly activated. When the cold metal block is brought into contact with the hot one, vibrations are transferred. In the end, the metal atoms in both blocks vibrate similarly about their resting positions, but on average not as strongly as the atoms in the hot block had moved before. The total amount of energy remains the same. However, the energy is now distributed over many more degrees of freedom. A constant energy balance could also be achieved if the hot piece of metal would get hotter and the cold piece would get colder by the same amount. But this process does not

◘ **Table 4.2** Experimental or quantum mechanically determined association energies in the gas phase

Dimer	Binding energy in kJ/mol
$CH_4 \cdots CH_4$	−2
$C_6H_6 \cdots C_6H_6$	−10
$H_2O \cdots H_2O$	−22
$NH_3 \cdots NH_3$	−18
$Na^+ \cdots H_2O$	−90
$NH_4^+ \cdots CH_3COO^-$	< −400
$Na^+ \cdots Cl^-$	< −400

happen. There is no increased distribution of energy over more degrees of freedom, and the system does not enter a more disordered state. However, this is a prerequisite for spontaneous processes. To quantify this finding, **entropy, S**, was introduced. It is a measure of the order of a system. It estimates over how many degrees of freedom a given amount of energy is distributed in a system and how much it will increase during a spontaneous process. In the case of protein–ligand complexes, a degree of freedom can be, for example, a certain vibration of the system or a rotation of individual groups against each other. A highly ordered system, where energy is trapped in only a few degrees of freedom, has a low entropy content. Increasing disorder increases entropy.

These aspects are particularly important for multicomponent systems, such as protein–ligand complexes, which also include the surrounding solvent molecules. To fully describe such a system, we need to consider not only the energy terms involved in the formation of noncovalent interactions between the two binding partners, but also how the energy is distributed over the multiple degrees of freedom of the formed protein–ligand complex. Therefore, we need a thermodynamic property that takes into account not only the energy contributions, but also how the energy is distributed over the system. In this way, the formed protein–ligand complex migrates to a more disordered state, otherwise its formation would not occur spontaneously.

The Gibbs free energy (ΔG) is the appropriate property to describe the formation of such complexes. Another important thermodynamic relationship is that the Gibbs free energy consists of two additive terms, an **enthalpic term ΔH** and an **entropic term $-T\Delta S$**. The latter is temperature weighted. It takes into account not only the energy balance of the process, but also the changes in entropy (Eq. 4.4).

$$\Delta G = \Delta H - T\Delta S \quad (4.4)$$

The **entropic component is temperature weighted** and, thus, receives the dimension of an energy. It makes a big difference whether the entropy in a system is changed at low temperature, where all particles are in a largely ordered state, or at high temperature, where the disorder is already very high. Because of the negative sign, an increase in entropy causes a decrease in ΔG and therefore an increase in binding affinity. When discussing the thermodynamic aspects of binding profiles, we always have to consider ΔH and $-T\Delta S$ in addition to ΔG. If the enthalpic and entropic contributions are both negative, this will improve affinity and the corresponding process is called **exergonic**. If one of them is positive, this will be detrimental to ligand binding. However, the sum of both quantities must be negative for the binding reaction to reach equilibrium. Otherwise, the complex formation will not occur and the process is called **endergonic**.

In biology, many processes require an **energy source** because they would be endergonic and, thus, not possible if considered alone. However, an overall exergonic balance is required for these processes to occur spontaneously for thermodynamic reasons. Examples are muscle contraction, transport of substances against concentration gradients, or the synthesis of many biomolecules. Such processes must be coupled to a reaction that provides the required Gibbs free energy and gives the overall process an exergonic inventory. Of central importance in this context is the **hydrolysis of adenosine triphosphate (ATP)** to adenosine diphosphate (ADP) and adenosine monophosphate (AMP). The triphosphate unit is high in energy and its hydrolysis releases a large amount of Gibbs free energy. Therefore, ATP is continuously produced by the organism and then hydrolyzed, usually in a specific ATPase, to allow the biological processes of interest to occur spontaneously.

Finally, in Eq. 4.4 the term "enthalpy" is used. Why is the term "enthalpy" used instead of the more common term "energy"? In chemistry and biology, processes take place as so-called open systems under atmospheric pressure. Since the volume of the environment is enormous, it can be assumed that the external pressure remains unchanged even in processes in which gas is produced. Therefore, these processes are considered to be under constant pressure conditions. Nevertheless, a gas produced during a reaction must first find its place among the surrounding particles in the air. Therefore, some energy must be spent and reduces the maximum possible energy amount that can be transferred by the system (so-called internal energy, ΔU). Accordingly, the energy reduced by this pressure–volume work is called **enthalpy (ΔH)**. It is therefore the energy converted during a process corrected for the pressure–volume work. In biological processes, the emission of a gas rarely plays a role. Therefore, internal energy and enthalpy are usually equal here (Note: in German usage, ΔG, which also refers to constant pressure conditions, is called a "free enthalpy." In English, the term "Gibbs free energy" is used instead).

4.4 Important Types of Protein–Ligand Interactions

Organic molecules can bind to proteins by forming noncovalent interactions between the ligand and the protein. In some cases, covalent bonds are also formed. For example, a chemically modified product of omeprazole reacts with its protein target and forms a covalent bond to the thiol group of a cysteine residue (Sects. 9.5 and 30.9). In this section, we will limit ourselves to ligands that bind to the protein by forming noncovalent interactions. For the following discussion, it is helpful to classify protein–ligand interactions into different categories. The different types of interactions are summarized in ◘ Fig. 4.3.

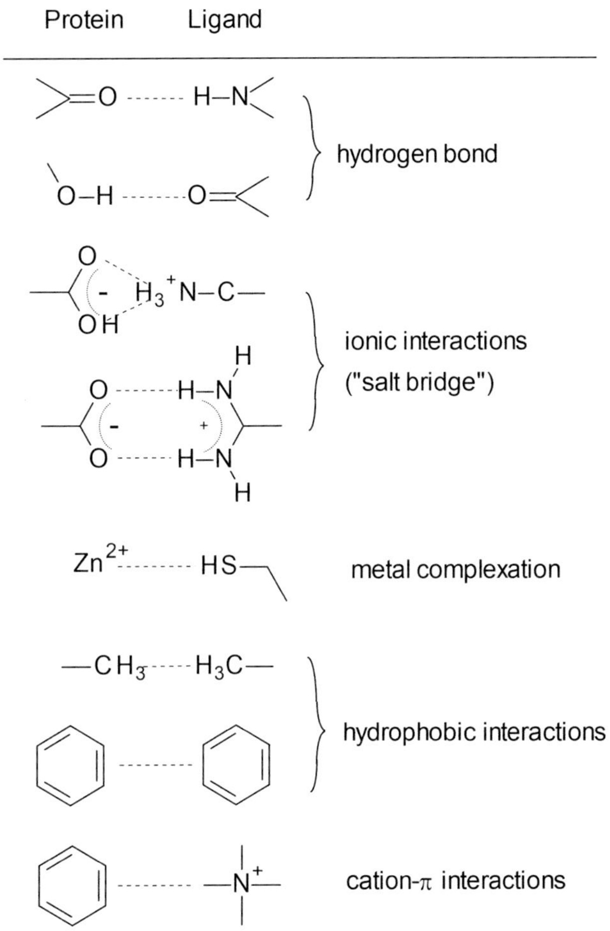

Fig. 4.3 Frequently occurring protein–ligand interactions. Important polar interactions are hydrogen bonds and ionic interactions. Metalloproteases contain, for example, zinc ions as a cofactor, the interaction of which with a ligand often yields important contributions to the binding affinity. Nonpolar parts of the protein and ligand contribute hydrophobic interactions. Because of the particular electron distribution in aromatic rings, the interaction between unsaturated ring systems is particularly large

Hydrogen bonds (H-bonds) between proteins and ligands are frequently observed. The proton-bearing partner in a biological system is usually an NH or OH group, termed the **hydrogen-bond donor**. The opposite group is an electronegative atom with a partial negative charge and is termed the **hydrogen-bond acceptor**. Examples of hydrogen-bond acceptors are oxygen and nitrogen atoms. Hydrogen bonds are predominantly electrostatic interactions. They achieve their extraordinary strength because the hydrogen atom of the donor group is bonded to a strongly electronegative atom, causing the electron density of the hydrogen atom to shift to the neighboring atom. The sphere of influence of the hydrogen atom becomes effectively smaller. This allows the acceptor to come closer to the proton than the sum of the van der Waals radii would allow. The electrostatic attraction between the partners thus becomes greater. The atoms of a hydrogen bond, for example of N–H···O=C, assume an almost linear arrangement to each other. The distance N···O is between 2.5 and 3.2 Å. The angle N···H···O is nearly always larger than 150°. Looking from the opposite side, a much larger variation between 100 and 180° is observed for the angle C=O···H.

It is often found that charged groups of the ligand bind to oppositely charged groups on the protein and they frequently overlap with a hydrogen bond. Then they are called **charge-assisted hydrogen bonds**. If both bear a formal charge, they are also named **ionic interactions** (also known as **salt bridges**) and they are particularly strong when the two groups are separated by only 2.7–3.0 Å. We will see that in many protein–ligand complexes the association is largely determined by such charge-assisted or ionic interactions. A few proteins contain metal ions as cofactors, such as Zn^{2+} in metalloproteases (Chap. 25). It is often the attractive interactions between the metal ion and the opposite charge on the ligand functional groups that contribute significantly to the affinity in these structures. In addition, there are a few groups that are particularly well suited to form complexes with transition metals. These include R–SH thiols, R–CONHOH hydroxamic acids, carbonic acid groups, and many nitrogen-containing heterocycles.

Whether the charge can increase the affinity contribution of hydrogen bonds depends strongly on the **protonation state of the involved functional groups**. Drugs are usually weak acids or bases, which means they contain so-called titratable groups (Sect. 19.4). Whether these groups, e.g., a carbonic acid, an acidic sulfonamide, or a nitrogen-containing heterocycle, can release or accept a proton and transform into a charged state depends strongly on the local pH conditions and the polarity and the distribution of electrons in a molecule. The same can apply for the functional groups of the acidic or basic amino acid residues. These groups can then form charge-assisted hydrogen bonds that contribute more to the binding affinity (Sect. 4.8).

The **pK_a value** is used to estimate whether a group is in the protonated or deprotonated state. It indicates the pH at which the two forms in equilibrium are present in equal amounts. The situation can become more complicated because the **pK_a value can be shifted by the local environment**. The pK_a value is in fact defined for an aqueous medium, but we also use it for solvent mixtures and even transfer it to the environment in a protein binding pocket. In a hydrophobic environment, adopting a charged state is less favorable for acidic and basic groups; this means a shift to a less acidic or basic character is the result. If an already protonated, positively charged group in the ligand faces an amino acid of the protein with the same charge, its protonated state becomes even more difficult to accomplish. Formally, the group will therefore exhibit less basic behavior. The opposite is the case when putatively positively charged basic groups bind in a negatively charged protein envi-

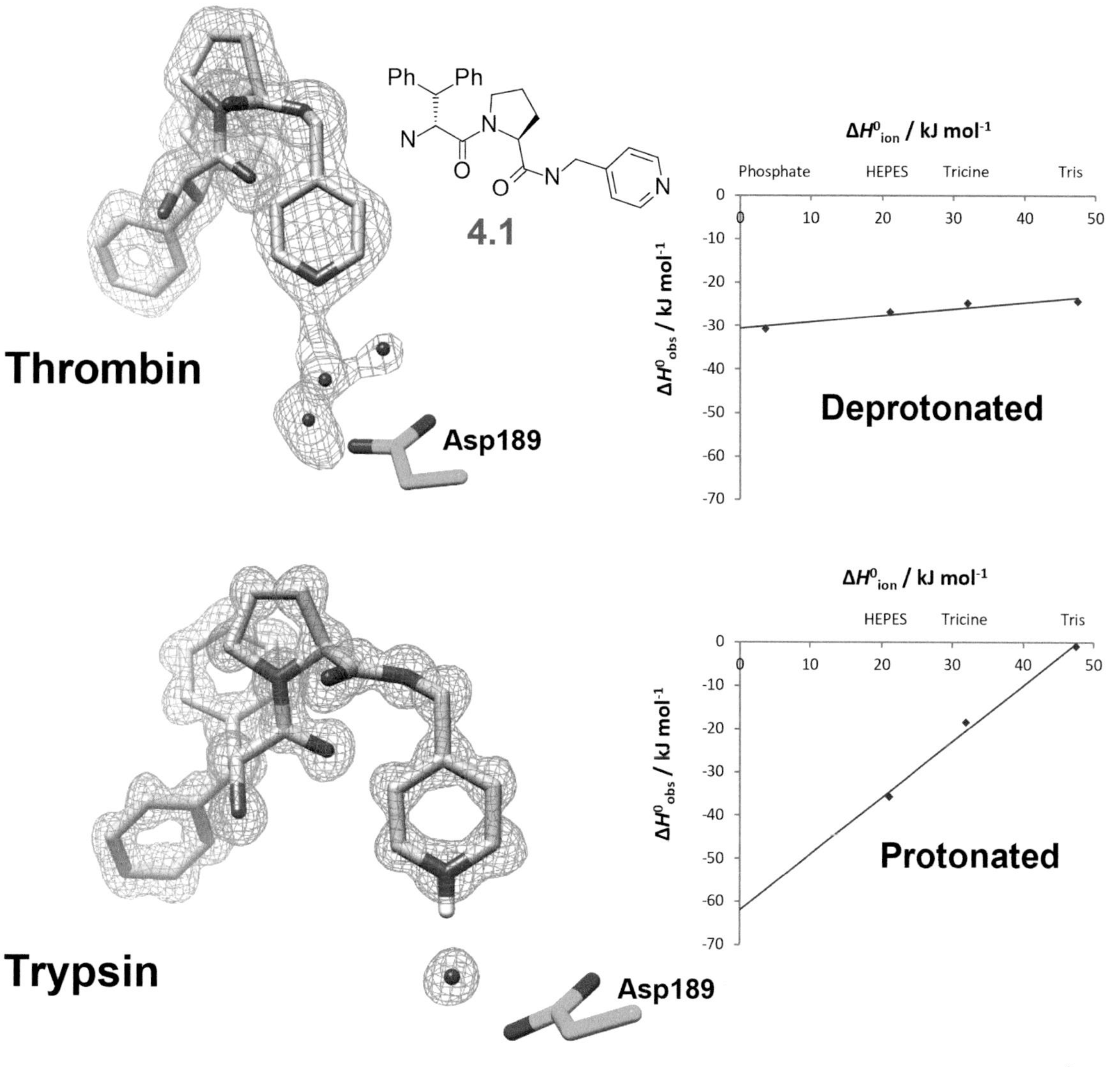

Fig. 4.4 Ligand **4.1** binds to thrombin and trypsin with different protonation states. In the case of trypsin, the interaction is mediated by an ordered water molecule; in the case of thrombin, the water molecule is distributed over three positions. The differences in the protonation states can be determined by isothermal titration calorimetry from several buffers, which exhibit different heat of ionization (ΔH^0_{ion}). In addition to the molecular scaffolds, the electron density around the ligand is shown (*green* chicken wire mesh), which determines the placement of the molecules in crystallography (Sect. 13.5)

ronment. Here, the charged state is formed even more easily, corresponding to a stronger basic character. The same principle is true for acidic groups only with opposite signs. Here, a positively charged protein environment makes an acidic group appear more acidic, whereas a negatively charged environment will make an acidic group appear less acidic. In this way, the protein environment can induce a significant pK_a shift of the titratable groups of the ligand. Uncharged H-bonds can become charge-assisted contacts that contribute significantly more to binding affinity (Sect. 21.9). Electrostatic calculations can be used to estimate the pK_a shift during complex formation (Sect. 15.4).

Khang Ngo in Marburg examined the following example that illustrates these differences (Fig. 4.4). Trypsin and thrombin are structurally very similar serine proteases (Sect. 23.4). They have almost identical binding pockets into which ligands bind (the so-called S_1 pocket, Sects. 14.5 and 23.3). An aspartate residue at position 189 (Asp 189) in this pocket is important for ligand binding. This occurs by formation of a hydrogen bond between the ligand and Asp 189 which is mediated via a water molecule. It is interesting that ligand **4.1** binds to thrombin with its pyridine ring in the deprotonated state, and to trypsin in protonated form. In both cases, a hydrogen bond is formed with the aspartate mediated via a water molecule. In thrombin, the water molecule is spatially scattered over three sites. In trypsin, the water molecule occupies only one position, indicating a more fixed geometry. Differences in protonation states can be measured by **isothermal titration calorimetry** (Sect. 7.7), since different heat signals are obtained when titrations are

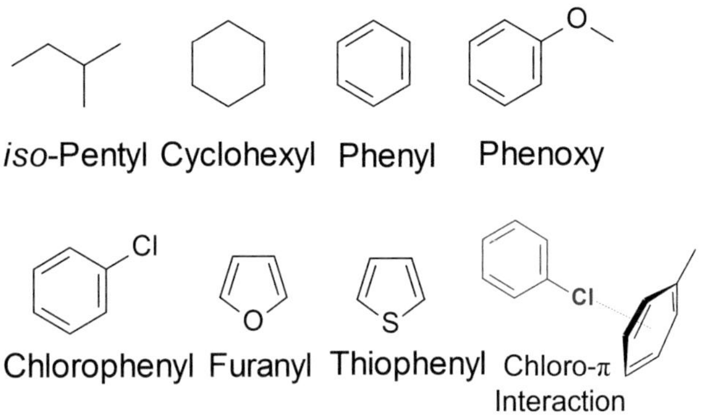

Fig. 4.5 Typical lipophilic groups in ligands are aliphatic and aromatic hydrocarbons, halogen substituents, as well as nonpolar heterocycles such as furan and thiophene. Halogens, such as chlorine or bromine, have a positively polarized tip (so-called σ-hole) on the side facing away from the aromatic ring in extension of the C–Cl bond, with which they can enter into an attractive electrostatic interaction with the π-electron system of a neighboring aromatic system

performed in buffers with different heats of ionization. If binding is accompanied by a change in protonation state, the titrations allow quantification of the molar amount of protons transferred during the binding reaction. For thrombin, there is hardly any buffer dependence, while for trypsin there is a strong effect (Fig. 4.4). In thrombin, a sodium ion is found next to the aspartate residue. This ion is absent in trypsin. Most likely, the positively charged sodium ion attenuates the charge on the adjacent acidic aspartate residue in thrombin to such an extent that the induced pK_a shift on the pyridine moiety is insufficient to reach a protonated state. In the geometrically analogous trypsin, however, the polarization effect is strong enough for protonation to occur.

Hydrophobic interactions are formed by the close proximity of nonpolar amino acid side chains of the protein to lipophilic groups on the ligand. Lipophilic groups include aliphatic or aromatic hydrocarbon groups, as well as halogen substituents (e.g., a chlorine) and many heterocycles such as thiophene and furan (Fig. 4.5). All areas that cannot form H-bonds or other polar interactions count as lipophilic parts of the surface of a protein and ligand. Unlike hydrogen bonds, hydrophobic interactions are not directional. The relative orientation of the lipophilic groups to one another does not matter. An exception is the interactions formed between aromatic rings, for which there is a preferred relative orientation. Halogens, such as chlorine or bromine, have a positively polarized tip on the side opposite to their covalent attachment. If they face with this tip, an aromatic ring, e.g., of a Phe of Tyr residue, an attractive electrostatic interaction will be formed (Fig. 4.5).

It has been shown that for ligands with large lipophilic groups, hydrophobic interactions often make a significant contribution to the binding affinity. However, the influence of direct attractive forces between the lipophilic groups is small. Hydrophobic interactions usually involve the displacement, or more precisely, the **release of water molecules from the lipophilic environment** of the binding pocket. In addition, the ligand with its lipophilic substituents leaves the bulk water phase in the vicinity of the protein. The solvent "cave" in which the ligand was hosted in water collapses. At first glance, this release of water molecules from hydrophobic surfaces into the bulk solvent increases the disorder of the system and, hence, has a favorable entropic contribution to the change in the free energy of binding. The role of water molecules is discussed in Sect. 4.6. Another important interaction should be mentioned here. Quaternary amines bind particularly well in binding pockets formed by the aromatic side chains of a protein. This contact is largely based on the polarization interaction between the positive charge and the electronic π-system of the aromatic rings.

4.5 The Strength of Protein–Ligand Interactions

When evaluating the strength of protein–ligand interactions, it is useful to first consider the noncovalent interactions between isolated small molecules. Information about these interactions is available from quantum mechanical calculations (Sect. 15.5) and spectroscopic studies. In this way, molecular pairs can be studied experimentally in the gas phase. The association energies obtained for the molecules give an indication of the strength of the direct interactions. Of course, the influence of effects resulting from the release of the solvent water (desolvation) is missing in such experiments. Some of these data are summarized in Table 4.2.

The results show that **electrostatic interactions** are the dominant energetic factor. The interaction between a cation and an anion in vacuum is more than −400 kJ/mol. This is equivalent to the strength of a covalent bond! This amount is enormous compared to the typical protein–ligand interactions in water, which were discussed in Sect. 4.4. The binding of an ion pair in the gas phase is, therefore, much larger than the typical strength of a protein–ligand interaction in water. Two water molecules bind with −22 kJ/mol. This interaction is also predominantly electrostatic in nature, as the large dipole moment of a water molecule is responsible for the strong binding. Interactions between small, nonpolar molecules are much weaker. Two methane molecules bind with about −2 kJ/mol. This is less than 10% of the $H_2O\cdots H_2O$ interaction. Correspondingly, methane boils at 90 K, while water is a liquid at room temperature. The direct interactions between polar groups are, therefore, orders of magnitude stronger than those between nonpolar groups.

4.6 Blame It All on Water!

The data presented in the previous section might suggest that protein–ligand interactions are mainly determined by H-bonds and ionic interactions. It is all the more surprising that the acetate ion CH_3COO^- does not form a dimer with the guanidinium ion $H_2NC(=NH_2^+)NH_2$ in water. Analogously, amides do not associate at all in water, although hydrogen bonds between two amide groups almost always occur inside protein structures. How can this be? The answer is that water is to blame for all of it!

All biochemical reactions take place in water. Indeed, **aqueous solvation** is an absolute requirement! How is a ligand solvated prior to protein binding? Water molecules form a shell around the ligand and the protein. This shell must be shed as the ligand enters the protein binding pocket. It involves breaking some of the hydrogen bonds formed by the polar functional groups of the ligand to the water molecules. At the same time, the "cavity" in which the ligand was trapped in the bulk water phase before binding to the protein will collapse. In consideration of this, we should not forget what the structure of pure water looks like. The structure of ice can serve as a first model. In fact, liquid water can transiently adopt the ice structure, although in the temperature interval 0–100 °C there are considerable modifications between the local volumes and their sizes where this occurs. In crystalline water, each water molecule is tetrahedrally surrounded by four other water molecules, resulting in a spatial structure similar to that of diamond. The individual water molecules are connected to one another by hydrogen bonds. However, the distribution of these H-bonds is not strictly ordered and static. Instead, a highly dynamic network is formed. The individual water molecules can spatially rotate, so that, on average, **each water molecule participates in four equivalent H-bonds**, twice as a donor and twice as an acceptor. Such a water structure is, thus, highly disordered, which will have a huge impact on the entropy contribution when structural changes occur. If this dynamic network is perturbed, for example by the introduction of a hydrophobic group of a dissolved drug molecule, the water molecules in the immediate vicinity of the hydrophobic group will no longer be able to establish this dynamic hydrogen bonding network. Hence, they can no further move and rotate randomly with the other water molecules as they can in the bulk water phase. They must choose one of the neighboring water molecules as their sole interaction partner. This automatically reduces the degree of disorder and leads to an unfavorable change in entropy. If a ligand is released from the water phase on its way into the protein pocket, these previously frozen degrees of freedom will be activated again, providing an entropically favorable contribution.

When the ligand enters a supposedly uncomplexed binding pocket of a protein, it almost always **encounters a cluster of water molecules** that must be displaced during the binding process. However, the local density of water molecules can vary greatly, from a pocket that is actually empty to one that is filled with water molecules similar in density to the bulk water phase. Some of the water molecules form hydrogen bonds with the protein and are in a spatially fixed, energetically favorable arrangement. Other water molecules are in contact with lipophilic regions of the protein surface and cannot form a perfect hydrogen bonding network. At most, they form a local cluster of interconnected water molecules. The exchange of these structurally very different water molecules will be associated with very different desolvation costs. An extreme example is the S_1' pocket of the metalloprotease thermolysin (Sect. 25.3). In this pocket, the enzyme binds its substrates via the hydrophobic side chains of the amino acids leucine or isoleucine. If the butyl moiety of these side chains is replaced by those of the smaller amino acids such as glycine, alanine, or valine, the affinity for substrates with these amino acids decreases dramatically. The transition from the glycine derivative to the leucine derivative represents a 41,000-fold increase in affinity! Any medicinal chemist would dream of achieving this increase in affinity with only four carbon atoms of an aliphatic substituent. Experimentally, Stefan Krimmer was able to show that the S_1' pocket of uncomplexed thermolysin is practically free of water. Thus, binding of the substrate does not require **desolvation of the pocket**. The result is an extremely potent, almost exclusively enthalpically preferred binding. The enzyme needs this trick to efficiently discriminate between substrates with small and large aliphatic residues of a homologous series. This example shows how useful it can be for any drug designer to study the substrates between which their target proteins can discriminate.

The displacement of fixed water molecules usually results in an entropic binding contribution, since the released hydrogen bonds of the displaced water molecules are replaced by comparably strong hydrogen bonds to the incoming ligand (◘ Fig. 4.6). The enthalpic inventory is, thus, to a first approximation balanced. However, the released water molecules have many more degrees of freedom in the bulk water phase, which implies an entropic advantage. This consideration is even more valid when we focus on hydrophobic regions of a ligand. We have already learned that release from the original cavity in the water phase is entropically favored. Now we add the entropic contribution for the displacement of fixed water molecules from the pocket (◘ Fig. 4.6). Based on these considerations, it was suggested that the hydrophobic binding contribution is essentially entropic in nature (the so-called "hydrophobic effect"). In the meantime, this picture no longer seems to hold. Hydrophobic binding can also be enthalpic. This is again due to the properties of the water molecules released from the protein during ligand binding (◘ Fig. 4.7). If one encounters a largely empty water-free pocket, or a pocket filled with only a few isolated water molecules that are already highly disordered in the protein-bound state prior to their displacement, there is hardly any entropic gain to be achieved upon their release. In such cases, the hydrophobic effect may have a preferential enthalpic signature.

However, water can be critical to the energy balance at a completely different point in the binding process. As mentioned above, a ligand is solvated in water before binding to the protein. In the water phase, it can adopt several conformations. Barbara Wienen-Schmidt and Tobias Hüfner have studied the binding of a series of ligands to protein kinase A. Surprisingly, the ligand in the series with the most rotatable bonds shows the most entropically favorable binding signature. Intuitively, one would expect the opposite, since a conformationally flexible ligand that is locked into a single conformation in the protein binding pocket should pay a high entropic cost for the loss of these degrees of freedom on protein binding. To investigate this phenomenon, the first aspect to check was whether the protein undergoes changes that are associated with a loss of entropy. However, no evidence for this could be found. It was then necessary to rule out the possibility that a change in the solvation water inventory of the binding process could explain the entropic advantage. Again, this could not explain the observed entropic effect. Finally, it was investigated whether the more flexible ligand exhibits any peculiarities in aqueous solution prior to binding. It was observed that the conformation of the unbound ligand entraps a water molecule so tightly that it can hardly exchange with the surrounding bulk water. Evidence for this behavior was provided by NMR studies (Sect. 13.7) and molecular dynamics simulations (Sect. 15.8). Upon binding of this ligand to the protein, the tightly bound water molecule had to be released, which was accompanied by a significant increase in entropy. In the end, this contribution was so large that it became determinant for the entire binding profile.

Water can still make an important contribution at the final step of the protein–ligand complex formation. When a new protein–ligand complex forms, the ligand adopts its binding pose in the protein pocket. In most cases, substituents of the bound ligand protrude into the surrounding water forming part of the surface of the newly generated complex. Both binding partners create a new, commonly exposed surface. Water molecules cluster around this surface and generate a new solvate shell, which can display a high order. The network of these water molecules can be used to enhance ligand binding through a suitable choice of side chain. In particular, when fused ring structures of water molecules are formed, this improves protein binding. For example, Stefan Krimmer and Jonathan Cramer were able to increase the binding affinity through tailored design by a factor of 50 for the thermolysin ligands shown in ◘ Fig. 4.8.

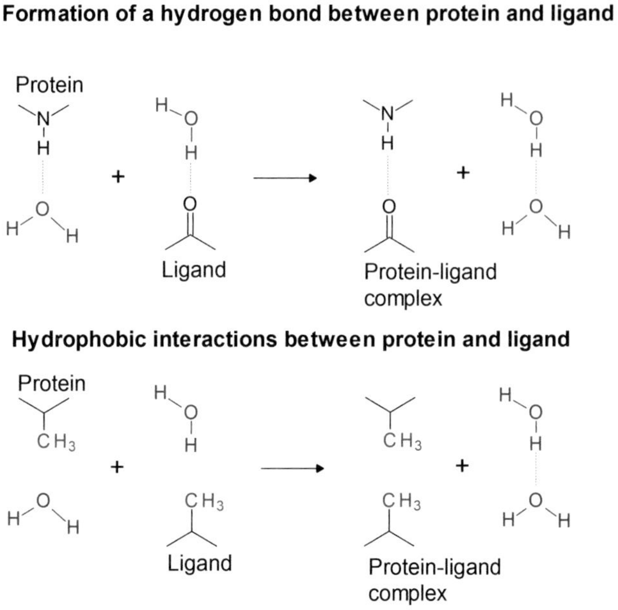

◘ **Fig. 4.6** Influence of water molecules on the strength of protein–ligand interactions. *Top row* The formation of an H-bond between protein and ligand requires the displacement of water molecules, which themselves form H-bonds to the protein or ligand. The H-bond inventory, meaning the number of H-bonds present before and after the binding, is balanced in this case. *Bottom row* When a hydrophobic contact is formed, water molecules are released from an environment that is unfavorable for them. In the bulk water phase, they can form H-bonds among one another

4.7 Thermodynamic Contributions to the Formation of Protein–Ligand Complexes

Having considered the individual factors involved in ligand binding, let us now turn to the successive steps of complex formation. As mentioned above, the binding affinity is crucial for assessing the strength of a protein–ligand complex. As outlined in Sect. 4.3, it can be described in terms of the binding constant (Eq. 4.1) or the change in Gibbs free energy of binding (Eq. 4.3). This is based on equilibrium thermodynamics. However, biological systems are so-called open systems for which a "steady state" can be assumed as a first estimate. Of course, application of equilibrium thermodynamics is only a crude approximation. However, describing open systems would further complicate the already very complex relationships. It is crucial for our considerations that besides purely energetic contributions, an entropic component must always be taken into account.

◘ Fig. 4.9 summarizes the steps involved in the formation of a complex between a ligand and its protein. At room or body temperature, protein and ligand can

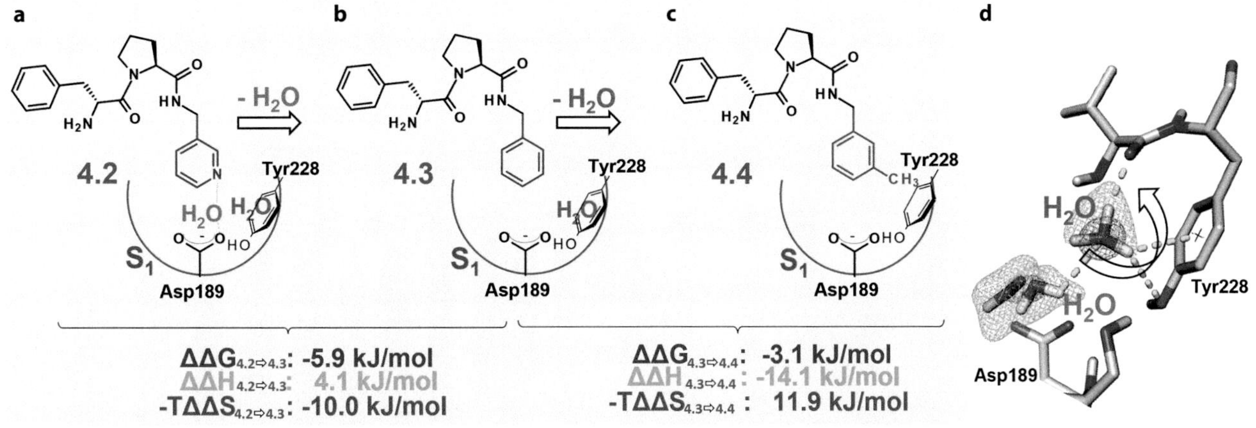

Fig. 4.7 Example of the displacement of two water molecules from the S_1 pocket of thrombin upon binding of ligands **4.2**, **4.3**, and **4.4**. Ligand **4.2** binds and two water molecules are found in the pocket (**a**). One water molecule (*orange*) connects **4.2** to Asp 189 via a hydrogen bond and is in a fixed position. The second water molecule (*green*) is dynamic and is located above Tyr 228 (compare (**d**), results of a neutron scattering study on the related trypsin). Comparing the binding of **4.2** (**a**) with that of **4.3** (**b**), the fixed *orange* water molecule is displaced. In the next step to **4.4** (**c**), the mobile *green* water molecule also leaves the pocket. The change in thermodynamic properties between the three states going from **4.2** (**a**) to **4.3** (**b**), and to **4.4** (**c**) is summarized by the relative differences $\Delta\Delta G$, $\Delta\Delta H$, and $T\Delta\Delta S$. Displacement of the *orange* water molecule leads to a strong entropic contribution (negative value, indicated here and in the following figures as relative difference ($-T\Delta\Delta S_{4.2\Rightarrow 4.3}$)). When **4.4** is bound, compared with **4.3**, the disordered *green* water molecule also leaves the S_1 pocket (**c**). However, in this case its displacement now proves to be an entropically unfavorable process (positive value for $-T\Delta\Delta S_{4.3\Rightarrow 4.4}$)

move in all spatial directions. In simple terms, during complex formation, **two independent particles become one common particle**. This reduction in degrees of freedom is expressed as a loss of affinity, ΔG, of approximately 16–20 kJ/mol. This contribution must first be produced by the newly formed interactions in the complex. From the considerations in Sect. 4.6, it is clear that this system, which at first glance appears to be binary, must be extended by a large number of water molecules. The water molecules of the solvation shell are also mobile, diffusing back and forth. Some of them are locally fixed for extended periods of time, especially if they are bound to the ligand or protein by multiple H-bonds. Such water molecules can often be identified during X-ray structure determination of a protein. Other water molecules are freely mobile and, therefore, much more difficult to detect in an X-ray structure determination. As described in Sect. 13.5, hydrogen atoms cannot usually be spatially resolved in X-ray structures of protein–ligand complexes. However, when neutron diffraction is used for structural analysis, H-atoms can be characterized with a high degree of confidence. A study by Johannes Schiebel on ligand complexes of trypsin showed that water molecules in the binding pocket exhibit different degrees of order depending on the bound ligand (Fig. 4.10). The **dynamic properties of these water molecules** affect the enthalpy/entropy profile of the ligand binding.

Conformationally flexible ligands can adopt multiple conformations in aqueous solution, which, as described above, can also efficiently entrap water molecules. In the binding pocket of the protein, the **conformational freedom of the ligand** is significantly limited. The conformation adopted at the protein site may differ from that in solution. In addition, the ligand must be desolvated. **Desolvation of a charged ligand** requires significantly more energy than desolvation of an uncharged ligand. The price to be paid is an unfavorable enthalpic binding contribution.

As described above, the incorporation of the ligand into the protein requires the **displacement of water molecules**. The associated thermodynamic signature depends strongly on the properties of the water molecules prior to their displacement. It should be noted, however, that ligand insertion can change the properties of all water molecules remaining in the pocket (Fig. 4.10). In some cases, the **water molecules** participate in ligand binding and **mediate contacts** between the binding partners. It is also important to remember that proteins can undergo conformational changes during binding, which also contribute to the energy balance. Sometimes binding opens pockets that were not present in the uncomplexed structure. Two extreme cases can be distinguished. Either the protein adopts several conformations on average over time, including the geometry of the ligand-bound state. The ligand selects and stabilizes this geometry during binding (**conformational selection**). However, there are also situations in which the desired conformation is so energetically unfavorable that it practically does not occur in the ensemble of possible conformations of the uncomplexed protein. Only when the ligand docks onto the protein surface does the pocket temporarily unlock and the ligand occupies the opening region. This **induced fit** represents the other extreme for binding processes that occur under structural adaptation of the protein.

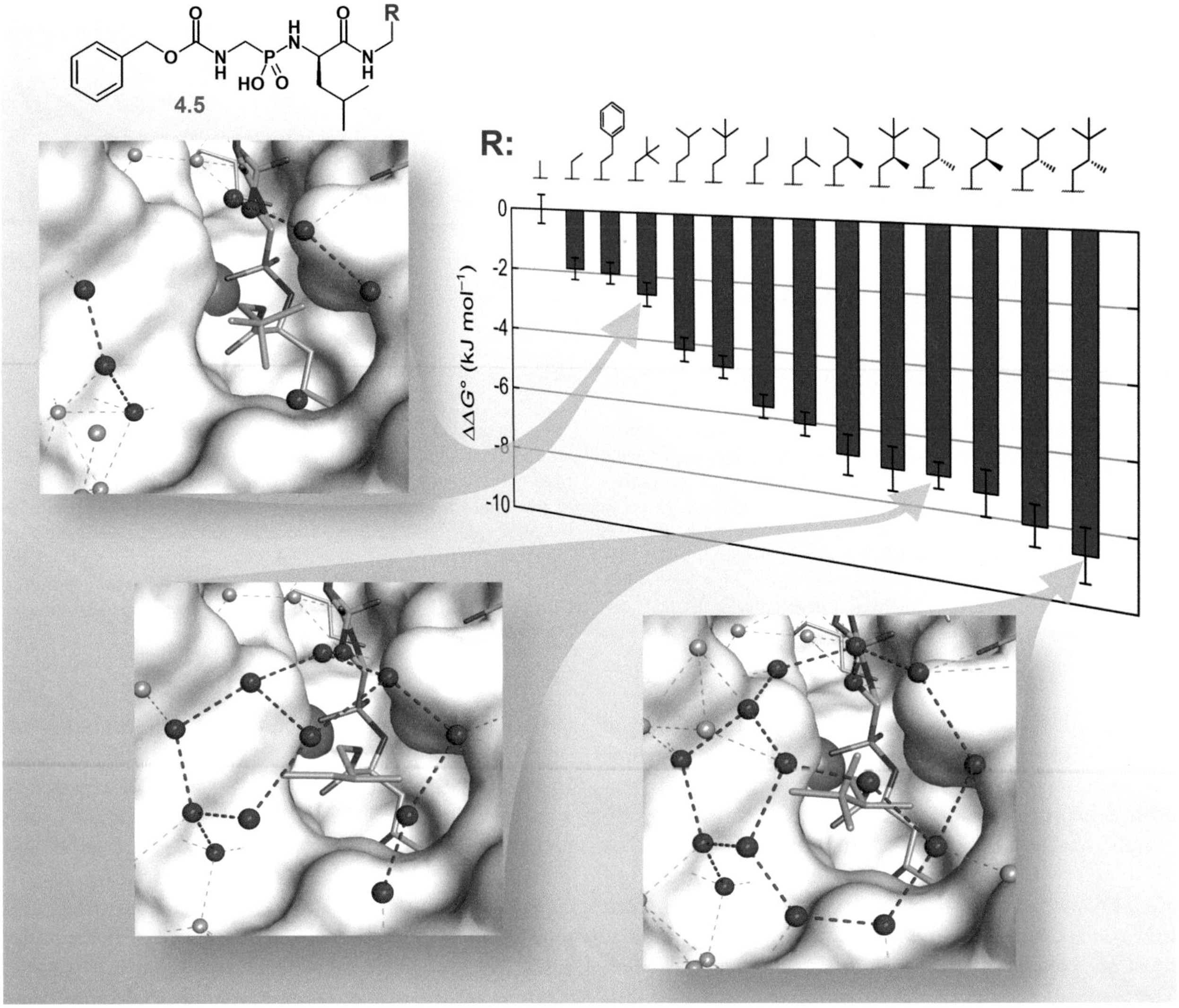

Fig. 4.8 Binding of a series of ligands with scaffold **4.5** to the metalloprotease thermolysin. The ligands place their side chain **R** in the bowl-shaped S_2' pocket (indicated by the *gray surface*), which is open to the surrounding solvent, and form a new surface of the complex together with the protein. It is solvated by water molecules. In the figure, the surface water molecules (*red spheres*) in contact with the protein and the hydrophobic side chain are shown for three singled-out complexes. Depending on their geometry, an increasingly well-adapted network of water molecules wraps around the ligand like a hood. In particular, the formation of ring structures can improve the strength of binding to the protein. Within the series, the binding affinity (expressed as a relative change $\Delta\Delta G$) of the ligands for thermolysin increases by a factor of 50, enhancing the inhibitory potency of the ligands. (Figure is taken from Krimmer et al., J. Med. Chem., **59** (23), 10530–10548 (2016), courtesy of the American Chemical Society)

4.8 What Is the Contribution of a Hydrogen Bond to the Strength of Protein–Ligand Interactions?

Now that we have qualitatively gone through the individual steps of protein–ligand complex formation, we turn to the question of how large the contribution of a particular interaction, e.g., a **hydrogen bond**, actually is to the binding affinity. This question can be answered experimentally by comparing two protein–ligand complexes that differ in their final composition by only one hydrogen bond. Such a comparison can be made, for example, by using protein mutants in which an amino acid that forms a hydrogen bond with the ligand is replaced by another amino acid that is unable to form this interaction.

In the enzyme aldose reductase (Sect. 27.4), the amino acid Leu 300 has been replaced by proline, thereby losing its ability to form an H-bond with the inhibitor fidarestat **4.6** (Fig. 4.11a). This was confirmed crystallographically. The loss of the H-bond is accompanied by a drop in affinity of 7.8 kJ/mol, and this loss is largely determined by enthalpy. In another experiment with the enzyme, Tyr 48 was replaced by Phe, again with the loss of an H-bond, here to the charged carboxylate group of the ligand IDD594 (**4.7**). Thus, a charge-assisted H-bond is lost in this case. The affinity loss is somewhat larger with

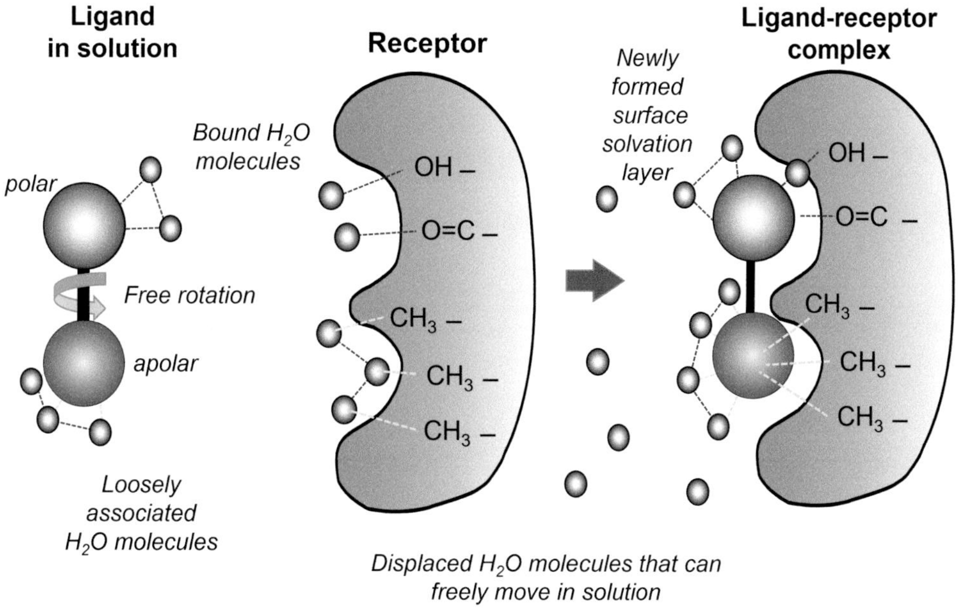

Fig. 4.9 Illustration of thermodynamic contributions to the change in Gibbs free energy ΔG. Before binding, the ligand can move freely in water. It has a certain translational and rotational entropy. In addition, the ligand is usually flexible and adopts different conformations in solution. Protein and ligand are solvated, forming H-bonds to water molecules (*blue spheres*). Some water molecules are in loose contact with the protein or ligand, while others are forming strong H-bonds. Translational and rotational degrees of freedom are lost during binding. The associated decrease in entropy is unfavorable for binding. In addition, the protein and ligand must shed some of their hydrate shell, also a process unfavorable for binding. The binding of the ligand leads to the formation of new direct interactions with the protein and releases water molecules from the protein pocket. Both are contributions that favor binding. In part, water molecules can mediate binding to the protein. The complex formed establishes a new surface that is re-solvated. H-bonds are shown as *blue dashed lines*, hydrophobic contacts analogously in *yellow*. (▶ https://sn.pub/iyZd6R)

8.5 kJ/mol, but this is partitioned into a large unfavorable enthalpic contribution and an opposite entropically favorable component. The strongly divergent profiles show that for charged groups additional contributions from stronger desolvation, electrostatic interactions and changes in the solvate structure have to be considered. The effects are even larger for a salt bridge, e.g., when an amide function is added to the thrombin inhibitor **4.8** (Sect. 23.4) that has an unsubstituted phenyl ring to produce **4.9**. The binding mode is retained and the added group forms a salt bridge with Asp 189 in the enzyme's S_1 pocket. The change in Gibbs free energy for binding of −14.4 kJ/mol demonstrates a substantially increased affinity. This improvement is due to a strong negative exothermic enthalpy contribution, but is partly compensated by an unfavorable entropic contribution. These and other examples show that the ΔG contribution of a hydrogen bond can vary between about −5 and −15 kJ/mol, largely depending on the given local charge conditions. The partitioning into enthalpy and entropy can extend over an even larger energy range, indicating that both can compensate their contributions to some extent.

It is also interesting to determine the influence of a water molecule on the binding profile, which mediates an interaction between protein and ligand. For this purpose, the complex of fidarestat **4.6** in aldose reductase is again compared with the analogous complex of sorbinil **4.10**. The two inhibitors differ in the attached carboxamide group. In fidarestat **4.6**, this group forms a direct hydrogen bond to the NH function of the amide group of Leu 300 (◘ Fig. 4.11a). The replacement of leucine for proline is accompanied by the loss of an H-bond. This results in a change in Gibbs free energy amounting to a decrease of 7.8 kJ/mol. Sorbinil **4.10** lacks the carboxamide groups (◘ Fig. 4.11d). Interestingly, the free enthalpy of binding for the Leu 300 → Pro exchange now remains almost the same. Since sorbinil lacks the group to form an H-bond with the NH group of Leu 300, the removal of the NH function in the proline variant is hardly noticeable. This explains the almost unchanged binding free energy. Nevertheless, binding to the wild-type enzyme is enthalpically more favorable, but entropically more "expensive" than for the mutant. The crystal structure indicates a water molecule that mediates an H-bond between the ether group of sorbinil and the NH function of Leu 300 (◘ Fig. 4.11d). This results in an enthalpy gain of −5.1 kJ/mol. At the same time, however, entrapping a water molecule is entropically unfavorable. This contribution of 5.9 kJ/mol just compensates for the enthalpic gain, leaving virtually no affinity benefit in ΔG

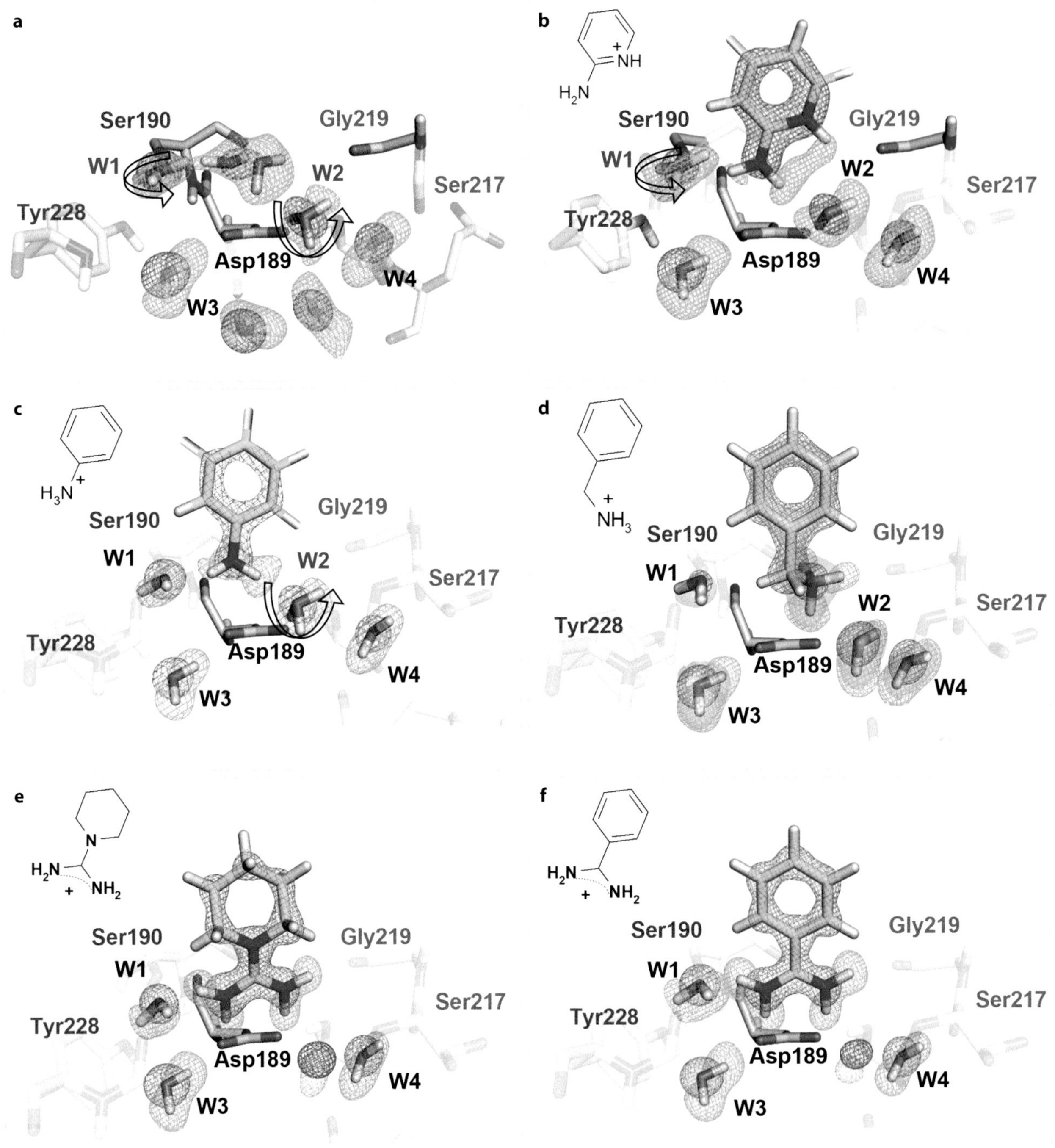

Fig. 4.10 Experimentally determined water structure in the S_1 binding pocket of trypsin before ligand binding (**a**), after binding of 2-aminopyridine (**b**), aniline (**c**), benzylamine (**d**), *N*-amidinopiperidine (**e**), and benzamidine (**f**). In the uncomplexed structure, five water molecules are found surrounding the aspartate residue (Asp 189), of which W1 and W2 are scattered over two orientations. They indicate dynamic properties of these two water molecules in the pocket (dynamic water molecules are labeled in *blue*, compare *arrows*). The binding of 2-aminopyridine (**b**) and aniline (**c**) displaces two of the five water molecules in the S_1 pocket. In the 2-aminopyridine complex, W1 remains disordered and W2 adopts an ordered arrangement. In the aniline complex, W2 remains dynamic and W1 adopts an ordered geometry. In the complex with benzylamine (**d**), both waters are ordered. In the case of *N*-amidinopiperidine (**e**) and benzamidine (**f**), W1 is ordered and W2 is displaced from the complex. The results are based on structural data determined by combining neutron (density in *green*) and X-ray scattering experiments (density in *orange*)

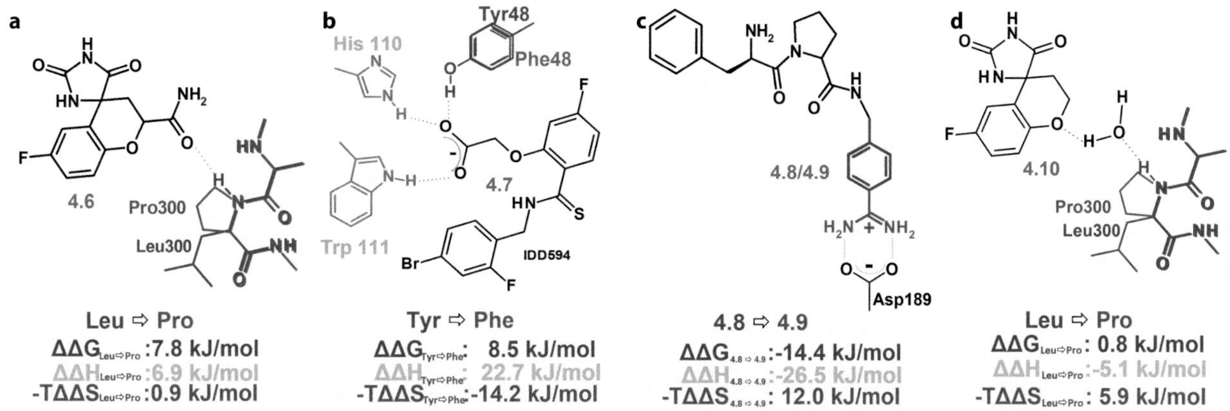

Fig. 4.11 Fidarestat **4.6** (**a**) forms a hydrogen bond with its carboxamide group to the NH group of Leu 300. By exchanging leucine for proline, the H-bond can no longer be formed. This leads to a $\Delta\Delta G_{\mathrm{Leu}\Rightarrow\mathrm{Pro}}$ loss of 7.8 kJ/mol, which is essentially being paid for by an enthalpic price ($\Delta\Delta H_{\mathrm{Leu}\Rightarrow\mathrm{Pro}}$: 6.9 kJ/mol). The inhibitor IDD594 **4.7** (**b**) binds to aldose reductase via three H-bonds with its carboxylate group. If Tyr 48 is exchanged for Phe 48, a charge-assisted contact is lost at this site. The loss in $\Delta\Delta G_{\mathrm{Tyr}\Rightarrow\mathrm{Phe}}$ is slightly larger than in the first example. However, it partitions into a huge enthalpic loss and smaller entropic gain. (**c**) In this example, the ligand **4.8** with a phenyl ring was modified into the benzamidine analogue **4.9**. A salt bridge to Asp 189 in thrombin is additionally formed, resulting in a strong gain in $\Delta\Delta G_{4.8\Rightarrow 4.9}$. This is governed by an enthalpic gain; entropy is opposing. In sorbinil **4.10** (**d**), the carboxamide group is lacking compared to fidarestat **4.6**. Considering again the exchange **Leu ⇨ Pro** in aldose reductase, this leaves the free binding energy $\Delta\Delta G_{\mathrm{Leu}\Rightarrow\mathrm{Pro}}$ virtually unchanged. However, sorbinil binds to the wildtype enzyme (leucine) enthalpically more favorably and entropically less favorably than to the proline variant. An interstitial water molecule mediates an H-bond between sorbinil and Leu 300, yielding an enthalpy advantage of −5.1 kJ/mol here. Simultaneously, however, the entrapment of a water molecule is entropically unfavorable ($-T\Delta\Delta S_{\mathrm{Leu}\Rightarrow\mathrm{Pro}}$: 5.9 kJ/mol) and virtually compensates the entire enthalpic advantage

in the inventory. The proline mutant cannot establish a water-mediated contact with sorbinil due to the lack of an NH function. Therefore, the enthalpic gain due to the H-bond is missing. However, there is also no entropic loss because no water molecule is entrapped.

The three-dimensional structure has been determined for a large number of protein–ligand complexes. Many of these complexes form hydrogen bonds between protein and ligand. The whole issue concerning the **contribution of a hydrogen bond to binding affinity** becomes evident in Fig. 4.12. Here, for a random selection of 80 protein–ligand complexes, the experimentally determined binding constants (logarithmic scale) are plotted against the number of hydrogen bonds. For a given number of hydrogen bonds, the measured binding constants cover a considerable range. Thus, the contribution of an H-bond is by no means constant, but varies considerably. Due to unfavorable desolvation effects, the contribution of an H-bond can even decrease the binding affinity.

A further important contribution to a **charge-assisted hydrogen bond** is where it is formed in the protein complex. In Fig. 4.11c, a salt bridge makes a large contribution to the transition from **4.8** to **4.9**. Deep in the S_1 pocket of thrombin, the benzamidine group of **4.9** forms a salt bridge with the carboxylate group of Asp 189 (Fig. 4.13, *left*). In the enzyme tRNA-guanine transglycosylase (Sect. 21.9), the binding mode of inhibitor **4.11** was characterized (Fig. 4.13, *right*). It carries a terminal phenyl ring, but is spatially close to the amino acid Arg 286 as seen in the crystal structure. To involve this charged amino acid in a salt bridge, **4.11** was terminally carboxylated to **4.12**. Although the crystal structure shows that the proposed salt bridge is indeed formed, its contribution to the affinity increase of **4.12** over **4.11** is

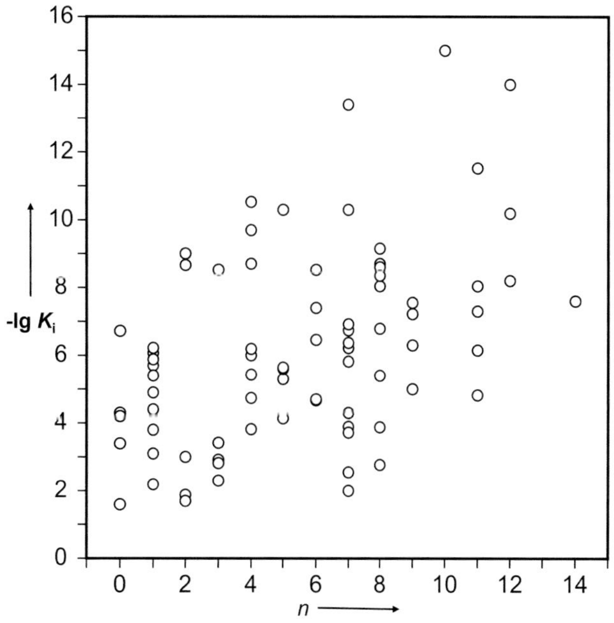

Fig. 4.12 A logarithmic plot of the binding constant of 80 protein–ligand complexes studied crystallographically against the number of hydrogen bonds formed between protein and ligand shows that there is no simple and direct correlation between the two properties

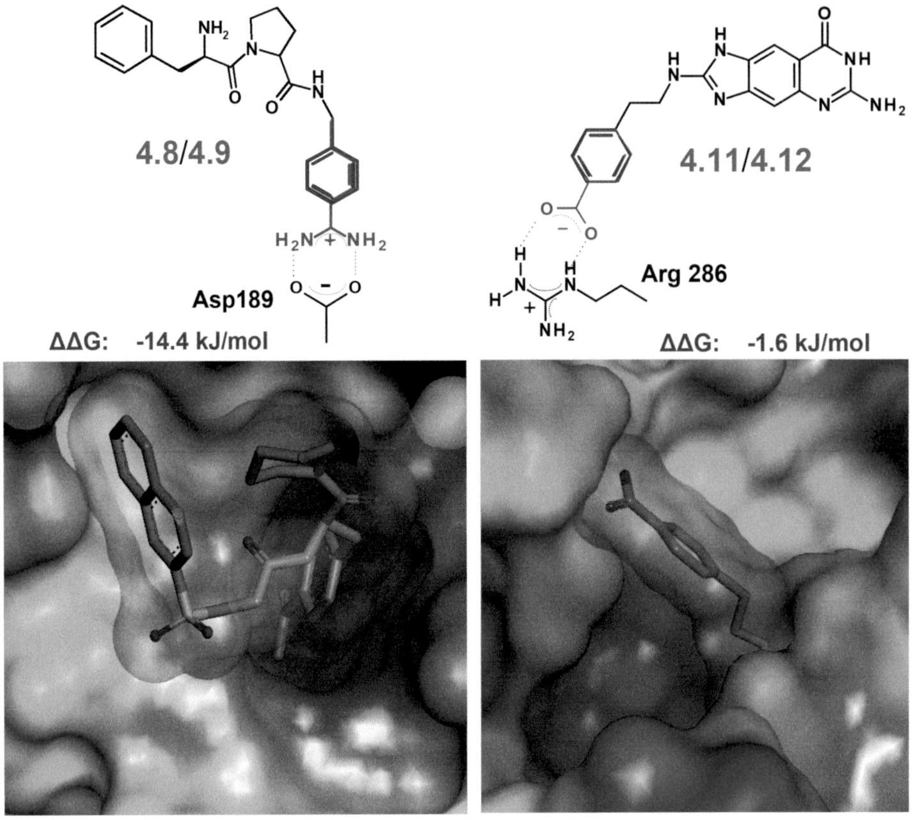

Fig. 4.13 For the contribution of a salt bridge to binding affinity, the environment in which this salt bridge is formed is crucial. In thrombin (*left*), the introduction of a benzamidine group **4.8**⇨**4.9** deeply buried in the S_1 pocket leads to the formation of a salt bridge with Asp 189. The affinity gain is very large (see Fig. 4.11c). In the tRNA-guanine transglycosylase (*right*), the addition of a carboxylate group to the terminal phenyl ring **4.11**⇨**4.12** does not enhance the affinity of the nanomolar binding inhibitor **4.11**. The crystal structure proves that the salt bridge to Arg 286 is formed geometrically. However, since it remains largely exposed to the surrounding solvent, its contribution is only about one tenth of the buried salt bridge in thrombin. (▶ https://sn.pub/BJz11h)

very small. The introduced salt bridge remains largely exposed to the surrounding solvent water, which reduces its impact on the affinity of the interaction.

An impressive example of the importance of hydrogen bonds is provided by the inhibitors **4.13** of the metalloprotease thermolysin, synthesized in the research group of Paul Bartlett. A phosphonamide $-PO_2HN-$ was replaced by a phosphinate $-PO_2CH_2-$ or a phosphonate $-PO_2O-$. The results of these exchanges are summarized in Table 4.3. Although the X-ray structure shows that the NH group forms an H-bond to the carbonyl oxygen of Ala 113, it can be replaced by a CH_2 group without loss of binding affinity. This result may be surprising at first glance, but it can be understood by comparing the number of hydrogen bonds before and after ligand binding for the phosphonamide and phosphinate, analogous to Fig. 4.6. The ligand forms an H-bond to water molecules via its NH function before binding to the protein in aqueous solution. The ligand has to abandon this H-bond when it enters the binding pocket, but there it forms a comparably strong interaction with Ala 133(C=O). Thus, the ligand loses one hydrogen bond but gains another; hence, the inventory is balanced. In the phosphinate, the $-CH_2-$ group cannot form an H-bond in water prior to protein binding. No H-bond is formed in the binding pocket either. Again, the inventory is balanced. In both cases, the number of H-bonds remains the same. If the NH group is replaced by an oxygen atom, the binding affinity decreases by a factor of 1000. In water, the oxygen atom that replaces the NH group can form a hydrogen bond with the bulk water. In the protein–ligand complex of the phosphonate $-PO_2O-$, the electronegative oxygen atom is exactly opposite the oxygen of the carbonyl group of Ala 113. Two acceptor groups face each other. A hydrogen bond cannot be formed here. The **hydrogen bond inventory** remains unbalanced. In addition, the two groups repel each other, resulting in weaker binding.

▫ **Table 4.3** Binding constants K_i for the thermolysin inhibitors **4.13**, which contain either a phosphonamide (X = –NH–), a phosphonate (X = –O–), or a phosphinate (X = –CH_2–) group. The phosphonamide group –PO_2NH– complexes the zinc ion and simultaneously forms an H-bond with Ala 113

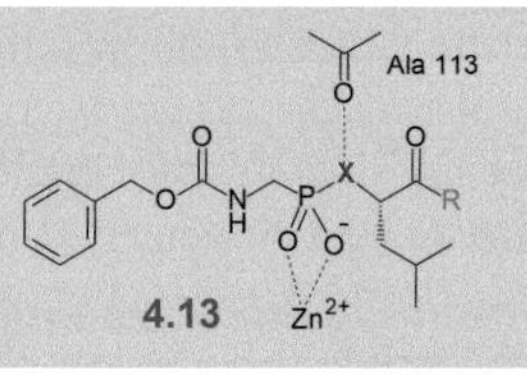

R	Binding constant K_i in µM		
	X = –NH–	–O–	–CH_2–
OH	0.76	660	1.4
Gly–OH	0.27	230	0.3
Phe–OH	0.08	53	0.07
Ala–OH	0.02	13	0.02
Leu–OH	0.01	9	0.01

A similar case is illustrated in ▫ Table 4.4. Here the binding affinity of three thrombin inhibitors **4.14** that were synthesized at Eli Lilly are compared with each other. The amine (X = –NH–) can form an H-bond with Gly 219 and binds the most strongly. The ether (X = –O–) binds 5000-times weaker because of an electrostatic repulsion between the ether oxygen atom and the carbonyl group of the protein. The aliphatic compound (X = –CH_2–) shows remarkable binding compared to X = –NH– that is merely reduced by a factor of eight (thrombin) and two (trypsin).

▫ **Table 4.4** Binding of **4.14** to the serine proteases thrombin and trypsin

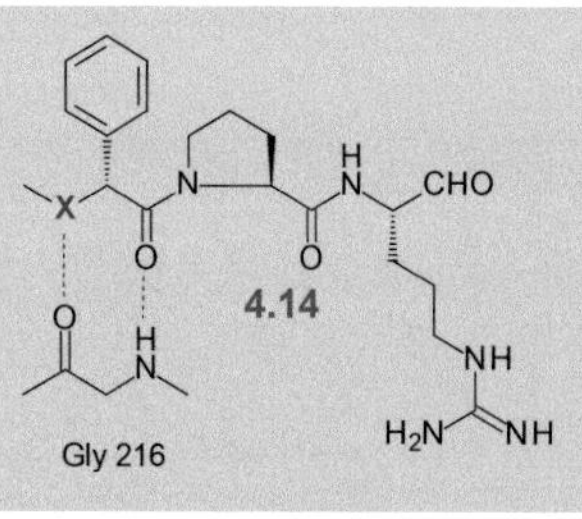

Enzyme	IC_{50} values in mg/mL		
	X = –NH–	–O–	–CH_2–
Thrombin	0.009	52	0.07
Trypsin	0.009	43	0.018

4.9 The Strength of Hydrophobic Protein–Ligand Interactions

We have seen that the direct attractive forces between lipophilic groups are much weaker than those between polar groups. Hydrophobic interactions are based on weak attractive forces in the immediate molecular environment (Sect. 4.4). They are formed in the protein environment as well as in the aqueous milieu. Therefore, they contribute little to the inventory. For **hydrophobic interactions**, however, it is mainly the displacement and rearrangement of the neighboring water molecules that counts. It has been shown in many experiments that their contribution to the binding affinity is, to a first approximation, proportional to the **size of the lipophilic surface** that is buried upon ligand binding and therefore no longer accessible to water. Typically, the contribution is found to be in the range of −50 to −200 J/mol per Å^2 of lipophilic contact

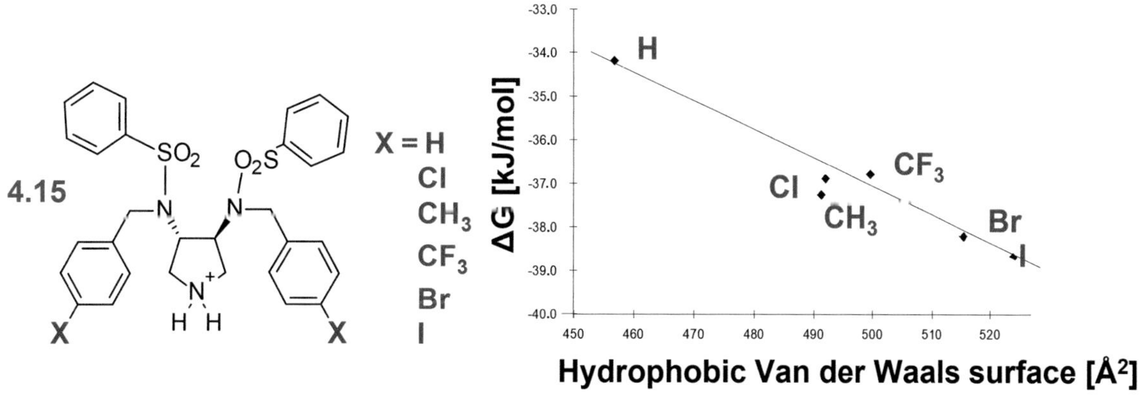

▫ **Fig. 4.14** The scaffold of the HIV protease inhibitor **4.15** was enlarged during the course of a lead structure optimization by adding hydrophobic groups to the aromatic *N*-benzyl group. An unchanged binding mode was evidenced crystallographically. The additional molecular volume improved the binding affinity in a linear manner by about −65 J/mol Å^2

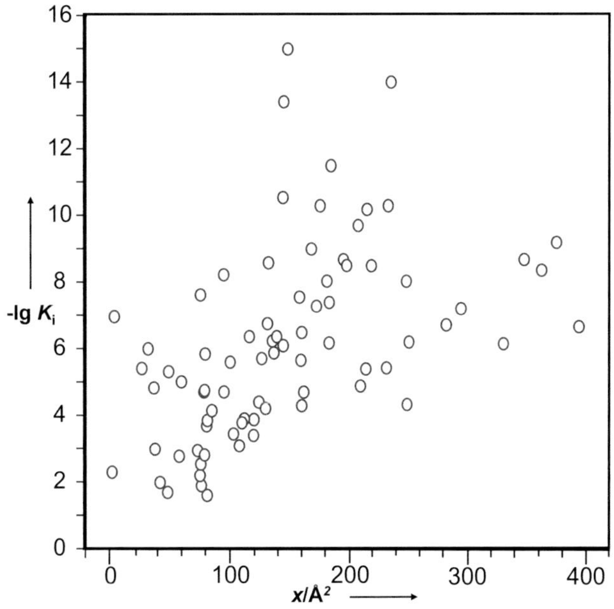

Fig. 4.15 In analogy to Fig. 4.12, a logarithmic plot of the binding constants K_i of the 80 crystallographically investigated protein–ligand complexes against the buried hydrophobic surface area shows that there is no simple function for this measure either

area. An example of this is retinol. It binds to the retinol binding protein (Fig. 4.1) with a binding constant of 190 nM exclusively through lipophilic contacts. This corresponds to a free energy of −39.8 kJ/mol. As a result of the binding, a lipophilic area of 250 Å^2 is buried. The contribution per Å^2 is −39,800/250 = −159 J/mol Å^2.

Six HIV protease inhibitors (Sect. 24.6) are listed in Fig. 4.14. During the course of a lead structure optimization, the hydrophobic surface of **4.15** was enlarged by adding hydrophobic groups. It could be confirmed crystallographically that the binding mode did not change. If the variations in the molecular volume in this series are plotted against the affinity, a linear relationship is obtained. The binding affinity increases by −65 J/mol Å^2.

In many cases, the hydrophobic interactions are a dominant contribution to the free energy of binding. In Fig. 4.15, the lipophilic surface area that is buried upon complex formation of the same 80 protein–ligand complexes as in Fig. 4.12 are shown together with their experimentally determined binding constants. Here too, the values are scattered over a broad range.

4.10 Binding and Mobility: Compensation of Enthalpy and Entropy

According to Eq. 4.4, enthalpy and entropy have a close physical relationship and their combined contributions give the Gibbs free energy of binding. Considering the formation of protein–ligand complexes, the ΔG of weakly binding millimolar complexes and strongly binding nanomolar complexes are in the range of about 35–55 kJ/mol. Lead optimization (Chap. 8) usually covers an even smaller range. Typically, the binding constants are improved by 5–6 orders of magnitude, corresponding to 25–30 kJ/mol. When functional groups are exchanged in a lead structure, the change in enthalpy ΔH usually varies over a much larger range. If the variation of ΔG is much smaller for this transformation, the change in enthalpy ΔH has to be compensated by a change in entropy $-T\Delta S$ in opposite direction, simply because of numerical reasons. Only in this way can the large variations in the two properties lead to the result that ΔG remains in a small window. This leads to an important question: Is there a connection that causes the opposing **enthalpy and entropy to partially compensate** during optimization? If there is compensation, how can both quantities be optimized without canceling each other out, so that ΔG remains unchanged?

Entropic optimization is aimed at increasing the hydrophobic surface area buried upon binding. This very descriptive quantity expresses that enlarged ligands displace an increasing number of water molecules upon binding. In particular, if these displaced water molecules were previously well fixed in the binding pocket, an entropically favorable signal will result.

The entropic contribution to binding of a ligand can be enhanced by synthesizing compounds that have reduced degrees of freedom around bonds, while maintaining a geometry that fits into the protein binding pocket. For example, in the peptide-like thrombin inhibitor **4.16**, a substituted glycine residue in the center was exchanged for the more rigid proline to form **4.17**. Crystallographically, it was found that the binding mode of both ligands remained unchanged. The rigidized ligand **4.17** is more potent by −10.8 kJ/mol (about two orders of magnitude in binding constant!), which can be attributed to a favorable entropic binding contribution. Molecular dynamics simulations of both ligands in aqueous medium prior to protein binding show that the accessible conformational space of the proline derivative **4.17** is significantly restricted compared to the open-chain glycine derivative **4.16**. This has a strong impact on the number of degrees of freedom that are sacrificed during binding.

The two thrombin inhibitors **4.18** and **4.19** were synthesized with the same goal. In their case, the rigidification should be realized through an intramolecular hydrogen bond to the hydroxymethylene and aminomethylene anchors on the phenyl ring, respectively. First, the two compounds were shown to adopt an identical binding mode in the thrombin binding pocket. The observed geometry also corresponds to the binding mode adopted by inhibitors that completely lack the anchor. But to what can the increase in affinity from **4.18** to **4.19** then be attributed, since both compounds form the desired stabilizing H-bond in the binding pocket? The difference is to be found in the rigidification of the ligands in aque-

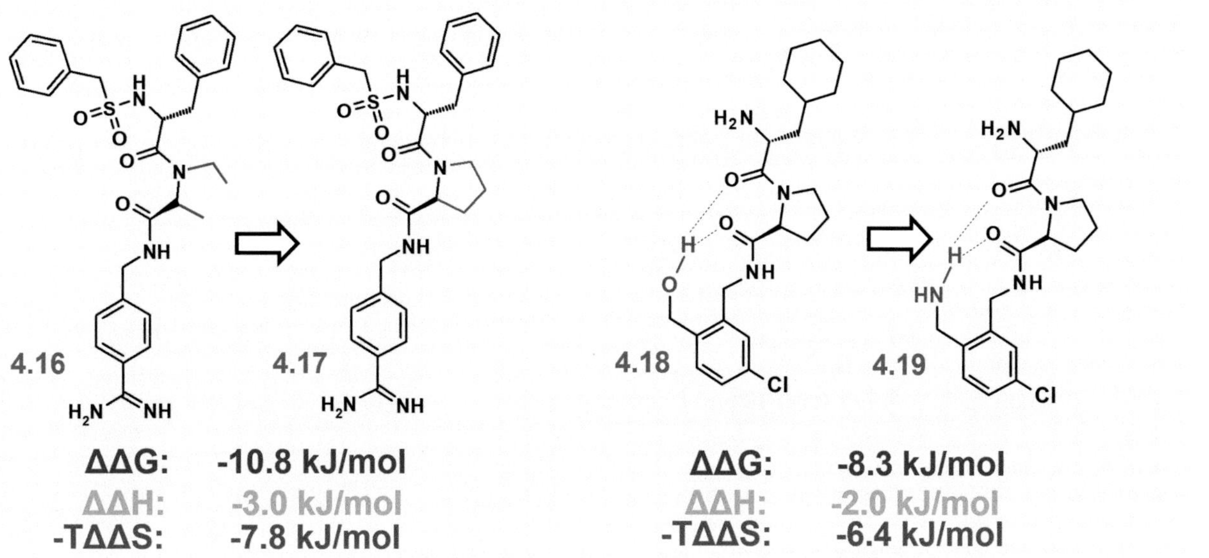

Fig. 4.16 Effect of rigidification or "pre-organization" of inhibitors on protein binding. *Left* The peptide-like thrombin inhibitors **4.16** and **4.17** differ only in the exchange of a central substituted glycine residue for proline. This conformationally restricts the molecular scaffold of **4.17** to a few conformers that are also adopted at the binding site. **4.17** therefore sacrifices fewer degrees of freedom upon binding than **4.16** and binds with a distinct entropic advantage. *Right* In thrombin inhibitors **4.18** and **4.19**, an anchor was introduced to stabilize the ligands in the bound state by an intramolecular H-bond. The binding mode adopted is similar to that present even in inhibitors that lack this anchor. However, in solution prior to binding, the intramolecular H-bond to the aminomethylene group is stable, whereas it is not in the case with the hydroxymethylene group. Therefore, only **4.19** is correctly "pre-organized" for protein binding. Here, rigidification results in the desired entropic gain. For **4.18**, this advantage is lacking as the compound adopts the required conformation only after accommodation in the protein binding pocket

ous solution before binding to the protein. Whereas the aminomethylene derivative **4.19** is stable with an intramolecular hydrogen bond even before protein binding, the H-bond in the hydroxymethylene analogue **4.18** is unstable. In the unbound state, there is hardly any formation of the conformation of this ligand taken up in the protein. Only if this is achieved does **rigidification of a ligand** lead to the expected entropic gain in affinity (see Fig. 4.16).

In order to enthalpically increase the binding affinity of a ligand to a protein, it is necessary to introduce additional polar interactions. However, this usually comes at the cost of the additional polar groups having to shed their water shell. This contribution to **ligand desolvation** must be provided. Fig. 4.11c compares thrombin inhibitors **4.8** and **4.9**. Here, the added amidino group in **4.9** allows a significant increase in affinity with a large increase in enthalpy. However, the desolvation of the charged inhibitor must be overcompensated. The cost of desolvation can be very high, as shown by the comparison of thrombin inhibitors **4.20** and **4.21** (Fig. 4.17). Crystallographically, the same binding mode is found for both. However, the affinity of **4.21** with the charged pyridinium group is 11 kJ/mol lower than that of the uncharged isostere **4.20**. This is due to the high enthalpic cost of shedding the solvate shell around the charged pyridinium derivative.

A further example of the importance of the contribution of the ligand dynamics of an interaction is demonstrated with the thrombin inhibitors **4.22** and **4.23** (Fig. 4.17). They differ only in the size of their terminal cycloalkyl residue, which was added to the parent scaffold to fill the hydrophobic S_3/S_4 pocket of the protein. Both inhibitors have virtually the same binding affinity for thrombin. However, their free binding energy decomposes very differently into contributions from enthalpy and entropy. The compound with the cyclopentyl substituent has an enthalpic advantage and an entropic disadvantage compared to the six-membered ring derivative.

What is the reason for this surprising effect of the **shift in the enthalpy/entropy partitioning**? The crystal structures of both derivatives with thrombin show an important difference with respect to the terminal cycloalkyl substituent. While the five-membered ring is well observed in the electron density (Sect. 13.5), virtually no density can be detected in the region where the six-membered ring should be found. Such an observation in a crystal structure indicates increased disorder of a particular structural building block in a protein–ligand complex. This disorder may be purely static, in which case the six-membered ring is scattered over many arrangements. Alternatively, for dynamic reasons, it may have a much higher residual mobility in the protein-bound state than the five-mem-

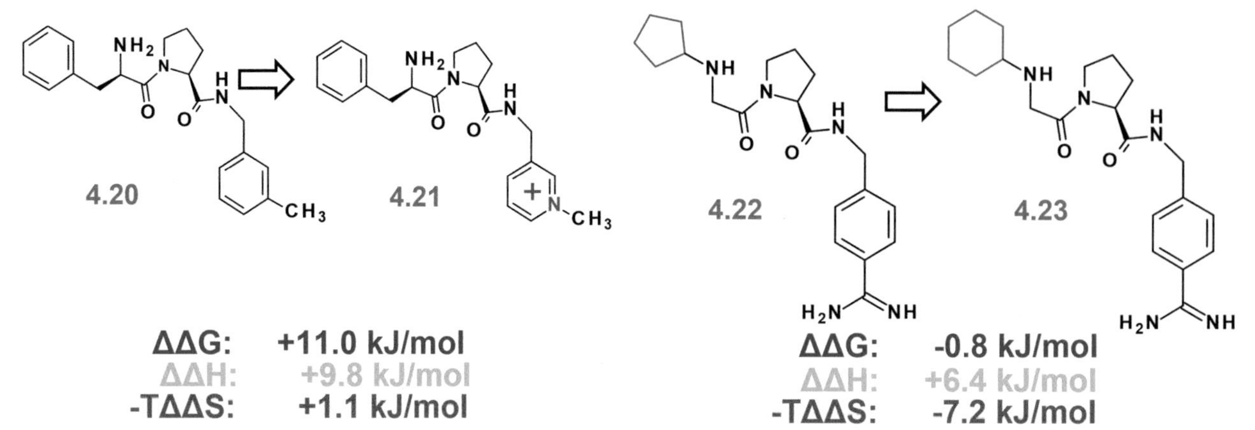

Fig. 4.17 *Left* Replacement of the *meta*-tolyl substituent in the thrombin inhibitor **4.20** with an isosteric methylpyridium group in **4.21** results in a significant loss of affinity, essentially at the price of enthalpy. The charged head group for the S_1 pocket of thrombin requires a very high price of desolvation but does not allow additional favorable interactions in the enzyme. *Right* The homologous ligands **4.22** and **4.23** bind equally strongly to thrombin, but the binding affinities partition quite differently into enthalpic and entropic contributions. **4.23** has much higher residual mobility in the binding pocket than **4.22**, giving this derivative an entropic advantage. But because of the poorer contacts to the protein on average, an enthalpic disadvantage results

bered ring derivative. Molecular dynamics simulations (Sect. 15.7) confirm this latter difference due to different residual mobility. In the case of the five-membered ring compound, the cyclopentyl moiety remains in a hydrophobic pocket and performs a so-called jump rotation from time to time. In this process, the planar ring swaps between two configurations, exchanging its top and bottom faces. But the overall position and occupancy of the ring in the pocket remains virtually unchanged. Ligand **4.22** does not form a hydrogen bond (Sect. 23.4) to the carbonyl group of Gly 216. The six-membered ring derivative **4.23** behaves quite differently. Here, the cyclohexyl substituent moves out of the binding pocket during the course of the simulation and returns back to it after some time. At the same time, **4.23** forms an intermediate hydrogen bond to Gly 216. Thus, **4.23** has a high residual mobility in the bound state and scatters over several spatially distinct arrangements.

This difference in the dynamic behavior of **4.22** and **4.23** explains their deviating thermodynamic profiles. The cyclopentyl derivative has an entropic disadvantage because it binds more tightly in the binding pocket. However, the uniform orientation gives this ligand an advantage in forming stable enthalpically favorable interactions with the protein. The situation is different for the six-membered ring derivative. Its reduced spatial fixation in the binding pocket is accompanied by a smaller loss of degrees of freedom in complex formation. This implies an entropic advantage. Enthalpically, however, this behavior is disadvantageous. Due to the intermediate leaving of the ligand from the binding pocket, interactions with the protein can only be formed with reduced strength.

What can we learn from this example? Even if ligands have a very similar chemical structure, their binding behavior can differ significantly. Their residual mobility in the binding pocket can be decisive for the thermodynamic binding contributions. Obviously, a **mutual compensation of enthalpy and entropy** leads to an almost unchanged free energy ΔG. This interplay between residual mobility in the binding pocket and the quality of the interactions formed clearly has consequences for the optimization process and is one explanation why enthalpy and entropy often compensate in optimization.

Medicinal chemists like to think in terms of standardized **group contributions** that the exchange of certain moieties and functional groups at a given scaffold might provide to the binding affinity. Statistical analyses of such group contributions have been performed and can be used as a set of rules for optimization strategies. Mostly, **these rules are considered as additive**. How much is gained by combining a particular group with another on a molecular scaffold to be optimized? However, care must be taken when applying such considerations. Small differences in binding behavior often cause these simple rules to break down.

As an example, consider the optimization of thrombin inhibitor **4.24** to **4.25** (Fig. 4.18). Two chemical modifications will be performed. In the first step, the terminal hydrophobic substituent at one end of the parent scaffold is increased from an *n*-propyl to a phenylethyl substituent. This results in a significant increase in the hydrophobic surface area of the molecule, but the binding affinity is only slightly improved by $\Delta\Delta G = -3.1$ kJ/mol. As a second subsequent optimization step, an amino group is introduced adjacent to the hydrophobic group to form a hydrogen bond to Gly 216. This adds another -15.5 kJ/mol, so the two modifications from **4.24** to **4.25** result in an affinity increase of $\Delta\Delta G = -18.6$ kJ/mol. Does the addi-

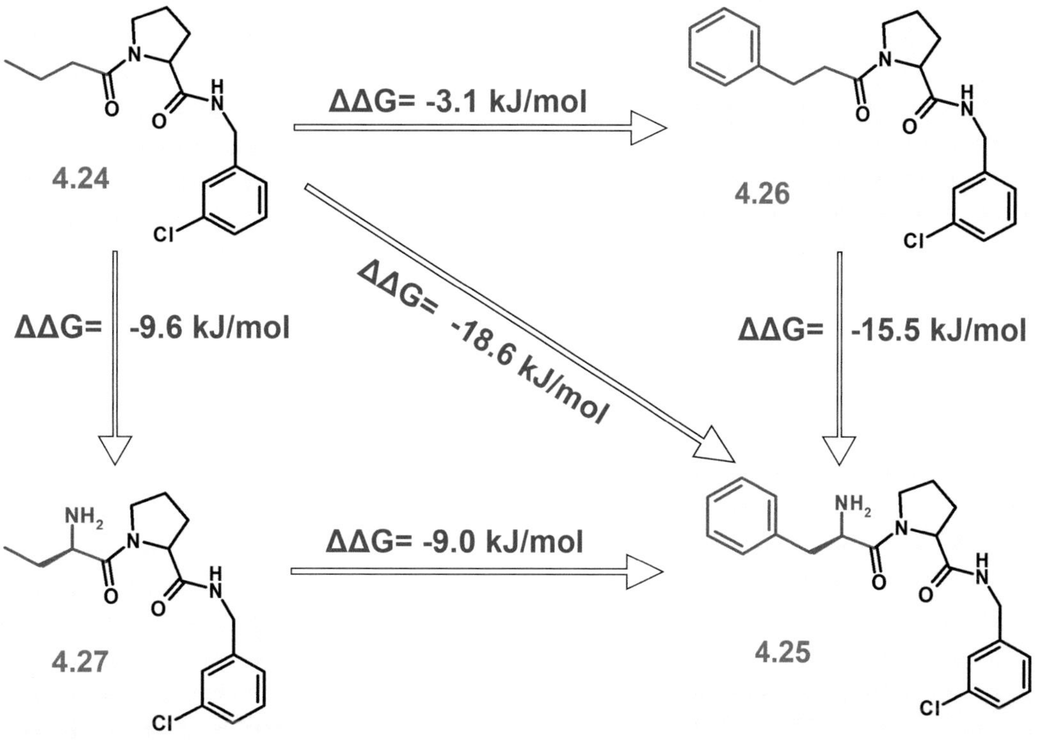

Fig. 4.18 Optimization of thrombin inhibitor **4.24** to **4.25** yields an affinity increase of $\Delta\Delta G = -18.6$ kJ/mol (*diagonal arrow*). This is achieved by increasing the hydrophobic side chain (*red*) from *n*-propyl to a phenylethyl residue and adding an amino group (*blue*). The changes can also be performed stepwise. Increasing the hydrophobic surface area to **4.26** improves the affinity by only −3.1 kJ/mol (*top arrow*). A major contribution of −15.5 kJ/mol is provided by the subsequently introduced amino group (*right arrow downwards*). The ligand forms a charge-assisted H-bond to the protein via this group. Reversing of the synthesis steps by adding the amino group first to **4.27** yields −9.6 kJ/mol (*left arrow downwards*), and the subsequent substitution of the hydrophobic portion enhances the affinity by a further −9 kJ/mol (*bottom arrow*). The reason why standardized group contributions cannot simply be added up here lies in the complex interplay of residual mobility, desolvation, and strength of the enthalpic interactions that are formed

tional amino group add that much to the affinity, and can we include this value in our list of rules for standardized group contributions? Of course, as a validation, the two modifications can be introduced in reverse order via intermediates **4.26** and **4.27**, respectively. If the reverse route is followed and the amino group is introduced first at **4.24** to **4.27**, the gain in $\Delta\Delta G = -9.6$ kJ/mol, a significantly smaller contribution by this group than along the first route. The subsequent increase of the hydrophobic surface from **4.27** to **4.25** now achieves an additional affinity gain of −9.0 kJ/mol. This now a significantly larger contribution for the hydrophobic group than along the first synthesis path.

This example shows that **simple additivity rules of standardized functional group contributions fail**. Instead cooperativity matters. As in the example with the five- and six-membered ring derivatives **4.22** and **4.23**, the balance between residual mobility, partial solvation of the binding pocket and the quality of the interactions formed has a decisive influence on the affinity increase. The interplay of **partially compensating enthalpic and entropic binding contributions** is responsible for this complex picture. The bulky phenylethyl substituent fixes **4.26** quite well in the binding pocket, whereby the ligand gains hydrophobic interactions and displaces water molecules from the protein pocket. However, it pays entropically for the loss of residual mobility compared to **4.24**, so the gain in affinity due to the increased size of the hydrophobic substituent is only moderate. However, once this price is paid, the additional charge-assisted H-bond of **4.26** ⇨ **4.25** provides a large amount of affinity gain. Following the reverse optimization strategy, the H-bond introduced by **4.24** ⇨ **4.27** now has to pay part of the price for the reduced residual mobility. Therefore, the contribution of the H bond is now much smaller. Once the ligand **4.27** is spatially fixed in the pocket, much more affinity can be gained by attaching the larger hydrophobic substituent to **4.25**. The price of the entropic loss of degrees of freedom has already been paid.

This example illustrates the dilemma that medicinal chemists often face when optimizing lead structures, as expectations of success on the way to the planned end product, which are usually based on estimates from standardized group contributions, can be disappointed already after the first synthesis steps. If we were to take the

4

path from **4.24** to **4.27**, we would likely be much more optimistic about achieving the optimization goal than if we were to go from **4.24** to **4.26**, because the first step in the latter synthesis route yields a much smaller affinity gain.

At first glance, **enthalpy/entropy compensation** appears to be an unavoidable curse that can easily thwart a medicinal chemist's optimization efforts. However, decomposing affinity as a free energy quantity into entropy and enthalpy provides deeper insight into the mechanisms that ultimately enable potency enhancement. However, it must never be forgotten that the entire process must be considered! Everything has to be taken into account, from the ligand and protein in the aqueous solution before binding, including all the water molecules involved, to the formed complex with its newly created solvation shell. Elaborate microcalorimetric analyses condense this complex process from a multitude of steps to three numerical values. In the end, they represent the thermodynamic profile of the entire binding event. Whether the result is a more **enthalpy- or entropy-driven binding** is determined by the overall balance. Michael Gilson and his group coined the term **enthalpy–entropy transduction** to describe the consequences of local perturbations in protein–ligand binding that, for example, mask the enthalpic binding of a ligand to its protein. The enthalpic signature of binding can be obscured by the thermodynamics of a global conformational change during complex formation. If it is induced as a consequence of the binding of the ligand, it may be accompanied by a change of the system under consideration to a conformational state of higher entropy. It is possible that such considerations help to explain why profiles with pronounced enthalpy–entropy compensation are observed experimentally in so many binding events. This masking could produce a misleading picture of the actual driving forces of binding. This makes it all the more important to always compare systems relative to each other by varying only one or a few parameters or properties.

4.11 Lessons for Drug Design

This chapter is not intended to leave the impression that protein–ligand interactions are too complicated for quantitative predictions of the strength of protein–ligand interactions. Rather, quantitative correlations can be established. All the more, some general rules or guidelines should be followed when optimizing a lead structure:

- Many strong protein–ligand interactions are characterized by **extensive lipophilic contacts**. Increasing the lipophilic contact area between the protein and the ligand often leads to an improvement in binding affinity largely as the result of entropic gains from the release of ordered water molecules from the surfaces. This means that the search for unoccupied lipophilic pockets in the protein should be the first step in the design and optimization of new ligands. However, this approach should not be taken too far, as a large increase in the overall **lipophilicity of a molecule** will increasingly reduce its water solubility.
- The binding profile for the release of a water molecule can range from dominantly entropic to dominantly enthalpic. The release of fixed water molecules usually correlates with an entropic signature, while the release of highly dynamic water molecules correlates with an enthalpic signature. Most likely, the whole range from more entropic to more enthalpic dominated signature for ΔG can be found, but with the drawback that the contribution of water displacement to the binding free energy of ligand binding is increasingly offset by mutually compensating effects. This is particularly the case when the water displacement profile is characterized by a balanced enthalpy/entropy signature. Neutron diffraction has shown that the residual mobility of water molecules involved in ligand binding may differ from ligand to ligand, making the estimation of the water signature even more complex.
- An increase in binding affinity due to additional H-bonds is not guaranteed. For the overall affinity contribution, an H-bond will only have a significant effect if the **interactions of the H-bonded groups** in the protein–ligand complex will be **stronger than in the aqueous environment** with the surrounding water molecules. In general, such an H-bond provides a much stronger affinity contribution when it is shielded from solvation and formed in a deep binding pocket. This is especially true when the binding partners carry a formal charge and **charge-assisted H-bonds** are formed. An optimal situation can be expected when the involved functional groups undergo a pK_a shift due to mutual polarization, transforming groups that were uncharged in aqueous solution prior to binding into charged groups in the protein binding pocket. This requires a sophisticated **adjustment of pK_a values** in the designed molecules.
- On the other hand, burial of polar atoms without saturating them with an H-bond almost always results in a loss of binding affinity. Ligand design must ensure that **polar ligand atoms find binding partners in the protein** when they are no longer accessible to water in the formed protein–ligand complex.
- **A ligand almost always displaces water molecules during protein binding**. There are protein binding pockets that are designed in such a way that they cannot be optimally solvated by water for geometric reasons. In these cases, a ligand may be able to form more H-bonds to the protein than water molecules can. The binding affinity of such ligands can be very high.
- However, hydrophobic binding pockets can also be filled with very few or even no water molecules in the uncomplexed state. In such pockets, hydrophobic groups achieve very strong enthalpic binding.
- **Rigid, properly "pre-organized" ligands can bind more tightly** than flexible ligands because the loss of inter-

nal degrees of freedom is less for rigid ligands. It is important to note that the "pre-organized" geometry must already be conformationally populated in aqueous solution prior to binding.

- When a ligand occupies binding sites that are open to the solvent, its substituents can contribute to the new surface of the formed protein–ligand complex. If they allow the **formation of an optimally structured solvent shell** with fused H-bonded rings of water molecules covering the ligand portions oriented towards the outside of the protein pocket, the ligand will gain binding affinity. In contrast, polar groups in this interface region can disrupt the solvation pattern and remain insufficiently solvated themselves. This will result in a loss of affinity.
- Placement of a titratable functional group of a ligand in a protein binding pocket can result in **local pK_a shifts of several log units**. Ideally, this leads to the formation of a **charge-assisted H-bond** to the protein, whereas this group is previously uncharged in the aqueous medium or during membrane passage. Its desolvation is more favorable, and the charge-assisted H-bond formed improves affinity. Charge-assisted H-bonds make a greater contribution to affinity when formed in deep pockets shielded from solvent access.
- By clever design, ligands in the complex can adopt a binding geometry in which they shield their charge-assisted H-bonds to the protein themselves from the surrounding solvent. This can greatly increase affinity.
- If two ligands that bind independently to the protein can be successfully merged or fused into one molecule without much perturbation of their binding modes, a strong gain in affinity can be expected. Such concepts can be applied especially in the early stages of fragment-based lead discovery (Sect. 7.9).

The relative contributions of changes in enthalpy and entropy to the binding affinity are of great importance. The combination of these terms results in ΔG which is the overall quantity that is optimized in a drug design project. This goal can be achieved by improving either the enthalpic or the entropic contribution. Ideally, both quantities are optimized simultaneously. However, this requires focusing on different parameters of the protein–ligand interactions (Sect. 8.8). An open question is whether a more **enthalpically or entropically driven binding will provide advantages** for the optimization of a particular drug. The strategy to be pursued will depend on whether tolerance to rapidly emerging resistance mutations (Sects. 24.5, 31.5, 32.5) is desired for the binding of the drug to be developed. In other cases, high target selectivity or broad binding promiscuity within a protein family may be desired for an optimal therapeutic effect (Sects. 25.6, 26.4, 27.5). Ultimately, these criteria determine whether a more enthalpically or entropically driven binding is the better choice in a given drug design project.

4.12 Synopsis

- Emil Fisher introduced the "lock-and-key" principle to describe the interaction of a small-molecule substrate and a macromolecular receptor. More than 50 years later, Koshland extended this picture by induced-fit considerations that allow both binding partners to change conformations and mutually adapt to each other to optimally interact.
- The cells are surrounded by a lipid double-layer membrane with polar head groups on the exterior and hydrophobic alkyl chains in the interior. This membrane is a barrier for polar substances, but sufficiently lipophilic compounds can penetrate and even pass through the membrane.
- The strength of protein–ligand interactions is measured by the binding constant, which quantifies the stability of a protein–ligand complex as a dissociation constant according to the law of mass action for complex formation.
- The binding constant is logarithmically related to the change in Gibbs free energy of binding. The change in free energy is composed of an enthalpic and entropic contribution. The enthalpic part summarizes all terms that relate to the interaction energy of the binding partners. The entropic part considers the ordering of the system and how its energy content is distributed over the degrees of freedom of the system.
- Protein–ligand complexes usually form through noncovalent interactions, predominantly through hydrogen bonds. The strength of hydrogen bonds strongly depends on the distributions of charges among the interacting functional groups. Whether a group is charged or not depends on its protonation state, which is defined by the pK_a value of the titratable groups involved in the protein–ligand interactions.
- Depending on the local environment in a binding pocket, the pK_a values of titratable groups can vary significantly and can, by this, transform a normal H-bond into a much stronger charge-assisted H-bond.
- Hydrophobic interactions form through the close proximity of nonpolar functional groups of the binding partners. As direct interactions, they are rather weak. Nevertheless, they can afford a significant contribution to binding affinity through the release of water molecules from either the lipophilic environment of the binding pocket or from the ligand surface next to a lipophilic surface patch.
- The strength of protein–ligand interactions is strongly influenced by the water environment. Both the protein binding pocket and the ligand are solvated before complex formation and functional groups of the protein and ligand will form H-bonds to water molecules. The total balance of the hydrogen-bond

inventory before and after complex formation matters for binding affinity considerations. Only if the number of newly formed hydrogen bonds in the complex is increased in number and/or strength compared to those previously formed in the bulk water phase will a net affinity increase result.

- The release of water molecules from hydrophobic surfaces can increase affinity by enthalpy and entropy. Release of fixed water molecules increases the degrees of freedom and boosts entropy. Replacement of highly disordered water molecules into the bulk water environment can contribute to an enthalpic gain.
- Entropic contributions to binding arise from an increase of the degrees of freedom of the protein–ligand–water system and, as a first approximation, correlate with the size of the hydrophobic surface buried in the formed complex.
- At the end of complex formation, the ligand forms a new surface with the protein which is solvated. If optimal water networks can form around the exposed parts of the ligand that participate in the shared surface, the bound ligand gains additional affinity.
- Correct rigidification ("pre-organization") of a ligand into the conformation required at the binding site can lead to a significant increase in affinity for entropic reasons. Importantly, however, the pre-organized conformation must be populated in aqueous solution prior to protein binding.
- Change in free energy variations are observed over a range of about 15–60 kJ/mol in protein–ligand complexes. Variations in enthalpy (ΔH) and entropy ($-T\Delta S$) can be much larger. This results from extensive enthalpy/entropy compensation. The entropy increases if the number of degrees of freedom in the system increases, previously fixed water molecules will be released from the binding pocket into the bulk water phase, or a high residual mobility will remain in the binding pocket. This usually has a detrimental effect on an increase in enthalpy, since the formation of strong interactions is then hardly possible in an efficient manner.
- The pronounced interdependence of enthalpy and entropy along with dynamic versus interaction–geometric phenomena causes simple additive rules about standardized functional group contributions to fail. Instead pronounced cooperative effects are in operation.

4

Bibliography and Further Reading

General Literature

T. E. Creighton, Proteins: Structures and Molecular properties, 2nd Ed., W.H. Freeman, New York (1992)

P. R. Andrews, Drug-Receptor Interactions, in H. Kubinyi, Ed., 3D-QSAR in Drug Design. Theory, Methods and Applications, Escom, Leiden, pp. 13–40 (1993)

Ajay and M. Murko, Computational Methods to Predict Binding Free Energy in Ligand-Receptor Complexes, J. Med. Chem., **38**, 4953–4967(1995)

P. R. Andrews, D.J. Craik and J.L. Martin, Functional Group Contributions to Drug-Receptor Interactions, J. Med. Chem., **27**, 1648–1657 (1984)

I. D. Kuntz, K. Chen, K. A. Sharp and P.A. Kollman, The Maximal Affinity of Ligands, Proc. Natl. Acad. Sci. USA, **96**, 9997–10002 (1999)

J. E. Ladbury, Just add water! The effect of water on the specificity of protein-ligand binding sites and its potential application to drug design, Chem. Biol., **3**, 973–980 (1996)

H. J. Böhm and G. Klebe, What Can We Learn from Molecular Recognition in Protein–Ligand Complexes for the Design of New Drugs?, Angew. Chem It. Ed. Engl., **35**, 2588–2614 (1996)

H. Gohlke and G. Klebe, Approaches to the Description and Prediction of the Binding Affinity of Small-Molecule Ligands to Macromolecular Receptors, Angew. Chem. It. Ed. Engl., **41**, 2644–2676 (2002)

H.-J. Böhm and G. Schneider, Eds., Protein-Ligand Interactions. From Molecular Recognition to Drug Design (Vol. 19, Methods and Principles in Medicinal Chemistry, R. Mannhold, H. Kubinyi and G. Folkers, Eds.), Wiley-VCH, Weinheim (2003)

S. G. Krimmer and G. Klebe, Thermodynamics of protein–ligand interactions as a reference for computational analysis: how to assess accuracy, reliability and relevance of experimental data, J. Comput. Aided Mol. Des., **29**, 867–883 (2015)

J. E. Ladbury, G. Klebe, E. Freire, Adding Calorimetric Data to Decision Making in Drug Development: A Hot Tip! Nat. Rev. Drug Discov., **9**, 23–27 (2010)

G. Klebe, Applying thermodynamic profiling in lead finding and optimization, Nat. Rev. Drug Discov., **14** 95–110 (2015)

G. Klebe, Broad-scale analysis of thermodynamic signatures in medicinal chemistry: Are enthalpy-favored binders the better development option? Drug Discov. Today, **24**, 943–948 (2019)

Special Literature

P. Ehrlich, Chemotherapeutics: Scientific Principles, Methods and Results. Lancet, **182**, 445–451 (1913)

F. W. Lichtenthaler, 100 Years "Schlüssel-Schloss-Prinzip": What Made Emil Fischer Use this Analogy? Angew. Chem. Int. Ed. Engl., **33**, 2353–2543 (1995)

R. P. Mason, D. G. Rhodes and L. G. Herbette, Reevaluating Equilibrium and Kinetic Binding Parameters for Lipophilic Drugs Based on a Structural Model for Drug Interaction with Biological Membranes, J. Med. Chem., **34**, 869–877 (1991)

D. E. Koshland, Application of a Theory of Enzyme Specificity to Protein Synthesis, Proc. Natl. Acad. Sci. USA, **44**, 98–104 (1958)

K. Ngo et al., Protein-Induced Change in Ligand Protonation during Trypsin and Thrombin Binding: Hint on Differences in Selectivity Determinants of Both Proteins? J. Med. Chem., **63**, 3274–3289 (2020)

J. Schiebel et al., Intriguing role of water in protein-ligand binding studied by neutron crystallography on trypsin complexes, Nature Comm., **9**, 3559 (2018)

B. Wienen-Schmidt, T. Hüfner et al., Paradoxically, Most Flexible Ligand Binds Most Entropy-Favored: Intriguing Impact of Ligand Flexibility and Solvation on Drug-Kinase Binding, J. Med. Chem., **61**, 5922–5933 (2018)

S. G. Krimmer, J. Cramer, et al., Rational Design of Thermodynamic and Kinetic Binding Profiles by Optimizing Surface Water Networks Coating Protein-Bound Ligands. J. Med. Chem., **59**, 10530–10548 (2016)

S. G. Krimmer, et al. How Nothing Boosts Affinity: Hydrophobic Ligand Binding to the Virtually Vacated S_1' Pocket of Thermolysin, J. Am. Chem. Soc., **139**, 10419–10431 (2017)

T. Petrova, H. Steuber et al., Factorizing Selectivity Determinants of Inhibitor Binding toward Aldose and Aldehyde Reductases: Structural and Thermodynamic Properties of the Aldose Reductase Mutant Leu300Pro-Fidarestat Complex, J. Med. Chem., **48**, 5659–5665 (2005)

B. Baum et al., More than a simple lipophilic contact: A detailed thermodynamic analysis of nonbasic residues in the S_1-pocket of Thrombin, J. Mol. Biol., **390**, 56–69 (2009)

M. Neeb et al., Occupying a Flat Subpocket in a tRNA-modifying Enzyme with Ordered or Disordered Sidechains: Favorable or Unfavorable for Binding? Bioorg. Med. Chem., **24**, 4900–4910 (2016)

B. P. Morgan, J. M. Scholtz, M. D. Ballinger, I. D. Zipkin and P. A. Bartlett, Differential Binding Energy: A Detailed Evaluation of the Influence of Hydrogen-Bonding and Hydrophobic Groups on the Inhibition of Thermolysin by Phosphorous-Containing Inhibitors, J. Am. Chem. Soc., **113**, 297–307 (1991)

E. Rühmann, et al., Boosting affinity by correct ligand preorganization for the S_2 pocket of thrombin: A study by ITC, MD and high resolution crystal structures, ChemMedChem, **11**, 309–319 (2016)

A. Sandner, et al., Strategies for Late-stage Optimization: Profiling Thermodynamics by Preorganization and Salt Bridge Shielding, J. Med. Chem., **62**, 9753–9771 (2019)

C. Gerlach, M. Smolinski, et al., Thermodynamic Inhibition Profile of a Cyclopentyl- and a Cyclohexyl Derivative Towards Thrombin: The Same, but for Deviating Reasons, Angew. Chem. Int. Ed., **46**, 8511–8514 (2007).

B. Baum, et al., Non-additivity of functional group contributions in protein-ligand binding: a comprehensive study by crystallography and isothermal titration, J. Mol. Biol., **397**, 1042–1054 (2010)

A. T. Fenley, H. S. Muddana and M. K. Gilson, Entropy–enthalpy transduction caused by conformational shifts can obscure the forces driving protein–ligand binding, Proc. Natl. Acad. Sci. USA, **109**, 20006–20011 (2012)

Optical Activity and Biological Effect

Contents

G. Klebe, *Drug Design*, https://doi.org/10.1007/978-3-662-68998-1_5

The three-dimensional shape of a molecule has a decisive influence on its biological activity. The **configuration** of a molecule is made up of the bonds between the atoms. Substances with an **asymmetric center** at the tetracoordinated carbon atom, which are considered here, are optically active and exist in two different forms. They are asymmetric and have a relationship to each other like an **image and its mirror image**. They are called chiral. It is impossible to change one form into the other without breaking and reforming bonds. **Chirality** is often unimportant to chemists because the image and mirror image behave exactly the same in a symmetrical environment. When they are placed in an asymmetric environment, such as the binding site of a protein, this is no longer true. The consequences of this for drug design and therapy are the subject of this chapter.

In the early 19th century, Jean Baptiste Biot observed that some quartz crystals rotated the plane of linearly polarized light to the right, while others rotated it to the left. Macroscopically, this optical activity is imprinted in the asymmetric, handed (enantiomorphic) shape of the crystals; they exist as left- and right-handed mirror images. A little later, Biot found that not only crystals, but also organic compounds such as turpentine or sugar solutions rotate polarized light in a particular direction.

5.1 Louis Pasteur Sorts Crystals

The decisive experiment was performed by the then 26-year-old Louis Pasteur in Paris in 1848. Several reports in the literature were inconsistent with his theory that there must be an obvious relationship between crystal forms and their optical properties. While carefully studying the sodium ammonium salt of optically inactive **tartaric acid**, he discovered that the crystals had different shapes. They had either right-handed or left-handed symmetry and could be sorted by hand. The crystals of enantiomers **5.1** and **5.2** (◘ Fig. 5.1) gave solutions that rotated the plane of linearly polarized light in opposite directions. This confirmed his suspicions. Before Pasteur could present his results to the Academy of Sciences, he had to repeat the experiment in public (!) in the presence of Biot at the Collège de France. He was lucky. His experiment was successful only because his solutions were allowed to evaporate slowly at room temperature. Above the critical temperature of 28 °C, a stoichiometric 1:1 mixture of the two enantiomeric forms, a racemate, would have crystallized in a homogeneous crystal form (Sect. 5.4).

A few years later, Pasteur made another important observation: mold contamination of a racemic solution

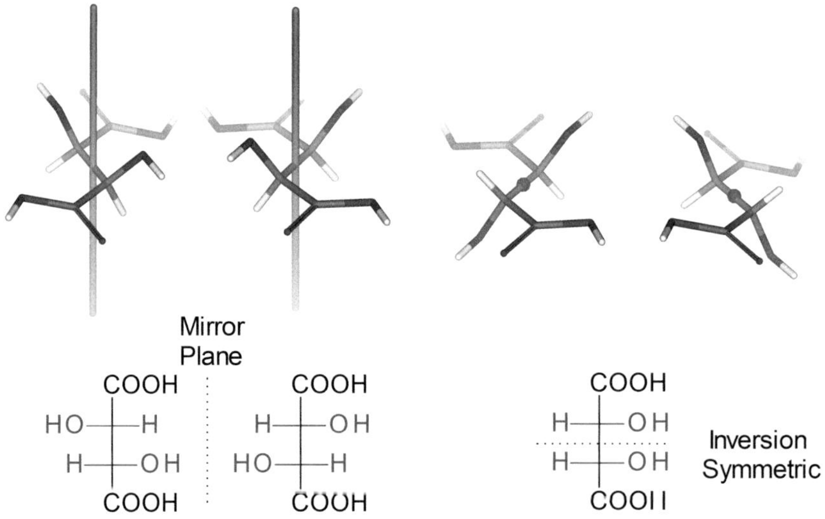

5.1 5.2 5.3

D-(-)-Tartaric acid L-(+)-Tartaric acid *meso*-Tartaric acid

◘ **Fig. 5.1** Optical isomerism in tartaric acid. The enantiomers (−)-tartaric acid **5.1** (m.p. 168–170 °C, $[\alpha]_D^{20} = -12°$) and (+)-tartaric acid **5.2** (m.p. 168–170 °C, $[\alpha]_D^{20} = +12°$) cannot be superimposed upon each other either in the plane of the paper or in 3D space. The sole symmetry element that they have is a twofold rotational axis (*orange axes*) that dissect the central C–C bond. Each mirror image rotates the plane of polarized light in opposite directions to the other. In contrast, *meso*-tartaric acid **5.3** (m.p. = 140 °C) has an inversion center of symmetry (the *purple center* on the central C–C bond). Solutions of *meso*-tartaric acid have no optical activity because the contribution from each stereogenic center compensates for the other. Racemic tartaric acid (m.p. = 206 °C, no rotation) is a 1:1 mixture of both enantiomers of tartaric acid **5.1** and **5.2**. Such mixtures are optically inactive and are called racemates (Lat. *racemus*, the grape–tartaric acid is found in grapes and wine). (▸ https://sn.pub/GCWqem)

of tartaric acid caused optical activity to develop. One enantiomer of tartaric acid is metabolized much faster than the other. He, thus, discovered two important methods for separating racemates into enantiomers. While mechanical sorting is limited to a few examples, **enzymatic kinetic resolution of enantiomers** has found wide application (Sect. 5.4).

5.2 Structural Basis of Optical Activity

An explanation for the optical isomerism was possible with the help of the **theory of the tetrahedral carbon atom** developed independently by Jacobus Henricus van't Hoff and Joseph-Achille Le Bel in 1874. When a carbon atom carries four different substituents, it creates an **asymmetric** or, as it is also called, **stereogenic center**. This property is not limited to carbon; nitrogen (in ammonium salts) or silicon atoms with four different substituents, phosphorus, for example, in phosphonic or phosphoric acid esters, or even sulfur atoms in sulfoxides (with two different substituents, oxygen, and the lone pair of electrons) can also be asymmetric. The spatial orientation of these compounds gives rise to two mirror-image isomers, each of which rotates polarized light in the opposite direction to the same degree. These forms are called **enantiomers** (formerly **antipodes**). Except for their optical activity, enantiomers are identical in all chemical and physicochemical properties, but only as long as they are in an achiral environment.

Compounds with two chiral centers configured as image and mirror image within the same molecule do not macroscopically exhibit optical activity; *meso*-tartaric acid **5.3** (◘ Fig. 5.1), an inversion-symmetrical molecule, exists as a racemic mixture of chiral conformers. Each conformer exists as an "internal" racemic mixture because the molecule has **inversion symmetry** in an energetically favored conformation. Its left part can be inverted to its right part by point reflection through the center of the central C–C bond. Therefore, the *R*,*S*- or *S*,*R*-*meso*-forms are identical and cannot be resolved into enantiomers (*R*,*S* assignment, see below).

Optical activity is also present in other forms of molecular asymmetry. One example is any regular or irregular tetrahedral orientation of different substituents on any skeleton other than a single carbon atom. Another case can be found in compounds where two groups are strongly rotationally hindered around a common bond. This results in an asymmetric center, which gives rise to optically active rotational isomers, called **atropisomers** (◘ Fig. 5.2).

The experimentally determined rotational value (+) or (−) (previously called *d* or *l*) is used to characterize enantiomeric compounds. The spatial configuration of a stereogenic center in a molecule is described as D or L (Lat. *dextro*, *levo*). This notation is based on the **Fischer convention** and is related to the absolute configuration of D- and L-glyceraldehyde, **5.7** and **5.8** (◘ Fig. 5.3). Most sugars, for instance glucose **5.9**, can be traced back to D-glyceraldehyde **5.7**, and the natural amino acids of proteins, for instance alanine **5.10**, can be traced back to L-glyceraldehyde **5.8**. For this reason, today the D/L nomenclature is still frequently applied to sugars and amino acids. The enantiomers of tartaric acid correspond to the D-(−) or L-(+) form.

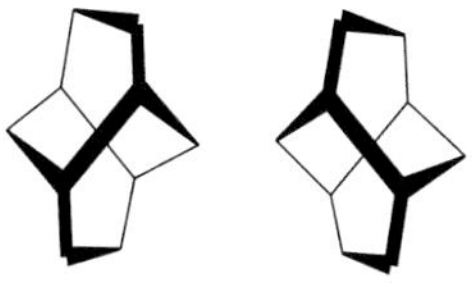

5.4
Twistane

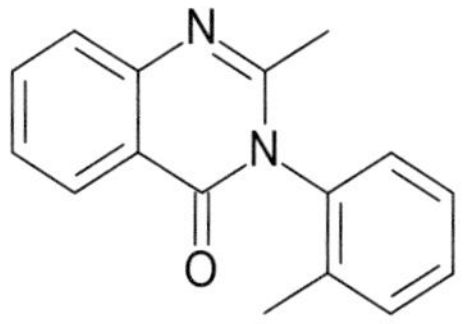

5.5 Methaqualone

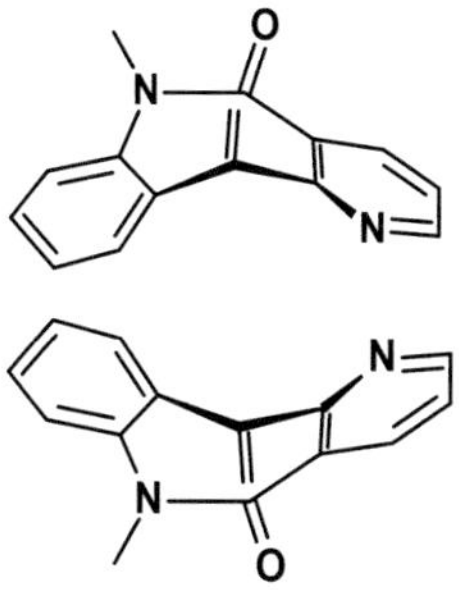

5.6

◘ **Fig. 5.2** Even molecules without stereogenic centers can form an image/mirror-image pair because of their spatial construction; an example is twistane **5.4**. If rotation around the bonds is limited, as in the case of the sedative methaqualone **5.5**, enantiomers will be separable (so-called atropisomers). In nonplanar fused ring systems like the dibenzocycloheptadiene derivative **5.6**, the enantiomeric separation depends on the barrier of inversion for the central ring system

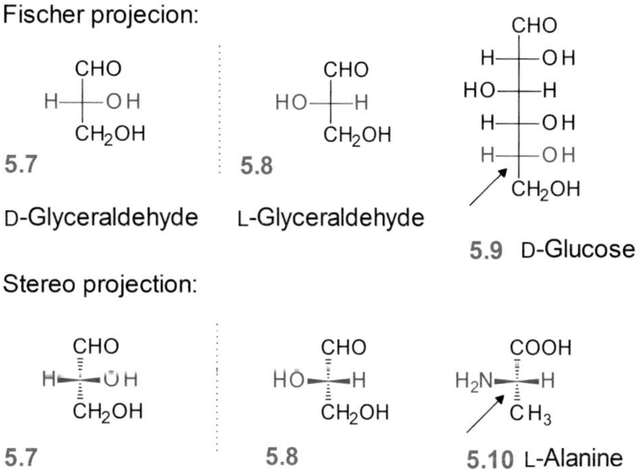

◘ **Fig. 5.3** The rotation (+ or −) and the Fischer assignment (D or L) is reported as part of the characterization of optically active compounds. To determine the Fischer assignment, the longest carbon chain is drawn vertically with the highest-oxidized carbon atom on top (e.g., **5.9**). The standard is set by the asymmetric carbon (*red*) of the D- and L-glyceraldehyde pair (**5.7** and **5.8**). With sugars (e.g., glucose **5.9**) or amino acids (e.g., alanine **5.10**), the carbon that is marked with the *arrow* decides whether the molecule is D or L

Cahn-Ingold-Prelog Rules

- Larger atomic numbers have higher priority over lower one e.g. I>Br>Cl>F>O>N>C>H;
- Free electron pairs always have the lowest priority;
- Larger atomic masses have priority, e.g. for isotopes D>H;
- In case the first sphere is identical (e.g. all C), the next sphere is consider:

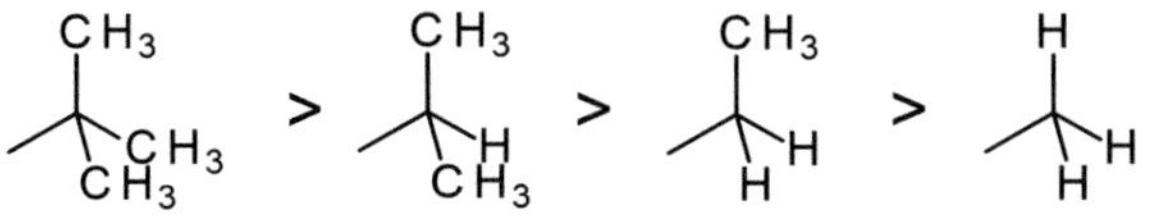

C[C+C+C] > C[C+C+H] > C[C+H+H] > C[H+H+H]

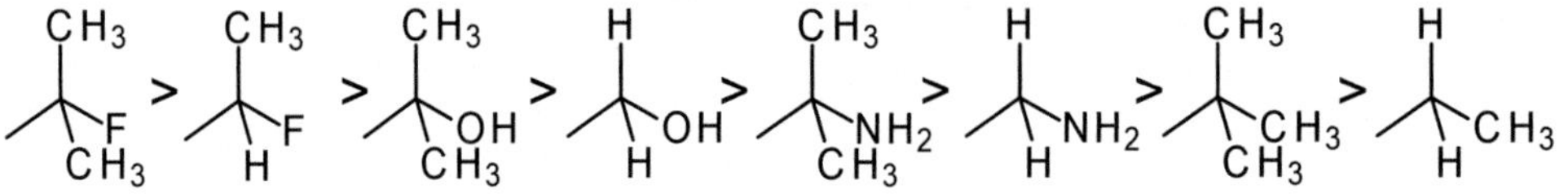

- Multiple bonds are formally considered as multiple single bonds, e.g. aldehyde: CHO = C [O + O + H] > CH_2OH = C [O + H + H];
- If the substituents are identical but chiral: (*R*) > (*S*) and (*R*,*R*) > (*R*,*S*), or (*S*,*S*) > (*S*,*R*);
- In case of differently configurated double bonds: Z > E
  Z=zusammen (*Engl. together)*, E=entgegen (*Engl. apart*) for the configuration of the double bond

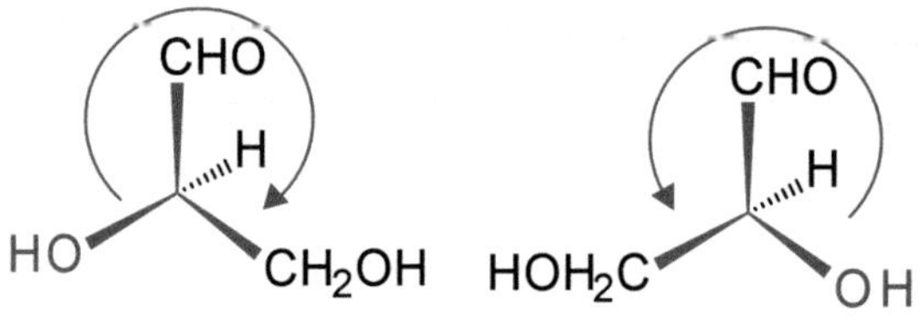

(*R*)-Glyceraldehyde **5.7** (*S*)-Glyceraldehyde **5.8**

Fig. 5.4 The *R*/*S* nomenclature that was proposed by R. S. Cahn, C. K. Ingold, and V. Prelog is unambiguous. Priority rules for each of the four different substituents on the tetrahedral stereogenic center were established. The substituent with the lowest priority is placed in the back, and the direction of remaining substituents determine the direction of rotation by decreasing priority

The **Cahn–Ingold–Prelog** rules allow an unambiguous stereochemical assignment (Fig. 5.4). By convention, the optical center is oriented so that the substituent with the lowest atomic number is at the back (e.g., a hydrogen atom or a lone pair of electrons). To use an intuitive explanatory model, we want to assign this substituent to the column of a steering wheel. Then the other substituents lie in the plane of the steering wheel. If these substituents are considered in descending order of atomic number and this order follows a rotation to the right, the stereogenic center will have an *R*-configuration; the opposite direction is the *S*-configuration (from the Latin: *rectus* and *sinister*). The only disadvantage of this nomenclature system is that the assignment of the stereocenter can change simply because of the atomic number, valency, or oxidation state. The homologous L-amino acids serine and cysteine, which are structurally stereochemical analogues differing only by the exchange of an oxygen for a sulfur atom, are classified as (*S*)-serine and (*R*)-cysteine.

If there is one stereogenic center in a molecule, there will be two enantiomers. Each additional symmetry-independent stereogenic center increases the number of enantiomers by a factor of 2. For n asymmetric centers, there are 2^n optical isomers. They occur as 2^{n-1} racemic mixtures because each has two isomers that are mirror images of each other. **Diastereomers** cannot be superimposed by translation and rotation in space or by creating a mirror image because the chirality of the stereocenters is different relative to one another. As a result, they have different physicochemical and chemical properties. All pairwise racemates of a diastereomeric mixture exist as a 1:1 mixture of enantiomers, but their relative proportions in the total composition can vary widely. Labeta-

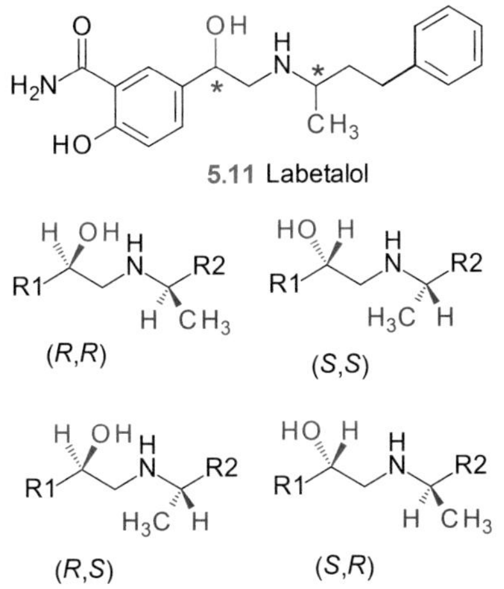

Fig. 5.5 Because it has two different asymmetric centers, labetalol **5.11** is a diastereomeric mixture of four different compounds with different activities on the same receptor. The antagonistic potency on the α_1 receptor of the (*R,R*)-, (*R,S*)-, (*S,R*)-, and (*S,S*)-isomers is: *S,R* >>*S,S*≈*R,R*>*R,S*; and on the β_1 receptor is: *R,R* >>*R,S*>*S,S*≈*S,R*; and on the β_2 receptor is: *R,R* >>*R,S* >>*S,S*≈*S,R*

lol **5.11** (Fig. 5.5) is just such a pair of diastereomers, consisting of two racemates, that is, two pairs of enantiomers. As a mixed antagonist, it acts on α-, β_1-, and β_2-adrenergic receptors (see Sect. 29.3). Due to the asymmetric architecture of biological macromolecules, the individual components of this mixture vary significantly in their qualitative and quantitative biological properties (Sects. 5.5, 5.6 and 5.7).

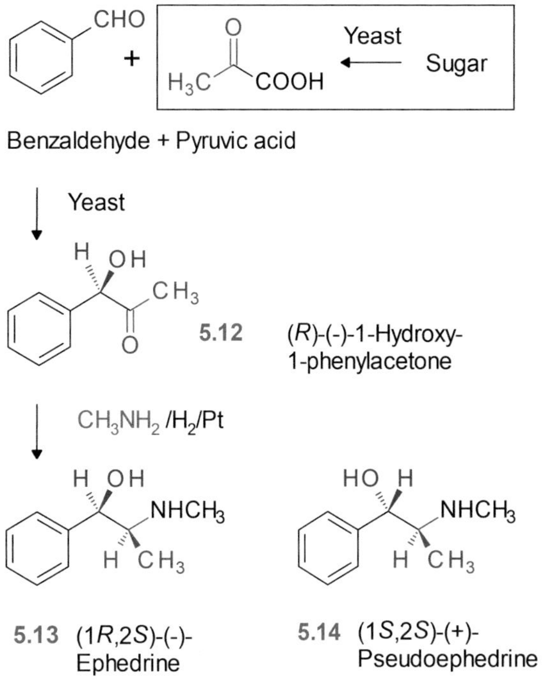

Fig. 5.6 The biotechnological production of ephedrine is accomplished by the fermentation of sugar with baker's yeast *Saccharomyces cerevisiae* to pyruvic acid. Pyruvic acid is coupled to benzaldehyde with decarboxylation to form (*R*)-(–)-1-hydroxy-1-phenylacetone **5.12**. Upon further chemical transformation (*1R,2S*)-( )-ephedrine **5.13** is obtained in optically pure form. (1*S*,2*S*)-(+)-pseudoephedrine **5.14** is a diastereomer of ephedrine. The configuration of one of the two chiral centers is different

5.3 The Isolation, Synthesis, and Biosynthesis of Enantiomers

Racemic acids and bases can often be separated by using other enantiomerically pure, optically active bases and acids because the formed diastereomeric salts have different solubilities. The chemical reaction of racemic acids, amines, and alcohols with optically active alcohols or acids results in diastereomeric reaction products. Because of their different properties, it is possible to separate them and finally isolate the desired optically active product by chemical cleavage.

Syntheses that do not start with optically active starting materials and do not use optically active auxiliaries always result in racemic mixtures, that is, an exact 50:50 mixture of the two enantiomers. Access to optically active compounds can be obtained by taking synthetic reaction components from the "**chiral pool**." All optically active natural products, their derivatives and degradation products that are available in an optically pure form can be used as easily accessible synthetic building blocks. Syntheses using chiral catalysts are particularly elegant. In most cases, the optimization of yield and enantiomeric purity, expressed as the **ee value** (ee = enantiomeric excess), requires considerable process development. The chromatographic separation of racemates on optically active solid supports is more suitable for analytical or semipreparative purposes.

Enzymatic and biotechnological techniques have become increasingly popular in recent years. Proteases, esterases, lipases, or hydantoinases react more or less selectively, preferentially with a distinctly different reaction rate; only one enantiomer of a racemic mixture is transformed to the product. The selectivity and yield of such a reaction can be optimized by the careful selection of the medium and other reaction conditions.

The production of optically pure ephedrine is an example of an industrial application of biotechnological synthesis that has been used for decades. This phytopharmacon is used in combination preparations for the adjuvant therapy of rhinitis, bronchitis, and asthma. The synthetic intermediate **5.12** (Fig. 5.6) is obtained from a mixture of benzaldehyde, sugar, and yeast. It is then transformed to (1*R*,2*S*)-(−)-ephedrine **5.13**, which is identical to the natural product in both of its optical centers. The C1 isomer (1*S*,2*S*)-(+)-pseudoephedrine **5.14** is a diastereomer of ephedrine. Its optical rotation, melting point, and biological properties differ from those of ephedrine.

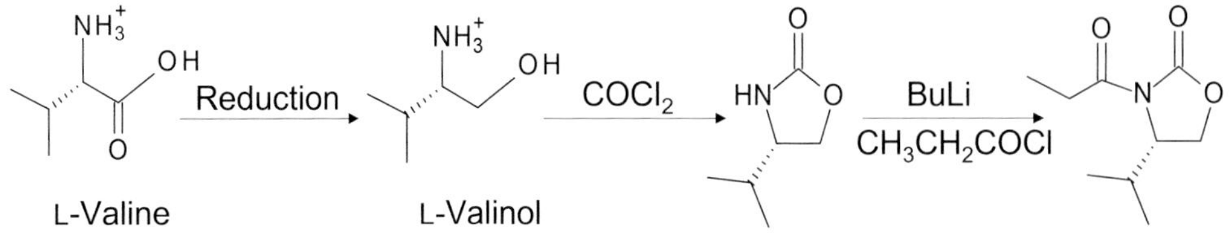

Fig. 5.7 Starting with enantiomerically pure valine, valinol is formed by reduction, which condenses with phosgene to an oxazolidinone. After abstraction of the amidic proton, the compound reacts with propionyl chloride to form the starting material for the Evans synthesis

Numerous other microbial syntheses provide optically pure products with or without the use of achiral, racemic, or enantiomerically pure starting materials. Of particular economic importance are the biotechnological syntheses of various antibiotics, especially penicillins and cephalosporins (Sects. 2.4 and 23.7). The biotechnological production of synthetic intermediates for chiral drugs is also becoming increasingly important.

In the last 35 years, organic synthesis has developed a variety of methods for the stereoselective preparation of compounds. Metal catalysts, which create a local chiral environment due to their spatial geometry, are mainly used for this purpose. In this environment, the conversion of one of the stereoisomeric reaction products is preferentially catalyzed. In addition, the conformational properties (Chap. 16) of the molecules to be reacted can also be used to steer a reaction preferentially in the direction of a particular product.

David Evans at Harvard University, Cambridge, USA, has proposed a procedure for the stereoselective introduction of alkyl groups in the α-position on carboxylic acids. It starts, for example, with an amino acid such as valine, which is reduced to the amino alcohol valinol and then condensed with phosgene to an oxazolidinone (Fig. 5.7). In the next step, a carboxylic acid, such as propionic acid, to which the alkyl group is to be added in the α-position, is coupled to the oxazolidinone via an amide bond. The conformational properties of the oxazolidinone ring are crucial for the further course of the reaction. The five-membered ring adopts a half-chair geometry. The conformer in which the large *iso*-propyl substituent is equatorial is energetically more favorable. One of the terminal methyl groups of the *iso*-propyl substituent is oriented towards the bottom of the ring plane of the heterocycle (Fig. 5.8). With a strong base such as lithium *iso*-propylamide, a proton is extracted from the reactant to form an enolate. The lithium ion plays an important role by forming a chelate ring with the enolate and the carbonyl oxygen, stabilizing a planar geometry in this part of the molecule. If, for example, benzyl bromide is used as a reagent, a benzyl group will be introduced stereoselectively. The newly formed C–C bond forms via an S_N2 reaction step. It proceeds through a trigonal–bipyramidal transition state. The new C–C bond to the enolic carbon is formed simultaneously with the displacement of the bromine atom as a bromide ion. This attack can only take place "from above," since the lower surface of the oxazolidinone ring is spatially shielded by the *iso*-propyl group (Fig. 5.8, *red semicircle*). In this way, only one stereoisomer is predominantly formed.

How can the benzyl group be introduced from the other side? The key is that now the upper face of the oxazolidinone ring must be sterically shielded. If a methyl and a phenyl group are attached to the two tetrahedral carbon atoms of the heterocycle with an upwards orientation, the larger phenyl substituent on the half-chair will assume the equatorial orientation. The methyl group will then be axially positioned towards the upper side of the ring. Now the attack can only occur from the bottom. The benzyl group introduced at the α-position is on the bottom side of the product. The other stereoisomer is formed. After hydrolysis of the amide bond, the enantioselective carboxylic acid is obtained. For the second reaction with one methyl and one phenyl group, another stereochemically uniform amino alcohol is required. (1*S*,2*R*)-Ephedrine can be used for this purpose. The example of the Evans synthesis with a chiral auxiliary is exemplary for many stereoselective syntheses that follow such a concept.

5.4 Lipases Separate Racemates

Enzymes are well suited to resolve racemates because of their asymmetric structure. This can occur either because one of the two enantiomeric substrates is more strongly bound and more rapidly converted. Alternatively, a chemical reaction may take place in the binding pocket of the protein with different efficiencies. Lipases are often used for **kinetic resolution of racemates** because they are stable in organic solvents due to their molecular composition and lipophilic surface. They belong to the large group of hydrolyzing enzymes (Chap. 23). A nucleophilic serine is present in the catalytic center that forms an acyl–enzyme complex upon hydrolysis of an amide or ester substrate. The protein is then itself converted to an ester through the OH group of the serine, the so-called acyl form is produced (Sect. 23.2). Such a complex can then react with another nucleophile, for instance an amine. The amine attacks the internal enzyme ester, the bond to the serine oxygen atom is broken, and a new amide bond is formed. If one employs the right- or left-handed form of an amine, one form will react preferentially. In this way, the racemate is resolved.

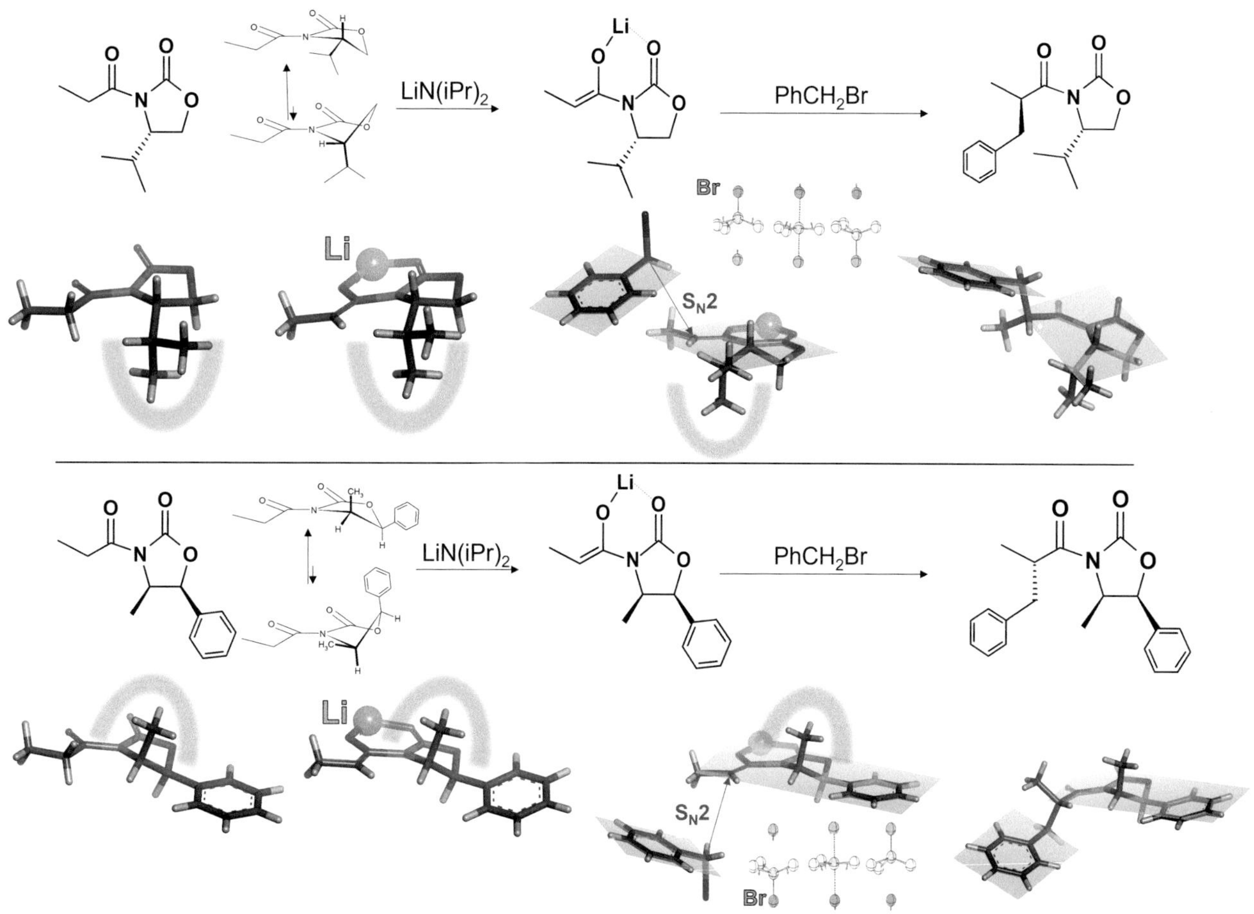

Fig. 5.8 Example of a stereoselective Evans synthesis for the formation of α-alkylated carboxylic acids using a chiral auxiliary. Stereochemically uniform oxazolidinones with either an *iso*-propyl (*upper row*) or methyl/phenyl substituent (*bottom row*) are converted into the corresponding enolates with LiN(iPr)$_2$. The oxazolidinone ring assumes a half-chair conformation. Here, the largest substituent (either *iso*-propyl or phenyl) is preferentially equatorially oriented. In an S$_N$2 reaction, the formed enolates are reacted with benzyl bromide. *Upper row* In the first case, the attack takes place from the sterically more accessible upper face. *Bottom row* In the second case, attack occurs from the lower face. In the product, therefore, the introduced side chain is either up (*R*-) or down (*S*-). Depending on the reaction conditions chosen, the Evans synthesis can achieve a diastereomeric excess of 93–98 (► https://sn.pub/kicyt3)

How does the enzyme distinguish between the two enantiomers of an amine? The reaction of (*R*)- and (*S*)-phenylethylamine **5.15** and **5.16** with the lipase *Candida antarctica* was carefully investigated (Fig. 5.9). The energy barrier for the faster-reacting *R*-form is lower than for the slower *S*-form. A detailed evaluation of the kinetic parameters showed that this is above all due to an enthalpic advantage of the *R*-amine. The *S*-amine has an entropic advantage. Altogether the enthalpic component is in excess so that the free energy difference (Δ*G*) favors the *R* form (Fig. 5.9). How is this discrimination to be understood? Structural **transition state analogues** have been synthesized. In the place of the unstable tetrahedral carbon atom intermediate, a phosphorus atom was introduced (**5.17** and **5.18**, Fig. 5.10). This trick gives a stable compound that is very similar to the transition state formed at the carbon atom. These analogues were synthesized with both enantiomeric amines, and complexes with the lipase were prepared. Marco Bocola managed to obtain a crystal structure of both. Interestingly the transition state analogue of the faster-reacting *R*-form fits well into the binding pocket (Fig. 5.10). On the other hand, the *S*-form demonstrated great residual mobility in the catalytic center.

Computer simulations of the molecular dynamics of both forms confirmed the picture: The *R*-analogue is well defined and temporally stable in a geometry that is ideal for the chemical reaction. In contrast, the *S*-analogue appears to be quite mobile. It is much less likely to remain in an arrangement that allows the catalytic reaction to occur in the lipase. Successful catalytic transformation of this substrate occurs much less frequently. The *R*-analogue, fixed as in a vise and waiting to be transformed, forms good enthalpic contacts with the enzyme. It practically takes a shape complementary to the enzyme pocket.

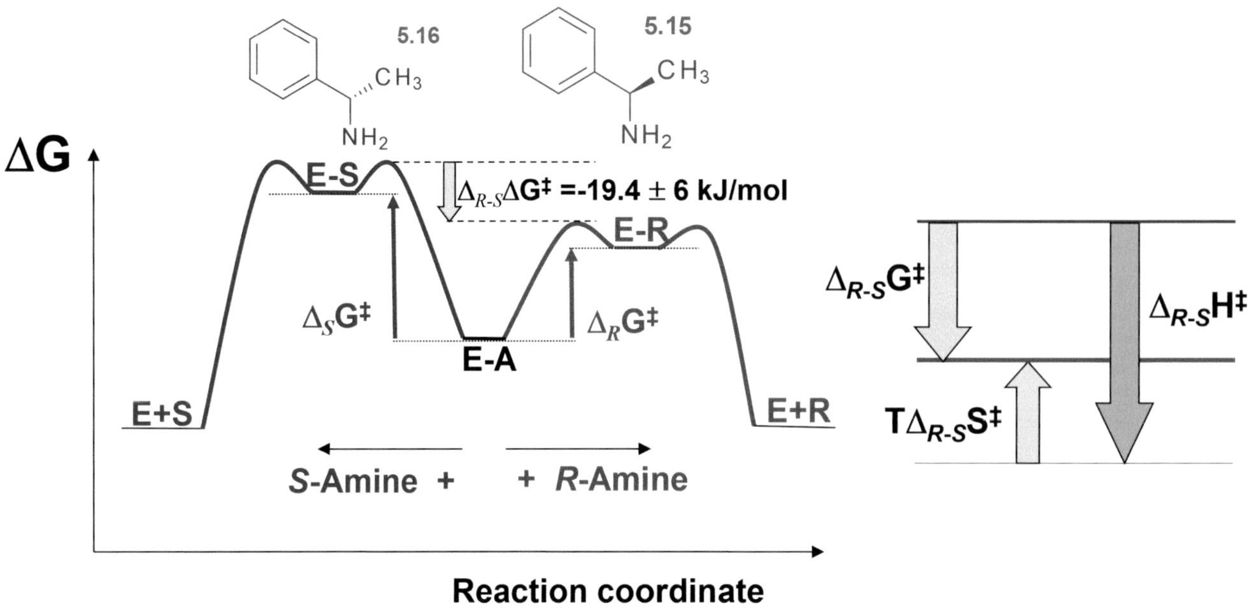

Fig. 5.9 The reaction of (*R*)- and (*S*)-phenylethylamine, **5.15** and **5.16**, with *Candida antarctica* lipase begins with the formation of an acyl–enzyme complex, E–A. The faster-reacting *R*-amine **5.15** (*red*) forms a lower-energy transition state that leads to the free enzyme and the *R*-amide (E+R). Analogously the *S*-amide (E+S) forms from the higher-energy E–S transition state (*blue*) from the *S*-amine **5.16**. Difference in $\Delta G^{\ddagger}$ is −19.4 kJ/mol and favors the *R* form. The $\Delta G^{\ddagger}$ difference is based on a combined enthalpic and entropic contribution in which the *R*-form is enthalpically favored, and entropically disfavored. The *S*-form is enthalpically disfavored but has an entropic advantage

This results in its great enthalpic advantage. Entropically, however, this fixation comes at a price. With its methyl group at the stereogenic center, the *R*-analogue becomes entangled in a small niche of the binding pocket. The *S*-analogue misses this opportunity because this crucial methyl group is oriented in the opposite direction. Therefore, it lacks this anchor for attachment to the binding pocket. It has a high degree of mobility in the catalytic center, so it does not lose as many degrees of freedom compared to the situation before enzyme binding. Entropically, this is favorable. But enthalpically, this substrate substrate cannot form good interactions. The complementary fit to the protein is rarely achieved. In the end, the enthalpic component predominates and the resulting Gibbs free energy difference leads to a much faster conversion of the *R*-amine. This is sufficient to produce practically only the *R*-amide in high yield. The lipase can also be immobilized onto a solid support and loaded into a glass column. After the acyl form is prepared on the column, a racemic mixture of the amine only needs to be poured onto the column. The *S*-amine and *R*-amide are then simply collected in a flask. If the solvent is well chosen, the amide will crystallize directly from the solution and can be mechanically separated whereas the amine will remain in solution.

Interestingly, the **enantiopreference** of the kinetic resolution is lost with increasing temperature or enlarging the enzyme pocket. Enlargement can be achieved by exchanging a tryptophan along the rim of the catalytic pocket for a histidine. The higher temperature or increased space in the binding pocket enhances the mobility of both substrates in the lipase pocket. The enthalpic advantage of the faster-reacting *R*-amine is lost. The entropic difference of both substrates levels out under these conditions.

This example shows at the molecular level how a lipase achieves kinetic resolution. With knowledge of the energetic parameters and structural information, an attempt can be made to tailor lipases for other transformations. Because of the importance of such reactions, the targeted design of enzyme catalysts has developed into an ever more important topic for the synthesis of chiral building blocks in new drugs.

5.5 Differences in the Activity of Enantiomers

Flora and fauna are known for their symmetry. Consider the face, the arms and legs, the ribs, or an orchid flower. The exceptions, such as the shell of a snail, are rare or, as in the case of the flounder, occur only under special evolutionary conditions. The internal organs of vertebrates are partially paired and partially asymmetrically oriented.

At the molecular level, there is no corresponding symmetry: optically active building blocks predominate. All specific interaction partners of biologically active molecules are chiral. Enzymes and receptors are composed of L-amino acids. Nucleic acids are built on a backbone of D-ribose or D-deoxyribose units. Most naturally occurring sugars have a D-configuration. Important vitamins,

5.18

5.17

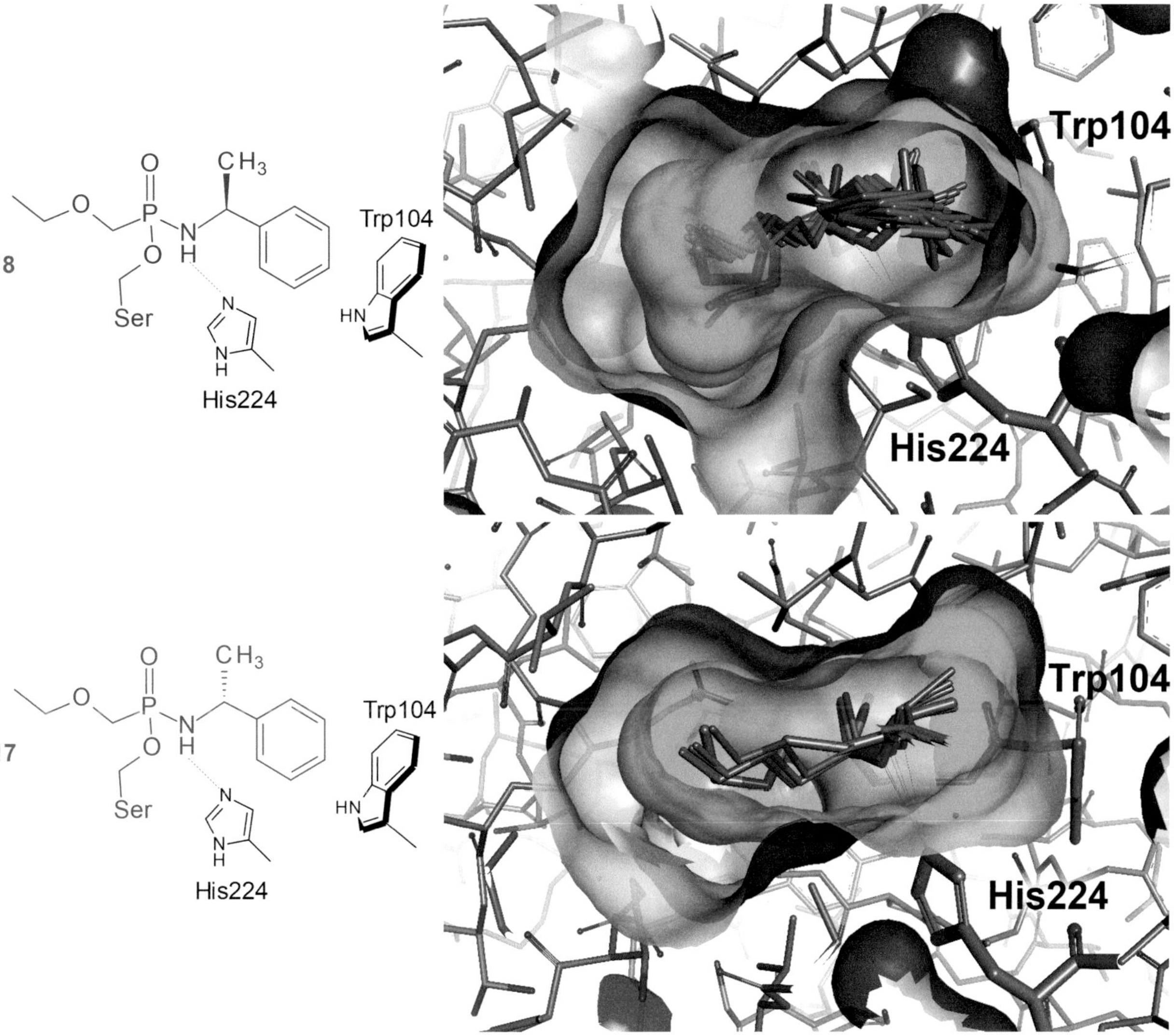

Fig. 5.10 *Upper row* Shown is a phosphorous transition state analogue **5.18** for the lipase with the *S*-amine. The crystal structure and MD (molecular dynamics) simulations indicate that it is less-rigidly fixed in the transition state and rarely adopts the geometry with an H-bond (*purple line*) to histidine (on the lower edge of the binding pocket) that is necessary for the reaction to occur. *Lower row* The relevant complex with the transition state analogue **5.17** of the faster-reacting *R*-amine is shown. This substrate is highly restricted in the binding pocket. Its methyl group (*above right*) is embedded in a small niche in the binding pocket. This substrate exclusively adopts the geometry with the H-bond to histidine. This orientation is required for a successful substrate reaction. Therefore, the *R*-amine **5.17** reacts with the enzyme faster. (▸ https://sn.pub/GLcA6h)

hormones, and second messengers exist in an optically homogeneous form. Accordingly, enantiomers of an optically active ligand are expected to have different effects. This has been demonstrated in many thousands of examples. In most cases, enantiomers show significant differences in potency and quality of action.

According to Everhardus J. Ariëns from the University of Nijmegen, the Netherlands, biologically active enantiomers are called **eutomers** and inactive enantiomers are called **distomers**. The quotient of the two affinities or effects is defined as the **eudismic ratio**, and the logarithm of this value is called the **eudismic index**. It should be noted that this value must be determined on extremely pure compounds. As little as 1% eutomer impurity in a completely inactive distomer can simulate 1% relative activity in the distomer!

The more the activities of enantiomers in a racemic pair differ, the more the eudismic ratio diverges from 1. Examples of this are given in compounds **5.19–5.22** (Fig. 5.11). A eudismic ratio of 500,000 was measured for an inhibitor of a chloride ion transporter. In this case, the chemists pulled out all the stops to purify the less potent enantiomer. Theoretically, a compound with nanomolar potency should yield even higher values.

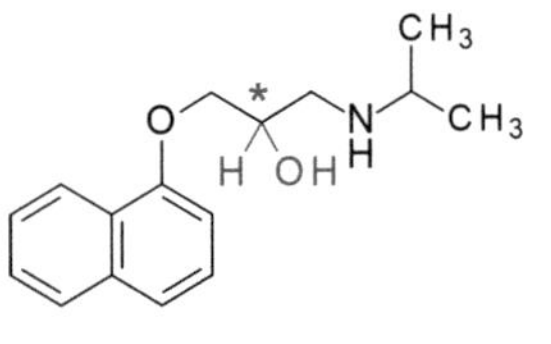

5.19 Propranolol

	Eudismic Ratio
β-Antagonism	100
Membrane effect	1

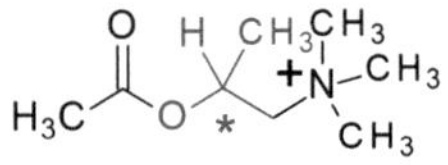

5.20 Methacholine

Cholinergic effect	320

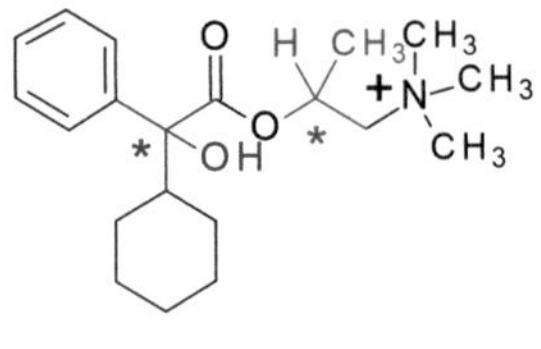

5.21 Anticholinergic agent

Ester group center	50 - 100
Amino alcohol center	2 - 4

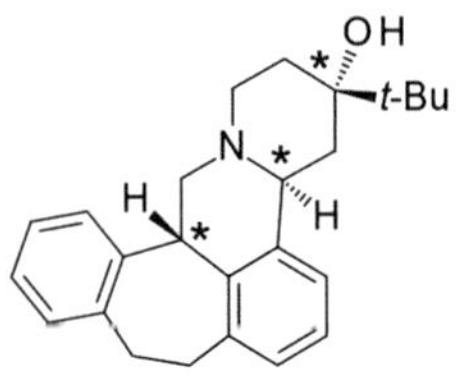

5.22 Butaclamol, (+) Enantiomer

α_1 Receptor	73
D_2 Receptor	1250
5-HT_1 Receptor	8
5-HT_2 Receptor	73
Muscarinic Receptor	0.5

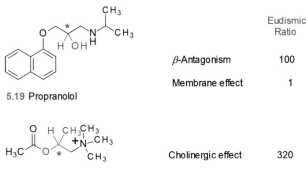

Fig. 5.11 Enantiomers have different biological effects. The eudismic ratio of propranolol **5.19** is 100 for β-antagonism, and for unspecific membrane interaction, it is, expectedly, 1. Identical partial structures can have entirely different eudismic ratios, for instance, compare the optical center of the alcohol moiety of the cholinergic compound methacholine **5.20**, with the identical center on the anticholinergic compound **5.21**. Compound **5.21** also proves that the eudismic ratio of different centers in a compound are independent from each other. The example butaclamol **5.22** also shows that the same substance can have different eudismic ratios on different receptors

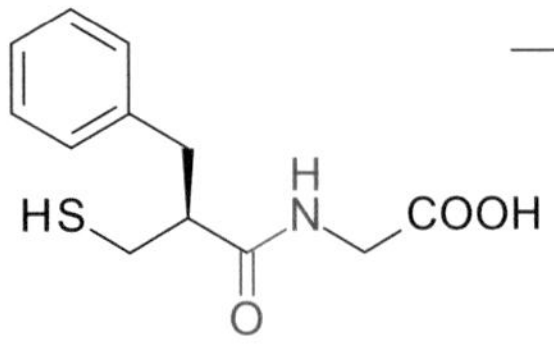

5.23 Thiorphan

Enzyme	K_i value in µM
NEP 24.11	0.0019
Thermolysin	1.8
ACE	0.14

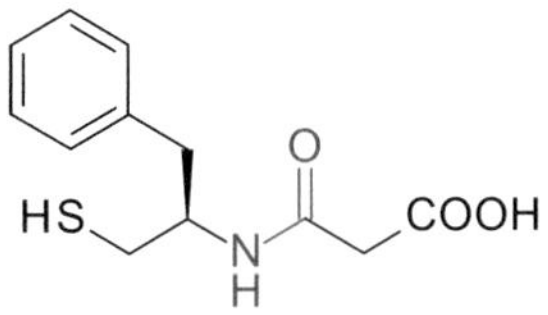

5.24 *retro*-Thiorphan

Enzyme	K_i value in µM
NEP 24.11	0.0023
Thermolysin	2.3
ACE	>10

Fig. 5.12 Thiorphan **5.23** inhibits the metabolism of enkephalins and contains a β-mercaptopropionic acid, the absolute configuration of which is analogous to L-phenylalanine. Application of the *retro–inverso* concept gives aminothiol **5.24**, the absolute configuration of which corresponds to D-phenylalanine due to incorporation in the opposite direction. The identical binding mode to the zinc protease was determined for both thiorphan **5.23** and *retro*-thiorphan **5.24**. Thermolysin and the neutral endopeptidase 24.11 (NEP 24.11, previously referred to as enkephalinase) are inhibited by both compounds to the same extent. On the other hand, angiotensin-converting enzyme (ACE), another zinc protease, discriminates decidedly between these substances

Some naturally occurring peptide antibiotics contain **D-amino acids**. This gives them better metabolic stability. For the same reason, D-amino acids are incorporated into many synthetic peptide molecules. In the best cases, a more potent and longer-acting analogue is obtained. Synthetic analogues of peptides with a ***retro–inverso*** **configuration** are a special case. In these molecules, the direction of the peptide chain or part of the peptide chain is reversed, that is, compared to the original peptide, the amino and carboxyl groups of individual amino acids are reversed. To maintain the relative configuration, D-amino acids or their analogues are used instead of L-amino acids. In this way, it is possible to deceive some enzymes or receptors; they bind the natural peptide and the retro–inverso peptide in the same way. This is true for thiorphan **5.23** and its *retro–inverso* analogue **5.24** for two enzymes, but not for a third (Fig. 5.12). In general, *retro–inverso* peptides are metabolically more stable than their original peptide analogues.

Enantiomers differ not only in the strength of their potency, but also in their biological qualities. These differences can manifest themselves as undesirable side effects of a stereoisomer, such as the chiral barbiturate **5.25** (Fig. 5.13). The most serious drug side effect of the last 70 years was the embryonic malformations caused by the sleeping pill thalidomide **5.26** (Contergan®); these were caused by one of the two enantiomers (Fig. 5.13). In the 1950s, thalidomide was claimed to be the best tolerated sleeping pill with the fewest side effects. It was introduced to the market in 1957 and was available in pharmacies as an over-the-counter drug. There was no concern that even women in the first months of their pregnancy were taking these sleeping pills. It was withdrawn from the market in 1961 because of its teratogenic effects. If drug testing had been what it is today, this catastrophe would certainly have been detected earlier and probably largely avoided. It could not have been prevented by administering only one of the two enantiomers. Both enantiomers racemize *in vitro*, meaning that one converts into the other even in the test tube. Accordingly, the effect was confirmed *in vivo*, after the administration of the supposedly safe enantiomer had led to teratogenic effects in an animal model.

The "other" enantiomer can also open up new therapeutic possibilities. The enantiomer of a synthetic opiate, such as propoxyphene **5.27** (Fig. 5.13), has weak an-

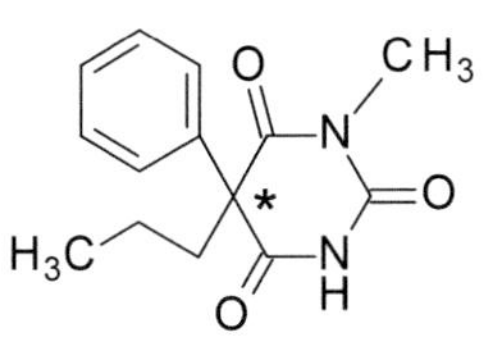

5.25 *N*-Methyl-phenyl-propylbarbituric acid

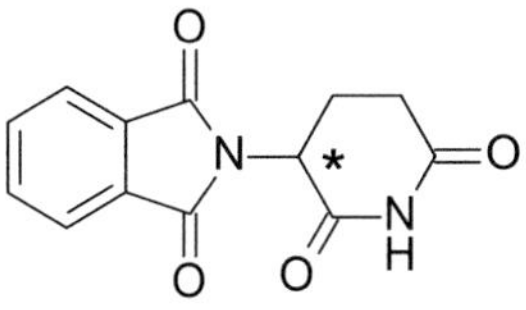

5.26 Thalidomide

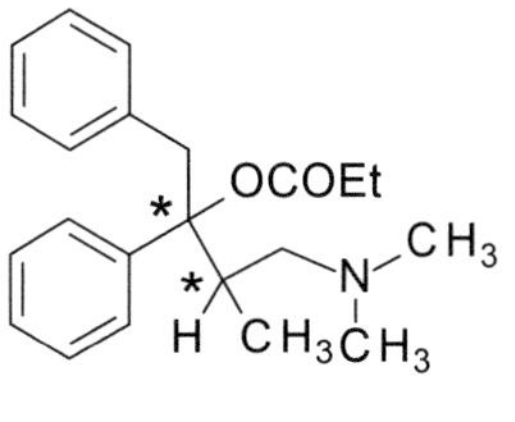

5.27 Propoxyphene

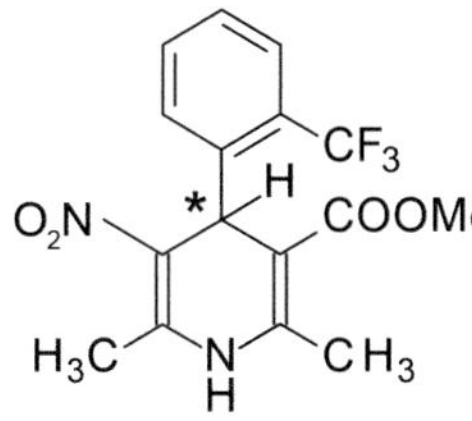

5.28 Bay K 8644

Fig. 5.13 Enantiomers also differ in their mode of action. The (*R*)-(−)-enantiomer of barbiturate **5.25** is a hypnotic agent, whereas the (*S*)-(+)-enantiomer causes seizures. In rats and mice, only the (*S*)-(−)-enantiomer of thalidomide **5.26** (Contergan®) is teratogenic, that is, it causes embryopathies. Thalidomide **5.26** racemizes *in vitro* as well as in rabbits. Therefore, even the (*R*)-(+)-enantiomer is teratogenic in rabbits. Propoxyphene **5.27** is a potent analgesic, the effect of which depends on the (2*S*,3*R*)-(+)-enantiomer, dextropropoxyphene. The (2*R*,3*S*)-(−)-enantiomer is a cough suppressant. The (*R*)-(+)-enantiomer of Bay K 8644 **5.28** is a weak calcium channel blocker. The (*S*)-(−)-enantiomer stabilizes calcium channels in the open form and is therefore an agonist, that is, a calcium channel opener (Sect. 30.4)

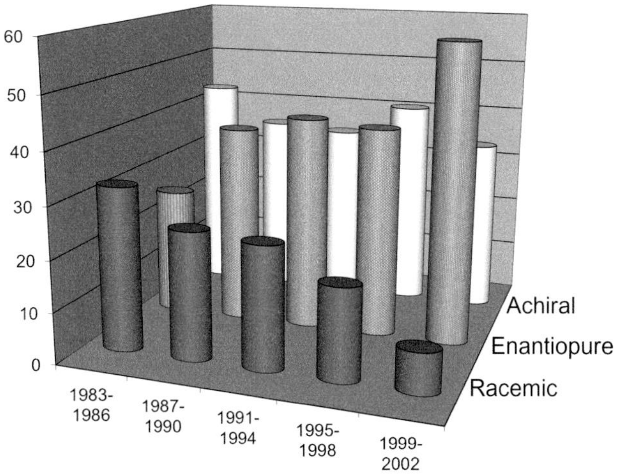

Fig. 5.14 The proportion of achiral, enantiomerically pure, and racemic drugs approved in the period from 1983–2002. In the meantime, the proportion of newly approved drugs has shifted decidedly in the direction of enantiomerically pure compounds

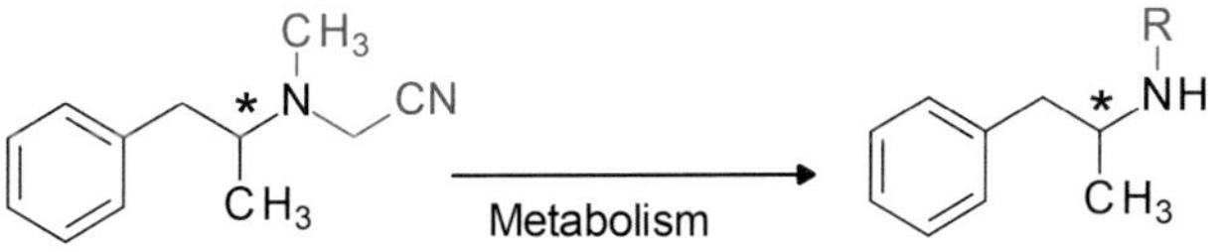

5.29

5.30 R = CH_3
5.31 R = H

Fig. 5.15 Upon metabolism of the monoamine oxidase inhibitor, selegiline **5.29**, which is used to treat Parkinson's disease, the more potent (*R*)-(−)-enantiomer is converted to methamphetamine **5.30** and amphetamine **5.31**. The less-active (*S*)-(+)-selegiline has less severe side effects because it is not metabolized to CNS-active stimulants

algesic and narcotic effects, but good antitussive properties. Enantiomers can also influence each other in their effects and even cancel each other out. In the case of the calcium channel ligand **5.28**, one enantiomer has a channel-opening effect, the other a channel-closing effect (see Sect. 30.4).

Between 1983 and 2002, 38% of all approved drugs were achiral, 39% were enantiomerically pure, and 23% were racemic or diastereomeric mixtures. The fact is that racemic mixtures of chiral drugs were much more readily accepted in earlier decades than they are today. This was certainly not due to stereophobia on the part of the chemical industry. Rather, it was due to a lack of understanding of stereospecificity and side effects, and perhaps also to economic considerations; kinetic resolution and/or enantiomerically pure syntheses are very expensive. You can certainly see that the proportion of **enantiomerically pure drugs** on the market is increasing (Fig. 5.14).

In the 1970s, Ariëns was the first to argue strongly against the use of racemic mixtures in therapy. In his view, **racemates** are compounds with 50% impurity. The inactive or less active enantiomer is called isomeric ballast and, despite its lack of effect, places an additional burden on the metabolism, since any drug-like compound must be chemically degraded in order to be eliminated from the body (Sect. 27.6). Ariëns used as an example the diastereomeric mixture labetalol **5.11** (Fig. 5.5), which is not a "mixed α,β-antagonist" but a mixture of four different drugs. The effect of this "combination" is a result of the effects of each enantiomer. In most cases, Ariën's criticism is fully justified. When designing and developing new drugs, it is important to ensure that the biological activity is as specific as possible and that side effects are minimized. Compound uniformity is usually easier to achieve for an enantiomer than for a racemate, which is a mixture of two substances, or even for a diastereomeric mixture.

Choosing the correct enantiomer can even reduce or prevent undesirable side effects of metabolites. Selegiline **5.29**, a monoamine oxidase inhibitor, is metabolized to the CNS active compounds methamphetamine **5.30** and amphetamine **5.31** (Fig. 5.15). Fortunately, the more active enantiomer of **5.29** forms the less active of these two metabolites! Using the correct enantiomer of the racemate will increase the desired effect and decrease the undesired CNS side effects.

There are also a few counter examples. The (−)-enantiomer of the calcium channel blocker verapamil (Sects. 2.6 and 30.4) is more effective than the (+)-enan-

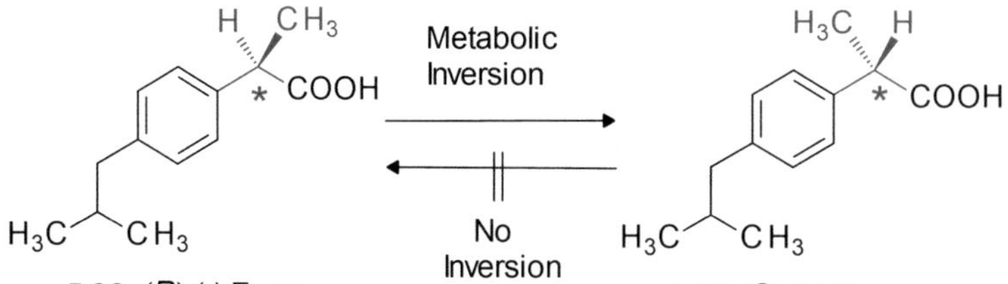

Fig. 5.16 The (*R*)-(−)-enantiomer of ibuprofen **5.32** undergoes a metabolic inversion of its stereocenter to form the (*S*)-(+)-enantiomer. As a cyclooxygenase inhibitor *in vitro*, the (*S*)-(+)-form is more potent than the (*R*)-(−)-form. The less-active form is converted to the more-active enantiomer *in vivo*. Therefore, both compounds exhibit equally anti-inflammatory properties in animal models

5

tiomer. The therapeutic spectrum of both enantiomers is practically identical. After oral administration, the (−)-enantiomer is quickly metabolized. Therefore, the (+)-enantiomer contributes substantially to the desired effect. In this case, it would not be economical to try to separate the racemic mixture.

Ibuprofen **5.32**, an anti-inflammatory drug of the arylpropionic acid class (Fig. 5.16 and Sect. 27.9), is a special case. The potency of the enantiomers are very different *in vitro*. *In vivo*, however, the inactive (*R*)-(−)-enantiomer is converted to a large extent to the (*S*)-(+)-enantiomer. The reverse reaction does not take place. Therefore, the racemate and each enantiomer are therapeutically identical, even at the same dose. Only the side-effect spectrum is different because the inversion of the (*R*)-(−)-enantiomer is not 100% complete.

Sometimes the cost of producing a pure enantiomer is not justified. In such cases, the efficacy and side effects of the two forms must be compared. Depending on the results, the continued use of the racemate or the development of an achiral analogue may be considered in special cases. In any case, these data must be complete before the drug can receive approval.

5.6 Image and Mirror Image: Why Is It Different for the Receptor?

Enantiomers and diastereomers have different biological properties because of the handedness of the proteins to which they bind. They occur in Nature in only one form. The amino acids with their chiral centers and the secondary structure elements (Sect. 14.2) with their helical orientation are responsible for these properties. If a protein is offered a left- or right-handed ligand, different binding modes can be expected, just as two right hands come together to shake hands more easily than a right hand and a left hand.

Only a few examples have been reported of protein–ligand complexes with both left- and right-handed ligands. This will only be possible if both enantiomers have sufficient affinity for the target protein, which means that they both bind to the protein strongly enough for an X-ray crystal structure to be determined.

The *R*- and *S*-enantiomers of the compound BX5633 (**5.33**) inhibit the serine protease trypsin (Sect. 23.3) to equal extents. They have a stereogenic center next to an acid group. The crystal structure determination explains this lack of discrimination. The inhibitor's acid group is oriented outside of the binding pocket so that no specific interaction is to be expected (Fig. 5.17). A stereopreference cannot exist.

Both enantiomers **5.34** and **5.35** bind to carbonic anhydrase II, a zinc hydrolase (Sect. 25.7). There is a difference of a factor of 100 in their affinities. As the X-ray structure with both enantiomers shows, they have similar binding modes (Fig. 5.18). All properties relating to the solvation of the ligands must be the same for both enantiomers. The difference in affinity is, therefore, only caused by differences in the binding mode of the ligands. The sulfonamide groups of both enantiomeric ligands bind almost identically to the catalytic zinc. In addition, the endocyclic SO_2 group forms very similar hydrogen bonds to Gln 92. The hydrophobic *iso*-butyl side chains are positioned in similar parts of the binding pockets. However, the six-membered ring must adopt a highly strained conformation in the case of the weaker binding enantiomer. The price of this strained conformation is a reduced binding affinity to the enzyme.

The enantiomeric agonists **5.36** and **5.37** bind to the ligand binding domain of the retinoic acid receptor with a factor of 1000 difference (Sect. 28.2). The receptor itself adopts the same geometry (Fig. 5.19). The alcohol function in the middle of the molecule is at the stereogenic center. In both cases, a hydrogen bond is formed with Met 272. As a result, the neighboring amide groups must take on deviating orientations in the binding pocket. On the "right" side, the tetralin moiety for both stereoisomers is found in similar spatial region. On the "left" side, the benzoic acid moiety of both enantiomers form a hydrogen-bond network with Arg 278, Ser 289, and Leu 233. The fluorine-substituted benzene ring adopts in both cases a 180° flipped orientation. These different orientations, together with the divergently oriented amide bond are responsible for the marked difference in the binding affinity of the mirror-image agonists.

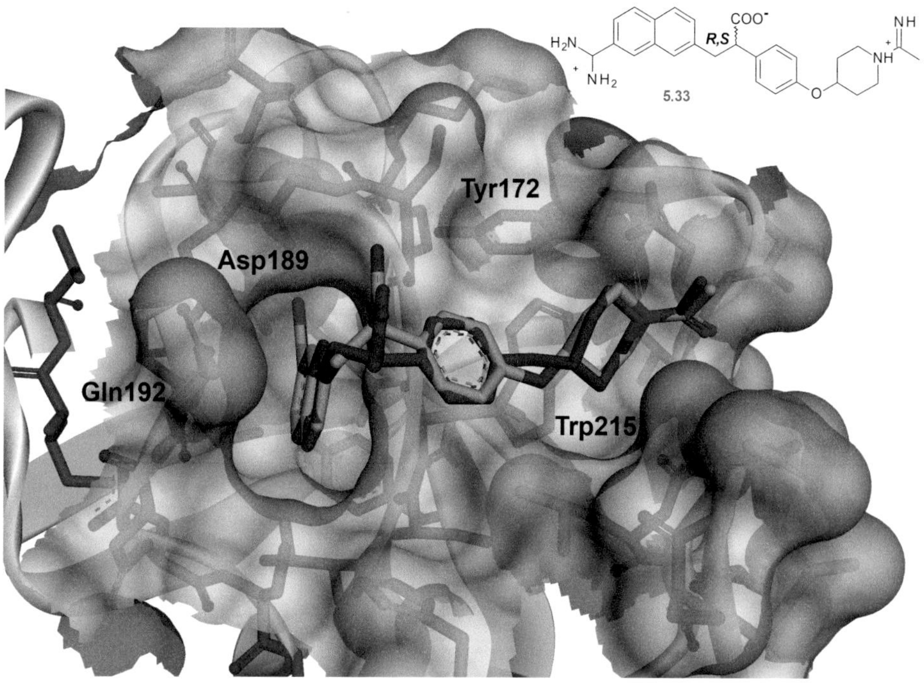

Fig. 5.17 The (*R*)- (*gray*) and (*S*)-enantiomers (*beige*) of the inhibitor BX5633 **5.33** bind with the same affinity to trypsin. Because the protein adopts practically the same geometry with both inhibitors, only one structure of the protein residues is shown. The crystal structure shows that both forms of **5.33** adopt almost identical binding modes. The acid function on the stereogenic center points out of the binding pocket and into the surrounding aqueous medium. Therefore, no stereochemical discrimination can take place. (▸ https://sn.pub/iAL3oo)

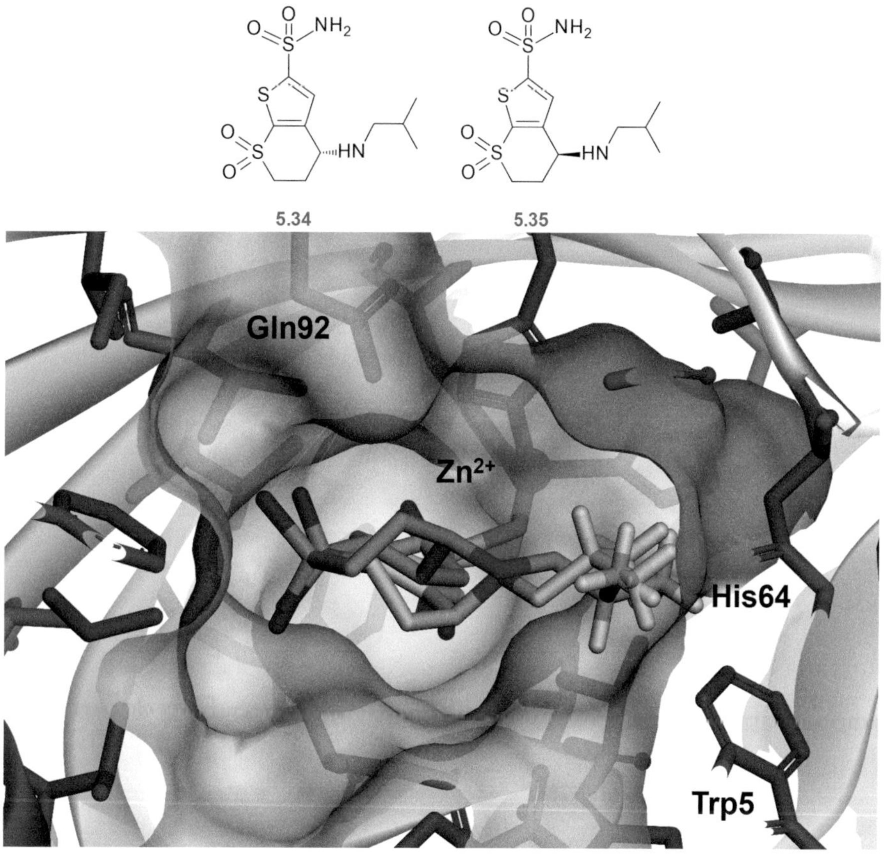

Fig. 5.18 The enantiomeric sulfonamides **5.34** (*gray*) and **5.35** (*yellow*) bind in a similar way to the enzyme carbonic anhydrase II. Because the protein adopts practically the same geometry with both inhibitors, only one structure of the protein residues is shown. The zinc ion in the catalytic center (*purple sphere*) is coordinated to the sulfonamide groups. The SO_2 groups in the six-membered ring form a hydrogen bond to Gln 92 (*green*). The hydrophobic *iso*-butylamino moieties on the chiral centers project into a hydrophobic pocket and fill this out to the same extent. In doing this, the six-membered ring must adopt a deviating conformation in both enantiomers. In one stereoisomer, this conformation is much more strained than in the other and causes a loss in binding affinity. (▸ https://sn.pub/YmklAy)

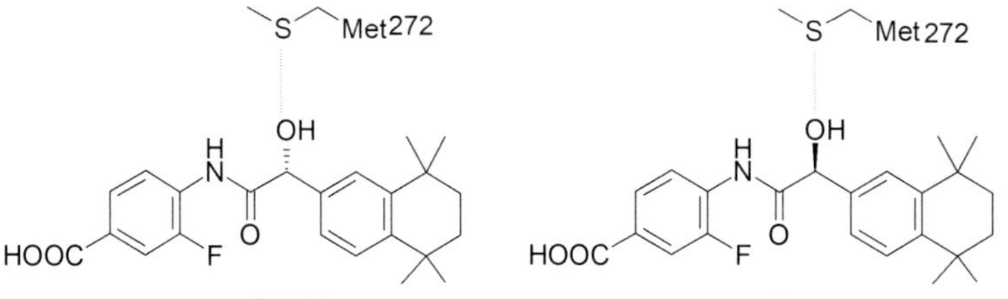

R-5.36 *S*-5.37

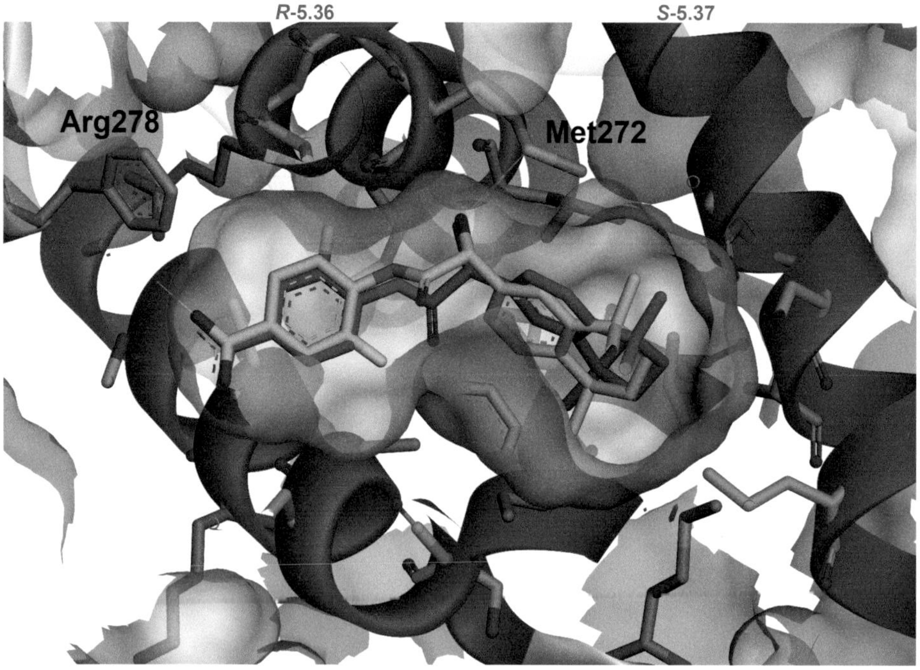

Fig. 5.19 Both enantiomers of the agonists **5.36** (*beige*) and **5.37** (*gray*) bind the retinoic acid receptor with 1000-fold difference in affinity. Because the protein adopts practically the same geometry with both ligands, only one structure of the protein residues is shown. Both ligands form H-bonds with their OH groups to the sulfur atom in Met 272. In doing so, the fluorine-substituted aromatic ring of the benzoic acid moiety on the left with its central amide bond has to adopt a deviating orientation. The tetrahydronaphthalene (tetralin) moiety, on the other hand, is positioned in the same way in both enantiomers. (► https://sn.pub/l3NCiE)

5

5.7 An Excursion into the World of Stereoisomers

Experience has taught us that if an enantiomer crystallizes with a particular auxiliary base or acid, the other enantiomer will crystallize with the mirror image of the auxiliary in the same way if identical reaction conditions are applied. Polypeptides composed of L-amino acids form right-handed helices, and polypeptides made of D-amino acids form left-handed helices.

Some naturally occurring peptides form ion channels in lipid bilayers. Their synthetic enantiomers can also do this. The more interesting question is: how does the mirror image of an enzyme behave? In 1992, Stephan Kent and coworkers prepared HIV protease (Sect. 24.3), a homodimer made up of 2 × 99 amino acids, entirely from D-amino acids. The naturally occurring protein was also prepared in parallel. The all-L-enzyme reacts only with L-peptide substrates and the all-D-enzyme reacts only with the all-D-enantiomer. The same is true for chiral inhibitors that block HIV protease. An achiral inhibitor, on the other hand, inhibits both enzymes in the same way.

Rubredoxin, an electron transport protein, was prepared as the all-D-protein for the sole purpose of mixing it with the naturally occurring all-L-protein and to make the racemate! If the effort involved is considered, this will certainly be an approach that takes some getting used to. The reward for the work was very high-quality crystals. The racemate crystallized in a centrosymmetric space group (Sect. 13.2), which allowed better resolution of the 3D structure than was possible with the natural, all-L-enantiomer and the phase determination is reduced to the assignment of "+" or "−".

What does a visit to the mirror-image world look like? Achiral drugs would have an identical potency and mode of action. On the other hand, many enantiomerically pure drugs would be useless. We would have to watch out for chiral barbiturates like **5.25**. They would rather cause a seizure than act as a sedative. In cases in which chiral antibiotics were used to treat bacterial infections, it would first have to be established whether the infecting bacteria came from the mirror-image world or the "normal" world. The administration of trimethoprim (Sect. 27.2) and a sulfonamide (both achiral) would help at any rate.

There would be tremendous nutritional problems. The carbohydrate and protein metabolism would not work anymore, nor would the absorption of monomers from the gastrointestinal tract. We would not be able to recognize some plants by their smell. (*R*)-Carvone smells of caraway seeds; (*S*)-carvone smells of spearmint. Our beloved sugar would have lost its sweet taste, and fruit juices and lemonade would taste sour. Coffee, tea, and Coca-Cola would retain their stimulatory effects because caffeine is achiral. Diet drinks would have to be sweetened with saccharine or cyclamate (both achiral) because aspartame is chiral.

Let us return to the normal world! But first, let us have a quick glass of vodka. It could also be cognac, whisky, or a dry red wine. The taste would be the same as in the normal world, or would it not? Despite the many hundred flavor components of wine, the exchange of a single chiral center could have the consequence that a connoisseur might no longer recognize the chateau. The euphoric effects would be the same, though this would not be the case for the hard, optically active drugs such as heroin, cocaine, or LSD.

5.8 Synopsis

- Compounds with an asymmetric or chiral center give rise to enantiomers, two isomeric forms that relate to each other like an image and mirror image and cannot be mutually transferred without breaking and reforming bonds.
- Enantiomers exhibit the same properties as long as they are found in a nonchiral environment. If exposed to the asymmetric environment such as a protein-binding site, they will experience different interactions and, thus, result in distinct biological properties.
- Chiral centers are mostly found at atoms carrying four different substituents, but also an overall handed scaffold can give rise to chirality. If n independent stereocenters are present, 2^n isomers (diastereomers) are produced occurring as 2^{n-1} racemic mixtures (pair of equally present enantiomers) as long as there is no internal inversion, mirror, or improper rotation symmetry present.
- Chiral centers are named according to the Cahn–Ingold–Prelog priority rules that bring the substituents in a unique sequence according to their atomic numbers. The substituent with lowest priority has to be oriented to the back and the direction of the remaining substituents determine *R*/*S* by the sense of rotation following decreasing priority.
- Enantiomers can be separated by fractional crystallization after being converted into diastereomeric salts with appropriate chiral auxiliaries. Enzymes such as lipases, esterases, or proteases can also be used for resolution because they transform one enantiomer faster than the other for steric and kinetic reasons.
- Most natural products are optically active and occur in just one form. Biologically active enantiomers are called eutomers, inactive ones distomers.
- Biological activities of enantiomers and diastereomers can vary greatly in strength and quality. Application of racemates has to be examined carefully for each individual case. Side effects, chemical stability, and deviating metabolism can have decisive influence on the activity profile.
- On the molecular level, the affinity discrimination of enantiomers is explained by deviating binding modes in the binding pocket of the target protein, thus, resulting in differences of the observed interaction pattern or strain of the adopted bound conformation.

Bibliography and Further Reading

General Literature

E. J. Ariëns, W. Soudijn and P. B. M. W. M. Timmermans, Stereochemistry and Biological Activity of Drugs, Blackwell Scientific Publishers, Oxford (1983)

D. F. Smith, Ed., CRC Handbook of Stereoisomers: Therapeutic Drugs, CRC Press, Boca Raton, Florida (1989)

B. Holmstedt, H. Frank and B. Testa, Chirality and Biological Activity, Alan R. Liss, Inc., New York (1990)

C. Brown, Ed., Chirality in Drug Design and Synthesis, Academic Press, London (1990)

M. Eichelbaum, B. Testa, A. Somogyi, Handbook of Experimental Pharmacology, Stereochemical Aspects of Drug Action and Disposition, Springer Verlag, Heidelberg (2002)

G. Klebe, Differences in Binding of Stereoisomers to Protein Active Sites, in Supramolecular Structure and Function 8, Ed. Greta Pifat-Mrzljak, Kluwer Academic/Plenum Pub., New York, pp. 31–53 (2004)

H. Caner, E. Groner and L. Levy, Trends in the Development of Chiral Drugs, Drug Discov. Today **9**, 105–110 (2004)

Special Literature

D. A. Evans, M. D. Ennis, and D. J. Mathre, Asymmetric Alkylation Reactions of Chiral Imide Enolates. A Practical Approach to the Enantioselective Synthesis of α-Substituted Carboxylic Acid Derivative, J. Am. Chem. Soc., **104**, 1737–1739 (1982)

D. A. Evans, Studies in Asymmetric Synthesis. The Development of Practical Chiral Enolate Synthons, Aldrichimica Acta, **15**, 23–32 (1982)

E. J. Ariëns et al., Stereoselectivity and Affinity in Molecular Pharmacology, Fortschritte der Arzneimittelforschung, **20**, 101–142 (1976)

E. J. Ariëns, Stereochemistry, a Basis for Sophisticated Nonsense in Pharmacokinetics and Clinical Pharmacology, Eur. J. Clin. Pharmacol., **26**, 663–668 (1984)

M. Bocola, M. T. Stubbs, C. Sotriffer, B. Hauer, T. Friedrich, K. Dittrich, G. Klebe, Structural and Energetic Determinants for Enantiopreferences in Kinetic Resolution of Lipases, Protein Eng., **16**, 319–322 (2003)

S. Mason, The Origin of Chirality in Nature, Trends Pharmacol. Sci., **7**, 20–23 (1986), and further contributions by additional authors, pp. 60–64, 112–116, 155–158, 200–205, 227–230 and 281–285.

E. J. Ariëns, Nonchiral, Homochiral and Composite Chiral Drugs, Trends Pharmacol. Sci., **14**, 68–75 (1993)

S. C. Stinson, Chiral Drugs, Chemical & Engineering News, 19. Sept 1994, p. 38–72, and Oct. 9, 1995, pp. 44–74

G. Jung, Proteins from the D-chiral world, Angew. Chem. Int. Ed. Engl., **31**, 1457–1459 (1992)

M. T. Stubbs, R. Huber, W. Bode, Crystal structures of Factor Xa specific Inhibitors in Complex with Trypsin: Structural Grounds for Inhibition of Factor Xa and Selectivity against Thrombin, FEBS Lett., **375**, 103–107 (1995)

J. Greer, J. W. Erickson, J. J. Baldwin, M. D. Varney, Application of the Three-dimensional Structures of Protein Target Molecules in Structure-based Drug Design. J. Med. Chem., **37**, 1035–1054 (1994)

B. P. Klaholz, A. Mitschler, M. Belema, C. Zusi, D. Moras, Enantiomer Discrimination Illustrated by High-resolution Crystal Structures of the Human Nuclear Receptor hRARγ, Proc. Natl. Acad. Sci. USA, **97**, 6322–6327 (2002)

R. C. Milton, S. C. Milton, S. B. Kent, Total chemical synthesis of a D-enzyme: the enantiomers of HIV-1 protease show reciprocal chiral substrate specificity, Science, **256**, 1445–1448 (1992)

L. E. Zawadzke, J. M. Berg, The structure of a centrosymmetric protein crystal, Proteins, **16**, 301–305 (1993)

The Search for the Lead Structure

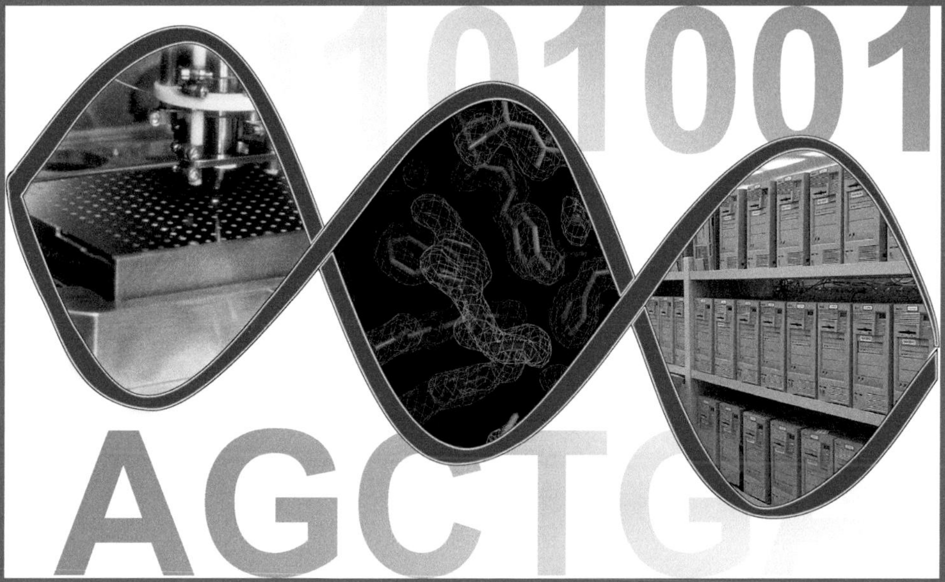

The starting point for the development of a new drug is the search for a suitable lead structure for a target protein. First, such a therapeutic target structure must be identified as a disease-relevant target in the genome or proteome of a cell. Genetic engineering techniques allow the production of this target structure. After setting up a high-throughput screening assay, thousands of test molecules are screened for binding to the target protein. The X-ray structure is elucidated and subsequently used to search for and optimize lead structures. Today, lead search and optimization is inconceivable without massive support from methods of bio- and chemoinformatics, molecular modeling, and computational chemistry (announcement poster of the author's group on the occasion of a 2003 conference in Rauischholzhausen, Marburg).

Contents

The Classical Search for Lead Structures

Contents

G. Klebe, *Drug Design*, https://doi.org/10.1007/978-3-662-68998-1_6

The starting point in the search for a new drug is the **lead structure**. Such a substance has already a desirable biological effect, but some specific characteristics are still inadequate for its therapeutic use. The definition of the term "lead structure" also means that analogues can be prepared by targeted chemical variations which produce compounds better than the lead structure in, for instance, their potency or selectivity. The goal is the optimization of all characteristics until a final substance is ready for therapeutic use.

The largest part of our pharmaceuticals originates directly or indirectly from natural products, that is, from plants, animals, or microbial sources, or from endogenous substances such as hormones and neurotransmitters. Only a few natural products have become drugs themselves. Examples include morphine, codeine, papaverine, digoxin, ephedrine, cyclosporine, and hirudin, the latter of which was isolated from leeches. Examples of endogenous drugs are the thyroid hormone T_3, insulin, coagulation factor VIII, erythropoietin, and further proteins for substitution therapy. Most naturally occurring compounds serve as lead structures. They are chemically manipulated with the goal of optimizing their desirable characteristics and minimizing their side effects (Chap. 8). Examples are found in the many natural products and endogenous receptor agonists that have been modified into selective agonists and antagonists (Sects. 6.2–6.4 and 6.6). Drugs are also derived from enzyme substrates (Chaps. 23–27) which can either be substrates for endogenous enzymes, for instance, that play a role in blood pressure regulation or inflammation, or they are substrates of enzymes from viruses, bacteria, or parasites, of which the metabolism should be specifically shut down.

Over the past 140 years, preparative organic chemistry has played a critical role not only in the systematic variation of lead structures, but also in their discovery. The search for new agents has yielded many drugs that bear no structural relationship to their endogenous counterparts. In other cases, the relationship between the biological effect and the mode of action was still unknown at the time of discovery. Some therapeutic principles that were subsequently pursued in drug development projects resulted from side effects observed during the administration of known drugs. This chapter summarizes how lead discovery has been successful in the time of classical drug research. Today, the screening of large compound libraries has become the focus of lead discovery and will be discussed in the next chapter.

6.1 How It Began: Hits by *In Vivo* Screening

The first example of discovering an active principle through testing occurred in the eighteenth century and is found in the effects of digitalis. The Scottish physician, William Withering, while working in England, was consulted by a patient who suffered from an extremely weak heart. After the doctor was unable to help him, the patient consulted a gypsy woman, who prescribed an herbal therapy. Impressed by the recovery of the patient, Withering sought out the woman and asked for the recipe. He received it in exchange for a handsome fee. The mixture contained an extract of the (poisonous) purple foxglove, *Digitalis purpurea*. The physician investigated the potency of different preparations of these plants in that he gave the medicines to 163 patients! With this experiment, he established that the best formulation was made up of the dried, powdered leaves. After the observation was made that a toxic dose is quickly reached, he recommended that diluted preparations be administered in repeated doses until the desired effect was achieved. Even though digitalis is still used today for congestive heart failure, no one would recommend that Withering's experimental "screening" technique be used to establish the therapeutic potential of a substance. This approach was neither ethical nor practical.

6.2 Lead Structures from Plants

The example of the previous section shows that Nature has furnished plants with highly potent substances. A plethora of secondary metabolites, for example, alkaloids, terpenes, flavones, and glycosides are also available. The contents of about a hundred different plant species have either directly or indirectly, in the form of analogues, found their way into human therapy. Traditional medicines use about 5000–10,000 of the several hundred thousand already known species from the rich plant kingdom. Morphine, caffeine, quinine, cocaine, ephedrine, coniine, atropine, and reserpine were already mentioned in Sect. 1.1. Further plant-based pharmaceuticals that are used in therapy or that have served as lead structures for the development of medicines are compounds **6.1–6.7** (◘ Fig. 6.1), and, in addition, emetine, pilocarpine, podophyllotoxin, and the vinca alkaloids vinblastine and vincristine.

Why do plants contain so many valuable therapeutic compounds? There is not a human-related answer because plants did not evolve so that they could become human medicines. The plants, however, had to respond to their environment, and a competition with other species occurred. The decisive disadvantage of being a plant is that it cannot run away! That is not a disadvantage when it comes to reproduction. Bees take care of the first part, and aerodynamic seeds help with the rest. Some plants have tasty fruits with robust seed bodies. In this way, predators later help to spread the undamaged seeds through their excretions. An effective protective mechanism of plants against, for instance, fungal infection and pests such as caterpillars, sheep, and cattle served as a se-

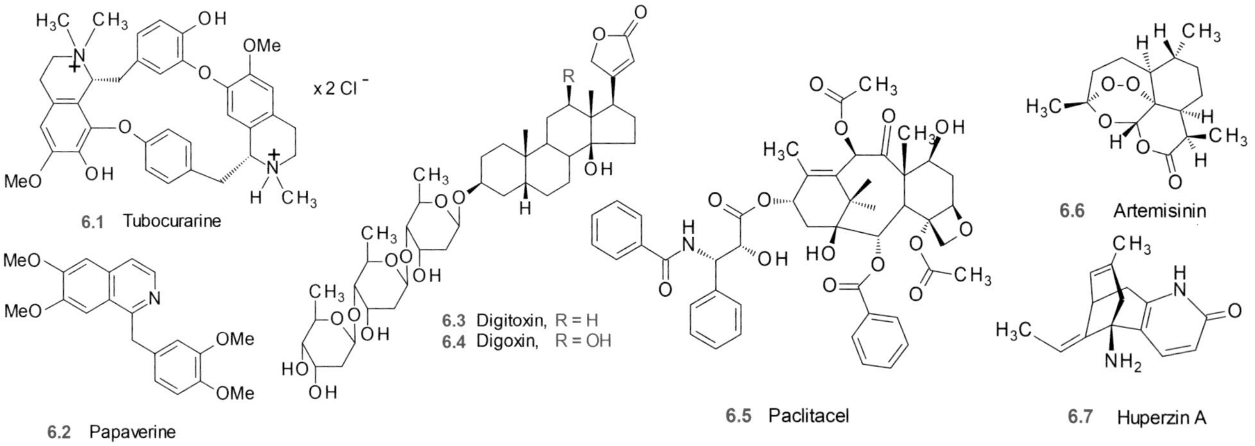

Fig. 6.1 Natural products from plants that have been introduced to therapy or have served as lead structures include, in addition to the substances introduced in Sect. 1.1, tubocurarine (curare) **6.1**, papaverine **6.2**, digitoxin **6.3**, digoxin **6.4**, and the related cardiac glycosides. Newer natural products from plants with great therapeutic potential include paclitaxel (Taxol®) **6.5** for tumor therapy, artemisinin **6.6** for malaria therapy (Sect. 3.3), and the acetylcholinesterase inhibitor huperzin A **6.7** for the potential treatment of Alzheimer's disease

lection advantage for some plants. The substances that offer an advantage have bitter taste, are spicy, or toxic. They exert their effects in that they interact with the enzymes or receptors of the "enemy." The stronger the effect, the better the protection. A successful principle of evolution is the development of defensive substances that do not kill, but cause an unpleasant experience for the predator, which in turn teaches the enemy to stay away. This is how butterflies survive, i.e., by accumulating poisonous plant-based substances in their bodies, and even some butterflies only imitate the appearance of these butterflies. After the first experience with the poisonous species, birds give both species a wide berth.

Plant substances have already undergone a selection process on biologically relevant proteins; during the course of evolution they have "seen" receptors and binding sites. In addition, the course of their biosynthesis takes place in the binding site of a protein, that is, they have functionality that mediates affinity to a protein. Certainly, there are many plant substances that coincidently have a biological effect in humans. Morphine contains a basic nitrogen, a phenolic hydroxyl group, an ether bridge, and a hydrophobic domain: a medicinal chemist would also choose such a mixture of functional groups, without the complicated ring structure, in the conception of an active substance. However, Nature has equipped its natural substances with the necessary solubility and bioavailability. If necessary, the metabolic stability has also been adapted accordingly. The correct adjustment of these properties is often difficult with optimizations that rely on synthesis products from classical synthetic chemistry performed in organic solvents.

The isolation of natural products from plants for lead discovery has experienced rather changing valuation in the last decades. Large pharmaceutical companies have repeatedly launched ambitious programs to elucidate the mechanism of action of traditional medicines, only to abandon the field in disappointment. These disappointments are a result of an unfavorable relationship between effort and reward. All too often only a toxin is isolated instead of a valuable lead structure, and all too often an already-known principle is found. Nonetheless, the search continues. Nature offers structural variation that the chemist can only dream of.

6.3 Lead Structures from Animal Venoms and Other Ingredients

In contrast to the plants, the evolution of animal venoms occurred with the objective of subduing prey or defending against an enemy. Many of these substances are proteins, peptides, and alkaloids. They function as potent poisons that can quickly lame or kill a victim. Because of this, many active substances from animals are unsuitable for therapy, but others, for the exact same reason, are interesting lead structures. Animal products offer many surprises, as illustrated in the following two examples.

Despite its simple structure, epibatidine **6.8** (Fig. 6.2, Sect. 30.6), which was isolated from the Ecuadorian poison dart frog *Epipedobates tricolor*, is a 100-fold more potent analgesic than morphine! It does not affect the opiate receptor, but rather it is an agonist at the nicotinic acetylcholine (nACh) receptor (Sect. 30.6). That comes as no surprise when its structural similarity to nicotine **6.9** is considered. Epibatidine has a binding constant of 0.04 nM on the nACh receptor, which is 50-fold stronger than nicotine. Unfortunately, its analgesic effects are coupled with a pronounced decrease in body temperature (hypothermia).

Dolastatin **6.10** (Fig. 6.2) was isolated from the wedge sea hare, *Dolabella auricularia*, a marine snail. It

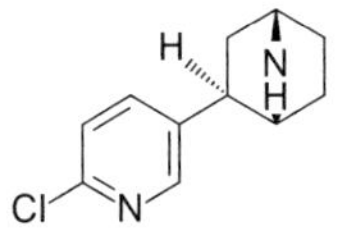

6.8 Epibatidine

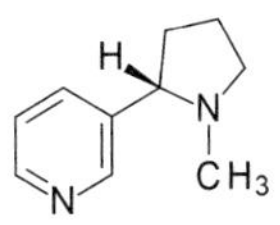

6.9 Nicotine

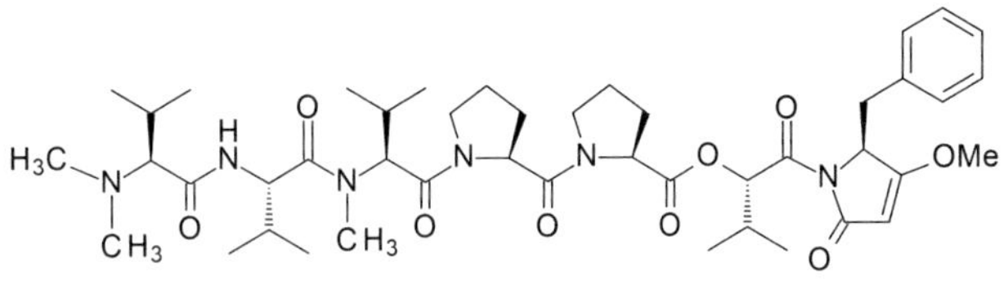

6.10 Dolastatin-15

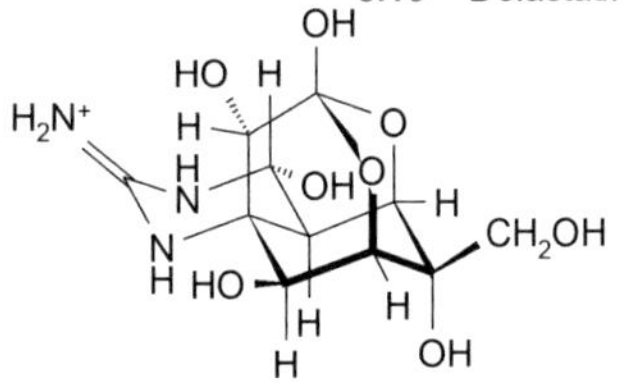

6.11 Tetrodotoxin

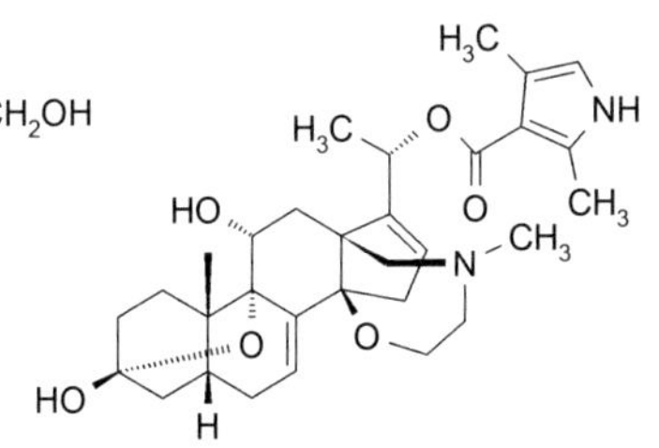

6.12 Batrachotoxin

Fig. 6.2 Epibatidine **6.8**, a non-opiate analgesic that binds 50-fold more potently to the nicotinic acetylcholine receptor than nicotine **6.9**, comes from a South American frog (Sect. 30.6). Dolastatin-15 **6.10**, which was isolated from a marine snail, is an interesting lead structure for cancer therapeutics. The toxin of the fugu fish, tetrodotoxin **6.11**, is not a lead structure but rather a sodium channel blocker for experimental (*in vitro*) use (Fig. 30.12). The steroid alkaloid batrachotoxin **6.12** is the most potent animal venom known. The LD_{50} value in mice, that is, the dose necessary to kill 50% of the experimental animals within 24 h, is 200 ng/kg

is an interesting lead structure for antitumor compounds. Synthetic analogues of **6.10** cause the complete remission of tumors in some animal models. The diversity of marine animals in particular has historically been a rich source of new and interesting lead structures and modes of action.

Other animal substances have gained importance in experimental pharmacology. Among them are the poison of the notorious fugu fish, tetrodotoxin **6.11**, and the steroid alkaloid batrachotoxin **6.12** from the skin of the Columbian poison dart frog (Fig. 6.2). Whereas tetrodotoxin specifically blocks sodium channels, batrachotoxin stabilizes sodium channels in the open form (Sect. 30.4).

Peptides from snake venom made a decisive contribution to the development of the antihypertensive angiotensin-converting enzyme inhibitors (Sect. 25.4). Research on the area of thrombin inhibitors in the past years have turned towards the active ingredient of leech saliva, hirudin. Aside from the direct use of hirudin, longer-acting derivatives, shorter peptides that only bind on the fibrinogen-binding site, and protein conjugates with other thrombin inhibitors have been derived from the structure.

Animal and human proteins as well as polymeric carbohydrates are extraordinarily important for substitution therapies. Insulin (isolated from the pig pancreas) is at the top of the list, followed by aprotinin, a protease inhibitor (isolated from cattle lungs), digestive enzymes, and the coagulation inhibitor heparin. Now that the possibility of producing insulin via gene technology is available, its isolation from animal organs has become less important. Other proteins, for example, the erythrocyte-stimulating hormone erythropoietin (Sect. 29.8), human growth hormone, tissue plasmin activator tPA, urokinase, and factor VIII, are all manufactured by using gene technology nowadays (Sect. 32.1). In this way, these proteins are available in practically unlimited quantities.

The protease ancrod, isolated from the venom of the Malayan pit viper *Agkistrodon rhodostoma*, cleaves fibrinogen, the precursor of fibrin, to products that can no longer aggregate. Thus, the viscosity and the coagulation ability of the blood is reduced (Sect. 23.4). An elevated thrombosis risk can be significantly reduced through this mechanism. To isolate the active component of this venom, several hundred snakes have to be "milked" regularly.

6.4 Lead Structures from Microbial Organisms

When speaking of active substances from microorganisms, antibiotics must be mentioned first. Substances such as streptomycin **6.14** can be used directly for therapy. The β-lactams penicillin and cephalosporin (Sects. 2.4 and 23.7) are highlighted as particularly valuable lead structures. Aside from oral bioavailability, the therapeutic goals were broad-spectrum activity and metabolic stability. Tetracycline **6.13** (Fig. 6.3) was also intensively structurally modified. Initially, all approved drug candidates with the tetracycline scaffold had been discovered and subsequently produced by chemical modification of the fermentation products. This severely limited the number of new structures. More recently, however, it has been possible to build a total synthesis platform technology that allows for completely different modifications, even to the basic scaffold. New clinical candidates have been discovered, including eravacycline, which was approved in 2018. Tetracyclines interfere with protein biosynthesis at the ribosome (Sect. 32.6). The new fully synthetic approach suggests a reassessment of other antibiotic classes based on natural products for which viable fully synthetic routes have not yet been developed. There may be underdeveloped resources with great potential for safer and more effective anti-infectives that we urgently need.

The immunosuppressants cyclosporine A (Sects. 10.1 and 22.8), FK 506, and rapamycin also originated from

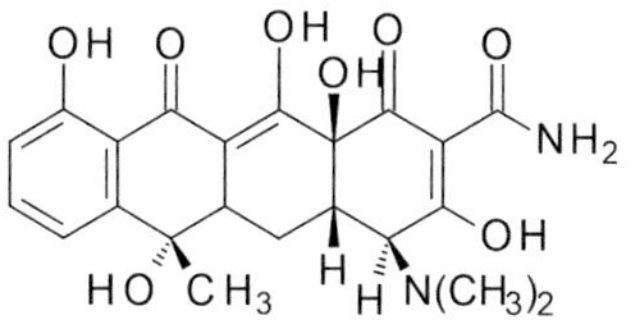

6.13 Tetracycline

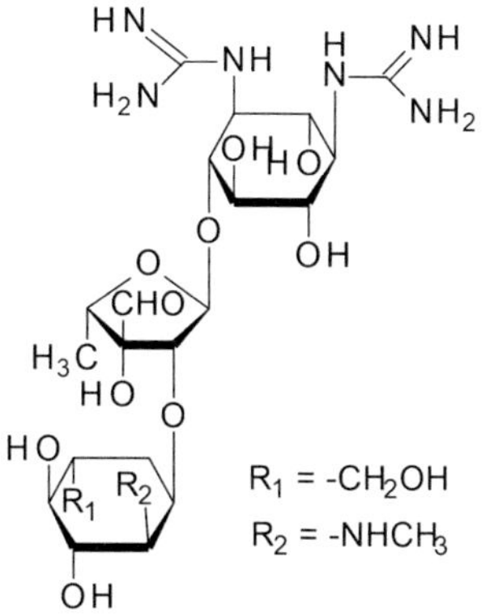

R_1 = -CH_2OH
R_2 = -$NHCH_3$

6.14 Streptomycin

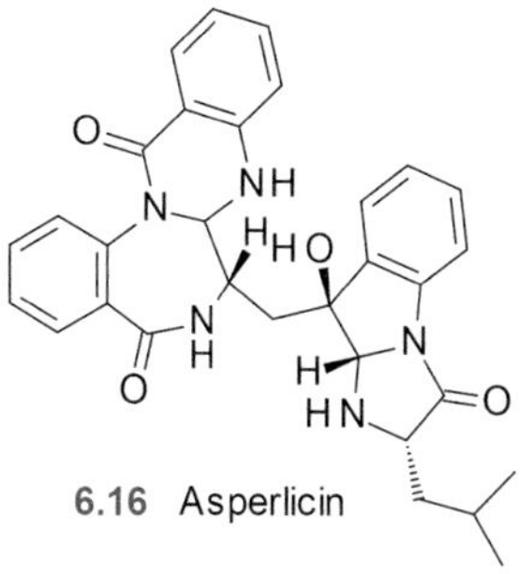

6.16 Asperlicin

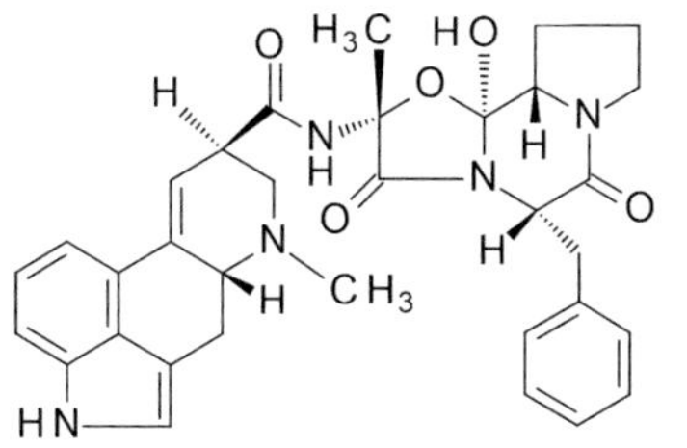

6.15 Ergotamine

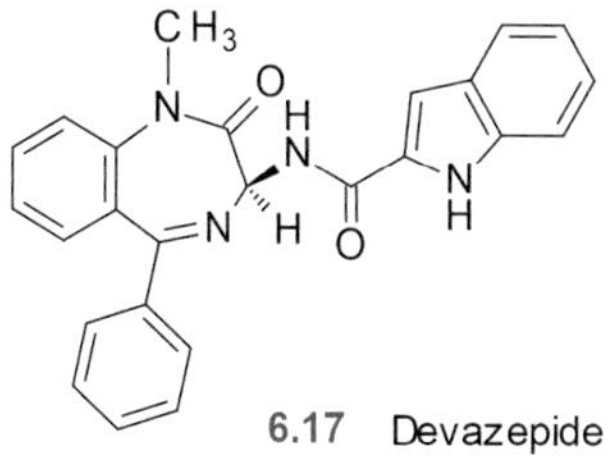

6.17 Devazepide

Fig. 6.3 Penicillins, cephalosporins (Sects. 2.4 and 23.7), and tetracycline **6.13** were important lead structures for even better antibiotics. In contrast, streptomycin **6.14** is used in therapy itself. Ergotamine **6.15** is a typical representative of the ergot alkaloids, from which a plethora of different drugs have been derived. Likewise, asperlicin **6.16** is a structurally complex microbial natural product. The 10,000-fold more potent derivative devazepide **6.17** was derived from it

microorganisms. Cyclosporine A is a convincing example of how difficult it is to predict the potential of a new therapeutic substance. Sandoz almost abandoned its development because of "lack of market potential." This decision would have had fatal consequences because a large portion of the success of transplantation surgery today can be attributed to this substance. Instead, cyclosporine became one of the company's best-selling products.

The fungus *Claviceps purpurea*, which grows in grain (ergot, *Secale cornutum*), contains a toxic alkaloid. For hundreds of years, the consumption of bread that had been made from contaminated flour was the cause of severe poisonings. The structures of these alkaloids, for example, ergotamine **6.15** (Fig. 6.3), were in large part elucidated at Sandoz. Their systematic modification led to active substances for many indications, e.g., for inducing contractions during labor, migraine therapy, perfusion disorders, and arterial hypertension. Today, they have little importance because of their limited therapeutic index. Another representative of this class is the hallucinogen lysergic acid diethylamide (Sect. 2.5), which was discovered by accident.

Lovastatin and some analogues (Sects. 9.2 and 27.3) are exceedingly important therapeutic substances that were isolated from microorganisms; they interfere in the biosyntheses of cholesterol. Cholecystokinin (CCK) is a peptide hormone that acts at a G protein-coupled receptor (Sect. 29.1). It induces multifaceted effects in the central nervous system and gastrointestinal tract. The nonpeptide CCK antagonist asperlicin **6.16** (IC_{50} = 1.4 μM) originated from extracts of *Aspergillus alliaceus*. After intensive structural variation, the much simpler devazepide **6.17** (IC_{50} = 80 pM) was designed, which has more than 10,000-fold higher affinity to the CCK receptor (Fig. 6.3). This antagonist is orally bioavailable and is an appetite stimulator.

The enzyme streptokinase for the dissolution of blood clots, and bacterial collagenase for wound treatment are examples of therapeutically important proteins that were isolated from microorganisms.

6.5 Dyes and Intermediates Lead to New Drugs

In 1903, Paul Ehrlich investigated hundreds of dyes in mice that had been infected with trypanosomes. The result of this research was Nagana Red, the first drug for *Trypanosoma crucei* infection, the causative agent of cattle trypanosomiasis. Other dyes followed, as did colorless compounds containing amide groups instead of azo groups. It was only after Ehrlich's death in 1916 that Bayer, after having investigated more than a thousand analogues, produced its wonder drug suramin (Germanin®) **6.18** (Fig. 6.4). The work in this area led to the discovery of the antibacterial sulfonamides in the 1930s (Sect. 2.3). Thousands, if not tens of thousands, of analogues were synthesized and tested. Many were introduced to the market. Depending on the structure, they cover an extraordinarily broad spectrum of different pharmacokinetic characteristics.

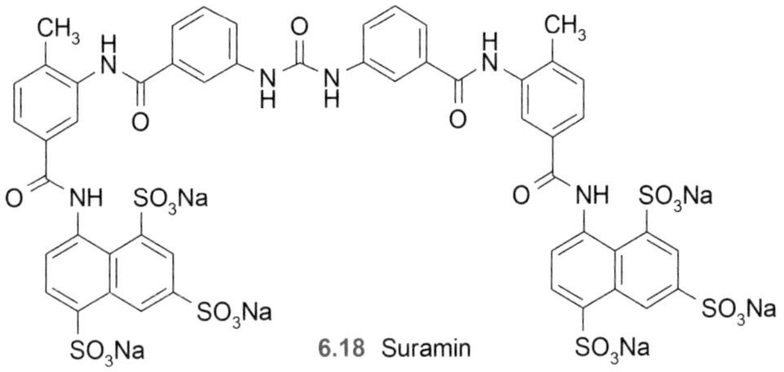

■ Fig. 6.4 Bayer's suramin **6.18**, which is also known as E 205 or Germanin®, had strategic importance for the colonies. An English engineer who was suffering from the African sleeping sickness (trypanosomiasis) and was near death despite aggressive treatment with diverse antimony and arsenic preparations, was cured after a few injections of this substance. The solvent for the preparation of the intravenous injection solution was rain water in the tropical clinical trials(!). After a short time, suramin was considered to be a "wonder drug." Despite the fact that the structure was kept secret, French researchers worked out their own synthesis within a short time. Suramin is still used for the treatment of trypanosomiasis because it has good efficacy and a long-lasting effect

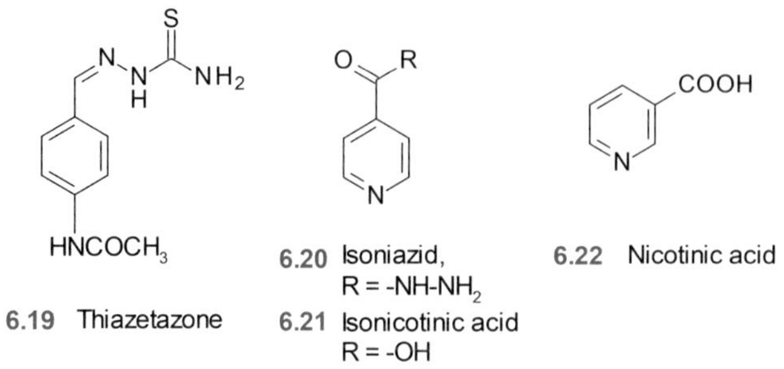

■ Fig. 6.5 Thiacetazone **6.19** and isoniazid **6.20** are tuberculostatics that originated as synthetic intermediates. Isoniazid penetrates the cell wall and irreversibly binds to the enzymatic cofactor NADH after radical formation. The originally accepted hypothesis that, upon metabolic degradation to isonicotinic acid **6.21**, it acts as an antimetabolite for nicotinic acid **6.22** proved to be incorrect

No actual biological activity was expected from synthetic intermediates. They were seen merely as starting material for the desired end product. Despite this, many intermediates were routinely tested for biological activity, and it was a good thing too!

Gerhard Domagk, the discoverer of sulfonamides (Sect. 2.3), investigated just such a synthetic intermediate in addition to the many end-target substances and found a surprisingly good effect against tuberculosis. Structural optimization afforded thiacetazone **6.19** (■ Fig. 6.5), which unfortunately turned out to be hepatotoxic. In the search for a follow-up substance, Bayer started a concerted program with 5000 compounds. In 1951, another synthetic intermediate showed surprisingly potent tuberculostatic activity. Isoniazid **6.20** (■ Fig. 6.5) was 15 times more active than the best antituberculosis antibiotic at the time, streptomycin **6.14** (■ Fig. 6.3). The discovery must have been "in the air." Two other research groups, both in the USA, simultaneously and independently discovered the effect of this substance, which, upon enzymatic radical formation, irreversibly binds to the cofactor NADH of a fatty-acid-synthesizing enzyme of the tuberculosis bacillus. The hypothesis that metabolic cleavage to isonicotinic acid **6.21**, which in turn exerts its effect by acting as an antimetabolite to nicotinic acid **6.22** (■ Fig. 6.5), was evidently wrong.

Inhibitors of the enzyme dihydrofolate reductase, for instance, methotrexate **6.23** (■ Fig. 6.6), are used in the treatment of leukemia (Sect. 27.2). During the investigation of analogues, a simple synthetic intermediate, mercaptopurine **6.24**, was tested. It showed efficacy, but was too toxic. Further development delivered azathioprine **6.25**, which releases mercaptopurine in the organism (■ Fig. 6.6). As an immunosuppressive, azathioprine was even better than the then-used corticosteroids (Sect. 28.5). Until the introduction of cyclosporine (Sect. 10.1), it was used in all organ transplantations. Another intermediate from this class, allopurinol **6.26** (■ Fig. 6.6), is a xanthine oxidase inhibitor. It is used for the treatment of gout.

6.6 Mimicry: How to Copy Endogenous Ligands

As of the middle of the nineteenth century, biological substances, enzyme substrates, neurotransmitters, and hormones were increasingly being used as archetypes for new medicines. The directed design of drugs from these lead structures led to the "golden age" of pharmaceutical research (Sect. 1.4).

The principal approach is demonstrated here using the example of enzyme inhibitors. Enzymes catalyze chemical reactions in that they stabilize the transition state of the reaction. In doing so, they decrease the activation energy, and the reaction can proceed at a lower temperature (Sect. 22.3). This specificity can be exploited particularly well for the optimization of enzyme inhibitors. By starting with knowledge of the reaction mechanism, substrate groups are assembled that are structurally analogous to the transition state (■ Fig. 6.7). They imitate it but do not lead to a product. In this way in a single step, through an entirely purposeful chemical transformation, a substrate can be converted into a potent and selective inhibitor.

The correct inhibitor binding geometry improves the affinity by several orders of magnitude. The two natural products pentostatine **6.29** and nebularine **6.30** (■ Fig. 6.8) are inhibitors of the enzymatic transformation of adenosine **6.27** to inosine **6.28** and impressive examples of transition state mimetics. The introduction of a hydroxyl group with the correct stereochemistry in-

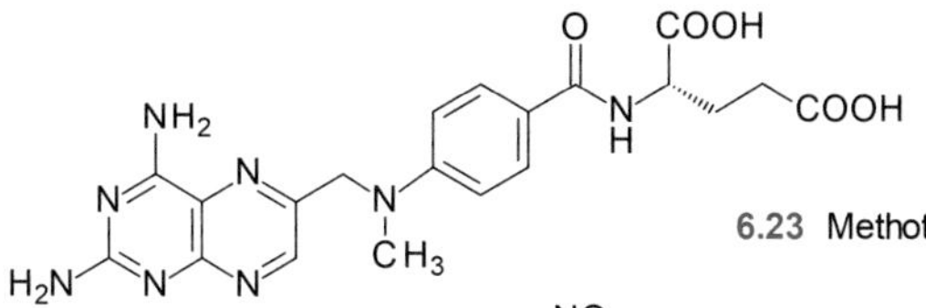

6.23 Methotrexate

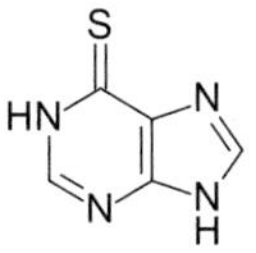

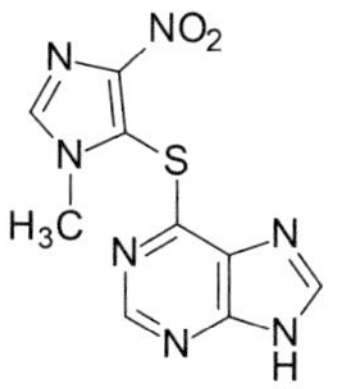

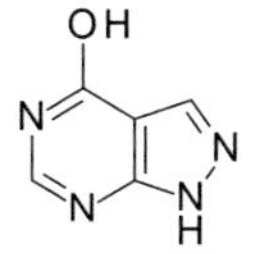

6.24 Mercaptopurine 6.25 Azathioprine 6.26 Allopurinol

■ **Fig. 6.6** Simple synthetic intermediates to methotrexate **6.23** turned out to be new drugs. Mercaptopurine **6.24** and azathioprine **6.25** are immunosuppressants, and allopurinol **6.26** is used to treat gout

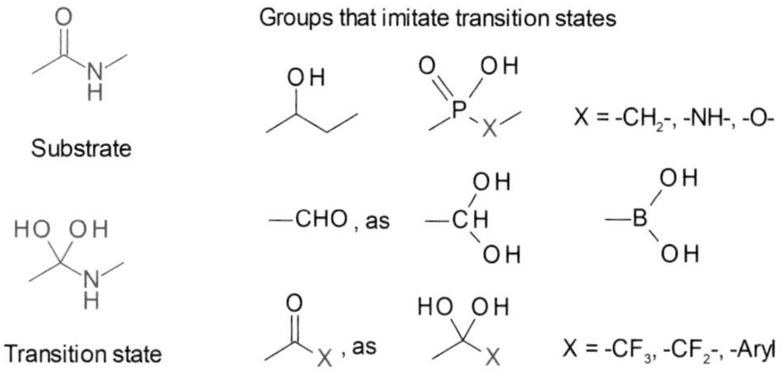

■ **Fig. 6.7** Examples of substrate, transition state, and groups that imitate the enzymatic transition state of an amide hydrolysis reaction. A few of the groups reversibly form covalent bonds to the serine in the catalytic pocket of a serine protease (see Sect. 23.2)

creased the affinity of the ligand to the enzyme by many orders of magnitude.

Never before was the search for new drugs as successful as it was in the two to three decades of the "golden age." Subsequently the success rate fell. Research became more expensive and laborious. How is this explainable? Because of the success during this period, many indication areas achieved a very high standard of care. That makes it difficult for modern research to be as successful as before, even with the use of superior tools. Other reasons include higher requirements for efficacy and safety.

6.7 Side Effects Indicate New Therapeutic Options

Many drugs came from the observation of side effects during clinical or practical use (see Sect. 2.8). The diuretic effects of mercury compounds were discovered purely by accident (Sect. 30.12). In 1919, physicians in the First Medical University Hospital in Vienna were testing a new treatment for syphilis. It was observed in a 21-year-old woman that her urine production increased from 200–500 mL a day to 1.2–2.0 L on the third day of treatment with the test substance. This result led to the development of the first effective diuretic (medicine to increase urine production). Fortunately, we are no longer dependent on extremely toxic mercury compounds for the therapy of venereal disease or as diuretics!

In 1948, it was observed in vulcanization factories that the antioxidant disulfiram **6.31** (■ Fig. 6.9) caused workers to become intolerant of alcoholic drinks. This discovery led to the use of the substance for the treatment of chronic alcoholism. The metabolic intermediate of ethanol, acetaldehyde, is not metabolized any further. This leads to generalized poisoning symptoms such as nausea, palpitations, and cold sweats. The effect is, however, difficult to control. Alcohol consumption after treatment has occasionally been fatal.

A classic example of the discovery of an important indication by observing side effects can be found in the sulfonamides. The sulfonamide diuretics and the oral antidiabetics (Sect. 30.2), drugs of choice to treat certain forms of diabetes, were found in this way (Sect. 8.4).

Iproniazid **6.32** (■ Fig. 6.9) is a derivative of isoniazid **6.20** (■ Fig. 6.5). In 1957, a tuberculosis patient noticed a distinctive improvement in mood, which led to its broad use for the treatment of chronic depression. The substance had to be withdrawn from the market a few years later due to severe side effects (Sect. 27.8).

Sweet clover has been used in Europe to feed livestock for hundreds of years. During its introduction in the 1920s to the USA and Canada, it was initially stored inappropriately, with disastrous consequences. Massive bleeding and fatalities in the cattle were attributed to the spoiled sweet clover (i.e., hemorrhagic sweet clover disease). The active substance, dicoumarol **6.33** (■ Fig. 6.9), was introduced into therapy in 1942, but its effects were unreliable. The Wisconsin Alumni Research Foundation investigated 150 analogues and produced warfarin **6.34**, which was sold as a rat poison. The name is derived from the company's acronym WARF, and the ending "arin" from coumarin. In 1951, an American soldier attempted suicide with a high dose of warfarin. Because he survived, a clinical trial was initiated. Despite the need for constant monitoring of coagulation levels, the use of warfarin was a standard therapy after myocardial infarction and stroke for many years.

Penicillamine **6.35** (■ Fig. 6.9) provides an example of an important indication extension. It was introduced for the treatment of Wilson's disease, an inherited metabolic disease that leads to copper accumulation in tissue. Because **6.35** forms complexes well, it is also appropriate for the treatment of heavy-metal poisonings. It was only later, after its practical use, that its much larger importance as a basis therapy for rheumatic disease was recognized. The mechanism of action remains largely unclear.

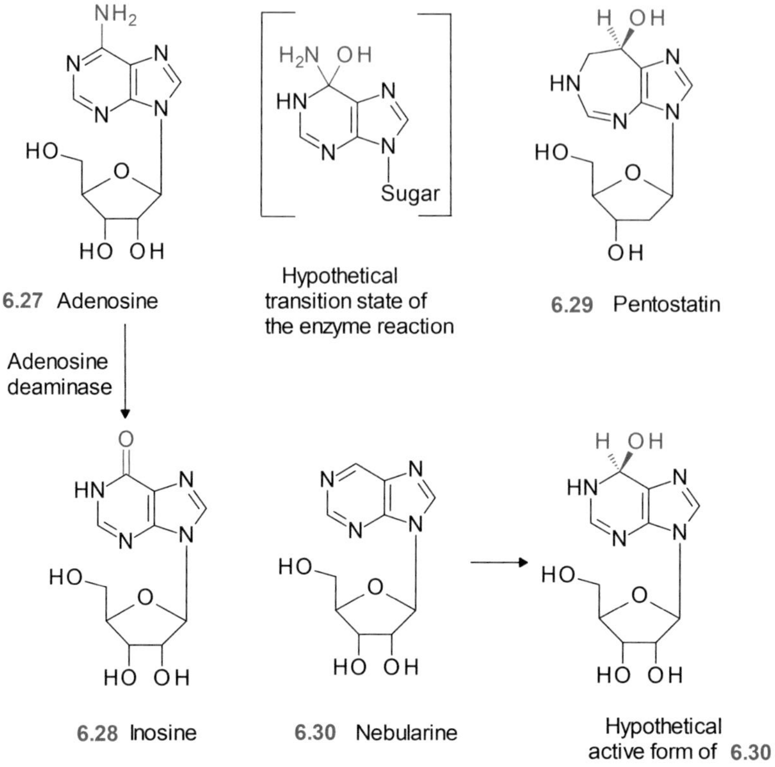

Fig. 6.8 Pentostatin **6.29** and nebularine **6.30** inhibit the enzymatic transformation of adenosine **6.27** to inosine **6.28**. The affinity of **6.29** is seven orders of magnitude more potent than the substrate adenosine (K_i = 2.5 pM), and the active form of **6.30** is even 10 orders of magnitude more potent (K_i = 0.3 pM). The structures of pentostatin as well as the active form of nebularine correspond to the transition state of the enzymatic reaction

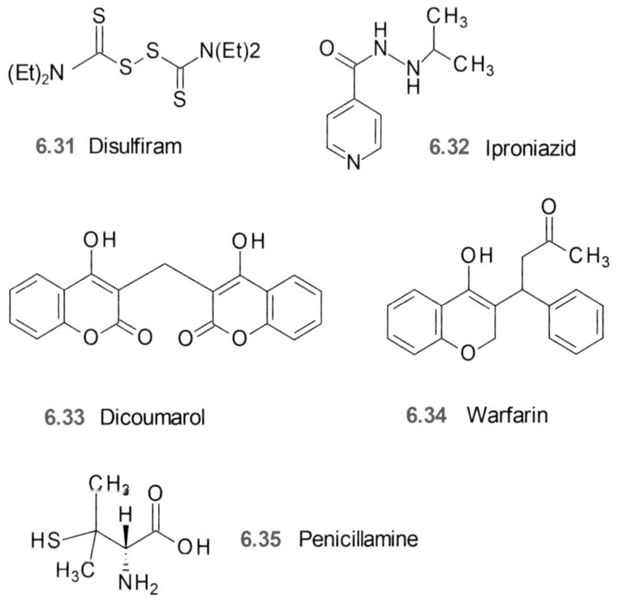

Fig. 6.9 Tetraethylthiuram disulfide **6.31** or disulfiram, better known as Antabuse®, is an aldehyde dehydrogenase inhibitor. The accumulation of the toxic acetaldehyde leads to nausea. Iproniazid **6.32**, a simple derivative of isoniazid **6.20** (Fig. 6.5), is a monoamine oxidase inhibitor (Sect. 27.8). It acts as an antidepressant by prolonging the effects of the biogenic amines. The rat poison warfarin **6.34** is derived from dicoumarol **6.33**. Its effect on blood clotting must be carefully monitored. Warfarin has long been a standard of therapy for diseases associated with an increased tendency for clotting, e.g., myocardial infarction or stroke. Penicillamine **6.35** is a complexation agent for heavy metals; it is used for—among other indications—the treatment of Wilson's disease, which is an inherited disease that leads to the accumulation of copper in the tissues. It was only later that its efficacy in chronic rheumatic diseases was discovered

6.8 From the Traditional Search to the Screening of Large Compound Libraries

The approaches described in the previous sections may still be important in industrial pharmaceutical research. However, because of the enormous costs associated with drug development, the search for original lead structures is now being approached in many different ways. Large sums of money are paid for novel therapeutic approaches, test models or three-dimensional (3D) structures of target proteins. This information can lead to a competitive advantage that may take time to realize, but must be zealously defended and exploited.

According to the principle of risk diversification and the maximal exploitation of all imaginable resources, today pharmaceutical companies subscribe to a strategy of broadly established screening of huge substance libraries of plant extracts, microbial fermentations, and synthetically prepared compounds. The last category comes from in-house chemistry as well as purchased compounds and combinatorial compound libraries (Chap. 11). Furthermore, a large part of the search for new lead structures takes nowadays place by computer methods.

The identification of therapeutically relevant target proteins plays an ever-increasing role for the discovery of new lead structures. The elucidation of the human genome (Sect. 12.3) has delivered the sequences of all human proteins. By comparing the expression pattern between diseased and healthy cells, it is possible to recognize particular proteins as a cause or consequence of a given pathology (Sect. 12.9). Should such a protein be detected, the next steps are certain. The therapeutic concept is tested on a genetically modified animal (Sect. 12.5), or the gene is silenced (Sect. 12.7), a molecular test system is established, and the 3D structure of the protein is elucidated. In parallel, all available techniques for lead structure search are employed. Because this process chain is being carried out with increasingly high throughput, the capacity for lead structure searching must be constantly extended.

Many companies try to simultaneously develop chemically unrelated lead structures for the same indication. The elaborateness of the animal models for the preclinical profiling and the preparations for clinical testing require so much labor and expense that it seems hardly justifiable to start such a program with only one compound class. Risk minimization and distribution are required for the search as well as the development of a medicine. Techniques that are used for the detection of new lead structures are presented in the next chapter.

6.9 Synopsis

- Many active substances originate from natural products found in plants, animals, and microbial sources. Their mode of action has been copied as an active principle for the development of drugs.
- Endogenous substances such as hormones and neurotransmitters also served as references for drug development.
- Only a few natural products became drugs themselves.
- Usually targeted chemical variations are required to optimize a lead for metabolic stability, half-life, or selectivity to be ready for therapeutic use.
- Plants contain many valuable therapeutic compounds that usually developed as an effective protective mechanism against various enemies.
- Nature offers a tremendous body of structural variations; however, ambitious programs to elucidate mechanisms of action of traditional medicines all too often only isolate toxins and discover already-known principles.
- Animals have developed venoms as aggressive or defense mechanisms that they use as predators or against enemies. They are mostly proteins, peptides, or alkaloids that either kill or lame a victim.
- Snake venoms served as references for the development of antihypertensive drugs; active principles to block blood clotting (e.g., by leeches or bats) were turned into active ingredients for anticoagulation drugs.
- Proteins for substitution therapy (such as insulin, erythropoietin, factor VII) are manufactured by gene technology.
- Microorganisms have provided leads for antibiotics (e.g., penicillins), which had to be optimized for oral availability, broad-spectrum activity, and metabolic stability.
- The immunosuppressant cyclosporine A (a cyclic peptide), ergotamine (a toxic alkaloid in ergot), lovastatin (an inhibitor of cholesterol biosynthesis), or streptokinase (used to dissolve blood clots) are successful drugs originating from microorganisms.
- Dyes and many synthetic intermediates produced in the chemical industry were investigated for biological effects and provided important compound classes such as the sulfonamides.
- Small but essential structural changes of endogenous ligands transform enzyme substrates, neurotransmitters, and hormones into successful drugs.
- Many drugs originated from clinical observations of side effects during practical use, for instance, the antidiabetic effect of sulfonyl ureas from the observation of side effects of sulfonamides.
- To exploit all imaginable resources to discover leads today, huge substance libraries of plant extracts, microbial fermentations, and libraries of synthetically prepared compounds are screened.

Bibliography and Further Reading

General Literature

A. Burger, A Guide to the Chemical Basis of Drug Design, John Wiley & Sons, New York (1983)

E. Verg, Meilensteine. 125 Jahre Bayer, 1863–1988, Bayer AG (1988)

W. Sneader, Chronology of Drug Introductions, in: Comprehensive Medicinal Chemistry, C. Hansch, P. G. Sammes and J. B. Taylor, Eds., Vol. 1, P. D. Kennewell, Ed., Pergamon Press, Oxford, pp. 7–80 (1990)

Special Literature

M. Suffness, Taxol: From Discovery to Therapeutic Use, Ann. Rep. Med. Chem., **28**, 305–14 (1993)

F. Liu, A. G. Myers, Development of a platform for the discovery and practical synthesis of new tetracycline antibiotics, Curr. Op. Chem. Biol., **32**, 48–57 (2016)

P. J. Hylands and L. J. Nisbet, The Search for Molecular Diversity (I): Natural Products, Ann. Rep. Med. Chem. **26**, 259–269 (1991)

M. S. Tempesta and S. R. King, Ethnobotany as a Source for New Drugs, Ann. Rep. Med. Chem., **29**, 325–330 (1994)

A. D. Buss and R. D. Waigh, Natural Products as Leads for New Pharmaceuticals, Burger's Medicinal Chemistry and Drug Discovery, M. Wolff, Ed., John Wiley & Sons, p. 983–1033 (1995)

G. R. Pettit *et al.*, Isolation of Dolastatins 10–15 from the Marine Mollusc *Dolabella Auricularia*, Tetrahedron, **41**, 9151–9170 (1993)

B. Badio et al., Epibatidine: Discovery and Definition as a Potent Analgesic and Nicotinic Agonist, Med. Chem. Res., **4**, 440–448 (1994)

Screening Technologies for Lead Structure Discovery

Contents

G. Klebe, *Drug Design*, https://doi.org/10.1007/978-3-662-68998-1_7

In the last chapter, examples were presented of how lead structures can be discovered through targeted searches, particularly using examples from nature or compounds with known modes of action. Even when a large number of natural products and synthetic compounds are available, it is not always easy to filter out the active molecules and assess their value for a given indication. This requires time-consuming and costly sorting or **screening** of huge compound libraries. Screening refers to the more or less specific biological testing of compounds. Although molecular test systems and cell culture models are used almost exclusively today, the cost of testing one compound ranges from US$1–5. Since millions of compounds are typically tested, a screening campaign can cost a lot of money!

The screening process can be divided into three phases. First there is an automatic **introductory screening**, which is usually carried out by robots and encompasses libraries of millions of compounds. The first substances that show an interaction are identified as "**hits**" that have to be validated by repeated testing. Next, a more detailed screening follows, with which the chemical space around the identified compounds is explored. The goal is to establish a structure–activity relationship (Chap. 18) and to improve the pharmacological and physicochemical properties (Chap. 19). Along the way, lead structures (so-called "**leads**") are discovered. Then in the last phase the **lead optimization** takes place through detailed biological testing, through which a **drug candidate** is selected for clinical testing (Chap. 8). How can we find appropriate hits from the enormous number of test candidates that have the potential to be developed into a medicine? The answer to this question is provided by automated screening for biological effects.

7.1 Screening for Biological Activity by HTS

A prerequisite for large-scale screening was the development of *in vitro* test systems to replace animal experiments. The first attempts were made with isolated enzymes and membrane homogenates for receptor binding studies. Later, gene technology (Sect. 12.6) provided sufficient quantities of pure proteins for the development of molecular test systems. This had the advantage that homogeneous protein samples, preferably those identical to the human proteins, could be tested.

In the mid-1990s, **automated test systems** with an extremely high capacity (***high-throughput screening***, **HTS**) led to a daunting boom. The discovery of candidates for drug development is now attempted by using the entire methodological repertoire of biochemistry in a test tube. Meanwhile, it is known how to reprogram cells and organisms so that the function of single genes is highlighted. The special trick with all of these test methods lies in translating the molecular effect into a macroscopically visible signal.

Despite the enormous effort that is associated with HTS, and the not-always-justifiable hit rate, HTS is here to stay in pharmaceutical research. There are always interesting lead structures to be found in this way (for example see Chaps. 23–32). A weakness may be the **limited diversity** of synthetic substances, compared with the **structural complexity** of plant and microbial metabolites. Another limitation of *in vitro* test systems is that neither the entire effect spectrum nor many other effects such as transport, distribution, metabolism, and excretion (Chap. 19) can be assessed. Furthermore, it must not be forgotten that hits from HTS campaigns must be validated. Unfortunately, assays designed for high throughput are susceptible and false-positive hits often emerge. It must also be clarified whether the detected substances actually bind specifically to the target structure as desired and switch off its function. In the meantime, so-called **phenotypic screening** has been established in parallel. It is also known as **high content screening**. Here, the effects of test substances on living cells all the way down to lower animals such as nematodes or zebrafish are investigated. The effects of the substance application on the phenotype of a cell, a specific tissue, or a living organism are recorded. How does the investigated sample change, e.g., its cell shape, cell growth or a cell function? Often in this screening, the molecular target is unknown; however, new targets are frequently discovered in this way. The capacities of high content screening are limited to lower throughput compared to target structure-oriented high throughput screening.

The composition of suitable screening libraries is exceedingly critical. Frequently molecules and test candidates are used that were prepared during the course of other drug development projects. As such, these molecules already have the size of a typical drug. Usually only modest, almost always micromolar binding to the tested macromolecular receptor is found. To improve the properties of such a hit, it must be structurally modified. As a general rule, this is accomplished by adding more chemical groups. This means that the molecular weight can quickly reach or exceed 500–600 Da, which is considered to be the upper threshold for good bioavailability (Sect. 9.1). The optimization of such a screening hit, therefore, means that the size must be reduced first, so that it can be increased again during a target-oriented optimization. Yet the size reduction often comes with a loss in binding affinity. Therefore, the **ligand efficiency** (**LE**) criterion was introduced to assess the optimization potential of a screening hit. It considers the number of nonhydrogen atoms (n) of a hit compound in relation to its binding affinity, expressed as $\Delta G/n$. Small compounds with good binding affinity relative to their size are considered particularly promising candidates for an optimization program. Two practical aspects of using LE

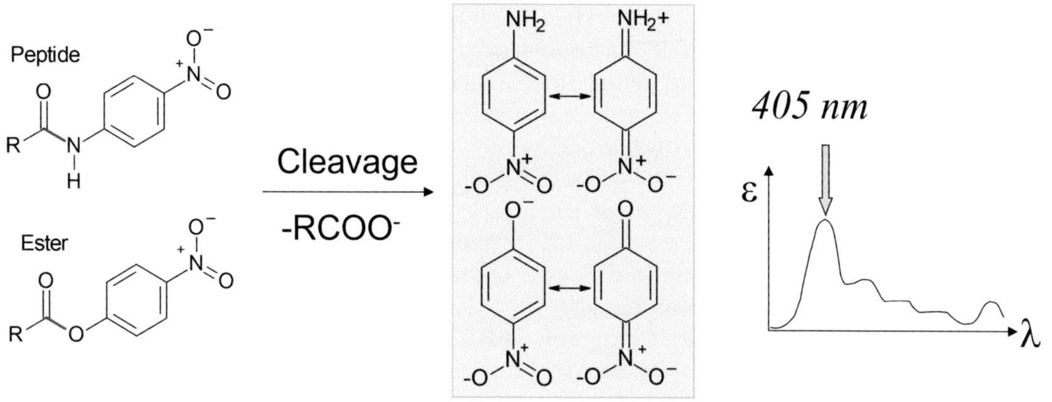

Fig. 7.1 A *p*-nitrophenolate or a *p*-nitroanilide group is added to the terminus of a natural protease or esterase substrate. The enzyme cleaves the *p*-nitrophenolate or *p*-nitroanilide, which becomes visible as a *yellow*-colored mesomerically stabilized anion (absorption maximum at 405 nm). If a competitive inhibitor is added along with the substrate to the enzyme, the cleavage reaction rate is suppressed depending on the binding strength. This is apparent by the more or less strong *yellow* color of the solution, which can be quantitatively measured

as a metric for optimization need to be considered. LE refers to ΔG, which is recommended to be measured in the SI unit kJ/mol. However, values based on the non-SI unit kcal/mol are more commonly used. Although the energy scales differ only by the factor of 4.184, care must be taken to ensure which scale is being referred to. Furthermore, an $LE_{kcal/mol}$ of 0.3–0.4 (or $LE_{kJ/mol}$ of 1.2–1.6) is considered a promising value. However, there are successful drugs in therapy that have an LE that is twice as high, and others that have an LE that is only half as high. Therefore, LE should only be used when optimizing one class of compounds for a given target.

7.2 Color Change Demonstrates Activity

Important target proteins for drug development are proteases and esterases, which are enzymes that cleave peptide and ester bonds (Chaps. 23, 24, 25). How can their enzymatic activity be visualized? In order to achieve this goal, synthetic substrates that are very similar to the natural substrate are prepared. For example, they carry a *para*-nitroanilide or a *para*-nitrophenolate group coupled to the peptide or ester bond to be cleaved (Fig. 7.1). When the enzyme degrades this substrate, yellow nitrophenolate or nitroanilide is released, and the absorption properties of the produced anion are a measurable change. This can be observed spectroscopically. If during screening, a compound acts as an inhibitor, the enzymatic cleavage of the synthetic substrate will be more or less suppressed, and the yellow color is minimized. In this way, the inhibition potency of test substances can be determined (Fig. 7.1).

A broad **palette of chromophoric reactions** that are suitable for the characterization of enzymatic activity have been developed. Many enzymes, for example, dehydrogenases, need NAD(P)H as a natural cofactor, which is subsequently oxidized to $NAD(P)^+$ (Sect. 27.1). Because the NAD(P)H starting material, in contrast to the product, absorbs at 340 nm, the progress of the enzymatic reaction can be monitored at this wavelength. As a variation, two enzymatic reactions can be coupled to each other. This route is of interest when the reaction in the enzyme of interest is difficult to observe, but the reaction forms a product that is a substrate suitable for spectroscopic monitoring. In this case, the activity of the enzyme of interest is recorded based on the conversion of the product of the previous reaction in the subsequent enzyme reaction. Although **absorption spectroscopic assays** are preferred for technical reasons, tests that are based on the reaction of radiolabeled compounds still play an important role. The activity of kinases is, for example, followed by using ^{32}P-labeled adenosine triphosphate. The terminal phosphate group of the labeled substrate is transferred to the phosphorylated protein by the kinase (Sect. 26.3). The incorporation rate serves as a measure of the kinase activity. Receptor-binding studies are carried out with a known radioactively labeled ligand. The assay investigates to what extent test compounds can displace the radioactively labeled ligand from the receptor-binding site. This type of test does not necessarily represent a functional assay though. Agonistic and antagonistic binding (Chaps. 28 and 29) must still be distinguished.

7.3 Getting Faster and Faster: More and More Compounds by Using Less and Less Material

Antibodies play an important role in assay development. The enormous specificity of antibody–antigen interactions can be exploited as a highly sensitive system (Sect. 32.3) for the detection of a specific molecular species. In classical immunoassays, either the release of a radioactively labeled substance is monitored (**radioim-**

munoassay, RIA) or an enzymatic reaction is induced (**enzyme-linked immunosorbent assay, ELISA**). In the latter case, for example, a specific antibody is used to find any unconverted substrate molecules left over from the enzyme reaction being tested for inhibition. Any antibody not bound to a substrate molecule is then removed. Since the antibody is also linked to another enzyme that can produce a chromogenic compound as reaction product, the amount of unreacted substrate molecules and, thus, the inhibition rate of the tested enzyme can be determined. The ELISA technique has a much wider range of application, mainly because radioactivity is best avoided as a measured quantity. Because they recognize only a single molecular species, immunoassays are not only highly specific but also versatile.

7

Screening techniques are optimized to be **automated** and **miniaturized**. Driven by the desire for higher capacity, these tests are hardly ever carried out in 96-well (8 × 12) microtiter plates anymore. The wells of these plates hold a reaction volume of about 0.3 mL. In the meantime, 384-well (16 × 24) microtiter plates are used or even 1536-well (32 × 48) plates, the volumes of which are only a few microliters per well. The **aggregation behavior** of **hydrophobic test compounds** poses a large problem. The aqueous buffer solutions that are used for these assays can cause these compounds to aggregate. This aggregation creates hydrophobic surfaces on which proteins can adsorb. This reduces the concentration of free protein, which can give the impression that the protein is well inhibited. Addition of **detergents** can reverse this effect. However, aggregation of test molecules and proteins on such aggregates, or even on the plastic material of the assay plates, can dramatically obscure assay results.

By using a sophisticated robot system, 100,000 assays a day can be carried out. This leads to an enormous flood of data to be evaluated. The reduced test volume has the advantage that much less material is consumed. Furthermore, the measurements can be carried out quickly. At the same time, the sample manipulation has become ever more difficult. One only has to consider the evaporation of such small amounts of solution, the enormously increasing **logistics** of comprehending so much data in parallel, or the reproducibility of the results, and the necessary sensitivity to measure weak signals with certainty to appreciate the difficulty.

In order to improve this last aspect, ever more sensitive detection procedures are used. **Fluorescence measuring techniques** are particularly sensitive. In the simplest case, a fluorescing substrate such as coumarin (Sect. 14.6) is incorporated in the place of *para*-nitroanilide. The protein–ligand binding can also be followed by **fluorescence anisotropy** (or polarization). For example, a known small molecule ligand for the target biomolecule to be screened is coupled to a fluorophore and excited with polarized light. The emitted fluorescence is also polarized. During the time that the excited molecule can freely diffuse in solution, the amount of induced polarization decreases. Because a small molecule can diffuse much faster than a large target biomolecule, the rate by which the polarization signal decreases depends on whether the small molecule is bound to the large biomolecule or not. In a screening assay, the fluorophore-labeled ligand is used as a reporter ligand and will be displaced by potential screening hits. Any difference in polarization decay is recorded and related to the binding event of a hit.

Even better sensitivity can be achieved with so-called **FRET measuring techniques** (**f**luorescence **r**esonance **e**nergy **t**ransfer). A resonance energy transfer can occur between **donor** and **acceptor fluorophores** of similar absorption if both are separated by no more than 50 Å. For example, to develop a phosphatase assay, a phosphorylated peptide substrate must be covalently linked to a donor fluorophore. This is added to the phosphatase to be tested, which has also been exposed to an inhibitor. Depending on the inhibitory potency of the test compound, the activity of the enzyme decreases and less substrate is cleaved. An antibody that binds to the unused phosphorylated substrate is added. The antibody is additionally labeled with an acceptor fluorophore whose absorption maximum overlaps with the emission spectrum of the donor fluorophore. If there is still a lot of phosphorylated substrate present, e.g., because the compound to be tested is a potent inhibitor, the spatial proximity between the donor and acceptor fluorophores will result in a strong FRET signal. Thus, the intensity of the FRET signal can be related to the activity of the tested enzyme.

In the meantime, progress in assay miniaturization allows the detection of single molecules. This is possible by using **fluorescence correlation spectroscopy** (FCS). A confocal laser microscope irradiates approximately a femtoliter of test solution. If a single fluorophore diffuses through the volume of interest, it will cause a time-resolved fluctuation in the fluorescence signal. An exact analysis of these signals delivers information about the concentration and diffusion constants. The diffusion velocity, on the other hand, depends on whether the fluorescence-marker-labeled substance is bound to a protein or not. If the proteins as well as the ligands are tagged with different markers, the association and dissociation can even be followed.

7.4 From Binding to Function: Testing in Entire Cells

The binding of a ligand to a protein does not tell us much about the associated biological function or the change in function that is induced. In the case of enzymes that catalyze a chemical transformation, it is often easier to relate an observed assay inhibition to a putative function. The correlation is less obvious with receptors and ion channels (Chaps. 28, 29, 30).

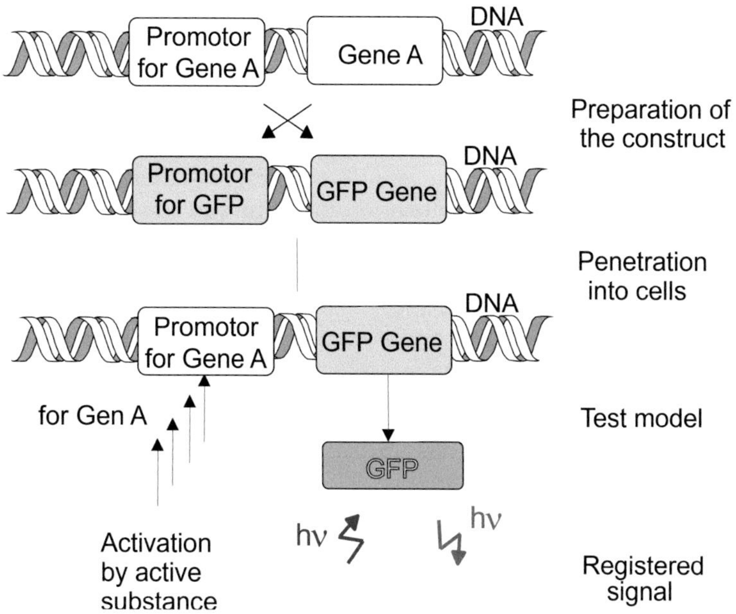

Fig. 7.2 Genes are controlled by promoters. Promoter-initiated gene activation leads to the synthesis of the relevant protein. By using green fluorescent protein (GFP), an easily observed assay can be constructed based on this principle. For this, the gene promoter that is activated by agonist binding is coupled to the GFP gene. Activation of the promoter then delivers not the original gene product, but rather the GFP. The presence of GFP is easily observed because of its fluorescence upon excitation with ultraviolet light

When **biochemical pathways** and **cell cycle regulation** are considered, it becomes even more complex to assign a particular function even in case of enzymes. Such correlations are not so easily reproduced in a test tube. Therefore, assays must also be developed to study function that allow the response of an entire cell to be measured upon ligand binding. It is possible to culture cells for many different tissues, which then allows the study of tissue-specific receptors.

Typically, the activity of ion channels can be investigated by using binding tests or radioactive assays. The so-called **patch–clamp technique** allows the influence of a drug candidate to be even better characterized. An electrode is placed on the surface of a cell and a voltage is applied, which creates a current. In this way, the opening or closing of individual ion channels can be recorded, especially when test molecules are added during these measurements. This method certainly does not have the capacity of high-throughput techniques. Rather, it is used to further investigate the function of hits from an initial prescreening. Fluorescence methods are often used for this first step. For example, in the case of calcium channels, the increase in intracellular calcium concentration can be observed using a dye that fluoresces sensitively in the presence of calcium ions.

Other tests employ the coupling to a **reporter gene**. Receptor stimulation initiates a signaling cascade that, for some receptors, leads to the transcription of gene products that are controlled by the relevant promoters (Sect. 28.1). If the sequence of the relevant gene is replaced with that of a reporter's, such as β-galactosidase, luciferase, or green fluorescent protein (GFP), then these proteins will be produced by the cell instead. This can subsequently be observed as an easily detectable signal (Fig. 7.2). As examples, if the produced β-galactosidase cleaves X-gal, a blue dye will be released, luciferase will develop an ATP-dependent chemiluminescence, and the green fluorescent protein will be detectable because of its own intrinsic fluorescence.

7.5 Back to Whole-Animal Models: Screening on Nematodes

Primary substance testing on animals, as it was once carried out, is ethically unjustifiable today. Furthermore, an animal model is not predictive for target-oriented optimization. Nevertheless, it does have advantages. The **reaction of an entire organism** to a substance is immediately transparent, the bioavailability is directly measured, and side effects as well as synergistic effects are immediately obvious. Back in 1963, Sydney Brenner recognized the complexity of molecular biology in that he emphasized the biochemical control of cellular development. He proposed that the pinworm (the nematode *Caenorhabditis elegans*) would be the simplest multicellular organism to investigate. This nematode normally lives in soil and feeds on bacteria. It is also easily culturable in microtiter plates and fed with *Escherichia coli* bacteria. It is a hermaphrodite, has a short lifespan, reproduces itself within three days, can be conserved in liquid nitrogen, is transparent, and homologous genes have been found in humans for 60–80% of its genes. The pinworm genome has been sequenced, and we now understand how to easily manipulate it. Because it is transparent, any internal changes can be easily observed so that, for instance, proteins can be tagged with fluorescence markers. Its 959 somatic cells form many different organs, including a nervous system with 302 neurons. Can substance testing be carried out in such a life form? The ethical threshold may be set lower in this case. But then, how predictive would any tests be? Can such an animal be used to predict mood changes, depression, or appetite and its relation to obesity? This is only possible if the causes of these diseases are known on the molecular level, for example, a defect caused by an altered serotonin-mediated signaling. In such a situation, the **worm** can serve **as a model**. A first step towards the discovery of a potential target is selective **gene silencing**. This is possible by using RNA interference (Sect. 12.7). If the pinworm (nematode) is exposed to a substance library, it will be possible to see a change in appearance or behavior. Is the life expectancy lengthened or shortened? These are indications that the compounds could interfere with the aging process or are toxic. If there are changes in muscle cells, perhaps it might be useful for neurodegenerative muscle disease. Aside from macroscopic changes in the body form, changes in the gene expression pattern can also be analyzed (Sect. 12.10). Are mutations in proteins apparent? Certainly, the worm does not have the same metabolic pathways as we do. Even its disease models only partially represent the pathophysiology that is seen in human disease. Nonetheless, direct testing of compounds on the pinworm seems to afford a new perspective for screening substance libraries. As an alternative, the fruit fly (*Drosophila melanogaster*) or the zebrafish (*Danio rerio*) are also available as test organisms. They help to test the validity of a therapeutic approach early in a program.

7.6 *In Silico* Screening of Virtual Compound Libraries

As described in the previous section, experimental high-throughput screening (HTS) has been automated with great effort. When fed with compounds from combinatorial chemistry (Chap. 11), several hundred thousand substances can be screened by using HTS. At first, it seemed that this would be the end of all rational structure-based techniques. In view of the enormous financial investment and the disappointingly low hit rate, the initial euphoria began to soberingly wane. Therefore, as an alternative, the technique of enumerating huge databases on the computer by fitting small molecules in a predefined binding pocket (docking, Sect. 20.8) was developed and referred to as **virtual screening**.

The unsatisfactory hit rate from HTS is attributed to the size, structural diversity, and poorly selected composition of the substance library with respect to the actual properties of the target protein. The recognition of **false-positive** and **false-negative** hits in biological systems causes large problems. Disappointing hit rates have been reported for the translation of initial hits into potential lead structures for lead optimization. This was all the more reason that virtual screening techniques were developed as a complementary and alternative method. The prerequisites for the successful use of these techniques are entirely different from those of the technology-driven HTS: virtual screening can only reasonably be applied if the factors that are responsible for a putative drug to bind to its target protein are well understood on the molecular level.

The starting point for this is the spatial structure of the target protein, which is usually determined by NMR spectroscopy, electron microscopy, or X-ray structure analysis (Chap. 13, ◘ Fig. 7.3). Models can be increasingly derived from structurally homologous proteins of known geometry (Sect. 20.5). It remains to be seen whether methods based on machine learning and artificial intelligence will provide sufficiently accurate models for virtual screening (Sect. 20.6). To successfully bind to a protein, the ligand must adopt a shape that is complementary to the binding pocket. Molecules are flexible and can change their shape through bond rotations that require very little energy (Chap. 16). In addition to adopting an appropriate conformation in space to accommodate the protein binding pocket, the functional groups of a potential ligand must find complementary functional groups of the protein in the binding pocket. Hydrogen bonds must be formed between ligand and protein, and hydrophobic molecular moieties must find their counterpart in the protein (Chap. 4). For this, the protein binding pocket is analyzed to highlight the regions that are essential for binding.

For a particular atom type, for instance, a hydrogen-bond donor or acceptor, the binding pocket is systematically scanned. By using computer graphics, it is possible to see where functional groups attached to a candidate ligand might be optimally placed (Sect. 17.10). The composite picture of all such placed atom types in the binding pocket that are indicated by this analysis reveals a spatial pattern of physicochemical properties that a ligand must meet to successfully bind to the protein ("**hot spots**" Sects. 17.1 and 17.10). With these criteria in hand, a molecular database can be searched that is composed of already-synthesized compounds or compounds that have been virtually assembled on the computer. To this end, programs have been developed that assemble huge libraries of virtual test molecules in the computer using previously programmed synthesis rules. In case a hit from the latter group is found, the compound can be subsequently synthesized. The search is divided into multiple filtering steps that become increasingly stringent and sophisticated with successive reduction of the search quantity. With the help of fast **docking programs** (Sect. 20.8), molecules are fitted into the binding pocket and a binding geometry is generated, from which the expected binding affinity can be estimated. This step is decisive, but unfortunately it is also the most difficult (Sect. 20.9). In Chap. 21, examples that were found by a virtual screening campaign are presented.

The evaluation of the generated binding geometries is accomplished with sufficient accuracy in about 70% of cases nowadays. An improvement in predictive power requires that we understand the ligand–protein recognition process better (Chap. 4). The role of water in the binding, the induced steric and dielectric adaptation, the flexible behavior and residual mobility of proteins and bound ligands, and the dynamic changes during complex formation are still poorly understood. The composition of the databases themselves plays a decisive role in the search's success. Enlarging the database alone is not enough. The enrichment of the compounds that could fulfill the requirements is crucial. Screening is often compared to the search for a **needle in a haystack**. When looking for such a needle, it is not helpful to simply double the size of the haystack! The haystack must be spiked with more promising needles. To achieve this, all available knowledge about the structure, function, and dynamic behavior of the target protein must be used to define the database search. Comparisons between proteins and protein binding pockets, especially among members of the same protein family, can offer decisive information (Sects. 20.3–20.6). In principle, all of the data that are needed about the composition of a suitable compound library for a virtual screening run are already intrinsically coded in the structure and geometric interaction properties of the binding pocket. It is only a question of applying it correctly. Another decisive criterion for a hit is an adequate pharmacokinetic profile so that satisfactory bioavailability can be achieved (Chap. 19).

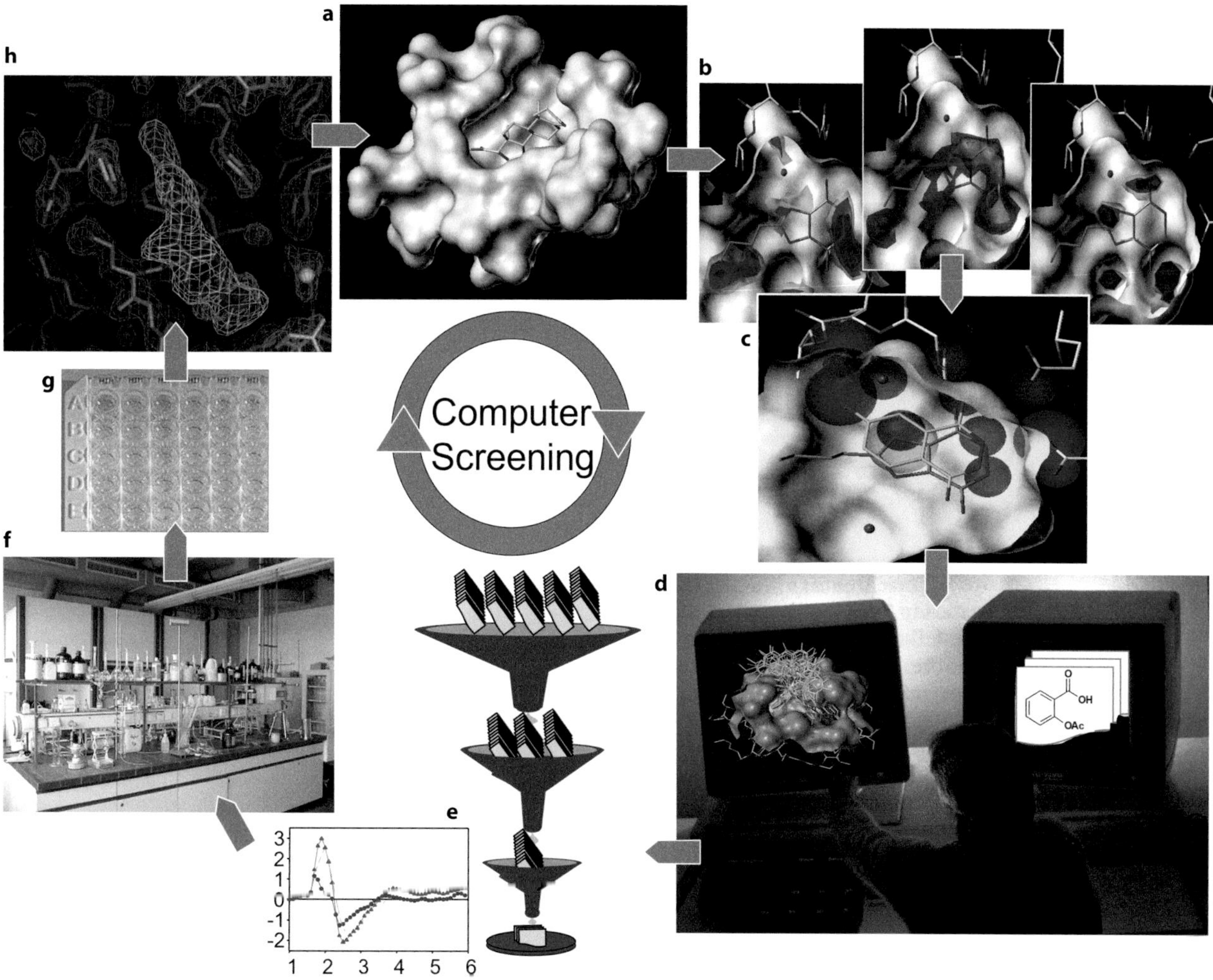

Fig. 7.3 The spatial structure of a protein is the starting point for virtual screening (**a**). The binding pocket is explored with a variety of different probe atoms, for instance, for hydrogen-bond acceptors or donors (**b**). Regions that are particularly favorable for such interacting groups are highlighted on the computer graphics. If the "hot spots" in these areas are summarized, a spatial pattern of properties that a potential ligand should have become apparent (**c**). This pattern is called "pharmacophore" and serves as the search criterion for a database retrieval (**d**). Potential ligands from a large database are filtered and energetically evaluated by docking (**e**). The discovered hits are either commercially available or synthesized in the laboratory (**f**). Next, biological testing takes place (**g**), and if the binding is successful, the lead structure will be crystallized with the protein. The subsequent structural determination (**h**) serves as a starting point for further design cycles (for an example see Fig. 02, ► https://sn.pub/4NppHn)

7.7 Biophysics Supports Screening

Surface plasmon resonance (**SPR**) is increasingly being used to screen for new lead structures. A target molecule is anchored to the gold-coated surface of a sensor chip. Polarized light (◘ Fig. 7.4, strictly speaking, transversely magnetically polarized light, that is, the direction of the magnetic field is perpendicular to the plane spanned by the incidence vector and the surface normal) is then irradiated from the bottom of a glass substrate. Changes in the refractive index, which can be followed by a shift in the angle of total internal reflection, are a measure of structural changes in the sensor surface. Changes in the masses of bound matter also contribute significantly to these changes. Free electrons in the deposited gold layer interact with the incident light to form PLASMA-like electron clouds (PLASMONS), which are nothing other than surface waves. These propagate parallel to the metal surface and, under certain conditions (i.e., wavelength and angle of incidence near total internal reflection), can be excited to resonance by the polarized light. This allows changes in the refractive index in the immediate vicinity (a few 100 nm) of the surface to be observed with very high sensitivity. For example, if a substance with a mass greater than about 100 Da binds to the biological target molecule, the resulting mass change can be registered on the gold surface.

7

There are two basic strategies for performing the measurements. In the first, the ligand is added to the chip with the immobilized protein in a cuvette. The binding equilibrium is established and the shifts in the reflectivity angle due to the mass effect are recorded. A dilution series is then performed. Depending on the binding affinity of the test ligand to the immobilized target protein, its binding behavior is evaluated from the resulting refractive index shifts. This equilibrium method has not become widely accepted. Instead, steady-state methods are used today. For this purpose, the test ligands are allowed to flow over the chip carrying the protein in a constant but, if necessary, adjustable liquid flow. Changes in the angle of reflection are observed and converted to changes in the refractive index. The test ligand is added to the liquid flow in a time-controlled manner. Again, concentration series must be performed. The latter procedure is faster to perform, but is dependent on the flow rates and, since chemical equilibrium can never be assured, assumes a "steady-state" situation. In particular, the diffusion rates must be strictly controlled. The SPR method works fast and a time course of the mass increase on the chip can be observed. Thus, in addition to stoichiometry, kinetic parameters of association or dissociation become available.

A prerequisite for using the method for screening is stable and reproducible immobilization of the compounds to be tested on the sensor chip. In the early days of SPR, the sensitivity required the binding of very large masses. Therefore, large libraries of low molecular weight test compounds were chemically anchored to the gold chips. The binding of a macromolecular receptor protein to the, thus, exposed test compounds could be easily detected by the enormous change in mass. Today, the surface plasmon resonance method has reached such a high sensitivity that even very small mass changes caused by small ligands can be detected. For this reason, today almost exclusively macromolecules are immobilized on the sensor chips and the low-molecular weight test substances are allowed to flow over these surfaces as possible binders. Since the test substances can also be "washed off" with the liquid flow, the "off kinetics" can be determined and the immobilized protein can be reused several thousand times for the measurement of different test ligands.

The concept of "**ligand efficiency**" was introduced in Sect. 7.1. To address the latter aspect, test libraries are increasingly being supplied with compounds that have a molecular weight of less than 250 Da. The term "chemical **fragment**" has become popular to refer to these search

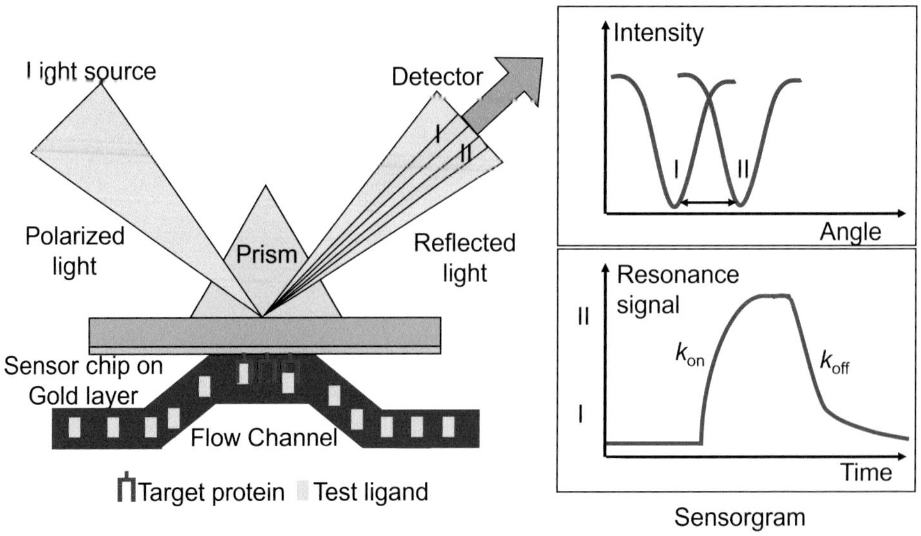

◘ **Fig. 7.4** Principle of surface plasmon resonance (SPR) using the continuous flow method. The method registers changes in the refractive index on the surface of a sensor chip (*green*). The extent of the mass change on the gold surface (*yellow*), which is caused by binding of a test ligand ( ) to a receptor (*red*) immobilized there, leads to a shift in the resonance angle of the reflected light (I and II). This not only measures the binding affinity, it also succeeds in determining kinetic parameters of association (k_{on}) and dissociation (k_{off})

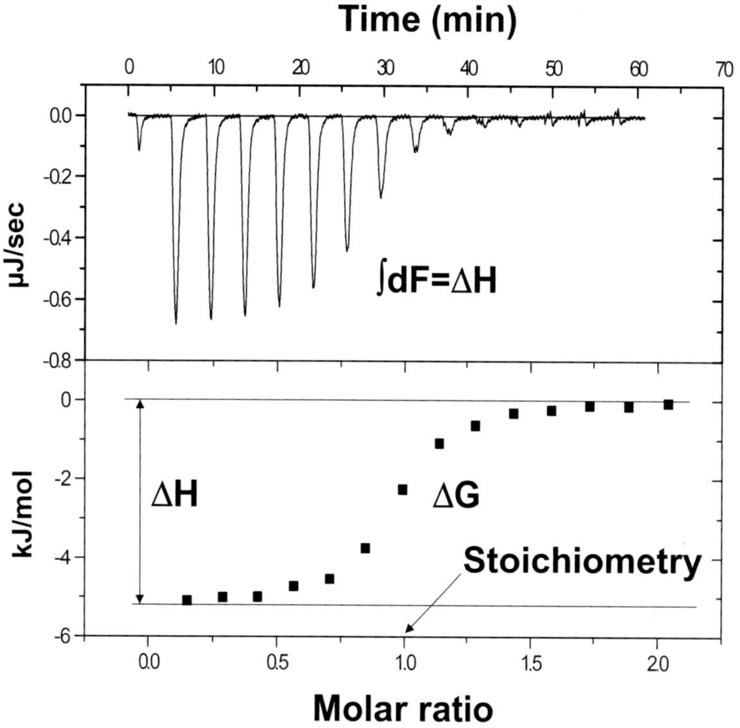

Fig. 7.5 In isothermal titration calorimetry, a solution with a ligand is added dropwise to a solution containing a protein. The binding to the protein leads to an exothermic or an endothermic reaction. The heat that evolves upon the addition of each drop is the area under the single signal peaks. The total integral of all signal peaks is the binding enthalpy ΔH. With increasing amount of ligand, the protein becomes saturated so that the signal intensity of the heat signal decreases. The binding constant (dissociation constant) can be derived from the shape of the curve, and the free energy ΔG can be obtained from the relationship $\Delta G = -RT \ln K_d$. The stoichiometry of the reaction is simultaneously obtained. The entropy is calculated by using the equation: $\Delta G = \Delta H - T\Delta S$

candidates. The term is somewhat unfortunate because the molecules are actually "complete" small molecules and not, as the term might suggest, simply a "fragment," that is an additional building block to be attached to a lead structure. In recent years, SPR has become the standard rapid method for fragment screening.

Proteins denature when heated. A "melting temperature" is defined when an **unfolding process** (Sect. 14.2) occurs. This temperature can be measured very sensitively with a thermal sensor. The binding of a ligand to a protein changes this melting point. As described in Sect. 7.3, fluorescence measurements are extremely sensitive indicators. This melting effect can be registered by allowing the unfolded proteins to interact with a fluorescent dye and detecting the change in a fluorescence signal. The temperature shift caused by ligand binding can be used as evidence of whether a ligand is bound to a protein or not. It has also been possible to construct quantitative binding assays using this effect. This highly sensitive **thermal shift method** is also suitable for the detection of weakly binding fragments.

Mass spectrometry (MS) has evolved significantly over the last few decades. By applying very gentle bombardment conditions, it is possible to detach single electrons from large biomacromolecules or even to generate negatively charged species. In the best case, the protein of interest can be detected in its intact form as a singly charged ion. The charged particles are then accelerated between charged, parallel-oriented condensator plates. The flow of charged particles can be bent by applying a magnetic field. The flight path of a given particle depends on its mass and charge. In this way, it is possible to separate and detect particles based on their **mass-to-charge ratio**. This principle has been refined with the most sophisticated technology and clever combination of electric and magnetic fields, so that it is now possible to detect individual mass differences of only a few daltons among even huge proteins. Using sophisticated experimental conditions, a given situation in solution, such as a protein–ligand complex, can be transferred to the gas phase without decomposition. There it is ionized and detected in the mass spectrometer (Sect. 21.10). With this technique, we have an assay that can detect the binding of very small ligands to proteins. It is even possible to tailor the decomposition of the complexes by varying the acceleration voltage. By registering the voltage at which decomposition occurs, the strength of the protein–ligand complex can be assessed. Since the decomposition occurs in the gas phase, information about the binding strength of such complexes in a water-free environment is available.

Recently, mass spectrometry has increasingly sought to capture spatial information about proteins in addition to their topological composition. By inserting chemically **reactive cross-linkers** between two functional groups of a protein or protein complex, distance information in its intact spatial structure can be captured and identified by MS. The positions of the cross-linked amino acids, together with the length of the cross-linker, reveal distance restraints in the 3D structure of a biomolecule. In addition, isotopic labeling by **hydrogen-deuterium exchange (HDX)** can be used to determine the spatial accessibility of main-chain amino groups. After certain time points and depending on the exchange conditions applied, fast/slow exchange in buried/accessible surface regions can be detected by digesting and analyzing the partially labeled protein. This provides insight into which sequence stretches have been involved in the formation of stable/labile contact interfaces of a protein complex.

Ligands can also be "**fished**" with proteins. This involves exposing a protein for which a ligand is being sought to an entire library of test compounds in aqueous solution. Those compounds from the library that bind to the protein are captured. The protein is then separated using a microfilter and the bound ligand is released by chemically denaturing the protein. The solution containing the released ligands is then processed and a **micro-HPLC separation** is performed. The chromatographically separated ligands are then subjected to highly sensitive analysis to determine which members of the original library have been fished out by the protein.

The binding of a ligand to a protein is a chemical reaction. As with all chemical reactions, a more or less pronounced **heat of reaction** can be observed. The process can either release heat (exothermic) or absorb heat (endothermic). This heat signal can be recorded to register the binding of a ligand to a protein. A highly sensitive **calorimeter** is required. When equipped with electronically controlled compensation heating, these devices can achieve amazing sensitivity. For example, such an instrument was built to study the activity of a butterfly being attracted by different pheromones. The heat generated by the wing beats was detected as a signal by such a calorimeter.

A dissolved ligand can be titrated by dropwise injection into the solution of a target protein in such a calorimeter. Each drop produces a heat signal. With increasing saturation of the protein, the heat signal decreases and a curve can be generated from which the **binding constant** of the ligand can be deduced (◘ Fig. 7.5). From the slope of the curve at the inflection point, the dissociation constant K_d can be determined. When all signals are integrated over the entire **titration**, the total heat of reaction for the binding event is determined. In this way, two different **ther-**

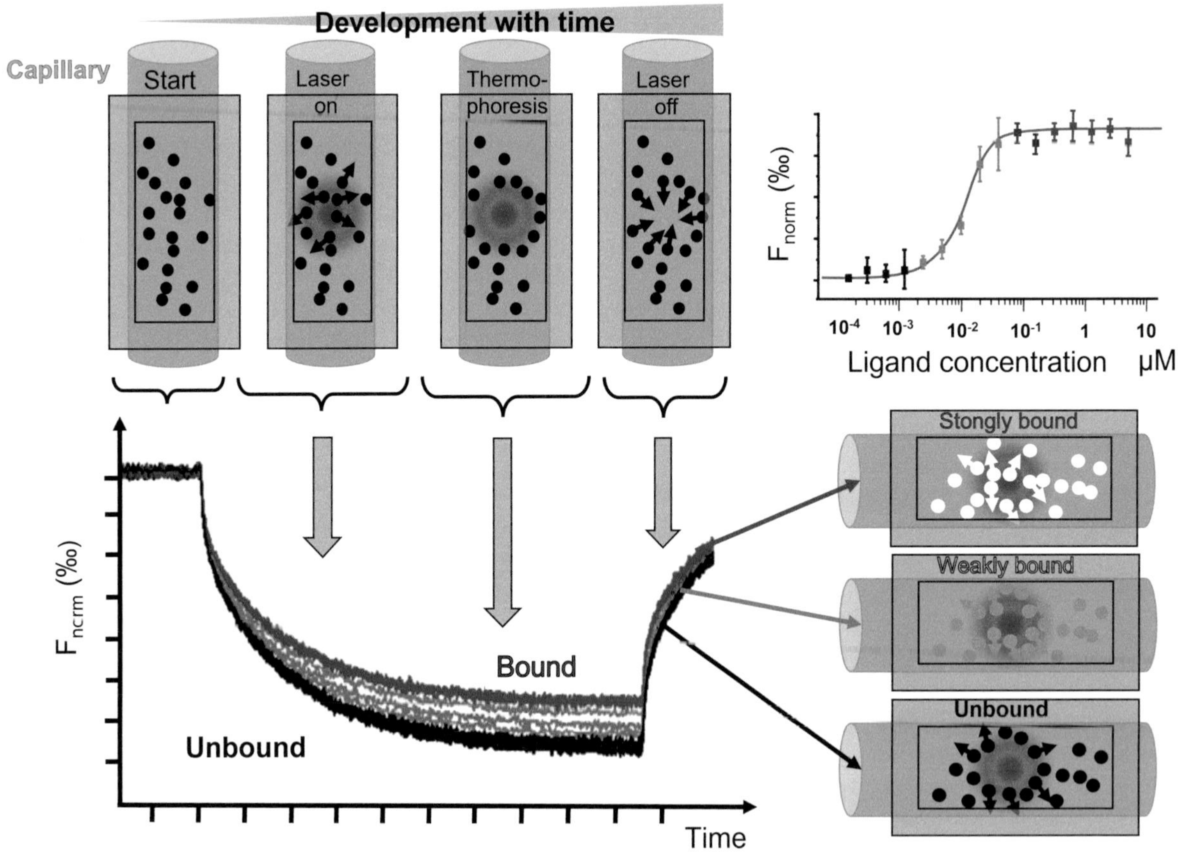

◘ **Fig. 7.6** The protein to be tested is placed in a capillary with a test ligand. Test series with several ligand concentrations are measured. The protein has previously been covalently labeled with a fluorophore. Now the infrared laser is switched on and generates a local temperature gradient in the region of the fluorescence observation window. The labeled molecules diffuse out of the heated region. Their migration is tracked by the decrease in fluorescence (F_{norm}). The thermophoretic migration properties depend on whether and how strongly a ligand is bound (concentration-dependent at equilibrium) to the target protein. At the end, the laser is turned off and the system returns to the initial state. Because the system is measured at different concentrations, the thermophoretic effect varies in strength (i.e., from the *black* to the *gray* and to the *red* decay curves) and can be converted into the shape of a binding curve (*top right*). (Figure after M. Jerabek-Willemsen et al., J. Mol. Struct., **1007**, 101–113 (2014))

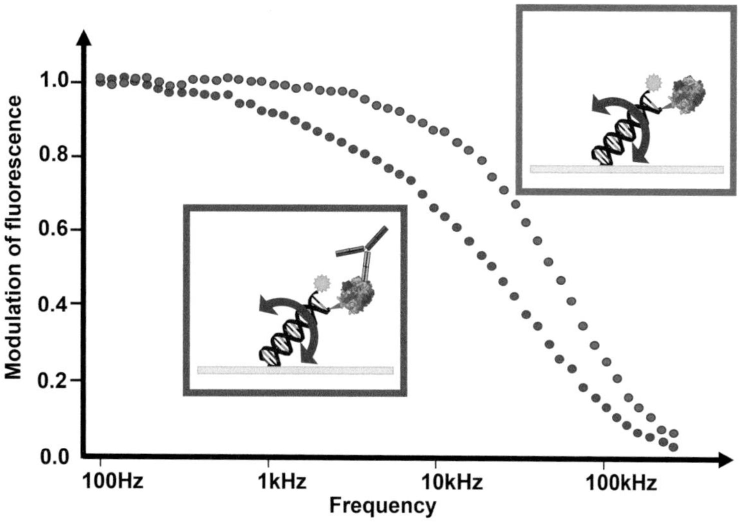

Fig. 7.7 In the switchSENSE method, a short DNA strand is immobilized on a gold surface (*yellow*). It carries a fluorophore at the end (*green*). A target molecule (here protein, *gray*) is attached to a complementary strand that hybridizes with the first strand. In an alternating electric field of varying frequency, the DNA double strand is set into an oscillating tilting motion (*purple arrow*). As the strand approaches the gold surface, the fluorescence is increasingly quenched periodically. Antibodies (schematically shown as Y) are added to the test solution. When one of these antibodies binds, the hydrodynamic frictional resistance changes. It differs whether the antibody binds to the protein (●) or not (●). In this way, binding can be detected and quantified. (Figure after J. Niemax et al., Laborwelt, **11**(5) 21–22 (2010))

modynamic binding characteristics are measured. The free energy ΔG is determined from the equilibrium constant and the enthalpy ΔH is given by the integrated heat signal (Sect. 4.3). Using Eq. 4.3, the entropy of binding can be calculated. It is important that in addition to proving the binding of the ligand to the protein, the most relevant thermodynamic parameters ΔG, ΔH, and ΔS can be evaluated in one experiment at one temperature.

The method also records the **stoichiometry** of the binding process. Many binding processes involve **changes in the protonation state** of the binding partners. This can be recorded with the isothermal titration calorimetry (ITC) method. To achieve this, the titration must be performed from different buffer solutions. If the buffer releases a proton during the binding process, its heat of ionization must be spent for this step. Conversely, if it captures a proton, the same energy will be released as a heat signal. Since different buffer compounds have different heat signatures for the protonation change, the molar amount of protons transferred can be inferred from this buffer contribution. In Sect. 4.4 (Fig. 4.4) such an example was presented.

Isothermal titration calorimetry is not a high-throughput method. Rather, it is used to analyze and characterize the binding process. Because of its importance, especially with respect to ligand optimization, the method will be discussed again in Sect. 8.8. Recently, the method has been further developed by Philippe Dumas in Strasbourg, France. **Kinetic data** can also be derived from the shape of the individual injection signals along the thermogram. The so-called **kinITC-ETC** (ETC: equilibration time curve) is used to evaluate how quickly the thermal signal returns to the baseline with each successive injection. This contains information about k_{off}. For a 1:1 binding process, the dissociation constant can be used to infer k_{on}, since it is the quotient of k_{off} and k_{on}.

A group of physicists led by Dieter Braun and Stefan Duhr at the University of Munich in Germany has developed a new biophysical method that is now being used worldwide for compound screening. **Microscale thermophoresis (MST)** is a powerful technique for quantifying biomolecular interactions. It is based on thermophoresis, the directional movement of molecules in a temperature gradient. This process is strongly dependent on molecular properties such as size, charge, hydration shell and molecular conformation of the moving particles. The technique is highly sensitive to virtually any change in these molecular properties. However, since these properties change primarily when a ligand binds to a protein, this approach can be used to quantify ligand binding. This is done by measuring the system at different ligand concentrations. Depending on the affinity and, thus, the percentage of the bound test ligand to the target protein in chemical equilibrium, the thermophoretic effect is different depending on the applied concentration. This can be converted into a binding curve (Fig. 7.6).

During an MST experiment, an infrared laser is used to abruptly create a local **temperature gradient** in the test solution (Fig. 7.6). Then, the directional movement of molecules through the temperature gradient is recorded and quantified. For this purpose, the protein of interest has been covalently labeled with a fluorophore, or the intrinsic fluorescence of, for example, the tryptophan residues in the protein can be used. By correlating the detected fluorescence, which changes its intensity strongly as a function of temperature and environment, with the variability of the thermophoretic effect, MST provides a robust and rapid way to study molecular interactions.

The development of new fluorescent dyes with improved environmental sensitivity has made the MST method even more sensitive. In particular, the temperature-dependent intensity change in the induced temperature gradient is exploited. As a result, it is now possible to perform the method in 384 microtiter plates. This opens the door to HTS; a plate can be screened in about 30 min.

Another relatively new method is the **switchSENSE technology** developed by Dynamic Biosensors, led by Ulrich Rant of Munich, Germany. In this technology, a short strand of DNA is anchored to a gold surface (◘ Fig. 7.7). On the opposite end, this strand carries a fluorescent probe. It can be hybridized to a second strand of complementary sequence. This second strand carries a chemical functionality on its side facing away from the gold surface so that a target molecule can be attached to it. Because the strands are made of DNA, they carry many negative charges. When an alternating electric field is applied across the gold surface, the DNA strands can be stimulated to oscillate towards and away from the gold surface. This oscillation can be tracked in real time because the fluorescence of the attached fluorophore is quenched as it approaches the metal layer, which has an increasing and decreasing positive charge in the alternating field. Thus, the position of the DNA strands in their switching motion can be observed by the intensity of the fluorescent light. The dynamics of the switching motion pull the molecule of interest through the solvent. Hydrodynamic friction and the charges on the target molecule influence the tilting motion. If a ligand is bound to the target molecule, or if a conformational change is induced for some reason, the switching motion can be slowed down. This is used as the measurement signal of the assay. This process can be analyzed very precisely by tuning the alternating field in a frequency-dependent manner. Thus, in addition to the binding affinity, binding kinetic parameters such as k_{on} and k_{off} become available. ◘ Fig. 7.7 shows an example of the binding of an antibody to a protein. The method can be extended in many ways to molecular systems, provided they can be chemically immobilized on a short DNA strand. For an assay developer, there are almost no limits to the measurement setups that can be created.

7

7.8 Screening by Using Nuclear Magnetic Resonance

The method of **NMR spectroscopy** is described in more detail in Sect. 13.7. Suffice it to note here that it is concerned with the orientation of the magnetic moments of the nuclei in a sample in a magnetic field. By applying a carefully chosen, spatially and temporally resolved sequence of electromagnetic fields, it is possible to specifically activate nuclei that are oriented within these magnetic fields. This can be done, for example, for one type of nucleus in a protein. If a solution of test ligands or a whole mixture of ligands is added to such a solution, protein binding can occur, provided that the ligands are suitable. Depending on their binding strength, they will remain on the magnetically saturated protein for a certain

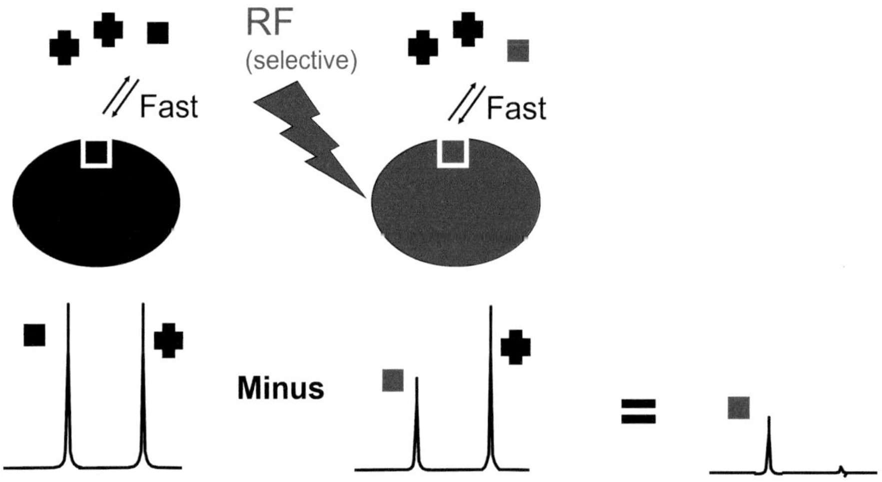

◘ **Fig. 7.8** To determine the saturation transfer difference (STD) with NMR spectroscopy, a library of test ligands (■, ✚) is added to a target protein (*ellipse*). Potential binders (here ■) reside for a finite time span bound to the protein. If the nuclear spin of one type of nucleus in the protein is selectively saturated (*red*) by using a suitable resonance frequency (RF), the protein magnetization will be transferred (nuclear Overhauser effect, see Sect. 13.7) to the ligand that was bound in the meantime (■). These ligands become apparent in that their spectrum is altered even though they are already dissociated from the protein. If the difference between the spectra in presence of the saturated and unsaturated protein is displayed, it will be possible to determine which ligands were bound immediately to the protein. Many variations and sophisticated experimental protocols have been developed for the principle of magnetization transfer

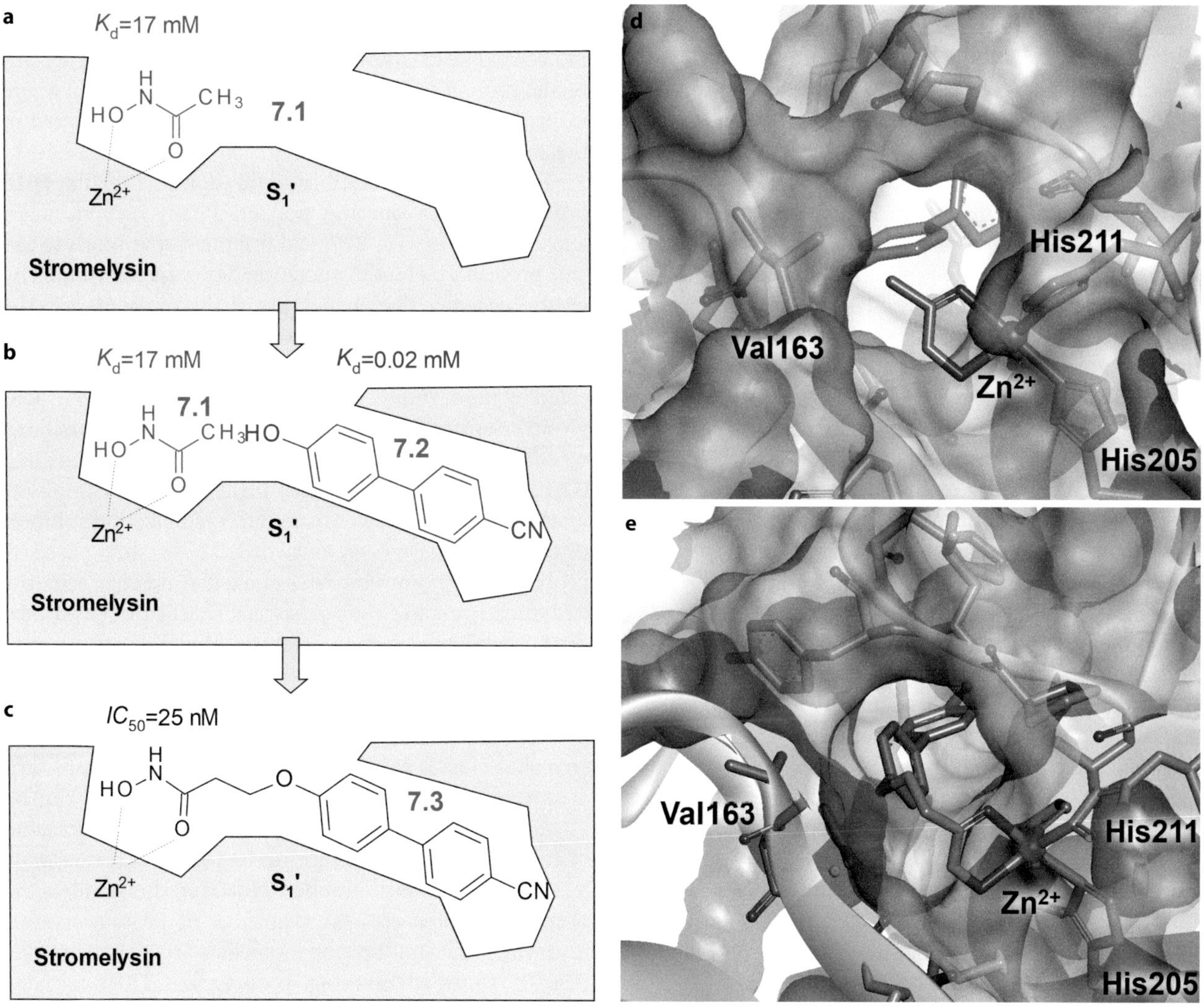

Fig. 7.9 In the *SAR by NMR* method, ligands with weak affinity to a protein, in this case stromelysin, are sought from a large complex mixture. ^{15}N-labeled protein is used and so-called ^{1}H–^{15}N HSQC spectra are measured. If a ligand such as acetohydroxamic acid **7.1** becomes apparent through a shift in the resonance of specific amino acids that protrude into the binding pocket, the binding geometry can be deduced (**a**, **d**). Later the binding site is saturated with these ligands. Further NMR measurements are carried out to identify ligands for neighboring binding positions. These are revealed by the shift in the resonances of neighboring amino acids. That is how 4-cyano-4′-hydroxybiphenyl **7.2** was discovered (**b**, **d**). A chemical coupling of both hits **7.1** and **7.2** with a –CH_2CH_2O– linker produced **7.3**, which is a nanomolar inhibitor of the protease stromelysin (**c**, **e**). (► https://sn.pub/zqmC0T)

amount of time. The magnetic signal is then transferred from the protein to the bound ligands. Upon dissociation, the changed magnetic properties of the temporarily bound ligands can be detected spectroscopically because the relaxation time of the transferred magnetization is now faster in the uncomplexed state. The solution is measured with and without the magnetized protein. The difference between the spectra is evaluated. Only ligands that have been bound to the protein in the interim and, thus, have undergone magnetization transfer will show signals. The so-called **saturation transfer difference (STD) spectrum** can be used to screen for potentially active ligands (■ Fig. 7.8). Many different variations and elaborate experimental protocols have been developed for the magnetization transfer principle described above. Even the use of so-called **reporter** or **spy ligands**, which have an easily measurable NMR signal, can be used. The resonance of fluorine atoms is particularly well suited. The ^{19}F nucleus is very sensitive to NMR detection and has a wide range of chemical shifts (see Sect. 13.7). In addition, there are usually only a few fluorine atoms present in the systems under investigation. This requires a fluorine-containing reporter ligand that binds to the protein, but the binding should not be too strong. The ligand should be easily released from the protein by the test ligand. This release is detectable as a change in the fluorine NMR spectrum and, thus, reveals the binding of the test ligand. As explained in detail in Sect. 13.7, the spatial structure of proteins can be determined by isotopic labeling and the measurement of mutually coupled NMR spectra (so-called multidimensional spectra). In this way, where a test ligand binds to a protein can be accurately determined by evaluating the specific resonance shifts of the labeled protein. However, this requires that all resonances be assigned to the atoms of the protein before the experiment begins. This is usually a rather time-consuming task. If the assignment is not yet available, it is still possible to qualitatively infer ligand binding from the observation of some signal shifts.

In the best case, it is even possible to see two ligands binding at the same time, or two different ligands binding at different non-overlapping positions in the binding pocket. Steven Fesik's research group at Abbott, USA, developed this method. It is known as **SAR by NMR** (SAR stands for structure–activity relationship) and is used for lead structure identification and optimization. This method was used to find a nanomolar inhibitor of the matrix metalloproteinase stromelysin (Sect. 25.6). First, a potent head group was sought that could bind to the zinc ion in the catalytic center of this protease. Such a molecule, acetohydroxamic acid **7.1**, was found with an admittedly weak but specific binding of K_d = 17 mM (■ Fig. 7.9). After the discovery of this ligand, the zinc binding site was saturated with this compound. Further NMR measurements focused on the search for a ligand capable of filling the adjacent S_1' binding pocket. A small library of heteroarylphenyl and biphenyl derivatives was used for this purpose. 4-Cyano-4′-hydroxybiphenyl **7.2** was identified as a hit. The right side of ■ Fig. 7.9e shows both ligands in the binding pocket. Evaluation of the structural data indicated that the hydroxylated phenyl ring binds near the methyl group of the acetohydroxamic acid. Therefore, linking the fragments was the next obvious step. An ethylenoxy group was used as a bridge and coupled to the cyanobiphenyl moiety. NMR spectroscopy confirmed this structural hypothesis and an inhibitor, **7.3**, with an affinity of 25 nM was produced.

7.9 Crystallographic Screening for Small Molecular Fragments

Crystal structure analysis provides the exact spatial position of a molecule in the binding pocket of a protein. Even the geometry of small, very weakly binding molecules is easily recognized. In structures that have a resolution better than 2–2.5 Å (Sect. 13.5), water molecules are usually still visible as discrete density maxima. Often, they indicate sites in the binding pocket that can be equally well occupied by polar functional groups of ligands (■ Fig. 7.10). In the early 1990s, Dagmar Ringe in the research group of Greg Petzko at Brandeis University, USA, intentionally exposed protein crystals to solvent molecules to allow the solvent to diffuse into the crystals. The solvent molecules can act as probes by populating binding regions of the protein pockets. As an example, the regions where isopropanol, acetonitrile, or acetone are encountered in thermolysin, a zinc protease, are shown in ■ Fig. 7.10.

Even phenol, a small organic molecule, manages to diffuse into the binding pocket. Phenylsuccinic acid, a lead structure with a typical fragment size, binds to the zinc protease. Its binding position has been determined by crystallography. The phenyl ring of this molecule is in the same position as that explored by phenol. One of the acid groups of the succinic acid is in the position that was indicated by the carbonyl carbon of acetone. The second acid group coordinates to the zinc ion and occupies positions where water molecules resided in the uncomplexed state (■ Fig. 7.10). There are many protein–ligand complexes in which small molecules from the crystallization solution or cryobuffer have been absorbed. These can be used as probes to map out a binding pocket. A creative scientist will directly exploit their positions for the design of new drug candidates. From there, it was obvious to use crystal structure analysis as a method to screen for the binding of small molecules or "fragments" (MW < 250 Da).

Crystal structure determination has become increasingly faster in recent years due to data acquisition at powerful synchrotrons, and data evaluation can be largely automated. As a result, **crystallographic screening of fragment** collections with up to 1000 test compounds has become standard practice. However, this requires the preparation of a very large number of protein crystals into which the small probe molecules are allowed to diffuse (the so-called "soaking" process, Sect. 13.9). Initially, attempts were made to soak the crystals with a "cocktail" of several test ligands at the same time in order to speed up the screening process. However, since this procedure leads to a large number of problematic artifacts and hardly interpretable electron density maps (Sect. 13.5), soaking with single compounds has been resumed. Often very high concentrations of fragments are used, and organic solvents such as dimethyl sulfoxide (DMSO), isopropanol, or dioxane are added to increase the solubility of the test compounds. Such conditions are often incompatible with the stability of the protein crystals, which are quite fragile. Under the very harsh conditions, they lose their diffraction power. It is, therefore, all the more important to develop strategies that **make the protein crystals more robust** against the highly concentrated fragment solutions containing additional organic solvents.

The hits discovered with the initial crystallographic fragment screening are often only very weak binders in the millimolar to weak micromolar affinity range. However, they can be optimized very effectively because the subsequent design process starts from very well-characterized binding modes (Sect. 20.7). This is also a crucial difference compared to hits from, for example, the HTS or the SPR methods. Whereas the latter methods require **hit**

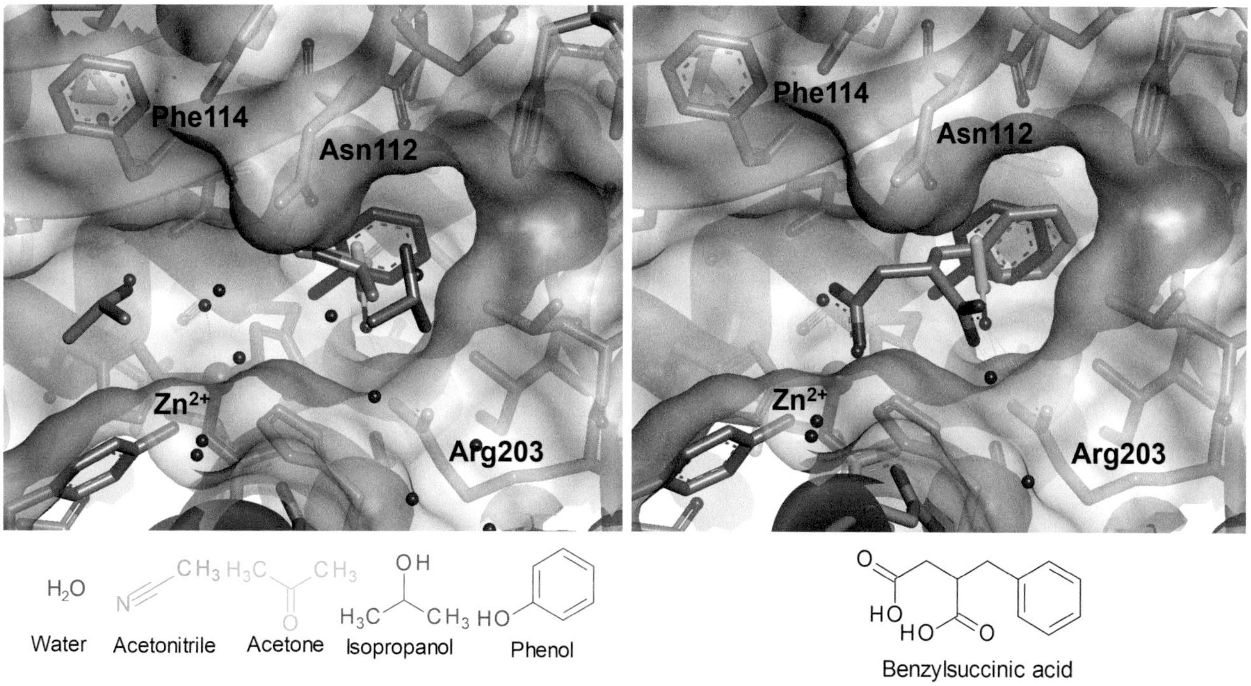

Fig. 7.10 It was possible to soak small probe molecules (so-called "fragments") into crystals of the protease thermolysin. *Left* Superposition of multiple structures in which water (*red spheres*), isopropanol (C-atoms are *gray*), acetone (C-atoms are *light blue*), acetonitrile (C-atoms are *green*), and phenol (C-atoms are *violet*) had penetrated the crystals. They describe potential positions for functional groups of putative ligands. The structure of benzylsuccinic acid, a weakly binding inhibitor of thermolysin, is also shown on the *right* side. This molecule coordinates with one of its acid groups to the catalytic zinc ion. Both oxygen atoms of the acid group displace two water molecules that are present in the noncomplexed structure. The other carboxylate group forms a salt bridge with the neighboring Arg 203. The oxygen of an acetone molecule was found at almost the same position. The phenyl ring of benzylsuccinic acid was found to occupy almost the same position as the phenol molecule in the fragment structure. Benzylsuccinic acid can be used as a starting structure for further optimization. (► https://sn.pub/7djPkK)

validation to verify that the binding really interferes with the mechanism and function of the target, crystallography directly shows where and how the hits have bound. One approach is to **chemically merge** two hits found in different regions of the binding pocket via a chemical linker, analogous to the SAR by NMR method presented in Sect. 7.6. Another, generally more successful, approach is to **chemically extend** fragment hits. This involves adding additional substituents to the original hit molecule based on the known crystal structure. After such a design candidate has been synthesized, a new crystal structure is determined. In this way, original hits that serve as seeds can be grown into enlarged ligands that bind more tightly to the protein in the binding pocket (Sect. 20.7 and Fig. 02, ► https://sn.pub/4NppHn).

Binding of low affinity fragments by biophysical methods is often difficult to characterize and reliably confirm. Different methods often result in **conflicting hit rates**. Therefore, an alternative strategy using **crystallography as the primary screening method** has been proposed. Since it reliably determines the binding geometries of fragments, hits are no longer selected according to their binding constants, which are very hard to determine with the necessary confidence. Rather, the binding poses found are evaluated for their relevance to interfering with protein function and for the possibility of synthetically extending the placed fragments using well-established chemistry. If promising candidate fragments emerge according to these criteria, they are chemically enlarged in a computer design session. They will then reach a size that allows reliable assay data to be collected. This approach typically results in high hit rates. If a structure is then determined from the best follow-up hits of this second round, single-digit micromolar compounds will usually be obtained. Since their binding mode is experimentally characterized, they can easily be further optimized in subsequent design cycles.

7.10 Tethered Ligands Explore Protein Surfaces

Ligands typically bind with very low affinity to **flat pockets** that are open to the surrounding solvent. Therefore, it is extremely difficult to detect their binding or even to obtain a crystal structure with a ligand bound to such a site.

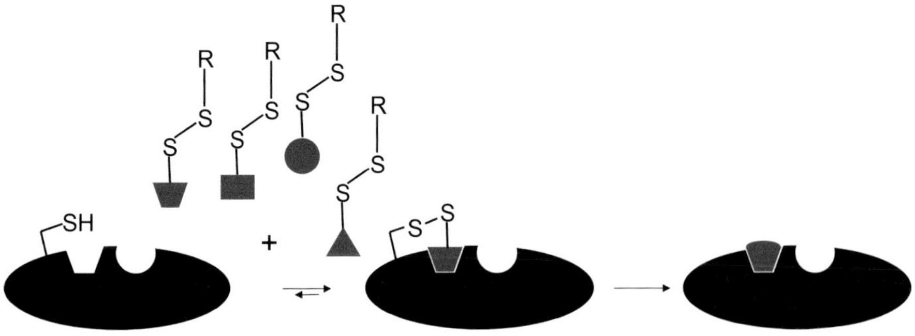

■ **Fig. 7.11** The thiol group of the exposed cysteine is used as an anchor group for the formation of disulfide bonds with ligand candidates from a compound library. There, suitable ligands react that are also able to interact with the surface region in the vicinity of the cysteine thiol. A crystal structure was determined from just such a covalently linked complex (■ Fig. 7.13). After optimization of the initially discovered hit, the disulfide anchor can be discarded and a noncovalent inhibitor can be developed

7

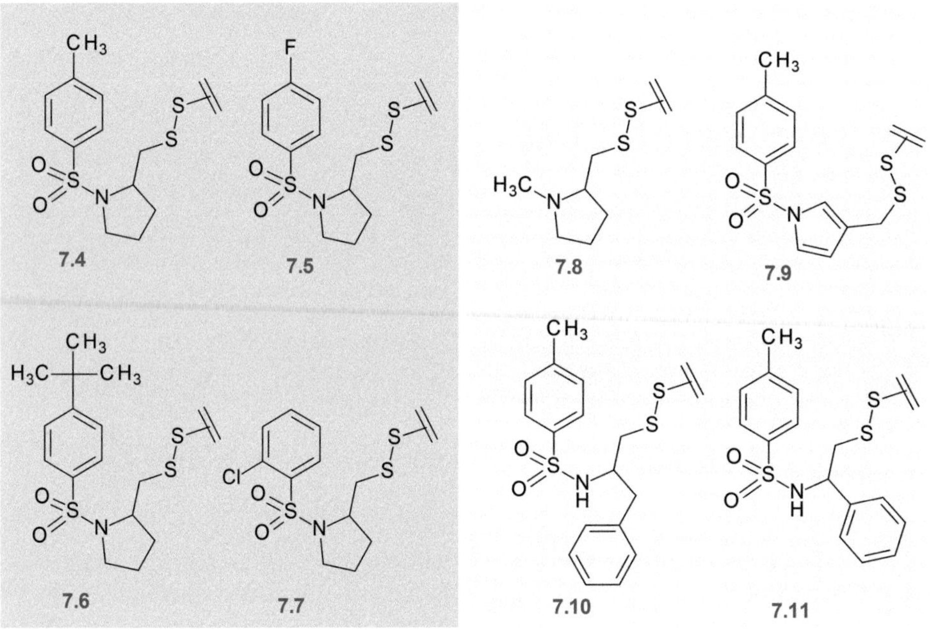

■ **Fig. 7.12** From a library of 1200 disulfides, the compounds on the side **7.4**–**7.7** (*left*) proved to be binders although structurally similar derivatives **7.8**–**7.11** (*right*) were synthesized but did not bind to the protein

James Wells and his colleagues at Sunesis in San Francisco, USA, came up with the idea of **attaching ligands to the surface of proteins** in order to record their binding to such pockets. Chemically, this means reacting with the exposed **thiol group** of a cysteine residue on the protein surface. Such a cysteine must be present in the native protein or can be introduced by mutagenesis (Sect. 12.2). Under appropriate reaction conditions, the ligand is anchored by a disulfide bond formed through the thiol group of the exposed cysteine (■ Fig. 7.11). Only those test candidates from the library that are able to interact with the surface and position themselves with a geometry near the cysteine thiol group that allows them to react with the sulfur atom will be bound. In essence, they explore the surrounding region, react with the cysteine, and remain coupled to the surface through the disulfide bridge. Successfully formed complexes are then detected by mass spectrometry. James Wells and Robert Strout chose thymidylate synthase as their first target. This enzyme plays an important rôle in the *de novo* synthesis of thymidine, an essential building block of DNA. Cells with a high division rate need this building block, and thus inhibition of this enzyme could be a potent anti-infective or antitumor agent (Sect. 27.2).

Thymidylate synthase has a cysteine residue at position 146, near the catalytic site. From a library of 1200 disulfides, compounds **7.4**–**7.7** proved to be binders, whereas the very similar derivatives **7.8**–**7.11** were not selected (■ Fig. 7.12). Accordingly, the phenylsulfonamide together with the proline moiety seemed to be essential for binding. Next, the disulfide anchor was removed, and the binding constant for *N*-tosyl-D-proline **7.12** was measured to be 1.1 mM (■ Fig. 7.13).

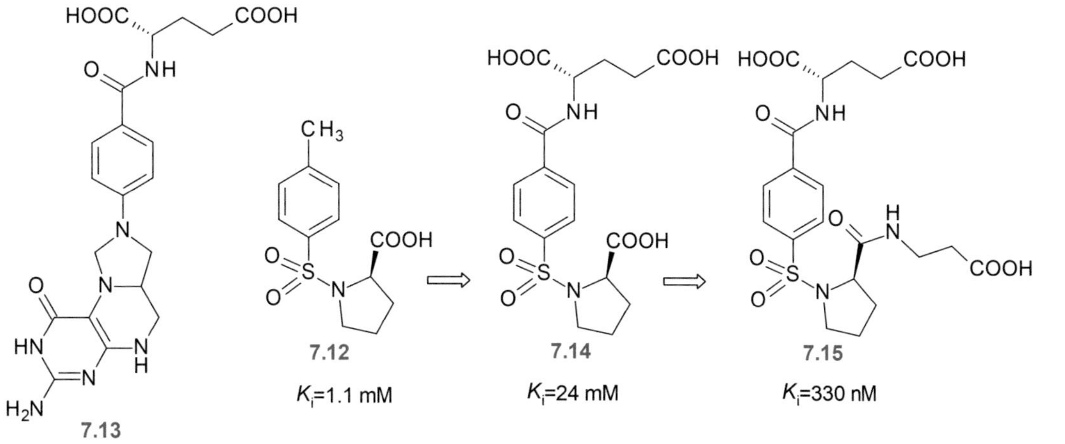

Fig. 7.13 The millimolar *N*-tosyl-D-proline **7.12** was optimized in two steps to the nanomolar inhibitor **7.15** by transferring the side chain from the natural cofactor methylenetetrahydrofolic acid **7.13**

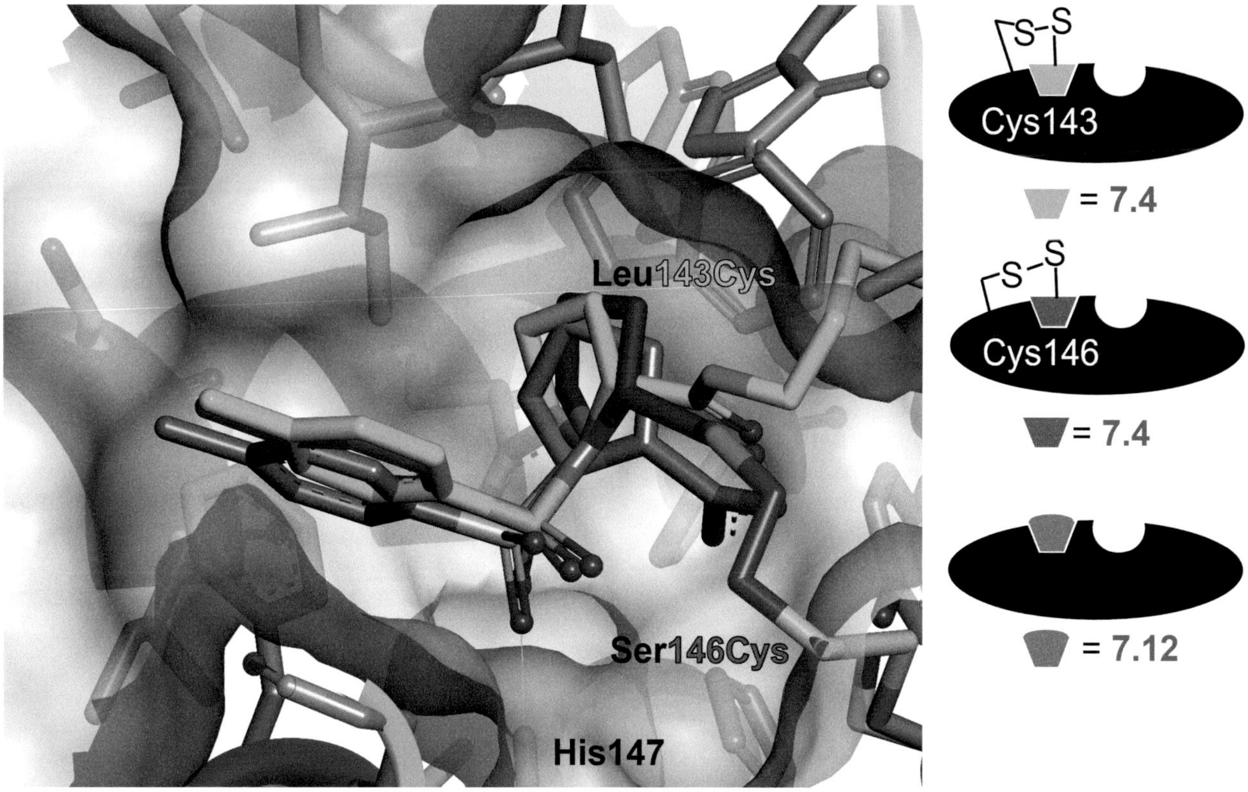

Fig. 7.14 Superpositioning of crystal structures of the enzyme thymidylate synthase with two tethered ligands, one bound to Cys 143 (C-atoms of ligand **7.4** are *green*) and the other to Cys 146 (C-atoms of ligand **7.4** are *violet*), both of which are *N*-tosyl-D-proline derivatives and which are covalently anchored through an S–S bridge. Upon cleavage of the disulfide anchor, the free *N*-tosyl-D-proline (C-atoms are *gray*, **7.12**) proved to be a ligand with an affinity of 1.1 mM. Its binding geometry is very similar to both of the covalently anchored derivatives. (▶ https://sn.pub/mhW8EO)

To further test the concept, Cys 146 was replaced with a serine residue (Fig. 7.14). Since no binding could be detected with this mutant, the neighboring His 147 was mutated to cysteine. However, this mutant was unable to fish out the *N*-tosylprolyl residue. In contrast, the mutant at position 143 was successful (Fig. 7.14). Here, a leucine is replaced by a cysteine. The subsequently determined crystal structures showed that the *N*-tosylprolyl moiety is accommodated almost identically in the two complexes with the covalent disulfide anchor, as it is bound in the complex without the S–S anchor (◘ Fig. 7.14). This is convincing evidence that the covalent attachment does not enforce the binding geometry. Rather, the method allows small, initially weakly binding ligands to be fished out of a large library. A nanomolar inhibitor **7.15** was developed from the initial millimolar hit **7.12** by transferring the side chain of the natural cofactor methylenetetrahydrofolic acid **7.13** in two steps (◘ Fig. 7.13).

The **tethering** method can be applied quite generally. It has been particularly successful in the search for ligands that disrupt the formation of **protein–protein surface contacts** (Sect. 29.8). A major advantage of this technique is that it does not require the development of an additional biochemical binding assay. Weakly binding ligands are covalently "tethered" and cannot be washed away as in the case of simple complex formation. In addition, the covalently bound chemical probes allow the adaptive capacity of the surface region to be explored.

7.11 Synopsis

- Large substance libraries are screened for biological effects to filter out active molecules and assess their value for a given indication.
- Three phases are distinguished: a broad automatic introductory screening for hits, a more detailed screening of chemical analogues around a hit to establish a first structure–activity relationship, and lead optimization to find candidates for clinical testing.
- A prerequisite for high-throughput screening was the development of *in vitro* test systems using pure proteins produced by gene technology along with the entire arsenal of biochemical methods in the test tube so that the function of single-gene products can be recorded.
- As a disadvantage, high-throughput screening does not assess the entire effect spectrum and ignores effects such as transport, distribution, metabolism, and excretion.
- Screening libraries are frequently assembled of molecules from other drug development projects; as such, they are rather inefficient with regard to their molecular size and their modest screening hit activity in the micromolar range. Small substances with high ligand efficiency and sufficient space for structural optimization are particularly promising.
- Enzymatic function and its inhibition can be recorded by the production of chromophoric reaction products.
- Radioactively labeled compounds or enzyme-linked immunosorbent assays are versatile techniques to record protein function on the molecular level.
- Progress in assay miniaturization calls for sophisticated robotic systems, ever-improving sensitivity of the read-out, including fluorescence measuring techniques, and reliable logistics to handle the enormous data flow.
- Aggregate formation of hydrophobic test compounds can exert significant influence on the assay read-out or even cause false-positive or -negative hits.
- Testing on cell-based assays is performed to study changes in cellular- or organism-related function beyond pure binding of a test compound to a given protein target.
- Primary animal testing in vertebrates has been abolished today for ethical reasons, but it is being increasingly replaced by whole-animal screening using nematodes as the simplest multicellular organism to record synergistic and side effects.
- As a complementary and alternative method, virtual computer screening has been developed to screen large compound libraries by docking ligand candidates into the known spatial structure of a target protein.
- Binding events are recorded by biophysical methods such as surface plasmon resonance, thermal stability shifting, mass spectrometry, fluorescence methods, microscale thermophoresis, or microcalorimetry. They are used to detect ligands as potential binders.
- NMR spectroscopy can be used to detect ligand binding by magnetization transfer. Multiple binders can be chemically linked to more strongly binding ligands according to the SAR by NMR technique.
- Exposure of small molecular probes and fragments to protein crystals allows for structural characterization of the binding modes of weakly binding fragments as a versatile starting point to lead optimization.
- Small-molecule fragments tethered to a protein through covalent attachment to the exposed thiol group of a cysteine residue allow the exploration of the binding properties of flat, solvent-exposed surface depressions and serve as a starting point to develop antagonists to perturb the protein–protein interface in complex formation.

Bibliography and Further Reading

General Literature

M.T.S. Stubbs and G. Klebe, in Die Pharmaindustrie, Eds. D. Fischer, J. Breitenbach, Spektrum Akademischer Verlag, Heidelberg, Berlin (2005)

L.M. Mayr and P. Fürst, The Future of High-Throughput Screening, J. Biomol. Screening, **13**, 443–448 (2008)

M. Vogtherr and K. Fiebig, NMR-Based Screening Methods for Lead Discovery pp. 183–202, in *Modern Methods of Drug Discovery*, Ed. A. Hillisch und R. Hilgenfeld, Birkhäusen Verlag (2003) ISBN: 376436081X

G. Klebe, Virtual Ligand Screening: Strategies, Perspectives and Limitations, Drug Discov. Today, **11**, 580–592 (2006)

T. L. Blundell, H. Jhoti and C. Abell, High-Throughput Crystallography for Lead Discovery in Drug Design, Nat. Rev. Drug Discov., **1**, 45–54 (2002)

P. J. Hajduk and J. Greer, A Decade of Fragment-based Drug Design: Strategic Advances and Lessons Learned. Nat. Rev. Drug Discov., **6**, 211–219 (2007)

M. M. Siegel, Early Discovery Drug Screening Using Mass Spectrometry, Current Topics in Medicinal Chemistry, **2**, 13–33 (2002)

A. K. Jones, S. D. Buckingham and D. B. Sattelle, Chemistry-to-Gene Screens in *Caenorhabitis Elegans*, Nat. Rev. Drug Discov., **4**, 321–330 (2005)

S. Löfås, Optimizing the Hit-to-Lead Process Using SPR Analysis. Assay Drug Dev. Technol., **2**, 407–415 (2004)

W. Jahnke and D. A. Erlanson, Fragment-based Approaches in Drug Discovery, Vol. 34 in Methods and Principles in Medicinal Chemistry, R. Mannhold, H. Kubinyi und G. Folkers, Eds., Wiley-VCH, Weinheim (2006)

D. A. Erlanson, S. W. Fesik, R. E. Hubbard, W. Jahnke and H. Jhoti, Twenty years on: the impact of fragments on drug discovery, Nat. Rev. Drug Discov., **15**, 605–619 (2016)

Special Literature

D. Cubrilovic, A. Biela, F. Sielaff, T. Steinmetzer, G. Klebe, R. Zenobi, Quantifying protein–ligand binding constants using electrospray ionization mass spectrometry: A systematic binding affinity study of a series of hydrophobically modified trypsin inhibitors, J. Am. Soc. Mass. Spectr., **10**, 1768–77 (2012)

L. Piersimoni, P. L. Kastritis, C. Arlt, A. Sinz, Cross-Linking Mass Spectrometry for Investigating Protein Conformations and Protein−Protein Interactions − A Method for All Seasons, Chem. Rev., **122**, 7500–7531 (2022)

P. Dumas et al., Extending ITC to Kinetics with kinITC, Methods in Enzymology, Vol. **567**, Chap. 7, A. Feig Ed. (2016)

M. Jerabek-Willemsen, T. Andre, R. Wanner, H. M. Roth, S. Duhr, P. Baaske, D. Breisrecher, MicroScale Thermophoresis: Interaction analysis and beyond. J. Mol. Struct., **1007**, 101–113 (2014)

A. J. Gupta, S. Duhr, and P. Baaske, Microscale Thermophoresis (MST). In Encyclopedia of Biophysics, G. Roberts and A. Watts, Eds. (Berlin, Heidelberg: Springer Berlin Heidelberg), pp. 1–5 (2018).

J. Niemax, R. Strasser, P. Hampel, U. Rant, Analyse von Biomolekülen mit aktiv bewegten Nano-Oberflächen, Laborwelt, **11**, 21–22 (2010)

P. J. Hajduk, G. Sheppard, D. G. Nettesheim, E. T. Olejniczak, S. B. Shuker, R. P. Meadows, D.H. Steinman, G. M. Carrera, Jr., P. A. Marcotte, J. Severin, K. Walter, H. Smith, E. Gubbins, R. Simmer, T. F. Holzman, D. W. Morgan, S. K. Davidsen, J. B. Summers and S. W. Fesik, Discovery of Potent Nonpeptide Inhibitors of Stromelysin Using SAR by NMR, J. Am. Chem. Soc.,**119**, 5818–5827 (1997)

C. Mattos and D. Ringe, Locating and characterizing binding sites on proteins, Nat. Biotechn., **14**, 595–599 (1996)

D. A. Erlanson, A. C. Braisted, D. R. Raphael, M. Randal, R. M. Stroud, E. M. Gordon and J. A. Wells, Site-directed Ligand Discovery, Proc. Natl. Acad. Sci. U.S.A., **97**, 9367–9372 (2000)

J. Müller et al., Magnet for the Needle in Haystack: "Crystal Structure First" Fragment Hits Unlock Active Chemical Matter Using Targeted Exploration of Vast Chemical Spaces, J. Med. Chem., **65**, 15663–15678 (2022)

Optimization of Lead Structures

Contents

G. Klebe, *Drug Design*, https://doi.org/10.1007/978-3-662-68998-1_8

A lead structure is the starting point on the way to a drug. The potency, specificity, and duration of effect must be optimized, and the side effects and toxicity must be minimized in a usually elaborate, iterative process. Every change in the chemical structure modulates the 3D structure of the molecule, its physicochemical properties, and the activity spectrum. The isosteric replacement of atoms or groups, the introduction of hydrophobic building blocks, the dissection of rings or the restriction of flexible molecular portions into cyclic structures, and the optimization of the substitution pattern are all possibilities to purposefully modify a putative lead structure.

Creativity and luck are always important prerequisites for success in pharmaceutical research. Nonetheless, there is a treasure chest of decades of accumulated experience that can be exceedingly supportive to the rational optimization process. Computer-aided methods can contribute to their full capability in this field in particular. Several general considerations and approaches to lead optimization are presented in the sections of this chapter. A discussion of the structure-based and computer-aided optimization of lead structures is presented in Chaps. 17 and 20. Examples for its application to different therapeutic areas are presented in Chaps. 21, 23–32.

8.1 Strategies for Drug Optimization

The optimization of active substances follows a process that is best characterized by the words of the philosopher Sir Karl Popper:

> "The truth is objective and absolute. But we can never be sure that we have found it. Our knowledge is always an assumed knowledge. Our theories are hypotheses. We test for the truth in that we exclude what is false." (Objective Knowledge, 1972)

Accordingly, the optimization of a compound's potency follows a working hypothesis, while an iterative process of trial and error refines the hypothesis. The accumulated data on the relationship between chemical structure and biological activity is used to design new structures. These are synthesized and tested, and a new working hypothesis is modified as appropriate. In negative cases, the hypothesis is discarded and a new one is formulated that better fits the biological data.

The following qualities are distinguished in the structure of the compound:

- The actual **pharmacophore** (Sects. 8.7 and 17.1) that is responsible for the specific binding and upon which only limited chemical modification can be carried out,
- The additional groups (**adhesion groups**) that improve affinity and biological activity,
- Further groups that do not influence the binding but rather the **lipophilicity** of the molecule and with it the transport and distribution in biological systems (Chap. 19), and
- The groups that must be cleaved or modified in the organism to **release** the **actual active form** (Chap. 9).

The most important steps in the optimization of lead structures are the systematic changes in the shape and form, that is, the three-dimensional structure, and/or the physicochemical properties. Single steps along this route are

- Changes in the lipophilicity and the electronic properties through the introduction or removal of hydrophobic or hydrophilic groups,
- Variations of substituents at aromatic or heteroaromatic rings,
- Introduction or elimination of heteroatoms in chains or rings,
- Changes in chain length of aliphatic groups or linkers,
- Introduction of space-filling substituents to stabilize a particular conformation,
- Changes in the ring size of alicyclic or heterocyclic rings,
- Incorporation of flexible partial structures in rings,
- Incorporation of branches or attachments to rings (rigidifying),
- Opening of rings,
- Elimination of chiral centers to simplify a structure,
- Addition of chiral centers to increase the selectivity, or
- Shifts of the thermodynamic binding profile and the drug's residence time at the target protein.

These processes are usually unidirectional in classical drug optimization, that is, the optimization takes place on one position of the molecule at a time, in one single direction. In the past, such unidirectional optimization has led to many disappointments because interdependent influences of the structural changes were neglected, or the optimal lipophilicity was exceeded. John Topliss developed a scheme for the variation of aromatic substituents that allows the biological activity to be optimized in a minimum number of steps (Sect. 8.3). The application of experimental design, simultaneously changing multiple parts of a molecule, and the evaluation of the results by using quantitative structure–activity relationships (Chap. 18) usually allows fast and effective optimization. In structure-based and computer-aided optimization, the 3D structure of the target protein and its complexes leads to directed structural variations of the active substances. Here again, the aspects of total lipophilicity and metabolism should not be neglected.

8.2 Isosteric Replacement of Atoms and Functional Groups

Isosteric replacement is the exchange of particular groups in a molecule for sterically and electronically related groups. If the biological effect is essentially maintained, the term **bioisosteric replacement** (◘ Fig. 8.1) can be used. In the simplest case, a single atom is exchanged, for instance, a Cl (lipophilic, weakly electron withdrawing) is replaced by a Br (same characteristics as Cl) or methyl (lipophilic, weakly electron donating), or an –O– (polar, H-bond acceptor) is exchanged for an –NH– (polar, H-bond donor) or a $-CH_2-$ (lipophilic, unable to form H-bonds) group. Furthermore, bioisosteric replacement also means the exchange of entire groups. For example, –COOH, an H-bond acceptor and donor, can be replaced with other groups that have the same or modified properties, for instance, with the similarly acidic tetrazole. Another example can be found in the exchange of a phenyl ring for a thiophene or a furan building block (◘ Fig. 8.1). The potential of isosteric replacement is illustrated in the exchange of all three iodine atoms of triiodothyronine T_3 **8.1** for alkyl groups to give 3,5-dimethyl-3′-isopropylthyronine **8.2**, which in turn retains impressive affinity and agonistic activity on the thyroid hormone receptor. In contrast to triiodothyronine, which is both iodinated and metabolized by a deiodinase, the alkyl groups of **8.2** are no longer metabolically cleavable.

Bioisosteric replacement is one of the most important strategies in pharmaceutical drug optimization. Nonetheless, surprises sometimes occur. The replacement of an ester for an amide group in local anesthetics (Sect. 3.4) expectedly improved the metabolic stability. In the case of acetylsalicylic acid **8.3** (◘ Fig. 8.2), this exchange cannot be made. An analogous exchange of the –COO– group for a –CONH– group results in a complete loss of activity because the amide can no longer acylate the cyclooxygenase enzyme (Sect. 27.9). In the case of *p*-aminobenzoic acid (R = –COOH, ◘ Fig. 8.2) the exchange of a carboxyl group for a sulfonamide group gives sulfanilamide **8.4** (R = $-SO_2NH_2$), which is an antimetabolite of *p*-aminobenzoic acid (Sect. 2.3).

A lead structure is rarely studied exclusively by one research group. Other companies adopt successful examples, at the very latest after the economic success of a new medicine. The goal of this so-called **"me-too" research** is to modify the competitor's lead structure to arrive at patent-free analogues that are more efficacious, more selective, or better tolerated. It must be accepted that even this form of competition has led to the therapeutically most valuable compounds in many therapeutic areas. On the one hand, a plentitude of duplicate work has been performed, while on the other hand, new analogues with improved properties have been produced and introduced to therapy which turned out to be successful in the long run. Penicillins of the third and fourth generation with broad-spectrum activity and metabolic stability, β-blockers with improved selectivity, and many other

Substituents: F, Cl, Br, I, CF_3, NO_2
Methyl, Ethyl, Isopropyl, Cyclopropyl, tert-Butyl,
-OH, -SH, $-NH_2$, OMe, $N(Me)_2$

Bridging Groups: $-CH_2-$, -NH-, -O-
$-COCH_2-$, CONH-, -COO-
>C=O, >C=S, >C=NH, >C=N-OH, >C=N-OAlkyl

Atoms and Groups in Rings: -CH=, -N=
$-CH_2-$, -NH-, -O-, -S-,
$-CH_2CH_2-$, CH_2-O-, -CH=CH-, -CH=N-

Larger Groups: $-NHCOCH_3$, $-SO_2CH_3$, -COOH, -CONHOH,

$-SO_2NH_2$,

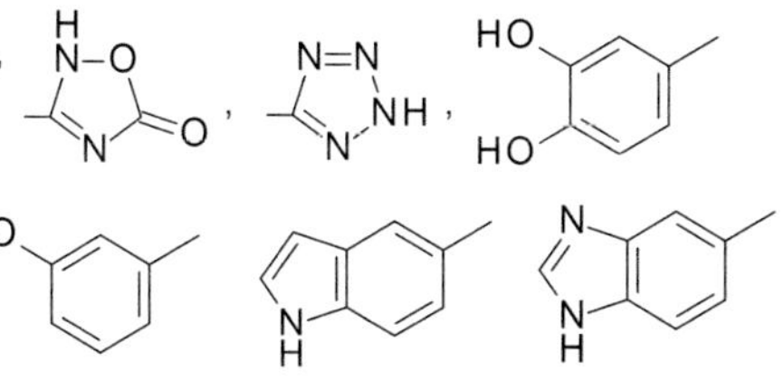

◘ **Fig. 8.1** A few possibilities for the isosteric replacement of atoms and/or groups

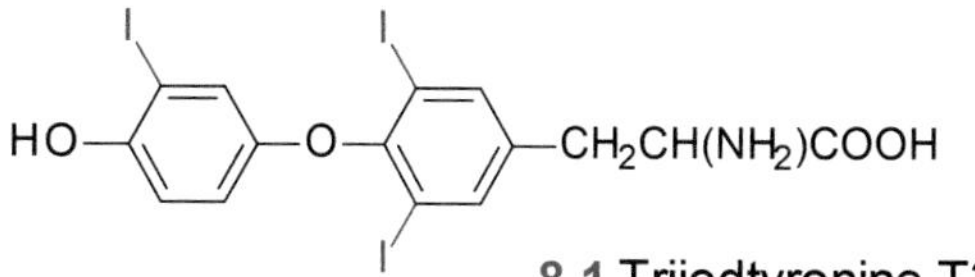

8.1 Triiodtyronine T3

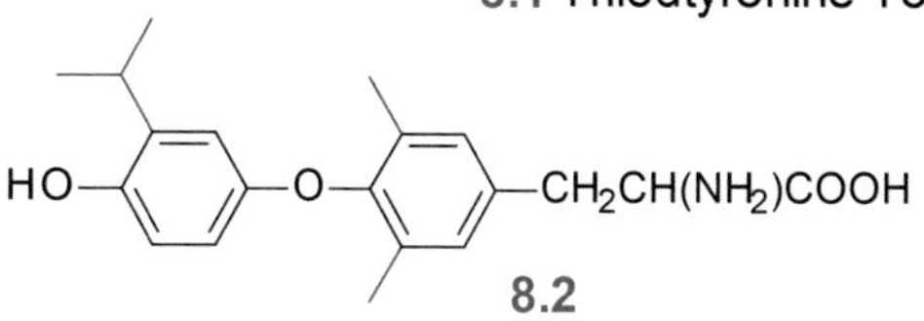

8.2

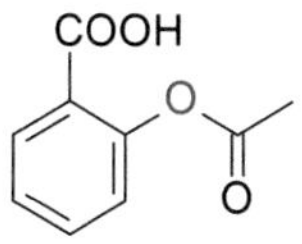

8.3 Acetylsalicylic acid

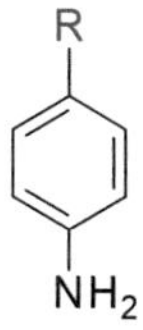

8.4 R = -COOH or $-SO_2NH_2$

◘ **Fig. 8.2** Isosteric replacement with retention, loss, and reversal of the biological activity. All three iodine atoms of the thyroid hormone thyroxine **8.1** can be replaced with alkyl groups and compound **8.2** is still active. In the case of acetylsalicylic acid **8.3**, the exchange of the $-OCOCH_3$ for an $NHCOCH_3$ group led to the loss of the acylating ability and, therefore, a nearly complete loss of the biological activity. The antimetabolite sulfanilamide **8.4** (R = SO_2NH_2) is derived from *p*-aminobenzoic acid **8.4** (R = COOH), which is a critical intermediate in the bacterial dihydrofolate synthesis; **8.4** (R = SO_2NH_2) is the result of the exchange of a carboxyl group for an isosteric sulfonamide group specific drugs would simply not exist if it were not for the much-disparaged "me-too" research.

8.3 Systematic Variation of Aromatic Substituents: The Topliss Trees

The goal of lead structure optimization has an impact on the planning of the relevant experimental series. If the biological consequences of structural changes are to be evaluated with minimal effort, careful **design** must precede the synthesis of the substances. Here, an almost unsolvable problem emerges in that, as a general rule, the exchange of a substituent or group leads to complex changes in multiple properties. The exchange of an ethyl group for a methyl group changes only the lipophilicity and size of the substituent. If a methyl group is exchanged for a chlorine atom, the polarizability, electronic properties, and moreover the metabolism are altered. Other substituents could then change the H-bond donor and acceptor properties as well as the ionization and dissociation.

In 1971, Paul Craig proposed the use of a simple diagram for the structural variation of aromatic substituents, with which the important characteristics of these substituents, for instance, lipophilicity and electronic properties, are plotted against each other. The selection of substituents from different quadrants of this diagram allows an evaluation of different combinations of properties. The concept can be extended to multiple dimensions, possibly with the aid of mathematical and statistical methods.

In 1972, John Topliss made a suggestion that went further, which would be called today an evolutionary strategy. One substituent at a time (e.g., hydrogen for chlorine) is exchanged in the optimization of the substitution pattern of an aromatic compound. The next compound is planned based on which of the first two compounds demonstrated better effects. If the new substituent improves the effect, a new substituent will be chosen that has the same physicochemical properties, in larger measure, or more of these substituents are added. If the new substituents impair the biological activity, then a substituent will be chosen that has the opposite physicochemical properties. If two different substituents produce the same effect, it should be evaluated whether changes in the physicochemical properties will influence the activity in the opposite direction. Despite its elegance, this strategy often fails for the mundane reason that it is too time consuming to take such a stepwise approach.

As a consequence of the work of Craig and Topliss, further design methods were developed. None of these methods should be interpreted too closely. Synthesis planning must be oriented on both the accessibility of the compounds as well as achieving the largest possible structural variation, that is, a diversity of physicochemical properties and 3D structure. Since the introduction of combinatorial chemistry (Chap. 11), the rational design of diverse substance libraries has taken on entirely new possibilities and perspectives.

8.4 Optimizing the Activity and Selectivity Profile

The structural variation of a lead structure influences not only the potency but also the **activity spectrum**. That can be thoroughly advantageous, but it also brings with it the risk that the selectivity can deteriorate. A simple rule of thumb is that enlarging the molecule, introducing optically active centers, and rigidification improves the selectivity, assuming that the activity is not entirely lost. On the other hand, removing a chiral center, establishing more flexibility, or reducing the size of the molecule usually results in unspecific and weaker activity.

Because of the sequencing of the human genome, the **gene family** to which a target protein belongs is known, as is the number of members of the gene family. By using gene technology, it is possible to construct single isoform test systems (assays). As a result, today pharmaceutical research is in a position to make a predictive **selectivity profile**. This has stimulated efforts to develop selective drugs. An interesting corollary to these efforts is the fact that the molecular weight of drugs has increased in the last few years as statistics have shown, thus, confirming the above-mentioned rule of thumb.

For drugs that are meant to act on neuroreceptors in the brain, polarity governs whether they can **cross the blood–brain barrier**. Polar compounds are unable to do this and act only in the periphery, for instance, on the circulatory system. Examples of this are adrenaline **8.5**

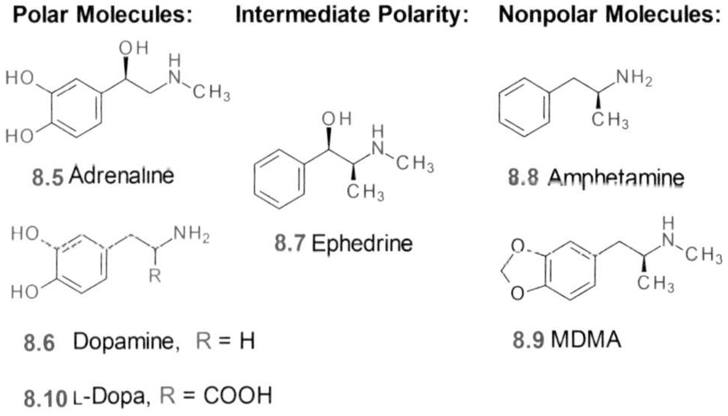

Fig. 8.3 The polar compounds adrenaline **8.5** and dopamine **8.6** are cardiovascularly active in the periphery after intravenous administration. Ephedrine **8.7** is more lipophilic and, therefore, shows both peripheral and central effects. The more nonpolar compound amphetamine **8.8** ("speed") has an overwhelmingly stimulatory effect in the CNS. 3,4-Methylenedioxymethamphetamine **8.9** (MDMA; "ecstasy") is hallucinogenic. Polar groups are *red* and neutral or lipophilic groups are *blue*

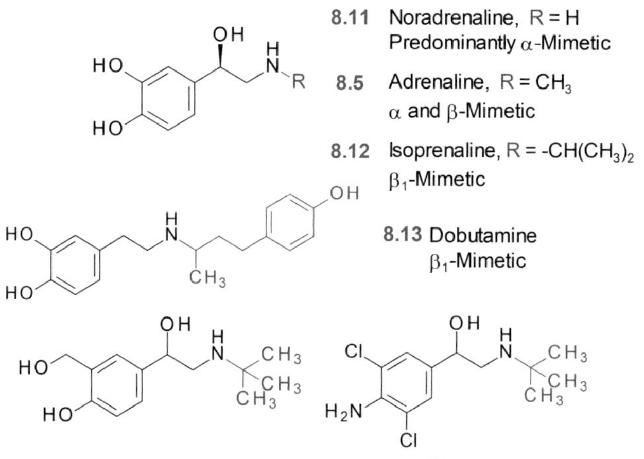

8.11 Noradrenaline, R = H
Predominantly α-Mimetic

8.5 Adrenaline, R = CH_3
α and β-Mimetic

8.12 Isoprenaline, R = $-CH(CH_3)_2$
β_1-Mimetic

8.13 Dobutamine
β_1-Mimetic

8.14 Salbutamol
β_2-Mimetic

8.15 Clenbuterol
β_2-Mimetic

Fig. 8.4 Noradrenaline **8.11**, adrenaline **8.5**, and isoprenaline **8.12** act to different extents on the α- and β-receptors. Selective β_1- and β_2-agonists, for instance, **8.13**, **8.14**, and **8.15**, act specifically as cardiac stimulants or bronchodilators

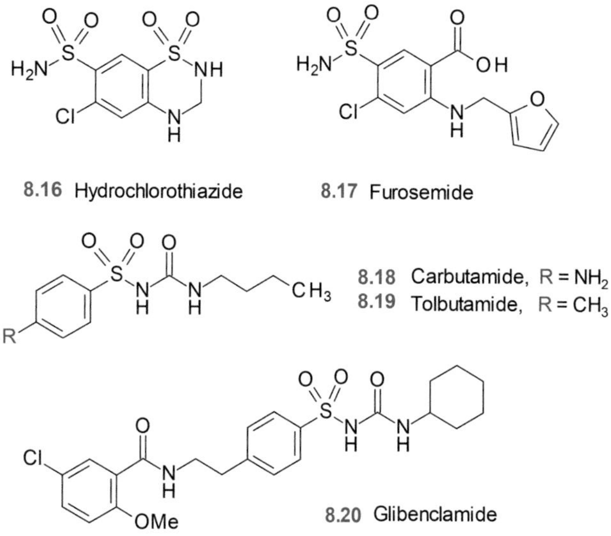

8.16 Hydrochlorothiazide

8.17 Furosemide

8.18 Carbutamide, R = NH_2
8.19 Tolbutamide, R = CH_3

8.20 Glibenclamide

Fig. 8.5 The sulfonamides hydrochlorothiazide **8.16**, furosemide **8.17**, and related diuretics are different from most antibacterial analogues because of the unsubstituted sulfonamide group. Carbutamide **8.18** and tolbutamide **8.19** were the first unspecific sulfonamides with hypoglycemic effects that were later replaced with specific hypoglycemics of the glibenclamide-type **8.20**

and dopamine **8.6** (Fig. 8.3). The stepwise removal or masking of polar groups brings the central effects into the foreground. Ephedrine **8.7** acts in the brain and in the periphery; it is centrally stimulating and raises the blood pressure. Amphetamine **8.8** ("speed") and the intoxicant MDMA **8.9** (the designer drug "ecstasy") are weak bases. Their relatively nonpolar neutral forms easily overcome the blood–brain barrier and their CNS effects dominate (Fig. 8.3).

There are exceptions even here. L-DOPA **8.10** (Fig. 8.3) is an extremely polar amino acid. It could never cross the blood–brain barrier by passive diffusion alone. Instead it is recognized by an amino acid transporter and actively transported over the membrane and into the brain. This simultaneously solves the problem of bringing dopamine **8.6**, which is used to treat Parkinson's disease, into the brain because L-DOPA is decarboxylated to dopamine there (Sects. 9.4 and 27.8).

The activity profile of the hormones and neurotransmitters noradrenaline and adrenaline and their synthetic analogues shows the decisive influence that even small changes in structure can have. Whereas noradrenaline **8.11** (Fig. 8.4) affects the α-adrenergic receptors, its *N*-methyl derivative adrenaline **8.5** (Fig. 8.3) acts on α- and β-receptors as a mixed α/β-agonist. This difference was used to enlarge the *N*-alkyl group to arrive at the specific β-agonist isoprenaline **8.2** (Fig. 8.4). Further differentiation of the effects could be achieved within the class of β-adrenergic substances. Dobutamine **8.13** is missing the alcoholic hydroxyl group of adrenaline. Despite its structural relationship to dopamine **8.6** (Fig. 8.3), it is a β_1-agonist with cardioselective effects. Specific β_2-agonists, for instance salbutamol **8.14** and clenbuterol **8.15** (Fig. 8.4) are used to treat asthma because they are bronchodilators without the cardiostimulatory effects of the unspecific β-agonists (Sect. 29.3).

The **sulfonamides** are a prime example for the **targeted optimization** of lead structures in different therapeutic indications. From the first antibacterial examples, diuretics as well as hypoglycemics (antidiabetics) resulted. It had already been noticed in 1940 that sulfanilamide (Sect. 2.3) inhibits the enzyme carbonic anhydrase and, therefore, should lead to increased urine production (Sect. 25.7). Among other substances, hydrochlorothiazide **8.16**, furosemide **8.17** (Fig. 8.5), and structurally related compounds gained therapeutic importance. In the early 1940s, hypoglycemic effects of a few sulfonamides were clinically observed. The antibacterial and simultaneously hypoglycemic carbutamide **8.18** was introduced into therapy in 1955; the lipophilic and therefore more bioavailable tolbutamide **8.19** was introduced later. Systematic structural variation finally led to glibenclamide **8.20** (Fig. 8.5 and Sect. 30.2), which is much more potent and specific.

8.5 From Agonists to Antagonists

There is no general recipe for the transformation of an agonist into an antagonist. An example of this is found in the tedious route from the agonist histamine to the H_2 antagonist, as described in Sect. 3.5. There are, however, recognized principles that have proven to be of value. For example, the exchange of polar for nonpolar substituents or the introduction of large groups such as additional aromatic rings changes some receptor ag-

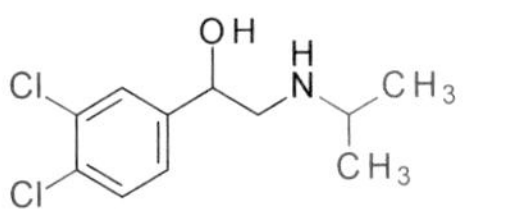

8.21 DCI

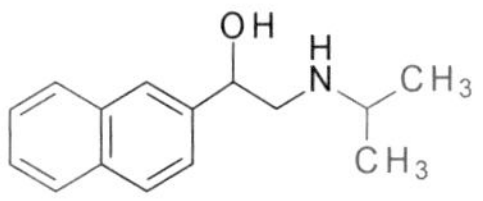

8.22 Pronethalol

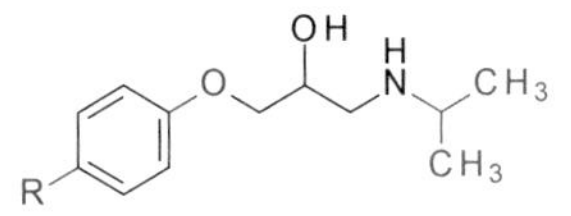

8.23 Practolol, R = -NHCOCH$_3$

8.24 Metoprolol, R = -CH$_2$CH$_2$OMe

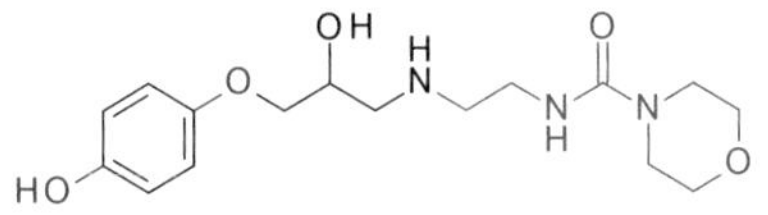

8.25 Xamoterol

Fig. 8.6 3,4-Dichloroisoprenaline **8.21** (DCI) and pronethalol **8.22**, the first unspecific β-blockers, were derived from isoprenaline **8.12**. Practolol **8.23** and metoprolol **8.24** are specific β_1-agonists. Xamoterol **8.25** is a partial β_1-agonist, a combined agonist and antagonist

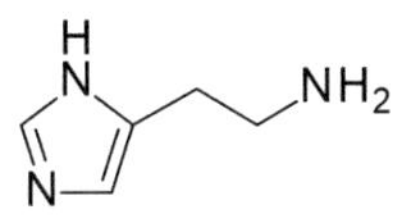

8.26 Histamine
H-Agonist

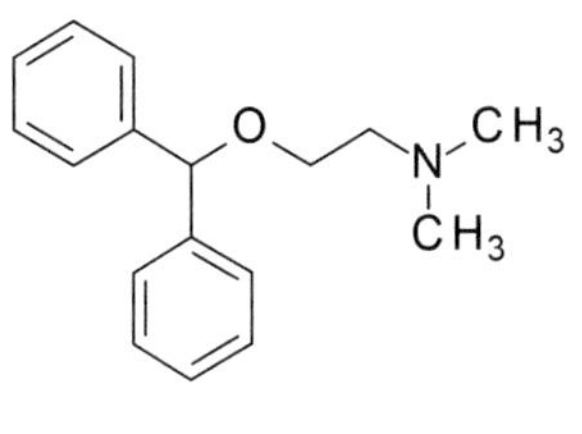

8.27 Diphenhydramine
Nonpolarer H$_1$-Antagonist
(sedating)

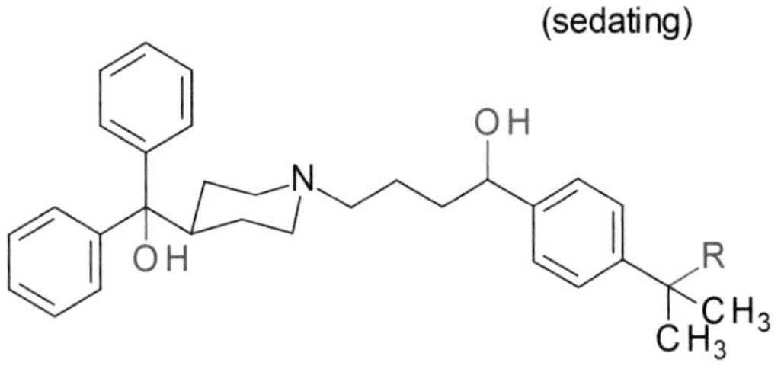

8.28 Terfenadine, R = CH$_3$
Polar H$_1$-Antagonist (nonsedating)
Fexofenadine, Active Metabolite: R = -COOH

Fig. 8.7 By starting with histamine **8.26** and introducing large hydrophobic groups, H$_1$-antagonists, for instance, diphenhydramine **8.27**, were obtained. The nonsedating terfenadine **8.28** (R = CH$_3$) crosses the blood brain barrier but is immediately expelled by a transporter. In the meantime, the active metabolite, fexofenadine with R = COOH, is on the market

onists to antagonists. The exchange of both phenolic hydroxyl groups in isoprenaline **8.12** for two chlorine atoms (DCI, **8.21**) or additional aromatic rings (pronethalol, **8.22**) delivered the first β-adrenergic antagonists, the so-called β-blockers. The introduction of an oxygen atom in the side chain, and further structural optimization afforded the first β_1-selective antagonists, for example, practolol **8.23** and metoprolol **8.24**. The β_1-selective partial agonist xamoterol **8.25** is a blocker as well as an agonist (Fig. 8.6). It occupies β_1-receptors and displays a moderately stimulating effect. By occupying the

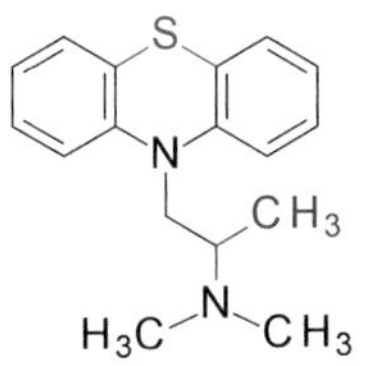

8.29 Promethazine
H$_1$-Antagonist

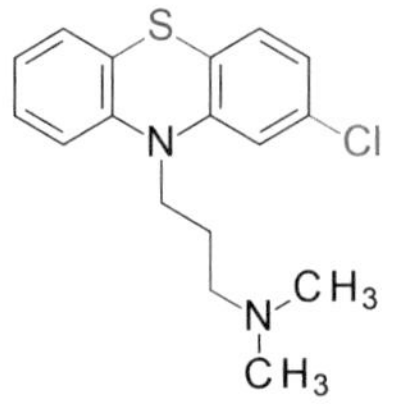

8.30 Chlorpromazine
Neuroleptic

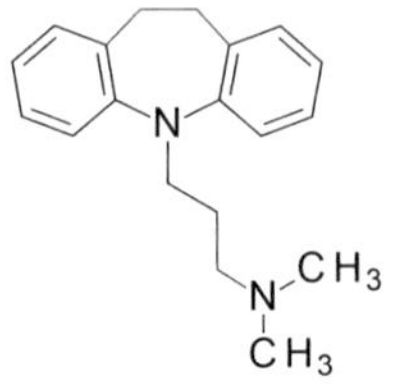

8.31 Imipramine
Antidepressant

Fig. 8.8 Closely related structures of active substances can have very different qualitative activity. Chlorpromazine **8.30**, a dopamine antagonist with neuroleptic activity, and imipramine **8.31**, a dopamine transporter inhibitor with antidepressant activity, are both derived from promethazine **8.29**, an H$_1$-antagonist with antiallergic activity

receptor, it protects it from an excessive response upon elevated adrenaline release, for instance, from exercise or stress.

Analogously, the exchange of the imidazole ring of histamine **8.26** for large hydrophobic groups led to the first H$_1$-antagonists, for instance, diphenhydramine **8.27** (Fig. 8.7). Sedation is the most troublesome side effect of the classic H$_1$-antagonists, which are used to treat allergies. The nonsedating terfenadine **8.28** (R = H) can cross the blood–brain barrier because of its high lipophilicity, but is immediately expelled by a transporter. Because of its cardiotoxicity, terfenadine has been withdrawn from the market and replaced by its active metabolite fexofenadine **8.28** (R = COOH).

The sedating side effects of antihistamines also led to neuroleptics and antidepressants (Sect. 1.6). Here, however, the limits of rational drug optimization are apparent. Promethazine **8.29** is an antihistamine with antiallergic action and sedating side effects. The neuroleptic chlorpromazine **8.30** is a central depressant and therefore an antipsychotic; the extraordinarily similar structure of imipramine **8.31** acts, on the other hand, as a stimulant and is an antidepressant (Fig. 8.8). All three substances have different mechanisms of action. The introduction of additional aromatic rings to other receptor agonists, for instance, to the neurotransmitters acetylcholine and dopamine, has led to antagonists.

8.6 Optimizing Bioavailability and Duration of Action

The **absorption** of the majority of pharmaceuticals depends only on their lipophilicity. The more polar the drug, the more poorly it can penetrate the lipid membrane, and the lower the absorption (Sect. 19.6). Increasing the lipophilicity improves the absorption. Extremely lipophilic compounds are insoluble in water, and their absorption is too slow. Lipophilic acids and bases will offer advantages here, if their acidity constant is not too far away from the neutral point, pH 7. In their ionized form, they are highly water soluble, while in their neutral form, with which they are in equilibrium, they are lipophilic and membrane permeable. These correlations are discussed in Sect. 19.5. The molecular size influences the bioavailability insofar that substances with a molecular weight above 500–600 Da are captured by the liver solely because of their molecular size, and they are quickly excreted with the bile. Aside from this, there are substances that penetrate the membrane regardless of their polarity. These are taken up into the cell or are eliminated from the cell by transporters (Sect. 30.10). Among these are structural analogues of amino acids and nucleosides.

Classical strategies to extend the **duration of action** are the conversion of free hydroxyl groups to ethers (see Sect. 9.2), the replacement of esters with amides, and the replacement of metabolically labile amide groups with isosteres. In a few cases, such structural changes are associated with a reduction in potency, which is more than compensated for by a longer duration of action. In the case of peptides, the replacement of L-amino acids with D-amino acids, the reversal of the direction of amide groups, and the replacement of larger structural elements with peptidomimetic groups (Sect. 10.4) have all proven successful.

The metabolism of aliphatic amino groups can be suppressed with alkyl substitution or branching at the α-carbon. Secondary alcohols can be converted to the more bioavailable tertiary alcohols by introducing an ethinyl group at the same carbon atom (Sect. 28.5). The introduction of an isosteric fluorine atom in the *para*-position as a replacement for hydrogen atoms prevents hydroxylation in this position. If steric considerations do not play a role, the *para*-position can also be blocked with a larger group, such as a chlorine atom or a methoxy group. In the hydroxylated 3- and 4-position of the neurotransmitters dopamine, adrenaline, and noradrenaline, the conversion to the monohydroxylated analogues 3,5-dihydroxy compounds or to the NH-isosteric indole group (◻ Fig. 8.1, Sect. 8.2) led to metabolically more stable and, therefore, longer-acting compounds.

8.7 Variations of the Spatial Pharmacophore

Rational design is characterized by the fact that the common feature of all active compounds and the differences to less potent or inactive analogues can be derived from the spatial structure of the pharmacophore. A **pharmacophore** (◻ Fig. 8.9) is defined as a special arrangement of particular functionalities that are common to more than one drug and form the basis of the biological activity (Sect. 17.1).

During the course of rational optimization, the molecular scaffold and the substituents at a pharmacophore are changed to maintain the principle function, while arriving at higher potency or better selectivity. Many computer methods have been developed to generate ideas for the spatial isomorphic replacement of ligand scaffolds. By considering the conformational aspects of the molecules (Chap. 16), they scan databases to find possible candidates that, despite a different parent scaffold, can place the side chains and interacting groups in the same spatial orientation. Examples of such approaches are presented in Sect. 10.8 and Chap. 17. However, a more indirect route via the protein structure has also been explored. Starting from the spatial structure of a protein–ligand complex, the part of the binding pocket for which a new building block in a ligand is sought is cut out. The shape and interaction properties of the extracted pocket are then compared with the database of all known protein–ligand complexes (Sect. 20.4). If a pocket section similar to the search pocket is found, the ligand binding to this pocket is of interest. The structure of the building block occupying the discovered pocket may provide a new idea for isosteric replacement of the ligand to be modified.

A different strategy that also considers the pharmacophore can be successful. In this approach, the pharmacophore is retained and only those groups that affect the **pharmacokinetic properties** (i.e., the transport, distribution, metabolism, and excretion of a molecule) are modified. An efficient and pragmatic strategy is important. For this, it is essential that not too many changes are made at the same time, and the changes should not be too biased. With little synthetic effort, a broad spectrum of physicochemical properties and spatial arrangements should be covered.

In the meantime, it has been established that binding to human **plasma proteins** such as serum albumin and the acidic κ_1-glycoprotein is of decisive importance for the **transport** and pharmacokinetic properties of a drug. Therefore, binding to these proteins is considered even in the early phase of drug development (Chap. 19). On the other hand, binding to the **hERG ion channels** (so-called "antitarget") is avoided because blocking these channels can lead to arrhythmias (Sect. 30.3). Drug metabolism is in itself a very important theme and must be considered in earlier phases of development. Cytochrome P450 en-

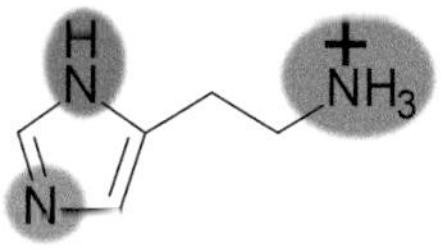

8.26 Histamine
(Positively charged form at pH = 7)

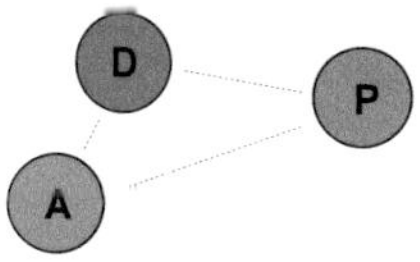

Pharmacophore

◻ **Fig. 8.9** The active substance histamine **8.26** and the corresponding pharmacophore assigned to it (*A* acceptor, *D* donor, *P* positively charged group) zymes are responsible for the vast majority of chemical transformations that occur on xenobiotics (Sect. 27.6). To be able to predict the behavior of drug candidates at this stage of the development process, the expected interactions with these metabolic enzymes are evaluated in an early phase of optimization. The expression of P450 enzymes can also be induced by xenobiotics. The trigger for this could be the binding to a transcription factor like the pregnane X receptor (PXR; Sect. 28.7). Drug candidates binding to this transcription factor can be evaluated early in their development to avoid this undesirable enhanced metabolism.

8.8 Optimizing Affinity, Enthalpy, and Entropy of Binding and Binding Kinetics

Generally, the **binding affinity** to a target protein is primarily improved during the course of optimization. If multiple candidates are available, the **ligand efficiency** (Sect. 7.1) in addition to the synthetic accessibility will lead the way. Small, potent lead structures offer legitimate hope that they can be well optimized. Very small compounds that have nanomolar affinity, despite their low molecular weight, can be problematic. Most of the time in such a molecule an optimal interaction pattern is already established. It is then almost impossible to transfer this pattern to another molecular scaffold. Medicinal chemists have established a **set of empirical rules** for standard functional group contributions to binding affinity (Sect. 4.10). According to these rules, it is possible to estimate how much a particular group, if correctly placed, will contribute to the binding affinity. It should be noted, however, that the application of these rules assumes additivity of the group contributions. However, this assumption is often too simplistic and cooperative effects determine the actual affinity gain to be achieved.

It was shown in Sect. 4.10 that the affinity is a combination of the enthalpic and entropic contributions. Usually one starts with a lead structure that has a binding affinity in the micromolar range. Expressed as the Gibbs free energy $\boldsymbol{\Delta G}$, this is usually about 30 kJ/mol. An increase in the binding affinity of 4–5 orders of magnitude causes an improvement in ΔG of another 20–30 kJ/mol. In order to optimize the most appropriate property of a lead structure, where and how should the screw be turned? Does it make more sense to improve the binding enthalpy, or is one better advised to improve the binding entropy? Given the enthalpy/entropy compensation described in Sect. 4.10, is it even possible to attempt optimization of both values independently? The prerequisite for using such a concept in the optimization is the determination of both values of a lead structure. Does this help in the choice of the right candidate for optimization? In the case that the thermodynamic binding profiles of multiple alternative lead candidates are known, should **enthalpically** or **entropically driven** binders be chosen for optimization? It is very interesting to compare the thermodynamic signatures of multiple generations of marketed products. The binding profiles for HIV protease inhibitors (Sect. 24.3) and HMG-CoA inhibitors (Sect. 27.3) are displayed in ◘ Fig. 8.10. Notably, it has been successful to shift the profile from initially strongly entropically driven binders to enthalpically driven ones. This observation suggests that it is initially simpler to optimize a substance's entropic binding contribution than its enthalpic contribution. Most of the time this

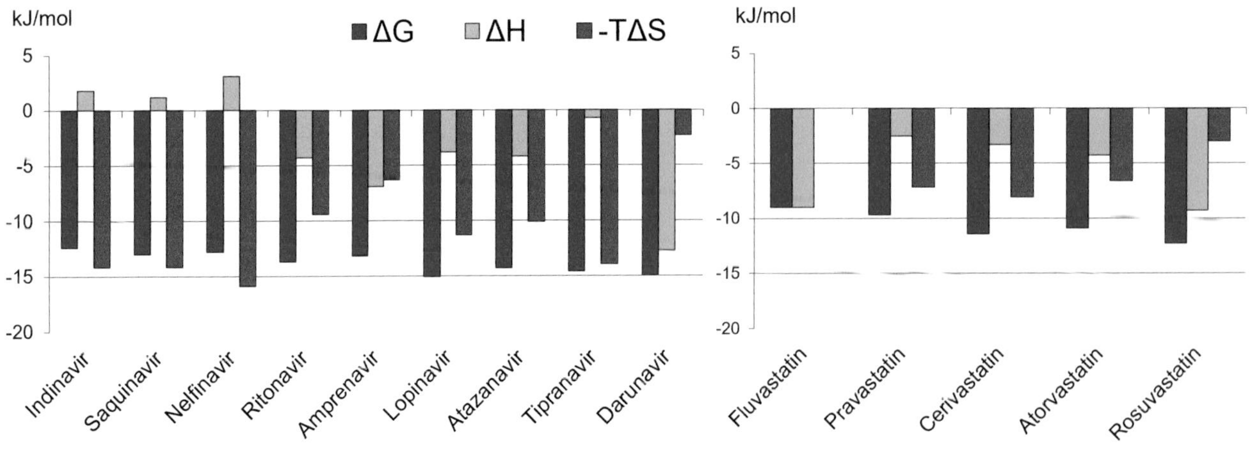

◘ **Fig. 8.10** Between 1995 and 2006, the profile of multiple development generations of HIV protease inhibitors (*left*; for formulas see ◘ Fig. 24.15) and statins as HMG-CoA inhibitors (*right*; for formulas see ◘ Fig. 27.13) could be optimized for their thermodynamic signatures, that is, the extent to which they are driven by entropy or enthalpy. The free energy ΔG is shown in *blue*, the enthalpy ΔH in *green*, and the entropic contribution $-T\Delta S$ in *red*. The more negative the column becomes, the stronger the binding affinity and the more the profile is determined by enthalpy or entropy. The initially developed compounds such as indinavir, saquinavir, nelfinavir, and pravastatin were predominantly entropic binders; in contrast, the newer derivatives such as darunavir or rosuvastatin have an improved enthalpic profile

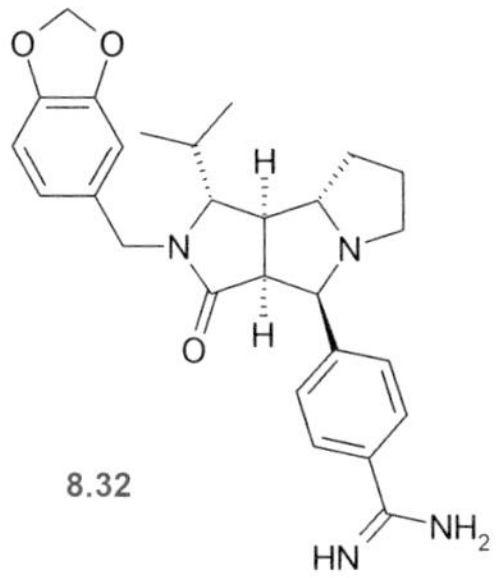

8.32

ΔG:	-42.3 kJ/mol
ΔH:	-6.2 kJ/mol
-TΔS:	-36.1 kJ/mol

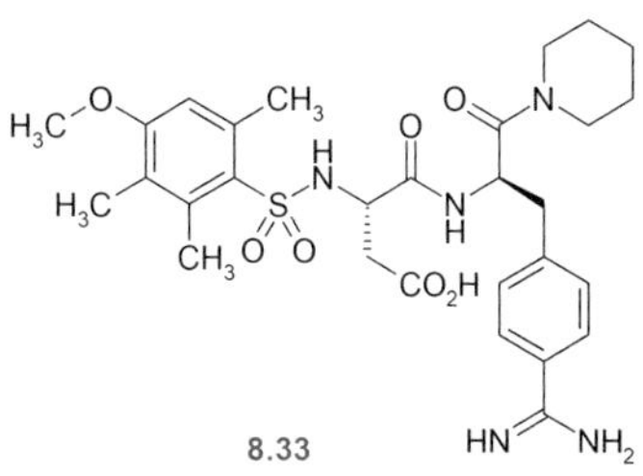

8.33

ΔG:	-49.2 kJ/mol
ΔH:	-48.5 kJ/mol
-TΔS:	-0.7 kJ/mol

Fig. 8.11 The rigid thrombin inhibitor **8.32** only exhibits a small number of rotatable bonds. It has an optimal shape complementarity to the binding pocket of thrombin. Its binding is, for the most part, entropically driven. On the other hand, the considerably more flexible ligand **8.33** has a higher enthalpic binding contribution

can be seen in the first lead structure upon which an enlargement of the hydrophobic surface area leads to better binding. The affinity that is gained is mostly explained by the displacement of ordered water molecules from the binding site (Sect. 4.6). Such contributions are assumed to be entropically favorable. A strategy of introducing rigid rings can also be pursued. In doing so, the compound loses degrees of freedom. If the geometry of the bound state is correctly frozen and if this state is also populated in solution prior to protein binding, the affinity will improve for entropic reasons. An example of this is the binding of the largely rigid thrombin inhibitor **8.32**, which binds in an almost exclusively entropically driven manner to the protein (Fig. 8.11). In contrast, the decidedly more flexible ligand **8.33** displays a large enthalpic binding contribution. Compound **8.32** represents the result of an optimization that led to a substance with single-digit nanomolar binding and an optimal shape complementarity for the binding pocket of thrombin.

As it seems, in general there are applicable concepts for the entropy-driven optimization. If one can "always win entropically," then for pragmatic reasons enthalpically favored lead structures should be preferred as a starting point for optimization.

However, caution is called for here. Why a ligand has a particular thermodynamic profile must be clarified first. The inhibitors **8.34** and **8.35** were discovered in a virtual screen as aldose reductase inhibitors (Fig. 8.12). The chemical structures of both ligands are very similar. Nevertheless, one is an enthalpically driven binder, and the other is an entropically driven binder. The crystal structures of both ligands with the protein delivered the reason: the enthalpically preferred inhibitor **8.34** traps a water molecule, which mediates binding between the ligand and the protein, whereas the other one does not. The incorporation of a water molecule is entropically disfavored, and therefore the profile appears to be that of an enthalpic binder. A resistance profile for inhibitors against mutants of the viral HIV protease was investigated in the research group of Ernesto Freire at Johns Hopkins University in Baltimore, USA (Sect. 24.5). Interestingly, the result was that resistance to the entropically favored inhibitors could be developed much faster than to inhibitors with enthalpic advantages. This observation indicates that it is worthwhile to concentrate on enthalpically favored binders in cases in which resistance can be expected to develop. In the investigated example, the enthalpically driven binder **8.33** has a less-rigid scaffold (Fig. 8.11). This would allow **8.33** to more easily elude changes that are caused by resistance mutations. It is much more difficult for rigid ligands that bind for entropic reasons to adapt to such steric modifications. On the other hand, an entropic binder may also experience an advantage in escaping resistance. If a ligand is entropically favored when binding in multiple binding modes, perhaps even with high residual mobility in the binding pocket, this may prove to be advantageous! If the protein tries to restrict the shape of its binding pocket against a given inhibitor by resistance mutation, a flexible ligand interacting with multiple binding modes will still find other orientations to continue to bind well (cf. example in Fig. 32.12). Again, it becomes apparent that such a concept can only be applied in a meaningful way if it is exactly understood why a certain binding profile prevails.

If it is clear that a lead structure is an enthalpically driven binder, and superimposed effects such as the entrapment of water molecules have not distorted the profile, how will the binding of an enthalpically driven binder be optimized? Let us remember from Sects. 4.5 and 4.8: hydrogen bonds, electrostatic interactions, and van der Waals contacts determine the contributions to binding enthalpy. However, a change in such an interaction property of a molecule is often coupled with a **compensation of enthalpy and entropy**. In the worst case scenario, it could well be that ΔG and, thus, binding affinity do not change at all! The optimization process can be compared to the challenge of repeatedly getting around the inherent enthalpy/entropy compensation. Enthalpically favorable hydrogen bonds should have an optimal geometry and should not induce severe structural changes in the protein environment. Otherwise this can lead to an entropic compensation by causing a shift in the degrees of freedom of the dynamic system. It seems to be more favorable to strengthen the hydrogen bonds in structurally rigid regions of the binding pocket. There, enthalpy is more likely to be gained because the compensatory shift in dynamic parameters is less likely. Furthermore, introduced hydrogen bonds should not reduce the degree of desolvation of a bound ligand by exposing a portion of its hydrophobic groups to the surrounding solvent as a result of small distortions in the binding geometry. It is also important that the local water struc-

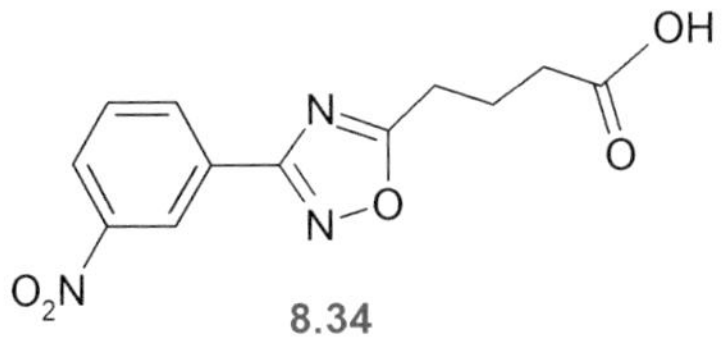

8.34

ΔG: -35.4 kJ/mol
ΔH: -25.6 kJ/mol
-TΔS: -9.8 kJ/mol

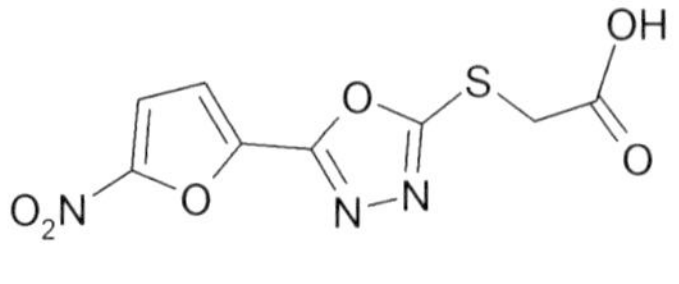

8.35

ΔG: -31.3 kJ/mol
ΔH: -8.7 kJ/mol
-TΔS: -22.6 kJ/mol

Fig. 8.12 Compounds **8.34** and **8.35** were discovered in a virtual screening as lead structure for the inhibition of aldose reductase. Although they are structurally similar, **8.34** is a stronger enthalpic binder and **8.35** is a more entropic binder. The subsequent crystal structure analyses of the complexes with the reductase showed that **8.34** traps a water molecule upon binding, whereas this was not observed with **8.35**. Because the entrapment of a water molecule is entropically unfavorable, the binding of **8.34** is overall enthalpically preferred

ture in the binding pocket is not unnecessarily modified to a large extent.

Another essential question concerns the **binding kinetics** that an optimal ligand should have. This aspect can only be answered with regard to the properties of the target structure to be modulated by a drug. In principle, it must be taken into account that the lifetimes of various proteins in humans vary considerably. The rate of resynthesis after degradation of a protein varies greatly from protein to protein. This is even further complicated by the fact that between individuals of our species, even the resynthesis rates of the same protein can vary.

The binding kinetics determine how quickly or slowly a drug binds to its target, how long it remains there in a bound state (so-called **residence time** RT = $1/k_{off}$) and how quickly it dissociates from there again. A long residence time can be advantageous for therapeutic interventions where prolonged drug binding to the target protein is desired. It guarantees that binding persists over a longer period of time and, thus, ensures a more sustained pharmacological effect. It can even extend beyond the time window in which, due to much faster pharmacokinetics, the local drug concentration has already decreased below the K_d for the target, since the drug is still effective as it remains in the bound state. For example, among the antihypertensive sartans (Sect. 29.5, Fig. 29.7), candesartan (**29.25**), which has a 30-fold longer residence time, appears to have a clinically proven better protective effect than losartan (**29.20**), which has a shorter residence time.

Conversely, if there is a risk of drug-related toxicity, it may be desirable not to set the target binding too long. On the one hand, the residence time should be maximized to achieve the therapeutic effect as efficiently as possible, while on the other hand, sufficiently rapid elimination must minimize unwanted side effects and off-target binding. If the latter aspect does not play a decisive role, as will be the case with most anti-infectives, the drug should hit the target as hard and as long as possible (e.g., by irreversible inhibition), because then a lower dose may be sufficient for efficacy. In contrast, if there is a risk of target-based toxicity, the mechanism of action should be tailored so that toxicity is minimized but efficacy is still maintained. In this case, faster dissociation of the

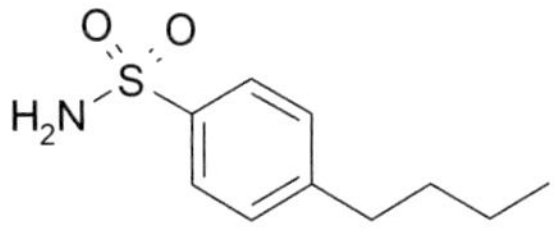

8.36

ΔG: -42.6 kJ/mol
ΔH: -44.2 kJ/mol
-TΔS: 1.6 kJ/mol
$k_{on} = 1.46 \cdot 10^5\ M^{-1}\ s^{-1}$
$k_{off} = 0.0491\ s^{-1}$

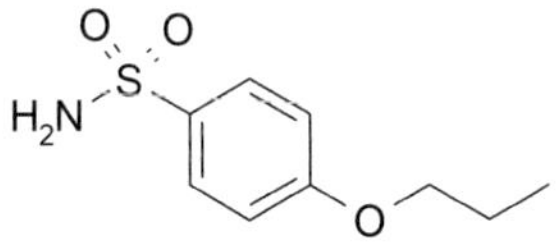

8.37

ΔG: -42.6 kJ/mol
ΔH: -49.8 kJ/mol
-TΔS: 7.2 kJ/mol
$k_{on} = 6.0 \cdot 10^4\ M^{-1}\ s^{-1}$
$k_{off} = 0.0221\ s^{-1}$

Fig. 8.13 The chemically similar inhibitors **8.36** and **8.37** bind with the same Gibbs free energy to carbonic anhydrase II, the second one binding slightly more enthalpically, and the first slightly more entropically. In terms of kinetics, **8.36** binds more rapidly to the enzyme, whereas **8.37** dissociates more slowly

drug may be advantageous (compare COX inhibitors (Sect. 27.9, Fig. 27.40) acetylsalicylic acid (**27.94**), which is an irreversible binder, or ibuprofen (**27.95**) and indomethacin (**27.98**), which are both reversible inhibitors with rapid dissociation kinetics).

Such optimized efficacy adapted to the therapeutic indication cannot be achieved by considering affinity alone. A rapid deactivation of a target by accelerated on kinetics can also be essential, e.g., in acute medicine. In Sect. 7.7, methods for determining binding kinetics have been presented. In the case of 1:1 binding, the thermodynamic equilibrium quantity "binding affinity" is expressed by the relative ratio of the binding kinetic dissociation rate (k_{off}) divided by the association rate (k_{on}). When discussing kinetic phenomena, however, it should be borne in mind that the binding of a ligand to its protein usually does not occur in a single step, but instead proceeds via many intermediate steps. However, only one of these steps on the way to the binding pocket is rate-determining, and in most cases, it is not the same step for the association as for the dissociation. We are still far from having conclusive concepts for establishing detailed structure–kinetics relationships. Preliminary evidence suggests that electrostatics or conformational adaptations of the ligand or the protein are important for the kinetics. Also, how quickly the local concentration of a ligand increases at the site of the target protein can determine the association rate. Furthermore, the dehydration of the previously uncomplexed binding pocket or the shedding of the ligand's solvation shell may be important. Conversely, the rewetting of functional groups of a ligand with water molecules can become a time-determining step for dissociation.

It has been observed that structurally highly similar ligands may well have strongly divergent kinetic binding profiles. ◘ Fig. 8.13 shows two carbonic anhydrase inhibitors with the same affinity, **8.36** and **8.37**, which have different on- and off-kinetics despite similar chain lengths of their substituents. **8.36** binds faster than **8.37** to human carbonic anhydrase, whereas **8.37** dissociates again more slowly.

Helena Danielson's group in Uppsala, Sweden, showed that therapeutically used HIV protease inhibitors have different binding profiles with respect to the development of resistance to mutated variants of the protease. It became clear that increased resistance to an active compound occurs when it is found to dissociate at an increased rate. This is a crucial criterion for guiding drug optimization in the desired direction.

8.9 Synopsis

- A lead structure is only the starting point on the way to a drug; potency, specificity, and duration of action have to be optimized concurrently to minimize side effects and toxicity.
- The structure of an active substance is determined by its pharmacophore, which is responsible for target binding. Its adhesion groups enhance potency and biological activity, its lipophilicity is responsible for transport and distribution, and groups to be cleaved or modified release the active form.
- Multiple concepts to modify the chemical structure of a lead can be planned; however, optimization is multifactorial due to highly correlated influences of the attempted changes.
- Bioisosteric functional group replacement attempts the exchange of groups on a given scaffold for sterically and electronically related groups that maintain activity but improve other drug properties.
- Me-too research follows the goal of modifying the competitor's lead structures to arrive at patent-free analogues with improved properties.
- Assuming unchanged activity, enlarging a molecule, adding chiral centers, and rigidification usually improve selectivity, whereas removing chiral centers, allowing more flexibility, and reducing the size make a drug less selective.
- The activity spectrum of a substance can be tailored even by the smallest structural changes that modulate affinity, transportation, distribution, or metabolism. Therefore, a particular compound class can show activity in very different therapeutic indications.
- Transforming agonists to antagonists does not follow clear-cut rules; however, increasing the size and the attachment of hydrophobic groups such as aromatic rings often shift the profile.
- The more polar a drug, the more poorly it can penetrate lipid membranes, and the lower the absorption is. On the other hand, special transporters can assist penetration.
- Extension of the duration of action is mostly achieved by replacement of metabolically labile groups with more stable isosteres, the introduction of more branching groups, blockage of metabolically labile positions at aromatic rings by F or Cl, or by exchanging L- for D-amino acids concurrently with the inversion of amide groups.
- Molecular databases can be screened to detect other scaffolds or substitution patterns that represent a given pharmacophore in an alternative fashion.
- In the early phase of drug development, undesired binding to plasma proteins, antitargets such as the hERG ion channel or preferred binding, inhibition, or activation of transcription factors or metabolizing cytochrome P450 enzymes are examined and possibly avoided.
- Proper adjustment of the thermodynamic binding profile can be essential for the optimization of binding affinity and to endow a drug with the required target-specific properties. Similarly, the interaction kinetics determining binding on and off rates or residence times are of decisive importance to develop drugs with, for example, an optimal resistance profile.

Bibliography and Further Reading

General Literature

W. Sneader, Drug Discovery: The Evolution of Modern Medicines, John Wiley & Sons, New York (1985)

J. B. Taylor and D. J. Triggle, Eds., Comprehensive Medicinal Chemistry II, Elsevier, Oxford (2007)

C. G. Wermuth, Ed., "The Practice of Medicinal Chemistry", 3rd Edition, Elsevier-Academic Press, New York (2008)

M. A. M. Subbaiah, N. A. Meanwell, Bioisosteres of the Phenyl Ring: Recent Strategic Applications in Lead Optimization and Drug Design, J. Med. Chem., **64**, 14046–14128 (2021)

Special Literature

C. Hansch, Bioisosterism, Intra-Science Chem. Rept., **8**, 17–25 (1974)

C. W. Thornber, Isosterism and Molecular Modification in Drug Design, Chem. Soc. Rev., **8**, 563–580 (1979)

C. A. Lipinski, Bioisosterism in Drug Design, Ann. Rep. Med. Chem., **21**, 283–291 (1986)

A. Burger, Isosterism and Bioisosterism in Drug Design, Fortschr. Arzneimittelforsch., **37**, 287–371 (1991)

H. Steuber, A. Heine and G. Klebe, Structural and Thermodynamic Study on Aldose Reductase: Nitro-substituted Inhibitors with Strong Enthalpic Binding Contribution, J. Mol. Biol., **368**, 618–638 (2007)

H. Ohtaka and E. Freire, Adaptive Inhibitors of the HIV-1 Protease, Prog. Biophys. and Mol. Biol., **88**, 193–208 (2005)

G. Klebe, Broad-scale analysis of thermodynamic signatures in medicinal chemistry: are enthalpy-favored binders the better development option? Drug Discov. Today, **24**, 943–948 (2019)

C. F. Shuman, P-O. Markgren, M. Hämäläinen, U. H. Danielson, Elucidation of HIV-1 Protease Resistance by Characterization of Interaction Kinetics between Inhibitors and Enzyme Variants, Antiviral Research, **58**, 235–242 (2003)

R. A. Copeland, D. L. Pompliano and T. D. Meek, Drug–target Residence Time and its Implications for Lead Optimization, Nat. Rev. Drug Discov., **5**, 730–740 (2006)

G. Klebe, The Use of Thermodynamic and Kinetic Data in Drug Discovery: Decisive Insight or Increasing the Puzzlement? ChemMedChem, **10**, 229–231 (2015)

D. C. Swinney, Opportunities to minimise risk in drug discovery and development, Expert Opin. Drug Discov., **1**, 627–633 (2006)

S. Glöckner, K. Ngo et al, Conformational changes in alkyl chains determine the thermodynamic and kinetic binding profiles of Carbonic Anhydrase Inhibitors, ACS Chem. Biol., **15**, 675–685 (2020)

A. Alsadhan, J. Cheung, et al, Variable Bruton Tyrosine Kinase (BTK) Resynthesis across Patients with Chronic Lymphocytic Leukemia (CLL) on Acalabrutinib Therapy Affect Target Occupancy and Reactivation of B-Cell Receptor (BCR) Signaling, Blood, **132**, 4401–4401 (2018)

Designing Prodrugs

Contents

G. Klebe, *Drug Design,* https://doi.org/10.1007/978-3-662-68998-1_9

After optimization of a lead structure, there may still be problems. Many substances lack important characteristics that are required for therapy in humans, for instance, adequate bioavailability, duration of action and metabolic stability, the ability to penetrate the blood–brain barrier, selectivity, or good tolerability. Often it proves impossible to address or improve these properties through structural variations. A solution to this problem can be found through special preparations, for instance, to be used for poorly water soluble substances, or via a derivatization to a **prodrug**. This term refers to a nonactive or poorly active precursor or derivative of an active molecule. In the organism, this form is converted to the actual active substance. In most cases, this is achieved by enzymatic reactions; in a few cases, it happens by spontaneous chemical decomposition.

Aside from this, the metabolites of some drugs also show favorable therapeutic properties. In some cases, this has led to new and improved drugs; in other cases, the original substance was retained as a prodrug.

9.1 Foundations of Drug Metabolism

Multiple factors are crucial for the absorption, bioavailability, and duration of action of an active substance. The most important are the solubility and lipophilicity of the drug, which are of nearly equal importance, followed by the molecular size and the metabolic stability. The terms absorption and bioavailability have very different meanings. **Absorption** refers to the amount of active substance that is taken up by the entire gastrointestinal tract. The **bioavailability** refers to just the portion of the active substance that is available in the circulation after the first pass through the liver.

After oral administration, the metabolism of the substance by enzymes begins. Ester and amide bonds are hydrolyzed, often already in the stomach and intestines, or by passage through the stomach and intestinal wall. The entire blood volume that flows through the intestines goes first to the liver via the portal vein (◘ Fig. 9.1). This passage is called "first pass." Because of its rich spectrum of hydrolyzing, oxidizing, reducing, and conjugating enzymes, the liver is the main site of drug degradation, that is, **metabolism**. A drug can have poor bioavailability despite good absorption because of fast and pronounced metabolism in the liver. For many substances, the first pass is already "the end of the road" They are well absorbed, but are immediately metabolized or excreted in the bile. The "**first-pass effect**" refers to cases of successful and extensive metabolism in the very first passage. Lipophilic active substances and those with a molecular weight of more than 500–600 daltons (Da) are susceptible to particularly intense first-pass effects. Of course, blood flows continuously through the liver, and metabolism carries on. The substances are no

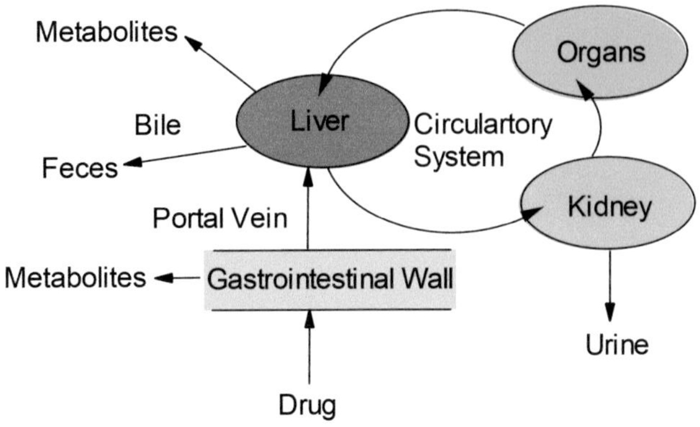

◘ **Fig. 9.1** Schematic sketch of the "life cycle" of a drug after oral administration. The drug is already metabolized during the passage through the stomach or intestinal wall, and above all, during the first pass through the liver. Lipophilic drugs and substances with a molecular weight of more than 500–600 Da are excreted with the bile. Polar substances and conjugated and/or metabolic products (metabolites) are excreted by the kidneys

longer in the blood stream at as high a concentration as they were before the first liver passage because they have been distributed to the tissue. In general, the hydrolytic cleavage of ester or amide groups leads to **highly water soluble metabolites** that can be excreted by the kidneys. **Conjugation**, that is, the coupling of the substance with native polar compounds, for instance, with sulfate groups, the amino acid glycine, or the glucose oxidation product glucuronic acid, leads to easily excreted products. In humans, conjugation is of great importance. It is more critical if the substance has neither easily degradable functional groups nor conjugation positions. Nonetheless, humans have enzymes that can metabolize xenobiotics. Among these, the cytochrome P450 isoenzymes are particularly important because they are able to chemically change a molecule oxidatively at various positions (Sect. 27.6). Usually this leads to better water solubility and therefore better-excretable substances. Since it is not possible to predict what properties the metabolites of these biotransformations will have, toxic compounds with mutagenic or carcinogenic properties may occasionally be formed.

Evolution has had time over millions of years to hone the degradation and excretion of foreign substances. For some compounds, however, the system can fail. Instead of detoxifying, the opposite happens, i.e., "poisoning." The carcinogenic effect of polycyclic hydrocarbons is attributed to an oxidative assault, just as are bone marrow damage and leukemia which are caused by benzene **9.1**. The simplest alkyl homologue of benzene, toluene **9.2** is less toxic for the sole reason that it can be oxidized to benzoic acid **9.3**, which, after conjugation with the amino acid glycine, can be excreted as hippuric acid **9.4** (◘ Fig. 9.2). There are even more conjugation possibilities available for the benzoic acid intermediate.

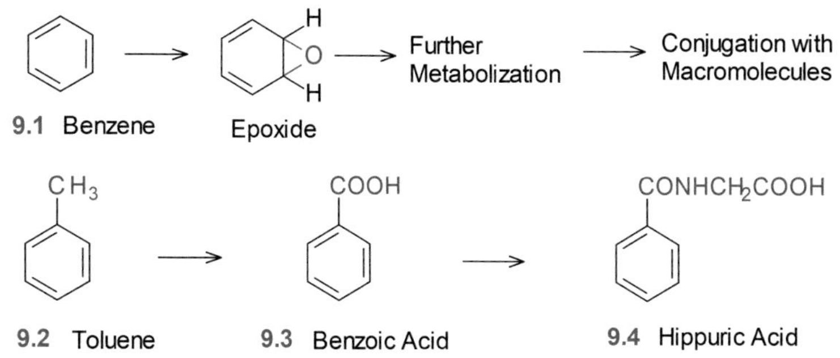

Fig. 9.2 The oxidation of benzene **9.1** leads to a reactive and toxic intermediate. In contrast, the oxidation of toluene **9.2** affords benzoic acid **9.3**, which can be excreted by the kidney as its nontoxic glycine conjugate **9.4**

One can speculate as to why no multienzyme complexes have evolved to immediately convert toxic intermediates into polar, nontoxic metabolites. In any case, it is an almost unsolvable problem because the properties of the metabolites would have to be predicted for each xenobiotic. A modification that leads to improved water solubility in one compound can cause a mutagenic effect in another. For their own protection, humans have, in fact, mechanisms for trapping reactive metabolites. Here glutathione and glutathione transferase must be mentioned because they detoxify electrophiles particularly well (Sect. 27.7). Perhaps toxic or carcinogenic effects were not a particularly decisive theme for evolution until now. Tumors play a secondary role for most animals because of their short lifespan. Up until just a few generations ago, war and infectious diseases were the primary causes of death in humans. It has only been in recent times that the average life expectancy has increased. In the sense of evolution, aging individuals play only a secondary role. Once reproduction is complete, the parents are only necessary for the care of their young until early adulthood. One only needs to think of female spiders that consider their mates to be nothing more than their next prey immediately after copulation!

From the above-described examples of toxic chemicals, the wrong conclusion should not be drawn that only human-made substances can cause cancer. This is true for a few natural products as well, for instance, aflatoxins. These microbial secondary metabolites, which form in spoiled nuts and other foodstuffs, are potent carcinogens. Certain alkaloids, for example, from the Spurge family (*Euphorbiaceae*) are also strongly cancer-promoting substances; they are so-called tumor promoters.

The principle of *nil nocere* (*Lat.* do not harm) is strictly applied to medicines, and only slowly have these standards been applied to other materials in our environment. For the testing and development of active compounds, this means that particularly rigorous tests for carcinogenic, mutagenic, and teratogenic effects must be conducted. The well-founded suspicion alone that a compound or one of its possible metabolites displays such effects leads to the consequence that the compound is not further developed.

9.2 Esters Are Ideal Prodrugs

Establishing sufficient water solubility in substances that are simultaneously suitable for passive transport across membranes is a special challenge in pharmaceutical optimization. Nowadays attention is paid to the correct balance of these parameters already in the early phase of development (Chap. 19). If it is not possible to achieve this optimum with the actual active substance, esters will often be produced as suitable prodrugs. Esters are easily cleaved by ubiquitously occurring esterases. The improved lipophilicity helps with the passive transport through **diffusion over membrane barriers**, as found in the intestines and, above all, the **blood–brain barrier**. One prodrug that has sadly achieved infamy is heroin **9.5** (Fig. 9.3), the diacetyl ester of morphine (Sect. 3.3). Because of its markedly increased lipophilicity, heroin penetrates the blood–brain barrier quickly. The pharmacologist Heinrich Dreser, who tested acetylsalicylic acid at Bayer, introduced heroin to therapy in 1898 as a pain and cough medicine because of its minimal respiratory depression. But heroin belongs to the substances with the highest addictive potential. Its abuse is an enormous social problem in many countries. It is used therapeutically in exceptional cases, for instance, for pain therapy in cancer patients, particularly those, who have exhausted other therapeutic options.

Many other prodrugs are also esters. The transformation from an acid or alcohol group to an ester usually leads to a better-absorbable product. The formerly used antilipidemic clofibrate **9.6** (Sect. 28.6) is just such an example of a bioavailable ester of a biologically active free acid **9.7**. The angiotensin-converting enzyme inhibitor enalapril **9.8** (Sect. 25.4) and its analogues are also prodrugs. The free acid **9.9** is not absorbed, but it is the active form *in vitro* (Fig. 9.3). The diester is chemically unstable and quickly forms the inactive diketopiperazine **9.10**. It is essential that only one of the acid groups is esterified to prevent the formation of this side product. The monoester **9.8** is "interpreted" as a dipeptide and is transported through the cell membrane by an oligopeptide transporter. The β-lactam antibiotics (Sect. 23.7) are also taken up by this transporter.

Hydroxymethylglutaryl coenzyme A **9.11** (HMG-CoA) is enzymatically reduced to mevalonic acid **9.12** in the biosynthesis of cholesterol (◘ Fig. 9.4). The antilipidemic lovastatin **9.13** (Sect. 27.3) prevents this reaction by inhibiting HMG-CoA reductase. It contains a lactone ring, which is transformed to its active form **9.14** by hydrolysis. This form is structurally very similar to the product of the enzymatic reaction, mevalonic acid **9.12**.

Other ester prodrugs were developed for depot formulations to achieve a longer duration of action after subcutaneous or intramuscular administration.

The phenolic hydroxyl group of bambuterol **9.15** is masked as a carbamate. Terbutaline **9.16** (◘ Fig. 9.5) is formed from this prodrug after hydrolysis by unspecific cholinesterases (Sect. 23.7). By using this prodrug strategy, it was possible to make a long-acting bronchospasmolytic that only needs to be administered once daily in contrast to the actual active substance, which must be administered three times daily.

Occasionally, a prodrug can be used to improve the taste, for instance, in the case of the extremely bitter chloramphenicol **9.17**. By converting it to the palmitate **9.18** (◘ Fig. 9.5), there is a marked reduction in the water solubility, but the substance no longer tastes bitter. The concomitant reduction in the absorption is of no consequence. The substance is hydrolyzed to the highly soluble and easily absorbed chloramphenicol in the duodenum by the pancreatic lipase enzymes.

The glucoside salicin (Sect. 3.1) represents a true prodrug that after hydrolysis and oxidation is converted to the anti-inflammatory salicylic acid. In contrast, acetylsalicylic acid (ASA) is a mixed type, as part of it reacts covalently with the enzyme and the remaining part occupies the active site non-covalently. It has its own activity through the irreversible inhibition of cyclooxygenase, above all as a coagulation-inhibiting substance. On the other hand, ASA has a prodrug character because the metabolic release of salicylic acid contributes a small part to the anti-inflammatory effect (Sect. 27.9). Furthermore, ASA is less irritating to the mucous membranes and tastes less unpleasant than salicylic acid. For a drug with a molecular weight of 180 Da, this combination of favorable characteristic in one structure is a proud achievement.

Esterification can also help with inadequate water solubility of an active substance. For this, esters with phosphoric acid or hemiesters with dicarboxylic acids such as succinic acid are formed. The added groups carry a charge and increase the water solubility of the active substance. In the organism, the esters are easily hydrolyzed again. The anticonvulsive compound phenytoin could be converted to a more hydrophilic phosphate prodrug **9.19** (◘ Fig. 9.5), which is easily hydrolyzed by phosphatases (Sect. 26.7). If a terminal sulfonamide group, as found in the prodrug of celecoxib (**9.20**, **9.21** ◘ Fig. 9.5), is acylated, water soluble salts are more easily formed. The acyl group is also easily hydrolyzed in the intestines.

Esterification with polyethylene glycol (PEG) can also be used to enhance solubility. This very water soluble polymer has been coupled through an ester group to the natural product paclitaxel (Sect. 6.2, 6.5). As PEG–paclitaxel, this compound can be used as an intravenous chemotherapeutic.

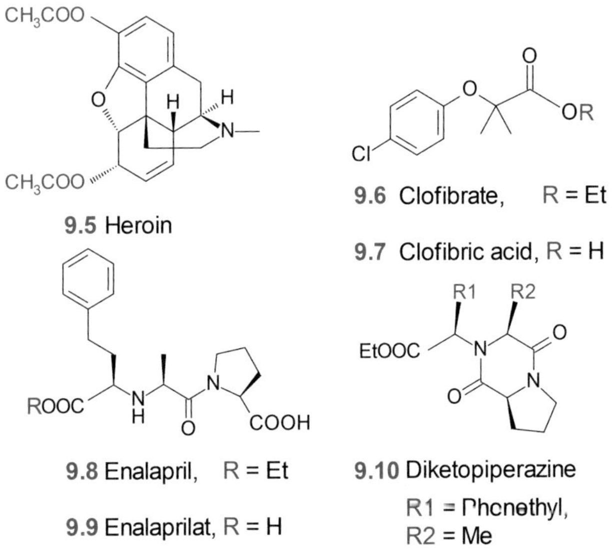

◘ **Fig. 9.3** Heroin **9.5**, the diacetyl derivative of morphine acts reliably and quickly, "heroically." Like morphine, it is slowly and inefficiently absorbed, but after intravenous administration it crosses the blood–brain barrier 100 times faster than morphine. There, the ester is converted by the enzyme pseudocholinesterase to morphine, which can no longer leave the brain because of its higher polarity. The cholesterol-lowering drug clofibrate **9.6** is a prodrug of the actual active compound, the free acid **9.7**. The antihypertensive enalapril **9.8** is also a prodrug of the active compound **9.9**. Here, the high lipophilicity is not responsible nor the absorption, rather it is actively transported by binding to a dipeptide transporter. The diester of enalapril is unsuitable as a drug because it spontaneously forms the inactive diketopiperazine **9.10**

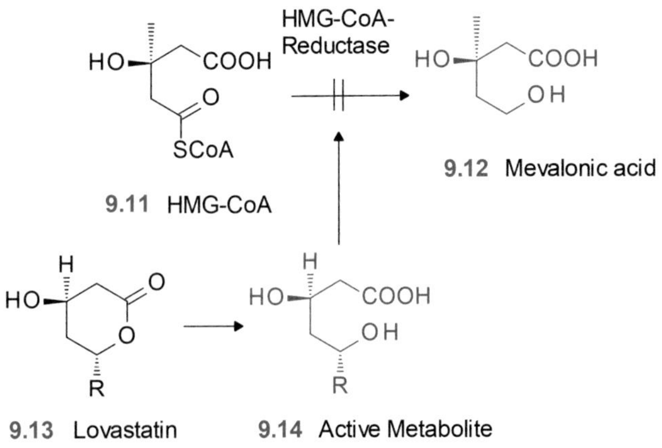

◘ **Fig. 9.4** The enzymatic reduction of hydroxymethylglutaryl coenzyme A **9.11** (HMG-CoA) to mevalonic acid **9.12** is inhibited by the lactone-ring-opened active metabolite **9.14** of lovastatin **9.13** (Sect. 27.3)

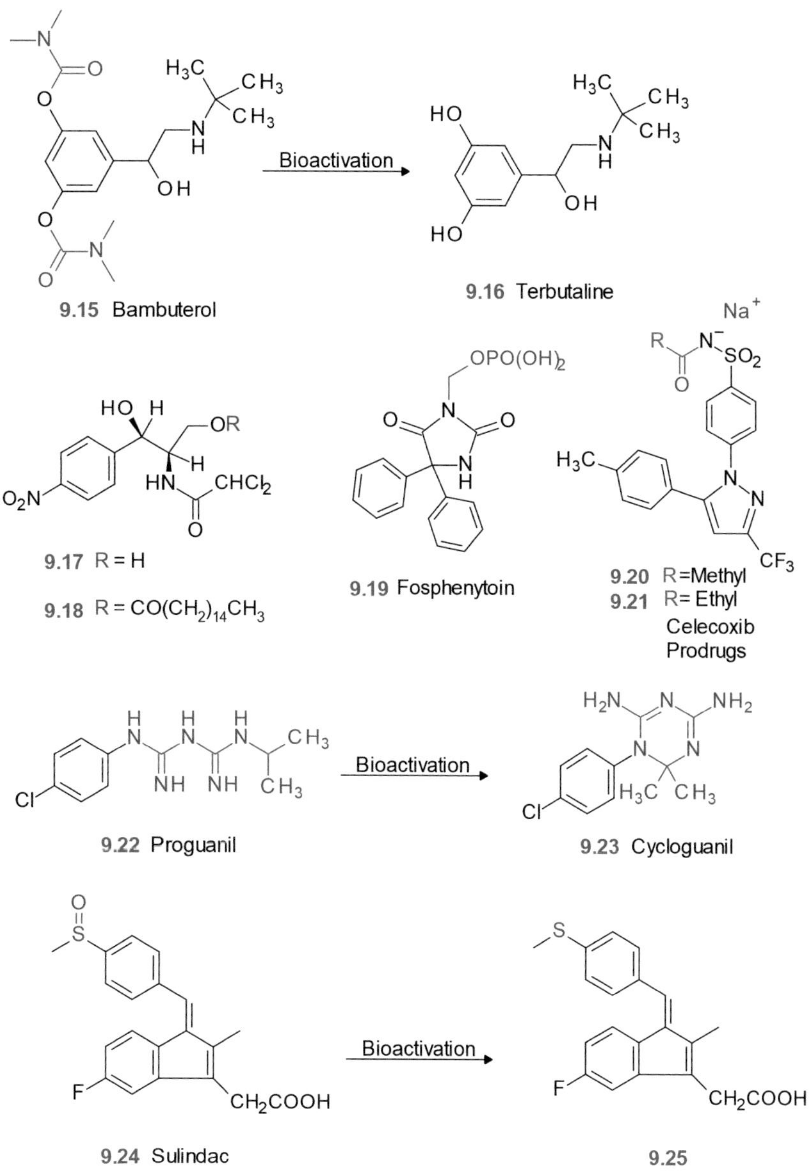

Fig. 9.5 Bambuterol **9.15** is a carbamate-masked prodrug of the bronchospasmolytic terbutaline **9.16**. It is transformed to the active compound slowly via hydrolysis. The prodrug **9.18** of chloramphenicol **9.17** masks only its extremely bitter taste. Phenytoin can be converted to a phosphoric acid ester **9.19**, which is significantly more water soluble. The cyclooxygenase inhibitor celecoxib can be converted to prodrugs (**9.20** and **9.21**) by adding acyl groups; these have markedly improved water solubility. The antimalarial cycloguanil **9.23** is formed by a metabolic cyclization of the inactive precursor proguanil **9.22**. The water solubility of the anti-inflammatory sulindac **9.24** is 100 times greater than its actual active form, the sulfide **9.25**. In addition to this reversible enzymatic reduction, an irreversible enzymatic oxidation to a biologically inactive sulfone also occurs

9.3 Chemically Well Wrapped: Multiple Prodrug Strategies

The antibacterial sulfonamide sulfamidochrysoidine (Sect. 2.3) is a prodrug. It is only after cleavage of the azo bond that the metabolic product, sulfanilamide, acts as an antimetabolite of *p*-aminobenzoic acid, which is critical for microorganisms. Additional prodrugs are proguanil **9.22**, which is converted to cycloguanil **9.23** (Sect. 27.2), and the anti-inflammatory sulindac **9.24**, which is metabolically converted to the active sulfide **9.25** (Fig. 9.5).

Amidines are used as building blocks in thrombin inhibitors and antagonists of the integrin receptor $\alpha_{IIb}\beta_3$ (Sect. 31.2). These strongly basic groups are detrimental for good bioavailability. Through oxidation to the corresponding amidoximes, a less basic group is formed that is not protonated under physiological conditions. Reductases, which are present in the liver, kidney, lung, and brain, release the original amidine structure. This concept, together with the esterification of the terminal acid function, was applied in a double-prodrug strategy for the thrombin inhibitor ximelagatran **9.26** (Sect. 23.4) and the receptor antagonist sibrafiban **9.27** (Sect. 31.3, Fig. 9.6).

The bombing of an allied ship that was docked in an Italian harbor in 1943 with 100 tonnes of mustard gas **9.28** (bis-*β*-chloroethylsulfide, Fig. 9.7) led to the observation that many of those who were poisoned experienced a severe reduction in their white blood cell counts. This severe toxicity for cells that quickly divide could be used for killing tumor cells. The cytotoxic effect arises from multiple alkylations of DNA. Consequently, replication and subsequent cell division are affected. A purposeful search for analogues of mustard gas with less

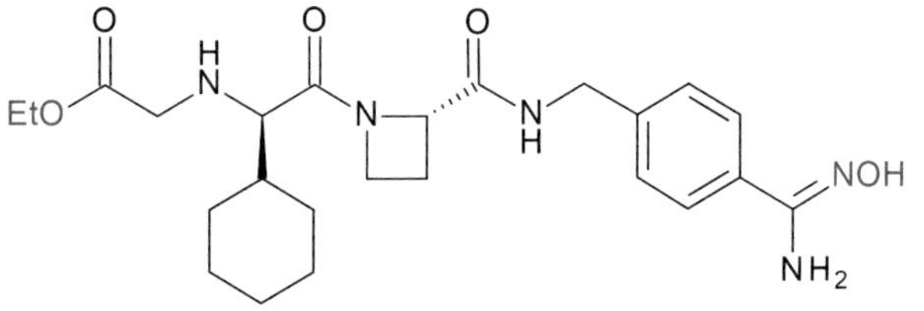

9.26 Ximelagatran

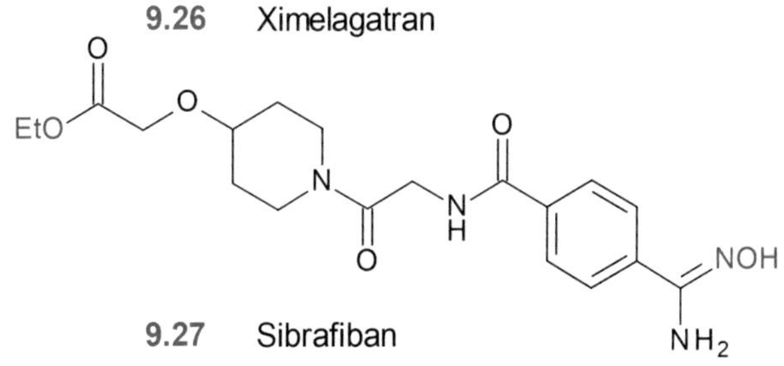

9.27 Sibrafiban

Fig. 9.6 Ximelagatran **9.26** and sibrafiban **9.27** were developed to improve oral bioavailability and contain both an uncharged amidoxime group and an ester function as a double prodrug

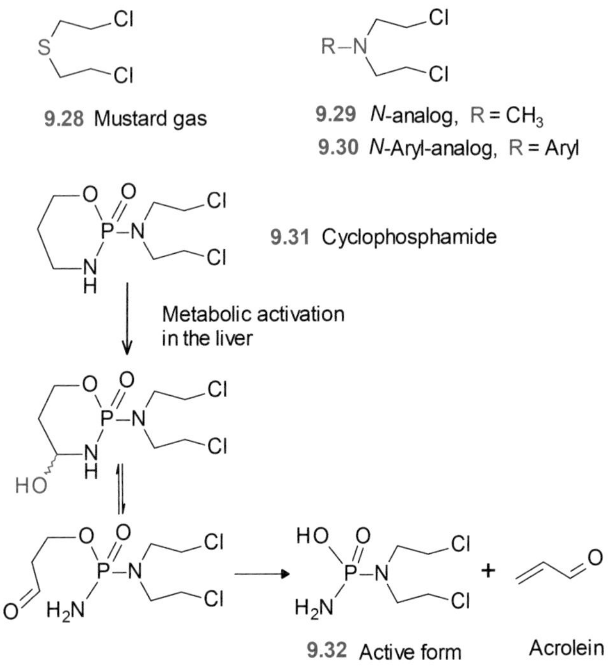

Fig. 9.7 The cytostatic *N*-methyl and *N*-aryl compounds **9.29** and **9.30** are derived from mustard gas **9.28**. The first step in the activation of the prodrug cyclophosphamide **9.31** is a metabolic hydroxylation of the carbon next to the nitrogen atom. The biologically active agent **9.32** and the toxic side product acrolein come from a labile intermediate that is formed by enzymatic degradation and spontaneous decomposition

toxicity led via *N*-derivative **9.29** to the aromatic-substituted derivative **9.30**, which still had inadequate tolerability and tumor specificity. Tumor cells are especially rich in phosphatases. Because of this, H. Arnold at the German company Chemie Grünenthal reasoned that phosphoric acid derviatives (**9.29** or **9.30**) of *N*-lost (mustard gas) might be suitable for a tumor-specific therapy. The most interesting compound was cyclophosphamide **9.31**, a substance that can cause the complete disappearance of tumors in animal experiments. The originally assumed mechanism was not correct because the substance is inactive *in vitro* in cell cultures of tumors. The metabolic activation occurs outside the tumor in the liver through oxidation (Fig. 9.7).

In the case of the cancer therapeutic 5-fluorouracil **9.33**, activation occurs through tumor-specific enzymes. The triple-prodrug capecitabin **9.34** is initially activated to **9.35** by a carboxylesterase in the liver (Fig. 9.8). Then cytidine deaminase cleaves an amino group to give **9.36** in the liver as well as in the tumor. Finally, thymidine phosphorylase releases the active substance **9.33** in the tumor cell. There, the compound unleashes its effect by blocking thymidylate synthase, an enzyme that plays an important role in thymine biosynthesis (Sect. 27.2) in that it delivers building blocks for DNA synthesis. Because cancer cells divide more quickly than healthy cells, they are more dependent on the activity of thymidylate synthase.

9.4 L-DOPA Therapy: A Clever Prodrug Concept

The neurotransmitters dopamine and acetylcholine fulfill different tasks in particular parts of the central nervous system. Parkinson's disease, also called "shaking palsy," is a result of the degeneration of dopamine-producing cells in the *Substantia nigra* in the midbrain. The ensuing disproportion between the dopaminergic and cholinergic nerve impulses leads to episodic chronic movement disorders such as rigidity, tremor, shaking, and an inability to move normally. Similar side effects are caused by substances that block the dopamine receptors, for instance, the tricyclic neuroleptics (Sect. 1.6). Intravenous administration of dopamine **9.37** (Fig. 9.9) does not lead to the desired effect because the substance cannot penetrate the blood–brain barrier. Because of its purely peripheral effect, undesirable cardiovascular and circulatory side effects are observed, for example, an increase in heart rate and blood pressure.

The desired equilibrium in the brain should also be established by suppressing the cholinergic system. This route is also taken by giving anticholinergics, that is, antagonists to the cholinergic receptors. The administration of the amino acid L-DOPA **9.38** (Fig. 9.9) is a more elegant possibility for dopamine substitutions. This metabolic precursor of dopamine is an orally bioavailable, central nervous system (CNS)-effective medicine. It is even more polar than dopamine and can neither be absorbed from the gastrointestinal tract nor can it cross the blood–brain barrier just by passive diffusion. Because it is an amino acid, it uses an amino acid transporter.

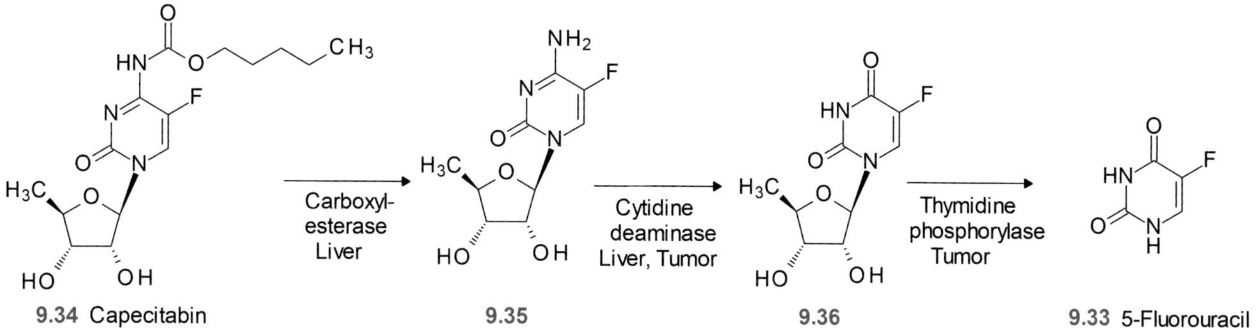

Fig. 9.8 The triple-prodrug capecitabin **9.34** is activated to **9.35** by a carboxylesterase in the liver, then it is transformed into **9.36** by a cytidine deaminase in the tumor, and a thymidine phosphorylase produces the cancer therapeutic 5-fluorouracil **9.33**

With this, the first goal, CNS activity, is achieved. Oral L-DOPA administration however, still presents too many side effects in the peripheral nervous system. Furthermore, L-DOPA is very short acting as dopamine is quickly metabolized in the brain. Therefore, one must try to prevent the metabolism of the substance, while simultaneously reducing its concentration in the periphery. The combination of L-DOPA with the peripheral decarboxylase inhibitor benserazide **9.39** and the CNS-effective monoamine oxidase inhibitor selegilin **9.40** (Sect. 27.8)

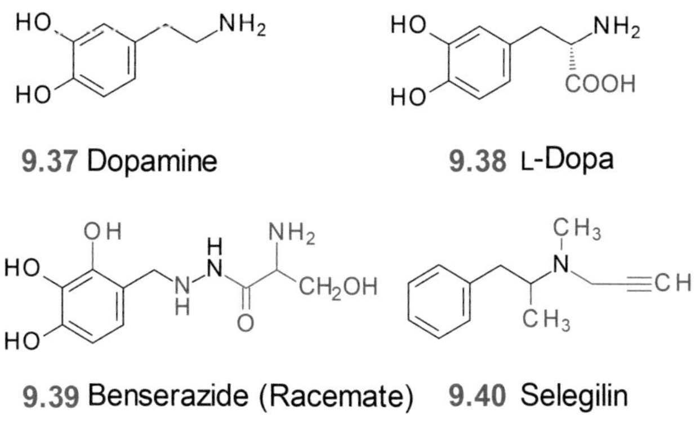

Fig. 9.9 Because dopamine **9.37** cannot enter the central nervous system, the metabolic precursor L-DOPA **9.38** is used. To reduce the cardiovascular effects of dopamine, L-DOPA is combined with a peripherally active decarboxylase inhibitor benserazide **9.39**. The administration of a monoamine oxidase inhibitor, for example, selegilin **9.40**, prevents the rapid degradation of dopamine

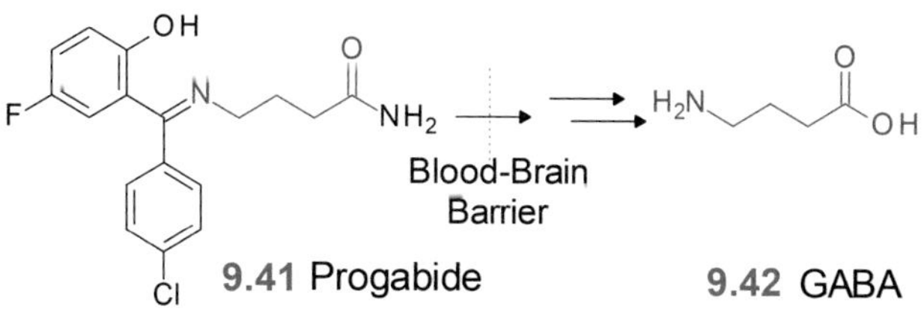

Fig. 9.10 Because it is a lipophilic neutral molecule, progabide **9.41** can cross the blood–brain barrier. It is transformed into the neurotransmitter γ-aminobutyric acid (GABA) **9.42** upon metabolic release of the amino and carboxyl groups

largely solves this problem. The peripheral side effects are reduced and the CNS effects are extended (Fig. 9.9). Despite this masterpiece of drug design, which has led to significant therapeutic progress, the metabolically produced dopamine still acts in too many places. Aside from the residual peripheral side effects, sudden changes between excessive movement, normal movement, and rigidity, insomnia, agitation, and hallucinations are all manifestations of the generalized CNS activity.

It has been speculated in conjunction with this observation whether, in addition to endogenous and genetic factors, environmental factors, for example, the metabolic transformation of structurally analogous foreign substances, might be responsible for triggering Parkinson's disease.

9.5 Drug Targeting, Trojan Horses, and Pro-prodrugs

The design of active substances that exert their effect only in, or overwhelmingly in, one particular organ is called **drug targeting**. Aside from general principles, for example optimal lipophilicity as a prerequisite for crossing the blood–brain barrier, specific metabolic transformations are used. The Parkinson's disease drug L-DOPA, which was introduced in the previous section, is such a prodrug. The anticonvulsive medicine progabide **9.41** is a double prodrug because both functional groups of the neurotransmitter are masked. After crossing the blood–brain barrier and release of the amino and carboxyl groups, the actual active compound, γ-aminobutyric acid (GABA, Fig. 9.10), is formed.

The ability of the blood–brain barrier to exclude polar substances can also be used as a prodrug concept. For this, an active compound with a metabolically labile group can be coupled to a dihydropyridine. The neutral conjugate **9.43** can cross the blood–brain barrier. Oxidation leads to a permanently charged compound **9.44**, which can no longer leave the brain. Upon metabolic cleavage, the free active compound is released *in situ*

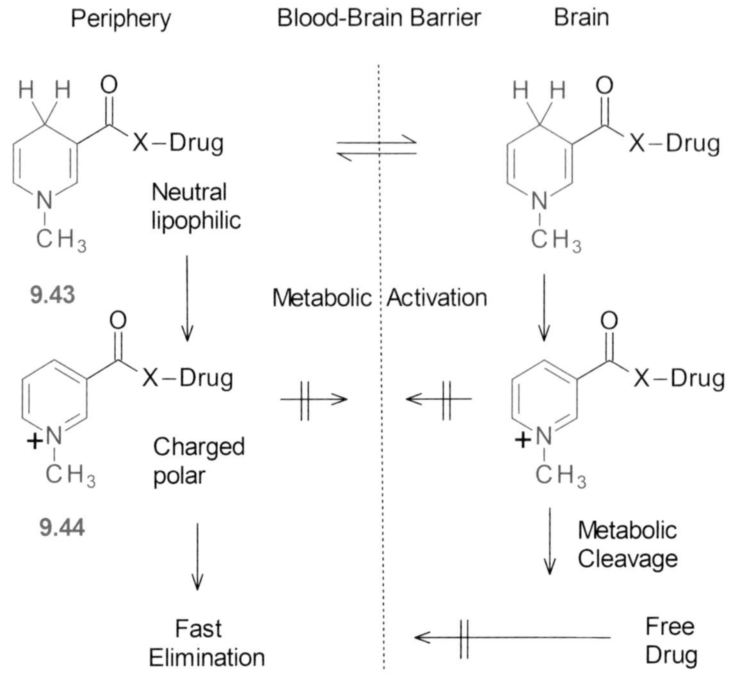

Fig. 9.11 Drug targeting in the brain is accomplished with a drug–dihydropyridine conjugate **9.43**. This substance can easily enter the central nervous system. Metabolic oxidation leads to a permanently charged pyridine **9.44**, which cannot cross the blood–brain barrier. The active compound is released in the brain, and the polar conjugate is quickly excreted from the periphery

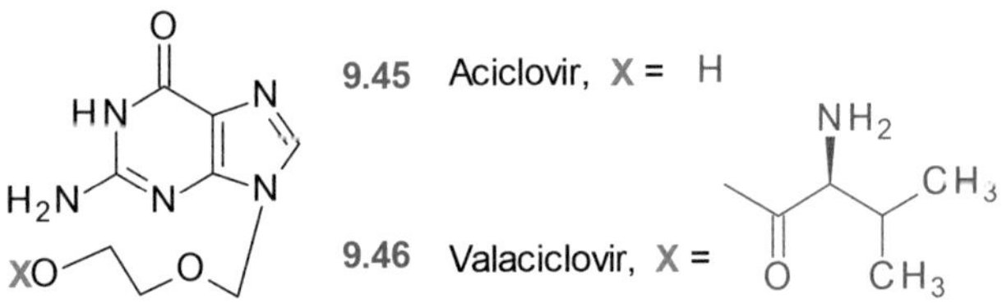

Fig. 9.12 Aciclovir **9.45** is a Trojan horse. An enzymatic phosphorylation of its hydroxyl group by a viral kinase affords its monophosphorylated form in virus-infected cells only, which is then transformed to the triphosphate derivative by the cellular kinases. Valaciclovir **9.46** is a pro-prodrug because it is first transformed to aciclovir by hydrolysis and subsequently activated

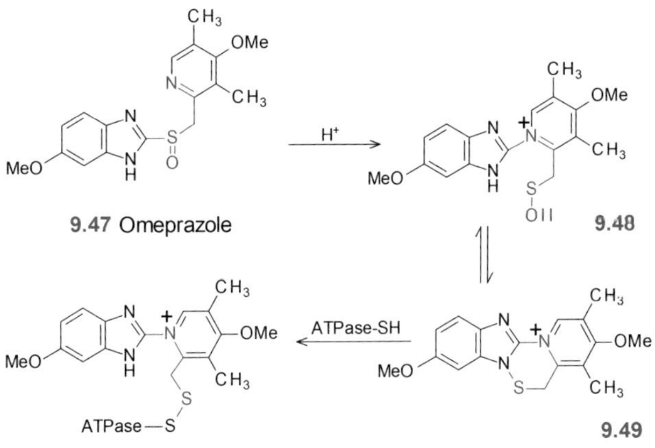

Fig. 9.13 In acidic milieu, omeprazole **9.47** is rearranged to a sulfenic acid **9.48**, which is in equilibrium with a cyclic sulfenamide **9.49**. This reacts irreversibly with the thiol group of a cysteine on H^+/K^+-ATPase, the so-called proton pump

(Fig. 9.11). If oxidation takes place in the periphery, the highly water soluble complex is excreted before the actual active substance is released. This example should show how, in principle, targeting to a specific tissue can be achieved when the specifically given chemical conditions are taken into account.

Several analogues of nucleobases and nucleosides are **Trojan horses**. The antiherpes medicine aciclovir **9.45** enters the cell as its inactive form. The first monophosphorylation occurs only in virus-infected cells by a virus-specific thymidine kinase. Next cellular kinases carry out the formation of the triphosphate, the actual active substance. Because of this aciclovir acts as a targeted antiviral. The compound is, however, poorly absorbed. The more suitable valaciclovir **9.46** (Fig. 9.12) can be considered to be a **pro-prodrug**. In the organism, it is initially hydrolyzed to aciclovir and then transformed into the active form by the viral enzyme. Valaciclovir is more lipophilic than aciclovir, but despite this, it is more soluble in water and has a bioavailability of approximately 55%.

Omeprazole **9.47** is the prodrug of an irreversible inhibitor of the H^+/K^+-ATPase, the so-called proton pump (Sect. 30.9). Only under strongly acidic conditions, in the acid-producing cells of the stomach, is it transformed into the sulfenic acid **9.48**, which is in equilibrium with the cyclic sulfenamide **9.49** (Fig. 9.13). This reacts irreversibly with an SH group of a cysteine residue of the enzyme to form a disulfide. Omeprazole is more effective than the H_2-antagonists (Sect. 3.5) because it blocks not only the histamine-induced acid secretions but rather all forms of acid secretions.

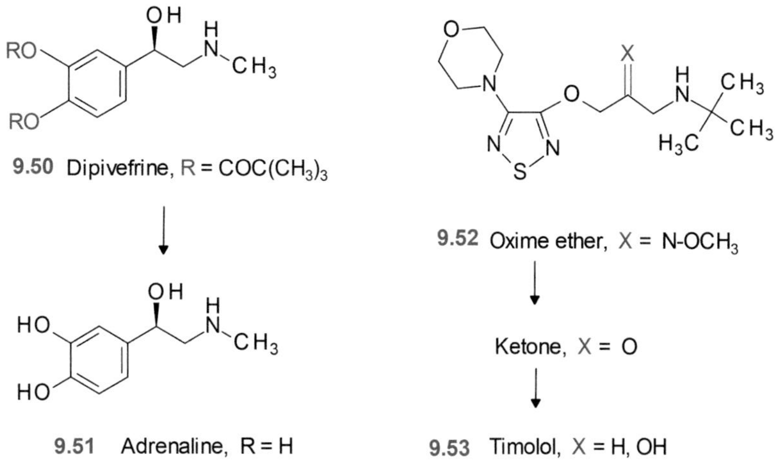

Fig. 9.14 The metabolic peculiarities of the eye are exploited for drug targeting in glaucoma therapy. After penetrating the cornea, the bis-pivaloyl ester, dipivefrine **9.50** of adrenaline **9.51** is hydrolyzed 20 times faster than it is in the periphery. The oxime ether of timolol **9.52** is metabolized through the ketone to the active form, timolol **9.53**, only in the eye

The different metabolic activity in different tissues can be used to achieve a selective effect in one specific organ. In principle, adrenaline (Sect. 1.4) as well as some β-blockers are suitable for the treatment of glaucoma, because they can normalize elevated intraocular pressure. However, they have substantial undesirable side effects on the heart function and circulation. This can be avoided by the administration of prodrugs that are metabolized more quickly in the eye, or only in the eye, for example, a particularly robust ester **9.50** of adrenaline **9.51**, or a ketone–oxime ether **9.52** of timolol **9.53** (Fig. 9.14).

The area of **drug targeting** has developed into an exciting field in recent years. Aside from the above-described prodrugs that release active compounds in the target area, the concept of **antibody–conjugate drugs** has been pursued especially for the development of novel cancer therapeutics. Another approach is the coupling of drugs to a cell-specific recognition sequence. The goal of this work is to trick the membrane transporters of very specific cells so that the drug–conjugate gains entry. Tumor therapeutics that were derived from *N*-lost were introduced in Sect. 9.3. These cytotoxic alkylating compounds, however, are very reactive and should only be activated in the desired target tissue. For this, the following strategies were developed. Aromatic *N*-lost derivative **9.55** (Fig. 9.15) is released from prodrug **9.54** by specific peptide cleavage with carboxypeptidase G2, an enzyme that only exists in bacteria. This enzyme was coupled to a monoclonal antibody (Sect. 32.3) that specifically recognizes human colorectal cancer cells. With this, the bacterial enzyme that "arms" the cancer drug is brought in the immediate vicinity of the cancer cell. In the future, this antibody-guided enzyme-activated prodrug therapy could make cancer therapy more tolerable and less toxic by releasing the active substance locally and in a distinctly more targeted way (see antibody–drug conjugates, Sect. 32.3).

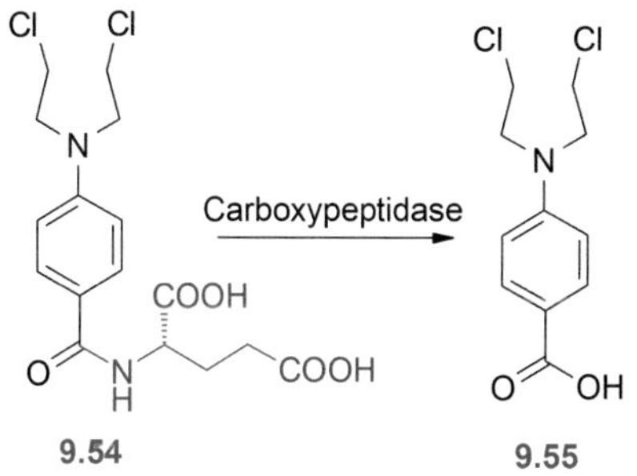

Fig. 9.15 The highly reactive cancer therapeutic derivative **9.55** is released from prodrug **9.54**, which is activated by a specific carboxypeptidase. The carboxypeptidase is bound to an antibody that is targeted to the cancer cell

9.6 Synopsis

- If it is impossible to achieve sufficient bioavailability, duration of action, membrane penetration, or metabolic stability by chemical modifications, a prodrug can be developed that corresponds to a nonactive or poorly active precursor or derivative that is converted in the organism to its active form.
- After absorption, a drug is transported to the liver and exposed to degrading enzymes that make it more water soluble for excretion. The amount of the drug that survives this first liver pass is referred to as the bioavailable portion and can be distributed in the organism.
- Esters are often used as prodrugs to mask polar acid groups; they are cleaved by ubiquitously present esterases.
- A large variety of chemical modifications have been applied to modulate the physicochemical properties of drug molecules; however, they require special enzymes in the targeted cells or organs for metabolic activation.

- L-DOPA, an amino acid analogue of dopamine, is delivered to the brain via an amino acid transporter and rapidly decarboxylated. To avoid side effects in the periphery, a combination with polar decarboxylase inhibitors is advisable.
- Drug targeting to particular organs or cells exploits specific metabolic transformations only present in these compartments of the body.
- Antibody-conjugate drugs are specifically delivered to those compartments or organs that present the antibody-specific recognition site on the surface of disease-related cells. To trick membrane transporters, drugs can be coupled to cell-specific recognition sequences and, thus, gain entry to the cells.

Bibliography and Further Reading

General Literature

H. Bundgaard, Eds., Design of Prodrugs, Elsevier, Amsterdam (1985)

N. Bodor, Prodrugs and Site-Specific Chemical Delivery Systems, Ann. Rep. Med. Chem., **22**, 303–313 (1987)

H. Bundgaard, Design and Application of Prodrugs, in: A Textbook of Drug Design and Development, P. Krogsgaard-Larsen and H. Bundgaard, Eds., Harwood Academic Publishers, Chur, pp. 113–191 (1991)

G. G. Gibson, Introduction to Drug Metabolism, Blackie, London (1994)

R. B. Silverman, M. W. Holladay, The Organic Chemistry of Drug Design and Drug Action, 3rd edn., Academic Press, (2014), Chapter 8, Drug Metabolism, and Chapter 9, Prodrugs and Drug Delivery Systems

L. P. Balant and E. Doelker, Metabolic Considerations in Prodrug Design, in: Burger's Medicinal Chemistry, M. E. Wolff, Eds., 5th edn, Vol. I, John Wiley & Sons, New York, pp. 949–982 (1995)

P. Ettmayer, G. L. Amidou, B. Clement and B. Testa, Lessons Learned from Marketed and Investigational Prodrugs, J. Med. Chem., **47**, 2394–2404 (2004)

B. Testa and J. M. Mayer, Hydrolysis in Drug and Prodrug Metabolism—Chemistry, Biochemistry and Enzymology, Wiley-VHCA, Zurich (2003)

K. Beaumont, R. Webster, I. Gardner and K. Dack, Design of ester prodrugs to enhance oral absorption of poorly permeable compounds: challenges to the discovery scientist, Curr. Drug Metab., **4**, 461–485 (2003)

B. Testa, Prodrug and Soft Drug Design, Comprehensive Medicinal Chemistry II, J. B. Taylor and D. J. Triggle, Eds., vol. 5, Elsevier, Oxford, pp. 1009–1041 (2007)

V. J. Stella, R. T. Borchardt, M. J. Hageman, R. Oliyai, H. Maag and J. W. Tilley, Eds., Prodrugs: Challenges and Rewards. 2 vols., Springer, New York (2007)

J. Rautio, Prodrugs and targeted delivery—towards better ADME properties. In: R. Mannhold, H. Kubinyi, G. Folkers, Eds., Methods and principles in medicinal chemistry, vol. 47, Wiley-VCH, Weinheim (2012)

Special Literature

M. E. Brewster, E. Pop and N. Bodor, Chemical Approaches to Brain-Targeting of Biologically Active Compounds, in: Drug Design for Neuroscience, A. P. Kozikowski, Eds., Raven Press, New York (1993)

N. Bodor and P. Buchwald, Ophthalmic Drug Design Based on the Metabolic Activity of the Eye: Soft Drugs and Chemical Delivery Systems, The AAPS Journal, **7**, E820–833 (2005)

M. P. Napier, S. K. Sharma et al Antibody-directed Enzyme Prodrug Therapy: Efficacy and Mechanism of Action in Colorectal Carcinoma. Clin. Cancer Res., **6**, 765–772 (2000)

9

Peptidomimetics

Contents

G. Klebe, *Drug Design,* https://doi.org/10.1007/978-3-662-68998-1_10

10

Peptides are open-chain polymers made up of **amino acids** (◘ Fig. 10.1). The main chain is constructed of alternating amide groups –CONH– and aliphatic carbon atoms, which are labeled C_α. The side chains branch from the main chain at the C_α atom. The amide group is barely flexible (Sect. 14.1). In contrast, rotation around the C_α–C_β bond is possible. The side chains are flexible as well. Because of this, each amino acid can take on multiple conformations. As a consequence, peptides are very flexible molecules with many rotatable bonds and a multitude of possibilities to adopt different spatial configurations. Formally, there is no difference between the construction of peptides and proteins. Nonetheless, oligomers of amino acids up to a size of 30–50 monomer building blocks are called peptides, and the term protein is preferred for any members of this substance class that are above this limit.

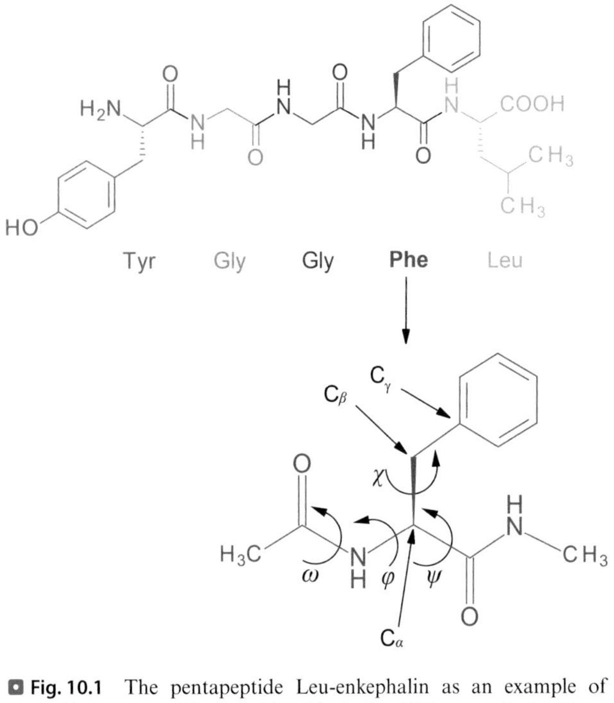

◘ **Fig. 10.1** The pentapeptide Leu-enkephalin as an example of a peptide structure. The left side with the free NH_2 group is the *N*-terminus, and the other is the *C*-terminus. Each amino acid contributes three nonhydrogen atoms to the peptide chain. Nature almost exclusively uses the 20 proteinogenic L amino acids for the construction of peptides (see page IX). Depending on the functional groups in the side chains, the distinction is made between hydrophilic acidic and basic amino acids and those with hydrophobic aliphatic and aromatic side chains. The amino acids are abbreviated with three letter codes. A one letter code is also used. The definition of the torsion angles $_\omega$, ϕ, ψ, and χ is shown in the example of the amino acid phenylalanine. The angle ω is practically always close to 180°. The spatial course of the peptide backbone is determined by the ϕ and ψ angles (see Sect. 14.2). The tetrahedral carbon atom in the chain is called C_α, the first atom in the side chain is named C_β, and the following atom is given the index γ

10.1 Therapeutic Relevance of Peptides

Peptides are responsible for numerous biological functions in humans, e.g., enzyme substrates and hormones. A few important examples are summarized in ◘ Table 10.1. Accordingly, peptides are interesting for therapeutic purposes, and in fact, several important drugs are peptides (◘ Fig. 10.2).

The use of peptides as drugs is significantly limited by several factors:

- Peptides are poorly absorbed after oral administration; this is mostly because of their high molecular weight and pronounced polarity.
- Peptides are easily degraded by proteases in the gastrointestinal tract and are, therefore, metabolically unstable.
- The body is able to very quickly excrete peptides via the liver and kidneys.

◘ **Table 10.1** Several important peptide hormones

Peptide	Function
Leu-Enkephalin, Met-Enkephalin	Opiate receptor ligands, analgesics
Fibrinogen	Platelet aggregation
Angiotensin II	Increases blood pressure
Endothelin	Increases blood pressure (among other actions)
Neuropeptide Y	Increases blood pressure (among other actions)
Substance P	Bronchoconstriction and pain mediation

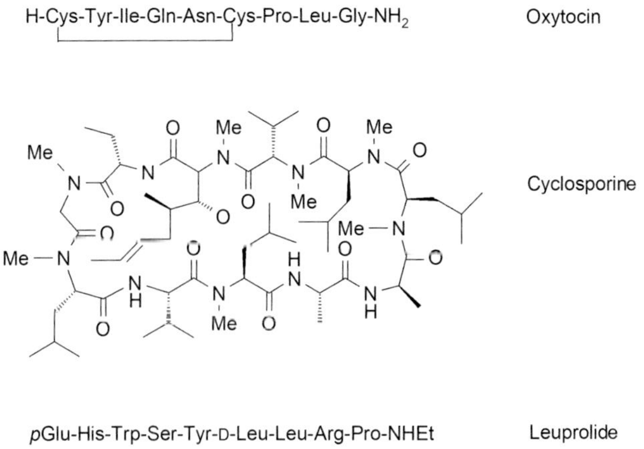

◘ **Fig. 10.2** Peptides applied as drugs: Oxytocin is used to induce and strengthen contractions during labor. The immunosuppressive cyclosporine prevents organ rejection after transplantation. Leuprolide (*p*Glu = *pyro*-glutamate) is an analogue of luteinizing hormone releasing hormone (LHRH), one of the hypothalamic hormones that, via luteinizing hormone (LH), controls the synthesis of male and female sexual hormones. Leuprolide is used to treat advanced-stage prostate cancer

Because peptides are involved in so many biological functions in our bodies, there is tremendous interest in finding active substances that do not have the above-mentioned detrimental properties, but that bind to the same receptors analogously to peptides or block enzymes that transform peptide substrates. A stepwise approach is taken in the search for such compounds. Peptide structures are replaced with isosteric building blocks so that the molecular recognition properties of the peptide remain, but the undesirable characteristics are reduced. Such **peptidomimetics** should have the following qualities:

- Few or no cleavable amide bonds to improve metabolic stability,
- Reduced molecular weight to improve oral bioavailability, and
- The same spatial orientation of groups responsible for strong binding to the receptor or enzyme as in the peptide.

Bacteria are the true masters of constructing peptide structures that frequently achieve the desired metabolic stability. They incorporate amino acids that do not belong to the typical 20 residues that are usually used for the construction of proteins. Stereochemically inverted amino acids are also employed, and many of these structures have a cyclic architecture. Due to an exceptional spatial structure, as e.g. in the lasso peptides, these peptides achieve very good proteolytic stability. Microorganisms have even evolved a dedicated synthesis machinery for this: the **nonribosomal peptide synthesis** (Sect. 32.7). This system of modular, coupled enzymes works like an assembly line. Depending on the desired product, different enzymatic functional units are lined up, one after the other, to successively assemble the amino acids cyclizing the product in the final step. The exchange of an enzymatic synthesis unit causes other amino acids to be incorporated into the otherwise unchanged peptide. Even ester bonds can be constructed with a very similar multienzyme complex. Many lead structures all the way to complete drugs can be derived from these original bacterial peptides, such as cyclosporine in ◘ Fig. 10.1, which is a very important immunosuppressant. A large number of macrolide antibiotics (Sect. 32.7) are also synthesized in this way.

Recently, a so-called chemoenzymatic synthetic strategy was developed for the construction of such macrolides. As discussed in Sect. 11.6, linear oligopeptides can easily be synthesized by using the Merrifield synthesis. Nonproteinogenic amino acids with L- and D-configurations can also be used to generate high combinatorial diversity. It is very difficult to cyclize these linear oligopeptides to the desired macrocycle by using synthetic chemistry methods. Here, the nonribosomal peptide synthetic machinery is of service. The synthetically prepared peptides are then funneled into the enzymatic process chain and the cyclization domain from the bacteria catalyzes the ring closure of the peptide: a perfect symbiosis between synthetic chemistry and enzyme biology!

10.2 Designing Peptidomimetics

In the beginning of the 1980s, there was only one generally accepted example for a low-molecular-weight active substance that takes over the function of an endogenous peptide: the opiate. It is assumed that morphine **10.1** is a **mimetic** of the endogenous peptide β-endorphin **10.2** (◘ Fig. 10.3). A comparison of both structures makes it immediately clear that morphine cannot possibly simulate all of the functional groups of the peptide. Obviously not all are necessary for the biological activity. This underscores the suspicion that other peptides also bind to receptors with only a few functional groups. If this hypothesis is true, it should be possible to identify the essential functional groups and find a small organic molecule that has the necessary functional groups in the correct relative orientation.

The starting point for the design of peptidomimetics is the identification of the biologically active peptide, the function of which is to be imitated. In the first step, single amino acids are excluded to determine whether a portion of the peptide retains sufficient activity. Next the importance of the individual side chains is investigated. In a so-called **alanine scan** (Sect. 10.7), each amino acid is successively replaced with alanine. A severe loss of activity is an indication that the removed side chain is important. Until now only peptides made up of the 20 amino acids found in proteins have been investigated. In the next step, structural elements are introduced that do not occur in the 20 **proteinogenic amino acids**. In principle, the following are possibilities for peptide structure modification:

- The use of D- instead of L-amino acids,
- Modifications of the side chain of amino acids,
- Changes on the peptide main chain,
- Cyclization to stabilize the conformation, and
- The use of templates that enforce a particular secondary structure, or that allow the attachment of side chains in a defined spatial orientation.

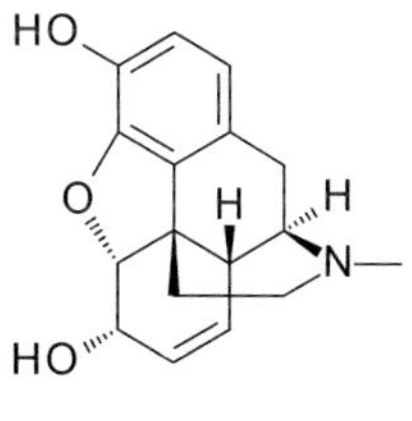

Tyr-Gly-Gly-Phe-Met-Thr-Ser-Glu-Lys-Ser-Gln-Thr-Pro-Leu-Val-Thr-Leu-Phe-Lys-Asn-Ala-Ile-Ile-Lys-Asn-Ala-Tyr-Lys-Lys-Gly-Glu

10.1 Morphine **10.2** β-Endorphin

◘ **Fig. 10.3** Morphine **10.1** is a peptidomimetic for the endogenous peptide β-endorphin **10.2** and the enkephalins (Sect. 1.4). It binds as an agonist to the opiate receptor

10.3 First Step to Variation: Modifying Side Chains

An improvement in a peptide's binding properties can often be achieved by using other side chains. For instance, in ◘ Fig. 10.4 a few analogues of the amino acid phenylalanine are shown that could be used as possible replacements. An increase in the binding affinity can be achieved if nonproteinogenic amino acids fill the binding pocket more completely. Rigid analogues will lead to improved binding if the **biologically active conformation**, the one that is adopted in or at the receptor site, is immobilized.

The introduction of nonproteinogenic amino acids can increase the metabolic stability. The hydroxylation of aromatic side chains can be suppressed by using a substituent, for example, a fluorine atom or a methoxy group, especially in the *para*-position. Stability to cleavage by the digestive enzyme chymotrypsin can be improved by adding substituents to the C_β-atom because the modified side chain no longer fits into the active site of this protease. A peptide's proteolytic stability can also be improved by exchanging L- for D-amino acids. As described above, bacteria have already recognized this trick. Distributing D-amino acids randomly in the peptide can furnish active substances with astonishing metabolic stability.

10.4 A More Courageous Step: Modifying the Main Chain

An important step in the design of peptidomimetics is the **replacement of amide bonds** in the main chain. A few commonly used groups are summarized in ◘ Fig. 10.5. It can be difficult or even impossible to find replacements for amide bonds, which form hydrogen bonds to the protein with both the C=O and NH groups, that do not significantly reduce binding affinity. If the amides only bridge functional groups to one another and do not form hydrogen bonds to the protein, then a large palette of different replacement groups is available. Substitution at the amide nitrogen atom leads to metabolic stabilization because proteases can hardly cleave ***N*-methylated amide bonds**. If *N*-methylation of an amide group of the main chain leads to a loss of affinity, several explanations are possible. One is that the *N*-methylated compound can no longer form hydrogen bonds, and an essential H-bond involving the NH group is lost. It is also possible that an undesired conformational change may have occurred as a result of the additional methyl group, or that the methyl group may sterically block binding to the protein. On the other hand, an improvement in binding as a result of *N*-methylation may indicate that the biologically active conformation is stabilized. At room temperature, an amide bond is almost exclusively in the *trans*-geometry. Therefore, it can be substituted by an ester bond that assumes the same geometry. However, the H-bond-donating properties of the amide group are lost.

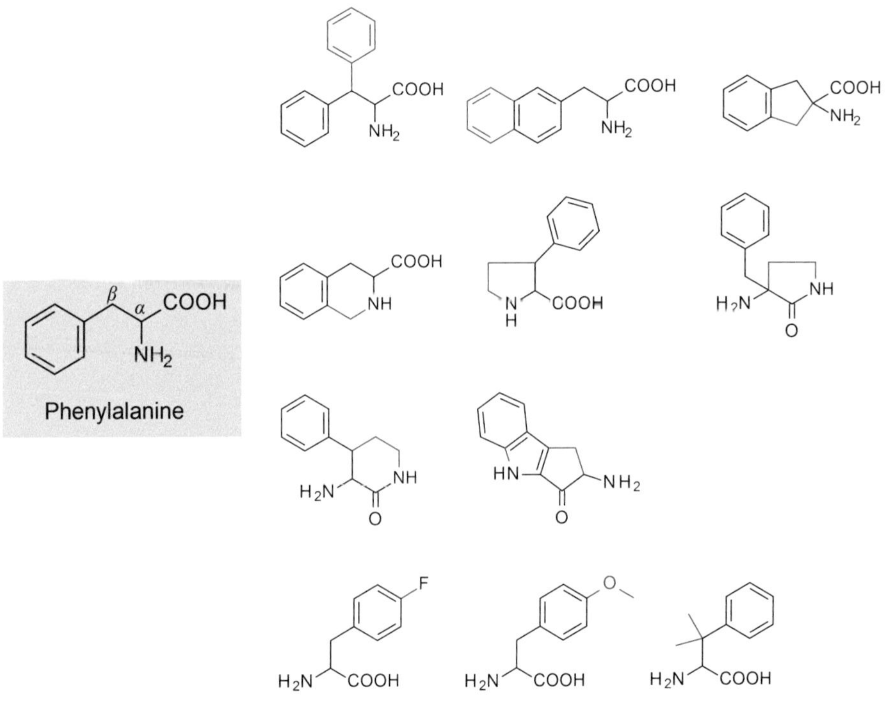

◘ **Fig. 10.4** Sterically demanding, conformationally fixed, or metabolically stable analogues of the amino acid phenylalanine; the structural enhancements are indicated in *red*

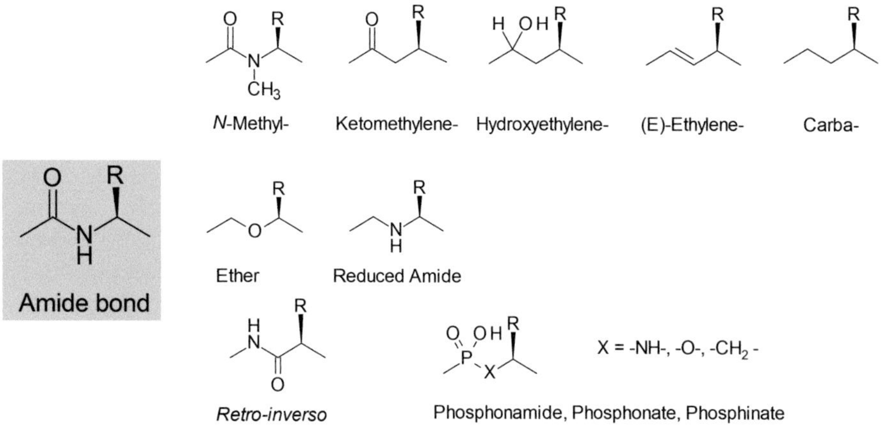

Fig. 10.5 Different functional groups that can serve as a replacement for amide bonds in peptidomimetics

An *N*-methyl substitution improves the stability of the 180°-rotated conformation of the amide. In the case of **proline**, the only proteinogenic amino acid with an *N*-alkyl substitution, both the *cis* and *trans* amide configuration can be found. The exchange for a 1,5-disubstituted tetrazole can replace the *cis*-orientation of a proline. In addition, *trans*-configured double bonds imitate the geometry of an amide bond well. The polar characteristics, however, are lost. To a certain extent, this can be compensated if the double bond is substituted with fluorine. The reduction of an amide or an isosteric ester bond means the loss of the carbonyl group and leads to increased flexibility. If the carbonyl group is exchanged for an –S=O, $-SO_2$, or $-PO_2$ group, the H-bond-accepting characteristics will be amplified; however, a geometry change comes with the bargain. The exchange of an amide for a **thioamide** results in a weakening of the H-bond-accepting properties and can serve as a test of the possible importance of H-bonds to carbonyl groups in the peptide backbone. Nonetheless, a measure of caution is warranted because the desolvation of a thiocarbonyl group is less difficult than that of a carbonyl group. This overlaps with the observed affinity and can mask the effect of the loss of the H-bond. The **retro–inverso exchange** of an amide bond can lead to marked improvement in the proteolytic stability without losing the binding qualities (Sect. 5.5).

An entirely different concept is the incorporation of β-amino acids (Sect. 31.7). In contrast to the proteinogenic α-amino acids, these residues have four chain members per monomer unit. The amide bonds are separated by two aliphatic carbon atoms. Peptides that are made from these amino acids also show secondary structural characteristics (Sects. 10.5 and 14.2). They have already successfully been incorporated into naturally occurring peptides as mimetics and can simulate peptide–protein interactions. Because of the altered sequence of amide bonds, they are stable to proteolytic degradation (for an example, see Fig. 31.23).

If the cleavable bond of a protease substrate is replaced with an isosteric, noncleavable group, a **substrate** can be converted to an **inhibitor** (Sect. 6.6). If the newly introduced group forms particularly favorable interactions with the active site of an enzyme, an exceedingly potent enzyme inhibitor can result. An example is found in the ketomethylene group in serine and cysteine protease inhibitors as a possible replacement for the amide bond that is destined for cleavage (Chap. 23). The hydroxyethylene group is especially suitable for aspartic protease inhibitors (Chap. 24). Phosphonamides, phosphonates, and phosphinates are often strong inhibitors of metalloproteases (Chap. 25).

10.5 Rigidifying the Backbone by Fixing Conformations

An important aspect in the design of peptidomimetics is the **peptide conformation**. Peptides are flexible molecules and can take on different conformations. It is known, however, that certain conformations are preferably adopted in proteins and in some peptides. Among these are the two most important secondary structural elements: the α-helix and the β-sheet (Sect. 14.2). Furthermore, there are loops and turns at the ends of these secondary structural elements that also adopt preferred patterns, particularly the **β-turn** (Fig. 10.6).

A β-turn is formed when a hydrogen bond exists between the carbonyl group of the amino acid i and the NH group of the amino acid i + 3. It is obvious that such hydrogen bonds can only form for certain combinations of the torsion angles ϕ and ψ, which are determined by the amino acids in the i + 1 and i + 2 positions (Fig. 14.6, Sect. 14.2).

β-Turns are especially interesting because many peptides bind to proteins in a **β-turn conformation**. Let us assume that the backbone of the peptide only serves to position the side chains for optimal receptor interactions. Then it should be possible to replace the peptide chain with a completely different scaffold to which functional groups are attached that assume the same spatial orientation as the amino acid side chains.

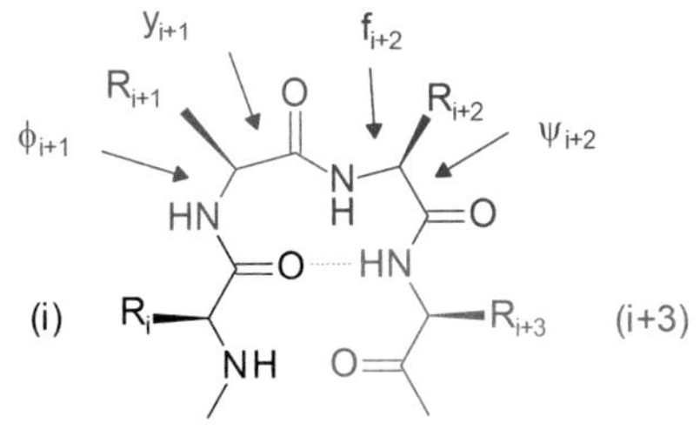

■ **Fig. 10.6** A β-turn is a peptide conformation in which a hydrogen bond is formed between the amino acids i and i + 3. Particular ranges for the values of the torsion angles ϕ_{i+1}, ψ_{i+1}, ϕ_{i+2}, and ψ_{i+2} are characteristic for the β-turn

If a β-turn-configured peptide binds to a receptor, then a rigid analogue that "freezes" the β-turn conformation should lead to improved binding. The simplest way to fix a β-turn is the incorporation of the necessary sequence in a small **cyclic peptide**. It is known from experimental structure determination that cyclic penta- and hexapeptides almost always contain a β-turn. The conformation of these peptides were investigated at length in
10
the research group of Horst Kessler at the Universities of Frankfurt and Munich, Germany. It could be shown that the position of a β-turn in a sequence can be controlled. Proline as well as D-amino acids prefer the i + 1 position in these loops. The introduction of D-amino acids supports the formation of a β-turn above other possible conformations (cf. Sect. 31.2).

A β-turn can also be forced by a **nonpeptide template**. Numerous β-turn mimetics have been proposed for this (■ Fig. 10.7). A part of the structures serves as a template on which two peptide chains can be forced in an antiparallel orientation. However, substitution by introduction of the R_2 and R_3 side chains is synthetically difficult. Benzodiazepines are interesting scaffolds onto which all four side chains R_1–R_4 can be coupled. Other peptide conformations can also be fixed by the introduction of rigid groups. A few examples of conformation-stabilizing ring systems are displayed in ■ Fig. 10.8.

An especially convincing example of a **scaffold mimic** is the design of a thyrotropin-releasing hormone (TRH) mimetic by Gary Olson and his colleagues at Roche in Nutley, New Jersey, USA. TRH is the tripeptide pGlu–His–Pro–NH_2 **10.3**. The approach is shown in ■ Fig. 10.9. After deducing a pharmacophore hypothesis, a rigid scaffold molecule was sought upon which the side chains could be appended in the correct relative orientation. Cyclohexane was chosen as a scaffold. Compound **10.4** is a potent TRH receptor ligand. The substance acts as an agonist and elicits the same effects as TRH. An improvement in cognitive function could be seen in animal experiments after the administration of **10.4**.

■ **Fig. 10.7** Typical β-turn mimics. The amino acids are added onto the template at the *red*-colored positions

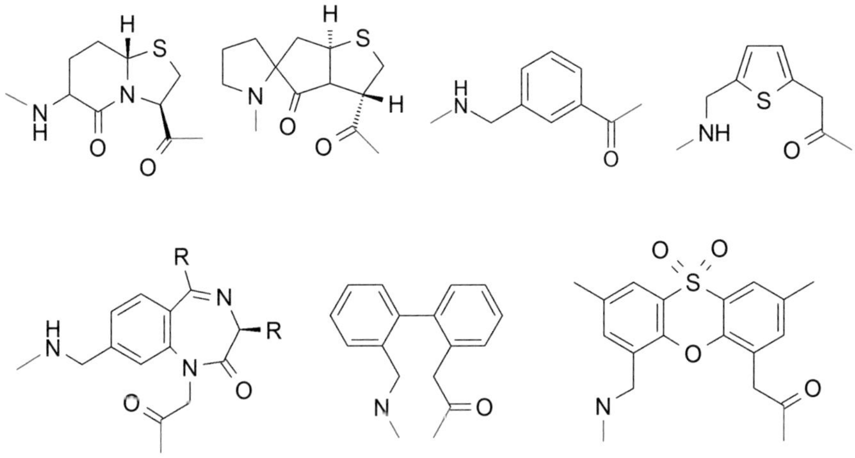

■ **Fig. 10.8** The illustrated rings replace one or two amino acids and force a particular conformation

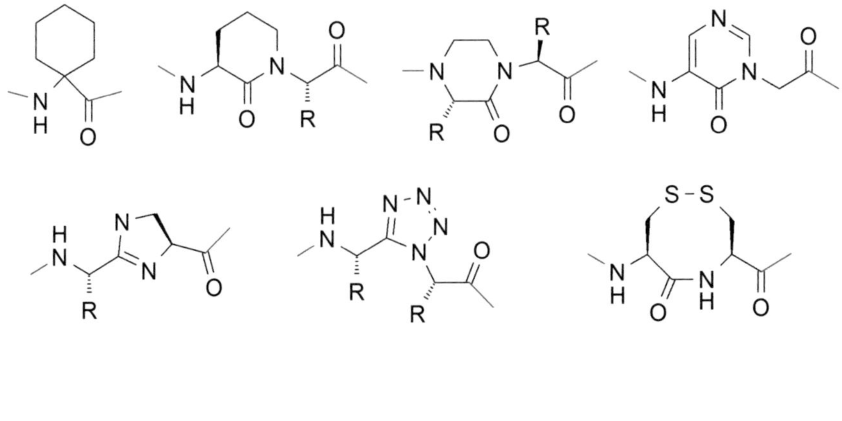

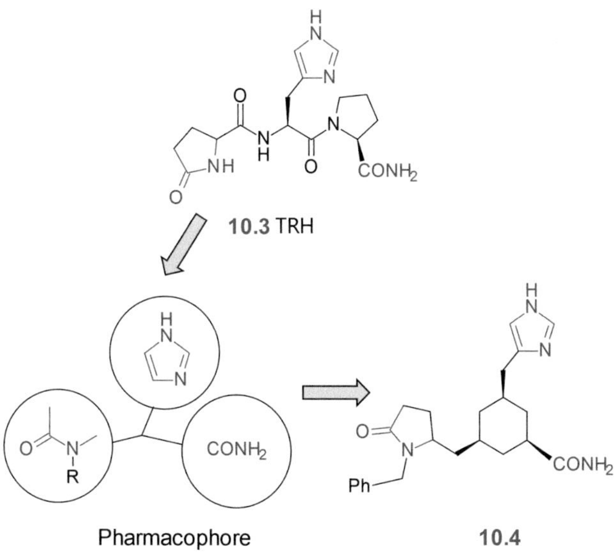

Fig. 10.9 By starting with the structure of tripeptide TRH **10.3** and a hypothesis for the functional groups that are essential for binding, the nonpeptidic molecule **10.4** was designed, which also binds to the TRH receptor

10.6 Peptidomimetics to Interfere with Protein–Protein Interactions

Proteins communicate with one another and transmit information and signals by forming complexes with each other via commonly shared surfaces. The area of the shared contact surface usually extends over more than a thousand square Ångstroms ($Å^2$). This is a large value when compared to the surface that a small organic molecule of typical drug size occupies upon binding. Furthermore, the **contact area between two proteins** is, as a general rule, not very jagged. It hardly resembles the deep binding pockets in enzymes that can host small ligands. Nevertheless, it would open entirely new perspectives for drug therapy if such protein–protein contact surfaces could be blocked with low molecular weight compounds. At first glance, this task seems almost impossible. How can a small molecule bind to a flat, barely structured protein surface with an interaction that is strong enough not to be "washed away" when the **protein–protein contact** forms? Furthermore, there is the problem that amino acid residues on the convex surface of a protein have in general much more space to flexibly adapt their conformation. A statistical analysis of the amino acid composition across the contact surfaces in protein complexes showed a preference for aromatic residues, aspartate, arginine and the aliphatic residues proline and isoleucine. The selective exchange of amino acids in the contact surface also showed that there are a few protruding residues that dominate the interaction (so-called "**hot spots**," Sect. 17.10). The search for possible binding sites of a small molecule that can compete with the formation of the protein–protein interface starts with a detailed analysis of the complementary geometry to the contacting surfaces. Are there clustered areas with charged residues or does a structural element such as a β-turn or α-helix penetrate a little more deeply into the opposite contact surface? Next, the peptide sequence that corresponds to the contact surface is synthesized. This can be portions that preferably adopt a helical structure or that can be fixed in a turn pattern such as a cyclopeptide. If an active peptide is found, it must be structurally characterized in complex with the opposite contact surface.

The complex of the BCL-X_L (B-cell lymphoma) protein with a 16-residue peptide that was cut from the BAK protein is shown in Fig. 10.10. BCL-X_L belongs to the proteins that prevent programmed cell death (apoptosis). Its function is regulated by binding to pro- and antiapoptotic factors such as BAK. Inhibitors of this contact formation might, therefore, deliver potential drugs for an anticancer therapy. The binding of the helical peptide takes place in a stretched-out groove. Small molecules have been discovered that fill this crevice (Fig. 10.11). The group of Andrew Hamilton at Yale University, New Haven, USA has been searching for a basic scaffold that can imitate the characteristics of a helix and simultaneously hold the side chains on one side. Terphenyl derivatives **10.5**–**10.7** were found that can arrange the side chains in a staggered conformation analogous to a helix. An alanine scan along the BAK peptide showed that four hydrophobic residues (Val 74, Leu 78, Ile 81, and Ile 85) are essential for binding. In addition, Asp 83 forms a salt bridge to BCL-X_L. The terphenyl scaffold was, therefore, furnished with an acidic group at the end and decorated with alkyl and aryl residues in the *ortho*-positions. Compound **10.6** binds to the BCL-X_L protein with an affinity of 114 nM.

A different approach was taken at AbbVie. Small molecules that interact with the BCL protein were sought by NMR spectroscopy (Sect. 7.8). The millimolar inhibitors *para*-fluorobiphenylcarboxylic acid **10.8** (Fig. 10.11) and 1-hydroxytetraline **10.9** were discovered. Both bind to distinct but neighboring positions. They replace Asp 83 and Leu 78 of the binding domain of the BAK peptide, and **10.9** occupies the Ile 85 position. From the two discovered fragments, the scientists at AbbVie developed compound **10.10**, which had two-digit nanomolar affinity for the protein. Further optimization led to **10.11**, a highly potent antagonist that blocks the entire family of antiapoptotic BCL-2 proteins. The synergistic effect of ABT-737 together with radiation and chemotherapy was demonstrated in animal experiments. Unfortunately, the bioavailability of ABT-737 was insufficient. Nevertheless, the clinical trials were also discontinued for the structurally similar follow-up compound navitoclax (ABT-263) due to severe side effects.

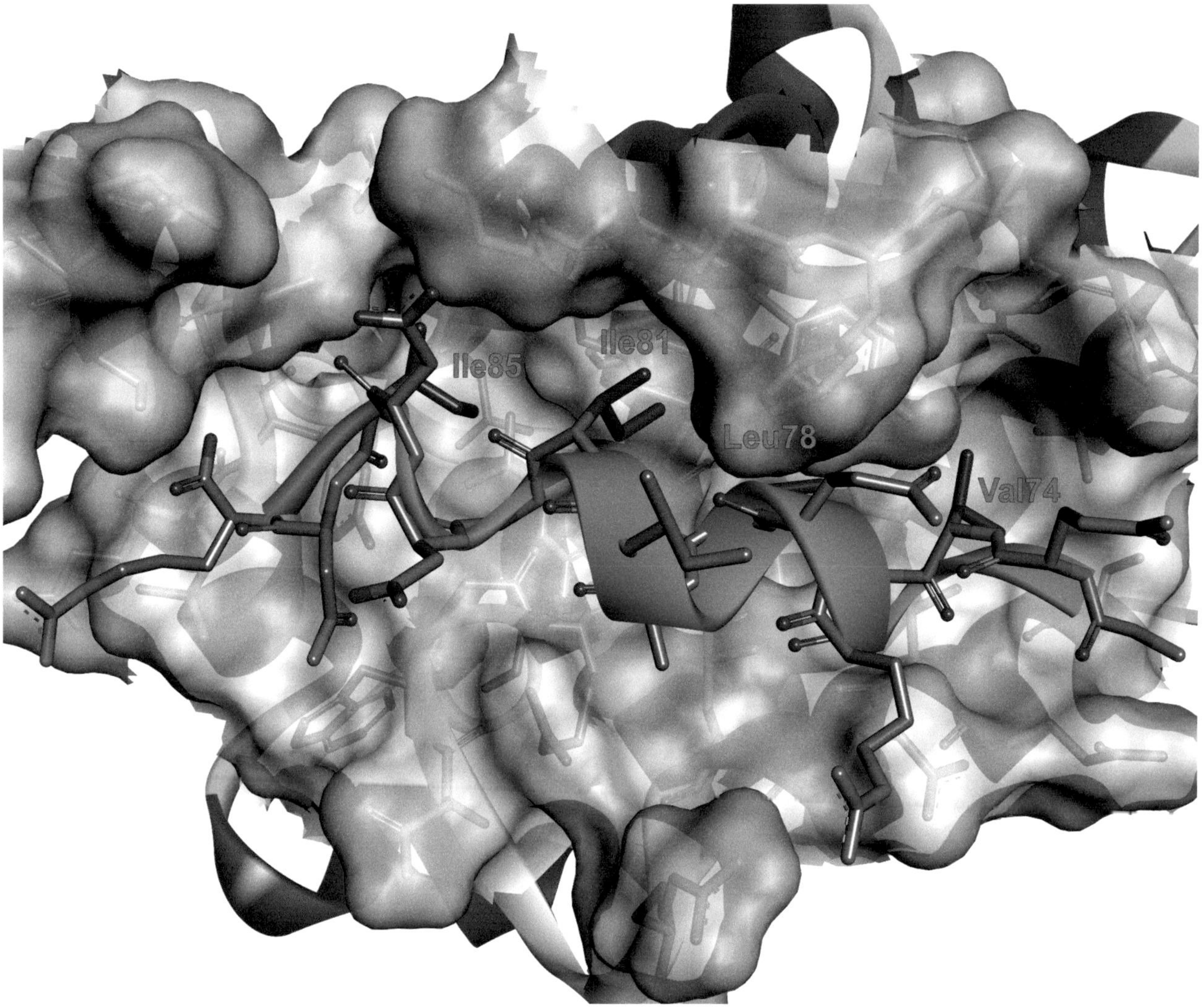

Fig. 10.10 The NMR spectroscopic structure of the BCL-X_L protein with the α-helical, 16-membered peptide fragment from the BAK protein (*blue–gray*). The peptide binds in a deep groove with the amino acids Ile 85, Ile 81, Leu 78, Val 74 (from *left* to *right*, side chains are in *magenta*). The surface of the BCL protein is shown in *white*, the contact surface of the hydrophobic amino acids of the peptide all protrude into the cleft. (► https://sn.pub/ooSzpg)

An analogous case was studied with the **MDM2** protein at Roche. MDM2 is overexpressed in many tumors. It binds to the tumor-suppressor protein p53, which protects cells from converting to a malignant state. It is, therefore, the protein that is most often inactivated during carcinogenesis. Inhibition of complex formation between the overexpressed MDM2 protein and p53 could, thus, represent an approach to a possible cancer therapy. Here too, an α-helical p53 peptide stretch binds to a hydrophobic groove on the MDM2 protein. A *cis*-imidazoline with an affinity of 100–300 nM was found in screening. The cocrystal structure was accomplished with **10.12** (◘ Fig. 10.11). The imidazoline scaffold imitates the side of an α-helix of the peptide from the p53 protein. The two *p*-bromophenyl rings replace a Trp and a Leu. The ethyl ether group on the third aromatic ring orients in the pocket that is filled with a phenylalanine in the peptide. The MDM2 protein is blocked through this competitive binding, and the level of free p53 increases. Through this, the p53 pathway in cancer cells is activated, and the cell cycle comes to a complete stop. The cell may go into programmed cell death. The tumor growth inhibition has already been demonstrated in animal models.

Another large class of proteins that is controlled by contacts with other proteins is the integrins. Numerous low molecular weight inhibitors have been discovered for this class. An example for the successful design of

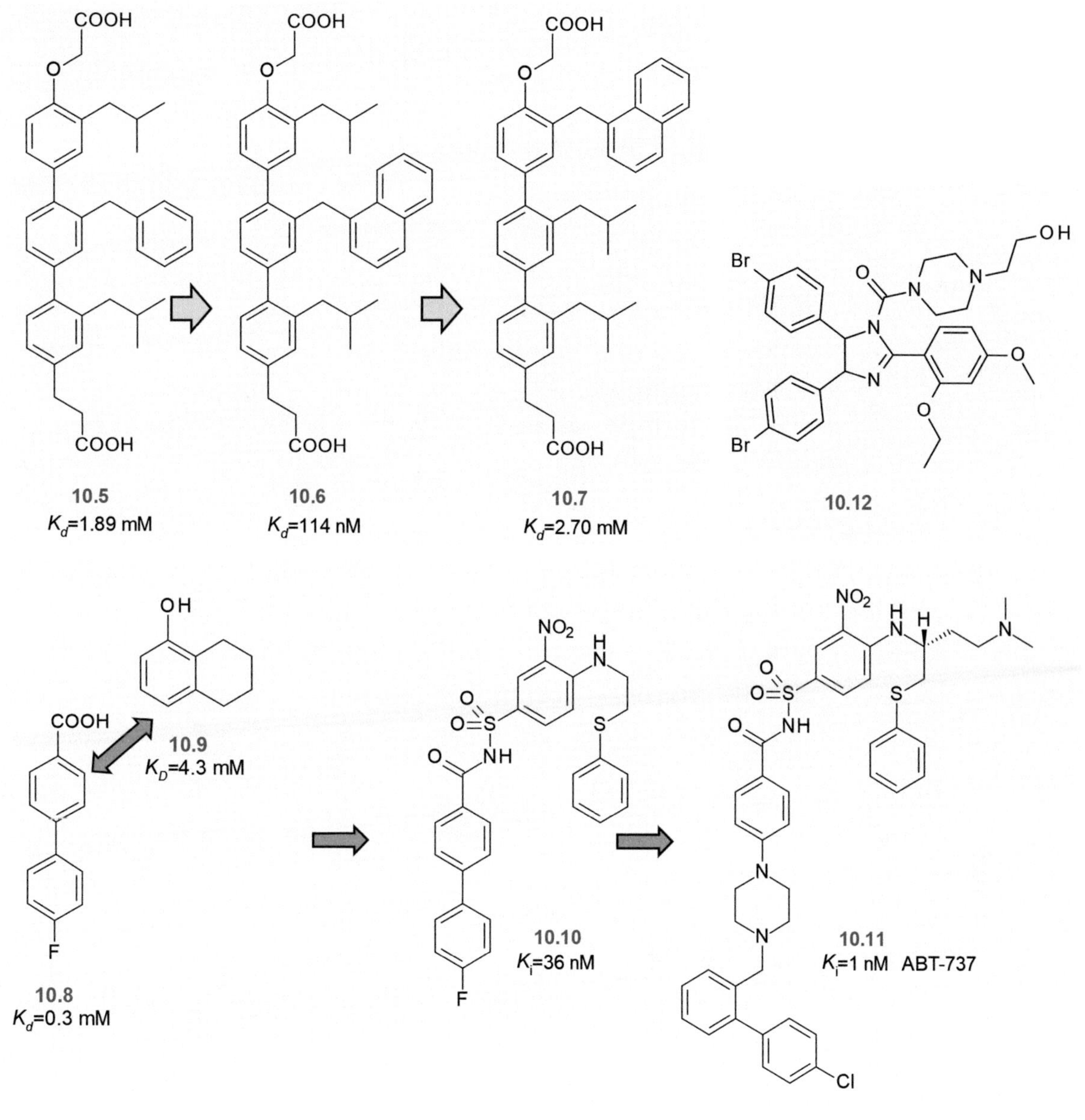

Fig. 10.11 Different inhibitors of protein–protein contacts that imitate the α-helical structural building blocks in the contact surface. The terphenyl derivatives **10.5**–**10.7** bind to the BCL-X_L protein in a pronounced crevice and block the binding site of a helix. The small fragments **10.8** and **10.9**, which led to the development of inhibitors **10.10** and **10.11** were discovered in the same area during NMR spectroscopic screening. Compound **10.12** is a different helix mimetic that prevents the interaction between the MDM2 and p53 proteins

antagonists by starting from cyclic peptides is presented in Sect. 31.2. Many G-protein-coupled receptors (Sect. 29.1) are controlled by endogenous peptides or proteins. For this, the peptide or protein bind to the receptor. The replacement of the peptide sequences with an organic molecule that imitates the binding of the natural ligand has also been attempted. An example of the design of such an active compound is given in Sects. 29.5 and 29.6. An example of how small fragments can lead the way to ligands that interfere with the formation of protein–protein interfaces is shown in Sect. 21.15.

10.7 Tracing Selective NK Receptor Antagonists by Ala Scan

Tachykinins are neuropeptides that all contain the same lipophilic *C*-terminus: –Phe–X–Gly–Leu–Met–NH_2. A well-investigated representative of the tachykinins is substance P, Arg–Pro–Lys–Pro–Gln–Gln–Phe–Phe–Gly–Leu–Met–NH_2 (**10.13**, Table 10.2). Tachykinins bind to at least three different tachykinin receptors, the NK_1, NK_2, and NK_3 receptors. All three belong to the class of G-protein-coupled receptors (Sect. 29.1). They mediate a variety of biological effects, for example, bron-

Table 10.2 The rational design of NK_2 receptor ligands

	No.	Structure	
Substance P	**10.13**	Arg–Pro–Lys–Pro–Gln–Gln–Phe–Phe–Gly–Leu–Met–NH_2	295
Minimal fragment	**10.14**	Leu–Gln–Met–Trp–Phe–Gly–NH_2	11.7
Ala scan	**10.15**	**Ala**–Gln–Met–Trp–Phe–Gly–NH_2	40
	10.16	Leu–**Ala**–Met–Trp–Phe–Gly–NH_2	138
	10.17	Leu–Gln–**Ala**–Trp–Phe–Gly–NH_2	156
	10.18	Leu–Gln–Met–**Ala**–Phe–Gly–NH_2	> 10,000
	10.19	Leu–Gln–Met–Trp–**Ala**–Gly–NH_2	8300
	10.20	Leu–Gln–Met–Trp–Phe–**Ala**–NH_2	28
	10.21	Leu–Gln–Met–Trp–Phe–NH_2	200
Dipeptide	**10.22**	Z–Trp–Phe–NH_2	2700
Immobilization of the biologically active conformation	**10.23**	Z–Trp–(*R,S*)–(α–Me)Phe–NH_2	327
N-Terminal optimization	**10.24**	(2,3-di-OCH_3)$C_6H_3CH_2OCO$-Trp-(*R,S*)-(α-Me)Phe-NH_2	37.6
Stereochemical optimization	**10.25**	(2,3-di-OCH_3)$C_6H_3CH_2OCO$-Trp-(*R*)-(α-Me)Phe-NH_2	10,000
	10.26	(2,3-di-OCH_3)$C_6H_3CH_2OCO$-Trp-(*S*)-(α-Me)Phe-NH_2	17.2
Addition of amino acid	**10.27**	(2,3-di-OCH_3)$C_6H_3CH_2OCO$-Trp-(*S*)-(α-Me)Phe-Gly-NH_2	1.4

Z = benzyloxycarbonyl group

10

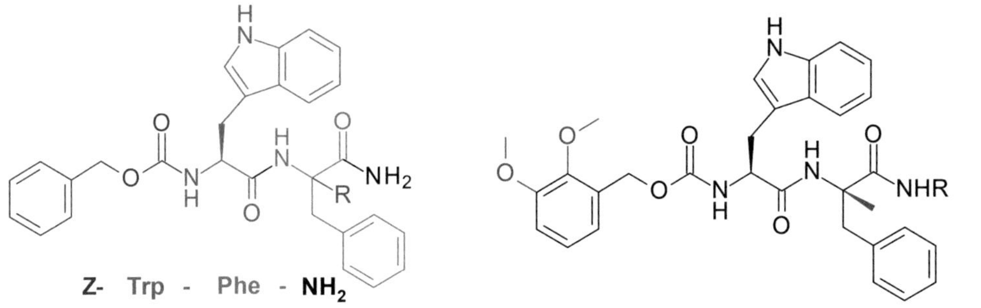

10.22 R = H K_i = 2700 nM

10.23 R = CH_3 K_i = 327 nM

10.26 R = H K_i = 17.2 nM

10.27 R = CH_2CONH_2 K_i = 1.4 nM

Fig. 10.12 Important intermediates on the way from the dipeptide **10.22** to the potent NK_2 receptor antagonists **10.27**

choconstriction or pain transmission. Consequently, a receptor antagonist could be helpful for the treatment of asthma as well as to fight pain.

The study conducted at Parke-Davis in Cambridge, UK, to develop an NK_2 receptor antagonist is a classic example of the conversion of a **peptide** into a **peptidomimetic** (Table 10.2 and Fig. 10.12). A compound was sought that would bind to the same receptor as substance P. The starting point of the work was a hexapeptide, Leu–Gln–Met–Trp–Phe–Gly–NH_2 (**10.14**), known from the literature to bind to the NK_2 receptor with an affinity of 11.7 nM. In a first step, each amino acid was systematically replaced by alanine (**10.15**–**10.20**). At a few positions, alanine substitution resulted in only a small decrease in binding affinity. For example, the *N*-terminal leucine could be replaced by an alanine (**10.15**). It was concluded that the Leu side chain may be of minor importance for receptor binding. However, the compound in which tryptophan or phenylalanine was replaced by alanine showed very little affinity for the NK_2 receptor. This was clear evidence that these two amino acids are essential for binding. Removal of the *C*-terminal amino acid glycine (**10.21**) decreased the affinity by a factor of 7. Apparently, this amino acid also has

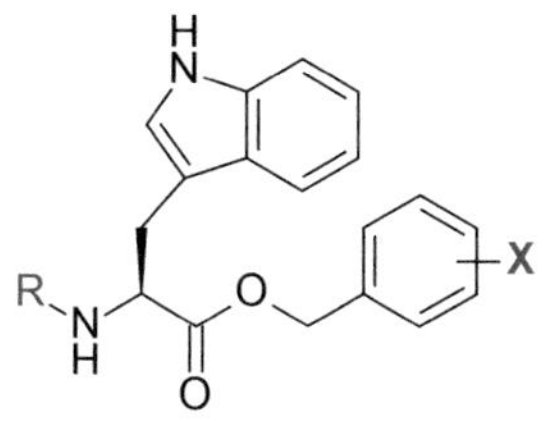

10.28 R = Et, X = H IC_{50} = 3800 nM

10.29 R = H, X = H IC_{50} >10000 nM

10.30 R = H, X = 3,5-di-CH_3 IC_{50} = 1533 nM

10.31 R = Ac, X = 3,5-di-CH_3 IC_{50} = 67 nM

10.32 R = Ac, X = 3,5-di-CF_3 IC_{50} = 1.6 nM

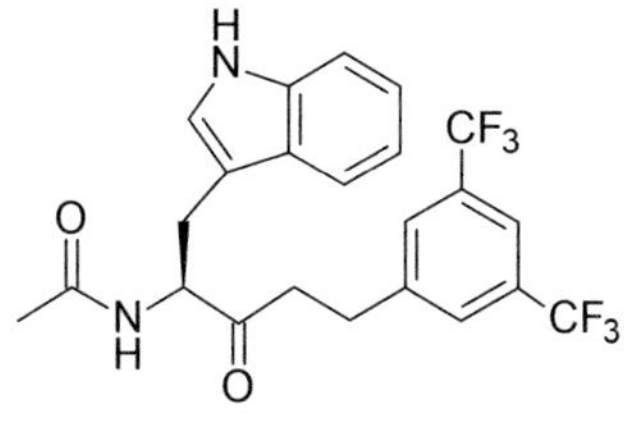

10.33 IC_{50} = 3 nM

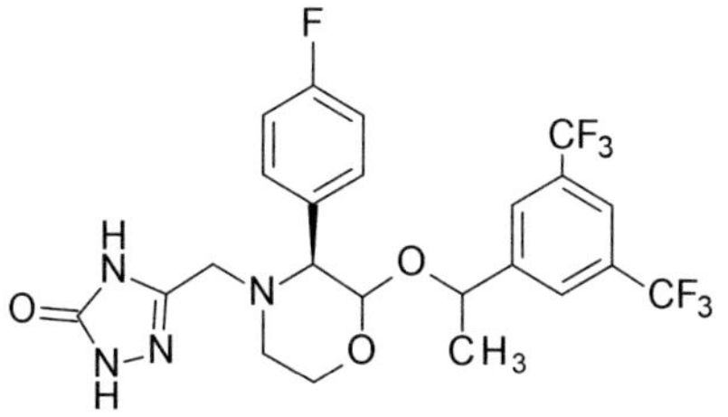

10.34 Aprepitant

Fig. 10.13 The optimization of lead structure **10.28**, which was found by screening, to selective NK_1 receptor antagonists **10.32** and **10.33**. In contrast to the metabolically labile benzyl esters **10.28**–**10.32**, ketone **10.33** is also active in animal experiments. The first NK_1 receptor antagonist aprepitant **10.34** was successfully brought to the market by MSD for the prevention of acute emesis

some importance for receptor binding. Testing of several *N*-terminal protected dipeptides led to Z–Trp–Phe–NH_2 (**10.22**, K_i = 2700 nM, Z = benzyloxycarbonyl-) as the lead structure for further work. This completed the first phase of the project. As a dipeptide, **10.22** was an interesting lead structure for subsequent work.

In the next step, additional methyl groups were introduced at different positions of the molecule. This restricted the **number of possible conformations**. A decrease in binding affinity was observed for many of the conformationally restricted compounds. A methyl group on the C_α-atom of phenylalanine increased the binding affinity by a factor of 8 (**10.23**, K_i = 327 nM). A possible explanation for this finding is that the conformation adopted at the receptor site is stabilized by the additional methyl group. The *N*-terminal part of the molecule was then varied. Replacement of the terminal phenyl ring with a 2,3-dimethoxyphenyl group further increased the binding affinity by a factor of 10 (**10.24**, K_i = 37.6 nM). This value corresponds to racemic α-methylphenylalanine. The enantiomerically pure compound **10.26** with this moiety in the *S*-configuration binds with a K_i of 17.2 nM. The reintroduction of the *C*-terminal glycine finally led to the highly potent compound **10.27** (K_i = 1.4 nM).

Independent of the work at Parke-Davis, lead structure **10.28** was optimized to the NK_1-specific receptor antagonists **10.32** and **10.33** at Merck Sharp & Dohme (MSD). Although **10.28**–**10.32** were only effective *in vitro*, **10.33** is also active *in vivo* because of its higher metabolic stability (Fig. 10.13). MSD was finally successful with the structurally related aprepitant **10.34**. The compound was introduced as a medicine to prevent acute emesis (vomiting) during highly nausea-inducing chemotherapy.

10.8 CAVEAT: Idea Generator for the Design of Peptidomimetics

In the previous sections, it was often highlighted that the side chains of the amino acids are responsible for the binding to receptors. Usually the main chain merely plays the role of a scaffold that serves to bring the side chains into the necessary spatial alignment for binding. Thus, a rigid, nonpeptidic scaffold to which the side chains could be attached in the same spatial orientation should be suitable for designing molecules with properties similar to those of peptides. This idea has been implemented in a computer program in Paul Bartlett's group at the University of California at Berkeley, USA. The program, **CAVEAT**, allows the search for rigid molecules that mimic a particular segment of a peptide backbone. To do this, the bonds on the peptide backbone are described by vectors (Fig. 10.14). The 3D structure of the peptide for the peptidomimetic being sought must be known as a prerequisite. The orientation of the side chains is determined by the binding vectors C_α–C_β. The relative orientation of, for instance, three amino acid side chains is found by the position of the relevant C_α–C_β binding vectors. This spatial pattern of vectors is used to search a 3D database of molecular scaffolds containing three substitutable bonds oriented analogously to the three C_α–C_β vectors. The result is a list of rigid, usually cyclic molecular scaffolds, the free positions of which can be coupled to the amino acid side chains.

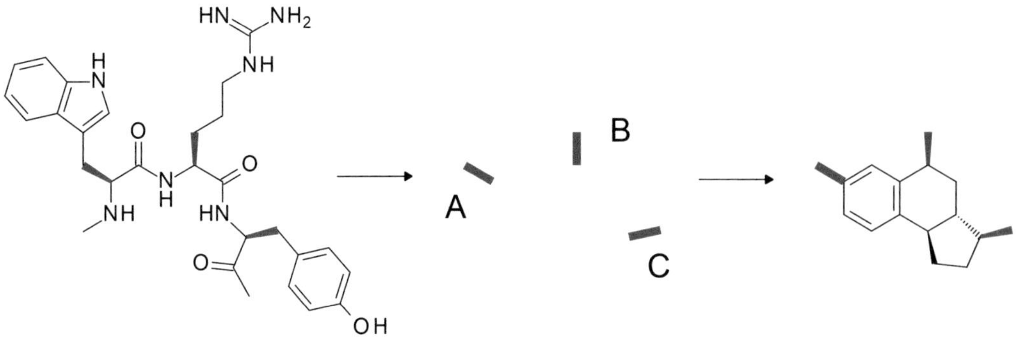

Fig. 10.14 Concept of a 3D search for scaffold mimetics with the program CAVEAT. First, the relative orientation of the biologically relevant side chains in the peptide lead structure is defined by the C_α–C_β bond vectors. In this example, the three amino acids Trp, Arg, and Tyr are taken as essential. The three vectors A, B, and C are the crucial information used to search the 3D database for rigid scaffold structures that bear substitutable bonds in the same relative orientation. A list of cyclic structures that represent possible templates for peptidomimetics is the result

10.9 Design of Peptidomimetics: *Quo Vadis*?

In this chapter, the systematic approach to the design of peptidomimetics has been described. The approaches have proven themselves in many cases and have led to many attractive drugs. Nevertheless, there are also difficulties. The first problem is the stepwise approach. A peptide is systematically modified, and the synthesized structures serve only to identify the essential functional groups. The synthesis of the many resultant derivatives, that is, practically all in which an amide group was replaced by one of the structures in Fig. 10.4, is laborious. Furthermore, these compounds only serve as tools because most modified peptides have high molecular weights, and this can result in poor oral bioavailability.

In the past, many new nonpeptidic active substances, especially as receptor antagonists, were found in high-throughput screening, and these could frequently be developed into clinical candidates in a relatively short time. These successes have pushed rational approaches to peptidomimetic design from the forefront. Nevertheless, the design of peptidomimetics remains an important area of research in drug design. The terphenyl scaffold helix mimetics are an example of this. The peptidic nature of many of the enzyme inhibitors presented in Chaps. 23, 24, 25 is still evident. Here, the **peptidic substrate** was clearly the inspiration for the **design of a mimetic**. Therefore, peptidomimetic concepts continue to play an important role in lead optimization.

10.10 Synopsis

- Peptides are open-chain polymeric molecules made up of amino acids that are mutually linked by amide bonds. Side chains branch from the main chain at the C_α atoms and show a high degree of flexibility. If such a polymer contains up to 30–50 amino acids, it is called a peptide; beyond this limit, it is called a protein.
- Peptides are responsible for many biological functions; their applicability as drugs is limited due to size, polarity, and poor proteolytic stability.
- Due to their multiple functions, peptides can be mimicked by smaller—similarly binding—and metabolically stable peptidomimetics.
- Peptidomimetic design starts with the identification of the minimal peptide sequence responsible for a biological effect, followed by successive replacement of each amino acid in the chain with alanine to detect the side chains responsible for activity. Finally, individual amino acids are replaced by nonproteinogenic ones or similar chemical building blocks.
- Multiple surrogates for amino acid side chains have been developed and can be tested to reveal better binding and conformationally more stable peptidomimetics. If not involved in direct binding, main-chain amide bonds can be replaced by a large variety of substitutes that achieve a similar geometry.
- Peptides are flexible and adopt multiple conformations. If a particular fold is adopted to correctly orient interacting side chains, the peptide backbone can be replaced by an entirely different scaffold that correctly positions the essential interacting groups.
- Peptides fold upon themselves through particular turn patterns. These turns stabilize a required conformation and can be chemically replaced by rigid structural surrogates that freeze a given turn conformation.
- Proteins communicate with one another through the formation of large, mutually shared surface patches. Small molecules designed to bind to such flat surfaces can antagonize complex formation and interfere with protein–protein communication.
- Design of small molecules to block protein–protein interfaces exploits depressions on the surface that accommodate spatial patterns such as turns or helical

portions of the penetrating contact surface of the binding partner protein.

- Peptides bind to receptors mostly via side chains, and the backbone provides the scaffold for their attachment. Computer programs can be used to screen structural databases to retrieve alternative scaffolds that are able to orient substituents in very similar fashion.

Bibliography and Further Reading

General Literature

A. Giannis and T. Kolter, Peptidomimetics for receptor ligands—discovery, development, and medical perspectives. Angew. Chem. Int. Ed. Engl., **32**,1244–1267 (1993)

J. Gante, Peptidomimetics—tailored enzyme inhibitors, Angew. Chem. Int. Ed. Engl., **33**, 1699–1701 (1994)

J.-M. Ahn, N. A. Boyle, M. T. MacDonald and K. D. Janda Peptidomimetics and Peptide Backbone Modifications, Mini Rev. Med. Chem., **2**, 463–473 (2002)

M. A. Marahiel, Working outside the protein-synthesis rules: insights into non-ribosomal peptide synthesis. J. Pept. Sci., **15**, 799–807(2009)

R. Hirschmann, Medicinal chemistry in the golden age of biology: lessons from steroid and peptide research. Angew. Chem. Int. Ed. Engl., **30**,1278–1301 (1991)

J. J. Perez, Designing Peptidomimetics, Curr. Top. Med. Chem., **18**, 566–590 (2018)

Special Literature

G. L. Olson, D. R. Bolin, M. P. Bonner et al., Concepts and Progress in the Development of Peptide Mimetics, J. Med. Chem. **36**, 3039–3049 (1993)

W. Howson, Rational Design of Tachykinin Receptor Antagonists, Drug News & Perspectives, **8**, 97–103 (1995)

A. M. McLeod, K. J. Merchant, M. A. Cascieri et al., *N*-Acyl-L-tryptophan Benzyl Esters: Potent Substance P Receptor Antagonists, J. Med. Chem., **36**, 2044–2045 (1993)

K. J. Merchant, R. T. Lewis and A. M. MacLeod, Synthesis of Homochiral Ketones Derived from L-Tryptophan: Potent Substance P Receptor Antagonists, Tetrahedron Letters, **35**, 4205–4208 (1994)

T. Oltersdorf et al. An inhibitor of Bcl-2 family proteins induces regression of solid tumours, Nature, **435**, 677–681 (2005)

B. Vu et al., Discovery of RG7112: A Small-Molecule MDM2 Inhibitor in Clinical Development, ACS Med. Chem. Lett., **4**, 466–469 (2013)

G. Lauri and P. A. Bartlett, CAVEAT: A Program to Facilitate the Design of Organic Molecules, J. Comput.-Aided Mol. Design, **8**, 51–66 (1994)

G. Lelais and D. Seebach, β^2-Amino Acids-Synthesis, Occurrence in Natural Products, and Components of β-Peptides, Biopolymers, **76**, 206–243 (2004)

Experimental and Theoretical Methods

A prerequisite for the 3D structure determination of a protein by the method of X-ray crystallography is the availability of a crystal (Chap. 13). The figure shows a set of crystals of a complex of protein kinase A, which was used to elucidate the structure and reaction mechanism of this class of enzymes (Chap. 26). (Courtesy of Dr. Dirk Bossenmeyer, German Cancer Research Center, Heidelberg).

Contents

Combinatorics: Chemistry with Big Numbers

Contents

G. Klebe, *Drug Design*, https://doi.org/10.1007/978-3-662-68998-1_11

The search for new lead structures and the optimization of their activity profile by systematic modification are among the most time- and cost-demanding steps in drug research. The optimization of a small organic molecule can serve as an example. Even if the number of different groups per position is limited to relatively few, several million structures are possible as exemplarily shown in the case of the multisubstituted tetrahydroisoquinoline carboxylic acid amide **11.1** (◘ Fig. 11.1). The combinatorial explosion of all imaginable substitution possibilities can no longer be realized with classical chemical techniques. The diversity increases even more when the different stereoisomers are considered. Their number is, thus, on the order of magnitude of all chemical structures recorded in *Chemical Abstracts* (160 million compounds, of which 68 million are protein and nucleic acid sequences) or in *Beilstein* (in *Reaxys*, 118 million compounds).

In the days when compounds were tested on whole animals or in complex *in vitro* pharmacological models, biological testing was the rate-determining step. The introduction of molecular test models, such as enzyme or receptor binding assays, and extensive automation of screening has fundamentally changed this situation. The testing of many thousands of compounds per day is technically unproblematic (Sect. 7.3). To fully exploit the capacity of these methods, the synthesis of thousands or even tens or hundreds of thousands of different molecules is desirable. The strategy can then shift either to **automated parallel synthesis** to cover a large number of single compounds, or to the simultaneous production of compound mixtures using **combinatorial chemistry**.

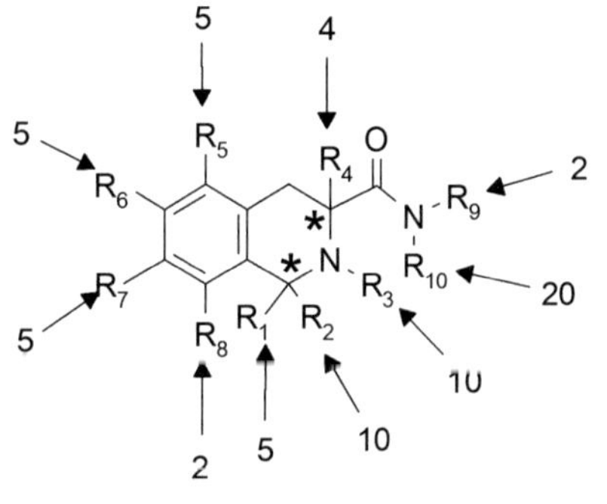

11.1

◘ **Fig. 11.1** The tetrahydroisoquinoline carboxylic acid amide **11.1** is to be substituted in 10 positions. The groups in these positions encompass a multiplicity of a total of 68 building blocks (R_1–R_{10} = 5, 10, 10, 4, 5, 5, 5, 2, 2, 20 groups). Twenty million compounds can be constructed in this way. If the structural diversity that results from the two stereocenters (*) is considered, this number increases again by a factor of 4

11.1 How Nature Produces Chemical Multiplicity

Nature has shown a way to achieve combinatorial diversity with the nucleic acids and with proteins. A 600-base-pair DNA sequence codes a protein with 200 amino acids. From the "pool" of four nucleic acids that code for the 20 proteinogenic amino acids in triplet sequences, 4^{600} (a number with 360 digits!) different DNA sequences are possible. This translates to 20^{200} (a number with 260 digits!) different amino acid sequences for the resulting protein. Short peptides with enormous structural variety can be constructed with just the **20 proteinogenic amino acids**. If instead of amino acid A, a manageable number of modified amino acids M is used, the number of possible analogues will increase even more (◘ Table 11.1).

Peptides play an important role in biological systems. They are found as protein ligands in the free form or as simple derivatives. Peptide sequences on the surface of proteins determine their recognition by a receptor. For this selective recognition, Nature exploits the full combinatorial diversity of the variable sequences in the surface regions (epitopes) of proteins. These principles of Nature can be used to generate vast libraries of compounds with widely varying compositions.

◘ **Table 11.1** A total of 400 dipeptides, 8000 tripeptides, 160,000 tetrapeptides, and 64 million hexapeptides can be generated from the 20 proteinogenic amino acids, **A**. If the palette is expanded to 100 modified, nonproteinogenic amino acids, **M**, the combinatorial diversity increases dramatically

Compounds	Number
Natural amino acids, **A**	20
Dipeptides, **A**–**A**	400
Tripeptides, **A**–**A**–**A**	8000
Tetrapeptides, **A**–**A**–**A**–**A**	160,000
Hexapeptides, **A**–**A**–**A**–**A**–**A**–**A**	64,000,000
Modified amino acids, **M**	100 (for example)
Modified hexapeptides, **M**–**M**–**M**–**M**–**M**–**M**	1,000,000,000,000
Number of known compounds	> 33,000,000

11.2 Protein Biosynthesis as a Tool to Build Compound Libraries

How can the biochemical synthesis machinery be used as a vehicle to generate a multiplicity of peptide sequences? It is possible to connect short sequences to a carrier protein so that they are exposed on the surface and can interact with the target protein in a molecular test system. The test system is constructed in a way that the binding to the target protein is monitored with an easily registered signal, for instance, a fluorescence signal or a colorimetric reaction (Sect. 7.2).

To use protein biosynthesis to construct such a library, the information about the randomly assembled peptides must be added to the "genetic make-up" of a DNA molecule. This molecule encodes the sequence of the protein on whose surface the library will be presented and, in addition, the randomly assembled double-stranded DNA sequences of individual members of the peptide library. The information of the latter is inserted into the DNA at an appropriate position. After producing a large number of identical copies (cloning), the resulting genes can be expressed. This produces a large population of proteins that carry the randomly assembled peptide sequences in a very specific region, usually at the beginning or end of the polymer sequence. The resulting proteins are then examined in a molecular test system. The distribution of the 20 proteinogenic amino acids over the variable sequence section is not entirely homogenous. That is because some amino acids are coded with a single triplet sequence (codons), and others are represented with up to six different codons (Sect. 32.7, ◘ Fig. 32.16). Because of this, biased libraries are inevitably formed.

The **bacteriophage M13** is an extremely popular expression system. M13 is a virus that infects *Escherichia coli* strains well. The virus carries six proteins on its coat. Two of these coat proteins allow randomly **assembled protein sections** to be added to their ends. Using this M13 system, a library of 20 million modified 15-mer peptides was generated. Their binding to the protein streptavidin was tested. A total of 58 candidates were identified as binding partners. They all shared the sequence segment –His–Pro–Gln–. The crystal structure of one of these oligopeptides complexed with streptavidin was successfully determined. The peptide occupies with its His–Pro–Gln segment the binding pocket normally populated by biotin. This demonstrates that such a strategy can be used to find selectively binding peptide sequences.

The biochemical approach to generating and presenting compound libraries has the overwhelming advantage that the high-capacity **protein biosynthesis** is exploited. Furthermore, the sophisticated protein and DNA synthesis techniques and analytical methods that have been developed for such substances (Sect. 11.7) can be used to characterize screening hits. But it also has disadvantages. The molecular diversity is limited to the **20 proteinogenic L-amino acids**, and only peptides result as lead structures. These are often the starting point for the development of a drug. However, we wish to move away from metabolically unstable, poorly bioavailable peptides. Therefore, structures are sought using classical organic molecular scaffolds. At least peptidomimetics or peptides with metabolically stable nonproteinogenic amino acids are desired. Unfortunately, the step away from peptides to alternative scaffolds that retain biological activity is not trivial (Chap. 10).

11.3 Organic Chemistry from a Different Angle: Random-Guided Synthesis of Compound Mixtures

Organic preparative methods were devised as an alternative to the biological approaches to generate compound libraries. Simple access to a compound library is gained by starting with reactive molecular building blocks, such as oligofunctional acid chlorides (**11.2**–**11.4**, ◘ Fig. 11.2). These components are simultaneously reacted with numerous reagents, for example, amines or amino acids. **A mixture of many products** is formed in an uncontrolled manner. Contrary to the general academic opinion that organic reactions should only deliver homogenous products, in this case as much product diversity as possible is desired. The advantage of this method is that it is easy to carry out and that automation is readily implemented. But this synthesis strategy also has disadvantages. The

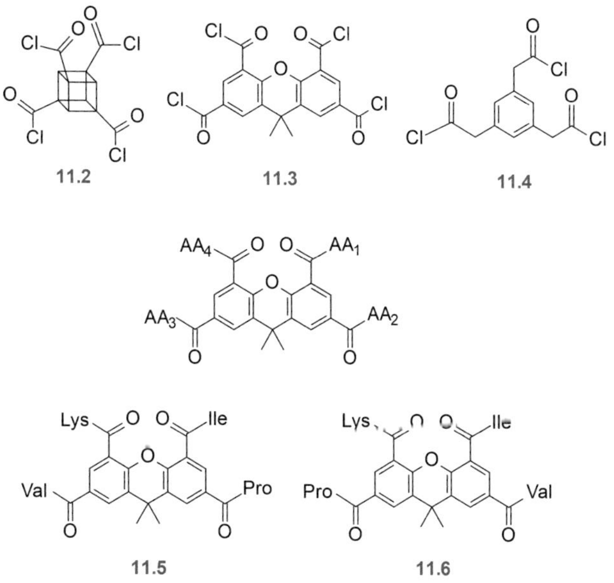

◘ **Fig. 11.2** The oligofunctional acid chlorides of the central building blocks cubane **11.2**, xanthene **11.3**, and benzene **11.4** are treated with protected amino acids (AA_1–AA_4). A xanthene-containing library inhibits the digestive enzyme trypsin. The active component of the library was deconvoluted and characterized by targeted resynthesis. In the end, isomers **11.5** and **11.6** remained as the most potent compounds. The derivative **11.5** inhibits trypsin with a K_i of 9.4 μM coupling partners have different reactivities. As a result, the products are not evenly distributed. The transformation of a particular functional group on the central building block can depend upon which components the central molecule has already reacted with and how this influences the other functional groups.

The thus-generated library is then tested. If binding to the target protein is found, the active substance in the mixture will be characterized, a task that is not particularly simple. On the one hand, sophisticated analytical techniques such as liquid chromatography coupled with NMR spectroscopy and mass spectrometry can be used. Moreover, an attempt can be made to "**deconvolute**" the library. For this, a targeted resynthesis of the library is carried out in which a partial library is prepared by using a defined selection of building blocks. This smaller library is then tested and the composition of the active mixture is determined. This strategy must be followed back to the level of single defined reaction products.

11.4 What Is Contained in Chemical Space?

At this point, the fundamental question must be asked: how many organic molecules are principally possible from which medicinal chemists can create their candidates? What does such an initially virtual chemical space contain? Much has been speculated about this question. Numbers between 10^{20} and 10^{200} possible molecules have been named. The last claim encompasses so many molecules that the entire mass of the universe would not be enough to synthesize at least one molecule of every compound! It is through the work of Jean-Louis Reymond's group at the University of Berne, Switzerland, that we now have a somewhat more solid idea about the principal composition of chemical space. Starting with mathematical graphs describing simple hydrocarbon scaffolds, molecules with up to 17 C, N, O, S, and halogen atoms have been generated on the computer. This selection covers a quite relevant molecular size (up to 350 Da), as 367 of today's approved drugs comprise ≤ 17 atoms. Combinatorially, heteroatoms and unsaturated bonds were scattered over the generated molecular graphs. Various filters, considering the chemical stability of the introduced functional groups, the strain of generated ring systems and the formation of tautomeric forms, resulted in the end in a database of 166,443,860,262 structures. Comparing this number with the known biologically active substances (about 2.5 million) that meet the criteria of molecules with 17 atoms, it seems that only a small fraction has been synthesized so far. It is interesting to notice that the number of entries increases exponentially with the square of the atomic number. For 15 active agents currently on the market with 14–17 atoms, several million isomers with the same empirical formula can be identified in each case. Molecules with small ring systems or nonaromatic heterocycles are encountered with greater extent in the systematically generated database. Since substances with these building blocks are more difficult to synthesize or often do not have the required stability, they occur much less frequently in the molecules synthesized to date. Also, acyclic substances can be discovered in the systematically generated database much less frequently than actually approved active substances can be found with this composition. Overall, the generated database entries show a higher proportion of polar compounds compared to the known active substances. It is also significant that many more molecules with a spatially bulkier structure appear in the systematically generated database. Candidates from medicinal chemistry often tend to have a spatially flat geometry. At this point, there is often a call for natural products as supposedly better candidates for drug development, since they usually have a "higher three-dimensionality" (often described as "escape out of the molecular flatland"). It is all the more interesting to see that the database of systematically generated molecules comprises a significantly higher number of stereogenic centers per molecule than is actually realized in the collection of known active substances of comparable size.

It is worthwhile to take a closer look at a smaller database comprising molecules of a size up to 11 non-H atoms. The average molecular mass in this database is 153 ± 7 Da. Molecules of this size fall into the range of typical fragments or "lead-like" molecules (Sect. 7.9). Exclusion criteria were proposed that emphasize promising candidates for drug development. The so-called "rule of three" leans on the "rule of five," which was established by Chris Lipinski at Pfizer (Sect. 19.7). If the database is filtered with these rules, approximately half of the entries will remain. Of these, ca. 15% are acyclic compounds, and about 43% contain one ring. It is very enlightening to see that only about 55% of the ring systems in the virtual database have been described in *Chemical Abstracts* or *Beilstein*. Comparison with a data collection of already-synthesized molecules of the same size makes clear where the chemical space has been only sketchily explored. It seems that very large gaps still exist! Over 99.8% of the entries in the virtual database are waiting to be synthesized. A comparison of the physicochemical properties of the molecules in both databases suggests that very broad areas still remain that until now have not been explored. If the chemical space is limited to compounds with 7, 8, or 9 atoms, it seems that the chemical space is well covered with already prepared molecules. Approximately 2/3 of the molecules with 10 or 11 atoms in the virtual database are chiral. In this group particularly, there are many candidates that meet the "lead-like" criteria. This is a real challenge for synthetic chemists. Chiral fused carbo- and heterocycles are difficult to make. Nevertheless, Nature has led the way: many biologically active natural products contain just these building blocks.

11.5 Compound Libraries on Solid Support: Complete Conversion and Easy Purification

An interesting variation to classical chemistry in solution is found in the synthesis of compound libraries on solid supports. Organic polymers, usually cross-linked polystyrenes, are used as carriers. This material is chemically modified so that it carries numerous reactive functional groups of a particular sort, for example, chloromethyl, carboxylate, or amino groups. Through these groups, the reaction product remains covalently attached to the insoluble polymer during the synthetic steps. Stepwise growth of the product is accomplished by coupling with appropriately protected building blocks (e.g., amino acids) and subsequent cleavage of these protecting groups. Large excess of reagents causes fast and nearly complete transformations. Unreacted starting materials can be removed by simple washing. After assembly of the target molecule, all protecting groups are removed. At the end of the synthesis, the product is either tested directly on the support or it is cleaved and its biological activity is tested in solution (Sect. 11.7).

The technique can be easily automated. In the beginning of the 1960s, Robert Bruce Merrifield developed **solid-phase synthesis** for peptides and small proteins (◘ Fig. 11.3). This earned him the 1984 Nobel Prize in Chemistry. At the beginning of the 1980s, the idea to use synthetic combinatorial principles for peptide synthesis emerged for the first time. H. Mario Geysen devised a multipin synthesis of peptides. By using a conventional Merrifield solid-phase synthesis, 96 different peptides or defined peptide mixtures were prepared in an 8 × 12 format on polymer pins. This concept was so revolutionary that the originally submitted manuscript was rejected for publication in 1984. The referees were too severely restricted by their traditional thinking. The absolute control of stoichiometry and yield were less in the foreground for Geysen, rather the creation of combinatorial diversity with minimal effort was more important. In this way, thousands of different peptides could be prepared weekly. Entire libraries of compounds could be prepared and tested. The new methods were originally used for "epitope mapping," that is, the structural probing of the surface of a protein with different antibodies (Sect. 32.1). This technique allows the recognition of areas in a polypeptide chain that are exposed to the surface of a protein. Later it served the search for optimal sequences of protease substrates (Sect. 14.6) and for the synthesis of biologically active peptides. In addition to the multipin method, high-efficiency methods have been established, for instance, the teabag method. Support beads are filled into teabags and dipped into solutions of protected amino acids with which their peptide sequence is to be elongated.

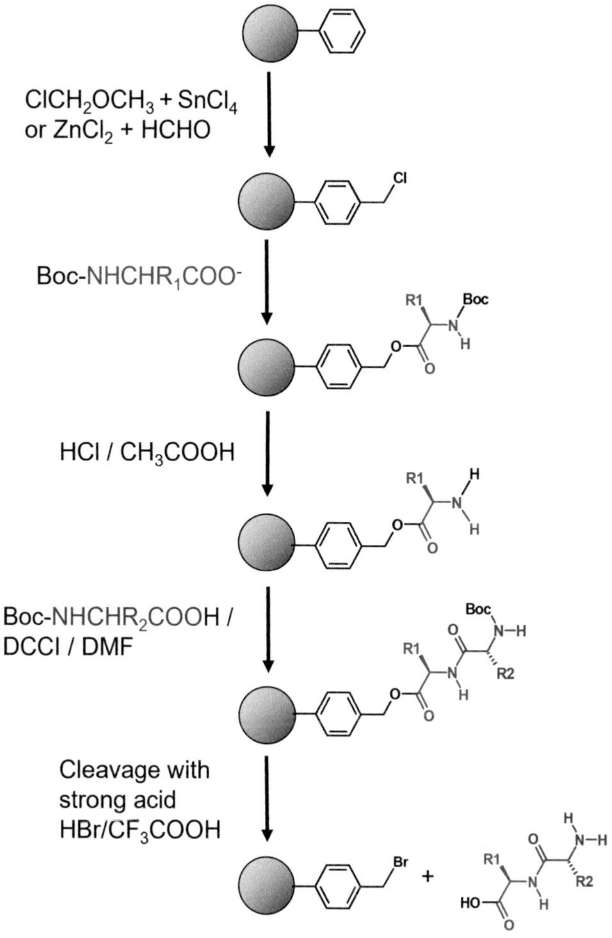

◘ **Fig. 11.3** The Merrifield peptide synthesis is assembled on a polymeric resin that is functionalized in an appropriate way. The first *N*-terminal-protected amino acid is coupled to the chloromethylene group (Boc = *tert*-butoxycarbonyl protecting group). Then the amino group is released, activated with dicyclohexylcarbodiimide (DCCI), and coupled with a second amino acid. The *N*-terminus of the resulting dipeptide can be deprotected and elongated. It can also be cleaved from the resin under strongly acidic conditions as a peptide

11.6 Compound Libraries on Solid Support Need Sophisticated Synthetic Strategies

A highly sophisticated synthetic strategy is required to build compound libraries. Hexapeptides are considered as an example. In principle, all 20 proteinogenic amino acids could be used and 20^6 = 64 million hexapeptides prepared and individually tested—an impossible undertaking. Therefore, intelligent strategies are needed to quickly identify biologically active sequences. As a consequence, an attempt is made to summarize the 64 million peptides in partial libraries. They contain constant amino acids in fixed positions. For example, all 400 partial libraries should be prepared for all possible hexapeptides with the form XXABXX (A, B = predefined amino acids, and X is any mixture of proteinogenic amino ac-

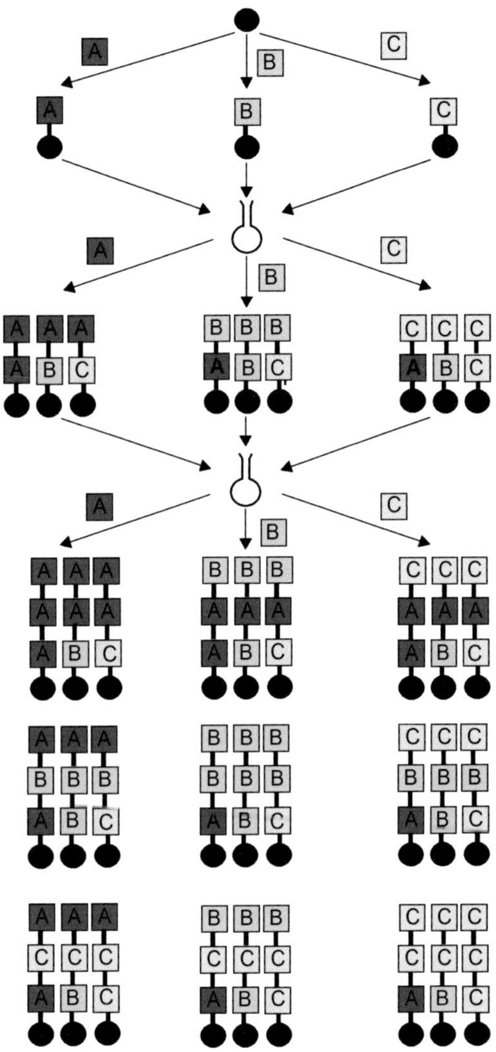

Fig. 11.4 The construction of a compound library according to the *split-and-combine* technique starts with a certain amount of resin beads. These are evenly distributed among *n* reaction vessels. Only three are considered here for the sake of simplicity. In the first flask reagent A (e.g., amino acid A) is coupled to the resin. Reagents B and C are analogously added to flask 2 and 3. In the next step, a dipeptide is constructed. To solve the problem of different reaction rates between the different amino acids A, B, and C, only one soluble reaction partner is added in excess to the mixture of solid-phase-bound starting materials. After the first reaction step, the resin, which is now loaded with an amino acid, is combined and mixed. It is again distributed between three (or more) reaction flasks. The next reaction is carried out. In the case of a peptide synthesis, amino acid A is added to flask 1, B to flask 2, and C to flask 3. The resin is combined and mixed thoroughly. In the meantime, all nine possible dipeptides are on the beads. After separating the beads again, the third step follows. In case the peptide chain is to be extended by another amino acid, amino acid A is added to flask 1, B to flask 2, and C to flask 3. Now all 27 imaginable sequential tripeptides are on the resin after three parallel reaction steps. A clearly identifiable compound is found on each resin bead. The library can be tested directly on the polymer or it can be tested in solution after cleavage from the support

ids). After testing these 400 substances, the biologically most active mixture is the starting point for the second round of synthesis. Another 400 libraries are generated, this time with the form XA(Aa1)(Aa2)BX. Aa1 and Aa2 are the amino acids from the most active mixture from the first testing. These are also subjected to the test. This identifies the "best" amino acids for positions 2 and 5. The strategy is followed step by step until the most active sequence is identified.

In a simpler procedure, the amino acids are varied in one position at a time. By starting with 20 libraries AXXXXX the most active amino acid (Aa1) is determined in the first position. The starting point for the next synthetic cycle is the most active mixture (Aa1) XXXXX. By varying the adjacent position, the second amino acid (Aa2) is ascertained. This is repeated and $6 \times 20 = 120$ hexapeptide libraries are prepared in the form of AXXXXX, (Aa1)AXXXX, ... (Aa1)(Aa2)(Aa3)(Aa4)(Aa5)A until the "best" amino acids in all positions are determined.

Another method allows the targeted construction of a library in a few working steps. The conceptual design of the synthesis ensures that a defined compound is produced on each polymeric support bead. This is achieved by using the so-called "**split-and-combine**" technique (Fig. 11.4). For example, it is possible to synthesize all 8000 possible tripeptides from the 20 proteinogenic amino acids in only 60 reaction steps. They are produced as 20 mixtures of 400 substances each. In the end, one definite compound is located on each polymer bead. The individual beads are available as a batch that is easily separated mechanically and individually tested.

11.7 Which Compound in the Solid Support Combinatorial Library Is Biologically Active?

The libraries generated on the solid support are biologically tested. This can be done directly on the polymer-immobilized compounds. As with the testing of bacteriophage libraries, there is a risk that the support material will interfere with the test, for example, by steric hindrance or unspecific interactions. Furthermore, it is important that the protein to be tested is in a soluble form. Membrane-bound receptors will, therefore, be excluded from the assay. Alternatively, the compound library can be cleaved from the resin. To do this, the resin must be coupled to the library component by a suitable "linker" that allows the library to be selectively released. This linker can be cleaved at low pH or photochemically with UV light. However, it must not interfere with the synthetic assembly of the library and must not be cleaved during the course of synthesis. Final cleavage from the resin must not destroy the products. Testing the cleaved products certainly correlates better with physiological conditions. Spreading the cleaved compounds over a large area or embedding them in a gel provides spatial separation so that compounds interacting with the test protein occur in local high concentrations. In this way, binding to insoluble proteins (e.g., membrane-bound receptors) can be tested. However, the advantage of mechanical manipulation of a polymer-bound library is lost upon release.

If biological activity is found in the assay, the next step is to determine which compound from the library is responsible. If the library is well defined by the synthesis protocol, then it is known which compounds have been tested. The active compounds are narrowed down by deconvolution and resynthesis of sublibraries. The **one-bead one-compound technique** produces only one defined compound on each resin bead. However, it is not known which compound it is. The characterization of the compound is attempted only after the activity is identified. There are many ways to do this: they can be tested on the resin by separating the relevant resin beads and analyzing the compounds. If the library consists of peptides or oligonucleotides, peptide sequencing is performed by Edman degradation (works even at 0.1 picomolar!), or polymerase chain reaction (Sect. 12.1) allows amplification and enrichment of oligonucleotides.

More sophisticated techniques are also used. During synthesis, the library is allowed to "grow" on several different linkers. The individual library compounds can be released from these linkers under different conditions (e.g., different pH values or photochemically at different wavelengths). First, the compound is cleaved from the first linker to perform the assay. Cleavage from the second linker is performed after mechanical separation of the desired resin beads. This method effectively "labels" the resin beads. The technique is, therefore, an elegant variation on library testing in the detached state. The different linker-bound compounds on the resin bead need not be identical. Thus, a test library of peptides can be linked to the resin bead by oligonucleotides used as labels. Halogenated aromatics have also been proposed as labels because they are easily identified by mass spectrometry, even in minute amounts. The labels can even be encoded with a binary code based on their sequence or the number of monomer building blocks.

In recent years, the new technology of **DNA-encoded chemical libraries** (**DECL**) has emerged. It starts with an oligonucleotide that is equipped with a chemical linker group. At this linker, molecules for biological testing are assembled in successive steps by chemical synthesis. Each individual synthesis step is documented by attaching a nucleotide to the original starting nucleotide. Comparable to a barcode, a DNA-encoded label is created. Many DNA-compatible reactions have been developed in order to link new synthesis building blocks to the growing test compounds. The technique does not necessarily rely on syntheses on a resin. It is also possible to create libraries purely in solution. This greatly simplifies the subsequent testing of the library in a biochemical activity assay with the target protein. Thus, the target protein can be immobilized on a solid support. The library then flows past the target protein on a column. Detected hits captured from the library are released from the protein again, e.g., by washing or with a known displacement ligand, and they are subsequently isolated. They are then analyzed by decoding the genetic information of the DNA label. PCR amplification (Sect. 12.1) is used for this purpose, followed by high-throughput DNA sequencing. In the meantime, however, techniques have been developed that do not require immobilization of the target proteins at all. Even testing in living cells has been reported. Extensive libraries have meanwhile been created. New clinical candidates have already been identified using this screening approach.

11.8 Combinatorial Libraries with Large Diversity: A Challenge for Synthetic Chemistry

Another aspect speaks for the last above-mentioned concept. In the meantime, a large number of organic reactions have been transferred to solid-phase synthesis. For each solid-phase synthesis, a special strategy, a specific linker, and a suitable cleavage method must be developed. Each single synthetic step must be compatible with the protecting groups, the polymer support, and the linker. However, a whole new dimension of chemical diversity is made available than is possible with peptides and nucleotides.

Careful design of the target molecules to be synthesized is indispensable for combinatorial chemistry. Limitations arise from the accessibility, that is, the development of an appropriate synthetic scheme, and furthermore from the desired structural **diversity** of the resulting library. Computer methods help to find a "reasonable" selection of synthetic components. How is the optimal composition obtained? This highly depends on what the constructed library should be tested for. A library can be developed for general-purpose screening. It should then be "**optimally diverse**." Their composition is based on general criteria such as molecular weight, overall lipophilicity, a balanced distribution of hydrogen-bond donor and acceptor groups, and the size of the hydrophobic surface. These criteria are important for the similarity or diversity of drug molecules (Chap. 17). However, the desired diversity of a library can also be considered in terms of its biological properties at a given receptor (**target-oriented**). Criteria that make molecules "similar" or "diverse" for one receptor are not necessarily identical for another receptor (Sect. 17.7). Thus, with respect to the wide range of proteins on which combinatorial libraries will be tested, there is no absolute measure of diversity. On the other hand, combinatorial chemistry plays an important role in establishing initial structure–activity relationships for a target protein. This requires very rapid chemical variations in different regions of a lead structure that has been found to be suitable. The design and synthesis of targeted compound libraries provides a rapid solution to this problem.

11.9 Nanomolar Ligands for G-Protein-Coupled Receptors

The following two sections present examples of how combinatorial libraries can be designed and successfully synthesized. Chemists at the company Chiron synthesized a library of trimeric *N*-substituted oligoglycines (peptoids) by using the *split-and-combine* method (Fig. 11.5). In their design of the nitrogen substituents, the scientists had G-protein-coupled receptors in mind. These receptors are the targets of many neurotransmitters and hormones. In the construction of their peptoids, they combined at least one aromatic group and a side chain with an H-bond donor in the form of a hydroxyl group (Fig. 11.5, groups A and O). Furthermore, a basic nitrogen atom is present in the molecules with X=H. These groups match those also found in neurotransmitters and hormones. For the remaining third substituents (group D), the substituent composition was chosen to be as diverse as possible. From these groups, a peptoid library of approximately 5000 di- and tripeptoids was prepared.

Different mixtures were tested on the adrenergic receptors. The H–ODA–NH_2 partial library was identified as the most active one. It served as a starting point for the stepwise deconvolution of the library. Partial libraries were resynthesized, first by keeping the hydroxy side chain from group O constant, then the members of the diverse group D, and finally the aromatic substituent from subset A. In the end, **11.7** remained as a nanomolar ligand (Fig. 11.6).

The same peptoid library was tested on another GPCR, the opiate receptor. In this case the most active partial library H–ADO–NH_2 was found in the first step. The relevant deconvolution through resynthesis delivered **11.8** as a nanomolar ligand. The molecule has a *p*-hydroxyphenelethyl moiety and a diphenylmethane group on both ends of the tripeptoid. It is known from detailed studies on Met-enkephalin **11.9** that the amino acids tyrosine and phenylalanine are essential for the activity. There are analogous groups for both moieties on the tripeptoid (Fig. 11.6).

11

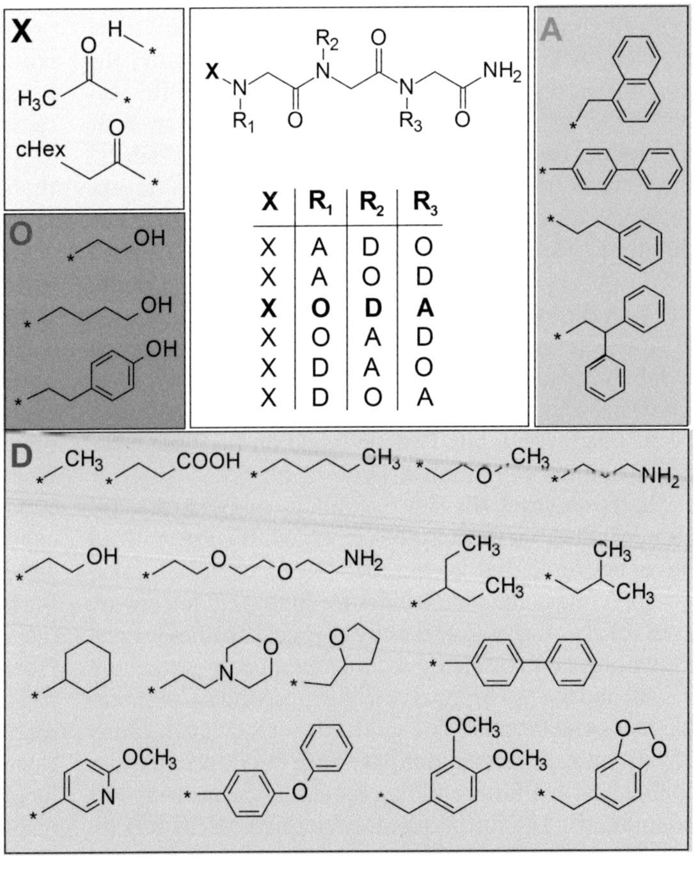

X	R_1	R_2	R_3
X	A	D	O
X	A	O	D
X	**O**	**D**	**A**
X	O	A	D
X	D	A	O
X	D	O	A

Fig. 11.5 Peptoids are oligoglycines that have substitutions at the nitrogen. A library of di- and tripeptoids was constructed according to the *split-and-combine* technique. Three **X** groups were added to the *N*-terminus. Three groups **O** with a hydroxy function, four groups **A** with an aromatic ring, and 17 groups **D** with diverse groups were used as nitrogen side chains. Eighteen mixtures (six permutations of **A**, **O**, and **D** with three end groups) gave ca. 5000 di- and tripeptides. The H–ODA–NH_2 library showed activity on the α-adrenergic receptor. First, the hydroxy groups **O** were deconvoluted. The compounds with *p*-hydroxyphenethyl groups were the most active ones. In the next synthesis round, 17 partial libraries were composed with this **O** group held constant, and defined groups were used from the diverse **D** group. Compounds with a diphenyl or diphenyl ether group were particularly active. With these groups in the D position, the work was continued. Deconvolution of the aromatic side chains A in the last position resulted in eight individual compounds

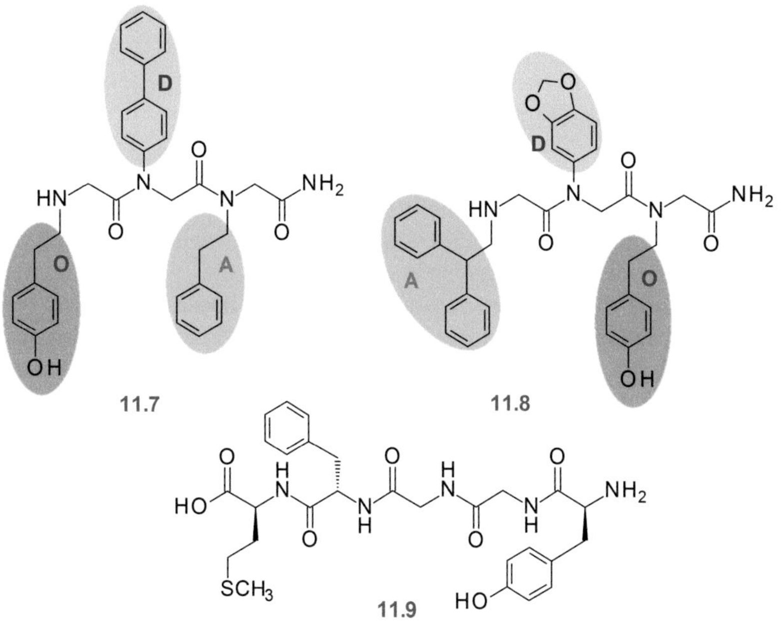

11.7 11.8 11.9

Fig. 11.6 The derivative **11.7** is the most potent compound from the H–ODA–NH_2 library with a $K_i = 5$ nM on the α-adrenergic receptor. Testing on the opiate receptor gave compound **11.8** as the candidate with highest affinity ($K_i = 6$ nM) from the H–ADO–NH_2 library after deconvolution. Met-enkephalin **11.9** is a potent opiate receptor ligand. The relationship between the *p*-hydroxyphenyl group in **11.8** and the tyrosine side chain in **11.9**, and a phenyl portion in the diphenylmethane groups of **11.8** and the benzyl groups of phenylalanine in **11.9** is obvious. Tyr and Phe are essential for the activity of Met-enkephalin

11.10 More Potent than Captopril: A Hit from a Combinatorial Library of Substituted Pyrrolidines

The Affymax company prepared a library of ca. 500 differently substituted pyrrolidines by 1,3-dipolar cycloaddition. In the first step, the resin was loaded with protected amino acids (Gly, Ala, Leu, and Phe; Fig. 11.7). Then the transformation to an imine was made with four different aromatic aldehydes. Cycloaddition with five different alkenes led to five-membered-ring heterocycles. In the last step, the pyrrolidines were *N*-substituted with three different thiols.

This last step was done in view of testing these ligands on the angiotensin-converting enzyme (ACE, Sect. 25.4). Inhibitors of this enzyme contain a functionalized proline residue at their *C*-terminus. The iterative deconvolution of the library afforded **11.10** as a potent ACE inhibitor (Fig. 11.7; $K_i = 160$ pM). It is distinctly a stronger binder than the marketed product captopril and belongs to the most potent thiol-containing ACE inhibitors.

11.11 Parallel or Combinatorial, in Solution or on a Solid Support?

While solid-supported combinatorial chemistry has enabled the automated synthesis of a large number of molecules, it has also posed problems. The difficulties of testing on resins or deconvolution and resynthesis of libraries have already been mentioned. Labeling is an elegant but laborious alternative. Another way to avoid deconvolution of a library but still take advantage of combinatorial chemistry is parallel synthesis in spatially separated reaction vessels. Throughout the entire reaction sequence, it is clear which reactant and product are in each vessel. There is no need for laborious deconvolution. At first glance, this strategy seems impractical. How can a thousand reaction components be reasonably transformed into a thousand reaction vessels? The reaction vessels should not be considered in the classical sense of organic chemistry. Instead, miniaturized reaction "automats" are developed, in which all reaction steps are carried out in parallel. Alternatively, methods have been developed in which the resin beads are filled into many small reaction capsules. These are open to the solution phase for compound transport, but the beads are mechanically enclosed. Each capsule has a label that can be read by a radio transmitter. All the capsules are then placed in a conventional round-bottomed flask and the usual chemistry is carried out. The capsules can be mechanically separated and brought into contact with different reagents. Which reaction sequence is performed on which capsule is tracked by the registration system with the radio transmitter. In this way, one molecule per reaction capsule can be produced by combinatorial principles, virtually as in parallel synthesis. The individual compounds are then available for testing

Synthesis on a solid support has disadvantages compared to solution phase chemistry. In general, the transformations are slower and the analysis to follow the reactions is much more complicated. Coupling to the solid support requires a suitable linker. Such a linker should be removed from the library prior to testing. Most importantly, removal of the linker (**"traceless linker"**) should not leave any functional groups in the library that could unintentionally be part of the pharmacophore. The chemistry used to attach and remove the linker must be compatible with all other reactions in the synthesis of the solid-supported library. This can limit the chemistry that can be applied. In preparative chemistry, molecules are preferably constructed using a **convergent synthesis strategy**. For this, a synthesis strategy is developed in which the components of the final product are prepared in separate steps, each in parallel. In the subsequent reaction steps, the previously prepared components are brought together and coupled to form the final product. Such a strategy is more efficient and leads to higher yields than a **linear synthetic route**.

However, a convergent strategy cannot be implemented by sequentially building on a resin. Therefore, for some syntheses, the tables have been turned. The prepared libraries are not bound to the solid support, but to the reagents with which they are treated. The advantage of carrying out reactions on the solid support remains. Good mechanical separation of reaction components, ease of working with large excess of reagents, and automated reactions are part of this technique. An advantage is that convergent syntheses are now possible. Even toxic reagents can be used, as their separation is ensured by their firm adhesion to a solid support. The usual analytical methods typically applied to the solution phase can also be used.

Some reactions, especially ring closure reactions or condensations, compete with intermolecular transformations. To avoid this, highly dilute solution conditions are used. When a solid supported reactant is used, the local concentration of the reactant is reduced because it is fixed to the solid support and spatially separated. Reactions that occur over a trapped reaction product can be simplified if the trapping reagent is coupled to a solid support. Mechanical filtration is sufficient to separate the trapped components. Similarly, products can be separated and purified by trapping them on a solid support. Acids and bases can be separated for purification by treatment with an immobilized amine or sulfonic acid. Metal complex formation or hydrophobic adhesion groups are now being used for the purification of combinatorially produced compound libraries.

How will combinatorial chemistry further evolve? The miniaturization of reaction vessels and synthetic automats seems to be a promising perspective. The **lab-on-a-chip concept** is already widely used for bioanalytical methods. Small reaction volumes, integrated separation columns, miniaturized valves and pumps controlled by piezo elements are integrated on small chip cards. It remains to be seen whether such serial reaction automats are the laboratories of the future.

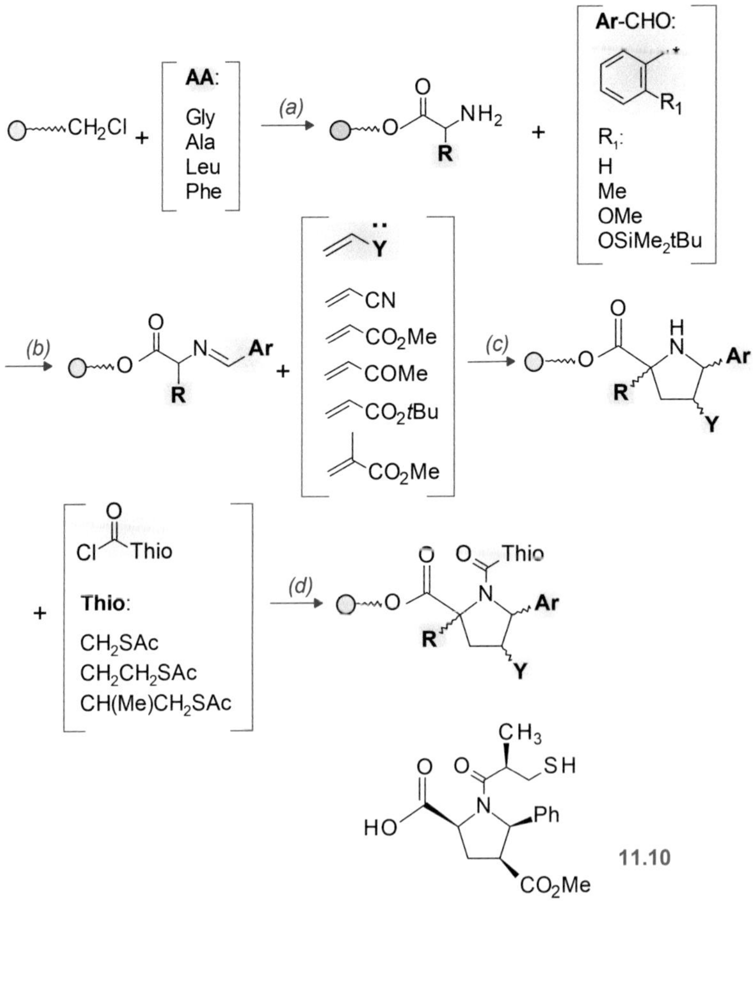

11.10

Fig. 11.7 The amino acids AA=Gly, Ala, Leu, or Phe are coupled to the support resin (**a**). Next, they are transformed to imines with four different aromatic aldehydes (**Ar**–CHO); (**b**), which react with alkenes under 1,3-dipolar cycloaddition conditions to give pyrrolidines (**c**). In the last step, the free NH proton on the heterocycle is treated with different thiol compounds (Thio-COCl); (**d**). With the help of the *split-and-combine* technique the library is cleaved from the polymer with release of an acid function. Its ability to inhibit the angiotensin-converting enzyme was tested. By resynthesis and renewed testing, the library was deconvoluted to the active compound. In doing so, compound **11.10** was identified as a high-affinity inhibitor

11.12 The Protein Finds Its Own Optimal Ligand: Click Chemistry and Dynamic Combinatorial Chemistry

Could a protein simply produce its own best inhibitor? This would have to be synthesized with an ideal geometry and optimal interactions directly in the binding pocket of the target enzyme or bound there immediately after synthesis. Which chemical reactions are suitable for such a concept? They must be reactions that can be performed in an aqueous medium. They should be orthogonal to the chemistry used by the biological systems themselves. In this way, unwanted cross reactions can be avoided. They must proceed with high fidelity, enthalpically driven, fast, and with nearly complete turnover. Such reactions have been developed under the term "**click chemistry**" in the groups of Barry Sharpless in La Jolla, California, USA, and Morten Meldal at the University of Copenhagen, Denmark. Carolyn Bertozzi, now at Stanford University in California, USA, was able to use these reactions in living organisms. For this work, the three scientists were awarded the Nobel Prize in Chemistry in 2022.

Cycloadditions of unsaturated compounds (1,3-dipolar cycloadditions, Diels–Alder reactions); nucleophilic substitutions, especially ring-opening reactions; nonaldol-like carbonyl reactions; and additions to C–C multiple bonds fulfill these requirements. These can be applied using combinatorial principles. The 1,3-dipolar cycloaddition (**Huisgen reaction**) is particularly well suited for the construction of five-membered triazole and tetrazole heterocycles (■ Fig. 11.8). 1,4-Disubstituted 1,2,3-triazoles can be regiospecifically prepared by the reaction of an azide and an alkyne in the presence of Cu(I) salts at room temperature. 1,5-Disubstituted triazoles are formed when copper ions are excluded or other ions such as ruthenium are added. The reaction proceeds over a wide pH range between 4 and 12. The reaction type can be extended to tetrazoles. This requires nitriles as dipolarophile reactants in the presence of zinc ions.

An example of the use of click chemistry is the search for an inhibitor of the target protein acetylcholinesterase (AChE, Sect. 23.7). The enzyme was added to a mixture of potential azide and alkyne reactants for the Huisgen reaction. Of the many possible reaction products, the enzyme captures two femtomolar inhibitors (**11.11**, **11.12**)! Provided with a phenylphenanthridine moiety at one end and a tacrine head group at the shallow entrance, azide and alkyne react in the center of the hose-like binding pocket to form a triazole (■ Fig. 11.9). Only a few products are formed. They are determined by the possible placement of the starting materials. Crystal structures have been determined for two potent products. The newly generated triazole ring forms a water-mediated H-bond to the catalytically active Ser 203 of the protein. It appears that the triazole ring is not formed exclusively as an entropically favored linkage product. Presumably, the polar interaction with Ser 203 exerts a directional influence.

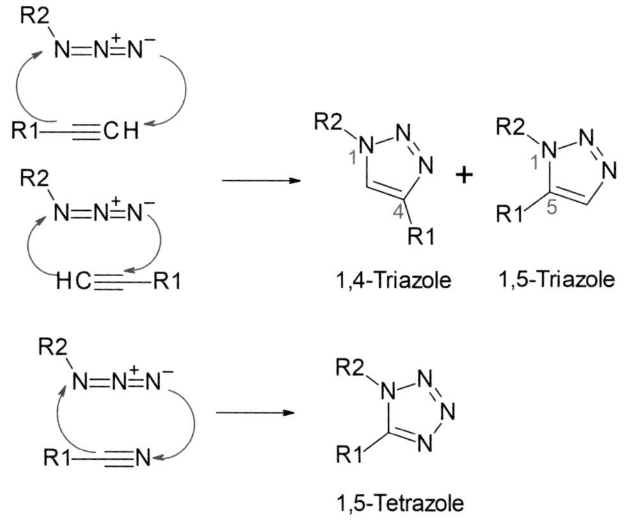

■ **Fig. 11.8** The 1,3-dipolar cycloaddition (Huisgen reaction) is a typical click chemistry reaction and leads to five-membered triazole and tetrazole heterocycles. In the presence of Cu(I) salts, azides and alkynes react regiospecifically at room temperature to form 1,4-disubstituted 1,2,3-triazoles, in the absence of copper but with ruthenium ions, 1,5-disubstituted products are formed. If a nitrile is used instead of the alkyne component, and the reaction is catalyzed by zinc ions, 1,5-disubstituted tetrazoles will be obtained as products

Jean-Marie Lehn's research group in Strasbourg, France, took a different approach. They developed "**dynamic combinatorial chemistry**" through the spontaneous construction of molecules from suitable starting materials via reversible chemical reactions (■ Fig. 11.10). All conceivable combinatorial products are formed from a mixture of different building blocks. A dynamic exchange equilibrium is established between them. The target receptor (e.g., a protein) is added to such an equilibrium system. In this way, the mixture components with the best protein-binding properties have an advantage, as the protein captures the best binders and shifts the equilibrium. This leads to a self-perpetuating selection of the ligands that fit best into the binding pocket. In this way, the added protein is effectively seeking its own best inhibitor.

A suitable reversible equilibrium reaction is the formation of acylhydrazones from a hydrazide and an aldehyde under acidic conditions (■ Fig. 11.11). Inspired by an earlier fragment screening on the aspartic protease endothiapepsin (Sect. 7.9), Milon Mondal and Anna Hirsch at the University of Gronningen in the Netherlands added this enzyme to a mixture of several acylhydrazides and aldehydes. STD NMR spectroscopy (Sect. 7.8) followed the formation of the products selected by the enzyme. Two micromolar inhibitors, **11.13** and **11.14**, were crystallographically characterized as inhibitors of the catalytic center of the protease. Based on these structures, a second round of design and dynamic

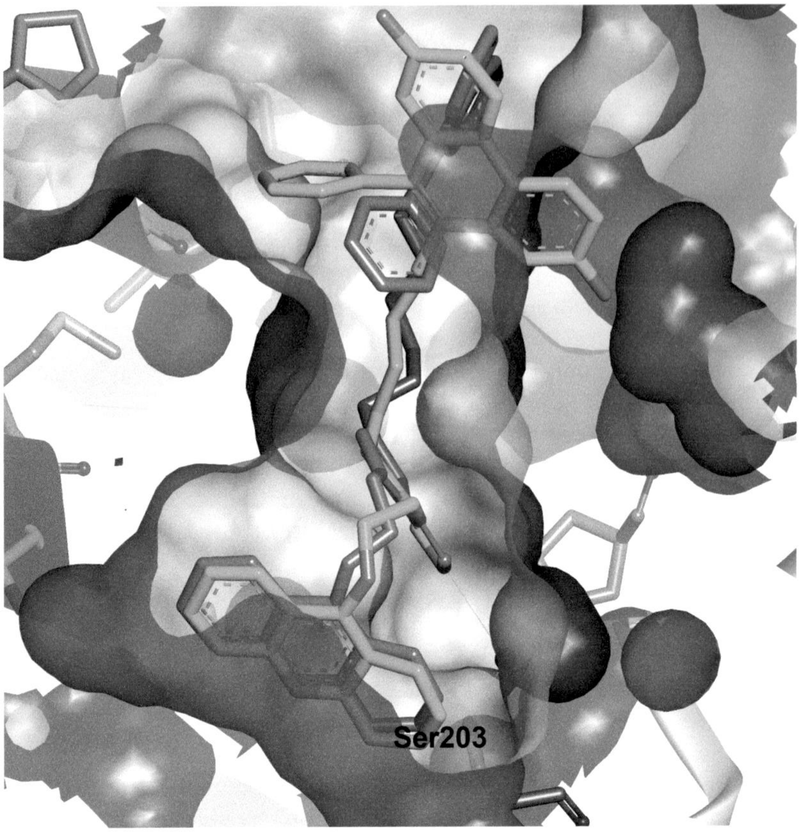

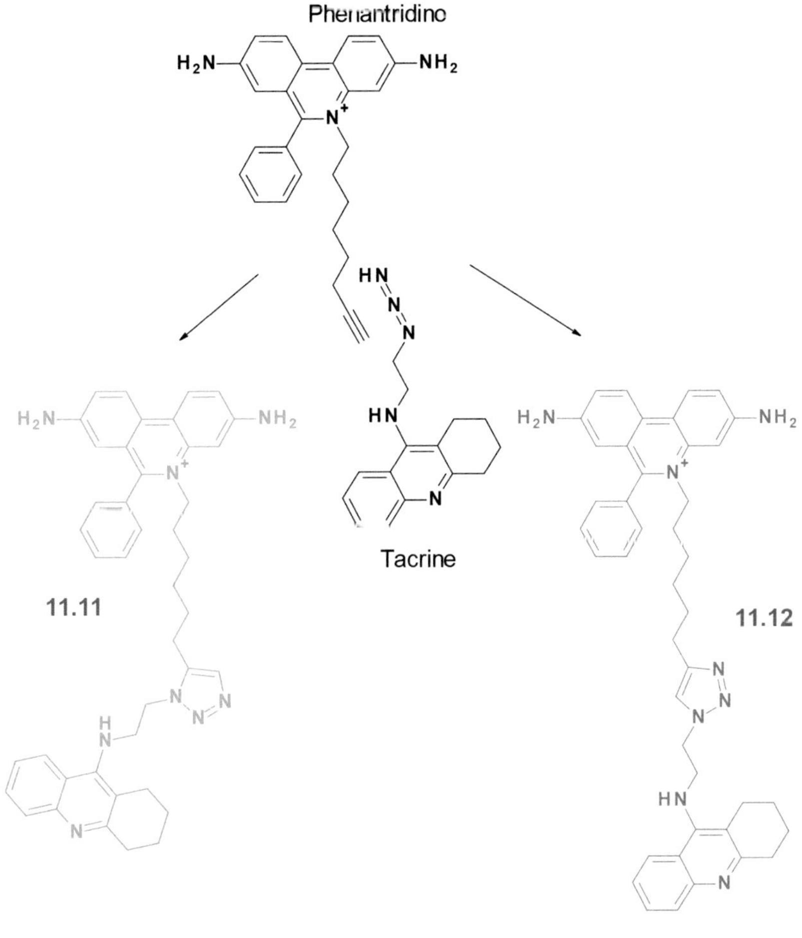

Fig. 11.9 The library produced from alkynes bearing an acetylcholinesterase (AChE)-suitable tacrine side chain and AChE custom-made phenylphenanthridine-substituted azides. In the presence of AChE the products **11.11** (*green*) and **11.12** (*gray*) are formed, which proved to be potent enzyme inhibitors. They differ in the topology on the five-membered ring. Crystal structure determinations were accomplished with both inhibitors. The surface around the protein is shown with the bound ligand **11.12**. Both ligands occupy the hose-like binding pocket of AChE. Compound **11.11** binds via a water molecule (*red sphere*) to the hydroxy function of Ser 203. (▶ https://sn.pub/vekrXl)

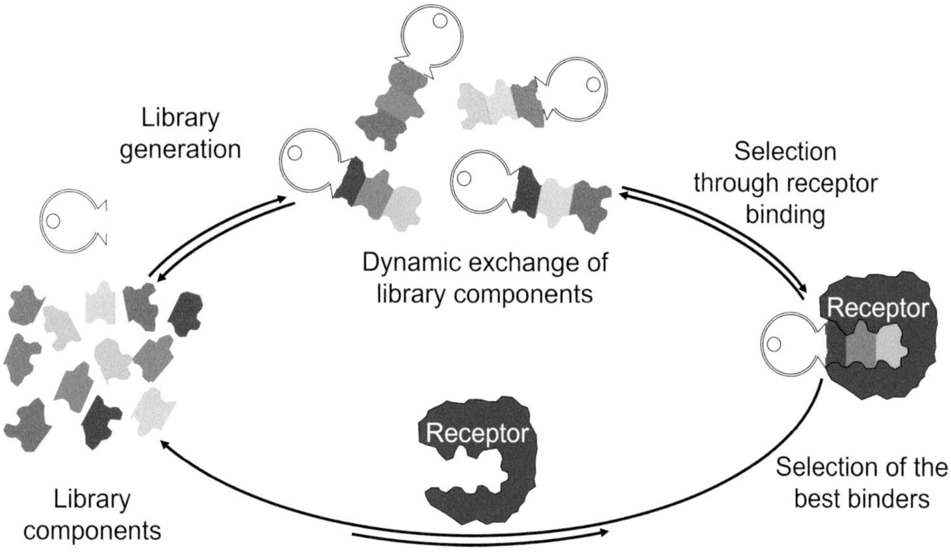

Fig. 11.10 A mixture of different library components that interact under equilibrium conditions in dynamic combinatorial chemistry is furnished. Numerous products can form in the equilibria. They represent potential "keys" that can fit in the "lock" of the target protein. The added receptor protein binds to the best-fitting ligands from the compound mixture and shifts the equilibrium in the direction of increased formation of this product. It is then removed from the equilibrium by protein binding. (Redrawn after O. Ramström and J.-M. Lehn, Nat. Rev. Drug Discov., **1**, 27–36 (2002))

combinatorial synthesis followed. It was now performed with a dialdehyde. A 54 nM inhibitor (**11.15**) was isolated from the reaction mixture.

Click chemistry and dynamic combinatorial chemistry in the presence of the target protein offer the perspective of very rapidly obtaining new potent compounds. Admittedly, the number of reactions suitable for this method is still limited, but further chemical transformations are being searched for intensively. If it is possible to elucidate the structures of the most potent products with the target protein, these can be modified by targeted design into structural analogs that also meet the criteria we demand of typical drug molecules.

11.13 Synopsis

- As a consequence of the tremendous acceleration of automated compound screening for biological activity, the number of compounds required for testing has significantly increased and stimulated the development of automated parallel synthesis and combinatorial chemistry.
- Nature produces an enormous chemical multiplicity by combining either amino acids or nucleic acids to reveal polymers that fold into 3D arrangements.
- The chemical space of organic molecules with up to 25 nonhydrogen atoms and that satisfy the requirements of drug-likeness has been estimated to host about 10^{27} imaginable candidates. A database systematically generated on the computer and containing up to 17 C, N, O, S, and halogen atoms, comprises 166.4 billion molecules.
- Chemical reactions on a solid support, usually organic polymer resins such as cross-linked polystyrene, follow a stepwise synthesis strategy to build up molecules sequentially on the solid phase. Complete conversions and easy purification can be achieved, and product release from the solid phase is accomplished as the final step.
- Sophisticated synthetic strategies have been developed to generate multiple products on the solid support from reagent mixtures in a limited number of reaction steps. Elaborate protocols have been established to keep track of product formation that also use elaborate chemical labeling techniques. In particular, DNA-encoded libraries are increasingly being used as this technique does not necessarily require syntheses on a resin. This greatly simplifies the subsequent testing of the library in biochemical activity assays.
- The biological activity testing of compound libraries generated by combinatorial chemistry on a solid support requires sophisticated protocols to detach and deconvolute the library.
- The design and selection of building blocks used for library synthesis are purpose-oriented and consider the properties of the target(s) at which the library is subsequently screened.
- Multiple protocols have been developed, either for combinatorial chemistry or parallel synthesis that immobilize either the library substrates on the solid

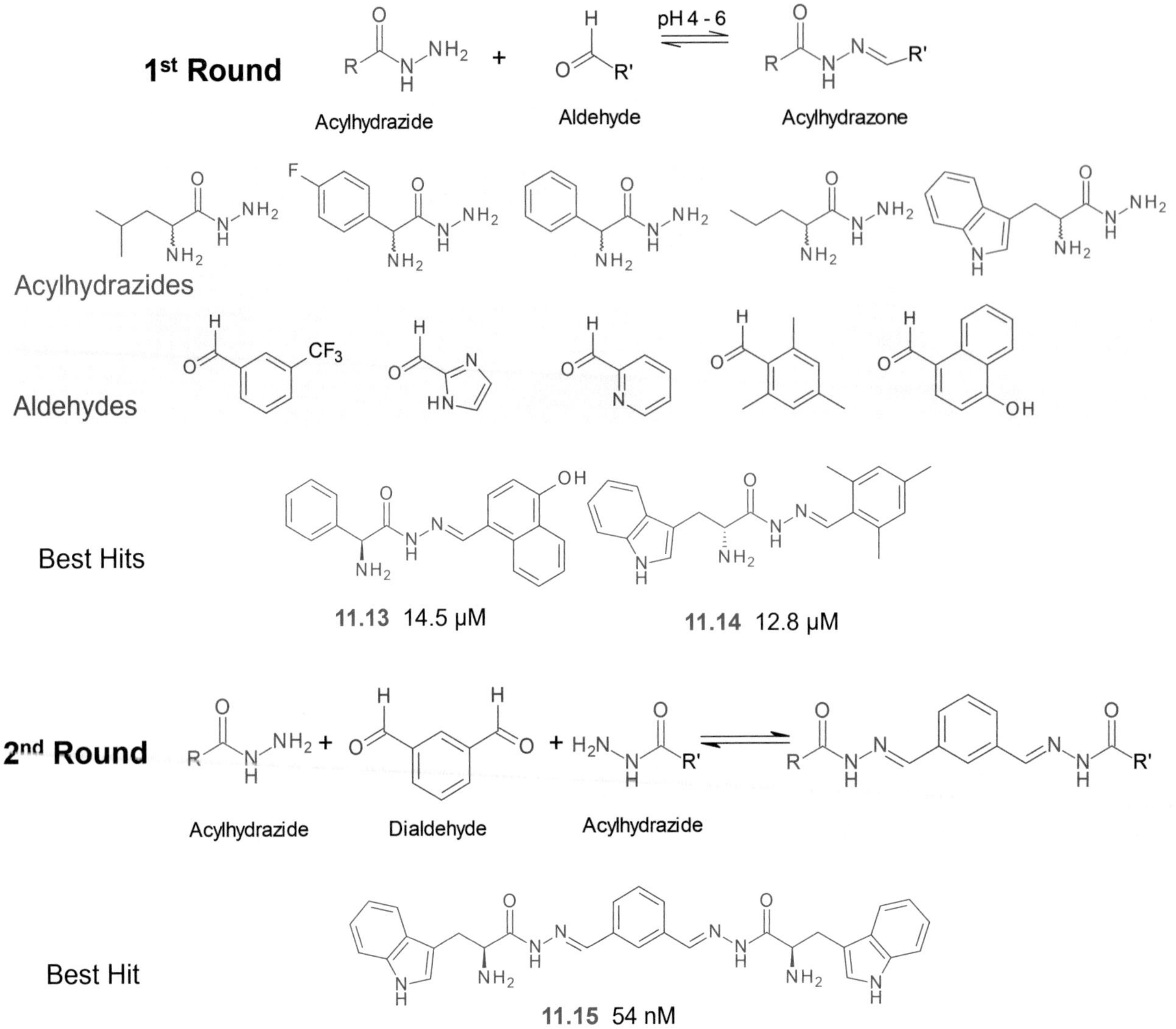

11

Fig. 11.11 Dynamic combinatorial chemistry was used to convert a mixture of acylhydrazides (*red*) and aldehydes (*blue*) to acylhydrazones in acidic conditions (1st round). From the mixture, **11.13** and **11.14** emerged as micromolar inhibitors of the aspartic protease endothiapepsin. **11.13** binds to the two aspartates of the enzyme with its hydrazone NH function mediated by a water molecule. **11.14** uses its free amino group for a direct interaction to the two residues. In a 2nd round, different acylhydrazides were again reacted but now with an aromatic dialdehyde. This mixture yielded **11.15** as a double-digit nanomolar inhibitor. It uses one of its two amino groups to bind to the aspartates

phase, or the reagents are immobilized and the library is developed in the solution phase.

- The target protein can be added to a mixture of reagents in click chemistry and dynamic combinatorial chemistry. From a large variety of possible reaction products, the protein binding pocket selects the best binder as a potent inhibitor or antagonist of the target protein.

Bibliography and Further Reading

General Literature

G. Jung and A. G. Beck-Sickinger, Multiple peptide synthesis methods and their applications. New synthetic methods. Angew. Chem. Int. Ed. Engl., **31**, 367–383 (1992)

W. H. Moos, G. D. Green and M. R. Pavia, Recent Advances in the Generation of Molecular Diversity, Ann. Rep. Med. Chem., **28**, 315–324 (1993)

L. Weber, The application of multi-component reactions in drug discovery. Curr. Med. Chem., **9**, 2085–2093 (2002)

R. M. Baum, Combinatorial Approaches Provide Fresh Leads for Medicinal Chemistry, Chemical & Engineering News, **72**(6), 20–26 (1994)

M. A. Gallop, R. W. Barrett, W. J. Dower, S. P. A. Fodor and E. M. Gordon, Applications of Combinatorial Technologies to Drug Discovery. 1. Background and Peptide Combinatorial Libraries, J. Med. Chem., **37**, 1233–1251 (1994)

E. M. Gordon, R. W. Barrett, W. J. Dower, S. P. A. Fodor and M. A. Gallop, Applications of Combinatorial Technologies to Drug Discovery. 2. Combinatorial Organic Synthesis, Library Screening Strategies, and Future Directions, J. Med. Chem., **37**, 1385–1401 (1994)

D. Madden, V. Krchnak, M. Lebl, Synthetic Combinatorial Libraries: Views on Techniques and Their Application, Persp. in Drug Discov. Design, **2**, 2690–285 (1994)

B. K. Kay, Biologically Displayed Random Peptides as Reagents in Mapping Protein-Protein Interactions, Persp. in Drug Discov. Design, **2**, 251–268 (1994)

F. Balkenhohl, C. von dem Bussche-Hünnefeld, A. Lansky, C. Zechel, Combinatorial synthesis of small organic molecules. Angew. Chem. Int. Ed. Engl., **35**, 2288–2337 (1996)

S. V. Ley and I. R. Baxendale, New Tools and Concepts for Modern Organic Synthesis, Nat. Rev. Drug Discov., **1**, 573–586 (2002)

H. C. Kolb, M. G. Finn and K. Barry Sharpless, Click chemistry: diverse chemical function from a few good reactions. Angew. Chem. Int. Ed. Engl., **40**, 2004–2021 (2001)

O. Ramström and J.-M. Lehn, Drug Discovery by Dynamic Combinatorial Libraries, Nat. Rev. Drug Discov., **1**, 27–36 (2002)

B. A. Bunin, The Combinatorial Index, Academic Press, San Diego (1998)

G. Jung, Combinatorial Chemistry, Wiley-VCH, Weinheim (1999)

P. Seneci, Solid-Phase Synthesis and Combinatorial Technologies, Wiley-Interscience, New York (2000)

A. G. Beck-Sickinger and P. Weber, Combinatorial Strategies in Biology and Chemistry, Wiley (2002)

K. C. Nicolaou, R. Hanko and W. Hartwig, Handbook of Combinatorial Chemistry. Drugs, Catalysts, Materials, Wiley-VCH, Weinheim (2002)

W. Bannwarth and B. Hinzen, Combinatorial Chemistry. From Theory to Application (Vol. 26 in Methods and Principles in Medicinal Chemistry, R. Mannhold, H. Kubinyi and G. Folkers, Eds.), Wiley-VCH, Weinheim (2006)

A. Gironda-Martínez, E. J. Donckele, F. Samain, D. Neri, DNA-Encoded Chemical Libraries: A Comprehensive Review with Successful Stories and Future Challenges, ACS Pharmacol. Transl. Sci., **4**, 1265–1279 (2021)

A. L. Satz et al., DNA-encoded chemical libraries, Nature Reviews Methods Primers, **2**, Article number: 3 (2022)

Special Literature

T. Carell, E. A. Wintner, A. J. Sutherland, J. Rebek, Y. M. Dunayevskiy and P. Vouros, New Promise in Combinatorial Chemistry: Synthesis, Characterization, and Screening of Small-Molecule Libraries in Solution, Chemistry & Biology, **2**, 171–183 (1995)

H. M. Geysen, R. Meloen and S. Barteling, Use of Peptide Synthesis to Probe Viral Antigens for Epitopes to a Resolution of a Single Amino Acid, Proc. Nat. Acad. Sci. USA, **81**, 3998–4002 (1984)

T. Fink and J.-L. Reymond, Virtual Exploration of the Chemical Universe up to 11 Atoms of C, N, O, F: Assembly of 26.4 Million Structures (110.9 Million Stereoisomers) and Analysis for New Ring Systems, Stereochemistry, Physicochemical Properties, Compound Classes, and Drug Discovery, J. Chem. Inf. Model., **47**, 342–353 (2007)

L. Ruddigkeit, R. van Deursen, L. C. Blum and J.-L. Reymond, Enumeration of 166 Billion Organic Small Molecules in the Chemical Universe Database GDB-17, J. Chem. Inf. Model., **52**, 2864–2875 (2012)

C. T. Dooley, N. N. Chung, P. W. Schiller and R. A. Houghton, Acetalins: Opioid Receptor Antagonists Determined Through the Use of Synthetic Peptide Combinatorial Libraries, Proc. Nat. Acad. Sci., USA, **90**, 10811–10815 (1993)

R. N. Zuckermann, E. J. Martin, D. C. Spellmeyer *et al.*, Discovery of Nanomolar Ligands for 7-Transmembrane G-Protein-Coupled Receptors from a Diverse N-(Substituted)glycine Peptoid Library, J. Med. Chem., **37**, 2678–2685 (1994)

M. M. Murphy, J. R. Schullek, E. M. Gordon and M. A. Gallop, Combinatorial Organic Synthesis of Highly Functionalized Pyrrolidines: Identification of a Potent Angiotensin Converting Enzyme Inhibitor from a Mercaptoracyl Proline Library, J. Am. Chem. Soc., **117**, 7029–7030 (1995)

Y. Bourne, H. C. Kolb, Z. Radic, K. B. Sharpless, P. Taylor and P. Marchot, Freeze-frame Inhibitor Captures Acetylcholinesterase in a Unique Conformation, Proc. Nat. Acad. Sci., USA, **110**, 1449–1454 (2004)

M. Mondal, N. Radeva, H. Köster, A. Park, C. Potamitis, M. Zervou, G. Klebe and A. K. H. Hirsch, Structure-based design of inhibitors of the aspartic protease endothiapepsin by exploiting dynamic combinatorial chemistry, Angew. Chem. Int. Ed., **53**, 3259–3263 (2014)

M. Mondal, N. Radeva, H. Fanlo-Virgós, S. Otto, G. Klebe and A. K. H. Hirsch, Fragment-Linking and -Optimization of Inhibitors of the Aspartic Protease Endothiapepsin: Fragment-Based Drug Design Facilitated by Dynamic Combinatorial Chemistry, Angew. Chem. Int. Ed. Engl., **55**, 9422–26 (2016)

Gene Technology in Drug Research

Contents

G. Klebe, *Drug Design*, https://doi.org/10.1007/978-3-662-68998-1_12

Engineers and writers have predicted many developments in science and technology. Among other sophisticated machines, Leonardo da Vinci described the principle of the helicopter. In the early 1820s, Charles Babbage designed an automatic calculating machine that was far ahead of its time. More than 160 years later, the mechanical precursor of a programmable computer was actually built, and it worked! Jules Verne described submarines and a trip to the moon, and Hans Dominik described creating energy by splitting the atom. All these visions became reality. Only one application of gene technology, the most groundbreaking invention of our time, was foreseen: the cloning of two genetically identical individuals in Aldous Huxley's *Brave New World* in 1932. Mammalian cloning has already been carried out. When Dolly, the first cloned sheep, was born in Scotland on July 5, 1996, it was hailed as a breakthrough in genetic research. Cloned mice, cattle, pigs, horses and, most recently, monkeys have followed. It is to be hoped that researchers will respect ethical boundaries and not make use of Huxley's idea of human cloning, despite its tangible feasibility.

Genetic engineering makes it possible to introduce new genes into a cell, multiply them, and exchange or remove them. When removed or altered, the cell can no longer produce the original protein derived from that gene. With the introduction of a new gene and a clever choice of method, the cell produces a foreign product, either an intentionally modified protein or a completely new one. For many diseases, the molecular cause is known to be the absence of a protein or a genetically caused mutation in a protein. These are only a few of the more common examples:

- Diabetes as a result of insulin deficiency,
- Particular, hereditary cancer forms (e.g., familial colon cancer, malignant melanoma),
- Chorea Huntington, a chronic form of brain atrophy,
- Sickle cell anemia, a genetic disease producing malformed red blood cells (Sect. 12.13), and
- Bleeding disorders that are caused by the absence of particular coagulation factors (see Sect. 12.13).

The possibility of purposefully producing arbitrary proteins has yielded the following **main applications** of **gene technology**:

- The identification of genes and proteins that could play a role in the treatment of a disease,
- The development of animal models to test a therapeutic principle,
- The production of proteins for therapies in which a particular protein is missing,
- The manufacture of monoclonal antibodies and vaccines,
- The manufacture of proteins for molecular test systems, and the determination of the 3D structures of enzymes and receptor proteins,
- The generation of proteins of which a targeted mutagenesis has been undertaken to exchange one or more amino acids for the elucidation of the mode of action of enzymes and for the characterization of receptor binding sites,
- Somatic individual gene therapy for specific patients, and
- Immunotherapy using genetically engineered T-cells with antigen-specific receptors.

Other application possibilities, for example, the manipulation of the human germline, or genetic changes in crops to achieve herbicide resistance, or to prolong the shelf life of fruits, are only briefly mentioned here.

12.1 The History and Basics of Gene Technology

The foundations of gene technology were first established in the middle of the twentieth century. It all started in 1953 after James Watson and Francis Crick in Cambridge (England) became aware of Rosalind Franklin's X-ray data; they correctly interpreted the available knowledge about **deoxyribonucleic acid** (**DNA**) and proposed the structure of a **double helix** as the three-dimensional geometry for the genetic material of all living beings. Immediate indications were obtained from the structure about the mechanism of our hereditary transfer and about the genetic code for the biosynthesis of proteins. A few years later, Werner Arber discovered enzymes that attack specific sites on the double helix and cleave DNA in a sequence-specific manner. These enzymes are called **restriction enzymes**. What was initially seen as a curiosity proved to be an exceedingly important discovery for gene technology. It is possible to selectively cleave DNA with these enzymes and to introduce new fragments. Next, the merging of new information with the original DNA, the **recombination** of the genetic constitution, is accomplished with **ligases** from special viruses called bacteriophages. The techniques for DNA sequencing have also made decisive progress. Soon afterwards, the amino acid sequence of a protein was no longer directly determined, but rather deduced from the analysis of the corresponding DNA. Today, sequencing is mainly done via cDNA, which is complementary to the RNA (Sect. 12.6).

In 1973, Stanley Cohen and Herbert Boyer managed to recombine the genome of a bacterium for the first time (◘ Fig. 12.1). Then, things happened one after the other: two years later the bacterial strain *Escherichia coli* K12, which is still used today, was developed. Part of its genetic constitution is missing, making it viable only under laboratory conditions. This bacterium can be genetically manipulated at will without fear of harm. The British scientists H. Williams-Smith and E. S. Anderson

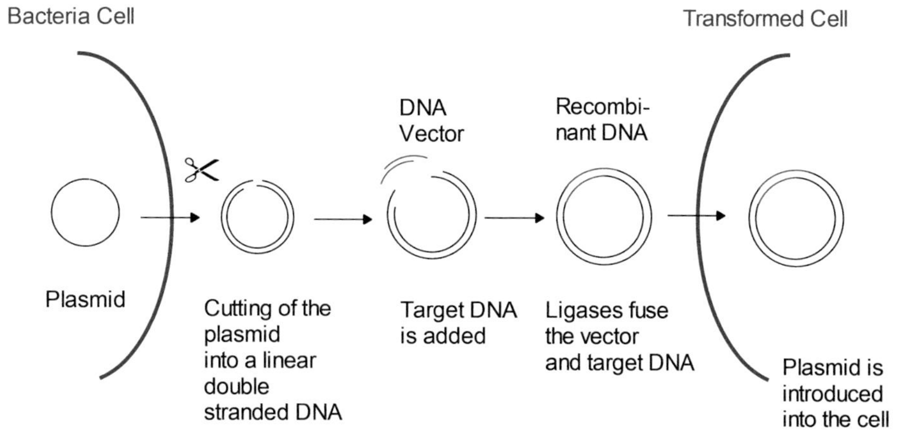

Fig. 12.1 The principle of gene technological recombination of hereditary information. Bacteria often contain additional genetic material in addition to their "chromosome" in the form of ring-shaped plasmids; these are used in gene technology as vectors to introduce foreign genes. Plasmids are removed from the cell and sequence-specifically cut with so-called restriction enzymes, which are isolated from bacteria. The target DNA that carries the desired gene, which was typically also treated with the same restriction enzyme, is bound to the overlapping single-stranded DNA ends *in vitro*. The DNA ends are coupled with the enzyme DNA ligase, and the modified, recombinant plasmid is brought into the bacterial cell. In addition to the DNA segment that is necessary for replication, plasmid vectors that are used in gene technology carry additional information that allows for the recognition and selection of the transformed cells (usually an antibiotic-resistance gene). In the presence of the selecting agent, only plasmid-containing cells grow

carried out independent experiments in which they took *Escherichia coli* K12 orally. They demonstrated that these bacteria survive only for a short time in the gastrointestinal tract and that the K12 gene, which confers antibiotic resistance for selection of the transformed cells, cannot be transferred to normal *Escherichia coli* found in the intestinal flora. At a conference in Asilomar, California, experts discussed the potential dangers of genetic engineering and defined various risk and safety classes. Genentech was founded in 1976. Its founder, Herbert Boyer, had to borrow US$ 500 as start-up capital! When the company went public in 1980, the value of his shares made him a millionaire within minutes. In 1982, Genentech introduced the first drug to be produced using gene technology: human insulin (Humulin®).

In 1983, Kary Mullis made a seminal contribution to gene technology when he developed the **polymerase chain reaction (PCR)** while working at Cetus, a Californian company founded in 1971. Double-stranded DNA is melted into its single strands by heating, then two short pieces of single-stranded DNA complementary to the regions at the beginning of the DNA, called **primers**, are added along with the four DNA nucleotides. A polymerase can now be used to synthesize new DNA in a test tube. This means that a new double strand is formed by starting with the primers (Fig. 12.2). A heat-stable DNA polymerase (originally from the bacterium *Thermus aquaticus*, endemic to the hot springs of Yellowstone National Park) is used for DNA synthesis. Each repetition of this step doubles the amount of DNA. Within a few hours, billions and trillions of DNA molecules can be produced from a single starting molecule. This amount is sufficient to sequence the DNA segment of interest.

PCR methods are applied diversely. The entire genetic information of an individual can be derived from a single DNA molecule. In medical diagnostics, this serves to provide evidence regarding genetic disorders, cancer, infectious diseases, and risk factors. PCR methods are also used to establish a genetic fingerprint in paternity tests and in forensic science.

New genetic information can be introduced not only into bacterial cells, but also into yeast, virus-infected insect cells, and even mammalian cells. However, as a first approximation, the more complex the organism, from bacteria to mammalian cells, the more difficult it is to produce proteins in these cells. On the other hand, insect and mammalian cells have the advantage of producing not only ordinary proteins, but also more complex ones (e.g., glycosylated proteins) in a functional form. Thus, in many cases, we are dependent on such organisms for the production of proteins.

12.2 Gene Technology: A Key Technology in Drug Design

The 1970s and 1980s were the golden age of receptor binding assays with membrane preparations. Radioactively labeled ligands were used to determine the specific binding of new drug candidates. The major receptors for hormones and neurotransmitters were known, and in some cases the difference between pre- and postsynaptic receptors. However, the different subtypes and

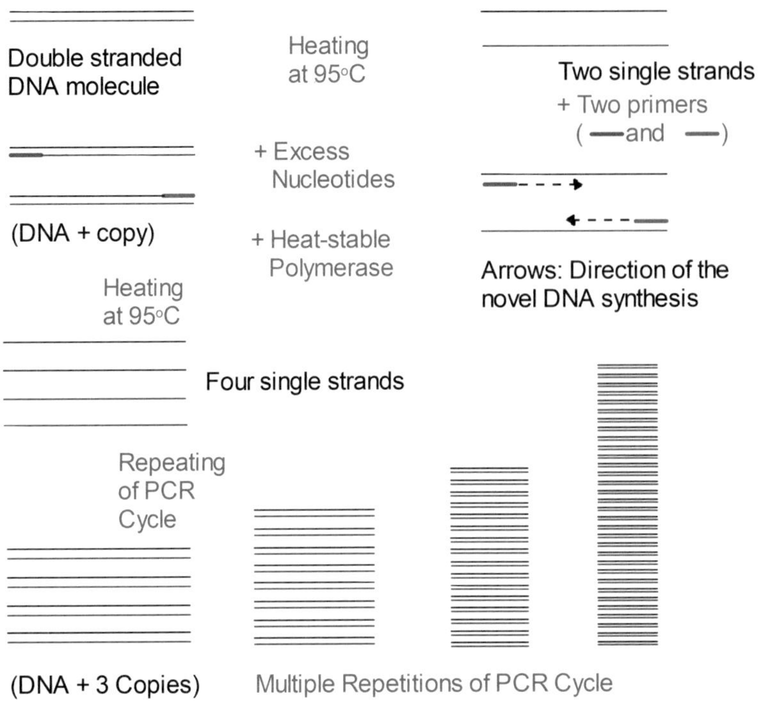

Fig. 12.2 The polymerase chain reaction (PCR) can make an unlimited number of identical copies of a DNA molecule. The DNA is heated to split the double-stranded DNA into complementary single strands. Synthetic oligonucleotides of about 20 bases, called primers, that are complementary to these DNA strands hybridize to the corresponding strand. Each primer must bind to one end of each strand of DNA. The primers define the boundaries of the amplified DNA. In addition, an excess of primers must be used because one pair of primers is required for each double strand of DNA in each cycle. The primers are needed to synthesize the new DNA in the presence of the DNA polymerase and an excess of the four different nucleotides. This occurs in the reverse direction (*dashed arrow*) due to the opposite orientation of the DNA strands and the specificity of the polymerase. The newly synthesized DNA segment can be several hundred to several thousand base pairs long. The result is two identical double-stranded DNA molecules. After heating, single strands are obtained and the above procedure is repeated. Because DNA polymerase is heat stable, it does not need to be added repeatedly. Each repetition of the above steps results in a doubling of the DNA molecule. Its number increases exponentially. Ten cycles lead to about 1000 DNA molecules, 20 to a million, and 30 to a billion. In this way, a single DNA molecule can be multiplied into a biochemically analyzable quantity

their amino acid sequences were not known. As a consequence, the results of these studies were quite imprecise.

Gene technology methods allow the production of homogeneous **recombinant proteins** in virtually unlimited quantities. They play an important role in the very first step of drug design: the **identification of a target protein**. Advances in methodology have led to the discovery of new receptors, some of which had unknown function or specificity. The next steps are to **test the therapeutic concept** in genetically modified animals. Another important contribution is the preparation of proteins for molecular test systems and the isolation of adequate material for the elucidation of the 3D protein structure (Chap. 13). With, perhaps, the exception of a few proteins that can be isolated from blood or other natural sources, the production of large quantities of proteins is dependent on gene technology. Today, the purification of proteins from animal or human blood is done rather reluctantly. The risk of transmitting viruses or infections is deemed to be too high.

Gene technology offers the possibility to selectively produce structural variants of proteins. The **generation of point mutations** (site-directed mutagenesis) allows particular properties in proteins to be improved, and the binding and catalytic properties of enzymes to be purposefully changed. Membrane-bound receptors can be probed position by position to establish which amino acids are responsible for the maintenance and stability of the 3D structure, the adoption of a particular conformation, or are of critical importance with respect to the binding of a ligand. Three-dimensional structural models of receptors can be generated in this way, or their relevance can be appraised.

In many cases, it has also proven worthwhile to introduce point mutations that change the surface properties of proteins and help to elucidate the **3D structure of proteins**. Sometimes the charge on individual amino acids or highly flexible amino acids that tend to cause disorder must be exchanged for the sake of protein crystallization. For proteins in which part of the sequence is membrane anchored, the membrane anchor, which would interfere with crystallization, is removed prior to the crystallization experiment. For soluble receptors, it has been found useful to remove individual domains, crystallize them, and determine their structure. Of course, such modified proteins must still perform their specific functions, such as ligand binding or DNA docking. Once the difficult crystallization step has been accomplished, the actual structure determination nowadays only takes a matter of a few weeks in most cases (Chap. 13). Membrane-bound proteins can be stabilized in their structure by point mutations so that they can be detached from the membrane. They can then be crystallized in special media, such as cubic lipid phases.

When considering the benefits of all this progress for humanity, we cannot avoid asking the question: where do the fears of broad sections of society about gene technology come from? It is not difficult to understand these reservations: with the use of gene technology, almost everything that is theoretically conceivable in the field of genetics becomes possible. However, people's trust in science is not as unshakable as it used to be before the atomic bomb. Now that the opportunities outweigh the risks, the sins of our forefathers have come back to haunt us. Too often in the past, scientists have underestimated potential risks and put their ethical concerns on hold. Scientists have still not managed to allay the public's fears. We must take these fears seriously and rebuild trust by acting responsibly.

12.3 Genome Projects Decipher Biological Constructions

The entire **human genome** is organized on 23 chromosomes. In 1990, the **Human Genome Organization (HUGO)**, with a budget of US$ 3 billion, began the then-ambitious task of sequencing the entire human genetic code from DNA within 15 years. By the end of 1993, the first annotated genome maps were available, and these were later refined. By 2001, the project had progressed to the point where the entire genome was published in *Science* and *Nature* by two parallel consortia.

The two competing consortia followed different strategies. The publicly funded international consortium chose the approach of setting progressively narrower parameters, the stepwise digestion of the genome, and the **systematic elucidation** of sequences for the complete genome analyses. In humans, this means that in addition to the 5% of DNA that corresponds to genes, the other 95% of sequenced DNA, the function of which was unknown, was classified with the somewhat derogatory term "junk DNA." It is now established that these regions play a pivotal role in the regulation of gene expression and host to a class of small non-coding RNA molecules, collectively termed microRNA. In 2024, Victor Ambros (UMass Chen Medical School, Worcester, USA) and Gary Ruvkun (Harvard Medical School, Boston, USA) were awarded the Nobel Prize in Medicine for their discovery of microRNA and the elucidation of their function in post-transcriptional gene regulation, particularly in the context of organismal development and function (Sect. 12.7). The second strategy, pursued by the privately funded consortium, used the **shotgun approach**. This was done by amplifying a longer strand of DNA and then cutting it into many small segments. After these segments were sequenced, the sequences were reconstructed into the original long strand of DNA using a powerful computer program. Of course, this can only work if the sequences of the cleaved segments overlap sufficiently. This technique proved to be much faster than the usual systematic sequencing methods. In particular, it benefited from the development of ever faster sequencing machines and powerful bioinformatics programs. In the end, it was not a disadvantage that the shotgun method required multiple sequencing of the genome due to the high redundancy of the method. Interestingly, the shotgun method was also used in the end by the international consortium that followed the systematic approach to elucidate local sequence regions. Since the initial intention of the private company was to patent the sequenced genome, the competition between the two initiatives was great. In March 2000, American President Bill Clinton declared that the human genome was not patentable and advocated its use by all for the common good.

How did it come that a competing private initiative started to sequence the genome? In spring 1995, Craig Venter and his group identified the entire genome for the bacterium *Haemophilus influenzae* using the shotgun method. The enormous number of 1,830,121 base pairs that code for 1749 genes was sequenced. The complete genomes of individual viruses were already known, but this was the

Table 12.1 Examples for the sequenced genomes of different organisms

Organism	Genome size[a]	Genes
HI virus [c]	9.2×10^3 [b]	9
HI-9.2 virus, Phage λ	4.85×10^4	70
Intestinal bacteria *Escherichia coli*	4.6×10^6	4800
Baker's yeast, *Saccharomyces cerevisiae*	2×10^7	6275
Pin worm, *Caenorhabditis elegans*	8×10^7	19,000
Wallcress, *Arabidopsis thaliana*	1×10^8	25,500
Fruit fly, *Drosophila melanogaster*	2×10^8	13,600
Green blow fish, *Tetraodon nigroviridis*	3.85×10^8	
Human, *Homo sapiens*	3.2×10^9	~21,500
Common newt, *Triturus vulgaris*	2.5×10^{10}	
Ethiopian lung fish, *Protopterus aethiopicus*	1.3×10^{10}	
Amoeba, *Amoeba dubia*	6.70×10^{10}	

[a] Number of base pairs
[b] Single-stranded RNA
[c] human immunodeficiency virus

decoding of the genetic information of a self-contained creature. The subsequent decoding of the sequence of 580,067 base pairs of the *Mycoplasma genitalium* genome by Venter's wife, Claire Fraser, took only four months.

Venter and his group worked with the shotgun method on the entire genome, the so-called "whole-genome shotgun sequencing." The statistical approach that was followed by Venter initially seemed so unusual and utopian that his application for a research grant from the American National Institutes of Health (NIH) was rejected. This brought about the founding of **T**he **I**nstitute for **G**enomic **R**esearch (TIGR) and the company Celera Genomics. There, Venter could pursue his research according to his ideas and plans. Finally, the success proved the feasibility of the proposed strategy.

Whose genome was actually sequenced? In both initiatives, the DNA of multiple individuals were mixed and the individual differences were purposefully calculated out. In this way the "consensus sequence" of the human genome was determined. But it did not stop with the human genome. The complete elucidation of baker's yeast *Saccharomyces cerevisiae*, and the common thale cress *Arabidopsis thaliana*, the rice plant *Oryza sativa*, the pinworm *Caenorhabditis elegans*, the fruit fly *Drosophila melanogaster*, the chimpanzee *Pan troglodytes*, the mouse *Mus musculus*, and many other organisms (◘ Table 12.1) has been accomplished. In the meantime, new ones emerge weekly. This raises new questions: how should this plethora of information be managed? How can the genetic information be translated into useful knowledge? The field of **bioinformatics** has been challenged. Computer programs for the intelligent comparison of sequences and the analysis of metabolic pathways and signaling cascades already existed. New initiatives were launched with the goal of determining the spatial structure of all or at least many sequences. The spatial structures of all real, naturally occurring proteins were slowly being elucidated. The crystal structures of all members of some protein families of the human genome have now been determined. In the meantime, structure predictions derived from amino acid sequences have become much more reliable (Sect. 20.6). Thus, it is only a matter of time until we will be able to place the sequence catalogs of very many genomes next to those with all the spatial structure maps.

12.4 What Is Contained in the Biological Space of the Human Proteome?

After the human genome was sequenced, the exciting question arose as to what gene products all these DNA sequences code for. The first thing to note is that the genome is not static but constantly changing. Only this way can the genetic variations that make up the diversity of all living creatures occur. In the course of evolution, the genetic constitution has expanded. Simple unicellular organisms without cell nuclei (prokaryotes) have a circular genome containing only coding genes. Single-celled organisms with a nucleus (eukaryotes), such as yeast, have a larger genome, of which about 20% consists of coding genes. Multicellular organisms, such as humans, have genomes that are several 100 times larger than that of yeast (◘ Table 12.1). However, the number of coding genes is not greater. In fact, some organisms, such as amoebae, have genomes that are 200 times larger than that of humans. Even the tiny water flea, with its 31,000 genes, dwarfs us numerically. Thus, the alleged masterpiece of creation does not necessarily have the largest genome. Obviously, only a small number of additional DNA sequences, which actually code for additional gene products, have accrued during the course of evolution. Many genes in higher organisms are similar to those in simpler species. If the number of coding genes has barely increased from unicellular organisms to humans, and even the gene products encoded are similar, what explains the massive increase in genome complexity in higher organisms? The answer does not lie in the diversity of required gene products, but rather in the finely tuned **regulation of gene expression** (Sect. 12.13).

In higher organisms, it is critical where and when specific gene products are synthesized. The 95% of human DNA that does not code for proteins contains numerous sequences and signals that control gene expression. Therefore, the total number of genes does not seem to increase in higher organisms, but rather **gene density** decreases. On average, there are 12 genes per million base pairs in the human genome, compared to 118 in the fruit fly, 197 in the nematode, and 221 in the common thale cress (*Arabidopsis thaliana*). Moreover, the human genome is highly scattered. It seems that it is not the number of genes, but rather how they are used and how their activation is regulated that determines the developmental state of an organism. It must also be considered that multicellular organisms also require a great deal of cell differentiation in the different organs. These processes must be reliably regulated and controlled. In addition, higher organisms achieve a much greater diversity in their protein composition through alternative splicing. **Posttranslational modification** after biosynthesis also plays a role. This is observed to a much lesser extent in prokaryotes, for example. After transcription from DNA to RNA, the splicing process cuts out parts of the RNA that do not code for proteins. In **alternative splicing**, a decision is made during splicing as to what is cut out and what is used for translation. Thus, one DNA sequence can code for several different proteins.

To date, one of the largest genomes of a prokaryote that has been found belongs to the pathogenic protozoa *Trichomonas vaginalis*. It consists of 160 million base pairs. In humans, this pathogen is usually transmitted through sexual intercourse and causes urinary tract infections. Its huge genome is disproportionately large in the cell. This could be an advantage for the pathogen because its large surface area makes it easier to adhere to the vaginal mucosa. In addition, the immune system has a tough time attacking and destroying such an oversized parasite. The genome of the soil bacterium *Sorangium cellosum* with 13 million bases and 10,000 genes is four times as large as the average genome of other bacteria. This may have something to do with the fact that this soil bacterium is able to perform special tasks that make its therapeutic use interesting. It is a versatile producer of complex natural products such as epothilones, which are potent chemotherapeutics that have great potential in the treatment of cancer.

According to estimates in 2016, the human genome consists of more than 3.088 billion bases. It contains approximately **21,500 protein-coding genes** and **several thousand RNA genes**. The former textbook knowledge that there is a gene product behind each DNA sequence needs to be expanded. It should not be overlooked that our genome contains many thousands of genes for noncoding RNA segments. The resulting RNA molecules perform important functions in our bodies. Particularly noteworthy are the large groups of tRNAs that serve as adapter molecules for reading and translating base-pair triplets in the genome into the correct amino acid sequence. In addition, it has been shown that the ribosome itself, the molecular machinery for protein synthesis, consists largely of RNA (Sect. 32.7). The spliceosome, the complex machinery for removing noncoding segments of the genome, contains RNA molecules called snRNAs. There are even more small RNA molecules (snoRNAs) that are responsible for processing and modifying other RNA molecules.

The number of protein-coding genes is, as mentioned, about 21,500, but we still do not know what functions all these proteins perform. Bioinformatics has contributed greatly to the classification of their biochemical function, that is, whether the protein is an enzyme (e.g., protease, kinase, or oxidoreductase) or a receptor, ion channel, or transporter. The function or class of protein to which a new sequence belongs can be determined by comparing it to previously annotated proteins. Multiple sequence comparisons within a protein family often reveal significant similarity. Information about spatial architecture and folding (Sect. 14.2) can be analyzed using relationships, because the spatial geometry of proteins is much more conserved than the sequential composition of the folded protein chain. Often, individual motifs or characteristic sequence segments reveal a particular biochemical function of a protein. Another tool in this detective *tour de force* of functional annotation has been the comparison of protein sequences across species.

Assigning a biochemical function to a protein sequence provides a first idea about its molecular function. It shows, for example, whether it acts as a catalyst to cleave a peptide sequence, carries out a metabolic reduction or, as a receptor, transmits a signal to the cell. What this regulation and control means for the organism remains to be resolved. It is also not known whether a particular protein causes a disease because of its defective function or its dysregulation. Correcting such a defect could lead to a successful pharmaceutical therapy.

In the *Science* publication from the Venter group in 2001, it was assumed that the genome coded for more than 26,500 proteins. At that time, a definitive function could not be assigned to 40% of the sequences. In the remaining part, about 10% were detected to be enzymes. Another 12% proved to be involved in signal transduction, and 13.5% are nucleic acid-binding proteins. The large remaining group was scattered across many different functions such as proteins of the cytoskeleton, surface receptors, ion channels, transporters, extracellular matrix proteins, immune system proteins, or chaperones. Seven years later, this picture could be refined. The largest protein family with more than 7000 members contains the zinc finger domain (Sect. 28.2). These proteins assume an important role in transcribing sequence segments of the DNA into RNA. Most zinc finger proteins belong to the group of transcription factors. Another large protein family contains the immunoglobulins. These domains (Sect. 32.1), which are

Table 12.2 Selected examples of protein families in the human genome (number of family members at the time of the study)

Protein superfamily	Number[a]
Zinc finger (C2H2 and C2HC)	7707
Protein kinase-like	876
G-Protein-coupled receptor-like	784
α/β-Hydrolases	151
Cysteine proteases	164
Trypsin-like serine proteases	155
Metalloprotease ("Zincins"), catalytic domains	132
FAD/NAD(P)-binding domains	79
Cytochrome P450	79
Integrin α, *N*-terminal domains	51
Cytokines	52
Cycl. Nucleotide-phosphodiesterase, catalytic domains	50
Caspase-like	39
Carbonic anhydrases	23
Aquaporin-like	20
Integrin domains	18
Aspartic proteases	16
ClC-chloride channel	16
Subtilisin-like	14

[a] Based on: http://hodgkin.mbu.iisc.ernet.in/~human/
For updated data see: https://www.proteinatlas.org/search/ (The Human Protein Atlas)

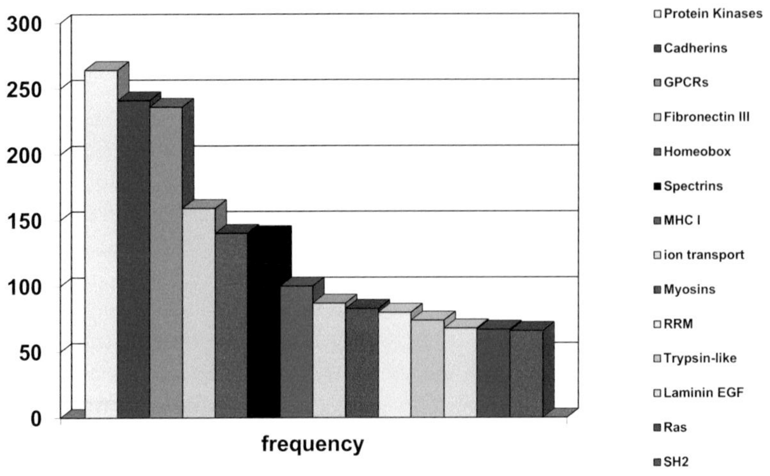

Fig. 12.3 The composition of protein families that are particularly often associated with human diseases (*GPCR* G-protein-coupled receptor; *Fibronectin* extracellular glycoproteins in tissue construction; *homeobox* proteins that influence the morphogenetic development; *spectrin* cytoskeletal proteins; *MHC I* major histocompatibility complex proteins that are involved in immune-recognition processes; *myosin* motor protein in muscle control; *RRM* RNA-recognition motif transcriptions factor; *trypsin-like* serine proteases; *laminin EGF* a growth factor in the extracellular matrix; *Ras* oncoprotein in tumorigenesis; *SH2* protein domains in the phosphorylation signal cascade)

12

constructed from β-pleated sheets, are present in antibodies. A few protein families are listed in Table 12.2 and are presented in more detail in Chaps. 23–32 of this book. It is interesting to note which protein family is frequently associated with which disease (Fig. 12.3). This list is headed by protein kinases (Chap. 26). It is, therefore, not surprising that drug research in the pharmaceutical industry has intensely focused for years on the modulation and inhibition of protein kinases. Next on the list are cadherins. These proteins are important for stabilizing cell–cell contacts. They play a role in embryonic morphogenesis, in signal transduction, and in the assembly of the cytoskeleton in cells. G-protein-coupled receptors, ion channels, trypsin-like serine proteases or RAS proteins are also prominent on the list of proteins potentially associated with disease, especially when genetically altered.

Finally, consider how the human genome differs from other eukaryotes. Of the more than 2200 protein families discovered in organisms with a nucleus, more than 1000 are missing in the human genome. Most of these families have specific functions in their respective organisms or can be explained phylogenetically. For example, venoms are found in snakes, scorpions, and insects. In plants, proteins are found that have a very specific function for the plant, such as nutrient storage in seeds or defense against the attack of pests. The proteins that are absent in humans usually perform biochemical functions that are irrelevant to our organism, or they perform a very specific task in lower eukaryotes.

12.5 Knock in, Knock out: Validation of Therapeutic Concepts

Molecular biology provides a wealth of information about how diseases develop and how their course can be influenced. This is the basis of the long road from the discovery to the development of a new drug. At the end of the process, it may be found that the result, although well planned, does not lead to the desired clinical success. It is therefore important to have an animal model that can be used to validate the therapeutic concept at an early stage. Classical test models are often not available because the disease in question does not occur in animals.

Since the 1980s, **transgenic animals** have been increasingly used in pharmacological research. Transgenic animals are animals in which a specific gene has been completely or partially turned off or replaced by a human gene. An animal in which the gene is completely knocked out corresponds to an animal in which the corresponding protein is absent or nonfunctional. A heterozygous animal in which the gene of only one parent is present corresponds to an animal in which the corresponding protein is only partially blocked. When the gene for an enzyme or receptor is affected, the effect of an inhibitor or antagonist can be simulated. The onset and progression of a disease, or the influence of protein inhibition on a disease, can be observed in such an animal. In this way, the relevance of a therapeutic concept can be established before an extremely long research and development process is launched. The increased production of a particular protein can be induced by the amplification of a gene. If the absence of a gene causes the overexpression of another gene, this will also become transparent. The gene product that is then produced in increased amounts can take over the missing function of the silenced product. In such a case, the planned therapeutic principle would only work if the function of the other gene product is also blocked. This question plays an important role in the inhibition of kinases (Sect. 26.2).

The **knock-out method** involves switching off a very specific gene. This technique was developed by Mario Capecchi at the University of Utah, USA, in 1987. The sequence of the gene to be knocked out must be known. A structurally homologous gene is created that is not functional, for example, because of the insertion of a stop signal. The gene is introduced into an animal and the intact gene is replaced at exactly the same location. This process is called **homologous recombination** or **gene targeting**. Mice are particularly well suited because the technology for manipulating their embryonic stem cells is particularly well established. A foreign gene, such as a human gene, can also be introduced. Mice are also well suited for this because their genome is surprisingly similar to the human genome.

To create a transgenic mouse, female mice are treated to produce a large number of egg cells. After fertilization, stem cells are extracted from the embryos at a very early stage, the blastocyst stage. They are cultured *in vitro* and the desired gene is injected into the cell. This procedure results in a low yield. A technique has been developed to differentiate transfected from nontransfected cells. The gene to be transferred is coupled to a gene that confers resistance to the cytotoxin neomycin. When cells are treated with neomycin, only transformed cells survive. The blastocytes are combined with blastocytes from other mice, and the altered embryos are carried to birth by mice. The offspring of the surrogate mothers are chimeric, meaning that they carry the genetic information of both the donor and acceptor mice. Here, mice with differently colored fur are selected so that the transformed mice are easily recognizable by their spotted fur.

Another method is to directly inject foreign DNA during an early embryonic stage. A disadvantage of random insertion of a gene is the possibility of destroying another gene, a lack of expression of the new gene, or multiple insertions. Animals from the first litter are bred to produce both genetically mixed, heterozygous animals and genetically homogeneous, homozygous animals. Sophisticated techniques can even selectively turn the new genes on and off.

In this way, transgenic animals are generated in which hereditary diseases such as cystic fibrosis, Crohn's disease, phenylketonuria, and others can be studied. Today, animal models also exist for diseases that have different or multiple causes, such as cancer, diabetes, rheumatoid arthritis, and Alzheimer's disease. Since 1988, when the U.S. Patent Office granted the **first patent for a transgenic mouse**, there has been controversy over whether a living creature can be patented at all. European patent law will prohibit patents if genetic modifications can lead to animal suffering. Exceptions are allowed only if a significant medical benefit is to be expected. In 1992, the first patent in Europe was approved for a genetically modified "cancer mouse." Since then, many similar patents have been filed, mostly on laboratory animals. But now farm animals such as cattle and pigs are also included. Because of their genetic similarity to humans, the European Patent Office declared patents on genetically modified primates invalid on ethical grounds in July 2020. Currently, patents on laboratory animals are not completely banned, but should be limited to a few exceptional cases.

12.6 Recombinant Proteins for Molecular Test Systems

Early on, pure or enriched enzymes were available for ***in vitro* assays**, but only in cases where the material was readily available, for example, human thrombin from blood. In other cases, animal material had to be used, with all the risks this entails, given the relevance to rational design (see Sect. 19.11). There are many proteins that cannot be isolated in sufficient quantities or in a homogeneous form. The sequence determination and production of such proteins is nowadays easy. The incredibly small amount of a few picomoles (1 pmol = 10^{-12} mol) is sufficient to determine the primary structure of a short protein segment. From the amino acid sequence determined in this way, the genetic code can be reconstructed into a gene. It should be noted that several base triplets can stand for a particular amino acid (so-called degenerate code, Sect. 32.7). A set of single-stranded oligonucleotides is synthesized that could theoretically cover the entire original peptide segment. These molecules can be used to find a complementary sequence in a **cDNA library**. cDNA (**complementary DNA**) is the DNA complementary to mRNA (**messenger RNA**). It is obtained from the mRNA, which contains only the sequence needed for protein biosynthesis, by reverse transcription with a reverse transcriptase (Sect. 32.5). Finally, the gene is produced in larger quantities using the PCR technique, and the amino acid sequence is determined from its base sequence, simply because polynucleotides are much easier to sequence.

Next, the gene is introduced into cells that are allowed to reproduce. In a few cases, there may be difficulties with this step. In bacteria, such as the intestinal bacterium *Escherichia coli*, or in yeast cells, only soluble proteins can be produced. Some proteins accumulate in **inclusion bodies**. They must be extracted, solubilized, and refolded under specific conditions. The gene segment for a small protein is often fused with the gene for another protein, and the fusion protein is then expressed. Often, the large protein conjugate formed in the cell is more soluble and/or better protected from metabolic degradation than small proteins. During purification, the nonessential part of the protein conjugate is cleaved off. Problems may arise if the protein is not folded correctly or if several chains (such as insulin) have to be linked by disulfide bridges. Larger proteins that require sugar groups to perform their function (glycosylation) must be produced in cells of higher organisms, such as mammalian cells. The production of complex proteins in **insect cells** has become particularly attractive. These cells are infected with a **baculovirus** that has incorporated the desired information into its genome. The virus encodes the target protein, and the insect cells provide for its production and subsequent glycosylation. Not only enzymes, but also receptors, ion channels, and entire signaling cascades can be produced in cells in this way.

12.7 Silencing Genes by RNA Interference

How genetically modified species can be created by intervening in the germline of an organism was described in Sect. 12.5. For example, such species may lack a particular gene and, therefore, a gene product, or a new gene may have been introduced. In this way, the function of certain genes can be studied in a living organism. This makes the consequences of blocking the corresponding gene product transparent before developing an effective drug. At the end of the 1990s, another technique that made it possible to silence genes without using mutagenesis to interfere with the genes of the organism was developed. This work was done by Andrew Fire and Craig Mello, who were awarded the Nobel Prize for their achievements in 2006.

Genes are stored on DNA. For gene expression, the coding parts of the genome are first transcribed onto mRNA. The ribosome then uses this transcribed information to convert the base sequence into a peptide sequence (Sect. 32.7). The idea of trapping the transcribed information on the single-stranded mRNA by adding a complementary single-stranded RNA was first proposed in the early 1980s. The two strands can be joined by "hybridization," which means that the single nucleotide strand is supplemented by the complementary strand to form a double strand. The resulting double-stranded RNA is no longer suitable as a template for protein biosynthesis. In practice, however, this antisense principle (Sect. 32.4) did not produce the expected breakthrough results. In some cases, genes were only weakly suppressed, and even the addition of the normal RNA strand could achieve a suppression. Fire and Mello suspected that neither the normal nor the antisense strand caused gene blockage, but rather the double-stranded form, which was present as an impurity. Further experiments confirmed their suspicions. Interestingly, even small amounts of double-stranded RNA were sufficient to render a large number of mRNA molecules unusable. In contrast, the use of antisense strands would require stoichiometric amounts. It was further shown that even short double-stranded RNA fragments of about 20 nucleotides were sufficient to silence entire mRNA gene sequences. Fire and Mello named this phenomenon **RNA interference**. What had happened? An enzyme called **dicer** chopped up the double-stranded RNA into pieces with a length of 21–23 nucleotides, which then resulted in blockade. The double-stranded RNA pieces are incorporated into the enzyme complex **RISC** (**R**NA-**i**nduced **s**ilencing **c**omplex) and separated into individual strands. One strand is released from the complex, while the other remains as a template for capturing other mRNA molecules.

The sequence of the captured strand allows the RISC complex to recognize and sequentially cleave all mRNAs with a complementary base sequence. They are then digested by enzymes in the cytoplasm. The cell selectively eliminates only those mRNAs that contain the sequence pattern complementary to the short RNA strands in the RISC complex. In practice, this gene blockade has proven to be simpler and more reliable than the antisense technique. RNA interference can, thus, be used to systematically silence discovered genes in order to draw conclusions about the resulting consequences for the organism. RNA interference is not only used for analytical purposes. There are already biotech companies that use small RNA fragments to silence disease-causing genes.

Another major problem in the development of **RNAi therapeutics** is how to get a 22-base RNA molecule into the cell where it is supposed to work. Highly charged molecules cannot cross the cell membrane. Therefore, a special delivery system is needed. Intensive research is underway to develop such systems, but the problem is far from being solved. A reliable and highly efficient system that can selectively deliver such polar and nuclease-sensitive molecules into the interior of the cell is likely to open up a completely new and currently unforeseeable perspective for disease therapy.

The goal is to construct delivery systems that can package the fragile and polar freight of RNA molecules and dock onto the cell. Once there, the coat of these carriers must fuse with the cell membrane or selectively penetrate to reach the interior of the cell. One concept is to package and compartmentalize RNA in polymers such as polyethyleneimine. The positive charges on the polymer backbone can bind and encapsulate a negatively charged polymer molecule such as RNA or DNA building blocks. Other systems attempt to make the RNA or DNA molecules bioavailable to the cell by encapsulating them in a membrane-like coating. This packaging in liposomes results in selective adhesion of the artificial cell to the membrane of the target cell, and then the liposome fuses with the target cell in an endocytosis-like process.

Another problem is the risk that small, **silencing RNA molecules (siRNAs)** could trigger an **immune response**. One solution to this dilemma is the chemical modification of siRNAs. The RNA molecules are modified in such a way that they still hybridize optimally to the targeted segment in the mRNA, but have improved properties in terms of transport, immunogenicity, and stability. The OH groups of the ribose building block of the nucleotides have been replaced by fluorine, methoxy, or hydrogen.

Certainly, siRNA research is still in its infancy. The potential of the methods seems impressive, since they use the principles of gene regulation that are applied in Nature. As described earlier, we have genes in our genome that encode microRNAs and that have long stretches of sequence complementarity. Structurally, they exist as double strands. They are cut by the dicer protein and can be used to interfere with RNA, resulting in an alternative means of gene regulation. For a broad therapeutic application of externally delivered RNA snips, these problems need to be overcome. In 2018, the company Alnylam launched the first RNAi therapeutic, patisiran, for the treatment of polyneuropathy in hereditary transthyretin-mediated amyloidosis. Worldwide, approximately 50,000 people suffer from this rare and extremely devastating disease. Patisiran successfully delivers siRNA to liver cells using lipid nanoparticles as a delivery vehicle.

12.8 PROTAC: How to force therapeutically untargetable proteins into targeted degradation

Traditional small molecule drug discovery for intracellular targets has focused on developing high-affinity ligands that target either an active, orthosteric, or allosteric site on a target protein to interfere with its physiological function. While this approach has been highly effective, it has left potential drug targets undruggable because their active sites are either too broad, formed as shallow pockets, or have flat, unstructured surfaces that provide very few sites with significant ability to bind small molecules.

Instead, the PROTAC (**Pro**teolysis-**Tar**geting **C**himeras) approach involves the targeted degradation of a given protein to down-regulate or completely shut off its biological function in the organism. It induces the selective proteolytic degradation of pathogenic and, therefore, unwanted proteins into amino acids by the cell's own proteolytic degradation machinery (proteasome, Sect. 23.8). To be degraded by the protein shredder, a protein must first be labeled. This is done by attaching **ubiquitin markers** to the protein (the so-called "**kiss of death**"). This requires several of these ubiquitin peptides, which are added one at a time. A bifunctional molecule ("chimera") is required to capture the protein to be degraded and to mark it for degradation. In **PROTAC molecules**, this is achieved by means of a chemical linker (e.g., a polyethylene glycol chain) connecting the two functional ends. On one side, it carries a decoy that binds the target protein to be degraded. On the other side, it carries another decoy that recruits an **E3 ubiquitin ligase** to do the labeling. The function of the E3 ligase, thus, determines which protein becomes the substrate for proteolytic degradation in the proteasome. Since the PROTAC molecule brings the protein to be degraded into spatial contact with the E3 ligase and, thus, with the "molecular machinery" for ubiquitin labeling (◘ Fig. 12.4), the protein to be degraded becomes a substrate for a ligase. Once the captured target protein has been labeled and targeted for degradation, the PROTAC molecule is released. Like a catalyst, it can then initiate the next cycle of degradation. Unlike a classical inhibitor, which must inactivate the target protein in stoichiometric amounts, PROTAC molecules are effective at lower dose levels. It appears that proteolytic degradation takes some time, but this does not necessarily reduce the success of the therapy.

A moderate affinity of the decoy to capture the target protein to be degraded is usually sufficient, making the design of such a decoy ligand easier. On the other hand, the ligand moiety that recruits the E3 ligase should be a potent binder with a slow off-rate. PROTAC molecules that attach even covalently to an E3 ligase may be superior because they reduce the three-body assembly kinetics to a two-body case, thereby increasing catalytic efficiency.

The attractiveness of this concept is that once a decoy for the protein to be degraded is found, the PROTAC

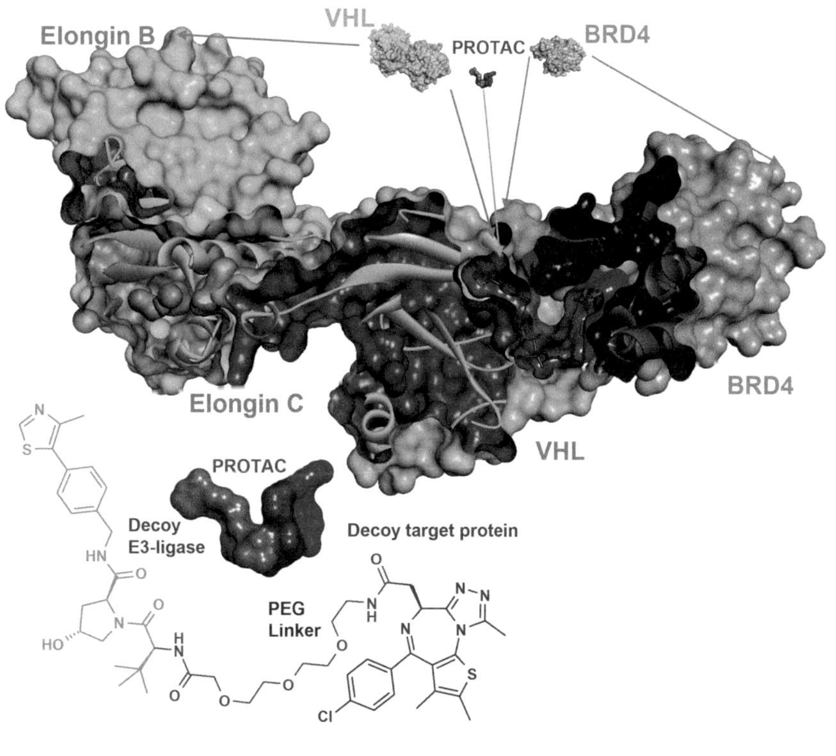

◘ **Fig. 12.4** Crystal structure of a complex of a VHL-E3 ligase (*light blue*, von Hippel–Lindau tumor suppressor with ubiquitin ligase E3 activity and two further domains, elongin B and C) and a target protein BRD4 (*magenta*, epigenetic target protein, a bromodomain that can read acylations on histone proteins, see Sect. 12.14) to be ubiquinylated. The PROTAC molecule can be seen in the center and links the two proteins via its decoy groups (*orange*/*blue*). Chemically, the two decoys are linked via a polyethylene glycol (PEG) inker (*purple*). The two decoys each fit highly specifically into the pockets of the ligase and the bromodomain, respectively. (▸ https://sn.pub/8AwGNS)

system can force any protein to be degraded in this way. This includes proteins previously thought to be therapeutically undruggable. Any small molecule ligand that binds specifically to the surface of the target protein can serve as a decoy, without the need to possess a functional role on the target protein.

It also needs an E3 ligase to capture the PROTAC molecule using its second decoy. Although there are over 600 E3 ligases in our genome, only a few can be captured using a small synthetic ligand as a decoy. One of these is thalidomide (**5.26**, Sect. 5.5). It binds to the cereblon protein complex (CRBN), which has several functions in the body. During embryonic development, for example, it forms the substrate-binding subunit of a ubiquitin E3 ligase that influences limb growth by regulating transcription. This has been implicated in the teratogenic effects observed with thalidomide. However, there are several other examples of such E3 ligands that can be used to synthesize a decoy in PROTAC molecules. ◘ Fig. 12.4 shows the structure of a ternary complex of another E3 ligase (VHL-E3 ligase), a PROTAC molecule with two specific decoy groups, and the epigenetic target protein BRD4.

Most PROTAC molecules described so far in the literature focus on epigenetic and oncogenic targets (Sect. 12.14), nuclear hormone receptors (Sect. 28.1), or kinases (Sect. 26.2). In 2019, the first PROTAC molecule entered clinical testing, in 2020, these trials provided the first clinical proof-of-concept against the estrogen (ER) and antrogen receptor (AR) as well-established cancer targets. PROTAC molecules have a large molecular weight and their physicochemical properties are outside the typical range of small molecule drugs (see Rule of Five, Sect. 19.7), which may pose challenges for drug delivery into the cells. Nevertheless, examples with sufficient oral bioavailability have been reported and first candidates have reached phase III clinical trials. Recent research extends to so-called bioPROTAC molecules, which use peptide-like decoys to lure target proteins for selective degradation. There is no doubt that the PROTAC concept has great potential and will hopefully prove itself in therapeutic applications.

12.9 Proteomics and Metabolomics

The approaches described in Sects. 12.5 and 12.7 aim to silence a disease-causing gene or a gene that plays an essential role in a disease. But how do we know whether a particular gene or gene product is involved in a disease process at all? Decisive indicators to answer this question can be extracted from the protein composition in the cell. This composition changes dynamically. It is called the **proteome** and reflects the totality of all proteins in a cell, or even in the entire organism, at a given point in time under well-defined conditions. Focusing on the protein pattern of a cell from a particular organ, important variables are the metabolic state, the developmental stage of the organism, the current stage in the cell cycle, and the surrounding temperature. Disease processes and drug therapy also alter this pattern. In the genome, all theoretically expressed proteins are encoded as static genetic information. In contrast, the proteome reflects the protein composition at a particular point in time. The difference between a butterfly's caterpillar and adult stages is a striking example of the difference between the genome and the proteome. The genome is the same for both, but the proteome is significantly altered, resulting in a completely different phenotype (caterpillar/butterfly).

The proteome can be used to compare the state of healthy cells, diseased cells, and cells under the influence of drug therapy. At first glance, this seems to be an extremely complex and almost impossible task. A cell contains thousands of proteins, many of which are modified after they are expressed. For example, the first amino acids of a sequence are cleaved (Sect. 25.9), phosphate groups are transferred (Sect. 26.3), sugar building blocks are added, disulfide bridges are coupled, prosthetic groups are added, and ubiquitin or prenyl groups are added (Sect. 26.11). In addition, alternative RNA splicing occurs as a mechanism of gene regulation, further increasing the diversity of the proteome from a comparatively small number of genes. All this dramatically increases the diversity of the protein composition, probably by a factor of 5–10 compared to the genome composition. Nevertheless, a sophisticated analytical method has been developed that makes it possible to analyze the proteome of a cell at a given point in time.

First, the proteins of a cell must be denatured in such a way that all modifying processes are abruptly stopped so that conclusions can be drawn about the cell contents. The cell lysate is then subjected to separation. Proteins contain many acidic and basic amino acids, so for each protein there is a well-defined pH value at which protonation or deprotonation reaches a state where the protein appears to be electrically neutral (**isoelectric point**). This pH is specific for each protein and depends on the amino acid composition. The protein mixture is added to a solid support (a polyacrylamide gel) such as is typically used in chromatography. A voltage is then applied. If the proteins carry a charge, they will migrate across the solid support in the direction of the oppositely charged pole. Thus, at some point during their migration across the gel, which is designed to provide a continuous **pH gradient** from one end to the other, the applied proteins will reach a point where their entire exterior appears uncharged. When this position is reached on the solid support, the proteins no longer migrate. The proteins are therefore separated according to their isoelectric point, a process known as isoelectric focusing. All proteins with the same isoelectric point migrate the same distance and appear as a mixture. The chromatography plate is then rotated by 90° and the proteins are separated once again but using a different principle. The proteins are thermally denatured and their charges are masked with sodium dodecyl sulfate, a highly charged anionic surfactant, so that

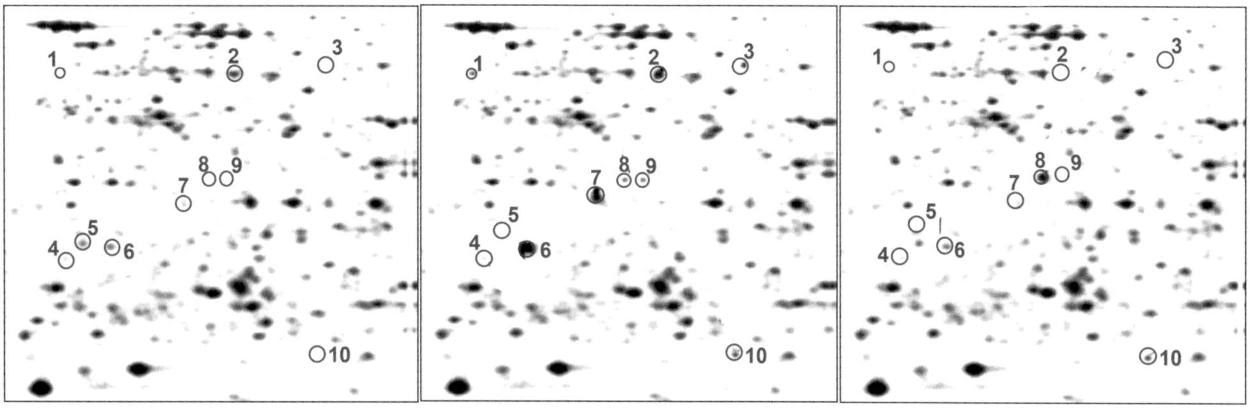

Fig. 12.5 2D-Gel electrophoresis for a cellular proteome analysis. *Left* Proteome of a normal cell. *Center* Proteome of a pathologically altered cell. *Right* Proteome of a pathologically altered cell after treatment with a drug. Changes in protein concentration are indicated by *red circles*. Above all, the proteins at positions 3, 6, and 7 are clearly up-regulated in the diseased state. A few of the pathological changes are corrected by the drug therapy, but new changes in the proteome (e.g., 2, 8, and 10) might be induced by side effects. (Figure from Lottspeich, Angew. Chem. Intl. Ed. Engl., **38**, 2476–2492 (1999) with the kind permission of the author and publisher)

they are all virtually equally charged on the outside (**SDS-PAGE**). When an electric field is applied, the denatured proteins migrate again. The speed of migration depends on the mass of the proteins. The direction of migration, which is perpendicular to the first isoelectric separation, results in the initial proteome being broadly distributed and well separated on the solid support.

Using this 2D electrophoresis, it is possible to separate many thousands of proteins. The amount and sequential composition of the separated proteins must be characterized. Many different staining and fluorescence techniques have been developed for quantitative analysis. They allow quantitative determinations, especially in comparison to the proteomes of analogous cells in a different state. Thus, a quantitative comparison of the protein composition in a diseased and a healthy state is possible. It is also possible to determine how the proteome changes under the influence of a drug (Fig. 12.5). But how can one find out what is contained in each individual protein spot on a 2D gel? The answer is to extract the proteins from the plate and digest them with trypsin. This protease (Sect. 23.3) cleaves the denatured proteins into small peptide fragments, which are then analyzed by mass spectrometry. Sophisticated technologies combined with computer analysis of precalculated protein fragmentation patterns allow the reconstruction and sequence characterization of proteins. Proteins in the proteome that are either up- or down-regulated due to a disease process can be detected in this way. However, whether the altered expression pattern causes or is a consequence of the pathological state remains to be determined by independent experiments.

As described above, the proteome of a cell can be altered by a drug. What are the interaction partners for a given drug? Will the induced effects always be the same if drugs of the same class are used? In the research group of Giulio Superti-Furga in Vienna, Austria, the properties of three kinase inhibitors developed for the treatment of chronic myeloid leukemia (Sect. 26.4) were investigated in detail. First, the drug had to be equipped with a chemically inert anchor group. It is certainly a challenging task to find the right position for an anchor on such a compound so that the mode of action is not significantly disturbed. Usually, several positions along the molecular scaffold have to be tried and tested. Finally, the drug is irreversibly covalently coupled to a chromatography column via the attached anchor group. The column is then loaded with the proteome of a cell lysate. Proteins that have affinity for the immobilized drug will adhere to the column. Finally, the binding partners detected in the **pull-down experiment** have to be released from the column, separated, and characterized analogously to the technique described above. The composition of all proteins with affinity to the tested drug is obtained. It is difficult to make quantitative statements about the affinity of the binding partners, especially since the amount of proteins and their composition in the lysate are highly variable. However, it is possible to construct a profile for each compound according to its protein interaction partners. This led to the surprising result that even drugs belonging to the same or similar substance classes and developed for the same therapeutic indication can exhibit **significantly different interaction profiles** in the cell. This is an impressive observation, the evaluation and application of which will require a great deal of research. We will see in the next section that this can explain the different efficacy, therapeutic deviations, and variable side effect profiles in patients.

Proteomic techniques can also be used in clinical diagnostics. Without the exact resolution of the analyte, significant changes can be detected in the form of a fingerprint of the distribution of molecular masses. Tumor diseases are revealed by changes in their protein composition. These can be detected at a very early stage, hopefully allowing curative treatment of the tumor.

Another technique analogous to proteomics is the analysis of metabolites produced in an organism. The term **metabolome** refers to all metabolites (e.g., metabolic degradation products) present at a given time. Metabolomics techniques attempt to quantify the metabolite composition and use this information to infer the status of a cell. This is especially true when a cell is exposed to foreign substances. If the metabolite profile is studied at a specific time point, especially in pathophysiological or genetically altered conditions, the term **metabolomics** is used. The goal of this technique is to draw conclusions about the molecular composition of cells from body fluids such as urine, serum, or cerebral spinal fluid. This can lead to improved and more sophisticated diagnostic procedures and, thus, to an easier early detection of diseases. It can also be used to characterize proteins for drug therapy or to analyze the overall effect of a drug on cellular events. It is hoped that these techniques will lead to a better understanding of the overall effects of drug use and ultimately to a higher standard of safety in therapy.

12.10 Expression Patterns on a Chip: Microarray Technology

The analysis of the genome, transcriptome, proteome or metabolome results in thousands of molecules that need to be characterized. This flood of data requires an immense measurement capacity. For this reason, the development of **microarray technology** was initiated in the late 1980s. Thousands of molecules, which are to be analyzed in parallel in an automated fashion, are attached to a support made of glass, silicon, gold, or nylon that is only a few centimeters in size (▫ Fig. 12.6). Very small quantities of biomolecules are required. This technique is now sufficiently mature to be used in routine analytical procedures. In addition to appropriate surface preparation, the art of reliable and standardized **immobilization of the molecules** required for precise analysis is critical for the success of the method. In addition to proteins and protein domains, antibodies, antigens and especially DNA, oligonucleotides and RNA can be immobilized. Proteins are often anchored by coexpress-

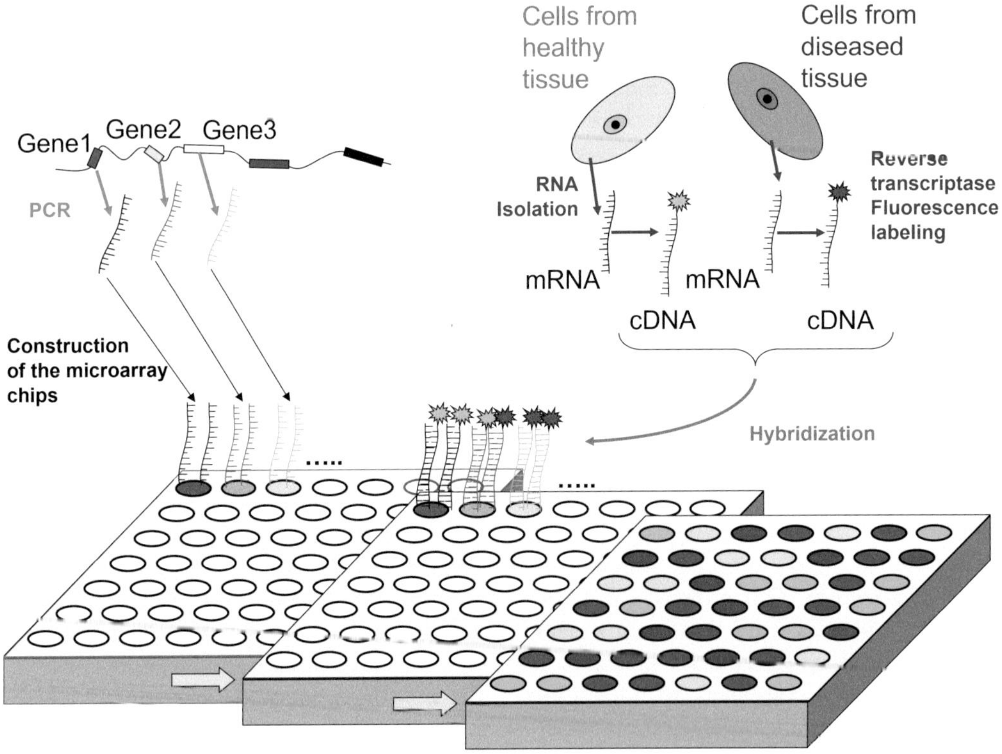

▫ **Fig. 12.6** Producing and testing an expression pattern using microarray technology. Individual gene segments are isolated from an organism and amplified by PCR (*top left*). They are then immobilized as single-stranded oligonucleotides on a microchip support (*bottom left*). In addition to the isolated and amplified DNA, synthetically produced DNA building blocks or cDNA molecules obtained by reverse transcription can also be immobilized on the support. One type of such probe molecule is placed at each position on the support. RNA molecules are isolated from the cells of healthy (*green*) and diseased tissue (*red*), transcribed into mRNA, and backtranscribed into cDNA (*top right*). The cDNA is labeled with a fluorescent dye. The test molecules are then added to the microarray plate in a single-stranded form, and if they are complementary, hybridization will occur (*bottom center*). Finally, the binding is analyzed under fluorescent light (*bottom right*). *Yellow* areas indicate that mRNA molecules from both healthy and diseased cells have bound. The mRNA that binds there is expressed in both healthy and diseased states. Areas that remain *dark* indicate that the mRNA is not upregulated in either the healthy or diseased state. Areas that fluoresce either only *green* or only *red* indicate a difference in the expression pattern between cells from healthy and diseased tissue ing the protein of interest coupled to an anchoring protein such as streptavidin as a fusion protein. The streptavidin anchor is attached to the surface via biotin. In addition, thiol group chemistry is used. Disulfide bridges are used to couple the thiol groups to the surface, which has previously been provided with appropriate reactive groups. Other strategies use amino groups, such as lysine, which are then coupled to a reactive aldehyde group on a solid support. To test the composition of an analyte, a soluble mixture is added to a premanufactured chip. If binding partners are found in this transformation, the components from the analyte solution will remain on the surface.

Such binding must be simple and detectable in a spatially resolved manner on the chip. Initially, staining and fluorescence were the methods of choice (◻ Fig. 12.6). Fluorescent dyes, such as green and red, are used because they can be easily excited and detected in a spatially resolved way. If mixed signals occur due to simultaneous red and green fluorescence, a yellow signal is obtained. In the meantime, surface plasmon resonance has become more significant (Sect. 7.7). As an alternative, the latter technique is used to detect binding. In addition, techniques similar to ELISA methods can be used (Sect. 7.3).

Microarrays are often used to analyze the **expression patterns** of biological systems. This is done by examining the transcriptome of a cell under different conditions, for example, in a diseased and healthy state. The first molecules to be successfully anchored on chips were single-stranded DNA oligonucleotides. To study the coding mRNA of a cell in a particular state, these molecules are transformed into a complementary DNA segment called cDNA using a reverse transcriptase (◻ Fig. 12.6). These cDNA molecules, or the fragmented sections of cDNA that are obtained, are immobilized on a chip and split into single strands. The cell lysate containing the single-stranded mRNA (transcriptome analysis) or the transcribed cDNA, prepared from it, is added to such a chip, and the complementary mRNA strand hybridizes with the oligonucleotide fragments anchored there. It is important that the samples to be analyzed are labeled with different fluorescent dyes depending on their origin. For example, the mRNA from a healthy cell is labeled green and that from a diseased cell is labeled red. After hybridization on the chip, there are areas that fluoresce green, red, or yellow when excited, and others that remain nonfluorescent. Areas that fluoresce yellow indicate that mRNA molecules from both healthy and diseased tissue are bound. Obviously, the mRNA that binds there is equally available in either the diseased and healthy states. Areas where no fluorescence is seen indicate that neither healthy nor diseased cells produced mRNA that is bound there. Areas that fluoresce either green or red are of interest because they indicate differences in expression patterns between healthy and diseased cells. In this way, gene products involved in a disease process can be discovered. If there is dysregulation, drug therapy can be used to try to correct the diseased state.

12.11 SNPs and Polymorphism: What Makes Us Different

What makes an individual organism of a particular species different and adds to the enriching diversity of a population? We are talking about the human genome, but there must be many interesting variations to make us all look different and have different personalities. Polymorphisms, or variations in the composition of the genome, cause the observed diversity in or shape the different **phenotypes** of a species. The most obvious phenotypic difference is the division into male and female individuals. Of course, this is not the only difference we recognize in the human species. Many sequence variations occur within a population at the genome level. If they occur in more than 1% of the population, they are called different **alleles**, otherwise they are attributed to mutations that have not yet become evolutionarily dominant. **Genetic polymorphisms** are observed, for example, as insertions or deletions in which at least one nucleotide is either partially or completely inserted or lost. The most common sequence variation, however, is single nucleotide exchanges. The term SNPs (pronounced "snips") is used here, which is an abbreviation for **s**ingle **n**ucleotide **p**olymorphism. Compared to the entire genome, polymorphisms represent a very small fraction. They are estimated to be 1% of the total genome, which is about three million bases. Of these, SNPs are the overwhelming portion, accounting for about 90%. Therefore, most of our genome is identical across the human species, even though there is enormous phenotypic diversity among us.

Within SNPs, coding and noncoding changes are distinguished according to whether the observed exchanges are translated into proteins or not. In the coding regions of the genome, a single nucleotide exchange can lead to an altered protein sequence. In Sect. 32.7, the translation process of a base triplet into a protein is introduced. If one base in a coding triplet is changed, the triplet can either be translated into the same amino acid, or it can lead to the incorporation of a different residue. This is due to the fact that some triplets code for the same amino acid. Incorporating a different amino acid into a protein can change its properties. For example, the amino acid composition of a glycosyltransferase determines the blood group we have. An example of how a change in the amino acid composition of a G protein-coupled receptor can affect our sense of smell is given in Sect. 29.7. Humans are divided into different alleles according to their ability to sense different intensities and qualities of smell.

However, it is not only SNPs in coding regions that lead to differences in our species. SNPs in noncoding regions of the genome can lead to changes in gene regulation. In the context of drug discovery and therapy, SNPs may be relevant even if they have no direct effect on the phenotype. Some SNPs are thought to confer susceptibility to disease or influence the cellular response to a drug. It should be noted that SNPs may also occur in the region of the binding site of a drug molecule, which is not necessarily identical to that of the natural substrate. SNPs then directly influence the affinity and binding profile of the drug. As a result, a drug may exert a stronger or weaker inhibition of protein function in patients with an observed SNP than in patients in whom this SNP is not present.

12.12 The Personal Genome: Access to an Individualized Therapy?

Genome sequencing and the analysis of SNPs and polymorphisms have impressively uncovered the source of **disease predisposition** and why drugs have **attenuated tolerability** and different **side effect profiles**. It has provided an explanation for why undesirably large variations in drug efficacy can occur between patients. All the more reason to ask whether the sequencing of the individual genome of each person would provide options for individual and **personalized therapy**. This is by no means a utopian idea, since it is now possible to sequence the entire genome of an individual person in a few hours at a cost of less than US$ 1000.

12

In medicine, it is well known that the blood group of the donor and the recipient must match for blood transfusions. Genome analysis would facilitate the search for a matching donor organ for transplantation. A particularly high density of SNPs has been found in the genome, especially in regions coding for proteins that present antigens to the immune system on their surface to stimulate an immune response (Sects. 31.7 and 32.3). SNP analysis of each individual could indicate the likelihood of developing a particular disease. Early detection of this risk and possible lifestyle changes could be better than any therapy. Today, high-resolution DNA chips (Sect. 12.10) allow the simultaneous detection of more than **500,000 genetic SNP markers**. Discovered SNPs can indicate an increased predisposition to developing Alzheimer's disease in aging, for example. A simple screening of an individual's DNA sequence could reveal a predisposition to a particular disease pattern.

Craig Venter, whose company sequenced the human genome using the mRNA shotgun method, had analyzed and published his own genome. The gene analyses of these data revealed a tendency towards obesity and cardiovascular disease. His own father had died of a heart attack at the age of 59. Based on this analysis, Venter decided to take a lipid-lowering drug from the statin class as a preventive measure (Sect. 27.3). A physician could simply check the personal genome to see whether the patient has a SNPs pattern that would indicate an intolerance to a particular drug therapy. In addition, the physician could see what type of metabolizer category (Sect. 27.7) the patient belongs to. This could reduce intolerance to simultaneous treatment with several drugs and allow safe adjustment of individual doses. It may also help in choosing the right drug for a therapy, especially when several drugs with different modes of action for one indication are available.

The dream of developing "**personalized drug molecules**" for individual therapy will be difficult to realize for cost reasons. Just adding one more methyl group to a drug requires a full toxicological and pharmacological testing program to gain approval. It would consume millions of dollars in development costs. As always, however, the determination of an individual's genome and the elucidation of all **conceivable predispositions** to possible diseases has its downside. In the hands of the treating physician, this information is a blessing. But what would a future employer read into these data about the prospect of hiring an employee? Insurance companies could accept only risk-free clients on the basis of their genomic data—a frightening idea that an individual's genomic make-up would determine their insurance premium!

Regardless of our genetic differences and the possible consequences for drug therapy, we must not forget that our gastrointestinal tract is home to millions of microorganisms. This flora has a decisive influence on our well-being, our health stability, our metabolism, and also on our response to drug therapy. The individual **gastrointestinal flora** begins to build up at birth and is decisively influenced by the mother. It varies considerably with lifestyle, food culture, and exposure to the regional microbial landscape. In India, China, or Europe you will find a different microbial culture than in America, for example. Interestingly, it changes when persons move their homes between continents. Different microorganisms cause a different configuration of secondary metabolites and contribute to a shifted health balance. Presumably, these differences between individuals are as important as the genetic diversity that makes us different.

12.13 When Genetic Differences Turn into Disease

Genetic diseases are molecular in origin. One gene (allele) is altered, sometimes the two genes from both parents. Each of us carries a large number of such altered genes, which are the result of random base exchanges: the SNPs. The principle of **evolution** is based on these random mutations. If a mutation makes an individual more adaptable to the environment, the chances of survival and reproduction will increase. These genes are then reproduced with increased probability. In asexually reproducing species, horizontal gene transfer has an accelerated effect on evolution. Here, entire DNA fragments are exchanged between individuals or even species. In this sense, **crossover** plays an important role in sexual reproduction. In this case, adjacent gene sequences from both parents randomly crossover and form new combinations. Without mutation and crossover, all species would remain absolutely constant. In individual cases, many errors are produced as a mechanism of evolution. Some of these errors cause genetic diseases. In **sickle cell anemia**, a single amino acid in hemoglobin, which gives blood its red color, is exchanged, and a glutamic acid at position 5 of the beta chain of hemoglobin A (HbA) is replaced by a valine. The modified hemoglobin aggregates: it "sticks" together in the red blood cells. The cells collapse and take on a characteristic sickle shape. Homozygous carriers, meaning individuals in whom the "sick" gene is inherited from both parents, cannot survive. Heterozygous carriers, who carry one "sick" and one "healthy" gene, produce both normal and altered hemoglobin. These people have a shorter life expectancy, but usually reach reproductive maturity. In areas where malaria is endemic, there is a selection pressure for the genetic disease. Heterozygous carriers of sickle cell anemia are more resistant to malaria than healthy individuals (Sect. 3.2). Here we are witnessing Nature's great experiment. How will it end? Even humans intervene. If malaria is successfully treated, wild-type HbA carriers will no longer be at a disadvantage, and the evolutionary advantage of sickle cell anemia and the resulting selection pressure towards this disease will disappear. This genetic disease could become "extinct" after a few generations. On the other hand, if sickle cell anemia is treated either conventionally or by gene therapy, then these people would have perfectly normal "healthy" red blood cells. The malaria parasite would be able to reproduce well in these cells again. The protection against this disease would disappear, and the susceptibility of these people to malaria would rise to a normal risk level.

In addition to sickle cell disease, about four thousand other diseases and their molecular causes are known. Some, such as cystic fibrosis, phenylketonuria, and inherited coagulopathies, are relatively common. Many others are rare, sometimes described only once. In recent years, a multifactorial genetic cause has been identified for an increasing number of diseases, including diabetes, rheumatoid arthritis, some cancers, asthma, and Alzheimer's disease. The occurrence of these diseases is caused by, or at least facilitated by, the simultaneous occurrence of multiple genetic changes.

The mechanisms of evolution are also responsible for the development of resistance (Sect. 3.2). In this case, the selection pressure is exerted by a drug or an insecticide (e.g., to eradicate malaria-carrying mosquitoes). Because of their rapid reproduction, bacteria and viruses adapt quickly to a "hostile" environment. The true masters are retroviruses, which, due to their high mutation rate, can develop resistance very quickly and, thus, wipe out the success of a drug in one fell swoop (Sect. 24.5).

12.14 Epigenetics: Lifestyle and Environment Influence Gene Activity Like a Pen Leaves a Mark in the Book of Life

The development of an organism does not only depend on the type of hereditary information stored in the DNA that can be translated into gene products. It is just as important that certain genes are only transcribed in certain cells at certain times. Social factors and the environment also influence genes and change their behavior. Scientists have observed the following example in zebra finches. When a male zebra finch hears the song of another male, the gene *EGR-1* is read more strongly. The unknown song of a potential rival leads to much stronger activity in *EGR-1* than background bird singing that the finch has already heard. *EGR-1* is itself a key gene in gene regulation, so a change in the finch's social environment leads to many shifts in the bird's protein expression pattern. This response helps the male bird adapt to changing conditions in his environment, as the intrusion of a potential competitor into his territory can be of critical importance to him. The exciting question is how these changes are expressed at the molecular level.

Pluripotent embryonic stem cells can differentiate into many different cell types. For example, liver, brain and muscle cells have the same set of chromosomes, although they are fundamentally different in function. Obviously, one genotype gives rise to many different phenotypes. This is true for the different cell types of an organism at the same time, as well as for different developmental stages in an organism. Research on twins has yielded remarkable results in this regard. Comparative studies of monozygotic twins, who are genetically identical, show that with increasing age, and especially with different lifestyles, there are progressively larger differences in phenotype. Therefore, there must be mechanisms that lead to changes in the phenotype that are passed on without changes in the genotype. They regulate the transcription process and pass this feature on to daughter cells. This process is called **epigenetics**. It leads to the formation of an additional layer of information that regulates the reading of genes from the DNA.

The environment affects genes through the **epigenome**. Upbringing, childhood experiences, exposure to chemicals or intoxicants, and stress are all epigenetic regulatory influences that temporarily or even permanently alter gene activity. As the following example of agouti mice shows, such information can even be passed on to subsequent generations. Normally, these rodents are small, brown, slender, and very agile. Their genes contain the so-called *agouti* gene, which, when activated, causes the animals to become sick and their fur to turn yellow. Moreover, they become greedy and fat. The offspring of these diseased mice have exactly the same coloring and are just as fragile as their parents. The American molecular biologist Randy Jirtle at Duke University in Durham, North Carolina, fed pregnant agouti females a special diet rich in supplements

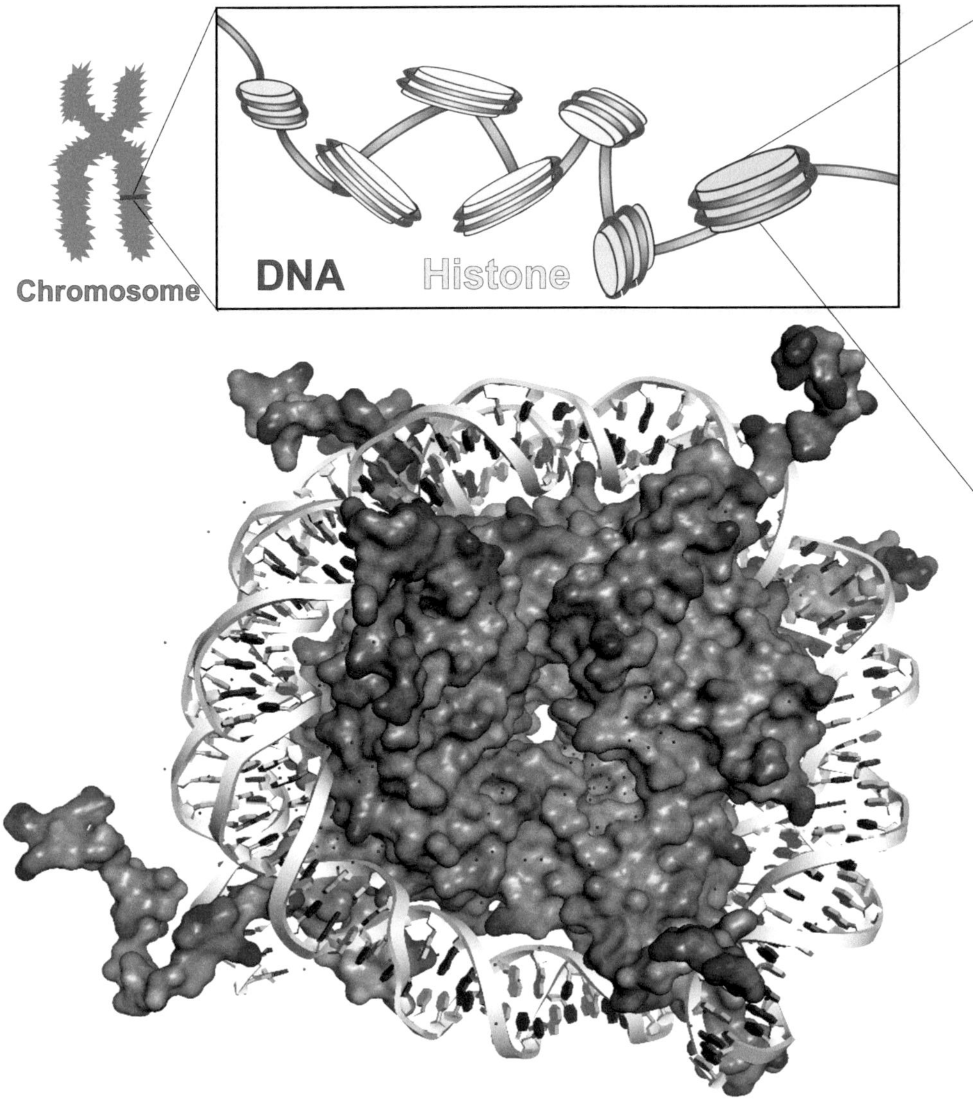

12

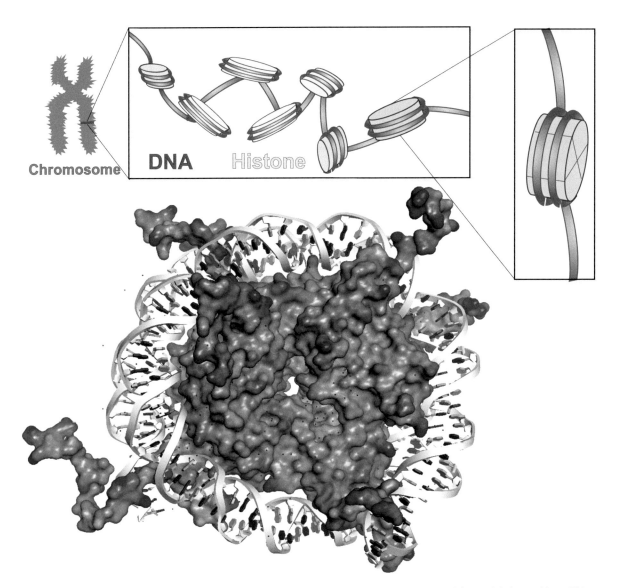

Fig. 12.7 The chromosomal DNA is shortened in its expansion by a factor of 10,000 to 50,000 by coiling itself onto the basic histone proteins, so that it fits into the cell nucleus as a chromosome. The individual histones serve as a kind of coil carrier (nucleosomes, diameter about 110 Å). Approximately 150 base pairs fit on such a coil. Many of them line up like pearls on a string and fold in space to form chromosomes. Histone proteins are made up of helices. In the figure, positively charged basic residues such as arginine and lysine are shown with *bluish* surfaces, while negatively charged acidic residues such as aspartate or glutamate are shown with *reddish* surfaces. The negative charges of the DNA backbone, which contains many phosphate groups, are compensated by the many arginine and lysine residues. This creates a strong electrostatic attraction. In order to read the DNA, this interaction must be loosened. For this purpose, individual Lys and Arg residues are acetylated, whereby the basic amino and guanidino groups shed their positive charge. This causes a decrease in the compact binding of the DNA to the histone coils. (► https://sn.pub/rpa861)

such as vitamin B12, folic acid, choline, and betaine. As a result, the majority of these females' offspring were brown, slender, and in excellent health. The *agouti* gene was switched off by the enriched diet without requiring any changes to the rodents' genome sequence.

At the molecular level, it is methylation and acetylation that carry the additional epigenetic information. In contrast to genetic changes, which cause mutations in the translated gene products, epigenetic changes have a strong dynamic component and, above all, are **reversible**. In the stretched state, there are more than two meters of DNA in the cell, and this is wound in a very compact form on small basic proteins: the **histones**. Lined up like pearls on a string, they form the **chromatin** that makes up the chromosomes in their most densely packed form (Fig. 12.7). Histones are the most highly conserved proteins in existence, for example, the 102-residue histone protein H4 from peas and cows differ in only two positions.

Epigenetic modifications modify DNA by transferring methyl groups to cytosine by **methyltransferases** (see

Sect. 26.10) to form 5-methylcytosine. Base pairing with guanine in the DNA is not affected by this modification, and the genetic code remains unchanged. When methylation occurs in a promoter region of DNA, it silences the corresponding gene. The methylation makes the DNA inaccessible to the reader, much like password-protected computer data. If the promoters in these gene segments are demethylated by **methylases**, translation into the corresponding protein will be possible again. A second epigenetic modification is the modification of histone proteins. For example, **histone acetyltransferases (HATs)** can enzymatically transfer acetyl groups to lysine residues of these basic proteins. The added acetyl groups neutralize the positive charge on the lysine residues (the so-called "histone tails"). As a result, they can no longer interact as efficiently with the negatively charged phosphate groups of DNA. Additional phosphate groups attached to serine, threonine, and tyrosine residues by **histone kinases** are even more repulsive. These changes result in less densely packed chromatin, making it easier to read DNA in certain regions. Transcription and gene expression are regulated in this way. On the other hand, the cleavage of acetyl groups by **histone deacetylases (HDACs)** or the methylation of lysine and arginine residues of histones by **histone methyltransferases (HMTs)** increases the packing density of chromatin and decreases the probability of DNA transcription in affected regions.

The many proteins that create, recognize, or reverse epigenetic changes are called **writers** (which make chemical modifications to DNA and histones), **readers** (which find and interpret these modifications), and **erasers** (which remove the chemical marks as enzymes). Misregulation of these enzymes is commonly associated with the development of various types of cancer. Since epigenetic processes are fundamentally reversible, there is a promising opportunity to correct the dysregulated function of these enzymes through drug therapy. In addition, attempts are being made to manipulate the function of the protein domains that read these modifications. For this reason, intensive research efforts are being devoted to finding ligands for the various methylases, methyltransferases, histone acetyltransferases, histone deacetylases, and reader domains. The **histone deacetylases** that have been most successfully studied to date are mechanistically closely related to metalloproteases (Sect. 25.10). It is hoped that inhibitors of these enzymes, which induce or suppress disease-causing epigenetic changes, can be developed into potent and safe drugs for human cancer therapy.

12.15 The Scope and Limitations of Gene Therapy

In September 1990, 4-year-old Ashanti DeSilva became the first patient to receive **gene therapy**. Both of her parents had deficient alleles for the enzyme adenosine deaminase. Because this enzyme is critical for the functioning of the immune system, the little girl suffered from a severe immunodeficiency that could not be treated with conventional therapies. As a therapy, the patient's white blood cells were repeatedly infected with a virus that carried the correct information for the missing enzyme. The patient, who had previously been hospitalized and was in constant danger of infection, became a person of perfectly normal health.

Gene therapy refers to any technology that introduces a gene into a patient's cell to **replace a defective or missing gene**. In principle, it is very simple. Viruses demonstrate it to us every day: they carry their own genetic information into a foreign cell and use it to encode a few key enzymes that are necessary for their own reproduction. For the rest, they use the biosynthetic machinery of the infected cell. **Retroviruses**, whose genetic information is encoded in RNA, translate this information into DNA and integrate it into the host's DNA. In gene therapy, a nucleic acid segment encoding the protein to be replaced in the patient is inserted into the genome of a virus. The **construct**, as these modified viral genes are called, is surrounded by the viral capsid and introduced into the patient's cells. This can be done either outside the body, such as in previously removed bone marrow or white blood cells, or inside the body, such as by injection into tumor tissue or a specific organ.

Adenoviruses, herpesviruses, or retroviruses are all well suited as gene carriers because these viruses incorporate their own genetic information into mammalian DNA. Although retroviruses transfer their genes only during cell division, adenoviruses can cause nondividing cells to incorporate and use foreign genetic information. Plasmids, DNA and liposomes, and pure DNA constructs are also being experimented with. The rates of transfer of the new information into cellular DNA are significantly higher than is the case with viruses.

There are now more than 1000 gene therapy clinical trials underway, most of them in the USA and mostly for tumor therapy. Cancer is not an inherited disease, but the genetic information that is passed from cell to cell creates a "local genetic disease." Oncogenes are a large group of proteins responsible for the development of cancer. Tumor suppressor genes encode proteins that interfere with the cell cycle and stop cells from dividing. The rapidly increasing knowledge of the molecular structure of these proteins has opened up many approaches to gene therapy of tumors.

Other diseases can also be targeted with gene therapy. The standard therapy for cardiovascular diseases, which are characterized by excessive growth of endothelial cells and consequent narrowing of blood vessels, is to dilate the vessels with a balloon catheter. This helps, but only temporarily. After a few months, the cells start to proliferate again and the blood flow in the downstream areas decreases threateningly. Gene therapy could be used here.

Adenoviruses can be released locally during the balloon catheter treatment. These carry the genetic information for a protein that inhibits cell division, the so-called retinoblastoma protein. The cells are then unable to proliferate.

In 2014, the first gene therapeutic drug, alipogentiparvovec, was approved for the treatment of hereditary lipoprotein lipase deficiency. This was followed in 2018 by voretigene neparvovec (Luxturna®), a treatment for a rare inherited retinal disease. Patients with hemophilia, who require continuous replacement of missing blood clotting factors, will soon benefit from a gene therapy medicine. It will hopefully enable them to produce the clotting factors in liver cells lifelong.

Will gene therapy replace traditional drug therapy? The answer is definitely no. The technique is very complex and each patient needs an individually tailored therapy. Moreover, in many cases, results have been disappointing and sometimes devastating. Gene therapy will take a firm place in the treatment of certain diseases, as it is a curative and not a symptomatic therapy. With increasing experience and better assessment of the possible risks, interventions in the human genome will become acceptable for some diseases, as they make it possible to eliminate the genetic defect once and for all for the individual and their offspring.

In recent years, genetic engineering, and with it the possibility of gene therapy, has gained new momentum from another direction. As early as in the 1980s, so-called **CRISPR** sequences (**c**lustered **r**egularly **i**nterspaced **s**hort **p**alindromic **r**epeats) were discovered in the genomes of various bacteria. These are repeated sequences of 33–47 nucleotides that are interrupted by variable regions of other nucleotides. It is now known that these sequences, in conjunction with the protein Cas, are used to specifically cut double-stranded DNA. This system, known as **CRISPR-Cas9**, serves as an adaptive immune defense for bacteria. For example, viral sequences introduced during a viral infection can be identified and "registered": in the event of a new infection with these viruses, CRISPR-Cas9 can cut the associated DNA and render it unusable.

Although this system originated from bacteria, it functions as precise **gene scissors** in almost all organisms and cells. The two scientists Emmanuelle Charpentier and Jennifer Doudna recognized the potential of this method for **gene editing**. It involves injecting a cell with the RNA that codes for the Cas9 protein and a recognition sequence. The cell uses this RNA to produce Cas9. This protein then finds the enclosed recognition RNA sequence. As a result, Cas9 then cuts double-stranded DNA according to the added RNA sequence fragment. Thus, it is only the added RNA recognition sequence that determines where Cas9 cuts the DNA. RNA fragments can be synthesized with any sequence, which theoretically means that the method can be used to cut at any arbitrary but predetermined position within a given genome.

After the DNA is cut, there are separate pieces of genetic material that are reassembled by cellular repair mechanisms. This step is often imprecise and produces short pieces of DNA (called "indels") that can be inserted or deleted at the cutting site. This renders the affected genes useless for further translation. Bacteria use this very same mechanism in their adaptive immune response to disable newly incorporated viral DNA.

However, gene editing uses a different step. If unbound DNA with loose ends is present in a cell, it is inserted without gaps into the cut DNA using the so-called **HDR system** (**h**omology-**d**irected **r**epair). In this way, the unbound DNA is introduced into the genome as a targeted modification. As early as 2015, there were reports about Chinese research groups using these genetic scissors to free human embryos from hereditary diseases.

More recently, virus-based gene therapy and gene scissors have been used in cancer therapy. In advanced cancer, the body's own gene-edited immune cells can be used for therapy. The T-cells of the immune system check whether a cell has been transformed into a cancer cell by degeneration (Sect. 31.7). If such cells are detected, they are normally eliminated by the immune system. Unfortunately, degenerated cells often know how to evade the immune response (Sect. 31.7). To overcome this, white blood cells are taken from the cancer patient and T-cells are isolated from them outside the body. These T-cells are genetically engineered to carry a chimeric antigen receptor that specifically recognizes degenerate cancer cells. This can be achieved either by introducing the modified DNA into the T-cells via viruses or by editing the genetic material of the T-cells with genetic scissors. The result is genetically modified CAR-T cells. They are then reinfused into the patient (therapy with, for example, Kymriah® or Yescarta®), where they recognize the cancer cells and initiate their killing. The CAR-T cells multiply in the body, providing long-term protection against the cancer.

The **CRISPR-Cas9 system** is currently the subject of heated debate and is being studied for its applicability, but also for its risks and dangers for therapy. The two scientists Emmanuelle Charpentier and Jennifer Doudna were awarded the 2020 Nobel Prize in Chemistry for their achievement. Certainly, molecular biology has never before had such precise and widely applicable genetic scissors. But is its use fully understood, with all the consequences for the modified organism? Will there be long-term effects, and how will the newly introduced gene be transmitted to the next generation? One can even ask whether an organism manipulated with these genetic scissors has been genetically modified at all. In fact, only a new variant of an already existing genetic setup has been created. This happens over and over again "naturally" during breeding. A new tool is available, its potential is undeniable. Only time will tell whether it will also bring the blessing of freeing us from hereditary diseases. In December 2023, the first CRISPR-Cas9 gene-editing therapy was approved. Casgevy, launched by Vertex Pharmaceuticals, is aimed at curing sickle cell disease. It does not correct the mutation that causes sickle cell disease (Sect. 12.13). Instead, Casgevy is designed to compensate for the loss of adult hemoglobin by inducing fetal hemoglobin, the main oxygen carrier in the fetus, which is normally switched off shortly after birth. The therapy is still very complex and expensive, but it represents a promising new treatment option.

Gene technology not only solves problems, but it also creates new ones. The technical barrier for creating a *homo perfectus* is as low as it has ever been in human history. The door to possible misuse is now wide open. We can only hope that ethics and common sense will prevent this from happening. Draconian legal regulations do more harm to the beneficial use of gene technology than to prevent its misuse. Those in positions of responsibility have recognized this and have created a framework within which gene technology can evolve for the benefit of humanity.

12.16 Synopsis

- Gene technology has developed into a key technology in modern drug research because it allows the production of pure proteins, the targeted mutagenesis to elucidate functional and mechanistic properties of proteins or to confirm and disprove binding modes, produces animal models by knocking in and out particular genes, and allows genes to be activated or silenced. With the PROTAC method, gene technology can be used to target proteins for cellular degradation. This makes individualized somatic gene therapy possible.
- The elucidation of the genetic code, the recombinant production of genes and gene products, and the polymerase chain reaction were milestones in the establishment of gene technology.
- The sequence analysis of the human genome revealed the constitution of our genes, the number of gene products, and many functional insights. Meanwhile, hundreds of genomes of other species have been sequenced, and the analysis of the genome of individuals is affordable.
- The human genome contains about 25,000 genes of which about 21,500 are translated into proteins. Some sequence segments are noncoding RNAs and they accomplish important functions in the organism (e.g., in the ribosome or spliceosome). About 95% of the genome contains numerous sequences and signals that control the regulation of the genome. A functional classification of the gene products has been accomplished for a significant portion of the genome.
- To study the relevance of blocking the function of a gene product, that is, a protein in a disease situation, a particular gene can be knocked out in an animal model, mostly in mice. Genes can also be knocked in. Such turning on and off of genes is of utmost importance in drug research because it provides decisive information about the relevance of a planned therapeutic intervention.
- *In vitro* models for drug screening could only be developed once proteins could be produced in pure form and high yield. Various expression systems from bacterial up to mammalian cells can be used for the production of foreign proteins, which are brought into cells via the corresponding coding DNA.
- Genes can be silenced by RNA interference. Therefore, small amounts of double-stranded RNA, usually produced by the enzyme dicer, are incorporated into the enzyme complex RISC. RISC uses one strand of the RNA dimer segments as a template to capture mRNA molecules with a complementary sequence and cleaves them sequentially. By doing this, mRNAs with particular sequences are eliminated.
- To copy this principle for therapy, one needs about 22-base RNA molecules that have to be transported across the membrane into cells, a difficult task with fragile and highly polar species. Furthermore, these molecules can cause unwanted immune responses. Chemical modifications of the RNA molecules are aimed at improvements in the transportation, immunogenicity, and stability properties.
- The PROTAC process (proteolysis-targeting chimeras) causes selective proteolytic degradation of pathogenic and, therefore, unwanted proteins into amino acids by the cell's own protein degradation machine.
- The proteome reflects the totality of all proteins in a cell at a given point in time under precisely defined conditions. Its composition changes dynamically and differs between healthy or diseased states or under the influence of therapeutic treatments.
- The proteome can be analyzed at any given time by 2D gel electrophoresis. This combines a separation by isoelectric focusing and SDS-PAGE analysis. Differences in expression patterns indicate the involvement of proteins in a disease situation. Back regulation under drug administration can indicate a possible therapeutic strategy.
- Pull-down experiments with immobilized drug molecules on a chromatographic solid support allow the trapping of proteins that show interaction with studied drug molecules. Interaction profiles for drug molecules in the cell can be determined.
- Biomolecules can be immobilized on microarray chips. In particular, RNA, DNA, and their oligonucleotides are anchored on these chips to extract com-

plementary RNA or DNA sequences from large mixtures. Suitable fluorescent labeling of the anchored decoy sequences or of the sequences to be "fished" makes it easy to automatically record the detection of binding. With this, the expression patterns of cells can be studied.

- Polymorphisms, particularly single nucleotide polymorphisms (SNPs), are variations in the composition of the genome of a species. These changes make individuals different, and some SNPs confer susceptibility or resistance to diseases or influence the cellular response to a drug.
- Differences in the individual genomes might be the key to a tailored individual and personalized drug therapy and can allow a susceptibility to a particular disease pattern to be recognized. Intolerance to a given drug therapy could become transparent or the classification of an individual into different metabolizer classes could be achieved.
- Genetic differences can be a reason for the development of diseases. In some cases, they are caused by single amino acid exchanges in one gene product (e.g., sickle cell anemia); in other cases multifactorial genetic causes are responsible for the disease development.
- Epigenetics do not alter the sequence of DNA, but regulate the transcription process by changing the accessibility of genes. Lifestyle, experience, and environment exert their effect on the genes through the epigenome.
- Methylations, phosphorylations, and acetylations transmit epigenetic information in a reversible manner. Either the bases of DNA are directly methylated or the packing density of stored DNA on the histone proteins is altered making it more or lesser accessible to DNA reader domains. The latter process modifies the charges of positively charged Lys and Arg residues involved in packing via the transfer of acetyl groups.
- The goal of gene therapy tries to replace a defective or missing gene in the cells of a patient. This would make it possible to eliminate the genetic disease for the individual and their offspring. A nucleic acid segment is inserted into the genome via viral carriers, and it codes for the protein that is to be substituted in the patient. Gene therapy attempts to replace defective or missing genes in a patient's body cells. Retroviruses are used for this purpose and a nucleic acid segment is introduced into their genome that codes for the protein to be substituted in the patient. Retroviruses transcribe this information into DNA and integrate it into the DNA of the cells in the patient. With the CRISPR-Cas9 method, highly precise gene scissors are also available, which in the future might make it easy to exchange diseased genes for healthy ones.

Bibliography and Further Reading

General Literature

B. R. Glick and J. J. Pasternak, Molecular Biotechnology: Principles and Applications of Recombinant DNA, ASM Press; 6th Edition (2022)

K. B. Mullis, F. Ferré and R. A. Gibbs, Eds., The Polymerase Chain Reaction, Birkhäuser, Boston (1994)

N. G. Cooper, Ed., The Human Genome Project. Deciphering the Blueprint of Heredity, University Science Books, Mill Valley, CA, USA (1994)

T. Strachan, The Human Genome, Bios Scientific, Oxford (1992)

G. M. Monastersky and J. M. Robel, Eds., Strategies in Transgenic Animal Science, Blackwell Science, Oxford (1995)

M. Békés, D. R. Langley, C. M. Crews, PROTAC targeted protein degraders: the past is prologue, Nat. Rev. Drug Discov., **21**, 181–200 (2022)

J. A. Wolff, Gene Therapeutics. Methods and Applications of Direct Gene Transfer, Birkhäuser, Boston (1994)

L. E. Post, Gene Therapy: Progress, New Directions, and Issues, Ann. Rep. Med. Chem., **30**, 219–226 (1995)

J. S. Kiely, Recent Advances in Antisense Technology, Ann. Rep. Med. Chem., **29**, 297–306 (1994)

S. B. Pandit, S. Balaji and N. Srinivasan, Structural and Functional Characterization of Gene Products Encoded in the Human Genome by Homology Detection, IUBMB Life, **56**, 317–331 (2004)

P. E. Slagboom and I. Meulenbelt, Organisation of the Human Genome and our Tools for Identifying Disease Genes, Biological Psychology, **61**, 11–31 (2002)

E. S. Lander, et al., Initial Sequencing and Analysis of the Human Genome, Nature, **409**, 860–921 (2001)

J. C. Venter, et al., The Sequence of the Human Genome, Science, **291**, 1304–1351 (2001)

S. L. Salzberg, Open question: How many genes do we have? BMC Biol., **16**, 94 (2018)

A. C. Lai and C. M. Crews, Induced protein degradation: An emerging drug discovery paradigm, Nat. Rev. Drug Discov., **16**, 101–114 (2017)

L. DeFrancesco, Life Technologies promises $ 1,000 genome. Nature Biotech., **30**, 126 (2012)

J.A. Doudna and E. Charpentier, Genome Editing. The new frontier of genome engineering with CRISPR-Cas9, Science, **346**, 1258096, 1–9 (2014)

Special Literature

K. B. Mullis, The unusual origin of the polymerase chain reaction, Scient. American, **262**(4), 56–61 (1990)

R. D. Fleischmann *et al.*, Whole Genome Random Sequencing and Assembly of *Haemophilus influenzae Rd*, Science, **269**, 496–512 (1995)

M. D. Adams *et al.*, Initial Assessment of Human Gene Diversity and Expression Patterns Based Upon 83 Million Nucleotides of cDNA Sequence, Nature, **377**, Suppl., 3–174 (1995)

L. Timmons, H. Tabara, C. C. Mello and A. Z. Fire, Inducible Systemic RNA Silencing in Caenorhabditis elegans, Mol Biol Cell., **14**, 2972–2983 (2003)

K. M. Sakamoto, K. B. Kim, A. Kumagai, F. Mercurio, C. M. Crews, R. J. Deshaies, Protacs: chimeric molecules that target proteins to the Skp1-Cullin-F box complex for ubiquitination and degradation, Proc Natl Acad Sci USA, **98**, 8554–8559 (2001)

C. Arnold, PROTAC protein degraders to drug the undruggable enter phase 3 trials, Nat Med, (2020), https://doi.org/10.1038/d41591-024-00072-8

M. W. Chang, E. Barr, J. Seltzer, Y.-Q. Jiang, G. J. Nabel, E. G. Nabel, M. S. Parmacek and J. M. Leiden, Cytostatic Gene Therapy for Vascular Proliferative Disorders with a Constitutively Active Form of the Retinoblastoma Gene Product, Science, **267**, 518–522 (1995)

C. Craig, Bristol-Myers to Pay $ 2.7 M for Transgenic Goats that Make Human Antibodies, BioWorld Today, **6**, 1 (1995)

M. S. Gadd et al., Structural basis of PROTAC cooperative recognition for selective protein degradation, Nat. Chem. Biol., **13**, 514–521 (2017)

U. Rix and G. Superti-Furga, Target profiling of small molecules by chemical proteomics, Nat. Chem. Biol., **5**, 616–624 (2009)

R. K. Seide and A. Giaccio, Patenting Animals, Chemistry & Industry, **16**, 656–659 (1995)

http://www.ensembl.org/Homo_sapiens/index.html (*Explore the Homo sapiens genome*) (Last accessed Nov. 16, 2024)

https://www.genecards.org/ (*The Human Gene Database with functional predictions*) (Last accessed Nov. 16, 2024)

https://www.proteinatlas.org/search/ (*The Human Protein Altas*) (Last accessed Nov. 29, 2024)

J. Wang, S. Yazdani, A. Han, M. Schapira, Structure-based view of the druggable genome, Drug Discov Today, **25**, 561–567 (2020)

S. Adhikari et al., A high-stringency blueprint of the human proteome, Nat Comm, **11**, 5301 (2020). https://doi.org/10.1038/s41467-020-19045-9

P. Amaral et al., The status of the human gene catalogue, Nature, **622**, 41–47 (2023)

J. M. Carlton et al., Draft Genome Sequence of the Sexually Transmitted Pathogen Trichomonas vaginalis, Science, **315**, 207–212 (2007)

S. Schneiker et al., Complete Genome Sequence of the Myxobacterium Sorangium cellulosum. Nature Biotech., **25**, 1281–1289 (2007)

Experimental Methods of Structure Determination

Contents

G. Klebe, *Drug Design*, https://doi.org/10.1007/978-3-662-68998-1_13

In this chapter, we will focus on experimental methods used to determine the structure of ligands and proteins. There are three main techniques that provide information about the three-dimensional structure of small organic molecules up to proteins: **crystal structure analysis**, **high-resolution NMR spectroscopy**, and **electron microscopy**. The first technique is the oldest. It goes back to an experiment performed by Max von Laue in 1912. Just 17 years earlier, Wilhelm Roentgen had discovered an electromagnetic radiation that was later named X-rays or "Roentgen rays" in German in his honor. Together with his collaborators Walter Friedrich and Paul Knipping, Laue demonstrated the wave nature of X-rays using a copper sulfate crystal. At the same time, they demonstrated the lattice structure of crystals. Just one year later, William Lawrence Bragg and his father William Henry Bragg reaped the rewards of these experiments. They determined the crystal structure of sodium chloride. The technique has evolved over the years. Today, the structures of proteins and nucleic acid complexes containing many thousands of amino acids and nucleosides have been determined.

NMR spectroscopy is likewise a relatively new technique. In 1945, the research group of Felix Bloch and Edward Purcell in the USA first observed the resonance absorption of hydrogen nuclei in a magnetic field. From this experiment, the technique has grown, largely due to advances in instrumentation, to the point where the structure of proteins with more than 800 amino acids can be determined. However, this requires the protein to be extensively labeled with different isotopes.

In recent years, **electron microscopy** has emerged as another very powerful imaging technique, providing direct images for structure determination of very large, often membrane-bound protein complexes and viruses. For this purpose, the samples must be "vitrified" (sealed in ice like a glass) in a process known as cryo-electron microscopy, so that they can then be studied as individual molecules at low temperatures. On the other hand, electron beams can also be used for crystallographic diffraction experiments to elucidate structures, similar to classical crystallography.

13.1 Crystals: Aesthetic on the Outside, Periodic on the Inside

The term "crystal" causes one to immediately think of well-formed minerals or sparkling gemstones with a magnificent cut. The association of crystals with the structures of the molecules that determine our lives only occur to us as a second thought. The crystal is typically associated with "dead" material. When Jack Dunitz took over his chair as professor of organic chemistry at the ETH in Zurich at the end of the 1950s, the famous natural product chemist Leopold Ruzicka dismissively told him that crystals are a "chemical graveyard." Nonetheless, Dunitz and his research group showed over many years that a crystal in no way belongs in a "graveyard," but rather is the key to understanding the structure, dynamics, and reactivity of molecules.

If a mineral is considered, the regular construction of the single crystals stands out. Even organic materials have the ability to form well-shaped crystals. One must only think of the fascinating crystals of candied sugar. Is this external regularity a representation of the inner structure? Before this question is answered, the way that crystals are obtained should be clarified. A mineralogist got it easy. Nature has already provided well-formed crystals over thousands or millions of years. Organic molecules and proteins rarely occur in Nature in a crystalline state. Conditions must be found under which they crystallize.

In general, crystals are grown from a solution. For simple organic substances, this can also be accomplished from melted material or by sublimation. Both crystallization methods are known from water when a lake freezes over or when beautiful hoarfrost crystals form. For **crystallization from solution**, a solvent is sought in which the com-

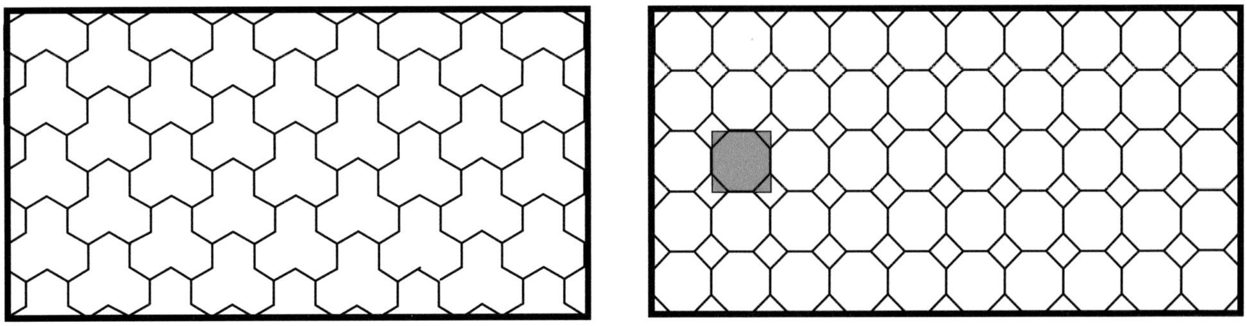

Fig. 13.1 Paving stones cover a surface without leaving holes (*left*). This is only possible if they are derived from a particular basic geometric pattern, for instance a parallelogram, rectangle, square, triangle, or hexagon. This basic pattern can by modulated by complementary bulges and recesses. A path cannot be covered without holes if equilateral pentagons or octagons are used. If an octagonal stone is combined with a square stone, however, the surface can be completely covered. It is immediately clear that if a square stone is cut along its two diagonals, two triangles result. Adding four such pieces an octagon can be amended to a square in this way (*right*)

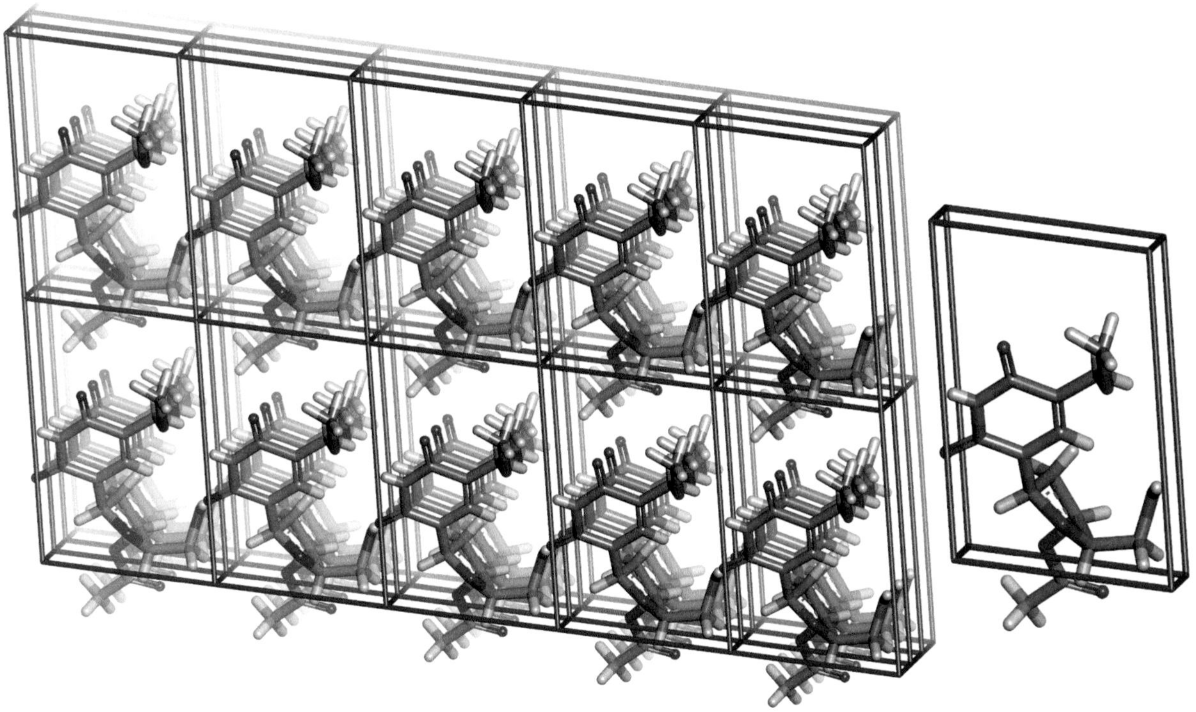

Fig. 13.2 In the simplest case, a molecular packing is generated by purely shifting the molecule in all three spatial directions. The generating unit, the elementary cell, is derived from an irregularly angled body, a parallelepiped (*right*, magnified image, *purple*). If a point near the molecule is picked out and all of the molecules in the crystal packing are connected via this point, a three-dimensional lattice will result. (► https://sn.pub/UDQaYm)

pound is sufficiently soluble. By changing the conditions, the saturation point of the solution is exceeded. If this is done slowly, small crystal nuclei will be formed which can grow into large crystals. As a rule, the solubility of the compound decreases as the temperature is lowered. The saturation point of the solution can be exceeded by changing the temperature. The solution can also be "thickened," meaning that some of the solvent is removed. Another possibility is to add a second solvent in which the compound is less soluble. If the ratio of the two solvents is chosen correctly, the saturation point can be approached slowly. For compounds with acidic or basic groups, pH conditions can be found at which the compound exists as a salt. Because of strong ionic interactions, salts often form better crystals. Organic compounds can also be "salted out." This is done by adding a salt, such as sodium chloride, to an aqueous solution of the compound. The salt "absorbs" the water molecules as it goes into solution. It is surrounded by a solvation shell of water molecules. This removes the water molecules from the surface of the organic compound, which is also surrounded by a solvent sphere. By this, the saturation point of the compound is gradually exceeded and crystallization begins.

Proteins are complex entities that, as a general rule, are only soluble in a buffered aqueous solution. Because of their amino acid composition, they carry charged ionic groups on their surfaces. Similarly, for proteins, it is also necessary to find conditions under which they associate in periodic arrays. This is achieved by slowly changing the amount of water in which the protein is dissolved. This can be done in both directions. Hydrophobic proteins begin to aggregate when the amount of water increases. Proteins that have stronger polar groups on their surfaces aggregate when the water molecules are removed from their surfaces. The adjustment of the local pH conditions to find an optimal value, the choice of a suitable salt for salting out, and the different temperatures are the conditions that need to be optimized. In addition to salts, surface-active substances (detergents) can also influence the solvation shell and support crystallization. Despite this, crystallization is a kind of fine art. The search for suitable conditions requires creativity and diligence. Today, however, the crystallization methods are so elaborate that the tedious work of setting up thousands of different test conditions is carried out by robots.

Sometimes considerable effort is invested into structure determination. In 1995, the crystallization and structure determination of HIV integrase, one of the key enzymes in the life cycle of the virus, was accomplished only after the 40th point mutation of the original protein.

These point mutations were made with the goal of altering the surface properties of the protein so that orderly aggregation into a crystal could occur.

Let us return to the original question of whether the orderly outwards appearance of a crystal is a reflection of its inner structure. Chemically, a crystal is homogeneous. The organic molecule or the protein is the basic building block. It is only when these building blocks are organized in a spatially ordered fashion that a periodic array emerges which optimally fills space. In everyday life, many solutions to these packing problems are easily seen, for example, sugar cubes that will only fit into the box if they are layered in the right direction, or paving stones that must be laid in a neat, periodic fashion to completely cover a sidewalk or a street without gaps (◻ Fig. 13.1).

A single paving stone, when correctly fitted to the next, represents a repeating unit in the lattice. A crystallographer refers to this unit as the **elementary unit cell**, and the orderly setting of one unit upon another in terms of periodic **translation**. In the simplest organic crystal structure, the elementary cell contains one molecule (◻ Fig. 13.2).

13.2 Just Like Wallpaper: Symmetries Govern Crystal Packings

The contents of an elementary cell can also be more complexly composed, for example, like a wallpaper pattern. A basic motif is repeated so that it fills the surface area. Crystallographers call the basic motif the **asymmetric unit**. In ◻ Fig. 13.3, this motif is a flower branch. Not all of the motifs can be generated simply by shifting the branch; some must be additionally mirror-reflected. A pair of image and mirror-image branches represent the elementary cell. The surface can now be filled with this building block by simply shifting it. In addition to reflecting, basic motifs can also be rotated. By using reflections and rotations, both so-called **symmetry operations**, the contents of the elementary cell is generated from the asymmetric unit. This cell is layered on itself in all three spatial directions in an orderly formed crystal lattice. Even as a three-dimensional entity, the elementary cell must take on a particular form to completely fill all of the space. If the basic types of elementary cells are combined with all of the possible symmetry operations, 230 possibilities result for the basic motif to fill the space. The crystallographer calls them the 230 **space groups**. For chiral molecules, and proteins belong to this group, mirror reflection does not occur. This reduces the number of possible space groups significantly. Therefore, proteins (and nearly all of the chiral organic molecules) crystallize in only 65 space groups.

13.3 Crystal Lattices Diffract X-Rays

Max von Laue used crystals to prove the **wave nature of X-rays** by **diffracting** them. For illustration, we shall consider a water wave. When a drop of rain strikes a puddle, circular waves propagate from the center outwards. The drop generates a so-called elementary wave upon submersion. If two drops that are separated by a particular distance simultaneously strike the water's surface, circular waves propagate outwardly from both submersion points. It is better to observe this experiment when the surface of the water is constantly "excited," for exam-

13

◻ **Fig. 13.3** An area can be covered not only by purely shifting an object, the asymmetric unit. Additional symmetry operations such as mirror reflection and rotation can also be used. This way multiple copies of the object are generated. In the presented case, the flower branch along with its mirror image makes up the unit (the elementary cell is outlined in *red*) that can be used to cover the surface simply by shifting it regularly

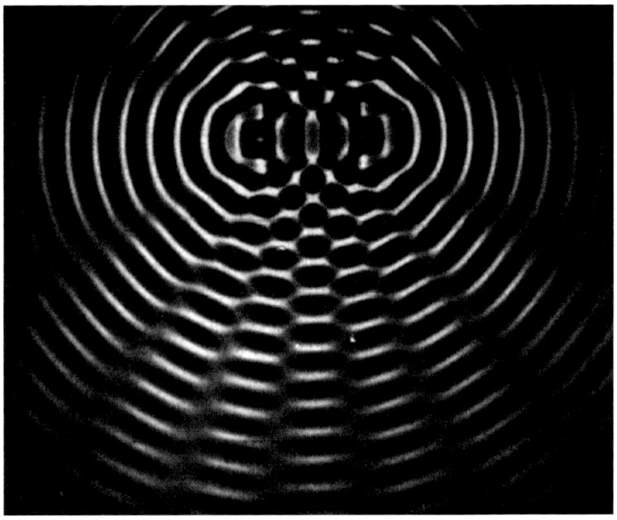

◻ **Fig. 13.4** Two raindrops strike the surface of the water and form circular, outwardly moving water waves. These superimpose on each other to give a band-like interference pattern. There are areas along these bands where the water surface is quiet. In other areas, it moves that much more strongly ple, by a constantly dripping tap. The circular outwardly spreading wave fronts meet each other at some point. What happens? A lamellar pattern forms, parts of the water's surface remain at rest and other parts seem to move vigorously (◘ Fig. 13.4). In the cross section, the water surface moves sinusoidally (◘ Fig. 13.5). How do two waves behave that collide and superimpose with each other? If the wave peak and another wave peak or the wave trough and another wave trough meet, the wave is amplified. If, on the other hand, a wave peak meets a trough, they cancel each other out. The water surface remains calm. The lamellar pattern of moving and still water surface between waves that are moving outwardly and inwardly is caused by this superimposition. It is called **interference**. The band density depends on the distance between the submersion points of the drops. The ensuing interference pattern, therefore, contains information about the relative position of the points from which the elementary waves were generated.

If parallel water waves (e.g., a wave front at the coast) collide with a barrier that has a small opening (e.g., a harbor entrance), semicircular waves spread outwards from the backside. If this barrier has two neighboring openings (double slit), semicircular waves develop behind each of the two openings. By this, the same picture as with the two raindrops is achieved (◘ Fig. 13.4). The two waves interfere with each other behind the double-slit barrier, and a **diffraction pattern** forms. The density of this pattern, that is, the progression of the bands, depends on the geometry of the double slit.

Formally, the diffraction sequence on the crystal lattice is analogous. The same principles are valid, but the superimposition is more complex. A very simple lattice shall be considered that has only one type of atoms. An X-ray beam runs as a parallel wave towards this crystal. It collides with an array of atoms and initiates an interaction that is comparable to that between the raindrop and the puddle. Each atom generates a spherical wave because of the interaction between the atom's electrons and the X-ray. The circular wave on the water's surface represents, therefore, the spherical wave in space. The spreading spherical waves superimpose on one another and form a wave that leaves the crystal in a changed direction (◘ Fig. 13.6). Formally seen, the incoming and outgoing waves have an angular relationship to each other that is equivalent to the reflection of the wave at a plane perpendicular to the considered array of atoms. Therefore, the diffraction of the three-dimensional crystal lattice can be treated formally as a reflection at a plane in the lattice.

Many parallel sets of such lattice planes can be inscribed on a crystal with differing relative separation from one another and relative occupation density with atoms (◘ Fig. 13.7). The reflected waves contain the information about the geometry (distance) and the relative occupancy (scattering power) in this plane. To record the diffraction properties of a crystal, each set of parallel planes of the crystal must be oriented in the X-ray beam so that a reflection is possible. This laborious work is taken over by a computer-controlled diffractometer.

13.4 Crystal Structure Analysis: Evaluating the Spatial Arrangement and Intensity of Diffraction Patterns

To demonstrate that different lattices indeed generate different diffraction patterns, a simple experiment should be considered. For this purpose, a laser pointer and different pinhole filters are needed. The pinhole filters can easily be made. A black-and-white printout of the periodic arrangements shown in ◘ Fig. 13.8 can be greatly reduced and transferred to high-resolution photographic film. These perforated pinhole masks represent two-dimensional periodic lattices. When the laser beam passes through these pinhole masks, a diffraction pattern is generated on a screen, which is shown in ◘ Fig. 13.8. The first two masks change the spacing and symmetry of the pinhole mask. This is clearly reflected in the diffraction images obtained. In the third and fourth masks, the repeating motif of three or five differently sized holes represents a molecule with two types of atoms. These motifs form a periodic lattice when arranged side by side. They have the same spacing or "dimension" as the first image on the left. When the diffraction images are compared, the intensity distribution of the light spots is different in the first, third, and fourth images. This intensity distribution contains the information about the spatial construction of the motif that created the lattice. This information is used to determine the crystal structure and the geometry of the molecule in the crystal.

The **reflections**, that is, the intensity of the individual light spots in the diffraction pattern, contain information about the spatial geometry of the molecule. There is a mathematical technique, the **Fourier transform**, that can be used to translate the diffraction pattern back to the generating motif expressed as a density in space. A Fourier transform is the superposition of many sine and cosine functions to describe the periodicity in space. The intensity of the diffracted reflections determines the contribution of the superimposed sine and cosine functions, as well as their relative phase. The importance of these aspects for the superposition of waves has already been emphasized in the explanation of interference (◘ Fig. 13.5). Unfortunately, this relative phasing information is lost in the diffraction experiment. The diffractometer records only the intensity of the reflections. This lack of information is known as **the phase problem in crystal structure determination**. The relative phases must be reconstructed for the individual reflections using computational methods and appropriate measurement conditions. Often large electron-rich elements (e.g., heavy metal ions) are embedded in the protein (e.g., by

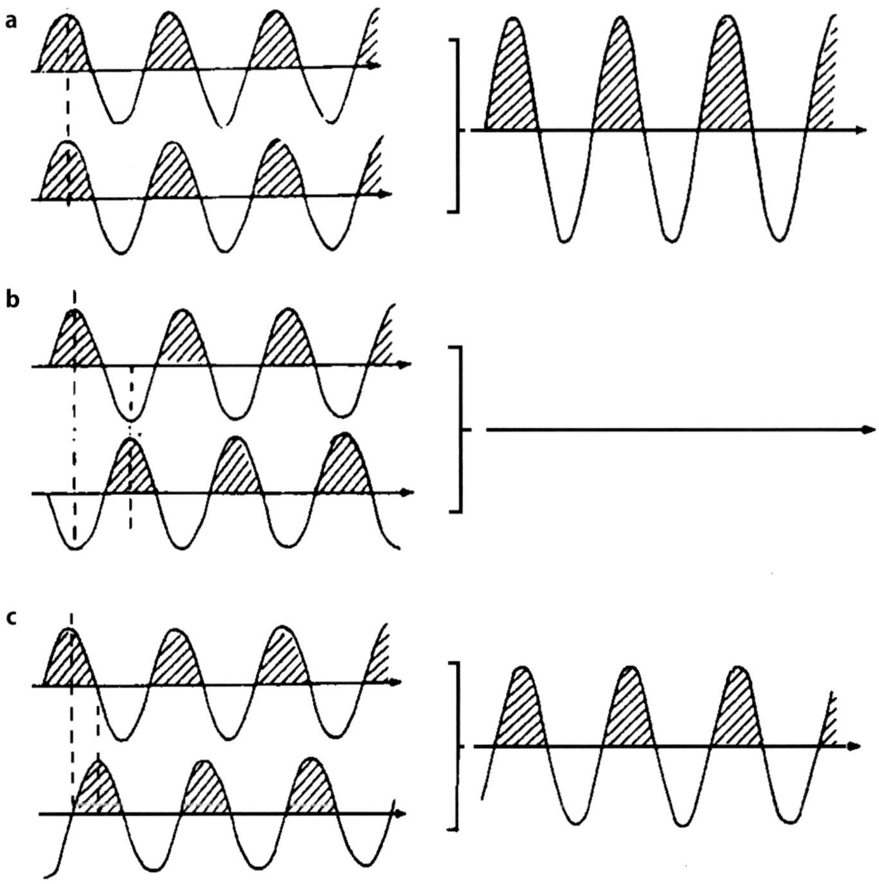

◻ Fig. 13.5 The waves are sinusoidal in cross-section. The distance between two wave peaks is called the wavelength. The height of the water wave at the summit is called the amplitude. The position at which the wave crosses the resting position determines the phase. (**a**) If two wave trains with the same phase meet, they will add to each other and the amplitude doubles. This situation is found in the places in ◻ Fig. 13.4 where the water's surface moves more strongly. (**b**) If there is a phase difference of exactly one half of a wavelength, the wave peaks will meet with the troughs. Both waves cancel each other out. This represents the parts of ◻ Fig. 13.4 where the water surface is very still. (**c**) Any other superimposed phase shift causes a wave, the amplitude of which is somewhere between the extremes in (**a**) and (**b**)

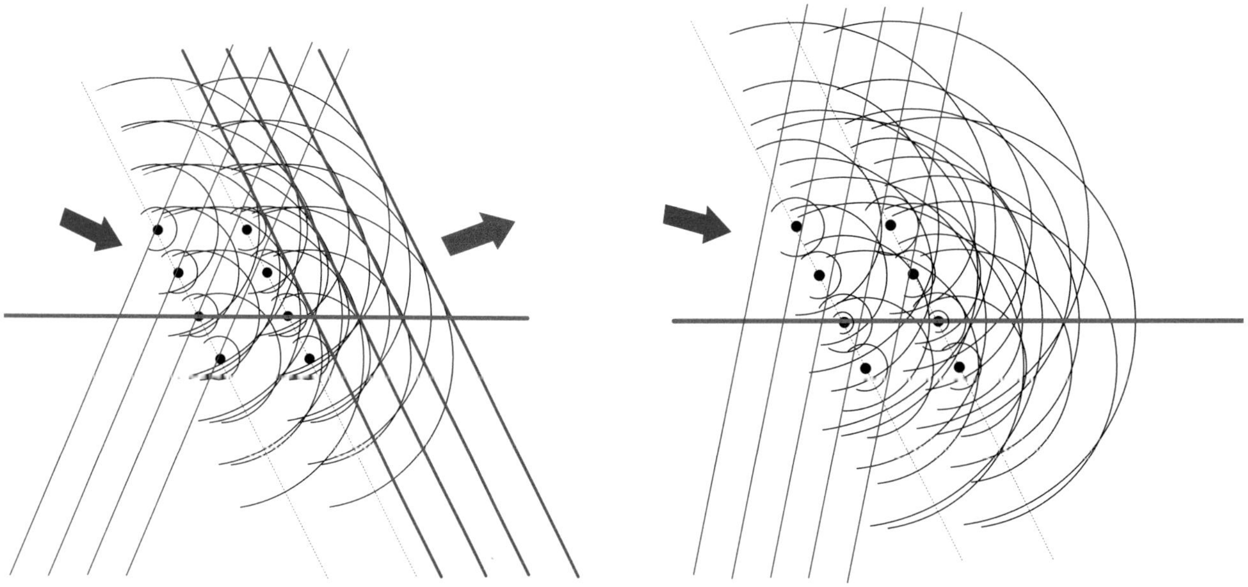

◻ Fig. 13.6 If a wave front (*blue*) in one plane meets with a row of atoms (*black points* on the *dotted lines*), each atom in this row will become the starting point for a three-dimensional circular wave. This is analogous to those created when the raindrop hits the surface of a puddle. The circular waves that formed from the back row of atoms superimpose upon one another just as in the case with the water waves (◻ Fig. 13.4). All circular waves are generated with the same phase in the indicated direction of the incoming wave (*left*). As a result of this superimposition, a new wave front forms (*red*) that leaves the crystal in an altered direction. Relative to the direction of the incoming wave, they have an angle that is formally a reflection of the incoming wave front on the atom row that is marked with the *green line*. If a different incoming direction is taken (*blue*), the circular waves will not be generated from the same place (*right*), that is, there is a phase difference between them. Their superimposition does not lead to a new wave front leaving the crystal

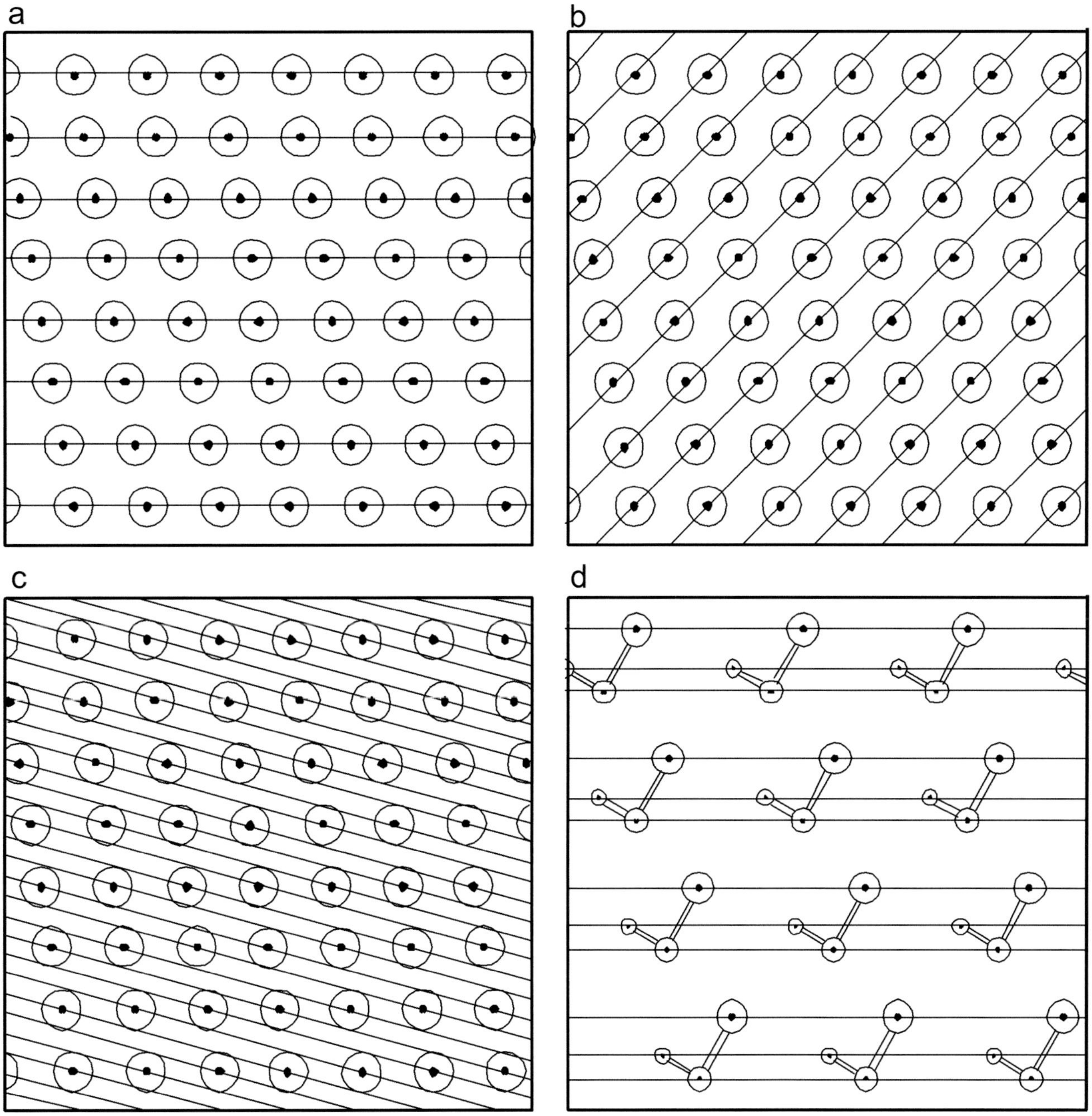

Fig. 13.7 A cluster of parallel planes can be laid through the atoms of a crystal lattice (**a**, **b**, **c**). In the first three images, one sort of atoms is present. The relative distance of the inscribed planes from one another and their atomic occupation density varies. Each of these planes can give rise to "reflections" in an X-ray diffraction experiment. For this, the crystal must be brought into the correct orientation for the incoming beam each time. The X-ray counter must be positioned in a way so that it captures the out-going X-ray beam. It is from this geometry that the spatial orientation of the cluster of planes in the crystal is determined. The occupation density of the atoms decides how "well" a particular array of planes reflects the incoming X-ray beam. This information is contained in the intensity (amplitude) of the outgoing wave. (**d**) Three different types of atoms in a molecular crystal have different spatial relationships to one another. A parallel cluster of planes can be placed through each type of atom in the molecule (here a triatomic molecule). The amplitude of the outgoing beam results as a superposition of three sets of waves reflected at these planes. Therefore, the observed intensity contains information about the relative distance of the atoms on the mutually parallel lattice planes and, therefore, their arrangement in the molecule

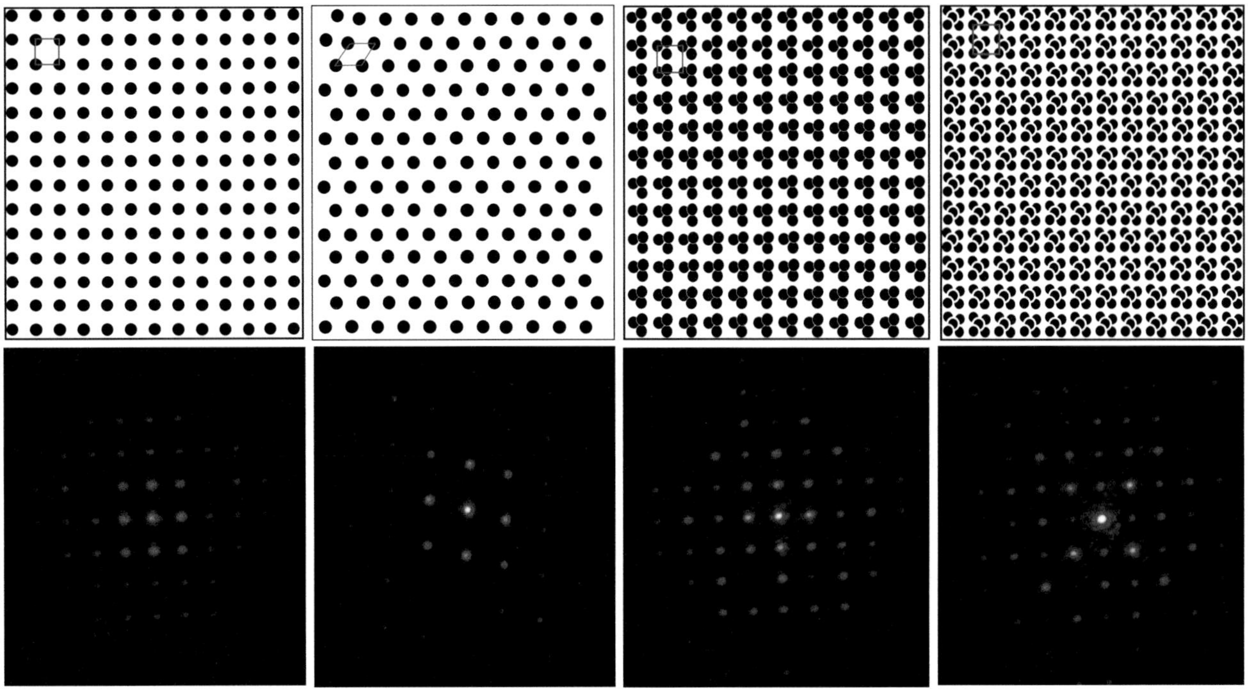

Fig. 13.8 A perforated pinhole mask can be used for a diffraction experiment with a laser pointer. For this the displayed hole patterns (*above*) must be brought to the size of the wavelength of laser light. The diffraction patterns *below* were generated from these masks. The holes in the *two left* masks are all of the same size, which is comparable to having only one type of atom. The hole pattern changes from wide-meshed squares to a skewed parallelogram (see *red* unit cells). The diffraction patterns reflect the symmetry and distance of the holes to one another. In the third and fourth masks on the *right*, the distance between the repeating units is identical to those of the first masks on the *left*. The composition of the motif in the repeating unit, however, varies. It is made up as clusters of multiple holes and can be compared to the different atoms in a molecule. The distance between the diffracted light reflections (*lower row*) is identical for the first, third, and fourth masks. The intensity of the diffracted radiation, however, varies from reflection to reflection. It contains information about the composition and the geometry of the individual motifs and, thus, the "molecules" which give rise to the diffraction patterns

13

coordination to histidine or cysteine). These **heavy atoms** dominate the diffraction pattern so that they betray their position in the crystal lattice. Another method takes advantage of **anomalous scattering**. This effect is based on the interaction of X-rays with the electrons of heavy atoms. As a result, a spherical wave propagating towards an atom is reflected with a phase shift. Simply stated, it is returned with a delay. This effect depends on the wavelength and can be used to determine the phase. The crystal is measured on a synchroton (a particle accelerator that also produces electromagnetic radiation in a broad wavelength range, including X-rays) and the diffraction experiment is performed at **several different wavelengths**. Anomalous scattering requires that a heavy atom is present in the protein structure. This is already the case for metalloproteins. But an alternative approach can also be taken. Proteins produced in a special expression system (Sect. 12.6) can be generated with **selenomethionine** instead of methionine. The heavier selenium acts as an anomalous scatterer in the diffraction experiment. Especially for small molecules, there are methods that allow a simple reconstruction of the phase information from probability considerations in the intensity distribution among different reflections, the so-called "**direct methods**." Such methods are being developed also for protein structure determination. Often an already solved, geometrically related protein structure can be used as a starting model for a structure determination (**molecular replacement method**). Since the structure prediction programs such as Alphafold or Rosettafold (Sect. 20.6) have meanwhile reached a convincing reliability, it is also possible to generate a starting model in this way. The model is translated and rotated in the elementary cell by computer simulations until a calculated diffraction pattern is obtained that matches the experimentally observed diffraction pattern of the unknown protein.

The phasing obtained at the beginning of the structural analysis with these methods is only approximate and must be "refined." For this purpose, the initial model is shifted and modified step by step until an optimal agreement with the experimental diffraction data has been achieved. Altogether the regeneration of the phasing information is not trivial. Even in the 1960s, phasing calculations kept one scientist busy for several years. The methodical progress and the increased performance of computers now allow this to be accomplished in a few minutes. Even today, however, this step can still be very challenging for proteins. It is becoming apparent, how-

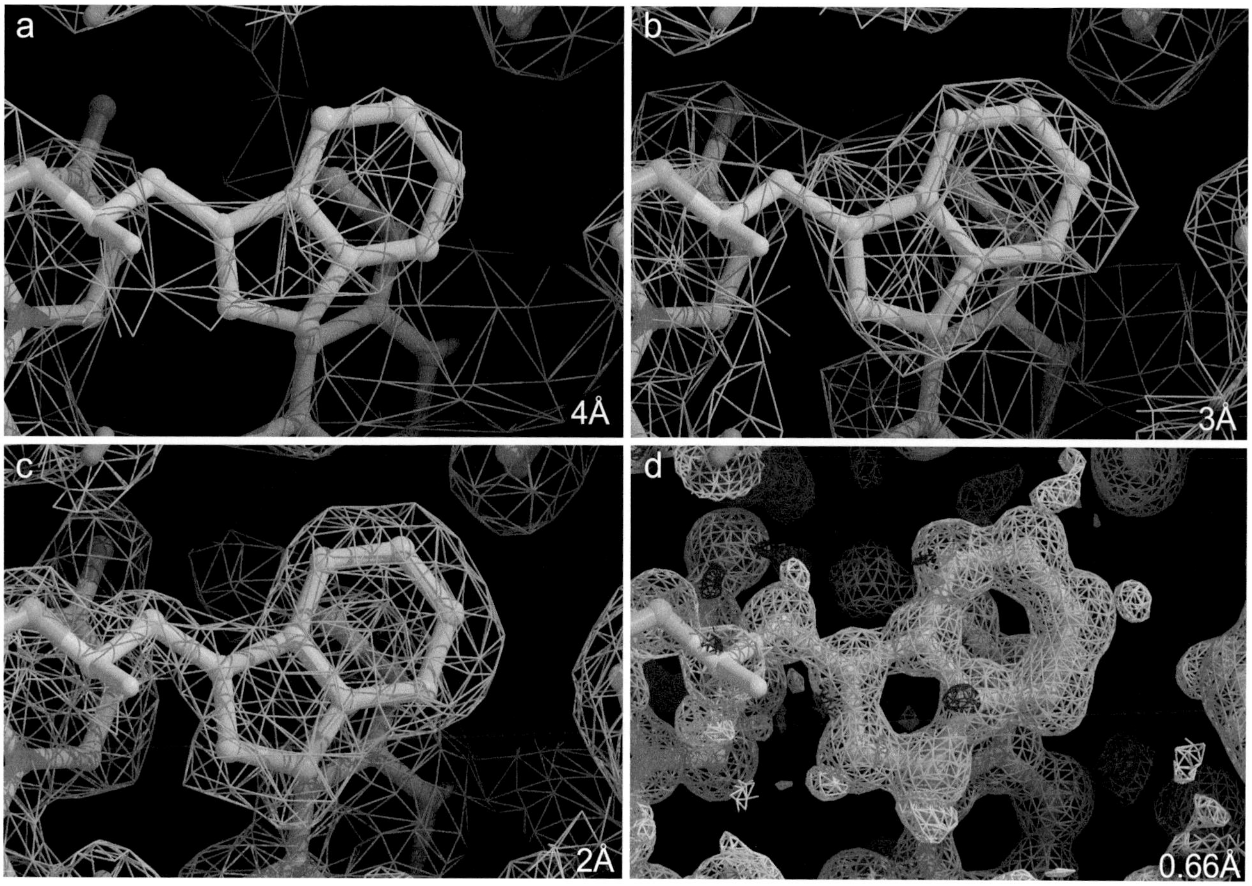

Fig. 13.9 View of a crystal structure of aldose reductase (Sect. 27.4). The electron density (the so-called $2F_0–F_c$ density at the 1σ level) is displayed as a *blue* mesh on the predefined contour level around a tryptophan residue. In (**a**), the diffraction data were obtained at a resolution of 4 Å, and a Fourier transform was used to calculate the electron density. The resolution increases from (**a**) 4 Å to (**b**) 3 Å, to (**c**) 2 Å, and to (**d**) 0.66 Å. The resolution in the last-shown contour density is so high that hydrogen atoms can be recognized as single density peaks in the difference density map (positive is *yellow*, negative is *violet* of the so-called $F_0–F_c$ difference density at the 2σ level). The electron density is so clearly structured at 2 Å (**c**) that it is simple to fit the indole building block in place. At 4-Å resolution (**a**), this assignment is problematic and can easily lead to errors

ever, that the structure determination of medium-sized proteins is becoming routine. Historically, the time span from crystallization to structure determination could be quite long. Urease is certainly a curiosity. It was the first protein to be successfully crystallized. James B. Sumner accomplished this back in 1926. Its 3D structure, however, was first elucidated in 1995, that is, 70 years later!

13.5 Diffraction Power and Resolution Determine the Accuracy of a Crystal Structure

A picture of the contents of the unit cell is the result of the Fourier transform. It is portrayed in terms of the electron density in space (Fig. 13.9). To which detail this electron density can be determined depends on the number of different wave fronts which are superimposed with correct amplitude and phase. The number of wave fronts is equivalent to the number of observed reflections (see above). In the diffraction images with the laser beam (Fig. 13.8), it can be seen that the intensities decrease significantly towards the rim. The maximum achievable resolution is, therefore, determined by the ultimately measurable reflection still observed at the rim. It is generated by the array of planes with the smallest mutual distance in the lattice that can still be observed in the diffraction experiment. In Fig. 13.7a, the spacing between a set of planes is large and the occupation density with atoms is high. Therefore, the observed reflection will be strong. In Fig. 13.7b, the planes are closer together, the occupation density with atoms, however, is lower, and the generated reflection intensity becomes weaker. In Fig. 13.7c, the planes of an array are very close to one another, and the occupation with atoms is strongly thinned out. Consequently, only a very weak reflection is expected here, possibly reaching the resolution limit.

For small organic molecules, this resolution is easily achieved in that the atoms are visible as distinct maxima in the electron density. If the crystal's quality is diminished due to lattice defects or disorder, the resolution is poorer. The resolution obtained from protein crystals is usually between 1.5 and 3 Å. This means, in the best case, a resolution is achieved that is on the order of magnitude of a bond length. The upper limit is in the range of the cross section of a benzene ring. Recently, resolutions of less than 1 Å have been increasingly achieved (◘ Fig. 13.9). In such cases, many details can be seen, for example, individual hydrogen atoms or multiple spatial arrangements of side chains.

At higher resolution, the electron density maxima are directly assigned to the atoms in the molecule (◘ Fig. 13.10). In the beginning this assignment is crude; the phases used in the Fourier transform are only approximate. The position of the detected maxima must still be optimized. This is defined as "**refinement of the structure**." For this, the experimentally observed diffraction pattern is compared with the diffraction pattern that is calculated from the atomic positions of the preliminary model. Such an initial model has to be optimized in a least squares refinement, which is an iterative multistep process. Structural parameters such as atomic coordinates and parameters describing thermal motion are iteratively modulated in small steps, and at each refinement cycle the achieved agreement is quantified by the so-called R-factor. It determines the agreement between the diffraction data calculated from the obtained model with those from the original X-ray diffraction experiment. R-values range from zero for perfect agreement between the calculated and observed diffraction patterns to about 0.6 for a set of measured diffraction data compared to a set of random data. In protein crystallography, an R-factor in the range of 0.2 is considered a desirable target for data with a resolution of about 2.5 Å. For small organic molecules, refinement typically results in values of $R < 0.05$.

If the measurement is very accurate, the density of a "pseudomolecule" with spherical atoms can be subtracted from the observed electron densities at the end of the structure determination. What remains is the electron distribution of the bonds between and lone pairs at the atoms in the molecule. This is, however, only possible with extremely high-resolution measurements. At lower resolution, as is the case in moderately resolved protein structure determinations, a direct assignment of the atoms of the protein to the electron density maxima cannot be made (◘ Fig. 13.10). More commonly, the course of the chains is fitted to the electron density. Because proteins are constructed from 20 different amino acids that prefer to take on typical geometries, the interpretation of the electron density is simplified (◘ Fig. 13.10e, f).

Electrons scatter X-rays. Therefore, the number of electrons around an atom determines how well it is detectable in the resulting density. Hydrogen atoms have only one electron in their shell. As a result, they are often undetectable or inaccurately located in the electron density. Hydrogen atoms can usually be identified in the density of small molecule crystal structures, but this will only be possible in protein structures if the resolution is less than 1 Å. This is unproblematic as long as the hydrogen atoms are in positions that correspond to spatially fixed positions on a rigid molecular scaffold, e.g., hydrogen atoms on phenyl rings. It is more difficult when the hydrogen atom is on a conformationally flexible group or groups that can be protonated or deprotonated. It would be ideal to know whether a carboxyl group is ionized or exists as a free acid and in which direction the hydrogen atom is oriented. This information can only be gleaned indirectly from the structure of the protein, through an accurate analysis of the spatial orientation and interaction geometry of the surrounding hydrogen bonding partners.

The **accuracy of structure determination** depends on the resolution of the data obtained from a crystal. Even if the structure of the protein is displayed on the computer screen like that of a small organic molecule, its geometry is determined much less accurately. The margins of error for small molecule determinations are about 0.01 Å for bond lengths, 0.1° for bond angles, and 1–2° for dihedral angles (Chap. 16). For protein structures, the errors are much larger and difficult to quantify. They depend on how the structure has been refined. Mostly the electron density does not allow the resolution of single atoms. Therefore, amino acids are placed in the electron density with idealized bond lengths and angles. Their geometry is left at the predefined knowledge-based values for subsequent refinement. The assignment of atom types for the placement of side chains is partly based on assumptions. Knowledge-based values are used or an attempt is made to keep the hydrogen-bonding network consistent. These aspects must be considered when evaluating the accuracy of a protein structure. The result of crystal structure determination is a **spatially and temporally averaged image** of an "average" molecule representing the entire crystal. It is often found that the electron density in some regions indicates only a reduced occupancy of a side chain or part of a bound ligand. In addition, alternative orientations (conformations) may be seen. Sometimes the electron density of entire regions is missing. This is indicative of spatial "**disorder**" and argues for a distribution over multiple orientations in the crystal. When describing the diffraction phenomena, we had seen that lattice planes, and thus the atoms arranged there, will contribute to the diffraction pattern only if they are found periodically as an array of planes at the same location over the whole crystal. This is not the case with disorder. The strict periodicity is lost and with it the contribution to the diffraction pattern. Disorder in the crystal can also be **dynamic**, that is, the corresponding group moves back and forth between two or more arrangements as a thermal motion. Alternatively, the disorder can be **static**, meaning that several orientations exist side by side in a crystal, but are

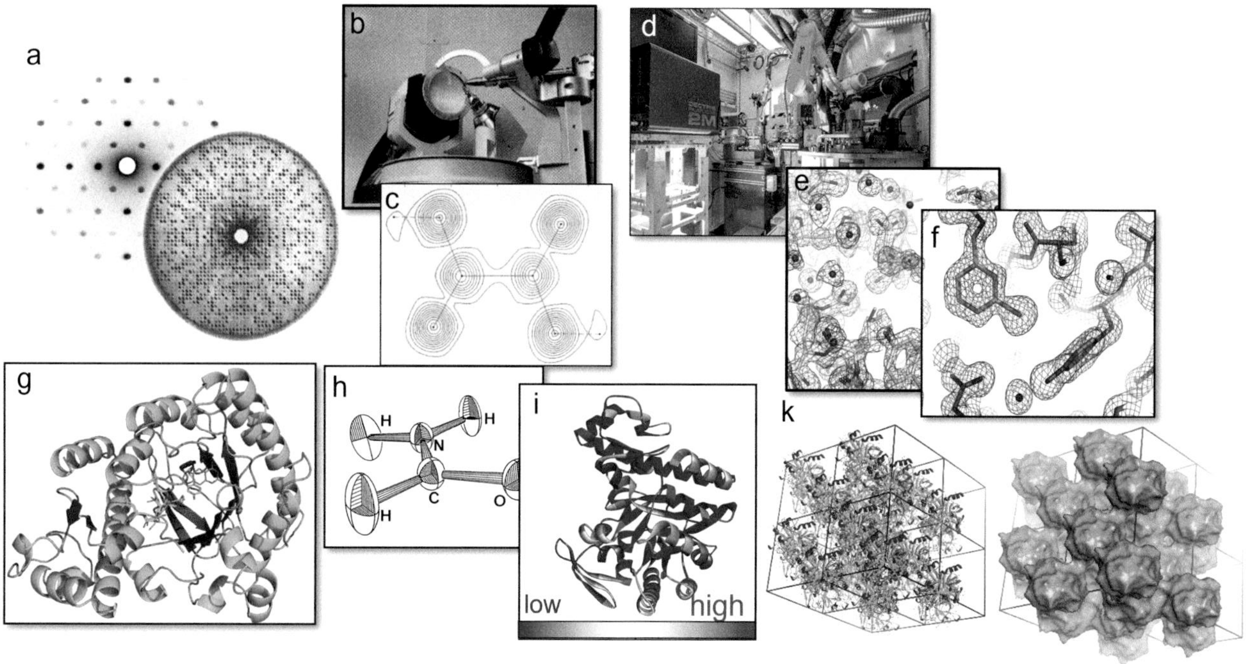

Fig. 13.10 The crystal structure determination of organic molecules and proteins requires crystals with an edge length of approx. 0.01–0.3 mm. (**a**) In the X-ray beam, a diffraction pattern is obtained (compare ◻ Fig. 13.8), which in the past was registered on a photographic plate, today with an area detector on a diffractometer. The diffraction pattern of protein crystals shows a much denser reflection pattern. (**b**) Structures of small molecules are usually measured in the laboratory on an automated diffractometer. (**c**) Electron density indicates the positions of individual atoms. (**d**) Data from protein crystals are nowadays collected almost exclusively at synchrotron sources. (**e**) With approximated phases, a Fourier transform is performed and the electron density in space is obtained, which is contoured according to a predefined electron density level. (**c, f**) After structure refinement, the density is interpreted and a model of the diffracting molecule is fitted. (**g**) Because of the complexity of their structure, proteins are usually represented by a ribbon model that depicts the course of their polymer chain in space. (**h**) The spatial blurring of the electron density is associated with thermal motion of the atoms. It is represented for small molecules by ellipsoids encompassing 50% of the atomic population probability. (**i**) For proteins, thermal motion is indicated as so-called B-factors using a color-coding scheme projected onto the folding pattern. *Red* indicates high thermal motion, while *blue* indicates low thermal motion. (**k**) The spatial arrangement of the molecules in the crystal lattice shows a dense packing of the protein molecules (here thrombin). However, large, at first glance "empty" channels exist in the crystal packing between the molecules. These channels are occupied by a large number of water molecules. Because of their extensive thermal motion and the resulting disorder, they are not detected in the electron density. However, small-molecule ligands can diffuse through these channels when the crystals are soaked with a solution of these ligands. (► https://sn.pub/m1wZeO)

randomly distributed. Because the structure is an averaged picture, these arrangements are randomly scattered throughout the crystal with different orientations. If part of the molecule is completely disordered, i.e., scattered in many orientations, the electron density is usually not visible. Today, diffraction data on protein crystals are mainly collected at synchrotron radiation sources. Only in rare cases are they still collected at an in-house facility using radiation from an X-ray tube. In such a tube, electrons are emitted from a filament and accelerated in a high-voltage electric field (1–100 kV). The electrons then collide with a metal anode. As the electrons decelerate on the anode material, X-rays are produced as characteristic bremsstrahlung. Electrons are expelled from the inner shell of the metal atoms and electrons from an outer shell take their place. The emitted radiation is therefore determined by the energy difference between the shells and is specific for the metal of the anode material. In a synchrotron, radiation is produced as electromagnetic waves when electrons accelerated to nearly the speed of light are forced to follow a curved path by magnetic fields. The radiation is emitted tangentially to the trajectory of the electrons, because physically a change in the direction of the velocity vector on the curved trajectory means an acceleration and thus leads to a special form of bremsstrahlung. In this way, a broad spectrum of wavelengths can be produced. If desired, monochromatic radiation can be produced using mirrors and crystals, where the strong reflection of a crystal (e.g. graphite) is used as the primary beam. The synchrotron beam is several orders of magnitude more intense than the beam from an X-ray tube. Measuring times of days on an instrument in one's own laboratory can be reduced to seconds. However, to minimize radiation damage to the crystal samples, it is

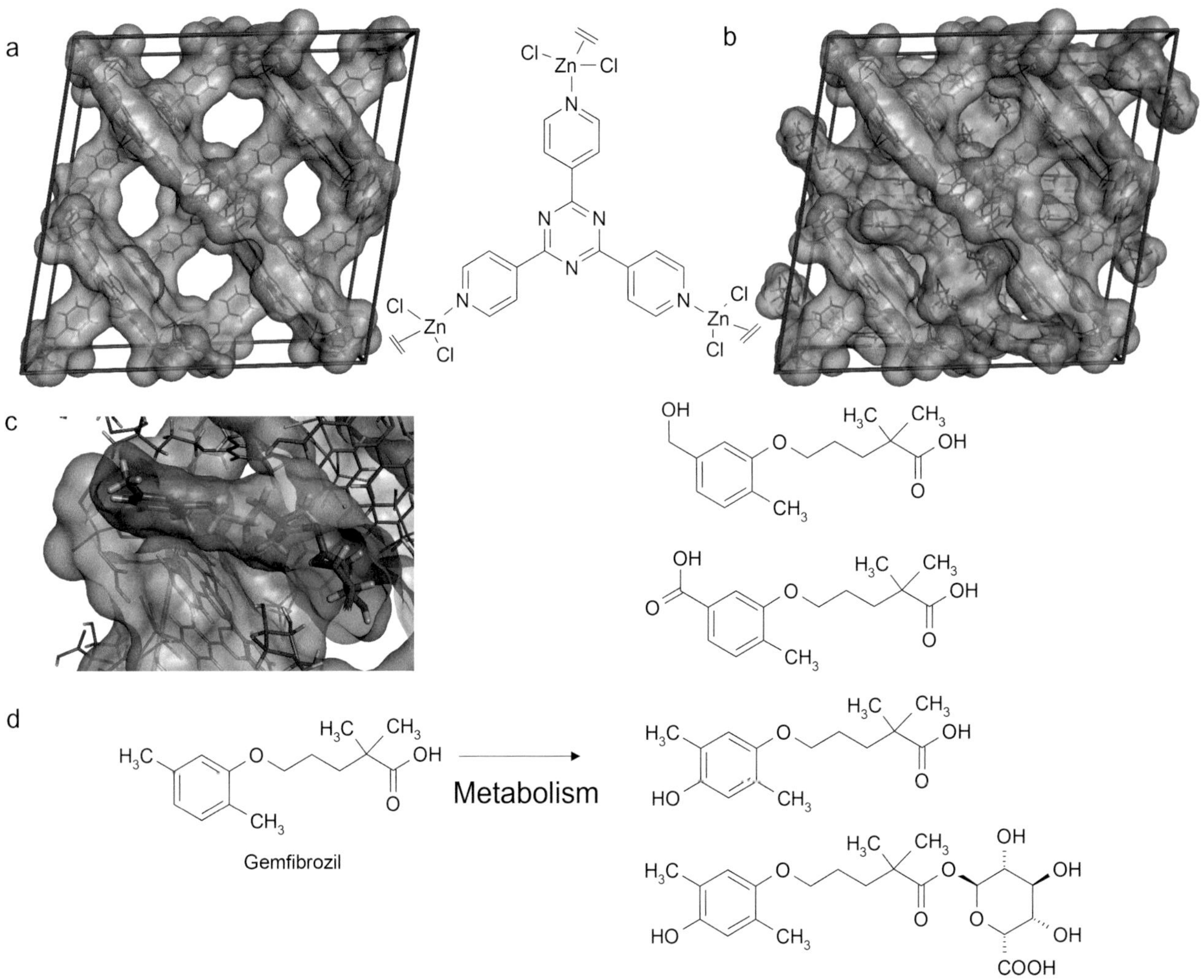

■ **Fig. 13.11** If zinc chloride is reacted with 2,4,6-tris(4-pyridyl)-1,3,5-triazine (*top center*), crystals of a porous structure with cavities are formed (**a**). A drop of solvent, in which an organic test molecule to be investigated is dissolved, is added to the crystals and the test molecule can diffuse into the huge cavities of the crystal lattice. If the process occurs slowly and in thermodynamic equilibrium, the test compound will place itself in a regular fashion in the empty cavities of the lattice (**b**). Its geometry can then be determined together with the lattice of the crystalline sponge (**c**). Only very small amounts of test substance are required for the method. For example, substances formed during drug metabolism can be characterized after their chromatographic separation (**d**). The molecule shown in (**c**) is the glucuronidated metabolite of gemfibrozil. (► https://sn.pub/aJLQMB)

necessary to measure the structures at about 100 K in a cryogenic nitrogen gas stream. At this temperature, many motions in the crystal are frozen and mostly static disorder is observed. Nevertheless, it has been shown that the structures determined correspond well to the situation at room or body temperature. These conclusions can be drawn by comparing the results with the analogous determination from NMR spectroscopy (Sect. 13.7) and molecular dynamics simulations (Sect. 15.7). Nevertheless, detailed studies of crystal structures collected at low and ambient temperatures have shown, as expected, that the conformer distributions of side-chain rotamers can differ at these different temperatures and this is reflected in the determined structures. It should be noted, however, that this is also where the largest differences to NMR structures and those from molecular dynamic (MD) simulations are observed.

Crystallography of small molecules is still the most powerful analytical method for characterizing the chemical composition and especially the stereochemistry of organic compounds. This method allows the absolute configuration of molecules to be determined with confidence on the basis of so-called **anomalous dispersion**. It requires the presence of an electron-rich atom that exhibits such an anomalous dispersion contribution in the molecule under investigation. For small molecules, phosphorus or sulfur atoms may be sufficient for these studies. For elements such as chlorine, bromine, or zinc, the effect is stronger. However, a prerequisite for all these crystallographic determinations is the growth of a single crystal. Unfortunately, this is often not trivial. What can be done if the compound under investigation will not crystallize? In many cases, there is simply not enough material to grow a crystal. A promising alternative has emerged in recent years: Diffusion in **crystalline sponges**! The concept dates back to Makoto Fujita's group at the University of Tokyo, Japan. For example, when zinc chloride is reacted with 2,4,6-tris(4-pyridyl)-1,3,5-triazine, crystals with a porous cavity structure are formed. ◘ Fig. 13.11a shows a section of the crystal packing of this structure, which contains large cavities. These crystals can be placed in a drop of organic solvent containing a solution of the substance to be studied. The substance can then diffuse into the cavities of this crystal lattice. If this diffusion is slow and in thermodynamic equilibrium, the guest molecules will be taken up and arranged regularly in the lattice. Everything else then proceeds as in a routine structure determination. The absolute configuration can also be determined in this sponge. The zinc atoms help in this process. They are the necessary anomalous scatterers in the crystal lattice. The method requires only very small amounts of substance and the crystallizability of the substance under investigation is not required. The method can greatly assist medicinal chemists in the analysis of their synthesis products.

However, this process can provide other extremely valuable assistance in drug development. Once a drug molecule has been developed to the stage of a clinical candidate, it is necessary to study in detail how the substance is chemically modified in the human body. In Sect. 27.6, we will learn how enzymes, particularly in the liver, chemically modify drugs, or "metabolize" them. This is the process of converting a drug molecule into a form that can be more easily eliminated from the body via the urine. Because the **metabolic process** creates active ingredients that are potentially new to the human body, it is important to determine exactly what substances are being formed. What is their stereochemistry and are they toxic to the body? But before these questions can be answered, the degradation products must first be structurally characterized. This is where the crystalline sponge method comes into play. So-called homogenates can be used to simulate the metabolism in the liver in the laboratory. Any metabolites formed are then separated by chromatography. Usually, only a very small amount of the substance is available for subsequent analysis. Mass spectrometry would be one option for analysis. However, this does not provide the exact topology and stereochemistry. If, on the other hand, the individual fractions are allowed to diffuse into the crystalline sponges, the chemical structure, including stereochemistry, can be obtained in the best case! ◘ Fig. 13.11d shows an example of the metabolic degradation of gemfibrozil, a lipid-lowering drug from the class of fibrates (Sect. 28.6). The four metabolites formed could be diffused into the crystalline sponge as guests. ◘ Fig. 13.11b, c shows the structure of the metabolite glucuronidated at the acid function. Only a tiny amount was needed. Crystallography performed on metabolites diffusing into crystalline sponges certainly has immense potential to massively assist in the difficult unraveling of drug metabolism.

13.6 Electron Microscopy: Topographic Images Reveal Macromolecular Structures

In addition to X-rays, **beams of electrons** and **neutrons** can also be used for diffraction experiments to determine the structures of molecules. Electron beams have the great advantage over X-rays and neutrons that they can be bent by magnets. Therefore, it is possible to build converging (convex) lenses for them, comparable to a light microscope, to show a magnified image. With X-rays, this is practically impossible or only very inefficient to realize. This leads to the described phase problem of X-ray crystallography. The operation that performs the task of a "converging lens" has to be replaced by a Fourier transform. Only then does the electron density of the molecules, and thus their spatial structure, become accessible with X-rays. In the electron microscope, a magnetic lens can be used to perform this Fourier transform directly and then even obtain an image of individual molecules as "shadows." However, for a long time this approach did not provide the necessary resolution to reveal the desired details in these structures.

In recent times, new developments—especially in the field of detectors for the registration of electron beams—have enabled **electron microscopy** to make a decisive breakthrough in the determination of the structure of huge macromolecular complexes. The method does not require crystallized proteins; single molecules can be studied. Particularly in the case of larger assemblies of huge protein complexes, crystallization is often the bottleneck for timely structure determination.

Cryo-electron microscopy (cryo-EM) examines molecules as individual particles, similar to the way a computer tomograph in medicine scans a patient from all sides. The intact protein samples are exposed to an electron beam in a high vacuum (◘ Fig. 13.12). Prior to this, they must be flash-frozen in a vitreous water environment. They are then exposed to the electron beam in many orientations at the temperature of liquid nitrogen. Thus "vitrified," thousands of projection images of the protein are taken. The next step is the reconstruction of the 3D structure of the studied molecule from these usually very noisy projections. This is done using a powerful computer programmed with sophisticated image-processing algorithms. The result is a composite

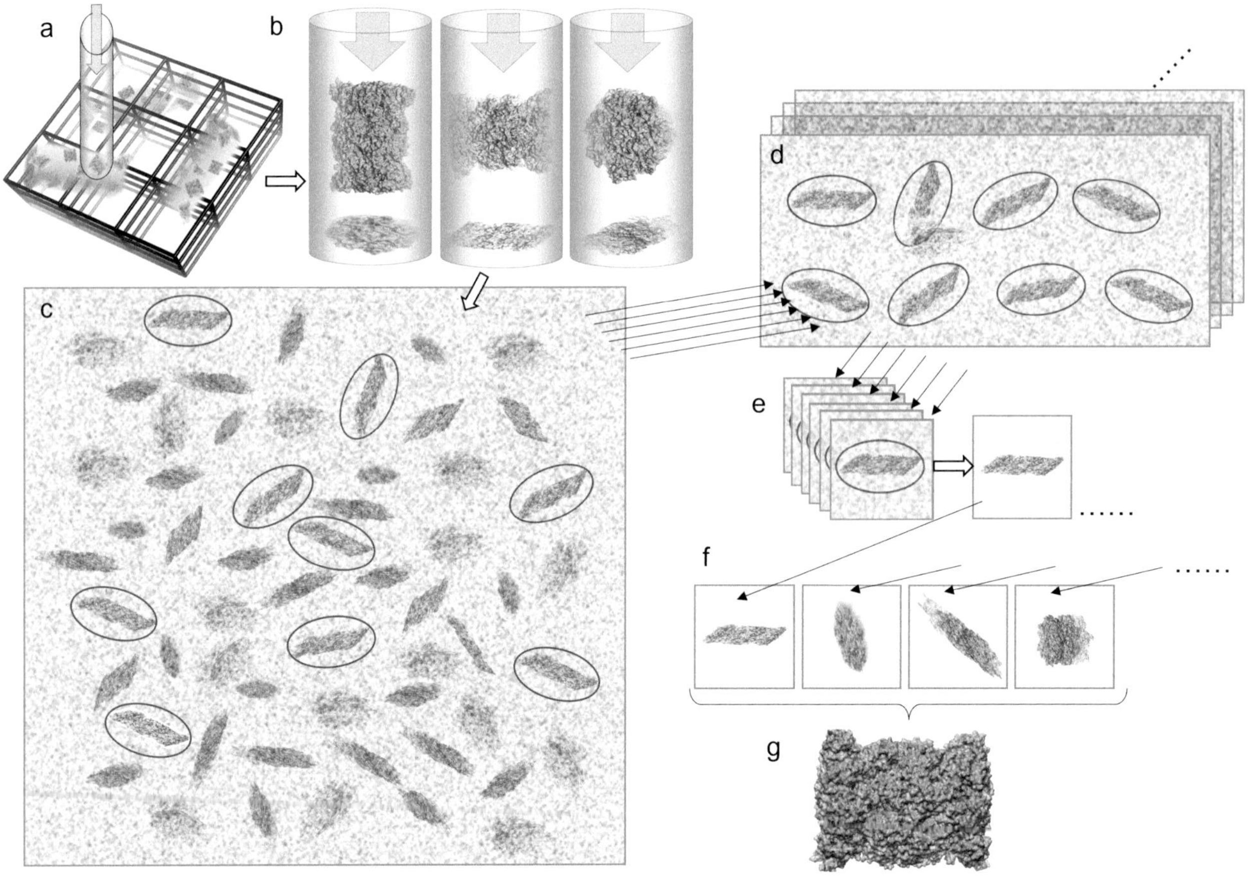

Fig. 13.12 Schematic representation of the workflow during structure determination with cryo-electron microscopy. Protein molecules are flash-frozen in a vitreous water coat and deposited on a graphite grid under high vacuum (**a**). An intense electron beam (sketched by *turquoise cylinder*) sweeps across the grid, creating many thousands of shadow images as projections of the molecules deposited on the grid (**b**). The extremely noisy shadow projections (**c**) are sorted in the computer in terms of similar orientations (**d**), compiled, and then averaged (**d**, **e**). As a result, the molecule under investigation emerges more and more strongly from the background noise (**e**). The same procedure is followed with the many thousands of projections (**f**), and an averaged spatial image of the protein molecule is generated from these images, similar to the way in which a computer tomograph is generated in medicine (**g**)

spatial image of the charge density of the microscopically observed object. Since electron beams have a much shorter wavelength than light waves, structures down to the 2–3 Å range can be resolved in favorable cases. By averaging a large number of measured shadow projections, the signal-to-noise ratio of the projections can be significantly improved. By iteratively evaluating individual projections, reconstructing the 3D structure, and subsequently improving the alignment and superposition of the raw shadow images, the representation of the molecular complex under investigation is continuously refined.

In 2017, Jacques Dubochet, Joachim Frank, and Richard Henderson were awarded the Nobel Prize in Chemistry for their contributions to the development of the cryo-EM technique. Dubochet helped the technique achieve a breakthrough by **cryopreparing the protein samples** in the form of a vitrified embedding in an amorphously solidified water shell about 100 nm thick. At the same time, this step allows the samples to be measured at low temperature, since the high-energy electron beam would destroy the proteins very quickly at room temperature. Joachim Frank developed the algorithms to discover and extract the recurring patterns from the noisy projected images of randomly distributed molecules, ultimately **reconstructing a model of the spatial structure** of the molecules under study. Richard Henderson helped cryo-EM make an experimental breakthrough. Step by step, he was able to increase the image resolution of the bacteriorhodopsin he studied (Chap. 29) until the folding pattern of this protein was finally visible.

Nowadays, under optimal conditions, complexes can be determined down to atomic details of their spatial structure. Only recently did Holger Stark's group at the MPI in Göttingen succeed in determining a cryo-EM structure of apoferritin with a resolution of 1.25 Å! This is certainly still the exception. When comparing the **resolution limits** of diffraction methods and cryo-EM, some caution is required because the resolutions are defined and calculated differently. Referring to the closest lattice plane distance, as in diffraction techniques, does not work in the cryo-EM technique. In order to reliably evaluate the quality of the data collected at the many projections in a three-dimensional volume, an evaluation in Fourier space is performed. This is done by dividing the dataset into two equal parts and calculating the Fourier shell correlation. From the inside to the outside, the data are processed in individual shells and it is determined how well the Fourier components of the transformed data correlate in both parts of the dataset. The shell in which the correlation falls below a predetermined threshold is specified as the resolution limit. The analysis is, therefore, more of a measure of the internal consistency of the two splitted datasets.

Cryo-EM is already the method of choice for studying multiprotein complexes, such as the spliceosome, and for addressing mechanistic questions about the function of individual proteins in a complex. Conventional structure determination in drug design typically requires resolutions of 2 Å or better. Structures must be able to be determined without significant effort on ligand series. In contrast, each additional EM structure virtually means a completely new structure determination. For these reasons, classical X-ray crystallography is likely to remain the workhorse for many years to come. One advantage of cryo-EM is certainly that the molecules are in a frozen water environment. This is certainly closer to physiological conditions than a crystalline assembly. In addition, the method is better at capturing different states of a protein. For example, if the protein exists as a monomer and a dimer side by side, this will be recorded. However, the computational cost of structure determination increases dramatically with increasing resolution.

An alternative, which was initially pursued intensively, is to use crystalline material for the electron microscopic electron microscopic studies. Since the averaged image of several aligned molecules is examined in a periodic crystalline array, a stronger signal is observed. The converging lens step (Fourier transform) is omitted and the reflections in the diffraction space are measured on the crystalline sample as in X-ray diffraction. Electrons penetrate only slightly into the crystalline sample material, but are scattered much more strongly by the molecules in the crystal. Therefore, much smaller crystals can be used, and even crystals that are wafer-thin in one direction and consist of only one or a few molecular layers are sufficient.

Despite the lower radiation exposure, the electrons still result in considerable destruction of the samples. It is important to remember that the crystals used are only about a billionth of the sample quantity of a crystal used for X-ray diffraction. The data for an X-ray structure can often be measured on one single crystal. The electron microscope, on the other hand, requires many hundreds of the tiny crystals, often only 5 µm in size. They are also flash-frozen under high vacuum and exposed directly to the electron beam. The images are also very noisy and have to be averaged over many images. To obtain detailed resolution perpendicular to the two-dimensional crystal plane, crystals must be measured in many orientations. A Fourier transform is used to obtain a charge density distribution of the molecules, similar to X-ray diffraction. Its interpretation or refinement is done in the same way as in X-ray experiments. The phases required for the Fourier transform can be determined in the electron microscope by direct imaging in the "converging lens mode." In the group of Tamir Gonen at the University of California, Los Angeles, USA, the idea was developed to minimize radiation damage by reducing the intensity of the electron beam. Thus, a greater number of reflections can be collected on individual crystals. In this way, the **micro-ED method** was successfully applied to a resolution range of about 1 Å by electron diffraction. The method can also be extended to the structure determination of small organic molecules. Since only very small crystals are required, the method works even with materials that look almost like an amorphous powder to the naked eye. To demonstrate its applicability, the Gonen group took finished drugs and crushed the tablets. The resulting powder was still crystalline enough to be used in an electron microscope to determine the structure of the drug molecules in the sample. It was also possible to find sufficiently large crystallites of the precipitant that remained in a flask after a substance was recovered following evaporation of the solvent. They still allowed the structure to be determined using the micro-ED method. This method may represent a breakthrough in crystallographic structure analysis, since in many cases the supposed bottleneck of single crystal growth may prove to be irrelevant.

Another type of radiation that can be used for diffraction experiments on biomolecules is a **beam of neutrons**. Neutrons are produced either in a reactor by nuclear fission of uranium isotopes or from a spallation source. In the latter case, heavy metals such as mercury emit out neutrons after they have been bombarded with high-energy electrons in a nuclear conversion reaction. A neutron beam behaves much like an X-ray beam in a diffraction experiment. However, the strength with which the chemical elements contribute to the scattering power in a reflection is completely different to X-rays. For example, hydrogen is a strong scatterer and the H isotope can be easily distinguished from the D isotope (deuterium). An element like sulfur is a weak scatterer, and the metal vanadium is practically invisible to neutrons. Therefore, this metal is often used as a material for sample containers for neutron diffraction experiments. For biological samples, neutrons have the incredible advantage of being very powerful at **visualizing geometries involving hydrogen atoms**. This is especially true for the study of hydrogen bonds and protonation states of acidic and basic groups. But also the orientation of water molecules and their dynamic behavior are visualized (■ Fig. 4.10). Why not do all structure determinations with neutrons? Unfortunately, the intensity of a neutron beam is much, much weaker than that of an X-ray beam, which means that the crystals required for neutron diffraction must be much larger, with edge lengths of several millimeters. Crystals of this size can be successfully grown from very few proteins.

13.7 Structures in Solution: The Resonance Experiment in NMR Spectroscopy

Many atomic nuclei have an angular momentum, or **spin**. The nuclei that occur in biological systems that have a nuclear spin are the hydrogen isotope ^{1}H, the carbon isotope ^{13}C, the nitrogen isotope ^{15}N, the fluorine isotope ^{19}F (in biology rarely found), and the phosphorus isotope ^{31}P (for the sake of simplicity, only nuclei with a spin of ½ will be considered here). Just as a top would, these nuclei rotate about their axes. As long as no magnetic field is applied, the tops orient in all possible spatial directions. In a magnetic field, they are forced into alignment (◘ Fig. 13.13). If a toy top is spun, it will move in the gravitation field. This field has, as the magnetic field, one preferred direction. If the alignment of the rotation axis of the top and the direction of the gravitation field, which is oriented towards the center of the Earth, are not exactly the same, the top will wobble. The end of the rotation axis performs a circular movement, an arc, with a very precise rotational velocity. It depends on the mass and geometry of the top. In physics this movement is known as **precession**.

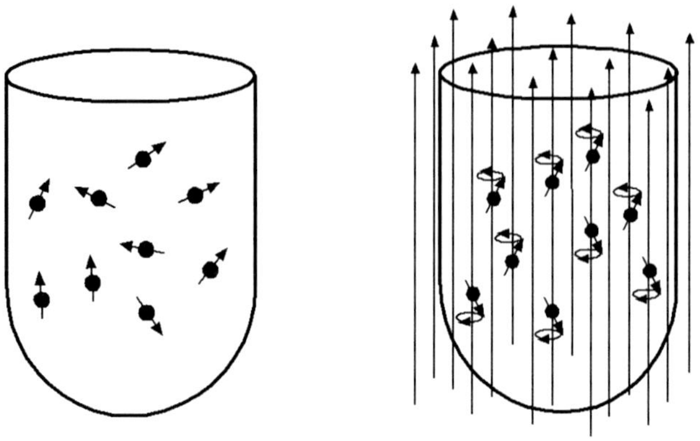

◘ **Fig. 13.13** Atomic nuclei with a rotational momentum behave like a spinning top. In the absence of an external magnetic field, they orient in all possible directions randomly (*left*). Upon application of a magnetic field, they orient their rotation axes parallel or antiparallel to the direction of the field (*right*). The precession movement is oriented in an arc around the applied field direction. The two orientations, parallel or antiparallel, with respect to the direction of the field are energetically different. Because of this, there is a small difference in occupancy between the two states. By applying an electromagnetic field with a frequency corresponding to the rotational movement of the precession of the axis of the top, the occupation can be inverted. This resonance absorption, the exact frequency of which depends on the type of nucleus and its immediate chemical environment, is registered with a spectrometer

13

Atomic nuclei with a spin behave in a very similar way. In contrast to the macroscopic top, they obey the laws of quantum mechanics. This means that the rotation axes that their precession movement takes on can only adopt very specific angles with respect to the applied field direction. The result for the ^{1}H, ^{13}C, ^{15}N, ^{19}F, and ^{31}P nuclei is that the rotation axis for the precession arc can only be parallel or antiparallel to the direction of the field (so-called spin ±½ particles). The orientation in the direction of the field is energetically somewhat more favorable than the rotation antiparallel to the direction of the field. Statistically, therefore, more nuclear spins in the substance sample will align with the direction of the field. If an additional magnetic field is applied to the outer magnetic field and its frequency corresponds to the precession frequency of the nuclear spin, the occupancy of "parallel" to "antiparallel" spinning nuclei can be reversed and a resonance absorption for the sample can be registered. After a particular time span, the original situation is restored (**relaxation**) for entropic reasons.

The rotational velocity of the top's axis for precession movements is characteristic for each type of nucleus. It further depends on, and is additionally modulated by, the composition of the chemical environment in which the nucleus resides. A carbon atom of a phenyl ring has a different resonance frequency than that of an aliphatic chain. The relative position of the resonance absorption in relation to a standard reference is also called the **chemical shift**. Furthermore, the individual nuclei can perceive the spin orientation of the neighboring nuclei. An alignment in the same direction as a neighboring nucleus is energetically different from that of an antiparallel orientation. This influence also modulates the rotational speed of the spin on the observed nucleus. The information transfer regarding the orientation or the magnetic state of the nuclei in the vicinity can be transmitted over several bonds. This transfer can even occur through space without any direct covalent connection.

To measure an **NMR spectrum** (nuclear magnetic resonance), a solution of the substance has to be placed in a strong magnetic field. In addition, a variable electromagnetic field is applied to the sample. The frequencies at which the nuclei in the sample have resonance, meaning when they flip from parallel to antiparallel, are recorded. The resulting spectrum discloses information about the composition and the chemical environment around the studied nuclei. It contains information about the spatial structure of the molecules under investigation. Based on the work of Richard Ernst, who received the Nobel Prize in Chemistry in 1991, multidimensional NMR techniques have been developed over the last 30 years. By applying suitable measurement conditions and selective electromagnetic fields, information about the mutual influence of resonance frequencies between individual nuclei is separated and analyzed. This bidirectional transfer of information about the magnetic state of neighboring nuclei is evident from the signal form in the multidimensional spectra, and it is registered in the form of cross peaks.

In 1965, Anet and Bourn performed decoupling experiments in which they saturated a particular hydrogen atom in the spectrum by irradiating it separately at the corresponding resonance frequency. They observed that the resonance intensity of a neighboring hydrogen atom, which was not covalently bonded to the saturated H atom, increased by 45%. This, of course, provides important information about the **spatial proximity** of the atoms. This so-called **nuclear Overhauser effect (NOE)** is often used in NMR experiments to elucidate the structure and conformation of biomolecules and their interactions.

Only the hydrogen isotope ^1H occurs in nearly 100% natural abundance. Therefore, it can be assumed that for statistical reasons, two ^1H nuclei will always be adjacent to each other in a molecule. In contrast, the ^{13}C and ^{15}N isotopes are scarce. As a result, statistically they are only very rarely found in the direct vicinity of one another. Data on the mutual influence of the magnetization of these nuclei are required for the spectra. Therefore, it is necessary to enrich the proteins with the appropriate isotopes. For this, bacteria are fed with isotopically labeled substrates such as glucose or ammonium chloride and will then produce proteins that are isotopically enriched. It is even necessary to produce deuterated proteins for the structural investigation of very large proteins. Today, by using numerous spectroscopic techniques, spectra from proteins of more than 800 amino acids have been successfully interpreted. The following questions can be addressed by NMR analysis:

- Which atomic nuclei occur in which chemical environment?
- What is in the immediate, covalently connected neighborhood of these nuclei? Information about the spatial orientation of atoms in the vicinity is also contained within these spectral parameters.
- Which geometric relationships are given between different segments of the polypeptide chain? This results from information transfer about magnetic states of nuclei that are not directly connected by covalent bonds.

NMR studies are usually performed in solution. In this way, dynamic processes can be observed by changes in the spectra. Fluorine atoms are isotopically pure as ^{19}F nuclei and they are rarely used by Nature. As a nucleus, it can be observed very well in the NMR spectrometer. Therefore, it is a useful probe to study the properties of biomolecules. Fluorine-containing amino acids can be introduced into proteins under special expression conditions. The ^{19}F resonances can then be used to study the structural and dynamic properties of proteins. An example is given in Sect. 21.14.

13.8 From Spectra to Structure: Distance Maps Evolve into Spatial Geometries

This last-mentioned observation, which results from the **nuclear Overhauser effect** (NOE), yields intramolecular distances of spatially neighboring but not directly covalently bound atoms. The entire connectivity, that is, a list of all covalent bonds within a molecule, and a list of the recorded intramolecular noncovalent NOE distances are applied to generate the structure for the molecule (◘ Fig. 13.14). For this purpose, so-called **distance–geometry calculations** are used to create the spatial coordinates of the atoms.

Often, distance geometry produces several equally good structural models that satisfy the experimentally

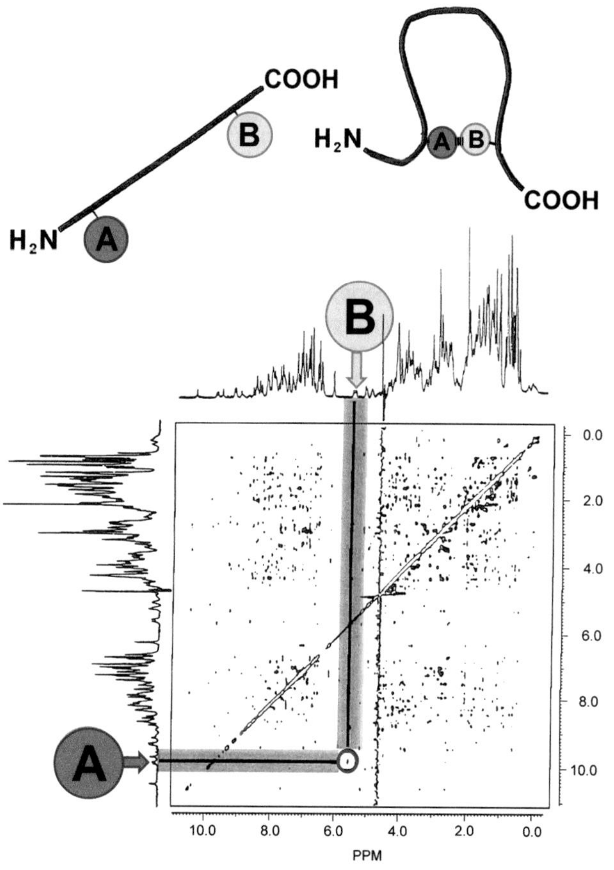

◘ **Fig. 13.14** A multidimensional NMR spectrum contains information about the spatial vicinity of atomic nuclei in a molecule (here, the trypsin inhibitor from bovine pancreas). It is expressed in cross peaks. This provides information about the distance between noncovalently bonded atoms in a molecule. The individual signals of the spectra are assigned to atoms in the molecule (e.g., **A** and **B**). The positions that these atoms have in the polypeptide chain are known from the sequence of the protein (*above left*). The intensity of the cross peak indicates which spatial distance is found between nuclei **A** and **B** in the folded polypeptide chain (*above right*). Just as was done for **A** and **B**, the many other cross peaks are evaluated and translated into distance conditions. With the totality of this distance information, it is possible to fold the polypeptide chain in space

solutions, was then determined in a crystallographic experiment. The binding constants were then determined from the occupancy data. They were in quantitative agreement with the inhibition constants determined in a functional assay in solution.

The diffraction data can be very quickly collected with even more intense, so-called white X-rays from a synchrotron source (the so-called **Laue technique**). With this experiment, it was possible to observe stable intermediates of enzyme reactions. Structural changes of the two-dimensional crystals of the acetylcholine receptor (Sect. 30.4) could be observed with electron microscopy after loading with the natural ligand. This and other experiments have proven that proteins exist in a crystal lattice that must be, at the very least, very similar to the biologically active form.

The **X-ray free-electron laser** (XFEL) is expected to provide further insights. It is a billion times more intense source of radiation for diffraction experiments. Tiny crystals (200 nm to 2 µm) are injected into the X-ray beam of such a source in the form of a continuous stream. Each of these crystals provides a diffraction pattern before bursting in the intense beam. However, the duration of the diffraction experiment is five powers of ten shorter than the time that elapses before the crystal explodes. Each crystal provides a limited diffraction pattern. However, since many crystals come into the reflection position in all possible orientations relative to the X-ray beam, a complete dataset can be collected in this way. Thus, a comprehensive structure determination (so-called **serial crystallography**) is possible. Due to the extremely short time between the injection of the crystalline samples into the beam and the measurement of the diffraction experiment, the method is ideally suited for the study of time-dependent phenomena. For example, light-dependent processes can be initiated by a laser pulse. In this way, the light-driven proton pump bacteriorhodopsin could be followed at work as it shuffles protons across the membrane. Like a movie, snapshots of the electron density distribution were recorded in steps of 16 ns to 1.7 ms. They show how retinal is rearranged, how amino acid residues are spatially shifted, and how water molecules accompany the process. To observe processes such as enzyme reactions, the tiny crystals are mixed with the reactants in the flow leading to the measuring beam. XFELs are currently being built at several locations around the world. They are not cheap to operate because they first require a kilometer-long linear accelerator for electrons as a radiation source. Over a very long distance using suitable alternating electrical fields the electrons are accelerated to almost the speed of light where relativistic effects dominate their behavior. They are injected as extremely short pulses and form individual particle bunches. Along specially arranged magnets, the electrons are then forced to follow a kind of slalom course, emitting laser-like X-ray pulses. These X-rays, traveling at the speed of light, interact with the electron bunches flying in front of them on the slalom course, by this slowing some of them down and accelerating others. In this way, a kind of "synchronization" of the electrons in the particle bunches is achieved along the further trajectory, and they emit their X-ray light in phase along the direction of travel, resulting in the pulsed, extremely intense, coherent and monochromatic laser-like beam. This beam hits the target, while the electron flow is deflected slightly in front of the target to avoid a collision with the sample. Unlike a synchrotron or a neutron reactor, where many experimental setups share the expensive radiation source, the XFEL allows only one experimental setup. However, it is hoped that this technology will enable completely new insights to be gained in the future, in particular into the dynamics and, thus, the detailed function of proteins.

13.10 Synopsis

- The most powerful methods to determine the spatial structure of molecules are X-ray crystallography and NMR spectroscopy. The former requires the biomolecules to be arranged in periodic arrays in a crystal, and the latter studies them in solution, usually in an isotopically labeled form.
- Crystals need special conditions to grow from saturated solutions. They spatially arrange in periodic arrays, and the molecules pack through translational symmetry in three dimensions. In addition to the pure shifting of basic motifs, usually one molecule that represents the asymmetric unit, symmetry operation such as mirror reflection, two-, three-, four-, and six-fold rotation or inversion can be applied.
- Crystal lattices diffract X-rays and the diffraction experiment can be understood as a three-dimensional interference of elementary spherical waves generated at the positions of the atoms in the lattice. The diffraction phenomenon at a 3D lattice can be treated formally as reflections at the multiple crystal planes in the lattice.
- Because the relative phases of the generated elementary spherical waves, superimposed in the various reflections, are not accessible by experiment, they must be regenerated by sophisticated phasing methods. Only then can a Fourier transform be calculated from the measured reflections that represents the spatial distribution of the electron density in the crystal. A model of the crystallized molecules is assigned to this electron density.
- The diffraction power and resolution of the crystals determine the accuracy of the resolved structure. For proteins, a resolution of 1.5–3 Å is usually achieved. At the lower end, molecular building blocks such as phenyl rings are well resolved, and individual water molecules are visible. At the upper limit, only the overall topology is determined, and the water molecules usually cannot be assigned.

- The crystal structure is an average structure over space and time. Enhanced B-factors give an estimate of the residual mobility of molecular portions in a molecule.
- Electron microscopy is an alternative method for determining the structure of very large, often membrane-bound proteins. One either performs diffraction experiments (micro-ED method) or collects thousands of shadow projections of individual molecules in the electron beam in microscope mode (cryo-EM method). In the case of diffraction, reflection data are collected from many thousands of tiny and wafer-thin crystals. Alternatively, the intensity of the beam can be reduced to collect larger diffraction datasets from the crystals. Micro-ED can also be applied to tiny crystals of low molecular weight compounds. Cryo-EM requires the inclusion of compound samples in frozen water droplets to collect many thousands of shadow projections of individual molecules. The projection images are then assembled into an averaged 3D structure in the computer.
- If only small substance samples are available or if crystallization fails, dissolved organic substances can be diffused into crystalline sponges. There they place themselves in cavities of the lattice. Structure determination succeeds by collecting diffraction data on the crystalline sponges with their absorbed "guest molecules."
- NMR spectroscopy records the resonance of magnetic nuclei such as ^{1}H, ^{13}C, or ^{15}N oriented in a strong magnetic field. The transition between parallel and antiparallel orientation of the nuclear spins can be induced by additional fields. Because the frequency at which these transitions take place depends on the chemical environment in a molecule, the spectral parameters contain information about the 3D structure of the molecules in solution.
- The large number of recorded spectral parameters, in particular information about the spatial neighborhood of atoms determined by the nuclear Overhauser effect (NOEs), can be translated into distance maps between individual magnetic nuclei. From this, the spatial structure of the protein can be folded in the computer. A distance geometry approach is used in combination with molecular dynamics simulations.
- It could be shown for many cases that the NMR structure of a protein in solution and the X-ray structure in a crystal largely coincide with each other. Differences are observed for the surface-exposed residues.
- Protein crystals contain up to 70% water and exhibit large water channels that pass through the crystal. Where appropriate, small-molecule ligands can diffuse through these water-containing channels to reach binding sites on the surface or in accessible binding pockets of the proteins. The binding modes of small-molecule ligands can be easily determined by using these soaking techniques.
- The significance of the architecture of proteins determined in a crystalline environment for biologically relevant conditions has been demonstrated. Examples are known of enzyme reactions that are usually carried out on the dissolved protein, but which also occur in a protein when it is in crystalline state.
- Structural data of proteins can also be collected with neutron radiation. In such determined structures, hydrogen atoms are revealed as strong scatterers. Protonation states of functional groups and the arrangement and dynamics of water molecules can be determined.
- Very strong synchrotron radiation from the X-ray laser can be used to resolve fast dynamic processes in protein crystals by serial crystallography.

Bibliography and Further Reading

General Literature

J. P. Glusker, K. N. Trueblood, Crystal Structure Analysis, A Primer, 2nd edn., Oxford Univ. Press, New York (1985)

J. P. Glusker, M. Lewis, M. Rossi, Crystal Structure Analysis for Chemists and Biologists, VCH, Weinheim (1994)

T. L. Blundell, L. N. Johnson, Protein Crystallography, Academic Press, London (1976)

J. D. Dunitz, X-Ray Analysis and the Structure of Organic Molecules, Cornell Univ. Press, Ithaca (1979)

J. Drenth, Principles of Protein X-ray Crystallography, Springer Verlag, Berlin (1994)

A. McPherson, Science in Pictures: Macromolecular Crystals, Scient. American, **260** (3), 62–69 (1989)

H. Friebolin, Basic One- and Two-Dimensional NMR Spectroscopy, Wiley-VCH, Weinheim (2010)

F. A. L. Anet, A. J. R. Bourn, P. Carter, S. Winstein Nuclear Magnetic Resonance Spectral Assignments from Nuclear Overhauser Effects, J. Am. Chem. Soc. **87**, 5250–5251 (1965)

K. Wüthrich, NMR of Proteins and Nucleic Acids, Wiley, New York (1986)

M. Pellecchia, I. Bertini, D. Cowburn et al., Perspectives on NMR in Drug Discovery: A Technique Comes of Age, Nat. Rev. Drug Discov., **7**, 738–745 (2008)

A. Watts, Solid-State NMR in Drug Design and Discovery for Membrane-embedded Targets, Nat. Rev. Drug Discov., **4**, 555–568 (2005)

M. Carroni, H. R. Saibil, Cryo electron microscopy to determine the structure of macromolecular complexes, Methods, **95**, 78–85 (2016)

M. Beckers, D. Mann, C. Sachse, Structural interpretation of cryo-EM image reconstructions, Prog. Biophys. Mol. Biol., **160**, 26–36 (2021)

G. Lander, A 3-minute introduction to CryoEM, https://www.youtube.com/watch?v=BJKkC0W-6Qk (Last accessed Nov. 17, 2024)

Special Literature

T. S. Koritsanszky and P. Coppens, Chemical Applications of X-ray Charge-Density Analysis, Chem. Rev., **101**, 1583–1627 (2001)

E. Keller, Röntgenstrukturanalyse von Molekülen I, II, Chemie in unserer Zeit, **16**, 71–88 and 116–123 (1982)

D. J. DeRosier, Turn-of-the-Century Electron Microscopy, Curr. Biol., **3**, 690–692 (1993)

M. A. Wear, D. Kan, A. Rabu and M. D. Walkinshaw, Experimental Determination of van der Waals Energies in a Biological System, Angew. Chem. Int. Ed., **46**, 6453–6456 (2007)

Y. Inokuma, S. Yoshioka, J. Ariyoshi, T. Arai, Y. Hitora, K. Takada, S. Matsunaga, K. Rissanen, M. Fujita, X-ray analysis on the nanogram to microgram scale using porous complexes, Nature, **495**, 461–466 (2013)

L. Rosenberger, C. v. Essen, A. Khutia, C. Kühn, K. Urbahns, K. Georgi, R. W. Hartmann, L. Badolo, Crystalline Sponges as a Sensitive and Fast Method for Metabolite Identification: Application to Gemfibrozil and its Phase I and II Metabolites, Drug Metab. Dispos., **48**, 587–593 (2020)

C. G. Jones, M. W. Martynowycz, J. Hattne, T. J. Fulton, B. M. Stoltz, J. A. Rodriguez, H. M. Nelson, T. Gonen, The CryoEM Method MicroED as a Powerful Tool for Small Molecule Structure Determination, ACS Cent. Sci., **4**, 1587–1592 (2018)

K. M. Yip, N. Fischer, E. Paknia, A. Chari, H. Stark, Atomic-resolution protein structure determination by cryo-EM, Nature, **587**, 157–161 (2020)

M. S. Doscher, F. M. Richards, The Activity of an Enzyme in the Crystalline State: Ribonuclease S, J. Biol. Chem., **238**, 2399–2405 (1963)

S. Wu, J. Dornan, G. Kontopidis, P. Taylor, M. D. Walkinshaw, The First Direct Determination of a Ligand Binding Constant in Protein Crystals, Angew. Chem. Int. Ed. Engl., **40**, 582–586 (2001)

A. R. Pearson, P. Mehrabi, Serial synchrotron crystallography for time-resolved structural biology, Curr. Op. Struct. Biol., **65**, 168–174(2020)

E. Nango et al. The three-dimensional movie of structural changes in bacteriorhodopsin, Science, **364**, 1552–1556 (2020), https://www.science.org/doi/10.1126/science.aah3497 Suppl. Mat.: aah3497s1, mp4, aah3497s2.mp4, aah3497s3.mp4

Three-Dimensional Structure of Biomolecules

Contents

G. Klebe, *Drug Design*, https://doi.org/10.1007/978-3-662-68998-1_14

Drug design focuses on the ligand, which is typically a small organic molecule with a molecular weight of less than 500 Da. It interacts with a macromolecular receptor and influences the properties of this receptor. On the other hand, the surrounding receptor can also determine the properties of the bound active ligand. Selective manipulation of these interactions requires an understanding of both the ligand and the receptor. After reviewing the methods used to determine the structure of biomolecules in the previous chapter, we will now look at what can be learned about the structural principles and properties of these macromolecules. Proteins consist of 20 basic building blocks, the proteinogenic amino acids (see page IX). A dipeptide is formed by linking two amino acids through an amide bond. Larger peptides and proteins are formed by the addition of many more amide bonds.

14.1 The Amide Bond: Backbone of Proteins

The simplest molecule with an amide bond is formamide **14.1**. Its structure is shown in ◘ Fig. 14.1. This bond occurs many hundreds of times in proteins, for example, over 50,000 times in the envelope of the rhinovirus (Sect. 31.6). The length of the bond between the carbon, oxygen, and nitrogen atoms can be determined from the crystal structure of formamide. The microwave spectrum of formamide in the gas phase also provides bond lengths, but different values are obtained. In the gas phase, formamide is "isolated," i.e., it does not "sense" any neighbors in its immediate vicinity. The C=O double bond is shorter and the C–N single bonds are longer than in crystalline formamide. In the crystal assembly, the individual formamide molecules are not "alone." They are connected to neighboring molecules by intermolecular hydrogen bonds. A hydrogen bond is a noncovalent interaction. It couples a functional group carrying a hydrogen atom (e.g., NH or OH) to an electronegative heteroatom (e.g., N, O; Sect. 4.4). Obviously, the involvement of a molecule in a network of hydrogen bonds causes a change in its geometry. The electron density between the atoms is shifted so that the C=O double bonds become longer and consequently weaker. At the same time, the C–N single bonds become shorter and stronger. Twisting the molecule away from planarity around this bond is, therefore, made much more difficult.

The amide bond is the fundamental building block of proteins. Every third bond in the polymer chain is an amide bond. They have planar geometry, meaning that a plane can be defined by their atoms, as we saw with formamide. The folding of the polymer chain and the resulting spatial structure of the protein is determined by the twist angle at the plane of the amide bonds against each other (◘ Fig. 14.2). The stiffness and planarity of the amide bonds determine the stability of the spatially folded protein. If one formulates a Lewis formula for such an amide bond embedded in a hydrogen bond network, the polar nature with the double-bond character of this bond becomes clear (◘ Fig. 14.1). In proteins, amide bonds practically only occur in the *trans* configuration (except for amide bonds adjacent to proline

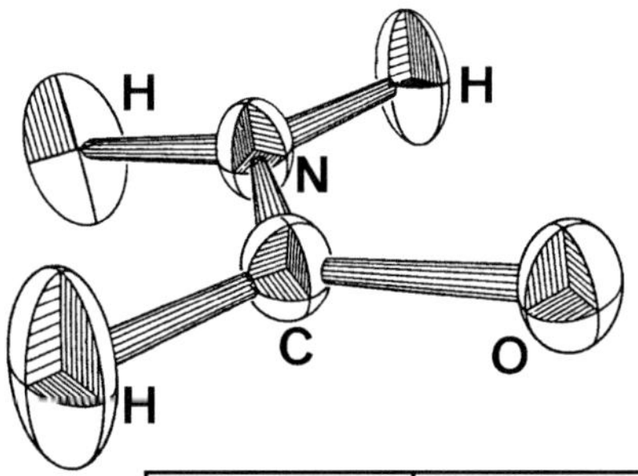

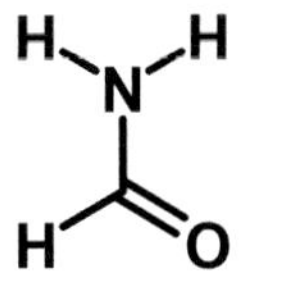

14.1

Formamide	Bond length in Å	
	C=O	C-N
Cystalline phase	1.241	1.318
Gas phase	1.219	1.352

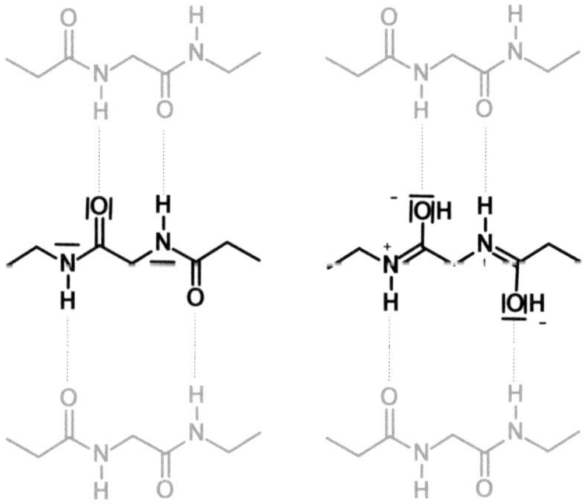

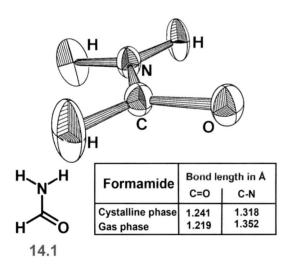

◘ **Fig. 14.1** Formamide **14.1** is the smallest molecule that has an amide group. Its molecular structure is shown on the *left*. Because of thermal motion in the solid state, the molecule carries out vibrational movements. Its electron density is, therefore, distributed over a larger area. This is described by using ellipsoids that encompass the 50% probability of occurrence of the atom (*left*). Two hydrogen bonds are incurred between the carbonyl group and the amide group of a neighboring molecule in the crystal packing. An extended H-bond network stabilizes the crystal structure and polarizes the amide group. The bond lengths (in Å) are different in the crystal structure and in the gas phase (Table). If we consider an amide bond in a protein, it will also be involved in hydrogen-bond contacts (*right*). It can be expressed as a Lewis formula, which indicates the polar nature and double bond character of the amide bond

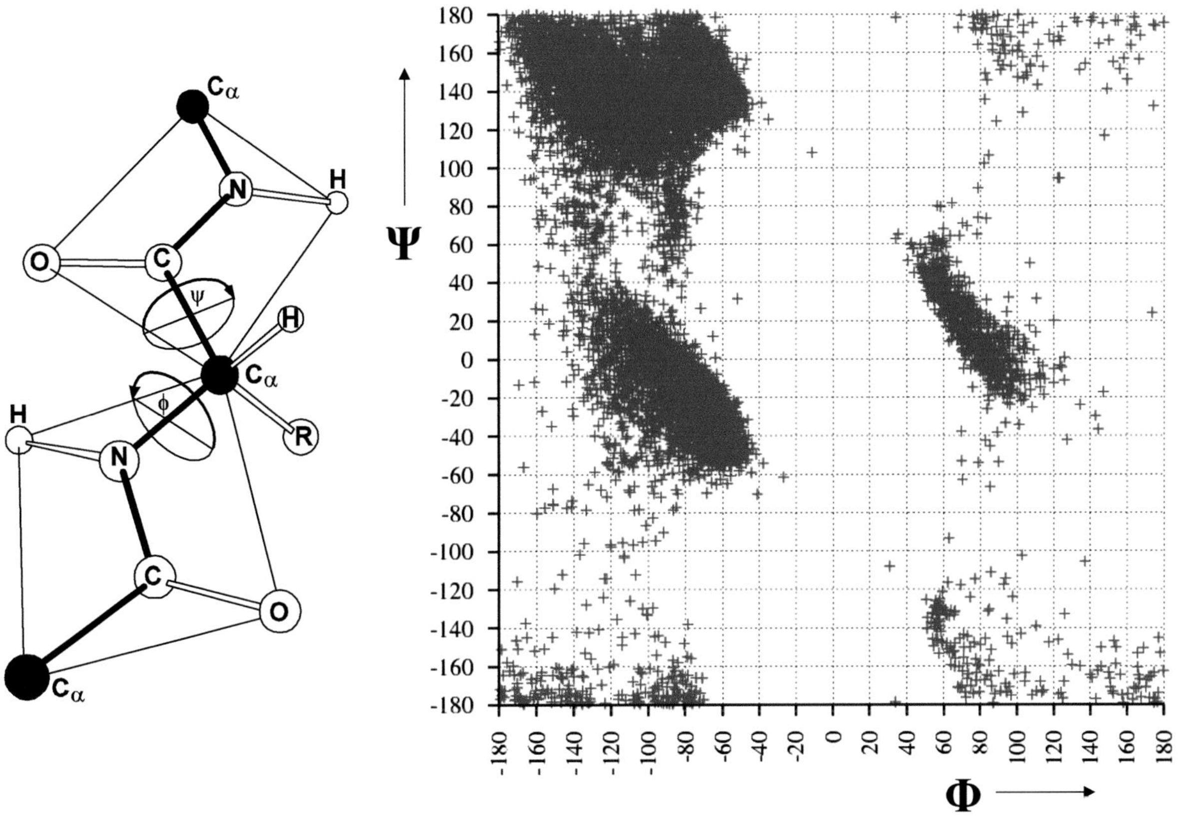

■ **Fig. 14.2** *Left* The spatial course of a polypeptide chain is determined by the relative orientation of the planar peptide bonds. The mutual twist of these planes against each other is measured on the basis of the two twisting or dihedral angles Φ and Ψ. These do not assume any value around the bond axes, but rather are limited to a few combinations of value ranges. *Right* In the diagram, a so-called Ramachandran plot, the values for both angles along the peptide chain are plotted. The angle combinations for an α-helix (■ Fig. 14.3) are found in the *middle left*, and those for a β-pleated sheet (■ Fig. 14.4) in the *top left*

residues, which show both configurations). The only remaining degrees of freedom for the polymer chain are the rotations about the entire amide bond planes. These rotations (Chap. 16) occur around the bonds that lie between the C_α carbon atoms. As shown in the comparison of the bond length between gaseous and crystalline formamide, the decisive additional stiffening of the amide bond is caused by its incorporation into a hydrogen-bonding network.

14.2 Proteins Fold in Space to Form α-Helices and β-Strands

Typically, the angles named Φ and Ψ are used to describe the two dihedral angles around the C_α carbon atom, and these angles usually take on value pairs from two ranges. These ranges are related to a helical or sheet-like course of the polymer chain (■ Fig. 14.2). In an α-helix with a right-handed turn, all CO and NH groups orient in the same direction (■ Fig. 14.3). Between them, a H-bond network is formed. Each amino acid in the helix is in contact with the next fourth amino acid in the sequence. This unidirectional orientation of the polar groups of the amide bonds in an α-helix has consequences for the electrostatic properties, and a significant dipole moment can build up along the helix (Sect. 15.5). At the tip of such helices, preferred binding sites for positively or negatively charged particles are formed (cf. the architecture of ion channels, Sects. 30.2 and 30.8). While a helix is composed of amino acids from a single contiguous segment of the peptide chain, amino acids from at least two sequence segments must come together to form a β-pleated sheet. Both strands can be bonded with each other in either a parallel or antiparallel orientation relative to the polymer chain (■ Fig. 14.4). This network exhibits a different progression of H-bonds for both orientations. The side chains alternate above and below the pleated sheet. The entire strand is slightly twisted upon itself. Because of this, a pleated sheet of multiple strands has a twist to it when viewed from the side (■ Fig. 14.5).

In addition to these two common secondary structures, there are other typical combinations of torsion angles. A polymer chain that folds into a globular structure in space must reverse its direction. This is accomplished in what is called the **turn or loop region**. Turns can be clas-

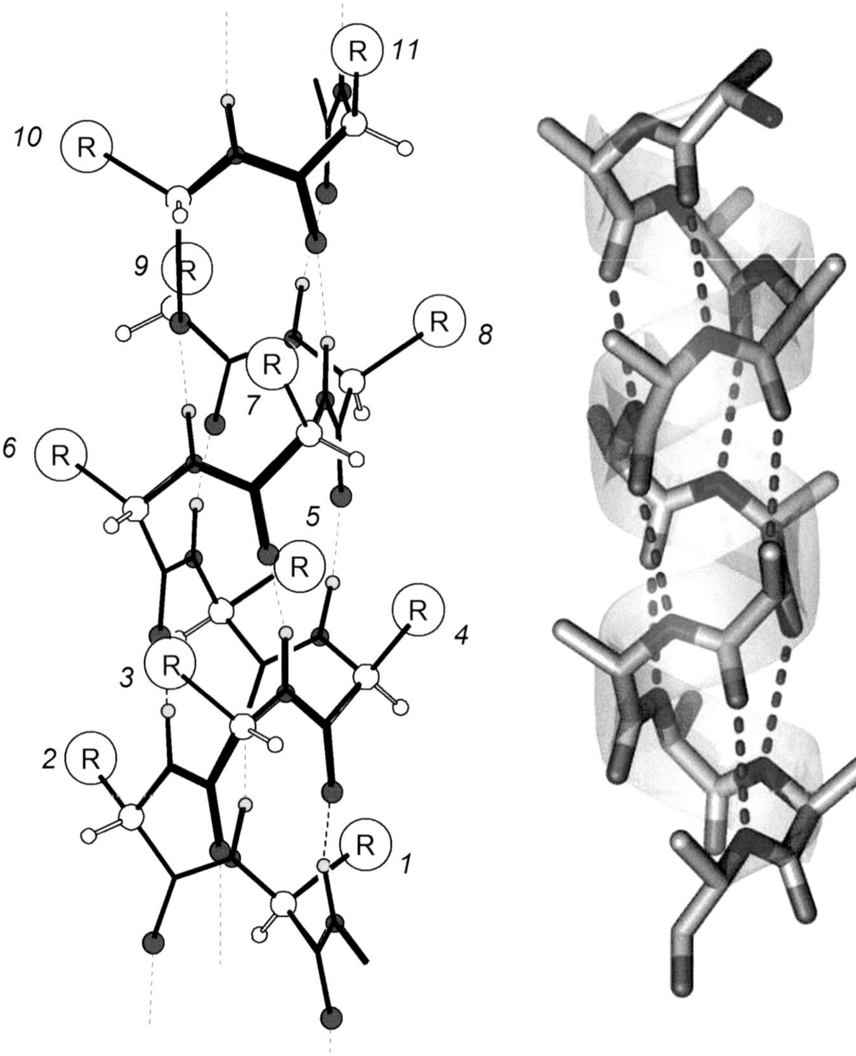

Fig. 14.3 The α-helix is a commonly found secondary structure. *Left* The polypeptide chain forms a right-handed spiral with a pitch of 7 Å, and 3.6 amino acids per turn. All carbonyl groups (oxygen atoms *red*) are oriented parallel to the helix axis in the same direction. The NH functionalities (nitrogen atoms *blue*, hydrogen atoms *cyan*) are oriented in the opposite direction. *Right* The groups form a pronounced hydrogen-bonding network (*violet dashed line*) between themselves. The side chains (R) on the C_α atoms are on the outside pointing away from the helix axis. This creates a typical groove pattern that spirals across the surface. This "ridge and groove" pattern determines the mutual packing of α-helices in proteins

sified according to the number of amino acids involved and the type of interaction that closes the loop. **Turns** that form a C=O···H–N hydrogen bond in the direction of the polymer chain, **inverse turns** with hydrogen bonds in the opposite direction, and **open turns** in the chain held together by van der Waals and polar interactions can be distinguished from one another (◘ Fig. 14.6). A total of 158 turn classes have been summarized in a comprehensive evaluation by Oliver Koch.

What force governs the organization of a protein? Amino acids have hydrophilic and hydrophobic side chains. Hydrophobic groups avoid aqueous environments (Sect. 4.2). During folding of the polymer chain in an aqueous medium, the hydrophobic amino acids aggregate to diminish their shared hydrophobic surface area. Therefore, the hydrophobic amino acids are predominantly found in the interior of a folded protein. The polar groups of the main chain amide bonds are saturated in the secondary structure by hydrogen bonds. The side chains of polar amino acids will only be found inside a protein if they can form a polar interaction with another surrounding amino acid. Otherwise, they are oriented towards the outside of the protein, protruding into the surrounding water environment. Proteins can also span a cell membrane. In the areas where they are in contact with the membrane, they have a large, contiguous hydrophobic surface (Sect. 14.7). The packing density inside the protein is of the same order of magnitude as that found in crystals of small organic molecules. The interactions that determine the molecular packing are virtually identical in both cases.

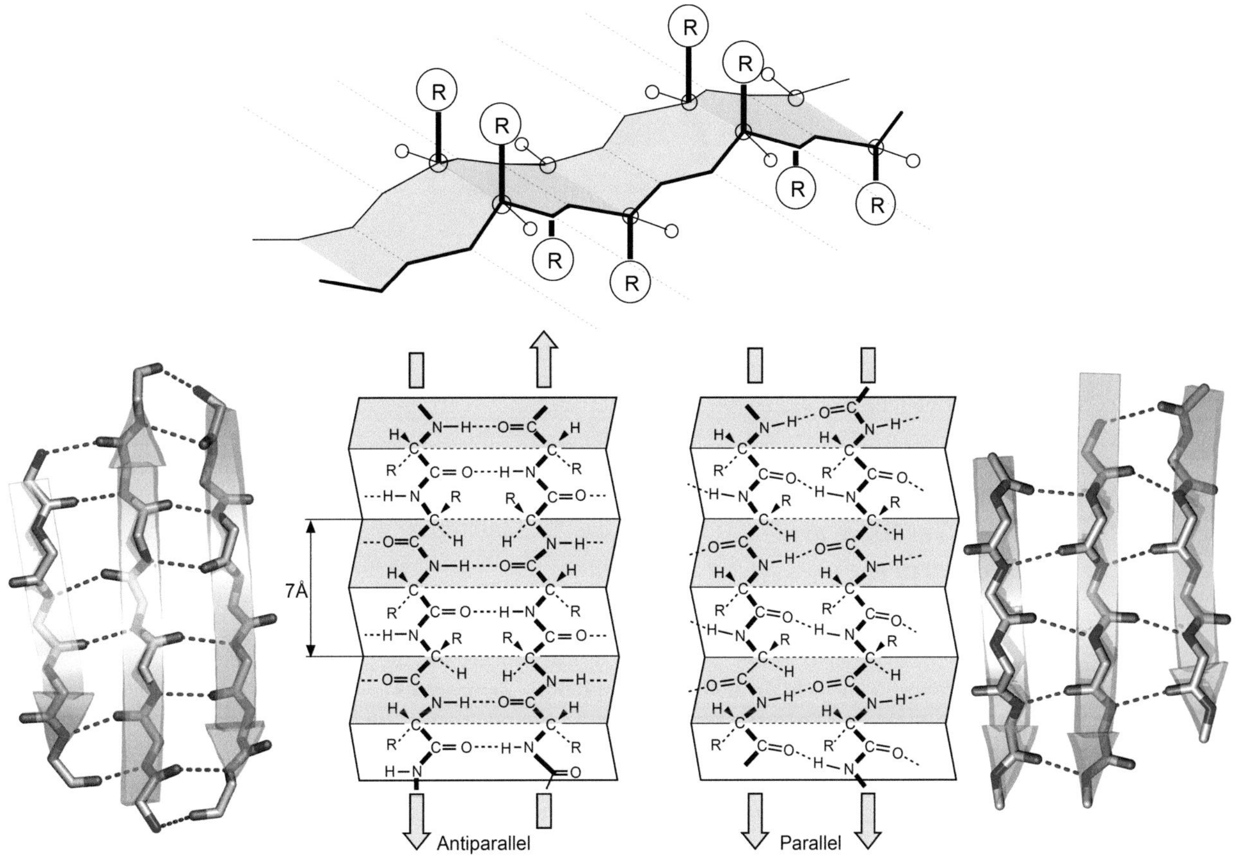

Fig. 14.4 A second important secondary structure, the β-strand is composed of multiple sections of the polymer chain that exist in a stretched conformation (*top*). The strands can run parallel or antiparallel. They are crosslinked to each other via hydrogen bonds (*violet*). The sheet-like structure displays a zigzag fold and is called a β-pleated sheet. The side chains (R) of the amino acids point away from it, alternating above and below the pleated sheet

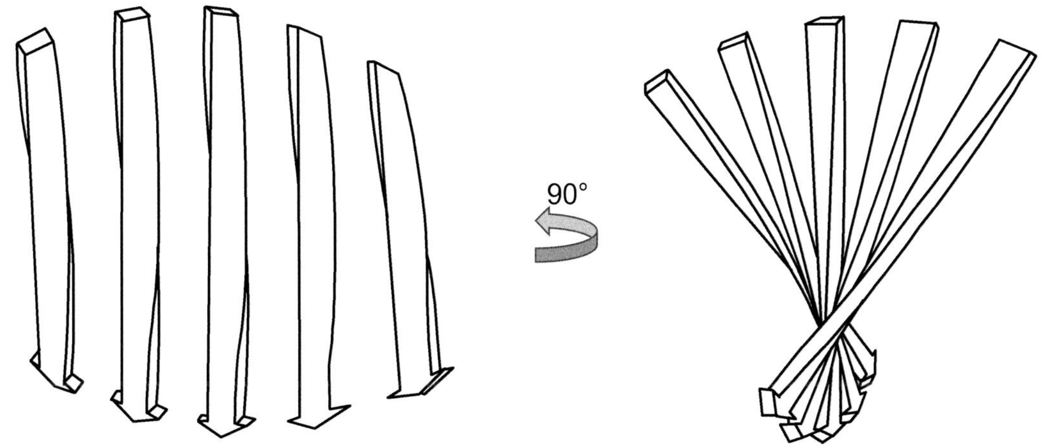

Fig. 14.5 Within a β-pleated sheet of multiple strands, here shown with a parallel orientation, a right-handed twist occurs. For simplification, the single β-strands are indicated with an *arrow*. The twist can be seen by the internal rotation of the *arrow*. The pleated sheet here is shown in two perpendicular views

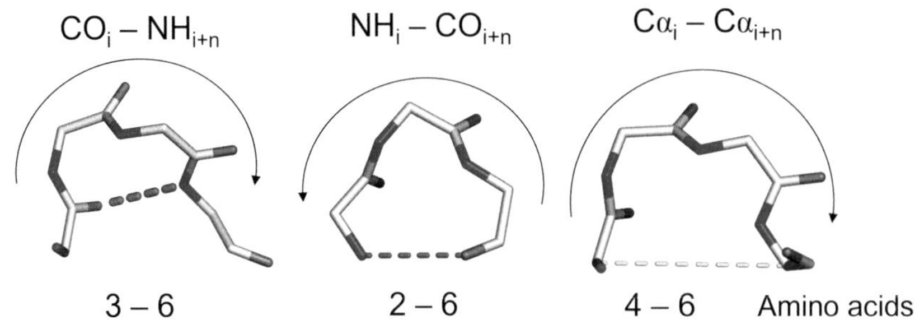

Fig. 14.6 The polymer chain of a globular protein reverses its direction in the loop or turn area. Numerous turn patterns have been found. They are made up of 2–6 amino acids. Normal turns (*left*) form a C=O···HN hydrogen bond (*violet*) in the direction of the polymer chain. This hydrogen bond has a different order in inverse turns (*center*). Another group of open turns (*right*) is held together by van der Waals contacts and polar interactions

14.3 From Secondary Structure Via Motifs and Domains to Tertiary and Quaternary Structure

Proteins organize their secondary structural segments into motifs. For example, the sequence of an α-helix, a β-strand, and another α-helix forms a **motif**. Multiple motifs fold into domains to yield the **tertiary structure** of a protein. Domains may be preferentially made up of helices, folded sheets, or a combination of both building blocks. Often the domain has a specific function. Many proteins consist of a single domain. Complex proteins can be made up of multiple domains. When a complex assembly of several separate polymer chains is formed (e.g., as in hemoglobin), it is called a **quaternary structure**.

Despite the enormous multiplicity that can be achieved by combining the 20 proteinogenic amino acids into sequences, there seems to be a rather limited number of folding possibilities for the domains. How many total folding patterns exist can only be speculated. Of all the crystal structures known today, about 1200 different folding patterns have been found. This number is mainly based on

14

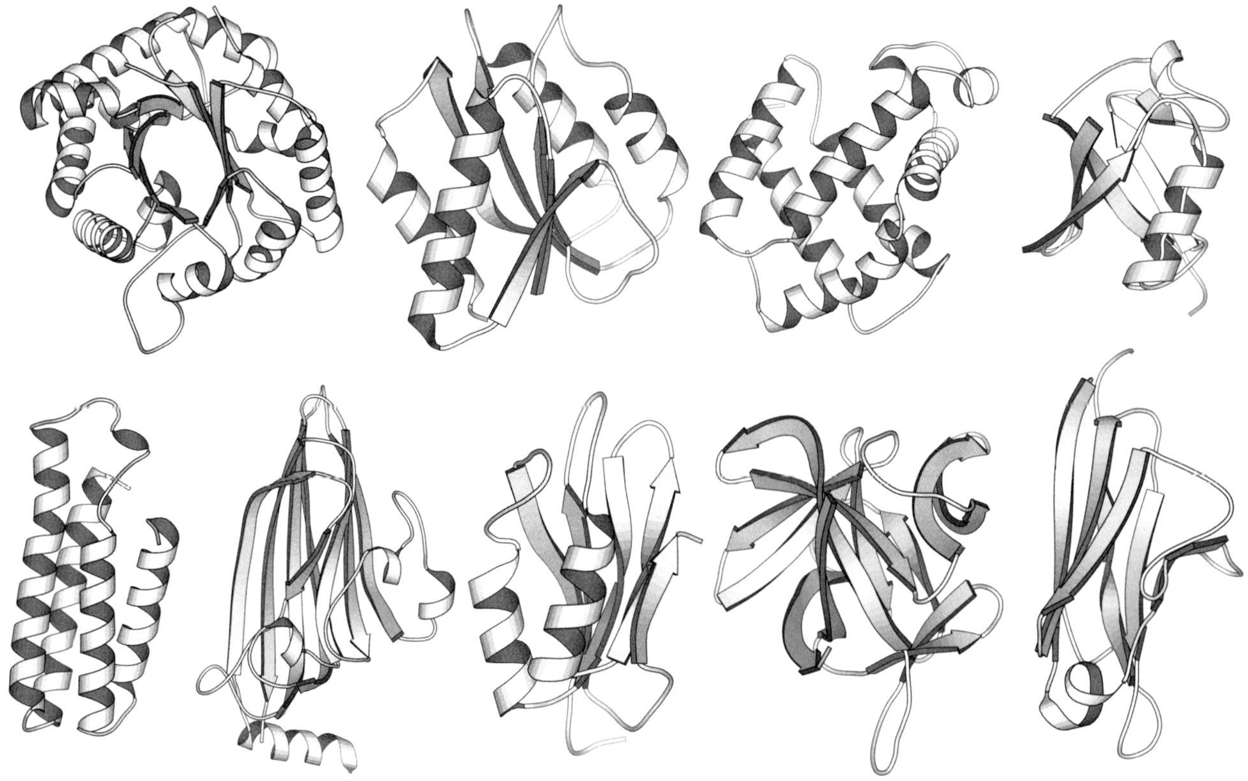

Fig. 14.7 The course of the polypeptide chains is symbolized with *spirals* for α-helices, with *arrows* for β-pleated sheets, and with *threads* for different turn segments. Approximately 30% of the structurally known proteins can be assigned to one of the nine shown folding classes. The first folding pattern (*upper left*) is a "TIM barrel," and the second one is an "open pleated-sheet" structure

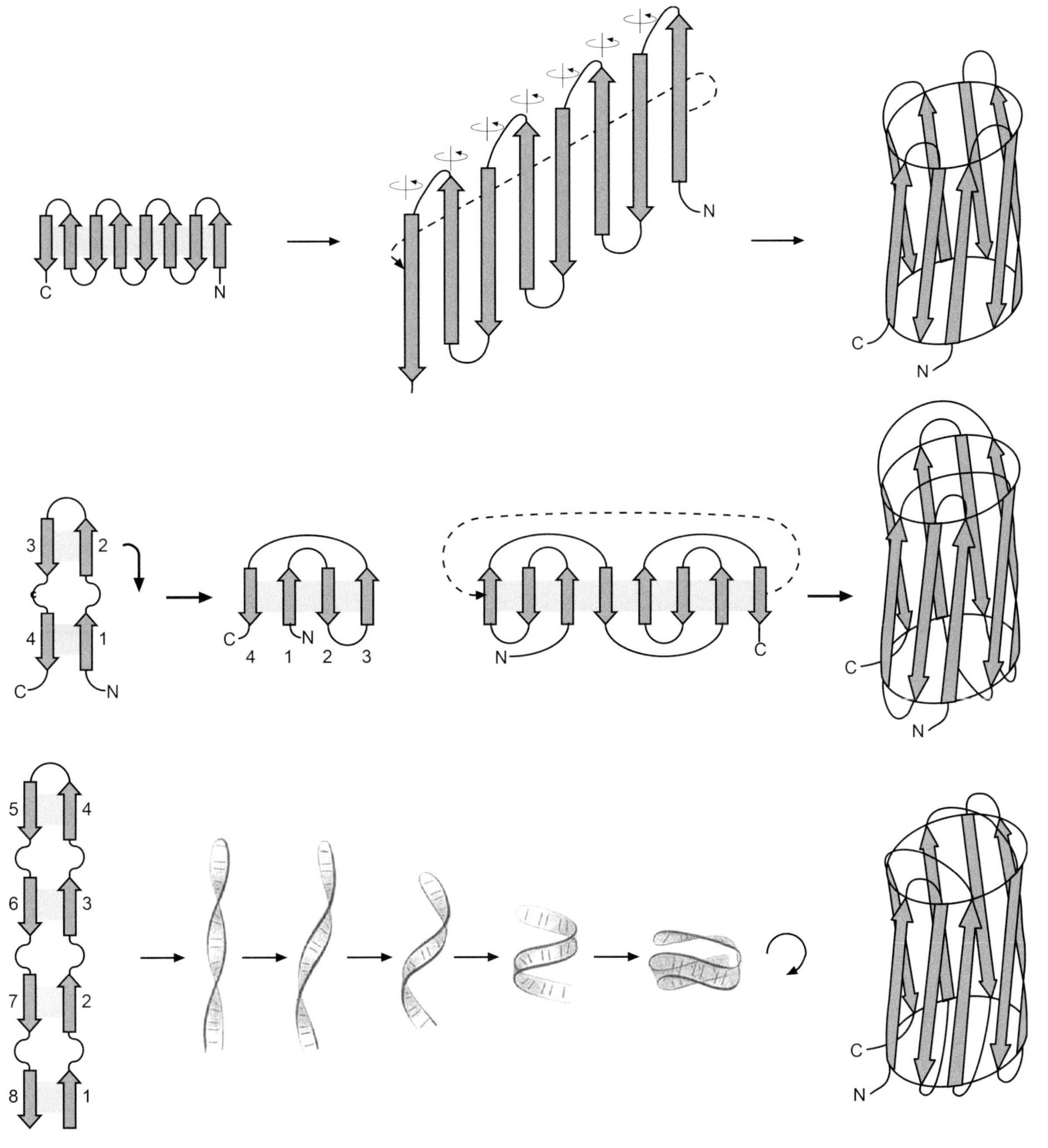

Fig. 14.8 The folding pattern of different β-barrel structures can be thought of as a polymer chain with eight separate β-strands (*arrows*). These are separated by loop areas. *Upper row* An up-and-down barrel forms when the folding of the polymer chain of eight β-strands follows a zigzag pattern. The antiparallel sections form hydrogen bonds between themselves that close up to form a cylinder. *Middle row* The four-β-strand polypeptide chains lie next to one another so that the first chain interacts with the fourth, and the second interacts with the third. Then the double strand folds and the first pair comes to lie next to the second. Because the course of the polymer chain is reminiscent of the engravings on Greek vases, the pattern is called a *Greek key*. Two such patterns can come together into a cylinder-like orientation and form a *Greek-key* barrel. *Bottom row* Another folding pattern is formed from a double strand that is placed together with an internal twist. The double strand wraps itself into a cylinder-like structure that is called a *jolly roll*

data from globular enzymes and transport proteins. Since hardly any new examples have been found in recent years despite intensive efforts, it can be assumed that there are perhaps 1500–2000 stable folding patterns used by Nature. Considering these numbers, it must be taken into account that there are many **intrinsically disordered proteins** that can adopt an ordered fold only when they are fulfilling their function. Very little is known about their structural organization. It remains to be seen whether new folding patterns can be found here. Using their computer programs, David Baker's group (University of Seattle, USA) was able to design proteins with novel folds and to predict the sequence that folds into that spatial structure. As a proof-of-concept, the group then synthesized these proteins and, as predicted, their designed architecture was different from all previously known proteins in Nature. This opens up the possibility of creating an enormous variety of new "artificial" proteins that can be used as drugs, vaccines, enzymes or sensors. For his achievements, David Baker was awarded the Nobel Prize in Chemistry in 2024.

About 30% of all folded proteins belong to one of the classes shown in ◘ Fig. 14.7. From the group of membrane-bound proteins, perhaps 1000 structures are known so far, although about 30% of all human proteins fall into this group. Based on such limited data, it seems difficult to make reliable predictions about possible additional folding classes.

Drug design focuses on the interaction of a ligand with a protein. As a result, the structural considerations of chemists are usually limited to the amino acid groups that protrude into the binding pocket. However, the folding pattern around the binding pocket influences the properties found there. For example, a helix oriented towards the binding pocket determines the local electrostatic potential. This can be exploited to design selective ligands that bind only to proteins of a particular folding class.

Despite the progress in structure determination techniques, it can happen that the structure analysis of an important protein fails, but the structure of a related protein, for example, can be solved. On this basis, a model of the desired protein can be built (Sect. 20.5). This requires information about the structure and folding principles of proteins. They allow us to understand which part of the protein stabilizes the scaffold, which parts determine the function, and which parts account for the differences between homologues.

A detailed discussion of these principles would go too far here. As an example, consider the folding pattern of the β-barrel. An extended sheet of multiple β-strands has an internal twist (see ◘ Fig. 14.5). For example, if eight such strands are lined up side by side, a cylinder can be formed. This barrel-like folding pattern of eight or more strands is often observed. Several variations of this folding pattern are shown in ◘ Fig. 14.8, which illustrate how and according to which principles a polypeptide can fold spatially.

A loop acts as a connecting element between the pleated sheet strands of the β-barrel in the example in ◘ Fig. 14.8. α-Helices can also serve as connecting elements (◘ Fig. 14.7). In the center a barrel-like structure forms

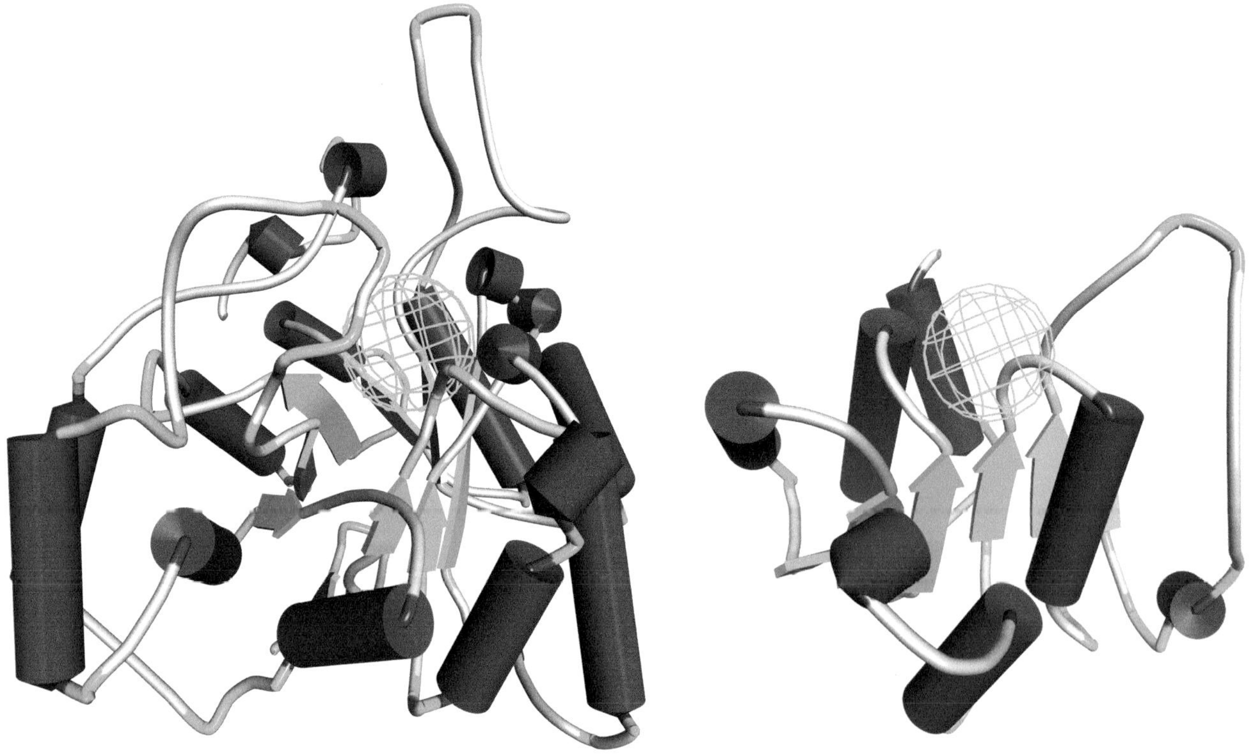

◘ Fig. 14.9 The amino acid groups that determine the folding pattern and those that carry the function are located in different regions of the protein. *Left* The catalytic site (*yellow spheres*), which binds and transforms substrates lies in a TIM-barrel-type structure (α-helices: *red cylinder*, β-strands: *light-blue arrows*) at the end of the barrel where one would expect to find a lid. The loops of the polymer chain that surround this "lid" (*gray* and *green threads*) carry the function-determining amino acids. *Right* The function-determining amino acids in the loop region occur in the open-pleated-sheet structure there, where the attached helices change from the top to the bottom of the pleated sheet. (► https://sn.pub/5hAz72)

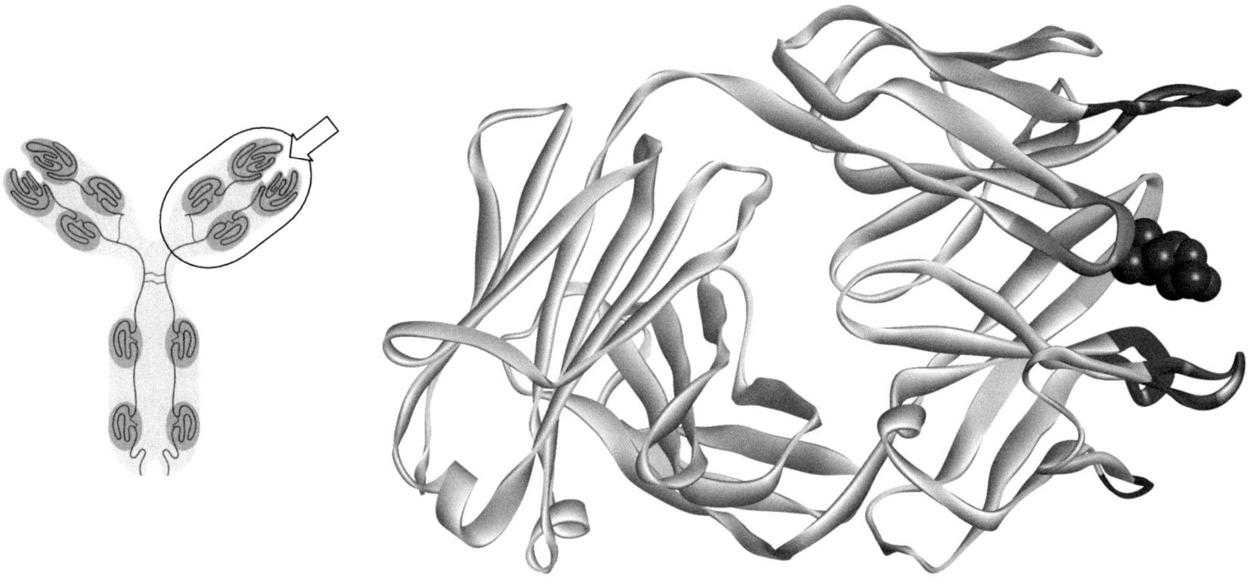

Fig. 14.10 Immunoglobulins form a highly specific binding pocket in which they recognize antigens. These are exogenous substances. The enormous structural diversity of these binding pockets is achieved by variations in the amino acids in the loop regions. *Left* The immunoglobulins have a Y-like shape, which is divided into a trunk (constant F_c domain) and two identical F_{ab} branches. *Right* The course of the polymer chain in these branches corresponds to the barrel type. The antigen binding site is found on the right-hand side. The displayed polypeptide chain covers the region circled in the schematic drawing on the left. At the right end (*colored*) are the loops responsible for recognition of exogenous substances. They grasp the antigen (*dark red molecule*) like the fingers of two hands. (▶ https://sn.pub/Soz3AF)

from β-strands, and on its outer surface the bridging α-helices are aligned. This folding pattern was first discovered in **t**riosephosphate **i**so**m**erase. It is therefore called a **TIM barrel** (◘ Fig. 14.7). Another important folding class that is made up of α-helical and β-pleated sheet segments comprise the open-sheet structures (◘ Fig. 14.7). In this class, the pleated sheet does not close to a cylinder but rather it remains open. Helices group above and below the sheet.

14.4 Are the Fold Structure and Biological Function of Proteins Correlated?

How does the structure of a protein relate to its function? For example, do all proteases have the same folding pattern? A large number of enzymes with distinctly different functions all belong to the TIM barrel type or open-sheet structure. There are many oxidases, isomerases, kinases, aldolases, synthases, dehydrogenases, or proteases that can be assigned to these two classes. Here, Nature started from a common origin and developed divergently. Consequently, the **function of a protein** is not necessarily coupled to a particular **folding pattern**. When the structure of the enzyme is further analyzed, it is found that the **catalytic sites of the proteins** of a folding class are located at the same position. This is found at the *C*-terminal end of the barrel in the TIM barrel structure and at the topological switch of the connecting helices from the top to the bottom of the open sheet structure (◘ Fig. 14.9). The function-determining amino acids are located in the loop region between adjacent folded sheets and helices. Why would Nature follow this principle of separating folding structure from function? The amino acids that enable the stable folding of a domain are separated from those that induce a specific function. This is a very efficient evolutionary strategy. Two regions have been optimized simultaneously:

- The stability of the protein scaffold in special folding patterns and
- The layout of the amino acid sequence to serve a special function.

By spatially separating and displacing the functional groups in the structurally less committed loop regions, the two tasks could be optimized in parallel. Changing a single amino acid in a secondary structural element could destabilize the entire folding pattern and stop folding. This will be avoided if the amino acid sequence to be functionally optimized is placed on a stable scaffold that does not interfere with the optimization.

A class of proteins that implements this principle to perfection are **immunoglobulins**. As antibodies, they recognize and bind to foreign substances called antigens. To remove an antigen, immunoglobulins with highly specific binding pockets and high affinity must be available within a few days. The substances recognized can be anything from small organic molecules to large proteins. To achieve this goal, it is estimated that approximately 10^{12} different

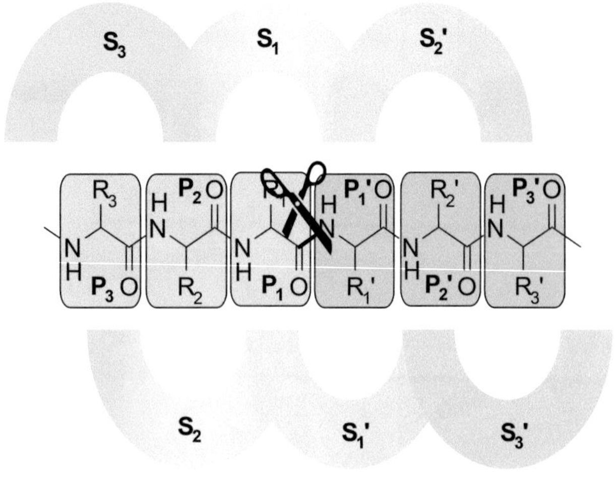

Fig. 14.11 The side chains of a peptide substrate and the binding pockets that they belong to are classified on the *N*-terminal side of the peptide as P_3, P_2, P_1... or S_3, S_2, S_1... (*left*); on the *C*-terminal side they are classified as P'_1, P'_2, P'_3... or S'_1, S'_2, S'_3... (*right*)

antigen sequences are produced. Considering that each antibody-producing cell makes just one specific immunoglobulin and that the human genome contains only about 10^5 genes, how can this difficult task of achieving such diversity be accomplished? It is solved by cells of the immune system using a combination of different variable gene segments and excessive amino acid exchange in these segments during lymphocyte maturation. In this way, variable loop regions are formed on a stable scaffold of barrel-shaped pleated sheet structures (Fig. 14.10). The therapeutic value of such biomolecules (called **biologicals**) has been recognized. Many **monoclonal antibodies** are nowadays available as therapeutics (Sect. 32.3).

14.5 Proteases Recognize and Cleave Substrates in Well-Tailored Pockets

Proteases cleave polypeptide chains during enzymatic degradation or release an active protein or peptide from an inactive precursor form. For this purpose, the enzymes possess a catalytic site at which cleavage takes place (Sect. 14.6 and Chaps. 23, 24, and 25). In order to specifically recognize a particular substrate, several binding pockets are located on their surface. These are structurally complementary to the side chains of the substrate, which are oriented around the catalytic sites. In 1967, Israel Schechter and Arieh Berger proposed a nomenclature to describe these pockets (Fig. 14.11). The positions of the amino acids of the peptide substrate are described as P_3, P_2, P_1–P'_1, P'_2, P'_3, and so on. Starting from the *N*-terminus, position P_1 is immediately before and position P'_1 is immediately after the cleavage site. The binding pocket of the enzyme for the side chain of amino acid P_1 is called S_1, and the same is valid for the other side chains. This very useful nomenclature is purely formal. Applying these labels to a particular enzyme does not mean that all of the pocket actually exists. Two binding pockets can appear as one large binding pocket in the 3D structure. The S_3 and S_4 binding pockets in the serine protease thrombin are actually just one large pocket (Sect. 23.3). It is also possible that a substrate amino acid does not have a complementary binding pocket in the enzyme. It then protrudes into the surrounding water environment.

14.6 From Substrate to Inhibitor: Screening of Substrate Libraries

Peptides are easily synthesized with enormous diversity (Sect. 11.5). If the peptide is attached to a probe that changes its color or fluorescence upon release (Sect. 7.2), the labeled peptide can be used to determine the substrate profile of the protease. To do this, a large library (Sect. 11.1) of these peptides is exposed to the protease, and the members that are well cleaved are identified. Fig. 14.12 shows the amino acid composition of a labeled tetrapeptide that is preferentially cleaved by the proteases trypsin, factor Xa, plasmin, and chymotrypsin. Peptides with basic groups such as arginine or lysine are preferentially cleaved by trypsin, plasmin, and factor Xa. Factor Xa almost exclusively converts peptides with arginine at the P_1 position. Chymotrypsin behaves quite differently. It prefers aromatic amino acids such as tyrosine, phenylalanine, and tryptophan at the P_1 position. The selectivity at positions P_2 to P_4 is not nearly as pronounced. Trypsin will convert tetrapeptides with branched groups at P_2 such as Phe, Tyr, Trp, Ile, or Val much more poorly if an arginine is at the P_1 position. Basic groups are also less favored. Trypsin shows virtually no selectivity at P_3 and P_4. Factor Xa has a special preference for the small glycine at position P_2, but there is hardly any difference for the groups at position P_3 for this enzyme. On the other hand, different groups are more strongly selected at the P_4 position. The substrate binding profile helps to reveal the selectivity characteristics of enzymes. They show the complementary properties in the binding pocket and help to inspire initial ideas for the design of possible inhibitors.

This concept was applied to cysteine proteases in the research group of Jonathan Ellman at the University of California at Berkley, USA. Substrate molecules were synthesized that carried a fluorescent marker at the end of an amide bond to be cleaved. Various organic building blocks were attached to the other side. When such a substrate molecule is cleaved by the protease, the organic moiety must be bound in the binding pocket of the enzyme. Therefore, the transformation indicates the binding of a test molecule. The method is ideal for screening purposes. A hit discovered in this way can easily be chemically converted from a substrate molecule to an inhibitor. For example, by replacing the cleaved amide bond with an aldehyde function, a cysteine protease inhibitor (Sect. 23.9, Fig. 23.27) that has very little in common with the peptide substrate can be designed.

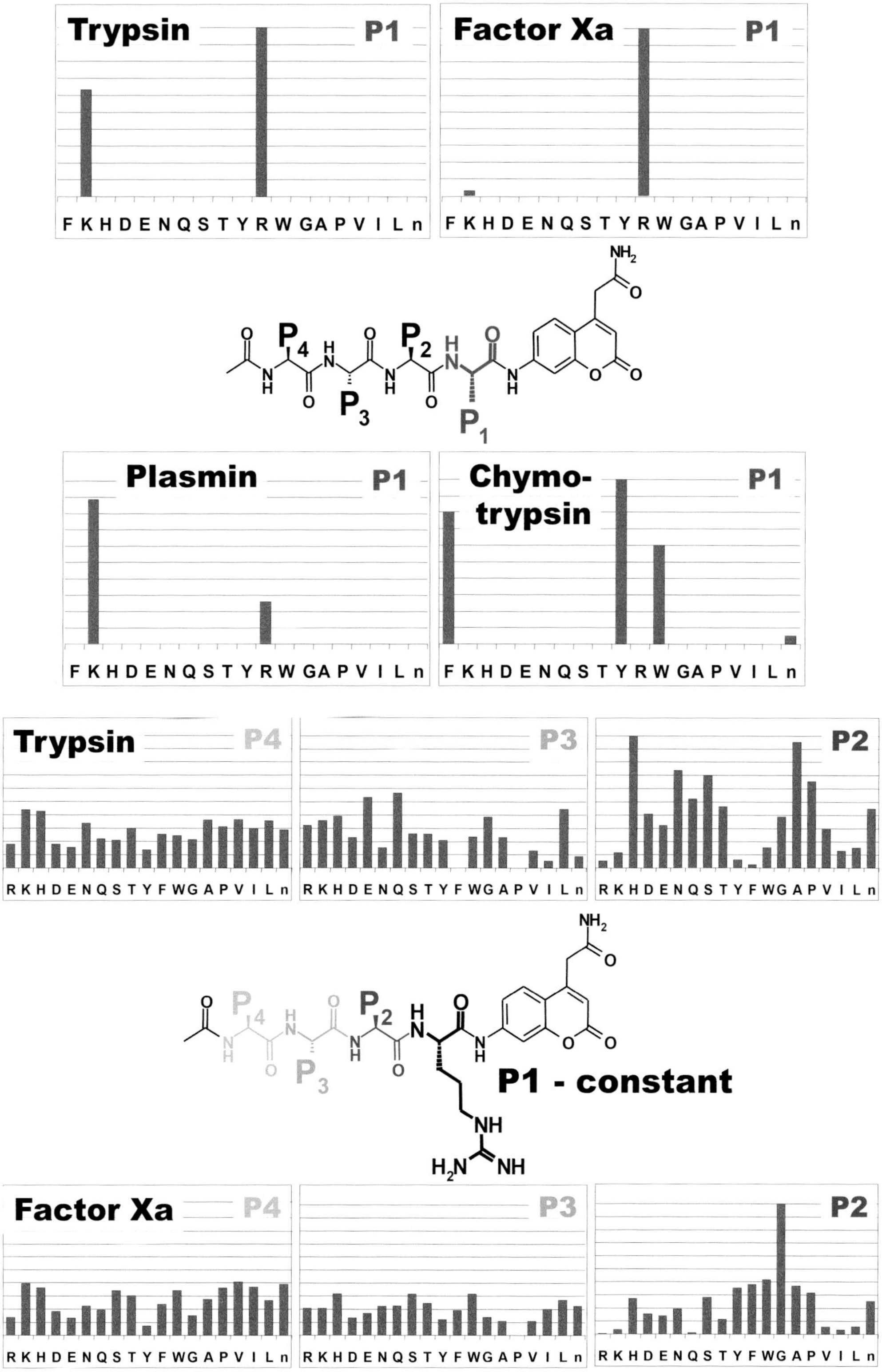

Fig. 14.12 *Upper rows* A tetrapeptide library, held constant in position P_2 to P_4, was varied at position 1 (*red*) with 19 amino acids (one-letter notation; *n* norleucine). It is cleaved by trypsin after arginine and lysine, by factor Xa after arginine, and by plasmin after lysine. Chymotrypsin prefers aromatic amino acids at position P_1. *Bottom rows* If arginine is held in position P_1 and the remaining three positions (*blue, green, orange*) are varied, trypsin shows practically no selectivity for the amino acids at P_2, P_3, and P_4. On the other hand, factor Xa prefers a glycine in position P_2

14.7 When Crystal Structures Learn to Move: From Static Structures to Dynamics and Reactivity

What information about the dynamics and reactivity of molecules can be extracted from a crystal structure? Molecular vibrations are visible even in the solid state. This is reflected in the blurriness of the electron density. When a molecule participates in a reaction, bonds are broken and new bonds are formed. The formation and cleavage of amide bonds is a central task in biochemical processes. Molecule **14.2** contains an amide and an ester group (◘ Fig. 14.13). When a crystal of this compound is exposed to thermal energy, a solid state reaction occurs to form **14.3**. The molecule is in a geometry in the initial crystal structure that is conducive for entering into the reaction pathway.

Information about changes in the geometric orientation of functional groups in the chemical reaction is critical to understanding the concomitant structural changes that occur. This knowledge is a prerequisite for the design of transition state analogue inhibitors (Sects. 6.6 and 22.3). With respect to the formation or cleavage of an amide bond, the question is: from which direction does the amino group attack the carbonyl carbon in the course of nucleophilic addition to form a new bond?

In the early 1970s, Hans-Beat Bürgi and Jack Dunitz at ETH Zurich in Switzerland began to extract information about the **geometric changes along such reaction steps** from crystal structures. Before the days of movies and television, people came up with creative ways to make pictures move, such as flip-books (◘ Fig. 14.14). They give the impression that a dynamic story is being told. Imagine that the pages of the book have fallen apart

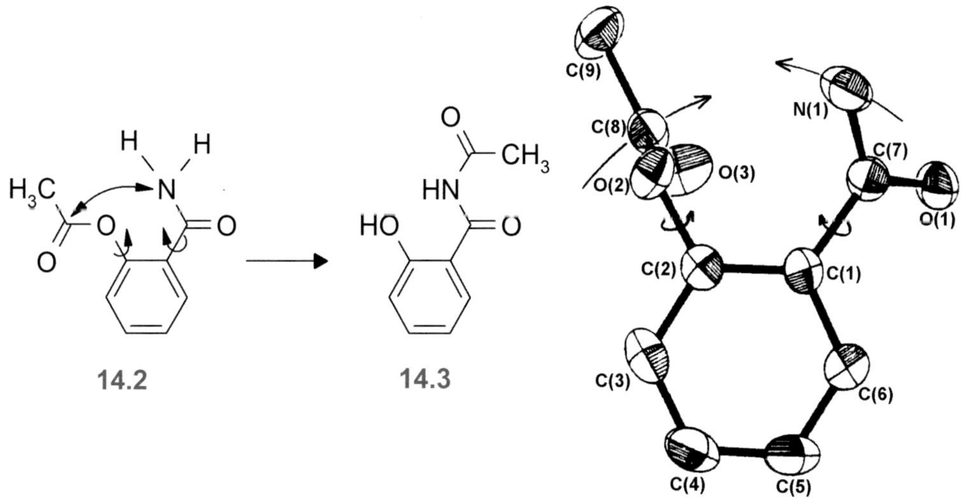

◘ **Fig. 14.13** If thermal energy is applied to a crystal of **14.2**, the carbonyl group of the ester function reacts with the amide NH_2 group and an imide bond is formed between N1 and C8 to give **14.3**. There must be an implicit vibrational motion that culminates in the reaction. At the same time, the ester bond between C(8) and O(2) is cleaved during these reaction steps

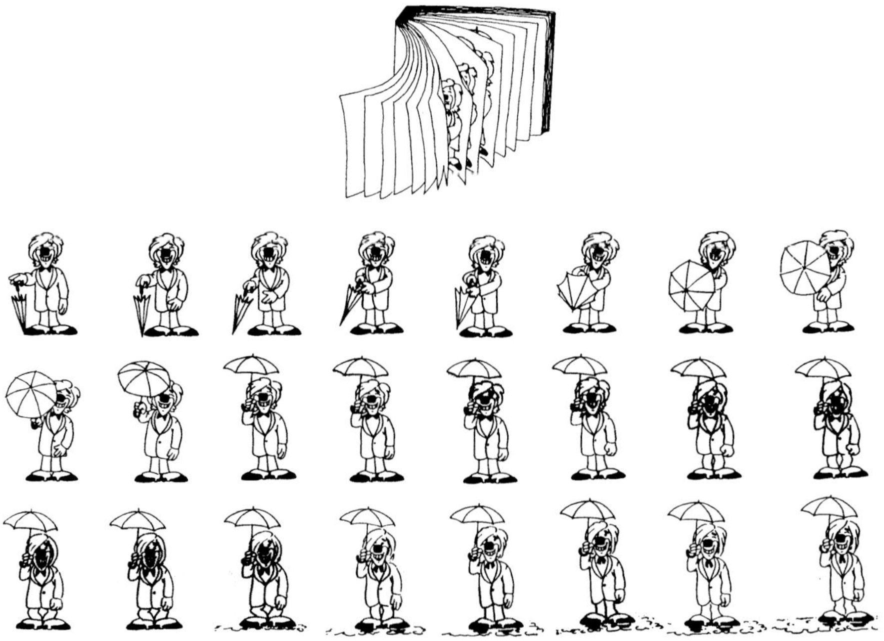

◘ **Fig. 14.14** A story is shown in static pictures in flip-books. If the different pages of this story flip pass the eyes quickly enough, the impression of a dynamic process will be given

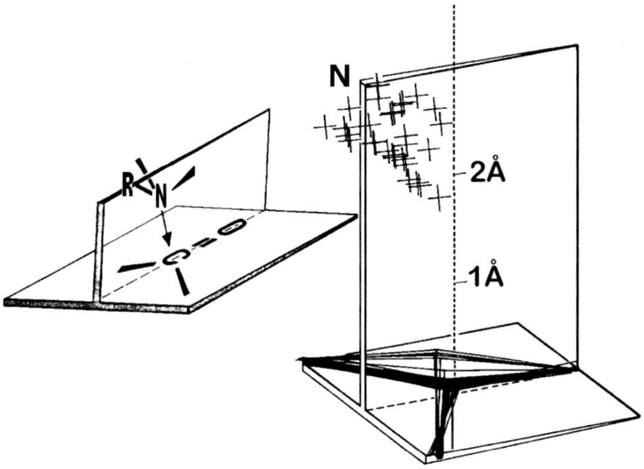

■ **Fig. 14.15** The formation or cleavage of an amide bond occurs by nucleophilic addition. A nucleophile, for instance, an oxygen or nitrogen atom, approaches the planar carbonyl carbon atom. During the reaction it rises out of the plane of the three neighbors and adopts a tetrahedral configuration. Examples were sought from low molecular weight crystal structures in which a nitrogen atom approaches a carbonyl group between a single bond and a van der Waals contact in the crystal packing. By superimposing these data, it is recognizable that the approach of the nucleophilic nitrogen towards the carbonyl group is "perfomed from back and behind." Simultaneous with this spatial approach, the carbon atom migrates out of the plane towards the nucleophile. The geometric requirements of this reaction step also determine the structural composition of the catalytic center of a variety of hydrolases (Sect. 22.3)

from frequent use and they are now in disorder. You need to put them back in order. In this case, ordering criteria are needed. A similar task is posed by organizing structural data to describe a reaction. From databases of known crystal structures (Sect. 13.9), particular crystal structures are sought in which an amino group is near a carbonyl group, as in the structure of **14.2**. Finally, they are put into a logical order (■ Fig. 14.15).

The systematic comparison of crystal structure data provides a preliminary information about structural molecular properties, for instance, about the preferred conformation (Sect. 16.4). The geometry of **noncovalent interactions** can also be evaluated in this way. The side chain of the amino acid histidine contains an imidazole ring with its two nitrogen atoms. In the neutral state, one of these nitrogen atoms is a hydrogen-bond acceptor and the other is a hydrogen-bond donor. There are hundreds of molecules with an imidazole ring in the database of low molecular weight crystal structures. In these structures, the imidazole ring actually has acceptor and donor interactions, usually with neighboring molecules. All of these structures are superimposed on one another based on their common imidazole ring (■ Fig. 14.16). It shows in which spatial direction the hydrogen-bonding partner of the imidazole nitrogen atoms are located. The task of estimating the possible interaction positions in the binding site of the protein for the functional groups of a ligand is undertaken in the course of *de novo* drug design (Chap. 20). In addition, this information is needed to compare the binding properties of molecules (Chap. 17) or to explore binding pockets for their preferred ligand binding sites (hot spot analysis). The **Isostar** database, assembled and maintained at the Cambridge Crystallographic Data Centre in England, contains many such contact geometries and spatial distributions.

14.8 Solutions to the Same Problem: Serine Proteases with Differing Folds Have Identical Function

It was shown in Sect. 14.4 that the amino acids that determine the folding and function of a protein occur in separate parts of the structure. However, for enzymes with the same function, Nature has found the same solution by using different spatial foldings.

The function and therapeutic significance of serine proteases will be discussed in more detail in Chap. 23. A set of three amino acids, the so-called catalytic triad, plays a key role in accelerating the hydrolysis of amide bonds by these enzymes. The two amino acids serine and histidine and one acidic amino acid residue, such as aspartic or glutamic acid, are found in a characteristic spatial orientation. They are defined by the narrow limits imposed by the reaction geometry required for nucleophilic addition (Sects. 14.7 and 23.2). Their composition is ideal for the cleavage of amide bonds.

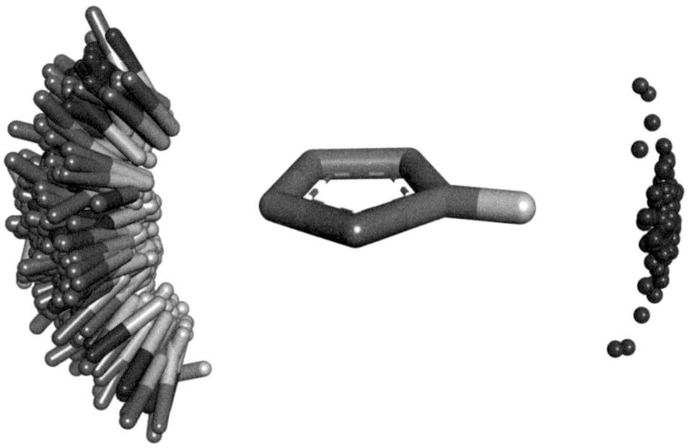

■ **Fig. 14.16** The crystal packing of low molecular weight compounds affords an overview of possible interaction geometries of hydrogen-bond donors (*left*) and acceptors (*right*) around the nitrogen atoms of an imidazole ring. Accordingly, all structures with an imidazole ring were sought in which at least one of the two nitrogen atoms participates in a hydrogen bond. The superposition of the structures shows where the positions of the interacting partners can be expected. (► https://sn.pub/G6cGKS)

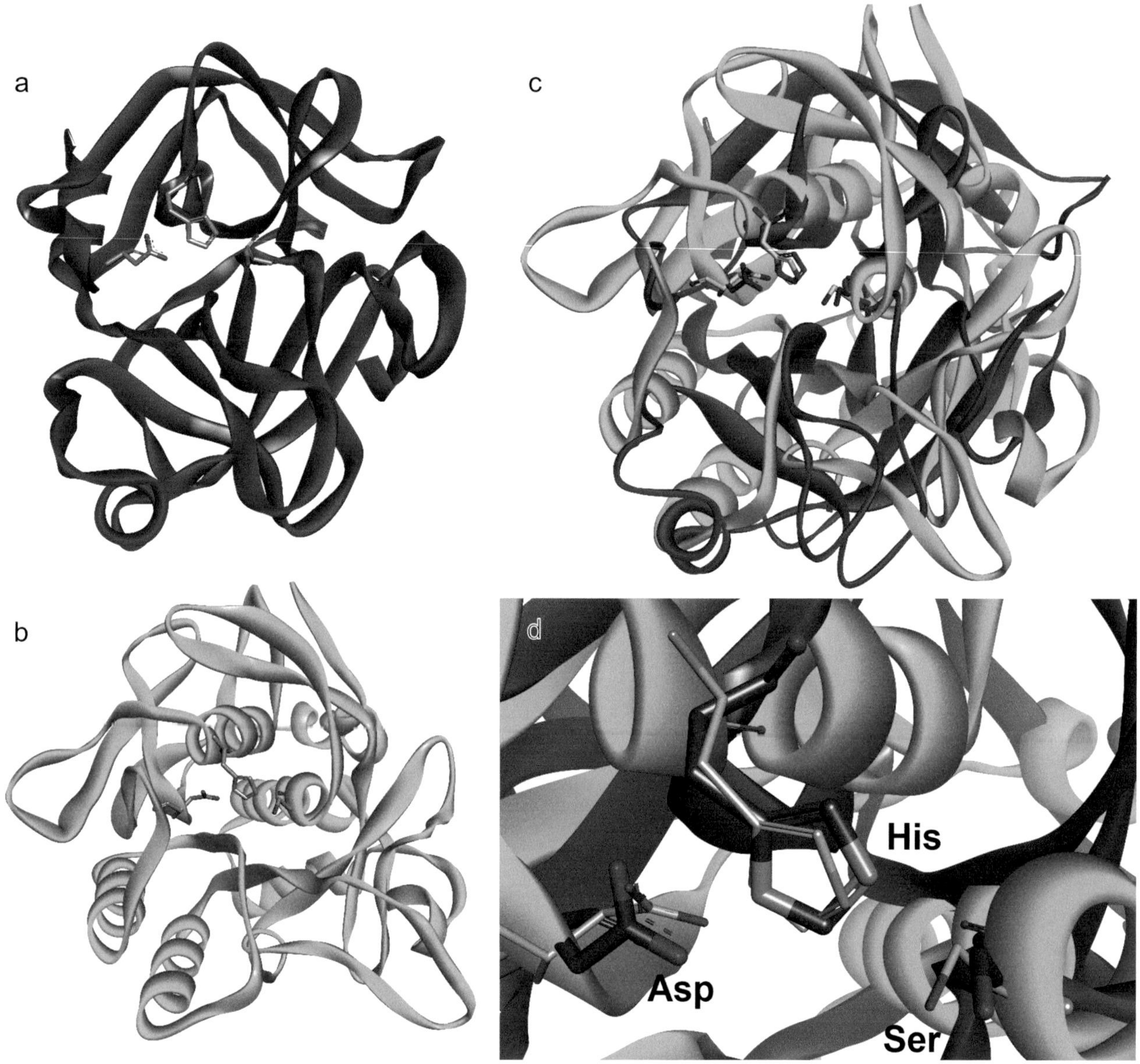

Fig. 14.17 Trypsin (**a**, *red*) and subtilisin (**b**, *green*) are serine proteases. They have the same catalytic triad of serine, histidine, and aspartic acid. These function-determining amino acids are, however, placed upon entirely different folding patterns In (**c**), the course of the chain of both proteins is superimposed upon one another. Despite this, the side chains of the amino acids of the catalytic triad are in the same spatial position (**d**). The course of the polymer chains are shown with *colored ribbons* that represent the spatial orientation of side chains of the three catalytic amino acids. (► https://sn.pub/9m77Ot)

The enzyme trypsin is composed of two barrel-like subunits (Fig. 14.17a). The catalytic site is located at the interface between these two subunits. Subtilisin is another serine protease that belongs to the class of open-pleated sheet structures. The catalytic triad occurs in a loop region at the edge of the folded sheet (Fig. 14.17b). If the amino acids involved in catalysis are removed from the protein and superimposed in space, the identical geometry of the triad becomes apparent. In addition to the enzymes mentioned, this catalytic triad is also found in lipases and esterases (Sect. 23.7), which also cleave peptide or ester bonds. Although they show divergent scaffold folding, the geometric orientation of their triads is once again identical.

14.9 DNA as a Target Structure of Drugs

Our genetic information is encoded in the DNA molecule. It is a thread-like molecule about 20 Å in diameter and up to 2 m long when stretched. It has the shape of a double helix (Fig. 14.18). On the outside, a polymer chain of sugar and phosphate building blocks wraps around the base pairs like a banister. The latter bases form a complementary pair at each step. The base pairs are linked together by a specific hydrogen-bonding pattern. A purine base (adenine A and guanine G) always interacts with a pyrimidine base (cytosine C and thymine T; in the related RNA molecule, thymine is replaced by uracil U; ◘ Figs. 14.18 and 14.19). The resulting spiral staircase has a pitch of 34 Å and reaches a full turn after ten steps. The two intertwined polymer strands form two grooves of different sizes on their surfaces (◘ Fig. 14.18). When the DNA is examined from the side along the steps at the major and minor grooves, the characteristics of the base pairs become visible. In the minor groove, there are three functionalities that determine the interactions with other molecules. In the major groove there are four such functionalities. Interestingly, the pattern that is read in the major groove is unambiguous because of the exposed properties for each base pair on a step. In the minor groove, only the overall difference between AT/TA or GC/CG can be distinguished (◘ Fig. 14.19), but not their spatial sequential topology.

The base pairs on three neighboring steps code for one amino acid each (see Sect. 32.7). In order to read this information unambiguously from the DNA, proteins that regulate gene expression (so-called transcription factors) read the information from the major groove, from the side (see Sect. 28.2). Only there is it possible to read the prescribed code (AT, TA, GC, CG) unambiguously. The DNA molecule is highly charged due to the many outward-facing phosphate groups. This charge is neutralized by the formation of ion pairs, mostly with

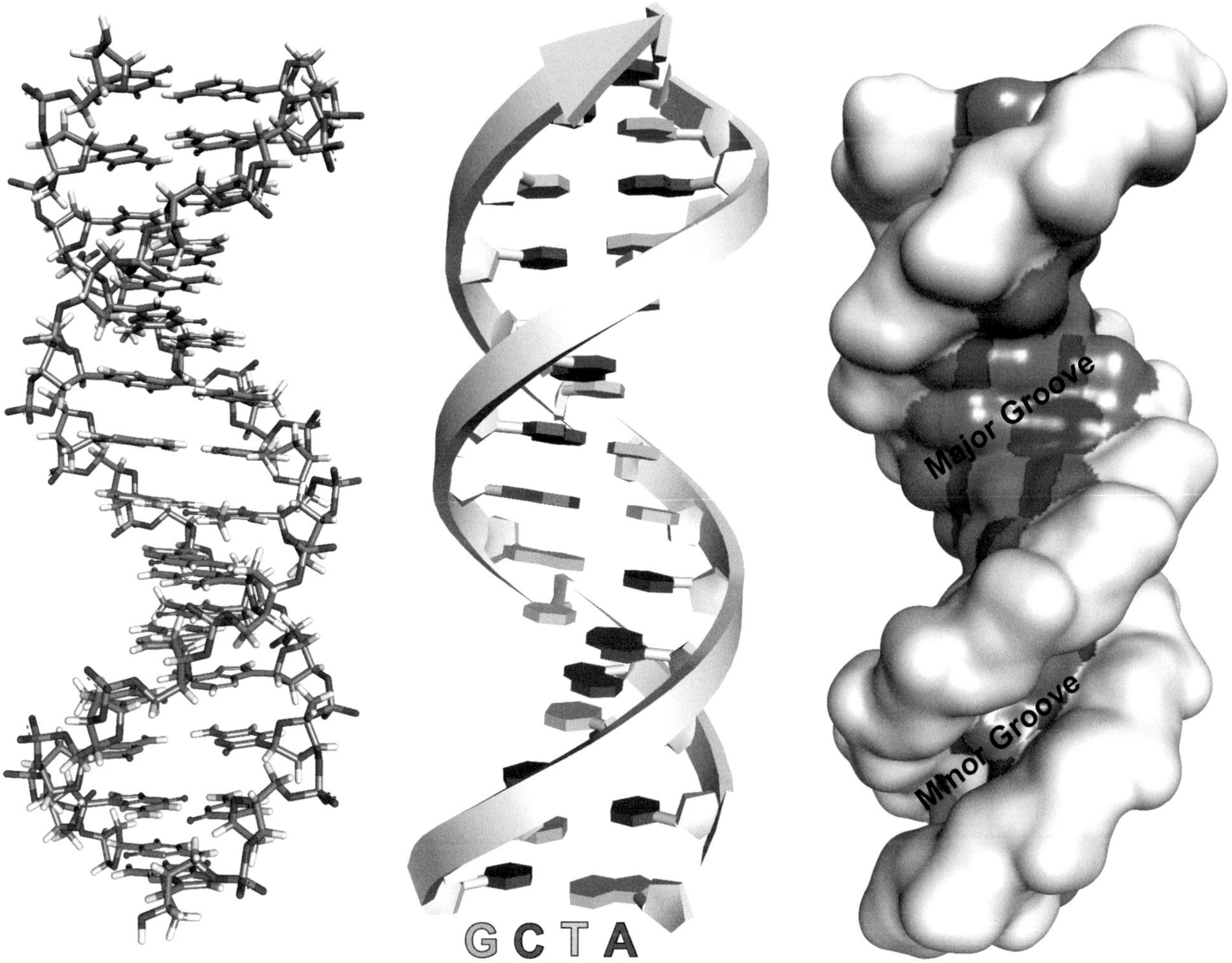

◘ **Fig. 14.18** The DNA molecule is made up of individual stair steps. One pair of bases forms each step. The sugar phosphate chain suspends the steps like a double banister. It forms a major and a minor groove on the surface. *Left* A segment of DNA with 14 base pairs. *Center* A schematic representation with the sugar–phosphate backbone as a *gray arrow*, guanine (*light green*), cytosine (*violet*), thymine (*cyan*), and adenine (*red*). *Right* A model of a DNA surface in which the size difference between the minor and major grooves is emphasized. The individual bases align according to their interaction properties (*blue* H-bond donor, *red* H-bond acceptor, *gray* hydrophobic contact). (► https://sn.pub/LGZoaa)

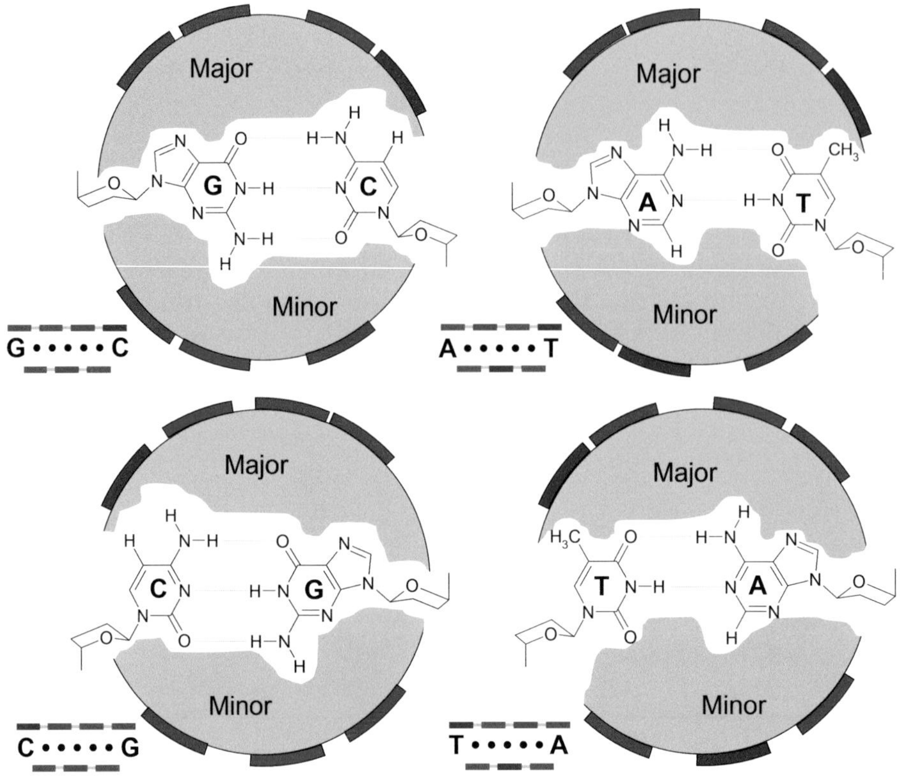

Fig. 14.19 The DNA base pairs of cytosine (C) with guanine (G) and thymine (T) with adenine (A) on the individual steps are formed by complementary hydrogen bonds. Each base carries a sugar–phosphate group that is coupled with the polymer chain. It affords a double-helical construction with a minor (*green*) and major (*yellow*) groove (cf. Fig. 14.18). If viewed from parallel to the steps, four groups can be seen in the major groove that possess either hydrogen-bond donor (*blue*), hydrogen-bond acceptor (*red*), or hydrophobic properties (*gray*). Three such groups are aligned in the minor groove. If an attempt is made to read the interaction pattern from this side, an GC or CG pair and an AT or TA pair will be recognized as identical. Here the orientation of the interaction pattern cannot be distinguished. In the major groove, on the other hand, the pattern of exposed interaction is unambiguous. Therefore, proteins read information about the DNA from the major groove

14

magnesium ions. Because of its important role in transmitting genetic information, several important drugs act on DNA. Two examples are briefly mentioned here. Cisplatin **14.4** is a reactive metal complex that can covalently bind to the nitrogen atoms of two nucleobases on two adjacent steps of the DNA by replacing both chlorine substituents (Fig. 14.20). This **cross-linking** distorts the DNA in such a way that the sequence information is no longer readable. Cisplatin and analogues such as carboplatin are used as potent chemotherapeutic agents in cancer therapy. Daunorubicin **14.5** is a representative with a slightly different mechanism of action, but it also prevents the reading of DNA base pairs. The planar molecular moiety of **14.5** slips largely between two adjacent base pairs, causing a structural distortion of the DNA (**intercalation**). This intravenously administered cytostatic drug is used as a combination therapy for the treatment of acute leukemias. Many natural products also use this intercalation mechanism for their antibacterial activity spectrum. Other pharmaceutical research approaches attempt to use segments of DNA itself for therapy. Such modified oligonucleotide therapeutics are discussed in Sect. 32.4.

14.10 Synopsis

- Every third bond in the polymer chain of a protein is an amide bond. It is the fundamental building block in the protein backbone and the mutual spatial arrangement of the sequential planar amide bonds determines the overall architecture of a protein.
- Typical arrangements involving the amide NH and C=O groups in hydrogen bonds lead to α-helical and β-pleated sheet structures. Reversal of the polymer chain in space is achieved in turns that can adopt a variety of distinct geometries.
- Helices, sheets, and turns, the secondary structural elements, assemble into motifs and domains to form the tertiary and quaternary structure of proteins.
- The function of a protein is not necessarily coupled to a particular folding pattern; however, the catalytic and ligand-functional sites within a folding class are found at the same position.
- Nature separates fold-stabilizing residues from function-carrying amino acids to keep the individual steps of the dual optimization problem separated.
- Proteases recognize peptide sequences specifically via the binding in well-tailored pockets on both sides of the cleavage site.
- Peptide libraries with an attached photometric or fluorescent label that can be cleaved by the protease reaction help to elucidate the substrate profile of different proteases.
- Structural arrangements of molecular portions found in multiple crystal structures can be arranged sequentially in a kinematic order to provide an idea of a dynamic process.
- The spatial arrangement of amino acid residues exerting a particular chemical transformation is highly

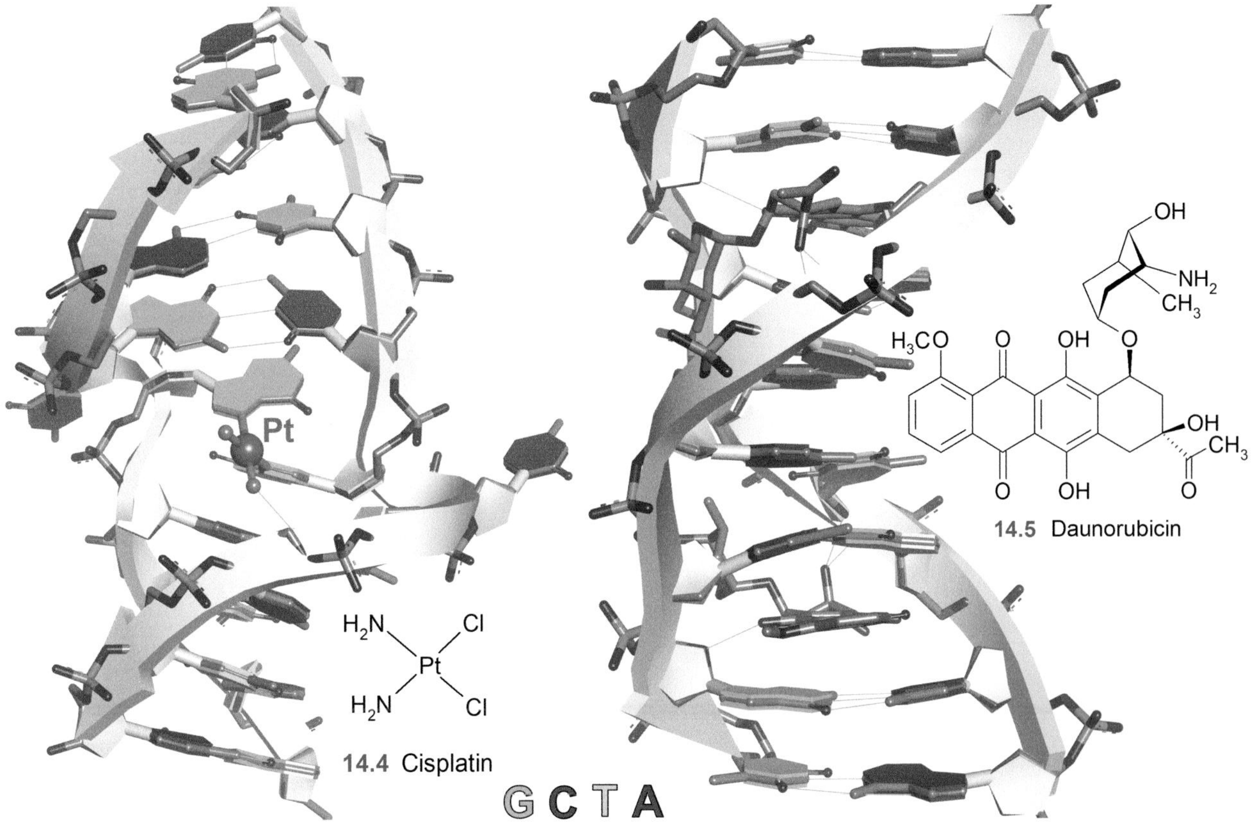

Fig. 14.20 Crystal structure of an oligomeric DNA segment after a reaction with cisplatin **14.4** (*left*) or intercalation with daunorubicin **14.5** (*right*). In both cases, the DNA molecule is severely distorted and the genetic information on the DNA cannot be read for cell division. Cisplatin reacts with the nitrogen atoms of two nucleobases (here guanine) of the DNA on neighboring steps with substitution of both chlorine atoms in a chemical reaction. With its planar tetracyclic ring system, daunorubicin intercalates between two neighboring base pairs by spreading the DNA along the helix axis. The compound's amino sugar accommodates in the DNA minor groove. (▸ https://sn.pub/PMGNV7)

conserved and can reside on protein architectures with similar geometry that are constructed from deviating folds.

- The DNA molecule encodes our genetic information and forms a double helix of two opposing strands wrapped like a handrail by sugar–phosphate polymer chains. They carry the complementary base pairs on the stacked steps of the double helix. The individual base pairs form a typical H-bonding pattern in the center of the helix.
- The typical H-bonding pattern of bases allows each single strand of DNA to be complementary to the second strand. When DNA is amplified, the double-stranded DNA can be completed from the single-stranded DNA. A small and a large groove are formed between the sugar–phosphate backbone. In the large groove, the coding of the base pairs on each step can be read from the side. By complexing bases on adjacent steps with metal ions or by intercalating planar agents between these steps, the DNA is geometrically distorted which will prevent reading of the genetic information, which consequently inhibits cell growth.

Bibliography and further reading

General Literature

C. Branden and J. Tooze, Introduction to Protein Structure, 2nd Ed., Garland Publ. Inc., New York (1999)

G. A. Jeffrey and W. Saenger, Hydrogen Bonding in Biological Structures, Springer Verlag, Berlin (1991)

H. B. Bürgi and J. D. Dunitz (Eds.), Structure Correlation, Vol. 1 and 2, VCH, Weinheim (1994)

G. E. Schulz and R. H. Schirmer, Principles of Protein Structure, Springer Edition, New York (1978)

Special Literature

I. Schechter, A. P. Berger et al., On the size of the active site in proteases, Biochem. Biophys. Res. Commun., **27**,157–162 (1967)

P. I. Lario and A. Vrielink, Atomic Resolution Density Maps Reveal Secondary Structure Dependent Differences in Electronic Distribution, J. Am. Chem. Soc., **125**, 12787–12794 (2003)

F. A. Allen, O. Kennard and R. Taylor, Systematic Analysis of Structural Data as a Research Technique in Organic Chemistry, Acc. Chem. Res., **16**, 146–153 (1983)

C. A. Orengo, D. T. Jones and J. M. Thornton, Protein Superfamilies and Domain Superfolds, Nature, **372**, 631–634 (1994)

K. Vyas, H. Monahar and K. Venkatesan, Thermally Induced O to N Acyl Migration in Salicylamides. Thermal Motion Analysis of the Reactants, J. Phys. Chem., **94**, 6069–6073 (1990)

G. Klebe, The Use of Composite Crystal-Field Environments in Molecular Recognition and the *De Novo* Design of Protein Ligands, J. Mol. Biol., **237**, 212–235 (1994)

W. J. L. Wood, A. W. Patterson, H. Tsuruoka, R. K. Jain, and J. A. Ellman, Substrate Activity Screening: A Fragment-Based Method for the Rapid Identification of Nonpeptidic Protease Inhibitors, J. Am. Chem. Soc., **127**, 15521–15527 (2005)

O. Koch and G. Klebe, Turns Revisited: A Uniform and Comprehensive Classification of Normal, Open, and Reverse Turn Families Minimizing Unassigned Random Chain Portions, Proteins: Struct., Funct. Bioinform., **74**, 353–367 (2008)

PDB Database: http://www.rcsb.org/pdb/home/home.do (Last accessed Nov. 17, 2024)

CSD Database: http://www.ccdc.cam.ac.uk/products/csd/ (Last accessed Nov. 17, 2024)

IsoStar Database: https://www.ccdc.cam.ac.uk/solutions/software/isostar/ (Last accessed Nov. 17, 2024)

Mogul, assessment of molecular conformations: https://www.ccdc.cam.ac.uk/solutions/software/mogul/ (Last accessed Nov. 17, 2024)

Molecular Modeling

Contents

G. Klebe, *Drug Design*, https://doi.org/10.1007/978-3-662-68998-1_15

Molecules are most commonly communicated in chemistry as two-dimensional molecular representations that show their topology. This formalism is well established and has proven enormously fruitful. The ability of a chemist to quickly grasp and intellectually process such structures should not be underestimated. However, the notation has its limitations. In particular, the three-dimensional shape of a molecule is not immediately apparent from the chemical formula. However, geometry is of great importance for the physical, chemical, and biological properties of drugs and thus for drug design. Therefore, 3D structure determination (Chap. 13) is of particular importance. Whenever possible, the experimentally determined 3D structure of the compound and the target protein is used to explain the mode of action and the structure–activity relationship. Notwithstanding this, there is often the problem that these structures are not always available. In these cases, the explanation of experimental results is limited to the structural consideration of generated models. In some cases where it is even not possible to determine an experimental structure, an attempt is made to calculate a model. Often the properties of molecules can be estimated more easily by calculations than by time-consuming measurements. In many cases, molecules can adopt several stable geometries. Through extensive simulations, it is possible to calculate these geometries and derive ideas about the dynamic transformations between different states. Computational models are, therefore, a crucial link to correlate experimental data with design ideas. Computations also allow scanning the structural property space around a given geometry and assessing its relevance for a given molecular property. The computational methods available for drug design, the assumptions and approximations on which they are based, and what can be deduced from them for the development of a novel drug will be the subject of this chapter.

15

15.1 3D Structural Models as Well-Established Tools in Chemistry

Three-dimensional structure models have been used since Jacobus H. van't Hoff and Joseph Le Bel. Emil Fischer reported in his book *Aus meinem Leben* about a vacation in Italy:

> "In the previous winter 1890/91 I was busy with the task of clarifying the configuration of sugar, without entirely achieving my goal. Then the thought came to me in Bordighera that the decision about the configuration of pentose has to do with its relation to trioxyglutaric acid. Unfortunately for lack of a model I could not tell to what extent such acids are possible according to theory and I therefore posed the question to Baeyer. He picks up such things with great enthusiasm, and directly constructed carbon atoms from balls of bread and toothpicks. But after many attempts he gave the cause up, ostensibly because it was too hard. Later in Würzburg after considering good models at length, I managed to find the conclusive solution."

Linus Pauling was the first to propose the α-helix as a secondary structure in proteins.

> "The key to Linus's success was his reliance on the simple laws of structural chemistry. The α-helix had not been found by only staring at X-ray pictures. The essential trick, instead, was to ask which atoms like to sit next to each other. In place of pencil and paper, the main working tools for this work were a set of molecular models superficially resembling the toys of preschool children."

These were the words Nobel Laureate James Watson used to describe Pauling's approach in his book *The Double Helix*. Pauling's success was also based on a sound knowledge of theoretical chemistry. For example, Pauling knew that an amide bond is rigid and flat, whereas his competitors, William Bragg, Max Perutz, and John Kendrew, mistakenly believed that it was flexible. James Watson and Francis Crick followed the same course as Pauling in their search for the structure of DNA:

> "We could thus see no reason why we should not solve the DNA problem in the same way [as Pauling]. All we had to do was build a set of molecular models and begin to play—with luck the structure would be a spiral."

Based on this background, the achievement of Watson and Crick seems even more impressive. They were awarded the Nobel Prize in 1962 for the elucidation of the double-helix structure of DNA. This example should underscore the importance of models in science. To end with a word from Francis Crick: "A good model is worth its weight in gold."

15.2 Strategies in Molecular Modeling

In contrast to the 1950s and 1960s, today's computers have impressive graphical and computational capabilities. Correspondingly, programs for working with molecular models are available. The new field of **molecular modeling** has emerged. This term encompasses the representation and manipulation of realistic three-dimensional molecular structures along with the calculation of their physicochemical properties. The main methods used in molecular modeling are summarized in ◘ Table 15.1.

In principle, molecular modeling can be approached from two directions. One approach is to extrapolate the geometry and physicochemical properties from known experimental data to the structure under investigation.

The other approach is to try to obtain as accurate a computational prediction as possible by starting from first principles. Quantum chemical methods and force field calculations are part of this strategy. In practice, both approaches are used in parallel and are increasingly coupled. If relevant experimentally determined structures are available, it would be stupid not to use them for model building. On the other hand, quantum chemical and molecular mechanical approaches are widely applicable and provide reliable results. Most importantly, models can be changed gradually, one parameter at a time. Experiments usually do not allow this. Instead, experimental investigations typically change several parameters at the same time, often unavoidably.

The construction of a structural model is achieved in three steps:

- Generation of a starting model,
- Optimization and analysis, and
- Work with the model.

It is advisable to stay as close as possible to experimental structures when creating the initial starting model.

Table 15.1 Overview of the most important molecular modeling approaches in pharmaceutical research

Technique	Objective
Interactive computer graphics	Display of 3D structures
Modeling small molecules	3D Structure generation (CONCORD, CORINA)
	Molecular mechanics—force fields
	Molecular dynamics
	Quantum mechanical techniques
	Conformational analysis
	Calculation of physicochemical properties
Comparing molecules	Superimposition of molecules according to their similarity
	Volume comparisons
	3D-QSAR (e.g., CoMFA methods)
Protein modeling	Sequence comparisons
	Protein homology modeling
	Protein-folding simulations
Modeling of protein–ligand interactions	Binding constant calculations
	Ligand docking
Ligand design	Searches in 3D databases
	Structure-based ligand design
	de novo design
	Virtual screening

This can be done by consulting the crystal structure of a compound. The Cambridge Crystallographic Database, which stores experimentally determined crystal structures of small molecules, is searched and the geometry of the resulting hits that most closely resembles the query molecule is used. The next step is to optimize the molecule using a force field calculation.

There are also off-the-shelf programs available for generating starting models, which translate a 2D structural formula into a 3D spatial structure according to the principle of a molecular construction kit. These "**electronic molecular construction kits**" store lists of bond lengths and angles, as well as preferred fragment geometries, and build molecules according to a sophisticated set of rules. In fractions of a second, they determine the 3D spatial structure for the 2D structural formula. The first programs in this field were **CONCORD** by Robert Pearlman in Austin, Texas, USA, and **CORINA** by Johann Gasteiger and Jens Sadowski at the University of Erlangen, Germany. Both programs can be used to generate 3D structures of small molecules. In addition, CCDC's program **Mogul** provides a comprehensive source of information for comparing and validating a computed molecular model with experimental data. However, these programs cannot build the 3D structure of a protein. More sophisticated techniques are required to predict the spatial structure of a protein from its sequence (Sect. 20.6).

15.3 Knowledge-Based Approaches

Perhaps the most widely used techniques in molecular modeling are knowledge-based approaches. They attempt to use the vast amount of knowledge accumulated from experimentally determined molecular structures, crystal packing, protein structures, protein sequences, and structure–activity relationships of protein–ligand complexes, etc. to efficiently solve a given problem. Basically, the computer program mimics the approach that a conscientious scientist would also take. First, as much experimental data as possible are collected and analyzed. Important sources of information are the Cambridge Crystallographic Database, which contains more than 1.25 million crystal structures of small molecules, and the Protein Databank (PDB), which contains more than 227,000 protein and DNA structures. Physicochemical properties are also available in databases. For example, the Beilstein database of nearly 10 million chemical structures contains pK_a values for more than 20,000 compounds. For amino acid residues in proteins, the PKAD database has been developed (▶ http://compbio.clemson.edu/PKAD-2/). The challenge lies in the extraction of the necessary data for the question at hand from the enormous plethora of electronically available information. Furthermore, it must be considered that the

$$E = E_{bond\ length} + E_{bond\ angle} + E_{torsion} + E_{noncovalent}$$

$$E = \frac{1}{2} \sum_{bonds} K_b (b - b_0)^2$$

$$+ \frac{1}{2} \sum_{bond\ angles} K_\Theta (\Theta - \Theta_0)^2$$

$$+ \frac{1}{2} \sum_{torsion\ angles} K_\Phi (1 + \cos(n\Phi - \delta)^2$$

$$+ \sum_{non\text{-}bonded\ atom\ pairs} (A_{ij}/r_{ij}^{12} - C_{ij}/r_{ij}^6 + q_i q_j / \varepsilon r_{ij})$$

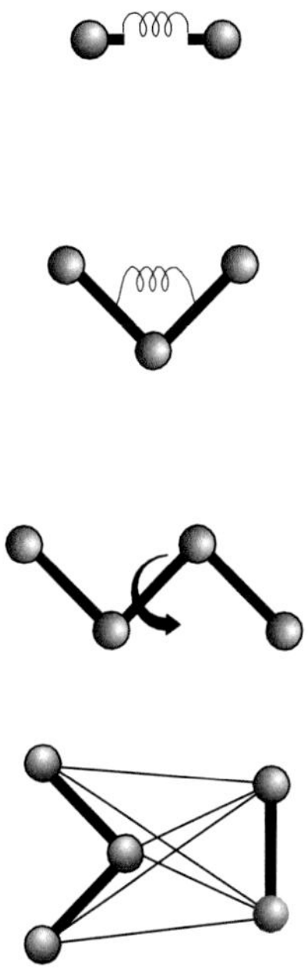

Fig. 15.1 E is the total energy of a molecule or complex of molecules. It is made up of several contributions. The first term describes the energy change when a chemical bond is stretched or compressed. In this example, it describes the so-called harmonic potential with the force constant K_b and the equilibrium bond length b_0 as a parameter. The second term describes the energy as a function of the bond angle Θ. Again, the harmonic potential is used with the force constants K_Θ and the equilibrium constant Θ_0. The third contribution describes the change in energy when the dihedral angle is changed, and the last term represents noncovalent interactions. The sum of three terms is used for this last contribution. The first term A_{ij} / r_{ij}^{12} is always positive and increases rapidly with decreasing distance. It describes the repulsion between atoms that come too close together. The contribution from $-C_{ij} / r_{ij}^6$ is always negative and tends to zero with increasing distance r_{ij}, but not as fast as the repulsive term. It describes attractive interactions, which are also called dispersion interactions. Other attractive interactions exist between polar molecules which are also proportional to $1/r_{ij}^6$ (for a description of the potentials see Sect. 18.12, Fig. 18.5). The last term $q_i q_j/\varepsilon r_{ij}$ describes the electrostatic interactions based on Coulomb's law, which works with a point charge model. The dielectric constant is ε. The noncovalent contribution to the total energy, without the electrostatic term, is called van der Waals energy

data come from different sources and could be partially erroneous or were measured and collected under barely comparable conditions.

The largest growth in electronically available data recently has occurred in the area of DNA sequences. Hundreds of genomes have been sequenced, and new ones are added weekly. The nearly endless number of sequences can only be conquered with intelligent searching protocols. Modeling of protein structures is now being conducted on a large scale with novel knowledge-based approaches using machine learning and artificial intelligence (Sect. 20.6). The generated models are collected in a database of computed structure models and contains more than a million models (▶ https://www.rcsb.org/news/6304ee57707ecd4f63b3d3db).

15.4 Force Field Methods

Force field methods, also known as molecular mechanics, are empirical techniques for calculating molecular geometries. The goal of a **force field calculation** is to determine an energetically favorable three-dimensional structure of a molecule or complex of molecules. The forces acting between the atoms are described in an analytical form, and appropriate parameters have to be assigned to the different terms. In principle, a large number of such analytical functions can be imagined. We will consider a simple form that has been used very often. It considers covalent and noncovalent forces. The central idea of molecular mechanics is the assumption that the bond lengths and angles take on values that are close to the standard values in molecules. Steric interactions, this means the repulsion of two atoms that are not directly bonded to each other, can cause some bond lengths and angles not to take their ideal values. These repulsive interactions are also known as van der Waals interactions. In the simplest case, the deviations from the ideal values are described by a parabolic potential (the so-called **"harmonic" potential**, which applies to the motion of a mass on a freely suspended spring). However, this form ignores the fact that very strong distortions lead to bond breaking. To better describe such strong distortions in the potential profile, a distance-dependent function with an exponential characteristic (e.g., **Morse potential**) is often used.

In 1946, three terms, **van der Waals interaction**, **bond stretching**, and **angular deformation**, were first proposed to be sufficient to calculate the structure and energy of molecules. At that time, however, performing such calculations was extremely difficult. It was not until the availability of computers increased that molecular mechanics calculations gained importance. In addition to the three terms originally proposed, a typical force field in use today contains at least one additional contribution that takes into account **rotations about the dihedral angles** (Fig. 15.1). In addition, many force fields use terms for **electrostatic interactions**. To do this, each atom must be assigned a partial charge. The sum of these charges gives the formal charge of the whole molecule. This is usually set to zero for uncharged particles.

Coulomb's law is used to describe the forces that occur between charges. This law states that the product of interacting charges is inversely related to the square of the distance between them, or, considering the potential, inversely related to the distance. The assignment of charges and the correct choice of the dielectric constant are critical to the correct treatment of electrostatic energy contributions. These values are in the denominator of Coulomb's law and can take values between $\varepsilon = 80$

for water and ε = 1 for vacuum. This dampens the electrostatic interactions in water very rapidly, while in a vacuum they tend to reach much further. Choosing the correct dielectric constant for force field calculations in proteins is very difficult. Many values between ε = 4 and ε = 20 have been tried. The constant is sometimes assumed to be environment dependent, so that larger values are chosen near the surface than for the interior of the protein. The van der Waals interactions are modeled by the Lennard–Jones potential. This interaction has an attractive term falling at a rate of $1/r^6$ and a repulsive term falling at a rate of $1/r^{12}$ (◘ Fig. 15.1). The result of the combination of these terms is a gradient that is very large near the atoms and approaches zero as the distance increases. In between, it passes through a potential energy minimum (◘ Fig. 18.5). In addition to A/r^6–C/r^{12} as the functional form, alternative distance dependencies with other potentials or exponential gradients have been used in force fields.

A force field is derived by calibration to experimental data or to the results of high-level quantum mechanical calculations. The 3D structures of small molecules and force constants derived from infrared and Raman spectroscopy are used. It is clear that different parameters must be used for a single bond between two carbon atoms than for a double bond. Therefore, several different atom types per element are defined in a force field. The crystal packing of small organic molecules can be consulted to parameterize nonbonded interactions. Amino acids and many functional groups of active compounds can exist in either a **protonated** or **deprotonated state** depending on the applied pH conditions (so-called **titratable groups**). The strength of the resulting interactions is strongly dependent on the charge state of the functional groups involved. The acidity or basicity of a given functional group is determined by its pK_a value. This indicates how easily a group accepts or releases a proton. This property, in turn, depends heavily on the partial charge that the group carries and what other charges are in the immediate vicinity of the group. Thus, the pK_a will change when a functional group is placed in a different environment. For example, carboxylic acids become more acidic when placed near a positive charge. On the other hand, their acidic nature will change if a partially negatively charged group is in the vicinity. This effect must be taken into account in a reliable force field calculation. An attempt can be made to predict the protonation state in protein–ligand complexes with such calculations. The contribution to the energy content of the complex is determined by evaluating all possible combinations of protonated states of titratable groups. In this way, the shift in pK_a values of functional groups can be estimated. To better describe the shift of charges in molecules, so-called **polarizable force fields** can be applied. They adjust the charge distribution to the local distances and take into account that charges are displaced or "polarized" in the molecule by attractive or repulsive effects. Certainly, these force fields give a more realistic picture. However, the disadvantage is that they are computationally much more demanding and require many more parameters to be adjusted.

The importance of water as a binding partner in the formation of protein–ligand complexes was emphasized in Sect. 4.6. The formation of a protein–ligand complex causes a change in the solvation conditions for the molecules involved. This has to be taken into account in the force field calculations. In order to do this, a force field is combined with estimates for the contribution of solvation. Approaches such as the **MM-PBSA** or **MM-GBSA** methods try to sum up these contributions over the local environment in a surface-dependent manner. Newer methods such as the **G**rid **I**nhomogeneous **S**olvation **T**heory (**GIST**) method can calculate thermodynamic contributions such as solvation enthalpy and entropy from the interaction contributions collected during a molecular dynamics simulation (Sect. 15.7). These values are then mapped onto a grid for subsequent graphical analysis.

The choice of a relevant initial starting geometry is important for any force field calculation. A force field calculation involves energy minimization. If one starts with an energetically unfavorable geometry, the force field will travel "downhill" to the next local minimum on the multidimensional energy surface (Sect. 16.2). If one starts with two different geometries, the structures obtained at the end of the minimization can be different, depending on which local minimum is reached. Many molecules, especially protein–ligand complexes, can adopt many energetically favorable conformations. It is, therefore, recommended to perform multiple force field calculations starting from different geometries. Molecular dynamics simulations, discussed below, also provide a solution to this problem of getting trapped in local minima.

15.5 Quantum Chemical Methods

In quantum mechanical approaches, the electronic structure of molecules is calculated using the **Schrödinger equation**. However, its mathematically closed solution is only possible for simple cases such as the hydrogen atom or the molecular ion of hydrogen, H_2^+. For molecules with more than one electron, approximate methods must be used to solve the quantum mechanical "many-body problem." The most commonly used approximation is the **Hartree–Fock method**. Here the many-body problem is reduced to several single-body problems. The sum of electron–electron interactions within a molecule is replaced by an effective field, which can be iteratively refined and optimized. This is where the popular name **SCF (self-consistent field)** comes from. In this model, each electron "sees," in addition to the potential of the nuclei, an averaged potential of the remaining electrons.

The state of each electron in a molecule is, thus, described by a **single-particle function** called the **atomic orbital (AO)** or, in a molecule, the **molecular orbital (MO)**. The wave function of the entire molecule is the antisymmetric product of the many orbitals considered. The Hartree–Fock equation is then obtained under the condition that the optimally chosen orbitals lead to a minimum overall energy. The main shortcoming of the Hartree–Fock approach, namely the **neglect of the electron correlation**, can be corrected by more sophisticated methods, which, however, increase the computational time considerably.

Quantum mechanical ***ab initio*** **calculations** allow the calculation of the molecular structure and electron density distribution as well as molecular properties without the assumptions necessary for force field calculations. In many cases, it is difficult to make a *priori* predictions based on the hybridization state of the atoms. For example, in the case of amines and sulfonamides, it is often impossible to predict whether the atoms bonded to the nitrogen in these compounds are all in the same plane or whether the nitrogen deviates with a pyramidal environment. In a force field calculation, this must be specified at the outset of the calculation by assigning which atom type to which atom in a molecule (i.e., in the above case, whether the nitrogen should be in a planar or pyramidal local geometry). Of course, if the wrong atom type is chosen, the resulting structure will be meaningless. As an advantage, quantum mechanical calculations do not require such assumptions.

Most currently applied force fields use a point-charge model to describe the electrostatic interactions. One option to derive the atomic charges is to calculate the electrostatic potential of a small molecule containing the group of interest using quantum mechanical methods. A set of partial charges is then assigned to the different nuclei in order to reproduce the quantum mechanical potential as accurately as possible. These charges can then be transferred to force field calculations for use in a large system.

15

Another important application of quantum mechanical calculations in drug design is the calculation of conformational energies of small molecules to calibrate force fields. The force fields developed for proteins and peptides are based on conformational energies calculated quantum-mechanically for small peptides.

In contrast to force field methods, quantum mechanical techniques are able to take into account the polarization of the electron density caused by the influence of neighboring groups. For example, the amide bond dipoles in an α-helix are all oriented in the same direction, so they add up to a significant total dipole moment (Sect. 14.2). As a consequence, such large and enhanced dipoles can polarize other groups located at the end of the helix. In fact, induced dipoles are incompletely described by standard force field methods (see polarizable force fields). This is not a problem for quantum mechanical methods. Another important area of application is chemical reactions, for which force fields are hardly parameterized, except for a few special cases. Here, quantum mechanical methods are the only option for a reliable theoretical description.

Quantum mechanical methods are considerably more elaborate than force field methods. The most accurate methods, which also consume the most computational time, are the so-called ***ab initio*** **methods**. However, these techniques quickly reach their limits for very large systems. Therefore, other less computationally intensive methods have been developed. In these so-called **semiempirical methods**, certain integrals, the determination of which represents the rate-determining step in *ab initio* methods, are replaced by adequate approximations that can be computed quickly. The resulting drastic reduction in computational time, at the expense of accuracy, allows the routine application of semiempirical calculations to active molecules and proteins. **Density functional theory** is another faster *ab initio* technique. In this method, the position-dependent electron density distribution is calculated in the ground state for a many-body system, avoiding the complete solution of the Schrödinger equation for a many-body system. All interesting properties are then derived from the electron density. For large protein–ligand systems, techniques have been developed that treat the interesting regions, such as the binding site or the catalytic reaction center, quantum mechanically. The surrounding regions are approximated with a faster force field method (**QM/MM methods**).

Recently, there has been a lot of discussion about new methods using **artificial intelligence** and **machine learning**. This is the attempt to transfer the way humans learn and think to computers. The idea is to give computer algorithms some intelligence. The goal is to use learning algorithms that allow the computer to find answers and solve problems on its own, without having to develop a new special program for each case. **Neural networks** are an important tool in this area. In 2024, the Nobel Prize in Physics was jointly awarded to John J. Hopfield (Princeton Univ., New Jersey, USA) and Geoffrey E. Hinton (Univ. of Toronto, Canada) for their fundamental discoveries and inventions that enable machine learning with artificial neural networks. Such methods are especially powerful for independent data analysis. Today's computers have the storage capacity and speed to crunch through vast amounts of data in a very short time and "remember" it all in a way that no human brain could capture, correlate, and evaluate in the same amount of time. In the field of drug development, such methods are not new. Data analysis methods and procedures have always been used to identify correlations in often high-dimensional data spaces. As the data have become larger and more complex, the algorithms have become better, and the computers have become faster. We will return to these quantitative assessments of structure–property correlations several times in Chaps. 18, 19, 20. Despite the excitement of the ever-increasing flood of data to be evaluated, it is important to remember that every evaluation, and ultimately every insight, is already contained in the data. If the data are not reliable, meaning that if they are too "noisy" due to bad and erroneous information, even the best machine learning or artificial intelligence algorithm will hardly bring any new insights to light.

But what can artificial intelligence contribute to the use of quantum chemistry in drug design? As mentioned above, quantum chemistry is characterized by high accuracy. But its computational cost is still prohibitive for many problems. Today, we are increasingly taking the approach of breaking molecules into small building blocks and performing sophisticated quantum chemical calculations on the resulting building blocks to determine their energy and geometry. When a large dataset of molecules is processed in this way and care is taken to ensure that the building blocks of interest are repeatedly embedded in different molecular environments, it is possible to generate a dataset with inherently redundant information. From such a dataset, one can then use artificial intelligence to infer parameters for a given force field. Typically, **neural networks** are used for this purpose. Force fields obtained in this way represent a real advance. They approach the accuracy of quantum chemical calculations, but can be computed as quickly as typical empirical force fields.

15.6 Computing and Analyzing Molecular Properties

The result of a molecular mechanics or quantum chemical calculation is initially a set of atomic coordinates that define the three-dimensional shape of the molecule. What can be done with this? An important application of the calculations is the determination of conformational energies: this is the relative energy of one molecular conformation compared to another (Sect. 16.1).

Two other molecular properties can be calculated: the shape and size of a molecule, along with its electronic characteristics. All current graphics programs have several ways of displaying such properties with the spatial structure of molecules. The most important ones are summarized in ◘ Fig. 15.2.

The most commonly used representation is a **line or stick representation** (Dreiding models), sometimes atoms are displayed as small spheres. Usually a color coding is used to represent the atoms: nitrogen is blue, oxygen is red, sulfur is yellow, fluorine is turquoise, chlorine is green, bromine is brown, and iodine is purple. Hydrogen atoms are shown in white, but are usually omitted for clarity. Carbon atoms are usually shown in black or gray. In most figures in this book, carbon atoms that belong to the protein are shown in orange, and carbon atoms that belong to the ligand are shown in gray, but sometimes in a different color is applied to distinguish them in different molecules. Another display option is the **space-filling model**, which shows van der Waals surfaces. In this representation, each atomic nucleus is represented by a sphere whose size corresponds to the van der Waals radius. Values for these radii are derived from crystal packing or from very accurate *ab initio* calculations. Such representations are also known as **CPK models** (named after the scientists **C**orey, **P**auling, and **K**oltun).

There are other ways to represent surfaces (◘ Fig. 15.3). The **solvent-accessible surface** has proven particularly valuable for proteins. The most commonly used way to display a protein in this book is a transparent opaque white surface. The van der Waals surfaces in ◘ Fig. 15.3 (*left*) give the impression that there is a gap at the location marked by the arrow. However, this gap is so narrow that no other atom can fit into it. Therefore, the solvent-accessible surface (◘ Fig. 15.3, *center*) is less misleading. It is created by rolling a sphere with a radius of 1.4 Å, the size of a water molecule, over the surface of the studied molecule. This surface appears much smoother. The depressions that are still present mean that small molecules—at least one water molecule—can actually fit in there. The **Lee–Richards surface** is less commonly used, but very helpful. It is chosen so that ligand atoms that come into contact with atoms of the protein under investigation will lie directly on this surface (◘ Fig. 15.3, *right*).

The surface can also be colored. For example, each atom type can be assigned a color, and then the color of the next closest atom is used in that part of the surface. Similar representations where the surface of the molecule is colored according to other properties, such as electrostatic or hydrophobic potential, are very instructive and often used.

15.7 Molecular Dynamics: Simulation of Molecular Motion

None of the processes we are interested in take place at 0 K, but rather at body temperature, which is about 310 K. It is, therefore, obvious that not only the potential energy but also the kinetic energy must be considered. Molecules also move at room temperature, which is close to body temperature (about 295 K). They diffuse and change shape by adopting different conformations. The flexibility and adaptability of both binding partners play an important role in protein–ligand interactions. A prerequisite for protein binding is that the ligand can adopt a conformation with a shape that fits into a protein–binding pocket. On the other hand, the protein is flexible to some extent. For example, side chains on the surface can adopt different conformations or entire domains can move relative to one another. The mutual adaptation of protein and ligand shapes plays an important role in the formation of protein–ligand complexes.

Molecular dynamics simulations (**MD**) are a theoretical method for describing these effects. Molecular dynamics simulations follow the motion of atoms and molecules under the influence of a selected force field. It

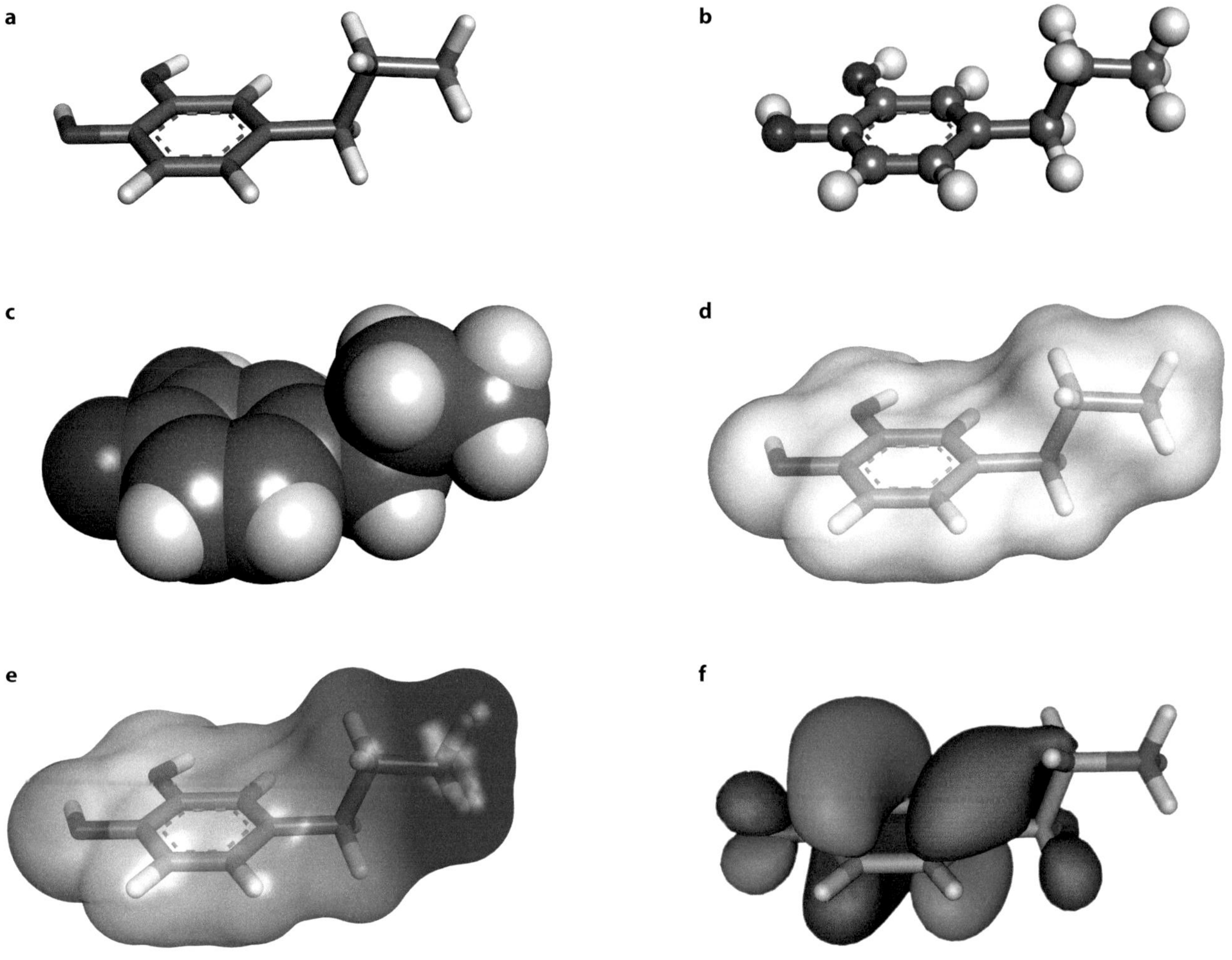

Fig. 15.2 Different computer graphics representations of dopamine (Sect. 1.4, formula **1.13**). Carbon atoms are colored *gray*, hydrogen atoms are *white*, nitrogen atoms are *blue*, and oxygen atoms are *red*. (**a**) Dreiding models. (**b**) Ball-and-stick models. (**c**) Space-filling models (CPK representation). (**d**) Solvent-accessible surface. (**e**) Electrostatic potential projected on the surface (positively charged areas are *blue*, negatively charged areas are *red*). (**f**) Highest-occupied molecular orbitals (HOMO), calculated for the uncharged dopamine molecule. The *blue* or *red* areas of the wave function indicate a different sign

15

is assumed that the interactions between particles obey the laws of classical mechanics. For this purpose, the **Newtonian equations of motion** are solved in parallel and stepwise for all particles simultaneously. It is usually assumed that the force between two particles is not affected by the individual spatial positions of the other particles.

In practice, a starting geometry is generated first (Fig. 15.4). If an experimentally determined structure, such as the crystal structure of a protein–ligand complex, is available, it is used as the starting point. To account for the surrounding water shell, the complex is immersed in a "**water bath**," meaning a large number of water molecules surround it. Furthermore, a sufficient number of ions is added to keep the whole system in an electrically neutral state. To avoid boundary effects on the "walls," a trick called "**periodic boundary conditions**" is applied to the water bath. When the simulated protein complex approaches such a wall and wants to leave the water bath, the computer will treat the process as if the complex had reentered the water bath from the opposite side. Formally, this eliminates the boundary regions of the water bath.

At the beginning of the actual simulation, each atom is assigned a **random starting velocity** with an arbitrary orientation. The velocities are chosen so that, on average, they correspond to the desired temperature (taken from a Boltzmann distribution). Then all forces from all surrounding atoms acting on a given atom are calculated. At fixed time intervals, the next position is calculated using Newton's equations of motion, and so on. The step size is typically one femtosecond (1 fs = 10^{-15} s). This small step size is necessary because many extremely fast processes occur at the molecular level. The evolution of the motion is followed for several nano- to microseconds and visualized in the form of a **trajectory**. Ten nanoseconds are sufficient to follow the movement of side chains and sometimes smaller movements of protein domains.

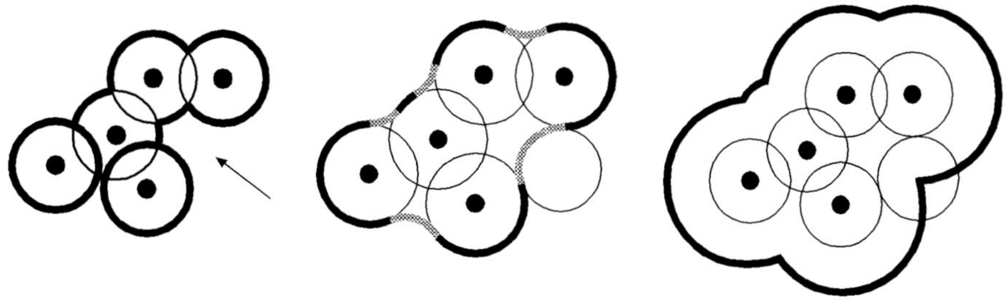

Fig. 15.3 Definitions of molecular surfaces. The van der Waals surface is shown on the *left*. The *arrow* marks a location where there is an irrelevant gap that is too small to accommodate even a single water molecule. *Center* Solvent-accessible surface created by rolling a water molecule across the surface. *Right* The Lee–Richards surface shows putative contact positions of atoms that will lie directly on the surface of the molecule being studied

However, this is not sufficient to describe conformational changes of protein domains or the diffusion of a drug molecule into the binding pocket. This requires significantly longer simulation times in the microsecond range. The folding of a protein is also difficult to follow with this technique. The time required for protein folding on the real time scale is between 20 ms and 1 h. The calculation of one time step (1 fs) still requires seconds of computing time even on the fastest computers. However, new algorithms and computers with more specialized architectures are being developed that will make such simulations possible in the foreseeable future.

Another important application of MD simulation is the **calculation of binding affinities**. In principle, the Gibbs free energy of binding ΔG can be calculated for a given system. Large systems such as protein–ligand complexes have many molecular degrees of freedom, e.g., groups can rotate around bonds and the molecules change shape in the process. Due to **Brownian molecular motion**, such a system is constantly in motion and can take on many different geometries, or as we say, configurations, over time. The frequency with which a particular configuration is assumed, on average over time, depends on its energy content. In low energy states, we encounter the system often, while in higher energy states we encounter it rather rarely. From the point of view of statistical thermodynamics, one now collects the energy contributions of all the configurations that the system assumes over time along the trajectory and summarizes them in the so-called **partition function**. One automatically obtains information about the distribution of the system over the many energetic states. This allows conclusions to be drawn about the **entropy content** of the system; this is how the energy is distributed over the degrees of freedom of the system. In the context of protein–ligand interactions, differences in free energies between protein complexes formed with different ligands are of particular interest. From this, it is possible to estimate which ligand is likely to have better binding properties compared to an alternative ligand. This is of great importance for the design of new drug candidates prior to their synthesis. Methods have been developed to **structurally morph one ligand into another** during the trajectory calculation. For example, a methyl group is transformed into a chlorine atom, and the force field parameters of a methyl group are gradually transformed into those of a chlorine atom during the course of the calculation. These free energy calculations provide an estimate of how much the binding mode and thermodynamic binding profile change during the transition from ligand A to ligand B. Meanwhile, methodological concepts have also been developed to completely eliminate certain groups at a given scaffold position or to create new groups from scratch at these positions. Such calculations are still quite computationally intensive, so they are used to design individual synthesis candidates. For broad screening of many thousands of compounds with docking methods (virtual screening, Sect. 7.6), where very large amounts of data have to be evaluated, simple empirical energy functions are still used to estimate binding affinities.

In Sect. 15.4, it was mentioned that the **search for local and global minima** of a molecule or a molecular complex can be approached with molecular dynamics simulations. Many approaches have been developed to speed up such calculations. One of them is **metadynamics**. It optimizes the sampling of the potential surface for minima by cleverly distorting the individual potentials during the calculation. This allows the simulation to leave local minima quickly. It was once casually expressed as follows: the method fills in potential minima during the simulation with previously calculated "computational sand." This "sand" is derived from arrangements that the simulation had already "visited" previously on the energy surface. In this way, the metadynamics bypasses arrangements that are visited most often during the course of a calculation. Thus, it is able to search a larger area of the energy surface of the system under investigation in a much shorter time.

Umbrella sampling is used to simulate changes in a molecular system along a predefined coordinate, such as the path of a ligand into the binding pocket of a protein. The system is forced along a path of interest by perturbing certain potentials. MD simulations are performed along this path in several time windows. The

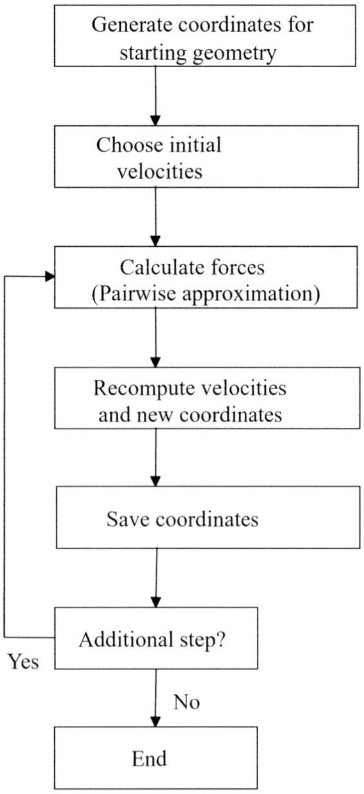

Fig. 15.4 Schematic of a molecular dynamics simulation. The starting geometry is either an experimentally determined structure or a geometry that was optimized with a force field. Randomly, each atom is assigned an appropriate starting velocity taken from a Boltzmann distribution. Then Newtonian equations of motion are stepwise solved beginning with these starting conditions and subsequently the coordinates are periodically saved

15

calculated properties in each time window are compiled and correlated along the entire path. This gives a rough idea of how and with what energy profile a ligand might bind to a protein.

15.8 Dynamics of a Flexible Protein in Water

Finally, an example of the application of molecular dynamics simulations to follow the motion of a molecule in solution is presented. A protein–ligand complex can be used to investigate which parts of a protein–binding pocket or a ligand in the complex remain rigid and which are flexible, and whether the shape of a binding pocket changes with time.

The enzyme aldose reductase has been shown to be a very flexible protein. It is able to adapt its binding pocket to the shape of a complexed ligand in many different ways. This property is related to the biological function of this protein. It reduces a very wide range of aldehyde substrates. Its exact function and role as a target structure for drug therapy is discussed in Sect. 27.4. Highly flexible and adaptable proteins present a special challenge to drug design. From the many crystal structure determinations, it has become clear that there are several parent conformations for aldose reductase that are most likely in dynamic equilibrium with one another. A binding ligand selects a conformation from this equilibrium that fits, and this conformation is stabilized upon binding. If a binding ligand selects a conformation from this equilibrium and stabilizes it by binding, this process is called **conformational selection**. However, it is also possible that the geometry of the open pocket is not determined until the ligand has bound at least in the vicinity of the pocket. This process is called **induced fit**. These two mechanisms are thought to play an important role in many flexible proteins, and they are particularly valid for structurally highly dynamic receptors such as GPCRs, which are discussed in Chap. 29.

Matthias Zentgraf performed extensive molecular dynamics simulations on aldose reductase. The resulting profile was consistent with multiple crystallographic structure determinations with this enzyme. Amino acids, which are repeatedly found in many protein–ligand complexes with modified geometries, were also shown to be very flexible in the MD simulations. When the trajectory of such simulations is evaluated, it is apparent that the protein flips between the above-mentioned **parent conformations**. In addition, many geometries appear that have only small but structurally critical variations from these parent conformations. For example, small areas of the binding pocket open up to accommodate an additional methyl group or a phenyl ring on a ligand. Such information can be used directly in the design of new inhibitors.

To obtain an overview of the flexibility of a protein, the variation of atomic positions from one simulation state to the next is calculated along a trajectory. As with a photographic film, these instantaneous images of the complex are called "snapshots." In particular, it becomes transparent when a protein fluctuates in one conformation for a certain amount of time before flipping to another geometry. As it progresses, it can either return to the original geometry or flip to a yet different parent geometry. Such an **orientation map** is shown in Fig. 15.5. From this map, it can be seen that the protein spends some time in several parent conformations. Superimposing representative snapshots from these clusters of parent conformations gives a very instructive picture of which groups in the binding pocket show increased flexibility. In this example, the side chains of two adjacent phenyl alanines (Phe 121 and Phe 122, Fig. 15.6) are particularly involved. These can swing out of the way to open a new, previously closed cavity in the binding pocket. In the context of drug design, such information can be translated into the design of new inhibitors that can occupy new binding pockets. In this way, improved affinity or selectivity for the target protein can be achieved.

15.9 Model and Simulation: Where Are the Differences?

To conclude this chapter, we will briefly compare and contrast the terms "model" and "simulation." Molecular models are used to address questions that are difficult or impossible to answer experimentally. What different conformations can a molecule assume? This question is currently difficult to resolve experimentally. Will a potential drug candidate fit into the binding pocket of a protein? This question is also difficult to answer experimentally in a predictive way without investing a lot of time and effort in actually doing the necessary experiments. The use of models is a fundamental part of every scientific discipline. In chemistry, models have always played a central role. Chaps. 23–31 show how models based on the crystal structures of protein–ligand complexes can make an important contribution to drug design, especially in the preselection of possible molecular candidates for synthesis.

The term "simulation" describes calculations with models. For a given mathematical model, several options or variable combinations can be quickly evaluated on the computer. Such studies can contribute significantly to a better understanding of the system. Along with theory and experiment, computer simulations have been called the third pillar of exact science.

However, beware of too high expectations in the field of drug design! It should not be overlooked that the performance of a reasonable simulation requires that the fundamental model is accurate and its limitations are well understood. In many areas of engineering, this requirement is well met, so that simulation plays an important role in the design of automobiles or computer chips. Unfortunately, chemistry is more complicated. Today's molecular models can be used to assemble and prioritize compounds for synthesis. They can also be used to design ligands with improved binding properties. However, current models are often not accurate enough to allow detailed simulations of protein–ligand complexes with sufficient accuracy to determine a binding affinity. This is mainly due to the need to correctly account for the behavior of the water molecules involved in the binding. In addition, polarization effects that alter the protonation states of functional groups of both the protein and the bound ligand are major problems that are still largely unresolved. Given the importance of molecular modeling in the field of rational drug design, this can only mean

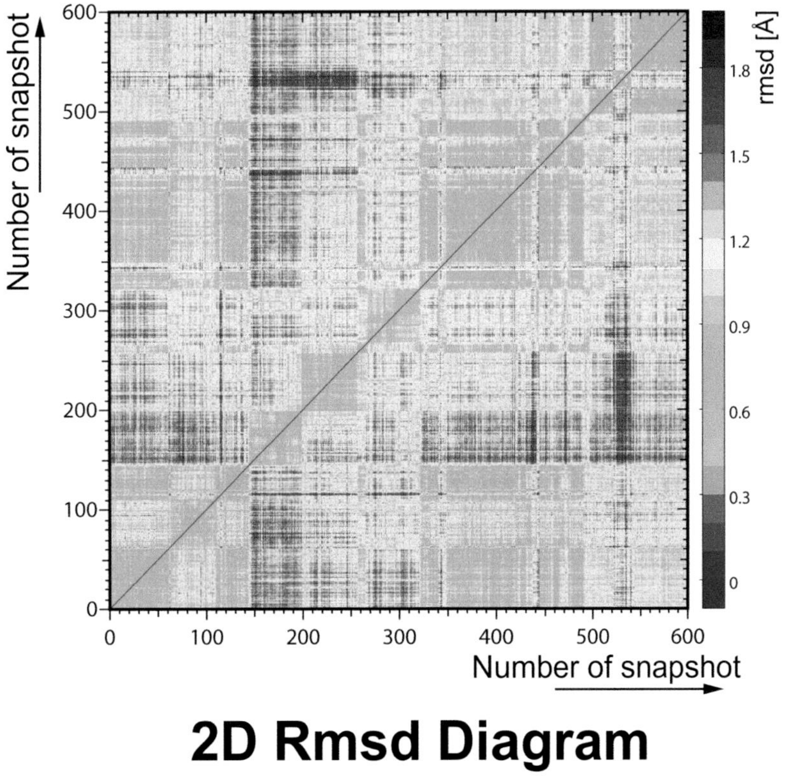

Fig. 15.5 The development with time of the spatial deviations of various snapshots along the simulation trajectory are visualized on this map. Large deviations are color-coded with *red*, medium-sized deviations with *green*, and small deviations are colored *blue*. *Green* delineated square areas are recognizable along the main diagonal. There the complex spends time near a parent conformation. The transition to the next square represents a flip to a new geometry. If sectors outside the main diagonal are colored increasingly *red*, the geometry will deviate more strongly from the previously adopted conformation. If an area outside the diagonal is reached that is *green*, the newly adopted geometry will not be very different from a state that the system had once reached. With such a map, it is possible to see which of the many parent conformations a complex oscillates between

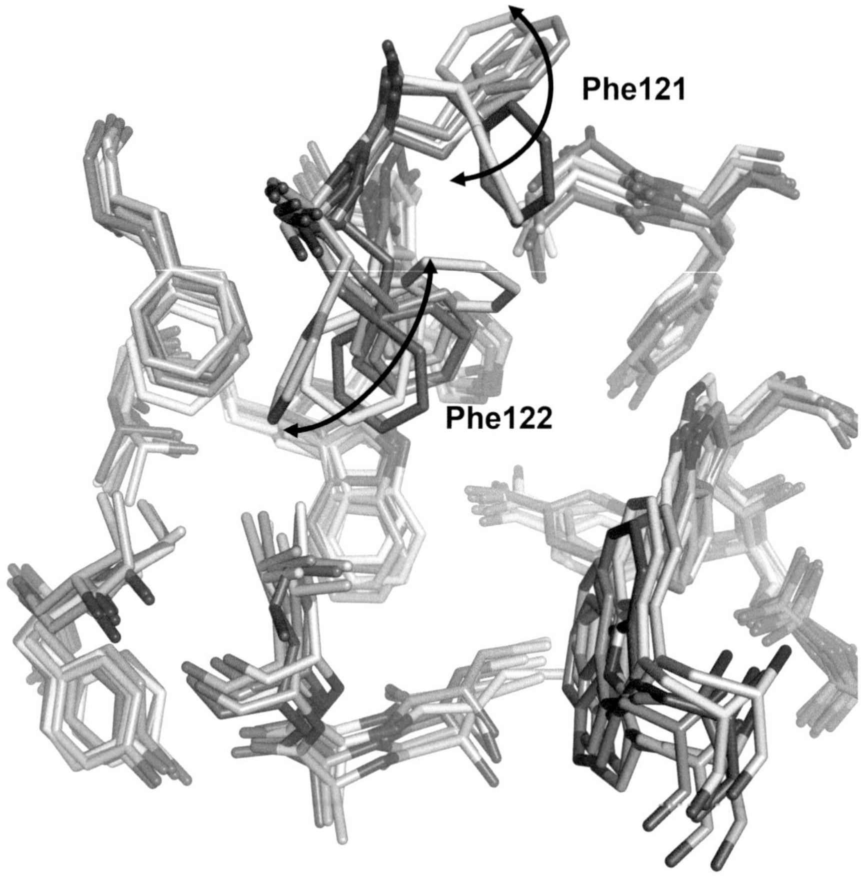

Fig. 15.6 Representative snapshots were taken from the different square area along the main diagonal in Fig. 15.5 and superimposed upon one another. It can be seen that above all else, the side chains of the phenyl alanines Phe 121 and Phe 122 can undergo extensive motion in the binding pocket. In doing so, they can also adopt conformations (e.g., the *light-blue* geometry) that open a new hydrophobic cavity in the binding pocket once had. With such a map, it is possible to see which of the many parent conformations a complex oscillates between

15

that more effort must be made to collect experimental evidence in order to develop improved models.

15.10 Synopsis

- Models have been and are still used in chemistry in general, but in particular in modern drug design. Computer graphics is a versatile tool to display structures and models along with various properties assigned and/or geometrically superimposed onto these molecules.
- Structures can be calculated by starting from first principles and by trying to regard physics as closely as possible. This is done with quantum mechanical calculations. Because these methods easily become elaborate and computationally intractable, an alternative is the use of empirical approaches. They are based on much simpler physics, normally classical mechanics, and treat molecules as a set of point charges in space interconnected by springs following harmonic potentials.
- Empirical approaches can only be used if enough experimental data are available to parameterize and calibrate the empirical concepts. Therefore, large databases assembling knowledge about molecular properties have been developed. Meanwhile, attempts are also being made to quantum-chemically calculate small building blocks of molecules in order to build up a large database, which can then be used to parameterize force fields using machine learning methods.
- Molecular mechanics to compute the geometry of molecules are based on empirical force fields. They comprise multiple energy terms that describe mutual interactions in the molecules either through bonds or through space. Particular potentials are used to describe the torsional barrier to rotations around single bonds. Furthermore, nonbonded interactions are handled by special potentials.
- The accuracy and required computational capacity of quantum chemical approaches depend on the sophistication of the basis sets of atomic or molecular orbitals used for the calculations. Parameterization of some parts of the calculations with empirical data can

significantly reduce the computational requirements. Density functional theory is a faster approach and works with electron density distributions instead of orbitals. Combinations of quantum chemical methods and force field approaches have been developed to handle large systems such as protein–ligand complexes.
- Properties such as charges can be displayed on the surface of molecules. Different types of surfaces have been defined such as the van der Waals surface or the solvent-accessible surface.
- Molecular dynamics simulations are normally based on potentials derived from empirical force fields. They consider the properties of a molecule under dynamic conditions by solving Newtonian equations of motion. As a result, the motion of a molecule can be evaluated over time by analyzing the so-called molecular trajectory.
- Molecular dynamics simulations can be used to study the flexibility of a protein next to its ligand-binding site. Such simulations can show multiple conformations of the protein that are competent to accommodate different ligands.
- Computer simulations allow the possible properties of molecules under different test conditions to be enumerated. They help to interpret results from experiments or help to predict properties of molecules to better plan the next experiments.

Bibliography and further reading

General Literature

K. B. Lipkowitz and D. B. Boyd, Eds., Reviews in Computational Chemistry, VCH, Weinheim (1990)
U. Burkert and N. L. Allinger, Molecular Mechanics, ACS Monograph 177, American Chemical Society, Washington (1982)
J. M. Goodfellow, Ed., Computer Modelling in Molecular Biology, VCH, Weinheim (1995)
A. Leach, Molecular Modelling: Principles and Applications, 2nd Ed. Prentice Hall (2001)
A. Hinchliffe, Molecular modelling for beginners. John Wiley & Sons (2003)
J. Gasteiger and T. Engel, Eds. Chemoinformatics: a textbook. John Wiley & Sons (2006)
G. Schneider and K.-H. Baringhaus, Molecular design: concepts and applications. Wiley-VCH (2008)
D. S. Sholl, J. A. Steckel, Density Functional Theory: A Practical Introduction, John Wiley & Sons, Inc. (2009)
W. Kauzmann, Quantum chemistry: an introduction. Elsevier (2013)
T. Tsuneda, Density Functional Theory in Quantum Chemistry. Springer Science & Business Media (2014)
S. J. Jang, Quantum Mechanics for Chemistry. Springer, Berlin (2023)

Special Literature

E. Fischer, Aus meinem Leben, Springer, Berlin, 1922, p.134
J. D. Watson, The double helix, Phoenix, London; originally published by Weidenfeld & Nicholson (1968)
D. J. Cram, The design of molecular hosts, guests, and their complexes, Angew. Chem. Int. Ed. Eng., **27**, 1009–1020 (1988)
B. Pullman, Molecular Modelling, With or Without Quantum Chemistry, in: Modelling of Molecular Structures and Properties, J. L. Rivail, Ed., Studies in Physical and Theoretical Chemistry, Vol. 71, pp. 1–15, Elsevier, Amsterdam (1990)
W. D. Cornell *et al.* , A Second Generation Force Field for the Simulation of Proteins, Nucleic Acids, and Organic Molecules, J. Am. Chem. Soc. **117** 5179–5197 (1995)
W. F. van Gunsteren and P. K. Weiner, Computer Simulations of Biomolecular Systems, ESCOM, Leiden (1989)
R.S. Pearlman, Rapid generation of high quality approximate 3D structures, Chem. Design Assoc. News, **2**, 1–7, (1987)
J. Gasteiger, C. Rudolph, and J. Sadowski, Automatic generation of 3D-atomic coordinates for organic molecules. Tetrahedron Comput. Methodol., **3**, 537–547 (1990)
C. Nguyen, M. K. Gilson, T. Young, Structure and Thermodynamics of Molecular Hydration via Grid Inhomogeneous Solvation Theory. (2011), arXiv:1108.4876. arXiv.org e-Print archive. https://arxiv.org/abs/1108.4876.
S. Ramsey, C. Nguyen, R. Salomon-Ferrer, R.C. Walker, M. K. Gilson, T. Kurtzman, T. Solvation Thermodynamic Mapping of Molecular Surfaces in AmberTools: GIST. J. Comput. Chem., **37**, 2029–2037 (2016)
Computed structure models of proteins: https://www.rcsb.org/news/6304ee57707ecd4f63b3d3db (Last accessed Nov. 17, 2024)

Conformational Analysis

Contents

G. Klebe, *Drug Design,* https://doi.org/10.1007/978-3-662-68998-1_16

16

Assembling a molecule with a modeling kit makes it clear that rotations about single bonds are easy to perform. The molecule takes on a different shape, or as chemists say, it is transformed into a different **conformation**. In a real molecule, the rotations around these bonds are not completely free. They are subject to a potential, and the molecule adopts certain energetically favorable arrangements as it rotates. *n*-Butane is the simplest case (◘ Fig. 16.1). The central **torsional** or **dihedral angle** determines the relative orientation of the two bonds to the methyl groups. When *n*-butane is rotated out of the arrangement with the two bonds to the methyl groups in a 180° orientation (*trans*), the methyl group on the "front" carbon and the hydrogen atom on the "back" carbon are directly coincident with each other at angles of rotation of 120 and 240°, called "**eclipsed**." In this geometry, they come closer to each other, so this arrangement is unfavorable for steric reasons. At an angle of rotation of 60 and 300°, the groups are again in a "**staggered**" geometry, which is an energetically more favorable situation. This arrangement is somewhat less favorable than the staggered *trans* orientation because of the spatial proximity of the methyl groups, which are now said to be "**gauche**" to each other. Finally, at 0 and 360° along the rotation path, an orientation is adopted in which both methyl groups are exactly behind each other. This is an even more unfavorable orientation.

16.1 Many Rotatable Bonds Create Large Conformational Multiplicity

Several energy maxima and minima can be passed through in the course of a full 360° rotation, depending on which atoms and groups are attached to the rotatable bond. They are at different energy levels relative to one another. The lowest minimum is called the **global minimum**, and the energetically higher minima are called **local minima**. Knowledge of these minima is important because molecules adopt geometries that correspond to such energy minima. To find these minima, calculations are necessary. One possible method is to systematically rotate all rotatable bonds, e.g., in 10° steps. At each step, the energy of the molecule is calculated using a force field. All minima found correspond to possible conformations of the molecule.

Most drug-like molecules have many single bonds and, therefore, exhibit more than one rotatable bond. For these bonds, multiple torsion angle values can be assumed. These values must be combined for all rotatable bonds in the molecule. The number of possible combinations increases multiplicatively. The molecule *n*-hexane has three rotatable bonds. If, analogous to *n*-butane, three local minima are assumed for each rotatable bond (±60, 180, and 300°), we can expect 3 × 3 × 3 = 27 minima. However, to systematically search for these minima in 10° steps, it would be necessary to evaluate

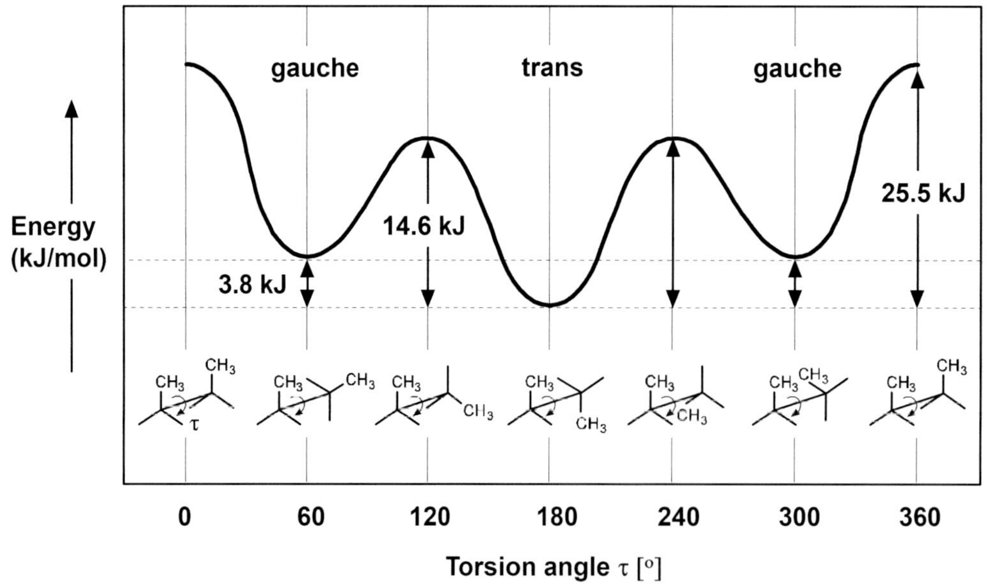

◘ **Fig. 16.1** Butane, $CH_3CH_2CH_2CH_3$, is made up of a linear chain of carbon atoms. If the terminal methyl groups are covering each other after rotation around the central C–C bond, the torsion angle about the central bond is 0°. At a 60° angle, the "back" methyl group is half way between the "front" methyl group and a hydrogen atom. This situation is called a "*gauche*" orientation. At 120°, a methyl group and a hydrogen atom are eclipsed to each other. At 180°, the terminal methyl groups are exactly opposite each other. Here, the energetically most favorable situation, the *trans* orientation, is achieved. From now on, the course of the rotation is mirror symmetrical, and ends in the starting position at 360°. The orientations at 120 and 240° are energetically less favorable than the 180° orientation by 14.6 kJ/mol. The *gauche* orientations at 60 and 300° are the least favorable ones and are 25.5 kJ/mol higher in energy. If a minimization method is applied that can only run "downhill," the three minima on the potential curve can be reached by starting, for example, at angles of 110, 130, and 250°

36 × 36 × 36 = 46,656 positions. In principle, the energy must be calculated for each of these positions. However, not all angle positions will lead to reasonable geometries. It can happen that parts of the molecule fold back upon itself and that these parts mutually superimpose. Such collisions can be identified by computer programs, and the corresponding geometry is discarded. It is also easy to imagine that with an increasing number of rotatable bonds, the number of local minima and adoptable geometries will increase dramatically in a systematic search.

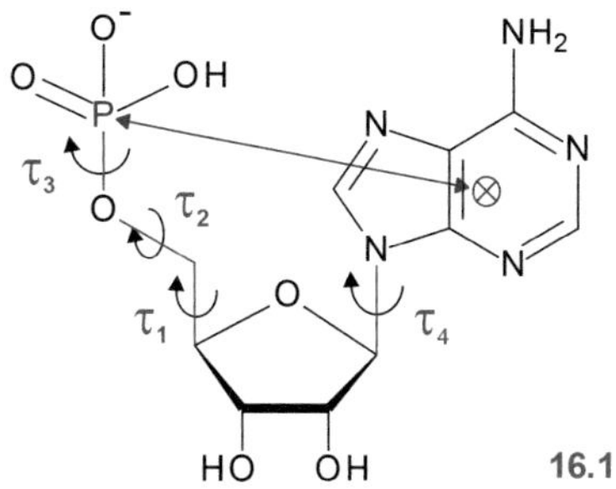

16.1

■ **Fig. 16.2** Adenosine monophosphate **16.1** exhibits the conformationally flexible ribose ring and four open-chain torsion angles, τ_1–τ_4. During the conformational analysis rotations are performed around these open-chain torsion angles. To get a rough description of the attained geometry, the distance between the phosphorus atom in the side chain and the adenine scaffold (⊗) is measured

16.2 Conformations Are the Local Energy Minima of a Molecule

It has been shown in the last chapter that the energy and geometry of a molecule can be calculated using a force field or a quantum mechanical method. In this way, all possible combinations of angles around the rotatable bonds in a molecule can be found that correspond to energetically favorable states. The mathematical method that is used to search for such a minimum geometry can only move downhill on the potential energy surface (Sect. 15.5). For this, the potential of *n*-butane should be considered again (■ Fig. 16.1). If an angle of 130° is used as a starting value, the minimization ends with a *trans* geometry. If an angle of 110° is started with, which is only 20° distant, the optimization will lead to a *gauche* orientation. By doing this, two of the three possibilities are detected. The third minimum that mirrors the *gauche* conformation will be reached if, for example, an angle of 350° is started from. In this way, all three conformations are found for the simplest possible case.

How are complex molecules to be approached? In principle, in exactly the same way. Because it is not known which torsion angles of the individual single bonds will give access to which potential minima, that is, stable conformations, the minimization must be started from numerous angles for each of the single bonds. From these values the minimization always goes "downhill." The minima on the potential surface are found in this way. The art is to efficiently define the starting points from which a given geometry is minimized. This is a very laborious task, particularly with large molecules. It is akin to a hiker in the mountains searching for the deepest valley.

Adenosine monophosphate **16.1** should serve as an example (■ Fig. 16.2). The analysis focuses on the five-membered ribose ring, the bond to the nitrogen in the adenine, and the three bonds of the sugar phosphate side chain. What conformations can this molecule assume? Rotations around the open chain bonds are performed in 10° steps. In the systematic search for the ribose ring, only orientations that allow the ring to close are considered. To get a rough overview of the hypothetically obtained geometries, the distance between the center of the adenine scaffold and the phosphorus atom is measured in each generated geometry. This distance falls between 4.5 and 9.3 Å for the more than 300,000 generated geometries. To estimate the energy content of the molecule in any geometry, its van der Waals energy (Chap. 15) is calculated. Such a calculation is very fast. The energies of the 300,000 geometries are found between 0 and 64 kJ/mol. The generated structures are not yet in local potential minima. To achieve this, each initial geometry must be minimized (cf. the potential energy curve of *n*-butane in ■ Fig. 16.1). The resulting conformations are then compared to see whether the same local minima are reached by starting from different points. With 300,000 starting geometries, this is quite a tedious task! It is akin to letting our hiker walk downhill from every cross-section on the map of the Alps to find the deepest valley. Hopefully he will be granted great longevity so that he lives long enough to see the results of the search! Can this search be made more effective?

16.3 How to Scan Conformational Space Efficiently?

Sometimes rolling the dice is better than systematic exploration! The hiker could randomly choose places in the mountains from which to descend into the next valley. With a little luck, he will find the deepest valley with much less effort. Such **Monte Carlo methods** are very popular in conformational analysis. The starting angles for the conformation search are chosen at random. The name of the method, which gives free rein to chance as in gambling, is an allusion to the casino in Monte Carlo. Another approach is **molecular dynamics**. The hiker would have to get into an airplane that flies between the mountains at high speed and changes direction at every obstacle. After a certain amount of time, the hiker jumps out of the plane and walks to the bottom of the valley. The higher the plane flies, the fewer mountain peaks it encounters and the faster it can traverse the mountains.

Molecular dynamics follows a molecular trajectory (Sect. 15.7) and stores the geometry at predefined time intervals to be used as starting points for energy minimization in a conformational analysis. By increasing the temperature (meaning flying higher), a larger area of conformational space can be searched in a shorter period of time.

16.4 Is It Necessary to Search the Entire Conformational Space?

So far, molecules have been considered in an isolated state. How does their flexibility change when they are placed in an environment such as the binding pocket of a protein? In principle, their conformational flexibility does not change. It could be that minima are found at different positions with different relative energies due to electrostatic and steric interactions with the binding pocket. This raises the question of whether the torsion angles for a ligand in a binding pocket need to be sought in all regions. If energy minima occur preferentially at certain torsion angles, it makes sense to limit the search to these angles. For example, the hiker might get the impression that the villages are mostly in valleys and hardly ever on peaks or slopes. Therefore, all villages would be worthwhile as starting points for his minimum search.

Ligands in the binding pocket of a protein are under the influence of directional interactions from the amino acids located there. Similar conditions exist for molecules in a crystal lattice. There, the environment is made up of identical copies of neighboring molecules (Chap. 13). These undergo directional interactions with the molecule, analogous to the amino acids in the binding pocket. Interestingly, the molecular packing density inside a protein is similar to that of organic molecules in a crystal lattice. As mentioned earlier, the crystal structures of many organic molecules are known and stored in a database. Unfortunately, experience has shown that the conformation of a flexible molecule in a crystal structure is often not identical or even similar to the geometry of the molecule in the binding pocket of a protein. The same is true for conformations found in solution.

16

Accordingly, the receptor-bound conformation of a molecule cannot be unambiguously deduced from its small-molecule crystal structure or from that in solution. Nevertheless, much can be learned from crystal structures. For example, it is not the whole molecule that should be considered, but rather individual torsion angles. The potential energy for the central torsion angle of *n*-butane is shown in ◘ Fig. 16.1. When the angles for multiple C–CH_2–CH_2–C fragments are extracted from a database of small-molecule crystal structures, they tend to cluster in areas where the potential energy curve shows local minima. Adenosine monophosphate **16.1** has four open-chain torsion angles τ_1–τ_4 (◘ Fig. 16.2). The bond between the ribose ring and the adenine backbone forms the torsion angle τ_4. Another fragment is the phosphate group with the oxygen and the attached carbon in the chain (τ_3). This fragment occurs in a large number of different structures in the database. A representative picture can be expected because this fragment will occur in very many different environments if enough crystal structures are considered. The results of such searches for the four torsion angles τ_1–τ_4 are shown in ◘ Fig. 16.4 as frequency distributions, so-called **histograms**. Experience has shown that for many torsion angles there are clearly preferred values. This is the case here for τ_1, τ_2, and τ_3. One might ask why this statistical evaluation is not better performed for ligands participating in crystallographically studied protein–ligand complexes. Unfortunately, the diversity of these data is still limited, and the data are usually not precise enough for the desired evaluation. Nevertheless, comparative studies have shown that the same torsion angles are preferentially found in protein–ligand complexes and small–molecule crystal structures (◘ Fig. 16.3). Meanwhile, the Cambridge Crystallographic Data Center has developed a tool called **Mogul** that provides easy access to the statistical frequencies found for the distribution of dihedral angles in torsion angle fragments found in crystal structures.

The experience that torsion angles prefer certain values can be used in the conformational search. The angle τ_4 between the ribose ring and the adenine backbone shows a broad distribution over many possible values (◘ Fig. 16.4). Unfortunately, the search cannot be narrowed down here. The situation is better for the other angles τ_1–τ_3. There are only certain values populated. If the systematic search is restricted to these areas and a search is performed in 10° steps around the mean value, only 6340 geometries would have to be generated. Almost the same distances between phosphorus and adenine are covered within the 5.9–9.3 Å range as in the unrestricted search. A van der Waals energy calculation on these geometries yields values between 0 and 16.3 kJ/mol. In contrast to the results of Sect. 16.2, all geometries corresponding to the energetically more unfavorable regions are discarded.

How can it be confirmed that this restricted search covers the part of the conformational space that includes the receptor-bound conformations? Adenosine monophosphate **16.1** often occurs as a substructure of cofactors in protein complexes, so there is enough information available about receptor-bound conformations for this particular example. They come from crystal structures of proteins with these cofactors bound. The distance range of 5.9–9.2 Å between the adenine backbone and the phosphorus in the receptor-bound structures covers the same range as was found in the exhaustive systematic search. It can, therefore, be assumed that enough geometries have been generated to

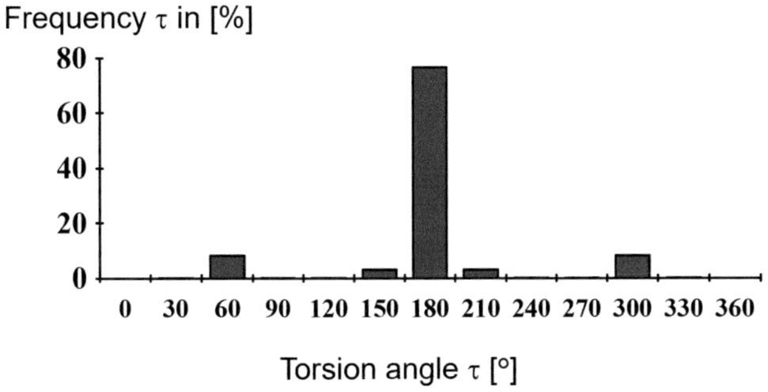

Fig. 16.3 A value distribution for the torsion angles with clusters at 60, 180, and 300° is derived from a database of small-molecule crystal structures for the C–CH_2–CH_2–C fragment. Most values are found at 180°. Torsion angles between 0 and 360° are entered as the relative frequency in percent. The maxima of the distribution are at the angles where the potential curve of *n*-butane (**Fig. 16.1**) shows its energy minima

satisfactorily populate the local minima of the bound state of adenosine monophosphate. Returning to the initial butane example (Fig. 16.1), this means that the starting points were well distributed so that all relevant minima were reached.

16.5 The Difficulty in Finding Local Minima Corresponding to the Receptor-Bound State

As already described, the local minima in a systematic conformational search are obtained by subjecting all generated geometries to a force field optimization. Problems can arise with this approach. To illustrate this, consider another molecule, citric acid **16.2**, in the binding pocket of citrate synthase. Its three carboxylate groups and the hydroxyl group form seven hydrogen bonds with three histidine and two arginine residues of the protein (Fig. 16.5). Considering the free citrate molecule not bound to the protein and minimizing its geometry in the isolated state, it assumes a conformation with internally saturated hydrogen bonds (Sect. 15.4). Of course, it is possible to start from a different geometry, but in all cases the minimization will result in conformations with intramolecular hydrogen bonds. Such hydrogen bonds rarely occur in the bound state of the protein. Therefore, the conformation obtained after minimization in

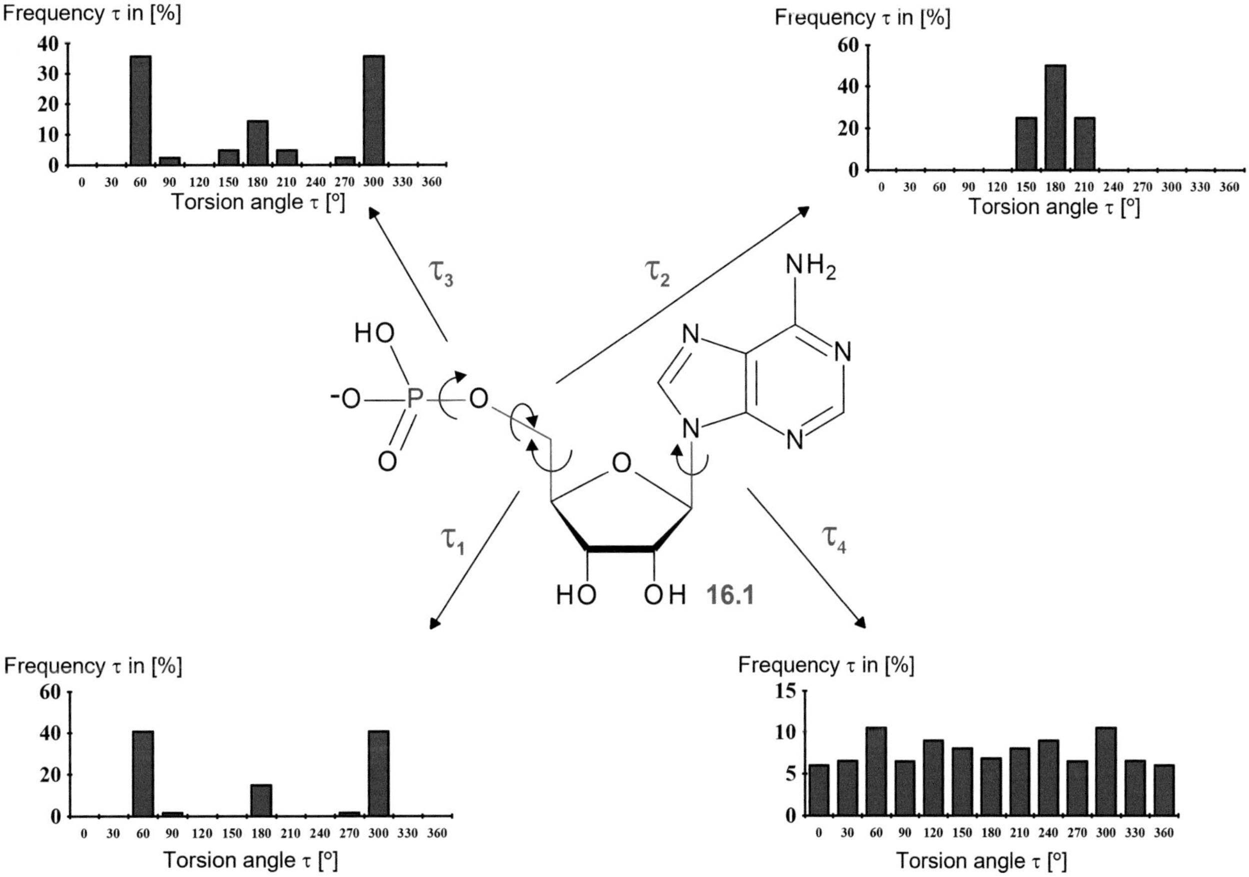

Fig. 16.4 The frequency distribution of the torsion angles of the open-chain bonds of adenosine monophosphate as found in the crystal structures of small organic molecules. The torsion angle histograms are constructed for fragments that are representative for corresponding portions of the test molecule. There are clearly preferred values for the angles τ_1–τ_3, but a broad distribution of all possible angles is found for τ_4. This knowledge is used in the conformational analyses and limits the search for τ_1–τ_3 to the preferred value ranges

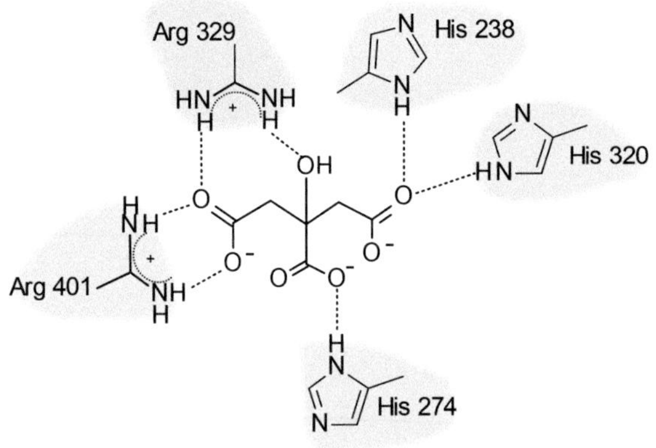

16.2

Fig. 16.5 Interactions between citric acid **16.2** and the enzyme citrate synthase. The molecule is bound by seven hydrogen bonds to three histidine and two arginine residues

the isolated state has little relevance to the conditions in the bound state of the protein.

In general, ligands rarely bind to proteins in a conformation that involves extensive formation of intramolecular hydrogen bonds. The polar H-bond forming groups are usually involved in interactions with the protein.

To avoid the problem of intramolecular H-bond formation, a minimization of the generated starting structure can be neglected and all geometries from the systematic search can be used for further comparison (Chap. 17). However, then a very large number of geometries would have to be examined. This would severely limit the scope of such comparisons for computational reasons. In addition, such results would probably describe rather distorted geometries. The force field responsible for the formation of intramolecular H-bonds could be neglected. But how reliable would such an artificially simplified force field be?

16.6 An Effective Search for Relevant Conformations by Using a Knowledge-Based Approach

A knowledge-based approach first analyzes the experimentally determined conformations and generates only those conformations for new molecules that are consistent with the experimental knowledge base. In this way, many geometries are never generated in the first place. The example of adenosine monophosphate **16.1** is used again. The approach recognizes a flexible five-membered ring and four open-chain rotatable bonds. Energetically favorable conformations of the ring are selected from a database. This database contains a large number of different ring systems, as they can be found, for example, in the crystal structures of organic molecules. In this case, the approach suggests the five most energetically favorable ring conformations, two of which are actually found in the protein-bound cofactors. For the open-chain part of the molecule, the method is guided by the aforementioned frequency distribution of the dihedral angle (Fig. 16.4). The starting geometries are generated only in angle ranges where these distributions show significant frequencies. The distribution is still rather coarse. In a final step, the generated geometries are optimized by readjusting the torsion angles. Clashes between noncovalently bonded atoms are avoided. At the same time, the adjusted dihedral angles are kept as close as possible to the preferred values. Such a so-called knowledge-based approach requires relatively few representative conformations. They are fairly evenly distributed in the part of conformational space relevant for receptor-bound conformations (Fig. 16.6).

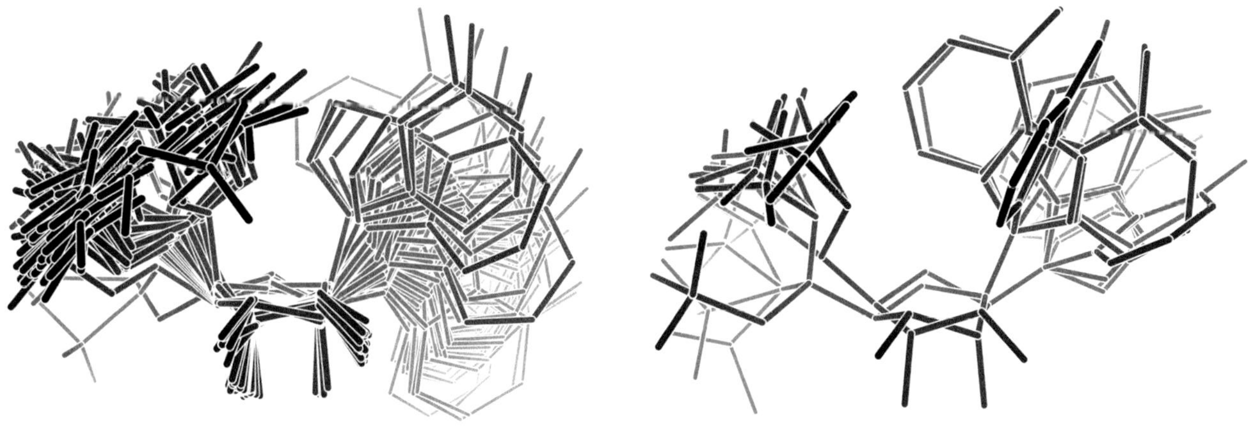

Fig. 16.6 *Left* Eighty-one conformers from experimentally determined protein–ligand complexes are superimposed upon one another to illustrate the areas in space that adenosine monophosphate **16.1** can adopt in a protein-bound state. The ribose ring is located in the center, for which two ring conformations occur. In each case, the possible orientations of the adenine ring are shown on the right side, and the conformations of the flexible phosphate chain are shown on the left side. *Right* Similar coverage of the conformational space is achieved with a manageable number of only 14 conformations generated by a knowledge-based approach

16.7 What Is the Outcome of a Conformational Search?

Many drug-like molecules are flexible. They can adopt markedly different conformations depending on their environment. Usually, the receptor-bound geometry is not the energetically most favorable conformation found for the isolated state, but it will be in an energetically favorable range. For conformational analysis, this means that it is not necessarily the deepest minimum that is sought. Rather, it should be the "relevant" minimum corresponding to the bound state. There will only be a chance to find it, if the criteria for the search are known. There is no difference in the difficulty of finding the energetically most favorable conformation or the one that "fits" the binding site best. An important tool in the search for novel lead structures is the docking of candidate molecules into the binding pocket of a given protein. Programs that follow this approach must be able to handle the conformation problem. A variety of methods have been developed that allow efficient docking searches on computer clusters, especially for molecules of drug-like size (Sects. 7.6 and 20.8).

16.8 Synopsis

- Drug-like molecules exhibit multiple rotatable bonds. Rotations around these bonds drive the molecules into different conformations that correspond to local minima on the energy surface of the molecule.
- The receptor-bound conformation of a drug-like molecule is the starting point for any drug-design considerations. Therefore, many methods have been developed to perform conformational analyses. Systematic searches by incremental rotations about each single bond torsion angle will produce a huge amount of geometries that need to be optimized to the local minima on the energy surface.
- The conformation of a drug-like molecule frequently changes with the environment. Usually the conformation in the protein-bound state differs from that in solution, in the gas phase, or in the small-molecule crystal structure.
- Considering torsional fragments in small molecules and analyzing them across databases of crystal structures by statistical means reveals clear-cut torsional preferences for many examples. Such knowledge can be exploited to perform a conformational search more efficiently. Not all values around a rotatable bond have to be tested, and the search can be limited to the ranges that are known to be preferred.
- A further obstacle in the conformational search of the protein-bound conformation of a drug-like molecule is that the molecule will interact with its environment. This environment, which is usually the protein's binding pocket, is often polar and will involve the bound ligand in multiple hydrogen bonds.
- Using a knowledge base on torsional preferences of small organic molecules can significantly enhance the conformational search, particularly during docking, in molecular comparisons, or in database searches based on predefined pharmacophores.

Bibliography and Further Reading

General Literature

J. Dale, Stereochemistry and Conformational Analysis, VCH, Weinheim, New York (1978)

K. Rasmussen, Potential Energy Functions in Conformational Analysis, Lecture Notes in Chemistry, Vol. 37, Springer (1985)

A. Leach, Molecular modelling: principles and applications, 2nd edn. Prentice Hall, Englewood Cliffs (2001)

Special Literature

G. Klebe, Structure Correlation and Ligand/Receptor Interactions, in: Structure Correlation, H. B. Bürgi and J. D. Dunitz, Eds., VCH, Weinheim, p. 543–603 (1994)

G. R. Marshall and C. B. Naylor, Use of Molecular Graphics for Structural Analysis of Small Molecules, in: Comprehensive Medicinal Chemistry, C. Hansch, P.G. Sammes and J.B. Taylor, Eds., Vol. 4, Pergamon Press Oxford, p. 431–458 (1990)

G. Klebe and T. Mietzner, A Fast and Efficient Method to Generate Biologically Relevant Conformations, J. Comput.-Aided Mol. Design, **8**, 583–606 (1994)

H. J. Böhm and G. Klebe, What can we learn from molecular recognition in protein–ligand complexes for the design of new drugs? Angew. Chem. Intl. Ed. Eng., **35**, 2588–2614 (1996)

G. Klebe, Toward a More Efficient Handling of Conformational Flexibility in Computer-Assisted Modelling of Drug Molecules, Persp. Drug Design and Discov., **3**, 85–105 (1995)

B. Stegemann, G. Klebe, Cofactor-binding sites in proteins of deviating sequence: Comparative analysis and clustering in torsion angle, cavity, and fold space. Proteins, **80**, 626–648 (2011)

I. J. Bruno, J. C. Cole, M. Kessler, Jie Luo, W. D. S. Motherwell, L. H. Purkis, B. R. Smith, R. Taylor, R. I. Cooper, S. E. Harris and A. G. Orpen, Retrieval of Crystallographically-Derived Molecular Geometry Information, J. Chem. Inf. Comput. Sci., **44**, 2133–2144 (2004)

S. J. Cottrell, T. S. G. Olsson, R. Taylor, J. C. Cole and J. W. Liebeschuetz, Validating and Understanding Ring Conformations Using Small Molecule Crystallographic Data, J. Chem. Inf. Model., **52**, 956–962 (2012)

Mogul, assessment of molecular conformations: https://www.ccdc.cam.ac.uk/solutions/software/mogul/ (Last accessed Nov. 18, 2024)

Quantitative Structure-Activity Relationships and Design Approaches

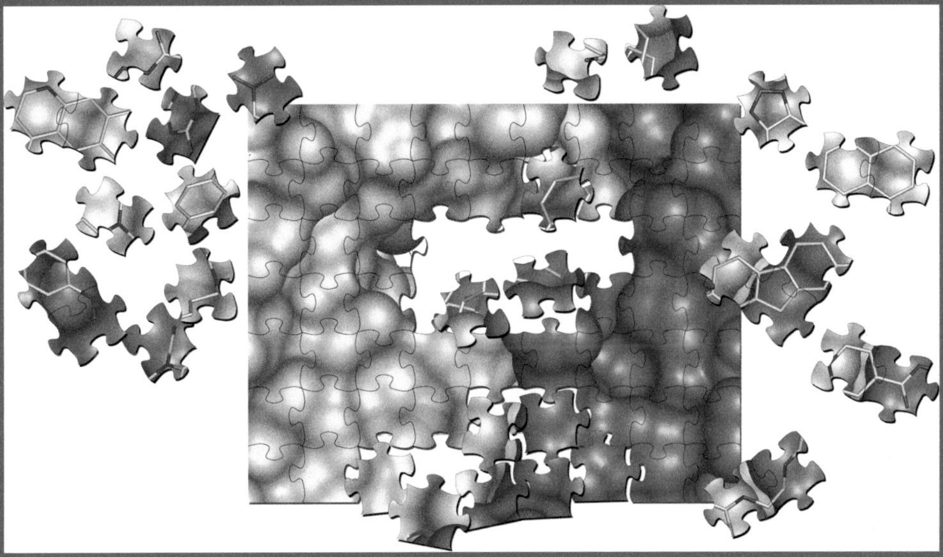

Today, drug design is supported by numerous computational approaches that, like the pieces of a puzzle, all make their contributions from the first design hypothesis to a clinical drug candidate (announcement poster from the author's working group on the occasion of a conference in 2005, Rauischholzhausen, Marburg, Germany).

Contents

Pharmacophore Hypotheses and Molecular Comparisons

Contents

G. Klebe, *Drug Design*, https://doi.org/10.1007/978-3-662-68998-1_17

Emil Fischer's lock-and-key principle (Sect. 4.1) demonstrates the specific interaction of an active compound with its receptor. In the case of the key, the spikes and notches on its beard interact with the pins of the lock to open it. With active substances, it is a specific part of the molecule that interacts with the amino acids in the binding pocket. In drug design, similar molecules are often compared to generate ideas for new structures. This chapter summarizes the criteria that make such comparisons possible. These criteria can also be used to search databases for alternative molecules that may bind to the protein in the same way.

17.1 The Pharmacophore Anchors a Drug Molecule in the Binding Pocket

The structure of the binding pocket determines which functional groups are necessary for ligand binding. The spatial orientation of these functional groups in ligands is referred to as the **pharmacophore** (Sect. 8.7, ◘ Fig. 8.9). Because of its importance in drug design and model hypothesis in medicinal chemistry, an official IUPAC definition has been established by Camille G. Wermuth (◘ Table 17.1). The interacting groups that a ligand must possess in order to interact successfully with a protein define the pharmacophore in space and are independent of the specific molecular scaffold to which they are attached. Hydrogen bonding groups or hydrophobic moieties are considered. A more detailed examination distinguishes between positively and negatively charged groups in a molecule. When derived from a set of similarly binding ligands, this generalized description is called a **ligand-based pharmacophore**. On the other hand, the protein structure can also be used as a starting point. This is done by analyzing which amino acid functional groups are located in the binding pocket. They define the properties with which a ligand can bind to them. In this sense, the protein structure determines how the pharmacophore of a ligand must be shaped to successfully bind to the protein. This description is called the **protein-based pharmacophore**. In contrast to the lock-and-key image, ligands and proteins are flexible. In ligands, the functional groups of the pharmacophore must be oriented towards the corresponding counter groups in the protein. Therefore, detailed knowledge of the conformational properties of the ligand is essential. Only then can it be predicted whether a ligand can potentially adopt a geometry that satisfies the interactions with the protein. On the side of the receptor, the geometry of the binding pocket can adapt to the shape of the ligand, similar to how a glove fits the hand of its wearer (*induced fit*, Sect. 4.1). In fact, binding pockets are found in the interior or in buried grooves on the surface of proteins, and it is there that the small but crucial conformational changes of the protein take place. An example of the adaptability of a protein is presented in Sect. 15.8. An attempt is made to describe these induced-fit adaptations, or the selection of binding-competent conformers of the protein, using molecular dynamics simulations.

17.2 Structural Superposition of Drug Molecules

For the moment, we will limit ourselves to examples where the receptor structure is unknown. All the effects of ligand binding on the protein are, therefore, neglected. An example should illustrate this. The fruit of the shrub *Anamirta cocculus*, the fish berry, contains the terpene picrotoxinin **17.1**, which causes convulsions. This compound acts on the chloride channel by blocking it at a narrow, constricted site like a plug (Sect. 30.7). Because of its central stimulatory effect, it has been used in the past as an antidote to sleeping pill overdoses. Due to its high toxicity, it is no longer of any therapeutic importance. The structure of picrotoxinin has been determined by crystallography (◘ Fig. 17.1).

Synthetic modifications of the cyclic core structure have led to active and inactive derivatives (◘ Fig. 17.2). The three-dimensional structure of each derivative can be constructed on the computer from the crystal structure of the parent compound. They are superimposed upon one another to identify their structural differences. The parts of the molecules that are considered equivalent in a li-

17

◘ **Table 17.1** Official IUPAC definition of a pharmacophore. (C. G. Wermuth et al. Pure Appl. Chem., **70**, 1129–1143 (1998))

– A pharmacophore is the ensemble of steric and electronic features that is necessary to ensure the optimal supramolecular interactions with a specific biological target structure and to trigger (or to block) its biological response.
– A pharmacophore does not represent a real molecule or a real association of functional groups, but a purely abstract concept that accounts for the common molecular interaction capacities of a group of compounds towards their target structure.
– A pharmacophore can be considered as the largest common denominator shared by a set of active molecules. This definition discards a misuse often found in the medicinal chemistry literature, which consists of naming as pharmacophores simple chemical functionalities such as guanidines, sulfonamides, or dihydroimidazoles (formerly imidazolines), or typical structural skeletons such as flavones, phenothiazines, prostaglandins, or steroids.
– A pharmacophore is defined by pharmacophoric descriptors, including H-bonding, hydrophobic, and electrostatic interaction sites, defined by atoms, ring centers, and virtual points.

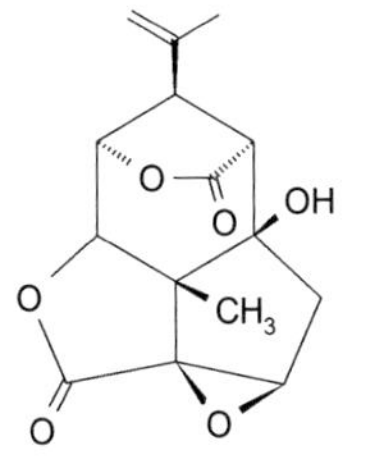

17.1

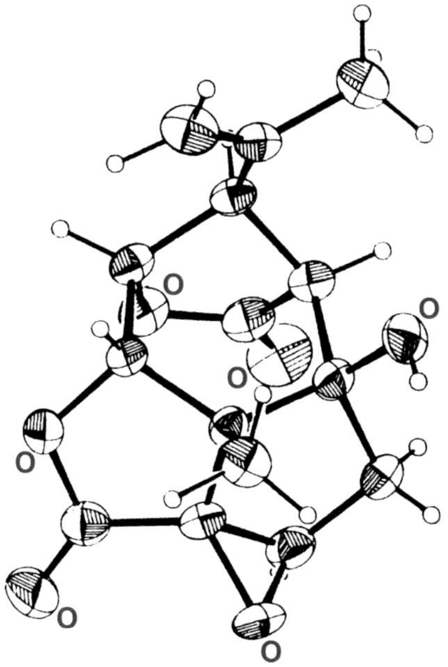

Fig. 17.1 Picrotoxinin **17.1** is responsible for the centrally stimulating effect of the extracts of fish berries. Its structure and spatial architecture were determined by X-ray structure analysis. The molecule inhibits the chloride channel and blocks the channel so that no more ions can pass (Sect. 30.7)

gand-based pharmacophore model are overlaid for this superposition. The superposition of all active and inactive derivatives along with the common volumes of both classes is shown in Fig. 17.3. The difference between the two volumes is computed. It describes those areas in space that are occupied only by the inactive derivatives. Since the structure of the chloride channel was determined many years later, it can be verified that the highlighted difference volume would actually extend into the occupied structural space of the protein. In Fig. 17.3, the lower part shows the binding site of picrotoxin in the channel. When the superimposed structures are placed in the channel, steric conflicts for the inactive derivatives are indeed indicated by the differential volume.

17.3 Logical Operations with Molecular Volumes

What information can be extracted from such comparative volumes? The working hypothesis is that a molecule can only be bound if its size does not exceed the maximum available space. What is the maximum available space? To get an idea, the common volumes of all active derivatives are considered and compared with the volumes of all inactive derivatives. A possible explanation for the lack of activity of a molecule could then be that the area in the binding pocket that the molecule would likely occupy is already occupied by the protein.

Volume comparisons between active and inactive derivatives provide information about the possible shape of the receptor pocket. Such comparisons can be very useful in drug design. Once the "forbidden" volume area has been determined for a compound class, it can be verified prior to synthesis whether a compound really leaves the "forbidden" area unoccupied.

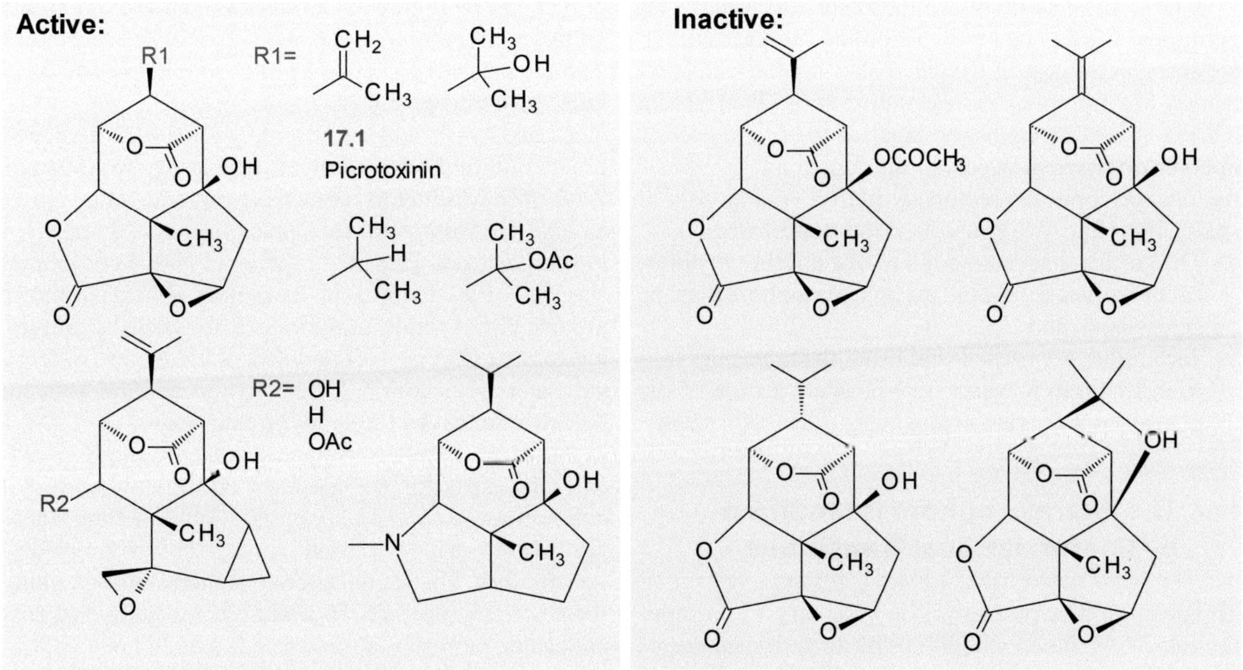

Fig. 17.2 By starting with picrotoxinin **17.1**, active and inactive derivatives were synthesized.

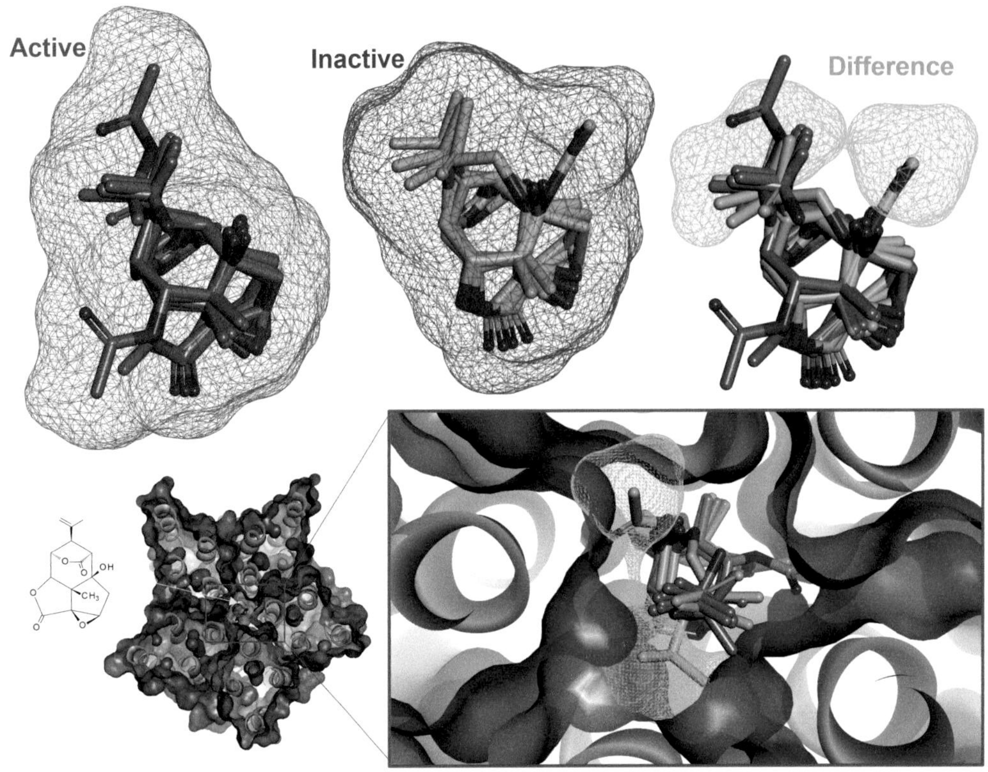

Fig. 17.3 Superposition of the spatial structure of active (*orange*) and inactive (*cyan*) derivatives of picrotoxinin. The united volumes around the active derivatives are shown by the *red mesh*. The total volume around all inactive derivatives is shown in *blue*. A difference is formed between the two volumes. The remaining volume (*green*) shows areas that are only occupied by inactive derivatives. An explanation for the lack of activity of these derivatives can be that they try to occupy volume areas that are already occupied by the receptor protein. This spatial clash does not occur with the active derivatives. Comparing the geometry of the protein environment of the chloride channel (*bottom left*, overview) shows that the *green* contoured volume indeed indicates steric conflicts with spatially occupied regions of the protein (detailed view, *bottom right*). The comparison is based on a cryo-EM structure of the complex of the channel with picrotoxinin (Sect. 30.7) which was determined many years later

Due to the rigidity of the molecule, it is easy to superimpose the analogues of picrotoxin on one another. However, in the case of flexible molecules, the transition from a 2D molecular representation to a 3D structure (Chaps. 15 and 16) cannot be expected to yield molecules in conformations in which all the functional groups of the pharmacophore are already placed analogously in space. Therefore, two issues need to be addressed:

- The groups that correspond to one another in different molecules and define the pharmacophore must be determined, and
- Techniques are needed that bring the molecules into conformations in which the equivalent groups of the pharmacophores are analogously oriented in space.

17.4 The Pharmacophore is Modified by Conformational Transitions

To solve the first problem, it is necessary to consider the role of the functional groups of an active substance that make contact with the receptor. They must form hydrogen bonds and hydrophobic interactions with the protein. In this context, the similarity of the functional groups means that they can form analogous interactions with the protein. To define a pharmacophore in space, at least three interacting groups are required. This can be illustrated by considering how many fingers are needed to hold a randomly shaped object (e.g., a potato) in space. With only two fingers, the object can still rotate about an axis. However, with three anchor points, its position in space is fixed. Practical experience with a compound class is often helpful in assigning pharmacophoric groups. For example, inhibitors of angiotensin-converting enzyme (Fig. 17.4 and Sect. 25.5) require a terminal carboxylate group, a carbonyl group, and a group that coordinates to the catalytic zinc ion.

How can it be determined whether a common orientation exists for the assumed equivalent groups in different molecules? In a computational method, these groups are assigned "virtual" springs that are coupled to one another. The spatial overlap is increased by pulling these springs together. To avoid a completely distorted molecular geometry, a force field is simultaneously considered for each molecule (Chap. 15). As an example, the steroid **17.2** and three different inhibitors **17.3**–**17.5**

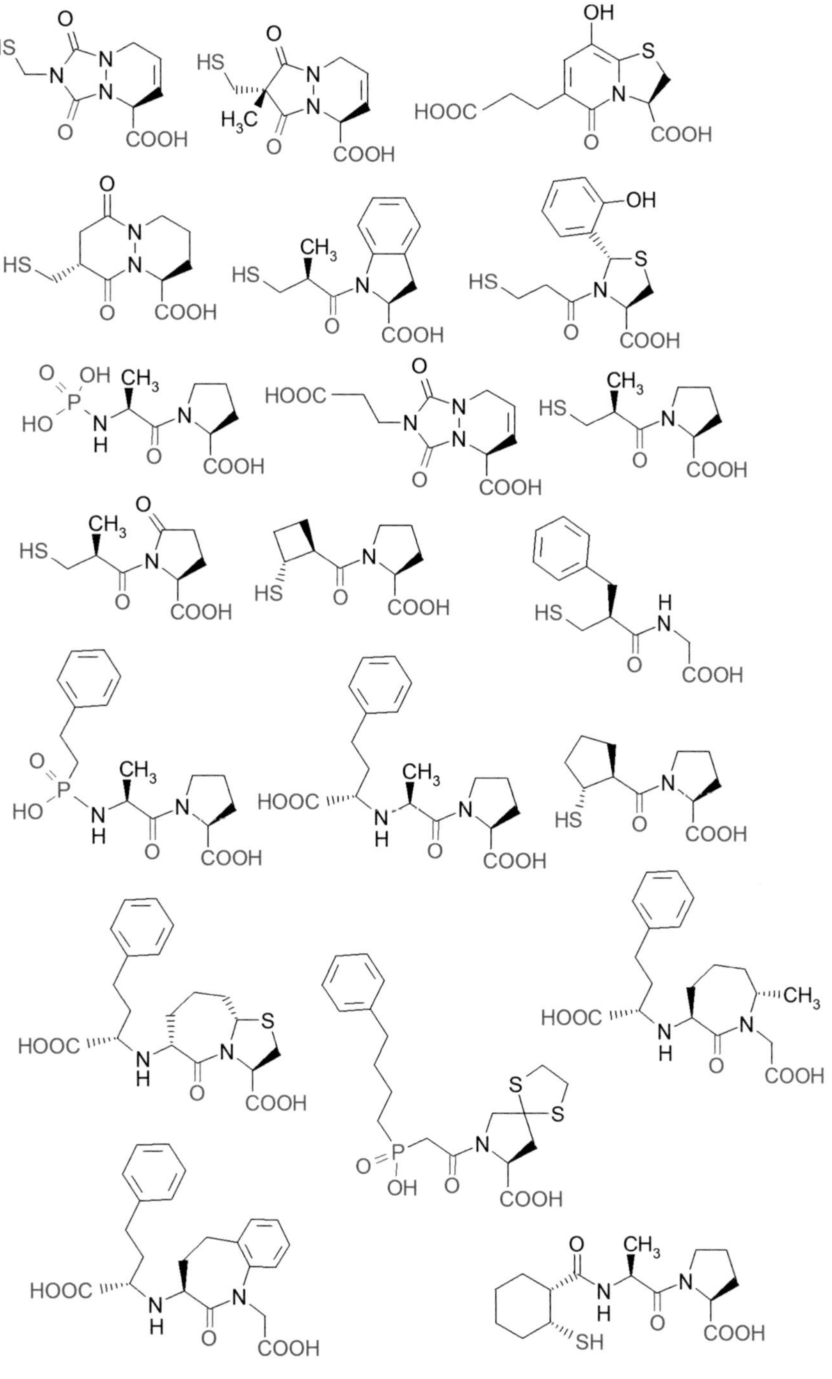

■ **Fig. 17.4** Inhibitors of the angiotensin-converting enzyme. A pharmacophore that consists of a terminal carboxylate group, a carbonyl group, and a group that coordinates to the catalytic zinc ion is necessary for binding to the enzyme. The latter function is assumed by a thiol, a phosphoric, phosphonic, or carboxylic acid group. The individual derivatives possess conformational flexibility in different areas

(■ Fig. 17.5) are considered. They are ligands of an enzyme in ergosterol biosynthesis. Spring forces are applied between the atoms marked with the same numbers. The minimization of these forces together with the individual force fields of the four molecules leads to the superposition shown in ■ Fig. 17.5.

Unfortunately, the resulting solution depends on the starting conditions. If the molecules are oriented differently in space at the beginning of the calculation, or if they start from different conformations, different superpositions can result. At first glance, this argument may seem rather implausible. It should be kept in mind that molecules are not only considered under the influence

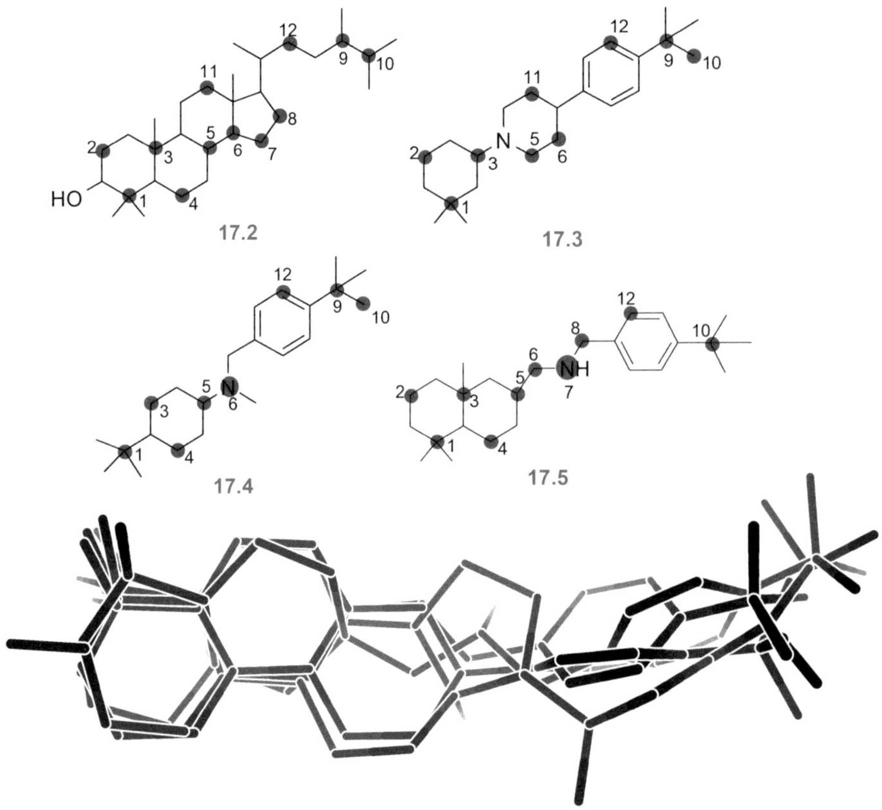

Fig. 17.5 "Virtual" springs are coupled to the atoms that are marked with numbers around the steroid **17.2** and the three derivatives **17.3**–**17.5**. The structural superposition (*bottom*) that is shown is determined by the forces of these springs and the simultaneous consideration of the individual molecular force fields

of "virtual" spring forces, but also under their own force fields. The multiple minima problem encountered in molecular force field calculations was already mentioned in Chap. 16. It also plays an important role here. The hiker from the last chapter should help to explain this issue. He stands on a mountaintop and wants to descend into the deepest valley possible. At the same time, he feels an "extra force" because he is very thirsty. He wants to meet his friends in a pub. The friends come from different peaks in the mountains. He sees a pub in each valley. But which one should he choose? For a common meeting place, he would also accept a less deep valley. At the beginning of his hike, he looks for the steepest descent to get down quickly. After a while, the other valleys fall out of sight. When he finally arrives at another pub, he has no longer the energy to look for a different one. If he would have started from a different peak, he might have found a similarly deep valley, but he would have found the pub of his choice and met his friends at the same time. The problem of choosing starting conditions for molecular comparisons with "virtual" spring forces is similar. How can you verify that the best possible solution has been found? Only an experiment can help. This requires the synthesis of molecules that are conformationally rigid in certain regions due to the insertion of rings. They impart a fixed spatial arrangement of the pharmacophore. If these molecules also have activity, their rigid geometry indicates a correct selection of the pharmacophore (see Sect. 17.9).

17.5 Systematic Conformational Search and Pharmacophore Hypothesis: The "Active Analog Approach"

The last chapter focused on conformational analysis. Could the techniques described there, such as the systematic rotation around certain bonds, be used to search for the spatial arrangement of the pharmacophore? Garland Marshall at Washington University, St. Louis, USA, developed such a technique, called the "**active analog approach**," in the late 1970s. First, putative pharmacophoric groups must be assigned to all molecules in a dataset. Then, the expected equivalence of the groups must be defined, meaning which groups are equivalent to which other groups. A systematic conformational search is then performed on the first compound in the dataset. During the search, the distances between the functional groups of the putative pharmacophore are determined for each specified geometry. These distances are stored. Since molecules cannot adopt just any geometry, the distances will fall into certain intervals. The same procedure is used for the second molecule in the dataset. In principle, only the distance ranges specified by the first molecule need to be searched. It may be that all the distances found with the second molecule were already found with the first. It is also possible that certain distances are excluded by the second molecule, thus, narrowing the "allowed" distance ranges. All molecules in the dataset are analyzed one after the other in this way.

If the conformational flexibility of the molecules is restricted in different regions of their scaffolds, there will be a chance that only one or a few distance ranges remain for the functional groups of the pharmacophore. This leads to the possible binding geometries of the pharmacophore groups in the ligands. Finally, geometry optimization can be performed, for which the "virtual" spring method is appropriate. This method is ideal at this stage because the desired solution is already very close. It is easy to imagine that the order in which the molecules are examined is crucial for the efficiency of the "active analog" procedure. It is best to start with the most rigid molecule in the dataset. With a little luck, this will already restrict a large part of the accessible conformational space. The list of possible distances will then remain rather small. By consistently using such constraints, Garland Marshall and his research group were able to propose in 1987 a model for the receptor-bound conformation of the ACE inhibitors shown in ◘ Fig. 17.4. What could be more rewarding than being able to personally validate this model years later and find that it was correct within a surprisingly small margin of error! The validation became possible because the crystal structure of the enzyme with several inhibitors from this dataset had been successfully determined in 2003 (see Sect. 25.5).

17.6 Molecular Recognition Properties and the Similarity of Molecules

The question is whether the concepts presented in the previous sections for representing the properties of molecules have really been adequately considered in the comparisons attempted. It is not easy to decide which functional groups correspond to the individual "anchor points" of a pharmacophore. Analogous functional groups must be oriented in a similar spatial direction in all molecules. In the case of the ACE inhibitors (◘ Fig. 17.4), conflicts arise even in the assignment of functional groups. Some analogues carry two carboxylate groups, which must be unambiguously assigned to the pharmacophore before they can be compared with other inhibitors.

The binding of small-molecule ligands to a protein is a mutual, targeted recognition process. Both partners must fit together to form a strong interaction. Parts of the ligand with complementary recognition properties determine the binding to the receptor. The term "recognition properties" refers to all properties that contribute to the specific interaction between molecules. Until now, only those properties and similarities that can be read directly from the molecular scaffold have been considered. But is that enough? What would the world be like if we could only recognize each other by our bones? Not even male and female would be immediately distinguishable! All the stimuli of interpersonal relationships conveyed by personal appearance and charisma would be lost. So far, molecules have been considered only in terms of their "molecular skeletons." Why should it be possible to describe ligand–receptor interactions at this level? Molecules also recognize each other by their shape and surface, as well as by the properties they transmit to their immediate environment and use to make contacts. The following example should illustrate this point. Methotrexate **17.6** (MTX) and dihydrofolate **17.7** (DHF) bind to the enzyme dihydrofolate reductase (◘ Fig. 17.6 and Sect. 27.2). The side chains of the two molecules are almost identical, but the heterocyclic moieties are different. It is known from NMR spectroscopic studies that the protonated form of MTX binds to the protein. Considering the chemical formulas, it is tempting to superimpose the two heterocycles directly. Perfect scaffold equivalence is achieved, and the heteroatoms in both molecules fall on top of each other. The receptor, however, does not care about an apparent equivalence of the molecular skeletons. Much more important is the interaction with the surface of the molecule. Polar molecules such as MTX or DHF will be bound to the protein by hydrogen bonds. The arrows in ◘ Fig. 17.6 characterize the H-bond donor and acceptor groups. The arrows point towards the molecule when an acceptor site is exposed and away from the molecule when a donor site is exposed. Initially, the molecules are oriented in space so that they correspond with respect to a direct atom–atom matching. For the moment, the basic molecular skeleton should be ignored and only the distribution of H-bond donor and acceptor groups should be considered (◘ Fig. 17.6, *bottom left*). The obtained equivalence is not very convincing. Therefore, another variant is considered in which the heterocycle of DHF is flipped along the bond between the heterocycle and the side chain. The spatial overlap of the two molecules is no longer optimal, but the pattern of exposed donor and acceptor groups for both molecules now shows a much better match (◘ Fig. 17.6, *bottom right*). In a different conformation, the molecules now present themselves with altered but much better matching molecular recognition properties. Even a trained eye can hardly read these differences from the structural formulas alone, even in a case as clear as the one presented here.

Models are fine, but are they actually correct? This can only be answered by experiment. Fortunately, in the present case, crystal structures are available for both ligands complexed with dihydrofolate reductase (DHFR). The observed binding geometries are shown

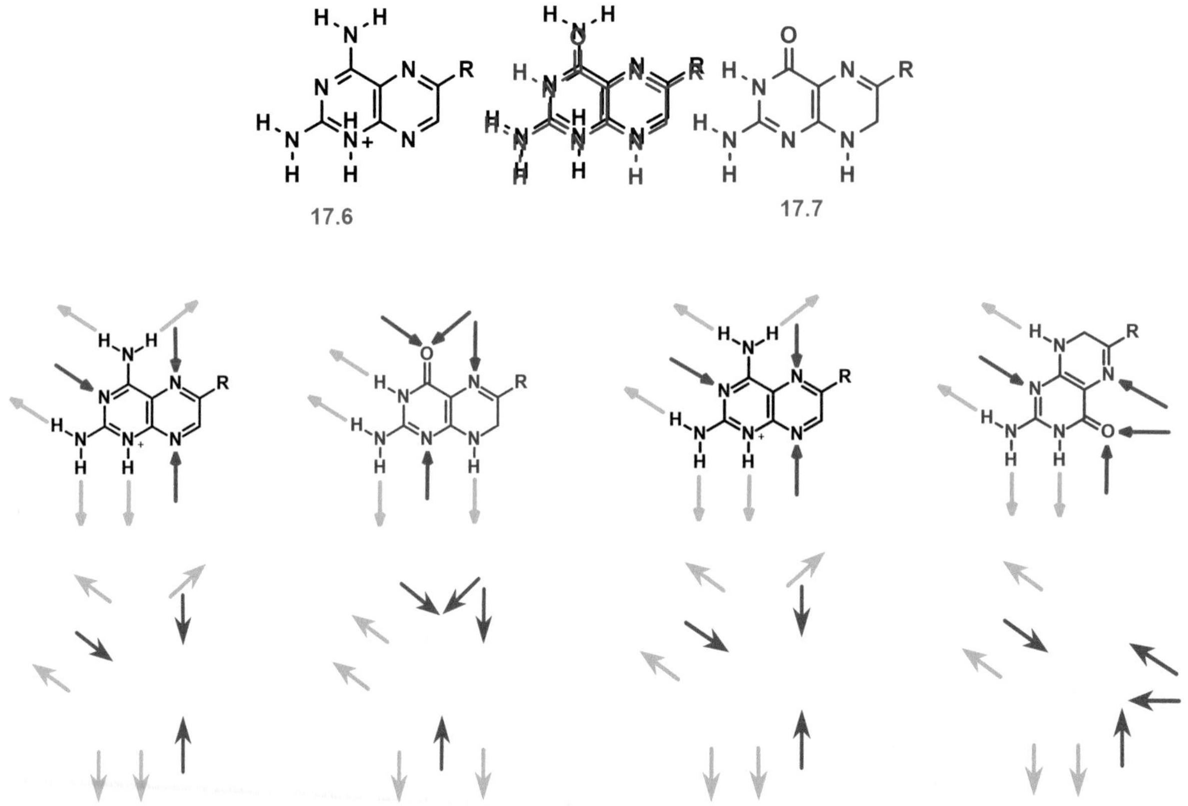

Fig. 17.6 Methotrexate **17.6** and dihydrofolate **17.7** are ligands of dihydrofolate reductase. The side chain R (see Sect. 27.2, Fig. 27.9) is identical for both except for a methyl group on the nitrogen atom. The heterocycles are different. *Upper row* Intuitively, superposition of both heterocycles directly upon each other when comparing the structures appear reasonable. Heteroatoms match pair-wise to one another. *Bottom row* Arrows are distributed around the molecules to compare the hydrogen-bonding properties. They are pointed to the molecule when an acceptor (*red*) is present and they point away for donor groups (*green*). If the molecular skeletons are masked out, and the distribution of H-bond donor and acceptor groups is concentrated upon, the atom–atom overlap obtained via the direct superposition of the rings will show rather unconvincing equivalence (*left*). Instead, if the heterocycle in **17.7** is flipped about the bond between the heterocycle and the side chain R, the pattern of donor and acceptor groups that will be obtained now exhibits convincing equivalence (*right*)

in Fig. 17.7. One aspartate and two carbonyl groups in the main chain and two water molecules are responsible for recognition in the binding pocket. The water molecules mediate hydrogen bonds between the ligand and the protein. The experimentally determined binding geometries show that the described considerations about the similarity of the hydrogen bonding properties led to the right conclusions. A surprising and apparently "non-equivalent" orientation of the two ligands in the binding pocket is easily explained. The properties responsible for the mutual recognition process must be compared. These are the only features that count in the comparison! It is noteworthy that the experimental confirmation of the considerations described above came eight years after the working hypothesis was proposed. This is a nice example of the performance of the model hypothesis.

Other properties besides hydrogen bonding can serve as additional criteria to define similarities in the molecular recognition process. The electrostatic potential (Chap. 15) calculated for the heterocyclic ring systems of DHF and MTX (Fig. 17.7) suggests very similar conclusions. In addition to the aforementioned H-bonding properties and electrostatic potentials, steric space filling and the distribution of hydrophobic properties on the surface of both ligands play an important role. When molecules are superimposed to predict their putative geometries in the binding pocket, their conformational flexibility must also be taken into account.

17.7 Automated Molecular Comparisons and Superpositioning Based on Recognition Properties

Is it possible to consider all of the properties mentioned in the last section in a method for superimposing molecules for relative comparison? To accomplish this, a measure of similarity must be computed for all the properties. This measure must be related to a spatial distance function. Then the spatial superposition can be performed. At the same time, the maximum similarity of the selected features is sought. The program SEAL

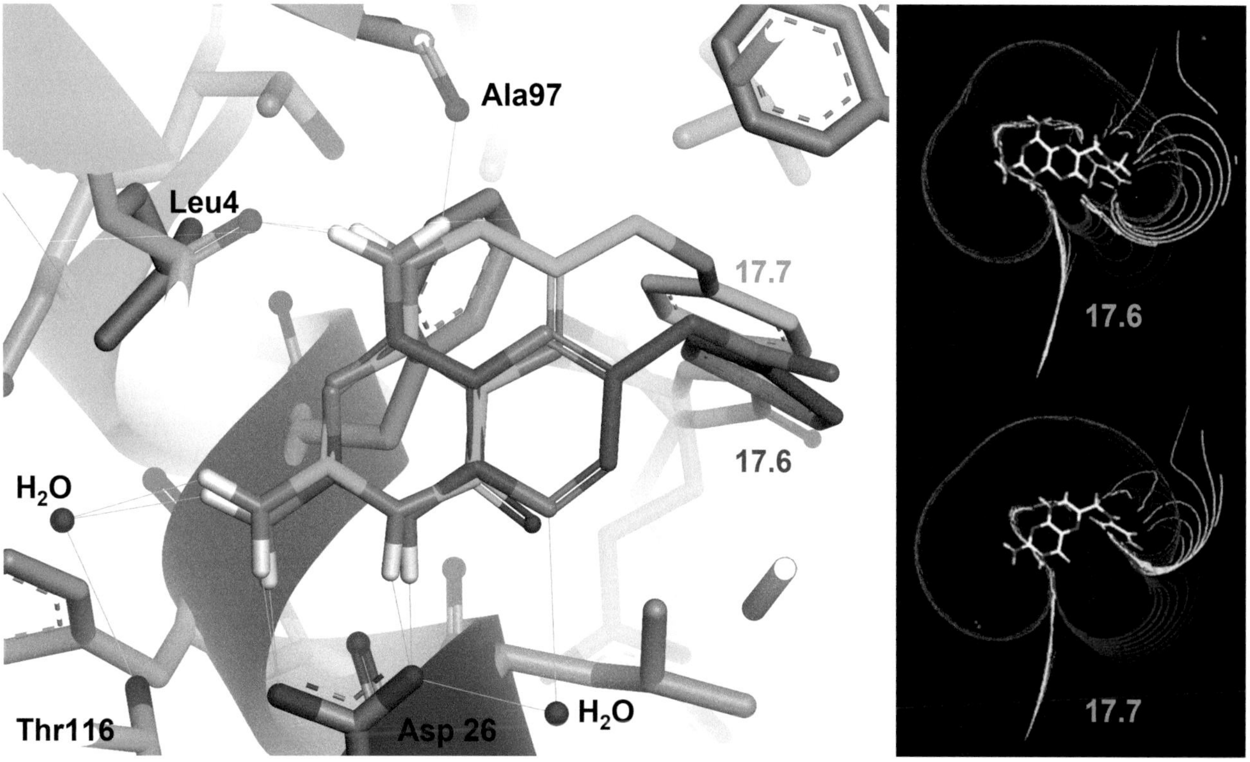

Fig. 17.7 Experimentally determined binding geometries of methotrexate (**17.6**, *gray* carbon atoms) and dihydrofolate (**17.7**, *green* carbon atoms) in dihydrofolate reductase. The heterocycles of the ligands are bound through H-bonds to the carboxylate or carbonyl group of an amino acid that is oriented into the binding pocket. Two water molecules (*red spheres*) mediate additional H-bonds between the ligands and the protein. The difference in the binding modes that is discussed in Fig. 17.6 is well approved. On the right-hand side, the electrostatic potentials around methotrexate (*top*) and dihydrofolate (*bottom*) are shown. The molecules are shown in a spatial orientation that was determined by crystal structure analysis. Considered qualitatively, the electrostatic potentials of both molecules have very similar spatial course in this orientation. (▶ https://sn.pub/UxHUJF)

by Simon Kearsley and Graham Smith at Merck Sharp & Dohme determines the spatial similarity of different properties distributed over the molecular scaffold. It simultaneously ranks the similarity with respect to the overlap volume of the molecules determined during superposition. In this way, the superposition of MTX and DHF is correctly predicted according to the experiment. The conformational flexibility is also considered in this analysis. For this purpose, precalculated conformers can be taken and successively compared with one another. This is implemented in the program ROCS by Anthony Nicholls at OpenEye, USA. Christian Lemmen at GMD in St. Augustin, Germany, has taken a slightly different approach in the program FlexS. First, a reference ligand is represented by a set of property-loaded Gaussian functions. In effect, the molecule is described as a density distribution of pharmacophore properties in space. The candidate molecule to be superimposed on the reference molecule is then taken. It is broken down into fragments. First, a central base fragment is placed on the reference so that its description by Gaussian functions overlaps the reference as much as possible. Then the other fragments are added to the base fragment until the complete ligand is recovered at the end. During this incremental attachment procedure, care is taken to ensure that the fragments added to the base fragment also fit optimally into the Gaussian description of the reference. At the same time, the conformational flexibility of a ligand is taken into account when adding subsequent fragments.

A complication arises when analyzing the similarity of molecules using these methods. It is assumed that the relevant properties defining the similarity have been found. However, the question arises as to what is accepted as "sufficiently" similar for a comparable effect at a receptor. There is a toy called a "shape sorter" in which children try to fit blocks of different shapes into a box through holes which are prepunched for them. There is a matching hole for each block shape, cube, cuboid, and cylinder with a circular or elliptical cross section. In a **similarity analysis**, one is tempted to group cubes and cuboids, or "circular" and "elliptical" cylinders, as related because of their shape. When you try to fit the pieces through the holes in the box, you may find that the cube fits not only through the rectangular hole, but also, perhaps with some effort, through the hole for the elliptical cylinder. The cube is only slightly too large to fit through the square hole and the hole for the circular cylinder. So are the cube and the elliptical cylinder or the cube and the circular cylinder more similar? The **measure of similarity** applied to the molecules is calibrated to the receptor into which the molecules are supposed to fit. Thus, it is always a **relative measure**!

Thiorphan and *retro*-thiorphan (Sect. 5.5, formulas **5.23** and **5.24**) differ only in the spatial sequence of the amide bond. They bind with almost identical affinity to the zinc protease thermolysin, and NEP 24.11. Therefore, they would be classified as very similar. The zinc protease ACE binds thiorphan by at least a factor of 100 times more strongly than *retro*-thiorphan (Sect. 5.5, ◘ Fig. 5.12). Relative to this enzyme, both substances must be considered dissimilar. Another extreme is seen in the oligopeptide-binding protein A (Sect. 4.1). It binds any tri- to pentapeptide containing a central Lys–Xxx–Lys unit (Xxx: any amino acid) with almost the same affinity. In principle, only information about the shape of the binding site is needed for a similarity analysis. Only then can the requirements be adequately defined. However, in a drug design project, the structure of the receptor may still be unknown; thus there is no choice: only hypotheses and their experimental testing in incremental steps can approximate the structural requirements of the receptor.

17.8 Rigid Analogues Trace the Biologically Active Conformation

The concepts in Chap. 16 showed that an enormously large number of conformers can be easily generated for many drug-like molecules. If a comparison of all conformers is desired, the undertaking will quickly become computationally intensive. How can we obtain a relevant image of the bound conformations? Either a compound in the dataset is highly rigid and constrains the putative arrangements of the pharmacophore in space, or the molecules under consideration are rigid in different regions of their molecular scaffold. ◘ Fig. 17.8 shows the structural superposition of the steroid **17.2** with the inhibitors **17.3**–**17.5** described above. This result was obtained from a similarity analysis with multiple conformers and the superposition is very similar to the calculation with the "virtual" spring forces. It has, however, a decisive advantage: we do not need preconceived definitions of equivalent centers between which the spring forces are applied. These equivalences arise automatically through a similarity comparison of the properties that are distributed over the molecules.

17.9 If Rigid Analogues are Lacking: Model Compounds Elucidate the Active Conformation

In the last example, a largely rigid reference compound was provided. How to proceed if no such reference compound is known? Only experiment can help. Rigid analogues have to be synthesized. These are tested for biological activity. If they still show affinity for the receptor, it can be assumed that the active conformation has been frozen.

The following example illustrates how the receptor-bound conformation can be studied by synthesizing rigid model compounds. The calcium channel blocker nifedipine **17.8** (Sects. 2.5 and 30.4) contains several rotatable bonds (◘ Fig. 17.9). It can, therefore, adopt many conformations. What is the orientation of the phenyl ring relative to the dihydropyridine ring? This question was elegantly answered by Wolfgang Seidel at Bayer through the synthesis and crystal structure determination of the cyclized derivatives **17.9**. An additional lactone ring changes the biological activity of the derivative depending on the ring size. In compounds with a six-membered lactone, the phenyl and dihydropyridine rings are almost in the same plane. Conversely, in the twelve-membered ring derivative, the phenyl ring is perpendicular to the dihydropyridine ring. The affinity of this compound is about five orders of magnitude higher than that of the six-membered lactone derivative. Therefore, it was hypothesized that nifedipine acts in a conformation in which the phenyl and dihydropyridine rings are perpen-

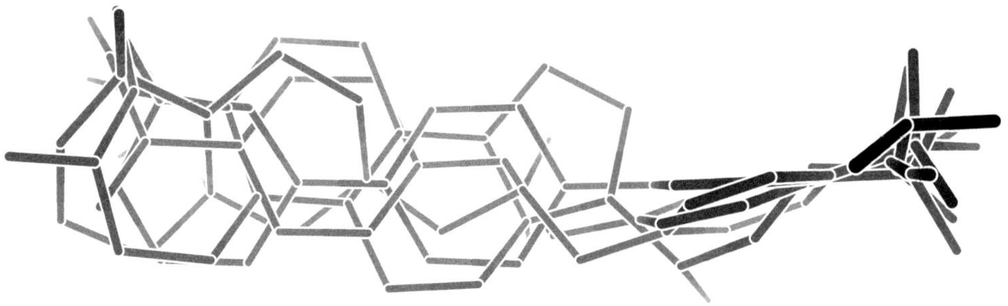

◘ **Fig. 17.8** Superposition of the steroid **17.2** and three inhibitors **17.3**–**17.5** according to a spatial comparison of their molecular properties. In contrast to methods with "virtual" spring forces, this method does not require a predefined equivalence of molecular groups. It is automatically generated by the similarity comparison of many different conformations

17

dicular to each other. Many years later, this hypothesis was shown to be correct. A cryo-EM structure of the calcium channel with bound nifedipine demonstrates the perpendicularity (Sect. 30.4).

After this question has been answered, more compounds can be designed. A relevant superposition that corresponds to the conditions in the protein's binding pocket will be possible. Such superpositions have gained a decisive meaning in the context of 3D structure–activity relationships (Chap. 18).

17.10 The Protein Defines the Pharmacophore: "Hot Spot" Analysis of the Binding Pocket

It was described in Sect. 17.1 that a pharmacophore can also be deduced from the protein structure. The **computer program GRID** from Peter Goodford is a tool that is often used for this purpose. It calculates favorable positions in protein-binding pockets for functional groups of a potential ligand. This could be a carboxylate group, a hydroxy group, or an aliphatic carbon atom. The potential function used for GRID has been calibrated for a variety of functional groups on the crystal structures of organic molecules. The result of a GRID calculation is a set of interaction energies at each point of intersection of a grid that is inscribed in the binding pocket. The energies are presented graphically, for example, by indicating the region of space where the interaction energy meets or exceeds a predefined threshold. They indicate **hot spots** for the placement of functional groups of a potential ligand. The areas in which the interactions with an aromatic carbon atom or a hydroxyl oxygen atom are favorable are shown for the enzyme thermolysin in ◘ Fig. 17.10. Such calculations are carried out with a set of different probes, for instance, a water molecule, an aromatic carbon, a hydrogen-bond acceptor or donor,

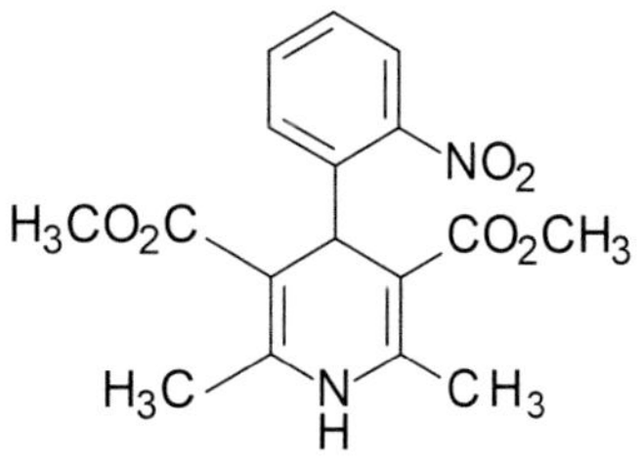

17.8 Nifedipine

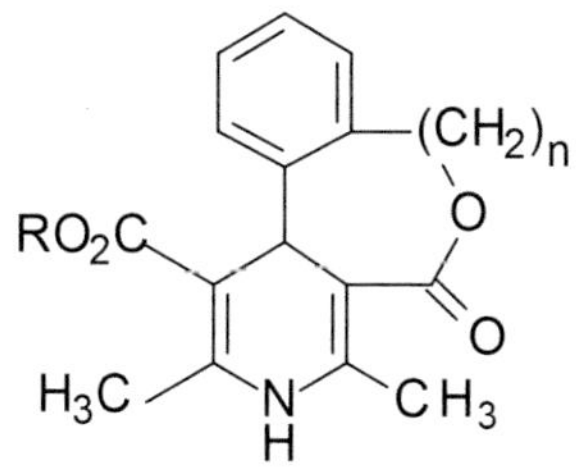

17.9

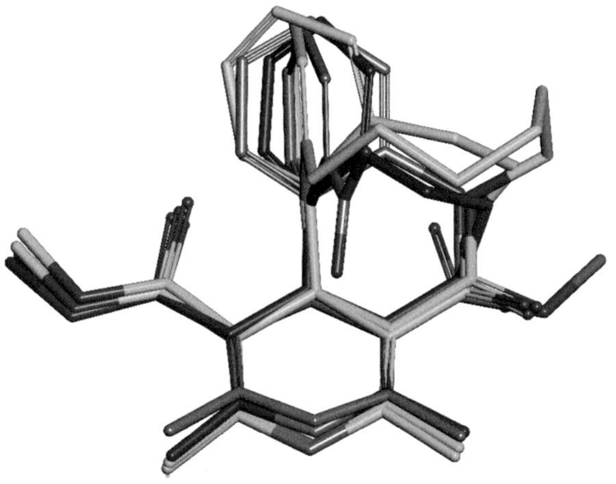

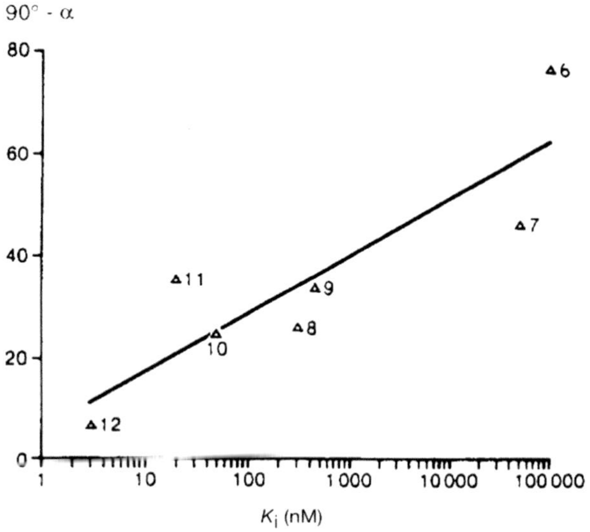

◘ **Fig. 17.9** The calcium channel blocker nifedipine **17.8** contains multiple rotatable bonds. The phenyl ring can coincide with a plane of the dihydropyridine ring or they orient perpendicular to each other. To distinguish between these possibilities, lactones with different ring size **17.9** were synthesized and their crystal structures were determined. The phenyl ring lies almost parallel to the dihydropyridine ring (α ≈ 0°) in the compound with the six-membered-ring lactone (*orange*). Upon increasing the ring size, the angle between the two rings grows so that a perpendicular orientation (α ≈ 80°) is achieved in the twelve-membered-ring derivative (*green*). The biological activity increases from virtually inactive, as in the six-membered ring, to almost five orders of magnitude higher for the twelve-membered-ring derivative. The bioactive conformation of nifedipine (*gray*), therefore, requires a perpendicular orientation of the two rings.
(► https://sn.pub/tNnF5T)

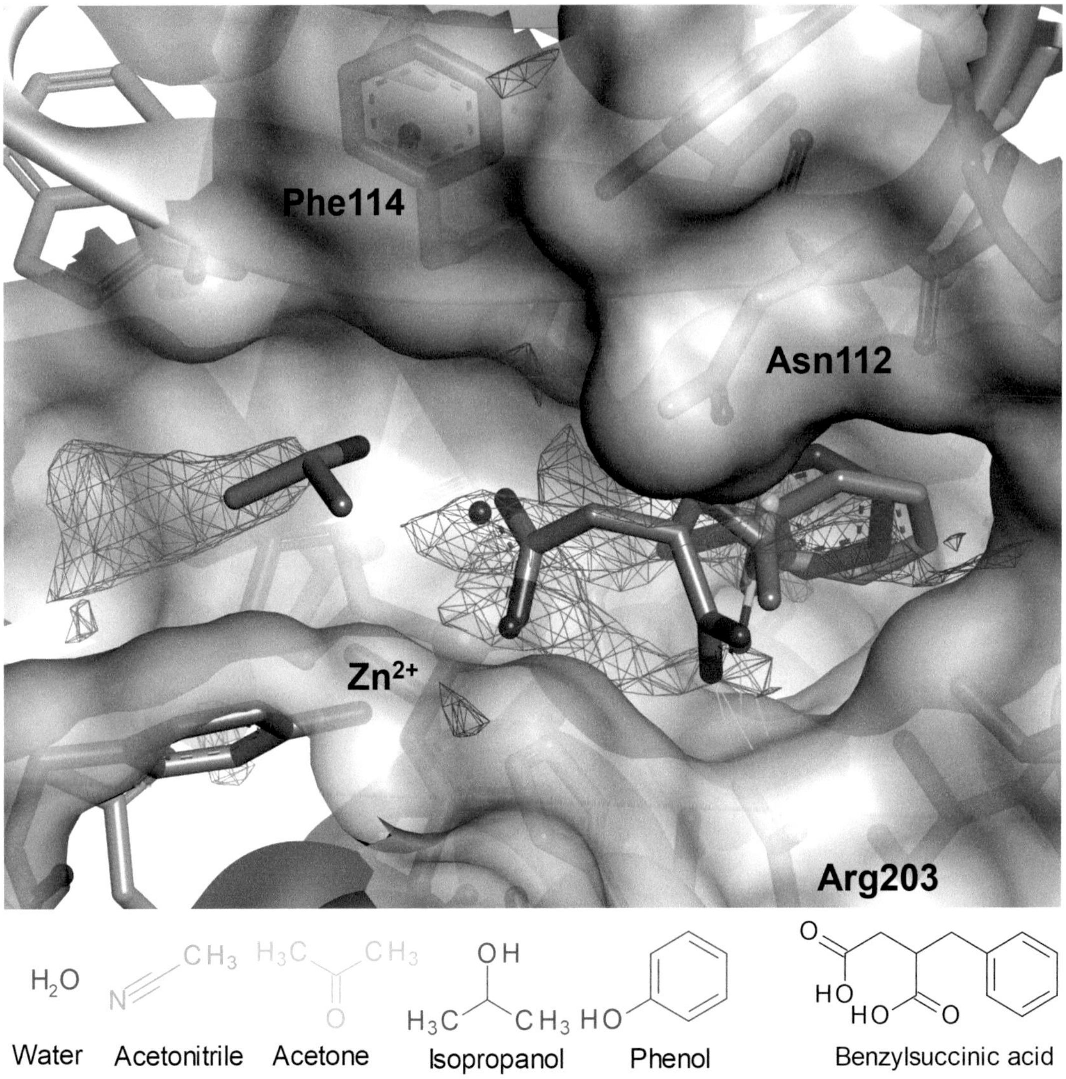

Fig. 17.10 An analysis of the binding pocket of thermolysin. Areas of favorable interactions were calculated for an aromatic carbon probe (*white*) and a hydroxyl oxygen atom (*red*). There are also fragments mentioned in Fig. 7.10 that could be determined by allowing the probe molecules to diffuse into the protein crystals. The calculated *hot spot* corresponds well with the positions that were crystallographically determined with molecular probes. (► https://sn.pub/m3N94q)

or a positively or negatively charged group. The results provide valuable information about the shape and electrostatic properties of the binding pocket.

Another approach to protein structure analysis is based on the idea that the physical nature of nonbinding interactions is identical in protein–ligand complexes and in the crystal packing of small organic molecules. The latter are particularly interesting for this purpose because the crystal structures of small organic molecules are regularly determined with great precision. More than 1.25 million crystal structures are stored in the Cambridge Database (Sect. 13.9). This collection is ideal for obtaining relevant and reliable data for ligand design through statistical analysis of crystal packing data (Sect. 14.7). Suppose there is a carboxylate group –COO– on the protein that protrudes into the binding pocket. Where must a partner group be positioned to form a favorable interaction? To answer this question, the Cambridge Database was first searched for compounds containing carboxylate groups. Then, for each of the retrieved acid groups, the position of the counter group forming an H-bond to the carboxylate was stored. Finally, the collective of all H-bonds found was superimposed by exactly mapping the carboxylate groups of all examples to one another. The distribution of H-bond donor groups (Fig. 17.11) provides a **composite picture** of the allowed range of **H-bonding geometries**. Such a distribution can then be superimposed on the protein structure by matching it to the carboxylate group of the residue in the

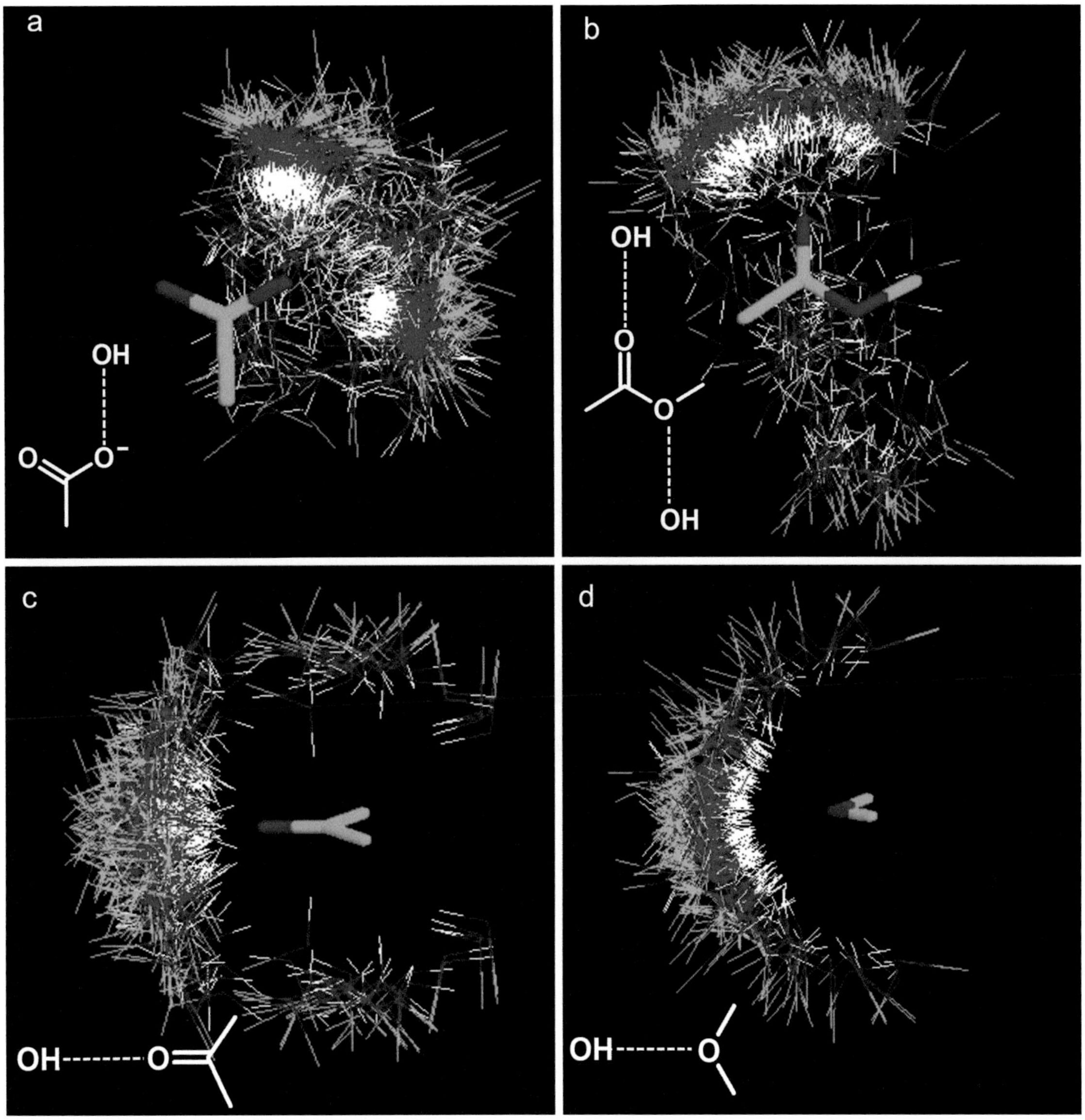

Fig. 17.11 Hydrogen-bonding geometries (carbon is *green*, oxygen is *red*, and hydrogen is *white*) around a carboxylate group (**a**), ester group (**b**), carbonyl group (**c**), and ether group (**d**). Structures with these central groups that form hydrogen bonds with OH-donor groups were extracted from the Cambridge Structural Database. These examples were superimposed based on the geometry of the central group. It is obvious that there is considerable variability in the interaction geometries, but also that preferred orientations are to be found. It is also shown that, for instance, the interaction pattern around an ester group (**b**) is not simply an additive superposition of the distribution around a carbonyl group (**c**) and an ether group (**d**)

protein. Areas where the distribution overlaps with other atoms in the protein are discarded. In this way, the energetically most favorable sites for a counter group in the binding pocket are identified. Fig. 17.12 compares these distributions with a protein–ligand complex. As expected, the hydrogen bond geometries found in the complex agree well with the range found in the crystal packing of organic molecules. A system of rules for nonbonding interactions in protein–ligand complexes was obtained from the statistical evaluations of all the groups found in proteins. These rules are compiled at the Cambridge Crystallographic Data Centre in the **Isostar database**. Once overlaid on the protein, they can be contoured to map binding hotspots using the **SuperStar** program.

Knowledge-based potentials are another approach to represent a protein-based pharmacophore. For this, the contact geometries in protein–ligand complexes are evaluated. A histographical distribution is generated showing how often a particular contact occurs between a group found in a ligand and in an amino acid of a protein. When such a statistical frequency distribution is related to a mean reference state, an energy function can be calculated from it. This function assumes that contacts, occurring more frequently than the average distribution, are

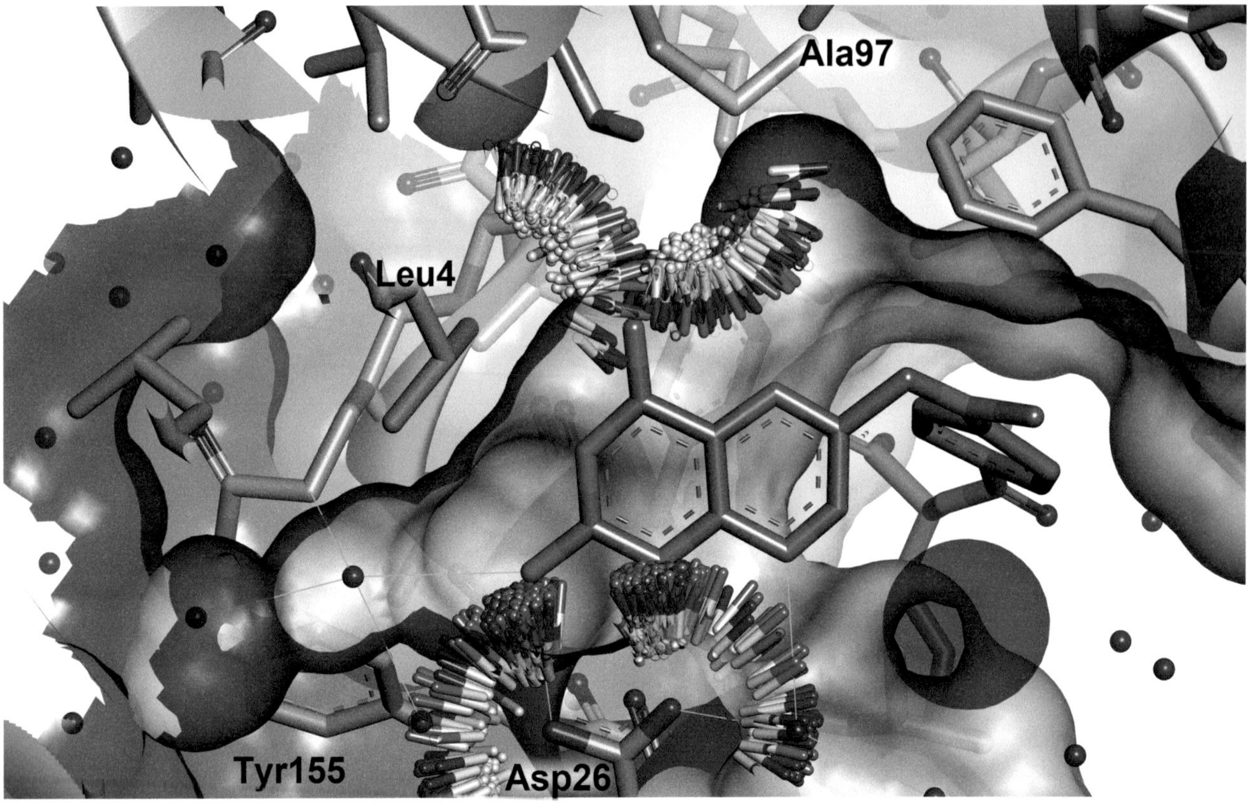

▣ **Fig. 17.12** The distribution of H-bond donor groups (carbon is *white*, oxygen is *red*, and nitrogen is *blue*) around a carboxylate group or a carbonyl group are superimposed with the 3D structure of the complex of methotrexate with dihydrofolate reductase (cf. ▣ Fig. 17.7). The distributions are imposed onto the acid group of Asp 26 and the carbonyl groups of Leu 4 and Ala 97. The hydrogen bonds formed between protein and ligand coincide geometrically with ranges often found in crystal packings of small organic molecules. (▸ https://sn.pub/rpQuWI)

energetically favorable. If they occur less frequently than the average, they will be considered unfavorable. These statistical potentials have been incorporated into the scoring function **DrugScore**, developed by Holger Gohlke in Marburg, Germany. They can also be used for graphical analysis of binding pockets and help to indicate the spatial distribution of **hot spots** for ligand binding.

17

The **MCSS method** was developed in the group of Martin Karplus at Harvard Medical School, Boston, USA. Several thousand random probe molecules such as acetone, water, methanol, or benzene were placed in a binding pocket for this. A computer simulation is started with which the single probe molecules are moved into optimal positions. They are driven by a calculation according to the underlying force field. The probe molecules experience the interaction with the protein, but they do not "see" one another. At the end of the calculation, a frequency distribution for the probe molecules is obtained. If this distribution is evaluated, a *hot spot* for an interaction with the protein can be highlighted. If the thus obtained *hot spots* are compiled into a composite picture, a protein-based pharmacophore will be obtained.

17.11 Searching for Pharmacophore Patterns in Databases Generates Ideas for Novel Lead Compounds

A pharmacophore can be used to search a database for promising candidates that can be accommodated in the binding pocket of a protein. The reference pharmacophore can either be derived from a set of superimposed ligands, or a reference protein can define its properties. How such a **database search** is performed and what is discovered depends on how much information is stored in the database itself. If only 2D structures are collected, all examples with a certain functional group or substructure can be retrieved. Based on the topology, different criteria are defined to determine the degree of similarity between molecules. If the definition of the pharmacophore is very general, e.g., an aromatic compound with an acidic group and a basic nitrogen atom, many hits will be found. However, the relative spatial distances between these groups are important. Such information is not considered when searching a 2D database. Matthias Rarey and Scott Dixon at the time at Smith Kline & Beecham developed the **Feature-Trees method**, which can search large databases using topological criteria. However, it does not compare the connectivity of chemical formulas. Rather, the database entries are first classified by the topological sequences of certain features, such as the presence of an H-bond donor group or a hydrophobic cyclic molecular building block. In this way, molecules can be compared and candidates with pharmacophore properties in a comparable topological connectivity can be found extremely quickly.

Databases that contain 3D molecular geometries allow the search for the spatial pattern of the pharmacophore. For example, the Cambridge Structural Database of crystal structures of small organic molecules (Sect. 13.9) can be used for such a search. Molecules are found with experimentally determined geometries that satisfy the pharmacophore. In the search for ligands for HIV protease (Sect. 24.3), a pharmacophore pattern was derived from the known crystal structure of the enzyme, and the Cambridge Database was searched for molecules that match this pattern. The results of this search are presented in Sect. 24.4 (◘ Fig. 24.16) in detail. It inspired the researchers at Dupont–Merck with first ideas that led to the development of an entirely new class of nonpeptidic HIV protease inhibitors.

Today, databases containing 3D structures of molecules generated from 2D structural formulas are commonly used alongside experimental structural databases. In other approaches, the spatial structure of candidate molecules is generated on the fly during the search (Sect. 15.2). Here, as with most entries in the Cambridge Database, each molecule exists in only one conformation. However, molecules can adopt many different conformations (Chap. 16). It is usually the exception rather than the rule that a flexible molecule will be stored in the "correct" conformation required for the search. Therefore, conformational flexibility must be taken into account during the search. An exhaustive search, such as the "active analog approach," would require too much computational time. Therefore, fast algorithms have been developed to identify whether certain pharmacophoric groups on the molecules could fall within predefined distance ranges. It is sufficient to estimate the minimum or maximum achievable distances. This concept has been realized, for example, in the program **UNITY** from the company Tripos. It is possible to start from a database that contains several precalculated conformers. In this case, it is very important that the distribution of the conformers in the conformational space is as representative as possible (Sect. 16.6). The individual conformers will then be checked to see whether they fit the defined pharmacophore pattern. This concept has been implemented in the database search engine **Catalyst** from Accelrys. A similar concept is followed by the program **LigandScout** of Thierry Langer in Vienna, Austria and Gerhard Wolber in Berlin, Germany.

Such database searches are not expected to immediately yield candidates for clinical testing. However, as a generator of ideas, they can lead drug discovery scientist to new lead structures and take his or her synthesis plans down completely different paths. Database searches are now widely used as part of virtual screening (Sect. 7.6). This involves screening proprietary collections of compounds or searching compilations of commercially available compounds. John Irwin and Brian Shoichet at UCSF in San Francisco, USA, have taken the initiative to continuously store commercially available compounds in the ***ZINC* database** and make them available for database searching. Preset filters help to extract the desired subset from the collection of several million compounds for the user's own search. Hits found in this way can be obtained commercially and tested experimentally in an assay. Many candidates for new lead structures have already been discovered through this **lead discovery by shopping** (for an example, see Sect. 21.7).

17.12 Synopsis

- The structure of the binding pocket determines which functional groups are necessary on the ligand side for successful protein binding. Either the ligand or the protein structure can be used as the starting point from which a pharmacophore is derived.
- The superposition of active and inactive small-molecule ligands from a series of related compounds can be used to define the allowed and forbidden areas in a hypothetical binding pocket. Logical operations of volume differences are indicative for the design of optimized ligands.
- Flexible molecules that can adopt different conformations present a special challenge in the mutual superpositions. The molecules must be energy-minimized as part of the superposition procedure or, alternatively, multiple conformations must be evaluated.
- Alternatively, a set of molecules can be superimposed by assigning pharmacophoric groups, and through systematic rotations about all open-chain single bonds a common alignment is found in the "active analog approach."
- Care must be taken to not be deceived by molecules that look similar with respect to their chemical formulas. Instead, the interacting functional groups are important for the molecular recognition at the binding pocket and not the scaffold itself. The role of water in the binding must not be underestimated.
- Molecular recognition properties can also be considered to mutually superimpose molecules.
- The synthesis of a structurally rigid analogue (or analogues) can help to define and validate the pharmacophore assignment and the determination of the biologically active conformation.

- Binding "hot spots" can be found by examining the protein by mapping the binding pocket with small molecules or molecular probes with different properties. These give some ideas as to what sort of molecule might successfully bind to the target protein.
- The Cambridge Database of crystal structures provides valuable insights into preferred interaction geometries and motifs. Such information is of high relevance for protein–ligand complexes because the forces that are responsible for crystal packing are the same as for nonbonding interactions between active substances and proteins.
- A variety of databases are available that can be screened by using a 3D pharmacophore as a search query. Usually, commercially available compounds are screened first. If they show activity on a certain protein of interest, they can be purchased and tested, and will hopefully provide a starting point for further lead discovery.

Bibliography and Further Reading

General Literature

T. Langer and R. D. Hoffmann, Pharmacophores and Pharmacophore Searches (Vol. 32 in Methods and Principles in Medicinal Chemistry, R. Mannhold, H. Kubinyi and G. Folkers, Eds.), Wiley-VCH, Weinheim (2006)

G. R. Marshall, Computer-Aided Drug Design, in: Computer-Aided Molecular Design, W. G. Richards, Ed., IBC Technical Services Ltd, London, pp. 91–104 (1989)

G. Klebe, Structural Alignment of Molecules, in: 3D-QSAR in Drug Design. Theory, Methods and Application, H. Kubinyi, Ed., ESCOM, Leiden, pp. 173–199 (1993)

Y. C. Martin, 3D Database Searching in Drug Design, J. Med. Chem., **35**, 2145–2154 (1992)

Special Literature

C. G. Wermuth, C. R. Ganellin, P. Lindberg, L. A. Mitscher, Glossary of terms used in medicinal chemistry (IUPAC Recommendations 1998), Pure Appl. Chem., **70**, 1129–1143 (1998)

17

W. E. Klunk, B. L. Kalman, J. A. Ferrendelli and D. F. Covey, Computer-Assisted Modeling of the Picrotoxinin and γ-Butyrolactone Receptor Site, Mol. Pharmacol., **23**, 511–518 (1983)

M. F. Mackay and M. Sadek, The Crystal and Molecular Structure of Picrotoxinin, Austr. J. Chem., **36**, 2111–2117 (1983)

G. R. Marshall, C. D. Barry, H. E. Bossard, R. A. Dammkoehler and D. A. Dunn, The Conformational Parameter in Drug Design: The Active Analog Approach, in: Computer-Assisted Drug Design, ACS Symp. Series 112, E. C. Olson and R. E. Christoffersen, Eds., Amer. Chem. Soc., Washington DC., pp. 205–226 (1979)

D. Mayer, C. B. Naylor, I. Motoc and G. R. Marshall, A Unique Geometry of the Active Site of Angiotensin-Converting Enzyme Consistent with Structure-Activity Studies, J. Comput.-Aided Mol. Design, **1**, 3–16 (1987)

D. J. Kuster and G. R. Marshall, Validated Ligand Mapping of ACE Active Site, J. Comput.-Aided Mol. Design, **19**, 609–615 (2005)

J. T. Bolin, D. J. Filman, D. A. Matthews, R. C. Hamlin and J. Kraut, Crystal Structure of *Escherichia coli* and *Lactobacillus casei* Dihydrofolate Reductase Refined at 1.7 Å Resolution, J. Biol. Chem., **257**, 13650–13662 (1982)

S. K. Kearsley and G. M. Smith, An Alternative Method for the Alignment of Molecular Structures: Maximizing Electrostatic and Steric Overlap, Tetrahedron Comput. Methodol., **3**, 615–633 (1990)

P. C. D. Hawkins, A. G. Skillman and A. Nicholls. Comparison of shape-matching and docking as virtual screening tools. J. Med. Chem., **50**, 74–82 (2007) https://www.eyesopen.com/rocs (Last accessed Nov. 18, 2024)

C. Lemmen, T. Lengauer and G. Klebe, FlexS: A Method for Fast Flexible Ligand Superposition, J. Med. Chem., **41**, 4502–4520 (1998) https://www.biosolveit.de/wp-content/uploads/2021/01/FlexS.pdf (Last accessed Nov. 18, 2024)

G. Klebe, T. Mietzner and F. Weber, Different Approaches Toward an Automatic Structural Alignment of Drug Molecules: Applications to Sterol Mimics, Thrombin and Thermolysin Inhibitors, J. Comput.-Aided Mol. Design, **8**, 751–778 (1995)

W. Seidel, H. Meyer, L. Born, S. Kazda and W. Dompert, Rigid Calcium Antagonists of the Nifedipine-Type: Geometric Requirements for the Dihydropyridine Receptor, in: QSAR as Strategies in the Design of Bioactive Compounds, J. K. Seydel, Ed., VCH, Weinheim, pp. 366–369 (1984)

P. Goodford, Drug design by the method of receptor fit, J. Med. Chem., **27**, 557–564 (1984) https://www.moldiscovery.com/software/grid/ (Last accessed Nov. 18, 2024)

I. J. Bruno, J. C. Cole, J. P. Lommerse, R. S. Rowland, R. Taylor and M. L. Verdonk, IsoStar: a library of information about nonbonded interactions, J. Comput.-Aided Mol. Design, **11**, 525–537 (1997)

IsoStar: https://www.ccdc.cam.ac.uk/solutions/software/isostar/ (Last accessed Nov. 18, 2024)

M. L. Verdonk, J. C. Cole, R. Taylor, SuperStar: A Knowledge-based Approach for Identifying Interaction Sites in Proteins, J. Mol. Biol., **289**, 1093–1108 (1999)

SuperStar: https://www.ccdc.cam.ac.uk/solutions/software/superstar/ (Last accessed Nov. 18, 2024)

G. Klebe, The Use of Composite Crystal-Field Environments in Molecular Recognition and the "De-Novo" Design of Protein Ligands, J. Mol. Biol., **237** 212–235 (1994)

R. Taylor and P. A. Wood, A Million Crystal Structures: The Whole Is Greater than the Sum of Its Parts, Chem. Rev., **119**, 9427–9477 (2019)

H. Gohlke, M. Hendlich and G. Klebe, Knowledge-based Scoring Function to Predict Protein-Ligand Interactions, J. Mol. Biol., **295**, 337–356 (2000) https://www.fz-juelich.de/en/ibg/ibg-4/expertise/databases-softwares-and-webservers-in-the-gohlke-group/drugscore (Last accessed Nov. 18, 2024)

A. Caflisch, A. Miranker, and M. Karplus, Multiple copy simultaneous search and construction of ligands in binding sites: application to inhibitors of HIV-1 aspartic proteinase, J. Med. Chem., **36**, 2142–2167 (1993)

D. Joseph-McCarthy, J. M. Hogle and M. Karplus, Use of the multiple copy simultaneous search (MCSS) method to design a new class of picornavirus capsid binding drugs, Proteins, Struct, Funct, Bioinform., **29**, 32–58 (1997)

M. Rarey and J. S. Dixon, Feature trees: A new molecular similarity measure based on tree matching, J Comput.-Aided Mol. Des., **12**, 471–490 (1998) https://www.biosolveit.de/wp-content/uploads/2022/03/FTrees.pdf (Last accessed Nov. 18, 2024)

G. Wolber and T. Langer, LigandScout: 3-D pharmacophores derived from protein-bound ligands and their use as virtual screening filters, J. Chem. Inf. Model., **45**, 160–169 (2005) https://ligandscout.software.informer.com/ (Last accessed Nov. 18, 2024)

J.J. Irwin and B.K. Shoichet, ZINC—A Free Database of Commercially Available Compounds for Virtual Screening, J. Chem. Inf. Model., **45**, 177–182 (2005) https://zinc15.docking.org/ (Last accessed Nov. 18, 2024)

Quantitative Structure–Activity Relationships

Contents

G. Klebe, *Drug Design*, https://doi.org/10.1007/978-3-662-68998-1_18

Quantitative structure–activity relationships, **QSAR** (usually pronounced [*'kyü:sar*]), attempt to describe and quantify the correlation between **chemical structure** and **biological activity**. The investigated substances should come from a chemically uniform series and must interact with the same biological target. They should also display the same mode of action. For example, structurally analogous inhibitors of a particular protein can be compared among themselves, but not different blood pressure lowering drugs that have diverse modes of action on different target proteins. The correlation between biological activity and physicochemical properties is always related to relative potency in a test model, but not to different modes of action.

The basis for quantitative correlations between chemical structure and biological effect is the entirely reasonable assumption that the differences in physicochemical properties are responsible for the relative potency of the interactions of the drug with biological macromolecules. In a first approximation, these are assumed to contribute additively to the affinity of a drug for its receptor. The concept of describing the biological activity of substances with mathematical models is derived from this approach.

For the system under investigation, it can be assumed that the simpler it is, the more likely it is that a quantitative structure–activity relationship can be derived. To a certain extent, this is true for *in vitro* systems, such as enzyme inhibition or receptor binding, where the assay records only the binding of a compound to a protein. The more complex the system, e.g., effects on the central nervous system of an animal after oral administration, the more different processes have to be considered. In this case, absorption, distribution, blood–brain barrier penetration, transport to the target tissue, metabolism, and excretion overlap with one another and with the actual effect at the receptor. In principle, an individual structure–activity relationship is required for each of these events. In order to establish valid and relevant models for each of these steps, appropriate test systems are needed to study the different steps separately. In favorable cases, it may be possible to characterize a complex multistep process by a single equation. This will only be feasible if one step, e.g., the penetration of the blood–brain barrier, dominates the entire structure–activity relationship.

18

18.1 How It All Began: Structure–Activity Relationships of Alkaloids

The South American arrow poison tubocurarine (Sect. 6.2) was the first therapeutic principle for which the exact mode of action was elucidated. In 1852, Claude Bernard realized that this quaternary alkaloid causes muscle paralysis, but that both the nerve and the muscle remain independently excitable. Curare must, therefore, act on the coupling between nerve and muscle. The Scottish pharmacologists Alexander Crum-Brown and Thomas Fraser studied in more detail whether the quaternization of the nitrogen atom of various alkaloids (▫ Fig. 18.1) influences their biological effects. In 1868, on the basis of very different effects observed before and after the transformation of alkaloids, they formulated a general equation to describe **structure–activity relationships** (Eq. 18.1).

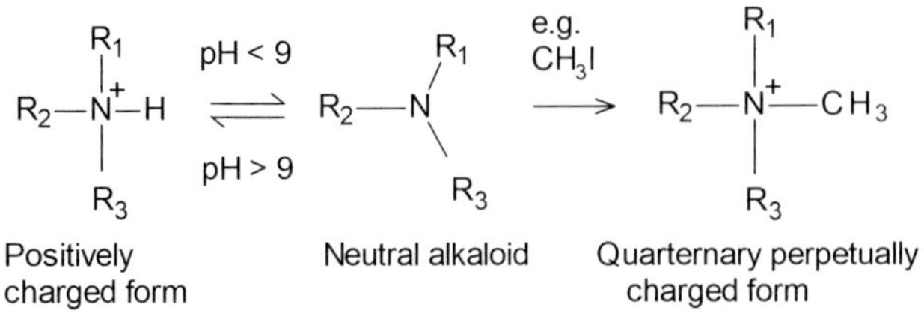

▫ **Fig. 18.1** The protonation of a tertiary amine depends on the pH value of the medium (*left*). On the other hand, the quaternization of a nitrogen atom leads to a permanently positively charged compound (*right*)

$$\Phi = f(C) \tag{18.1}$$

This equation is ingeniously simple, but it says only that Φ, the biological activity, is a function of *C*, the chemical structure. At that time, the tetrahedral structure of the carbon atom had not been elucidated, and the composition of many organic compounds, especially complex natural products, was completely unknown.

18.2 From Richet, Meyer, and Overton to Hammett and Hansch

In 1893, Charles Richet published a study on the toxicity of organic compounds. Comparing the water solubility of ethanol, diethyl ether, urethane, paraldehyde, amyl alcohol, and absinthe extract (!) to the lethal dose in the dog, he concluded *plus ils sont subles, moins ils sont toxiques*, that is, the better the solubility, the less the toxicity. This was the first evidence of a linear inverse relationship between water solubility and biological activity.

At the turn of the last century, the pharmacologist Hans Horst Meyer and the botanist Charles Ernest Overton independently established the **lipid theory of anesthesia**, which combines three important statements:

- All chemically unreactive substances that are lipophilic and can be distributed in biological systems have anesthetic effects.
- The biological effect occurs in nerve cells because fats play an important role in their function.
- The relative potency of anesthetics depends on their partition coefficient (Sect. 19.2) in a mixture of fats and water.

The work of Crum-Brown, Fraser and Richet or the contribution of Meyer and Overton can be considered as the origin of quantitative structure–activity relationships. In fact, after the formulation of the anesthetic theory, numerous other linear and later nonlinear dependencies on the lipophilicity, the "fat affinity" of drugs, were found. However, these were all relatively unspecific "membrane" effects.

In the mid-1930s, Louis P. Hammett formulated a relationship between the electronic properties of substituents and the reactivity of aromatic compounds. According to this relationship, the relative contributions of electron-withdrawing and electron-donating substituents to the electron density of the aromatic ring are always constant. They are determined by the electronic parameter of the substituent, the **Hammett constant**, σ. Electron-accepting substituents with positive σ values are, among others, the nitro group, the cyano group, and the halogens. Electron-donating substituents with negative σ values are hydroxyl and amino groups, the methoxy group, and alkyl substituents. Acceptor substituents enhance the acidity of benzoic acids and phenols, they reduce the basicity of anilines, and they accelerate the basic hydrolysis of benzoic ethers. Electron-donating substituents exert an opposite influence.

However, an individual reaction constant ϱ must be applied for each reaction type of aromatic compounds. By using Eq. 18.2, later generally called the Hammett equation, the equilibrium constant K for an arbitrary reaction can be calculated from ϱ and σ. R–X and R–H represent the relevant aromatic compounds substituted with the group X, or unsubstituted, respectively.

$$\varrho\sigma = \log K_{R-X} - \log K_{R-H} \tag{18.2}$$

Acceptor and donor substituents influence the electron density on the heteroatoms and reduce or increase the ability to form hydrogen bonds. This explains, among other things, the electronic influence of aromatic substituents on the biological activity of drug molecules. The Hammett equation has, therefore, been seen as a challenge for medicinal chemists and biologists to derive quantitative structure–activity relationships from this concept. Many groups have attempted to find relationships between biological activity and the Hammet constants σ, or between σ and/or ϱ-analogous substituents, and to derive test parameters for biological systems. Despite some interesting results, no general concept could be established.

It was Corwin Hansch and Toshio Fujita who published a paper in 1964 that laid the foundation for **quantitative structure–activity relationships**. In it they describe:

- The definition of a lipophilicity parameter π, analogous to the electronic term σ in the Hammett equation.
- The combination of several parameters in one model.
- The formulation of a parabolic model to describe nonlinear lipophilicity–activity relationships.

18.3 The Determination and Calculation of Lipophilicity

Corwin Hansch had previously studied the structure–activity relationship of phenoxyacetic acids, which have growth-stimulating effects in plants. In addition to their biological activity, he was particularly interested in their lipophilicity, which can be measured by the partition coefficient in an octanol/water system (Sect. 19.1). While analyzing the data, he realized that lipophilicity is an additive molecular parameter. The logarithm of the octanol/water **partition coefficient *P*** is given by the sum of the group contributions of each part of the molecule. Hansch defined a **lipophilicity parameter π** (Eq. 18.3) analogous to the Hammett equation. R–X and R–H have the same meaning as in Eq. 18.2. The absence of a reaction-specific ϱ term in Eq. 18.3 results from relating the π values to a single distribution system, the two-phase mixture of *n*-octanol and water.

$$\pi = \log P_{R-X} - \log P_{R-H} \tag{18.3}$$

n-Octanol was chosen for theoretical and practical reasons. It has a long aliphatic chain and a hydroxyl group that is an H-bond donor as well as an acceptor. Its structure, therefore, resembles the membrane lipids to some extent. It dissolves a large number of organic compounds, it has a low vapor pressure, but can nonetheless be easily removed. Its UV transparence over an extremely wide range is particularly advantageous.

With the help of the lipophilicity parameter π, the log *P* values of new compounds, and therefore their lipophilicity, can be calculated. For this, the lipophilicity of the basic scaffold and the π values of the substituents must be known. In this way, the biological activity can be correlated without the tedious experimental measurements of each individual partition coefficient. In addition to the π values of all important substituents, a very large number of experimentally determined octanol/water partition coefficients are available in the literature.

18.4 Lipophilicity and Biological Activity

Lipophilicity plays an overwhelming role in describing the dependence of biological effects on chemical structure and, therefore, explains many quantitative structure–activity relationships. This is easily understood because biological systems consist of aqueous phases separated by lipid membranes. The transport and distribution of small molecules in such systems must, therefore, depend on their lipophilicity. For polar substances, the lipid membrane is an insurmountable barrier. Only substances with moderate lipophilicity have a good chance of "migrating" into both the aqueous and lipid phases to reach the target tissue in adequate concentrations (Chap. 19).

Although soluble proteins carry predominantly polar amino acid residues on their surface, the more or less buried binding sites for ligands are composed of polar and nonpolar regions. The hydrophobic parts of the ligand bind to the hydrophobic parts of these pockets. The size of these hydrophobic surfaces is always limited. The size and shape of the lipophilic part of the ligand must fit the hydrophobic surfaces in the binding pocket. Since the natural ligands normally bound in these pockets are themselves sufficiently water soluble, the lipophilic regions in the binding pockets are of limited size. This fact is another explanation for the complex, generally nonlinear, lipophilicity–activity relationships.

Many linear and nonlinear lipophilicity–activity relationships describe relatively unspecific biological effects, such as anesthetic, bactericidal, fungicidal, and hemolytic effects. They will not be discussed further here. Other relationships describe the transport and distribution in a biological system. Such structure–activity relationships are discussed in Chap. 19.

Table 18.1 The biological activity of *meta*- and *para*-substituents of phenethylamines **18.1** (i.v. application in the rat; C in mol/kg rat)

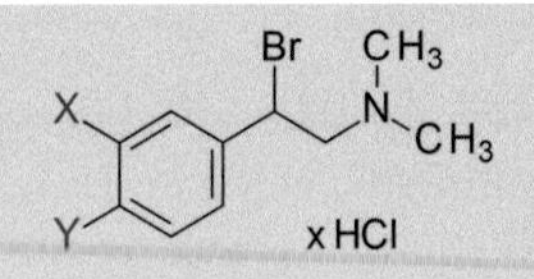

18.1

meta	*para*	log 1/*C*
H	H	7.46
H	F	8.16
H	Cl	8.68
H	Br	8.89
H	I	9.25
H	Me	9.30
F	H	7.52
Cl	H	8.16
Br	H	8.30
I	H	8.40
Me	H	8.46
Cl	F	8.19
Br	Cl	8.57
Me	F	8.82
Cl	Cl	8.89
Br	Cl	8.92
Me	Cl	8.96
Cl	Br	9.00
Br	Br	9.35
Me	Br	9.22
Me	Me	9.30
Br	Me	9.52

18.5 The Hansch Analysis and the Free–Wilson Model

In 1964 Corwin Hansch and Toshio Fujita derived a mathematical model more intuitively than theoretically that can quantitatively describe structure–activity relationships, the **Hansch analysis** (Eq. 18.4).

$$\log 1/C = -k_1(\log P)^2 + k_2 \log P + k_3\sigma + \ldots k \quad (18.4)$$

In Eq. 18.4, C is a molar concentration that produces a particular biological effect. When related to a series of substances, it is the **equieffective molar dose**. Log P is the logarithm of the octanol/water partition coefficient P, and σ is the Hammett constant. The square of the log P term allows the quantitative description of nonlinear lipophilicity–activity relationships. This term is omitted when the dependence is linear. Other terms such as polarizability and steric parameters can additionally occur.

The coefficients k_1, k_2, ... and k are determined using the method of **regression analysis**. The Hansch analysis, therefore, establishes a hypothetical model for quantitative relationships between biological activity and physicochemical parameters. Biological data are flawed, and the same is true for physicochemical properties. Despite this, the reliability of the latter parameters is usually greater than those of the biological data. The result of a calculation is judged by the squared differences between the measured biological data and the values that were calculated from the model. The sum must be as small as possible over all the compounds investigated. It is an important criterion for judging the quality of a model or for comparing different models of different quality.

The quantitative structure–activity relationship of the antiadrenergic effect of *N*,*N*-dimethyl β bromophenethylamines **18.1** (◘ Table 18.1) is considered as an example. According to their structure, these compounds more or less reverse the agonistic effect of an adrenaline dose. The value C is the dose of an antagonist that blocks the adrenaline effect by 50%. The data can be described using the Hansch model shown in ◘ Fig. 18.2.

The entire dataset can be described by a mathematical model using the derived equations. When bromine is cleaved, a carbocation is formed and the substances bind irreversibly to the adrenergic receptor. Accordingly, the σ^+ term is found in the Hansch equation (◘ Fig. 18.2), which describes this type of reaction particularly well. Lipophilic substituents increase the biological activity (positive π term) and electron withdrawing substitu-

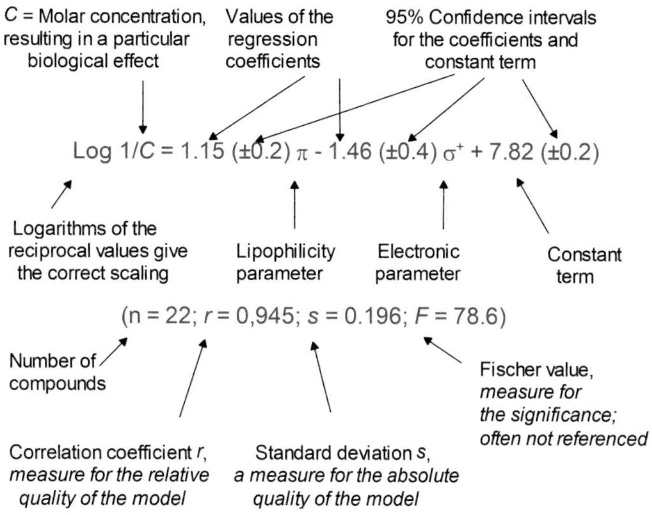

Fig. 18.2 A QSAR equation delivers individual parameters for a quantitative model for the prediction of biological activity, in this case from substituted *N*,*N*-dimethyl-β-bromophenethylamines (Table 18.1)

ents decrease it (negative σ^+ term). Therefore, lipophilic electron-donating substituents, such as large alkyl substituents, should be optimal for activity. Second, within certain limits, the effect of other compounds can be predicted. Interpolations (i.e., conclusions based on very similar substituents) are generally much more reliable than attempts at extrapolations (i.e., predictions made outside the parameter space, e.g., for considerably more lipophilic, more polar, or larger substituents). As a first approximation for the statistical parameters *r*, *s*, and *F* (Fig. 18.2), it can be said that the correlation coefficient *r* should have values close to 1.00, the standard deviation *s* should be as small as possible, and the *F* value should be as large as possible. The better these criteria are met, the better the quantitative model will be, in other words, the better the agreement between the experimental and calculated values.

Also in 1964, and independent of Hansch and Fujita, S. R. Free and J. W. Wilson developed a completely different model for structure–activity analysis. Since the original approach is confusingly formulated and difficult to use, only a variant, which was later proposed by Fujita and T. Ban, will be discussed here, the **Free–Wilson analysis**. The Free–Wilson analysis assumes that within a set of chemically related substances, a reference compound, usually the unsubstituted parent compound, *per se* makes a specific contribution μ to the biological effect. Each substituent on this scaffold makes an "additive and constitutive" contribution a_i to the biological activity (Fig. 18.3)—additive, because there is no consideration of structural variation at other positions in the molecule, and constitutive, because it matters where in the molecule the specific structural change is made. Despite these relatively simple assumptions, Free–Wilson analysis provides good quantitative models for many structure–activity relationships.

In contrast to the Hansch analysis, which compares properties, the Free–Wilson analysis is a real "structure–

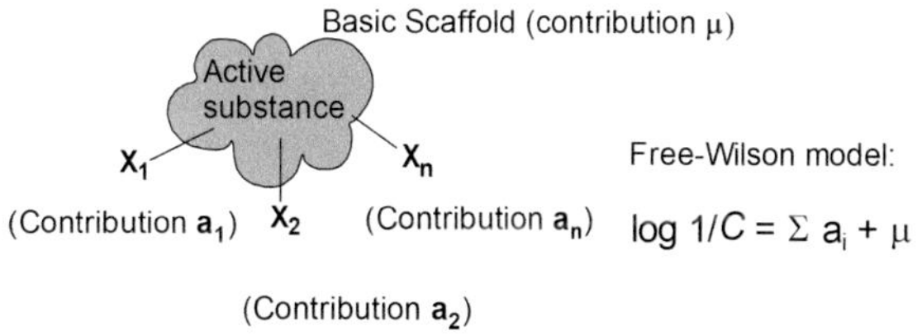

Fig. 18.3 The Free–Wilson analysis uses the additive nature of the group contributions to describe the biological activity. Accordingly, the biological activity in the displayed equation is made up of the activity μ of the basic scaffold and the constant group contributions a_i of the substituents X_i

activity analysis," because the parameter that codes for the structural information (1 for present, 0 for absent) correlates with biological effects. It is easily carried out, but the structures and the biological data must be known. Unfortunately, the Free–Wilson analysis also has disadvantages:

- The structural variation must be present on at least two different substitution sites, because otherwise there will not be enough degrees of freedom to use statistical methods.
- The usually large number of variables diminishes the predictive value and reliability of the analyses.
- Predictions are only possible for combinations of substituents that have already been considered in the analysis, and not for new substituents.

When the Free–Wilson analysis is applied to the above example of the antiadrenergic phenethylamine, the values for the scaffold and substituent contributions shown in Table 18.2 are obtained. At first glance, an increase in the values from F to Cl and from Br to I, i.e., the influence of lipophilicity, is obvious. Despite having almost the same lipophilicity, the methyl and chloro substituents are different. This is due to their different electronic properties. Differences in the *meta* and *para* positions on

Table 18.2 Free–Wilson group contributions for phenethylamines

Atom type Position	H	F	Cl	Br	I	Me
meta	0.00	−0.30	0.21	0.43	0.58	0.45
para	0.00	0.34	0.77	1.02	1.43	1.26

$\mu = 7.82$
($n = 22$; $r = 0.97$; $s = 0.19$)[a]

[a] For an explanation of these values see Fig. 18.2

the electronic influence can also be followed. Therefore, the Free–Wilson analysis indeed has advantages for the analysis of substituent effects.

18.6 Structure–Activity Relationships of Molecules in Space

As shown in the previous section, an attempt is made to correlate structure–activity relationships with compound-specific parameters. These parameters, such as volume, polarizability, or lipophilicity, are properties that are calculated or measured for the entire molecule or for specific groups of substituents. The **3D structure** of the molecules is only conditionally taken into account by these **descriptors**. Therefore, in the context of increasing knowledge of the spatial structure of protein–ligand complexes, QSAR methods focus on parameters that can be derived from the 3D structure. In general, the goal of these approaches is to calculate binding affinity. The techniques can also be used to describe other biological properties such as bioavailability, toxicity, or metabolic reactivity (Chap. 19). To distinguish them from the classical QSAR techniques described above, they are referred to as **3D-QSAR methods**.

Ideally, parameters that can be read directly from the 3D structure of a compound and used to infer its binding affinity would be desirable. However, the interplay between these parameters and activity is very complex and still far from being fully understood. In addition, there are many other biological systems to which one would like to apply 3D-QSAR methods, but the structures of the relevant target proteins are unknown. Many pharmacologically relevant receptors are membrane-bound and their structure determination has proven to be extremely difficult. However, the knowledge of their structures is a prerequisite for a reasonable estimation of the binding affinity of a ligand from the geometry of the formed complex (Chap. 4). Therefore, instead of trying to calculate the absolute values of the binding affinities from these incomplete data, we will focus on the **relative affinity differences** between compounds in a dataset. The gradual changes in the compound-specific parameters are then correlated with the biological data.

18.7 Structural Alignment as a Prerequisite for the Relative Comparison of Molecules

Assumptions about the spatial structure of molecules are already taken into account in classical QSAR techniques. Different positions of substituents, e.g., in the *meta* or *para* position of an aromatic ring, are often described by individual parameters. In this form, they are considered in the Hansch equation as well as in the Free–Wilson analysis (Sect. 18.5). Furthermore, in classical QSAR models, indicator variables are defined for different configurations of substituents, e.g., the configuration of stereoisomers. The use of these parameters assumes an analogous orientation of the molecules in a hypothetical binding pocket. For example, in a series of *ortho*-substituted derivatives, it is assumed that all *ortho* substituents are oriented towards the "same side." Structure–activity relationships that correlate biological activity with properties of the 3D structure require a **spatial superposition of the compounds**. This superposition should approximate the relative orientation in the binding pocket as closely as possible. Methods for calculating these spatial superpositions were discussed in Chap. 17.

18.8 Binding Affinities as Compound Properties

What compound-specific properties can be used to correlate the properties of the 3D structure with the binding affinity? As discussed in Chap. 4, **binding affinity** is composed of **enthalpic** and **entropic components**. The former includes everything that depends on direct energetic interactions. These are mainly of **steric** (van der Waals potentials, Sect. 15.4) or **electrostatic** (Coulomb potentials) nature. The second contribution focuses on the degree of order and the distribution of energy over the different degrees of freedom of the system under investigation. The ligands as well as the binding pockets of a protein are solvated by water molecules in the uncomplexed state. Upon complex formation, the enthalpic interactions to these water molecules are lost. They are replaced by direct interactions between the ligand and the protein. Since only relative differences between the molecules of a dataset are of interest, effects that are the same for all ligands are not considered. This includes all influences that affect the protein. This omission is certainly an oversimplification, since the protein changes its degree of solvation upon ligand binding and is polarized differently by ligands. Water molecules are displaced from the binding site. Ligand-induced adaptations of side chains in the binding pocket or changes in the rotational degrees of freedom of methyl groups and side chains (Sect. 4.10) are conceivable. These effects are either not considered or are assumed to be the same for all molecules in the dataset. This assumption is probably valid in many cases. However, many recent investigations clearly show that changes affecting the protein or the dynamics of the ligand are often not constant within a series of compounds. This is where the methods fail.

Initially, only the steric and electrostatic interactions of a compound in the binding pocket should be considered. How can these properties be compared for a set of ligands? A first approach has been the **hypothetical interaction models** developed by Hans-Dieter Höltje and Lemont B. Kier. A key assumption of these models was the selection and spatial positioning of amino acid side chains around the ligands. When the molecules are embedded in a lattice and systematically scanned with an interaction probe, these assumptions are no longer necessary. Richard Cramer and M. Milne proposed such a model in 1978 (DYLOMMS). It took another 10 years before the generally applicable **CoMFA** (*comparative molecular field analysis*) method was established. Despite many theoretical and practical shortcomings in its application, the method was quickly accepted. Today, it is applied in many different variations.

Before performing such an analysis in practice, some basic considerations should be made. Do steric and electrostatic interactions account for all contributions to ligand binding that ultimately lead to a correct **relative ranking of binding affinities**? As mentioned above, binding affinity is composed of enthalpic and entropic contributions. Sampling properties via probes to map interactions certainly provides a measure of how well a molecule can undergo energetically favorable interactions. But how well are the entropic contributions accounted for? A significant part of this is due to **solvation and desolvation processes** (Sect. 4.6). In the dissolved state, in the immediate vicinity of the hydrophobic surface portion of a ligand, the water structure must assume a more ordered state compared to the bulk water phase. The transfer of such a ligand from water to the protein-binding pocket, thus, requires that a certain number of water molecules in the water phase change to a significantly less ordered state. This increases the entropy of the system and favors the spontaneous occurrence of the binding event. The number of water molecules involved in this process depends on the size of the hydrophobic surface of the ligand. Furthermore, the displacement of bound water molecules out of the binding pocket by the ligand to be accommodated increases the disorder of the system under consideration and, thus, also the entropy of the system. In the approximation discussed above, it is assumed that these effects are the same for all molecules in the dataset and are not important in a relative comparison. In addition, rotational, translational, and internal conformational degrees of freedom are frozen. As a result, the entropy of the system decreases. For the affinities to be correctly considered, all these effects would have to be taken into account.

In 2019, Tobias Hüfner in Marburg performed MD simulations on enzyme complexes using the **GIST method** (Sect. 15.4) to predict binding affinities considering enthalpic and entropic solvation contributions. He made a very interesting observation. A set of crystal structures of different ligands with a target protein was used. All the structures were superimposed in their experimentally observed geometry. Thus, the problem of ligand alignment, which is essential for performing a CoMFA analysis, could be solved using experimental data. Tobias Hüfner then used different mathematical models to calculate the GIST contributions to the dataset. He also analyzed a dataset in which only the ligands were considered, but in their protein-bound conformations. In this way, the ligands are practically sampled for their potential interactions with water molecules using an MD simulation. The calculated solvation contributions have been deposited on a lattice surrounding all ligands. Accordingly, the result is very similar to a CoMFA analysis. Surprisingly, this simulation, performed with only the ligands, gave an excellent affinity prediction for the ligands. How can this be understood? It is possible that a significant part of the relative differences in binding affinities is already accounted for by the desolvation properties of the ligands. Contributions due to the properties of the protein and its desolvation seem to cancel each other out to a first approximation in the relative comparison. Certainly, there are water molecules in the binding pockets whose enthalpic and entropic properties are very different from those in a surrounding water phase. Apparently, a large part of these differences in the individual desolvation contributions required to displace the water molecules from the binding site are in turn compensated for by the individual functional groups of the ligands binding to these sites and achieving comparably graded binding contributions with the residues of the protein. Therefore, simply scanning the ligands with their different functional groups in the correct bound geometry with a water probe already provides a relevant picture to reasonably predict the relative differences in the affinity data. Perhaps this is a clue as to why comparative field analysis works so surprisingly well.

18.9 How Is a CoMFA Analysis Performed?

The most important and widely used 3D structure–activity method is the CoMFA method. The first step in performing a CoMFA study is to select a **dataset** of suitable compounds. This dataset should contain about 50–100 compounds with related overall geometry. It should also be ensured that all compounds bind to the same protein at the same site and that a binding affinity is known for all of them. The ligands must have a certain diversity of their structural variation. Their binding affinities should be spread over **at least three orders of magnitude**. Conformations are generated for all molecules (Chap. 16) and superimposed using one of the techniques discussed in Chap. 17. In general, one refers to the spatial structure of the target protein, if available, and fits the ligands of interest into the binding pocket. Of course, it will be optimals if a crystal structure with the protein is available for many, if not all, bound ligands. Finally, the superimposed molecules are embedded in a lattice (◘ Fig. 18.4) that surrounds them by a sufficiently large margin. The intersections of the lattice should have a spacing of 1 or 2 Å. A probe, this is an atom with the properties of hydrogen, carbon, or oxygen, or a particle with a formal charge, is placed at each of the lattice intersections. The interaction energies between this probe and each molecule in the dataset are calculated. The collective interaction contributions on the lattice are referred to as the **interaction field** of the molecule. This is where the name of the method comes from. Finally, the fields of the molecules in the dataset are compared. With a box size of 10–20 Å and a grid spacing of 1–2 Å, there are many thousands of field values per molecule in the dataset to be processed. This huge amount of data means that field evaluation can be computationally rather intensive.

18.10 Molecular Fields as Criteria of a Comparative Analysis

Steric and electrostatic interactions are described by a Lennard-Jones or Coulomb potential (◘ Fig. 18.5) in force fields (Sect. 15.4). As the distance between a probe and an atom of the molecule approaches zero, the **Lennard-Jones** and **Coulomb potentials** increase towards infinity. For like-charged particles, the Coulomb potential approaches infinity; for oppositely charged particles, it approaches negative infinity. These values reach extremely high field contributions at grid points near the surface or inside a molecule. They must be avoided in a CoMFA analysis. Therefore, the field contributions above and below a certain threshold are set to a predefined **cut-off value**. Following these procedures, a Lennard-Jones or Coulomb potential can be calculated. For example, aliphatic carbon atoms can be used as probes. These probes are given a positive or negative charge to study the electrostatic properties of the molecules. The program **GRID** by Peter Goodford was introduced in Sect. 17.10. With this program, molecular fields can be calculated for numerous probes describing different functional groups. For each predefined probe, there are regions in space where favorable or unfavorable interactions between the probe and the studied molecule are expected.

In addition, other fields can be defined besides those that probe the steric and electrostatic properties of molecules. It was discussed in Sect. 18.8 that the hydrophobic surface of a molecule is a measure of the entropic contribution, especially in the transition from the bulk water phase. In the group of Donald Abraham, at Virginia Commonwealth University, Richmond, USA, molecular fields have been developed which allow the hydrophobic properties of molecules (program **HINT**) to be studied. These are calculated using a very similar distance-dependent function. The resulting molecular field describes the lipophilicity distribution on the surface of a molecule.

18.11 3D-QSAR: Correlation of Molecular Fields with Biological Properties

Let us assume that multiple molecular fields have been calculated for each molecule in a dataset, and a correlation of their differences with binding affinity is attempted. How are these differences expressed? For this, we will consider three hypothetical examples of substituted phenyl derivatives.

- First, all substituents on the phenyl ring in a series of compounds should be varied so that increasingly larger field contributions occur in the vicinity of the substituent when scanned with a positively charged probe. If the binding affinities increase as the field contributions increase, this will be reflected in the quantitative analysis. They indicate that ligands with increasingly negatively charged groups in this region of the molecule lead to more potent compounds. This is explained by the fact that the more negatively charged groups interact better with the positively charged probe.
- The second example is a bit different. Now the substituents on the phenyl ring are given positive or negative partial charges. Their variation has no influence on the potency of the compounds. The quantitative analysis shows that the changes in the electrostatic field contributions have no correlation with the biological activity. A possible explanation could be that this effect and another property, e.g., the size of the substituents, cancel each other out. It could also be that the biological activity is influenced by other properties of the substituents, such as their hydrophobic character.
- In the third case, the electrostatic properties of the substituents that are important for binding to the receptor should not vary much at the examined position. There may be different substituents present, but they all have comparable partial charges. The model that analyzes the field contributions in the vicinity of these groups does not recognize differences and, therefore, does not find a correlation with binding affinity. It may be that a class of substituents at a particular posi-

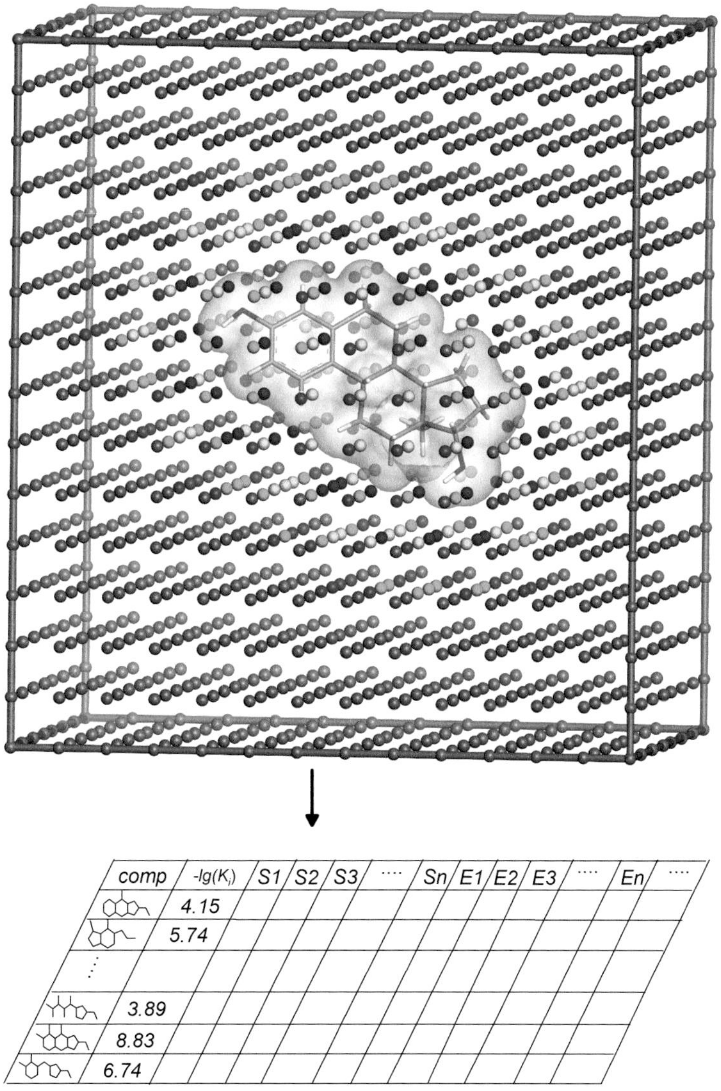

$$-\lg(K_i) = y + a\ S1 + b\ S2 + c\ S3 + \ldots + h\ Sn + k\ E1 + m\ E2 + n\ E3 \ldots + z\ En$$

Fig. 18.4 A grid is generated for the calculation of molecular fields that broadly encompasses a molecule. In this image, the grid points are color-coded with increasing distance from the ligand (*red < yellow < green < blue < gray*). The contributions from the chosen fields are calculated at all points of the lattice, which have a grid spacing of 1–2 Å. The field contributions at each point in the grid (S1, S2, … Sn, E1, E2, … En) are entered into a spreadsheet. The analysis is carried out for all molecules in the dataset. The binding affinities are incorporated into the spreadsheet as, for instance, −log (K_i). The field contributions are weighted with appropriate coefficients (a, b, … z) and using a special statistical method, the partial least squares (PLS) analysis, they are related to the affinity. A model is obtained in the form of an equation that indicates at which grid points and with what weights the different field contributions explain the biological activity. (► https://sn.pub/jndLSI)

tion on a molecular scaffold is actually very important for binding, but remains insignificant in the analysis. This has to do with the fact that a **QSAR** analysis only **makes relative comparisons within a dataset**. In other words, if a property is the same for all ligands in the dataset, then that property will not be able to explain any differences in the biological data.

These examples are still easy to handle. The question can be asked whether a tedious correlation method with a "detour" via molecular fields is really necessary. In practice, the situation is more complicated, especially when considering molecules with different scaffolds. The substituents do not fall exactly on top of one another in the molecular superposition. Their contribution must

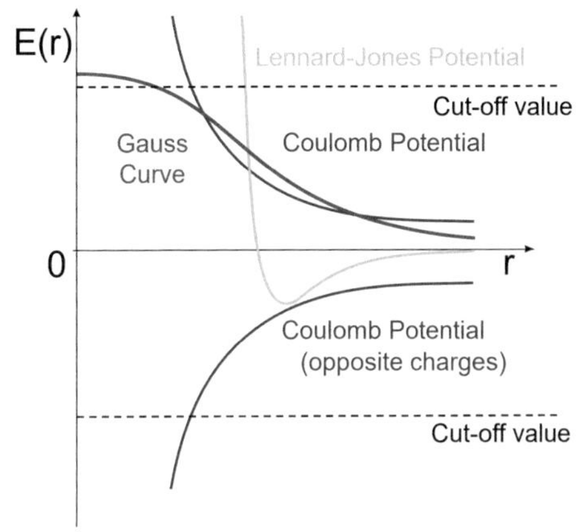

Fig. 18.5 The Lennard-Jones potential (*green*) is a model for describing the intermolecular interactions of two atoms without considering their charge. Negative potential values correspond to mutual attraction, positive values correspond to a repulsion of the particles. If a reciprocal distance becomes infinite, the potential will approach zero. Upon approach, it goes through a shallow minimum due to alternating polarization. At even shorter distance, it very steeply rises towards positive infinity because of atom–atom repulsions. The Coulomb potential (*blue*) considers only electrostatic interactions that formally reside as point charges on the atomic nuclei. It also approaches infinity when the distance disappears for like-charged particles. For oppositely charged atoms, negatively infinite values result. The hyperbolic form of the Coulomb potential is considerably less steep, so that the particles can still "feel" one another at larger distances. Boundary values are set for potentials in a CoMFA analysis. A Gaussian function, which takes the course of a bell-shaped curve (*red*, here only the right half of the "bell" is shown) describes the distance dependence of the interaction potential between the particles in the context of the CoMSIA model. As the distance disappears between the particles, the curve reaches its maximum value, which remains finite

be described as a field in space, and only as such can it be evaluated. In any case, these examples underline the importance of careful planning of the analysis. The structures of the dataset must be chosen in such a way that the variation of the substituents and their properties is maximized.

18.12 Results of a Comparative Molecular Field Analysis and Their Graphical Interpretation

18

When the full complexity of the field contributions is considered in terms of a multidimensional matrix, a simple regression analysis cannot be applied to extract the interdependence of the variables, e.g., the binding affinity. **Partial least squares (PLS)** analysis is a statistical method that extracts relevant and explanatory factors, called PLS vectors, from large amounts of data. In CoMFA analysis, these vectors describe the range of fields that best correlate with the experimentally determined affinity. The result is an equation analogous to the results of classical QSAR methods. This equation shows to what extent particular grid points in each of the individual fields contribute to the relative differences in the binding affinities. Depending on how many grid points are to be evaluated in the analysis, the statistical significance of the derived results must be strictly monitored. This significance is checked by a special test: the **cross validation**.

This is done by randomly removing one or more compounds from the dataset. A model is constructed with the remaining compounds of the dataset and the affinities of the removed compounds are predicted with this model. The removal of compounds is repeated several times, in the simplest case until all compounds have been removed consecutively one after the other (**leave-one-out** validation). The quality of the overall prediction obtained is a measure of the reliability and validity of the model. The obtained result is expressed by the **q^2 value**, which can be calculated from the square of the deviation from the predicted values. It takes values from $-\infty$ to $+1$. A value of $+1$ indicates a perfect model. All predictions exactly match the measured binding affinities. There is no deviation. A value of $q^2 = 0$ indicates that the predictions of the model are no better than no model at all; they are as good as the average of all affinities. If q^2 takes negative values, the model is worse than the average, which means it is worse than no model. Therefore, a model can only be trusted if q^2 is above 0.4–0.5.

A further step is required to **check the predictive power** of a trained model (one speaks of a "trained" model and, therefore, one refers to the dataset as a "training dataset"). This validation step requires a **test dataset of molecules** that are similar to the molecules in the training dataset, but were not used for the initial training. Binding affinities are predicted for these molecules. Only if a correlation coefficient, calculated analogously to q^2 with the training dataset, is of similar magnitude will the derived model have sufficient predictive power. This procedure is, therefore, a test of the model on independent data that were not used to derive the model.

The derived model can then be used to estimate the affinity of new compounds that have not yet been synthesized. The conformations of these compounds are generated and superimposed on the other structures. They must fall within the grid dimensions defined in the training set. Then their field contributions are calculated. Using the correlation derived by CoMFA for the training set, it is possible to calculate which grid points are predictive of the binding affinity of new compounds.

CoMFA techniques establish a correlation between activity data and molecular properties. From the relative comparison within a training set, a model can be derived that encompasses the properties of new molecules. Relevant predictions will only be obtained if the structural variations in the new molecule remain within the scope of the model. In other words, the model cannot make predictions about the influence of substituents that occur in regions where there were no structural variations in the training set. CoMFA models **interpolate** between the field contributions of molecules. **Extrapolation** to regions not covered by the dataset is not possible.

The results of a CoMFA analysis can be **evaluated graphically**, which is probably the most powerful aspect of the method for the medicinal chemist. From the model and the derived equation, it is known at which grid points field contributions are obtained that provide a significant explanation for the binding affinity. These contributions can be outlined for the different fields according to their importance. They indicate volume regions around the molecules in which changes in the field contributions run parallel or opposite to the affinity changes in the dataset. These **contour maps** are an important support in the design of new compounds (Sect. 18.14). They indicate where the properties of a lead structure need to be varied to increase affinity.

18.13 Scope, Limitations, and Possible Expansions of the CoMFA Analysis

Typically, only steric and electrostatic field contributions are evaluated in CoMFA analyses. A hydrophobic field can quantify the size of the hydrophobic surfaces and, therefore, partially accounts for the entropic contribution to affinity. Since CoMFA evaluations yield relevant models without the explicit use of hydrophobic fields, these field contributions must be at least partially included in the Lennard-Jones and Coulomb fields. The lipophilicity of a molecule increases when an uncharged, sterically demanding group is enlarged, e.g., from methyl to butyl. Here the changes in the steric field contributions can correctly reflect the lipophilic surface. A correlation with electrostatic properties is also possible. Hydrophobic moieties usually carry only small partial charges. Positively or negatively charged groups represent hydrophilic regions. In this way, the lipophilic and hydrophilic regions of the surface can be quantified by differences in charge.

The deviation that cannot be explained by a CoMFA model also includes, apart from experimental errors, all inadequately described binding contributions. These include structural adaptations of the protein that are not identical for all compounds in the dataset. Entropic contributions resulting from (*i*) conformational fixation of the drug molecules in the binding pocket, (*ii*) residual mobility of the ligands in the binding pocket, (*iii*) conformational changes on the side of the protein, (*iv*) or significant differences in the solvate structure remaining on the protein are all not considered in the fields. It should be noted, however, that such deviations will only be significant if they differ from one ligand to another in the dataset.

In addition to these shortcomings, the fields themselves cause some problems. Because of their mathematical functional form, they reach very large or very small values near the surface or inside the molecules (■ Fig. 18.5). Because the Lennard-Jones potential grows faster than the Coulomb potential as it approaches the atoms, they both reach an arbitrarily set cutoff (Sect. 18.10) at different distances from the molecules. The extremely steep Lennard-Jones potential can change its functional values from practically zero to the cut-off value within a distance of 2 Å, the commonly used grid spacing! These discontinuities and accordingly the lack of any variation within the cut-off regions near the ligand surfaces cause considerable problems in the evaluation. Moreover, they often produce "disrupted" and, therefore, difficult to interpret contour maps of the different fields.

The shortcomings of these fields have stimulated the search for alternative solutions. One method is to determine the **similarity of molecules** by their steric and physicochemical properties in space. These are then correlated with binding affinities. This is done in the **CoMSIA** method (**co**mparative **m**olecular **s**imilarity **i**ndices **a**nalysis). The molecules are superimposed similarly to the CoMFA method. How similar they are to one another is then measured in relation to a probe, for example, a carbon atom. For each molecule, the similarity to this probe is sampled at the intersections of a surrounding grid. The similarity measure between the probe and the molecule is defined as a distance-dependent function. A Gaussian function is chosen for this purpose (■ Fig. 18.5). Unlike the hyperbolic course of the potentials described above, the Gaussian bell curve does not tend to infinity for decreasing distance values. Therefore, no cutoff values need to be set. At any grid point, a similarity measure can be determined for a large number of properties. A prerequisite for the CoMSIA method is the description of the properties in terms of atomic values (e.g., partial charges, atomic volumes, lipophilicity, H-bonding properties). The same distance dependence is used for all properties. Property-specific similarity fields are obtained and correlated with the binding affinities. The interpretation of the field contributions is analogous to the CoMFA method. The main advantage of this method is the **ease of interpretation of the resulting contour maps**. If a particular property in a region of the superimposed molecules correlates significantly with binding affinity, that region will be highlighted and this information can be easily translated into the design of a new molecule. In contrast, the CoMFA method contours only regions outside the molecules where a property reveals changes in field contributions that positively or negatively affect affinity. However, by setting cut-off values, entire regions of these field contributions are hidden, especially near ligand surfaces (■ Fig. 18.5). Instead, the CoMSIA approach also contours at the atomic positions of the molecules that are responsible for the trends in the affinity changes. This gives medicinal chemists a much more intuitive picture of where to modify molecules in an optimization process.

3D-QSAR analyses were originally designed to establish structure–activity relationships in cases where the structure of the target protein was not available as a reference. Today, as more and more crystal structures of target proteins become available, the technique is increasingly used for cases where the protein reference is indeed known. The protein structure then helps to generate a reasonable and relevant superposition of the training compounds to be compared in their biologically active conformations. It seems paradoxical to use the information about the protein environment only to superimpose the molecules and then to disregard this valuable data in the comparative field analysis. Therefore, methods have been developed that take this information into account. Rebecca Wade's group at the EMBL and HITS in Heidelberg, Germany, has developed the **COMBINE method**. It uses a set of modeled protein–ligand complexes to calculate a data table. It contains the interaction energies between individual ligand atoms in the test molecules of the dataset and the amino acid residues and water molecules in the surrounding protein. The interpretation of this huge data table is achieved using a technique similar to CoMFA methods. The graphical interpretation of the correlation model obtained by COMBINE indicates which regions of the protein are critical in explaining the affinity differences in the ligand dataset. These are very valuable details, but they are of limited help for the direct design of improved molecules with higher affinity.

The variant **AFMoC** (**a**daptation of **f**ields for **mo**lecular **c**omparison), developed by Holger Gohlke in Marburg, allows the integration of information about the protein environment into the field-based model. The advantages of the intuitive evaluation of the field contributions with regard to the structural optimization of the ligands are not lost. To this end, the empirical scoring function **DrugScore** (Sect. 17.10) is first used to map onto the intersections of a CoMFA-like grid the values that a protein environment would sense as an interaction at each of the grid points if sampled with a particular atomic probe according to the functional form in DrugScore. The lattice is effectively "prepolarized" by the protein environment. The ligands of the training set are then placed on this lattice (using either a docking program or one of the superposition methods described). Whenever a ligand places an atom type on a region of the lattice where the protein considers that atom type to be advantageous, the field contribution is increased. Otherwise, the interaction contribution on the lattice is reduced. In this way, a data table analogous to the CoMFA method is created for the entire training dataset. This table is then evaluated to generate a QSAR equation. The individual contributions can be graphically visualized on the grid. They illustrate where certain atom types cause an increase or decrease in affinity, but now taking into account simultaneous information from the surrounding protein.

Similar field analyses can also be used to correlate and **predict selectivity differences** between ligands. Many enzymes exist as isoforms. They therefore have similarities in their binding pockets. As a consequence, ligands show graduated affinities or "selectivity profiles" for these isoforms. If a ligand is to be optimized to improve selectivity, the positions at which a change in a property leads to an improved profile must be known. A 3D QSAR model is constructed for each isoenzyme. Either the **difference of the affinity values** can be calculated and used for the model as the values to be predicted, or alternatively two correlation models can be constructed and the **field contributions at each grid point are subtracted** from one another. The models obtained from either approach can be interpreted graphically. Contour plots show where and how molecules should be modified to improve their selectivity for one or the other isoenzyme.

18.14 A Glimpse Behind the Scenes: Comparative Molecular Field Analysis of Carbonic Anhydrase Inhibitors

Today, comparative field analyses are part of the standard repertoire in drug discovery. As an example, the binding of inhibitors to carbonic anhydrase I (CAI) and II (CAII) will be examined. The biological function of these enzymes is described in detail in Sect. 25.7. The sequence identity of the two isoforms is 60%. The ligands in the training dataset are derived from the parent structures shown in ◘ Fig. 18.6. First, a superposition model is created by docking the ligands into the protein (◘ Fig. 18.7). The funnel-shaped binding pocket of the enzyme is occupied by ligands in a variety of ways. A good correlation model is obtained with the three methods, CoMFA, CoMSIA, and AFMoC. The models also achieve convincing predictive power on a test dataset independent of the training set.

The contours for the acceptor properties with respect to the inhibition of carbonic anhydrase II are shown in ◘ Fig. 18.8. Molecules in the dataset that exhibit an acceptor function in the areas marked in red have lower potency. On the other hand, an acceptor function in the blue area improves potency. Compound **18.2**, which has both acceptor functions of an SO_2 group oriented in the detrimental red area, is a weak CAII inhibitor. Moreover its NH group is in the blue region, which should be occupied by an acceptor. Compound **18.3**, which is about four orders of magnitude more potent, leaves the area that was occupied by the oxygen atoms in **18.2** empty, and orients its thiadiazole ring in the direction of the desirable acceptor function. It achieves considerably better inhibition of the target enzyme.

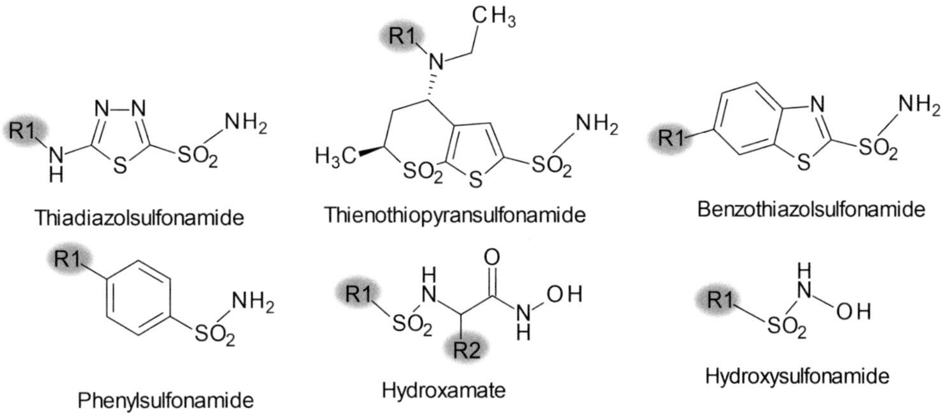

Fig. 18.6 The scaffolds of inhibitors that were used in different field analyses to establish affinity (pK_i[CAII]) and selectivity models (pK_i[CAII] − pK_i[CAI] = ΔpK_i[CAII − CAI]) to describe the inhibition of the carbonic anhydrase CAI and CAII. Different substituents were varied at the positions that are marked as R1 and R2

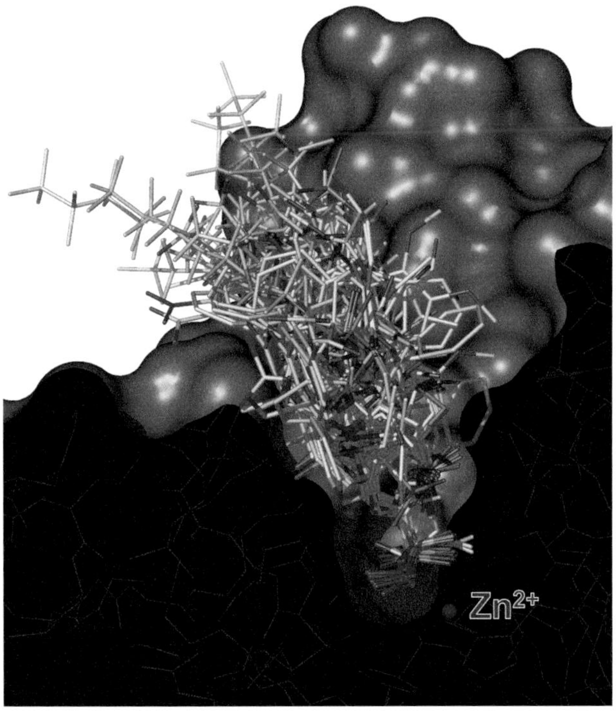

Fig. 18.7 The superposition of inhibitors from the dataset in the funnel-shaped binding pocket of carbonic anhydrase II; the zinc ion is shown as the *blue-gray sphere*, carbon atoms are *light yellow*, oxygen *red*, nitrogen *blue*, sulfur *orange*, and hydrogen *light cyan*

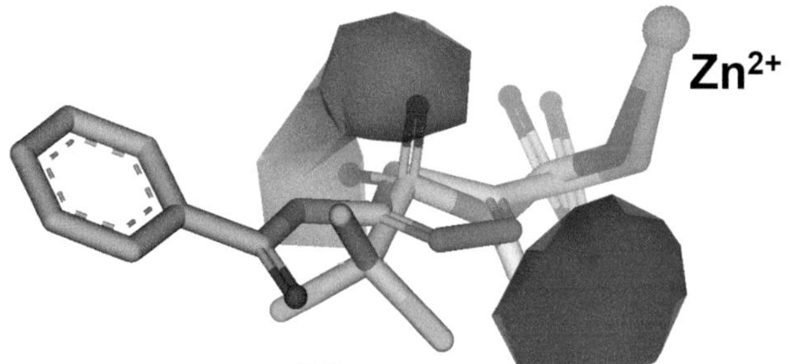

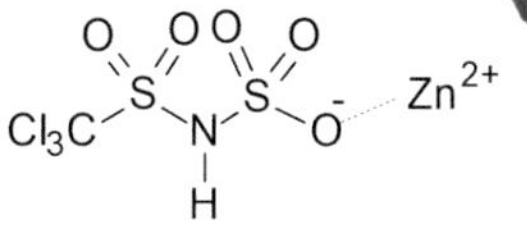

18.2 CA II pK = 4.7

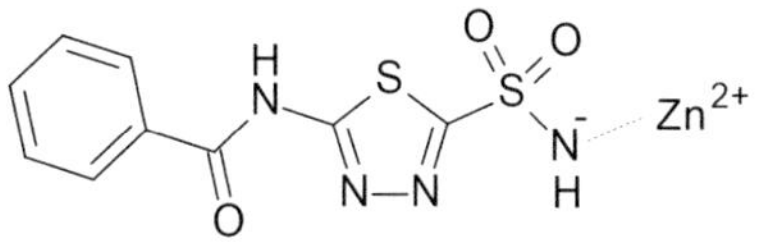

18.3 CA II pK_i = 8.7

Fig. 18.8 Contour map (CoMSIA) for the description of the binding contributions of H-bond acceptor properties. Inhibitors that occupy the *red*-contoured areas with H-bond acceptor groups do not inhibit carbonic anhydrase II (CAII) well; however, the occupancy of the *blue* areas with acceptor groups leads to increasing values. Both oxygen atoms of the sulfonamide group of **18.2** occupy the *red*-contoured area, which is unfavorable for acceptor properties. On the other hand, **18.3** leaves these areas unoccupied and places its basic nitrogen in the vicinity of the *blue*-contoured region, which is favorable for the occupancy by acceptor groups. This explains the markedly better inhibition of CAII by **18.3**

As with acceptor properties, contour maps can be generated for steric, electrostatic, hydrophobic, and hydrogen-bond donor properties. Their evaluation helps to identify where certain properties improve or decrease binding affinity. Such correlation analyses help the synthetic chemist plan the optimization of lead structures.

Contour maps based on a CoMFA analysis for steric properties that cause a difference in selectivity between CAI and CAII are shown in Fig. 18.9. Placing an inhibitor next to the green areas improves selectivity for CAI. On the other hand, occupancy next to the yellow areas improves the selectivity for CAII. Compound **18.4** binds unselectively with the same affinity to both isoforms, but **18.5** can clearly discriminate between the two. The model shown is derived purely from the correlation of ligand-binding data. The relative alignment of the molecules in the dataset is achieved by docking into

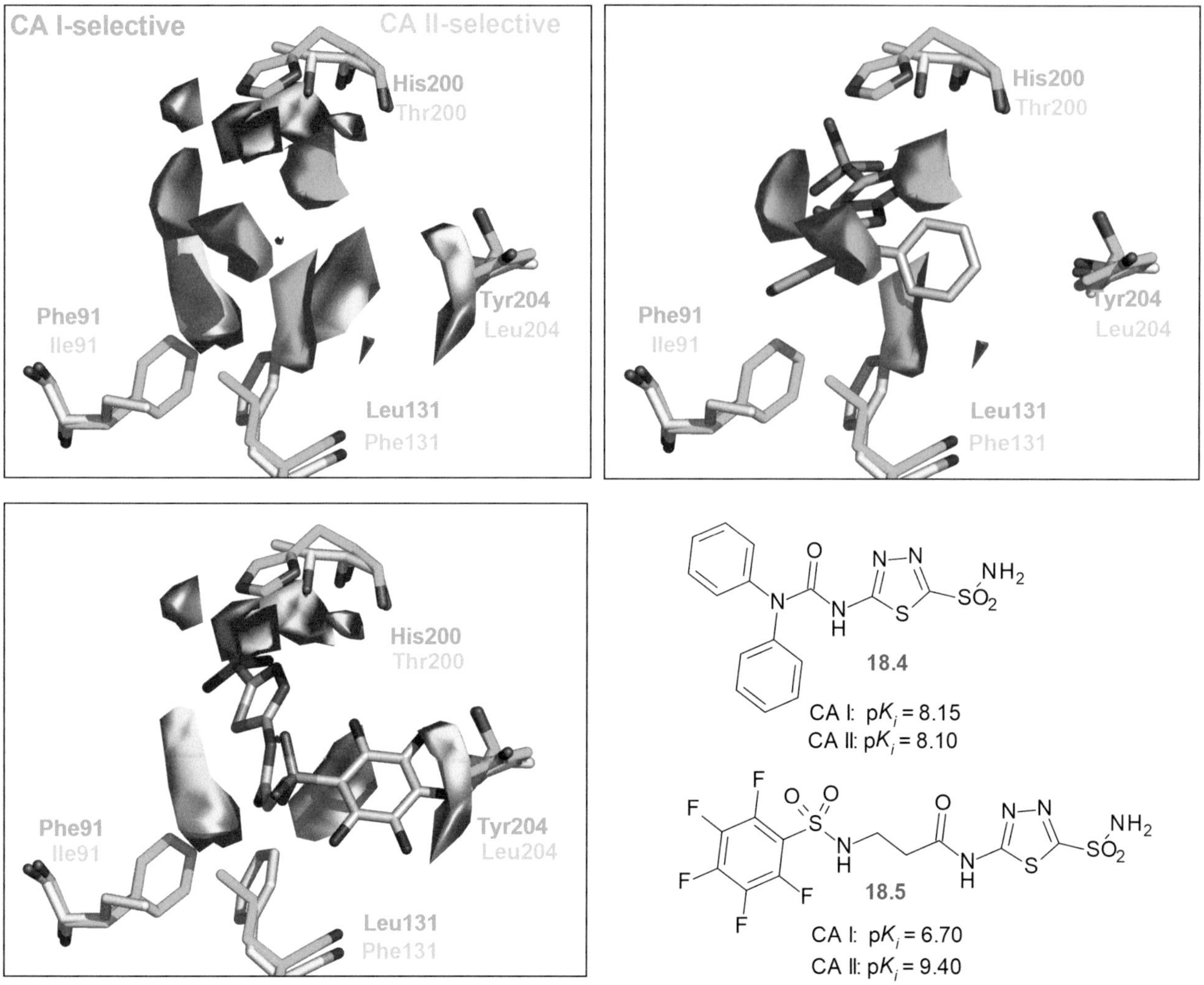

Fig. 18.9 *Top left* The selectivity can be improved with regard to carbonic anhydrase II (CAII) inhibition by sterically filling the region next to the *yellow*-contoured area. Filling the *green* area with sterically demanding group causes an increase in selectivity with regard to CAI. *Top left* and *top right* Compound **18.4** occupies virtually no area that is particularly selectivity discriminating; the compound is not isoenzyme specific. On the other hand, **18.5** occupies a *yellow*-contoured area neighboring position 204 (*bottom left*), which causes a selectivity enhancement for CAII. Compound **18.5** inhibits CAII decidedly more potently than CAI. All contour maps shown are based on data analysis by CoMFA; the adjacent protein residues were added to the images from corresponding crystal structures

the binding pocket of the protein. Therefore, the protein environment around this binding pocket should be examined more closely to see whether the derived contours are reasonable. Comparing the amino acid replacements between the two isoforms, it is apparent that CAI has two large residues, Phe 91 and Leu 131, which restrict the lower left portion of the binding pocket more than in CAII. The inhibitors have less space in CAI than in CAII. In fact, the comparative field analysis generates a yellow contour in this region (near position 91), the occupancy of which should be favorable for potent inhibition of CAII. CAII also provides a large space for inhibitors near position 204, which is occupied by the less crowded Leu 204 in CAII instead of Tyr 204 in CAI. A yellow contour is visible, indicating a favorable occupancy of this site. Inhibitor **18.5**, which is much more potent at CAII, orients its pentafluorophenyl group exactly in this region (Fig. 18.9, *right*). In the vicinity of position 131 (Leu 131/Phe 131), a yellow and a green region appear directly adjacent to each other but spatially separated, the occupancy of which is favorable for either CAI or CAII inhibitors. Compound **18.4**, which can hardly distinguish between the two isoforms, occupies the upper edge of both regions equally well. Moreover, it leaves virtually all regions unoccupied which, for steric reasons, should lead to a better inhibition of either CAI or CAII. This explains why this compound shows no particular selectivity.

Finally, the binding of the well-discriminating compound **18.6** should be considered (Fig. 18.10). The evaluation of the acceptor properties of the ligands in the training dataset shows that the occupancy of the red contour regions with H-bond acceptor groups shifts the

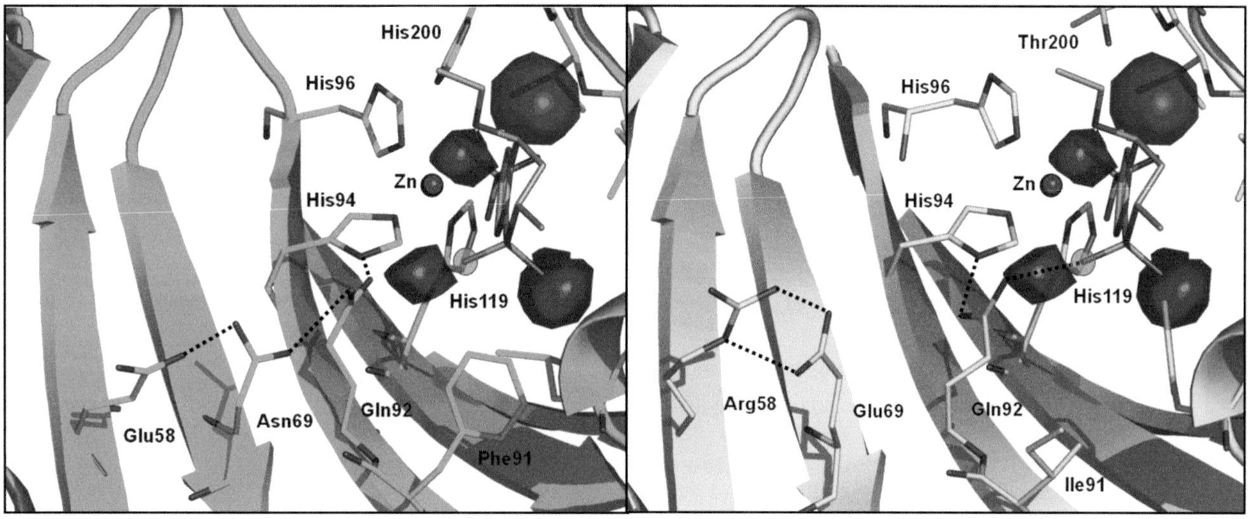

CA I CA I-selective CA II CA II-selective

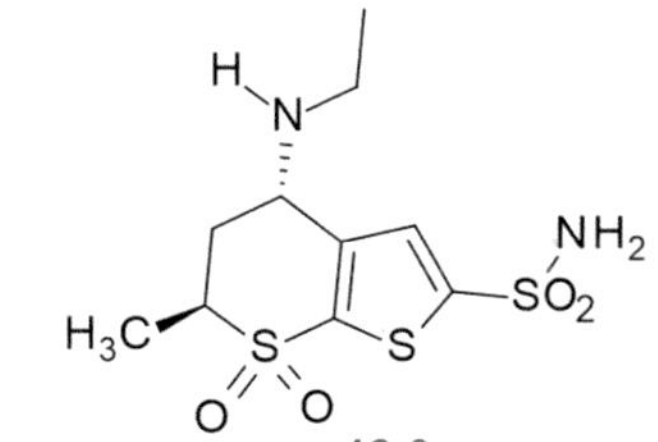

18.6

CA I: $pK_i = 4.30$
CA II: $pK_i = 8.05$

Fig. 18.10 Compound **18.6** inhibits carbonic anhydrase I (CAI; *left, green*) significantly less potently than CAII (*right, yellow*). The sulfone oxygen atom on the left hand side (*circled in magenta*) falls close to a *red* contoured area (occupancy with H-bond acceptor group favorable), the filling of which causes an increase in the selectivity for CAII binding. Interestingly, Gln 92 is found in this region in both isoforms. However, it is only in CAII that this group is available to accept an H-bond from the inhibitor that will contribute to binding affinity (*right, black dotted line*). The comparable residue Gln 92 in CAI is involved in a network of H-bonds to neighboring amino acids Asn 69 and Glu 58. Therefore, it is not available as a binding partner, and a decrease in the affinity for CAI is the consequence. The shown contour maps are based a CoMSIA analysis and superimposed onto crystal structures of CAI and CAII

selectivity in favor of CAII. Filling the blue contours with this property results in an increase in potency with respect to CAI. Compound **18.6** places its oxygen atoms of the endocyclic SO_2 group near the red CAII selective regions. Again, it is important to note that the model shown is derived purely from the correlation of ligand-binding data, and that in the following the information from the protein isoforms is used only to understand the obtained correlation model. The protein structures show a glutamine residue adjacent to position 92 in both CAI and CAII. This amino acid can accept an H-bond from the inhibitor via the NH_2 group of its carboxamide group. However, only CAII allows these structural conditions. Gln 92 is adjacent to Asn 69 and Glu 58 in CAI. The carboxamide group of Glu 92 forms a continuous H-bonding network with these residues and with His 94. Therefore, the NH group is no longer available for interactions with a bound inhibitor. This is reflected in the lower binding affinity of inhibitors that have an acceptor function at this position, such as **18.6**. The situation is completely different for CAII. The adjacent functional groups of Glu 69 and Arg 58 form an internal salt bridge. Therefore, they are not available as H-bonding partners for Gln 92. The carboxamide group of Gln 92 involves His 94 in an H-bond via its carboxamide CO group, and its NH_2 group is now available as an acceptor functionality to interact with a bound ligand. This results in significantly enhanced binding to CAII and is expressed as a selectivity advantage. It is important to remember that the applied CoMSIA analysis was trained only on ligand-binding data. Therefore, it did not know anything about the properties of the neighboring amino acids. Nevertheless, it produces a correlation model that can be easily explained by the unexpected binding behavior of the adjacent amino acids in the two isoforms.

18.15 Synopsis

- The concept of quantitative structure–activity relationships is not new. It was first described in the nineteenth century qualitatively, and later more quantitatively by Hansch and Fujita. It is an attempt to describe structure–activity relationships with mathematical models.

- Across a series of structurally closely related test compounds, the equieffective dose that induces a particular biological effect is related in a linear or squared dependence on the logarithm of the octanol/water partition coefficient and the Hammett constant, which describes the electronic properties of substituents at a given scaffold. A mathematical correlation model is computed by regression analysis.
- 3D QSAR methods have been developed to consider and correlate the spatial structure of active substances beyond molecular topology.
- The mutually aligned test molecules are embedded in a regularly spaced grid and their properties are explored with an interaction probe. The probe is placed systematically at all grid points and a molecular interaction field is computed around the aligned molecules by using a distance-dependent property potential.
- Usually, Lennard-Jones and Coulomb potentials are evaluated, and the generated data table for all molecules of the training dataset is correlated by the partial least squares technique.
- The derived CoMFA correlation model can be used to predict the biological properties of novel ligands not included in the training dataset. Strict criteria to monitor the statistical significance of the derived correlations must be met.
- Other property fields beyond Lennard-Jones and Coulomb potentials with mathematically different functional forms can be applied. With respect to the prediction of binding affinity, it has to be regarded that hydrophobic properties also implicitly reflect an entropic contribution to binding that is particularly difficult to regard in property fields.
- QSAR analysis only performs a relative comparison of molecules with regard to the considered biological property. Any dependence on a particular descriptor across a compound series can only be expected if the property related to this descriptor is varied in the series. QSAR methods only interpolate and never extrapolate beyond the scope of molecular properties reflected by the training set
- A number of alternative 3D QSAR approaches have been developed. They use different types of fields (e.g., for similarity analysis, CoMSIA) or try to incorporate protein information into the analysis (COMBINE and AFMoC).
- Comparative molecular field analyses can be evaluated graphically. Results are displayed as contours around the molecules and indicate where the change of a particular property runs either parallel or opposite to the changes in the biological property in the dataset.
- The graphical information can be directly translated into the design of modified molecules and, thus, support the medicinal chemist in optimizing a given lead structure in a systematic fashion.

Bibliography and Further Reading

General Literature

C. A. Ramsden, Eds., Quantitative Drug Design, Vol. 4: Comprehensive Medicinal Chemistry, C. Hansch, P. G. Sammes and J. B. Taylor, Eds., Pergamon Press, Oxford (1990)

H. Kubinyi, QSAR: Hansch Analysis and Related Approaches, VCH, Weinheim (1993)

H. van de Waterbeemd, Chemometric Methods in Molecular Design, VCH, Weinheim (1995)

H. van de Waterbeemd, Advanced Computer-Assisted Techniques in Drug Discovery, VCH, Weinheim (1995)

C. Hansch and A. Leo, Exploring QSAR. Fundamentals and Applications in Chemistry and Biology, 2 Volumes, American Chemical Society, Washington (1995)

H. Kubinyi, Ed., 3D-QSAR in Drug Design: Theory, Methods, and Applications, ESCOM, Leiden (1993)

H. Kubinyi, G. Folkers and Y.C. Martin, 3D QSAR in Drug Design, Vol. 1–3, Kluwer/ESCOM, Dordrecht, Boston, London (1998)

Special Literature

S. H. Unger and C. Hansch, On Model Building in Structure-Activity Relationships. A Reexamination of Adrenergic Blocking Activity of β-Halo-β-arylalkylamines, J. Med. Chem., **16**, 745–749 (1973)

J. M. Blaney, C. Hansch, C. Silipo and A. Vittoria, Structure-Activity Relationships of Dihydrofolate Reductase Inhibitors, Chem. Rev., **84**, 333–407 (1984)

C. Hansch and T. E. Klein, Quantitative Structure-Activity Relationships and Molecular Graphics in Evaluation of Enzyme-Ligand Interactions, Methods Enzymol.,**202**, 512–543 (1991)

H.-D. Höltje and L. B. Kier, Nature of Anionic or α-Site of Cholinesterase, J. Pharm. Sci., **64**, 418–420 (1975)

R. D. Cramer and M. Milne, Abstracts of the ACS Meeting, April 1979, COMP 44

R. D. Cramer, D. E. Patterson, and J. D. Bunce, Comparative Molecular Field Analysis (CoMFA). 1. Effect of Shape on Binding of Steroids to Carrier Proteins, J. Am. Chem. Soc., **110**, 5959–5967 (1988)

S. A. DePriest, D. Mayer, C. B. Naylor and G. R. Marshall, 3D-QSAR of Angiotensin-Converting Enzyme and Thermolysin Inhibitors: A Comparison of CoMFA Models Based on Deduced and Experimentally Determined Active Site Geometries, J. Am. Chem. Soc., **115**, 5372–5384 (1993)

P. J. Goodford, A Computational Procedure of Determining Energetically Favorable Binding Sites on Biologically Important Macromolecules, J. Med. Chem., **28**, 849–857 (1985)

G. E. Kellogg and D. J. Abraham, Key, Lock and Locksmith: Complementary Hydropathic Map Predictions of Drug Structure from a Known Receptor-Receptor Structure from Known Drugs, J. Mol. Graphics, **10**, 212–217 (1992)

T. Hüfner-Wulsdorf and G. Klebe, Protein-Ligand Complex Solvation Thermodynamics: Development, Parameterization and Testing of GIST-based Solvent Functionals, J. Chem. Inf. Model., **60**, 1409–1423 (2020)

G. Klebe, U. Abraham and T. Mietzner, Molecular Similarity Indices in a Comparative Analysis (CoMSIA) of Drug Molecules to Correlate and Predict Their Biological Activity, J. Med. Chem., **37**, 4130–4146 (1994)

A. R. Ortiz, M.T. Pisabarro, F. Gago and R.C. Wade, Prediction of Drug Binding Affinities by Comparative Binding Energy Analysis, J. Med. Chem., **38**, 2681–2691 (1995)

H. Gohlke and G. Klebe, DrugScore Meets CoMFA: Adaptation of Fields for Molecular Comparison (AFMoC) or How to Tailor Knowledge-based Pair-Potentials to a Particular Protein J. Med. Chem., **45**, 4153–4170 (2002)

A. Weber, M. Böhm, C. T. Supuran, A. Scozzafava, C. A. Sotriffer and G.Klebe, 3D QSAR Selectivity Analyses of Carbonic Anhydrase Inhibitors: Insights for the Design of Isozyme Selective Inhibitors, J. Chem. Inf. Model., **46**, 2737–2760 (2006)

A. Hillebrecht, C. T. Supuran and G. Klebe. Integrated Approach Using Protein and Ligand Information to Analyze Affinity and Selectivity Determining Features of Carbonic Anhydrase Isozymes, ChemMedChem, **1**, 839–853 (2006)

From In Vitro to In Vivo: Optimization of ADME and Toxicology Properties

Contents

G. Klebe, *Drug Design*, https://doi.org/10.1007/978-3-662-68998-1_19

The interaction between a compound and the binding site of a therapeutically relevant biological macromolecule is the critical prerequisite for its suitability as a drug. Another, no less important, requirement is the ability of the substance to make its way from the site of application through an often rather tortuous path to the target tissue and finally into the binding pocket of the macromolecular target. To do this, the compound must penetrate aqueous phases and lipid membranes. Depending on its water and lipid solubility, it will end up in different compartments of the biological system. It is also modified by metabolizing enzymes. After conjugation or degradation, it is finally excreted through the kidney, bile, and/or intestine (Sects. 9.1 and 27.6).

In contrast to the biological activity of a drug, which is called **pharmacodynamics**, the sum of all processes affecting **absorption**, **distribution**, **metabolism**, and **excretion**, the so-called **ADME** parameters, is called **pharmacokinetics**. Roughly speaking, pharmacodynamics can be thought of as "the effect of the substance on the organism" and pharmacokinetics as "the effect of the organism on the substance." In recent years, this clear separation of definitions has begun to fade. The term pharmacodynamics has also been expanded more and more to processes of pharmacokinetics. This is mainly due to the increasing knowledge that transporters or enzyme systems are responsible for properties such as absorption, distribution, or metabolism. A growing number of structures of the involved enzymes and transporters has been determined, allowing specific structure–activity relationships to be established for such systems (Sects. 27.6 and 30.10).

The pharmacokinetics of a given biological system and the time dependence of the absorption, distribution, and excretion processes are described by mathematical models. The pharmacokinetics of every pharmaceutical is scrupulously investigated and a dosing scheme is determined before entry into **clinical trials**, especially during clinical phases I and II, which evaluate tolerability and efficacy in humans. The isolation and structural elucidation of metabolites formed in humans helps to identify the animal model that most closely resembles humans in terms of metabolic properties. These species are then used for toxicology studies, which are chosen to investigate possible teratogenic effects, and long-term studies to investigate possible carcinogenic effects. In parallel, individual metabolites of a pharmaceutical are investigated for their toxic side effects.

19

In the context of the rational design of new active substances, a substantial problem arises from the pharmacokinetic parameters and the toxicity: these investigations are only carried out for very few compounds because of the enormous experimental effort and the high costs, and only for those compounds that are intended for clinical development. This approach comes with a serious danger: inadequate pharmacokinetic properties are only recognized in very late development stages, and only then after considerable sums have already been invested in the development of a new pharmaceutical. In the mid-1990s, a study showed that many unsuccessful development campaigns failed due to **unsatisfactory pharmacokinetics and intolerable toxicity**. For these reasons, the search for ***in vitro* models to predict ADME toxicity** has intensified over the last 30 years. Rather than studying the pharmacokinetics of a single compounds in detail, the dependence of various pharmacokinetic parameters on the properties of many different compounds is nowadays investigated. This provides a better understanding of the relationship between chemical structure and pharmacokinetics. At the same time, it leads to the establishment of general rules and numerous computer models that are now used at an early stage in the design of new drugs.

19.1 Rate Constants of Compound Transport

The distribution of a substance into phases of different lipophilicities is measured as the partition coefficient P (Sect. 18.3). This definition is valid for systems at equilibrium. The distribution between the **water** and **octanol phases** is considered as a model system. The ratio of the concentration of the nonionized form of an investigated compound in the two phases is evaluated. In addition, the pH value is adjusted during the measurement so that the investigated compound overwhelmingly occurs in its nonionized form. As a general rule, log P, the logarithm of this value is used.

$$\log P_{(\text{octanol/water})} = \log \frac{\text{concentration(dissolved compound)}_{\text{octanol}}}{\text{concentration(dissolved compound)}_{\text{nonionized in water}}}$$

Biological systems are open systems that are kinetically controlled. They can be temporarily found in a dynamic equilibrium. This condition can be compared to a chromatographic process in which a substance is in constant exchange between the solid support and the mobile phase. Locally, equilibria occur that are disrupted by the continuous progression of the mobile phase. In contrast to the relatively simple conditions in chromatography, there are a plethora of different phases in biological systems. A drug is distributed throughout all of these phases. Furthermore, metabolic processes are running in parallel that lead to different metabolites.

To analyze these dynamic equilibria, the **kinetic equilibrium constants of the substance transport** from the aqueous phases into the lipid phases and in the reverse direction must be known. It is astonishing that such fundamental experimental investigations on organic substances were first carried out by Bernard Lippold in the mid-1970s, and later also by Han van de Waterbeemd. Lippold used a three-phase system: water/*n*-octanol/water (■ Fig. 19.1). After adding the substance to one of the two aqueous phases, the time dependence of the substance concentration in the different phases was measured. From this, the equilibrium constant k_1 for the transport from water to the octanol phase and the rate constant k_2 in the opposite direction can be calculated.

In addition to the partition coefficient P, which is described in Eq. 19.1, a very simple correlation has been shown for the dependence of k_1 and k_2 (Eq. 19.2); β and c are constants that depend on the system and not on the structures of the substances.

$$P = k_1 k_2 \tag{19.1}$$

$$k_2 = -\beta k_1 + c \tag{19.2}$$

The dependence of the rate constants k_1 and k_2 on the partition coefficient P results from the combination of both equations (Eqs. 19.3 and 19.4).

$$\log k_1 = \log P - \log(\beta P + 1) + \text{constant} \tag{19.3}$$

$$\log k_2 = -\log(\beta P + 1) + \text{constant} \tag{19.4}$$

The experimental k values for 20 different sulfonamides and 15 further substances that were experimentally determined by Han van de Waterbeemd are shown in ■ Fig. 19.2. Among the latter are neutral, acidic, basic, and even quaternary charged compounds with very different molecular weights. The characteristic curve shows that the rate constant k_1 for the transfer from the aqueous to the organic phase depends on the partition coefficient P for relatively polar substances. It is thermodynamically controlled, which means that it increases with increasing lipophilicity. However, there is a point at which the diffusion of the substance is limited by k_1 at the maximally achievable value. More lipophilic substances cannot simply penetrate the organic phase faster. The same applies in the opposite direction, where the diffusion from the organic phase into the aqueous phase is described by k_2. In both cases, the chemical structure plays a role in determining the value of the partition coefficient P. Since the rate constants are limited by diffusion, there must be an apparent dependence on the molecular size in this area. According to Fick's law of diffusion, the diffusion should be proportional to the radius of the particle, parallel to the cube root of the volume, as a first approximation. Due to the relatively low variability of the molecular size of organic drugs and their conformational flexibility, this effect is likely to be lost in the noise of experimental error. In addition, it should not be forgotten that the octanol/water system discussed is very simple and only slightly approximates the complex structural relationships of real membrane systems. For this reason, more relevant models, such as the so-called **PAMPA** or **Caco-2 models**, are increasingly being used to obtain experimental distribution data (Sect. 19.6). Here, more complex relationships are indicated. Obviously, how a compound is distributed and structurally oriented in the vicinity of membrane structures is important. At the same time, these properties influence how the penetration, and therefore the distribution, is to be described.

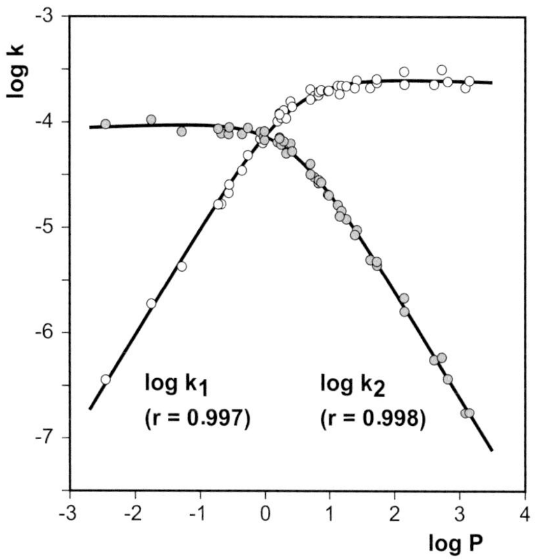

■ **Fig. 19.2** Experimentally determined rate constants k_1 and k_2 for the transport of 20 sulfonamides and 15 further chemically different substances with molecular weights between 100 and 500 Da. The curves and correlation coefficients r correspond to the fitting of the data with Eqs. 19.3 and 19.4

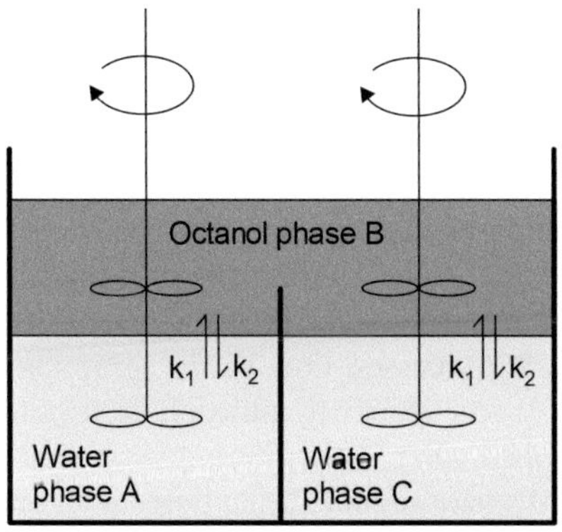

■ **Fig. 19.1** Three-compartment system for the determination of the rate constants k_1 and k_2. At the beginning of the experiment the substance is dissolved in aqueous phase A. Next the substance concentration is measured in phases A, B, and C after different times until an equilibrium is established between the individual phases

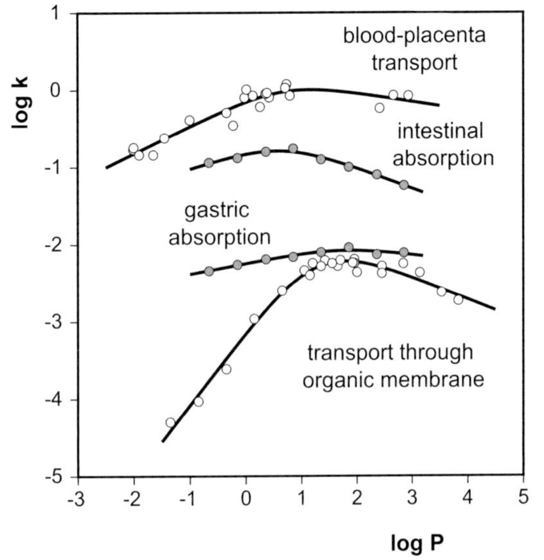

Fig. 19.3 The rate constant k for the transport of drugs depends nonlinearly on lipophilicity. This is valid for simple *in vitro* models as well as for biological systems. The bottom curve describes the log k values of the transport of barbiturates in an *in vitro* absorption model from an aqueous phase, through an organic membrane into another aqueous phase. Both curves in the center (*gray dots*) describe the dependence of the absorption rate constants k on the lipophilicity for the absorption of homologous carbamates from the stomach (gastric absorption) or the gut (intestinal absorption) of rats. The top curve was determined for the entry of different drugs into the placenta from the circulation. In all cases an increase in log k dependent upon log P is seen, until a more-or-less-pronounced maximum for substances with moderate lipophilicity. For very nonpolar substances, this curve falls, and in rare cases a plateau is reached. The curves for gastric and intestinal absorption and for the penetration into the placenta run flatter than the curve for the *in vitro* transport of barbiturates (*below*), because here no lipid barrier is present

19.2 Absorption of Organic Molecules: Model and Experimental Data

The rate constant, k, for the penetration through the lipid membrane from the aqueous phase is described by another equation, Eq. 19.5. Here, the rate constants k_1 and k_2 also describe the entry into the organic phase and the transport in the opposite direction, respectively.

$$\log k = \log k_1 + \log k_2 + \text{constant} \qquad (19.5)$$

In the first approximation, this equation should also describe transportation processes in multicompartment systems. Model calculations on arbitrary, complex systems show that this is indeed the case. They confirm that there is bilinear dependence of the transport in different phases on the total lipophilicity of a substance. For multiple groups of drugs, for example, barbiturates, this was demonstrated experimentally in simple *in vitro* model systems (Fig. 19.3, bottom). The log k values increase linearly upon penetration through an organic membrane, which correlates with the increase of k_1 with constant k_2. After passing through a maximum, they decrease with a constant k_1 value and decreasing k_2 value. This dependence was quantitatively summarized by Hugo Kubinyi in the so-called **bilinear model** (Eq. 19.6); a, b, β, and c are constants, by which the nonlinear regression analysis is ascertained.

$$\log k = a \log P - b \log(\beta P + 1) + c \qquad (19.6)$$

Entirely analogous dependencies are observed with the absorption of compounds, that is, out of the stomach or intestines (Fig. 19.3, center). Active substances that should be orally available should not be either very polar or very nonpolar. Substances with intermediate lipophilicity can cross the blood–placenta barrier more easily than very polar or very nonpolar compounds (Fig. 19.3, top). A nonlinear dependence on the lipophilicity for substance penetration through the blood–brain barrier is particularly pronounced (Fig. 19.4). The optimum for this barrier is in the range of log P = 1.5–2.5. For central nervous system (CNS)-active substances, an optimal lipophilicity around log P = 2 should be aimed for in order to facilitate penetration across the blood–brain barrier.

19.3 The Role of Hydrogen Bonds

The simple concept about the dependence of absorption on the octanol/water partition coefficients outlined above has been questioned in recent years. While octanol is a relevant model for lipid membranes in many respects (Sect. 4.2), it can only incompletely model the influence of hydrogen bonding. After equilibration in the octanol/water system, the organic phase contains considerable amounts of water, corresponding to a molar ratio of octanol:water = 4:1. Substances with polar, solvated groups, therefore, do not have to completely release their water solvation shell upon entering the octanol phase. Entry into a biological membrane is obviously different. Apart from the dependence on lipophilicity, even poorer membrane penetration is observed for substances that can form an increasing number of hydrogen bonds. Similarly, a ligand must release its water shell before it can be accommodated in the binding site of a protein.

The water/cyclohexane system is more suitable for describing such processes. Due to the nonpolar nature of this hydrocarbon, the drug molecule cannot take its water shell with it when transitioning from water to cyclohexane. Many years ago, P. Seiler derived an increment I_H (Eq. 19.7) from the differences in partition coefficients in cyclohexane/water (loss of water shell) and octanol/water (no loss of water shell) for different functional groups. These I_H values characterize the tendency of the groups to form hydrogen bonds.

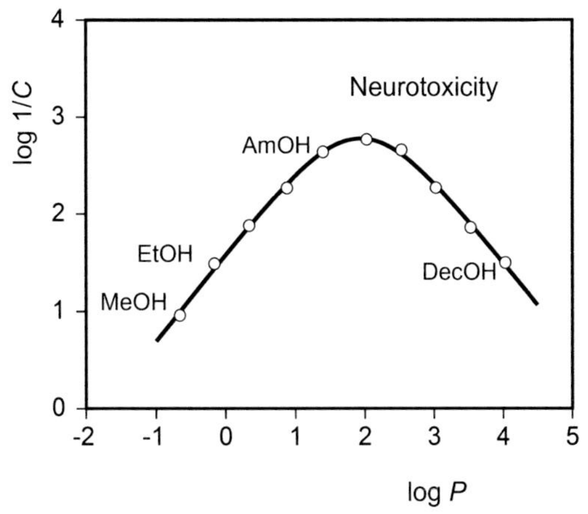

Fig. 19.4 The neurotoxicity (C = molar dose that induces a specific toxic effect) of homologous primary alcohols in the rat is a measure of their ability to cross the blood–brain barrier. Polar substances remain predominantly in the circulation. In contrast, substances with moderate lipophilicity easily reach the central nervous system. Accordingly, neither methanol (MeOH) nor ethanol (EtOH) exhibit pronounced neurotoxicity. The high general toxicity of methanol (blindness) is not due to its own action, but rather to the highly toxic metabolites formaldehyde and formic acid (acidosis). Short-chain alcohols such as amyl alcohol (AmOH) are much more neurotoxic. The highly lipophilic decanol (DecOH) shows low toxicity

$$\log P_{\text{cyclohexane}} + \sum IH = 1.00 \log P_{\text{octanol}} + 0.16 \quad (19.7)$$

Seiler's concept remained largely ignored. In 1988, Robin Ganellin and coworkers described the CNS bioavailability of various substances, that is, their ability to cross the blood–brain barrier, as a linear function of a Δlog *P* value. This Δlog *P* value is the difference between the log *P* values in the cyclohexane/water and octanol/water systems. The bioavailability of peptides also runs in a first approximation parallel to the Δlog *P* value or the number of groups potentially involved in hydrogen bonding. In fact, methylation of all NH groups of a peptide scaffold can provide compounds with good bioavailability. The requirements for good membrane penetration are similar to those for high affinity at the binding site (Chap. 4). Here again, the need to release relatively tightly bound water molecules can have a detrimental effect on binding affinity. It should not be forgotten that more complex molecules can easily undergo conformational transitions to a geometry that buries part of the hydrogen bonds and, thus, their polarity intramolecularly (cf. cyclic peptidomimetics such as cyclosporine, Sect. 10.1, Fig. 10.2).

Several other partitioning systems, such as heptane/ethylene glycol, have been proposed as alternatives to the octanol/water or cyclohexane/water systems for simulating penetration through a lipid membrane. However, even these systems cannot correctly reflect the architecture of membranes with an inner lipophilic zone and a polar, negatively charged outer rim. Another option is the determination of the membrane/water partition coefficient, which is rather laborious experimentally. For this, artificial membranes or liposomes are used as models.

19.4 Distribution Equilibria of Acids and Bases

Many drugs are acids (HA) or bases (B). They exist in two forms through dissociation (Eq. 19.8) or protonation (Eq. 19.9); one is usually a nonpolar neutral form and the other is a polar ionic form. The values of the partition coefficients of the ionic species are generally three to five orders of magnitude smaller than those of the corresponding neutral molecule.

$$HA + H_2O \leftrightarrow A^- + H_3O^+ \quad (19.8)$$

$$B + H_3O^+ \leftrightarrow BH^+ + H_2O \quad (19.9)$$

The partition equilibrium of an acid and its anion in a two-phase system depends on the pK_a and pH of the aqueous phase and the partition coefficients P_u and P_i of the substance (Fig. 19.5). All components in each phase must be in equilibrium with one another for the total system to be in equilibrium. The dependence of the partition coefficient *P* on pH, the **pH partition profile**, is usually sigmoidal (i.e., S-shaped). Plateaus are observed for the uncharged neutral form and for pH values where so little of the neutral form is present that only the transfer of the charged species into the organic phase determines the measured partition coefficient (Fig. 19.6). The charged species enters the organic phase as an ion pair together with a counterion. The counterion is either the corresponding ion of the salt or the excess of ions in the aqueous buffer. The partition coefficient of the **ion pair** depends on the lipophilicity of the counterion. The tetrabutylammonium salt of salicylic acid has only a slightly lower partition coefficient than the neutral form of salicylic acid. In contrast, the sodium salt of salicylic acid has absolutely no tendency to partition into the organic phase. Amino acids and other mixed acidic and basic compounds yield pH partition profiles with a maximum between the pK_a values of the two ionizable groups (Fig. 19.6), this means when the **zwitterionic form** is present.

By knowing the log *P* value of the neutral form and the pK_a value, the partition coefficient of a substance at neutral pH can be calculated. These concepts allow the estimation of the absorption and distribution properties of new compounds. Of course, these considerations are only valid for drugs for which there is no transporter to facilitate membrane penetration (Sects. 22.7 and 30.10).

Because of their importance, pK_a values are now routinely measured by potentiometric titration in pharmaceutical research. However, it is often overlooked that the definition of the pK_a of acids and bases is only valid for aqueous solutions. The addition of an organic solvent, which changes the dielectric constant, shifts this value (Sect. 4.4). This is even more true for the binding site of a protein or the interior of a membrane. In some cases, experimental values have been determined by NMR spectroscopy and isothermal titration calorimetry.

19.5 Absorption Profiles of Acids and Bases

For example, the absorption of an active substance from the intestine into the blood should depend on the pH of the surrounding medium and the pK_a of the substance, just like the distribution between an aqueous buffer system and an organic phase. Absorption should, therefore, follow very similar profiles as the distribution. In the 1950s, Brodie, Hogben, and Schanker formulated the **pH partition theory** to describe this effect. It states that the dependence of the absorption profile on the pH value, the **pH–absorption profile**, is identical to the pH–partition profile (Sect. 19.4). This theory was confirmed by, among other things, the investigation of the rate constant of absorption of a few acids and phenols from the colon of the rat at pH 6.8. The neutral forms of the strong acids 5-nitrosalicylic acid (pK_a = 2.3), salicylic acid (pK_a = 3.0), *m*-nitrobenzoic acid (pK_a = 3.4), and benzoic acid (pK_a = 4.2) display comparable lipophilicity with log *P* values between 1.8 and 2.3. Under experimental conditions near neutral pH, they are largely dissociated. Less than 0.1% of the compounds in the neutral form. Therefore, they are distinctly more slowly absorbed than the comparably lipophilic, weakly acidic phenols *p*-hydroxypropiophenone (pK_a = 7.8) and *m*-nitrophenol (pK_a = 8.2), which are more than 90% in their neutral form at pH 6.8.

Neutral forms can diffuse through membranes; charged forms are highly soluble in water. An equilibrium between the two forms is quickly established in an aqueous medium and also at the phase boundaries. If the pK_a values of the substances are not more than 2–3 units

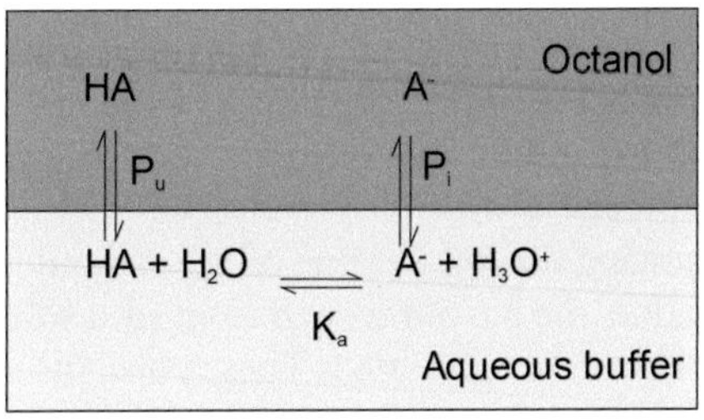

Fig. 19.5 Two-phase system with partition and dissociation equilibria for an acid HA (Eq. 19.8). K_a is the dissociation constant, P_u and P_i are the partition coefficients of the undissociated and ionic forms, that is, neutral and charged species, respectively. Because there is usually a difference of several orders of magnitude between the P_u and P_i values, in many cases the P_i value can be neglected. This leads to considerable simplification of the corresponding mathematical models

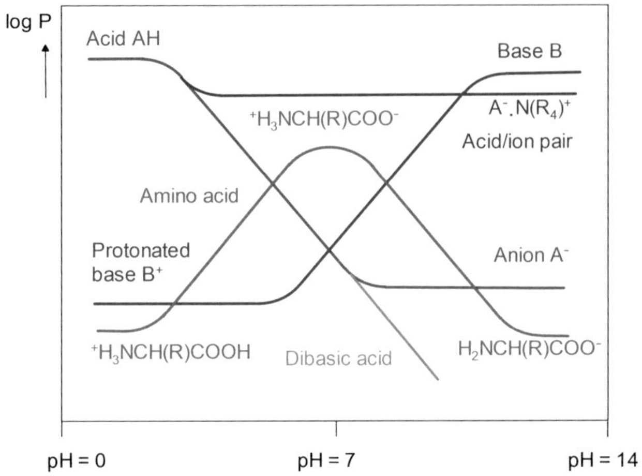

Fig. 19.6 The pH dependence of the distribution equilibrium of acids and bases, the so-called pH distribution profile, follows simple rules. Typically when an acid (*red*) or a base (*blue*) is present, sigmoidal, that is, S-shaped, curves are observed. For a dibasic acid, for example, oxalic acid, the decrease in the partition coefficient continues with increasing pH values (*violet*). In the presence of lipophilic counterions, for example, the tetrabutylammonium salt of salicylic acid, the ion pair displays a very high partition coefficient (*magenta*). Amino acids with neutral side chains carry one basic amino group and an acidic carboxyl group (*green*). Accordingly, they have a maximum partition coefficient at the neutral point. Here, the majority of the substance is indeed present as a zwitterion, but a larger fraction is in the neutral form than at lower or higher pH values

from the neutral value of pH 7, the neutral form will be present in the aqueous phase in a quite sufficient concentration of about 0.1–1%. It penetrates the membrane. In the aqueous phase, it is immediately regenerated by the dissociation equilibrium. In a biological system, the distribution of such substances is rapid and effective (Fig. 19.7), and it is even better the closer the pK_a is to neutral pH 7. This also explains why so many drugs are organic acids or bases. Because of the strongly deviating pH values in the stomach and intestine, there is a point in the gastrointestinal tract where a neutral substance, an acid or a base, can be well absorbed. If the pK_a values are too far away from the physiological pH values, for example, amidines or guanidines with extremely high pK_a values, absorption can become problematic. This is also true for zwitterionic compounds, such as amino acids, and compounds with multiple acidic or basic groups in the molecule. Because of the large volume available for distribution, diffusion is predominantly from the gastrointestinal tract into the blood or tissues, and only to a negligible extent in the opposite direction (Fig. 19.7).

The absorption of strongly acidic compounds outside the range in which the compound exists as a neutral molecule, runs in first approximation parallel to the difference pH − pK_a, and for bases the difference is pK_a − pH. There are exceptions to this approximation. Highly lipophilic compounds require a more detailed description of the pH–absorption profile. The neutral forms of these

19

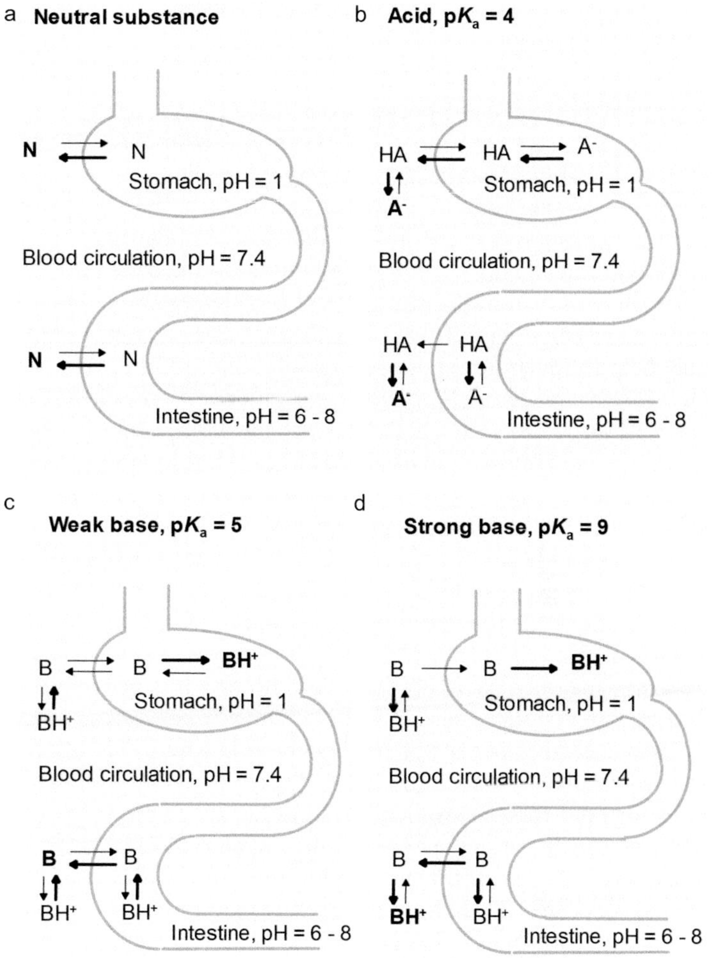

Fig. 19.7 **a** A moderately polar neutral substance N is absorbed very well from the stomach as well as from the intestines. It is quickly distributed in the circulation so that back-transport does not play a notable role. **b** An organic acid HA ($pK_a = 4$) is absorbed well from the stomach, as long as it is not too polar, because it exists there overwhelmingly in the neutral form. The absorption is facilitated by the fact that the free acid is in considerably lower concentration in the blood than in the stomach. The formation of an anion shifts the concentration gradient in this direction. The absorption is slower from the gut because there the equilibrium lies overwhelmingly on the side of the ionized form. **c** A weak base ($pK_a = 5$) is relatively poorly absorbed from the stomach because it is predominantly present in its polar, protonated form. It is well absorbed in the intestine, where it exists in its neutral form. **d** A strong base with a $pK_a = 9$ cannot be absorbed via the stomach. The equilibrium indeed lies heavily on the side of the protonated form in the intestines, but the nonpolar form is available in adequate quantities. Therefore, the substance can be absorbed. When a substance reaches a pK_a value of more than 11, the concentration of the neutral, bioavailable form is too low for good absorption to take place

substances enter the lipid phase as soon as they come near the membranes. The neutral molecule is being constantly removed from the dissociation equilibrium, which is established in the aqueous phase. However, it is very quickly replenished by this equilibrium. In equilibrium, there is a continuous transport of substance from the aqueous phase into the membrane. The small amounts of the uncharged neutral form are the door through which the entire process takes place. The rate of transition into the lipid layer does not depend on the (often very low) concentration of the neutral form, but rather on

- The total concentration of the compound,
- The rate constants of the dissociation equilibrium, and
- The diffusion constant of the compound.

Accordingly, a **shift in the pH–absorption profile** is observed in biological systems for lipophilic acids and bases relative to the **pH–partition profile**, which is referred to as **pH shift**. This always occurs in the direction towards the neutral point, which means to higher pH values for acids and to lower pH values for bases. The greater the lipophilicity of an acid or base, the greater the observed shift in the absorption profile. The log P value and the pK_a values cannot be considered separately when assessing how well a substance is absorbed. Their combination is crucial. For the design of new drugs, this means that a substance with an unfavorable partition behavior, which means with a pK_a value that is too high or too low, can be favorably modified in the desired direction by increasing its lipophilicity. To describe the pH dependency of the distribution equilibrium, a **distribution coefficient *D*** was introduced as a supplement to the partition coefficient *P*. For this, the ratio of the sum of all concentrations of ionized and nonionized forms of an investigated compound in the two phases are considered. The pH value is adjusted for measurement in a buffer solution so that the addition of the investigated compound does not shift the pH. Usually log *D*, logarithm of the distribution coefficient, is used here (Eq. 19.10).

$$\log D_{(\text{Octanol/Buffer})} = \log \frac{\text{Sum}(\text{Conc.}(\text{Sovated Substance})_{\text{Octanol, ionized/nonionized forms}})}{\text{Sum}(\text{Conc.}(\text{Solvated Substance})_{\text{Buffer, ionized/nonionized forms}})} \quad (19.10)$$

19.6 What Is the Optimal Lipophilicity of a Drug?

Lipophilicity plays an important role in the assessment of a drug's therapeutic potential. This applies to absorption, distribution, metabolism, and excretion. With the exception of substances that are absorbed via a transporter, absorption is usually better when compounds are more lipophilic. This advantage is limited by the **solubility** in aqueous phases, which decreases significantly with increasing lipophilicity. The following comparison may help to illustrate the low water solubility of some hydrophobic drugs. The neuroleptic chlorpromazine (**19.1**, ◘ Fig. 19.8) with log $P = 5.4$ has a water solubility of 2.55 mg/L at 24 °C in its neutral form. Marble with 14 mg/L and sand with 10 mg/L are much more soluble in water. Due to its basic nature ($pK_a = 9.3$), chlorpromazine can be converted to its hydrochloride, which is much more water soluble at about 10 g/L. However, it does not approach the solubility of common salt, which is 350 g/L.

These factors depend on the intermolecular interactions in the crystalline solid and can vary greatly between different polymorphic crystal modifications in which the drug molecule has been crystallized. Therefore, correlations for predicting bioavailability consider the melting point as another simple parameter to estimate the stability of a polymorphic form, in addition to lipophilicity and solubility. Furthermore, computational methods are used to investigate whether additional polymorphic forms may exist with more suitable properties. In addition to solubility, dissolution kinetics is important for galenic formulations, namely the final drug product. It determines the amount of substance that is dissolved

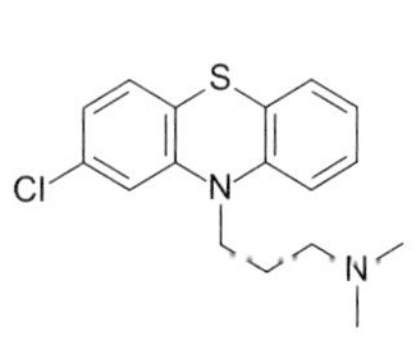

19.1 Chlorpromazine

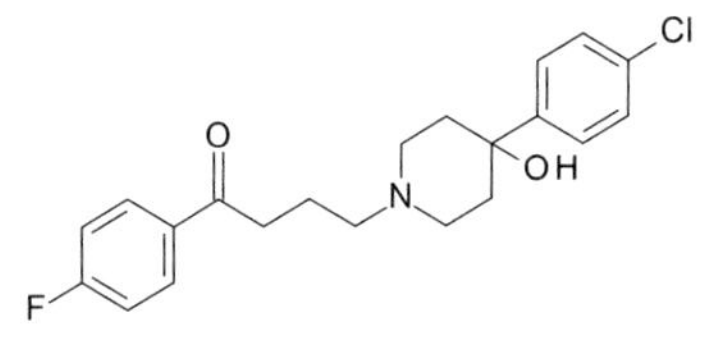

19.2 Haloperidol

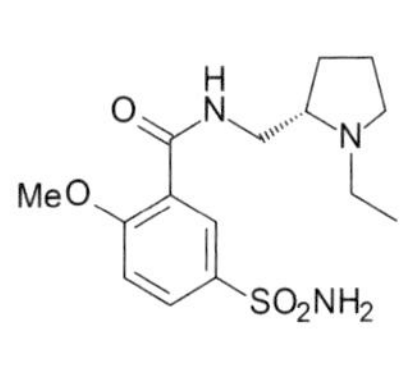

19.3 Sulpiride

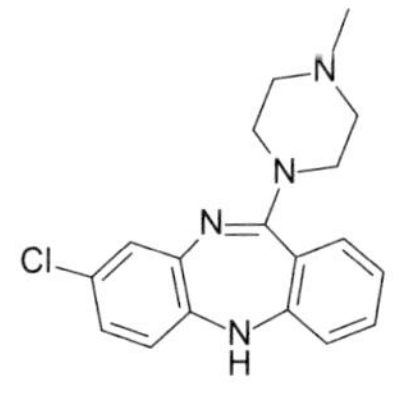

19.4 Clozapine

◘ **Fig. 19.8** Chlorpromazine **19.1**, haloperidol **19.2**, and sulpiride **19.3** are neuroleptics with typical side effects that are associated with dopamine antagonists. Clozapine **19.4** is different from these substances in its binding profile on the dopamine receptors (◘ Table 19.2) as well as in its side effects

during the gastrointestinal passage. This amount can be influenced by several factors, such as, the following:

- Increasing the surface area by grinding the crystals into miniscule particles (micronization),
- Growing a modified crystal with better solubility properties,
- Crystallization under special conditions to afford a more uniform (usually smaller) size, or crystals with lattice defects,
- Changing the salt form,
- Adding solubility-mediating additives, and
- Embedding the drug as amorphic solid solutions of easily dissolvable polymers.

Because of its importance, high-throughput solubility measurement techniques have been established in recent years. As mentioned above, large-scale computer simulations are nowadays used to predict the polymorphic crystal forms of new drug molecules in order to assess whether there might be a different crystal form that could positively influence dissolution behavior, solubilization kinetics and, thus, bioavailability.

Cell cultures are also increasingly used as *in vitro* models to study drug absorption. A thin layer of cells from human colon carcinomas (so-called **Caco-2**, HT29 or MFCH cell lines) is grown in a two-chamber system. Drug transport can be monitored from either the apical or basolateral side. Since these cells also express transporters, the involvement of specific transport mechanisms can also be studied. These models are less suitable for studying the possible consequences of drug metabolism because the metabolizing enzymes (Sect. 27.6) are expressed at reduced levels in these cells.

In vitro models have also been developed to study blood–brain barrier penetration. However, these models are relatively labor-intensive and the results can often only be compared within a series of structurally related compounds. Artificial membrane assay systems (**PAMPA**, from **p**arallel **a**rtificial **m**embrane **p**ermeability **a**ssay) can be constructed to allow high-throughput screening. In addition, the penetration behavior in liposomes can be evaluated by surface plasmon resonance.

When experimentally determining the absorption of various substances, results obtained with saturated solutions of the substances should not be compared with results obtained with solutions of constant concentration. In the first case, the absorption rate of highly lipophilic substances decreases linearly due to the decrease in solubility with lipophilicity. In the second case, the absorption rate often remains at a more or less constant value, even for highly lipophilic substances. A comparison of such different experimental conditions is likely to lead to erroneous conclusions. Further confusion arises when the terms **absorption and bioavailability** are used incorrectly (Sect. 9.1). Absorption of a compound may be excellent, but bioavailability may be poor. Lipophilic compounds and substances with a molecular weight greater than 500–600 Da are often well absorbed but suffer from very rapid biliary elimination. This usually occurs during the first liver passage (first-pass effect, Sect. 9.1) immediately after absorption from the intestine. To achieve good bioavailability, the lipophilicity must not be too high. The route of excretion also depends on lipophilicity. In general, extremely lipophilic substances are metabolized more rapidly, but they are also of greater toxicological concern. Hydrophilic substances and polar metabolites, even after conjugation with polar groups, are excreted via the kidneys. Lipophilic substances are usually excreted by the liver and then by the intestine. Such substances often undergo oxidative metabolism, with the potential for the formation of toxic metabolites.

Substances that interact with membrane-bound receptors or ion channels can often reach their targets more easily if they are enriched in the surrounding membrane. To achieve this, the substances should be lipophilic or carry a large lipophilic group with which they can be anchored in the membrane (Sect. 4.2, ◘ Fig. 4.2).

19.7 Computer Models and Rules to Predict ADME Parameters

In addition to setting up appropriate test systems to systematically record parameters that determine pharmacokinetic properties, much effort has been devoted to establishing rules and computer models to predict favorable ADME properties. First and foremost is the **Rule of Five**, developed by Chris Lipinski at Pfizer. It states that an active substance should not violate more than two of the five criteria listed in ◘ Table 19.1. These simple rules are derived from experience and are often used to preselect compounds for screening. Tudor Oprea (Albuquerque, New Mexico, USA) has further refined these rules and extended them to include the occurrence of certain structural building blocks, such as the maximum number of rings of a certain size. Programs such as CLOGP, or ACD/pKa and Pallas/pKa have been developed to estimate lipophilicity and pK_a values. Meanwhile, the simple Ro5 has come under increasing criticism as more and more examples of successful drugs that do not obey the rule have become known. These so-called "beyond Ro5" drugs typically bind to flat and groove-shaped binding sites. Chemically, they often belong to the class of macrocycles. Solvation enthalpies are used to attempt to predict solubility. Predictions of permeability, absorption, and bioavailability are based on empirical correlation models. Experimental observations are related to the chemical structure of the investigated molecules. The methods used are derived from the QSAR models presented in Chap. 18. The properties to be predicted are described by models based on intuitively chosen or more or less obvious descriptors. Usually, molecular parameters are used, which are often derived from molecular surface contributions and may be determinant for the target properties. In addition to routine regression analysis, more recent mathematical models such as neural networks, nearest neighbor classifiers, decision trees, or machine learning techniques such as support vector machines are applied.

◘ **Table 19.1** Criteria for the *Rule of Five*

Molecular weight	≤ 500 Da
Octanol–water partition coefficient	log P ≤ **5**
Number of H-bond donor groups	not more than **5**
Number of H-bond acceptor groups	not more than **2 • 5 = 10**

In addition to the easy-to-evaluate Rule of Five, the following criteria should be considered for rational design: substances that act in the periphery, such as cardiovascular drugs, should be relatively polar. Of course, a certain degree of minimal lipophilicity is necessary for their absorption. Because of the risk of central side effects or the formation of toxic metabolites, this lipophilicity should not be exceeded too much. As a general rule, it is better to be a little less potent than to have all the other problems! A good therapeutic window is much more valuable than a picomolar affinity to a protein. Substances that act on membrane-bound proteins and substances that act in the central nervous system should have a moderate to high log P value of > 1. To avoid the development of toxic metabolites, the inclusion of the following is recommendable:

- Easily conjugated groups, for example hydroxyl, amino, or carboxyl groups,
- Preconceived metabolic cleavage points such as ester or amide bonds, and
- Oxidizable groups that lead to nontoxic and easily excretable metabolites, for example, methyl groups.

Of course, this strategy should not be exaggerated, otherwise the substances will be excreted too quickly. The **biological half-life** is then reduced to a value that makes therapeutic administration in humans impossible.

The structural consideration of properties that lead to optimal bioavailability, adequate biological half-life, and nontoxic metabolites is a problem in the search for new drugs. Structure-based drug design initially focuses on the fit of a ligand to its binding site. Often, aspects related to pharmacokinetics and metabolism are not adequately considered at this stage. Disappointments at the end of a successful optimization in the preclinical phase, or at the latest in the clinic, punish such a one-sided approach. As the spatial structures of **transporters, channels, and metabolic enzymes** become increasingly available, structure-based design concepts can be used to test the cross-reactivity of proposed or developed ligands on these target structures.

Binding to the potassium-ion-transporting **hERG ion channels** leads to their blockage. This can lead to life-threatening cardiac arrhythmias (Sect. 30.3). For this reason, QSAR models have been developed to screen molecules for potential hERG channel binding. Methods have been developed to dock ligands directly into the spatial structure of the channel. Another system that has recently been structurally characterized is the membrane-bound **glycoprotein GP170**. It is a transporter capable of expelling drugs from the cell (Sect. 30.10). It is desirable to avoid interactions with this protein as much as possible. Another large family of enzymes worthy of attention are the **cytochrome P450 metabolic enzymes** (Sect. 27.6). Here one tries to estimate how drugs interact with these proteins and how they are metabolized. This opens up a wide field for structure-based design.

19.8 From *In Vitro* to *In Vivo* Activity

Compounds are first tested in simple *in vitro* models, such as enzyme inhibition or receptor binding, in cell culture and later in organs and animal models. In general, the simplest model where the results are predictive of the expected effect in animals or humans is chosen. To do this, it is necessary to derive quantitative relationships between the different test models, known as **activity–activity relationships**. These describe the relationship between biological activity, e.g., between *in vitro* and *in vivo* data. In the best case, they even allow extrapolation to therapeutic effects in humans from the determination of affinity in a binding or inhibition assay.

Confirmation of a correlation between a simple test model and a therapeutic effect is often more important than the derivation of a structure–activity relationship. Once the relevant quantitative relationship has been established, inexpensive and rapid assays can be used instead of time-consuming and costly animal experiments. This significantly **reduces the number of animal experiments**. But that is not the only benefit. The use of automated molecular testing systems allows reliable characterization and standardization of compound profiles.

19.9 Compartmentalization: Natural Ligands Are Often Unspecific

Prior to biological testing of a compound, the following questions must be answered: What is the therapeutic goal to be achieved and how will it be achieved? Therapeutic concepts are derived from the pathophysiology of the disease mechanism. Regulatory intervention with drugs should restore the original physiological state as far as possible. To mimic the natural ligands of enzymes and receptors, the drug must have sufficient specificity and be able to clearly access the target site.

Nature operates on two orthogonal principles with respect to endogenous substances: specificity of action and usually highly **pronounced spatial compartmentalization**. Hormones act predominantly systemically, which means they are released at one site in the body and transported by the bloodstream to another, completely different site. There they exert their effect. Other substances, such as neurotransmitters, act strictly locally. In the context of the picture of lock and key (Sect. 4.1), Nature prefers to have a master key that can act on different locks. It acts only at the site where it is produced and is removed once it has served its purpose. Neurotransmitters are synthesized in nerve cells, stored, and released when the cell is stimulated at the synaptic cleft (Sect. 22.5). There they bind to specific receptors and stimulate the neighboring nerve cell. The effect is quickly dissipated after reuptake into the cell or after degradation, for example, by monoamine oxidases (amines), esterases (acetylcholine), or peptidases.

Nature's efficiency is demonstrated most impressively by the diversity with which small molecules such as adrenaline and noradrenaline (Sect. 1.4) can be used as hormones and neurotransmitters. There is a plethora of different receptors and receptor subtypes for these substances, allowing the same molecule to have completely different effects. The amino acid sequence of a particular receptor, and thus its binding site, can be altered relatively easily at the gene level. The evolution of complex biosynthetic pathways for nonpeptidic ligands, often involving multiple enzyme-catalyzed steps, is much more complicated. Accordingly, almost all neurotransmitters and many hormones are derived in a simple way from the central intermediates of the metabolism of, for example, amino acids. On the other hand, the steroid hormones (Sect. 28.3) demonstrate that Nature can achieve very different effects with a set of chemically similar structures and evolutionarily and structurally related receptors, such as the estrogens, gestagens, androgens, glucocorticoid steroids, and mineralocorticoid steroids.

Often, the spatial distribution of biosynthesis or the release of a receptor ligand or the distribution from membrane-bound receptors or enzymes plays a decisive role in the specificity of an effect. Different effects of the same ligand can be achieved by locally restricted release or by the presence of different receptors. The differentiation is not only between certain organs or areas, but also between individual cells and cell compartments. For example, the concentration of dopamine in different regions of the rat brain has been determined. While in some regions, such as the caudate nucleus (lat.: *Nucleus caudatus*), an important synaptic site for the motor and olfactory systems, concentrations of up to 100 ng dopamine per mg protein are reached, while most other areas of the brain only contain between 0.2 and 10 ng/mg. Even in the *Substantia nigra* of the midbrain, dopamine levels are only 5–6 ng/mg. Degeneration of dopaminergic neurons in this area leads to Parkinson's disease in humans. It is known from labeling experiments that the distribution and population density of receptor subtypes in distinct areas of the brain and other tissues can be very different.

19.10 Specificity and Selectivity of Drug Interactions

How specific should a drug act? There is no absolute answer to this question. Since drugs are almost always administered orally or intravenously, they act systemically, which means on the whole organism. The lack of restriction to a specific organ or compartment must be compensated for by a greater specificity. In any case, the drug must be as specific as necessary to achieve a successful therapy with tolerable side effects.

In the case of enzyme inhibitors, substances that are specific enough to inhibit only one particular enzyme are preferred. Nonspecific inhibitors that simultaneously inhibit several serine or metalloproteases would have a devastating effect on an organism. For example, a thrombin inhibitor designed to reduce an increased risk of thrombosis should not also act as an inhibitor of the closely related plasmin, which causes fibrinolysis and leads to the dissolution of blood clots that have already formed. The situation with kinase inhibitors (Sect. 26.3) is somewhat different. Because of the similarity among kinases, one member of the family can easily do the work of another related kinase that has been blocked. In doing so, it will reduce the therapeutic effect to zero. Here, a broad-spectrum kinase inhibitor that can simultaneously block an entire family of proteins may be desirable. A broad-spectrum action that inhibits multiple isoenzymes of a parasite equally well may also be advantageous for antibacterial or antiparasitic compounds (e.g., plasmapepsins, Sect. 24.7).

Receptor agonists and antagonists should also be highly selective. β-Agonists used to treat asthma (Sect. 29.3) must be β_2-specific so that they do not induce an unwanted increase in heart rate or blood pressure. Often, a single active agent cannot achieve the desired therapeutic response. The simultaneous use of several drugs is often indicated for the treatment of arterial hypertension (Sect. 22.10). More complex, multifactorial disease processes must be treated by targeting multiple mechanisms. Due to the low dosage of the different components, the nonspecific side effects of the individual components fade into the background.

Specificity is critical to the efficacy of drugs acting in the CNS. Advances in genetic engineering have given us an explosion of knowledge about receptors, but also a dilemma. We know the **exact receptor profile** of established compounds. We know what specificity must be achieved to mimic a particular type of effect. However, in many cases we do not know what that profile of a compound should look like to achieve a **better therapeutic effect**. An example should illustrate this point. Neuroleptics and many antidepressants (Sect. 1.6) act on neuroreceptors. The classical neuroleptics chlorpromazine **19.1** and haloperidol **19.2** (see Sect. 19.9), used in the treatment of schizophrenia, are relatively unspecific dopamine receptor antagonists (◘ Table 19.2). The mixed neuroleptic/antidepressant sulpiride **19.3** acts simultaneously on D_2 and D_3 receptors. All of these drugs have side effects on the musculoskeletal system, as seen in Parkinson's disease (see Sect. 9.4), which is caused by a lack of dopamine. Because of their mode of action, it was assumed that the side effects of neuroleptics were inevitable consequences of the antagonism of dopamine receptors. Then an atypical neuroleptic, clozapine **19.4**, was introduced (◘ Fig. 19.8). It does not have the side effects described above. Today we know that clozapine, unlike the other neuroleptics, is much more potent at the D_4 receptor than at the D_2 and D_3 receptors (◘ Table 19.2). However, at the concentrations at which clozapine acts on the D_4 receptor and which can be detected in the cerebrospinal fluid of treated patients, clozapine also binds to certain serotonin and muscarinic receptors, in some cases with higher affinity. It is, therefore, possible that the antago-

◘ **Table 19.2** The natural neurotransmitter dopamine binds with higher affinity to dopamine receptors of the D_1-type. The classic neuroleptics chlorpromazine **19.1**, haloperidol **19.2**, and (*S*)-sulpiride **19.3** are different from clozapine **19.4** (◘ Fig. 19.8) in one point: they have no comparable selectivity for the D_4 receptor

	Binding to	**the dopamine receptors, K_i in nM**			
Substance	D_1-Type		D_2-Type		
	D_1	D_5	D_2	D_3	D_4
Dopamine	0.9	< 0.9	7	4	30
Chlorpromazine **19.1**	30	130	3	4	35
Haloperidol **19.2**	80	100	1.2	7	2.3
(*S*)-Sulpiride **19.3**	45,000	77,000	25	13	1000
Clozapine **19.4**	170	30	230	170	21

nistic effect of clozapine at these receptors is responsible for its atypical action profile.

Many drugs are classified as "**dirty drugs**" because of their multifaceted action on many completely different receptors. From a pharmacologist's point of view, such a characterization is appropriate. A general statement about the therapeutic value cannot be derived from this. It may well be that many dirty drugs are optimal for therapy because of their balanced action on multiple receptors. Recently, these compounds have been termed "rich in pharmacology" and they define a "**polypharmacology**." The suitability or unsuitability of a drug is decided only in clinical trials and later by the experience gained from broad application in patients.

The differences between enzymes and receptors in distinct species also offer the chance to achieve desired selectivity therapeutically. Species differences play a role when an unwanted organism is to be killed, for example with antibiotics, antifungals, antivirals, and antiparasitic drugs. To avoid side effects in humans, the metabolic pathways of the bacteria, fungi, viruses, or parasites are targeted either by adequate selectivity or by selecting a site of action that is not present in higher organisms (see Sects. 23.7, 24.3, 27.2, or 30.11).

19.11 Of Mice and Men: The Value of Animal Models

Quantitative activity–activity relationships are used to extrapolate from animals to humans, but they are also valuable for comparing different biological models. From the vast number of examples described in the literature, some typical relationships will be mentioned.

Even before the characterization of the different dopamine receptors (Sect. 19.10, ◘ Table 19.2), 25 clinically used neuroleptics were investigated to unravel correlations between the results of *in vitro* models, animal experiments, and the potency of these substances in humans. Two radioactively labeled ligands, dopamine and haloperidol **19.2** (Sect. 19.10, ◘ Fig. 19.9), one of which prefers the D_1-type and the other the D_2-type dopamine receptor, were used to characterize binding. It was shown that the average clinical dose correlated significantly with the displacement of the D_2-type ligand haloperidol **19.2**. Significantly higher concentrations were required to displace the D_1-type ligand dopamine. There is virtually no correlation with these data. Not only clinical efficacy, but also data from animal models used to test neuroleptic effects correlate better with the displacement of haloperidol than with dopamine (◘ Table 19.3). Retrospectively, the results suffer from a lack of ligand specificity for a single receptor, and the preparations are affected by receptor heterogeneity because the presence of different receptor subtypes was not standardized in the calf brain homogenates used. All compounds were tested with *dirty ligands* in *dirty test models*. The profile of the compounds can only be unambiguously assigned by using uniform receptor subtypes produced by gene technology (see ◘ Table 19.2).

In many cases, the relationship between different experimental models is strongly dependent on the species used. Studies of isolated arteries and veins from rabbit, sheep, pig, and human lungs indicate that rabbit and

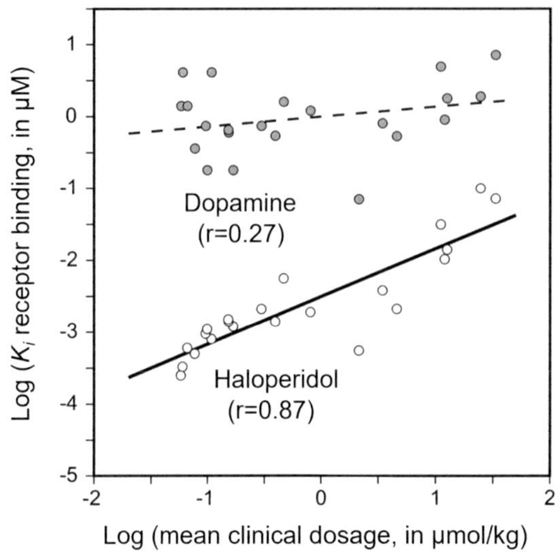

◘ **Fig. 19.9** The agonist dopamine preferably binds to the D_1-type of dopamine receptors (◘ Table 19.2). It was clear very early, however, from binding studies on membrane homogenates that the potency of clinically used neuroleptics correlated with the displacement of haloperidol (r = 0.87) rather than with dopamine binding (r = 0.27)

◘ **Table 19.3** Correlation of the clinical efficacy (◘ Fig. 19.10) of 25 different neuroleptics and their potency in different animal models that are typically used for the evaluation of neuroleptic effects with the displacement of dopamine or haloperidol **19.2**. The clinical data and the results of the animal models correlate conspicuously better with the displacement of the D_2-type ligand haloperidol than with the displacement of the D_1-type ligand dopamine (r = correlation coefficient)

Model	Correlation with dopamine displacement (r)	Correlation with haloperidol displacement (r)
Mean clinical dose in humans	0.27	0.87
Inhibition of the stereotypical behavior after administration of apomorphine (rat)	0.46	0.94
Inhibition of the stereotypical behavior after administration of amphetamine (rat)	0.41	0.92
Protection from apomorphine-induced emesis (dog)	0.22	0.93

19

Table 19.4 Binding of substance P and displacement by the antagonist CP 96 345 **19.5** (tested as a racemate) on cells of different origins

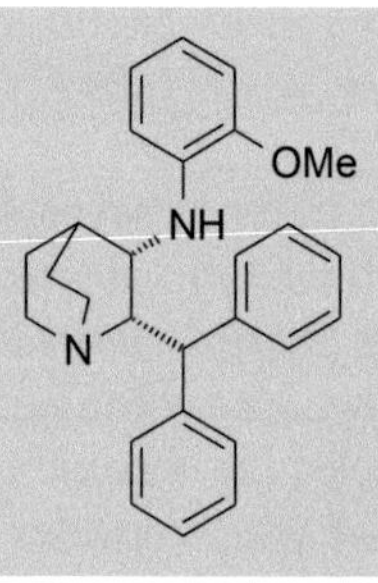

19.5 CP 96 345

System	Binding of substance P, IC_{50} in nM	Displacement of substance P by 19.5, IC_{50} in nM
Human cell line U373	0.13	0.40
Human cell line IM9	0.22	0.35
Guinea pig brain	0.07	0.32
Guinea pig lung	0.04	0.34
Rabbit brain	0.16	0.54
Mouse brain	0.19	32
Rat brain	0.20	35
Chicken brain	0.26	156

Table 19.5 Inhibition of the renins of humans and other animal species by remikiren **19.6** and aliskiren **19.7**

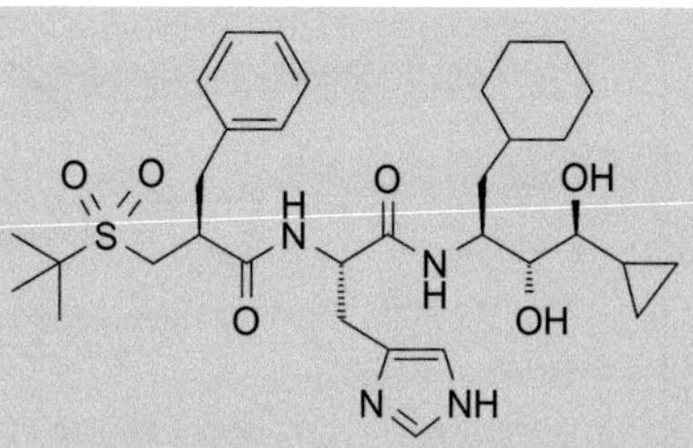

19.6 Remikiren

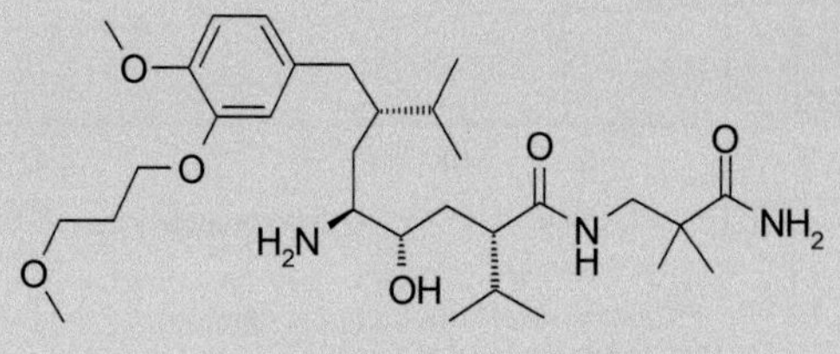

19.7 Aliskiren

Renin from:	IC_{50} in nM, Remikiren	IC_{50} in nM, Aliskiren
Human	0.8	0.6
Monkey	1.0	1.72
Dog	107	7
Rat	3600	80

human vascular preparations are similarly sensitive to noradrenaline. Sheep and pig arteries are much less sensitive. Isolated pig veins cannot be stimulated at all with comparable doses of noradrenaline. The experimental results are even more heterogeneous and difficult to interpret with acetylcholine stimulation. It should not be forgotten that the metabolism of humans and animals differs and influences the test results.

Tachykinins are short peptides that trigger a variety of physiological and pathological processes. Their central role in pain and asthma is well established. They act via the NK1, NK2, and NK3 receptor subtypes, which also bind specifically to the three peptide agonists substance P, neurokinin A, and neurokinin B (Sect. 10.7). CP 96 345, **19.5**, a nonpeptide NK1 antagonist, displaces substance P with high affinity in two human cell culture models and in guinea pig and rabbit membrane preparations. In membrane preparations from mouse, rat, and chicken brain, to which substance P binds with quite comparable affinities, **19.5** has IC_{50} values that are 60–500 times higher (Table 19.4). It has been suggested based on sequence-specific point mutations that the agonist substance P and the antagonist CP 96 345 may bind to different regions of the receptor (see Sect. 29.6).

The differences between humans and individual animal species are not surprising considering that the amino acid sequence of the receptor proteins usually differs at multiple positions. The use of human proteins in molecular test systems is just as critical to the relevance of the results obtained as it is for the determination of the 3D structures (Chaps. 13, 14). This can be seen very clearly in the results for the aspartic protease renin (Sect. 24.2). The inhibitors remikiren **19.6** and aliskiren **19.7** were tested on renin from different species. The renins of two primate species and humans were inhibited at very low concentrations. In contrast, renin from rat and dog, two of the most commonly used species in cardiovascular pharmacology, was inhibited only at significantly higher concentrations (Table 19.5). Remikiren would have been found in a classic antihypertensive test using these animal models, but would probably have been rated as much too weak. A comparison of the crystal structure analyses of murine and human renin also reveals a conserved binding mode in the main chain of the peptide inhibitors that is common to the other aspartic proteases. However, there are subtle differences at the rim of the binding pocket that can be attributed to sequence differences between the two species.

More than 90% of the amino acid sequences of the 5-HT_{1B} and 5-$HT1D_{\beta}$ subtypes of the human and rat serotonin receptors are identical. When the relationships between the individual amino acids are considered, the homology is as high as 95%. Despite these similarities,

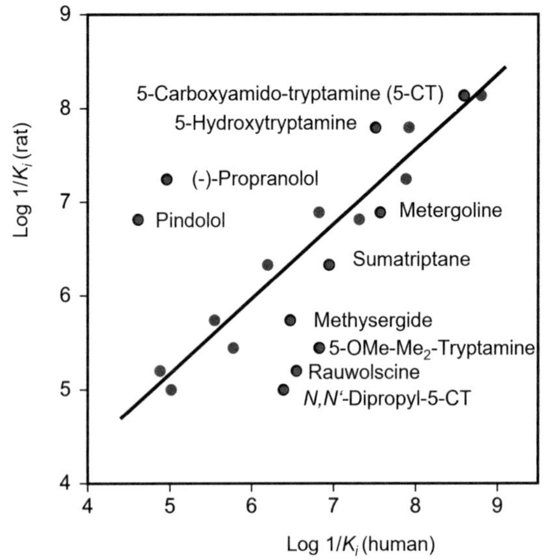

Fig. 19.10 Different serotonin receptor ligands and the β-blockers propranolol and pindolol show very different binding affinities on the highly similar 5-HT receptors from rats and humans. The *red circles* refer to the wild-type human receptor. They are irregularly scattered over the diagram (correlation coefficient $r = 0.27$). If one amino acid in the human receptor is exchanged for the corresponding amino acid in the rat receptor, the binding profile changes. Relative to the affinity of the ligands, the human receptor becomes very close to the rat receptor. The *gray circles* refer to this Asn 355 mutant (correlation coefficient now $r = 0.98$)

a number of drugs bind to these two receptors with very different affinities. The difference is due to a single amino acid: the exchange of threonine 355 for an asparagine (Fig. 19.10). This mutation converts the human receptor into the rat receptor in terms of affinity! After the exchange of this amino acid, the β-blockers propranolol and pindolol (Sect. 29.3, Fig. 29.1) bind with approximately three orders of magnitude higher affinity. The affinities of many other ligands are significantly reduced.

19.12 Toxicity and Adverse Effects

One of the most difficult tasks in preclinical research is **to estimate** the **toxicity** of a substance, especially human toxicity, from data obtained in other species. Such considerations must be made in order to assess the potential danger of the substance before it is introduced into the clinic. Are there any drugs without toxicity and without side effects? Paracelsus recognized in the sixteenth century that:

> "Everything is poison and nothing is without poison, it is the dose alone that makes a thing nonpoisonous."

Friedrich Schiller had his Fiesko say:

> "A desperate evil needs a bold remedy [medicine]."

And the pharmacologist Gustav Kuschinski formulated:

> "Whenever it is proclaimed that a substance has no side effects, the urgent suspicion ensues that there is also no main effect."

Determination of acute toxicity in several species and determination of chronic toxicity in at least two species is routine prior to entering phase I clinical trials, which are tolerability studies in healthy volunteers. It is common practice to select species for chronic toxicity studies primarily on the basis of which species most closely resembles humans in pharmacokinetics and metabolism for the substance of interest.

Cats and guinea pigs are extremely sensitive to cardiac glycosides. For this reason, they have traditionally been used as models for the effect on humans. Rats are much less sensitive. The hallucinogen lysergic acid diethylamide (LSD **2.21**, Sect. 2.5, Fig. 2.8) shows markedly different toxicity in several species. An experiment to test the hallucinogenic effects of LSD on an elephant ended in a disaster. A hallucinogenic but nontoxic dose was desired. Despite careful dose estimation, the elephant died within minutes of being given 0.3 g of LSD (equivalent to 0.06 mg/kg). Compared to the mouse, which is relatively insensitive (Table 19.6), the elephant was at least 1000 times more sensitive. This experiment has not been repeated! The discoverer of LSD, Albert Hofmann, took 0.25 mg of LSD in his first controlled self-experiment. In retrospect, his choice of dose was about three to five times the effective dose in humans. With about 0.0035 mg/kg he was well below the dose that killed the elephant. Nevertheless, it can be assumed that LSD is less toxic for humans than for elephants. Direct fatalities due to LSD are not known, only mortality as a result of accidents or suicides while in a psychotic state.

The **toxicity** of the poisons that end up in our environment is studied in great detail. Chlorinated dibenzodioxins and furans are formed by the uncontrolled chemical decomposition of the corresponding substituted chlorophenols. The Seveso accident is attributed to such an incident. Toxic chlorinated dioxins and furans are also

Table 19.6 Acute toxicity of lysergic acid diethyl amide (LSD, **2.21**, Sect. 2.5, Fig. 2.8) in different species and in humans (LD_{50} = dose that was lethal for 50% of the animals)

Species	Toxicity, LD_{50} (in mg/kg)
Mouse	50–60
Rat	16.5
Rabbit	0.3
Elephant	≪ 0.06
Human	≫ 0.003

formed during many combustion processes. **Tetrachlorodibenzodioxin 19.8** (TCDD, "Seveso Dioxin") is one of the best-studied substances in terms of its toxicity. Here, too, different species react differently (◘ Table 19.7). There is a difference of three orders of magnitude regarding toxicity between the two relatively closely related species, the hamster and the guinea pig. It is, therefore, difficult to draw conclusions about toxicity in humans. If extrapolated between primates and humans, TCDD would be classified as relatively nontoxic. In the context of humans, the definition of an acute LD_{50} is completely inappropriate. In order to exclude one fatality per one million people, an "$LD_{0.00001}$" must be determined or calculated. Because of its pronounced mutagenic effects, long-term damage is the primary concern with TCDD. In this case, it is questionable whether an absolute no-effect level, i.e., the lowest ineffective dose, can be defined. The assessment of the hazard potential of environmentally relevant chemicals looks quite different when compared with toxic natural products, natural radioactivity, cosmic radiation, etc., or even with socially tolerated substances of abuse such as alcohol and nicotine. This puts into perspective some things that are very controversially discussed in public forums.

When discussing structure–activity relationships, it is important to recognize the difficulty of using *in vitro* studies to estimate the mutagenicity and carcinogenicity of a substance. While such tests provide valuable indications that must be carefully reviewed, they are not conclusive in either a positive or negative sense in individual cases. It is extremely difficult to provide theoretical models for **estimating toxicity and carcinogenicity** with sufficient reliability and predictive power. The mechanisms responsible for the effects are too diverse and multifaceted, and the chemical structures and structure–activity relationships applicable to different classes of substances are too different.

◘ **Table 19.7** Acute toxicity of tetrachlorodibenzodioxin **19.8** in different animal species

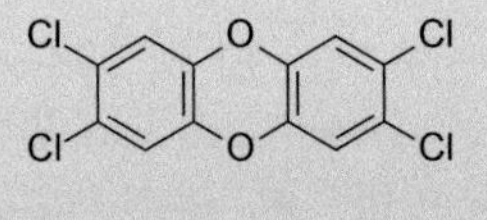

19.8 2,3,7,8-Tetrachlorodibenzodioxine

Species	Toxicity (LD_{50} in µg/kg)
Mouse	114–280
Rat	22–320
Hamster	1150–5000
Guinea pig	0.5–2.5
Mink	4
Rabbit	115–275
Dog	> 100–< 3000
Monkey	< 70
Human	?

Today, testing for toxic, carcinogenic, and teratogenic side effects has reached a high standard. The pharmaceutical disasters of earlier decades, such as the following, would be almost impossible with today's standards:
- Early childhood brain damage and death of many premature and mature newborns by the sulfonamides in the late 1930s,
- Over 100 fatalities in the USA because of the use of diethylene glycol as a solvent for sulfanilamide (this incident led to the foundation of the Food and Drug Administration, FDA),
- The SMON (subacute myelo-optic neuropathy) illness of thousands of Japanese, caused by the prolonged and too-frequent use of an antidiarrheal medicine, and
- The severe birth defects of approximately 10,000 children worldwide that were caused by thalidomide (Contergan®) in the late 1950s.

However, criminal intrigue and the uncontrolled distribution of fake drugs from internet-based vendors, or the unscrupulous pursuit of economic gain, can still cause such disasters today. A case in point is the melamine-contaminated infant formula (melamine makes the protein content of inferior or diluted milk appear higher) in China in September 2008, which sickened many thousands of toddlers and babies and even caused some deaths.

In addition to the much stricter testing guidelines for medicines that are now in place in most countries, there is a reporting system that records and investigates adverse drug reactions. The slightest suspicion of a causal relationship can result in anything from public announcement or warning all the way to withdrawal of the marketing license.

A complication in estimating toxicity is the formation of toxic and especially reactive metabolites, even at low levels. As discussed in Sects. 9.1 and 19.6, an ideal drug should contain predetermined cleavage and/or conjugation sites in addition to finely tuned pharmacodynamics and pharmacokinetics. The more these requirements are met, the lower the risk that the compound will exert toxic effects.

Some toxicity studies suffer from the fact that the results extrapolated to humans reflect a higher toxicity than is actually the case due to the unphysiologically high doses used in the studies. On the other hand, even the most comprehensive study cannot eliminate the risk of serious adverse events occurring in extremely rare cases once the drug is in widespread use. An adverse event rate of 1 in 10,000 or less may go undetected in even the most carefully conducted preclinical and clinical studies.

Toxic side effects in humans are particularly common after **chronic drug abuse**. The lifetime consumption of large quantities of painkillers adds up to kilograms. In the case of phenacetin (Sect. 2.1), this resulted in an effective and generally well-tolerated drug having to be withdrawn from the market due to kidney damage caused by inappropriate (abusive) use.

Gradually, humanity is becoming more concerned about what happens to the products we release into the environment. Pharmaceuticals are for consumption by patients. However, much of what goes into the body also comes out of the body. Many **pharmaceuticals enter the environment** through wastewater. Often they remain unchanged, but in some cases they are metabolically modified. Once in the environment, they can be taken up by other organisms and have similar or different effects. In general, these effects are unknown and unstudied. In spite of high dilutions, active substances can build up locally to high concentrations or exceed therapeutically critical limits through specific accumulation. In the case of antibiotics, the **emergence of resistant bacteria** is known to occur in inadequately treated wastewater, e.g., after carelessly operated pharmaceutical factories, hospitals, or slaughterhouses. Hormones and contraceptives can interfere with the reproduction of fish populations. Lipid-lowering drugs such as statins (Sect. 27.3) interfere with cholesterol biosynthesis, a pathway that is essential for fundamental processes in Nature. The painkiller diclofenac (Sect. 27.9) has been found to have a devastating effect on birds and fish, causing kidney damage. We need to develop a more responsible approach to the **disposal of excreted pharmaceuticals**. Presumably, the additional requirement that a drug be toxicologically safe as a degradation product in the environment would raise the bar for an optimally effective therapeutic to heights that are almost impossible to achieve. Nevertheless, there seems to be an urgent need to think more carefully about the fate of excreted drugs. Stricter requirements, especially for the environmentally sound disposal of unused medicines, seem more than necessary.

19.13 Animal Protection and Alternative Test Models

19

As early as 1780, the philosopher Jeremy Bentham discussed the rights of animals. The first mass **protests against animal experimentation** took place almost 150 years ago. In 1875, dedicated animal rights activist Frances Power Cobbe founded the first Society Against Vivisection in England, and a year later the demand for anesthesia in animal experimentation led to the first Animal Welfare Act. In Germany in 1879 the *Internationale Gesellschaft zur Bekämpfung der Wissenschaftlichen Thierfolter* (International Society for the Abatement of Scientific Animal Torture) was founded, followed in 1883 by the American Antivivisection Society. A new militant form of protest against animal experimentation, complete with the violent liberation of laboratory animals and attacks on scientists, emerged in the 1970s. Peter Singer's book *Animal Liberation* was published in 1975 and became the bible of animal rights activists. The often-quoted story of animal trappers selling their prey to the pharmaceutical industry was a fantasy even in the early days of drug discovery. Every pharmacologist knows that any results obtained from such diverse animals would be completely useless without any knowledge of their health history.

More in parallel with the development of the **animal welfare movement** than inspired by it, alternative methods for pharmaceutical research were introduced in the 1960s, consisting mainly of binding studies on membrane homogenates and cell culture studies. The **number of animal experiments** has been significantly **reduced** in recent decades due to the economic motivation arising from the enormous costs of breeding and maintaining experimental animals, but also because of the rapid progress in gene technology. As explained in Sect. 7.5, models using lower animals such as pinworms, fruit flies, or zebra fish are increasingly being used for screening. Here, the ethical threshold for animal testing is certainly lower.

More than 50% of all laboratory animals are used for drug testing, with 12–15% each for basic research, investigation of medical methods, and detection of environmental hazards. About half of all laboratory animals are mice. The rest are rats and other rodents, and a small proportion are fish and birds. Only about 1.5% of the total number are cats, dogs, pigs, and other animals. Much of the testing on the latter species is **chronic toxicity testing**, which is required by law.

The decline in these numbers is remarkable because pharmaceutical companies are investigating the biological activity of more compounds than ever before. Each year, tens or even hundreds of thousands of compounds are meticulously characterized, usually in automated ***in vitro*** **assays**. Only a few of these compounds are ever tested in animals. The number of animals used must also be seen in the context of **regulatory requirements** to demonstrate the efficacy and safety of new drugs, which are increasing rather than decreasing. The vast majority of these tests must still be performed on animals.

19.14 Synopsis

- Apart from potent and selective binding to a target protein, a successful drug candidate must exhibit favorable pharmacokinetics. This comprises all processes that affect the absorption, distribution, metabolism, and excretion along with minor toxic side effects.

- Due to high costs and enormous experimental effort, full pharmacokinetics and toxicity studies can only be carried out on a few drug development candidates. A plethora of test methods have been developed to relate chemical structure with ADME and toxicology properties in the lead-optimization phase to reduce the chances of failure due to insufficient pharmacokinetics at a late stage.
- An active substance has to penetrate multiple lipid membrane barriers and aqueous compartments on its way from the site of application to the locus of the target protein. To achieve sufficient distribution, adequate lipophilicity must be present. This is described by the partition coefficient between the lipid and aqueous phases. In the simplest model, the distribution between octanol and water is measured.
- Rather sophisticated models have been established to relate chemical structure with penetration properties. Considerations about the release of the water solvation shell around a drug molecule and its potential to form hydrogen bonds upon crossing lipid membranes are particularly important.
- Many drugs are either weak acids or bases. Depending on the pH used, they exist through dissociation equilibria in either a more lipophilic neutral or more polar ionized form. Membrane penetration of such species will, therefore, depend very much on the local pH conditions.
- Because of the progressively changing pH conditions in the stomach and intestines, appropriate pH conditions exist at some place along the gastrointestinal tract that allow sufficient penetration of the neutral form of weakly acidic or basic drug molecules.
- Because of established equilibria, small amounts of the neutral form of an acidic or basic drug molecule are the intermediate over which membrane penetration occurs. Constant removal of the neutral species from the aqueous phase into the membrane is quickly replenished by the dissociation equilibrium.
- The adjustment of the lipophilicity of a drug is crucial for pharmacokinetics. Usually the more lipophilic a compound is, the better it will be absorbed; however, limited solubility in the aqueous phase restricts lipophilicity. Relevant test models have been developed by using thin layers of human colon cells. These also allow the absorption by transporters to be studied.
- Active substances are initially tested in simple *in vitro* test models. Testing is gradually moved into animal models via cellular assays. Relevant activity–activity relationships must be established to correlate response in animal models. At best, results from appropriate *in vivo* testing allow the therapeutic effects in humans to be predicted. They standardize test data and reduce the number of animal experiments that are required.
- Nature works with two orthogonal principles upon release of its native substances: the specificity of the biological effect and a pronounced spatial compartmentalization. Some compounds are highly specific and travel long ways through the organism to exert their action. Others are locally synthesized and stimulate their target protein in the immediate vicinity. Here, high specificity and selectivity are not required.
- Drugs administered orally or intravenously act systemically on the entire organism; no organ- or cell-specific compartmentalization can be achieved. This has to be compensated for by sufficient specificity and selectivity. Whether high isoform selectivity or protein-family-wide promiscuity is required depends very much on the mode of action and biological function of the target protein.
- Prior to administration in humans, clinical candidates are tested in animals. To draw conclusions about humans from animals, it must be considered that test models strongly depend on the species used. Even metabolism can be very different in humans and various animal species.
- Deviating therapeutic responses in animals and humans are also related to small differences in the amino acid composition of the target proteins in various species.
- Estimates of human toxicity must be made from data obtained in other species. Chronic toxicity is routinely determined in two species and must be evaluated in the animal species with the closest pharmacokinetic and metabolic similarity to humans.
- Different species, however, react differently to active substances and frequently show several orders of magnitude differences in toxicity.
- Today, testing for toxic, carcinogenic, and teratogenic adverse effects has reached a high standard. The pharmaceutical catastrophes of earlier decades, which were mostly caused by these adverse effects, will hopefully be almost impossible nowadays.
- Even the most comprehensive toxicity studies cannot eliminate the risk of severe adverse effects occurring in extremely rare cases once a new drug is administered broadly.
- The animal experiments of the past have been replaced by more conclusive binding studies on membrane homogenates and cell cultures. Whole-animal testing has shifted towards lower animals such as pinworms, fruit flies, or zebra fish. The toxicity studies required by law make up a large part of the animal experiments that are done today.

Bibliography and Further Reading

General Literature

H. Kubinyi, Lipophilicity and Drug Activity, Progr. Drug Res., **23**, 97–198 (1979)

J. K. Seydel and K.-J. Schaper, Quantitative Structure–Pharmacokinetic Relationships and Drug Design, Pharmac. Ther., **15**, 131–182 (1982)

J. M. Mayer and H. van de Waterbeemd, Development of Quantitative Structure–Pharmacokinetic Relationships, Environ. Health Perspect., **61**, 295–306 (1985)

J. C. Dearden, Molecular Structure and Drug Transport, in: Quantitative Drug Design, C. A. Ramsden, Eds., Vol. 4: Comprehensive Medicinal Chemistry, C. Hansch, P. G. Sammes and J. B. Taylor, Eds., Pergamon Press, Oxford, p. 375–411 (1990)

C. A. M. Hogben, D. J. Tocco, B. B. Brodie, and L. S. Schanker, On the Mechanism of Intestinal Absorption of Drugs, J. Pharmacol. Exp. Therap., **125**, 275–282 (1959)

H. Kubinyi, QSAR: Hansch Analysis and Related Approaches, VCH, Weinheim (1993)

C. Hansch and A. Leo, Exploring QSAR. Fundamentals and Applications in Chemistry and Biology, Vol. 1, American Chemical Society, Washington (1995)

R. L. Lipnick, Selectivity, in: General Principles, P. D. Kennewell, Eds., Vol. 1: Comprehensive Medicinal Chemistry, C. Hansch, P. G. Sammes and J. B. Taylor, Eds., Pergamon Press, Oxford, p. 239–247 (1990)

H. Kubinyi, Lock and Key in the Real World: Concluding Remarks, Pharmac. Acta Helv., **69**, 259–269 (1995)

C. A. Reinhardt, Hrsg., Alternatives to Animal Testing, VCH, Weinheim (1994)

R. Mannhold, Ed. Molecular Drug Properties, Wiley-VCH, Weinheim (2008)

D. A. Smith, H. van der Waterbeemd and D. K. Walker, Pharmacokinetics and Metabolism in Drug Design, Wiley-VCH, Weinheim (2006)

B. Testa and H. van der Waterbeemd, (Eds)., ADME-Tox Approaches, Vol. 5 of Comprehensive Medicinal Chemistry II, Elsevier (2007)

G. Orive, U. Lertxundi, T. Brodin, P. Manning, Greening the pharmacy. New measures and research are needed to limit the ecological impact of pharmaceuticals, Science, **377**, 259–260 (2022)

Special Literature

B. C. Lippold and G. F. Schneider, Zur Optimierung der Verfügbarkeit homologer quartärer Ammoniumverbindungen, 2. Mitteilung: *In-vitro*-Versuche zur Verteilung von Benzilsäureestern homologer Dimethyl-(2-hydroxyäthyl)-alkylammoniumbromide, Arzneim.-Forsch., **25**, 843–852 (1974)

H. Kubinyi, Drug Partitioning: Relationships between Forward and Reverse Rate Constants and Partition Coefficient, J. Pharm. Sci., **67**, 262–263 (1978)

R. C. Young, R. C. Mitchell, T. H. Brown, C. R. Ganellin, R. Griffiths, M. Jones, K. K. Rana, D. Saunders, I. R. Smith, N. E. Sore, T. J. Wilks, Development of a new physicochemical model for brain penetration and its application to the design of centrally acting H_2 receptor histamine antagonists, J. Med. Chem., **31**, 656–671 (1988)

H. van de Waterbeemd, P. van Bakel and A. Jansen, Transport in Quantitative Structure–Activity Relationships VI: Relationship between Transport Rate Constants and Partition Coefficients, J. Pharm. Sci., **70**, 1081–1082 (1981)

C. Hansch, J. P. Björkroth and A. Leo, Hydrophobicity and Central Nervous System Agents: On the Principle of Minimal Hydrophobicity in Drug Design, J. Pharm. Sci., **76**, 663–687 (1987)

H. van de Waterbeemd and M. Kansy, Hydrogen-Bonding Capacity and Brain Penetration, Chimia, **46**, 299–303 (1992)

C. A. Lipinski, Lead- and drug-like compounds: the rule-of-five revolution, Drug Discov. Today, Technol., **1**, 337–341 (2004)

L. Z. Benet, C. M. Hosey, O. Ursu, T. I. Oprea, BDDCS, the Rule of 5 and drugability, Adv. Drug Deliv. Reviews, **101**, 89–98, (2016)

D. C. Doak, J. Zheng, D. Dobritzsch, J. Kihlberg, How Beyond Rule of 5 Drugs and Clinical Candidates Bind to Their Targets, J. Med. Chem., **59**, 2312–2237, (2016)

D. G. Jimenez, V. Poongavanam, J. Kihlberg, Macrocycles in Drug Discovery_Learning from the Past for the Future, J. Med. Chem., **66**, 5377–5396, (2023)

A. Tsuji, E. Miyamoto, N. Hashimoto and T. Yamana, GI Absorption of β-Lactam Antibiotics II: Deviation from pH-Partition Hypothesis in Penicillin Absorption through *In Situ* and *In Vitro* Lipoidal Barriers, J. Pharm. Sci., **67**, 1705–1711 (1978)

P. Seeman and H. H. M. Van Tol, Dopamine Receptor Pharmacology, Trends Pharm. Sci., **15**, 264–270 (1994)

B. D. Gitter *et al.*, Species Differences in Affinitites of Non-Peptide Antagonists for Substance P Receptors, Eur. J. Pharmacol., **197**, 237–238 (1991)

J.-P. Clozel and W. Fischli, Discovery of Remikiren as the First Orally Active Renin Inhibitor, Arzneim.-Forsch., **43**, 260–262 (1993)

V. Dhanaraj *et al.*, X-Ray Analyses of Peptide–Inhibitor Complexes Define the Structural Basis of Specificity for Human and Mouse Renins, Nature, **357**, 466–472 (1992)

E. M. Parker, D. A. Grisel, L. G. Iben and R. S. Shapiro, A Single Amino Acid Difference Accounts for the Pharmacological Distinctions Between the Rat and Human 5-Hydroxytryptamine-1B Receptors, J. Neurochem., **60**, 380–383 (1993)

D. J. Hanson, Dioxin Toxicity: New Studies Prompt Debate, Regulatory Action, Chem. Eng. News, **69**,7–14 (1991)

M. J. Matfield, Animal Liberation or Animal Research? Trends Pharm. Sci., **12**, 411–415 (1991)

CLOGP program: https://www.daylight.com/products/pcmodels.html (Last accessed Nov. 19, 2024)

ACD/pKa program: https://www.acdlabs.com/products/percepta-platform/physchem-suite/pka/ Last accessed Nov. 19, 2024)

Pallas/pKa program: http://www.ccl.net/ccl/pallas.html (Last accessed Nov. 19, 2024)

Protein Modeling and Structure-Based Drug Design

Contents

G. Klebe, *Drug Design*, https://doi.org/10.1007/978-3-662-68998-1_20

Structure-based drug design focuses on the search, design, and optimization of a small molecule that fits well into the binding pocket of a target protein to form energetically favorable interactions. First, a detailed analysis of the target protein is performed. All information about its structure and that of related proteins is evaluated. Next, the properties of the binding pocket are thoroughly explored, looking for areas where optimal binding is expected. Experimental and computational methods are used to discover a lead structure from a screening library (Chap. 7). Alternatively, approaches are used that start with a small-molecule "seed" (or "fragment") in the binding pocket that is then allowed to "grow" into a potent ligand using stepwise iterative design. This approach uses fast **docking** techniques that suggest relevant binding geometries. The geometries are evaluated with a **scoring function** that estimates whether they are energetically favorable.

However, a crucial prerequisite for the use of structure-based drug design is the knowledge of the spatial structure of the target protein. Impressive progress in the field of protein structure determination (Chaps. 13 and 14) has led to the fact that the 3D structures of almost all therapeutically relevant proteins are known today or will be determined at the beginning of a new project. Nevertheless, it should not be overlooked that for some target proteins of interest there is still no experimentally determined three-dimensional structure available, and in these cases models have to be built.

Thanks to the sequencing of the human genome, the blueprints for all the proteins of our species are now known at the sequence level. The genomes of many pathogens have also been sequenced, and new ones are being discovered weekly. How can this enormous advance in information be used to design new drugs? Unfortunately, the step from the primary structure, this means the amino acid sequence, to the 3D structure is very difficult and, until now, has usually been followed by experimental methods of structure determination (Chap. 13). Methods for reliable *ab initio* prediction of the spatial structure of proteins based on theoretical concepts have not yet been successfully developed. Increasingly, however, the situation arises where the structure of the protein of interest is unknown, but the structure of another related protein has been determined. In such a situation, a model of the unknown protein can be constructed from the spatial coordinates of the already characterized biopolymer. Recently, however, there has also been a breakthrough in the structural prediction of protein models from their primary sequence. Successful methods learn from the wealth of experimental structural data using artificial intelligence techniques and are, thus, trained to model protein architectures (Sect. 20.6).

20.1 Pioneering Studies in Structure-Based Drug Design

Considering that most of the structures of therapeutically relevant proteins have been determined in the last 20 years, it is even more impressive that the first work in structure-based drug design was already done in the 1970s. The pioneers in this field were Chris Beddell and Peter Goodford, who began developing methods for ligand design at the Wellcome Research Laboratories in 1973. Hemoglobin was chosen as the target protein because, at the time, it was the only example with a known 3D structure that had some relevance to pathophysiology. The goal of this work was to find a ligand that would exert an allosteric modulating effect, analogous to the natural ligand diphosphoglyceric acid **20.1** (DPG; ◘ Fig. 20.1). The hope was to find a therapeutic approach that could help homozygous patients with lethal sickle cell anemia (Sect. 12.13). DPG is synthesized in red blood cells. It binds to hemoglobin and reduces its affinity for oxygen. This allows oxygen absorbed in the lungs to be released to other tissues.

The portion of hemoglobin that binds to DPG contains a large number of positively charged amino acids (◘ Fig. 20.1). An optimal ligand should, therefore, contain a negatively charged group to form multiple salt bridges to hemoglobin, just as DPG does. However, such compounds cannot penetrate the membrane of a red blood cell. Therefore, the Wellcome group considered structures that interact with hemoglobin in other ways. Compounds were selected that contained reactive groups that could be attached to the amino groups of the lysines in the binding pocket or to the *N*-terminus. The idea was to design a compound containing two correctly spaced reactive groups that could form Schiff bases with two of these amino groups. Dibenzyl-4,4′-dialdehyde **20.2** (◘ Fig. 20.2) was chosen as the parent structure. The putative binding mode of this compound is shown in

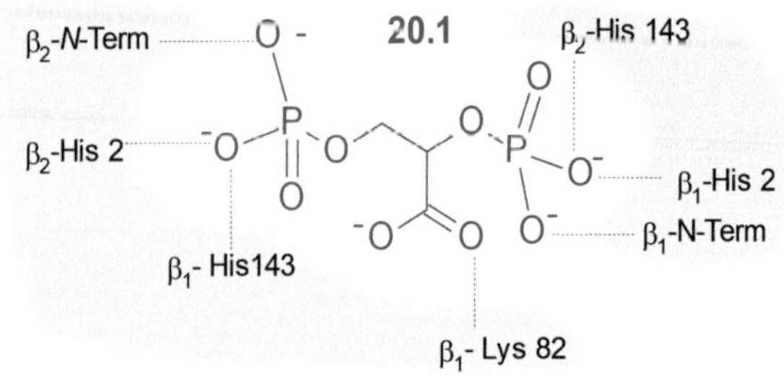

◘ **Fig. 20.1** Schematic binding mode of diphosphoglyceric acid **20.1** (DPG) to the allosteric binding site of hemoglobin. The ligand is bound through multiple charge-assisted hydrogen bonds (*N*-terminal amino groups, His 2, Lys 82, and His 143) from the β_1 and β_2 subunits

▫ Fig. 20.3. Compound **20.2** was synthesized but proved to be too insoluble for testing. Sufficient solubility was achieved by introducing an additional carboxyl group in **20.3**. In addition, this compound with its carboxyl group should provide an additional favorable interaction with the lysine side chain of the protein. Compounds **20.4** and **20.5** are the bisulfite adducts of the corresponding aldehydes. These compounds were tested and indeed showed the desired allosteric effect. However, they bind to the oxy form of the protein and increase its oxygen affinity. They proved to be potent inhibitors of the erythrocyte deformation that occurs in sickle cell disease by stabilizing the oxy form. The deformation begins with the aggregation of the desoxy form. The targeted design of these dibenzyldialdehydes is the first example of rational, structure-guided protein–ligand design.

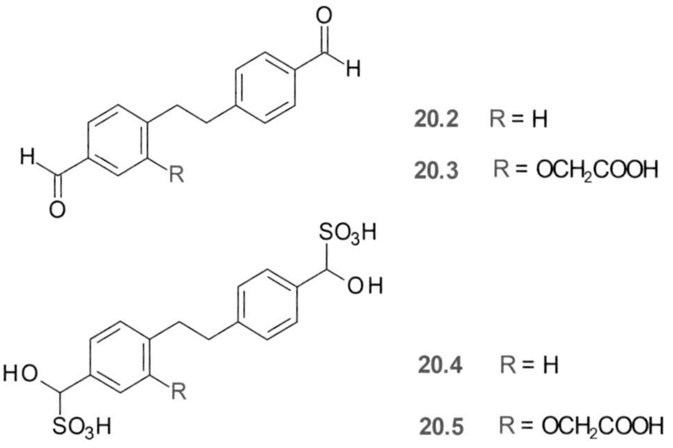

▫ **Fig. 20.2** Structures of the diphosphoglyceric acid competitive hemoglobin ligands **20.2**–**20.5** that were developed by Beddell and Goodford

20.2 Strategies in Structure-Based Drug Design

In order to design a ligand for a protein with a known 3D structure, it is necessary to analyze the structure of the protein. What does the binding pocket of the protein look like? Where are the hot spots of binding, this means where can functional groups of a ligand bind particularly well to the protein? For such analyses, experimental data can be used or computer programs are available. They search the surface of a protein for suitable binding sites for different functional groups, such as a carbonyl group, a carboxylic acid function, or an amino group. A selection of experimental methods for finding hot spots using X-ray structural analysis and NMR spectroscopy were presented in Sects. 7.8 and 7.9, and some computational methods are described in Sect. 17.10.

To design new drugs, an attempt is made to find new ideas for potential ligands, either by using docking methods through **virtual screening** (Sect. 7.6). Or, alternatively, a molecule that has been discovered using one of the techniques described in Chap. 7 can be modified and progressively increased in size in a stepwise fashion. Modification of a known structure has the advantage that potent and selective protein ligands can be obtained relatively quickly. In addition, if the 3D structure of the protein is known, clear structure–activity relationships will usually emerge. However, there is a risk that the proposed structures will remain close to the original lead structure. For example, when an initial 3D structure of an enzyme complexed with a peptidic ligand is solved, the resulting design proposals are often very similar to peptides (Chaps. 23, 24, 25). The path to an orally available drug can then be quite long (Chap. 10). Another approach is ***de novo*** **design**, or **fragment-based lead**

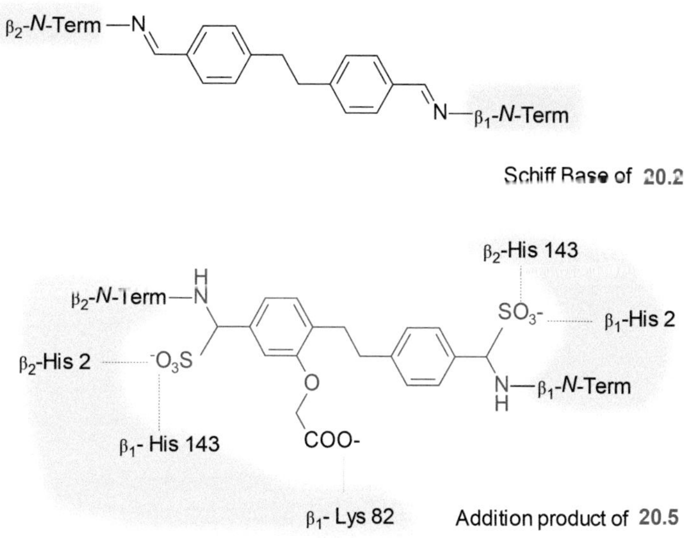

▫ **Fig. 20.3** Postulated binding mode of the hemoglobin ligands **20.2** and **20.5** after chemical reaction to the Schiff base or the bisulfite addition product. It is assumed for both compounds that they bind covalently to the β_1 and β_2 subunits of hemoglobin through their *N*-terminal amino acids. Compound **20.5** should also be able to form a hydrogen bond with its charged groups to the side chains of amino acids His 2 and His 143 of the β_1 and β_2 subunits, as well as Lys 82 of the β_1 subunit

discovery. These methods can lead to completely novel, nonpeptidic structures.

An essential requirement for the success of structure-based drug design is an iterative approach, as illustrated in Sect. 7.6 and ◘ Fig. 7.3. Further examples of this approach are given later in this chapter. In all cases, however, the existence of a 3D structure of the protein is the prerequisite and starting point for the structure-based design of a ligand, which is then synthesized and tested. How do you get such a **3D structure of the protein**, and does it have to be determined experimentally in every single case (Chap. 13)?

20.3 Search Tools for Databases of Experimentally Determined Protein Complexes

The number of experimentally determined protein structures has grown exponentially in the last years. In 1988, 200 3D structures were found in the **Protein Data Bank (PDB)**, in the meantime there are more than 227,000 entries, mainly from proteins and protein–ligand complexes. This rapid growth of known spatial protein structures is stimulating the development of methods to use this structural information for the design of new active compounds. Most of the available examples are still predominantly globular, water-soluble enzymes. However, the number of novel membrane-bound proteins is steadily increasing. To really exploit this wealth of structures, **database tools** are needed that can retrieve, correlate, and analyze structures and structural motifs. There are many programs that can compare the sequence and folded structure of proteins. The Relibase database was one of the first tools developed specifically for the analysis of protein–ligand complexes. Such a database can be used to search for sequence patterns in proteins and also to compare the connectivity of bound small-molecule ligands. The database automatically superimposes proteins using an iterative process to find optimal superposition, especially of binding pocket regions. Structures aligned in this way can be systematically evaluated. Which amino acids are involved in ligand interactions? What functional groups do the ligands use to interact with the amino acids of the protein? Which residues in the binding pocket occur repeatedly with identical geometry or are highly flexible? The water structure at the protein–ligand interface can be studied in detail. Surprisingly, a statistical analysis revealed that in about two thirds of all protein–ligand complexes, at least one crystallographically determined water molecule is involved in the binding of a ligand. This underscores the importance of including water molecules in modeling efforts. However, it is at this point that concepts for the treatment of water need to be significantly improved and extended.

20.4 Comparison of Protein-Binding Pockets

Another important question concerns the shape and composition of the binding pocket. Are there other proteins with similar amino acid compositions in which the pocket has an analogous shape? The actual amino acids are less important here. Rather, it is the analogous physicochemical properties of the exposed groups, such as hydrogen bond donors or acceptors, that are oriented towards the binding pocket. Programs that enable these comparisons describe the shape and surface of protein pockets along with the exposed properties. The function of proteins is often coupled to the recognition and binding of small-molecule ligands or segments of peptide sequences (e.g., proteases). Once bound, these molecules are chemically modified in the case of enzymes. In the case of receptors, the ligands are able to induce an effect within the receptor, for example, stabilizing an active or inactive conformation of the protein or changing its dynamic behavior. In this way, a signal is transmitted. The discovery of similarities in binding pockets can lead to the discovery of functional similarities between proteins. This is independent of whether there is sequence or folding homology between the proteins. There is also a chance of finding unexpected cross-reactivity through similarities in the shape and properties of binding pockets. Such unexpected binding is often the cause of adverse effects. By evaluating similarities and differences in such pockets, it is also possible to identify how ligands should be modified to achieve the desired selectivity for the given target protein. Valuable ideas for the design of new or modified protein ligands can be generated by studying and comparing bound ligands or ligand building blocks in similar pockets. This provides valuable ideas for isosteric replacements in the structure-based optimization of initial lead structures. The **Cavbase search engine**, implemented in the Relibase database, enables such pocket comparisons. Ruben Abagyan's group at UCSD in San Diego, USA, has developed **Pocketome, a comprehensive encyclopedia** that can be used to search for related binding pockets in protein families. The recent tool **SiteMine** for binding site comparisons has been developed by Matthias Rarey's group at the Hamburg university in Germany.

20.5 High Sequence Identity Facilitates Model Generation

An indispensable prerequisite for the use of the method arsenal of structure-based drug design is the existence of a spatial structure. A crystal structure cannot always be obtained. Under what conditions can a model of an unknown protein be constructed from a given sequence?

```
(a)  NEGDAAKGEKEF-NKCKACHMIQAPDGTDIKGGKTGPNLY
(b)  -EGDAAAGEKVS-KKCLACHTFDQGGAN-----KVGPNLF
(c)  --GDVAKGKKTFVQKCAQCHTVENGGKH-----KVGPNLW

(a)  GVVGRKIASEEGFKYGEGILEVAEKNPDLTWTEANLIEYV
(b)  GVFENTAAHKDNYAYSESYTEMKAK--GLTWTEANLAAYV
(c)  GLFGRKTGQAEGYSYTDA-----NKSKGIVWNNDTIMEYI

(a)  TDPKPLYKKMTDDKGAKTKMTFKMGKNQADVVAFLAQBBP
(b)  KDPKAFVLEKSGDPKAKSKMTFKLTKDD--------EIEN
(c)  ENPKKYI--------PGTKMIFAGIKKKGER-------QD

(a)  BAGZGZAAGAGSBSZ
(b)  VIAYLK------TLK
(c)  LVAYLKSATS
```

Fig. 20.4 The primary sequences of three cytochrome C proteins arranged using the typical one-letter code are shown from **(a)** the denitrifying bacterium *Paracoccus denitrificans* (134 amino acids), **(b)** the proteobacterium *Rhodospirillum rubrum* (112 amino acids), and **(c)** the mitochondria of a tuna fish (103 amino acids). The proteins vary in their length and composition. The sequence comparison shows the alignment with the best agreement. Invariable or conserved positions in the sequence are marked in *bold*. *Dashes* stand for areas in which other proteins carry additional amino acids (*insertions*). The *red bars* underscoring the sequences show preferred helical areas. *A* Ala, *C* Cys, *D* Asp, *E* Glu, *F* Phe, *G* Gly, *H* His, *I* Ile, *K* Lys, *L* Leu, *M* Met, *N* Asn, *P* Pro, *Q* Gln, *R* Arg, *S* Ser, *T* Thr, *V* Val, *W* Trp, *Y* Tyr

Proteins with similar functions from different species differ in their amino acid sequences. As the distance up the **phylogenetic tree** increases, these differences become more pronounced. Consider the example of cytochrome C (Fig. 20.4). This widely distributed protein in mitochondria plays a central role in the respiratory chain. It consists of a polypeptide chain of about 100 ± 20 amino acids. Three cytochromes are shown in Fig. 20.5 which, despite their different peptide chain lengths and compositions, have a very **similar folding pattern**. The proteins from the phylogenetically related species human and chimpanzee have 100% sequence identity. In contrast, the yeast enzyme has only 45% identity with that of these mammals. If the homology is very high and only a few mutations are present, model construction will be relatively easy. When sequence identity is greater than 90%, models can be constructed with uncertainties approaching the error margins of experimental structure determinations (Sect. 13.5). As sequence identity further decreases, model building becomes less accurate. At 50%, the average coordinate error can be a few angstroms. Below an identity of 25–30%, recognition of structural relationships becomes very problematic.

The vast majority of sequence differences between homologous proteins are located on the surface of the protein in loop regions that are not critical for the folding of the protein backbone (Sect. 14.4). Exchanges in the interior of the protein can have a much greater effect on its architecture. They are usually limited to amino acids of similar volume and very similar physicochemical properties, such as the exchange of a leucine for an isoleucine. Often the exchange of one amino acid is coupled with the complementary exchange of one or more other amino acids in the immediate vicinity. This is especially true when polar amino acids are exchanged inside the protein, which are internally saturated, e.g., by salt bridges. In the newly mutated protein variants, these amino acids form a stable orientation. Since the spatial proximity of amino acid residues in the fold does not necessarily correspond to their sequential proximity in the protein chain, the recognition of such structural relationships is considerably complicated. Mutations in the protein core can lead to expansion, spatial shifts, or twisting of the structural building blocks of the protein.

If the identity is very high, only a few amino acid side chains need to be exchanged. The conformations of the involved side chains can be deduced from a comparison with the structurally resolved proteins showing these amino acids in a similar environment. With decreasing identity, **insertions and deletions in loop regions**, i.e., an expansion or contraction of the polypeptide chain, must be considered. To predict the conformations of these loops during model building, libraries of known protein structures have been compiled. Based on length and sequence, these loops are classified into conformational families. They can be retrieved by the computer and support the construction of the spatial arrangement of a modified loop. The validation of the relevance of these protein models follows empirical rules. It is checked whether the constructed geometry agrees with experimental evi-

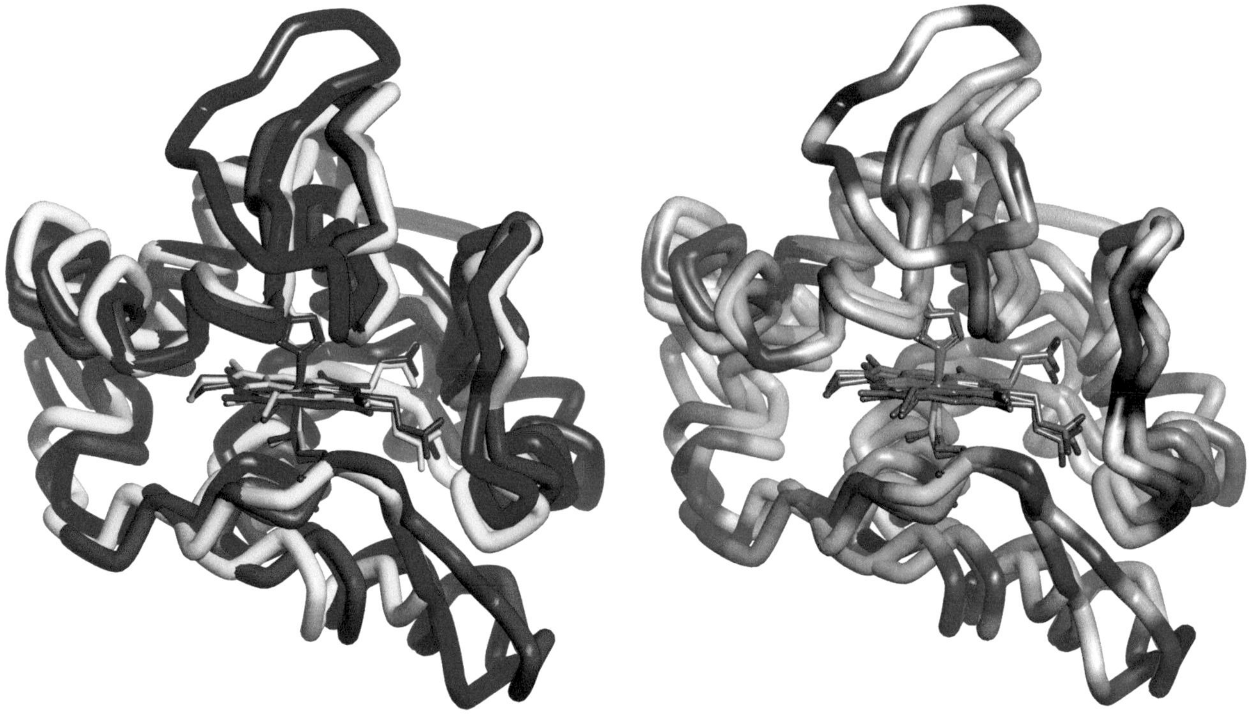

Fig. 20.5 *Left* Superposition of the folded structures of the three cytochrome C proteins from Fig. 20.4 based on a ribbon model: *Paracoccus denitrificans* in *blue*, *Rhodospirillum rubrum* in *red*, and tuna fish in *yellow*. The cytochromes bind via a histidine and a methionine to an iron–heme center. The structures were determined by X-ray crystallography. Structural deviations occur particularly in the loop regions. *Right* The same superposition is shown, only here the individual amino acids are color-coded. The same colors in all three ribbon models show identical amino acids at different positions (color coding: Ala: *light gray*, Val: *chartreuse*, Gly: *white*, Ile: *bright green*, Leu: *olive green*, Pro: *pink*, Phe: *violet*, Tyr: *dark purple*, Trp: *light violet*, Asp: *dark red*, Glu: *wine red*, Asn: *turquoise*, Gln: *cyan*, Lys: *blue*, His: *light blue*, Arg: *medium blue*, Ser: *light orange*, Thr: *dark orange*, Cys: *light yellow*, Met: *dark yellow*). (▸ https://sn.pub/UmXxlS)

dence. For example, it must be ensured that hydrophobic groups are oriented inwards and hydrophilic groups are oriented mainly outwards. The contact between amino acid groups is checked, and the chosen torsion angles are compared with those typically observed.

20.6 Secondary Structure Prediction and Amino Acid Replacement Propensities Support Model Building at Low Sequence Identity

When the sequence identity between the known and modeled protein falls below 30%, determining structural homology becomes increasingly difficult. All additional information must be employed as a resource. An attempt is made to estimate where in the polymer chain of the modeled protein certain secondary structure elements are expected to occur (Sect. 14.2). When evaluating the frequency with which individual amino acids occur in helices, pleated sheets, or loops, significant differences are found. For example, proline is considered a "helix breaker." It occurs at most in the first turn of a helix; at other positions, it disrupts the geometry and induces a kink. To determine whether a particular sequence segment folds as a helix, pleated sheet, or loop, the information about positional preferences is evaluated for several neighboring amino acids in an overlapping fashion.

Once analyzed, the primary sequence is compared to a reference protein of known geometry. Since its 3D structure is known, the assignment of the sequence to the secondary structural elements is straightforward. If not only one, but several 3D structures of members of a homologous protein family are known, multiple sequence alignments can be used to construct a representative profile of the expected secondary structure. This profile then serves as a reference for the mutual comparison of the sequences of structurally known and unknown proteins.

For this purpose, in addition to the physicochemical properties of the amino acids, their local conformational properties, their main and side chain orientations, their accessibility for solvent molecules, or their involvement in hydrogen bonds are analyzed. At the same time, the probabilities with which such an amino acid substitution can occur at the DNA sequence level are taken into account (Sect. 32.7). These parameters can be easily determined for proteins with a known spatial structure. Comparing the structures within a set of homologous proteins yields probabilities for mutual amino acid substitutions. For example, unlike all other amino acids, glycine has no side chain (see page IX). Therefore, it can adopt conformations in the polymer chain that are sterically impossible for other amino acids. The polymer chain assumes such conformations in regions close to the protein surface, where it reverses its orientation. This is where the conformationally flexible glycines play an important role. Such glycines that are exposed to the solvent and have unusual torsion angles are largely conserved between the folding of homologous proteins. The conserved glycines can be searched for during the sequence alignment of a protein to be modeled. They thus provide anchor points for sequence alignment. Many similar rules have been established. They serve as criteria for the recognition of sequence segments of structural importance. They are then applied to the protein sequence to be modeled. Even with relatively low sequence identity, structural homologies between a primary sequence and a protein of known 3D structure can be detected in this way. They are used as criteria for homology modeling.

Before the breakthrough in successful experimental structure determination of G-protein-coupled receptors, the most important group of membrane receptors (Chap. 29), it was necessary to rely on their model building. Additional criteria were used for this purpose. The modeling had to ensure that the hydrophobic amino acids in the helical areas were embedded in the membrane and oriented towards the membrane environment. Meanwhile, the homology modeling programs have achieved a high degree of automation. At the Biozentrum in Basel, the **SWISS-MODEL** server which transforms submitted sequences fully automatically into 3D structures was set up. The program **Modeller** from the group of Andrej Šali in San Francisco is able to assemble protein models from the sequences of entire genomes *in silico*. Although the structures are certainly rough, and many may be incorrect, such an approach allows a search for similarities in the recognition determinants of proteins. In this way, possible interactions between proteins can be discovered, or commonalities in metabolic pathways become transparent.

The modeling of proteins gives good results especially when they have a high homology. This is given in regions that determine the folding scaffold. The binding pockets are mostly located in the loop regions (Sect. 14.4). This is where even homologous proteins differ greatly. Therefore, the model constructions do not achieve the desired accuracy in these regions. An improvement can be achieved if a ligand is already placed in the assumed binding region during model construction. Model and placement must be optimized in an iterative process using appropriate energy functions.

As mentioned above, the problem of predicting the structure of a protein from its amino acid sequence using theoretical concepts has not yet been successfully solved. Therefore, the development of methods for prediction using data from experimentally determined structures has been the subject of intensive research for many years. Frustratingly, progress has been very incremental. The breakthrough to significantly improve the quality of predictions came in 2020, when the researchers at the Google subsidiary **DeepMind**, who had already attracted attention with their artificial intelligence (AI)-based programs AlphaGo and AlphaZero for simulating board games, achieved surprising success. Under the leadership of John Jumper, they won the **CASP14** competition (Critical Assessment of Protein Structure Prediction) by a wide margin with their **machine learning-based program AlphaFold 2.0**. The competition, which has been held since 1994, is performed to compare approaches to protein structure prediction. Research groups are provided with amino acid sequences of proteins. Their structures have been experimentally determined but not yet published. For these sequences, the participants have to predict structures using their computational models. AlphaFold 2.0 is based on a **neural network** (deep learning). It is trained to learn structure predictions based on experimentally determined protein structures from the PDB database. It evaluates input data by incorporating multiple sequence comparisons, evolutionarily related sequences, and physical and geometric constraints such as the preferred spatial pairing of amino acid residues. It applies rules similar to those used in the homology modeling described above. The scientists at DeepMind use the "**attention mechanism**" known from transformer architectures in their neural network. This involves highlighting parts of the input data in the context of the overall environmental data, while attenuating others. For language, this would be words in the context and meaning of a sentence or text. In this way, the neural network automatically focuses on and weights the critical information. Similar approaches have revolutionized natural language processing in recent years and are the basis for significant performance improvements in current AI systems for language and translation programs. As AlphaFold iterates through the entire structure of the protein to be folded, it simultaneously understands how to locally improve each part of the structure through repeated application. In doing so, it takes into account information already learned from previous steps. But how AlphaFold ultimately achieves its impressive results is not easy to deduce. With AI approaches, it is by no means trivial to deduce the system's rules and decision trees after the fact. This is especially true for analyses in which the decisions of the neural network are self-organizing and in which humans no longer intervene during the learning process.

DeepMind left the other participants far behind in the CASP14 competition. But academic research quickly caught up. David Baker's group in Seattle has since integrated DeepMind's AI approaches into the **RoseTTA Fold** software, which now delivers equivalent results. DeepMind has released AlphaFold 2.0 for free via a cloud computing tool. It can also be installed locally on a computer. The free use appeared only logical, as DeepMind has trained its AI on the publicly available structures in the PDB database. With the release, huge databases are now filling up with structures predicted by AlphaFold 2.0. There are now more than a million models available in the PDB. Meanwhile, a new version, **AlphaFold 3**, with enhanced functionality has been released and published in Nature. It is described to be not limited to single-chain proteins, it can also predict the structures of protein complexes with DNA, RNA, post-translational modifications, and selected ligands and ions. Unfortunately, this time its publication was without freely accessible executables or source code and for noncommercial trial applications it can be accessed through a server. In the interest of free scientific exchange and despite the predictable high level of commercial interest, it would be greatly appreciated if DeepMind could consider a more open approach to the release of the tool. It is believed that only through frequent use and feedback from the community, who provided the required database to train the neural network, will the computer tool achieve further improvements.

It seems reasonable to conclude that AI programs have already made a notable impact on protein structure research, and it seems probable that they will also influence predictions in future drug design. It is important to remember that structures for targeted ligand design require high resolution and the modeled geometries must be accurate to less than an Angstrom. Accurately predicting folding patterns to within a few angstroms may be sufficient for some applications. However, drug design researchers have learned the hard way from a renin model derived from the related endothiapepsin that small deviations can mislead targeted ligand design (Sect. 24.2).

The tools have been met with considerable enthusiasm and were recognized in 2024 with the Nobel Prize in Chemistry, which was bestowed upon the two leading DeepMind scientists, John Jumper and Demis Hassabis. The long-standing achievements of David Baker's group were also acknowledged, particularly given that David Baker has succeeded in predicting entirely novel folded proteins with tailored properties, which were subsequently validated through experiment. This opens up new perspectives for countless applications in biochemistry, structural biology, and drug design.

20.7 Ligand Design: Seeding, Expanding, and Linking

The next step after the analysis of the binding pocket of the experimentally determined or modeled protein is the actual ligand design. There are several **computer-aided design approaches** available to suggest new protein ligands. A docking program can be used to place successively preselected ligands from a database into the binding pocket (◘ Fig. 20.6). Typically, the database is populated with molecular candidates that closely resemble common drug-like molecules. This method, which has become known as "**virtual screening**" (Sect. 7.6), can now screen many millions of candidate molecules in a few hours using parallel computers.

Another approach starts with a "seed" in the binding pocket. Starting from this point, the ligand gradually grows into the binding pocket. This is the principle followed by most ***de novo* design** programs. The placement of the first seed is critical. Such approaches are particularly successful when there is a specific hot spot in the binding pocket from which further optimization starts. Salt bridges to charged amino acids or coordination of metal ion centers are particularly well suited for this approach. This concept has been successfully applied, for example, to the serine proteases trypsin and thrombin (Sect. 23.4) and the zinc-containing carbonic anhydrases (Sect. 25.7).

Another approach starts with small **fragments** that are placed into the protein binding pocket. Increasingly, the results of experimental fragment screening (Sect. 7.9) are used to provide reliable starting points for further growth of these candidate molecules in the binding pocket. The approach is an ideal symbiosis of experimental work and subsequent computational design. It combines the strengths of both techniques. Another approach is to chemically link several of the fitting fragments together. This strategy has been successfully applied, for example, in the SAR-by-NMR method (Sect. 7.8). The cover of this book shows an example of how a nanomolar protein kinase A inhibitor was developed through several design steps, starting with the crystal structure of a millimolar phenol fragment (see ◘ Fig. 02 page XI, ▶ https://sn.pub/4NppHn).

20.8 Docking Ligands into Binding Pockets

Docking attempts to fit potential protein ligands into a binding pocket using a computer. A docking program takes one candidate at a time from a precompiled library of molecules. For each entry, a 3D structure is generated. If a flexible molecule is encountered, either multiple conformations will be generated and docked individually, or they will be generated on the fly during docking. The next step is to fit each molecule into the binding pocket. First, the structures that cannot bind to the protein are discarded. In addition, other structures that cause obvious problems, such as electrostatic repulsion with the protein in the assumed docking mode, are eliminated. Typically, a docking program generates several solutions. These are evaluated on the basis of the generated binding geometries and their affinity is estimated.

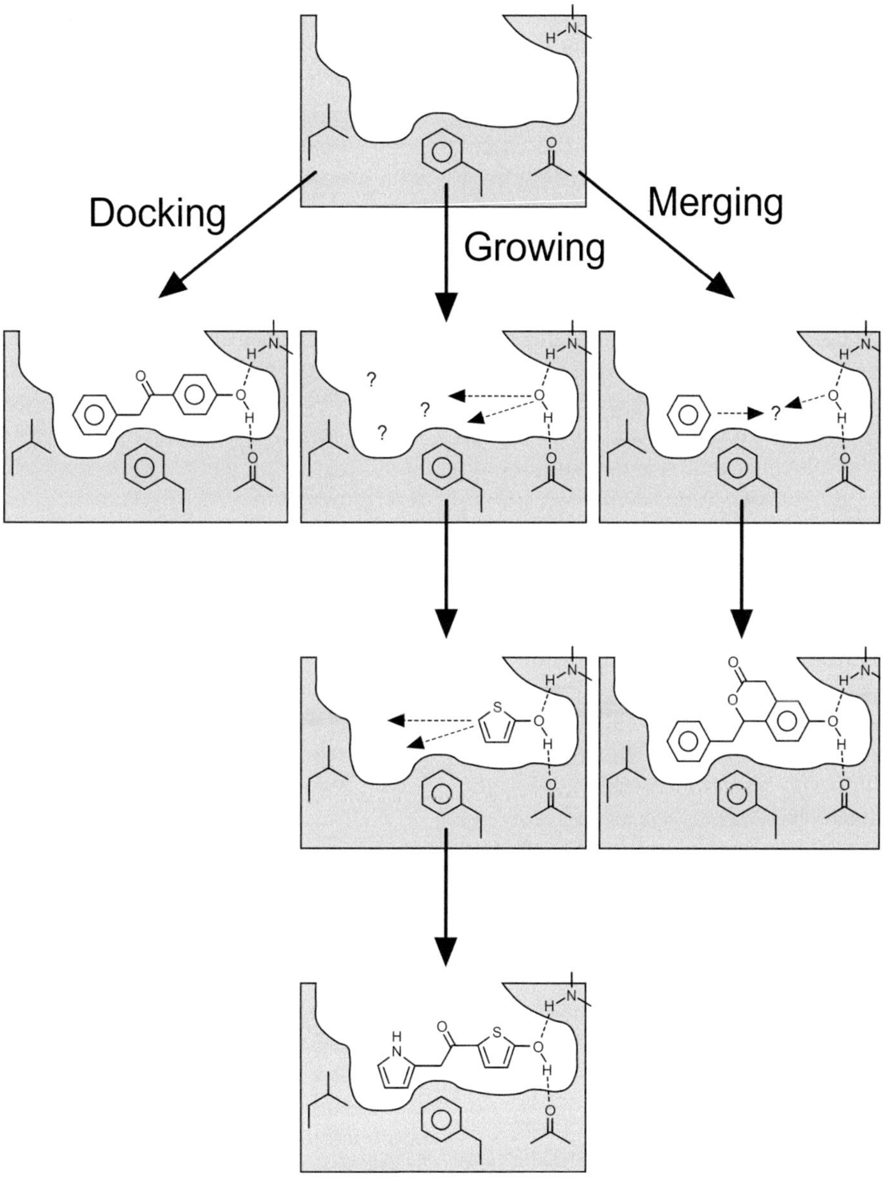

Fig. 20.6 Potential strategies for ligand design. The complete 3D structures of putative ligands are docked into the binding pocket (*left part of the image*). The construction of new molecules is sketched in the *middle* and *right* part of the picture. Basically there are two possibilities. A fragment can be placed as a seed and other groups can be attached step by step (*middle*). Alternatively, several small-molecule fragments can be placed in the binding pocket independently of one another and later linked or merged together (*right*)

Irwin Kuntz is a pioneer in the field of docking programs; **DOCK** was developed in his group at UCSF in San Francisco, USA. The original version in 1982 evaluated only the steric complementarity of ligands and proteins. The shape of the binding pocket was approximated by a set of different spheres so that the pocket was completely filled. A mathematical method was then used to place the test ligands on this distribution of spheres. Complementarity, a measure of direct protein–ligand contacts, served as a scoring function. Since the first version, DOCK has evolved considerably. The program now uses a force field for scoring and calculates the contributions for desolvation. Even the placement of the ligands is flexible by considering rotatable bonds. Another docking prototype was developed at the GMD in Bonn, Germany, by Matthias Rarey. The program **FlexX** was the first program that could quickly handle ligand flexibility during docking. It decomposes the test ligands into individual fragments and then uses an algorithm very similar to the positioning algorithm implemented in the *de novo* design program LUDI (Sect. 20.10). After placement of the first building block, the ligand is successively reconstructed in the binding pocket. Different conformers along the rotatable bonds are considered. The program maintains stored tables of preferred torsion angles, similar to those described in Sect. 16.6. The energetic evaluation of the placement is performed in this step. The program **AutoDock** from the group of Art Olson at Scripps in La Jolla, San Diego, USA, uses a lattice-based algorithm for placement. A force field function similar to that of the GRID program (Sect. 17.10) is used to place potential values on a grid embedded in the binding pocket. Starting from a random orientation, the ligand is moved over the grid until an optimum is found. In doing so, it "feels" the interaction potential with the protein. Since the potential has already been precalculated on the lattice, this evaluation is particularly fast. At the same time, twisting around rotatable bonds is performed. The program **GOLD**, developed by Gerrith Jones in the group of Peter Willett in Sheffield, England, also uses a lattice for placement. However, the interaction potentials are parameterized using crystal data. GOLD uses a genetic algorithm to optimize the geometry. Over time, a plethora of docking programs has been developed. All follow a slightly different strategy, but are based on the concepts described for the prototypes mentioned above. Some follow the idea that it is better to generate a well-distributed number of rigid ligand conformers and then dock them quickly as rigid bodies.

Today there are three main problems that impose limits on docking. One is the energetic evaluation of the generated geometries. This will be specially addressed in the next section. Another is that water plays a decisive role in ligand binding (Sect. 20.3). Even today no really convincing solution to the handling of water molecules during docking has been found. The third problem is the flexible adaptation of the protein (Sect. 15.8). Usually there are small adaptations on the side of the protein that slightly change the shape of the binding pocket. In fact, they are big enough to make the docking programs look for the proverbial red herrings (means: they lead to wrong tracks).

20.9 Scoring Functions: Ranking of Constructed Binding Geometries

A relevant scoring of the generated binding geometries is essential for all docking and *de novo* design approaches in structure-based drug design. Among the numerous geometrically plausible placements, only those that reasonably approximate the experimentally found situations must be filtered out. The enthalpic and entropic contributions, which determine the affinity of a ligand to its target protein, were described in Chap. 4. The goal of a **scoring function** is to quickly estimate the expected binding affinity from a given interaction geometry. Theoretically, a single geometry is not sufficient to solve this problem. Molar energies are determined by a finite set of conformations. They are distributed over an "ensemble" of multiple states. These states are differently populated according to their energy content. One group of methods tries to account for this fact in the calculations. This is the most theoretically sound approach. The energy contributions of the ensemble (usually taken from the **trajectories** of a molecular dynamics simulation, Sect. 15.7) are summed. The Gibbs free energy ΔG (Sect. 4.3) can be estimated from the resulting **partition function**. However, the computations required for such evaluations are very time consuming. As a result, they are usually not feasible for screening large amounts of data in a structure-based drug design session.

Instead, **regression-based scoring functions** are used as an alternative approach. Assuming that a particular state is predominantly populated, it may be justified to consider only one state (or conformation) in the scoring function. The enthalpy and entropy contributions that most likely determine the binding affinity are considered. The approach is reminiscent of setting up a QSAR equation (Sect. 18.2). The terms are combined into an energy function. Molecular descriptors that correctly reflect the contributions to these terms are sought. In doing so, the erroneous assumption is made that the individual contributions to the description of the free enthalpy are additive (Sect. 4.10). The individual terms of the equation are each given an adjustable weighting factor. As with QSAR equations, a mathematical technique is used to optimally fit these weighting factors to a training dataset. This set consists of crystallographically determined protein–ligand complexes for which experimental binding affinities are available.

A third approach follows a so-called **knowledge-based concept**. As already discussed in Sect. 17.10, the frequency of individual contact geometries in the crystal structures of protein–ligand complexes is evaluated. A kind of "normal distribution" is defined as a reference state. Then, all contacts that occur more frequently than average are classified as energetically favorable. All contacts occurring less frequently than the average are classified as unfavorable. These scores are then applied to all contacts present in a computationally generated protein–ligand complex and combined for a ranking. So far, a function derived in this way can be used for the relative energetic ranking of a set of ligands binding to the same reference protein. However, affinity prediction will also be successful here if training is performed against a reference dataset of known geometry and binding affinity in a manner analogous to regression-based scoring functions. The evaluation with the regression-based or knowledge-based function is very fast. In the meantime, a large number of scoring functions have been developed. So far, none of them has proven to be ideal and generally applicable. Therefore, in each case it has to be checked which function provides the best performance for the protein under investigation.

20.10 *De Novo* Design: From LUDI to the Automated Assembly of Novel Ligands

The first program for stepwise ***de novo* design** was **GROW** by Jeffrey Howe and Joseph Moon at Upjohn. It focuses

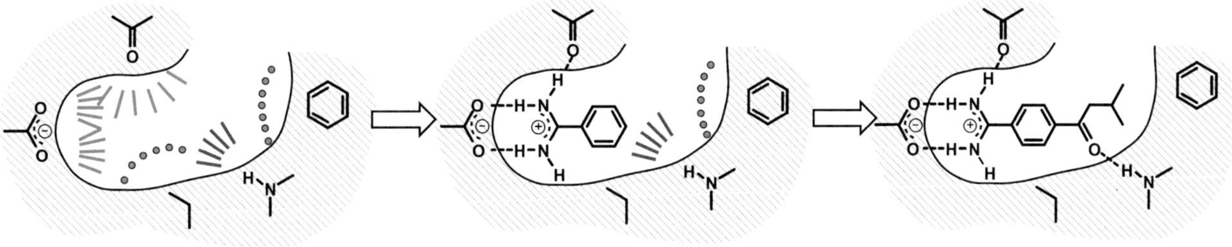

Fig. 20.7 Concept of the program LUDI for *de novo* design of protein ligands. In the first step, the interaction sites are determined (*left*). Donor sites are represented by *blue lines*, acceptor sites by *red lines*. The *green dots* symbolize lipophilic sites. Subsequently, small molecules from a database are fitted into the binding pocket in that they are matched with the interaction sites (*middle*). Finally LUDI can chemically link groups or parts of molecules to larger structures to match the remaining unsatisfied interaction sites and to fill the entire binding pocket (*right*)

on peptides as lead structures. An amide group is positioned in a favorable orientation in the binding pocket. The next amino acids are added to the starting amide group in a stepwise fashion. At each step, a large number of different conformations of all 20 proteinogenic amino acids are attached to the seed on the fly. The "best" solutions for each are followed. In this way, GROW constructs a peptide ligand of increasing length in the binding pocket.

In the early 1990s, Hans-Joachim Böhm developed the program **LUDI** at BASF, Ludwigshafen, Germany. The idea was to read small molecules or molecular fragments from a database with precalculated spatial geometries and position them in the binding pocket so that hydrogen bonds are formed with the protein and hydrophobic pockets are filled with nonpolar groups. As input, the program requires the coordinates of the protein and a library of 3D structures of fragments or drug-like molecules.

The precalculation of interaction sites is crucial. These are placed in the binding pocket around the amino acid residues in the form of fitting points or directional vectors (Fig. 20.7). The program uses rules derived from the nonbonded interactions found in the crystal packing of small organic molecules (Sects. 14.7 and 17.10). LUDI then extracts small molecules or molecular fragments from the 3D library. For each entry, an attempt is made to position it in the binding pocket of the protein so that as many of these interaction sites as possible are satisfied (Figs. 20.7 and 17.12). All successfully placed fragments are then ranked. The scoring function used takes into account the number and quality of H-bonds and ionic interactions formed, hydrophobic contact surfaces shared by the protein and ligand, and unfavorable contributions due to the number of rotatable bonds in the ligands. An example of the successful application of this program is described in Sect. 21.5.

As the first prototype, LUDI was the gold standard for many *de novo* design programs that were developed later. These approaches implemented enhanced scoring functions and improved fragment libraries. The programs were also taught synthesis rules so that the chemical accessibility of the generated molecules was taken into account. The search space of the programs has also been expanded to include multiple conformations and configurations.

A *de novo* design program is an idea generator. Its value is, of course, determined by the concepts that went into its development. On the other hand, its value also depends strongly on the user and how the suggestions of such a program are interpreted and used for further design.

20.11 The Feasibility of Designing Ligands *In Silico*

Certainly, many examples have been presented that demonstrate the power of *de novo* design, virtual screening, and docking. The example described in Chap. 21 was successful because of the use of such methods. However, it is critical that the computational methods are tightly integrated into an iterative process of synthesis, biological testing, and experimental structure determination. It is important to keep in mind that not all hits from computational screening are based on the correct assumptions, and not all hits are picked up for the right reasons.

The predictive power of available methods is still limited. The synthetic accessibility of a proposed molecule is often not sufficiently taken into account, the flexibility of the protein is frequently neglected, and the methods for estimating binding affinity are still too inaccurate. This is because the process and factors responsible for molecular recognition and ligand binding are still poorly understood. The correct description of solvation effects, the involvement of water molecules in the binding process, and the change of protonation states are major problems. The contribution of hydrogen bonding to binding affinity is still only an estimate, despite efforts to the contrary. In the case of lipophilic interactions, it can at least be assumed that the filling of an unoccupied lipophilic pocket with additional nonpolar substituents is in most cases accompanied by an increase in binding affinity.

How to account for the changes in binding entropy that contribute to the free energy of ligand binding is still insufficiently considered. Certainly, there is evidence that the oversimplified assumption that entropic contributions will be constant within a set of congeneric ligands is definitely not valid.

However, there are other fundamental limitations of this approach. The most important is that the technique is limited to optimizing direct interactions with the protein. Successful binding to a target protein is essential for any drug. However, to be suitable as a drug, additional requirements must be met. These include good selectivity, metabolic stability, adequate duration of action, low addictive potential, and negligible toxicity. Today, at least the selectivity of a compound towards members of a structurally related protein family can be estimated with some certainty.

Fully automated molecular design on a computer is not yet possible, nor is it likely to be in the long term. The methods of structure-based design are valuable as idea generators. The resulting suggestions must be checked and modified if necessary. Time will tell whether these methods will gradually approach the "holy grail" of drug design: the design of drug molecules from scratch.

20.12 Synopsis

- In structure-based drug design, attempts are made to design small-molecule ligands by docking them directly into the binding pocket of a target protein. This requires a 3D structure of the reference protein. The goal is to optimally fill the binding pocket with a ligand by satisfying nonbonded interactions with the functional groups of the binding-site residues.
- Structure-based design starts with a detailed analysis of the binding pocket to elucidate hot spots for putative interactions with the protein. Either experimental methods or computational tools can be used to perform an active site mapping with molecular probes or small solvent-like molecules.
- In an iterative process of structure determination, modeling of modified ligands, docking and screening, synthesis, and biological testing, the properties of small-molecule ligands are improved to optimize binding to the target protein.
- Databases have been developed to retrieve and compare structural information about the exponentially growing body of structural data on protein–ligand complexes. They allow comparison of binding poses, active-site interaction geometries, protein–ligand binding motifs, and the original solvation structures in the protein's binding pocket.
- Proteins can be compared in terms of their exposed binding pockets. The shape and the exposure of groups which have particular physicochemical properties in binding pockets are compared and help to design small-molecule ligands with the desired selectivity. Ideas for isosteric replacements on the ligand scaffold can also be generated in this way.
- If an experimentally determined structure of the target protein is unavailable, a homology model can be constructed by using a related protein of known architecture as a template. The accuracy and the success of such homology modeling depend strongly on the sequence homology with the template structure.
- Tools for secondary structure prediction and amino acid replacement propensities have been developed to improve the reliability of the sequence assignment of proteins to the 3D structure of the reference template. Recently, AI programs have been developed based on neural networks to generate 3D structures from sequence data. They were trained to learn structure predictions based on experimentally determined protein structures.
- Ligand design approaches screen large databases of candidate molecules by docking them into the binding pocket.
- Alternatively, *de novo* design starts with a small molecule seed or fragment and grow them into putative ligands in the pocket. Two nonoverlapping fragments can also be linked together to form a larger ligand with improved binding properties.
- The geometry of a constructed protein–ligand complex must be evaluated in terms of the expected binding affinity in all structure-based design strategies. A large variety of scoring functions are used to predict the binding affinity based on the geometry of the formed complex.

20

Bibliography and Further Reading

General Literature

C. Branden, J. Tooze, Introduction to Protein Structure, Garland Publishing, Inc. New York, 1991, 2nd Edition, 1999

T. J. P. Hubbard, A. M. Lesk, Modelling Protein Structures, in Computer Modelling in Molecular Biology, J. M. Goodfellow, Eds., VCH, Weinheim, 1995

C. Hutchins, J. Greer, Comparative Modeling of Proteins in the Design of Novel Renin Inhibitors, Crit. Reviews Biochem. Molec. Biol., **26**, 77–127 (1991)

P. Goodford, Drug Design by the Method of Receptor Fit, J. Med. Chem., **27**, 557–564 (1984)

J. Greer, J. W. Erickson, J. J. Baldwin, M. D. Varney, Application of the Three-Dimensional Structures of Protein Target Molecules in Structure-Based Drug Design, J. Med. Chem., **37**, 1035–1054 (1994)

C. R. Beddell, Ed., The Design of Drugs to Macromolecular Targets, Wiley, Chichester, 1992

I. D. Kuntz, Structure-Based Strategies for Drug Design and Discovery, Science **257**, 1078–1082 (1992)

S. Borman, New 3D Search and De Novo Design Techniques Aid Drug Development, Chem. & Eng. News, **10**, 18–26 (1992)

Y. C. Martin, 3D Database Searching in Drug Design, J. Med. Chem., **35**, 2145–2154 (1992)

I. D. Kuntz, E. C. Meng, B. K. Shoichet, Structure-Based Molecular Design, Acc. Chem. Res., **27**, 117–123 (1994)

H. J. Böhm, Ligand Design, in: 3D QSAR in Drug Design, H. Kubinyi, Eds., Escom, Leiden, 1993, pp. 386–405

K. Müller, Ed., De Novo Design, Persp. Drug Discov. Design, Vol. 3, Escom, Leiden, 1995

H. J. Böhm, G. Schneider, Molecular Recognition in Protein–Ligand Interactions (Vol. 19, in Methods and Principles in Medicinal Chemistry, R. Mannhold, H. Kubinyi und G. Folkers, Eds.), Wiley-VCH, Weinheim, 2006

G. Schneider, K. H. Baringhaus, Molecular Design, Wiley-VCH, Weinheim, 2008

M. Eguida, D. Rognan, Estimating the Similarity between Protein Pockets, Int. J. Mol. Sci., 23, 12462 (2022)

K. A. Carpenter, R. B. Altman, Databases of ligand-binding pockets and protein-ligand interactions, **23**, 1320–1338 (2024)

Special Literature

C. R. Bedell, P. J. Goodford, F. E. Norrington, S. Wilkinson, R. Wootton, Compounds Designed to Fit a Site of Known Structure in Human Haemoglobin, Brit. J. Pharmacol., **57**, 201–209 (1976)

M. Hendlich, A. Bergner, J. Günther, G. Klebe, Design and Development of Relibase—a Database for Comprehensive Analysis of Protein–Ligand Interactions, J. Mol. Biol., **326**, 607–620 (2003)

S. Schmitt, D. Kuhn, G. Klebe, A new method to detect related function among proteins independent of sequence and fold homology, J. Mol. Biol., **323**, 387–406 (2002)

A. Weber et al., Unexpected Nanomolar Inhibition of Carbonic Anhydrase by COX-2-Selective Celecoxib: New Pharmacological Opportunities due to Related Binding Site Recognition, J. Med. Chem., **47** 550–557 (2004)

C. Gerlach et al., KNOBLE: a knowledge-based approach for the design and synthesis of readily accessible small molecule chemical probes to test protein binding, Angew. Chem. Int. Ed. Engl., **46**, 9105–9109 (2007)

I. Kufareva, A. V. Ilatovskiy, R. Abagyan, Pocketome: An Encyclopedia of Small-molecule Binding Sites in 4D. Nucl. Acid. Res., **40**, D535–D540 (2012)

K. N. Allen, C. R. Bellamacina, X. Ding, C. J. Jeffery, C. Mattos, G. A. Petsko, and D. Ringe An Experimental Approach to Mapping the Binding Surfaces of Crystalline Proteins, J. Phys. Chem. **100**, 2605–2611 (1996)

J. Overington, M. S. Johnson, A. Sali and T. L. Blundell, Tertiary Structural Constraints on Protein Evolutionary Diversity: Templates, Key Residues and Structure Prediction, Proc. Royal Soc. Lond. **B 241**, 132–145 (1990)

M. S. Johnson, J. P. Overington, T. L. Blundell, Alignment and Searching for Common Protein Folds Using a Data Bank of Structural Templates, J. Mol. Biol., **231**, 735–752 (1993)

A. Waterhouse et al., SWISS-MODEL: Homology Modelling of Protein Structures and Complexes, Nucl. Acid. Res., **46**, W296–W303 (2018)

A. Sali and T. L. Blundell, Definition of General Topological Equivalence in Protein Structures, J. Mol. Biol., **212**, 403–428 (1990)

A. Sali and T. L. Blundell, Comparative Protein Modelling by Satisfaction of Spatial Restraints. J. Mol. Biol., **234**, 779–815 (1993)

J. Jumper et al., Highly accurate protein structure prediction with AlphaFold, Nature **596**, 583–589 (2021)

J. Abramson, J. Adler, J. Dunger et al. Accurate structure prediction of biomolecular interactions with AlphaFold 3. Nature 630, 493–500 (2024)

M. Baek et al. Accurate prediction of protein structures and interactions using a three-track neural network, Science, **373**, 871–876 (2021)

E. Callaway, It will change everything: DeepMind's AI makes gigantic leap in solving protein structures, Nature **588,** 203–204 (2020)

M. Hibert, S. Trumpp-Kallmeyer, J. Hoflack, A. Bruinvels, This is Not a G-Protein-Coupled Receptor, Trends Pharm. Sci., **14**, 7–12 (1993)

J. Hoflack, S. Trumpp-Kallmeyer and M. Hibert, Re-evaluation of Bacteriorhodopsin as a Model for G-Protein-Coupled Receptors, Trends Pharm. Sci., **15**, 7–9 (1994)

R. Henderson, J. M. Baldwin, T. A. Ceska, F. Zemlin, E. Beckmann, K. H. Downing, Model of the Structure of Bacteriorhodopsin Based on High-Resolution Electron Cryo-Microscopy, J. Mol. Biol., **213**, 899–929 (1990)

G. F. X. Schertler, C. Villa, R. Henderson, Projection Structure of Rhodopsin, Nature, **362**, 770–772 (1993)

J. Travis, Proteins and Organic Solvents Make an Eye-Opening Mix, Science, **262**, 1374 (1993)

C. S. Ring et al., Structure-based Inhibitor Design by Using Protein Models for the Development of Antiparasitic Agents, Proc. Natl. Acad. Sci., **90**, 3583–3587 (1993)

I. D. Kuntz, J. M. Blaney, S. J. Oatley, R. Langridge, T. E. Ferrin, A Geometric Approach to Macromolecule-Ligand Interactions, J. Mol. Biol., **161**, 269–288 (1982)

M. Rarey, B. Kramer, T. Lengauer, G. Klebe, A Fast Flexible Docking Method using an Incremental Construction Algorithm, J. Mol. Biol., **261** 470–489 (1996)

D. S. Goodsell, A. J. Olson, Automated Docking of Substrates to Proteins by Simulated Annealing, Proteins, **8**, 195–202 (1990)

G. Jones, P. Willett, R. C. Glen, A. R. Leach, R. Taylor, Development and Validation of a Genetic Algorithm for Flexible Docking, J. Mol. Biol., **267**, 727–748 (1997)

Pocketome: https://ngdc.cncb.ac.cn/databasecommons/database/id/501 (Last accessed Nov. 19, 2024)

SiteMine: https://uhh.de/naomi (Last accessed Nov. 19, 2024)

Explore Computed Structure Models Alongside PDB Data: https://www.rcsb.org/news/6304ee57707ccd4f63b3d3db (Last accessed Nov. 19, 2024)

SWISS-Model web server: https://swissmodel.expasy.org/ (Last accessed Nov. 19, 2024)

Modeller: https://salilab.org/modeller/ (Last accessed Nov. 19, 2024)

AlphaFold: https://neurosnap.ai/academic?gclid=CjwKCAiAqNSsBhAvEiwAn_tmxbN3V2G2yrU1LJABBGDpMRMSgu_2u6c-8v76nQEU-tDj1OMPXPFoQxoCX7gQAvD_BwE (Last accessed Nov. 19, 2024)

A Case Study: Structure-Based Inhibitor Design for tRNA-Guanine Transglycosylase

Contents

G. Klebe, *Drug Design*, https://doi.org/10.1007/978-3-662-68998-1_21

The numerous examples in the last part of this book (Chaps. 22–32), many of which have successfully led to marketed products, are preceded by an exemplary case study. The main purpose of this chapter is to show how the inhibition of a target protein can be achieved by different approaches and how initial ligands can be candidates for further medicinal chemistry optimization. Using the tRNA-modifying enzyme tRNA-guanine transglycosylase (TGT) as an example, we present the options for inhibitor development that result from the iterative application of several design cycles and the biophysical and crystallographic characterization of the structure-based design methods described in the previous chapters (Chaps. 7, 8, 13, 14, and 20). The example refers to work done by the research group of François Diederich at the ETH in Zurich, Switzerland, as well as to work performed by the author's research group at the University of Marburg, Germany. Because the work was accomplished in an academic environment, it was possible to use different tools of structure-based design and to pursue some of the more fundamental problems in the context of the project.

21.1 Shigellosis: Disease and Therapeutic Options

Shigella dysentery is a severe **diarrheal illness** that is caused by ***Shigella* bacteria**. These bacteria are ingested with contaminated water or food and adhere to epithelial cells in the intestinal mucosa. They are extremely contagious: 10–100 bacteria are enough to cause an infection. Worldwide, shigellosis represents a serious problem. Almost 270 million cases are reported annually, of which over a million are fatal. The disease is widespread in developing countries, but over half a million cases are also annually reported in industrialized countries. Above all, the disease flourishes under conditions of inadequate hygiene and poor water quality as is found in war, natural catastrophes, famine, and in refugee camps. Dysentery is a particular problem in Africa where it can occur concomitantly with AIDS.

As with any bacterial **infectious disease**, shigellosis can be treated with antibiotics. The infections that occur in industrialized countries are cured in this way. Unfortunately, *Shigella*, which is very similar to the *Escherichia coli* that naturally occurs in the intestinal flora, has a tendency to **become resistant to antibiotics** very quickly. Moreover, antibiotic therapy also kills the naturally occurring bacteria of the intestinal flora, and this also produces diarrheal symptoms and severe dehydration in the patients. This can lead to a life-threatening disruption of electrolyte homeostasis, particularly in small children. Therefore, specific therapeutic approaches that suppress the pathogenicity of *Shigella* are sought.

21.2 Blocking Pathogenesis on the Molecular Level

Shigella infect the epithelial cells of the colon. Once in contact with these cells, they secrete invasins through a **complex secretion apparatus**. The invasins create a pore in the host cell membrane. As this invasin pore remains connected to the secretion apparatus, a continuous channel is formed. Through this channel, further **virulence factors** are transported directly from the bacterial cytoplasm into the host cell cytoplasm. There, the virulence factors cause the bacteria to be engulfed by a specific type of endocytosis, that is through an internalization of the cell membrane, which completes the uptake of fluid and particles into the cell. Initially, the ingested bacterium is still enclosed in an envelope (a type of endosome), but this dissolves within a few minutes. The bacteria are then free to move around in the cytoplasm of the host cell and use the infected cell for further replication.

The genes for the components of the secretion apparatus as well as the genes for the invasins and other virulence factors are located on a very large plasmid. The expression of all these genes requires the **transcription factor VirF**, whose gene is also located on this plasmid. In order for the protein encoded by the *virF* gene to be effectively synthesized on the ribosome, a specifically modified tRNA base is required. The **tRNA** is a ribonucleic acid of about 80 nucleotides (◘ Fig. 32.18, Sect. 32.7). It is terminally loaded with an amino acid that is defined by its specific base triplet in the central loop, called the anticodon loop. When the gene information is translated from mRNA, a corresponding tRNA is bound to the ribosome for each amino acid encoded there, which is deposited as a base triplet. This tRNA carries the required amino acid so that the correct residue is incorporated into the nascent peptide chain of the resulting protein. The modified base required for the efficient biosynthesis of the VirF transcription factor is located at position 34 (the so-called **wobble position**) of certain tRNAs. They are loaded with the amino acid aspartate, asparagine, histidine, or tyrosine. If this modification is missing, only a small amount of VirF will be produced. *Shigella* bacteria then produce very little of the invasins needed to infect colon epithelial cells. Their pathogenicity is, therefore, greatly reduced.

Bacteria have enzymes that can make these changes in tRNA. In the first step, a guanine **21.1** is cut out of the tRNA molecule at position 34 and replaced with an altered base, **preQ$_1$ 21.2** (◘ Fig. 21.1). This step is catalyzed by the enzyme tRNA-guanine transglycosylase (TGT). The exchanged base in the tRNA is further modified in the next step of an enzymatic cascade to yield the base **queuine** as the final product. Inhibitors of the bacterial TGT, therefore, represent a specific therapeutic principle for selectively targeting the pathogenicity of *Shigella*. In contrast to therapy with broad-spectrum antibiotics, the bacteria are not killed, but the disease-causing infection of the epithelial cells is prevented. Higher developed eukaryotic organisms such as humans also possess such an enzyme. Unlike bacteria, which use a homodimeric enzyme, the eukaryotic enzyme is a heterodimer. In addition, higher organisms do not convert $preQ_1$ to the end product queuine, but incorporate queuine directly into tRNA.

21.3 The Crystal Structure of tRNA-Guanine Transglycosylase as a Starting Point

First, the crystal structure determination of TGT in complex with $preQ_1$ was determined from a related species. This species shows an exchange of a Phe for a Tyr in the active site, which is immaterial for substrate or ligand binding. Later, the structure complexed with a part of the tRNA was elucidated (◘ Fig. 21.2). According to these structures, **base exchange** occurs via the following **reaction pathway** (◘ Fig. 21.3). Initially, the tRNA with the covalently attached guanine binds to the enzyme. The base with its ribose moiety is pulled out of the tRNA molecule and is specifically recognized by Asp 102, Asp 156, Gln 203, Gly 230, and Leu 231. The reaction starts with a nucleophilic attack at carbon C1 of the ribose ring. The C1–N bond is cleaved, and a proton is transferred from the contact-mediating water to guanine. The released base, together with the formed hydroxide ion, leaves the binding pocket. Subsequently $preQ_1$ is taken up by the same binding site. For this, the peptide bond between Leu 231 and Ala 232 must flip over. The nitrogen atom at position 9 of $preQ_1$ then relays a proton to the nitrogen atom at position 3 and carries out a nucleophilic attack on the ribose, which is covalently attached to Asp 280. Once the new bond to the tRNA is formed, the chemically altered tRNA leaves the enzyme. Asp 102 is critically involved in the recognition process of the bound base.

21.4 A Functional Assay to Determine Binding Constants

The **base-exchange reaction** is accomplished in two steps. In principle, both steps can be blocked by inhibitors. This must be considered in a **functional assay**. In the first step, the unmodified tRNA is bound (◘ Fig. 21.4). Sufficiently large inhibitors could competitively prevent this step. After the tRNA is covalently attached to the enzyme, the guanine base is released and leaves the protein. Next, $preQ_1$ binds. A potential inhibitor can also compete with this uptake into the binding site, but must not be much larger than guanine or $preQ_1$. In this way small inhibitors display a different inhibition profile than structurally larger inhibitors.

Radioactively labeled guanine is used to measure inhibition. If this guanine is added to the tRNA, the TGT will catalyze its incorporation, and the tRNA molecule will become radioactively labeled. If the tRNA is separated at fixed intervals, and the incorporated radioactivity is measured, the reaction kinetics of the incorporation process and, therefore, the catalytic rate of the enzyme can be followed. If potential inhibitors are added, fewer TGT molecules will be available for the transformation,

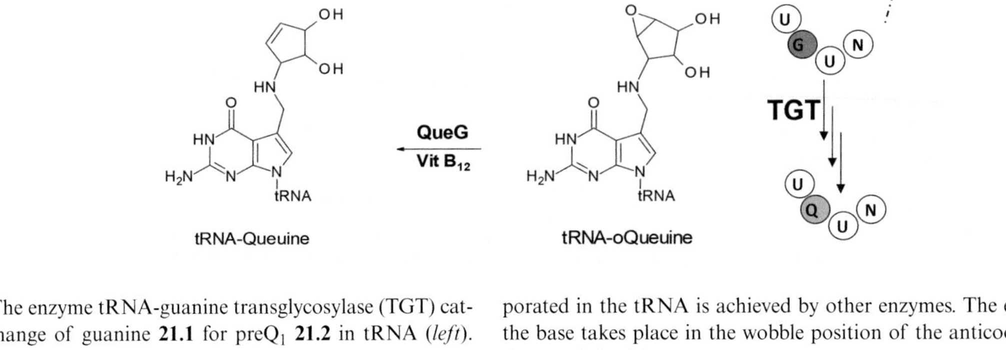

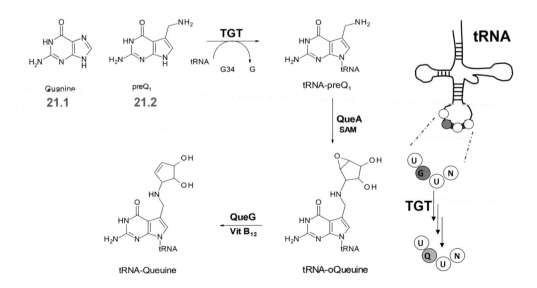

◘ **Fig. 21.1** The enzyme tRNA-guanine transglycosylase (TGT) catalyzes the exchange of guanine **21.1** for $preQ_1$ **21.2** in tRNA (*left*). Next, the further modification of this base to queuine, which is incorporated in the tRNA is achieved by other enzymes. The exchange of the base takes place in the wobble position of the anticodon loop of the tRNA (*right*)

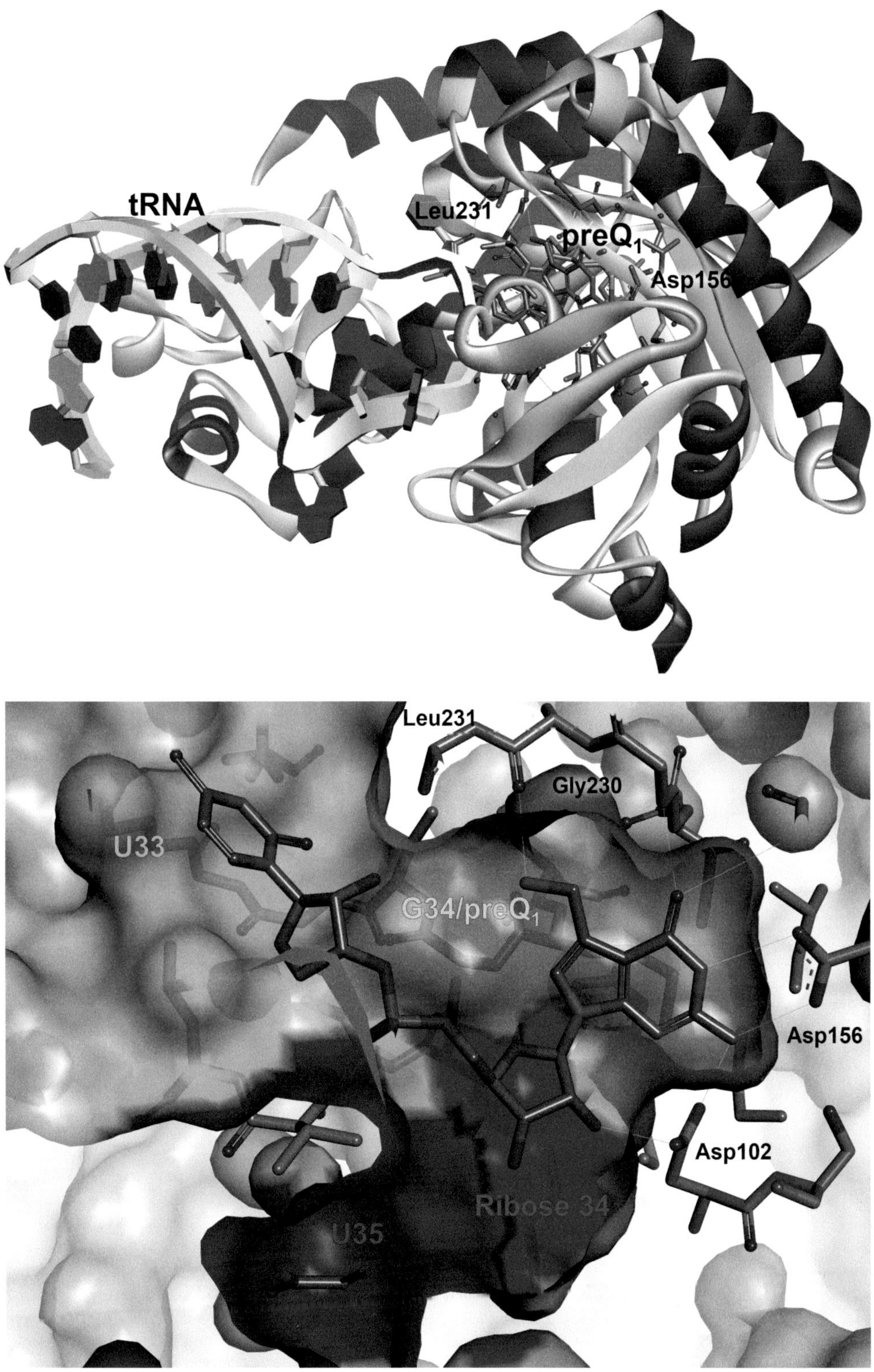

Fig. 21.2 The crystal structure of TGT with a portion of the tRNA. The protein adopts a TIM-barrel fold. The tRNA binds to the protein near the catalytic center with the bases U33, G34, and U35, and the base (*gray*) to be exchanged at position 34 is completely rotated out from the tRNA molecule (*upper part*). A view into the binding site is shown (*below*). The already-incorporated, modified base $preQ_1$ is held in place in the guanine-recognition pocket (*orange*) by Asp 102, Asp 156, Gly 230, and Leu 231. The ribose moiety is arranged in a small hydrophobic pocket (*blue*). Uracil 33, preceding guanine 34, lies in the *green*-colored part of the binding pocket, and the uracil 35 residue, following guanine 34, lies in the *red*-colored binding areas. (▶ https://sn.pub/TTqICn)

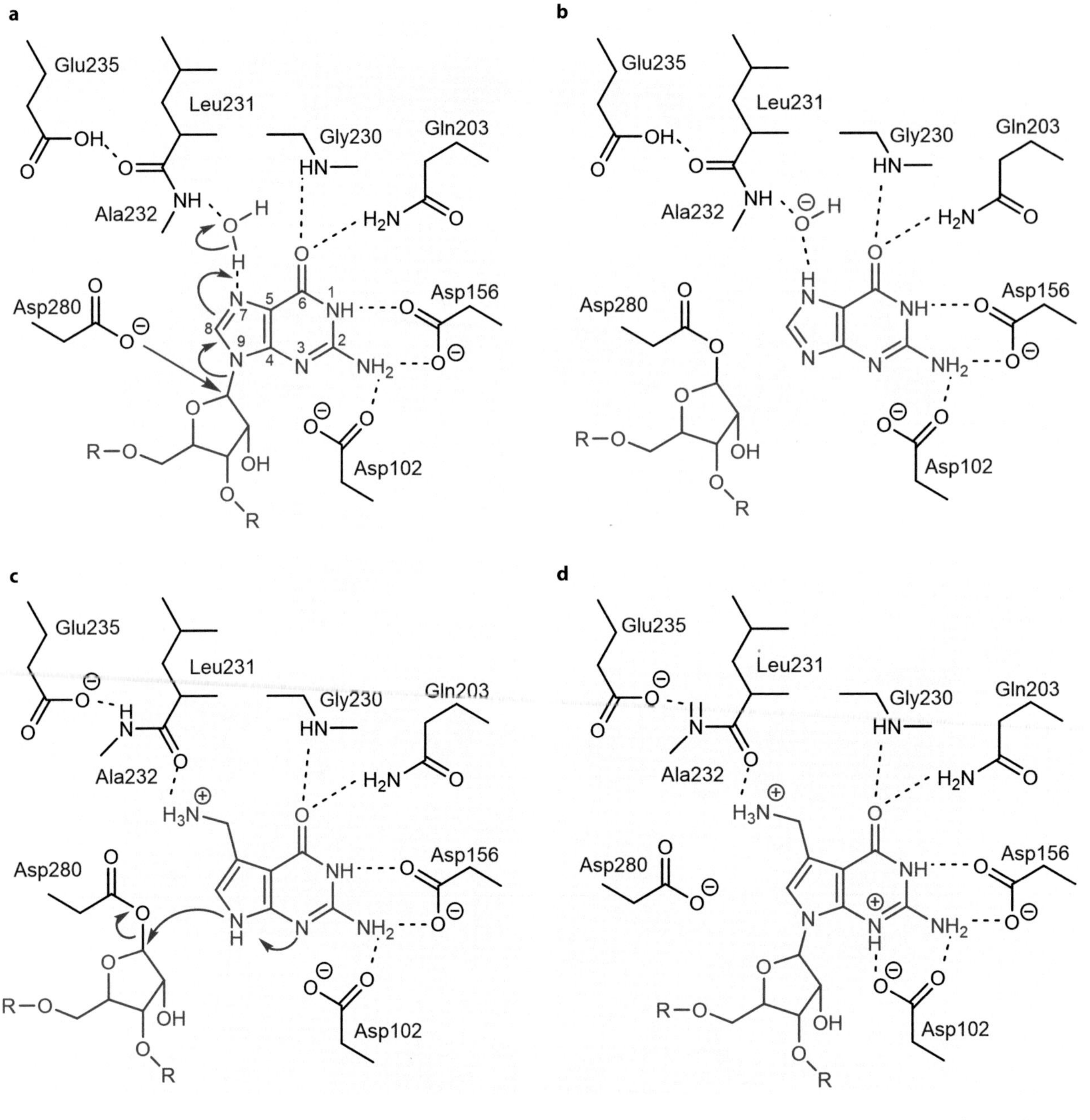

Fig. 21.3 Mechanism of the base exchange reaction in the transglycosylase. The tRNA with guanine 34 is bound and a water molecule mediates the contact with the nitrogen atom at the 7-position. Asp 280 attacks the C1 carbon of the ribose ring as a nucleophile (**a**). The C1–N bond is cleaved and guanine is released by accepting a proton from the adjacent water molecule (**b**). Together with the resulting hydroxide ion, it leaves the binding pocket. In the same binding site, $preQ_1$ is now incorporated by flipping the peptide bond between Leu 231 and Ala 232 over (**c**). The proton at position 9 of $preQ_1$ is transferred to the nitrogen atom at position 3. This creates a guanidium-like moiety that is stabilized by the carboxylate groups of Asp 102 and Asp 156. The deprotonated nitrogen atom at position 9 then nucleophilically attacks the ribose moiety which is covalently attached to Asp 280. A new bond to the tRNA is formed (**d**). This releases the modified tRNA, which leaves the enzyme. The incorporated $preQ_1$ base releases its proton into the surrounding solvent

and incorporation rate is reduced. This can be seen in the observed **enzyme kinetics**. Inhibition constants can be determined by detailed evaluation of the kinetics. It can also be determined separately whether inhibitors interact **competitively with the binding tRNA** or whether they also **compete with the exchange of the small base**. If a small inhibitor occupies the $preQ_1$ binding pocket after the tRNA is already bound and the guanine base has been removed, it will prevent the tRNA from further reaction. This is called noncompetitive inhibition. If the binding of such a small inhibitor occurs before the tRNA binds to the enzyme, it will act competitively with respect to the entire tRNA, just like a larger inhibitor. Overall, small inhibitors that fill only the guanine/$preQ_1$ binding

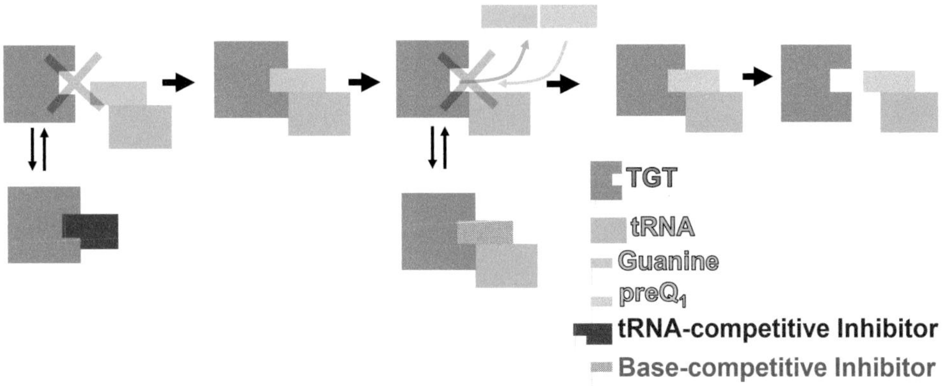

Fig. 21.4 The base-exchange reaction takes place in two steps. Inhibitors can compete with the binding of the complete tRNA (*left, dark gray*) as well as the exchange of the small nucleobase (*middle, light gray*)

pocket are "mixed" inhibitors. They inhibit the enzyme in two different ways and are, therefore, more efficient than larger inhibitors at the same binding strength.

21.5 LUDI Discovers the First Leads

In the beginning of the project, only the structure of the binary complex of TGT with $preQ_1$ was known. The two-step inhibition mechanism explained in the last section was also unknown at the time. During the course of the project Bernhard Stengl managed to clarify the details of this process. Ulrich Grädler used the binary TGT•$preQ_1$ structure as a reference and initiated a search for potential inhibitors with **LUDI** (a *de novo* design program; Sect. 20.10). He was able to find hits in a chemical catalog. The compounds listed in Fig. 21.5 were proposed. Among them, **21.3** proved to be a micromolar inhibitor. A crystal structure could be determined with this hit (Fig. 21.6). There was great delight when 4-aminophthalic acid hydrazide **21.3** was shown to bind to the enzyme exactly as LUDI had predicted.

Next, LUDI was consulted to predict further groups for the inhibitor that would fill in the as-yet unoccupied areas in the binding pocket. On the one hand, an expansion of the ring system by an additional aromatic ring was proposed. On the other hand, the placement of a nitrogen-containing heterocycle at the unoccupied interaction site near Asp 102 and Asp 280 was considered. Hans-Dieter Gerber synthesized derivatives **21.4**–**21.6** (Fig. 21.6). Compounds **21.4** and **21.5** achieved 10-times better inhibition of the enzyme in the assay than **21.3**. The results were quite different with the heterocyclic derivative **21.6**. It was significantly worse than the initial lead structure. Ulrich Grädler was able to solve the crystal structures with these inhibitors, which exhibited the expected binding mode. It was shown in the structure with **21.6** that the heterocycle falls very near the terminal amide

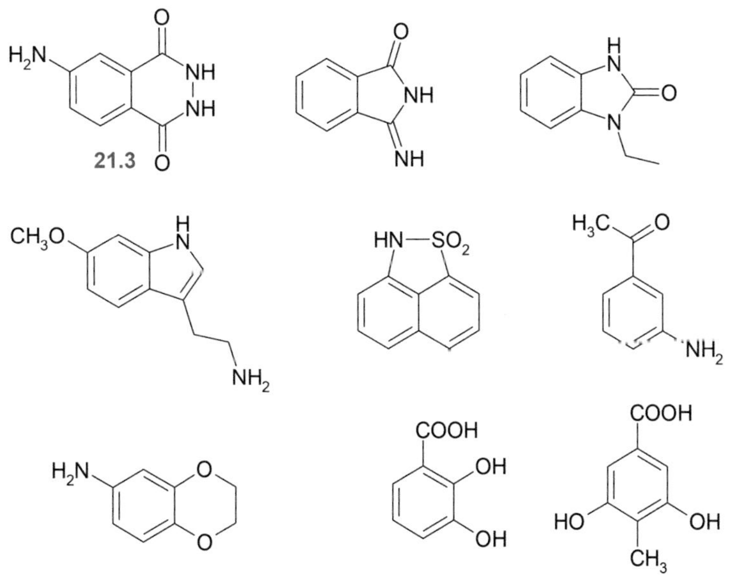

Fig. 21.5 Proposals for the first lead structures by LUDI. Among them, **21.3** proved to be a two-digit micromolar inhibitor

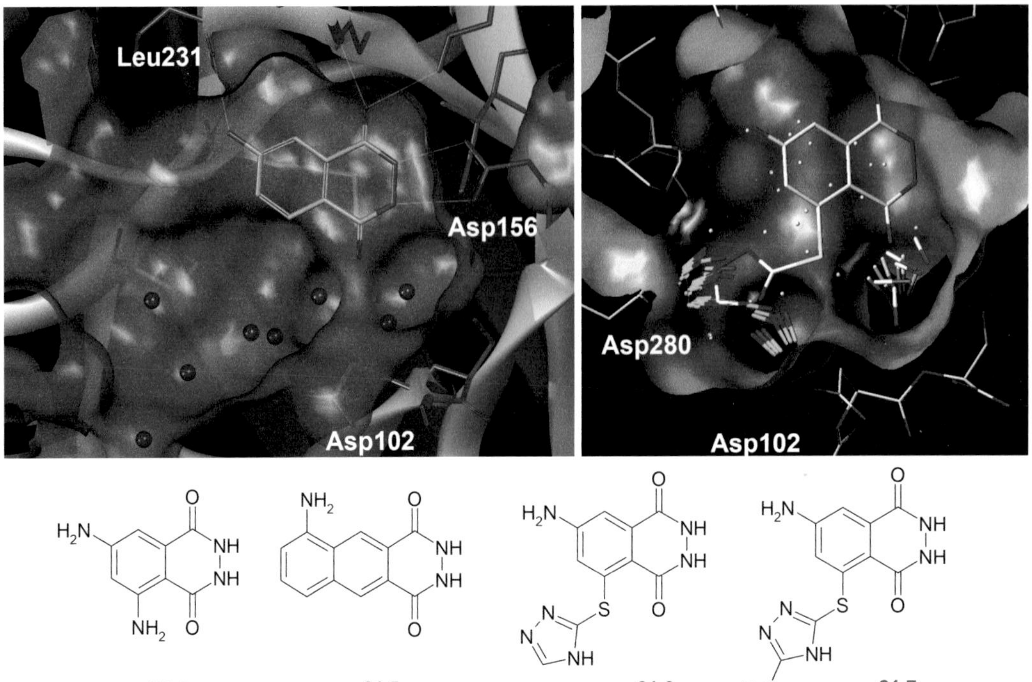

21.4 21.5 21.6 21.7

Fig. 21.6 Crystal structure of TGT with **21.3**, the first hit from LUDI. The agreement between the predicted (*left*) and the final experiment is almost perfect. LUDI indicated additional interaction centers in the lower part of the binding pocket that had not yet been used (*right*). Therefore, starting from **21.3**, the inhibitors **21.4** and **21.5**, which bind better by a factor of 10, were synthesized. The two derivatives **21.6** and **21.7**, which were extended with a heterocycle, occupy the still unused additional interaction sites even better. However, the two derivatives showed decreased binding affinity, presumably due to repulsive interactions with the two neighboring aspartate residues 102 and 280, since the heterocycles do not have the desired positive partial charge at the site of action. (▶ https://sn.pub/R6M5Uw)

group of Asn 70. It was then obvious that **21.6** needed an additional amino group to build an additional contact with the protein. This synthesis was accomplished, and the crystal structure with **21.7** in fact did show the expected binding mode with the additional H-bond. However, even this derivative was less potent than the original lead structure **21.3**. A more detailed analysis of the structural data showed that the appended heterocycle in **21.6** and **21.7** is disordered, and a hydrogen bond between the exocyclic amino group and the carbonyl group of Leu 231 is very long. The heterocycle was incorporated based on the idea that it would be beneficial to have a charged group that can also form hydrogen bonds to the two neighboring aspartate groups. These two groups were assumed to adopt a deprotonated state. Then a positive charge on the triazole group would be ideal for an interaction. But, which **protonation state** do these groups adopt? A pK_a measurement was carried out on a related model compound. A small-molecule crystal structure determination was undertaken on a crystal grown under the same buffer conditions as the protein complex was crystallized. Both experiments showed that the heterocycle exists without a charge, that is, both of the neighboring nitrogen atoms are deprotonated. Although it is not obligatory that the same protonation state is found in the protein's binding pocket, this model appears to be plausible to explain the decreasing binding affinity of **21.6** and **21.7**: An uncharged triazole ring between the two negatively charged aspartate groups must experience a repulsive interaction with at least one of the two acidic groups. This could reconcile the decreasing binding affinity, the observed disorder, and the elongated H-bond to the carbonyl group of Leu 231.

21.6 Surprise: A Flipped Amide Bond and a Water Molecule

Novo Nordisk kindly provided an additional compound, **21.8** that emulates the original interaction pattern of the initial lead structure (Fig. 21.7). Upon docking this derivative, however, it was shown that the distance between the polar nitrogen atom in the central pyridazinone ring and the carbonyl group in Leu 231 was too large. Nevertheless, the compound was a micromolar hit. The crystal structure that was determined with the related derivative **21.9** delivered an explanation. The **peptide bond**, which, **for mechanistic reasons, acts as a switch** between two conformations, takes on a different orienta-

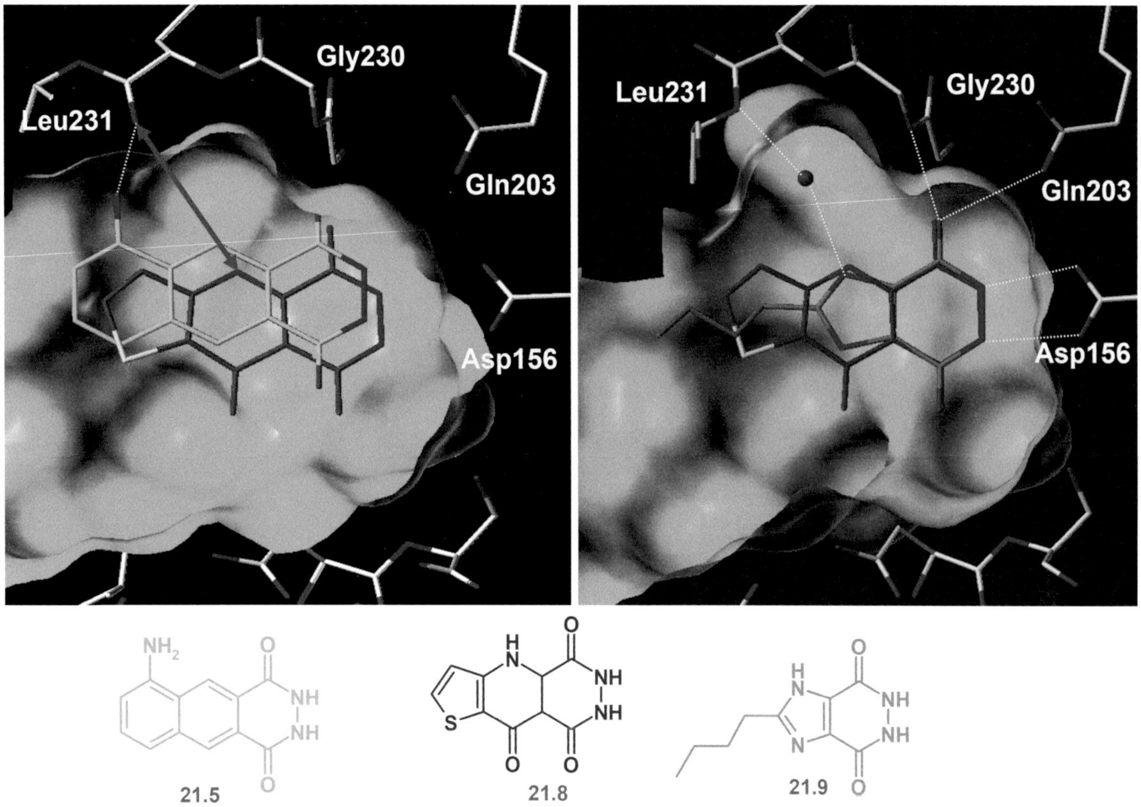

■ **Fig. 21.7** *Left* Analogue **21.8** should also emulate the interaction pattern of the original lead structure. If this derivative is placed in the binding pocket (*purple*), the distance between the polar nitrogen atoms in the central pyridazinone ring and the carbonyl group on Leu 231 seems to be too large for an H-bond. Nonetheless, **21.8** binds to the protein with micromolar affinity. *Right* The crystal structure that was determined with the very similar inhibitor **21.9** (*orange*) shows two surprises: The peptide bond rotates its orientation and now directs its NH group towards the binding pocket, and a water molecule (*red sphere*) mediates the interaction with the ligand!

tion! When flipped, the NH functional group is found in the binding pocket. The contact between this NH group and the polar nitrogen atom in the ligand is mediated by an **interstitial water molecule**. Because the details of the above-described enzymatic mechanism were not known at that time, the flipping of the peptide bond switch could not have been predicted. Furthermore, the incorporation of a water molecule was a big surprise. It underscores the importance of repeatedly determining crystal structures with newly found lead structures.

21.7 Hot Spot Analysis and Virtual Screening Open the Floodgate to New Ideas for Synthesis

How can multiple binding modes be made a virtue out of necessity? Ruth Brenk used the protein conformers in the structure with **21.3** as well as the geometry in the complex with **21.9** to carry out a **hot spot analysis** (Sect. 17.10). The result of this analysis is shown in ■ Fig. 21.8. A **virtual screening** (Sect. 7.6) was performed with the **generated pharmacophore** and this produced a plethora of alternative molecular scaffolds (■ Fig. 21.9) to occupy the guanine-binding site (■ Fig. 21.2). Many of the hits that were discovered in this way proved to be micromolar inhibitors. They afforded many new ideas for synthetic entry points to develop new inhibitors. Of these the pyridazinone (trione, **21.10**), pteridine (**21.11**), 6-amino-quinazolinone (**21.12**), and particularly the *lin*-benzoguanine scaffold (**21.13**), which was studied with the group at ETH Zurich, were investigated in detail.

Let us turn to the distribution of hot spots in the binding pocket. The new lead structures all interact at sites in the "upper part" of the binding pocket. However, an additional favorable binding area, capable of interacting with donor properties as well as hydrophobic moieties, is indicated in the "lower left part" of the binding site next to the two aspartic acid residues Asp 102 and Asp 280. These binding sites were not used in previous design. Considering the binding mode of the bound tRNA (■ Fig. 21.2), the ribose sugar moiety at position 34 is accommodated in this region. The hot spot analysis suggests a hydrophobic molecular fragment. A favorable site for an H-bond donor is located slightly above. This region corresponds to the binding site between the two aspartic acids, where the two heterocyclic derivatives **21.6** and **21.7** have already been placed. An-

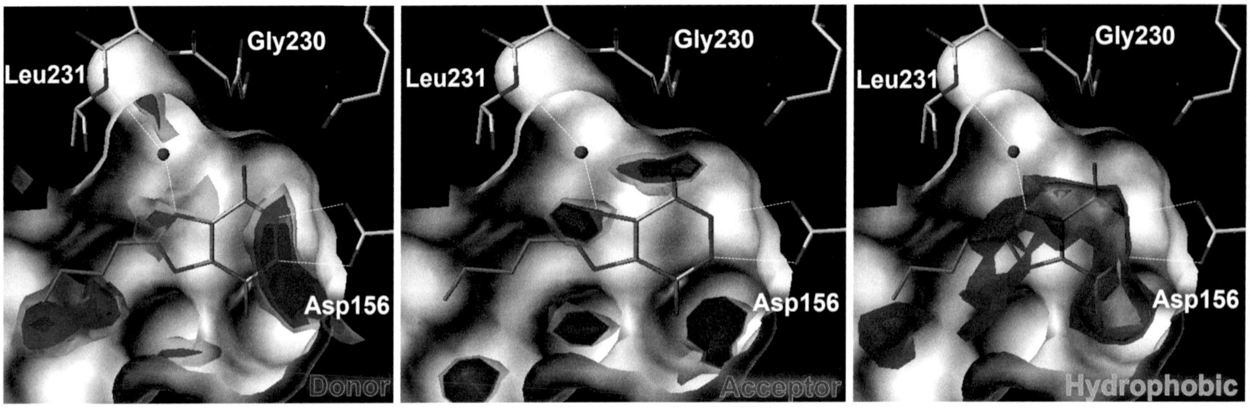

Fig. 21.8 Hot spot analysis shows preferred binding areas for a hydrogen-bond donor (*left*), acceptor (*center*), and a hydrophobic group (*right*). In addition, it was shown that the polar groups of **21.9** (cf. Fig. 21.7) fall into the preferred binding area. In the *bottom left* corner of the binding pocket (near the binding site of the ribose moiety Fig. 21.2, *blue*), other binding areas are indicated that were addressed in subsequent design steps

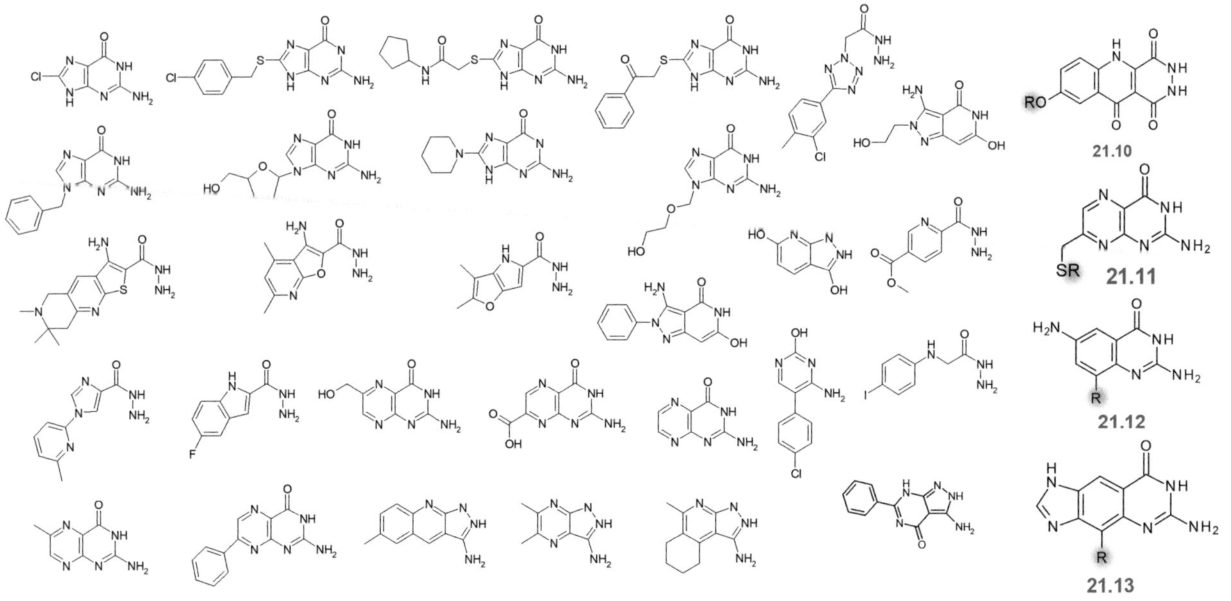

Fig. 21.9 A variety of suggestions from virtual screening, some examples of which were experimentally tested and found to be micromolar inhibitors. In particular, the pyridazinone (triones, **21.10**), the pteridine (**21.11**), the 6-aminoquinazolinone (**21.12**), and the *lin*-benzoguanine scaffolds (**21.13**) served as the first possible lead structures for further synthesis and optimization. By addition of suitable substituents R, numerous derivatives could be synthesized

other favorable area for an acceptor group is indicated at the rim of this pocket. The 2′ and 3′ hydroxyl groups of the tRNA ribose moiety are placed in this area.

21.8 The Filling of Hydrophobic Pockets and Interference with a Water Network

A golden rule in drug design is that the **occupancy of an empty hydrophobic pocket** with a lipophilic group leads to an increase in affinity (Sect. 4.9). Accordingly, inhibitors with such side chains were designed and led to the derivatives displayed in Fig. 21.10. Disappointingly, these showed only a modest improvement. In addition to the synthesis of the pteridines and aminoquinolinones developed in Marburg, the *lin*-benzoguanines **21.13** were advanced by Emanuel Meyer and Simone Hörner at the ETH in Zurich. But unfortunately, no really striking improvement in affinity could be found for any of the derivatives listed in Fig. 21.10. They occupy the small hydrophobic pocket between Val 45, Leu 68, and Asn 70, as planned and as shown in the crystal structure with **21.14**. Bernhard Stengl and Tina Ritschel took another

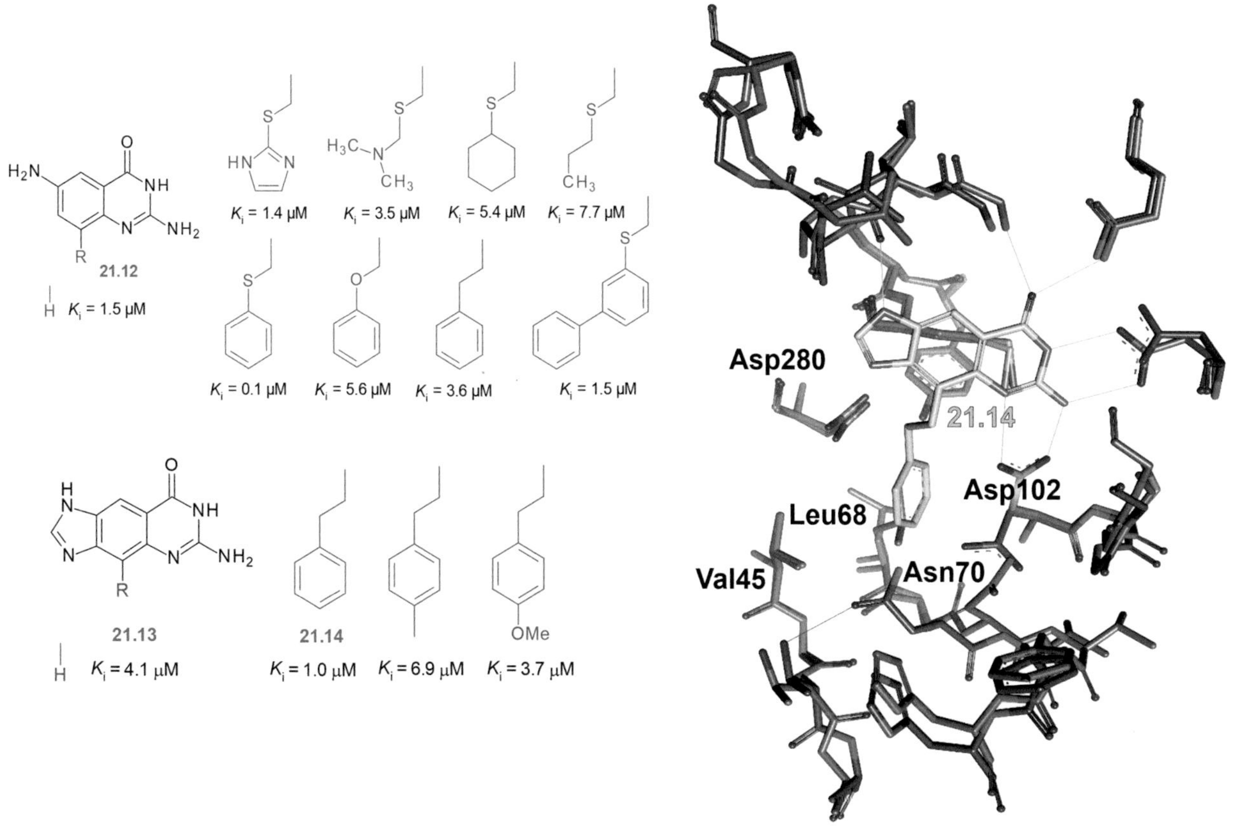

Fig. 21.10 From the 6-amino-quinazolinone scaffold **21.12**, the listed derivatives could be synthesized and tested by adding different substituents R (*red, upper left*). To our surprise, even the best compounds of this series remained in the single-digit micromolar range. As an alternative inhibitor scaffold, *lin*-benzoguanines **21.13** were provided with hydrophobic residues in the 4-position (*lower left*). Despite very good inhibition of the basic scaffold, the substituted derivatives failed to achieve a significant improvement in affinity. The crystal structure with **21.14** shows (*right*) that the indicated phenylethyl substituent binds into a small hydrophobic pocket of Val 45, Leu 68, and Asn 70. The ligand-bound structure is shown with *orange* carbon atoms and the ligand-free structure is displayed with *gray* carbon atoms. (▸ https://sn.pub/DNgZHZ)

look at individual derivatives. It was surprising that the small backbones without any side chains already showed single-digit micromolar binding. Adding another small substituent to the hydrophobic pocket initially led to a loss of binding affinity. This loss of affinity could only be compensated for by filling the hydrophobic pocket with an aromatic residue. A comparison of the arrangement of the water molecules in the different inhibitor structures was revealing. In the unsubstituted derivatives, several water molecules form a network between the two presumably charged aspartate residues 102 and 280 (Fig. 21.11, *left*). This network significantly contributes to the **residual solvation** of these two polar acid residues in the protein. Presumably, the water molecules in this region buffer the accumulation of negative charges on the two adjacent acid groups. All of the derivatives listed in Fig. 21.10 span this region of the **water network** with a hydrophobic linker in order to place their hydrophobic substituents in the small hydrophobic pocket at the end. In doing so, however, they inevitably destroy the water network. This has its price!

An affinity comparison between compounds **21.15** and **21.16** (Fig. 21.11) was striking. The derivative with a 7-dimethylamino group on the quinazolinone scaffold **21.15** lost binding affinity by a factor of more than 10 compared to the unsubstituted derivatives. Replacing one of the methyl groups with a benzyl group (**21.16**) partially restores the lost affinity. The crystal structure of this derivative shows that the benzyl group is not oriented towards the small hydrophobic pocket, but rather towards a pocket occupied by uracil 33 in the natural substrate (Fig. 21.2, *green* pocket). With this result, a new concept for further design was obvious. Under no circumstances should the water network between Asp 102 and Asp 280 be crossed by a hydrophobic linker. Furthermore, a hydrophobic group should be added facing the uracil 33 pocket to allow the ligand scaffold to grow into this pocket.

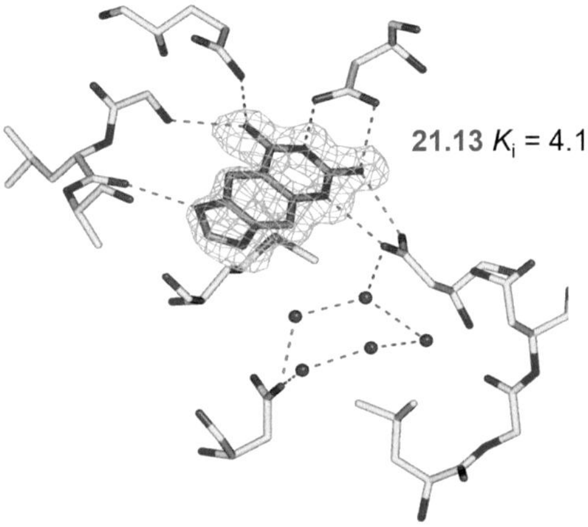

21.13 K_i = 4.1 µM

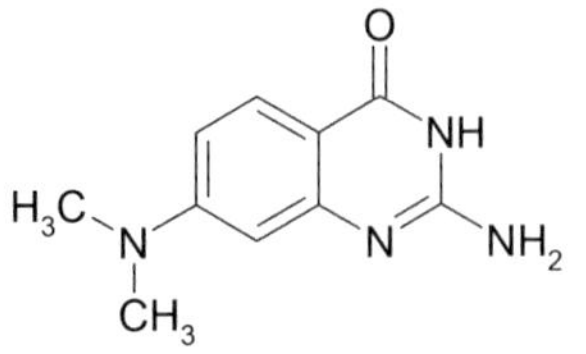

21.15 K_i = 31.0 µM

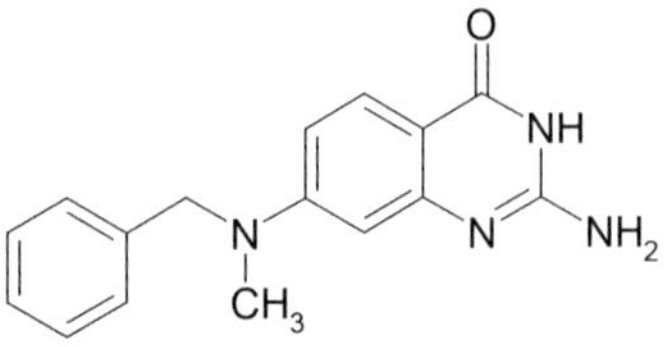

21.16 K_i = 7.6 µM

Fig. 21.11 *Left* The backbone of *lin*-benzoguanine **21.13** binds to the protein with 4.1 µM, leaving the water network (*red spheres*) between the two presumably negatively charged aspartates 102 and 280 intact. *Right* The binding affinity of the quinazolinone derivative with a 7-dimethylamino group (**21.15**) decreases by a factor of 10 compared with the unsubstituted derivative. However, when one of the two methyl groups is replaced by a benzyl group (**21.16**), an increase in activity is again obtained. The crystal structure with this derivative showed that the benzyl group is not oriented towards the small hydrophobic pocket (*blue* pocket in Fig. 21.2), but projects into the uracil 33 pocket (*green* pocket in Fig. 21.2)

21.9 With a Salt Bridge: Finally Nanomolar!

Synthetically, the desired modifications of the substituent were easier to achieve on the *lin*-benzoguanine parent scaffold. Unsubstituted *lin*-benzoguanine **21.13** displays a water network containing five distinct water molecules in the crystal structure with the enzyme (Fig. 21.11, *left*). It was decided not to change this network in the next design cycles, so that derivatization first concentrated on the 2-position and the study of the 4-position was postponed (see below). The sole attachment of a methyl group to the 2-position (**21.17**) improves affinity by a factor of 2.7 (Fig. 21.12).

If the methyl group is then exchanged for an amino (**21.18**) or methylamino group (**21.21**), the binding constant will dramatically improve into the two-digit nanomolar range. The introduction of an amino group in the 2-position of the *lin*-benzoguanine scaffold improves the affinity by a factor of 50! How can this surprising result be explained? The hydrogen bond to the carbonyl group of the main chain in Leu 231 was discussed in Sect. 21.5. This functional group is part of the peptide bond, which can flip its orientation like a switch. The *lin*-benzoguanine scaffold also forms a hydrogen bond to this carbonyl group of Leu 231, and the introduction of an amino group in the 2-position transforms the imidazole portion into a guanidine-like moiety (Fig. 21.13, *center*, *red*). Such a change increases the basicity of the scaffold and possibly alters its protonation. However, this is not the only guanidine-like group in the molecule. The aminopyrimidone moiety also contains such a group (Fig. 21.13, *center*, *blue*). Measurements of **pK_a values** in aqueous solution indicate that the aminoimidazole is the group with the higher pK_a value, so this group should be protonated first. This was experimentally verified by Manuel Neeb. As described in Sect. 4.4, isothermal titration calorimetry can be used to determine how many protons are transferred during ligand binding to the protein. The titration was carried out at pH 7.8. In total, **21.18** takes up one proton when it binds to the protein. But where does the proton go? In **21.17**, the guanidine moiety in the imidazole ring is missing; nevertheless, the ligand still takes up a proton (Fig. 21.13, *left*). In contrast, no proton is taken up when the guanidine group in the pyrimidone moiety is removed (**21.20**). These results suggest that the proton is taken up by the aminopyrimidone moiety, even though the pK_a of this group is lower by 1.3 pH units! At first sight, this seems to be very contradictory. To be sure, we replaced the two charged aspartates 102 and 156 in the binding pocket of the TGT with uncharged asparagines by mutagenesis (Fig. 21.13, *right*). Interestingly, this eliminates the uptake of a proton by **21.21**, again emphasizing that the effect occurs at the aminopyrimidone moiety. The observed protonation step is induced by the two adjacent and negatively charged Asp residues, which **locally shifts the pK_a values** quite strongly.

However, this does not explain the sudden increase in affinity due to the introduction of an amino group on the *lin*-benzoguanine scaffold in the 2-position. This observation would have been easily explained by the more basic nature of the aminoimidazole moiety, which would have resulted in a charge-assisted hydrogen bond to the carbonyl group of Leu 231. Further structural studies were performed at different pH values to rule out the formation of this charge-assisted interaction. Finally, any

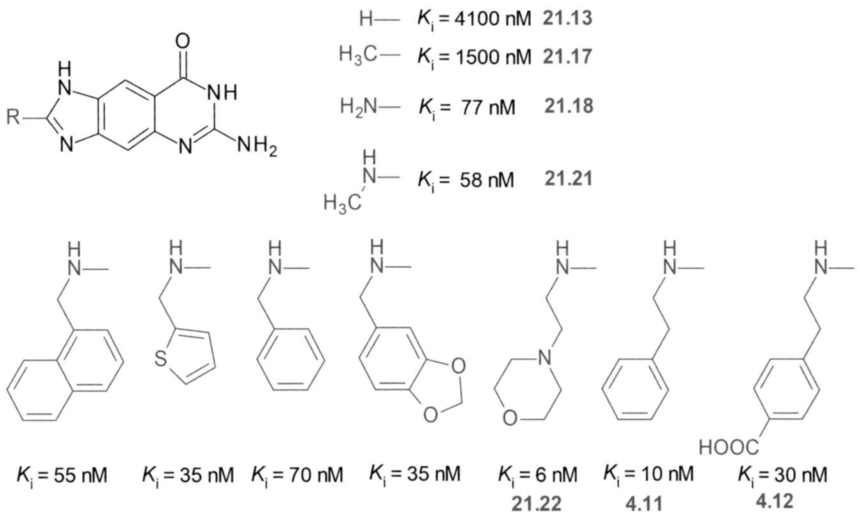

Fig. 21.12 Substitution of the *lin*-benzoguanine parent scaffold **21.13** in the 2-position leads to a significant improvement in binding affinity. In particular, the introduction of a 2-amino group (**21.13** ⇨ **21.18**) leads to a tremendous increase in binding affinity. Larger substituents are accommodated by the U33 pocket, but are mostly disordered there. The morpholino derivative **21.22** represents a single-digit nanomolar inhibitor. The attempt to introduce another affinity-increasing salt bridge via **4.11** to **4.12** has no effect, since the salt bridge remains exposed to the solvent (cf. Sect. 4.8, Fig. 4.13)

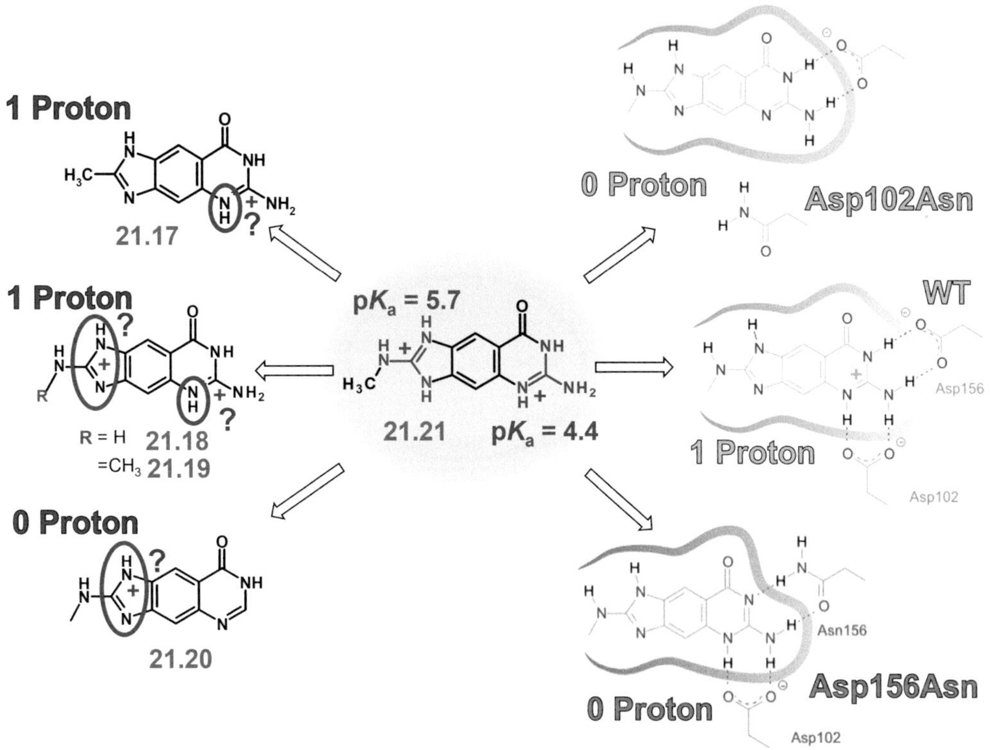

Fig. 21.13 The parent scaffold of the *lin*-benzoguanine derivative **21.21** (*center*) has two guanidine-like groups that account for the basic character of the compound. The aminoimidazole moiety (*red*) has a pK_a of 5.7 in water, while the aminopydimidone moiety (*blue*) has a pK_a of 4.4. Therefore, protonation should occur more readily at the imidazole moiety. ITC measurements (*left*) show the uptake of one proton upon binding to the wild-type TGT. For **21.17** (*top left*), which cannot accept a proton at the imidazole moiety, this still results in the uptake of one proton upon protein binding. In contrast, no proton is taken up upon binding of **21.20** (*bottom, left*). This suggests that despite the lower pK_a, the aminopyridmidone moiety is protonated. Mutagenesis and crystal structure determination confirm these data (*right*). Only the binding of **21.21** to the wild type, in which two negative charges occur in close proximity at Asp 102 and Asp 156, shows proton uptake. If Asp is replaced by Asn by mutagenesis, one negative charge on each side of the protein variants is lost, and **21.21** binds to the protein without picking up a proton

additional protonation effects on the side of the protein that could have disturbed the found protonation inventory could be ruled out. Thus, the observed increase in affinity could only be explained by the **absence of repulsive effects** in the **formation of the two parallel hydrogen bonds** (◘ Fig. 21.14). If the donor or acceptor groups are on the same side of the binding partners that come together to form the complex, no additional repulsive effects between the hydrogen atoms in the twinned H-bonds need to be overcome during complex formation (◘ Fig. 21.14, *left*). The repulsive effects are also present there, but they had to be overcome already during the synthesis of the individual components. Therefore, they are not important for the balance of complex formation. If, on the other hand, donor and acceptor groups arrange in alternate fashion on the partners, the repulsive effects occur during complex formation. Since these effects come at a cost during complex formation, they will weaken the strength of the H-bonds that are formed. This effect is surprisingly strong and it has been described for the first time in host–guest complexes. Interestingly, such effects also occur during H-bond formation between the nucleobases on the individual steps of DNA (Sect. 14.9). For the optimization of our TGT inhibitors, this was of course an extremely welcome additional affinity contribution!

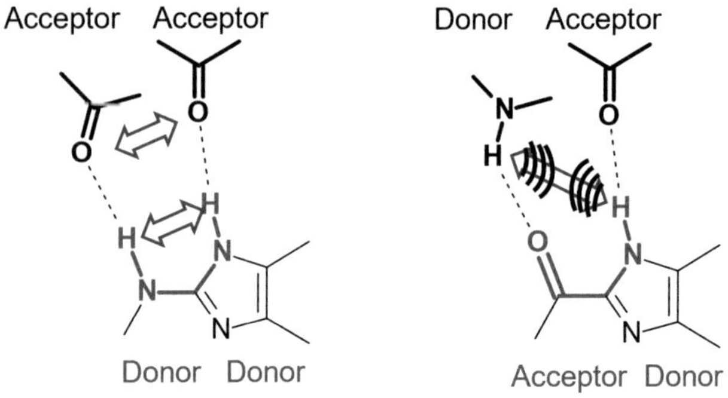

◘ **Fig. 21.14** When two binding partners come together to form two parallel hydrogen bonds, it is crucial for the contribution to the strength of these H-bonds whether the two donors and acceptors are on the same side (*left*) or on opposite sides (*right*) of the binding partners. *Left* There are no repulsive interactions of the H-atoms to overcome during complex formation. Any repulsive effects had to be exceeded during the synthesis of the binding partners and, therefore, do not have to be paid for during complex formation. *Right* These repulsive effects arise during complex formation once the H-bonds are formed. This weakens the affinity contribution to be achieved during complex formation

It has already been demonstrated that filling the uracil 33-binding pocket is associated with an improvement in the affinity. Therefore, groups were introduced onto the 2-amino group. However, a methylene group was used as a linker to keep the amino group electronically unconjugated to the added aromatic substituents. Of the synthesized derivatives, morpholine derivative **21.22** proved to be the strongest binder. It also has the best water solubility. Interestingly the added side chains in this area are not clearly visible in the electron density. They are probably in a disordered state in the binding pocket (◘ Fig. 21.17, *left*). This speaks **against a good enthalpic interaction** for these groups in this area, but this effect should be compensated for due to **entropic reasons** so that a good contribution to the free energy is achieved in the sum, and overall the binding affinity is improved. This situation is explained in an example in Sect. 4.10.

After optimization of the substituents at the 2-position, an enlargement of the molecular scaffold of the *lin*-benzoguanines at the 4-position was approached. Special attention was paid to the **perturbation of the polar interactions with the water network**. This network buffers the charges between Asp 102 and Asp 280 (◘ Fig. 21.11, *left*). Purely hydrophobic side chains had not yielded the desired increase in affinity (◘ Fig. 21.10). Therefore, a basic nitrogen atom was introduced into the **linker to actively participate in the water network**. As a result, single-digit nanomolar inhibitors have also been successfully developed (◘ Fig. 21.15). Crystal structures **21.23** and **21.24** (◘ Fig. 21.16, *top*) show that the water network is successfully incorporated and stabilized. The introduced polar nitrogen probably binds to Asp 280 in protonated form via a salt bridge. It also participates in the water network.

However, another phenomenon was observed, the significance of which only became clear as the project progressed. The two inhibitors differ by only one mem-

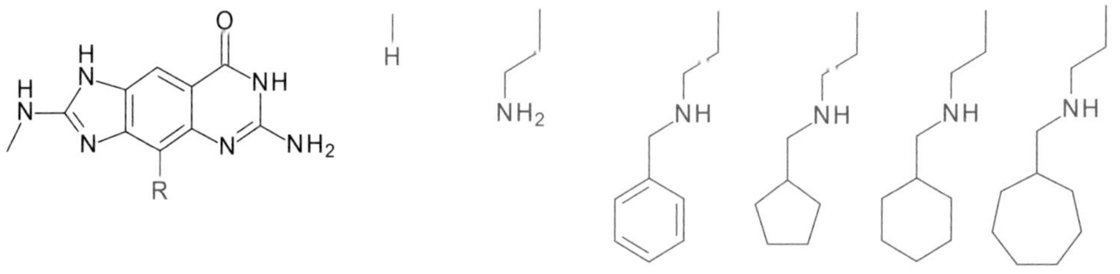

◘ **Fig. 21.15** Substitution of the *lin*-benzoguanine parent scaffold **21.21** in the 4-position by a basic nitrogen in the side chain leads to a significant improvement of binding affinity down to the nanomolar range. Terminal cycloaliphatic groups proved to be the best representatives of the series (**21.23**, **21.24**)

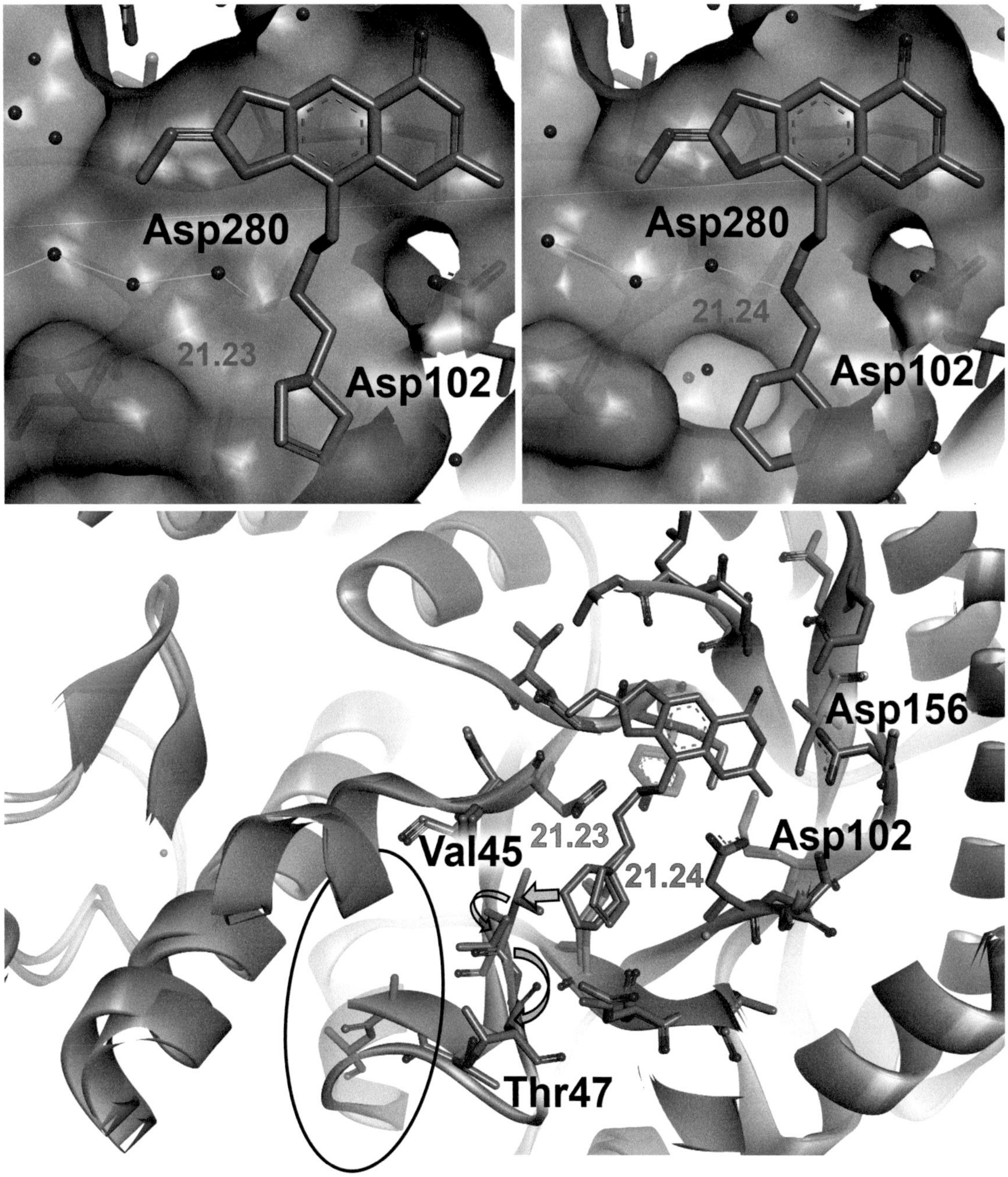

Fig. 21.16 The crystal structures with the 4-substituted *lin*-benzoguanine derivatives show that the basic nitrogen in the chain is incorporated into an H-bonding network with the water cluster and with Asp 280 (*top*). At first glance, **21.23** and **21.24** appear to adopt an almost identical binding mode. On closer inspection, however, there is a crucial difference that, as was shown later, is of great importance for the stability of the protein (*bottom*). The cyclohexyl ring (*blue*) in **21.24** occupies a slightly larger space than the five-membered ring (*ochre*) in **21.23**, displacing Val 45 from its position in the uncomplexed protein (see *red arrows*). The cyclopentyl ring in **21.23** is slightly smaller, so it does not cause this displacement. When Val 45 is dislocated, it continues like a series of dominoes, first on Thr 47 (*red arrows*) and then on the entire loop with the attached helix, which becomes disordered at the end. It can no longer be seen in an orderly fashion in the crystal structure (in the *blue* structure, this part is missing in the region of the *ellipse*, whereas it is clearly visible in the *ochre* structure). As will be shown later, this breakdown of the ordered structure of the loop–helix motif has a massive influence on the dimer stability of the enzyme. (▶ https://sn.pub/eM2mQk)

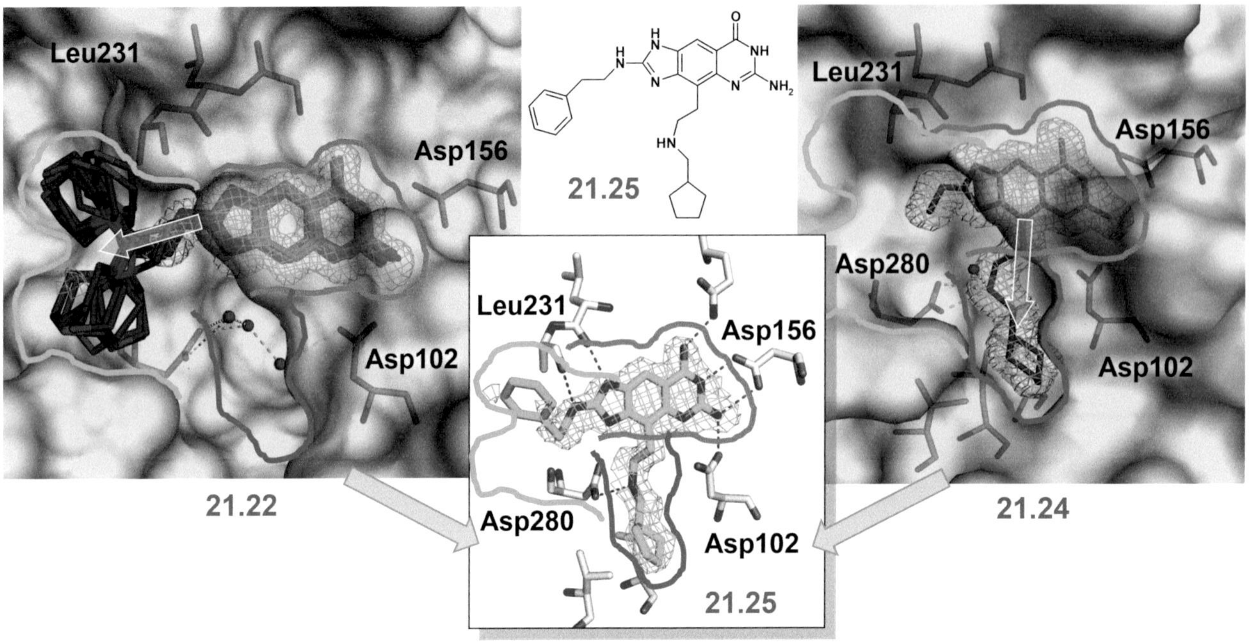

Fig. 21.17 The crystal structure of morpholino derivative **21.22** with a substituent in the 2-position (*left*) shows no well-defined difference electron density (*green mesh*) around the morpholino side chain in the region of the uracil 33 pocket (*green arrow*). This observation suggests a strong disorder of this substituent scattered across multiple spatial orientations. MD simulations confirm this hypothesis and indicate two possible placements of the side chain. By incorporating a basic nitrogen atom into the side chain at the 4-position of the *lin*-benzoguanine scaffold in **21.24**, the hydrogen-bond network between Asp 102 and Asp 280 can be actively incorporated and does not lead to a collapse of the binding affinity by disrupting the water network (*right*). **21.24** is clearly seen in the differential electron density. It forms H-bonds to Asp 280 and fills the small hydrophobic pocket (*blue arrow*). Combining the results of the 2- and 4-position substitutions in a derivative such as **21.25** yields subnanomolar inhibitors (*center*). Again, the 2-side chains showed increased mobility in the uracil 33 pocket

ber in the terminal cycloalkyl substituent, a five- or six-membered ring. The slightly smaller five-membered ring fits easily into the ribose-34 pocket (*blue* pocket, see Fig. 21.2, *bottom*), leaving the protein virtually unchanged from the uncomplexed structure. In contrast, the six-membered ring requires more space and exerts steric pressure on the side chain of Val 45 (Fig. 23.16, *bottom*, *red arrows*). This triggers a domino-like cascade of displacements. As a result, the following residues (see Thr 47) also shift their positions, and finally the entire loop–helix motif from sequence position 45 to 63 is disordered. Structurally, this section of the protein collapses. Much later, we realized that this loop–helix motif is crucial for the stability of the enzyme. As we will see in the next chapter, TGT can only perform its function as a dimer. Ligands can, therefore, interfere with the quaternary structure of the protein via their side chains. This serious effect is either induced by ligands or it is absent and, surprisingly, it is induced by ligands that differ by only one methylene group in their terminal ring.

21 However, for the design of inhibitors for the active pocket, it was important to see that the addition of substituents in both the 2- and 4-positions led to single-digit nanomolar inhibitors. As a further approach, Luzi Barandun and Florian Immekus attempted to add substituents at both positions and characterize them structurally (Fig. 21.17). An additional increase in potency was observed with compounds such as **21.25** in the subnanomolar range. Thus, the inhibitory effect was below the limit of reliable detection by the enzyme assay. From an affinity optimization point of view, the initial goal was achieved. However, the compounds proved to be very large and polar. Therefore, they did not have the desired properties for **sufficient bioavailability**. On the one hand, ligands with sugar moieties in the side chain attached in the 4-position were used for further design (cf. **21.27**, **21.28**, Fig. 21.20). On the other hand, a prodrug strategy was pursued with the smaller ligands (not described here).

21.10 Surprise: The Enzyme is Only Functional as a Dimer

Previous inhibitor design focused solely on blocking the catalytic center to eliminate the function of the enzyme. This ignored the fact that the protein must interact with the tRNA. In the determined crystal structures, there is always a pair of two protein molecules aligned with each other due to a twofold rotation axis in crystal packing (Fig. 21.18, *yellow box*). However, this does not necessarily indicate that TGT has to exist as a **dimer in**

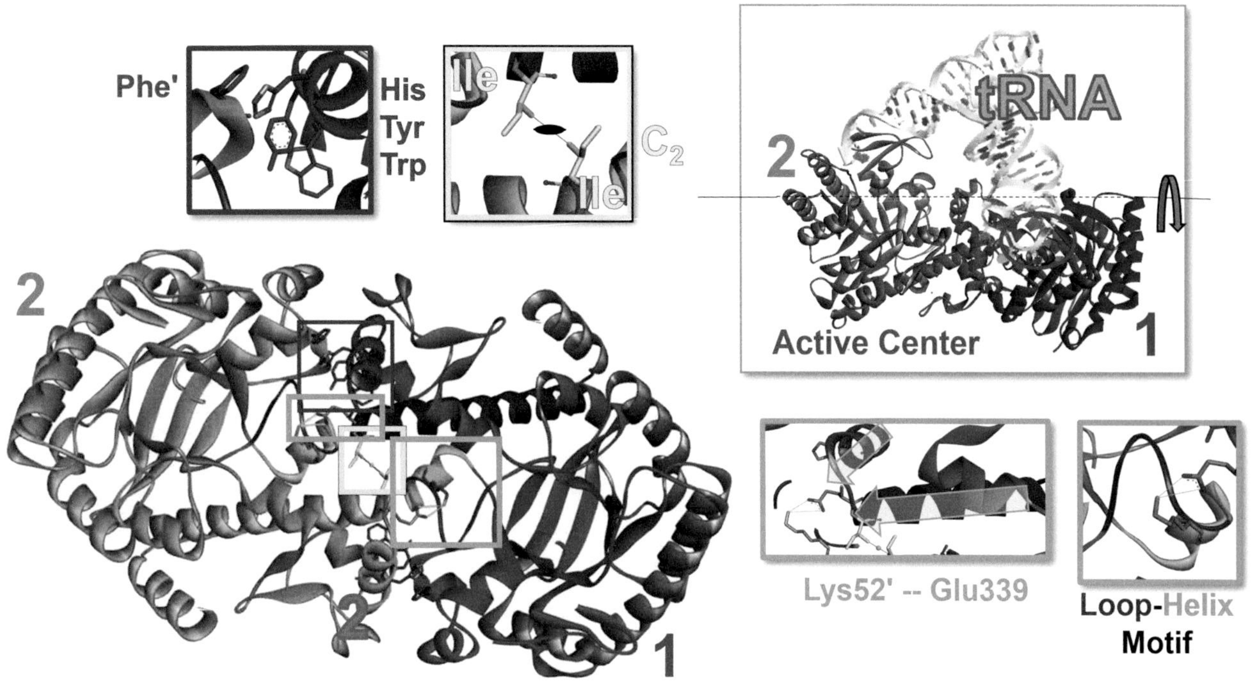

Fig. 21.18 The enzyme TGT forms a 2:1 complex with tRNA (*top right*). One monomer (*dark blue*, **1**) catalyzes the base exchange in the tRNA. The second monomer (*dark green*, **2**) holds the substrate in position for the reaction. The homodimeric arrangement has twofold *C2* symmetry and is located on a rotational axis in the crystal structure (*yellow box*). A cluster of four aromatic amino acids (*red box*, Trp 326, Tyr 330, His 333, and Phe 92′ of the neighboring monomer **2**) is crucial for the stability of the dimer. For reasons of symmetry, this cluster occurs twice, as do all the other patterns. A salt bridge of Glu 339···Lys 52′ spans the contact surface (*light blue box*). A motif consisting of a loop (*violet*) and a helix (*light green*), which occurs repeatedly with different geometries in the numerous crystal structures with different ligands (see Fig. 21.20, *left*), proves to be important for the formation of the dimer contact (*green box*). (► https://sn.pub/oZX57w)

solution in order to function. This picture changed when a structure of the enzyme with a 20-base RNA oligonucleotide became available. This nucleotide contains the anticodon loop where the base exchange takes place. In this crystal structure, the enzyme binds only one RNA molecule as a dimer, although each monomer unit contributes a complete catalytic center. These centers are both located on the same side of the homodimer. Therefore, for steric reasons, simultaneous binding of two tRNA substrates would be impossible. Obviously, **one monomer unit performs catalysis**, while the **second monomer is responsible for the correct positioning of the tRNA substrate** for the enzyme reaction (Fig. 21.18, *upper right inset*).

Next, we were interested in what determines the stability of the contact interface between the two monomer units. Each monomer consists of 385 amino acids. The question of stability can be investigated by targeted exchange (mutagenesis) of individual amino acids in the contact interface. This can lead to local destabilization of the protein contact. In this approach, we pursued the working hypothesis that the function of this enzyme can also be disrupted if the stability and, thus, the formation of the **dimer structure is blocked**. This would be an alternative strategy to inhibit the function of TGT. But which residues are crucial and how do they contribute to stabilization? The contact area of both monomers spans more than 1600 Å^2 and is formed by 43 amino acids with 10 hydrogen bonds each, 14 salt bridges, and 188 individual van der Waals contacts. Computer simulations can help at this point. Of course, it is much easier to exchange amino acids in the computer than experimentally by targeted mutagenesis. Stephan Jakobi first carried out such a computer screening to find the "**hot spots" of contact surface adhesion**. To do this, amino acids are replaced one after the other and MD simulations are run repeatedly. The contribution of each amino acid to the stability of the contact surface in the mutated variants is then analyzed. In this way, Stephan Jakobi discovered a cluster of four aromatic amino acids that turned out to be important for the stability of the dimer (Fig. 21.18, *red box*). In addition, a salt bridge from Glu 339 to Lys 52′ spans the contact surface (Fig. 21.18, *light blue box*). To differentiate between the two monomers, a dash (′) is always added as label to the second monomer in the following. In addition, the **loop–helix motif** mentioned above (Fig. 21.18, *green box*), which was repeatedly found in the crystal structures with deviating geometry,

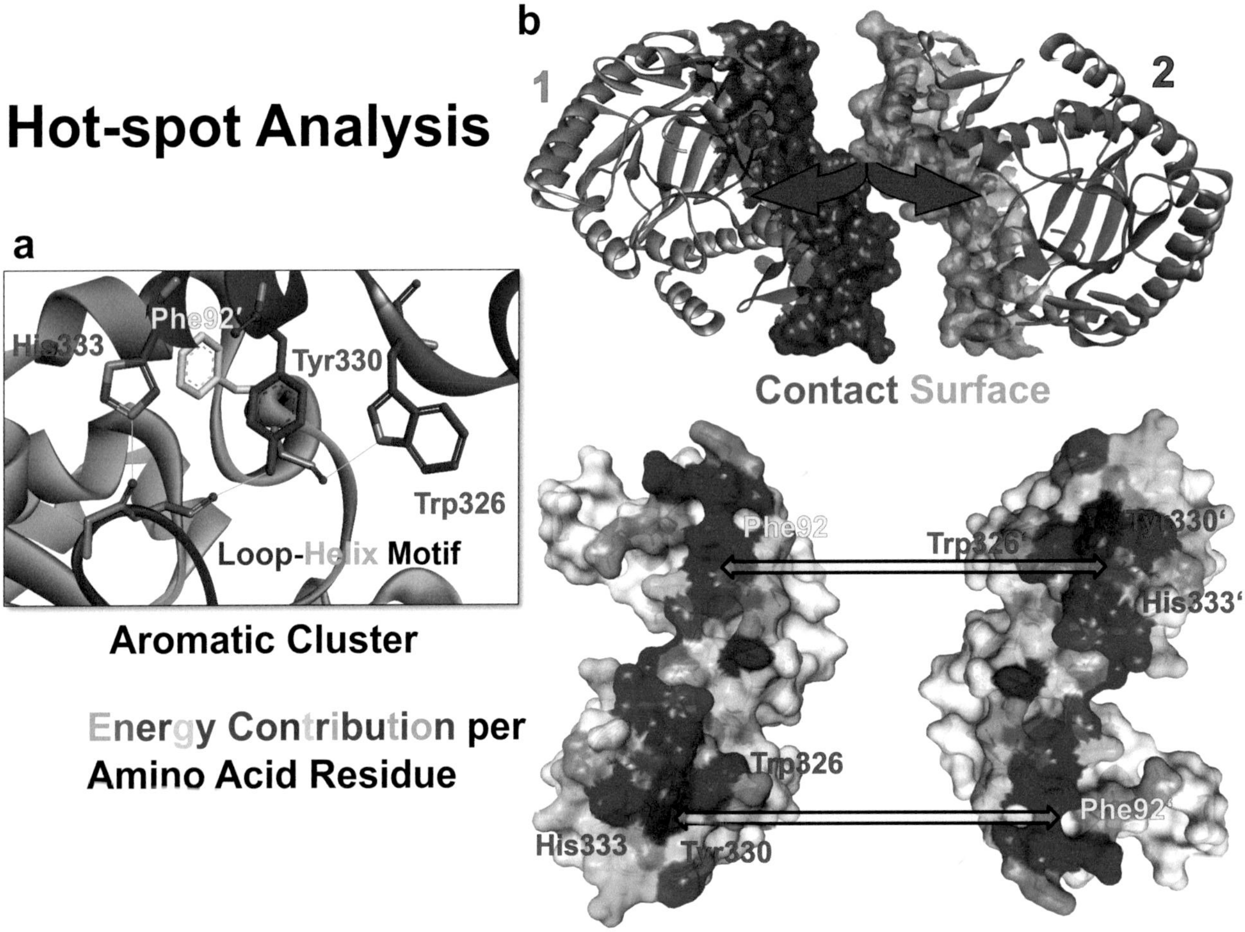

Fig. 21.19 Results of a hot spot analysis of the contact area covering more than 1600 Å^2. The stability contributions are summarized by a *bluish* color coding (*top right*, *red*/*green*, below unfolded in top view with color coding). The more intense and darker the *blue*, the stronger the contribution of a single amino acid to the contact (*bottom right*). A cluster of four aromatic amino acids (Trp 326, Tyr 330, His 333, and Phe 92′ (*yellow*) on the adjacent dimer) stands out. This cluster is shown in detail on the *left*. Note that the three residues Trp 326, Tyr 330, and His 333 each form a hydrogen bond (*green line*) across the interface contact surface to amino acids on the loop–helix motif of the neighboring monomer. (▶ https://sn.pub/og37OX)

stood out. This led to the assumption that this motif plays an important role in the formation of the dimer contact (Fig. 21.19).

But how can we experimentally determine whether and to what extent the dimer dissociates in solution? In collaboration with Sarah Cianférani's group in Strasbourg, France, we used native **nanoESI mass spectrometry**. In this method, the protein is transferred from an equilibrium solution into the gas phase and its mass is determined from the intact protein molecule. If two or more species are present, they can be identified by their different masses. Since the transfer to the gas phase is extremely fast, it is assumed that a correct representation of the mass ratios in solution is obtained.

21

21.11 Site-directed Mutagenesis: What Binds the Dimer Together

The first step was to successively exchange individual residues of the aromatic cluster: Trp 326 for glutamate, Tyr 330 for cysteine, His 333 for aspartate, and Phe 92′ on the other monomer also for cysteine. The choice of these residues was guided, on the one hand, by our calculations. We also wanted to introduce more polar residues to improve the solubility of the protein in the monomeric state. In the wild type, the contact surface is largely hydrophobic, so we expected a decrease in solubility when this surface is exposed in the monomeric state. Finally, when selecting the cysteines, we had in mind the idea of **introducing new attachment points** for subsequent chemical modification

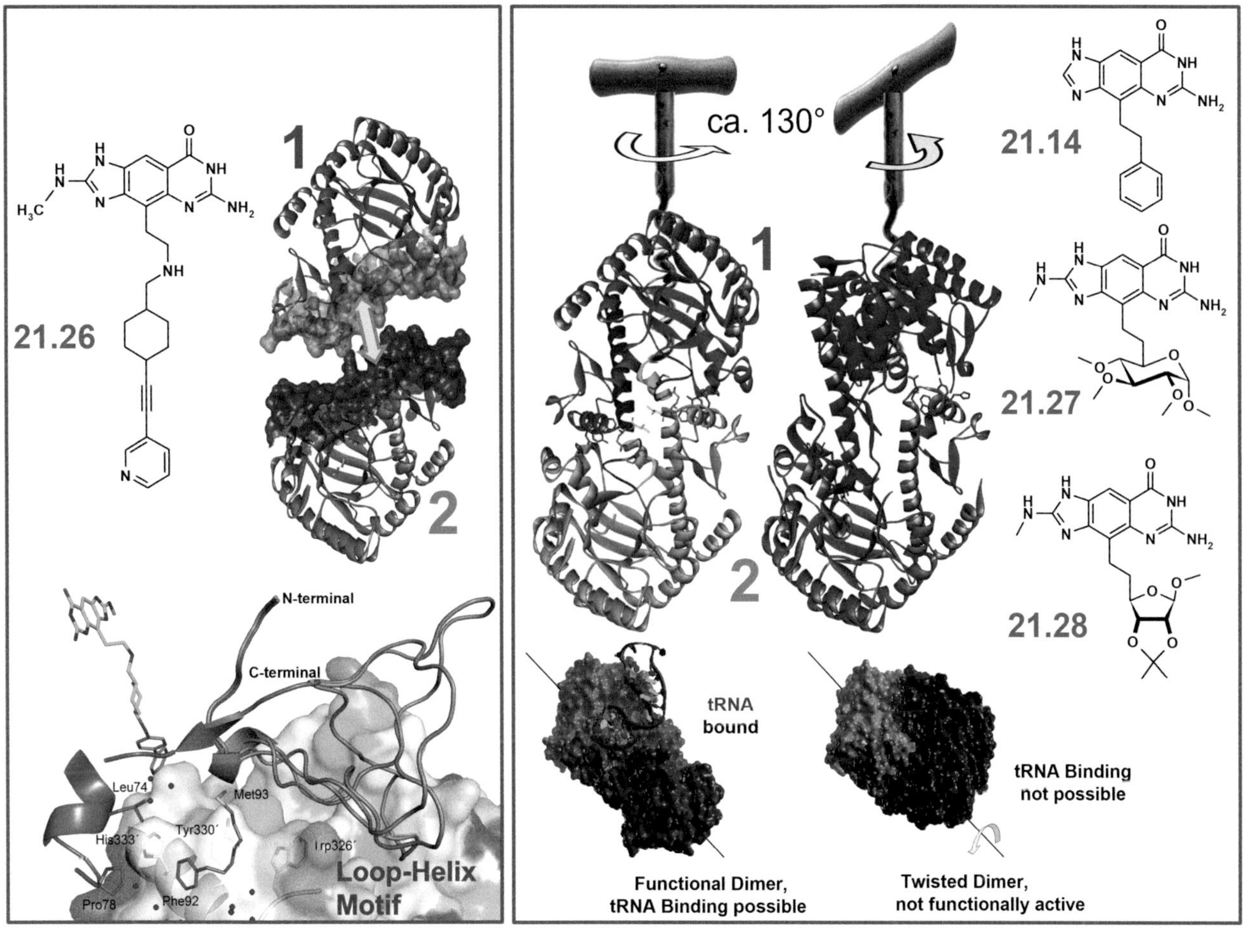

Fig. 21.20 *Left* Since the catalytic center and the flexible loop–helix motif (*bottom left*: *orange*, shown in different conformations) are close to the dimer contact surface, inhibitors of the catalytic center could be designed with long, spike-shaped substituents such as **21.26** that protrude into the region of the aromatic cluster and the Glu 339···Lys 52′ salt bridge. They disrupt the geometry in this region. Their binding increases the dissociation of the dimer into monomers. *Right* With ligands **21.14**, **21.27**, and **21.28**, we had the surprising result that two differently packed dimers crystallized side by side. In one, we found the usual functional dimer. In the second, the monomer units were packed together in an altered form, which formally corresponded to a rotation of one monomer unit with respect to the other one by about 130°. While the functional dimer is able to bind tRNA, the twisted dimer is unable to do so for steric reasons (*bottom right*). (▶ https://sn.pub/b30k5W)

to the dimer contact surface near the aromatic cluster. We planned to "tether" small fragments to the thiol groups of the cysteines via a chemical reaction (see Fragment Tethering Approach, Sect. 7.10). During the course of the project, the cysteines introduced in this way proved to be extremely instructive, as they opened the way to a modified strategy in the search for small ligands (Sect. 21.15). In a further step, we extended our mutagenesis to the salt bridge residues Glu 339···Lys 52′ and Trp 95. For this purpose, double mutants have also been created.

While the wild type featured hardly any monomer in solution, the monomer fraction increased in all variants. This even led to mutant variants that were largely monomeric in solution. In the crystal structures of these variants, however, we still found the *C2* symmetric homodimer with only some rearrangements in the region of the exchanged amino acids being observed. In addition, a few water molecules had crept into the structure in all cases. They probably indicate how the destabilization of the contact is initiated in aqueous solution. The fact that we found the homodimer again and again in the crystals can be understood as a consequence of chemical equilibrium. As the local concentration of protein in solution increases, as is the case at the surface of a growing crystal, equilibrium automatically shifts towards the dimer (Le Chatelier's principle). Therefore, only the dimer could be obtained from the crystallization solution. When we performed mass spectrometric experiments with solutions of increasing protein concentration, we also found an increase in the dimer fraction in solution. This underlines the concentration dependence of the dissociation in the monomer/dimer equilibrium.

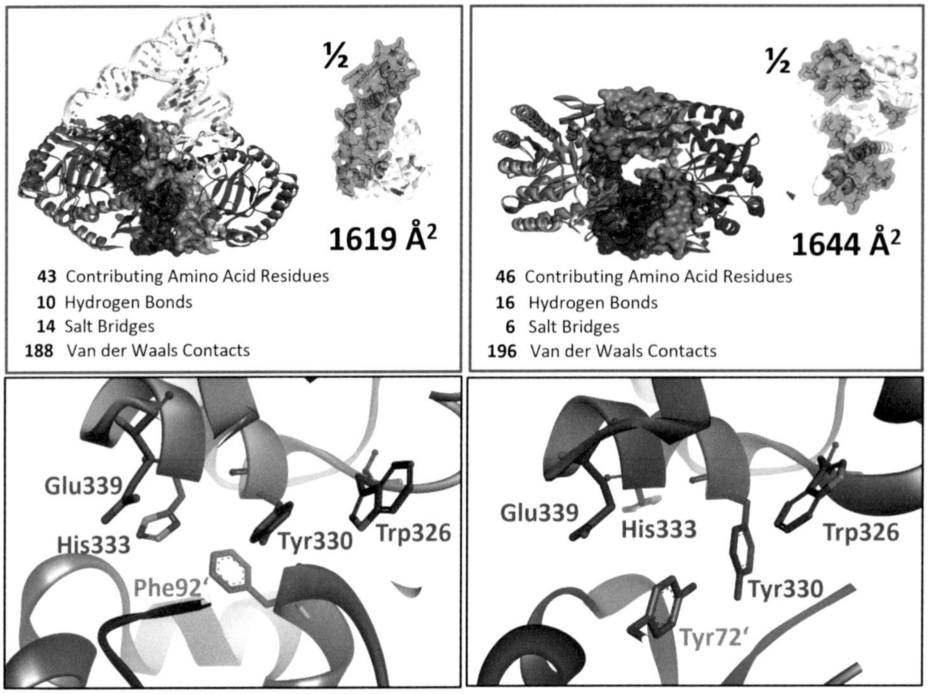

Fig. 21.21 *Left* The functionally active homodimer of TGT is shown in the *upper left*. It is able to recognize and catalytically convert a tRNA molecule. *Right* The new twisted homodimer packing induced by ligand binding is shown on the right. It locks the enzyme into a geometry that can no longer bind a tRNA molecule. The two dimers formally differ by a rotation of about 130° of the two structurally nearly unchanged monomer units with respect to each other. The new contact area in the twisted dimer differs significantly from that in the catalytically active dimer, although both are similar in size (in Å²) and compositional inventory. However, the contact area in the catalytically active dimer is contiguous. In the twisted dimer, the interface is divided into four separate sections (compare *red surface* fractions). An important component of the interface in the catalytically active dimer is the completion of the aromatic cluster of Trp 326, Tyr 330, and His 333 by the addition of Phe 92′ from the other dimer partner (*bottom left*). In the twisted state, the role of Phe 92′ is taken over by Tyr 72′ from the dimer partner, resulting in a comparable aromatic cluster (*bottom right*)

In the context of these stability studies of the contact surface interface, we were also interested in **how fast the homodimer exchanges its monomer subunits** with each other. For this purpose, we mixed a form of TGT still bearing the so-called Strep-tag®II from a purification step at the *N*-terminus with an equimolar amount of TGT without this tag. Using mass spectrometry, we then observed the appearance of a "heterodimeric" TGT that differed in mass and consisted of both a tagged and an untagged subunit. In the end, it was present at 50% alongside the two "pure" TGT forms (25% each). While virtually no "heterodimeric" TGT was found approximately two minutes after mixing the two "pure" forms, complete exchange equilibrium was not reached until more than 10 h had elapsed. Thus, the exchange process is very slow. In contrast, a destabilized variant, in which His 333 was exchanged for alanine, reached equilibrium much more rapidly, i.e., in less than ten minutes.

21.12 When Nothing Else Works: Chemical Poking at the Contact Interface

Obviously, all the mutagenesis did not help to obtain a crystal structure of the monomer. Although it became clear that the equilibrium and the kinetics of its establishment could be shifted towards monomerization by destabilizing the interface contact surface, what effect did this have on the structure? A structural feature of TGT is that the catalytically important preQ$_1$ binding site is spatially very close to the dimer interface contact. In collaboration with the group of François Diederich in Zurich, we succeeded in synthesizing inhibitors of the catalytic center with long needle-shaped substituents (e.g. **21.26**). This allows these **ligands to spike into the nearby dimer interface** (Fig. 21.20, *left*). As a result, these **spiking ligands** were able to achieve at least partial disruption of the TGT dimer. Mass spectrometry showed an increase to approximately 25% of the dissociated species. Optimization of these spiking ligands could lead to an alternative concept for drug development.

21.13 Only Serendipity Can Help: Different Crystal Form—New Dimer

While searching for new inhibitors addressing the $preQ_1$ binding site with sugar-like side chains, Frederik Ehrmann made a surprising observation: from the same crystallization sample, the enzyme with these inhibitors crystallized side by side in two different crystal forms! On the one hand, we found the well-known crystal form in space group C2. It showed the usual functional dimer with the dimer packing described above. In the other crystal form, now in space group $P2_1$, the dimer adopts a different mutual interface packing. Here, the structurally almost unchanged monomer units form a new contact surface interface in an orientation rotated by about 130° with respect to the other dimer form. We will call this form the **"twisted" homodimer** (◘ Fig. 21.21, *right*). Remarkably, the size and number of interface contacts were hardly reduced. However, what is critical to the disruption of enzyme function is that tRNA binding is no longer possible to the twisted dimer for steric reasons. Thus, this new dimer packing, apparently induced by the binding of our novel inhibitors, freezes the protein in an inactive state. As a result, it is no longer able to induce catalytic turnover. The inhibitor **21.14** had been found previously, but had only been studied structurally by crystal soaking. This method, which works with premanufactured crystals, naturally yielded the usual *C2* homodimer. Now, following a cocrystallization protocol where the ligand is added already to the crystallization solution, the twisted homodimer was found in the crystals in space group $P2_1$. Thus, we had to learn that the applied crystallization protocol used initially led to incomplete conclusions.

Since both crystal forms grow side by side from the same crystallization well, we assume that both forms coexist in solution and have very similar stability. However, crystallography only determines the end points of this assumed rotational transformation between the two forms. Therefore, the intriguing questions were: does this transformation also occur in solution, does it depend on the ligand used, and are there structural features that indicate what causes the transformation? If there is a substance that stabilizes the twisted, catalytically inactive form, it will also be a candidate for drug development as it will also block the biological function of TGT. Such a substance will indeed be able to keep our enzyme in a catalytically inactive state. Closely related to this is the question of why the enzyme is able to adopt such a state in the first place. Is this a state of self-regulation and is the enzyme "slowed down" in its catalytic activity by certain ligands that may be undesirable or present in too high a concentration?

21.14 Tracking the Dynamic Transformation with the Appropriate Spins

In solution, methods that observe the spins of magnetic nuclei and their coupling to each other are very powerful techniques for structure determination (see Sects. 7.8 and 13.7). While NMR spectroscopy detects spin couplings only over relatively short distances, the spins of unpaired electrons can couple over much longer distances. Thus, electron paramagnetic spin resonance (EPR) spectroscopy seemed ideally suited to study the behavior of TGT in solution. **EPR spectroscopy** requires unpaired electrons in the molecule. Since paramagnetic centers are absent in most biomolecules, spin labels must be introduced into the biological system in a site-specific manner. If two or more spin labels are present in a molecule, so-called pulsed electron–electron double resonance (**PELDOR** or **DEER**) experiments can be used to obtain precise information about changes in the distance between the spin labels in the range of 15–60 Å. This is exactly the range in which we expected the changes to occur during the transformation of the TGT between the two dimeric states. **Nitroxide spin labels** are most commonly used for this purpose. They can be selectively coupled via the thiol group of a cysteine residue in the form of a disulfide bridge (◘ Fig. 21.22). Site-specificity on the protein is achieved by mutating cysteine residues into the protein sequence at the desired positions. At the same time, undesired cysteines found in the wild-type sequence are replaced by, for example, alanine or serine. The search for the best spin-labeling sites on the TGT dimer in both of its forms was initially planned with the help of a computer simulation.

Dzung Nguyen from the group in Marburg and Dinar Abdullin from Olav Schiemann's group in Bonn, Germany, found positions 87 and 319 to be particularly suitable for spin labels. After the protein variants had been prepared and labeled, and the equilibria with the different ligands had been established in solution, the samples were shock-frozen at −196 °C and measured. This gave the result shown in ◘ Fig. 21.23.

Obviously, ligand **21.27** predominantly induces the twisted form of the enzyme. It can, therefore, be considered as a stabilizer of the inactive twisted form. Such a compound is expected to inhibit the enzyme by a completely different mechanism. Time-dependent measurements were also performed for this ligand using the PELDOR method. After only one hour of equilibration in solution, a significant amount (> 70%) of twisted dimer is present. After 24 h, the final equilibrium is reached with more than 85% of this form.

EPR spectroscopy is not the only way to study solution equilibrium. **^{19}F-NMR spectroscopy** is also a very sensitive method for tracking structural differences in

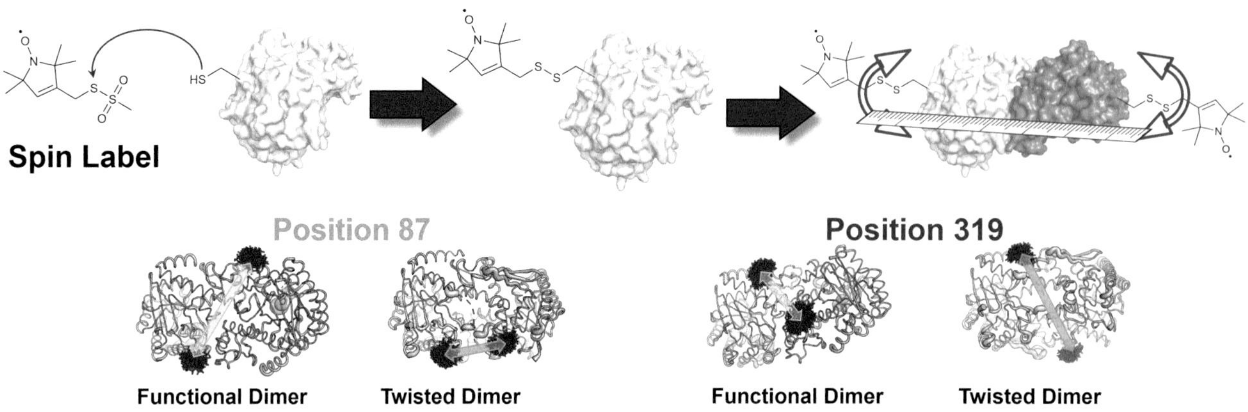

Fig. 21.22 A nitroxide spin label is attached via a disulfide bridge to the thiol group of an appropriately positioned cysteine residue. In the homodimer, the distance between the two labels can now be measured in the range of 15–60 Å. At positions 87 and 319, a glycine and a histidine, respectively, are exchanged for cysteines. In the functional (*yellow arrow*) or twisted (*green arrow*) dimer, the distance between the labels changes from about 55 to 25 and 30 to 60 Å, respectively. Because of the conformational flexibility of the nitroxide group at the protein surface, a distribution in a narrow distance range is detected (*dark blue* scattered distribution)

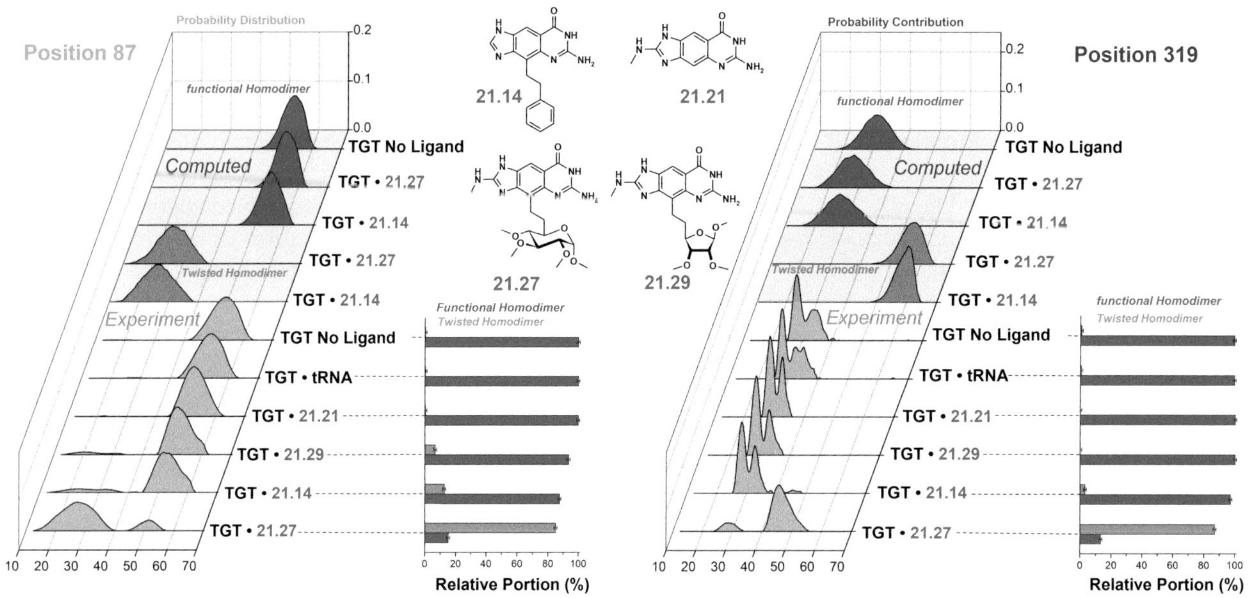

Fig. 21.23 *Left* For position 87, calculations based on the crystal structures of the functional and twisted homodimer give a distribution of about 55 and 25 Å with **21.14** and **21.27**, respectively. The experiment shows that TGT is present in the functional form without any bound ligand or with bound tRNA. Ligand **21.21** without a side chain also binds only to TGT in the functional dimer form. For ligands **21.29**, **21.14**, and especially **21.27**, the relative proportion of the twisted dimer population increases. For **21.27**, up to 85% is present in the twisted form. *Right* Position 319 confirms this finding. Here the calculated distance in the functional dimer is about 30 Å and increases to 60 Å in the twisted form. Experimentally, there is confirmation that **21.27** is a stabilizer of the twisted form. The distribution in this case is more complex and shows two maxima, suggesting two different conformational families for the attached spin labels

solution. Fluorine atoms are generally hardly present in biological structures. Therefore, they have to be incorporated artificially via **fluorine-labeled amino acids**. We have chosen a fluorinated tryptophan substituted at position 5. TGT contains four tryptophan residues per monomer unit. Of these, Trp 95 and Trp 326 are located near the dimer interface. The other two residues, Trp 178 and Trp 296, are more distant. Trp 178 is completely buried inside the protein, while Trp 296 is partially oriented towards the surface. To introduce the fluorinated residues, Andreas Nguyen used the expression of a tryptophan-auxotrophic cell line that cannot produce the aromatic amino acid itself. After adding the fluorinated amino acid in place of the natural one to the expression medium and starting protein expression, the ^{19}F-labeled amino acid is incorporated into the target protein by the cellular machinery. The correct incorporation was subsequently verified by both crystallography and mass spectrometry.

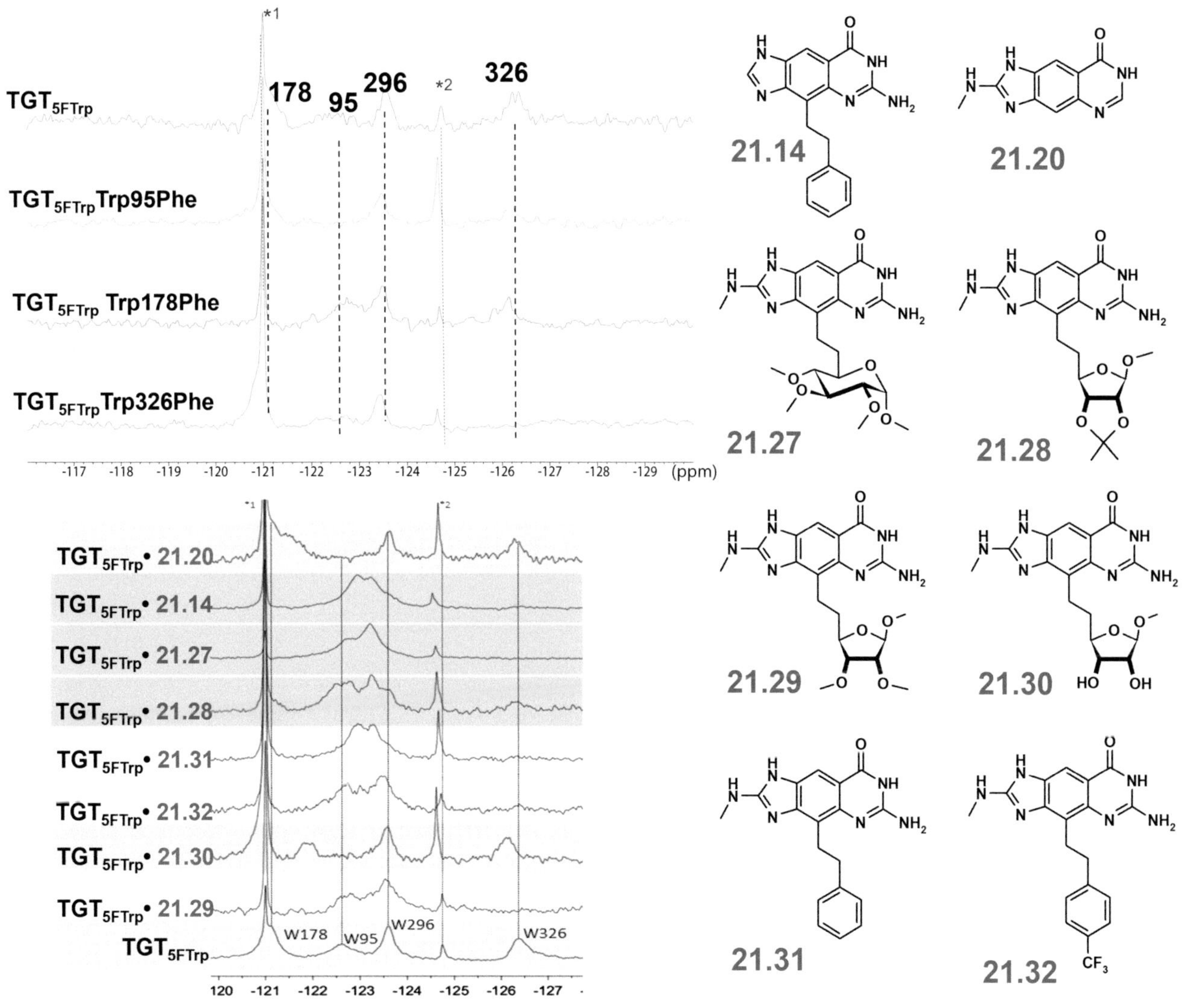

Fig. 21.24 *Upper left* ^{19}F-NMR spectra of $TGT_{5F\text{-}Trp}$ and three different Trp ⇨ Phe mutants of the wild type, Trp95Phe, Trp178Phe, and Trp326Phe variants. Two low molecular weight contaminants (fluoride ions and free 5F-Trp) are indicated by an *asterisk*. The absence of individual signals in the Phe variants assigns the resonances to the individual Trp residues. *Bottom left* ^{19}F-NMR spectra of $TGT_{5F\text{-}Trp}$ without ($TGT_{5F\text{-}Trp}$, lowest spectrum) and with the listed ligands **21.14**–**21.32**. For 5F-Trp178, 5F-Trp296, and 5F-Trp326 with ligands **21.20** and **21.30** (*green background*) hardly or only slightly changed resonance positions are observed compared to the uncomplexed protein. For both, no evidence for the formation of the twisted dimer could be detected crystallographically. For ligands **21.14**, **21.27**, and **21.28**, however, structures of both forms could be crystallized. With these ligands similar spectra are observed, which are strongly altered compared to the uncomplexed protein (*red background*). But also the spectra with the ligands **21.31**, **21.32**, and **21.29** show clear changes compared to the uncomplexed protein. Here, both forms are probably present in solution

We performed the NMR studies together with Michael Sattler's group at the TU Munich, Germany. As expected, the spectra indicated four different fluorine resonances. First, the different ^{19}F-labeled tryptophans had to be assigned to the signals. This was done by successively replacing the tryptophans with phenylalanine and observing which signals disappeared from the spectrum (◘ Fig. 21.24, *top*).

Subsequently, different ligands from the series **21.14**–**21.32** were titrated into a solution of the labeled $TGT_{5F\text{-}Trp}$ until maximally a three-fold molar excess of the ligands was present. Depending on the ligand used, different amounts of changes in the resonance positions were observed. Using the unliganded protein as a reference (◘ Fig. 21.24, bottom spectrum), which according to all previous results is present in solution as the functional *C2* symmetric homodimer, its spectrum shows similarities to that of the complex with **21.20**. Only Trp 95 undergoes a significant shift. The other residues, especially Trp 326 in the interface contact region, remain virtually unchanged. The situation is very similar for ligand **21.30**. Ligand **21.20** lacks a side chain and there is no evidence that this molecule triggers the transformation to the inactive, twisted form. Similarly, no transformation was observed for ligand **21.30**, despite its close similarity to **21.27** and **21.28**. According to the

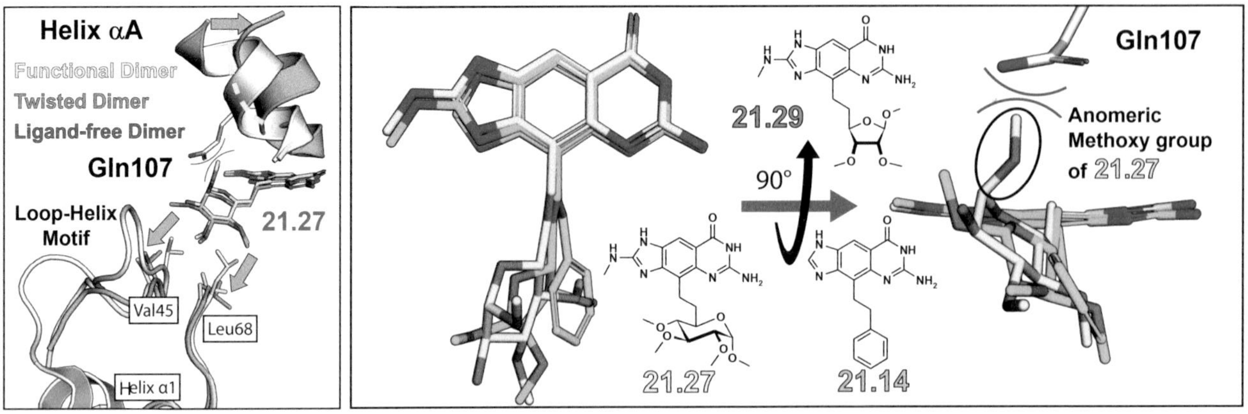

■ **Fig. 21.25** *Left* Binding mode of ligand **21.27**, which stabilizes the twisted form of TGT. In the ligand-free state, the homodimer adopts the geometry of the functional dimer. When the enzyme adopts the structure of the twisted dimer, the helix αA shifts from its position in the ligand-free state (*gray*) to the geometry of the twisted dimer (*ochre*). In the structure with **21.27**, corresponding to the geometry of the functional dimer, this helix is completely disordered (*yellow dashed line*). When compared to the geometry of the uncomplexed TGT (*gray*), we see that the anomeric methoxy group of **21.27** pushes against residue Gln 107 in the center of the helix. This triggers the transition from the functional to the twisted dimer. *Right* In the detailed view, this geometric conflict of **21.27** with residue Gln 107 is again highlighted. The ligands **21.14** and **21.29**, for which crystal structures in the geometry of the functional as well as the twisted dimer could be determined, indicate a smaller spatial demand of the 4-substituent in this region. This explains why, according to the PELDOR measurements, only **21.27** proves to be an excellent stabilizer of the twisted dimer arrangement

PELDOR measurements, ligand **21.27** is a very **strong stabilizer of the twisted form**. Its NMR spectrum shows clearly different resonance positions compared to those with **21.20** and **21.30**, especially for Trp 326. The same is true for **21.14** and **21.28**. For ligand **21.29**, both species seem to coexist in solution. This is also true for **21.31** and **21.32**, for which both the functional and the twisted species can be detected in equilibrium.

The results of the ^{19}F-NMR spectroscopy, thus, confirm the picture from the PELDOR studies. However, they differentiate it even further. Not only is **21.27** able to induce the twisted form effectively, but this raises the question of why some ligands induce the twist, while others do not. A comparison of the crystal structures of ligands **21.14**, **21.27**, and **21.29** provides further insight at this point (■ Fig. 21.25).

To transform into the twisted dimer geometry, the helix αA has to be shifted in its position (■ Fig. 21.25, *left*). In the TGT complex with the geometry of the functional dimer and the ligand **21.27** (*yellow*), the geometry of this helix is already disturbed to such an extent that its arrangement in the electron density can no longer be observed. The anomeric methoxy group on the six-membered ring sugar comes into spatial conflict with the side chain of Gln 107 in the center of the helix αA. This leads to steric stress that triggers the transformation to the twisted dimer geometry and subsequently stabilizes this geometry. For the other ligands that also trigger the transformation, this steric stress is weaker (■ Fig. 21.25, *right*). Thus, the **design concept for potent stabilizers** of the catalytically inactive twisted dimer form is in place. Ligands are needed that, on the one hand, exert efficient steric pressure on Gln 107 to induce transformation. At the same time, they must influence the conformation of the loop–helix motif. Finally, the barrier to return to the functional dimer geometry must be sufficiently high. It is also possible that the discovered rearrangement mechanism is triggered by substrates or products of TGT to bring the enzyme to a **resting state**. In any case, it is important for drug development that a completely new inhibition mechanism of TGT has been discovered with ligand **21.27**. The enzyme is blocked by putting it into a kind of "dormant" state.

21.15 When Sulfur Accidentally Oxidizes and Starts a Fragment Design Project in a New Arrangement

Despite our best efforts, we had not yet found a way to get an idea of the geometry of the contact interface surface in the monomeric state. By all accounts, the **loop–helix motif** seemed to be important in controlling monomerization. The many crystal structures that had been solved indicated that this motif had **great conformational flexibility** (■ Fig. 21.20, *left*). As mentioned above, cysteine residues were introduced to replace the amino acids of the aromatic cluster. By this, we also aimed to introduce as many potential attachment points for small fragments into the contact surface as possible (cf. Fragment Tethering Approach, Sect. 7.10). Fortunately, serendipity was once again our friend. Stephan Jakobi made an exciting discovery: after some time, a new crystal form had grown in one of his crystallization batches. It was the variant in which Tyr 330 had been replaced

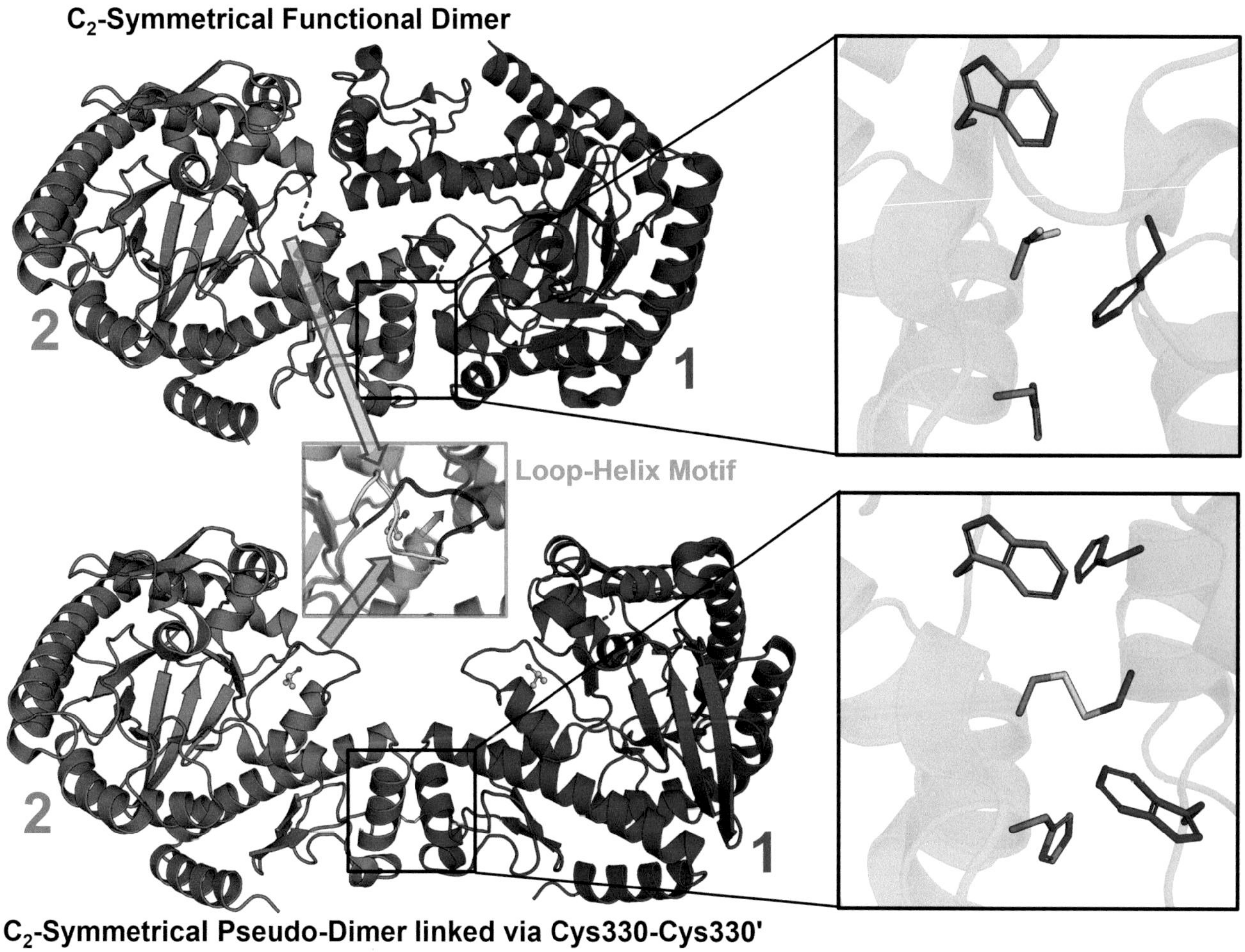

Fig. 21.26 *Upper row* Structure of the functionally active homodimer of TGT. In the aromatic cluster, Tyr 330 has been replaced by a cysteine. However, there is still a short contact with Phe 92′ of the opposite monomeric unit (see detailed view, *upper right*). *Bottom row* Oxygen from the air oxidizes the thiol function of the exposed Cys 330. A disulfide bridge is formed between Cys 330–Cys 330′. The enzyme is now "pseudomonomerized." The monomer units, which have hardly changed structurally, now pack against each other with altered geometry (detailed view, *bottom right*). The former contact surface is exposed. As a result, the flexible loop–helix motif is no longer embedded in the interface contact surface (*green box*). It takes a new course (*red*) that differs from the original geometry in the homodimer (*yellow*). With this geometry, the loop no longer fits into the packing of the former homodimer for steric reasons and, thus, blocks its formation. In order to disrupt the mutual recognition and binding of the two monomeric units, this new conformation of the loop–helix motif would have to be stabilized by a ligand. (▸ https://sn.pub/iYaYHJ)

by a cysteine. The new crystals had a hexagonal shape and not the usual monoclinic crystal habit! Structure determination solved the puzzle. The original homodimer was completely altered, although the geometry of the individual monomer units had hardly changed. Formally, the enzyme was no longer a dimer. Atmospheric oxygen had oxidized the sulfur at Cys 330, and as a consequence, forming a covalent disulfide bridge between the two former monomer units. This new S–S bridge was formed between Cys 330–Cys 330′. It coincides with a two-fold crystallographic rotation axis in the new crystal packing. Thus, it causes one monomer subunit to tilt towards the other (Fig. 21.26). The original contact area of over 1600 Å^2 shrinks to 537 Å^2. Most likely, the stability of the contact area in the functional dimer of the studied Tyr330Cys variant is significantly reduced and, in solution, the amount of dissociated monomer is, therefore, strongly increased. This enhances the probability of **oxidative disulfide bridge formation**. However, it was important for the further course of our project that in the new reversibly covalent "dimer" the original **contact interface surface** was largely exposed. For the first time, we had a "pseudomonomerized" TGT structure. To increase the likelihood of the formation of this disulfide-bridged form, we later examined the enzyme in which His 333 was also exchanged for alanine. As a result, the monomer fraction in solution equilibrium increases even further.

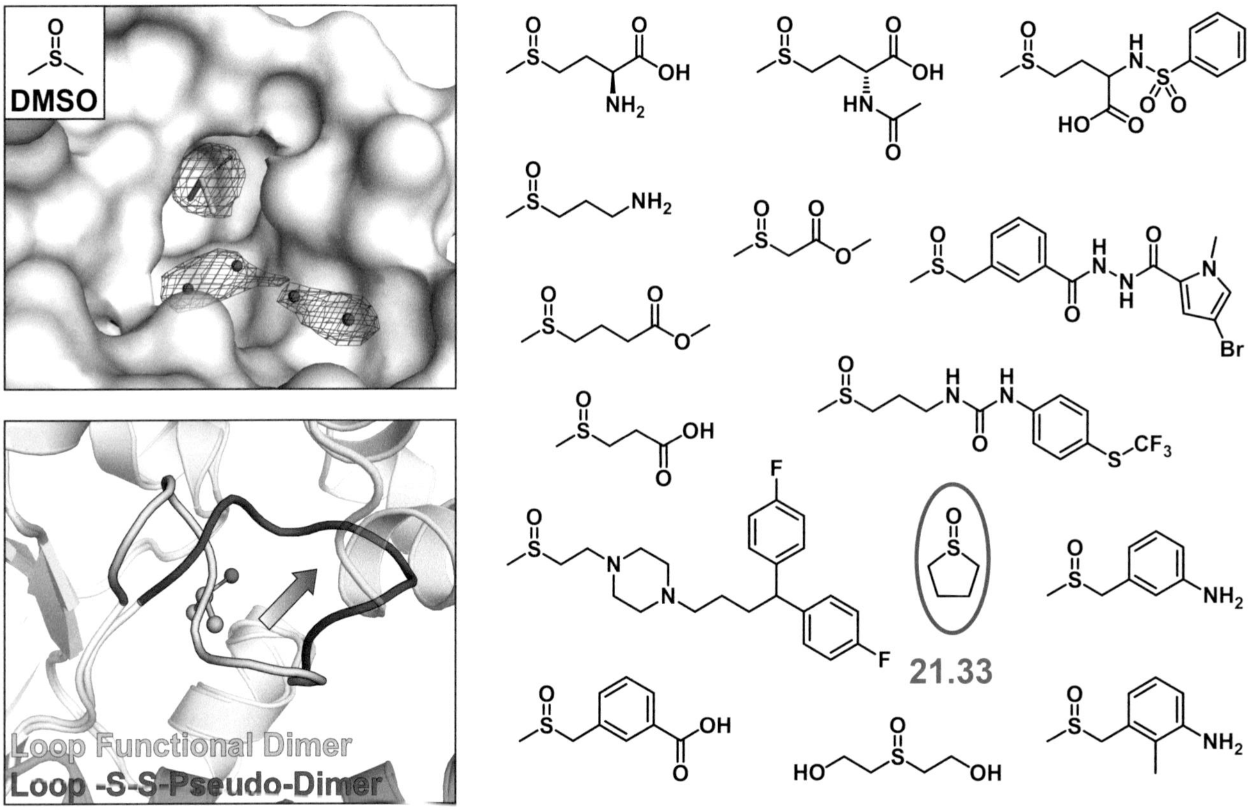

Fig. 21.27 A small pocket under the loop–helix motif in the crystal structure of the "pseudomonomerized" enzyme accommodates a DMSO molecule when transferred to the cryobuffer (*left*). In addition, four water molecules are found in this pocket. Subsequently, a series of commercially available sulfoxide fragments (*right*) were tested directly for their binding into the protein crystals. However, only **21.33** could be detected as a hit in the pocket

What did the new form look like? In Fig. 21.26, the dimer packing of the functional homodimer is shown at the top, and that of the pseudomonomer spread by the covalent disulfide bridge is shown at the bottom. The original aromatic cluster (*right*), in which Tyr 330 is replaced by cysteine, is rearranged to reveal the two covalently linked cysteines in its center. The old contact area is oriented into the crystal packing with virtually no direct contact to any neighboring molecule. Interestingly, the loop portion of the loop–helix motif presents a new geometry. Importantly, the loop adopts an arrangement that no longer matches the original dimer packing of the functional homodimer for steric reasons. Our hypothesis was that the loop also adopts this geometry in the monomeric state in solution. This idea was supported by MD simulations. Protein crystals are briefly immersed in a cryobuffer before being measured at low temperatures. This ensures that any remaining water in the crystals solidifies into a glassy state upon freezing. The used cryobuffer contained a small amount of the solvent DMSO, which was not present in the crystallization buffer. One molecule of this solvent diffused into a **small pocket under the loop** together with four water molecules (Fig. 21.27, *top left*). We took this observation as a strong indication that small molecules could be introduced as ligands into the pocket under the rearranged loop! This was the starting point for a **fragment-based lead structure search**.

A whole series of commercially available sulfoxides was tested by directly diffusing them into the protein crystals (Fig. 21.27). Surprisingly, besides DMSO, only the cyclic tetramethylene sulfoxide **21.33** was found to be a hit. Analysis of the crystal packing of the new hexagonal crystal form of the "pseudomonomerized" TGT suggested that the binding pocket of the fragments could only be reached through a very narrow solvent channel. Therefore, only very small fragments could enter the binding pocket. However, they were not prevented from successful binding by constrictions of the channel in front of the binding site. Dzung Nguyen, therefore, extended his fragment search to small analogues of **21.33** and was able to discover a total of six additional fragments as hits. Fig. 21.28 shows the binding mode of two of these derivatives. For **21.34**, it was even possible to accommodate two of these fragments in the pocket at the same time. They form specific contacts with the surrounding amino acids Gly 46, Thr 47, Pro 56, and Met 93, which establish hydrogen bonds with the residues

21

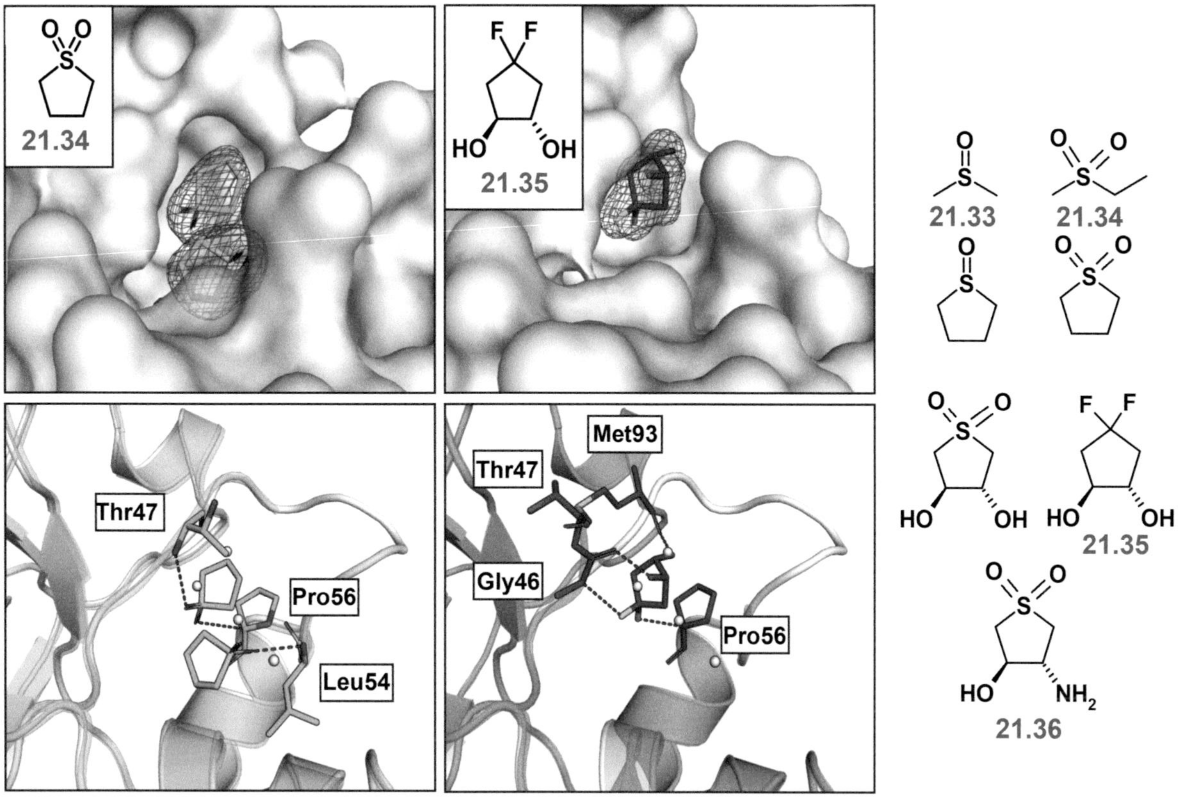

Fig. 21.28 Binding geometry of sulfolane **21.34**, which binds analogously to sulfoxide **21.33**, in the pocket below the loop–helix motif. A second copy of **21.34** also fits into the pocket and displaces some of the water molecules. The introduction of two *trans*-oriented hydroxy groups allows the binding to be improved by further contacts to the amino acids Gly 6, Thr 47, Pro 56, and Met 93. In the structure of the original functional homodimer, these residues are involved in H-bonds to the amino acids of the aromatic cluster. The sulfone group can be replaced by an isosteric difluoromethylene group. Fragment **21.35** and sulfolane **21.36** have been characterized by NMR spectroscopy as micromolar binders to TGT in solution

of the aromatic cluster in the functional dimer. Interestingly, the sulfone group in **21.34** can be replaced by an isosteric CF_2 group to form fragment **21.35**, which binds to the protein as the *R*,*R*-stereoisomer.

The key question was whether these fragments also bind to the enzyme in solution. To prove this, we used NMR spectroscopy. So-called DOSY experiments were performed, in which molecular diffusion coefficients are compared with one-dimensional chemical shifts. For the experiments, we used a TGT variant in which His 333 was replaced by aspartate. This variant is significantly weakened in its dimer stability, so that the monomer is present in solution at appreciable concentrations. This clearly favors the detection of the weak binding of a small fragment. Indeed, the binding of the racemic **21.35** was successfully detected, with a binding constant estimated to be about 90 µM. As another fragment, for which no crystal structure could be determined, the racemic *trans*-3-hydroxy-4-aminosulfolane **21.36** was detected with a binding constant in the same range.

The fragment approach described above shows that ligands can be found that bind to the interface contact surface of the original functional homodimer. They stabilize the conformationally flexible loop–helix motif in a geometry that no longer fits the original packing of the functional homodimer. As a design concept, it is therefore necessary to develop sufficiently potent ligands for the binding pocket under the loop–helix motif that bind to the monomeric form of TGT. They would then have to alter the geometry of the former interface contact surface to such an extent that the two monomeric units would no longer be able to recognize and interact with each other. As a result, the catalytically active homodimer can no longer form and the function of the enzyme would be blocked. With regard to the inhibition of the function of TGT, the observed fragment structures point to a further strategy for the development of putative drugs.

21.16 A Fragment Opens a Transient Pocket and Suggests the Design of Bacteria-specific Inhibitors

A critical aspect of any drug design project is to achieve **sufficient selectivity** of the developed drug candidates. With TGT, we have chosen an enzyme as target. Inhi-

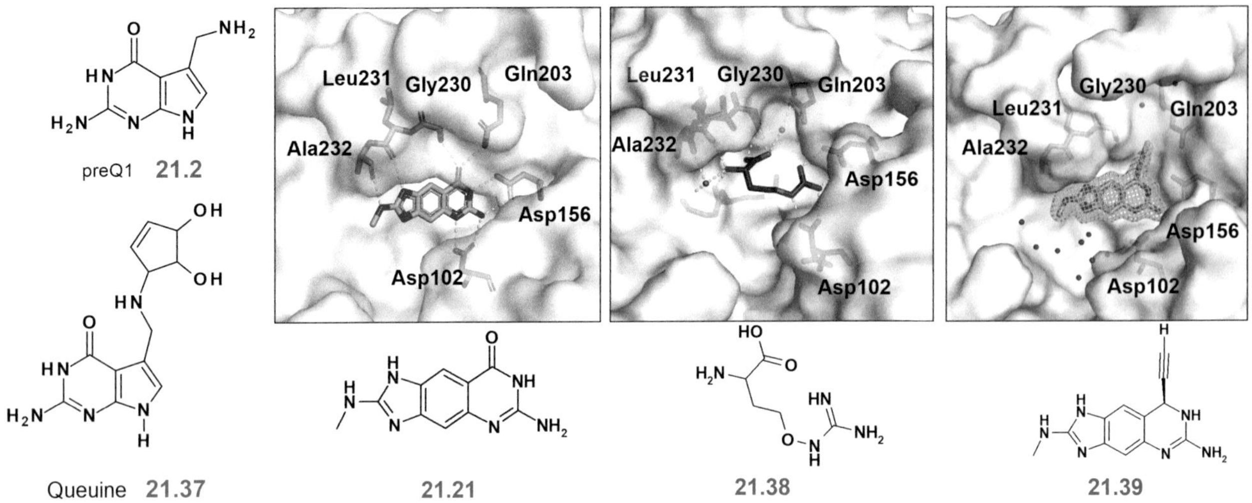

Fig. 21.29 Bacterial TGT incorporates preQ$_1$ **21.2** into tRNA as a substrate. Instead, the human enzyme uses queuine **21.37** as the substrate at this site. On the *right*, complexes with three ligands are shown: binding geometries of the parent scaffold of *lin*-benzoguanine **21.21** (*left, light blue*), of the fragment **21.38** (*center, purple*), and inhibitor **21.39** (*right, ochre*, electron density *outlined with a mesh*) extended with a propargyl substituent are shown. Compared to **21.21**, fragment **21.38** and the modified inhibitor **21.39** are able to open a small transient pocket (*marked in yellow*) near Gln 203 and Gly 230

bition of this enzyme represents a therapeutic concept for the treatment of *Shigella* dysentery. However, such an enzyme also exists in humans. It uses queuine **21.37**, a slightly larger substrate compared to the bacterial species (Fig. 21.29). It is similar to the final product that bacterial enzymes produce from the precursor preQ$_1$ **21.2**, which they then modify in several steps to produce queuine (Fig. 21.1). Despite this difference in size, the catalytic centers of the bacterial and human enzymes are related. The human enzyme also catalyzes as a dimer. However, it is a heterodimer consisting of two different monomeric units. This makes the contact area between the bacterial and human enzyme significantly different and should simplify the selectivity problem for inhibitors targeting the interface contact area. But what about active site inhibitors? Based on a structure determination with a fragment, Engi Hassaan observed an unexpected result. It gave us an idea of how to **selectively inhibit bacterial TGT**.

Fragment **21.38** has a marked similarity to the amino acid arginine. It binds to the catalytic center of TGT and interacts with Asp 156 in a manner similar to the *lin*-benzoguanine scaffold found in **21.21**. However, the **21.38** fragment is able to open a **transient pocket** at this site. The resulting pocket is filled with several water molecules. Its opening brings the amino acids Cys 158 and Val 233, which in the closed state are in almost van der Waals contact with each other, to a greater distance (Fig. 21.30). The previously buried Cys 158 becomes freely accessible. With inhibitor **21.39**, we have found a first candidate that pushes into this pocket with its propargyl group and comes close to the thiol group of Cys 158. Freely accessible thiol groups can be covalently attached to a suitable ligand by chemical reaction. Several examples of disulfide tethering (Sect. 7.10) have demonstrated this as a promising strategy for drug design. An irreversible covalent bond to Cys 158 with an inhibitor derived from **21.39** that binds to the active site would inhibit the function of TGT. It is interesting to note that a cysteine at this position is found only in the bacterial enzymes. The comparable TGTs from higher developed eukaryotic organisms do not have this amino acid in that position. This provides a promising structural site for inhibition that is unique to bacterial enzymes: an ideal prerequisite for the development of selective inhibitors!

21.17 Many Ways to a Smart Antibiotic Against Shigellosis

The development of so-called **"smart" anti-infectives** is particularly attractive because, when used therapeutically, they do not radically kill the entire bacterial culture in the intestinal flora. Instead, they use specific mechanisms to prevent bacteria from becoming pathogenic. Inhibiting TGT in *Shigella* could be such a successful concept. The enzyme which controls the production of the proteins that initiate the invasion process into the epithelial cells is blocked. However, there is still a long way to go before suitable candidates are ready for clinical testing. The goal of our academic research was to **identify potential concepts and starting points** for modulating the function of this versatile enzyme. The translation into a commercial drug candidate for potential therapeutic applications is still almost exclusively carried out by the pharmaceutical industry, where the necessary infrastruc-

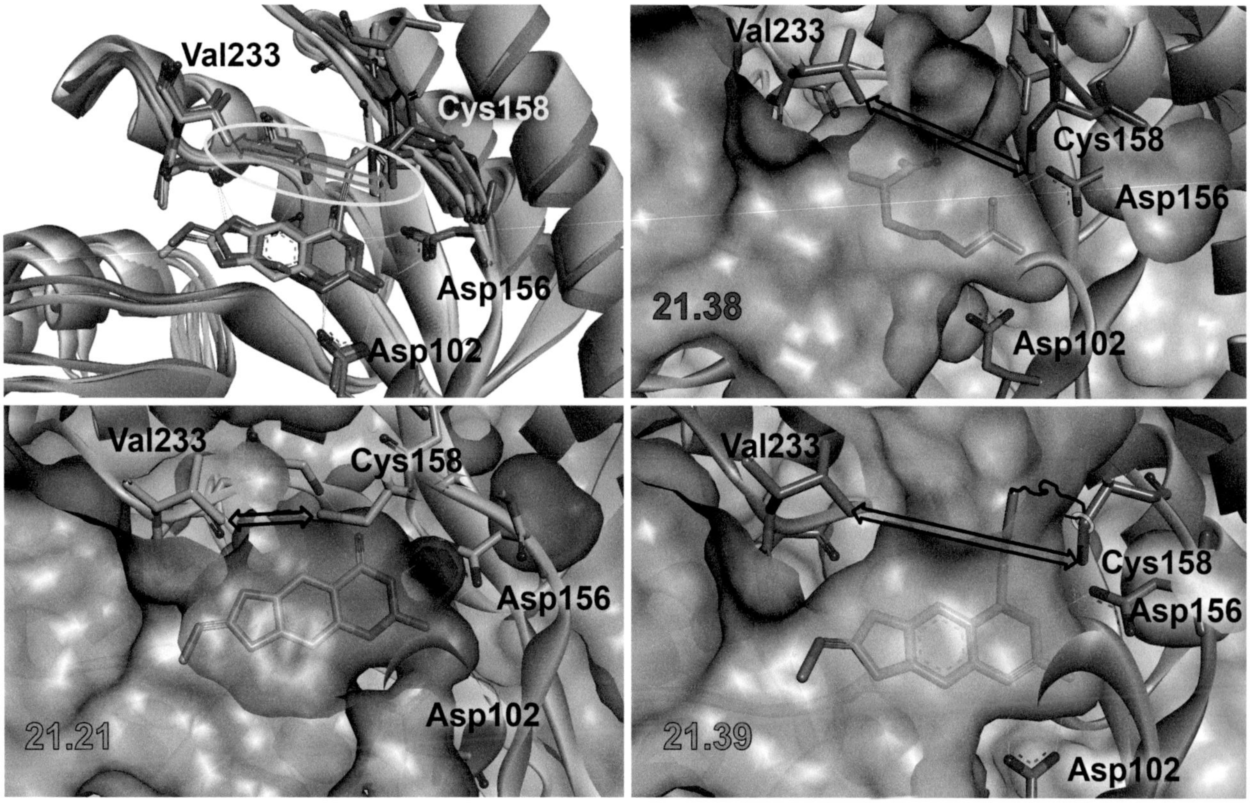

Fig. 21.30 *Upper left* An overlay of the binding modes of **21.21** (*light blue*), **21.38** (*purple*), and **21.39** (*ocher*) is shown with TGT. The latter two ligands open a transient pocket (*upper and lower right*) compared to **21.21** (*lower left*). This increases the distance between Cys 158 and Val 233. While in the complex with **21.21**, the thiol group of Cys 158 is buried by the van der Waals contact with Val 233, it becomes accessible to chemical attack by opening the transient pocket in the complexes with fragment **21.38** and extended inhibitor **21.39**. It is convenient to replace the propargyl group in **21.39** with a suitable chemically reactive group to form a covalent bond between the inhibitor and the thiol group of Cys 158. Only bacterial TGTs contain a cysteine at this position. This is a promising starting point for the development of selective inhibitors of bacterial TGTs. (▶ https://sn.pub/AMbWJF)

ture, financial resources, and organizational structures are available for such a project.

The purpose of this chapter was to show, using a selected example, how the in-depth study of the properties and functions of a target protein can provide **entry points for manipulating its biochemical function**. Several strategies can be proposed. One is the more classical **blocking of the catalytic center**. Model compounds down to subnanomolar inhibition have been developed. Due to their highly polar nature, they will probably have to be used as prodrugs. To achieve specificity for bacterial enzymes only, irreversible covalent inhibition of a cysteine near the catalytic center would be conceivable. Only in bacterial enzymes is this residue located next to a transient pocket. Ligands are needed that open the transient pocket and bind covalently to the then accessible thiol group of the cysteine. Since the protein binds tRNA exclusively as a homodimer, its function can also be disturbed by **disrupting the dimer structure**. On the one hand, inhibitors of the active site can be equipped with a molecular spike. This allows them to interfere and disrupt the contact surface between the monomer units. Ligands that bind under an exposed loop in the monomeric state can also block dimer formation. Their binding restructures the original monomer units in a way so that they no longer fit together for dimer formation. Since many enzymes have feedback regulatory mechanisms, the ligand-induced transformation of TGT into an inactive "resting state" may provide another strategy. By forming an alternatively packed dimer, the enzyme is unable to bind its substrate tRNA. Substances that stabilize this "dormant" state may, therefore, also inhibit the function of the target enzyme.

There seem to be many approaches to the desired goal. So, does TGT represent a special case that has led to so many different approaches to inhibiting it? Probably not, because proteins are "social," involved in many networks and, thus, participate in many mechanisms. It is more a question of time and thoroughness of characterization, regarding how many possible inhibitory mechanisms can be traced. In terms of drug design, TGT is a prime example. You can see how many different concepts are open to drug designers if they only "get to the heart" of the target structure for a long enough time! In the end, it has to be shown pharmacologically which of these concepts is the most suitable for therapeutic success.

21.18 Synopsis

- *Shigella* dysentery is a severe bacterial diarrheal illness. *Shigella* bacteria that are ingested with contaminated water or food adhere to epithelial cells in the intestinal mucosa. To gain entrance to these cells, the bacteria produce their own virulence factors, so-called invasins.
- The genes encoding the invasins will only be transcribed in sufficient amounts if the transcription factor VirF is present in sufficient quantities. A prerequisite for its efficient protein biosynthesis is the tRNA-modifying enzyme tRNA-guanine transglycosylase, which catalyzes the incorporation of the modified $preQ_1$ base into the wobble position of certain tRNAs.
- A functional assay recording the exchange of guanine by radioactively labeled guanine can determine the potency of ligands inhibiting the function of the target enzyme.
- The first hits were detected by using the *de novo* design program LUDI, and the predicted binding mode of a micromolar hit was confirmed by crystallography.
- The active site shows adaptations by flipping a peptide bond and mediating important interactions to the substrates through a water molecule.
- Virtual screening suggests a broad variety of basic scaffolds for inhibitor design. A *lin*-benzoguanine scaffold served as the most promising lead structure.
- Substitutions at the 2- and 4-positions of the *lin*-benzoguanine scaffold lead to very different increases in affinity. The addition of a 2-amino group leads to the presence of two basic centers in the molecule. Interestingly, the less basic center is protonated and, thus, charged by the local polarity in the binding pocket. This leads to salt bridge-like interactions with two aspartate residues. Nevertheless, the addition of the 2-amino group also results in an enormous increase in affinity. Here, the surprising increase in affinity can be attributed to the formation of two parallel H-bonds, which do not experience secondary repulsive effects and, thus, turn out to be very strong.
- Substitutions at the 4-position have to interfere with and partially replace a contiguous water network between two facing aspartic acids. They can potentially link the parent scaffold with substituents which fill a small hydrophobic pocket. A significant potency enhancement can be achieved only if the spacer linking the two portions contains polar atoms to cross the water network. These atoms can actively participate in the network.
- Several iterative cycles of design, crystal structure analyses, and inhibitor syntheses were required to develop the initial double-digit micromolar hits into subnanomolar inhibitors. The best potency is achieved by *lin*-benzoguanines carrying both a 2- and 4-substituent.
- Surprisingly, the enzyme is only functional as a homodimer. Thus, disruption of the dimer geometry may provide a further principle for inhibitor design. Computational analysis and subsequent targeted mutagenesis of individual residues in the dimer interface identified residues that contribute primarily to the stability of the contact area expanding over more than 1600 $Å^2$. H-bonds and a cluster of aromatic amino acids were found to be critical. The cluster is embedded in an environment of hydrophobic amino acids and prevents the entry of water molecules.
- Since the interfacial contact region is located near the catalytic center, active site inhibitors with very long, spike-like substituents were able to disrupt the dimer contact and convert the dimer, at least partially, to a monomer.
- With ligands carrying bulkier, partially sugar-containing substituents in the 4-position, a new crystal form was found by chance under the same crystallization conditions. The enzyme appears with a twisted arrangement of the structurally unchanged monomer units. A new interface contact area of almost the same size is formed. In this ligand-induced form, the monomer units pack together in such a way that the tRNA substrate can no longer be bound. Thus, ligands that induce and stabilize this twisted, catalytically inactive form are another concept for the development of inhibitors for the enzyme.
- Since crystallography only characterizes the endpoints of a possible dynamic rearrangement, it was possible to infer a dynamic transformation in solution by spin labeling using ESR spectroscopy. By introducing ^{19}F-labeled tryptophan, NMR spectroscopy was also able to detect the ligand-induced dynamic rearrangement process between the catalytically competent and the twisted inactive dimer forms in solution. One ligand proves to be an effective stabilizer of the twisted inactive form and can, therefore, be considered as a starting point for an alternative inhibitor design.
- Surprisingly, a cysteine residue introduced into the interface contact surface by mutagenesis led to the formation of a new crystal form under oxidative conditions, in which the former contact surface is structurally exposed by the altered dimer packing. The new packing was created by a disulfide bridge between the cysteine residues introduced into the dimer. The enzyme, thus, assumes a "pseudomonom-

erized" form. In the exposed packing, a loop motif assumes a geometry that is no longer consistent with the original packing of the homodimer.

- Below the loop, a small pocket forms in the "pseudomonomerized" form that can accommodate small fragments. The fragments can be diffused into crystals of the enzyme and, thus, structurally characterized. NMR spectroscopy provides evidence that these small ligands can also bind in solutions and interfere with the formation of the contact surface. They suggest another principle for inhibitor design by interfering with the functionally essential homodimer formation.
- A transient pocket that can form near the active site, points to another design concept for developing selective inhibitors for the bacterial enzyme. Such inhibitors must approach a cysteine that is unique to the bacterial enzymes by filling the transient pocket in order to undergo irreversible covalent inhibition via the thiol group of the cysteine.
- Several design concepts for the inhibition of the bacterial enzyme are emerging:
  - Inhibition of the active site by potent inhibitors;
  - Disruption of homodimer formation by active site inhibitors that perturb the interface contact surface with long, spike-like substituents;
  - Inhibitors that induce and stabilize formation of the twisted, functionally incompetent homodimer;
  - Small ligands that fix a loop in the contact surface in a geometry that blocks the formation of the original homodimer; and
  - Filling a transient pocket with inhibitors that irreversibly bind to the thiol group of a cysteine residue present only in the bacterial enzymes.

Bibliography and Original Papers

C. Romier, K. Reuter, D. Suck, D. and R. Ficner, Crystal structure of tRNA-guanine transglycosylase: RNA modification by base exchange, EMBO J., **15**, 2850–2857 (1996)

U. Grädler, H.-D. Gerber, D. A. M. Goodenough-Lashua, G. A. Garcia, R. Ficner, K. Reuter, M. T. Stubbs and G. Klebe. A New Target for Shigellosis: Rational Design and Crystallographic Studies of Inhibitors of tRNA-Guanine Transglycosylase J. Mol. Biol., **306**, 455–467 (2001)

E. A. Meyer, R. Brenk, R. K. Castellano, M. Furler, G. Klebe, F. Diederich. De Novo Design, Synthesis, and in Vitro Evaluation of Inhibitors for Prokaryotic tRNA-Guanine Transglycosylase (TGT): A Dramatic Sulfur Effect on Binding Affinity, ChemBioChem **2**, 250–253 (2002)

R. Brenk, L. Naerum, U. Grädler, H.-D. Gerber, G. A. Garcia, K. Reuter, M. T. Stubbs and G. Klebe. Virtual Screening for Submicromolar Leads of TGT based on a New Unexpected Binding Mode Detected by Crystal Structure Analysis J. Med. Chem., **46**, 1133–1143 (2003)

R. Brenk, M.T. Stubbs, A. Heine, K. Reuter, G. Klebe, Flexible adaptations in the structure of the tRNA modifying enzyme tRNA-guanine transglycosylase and its implications for substrate selectivity, reaction mechanism and structure-based drug design, ChemBioChem, **4**, 1066–1077 (2003)

R. Brenk, H-D. Gerber, J. Kittendorf, G.A. Garcia, K. Reuter, G. Klebe, From Hit to Lead: De Novo Design based on Virtual Screening Hits of Inhibitors of tRNA-guanine transglycosylase, a putative Target of Shigellosis Therapy, Helv. Chim. Acta Vol. **86**, 1435–1452 (2003)

W. Xie, X. Liu and R. H. Huang. Chemical Trapping and Crystal Structure of a Catalytic tRNA Guanine Transglycosylase Covalent Intermediate. Nat. Struct. Biol., **10**, 781–788 (2003)

E. A. Meyer, M. Furler, F. Diederich, R. Brenk, G. Klebe, Synthesis and In Vitro Evaluation of 2-aminoquinazolin-4(3H)-one-based Inhibitors for tRNA-Guanine Transglycosylase (TGT), Helv. Chim. Acta, **87**, 1333–1356 (2004)

R. Brenk, E. Meyer, K. Reuter, M.T. Stubbs, G.A. Garcia, F. Diederich, G. Klebe, Crystallographic study of inhibitors of tRNA-guanine transglycosylase suggests a new structure-based pharmacophore for virtual screening, J. Mol. Biol., **338**, 55–75 (2004)

B. Stengl, K. Reuter and G. Klebe. Mechanism and Substrate Specificity of tRNA–Guanine Transglycosylases (TGTs): tRNA Modifying Enzymes from the Three Different Kingdoms of Life Share a Common Mechanism. Chem-BioChem, **6**, 1926–1939 (2005)

E. A. Meyer, N. Donati, M. Guillot, B. Schweizer, F. Diederich, B. Stengl, R. Brenk, K. Reuter, G. Klebe, Synthesis, biological evaluation, and crystallographic studies of extended guanine-based (*lin*-benzoguanine) inhibitors for tRNA-guanine transglycosylase (TGT), Helv. Chim. Acta, **89**, 573–597(2006)

B. Stengl, E. A. Meyer, A. Heine, R. Brenk, F. Diederich and G. Klebe. Crystal Structures of tRNA-Guanine Transglycosylase (TGT) in Complex with Novel and Potent Inhibitors Unravel Pronounced Induced-fit Adaptations and Suggest Dimer Formation upon Substrate Binding. J. Mol. Biol., **370**, 492–511 (2007)

S. Hörtner, T. Ritschel, B. Stengl, C. Kramer, G. Klebe, F. Diederich. Potent inhibitors of tRNA-Guanine Transglycosylase, an Enzyme linked to the Pathogenicity of the Shigella Bacterium: Charge-assisted Hydrogen Bonding. Angew. Chem. Int. Ed., **46**, 8266–8269 (2007)

T. Ritschel, C. Atmanene, K. Reuter, A. Van Dorsselaer, S. Sanglier-Cianférani, G. Klebe, An Integrative Approach combining Noncovalent Mass Spectrometry, Enzyme Kinetics and X-ray Crystallography to Decipher Tgt Protein-Protein and Protein-RNA Interaction. J. Mol. Biol., **393**, 833–847 (2009)

P. C. Kohler, T. Ritschel, W. B. Schweizer, G. Klebe, F. Diederich, High-Affinity Inhibitors of tRNA–Guanine Transglycosylase Replacing the Function of a Structural Water Cluster, Chem. Eur. J., **15** 10809–10817 (2009)

T. Ritschel, P. C. Kohler, G. Neudert, A. Heine, F. Diederich, G. Klebe, How to Replace the Residual Solvation Shell of Polar Active-site Residues to Achieve Nanomolar Inhibition of tRNA-guanine transglycosylase, ChemMedChem, **4**, 2012–2023 (2009)

L. J. Barandun, F. Immekus, P. C. Kohler, S. Tonazzi, B. Wagner, S. Wendelspiess, T. Ritschel, A. Heine, M. Kansy, G. Klebe, F. Diederich, From *lin*-Benzoguanines to *lin*-Benzohypoxanthines as Ligands for *Zymomonas mobilis* tRNA-Guanine Transglycosylase: Replacement of Protein-Ligand Hydrogen Bonding by Importing Water Clusters, Chem. Eur. J., **18**, 9246–9257 (2012)

I. Biela, N. Tidten-Luksch, F. Immekus, S. Glinca, Tran Xuan Phong Nguyen, H.-D. Gerber, A. Heine, G. Klebe, K. Reuter, Investigation of Specificity Determinants in Bacterial tRNA-Guanine Transglycosylase Reveals Queuine, the Substrate of Its Eukaryotic Counterpart, as Inhibitor, PLoS ONE **8**, e64240 (2013)

F. Immekus, L.J. Barandun, M. Betz, F. Debaene, S. Petiot, S. Sanglier-Cianférani, K. Reuter, F. Diederich, G. Klebe, Launching Spiking Ligands into a Protein-Protein Interface: A Promising Strategy to Destabilize and Break Interface Formation in a tRNA Modifying Enzyme. ACS Chem. Biol., **8**, 1163–1178 (2013)

L. J. Barandun, F. Immekus, P. C. Kohler, T. Ritschel, A. Heine, P. Orlando, G. Klebe, F. Diederich, High-affinity Inhibitors of *Zymomonas mobilis* tRNA-Guanine Transglycosylase through Convergent Optimization. Acta Cryst., Sect., D **69**, 1798–1807 (2013)

S. Jakobi, T.X.P. Nguyen, F. Debaene, A. Metz, S. Sanglier-Cianférani, K. Reuter, G. Klebe, Hot-spot Analysis to Dissect the Functional Protein-Protein Interface of a tRNA-modifying Enzyme. Proteins, **82**, 2713–2732 (2014)

M. Neeb, P. Czodrowski, A. Heine, L. Jakob, S. Barandun, C. Hohn, F. Diederich, G. Klebe, Chasing protons: How isothermal titration calorimetry, mutagenesis, and p*Ka* calculations trace the locus of charge in ligand binding to a tRNA-binding enzyme. J. Med. Chem., **57**, 5554–5565 (2014)

M. Neeb, M. Betz, A. Heine, L. J. Barandun, C. Hohn, F. Diederich, G. Klebe, Beyond affinity: enthalpy–entropy factorization unravels complexity of a flat structure-activity relationship for inhibition of a tRNA-modifying enzyme, J. Med. Chem., **57**, 5566–5578 (2014)

S. Jakobi, T.X.P. Nguyen, F. Debaene, S. Cianférani, K. Reuter, G. Klebe, What Glues a Homodimer Together: Systematic Analysis of the Stabilizing Effect of an Aromatic Hot Spot in the Protein–Protein Interface of the tRNA-modifying Enzyme Tgt. ACS Chem. Biol., **10**, 1897–1907 (2015)

L.J. Barandun, F.R. Ehrmann, D. Zimmerli, F. Immekus, M. Giroud, C. Grünenfelder, W.B. Schweizer, B. Bernet, M. Betz, A. Heine, G. Klebe, F. Diederich, Replacement of Water Molecules in a Phosphate Binding Site by Furanoside-Appended *lin*-Benzoguanine Ligands of tRNA–Guanine Transglycosylase (TGT), Chem. Eur. J. 21, 126–135 (2015)

M. Neeb, C. Hohn, F. R. Ehrmann, A. Härtsch, A. Heine, F. Diederich, G. Klebe, Occupying a Flat Subpocket in a tRNA-modifying Enzyme with Ordered or Disordered Sidechains: Favorable or Unfavorable for Binding?, Bioorg. Med. Chem., **24**, 4900–4910 (2016)

F. R. Ehrmann, J. Kalim, T. Pfaffeneder, B. Bernet, C. Hohn, E. Schäfer, T. Botzanowski, S. Cianférani, A. Heine, K. Reuter, F. Diederich, G. Klebe, Swapping Interface Contacts in the Homodimeric tRNA Guanine Transglycosylase: An Option for Functional Regulation. Angew. Chem. Int. Ed. Engl., **57**, 10085–10090 (2018)

E. Hassaan, C. Hohn, F. R. Ehrmann, F. W. Goetzke, L. Movsisyan, T. Hüfner-Wulsdorf, M. Sebastiani, A. Härtsch, K. Reuter, F. Diederich, G. Klebe, Fragment Screening Hit Draws Attention to a Novel Transient Pocket Adjacent to the Recognition Site of the tRNA-Modifying Enzyme TGT. J. Med. Chem., **63**, 6802–6820 (2020)

A. Nguyen, D. Nguyen, T. Nguyen, Tran, M. Sebastiani, S. Dörr, O. Hernandez-Alba, F. Debaene, S. Cianférani, A. Heine, G. Klebe, K. Reuter, The importance of charge in perturbing the aromatic glue stabilizing the protein–protein interface of homodimeric tRNA-guanine transglycosylase, ACS Chem. Biol. **15**, 3021–3029(2020)

D. Nguyen, X. Xie, S. Jakobi, F. Terwesten, A. Metz, T. X. P. Nguyen, V. A. Palchykov, A. Heine, K. Reuter, G. Klebe, Targeting a Cryptic Pocket in a Protein–Protein Contact by Disulfide-Induced Rupture of a Homodimeric Interface. ACS Chem. Biol., **16**, 1090–1098 (2021)

D. Nguyen, D. Abdullin, C. A. Heubach, T. Pfaffeneder, A. Nguyen, A. Heine, K. Reuter, F. Diederich, O. Schiemann, G. Klebe, Unraveling a Ligand-induced Twist of a Homodimeric Enzyme by Pulsed Electron–electron Double Resonance. Angew. Chem. Int. Ed Engl., **60**, 23419–23426 (2021)

A. Nguyen, G. Gemmecker, C. A. Softley, L. D. Movsisyan, T. Pfaffeneder, A. Heine, K. Reuter, F. Diederich, M. Sattler, G. Klebe, ^{19}F-NMR unveils the ligand-induced conformation of a catalytically inactive twisted homodimer of tRNA-guanine transglycosylase. ACS Chem. Biol., **17**, 1745–1755 (2022)

M. Sebastiani, C. Behrens, S. Dörr, H. D. Gerber, R. Benazza, O. Hernandez-Alba, S. Cianférani, G. Klebe, A. Heine, K. Reuter, Structural and Biochemical Investigation of the Heterodimeric Murine tRNA-Guanine Transglycosylase, ACS Chem. Biol., **17**, 2229–2247 (2022)

Drugs and Drug Action: Sucesses of Structure-Based Design

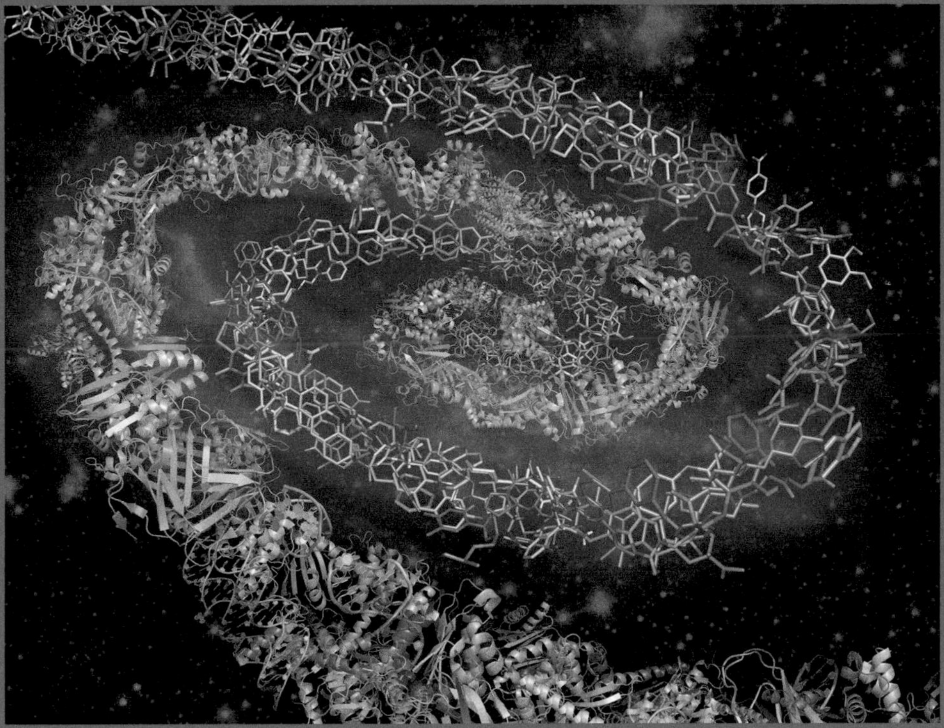

The design and development of a suitable small-molecule drug candidate for a given macromolecular target structure, selected from the universe of all possible proteins, means finding the most suitable compound from the chemical space of all conceivable pharmacologically relevant molecules. This is equivalent to the task of merging both the chemical and the biological space. The figure tries to symbolize the intersection of these two spaces by the spiral nebulae of protein and ligand structures merging into each other (announcement poster of the author's working group on the occasion of a conference in 2007, Rauischholzhausen, Marburg).

Contents

How Drugs Act: Concepts for Therapy

Contents

G. Klebe, *Drug Design*, https://doi.org/10.1007/978-3-662-68998-1_22

How many drug targets and modes of action are there? It has been estimated that currently available drugs act on approximately **600–700 targets**. Optimistic projections suggest that this number could be increased by a factor of 10. But this number is still small compared to the diversity of proteins that play a role in our organism. Our genome has been sequenced. We know that the number of our coding genes (about 21,500) is much smaller than originally thought (Sect. 12.3). However, the number of relevant proteins encoded by these genes is much larger because, among other reasons, versatile **posttranslational modifications** and **alternative splicing** cause the genetic information to be diversified into multiple protein variants. So our genome is mapped, but do we know the function of each gene? How can predictions about proteins and their functions and possible roles in pathophysiology be extracted from this flood of sequence information? Many of the proteins discovered in the genome can be assigned to protein families based on sequence comparisons. Nevertheless, a significant portion of our genetic information still awaits annotation. The first step has been taken, but what are the spatial structures of these proteins for which only sequences are known? Which ligands are recognized by these proteins and what is their biochemical role in our organism? The **biochemical function**, that is whether a protein is a protease, an ion channel, or a transporter, still does not provide information about the **systemic role** of the protein in the functional processes of a cell or an entire organism. The spatial structure of a protein is responsible for this function. For this reason, the spatial structures of the proteins in our genome are intensively studied or their geometries are predicted by computer methods. The goal is to map the structural space of all proteins as well as possible. Then it might be possible to find a spatially resolved and sufficiently homologous reference structure for each discovered sequence. Today, the structures of all members of a few gene families have already been determined. It is only a matter of time before we have the spatial structure of all relevant proteins. The road may be long and arduous, but it is clearly marked out. Will this revolutionize the market for potential drugs and enable completely new therapeutic approaches? The **chemical space** of all conceivable compounds and the **biological space** of all possible pathology-relevant proteins were discussed in Sects. 11.4 and 12.4. Drug design attempts to merge these two spaces. In the cross-section of both spaces, there are molecules to be found as candidates for potential drugs. This chapter attempts to provide an overview of the range of mechanisms of action and some general principles in the families of similar proteins and biomolecules. Details of the individual families can be found in Chaps. 23–32.

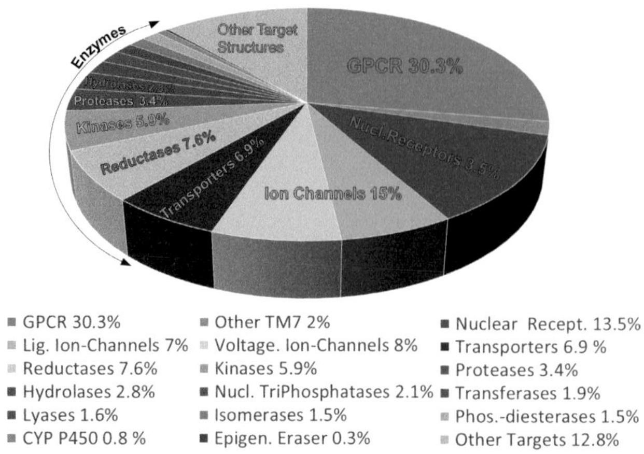

Fig. 22.1 Distribution of the target proteins for drugs that are on the market today

22.1 The Druggable Genome

Andrew Hopkins and Colin Groom first provided an **overview of the pharmaceutical market** in 2002. Later in 2017, a more comprehensive assessment of this market was published by a consortium of authors, suggesting some shifts over the past 15 years (Fig. 22.1). About one-third of the drugs on the market today inhibit enzymes. Another 30% affect the behavior of G-protein-coupled receptors (GPCRs). About 15% develop their therapeutic significance at ligand- or voltage-gated ion channels. Nearly 7% affect transporters. About 13.5% of drugs target nuclear hormone receptors. However, these market shares do not correspond to the frequency of these targets in our genome. For example, GPCRs represent only 2.3% of our genome if sensory GPCRs are excluded. GPCRs represent about 15% of the **"druggable" genome**, the part of the genome whose function could be modulated by drug therapy. In contrast, kinases represent more than 22% of the druggable genome, however, only 5.9% of currently used drugs act on these enzymes. Due to the long time span between current discovery research and market launch, the drug market is expected to continue to change in the coming years, especially as the proportion of macromolecular drugs, known as biologicals, has recently increased sharply. These are not included in the figures above.

In the chapters that follow, examples of individual target structures that represent potential drug therapy targets are presented. They are discussed on the basis of their key structural features because the structure of the target generally defines what is needed to qualify a molecule as an inhibitor, agonist, antagonist, or allosteric modulator. These principles serve as a general concept for the design of new drugs. In modern drug discovery, the target structure for which a new compound is being sought is usually known. In many historical examples of drug development, this was not initially the case. Today,

Table 22.1 Enzyme classification based on the four-digit number code

Class	Name	Biochemical function	Examples	Coenzymes
EC 1.x.x.x	Oxido-reductases	Catalyze redox reactions; transfer of H- and O-atoms or electrons between molecules	Dehydrogenases, Oxidases, Oxygenases, Hydroxylases	NAD^+, $NADP^+$, FAD, FMD, and Liponic acid
EC 2.x.x.x	Transferases	Transfer functional groups such as methyl, acyl, amino, or phosphate groups from one molecule to another	Phosphotransferases (including kinases) Aminotransferases	*S*-Adenosyl methionine, Biotin, cAMP, ATP, Thiamine pyrophosphate (TPP), Tetrahydrofolic acid
EC 3.x.x.x	Hydrolases	Hydrolytic cleavage of molecules	Esterases, Lipases, Phosphatases, and Peptidases	Not needed
EC 4.x.x.x	Lyases	Nonhydrolytic addition or cleavage of groups on molecules, particularly double bonds, cleavage of C–C, C–N, C–O, and C–S bonds	Decarboxylases, Aldolases, Synthases	TTP, Pyridoxal phosphate
EC 5.x.x.x	Isomerases	Intramolecular rearrangement and isomerization within a molecule	Racemases, Mutases	Glucose-1,6-bisphosphate, Vitamin B12
EC 6.x.x.x	Ligases	Coupling of two molecules by the formation of C–C, C–N, C–O, or C–S bonds by using ATP	Synthestases, Carboxylases	ATP, NAD^+

however, many modes of action are known. Peter Imming and his research group at the University of Halle, Germany, have compiled a summary of the modes of action for a broad collection of drugs in use today. In addition, Tudor Oprea's **WOMBAT database** at the University of New Mexico in Albuquerque, USA, provides quick access to functionally annotated drugs along with their characteristic properties. The **ChEMBL database** continues to be a treasure trove for correlating the chemical structure and biological action of a target structure.

22.2 Enzymes as Catalysts in Cellular Metabolism

All metabolic processes, biosynthetic pathways, and the regulation of important physiological processes are mediated by enzymes. Enzymes are **macromolecular biocatalysts** that enable complex chemical reactions to take place in an aqueous medium, usually at 37 °C and under normal pressure. During evolution, families of enzymes with analogous architecture and identical catalytic sites have evolved. Small differences in the structure of the binding sites result in quite different substrate specificities, making these enzymes either highly specific or highly promiscuous, depending on the function required.

Enzymes do not bind particularly strongly to their substrates and reaction products. The bound conformation of the ligand is often different from the energetically most favorable conformation in aqueous solution. An enzyme binds the substrate in a geometry that prepares it for the transition state of the reaction. In addition, polar groups can induce the necessary charge shifts. The enzyme stabilizes the transition state of a chemical reaction by the spatial arrangement and orientation of its reactive groups. At the same time, the enzyme lowers the activation energy of the reaction and allows for sometimes dramatic rate accelerations of chemical reactions. After dissociation of the product, the enzyme is available for the conversion of the next substrate molecule.

Enzymes are classified according to the reactions they catalyze. An international commission has divided enzymes into six classes, each of which is assigned a **four-digit code** (Table 22.1). The main class indicates the type of reaction catalyzed (redox reactions, transfer reactions, transfer of functional groups to water, cleavage and elimination reactions, isomerization of groups within the substrate, or condensation or linkage of molecular groups). The remaining numbers classify, for example, which group is transferred or whether the protein is regulated by cofactors. The **MEROPS database**, maintained by the Sanger Institute in Cambridge, England, provides quick access and a broad overview of proteases, their substrates, reaction mechanisms, and selectivities. Chaps. 23–27 review the major classes of enzymes for which drugs have been successfully developed. A seventh class has now been defined. It includes enzymes that catalyze the movement of ions or molecules across membranes or their separation within membranes. The reaction actually describes the transfer of these particles, and the subclasses differentiate the types of components transferred. The next subclass further differentiates the reaction processes that are the driving force for spatial transport.

22.3 How Do Enzymes Push Substrates Towards the Transition State?

To illustrate how an enzyme prepares its substrate for the **transition state of a reaction**, let us consider an example. The crystal structure of creatinase with its natural substrate creatine **22.1** and a very similar inhibitor, carbamoylsarcosine **22.2**, was determined in the research group of Robert Huber at the Max Planck Institute in Martinsried, Germany. The enzyme catalyzes the cleavage of creatine into urea and sarcosine (◘ Fig. 22.2). The central carbon of the C–N bond in the guanidinium moiety of creatine is nucleophilically attacked by a water molecule. All three C–N bonds in the guanidinium moiety have a double bond character and the group prefers a planar geometry due to electron delocalization. How does the enzyme manage to distort creatine towards the transition state of the reaction to prepare it for nucleophilic attack and bond cleavage? The zwitterionic creatine is bound by its guanidinium function through two glutamate residues that form two salt-bridge-like hydrogen bonds (◘ Figs. 22.3 and 22.4). The opposite acid function finds strongly polarizing binding partners in two arginine residues. In addition, a water molecule is found near the central imine-like carbon atom in the crystal structure. Next to it in the binding pocket is a histidine residue. This histidine orients the water molecule in exactly the right position and also supports the abstraction of a proton from this water molecule. This increases the

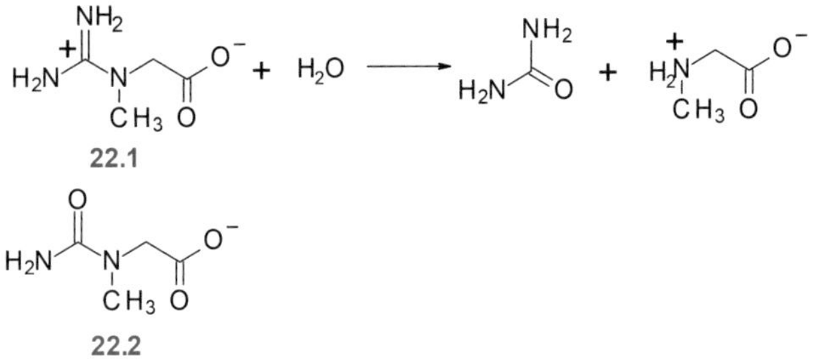

◘ **Fig. 22.2** The enzyme creatinase cleaves creatine **22.1** with water into urea and sarcosine. The structurally very similar molecule, carbamoylsarcosine **22.2**, is an inhibitor of this enzyme

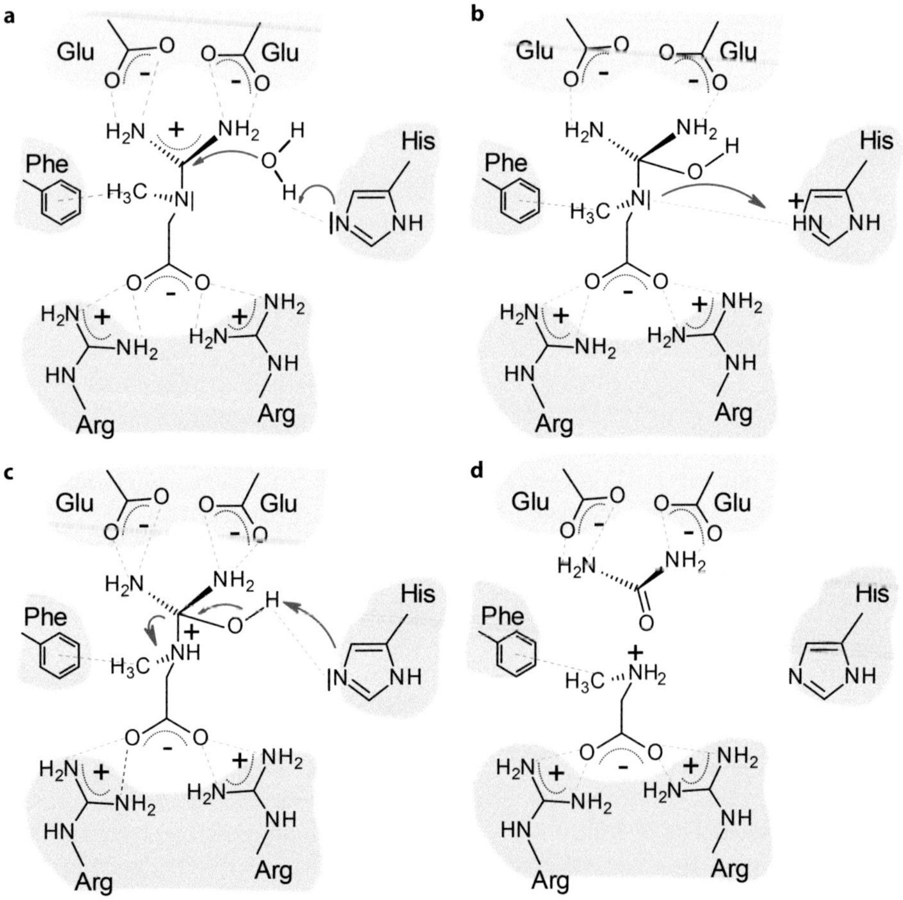

◘ **Fig. 22.3** **a** In the first step, a water molecule is polarized by a neighboring histidine so that a nucleophilic attack on the imine-like carbon is facilitated. **b** Then, the histidine transfers a proton to the central nitrogen atom. **c** The substrate reacts further in that a C=O double bond is formed and the C–N bond is cleaved. **d** The products urea and sarcosine leave the binding pocket

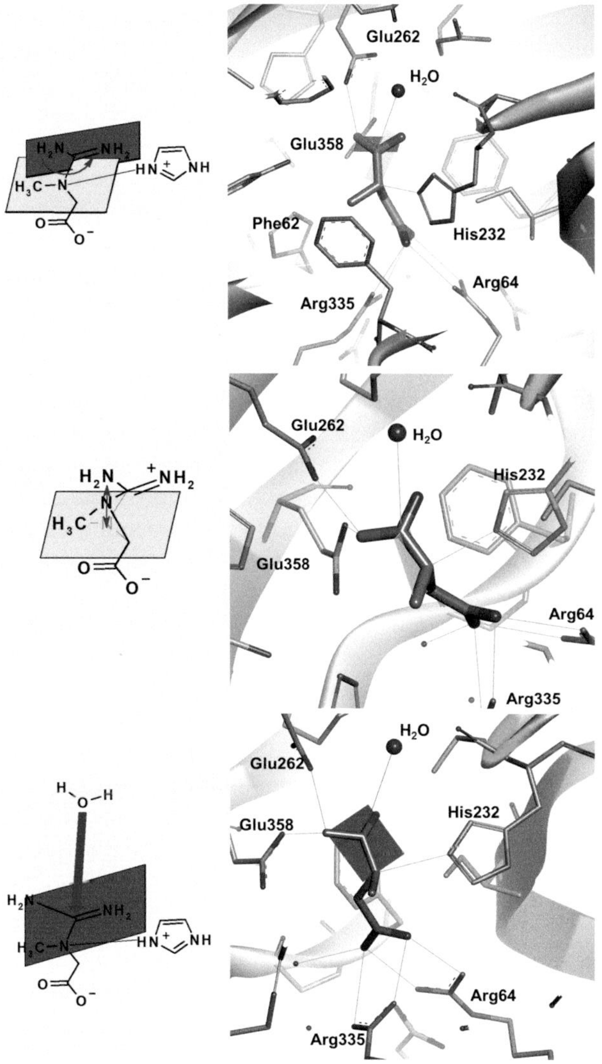

Fig. 22.4 *Upper row* The vice-like fixation of the creatine molecule by two glutamate and two arginine residues causes a twist in the guanidinium group, which is planar in the unbound state. This disrupts conjugation across the guanidinium moiety and weakens the C–N bond to be cleaved. The twist is indicated by the *red* and *yellow* planes passing through the atoms of the guanidinium group. *Middle row* The neighboring protonated histidine further polarizes the methyl-substituted nitrogen atom and involves it in a hydrogen bond. In the course of the reaction, the nitrogen atom takes on a pyramidal configuration, deviating from the plane (*yellow*) of its next three neighbors. *Bottom row* In the structure with the substrate-like inhibitor carbamoylsarcosine, a water molecule can be found at a position from which the nucleophilic attack on the substrate creatine is initiated. This occurs from above and diagonally behind the C=N bond. (► https://sn.pub/MXp0Uw)

nucleophilicity of the water molecule to form an OH^- group. The **vice-like fixation of the guanidine group** by the two glutamate residues causes a twisting of this building block, which is, as mentioned, planar in the unbound state. This disrupts the conjugation and significantly weakens the C–N bond to be cleaved. Nucleophilic attack occurs and a **tetrahedral transition state** is formed. At the same time, the now protonated histidine is able to polarize the methyl-substituted nitrogen atom and involves this atom in a hydrogen bond. This prepares the substrate for the transition state of the bond-breaking reaction step. After the proton is transferred from the histidine to the substrate, a positive charge is formed on the nitrogen atom of the bond to be cleaved. Histidine accepts a proton from the oxygen atom of the tetrahedral transition state as a C=O double bond is formed, and the central C–N bond is cleaved. The products then leave the binding pocket. In this way, the enzyme creates a **stereo-electronically complementary environment** for the cleavage reaction. Its polar groups position the water molecule correctly for nucleophilic attack, and histidine induces a **pyramidalization** of the nitrogen atom in the bond to be broken. At the same time, it serves as both a proton donor and acceptor during the reaction.

The crystal structure in Fig. 22.4 was determined together with carbamoylsarcosine. This molecule differs from the substrate creatine in that a nitrogen atom has been replaced by an oxygen atom. However, this part of the molecule does not carry a positive charge like creatine. The addition of the nucleophilic OH^- leads to decomposition and compensation of the charge in the guanidinium moiety of creatine. A similar attack on carbamoylsarcosine would lead to the formation of a negative charge next to the two negatively charged glutamates. This is energetically unfavorable. As a result, the cleavage reaction does not occur on this molecule, and the conversion is blocked. This example shows how perfectly the substrate and the enzyme must match. Small changes can drastically alter this system, turning a substrate molecule into an inhibitor of the desired transformation reaction.

22.4 Enzymes and Their Inhibitors

Enzymes can be organized into multienzyme complexes that carry out multiple reactions on a substrate sequentially. They can also form cascades in which one enzyme activates the inactive precursor of the next enzyme. This activation is passed on to the next enzyme, and the next, and so on. The coagulation cascade (Sect. 23.3) is activated by two independent pathways, each with several steps that ultimately merge into a common pathway. Thus, a small initiating event is amplified by several orders of magnitude. This is good for normal coagulation after injury, but in the context of a coagulopathy (i.e., a tendency to form clots too easily), it can have disastrous consequences!

Many inhibitors prevent the catalytic activity of an enzyme by occupying the site where the substrate binds site, the so-called orthosteric site. Such inhibitors are called **competitive inhibitors**. In addition, there are **allosteric inhibitors**, which bind to another position on the enzyme and cause a change in its three-dimensional structure or dynamic properties. This can prevent the enzyme from assuming the conformation necessary for catalysis and lead to a weakening of catalytic activity. Detailed studies of enzyme kinetics allow the distinction between competitive and noncompetitive inhibition. Depending on the type of interaction with the enzyme, **reversible and irreversible inhibitors** can be distinguished. In the case of reversible inhibitors, the binding to the enzyme must be strong enough to reliably prevent the conversion of the substrate. Some reversible inhibitors form a covalent bond to the catalytic center that is chemically labile and, therefore, fully reversible, such as a hemiacetal bond. Irreversible inhibitors react with the enzyme by forming a chemically stable bond. These inhibitors, or the reacting groups, cannot be removed and the enzyme remains inhibited for the remainder of the enzyme's life until it is degraded by the organism. In addition, there are naturally occurring protease inhibitors that bind reversibly but adhere so strongly that the complex is degraded before the inhibitor is released.

The rational design of an enzyme inhibitor usually starts with the structure of the substrate. A particularly successful approach is to mimic the transition state with a chemically analogous group that is not attacked by the enzyme. Many examples of the design of such inhibitors are given in Chaps. 23–27. In general, irreversible enzyme inhibitors play a minor role compared to reversible inhibitors, but important drugs such as acetylsalicylic acid (ASA, Sect. 27.9), omeprazole (Sect. 30.9), clopidogrel (an inhibitor of platelet aggregation), penicillins and cephalosporins (Sect. 23.7), and some monoamine oxidase inhibitors (Sect. 27.8) belong to this group.

22.5 Receptors as Target Structures for Drugs

Receptors are proteins or protein complexes that

- Mediate the information exchange between cells (membrane-bound receptors),
- Regulate hormone-controlled gene expression (soluble receptors or transcription factors), and
- Are coupled to ion channels and control the flow of ions into or out of a cell along a concentration gradient.

Important **membrane-bound receptors** include those for adrenaline, serotonin, dopamine, histamine, acetylcholine, adenosine, and thromboxane, and for peptides such as the enkephalins (opiate receptors), neurokinins, and endothelins for glycoproteins, as well as the group of sensory receptors. **Neurotransmitters** are the endogenous agonists of many membrane-bound receptors (Sect. 1.4). Nerve cells are connected to each other by synapses; these are zones where chemical information is transmitted by neurotransmitters. The **synaptic gap** is located between the transmitting cell (presynaptic neuron) and the receiving cell (postsynaptic neuron). Neurotransmitters are synthesized in the presynaptic neuron and stored in vesicles. Upon nerve stimulation, they are released into the synaptic cleft. There, by binding to a specific receptor on the postsynaptic neuron, they cause a change in membrane potential, thereby stimulating that cell. After reuptake into the cell, entrapment in vesicles, or degradation by, for example, the enzyme monoamine oxidase (amines), esterases (acetylcholine), and peptidases, or in glial cells by the activity of catechol-O-methyltransferase, the effect rapidly subsides (see ◘ Fig. 22.7).

Inside the cell, these receptors act on **G-proteins** (◘ Fig. 22.5), which derive their name from guanosine di- and triphosphate that they bind. All **G-protein-coupled receptors (GPCRs)** are identical in structure and function. They consist of a protein chain with seven hydrophobic segments that penetrate the cell membrane and anchor the receptor. These segments are connected by loops. To date, about 800 human GPCR sequences

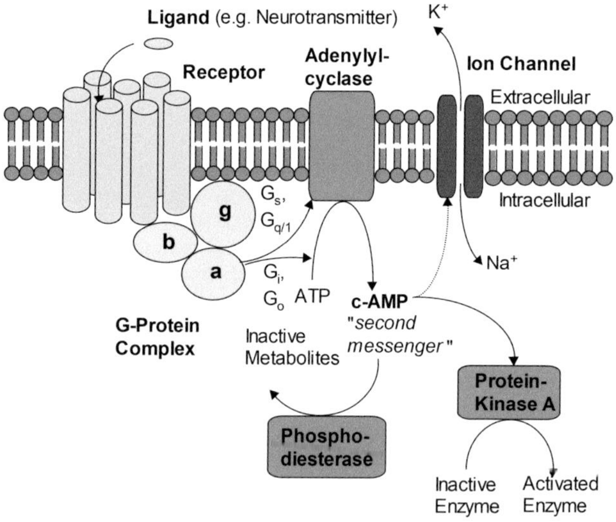

◘ **Fig. 22.5** Schematic representation of the structure and function of a G-protein-coupled receptor (GPCR). The seven cylinders symbolize the seven transmembrane helices. The extra- and intracellular loops that bind the helices are not shown. After binding an agonist, the α-subunit dissociates from the so-called G-protein complex. If a G_s or $G_{q/11}$ protein is present, then an enzyme is activated that generates an internal hormone, a "second messenger." For example, the membrane-bound enzyme adenylate cyclase generates cyclic adenosine monophosphate (cAMP) from adenosine triphosphate (ATP). This second messenger can further affect target proteins via protein kinase A, or open an ion channel. To avoid an overreaction, cAMP is constantly being degraded by the enzyme phosphodiesterase. $G_{i/0}$ proteins inhibit enzymes that form second messengers

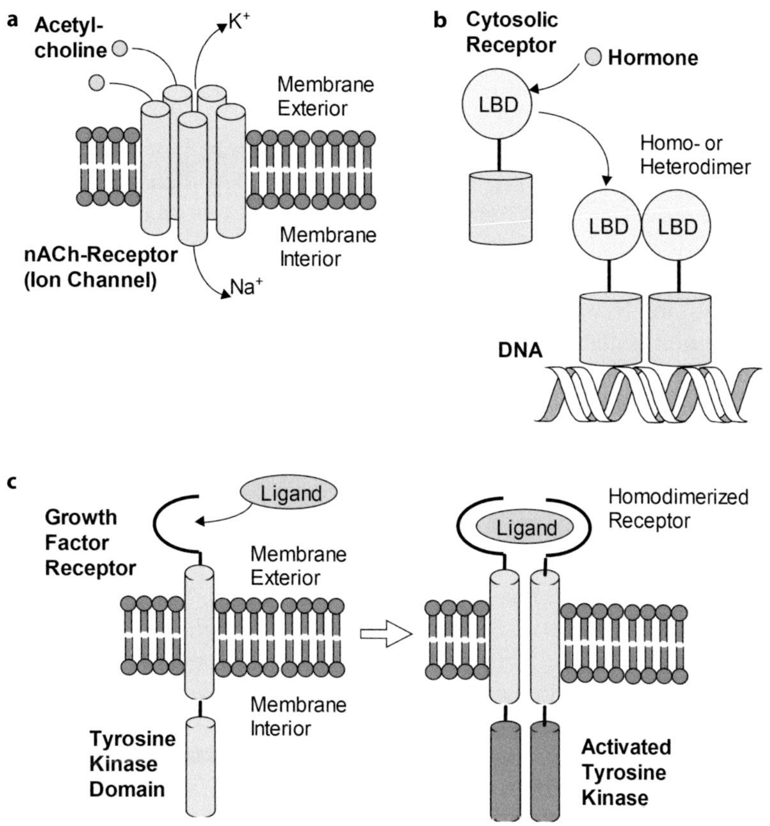

Fig. 22.6 **a** The nicotinic acetylcholine receptor (nAChR) is a ligand-gated ion channel (Sect. 30.4). Here the cylinders do not stand for segments but rather for five separate proteins, each of which has four transmembrane domains. After binding acetylcholine, the channel is quickly opened. For example, chloride channels use the same principle of construction (Sect. 30.7). **b** Soluble receptors dimerize after agonist docking to their ligand-binding domains (LBD). Here homodimers composed of two identical receptors as well as heterodimers of two different receptors can be formed. The so-called zinc fingers of the DNA-binding domains (DBD) recognize very specific sequences of DNA. A particular DNA segment is addressed by dimerizing two receptor units. **c** Membrane-bound receptors for growth factors and insulin also dimerize. Two receptors form a complex in the membrane and in doing so activate the intracellular domain of the receptor, in this case, a tyrosine kinase

are known, but new ones are still being discovered and characterized (Sect. 29.1).

After an **agonist** docks, the active conformation of the receptor is stabilized. **Antagonists** prevent agonists from docking, and **inverse agonists** stabilize the inactive conformation of the receptor.

The initiated receptor response proceeds along identical pathways, despite the different receptor types, and then branches off again. This economical principle of Nature is also used in other cases, such as the regulation of cell proliferation. The more or less pronounced effect specificity is achieved by

- The different structures of the agonists and receptors and the resulting activation of different G-proteins and effector proteins,
- The different receptor occupancy and density of different cells, and
- The location of the cells that produce and release the hormone or neurotransmitter. This is accomplished in very specific cells; neighboring cells or organs are not involved.

The picture of such receptors can be very complex. For example, there are two distinct groups of **acetylcholine receptors** that preferentially bind either muscarine, a toxin from the toadstool *Amanita muscaria*, or nicotine, the active ingredient in the tobacco plant *Nicotiana tabacum*. Unlike the muscarinic acetylcholine receptor, the nicotinic acetylcholine receptor (nAChR) is a ligand-gated ion channel (Sect. 30.5). It has a complex architecture of five protein chains located in the cell membrane (Fig. 22.6a). For the nAChR from the electric organ of the torpedo electric ray, a fish, cryo-electron microscopy images are available of the closed and open structure (after activation by acetylcholine) of the 290 kD nAChR protein complex (Sect. 30.5).

Many hormone receptors, such as those for thyroid hormone, sex hormones, the corticosteroids, and retinoic acid, are **soluble receptors** that can move freely in the cytosol, or cell fluid. After binding the agonist, the complex translocates to the nucleus. There it binds as a dimer to the signal sequences of the DNA, operator and repressor genes and induces or suppresses the new synthesis of specific proteins (Fig. 22.6b). These receptors are, therefore, also referred to as **transcription factors**.

All cytosolic hormone receptors or **nuclear receptors** are based on common structural principles (Sect. 28.2). They have domains with a DNA-binding site and a ligand-binding site. The DNA-binding site is highly conserved, i.e., its amino acid sequence varies very little between different receptors. It contains two "zinc fingers" consisting of two Zn^{2+} binding sites, which are highly conserved motifs that bind to very specific DNA segments known as recognition sequences. The ligand binding site is much more variable. Dimers, either of two identical receptors (homodimers) or of two different receptors (heterodimers), are formed to interact with DNA. Four zinc fingers in the dimer recognize a total of 12 base pairs of DNA.

Dimerization is also found in other classes of membrane-bound receptors that are not GPCRs. These include **receptors for growth factors** such as human growth hormone (hGH), epidermal growth factor (EGF), and insulin (Sect. 29.8). Upon binding to the factor, these receptors dimerize with the extracellular domains in the membrane. This activates **intracellular kinases** that are part of the receptor protein (Fig. 22.6c). In addition, there are receptors that must form complexes of more than two units to elicit a receptor response. These include a number of immunologically important receptors, as well as receptors for nerve growth factor (NGF) and **tumor necrosis factor** (TNF, Sect. 29.8).

Several examples of proteins that function as oligomers are presented in this section. In fact, **oligomerization** is also common in enzymes. There are many reasons why oligomerization is advantageous. On the one hand, there are functional requirements that require multiple adjacent domains, as described above. On the other hand, there may be mechanistic advantages, especially for enzymes. Individual domains of an oligomer are not necessarily independent. Their catalytic efficiency may depend on the current state of the other domains of the oligomer. This provides an additional means of regulating protein function. Oligomerization can also have another significance. The interior of a cell is highly crowded with proteins, ligands, substrates, and ions. It can be compared to a ticker-tape parade given to a winning football team: hectic pushing and shoving! One way to reduce this number without compromising catalytic productivity by sacrificing catalytic centers is to form oligomers.

22.6 Drugs Regulate Ion Channels: Our Extremely Fast Switches

Ion channels embedded in the cell membrane allow ions to enter or leave the cell along an electrochemical gradient when open. The opening or closing of the channel can be either voltage-, ligand-, or receptor-gated. All of these processes are extremely fast (Sect. 30.1).

Many drugs act on **voltage-gated ion channels** (Sect. 30.4). Local anesthetics and their derivatives are sodium channel blockers; they reduce the excitability of nerve cells. The fugu fish toxin, tetrodotoxin (Sect. 30.4), also blocks this channel. Other antiarrhythmic drugs block potassium channels. Substances that stabilize the K^+ channel in the open form, called K^+ channel openers, have vasodilatory and antihypertensive effects. The antidiabetic sulfonylureas are K^+ channel blockers that act on the insulin-producing cells of the pancreas (Sect. 30.2). Voltage-gated calcium channels are modulated by ligands such as the nifedipine-type antagonist or blockers such as the verapamil or diltiazem-type ligands (Sect. 30.4).

The nicotinic acetylcholine receptor (nAChR, Fig. 22.6a), glutamate receptors, and GABA receptors belong to the **Cys–loop family of ligand-gated ion channels** (Sects. 30.5, 30.6, and 30.7). The opening and closing of the channel is not caused by an electrical impulse, but by the binding of a ligand.

Benzodiazepine-type tranquilizers (Sect. 30.7) enhance the binding of the neurotransmitter γ-aminobutyric acid (GABA) to chloride channels. Prolonged opening of this channel results in an increased influx of chloride ions and a change in the response behavior of nerve cells. They bind between different domains of the channel. Barbiturates and inhaled anesthetics also act on GABA receptors, but at a different binding site between different domains.

22.7 Blocking Transporters and Water Channels

Transporters are proteins that affect the active uptake of molecules or ions into cells. They play a crucial role in the digestive process. Because amino acids and sugars cannot cross membranes on their own, they can only be absorbed in the digestive tract with the help of **transporters**.

Transporters are also extremely important for signal transmission in nerve cells. Once released, a **neurotransmitter** must be quickly removed from the synaptic cleft to prevent prolonged stimulation of the nerve cell. This is accomplished in part by metabolic degradation, but this is very wasteful for the releasing cell. Uptake (often incorrectly called reuptake) by a specific transporter is more economical. The neurotransmitter is stored in vesicles and made available for the next release.

Transporters work against concentration gradients. The transport process is relatively slow, much slower than an ion channel, and it requires energy. The amino acid sequence of the specific transporter is known for many neurotransmitters, amino acids, sugars, and nucleosides. Like G-protein-coupled receptors, transporters are divided into many families. Most have an even more complex structure with 12 transmembrane domains (Sect. 30.10).

A few drugs target the transporters directly and displace the natural ligands. The euphoric effects of cocaine (Sect. 3.4) are due to its binding to the dopamine transporter, which is responsible for the active transport and uptake of dopamine into neurons. A rapid onset of cocaine causes a delayed uptake of dopamine from the synaptic cleft, which is responsible for the typical physical and psychiatric effects. Some antidepressants are ligands for the noradrenaline and serotonin transporters (Sect. 1.4). They are bound but not transported into the cell. In contrast, some amino acid analogues are transported into neurons by transporters and act as neurotoxins. The complex interplay of neurotransmitters, enzymes, receptors, and transporters is illustrated in ◘ Fig. 22.7. Some anti-gout drugs bind to the uric acid transporter. They displace uric acid, inhibit its absorption from primary urine, and accelerate its excretion with urine. There are even specific transporters for bile acids.

In addition to the transporters described above, other members of this protein class are also important for the uptake or excretion of foreign substances into or from cells. Tumor cells often respond to therapeutic measures by developing multiple resistance to many structurally different substances (Sect. 30.10). The **glycoprotein GP 170**, also a transporter with 12 transmembrane domains, is responsible for this process.

In contrast to ion channels, **ion transporters** work against concentration gradients. This is an active process that requires energy (Sect. 30.9). Drugs can also affect this process. One example is drugs that increase urine production: diuretics. They inhibit different ion transporters. The Na^+/K^+ ATPase, a pump that exchanges sodium for potassium ions, is inhibited by cardiac glycosides, which are used to treat heart failure. Substances such as omeprazole (Sects. 9.5 and 30.9) inhibit the H^+/K^+ ATPase, the so-called **proton pump**. Nature uses special water channels to regulate water homeostasis and also to transport small, uncharged molecules such as glycerol or urea across the cell membrane quickly and selectively. Unlike transporters, and analogous to ion channels, these **aquaporins** allow water to flow along the osmotic gradient (Sect. 30.12). Ten isoforms with different permeabilities have been discovered in mammals. They are tetramers composed of six transmembrane helices. Each monomer unit forms a channel. The channels are partially made available for water homeostasis by the release from cytosolic vesicles or they can be activated by phosphorylation. Regulation of the water channels by drugs represents a concept for diuretic therapy, but the treatment of parasitic infections has also been discussed as an additional indication.

22.8 Modes of Action: A Never-Ending Story

Therapies for viral, bacterial and parasitic diseases attempt to target a specific pathogen. Different mechanisms are used, such as biosynthetic pathways that are not present in humans in identical form or that do not play an important role in humans. In this way, the risk of adverse effects can be minimized from the outset.

Antimetabolites are substances that are incorporated as a false substrate instead of the natural biological reagents, for example as enzyme cofactors or in DNA. One example is the sulfonamide sulfonamidochrysoidine. Its cleavage product sulfanilamide (Sect. 2.3) is similar to *p*-aminobenzoic acid, which is the starting material in the biosynthesis of an important bacterial cofactor, dihydrofolic acid. Only bacteria are affected. Humans are not dependent on this biosynthetic pathway. Like other

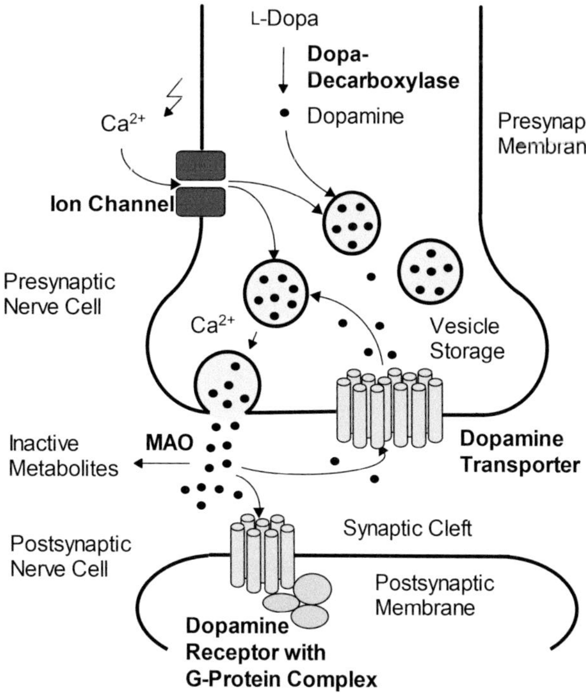

◘ **Fig. 22.7** Nerve signal transmission through neurotransmitters is based on a complex interplay of enzymes, receptors, ion channels, and transporters. Dopamine is produced by enzymatic decarboxylation of the amino acid L-DOPA. As with other neurotransmitters, it is stored in special vesicles. Upon electrical stimulation, Ca^{2+} ions flow into the cell. This causes the neurotransmitter to be released into the synaptic gap. The nerve impulse is conducted further by the interaction with the postsynaptic receptor. Finally, the uptake in the presynaptic cell is accomplished by a transporter and the neurotransmitter is stored in a vesicle again, or degraded by the enzyme monoamine oxidase (MAO). In the case of dopamine deficiency, dopamine can be released by enzymatic decarboxylation when the drug L-DOPA, an amino acid that enters the nerve cell through the amino acid transporter, is administered mammals, humans must obtain dihydrofolic acid from the diet. Some **antiviral and anticancer drugs** are **nucleoside analogues**. Depending on the type of structure, they use a modified base, a modified sugar, or both. All affect DNA or RNA synthesis. Acyclovir and a few other analogues enter cells as Trojan horses in an inactive form and are "armed" once inside the cell. Their activation is carried out by viral enzymes, and this process occurs only inside cells that have been infected by the virus (Sects. 9.5 and 32.5). Another mechanistic principle attempts to interfere with the translation process so that certain proteins are never produced by protein biosynthesis. The **translation of mRNA** is blocked by complexation with **antisense oligonucleotides** (Sect. 32.4). The resulting double-stranded mRNA cannot be read by the ribosome. This type of therapy can be used to treat exaggerated immune reactions, septic shock, arterial hypertension, pulmonary emphysema, or pancreatitis.

Many **antibiotics**, such as the penicillins and cephalosporins (Sect. 23.7), inhibit **bacterial cell wall biosynthesis**. In the latter process, they block the catalytic center of a transpeptidase, which has a similar mode of action to a serine hydrolase (Sect. 23.7). The antibiotic D-cycloserine, also an inhibitor of cell wall construction, penetrates into the interior of the bacteria using a D-alanine transporter. Other antibiotics are inhibitors of protein biosynthesis (Sects. 32.6 and 32.7). Tetracycline (Sect. 6.4), streptomycin (Sect. 6.4), and chloramphenicol (Sect. 9.2) also inhibit the protein synthesis machinery. They interact with the 30S or 50S subunit of the **ribosome** and block ribosomal peptide synthesis. The elucidation of the spatial structure of the ribosome has provided a basis for understanding the mechanism of action of a large number of macrolide antibiotics and has provided a perspective on the development of resistance mechanisms (Sect. 32.7). Antibacterial quinolone carboxylic acids inhibit **gyrase**. Gyrase is an enzyme that causes DNA to twist, thus, allowing DNA to be tightly packed in bacterial cells. Without this twisting, there is simply not enough space in the cell for the genetic material (Sect. 32.6) Polyene antibiotics are used to treat fungal infections. They form channels in the fungal cell membrane that cause a loss of intracellular ions and, consequently, cell death. Azoles inhibit the **biosynthesis of ergosterol**, which is essential for the formation of an intact cell membrane.

Alkylating agents play an important role in tumor therapy. Alkylation of **DNA** bases results in reading and writing errors that are much more severe in rapidly dividing tumor cells than in normal cells, but also have significant side effects (Sect. 12.14). Intercalating tumor therapeutics are planar molecules that slip between two base pairs of DNA (Sect. 14.9). The resulting disruption also leads to errors in cell division. Other DNA ligands bind to the minor or major groove on the outside of the double helix. **Paclitaxel** (Taxol, Sect. 6.1) and the **epothilones** are important agents in cancer therapy. They bind to tubulin, a protein that forms tube-like structures called microtubules. Since the formation of microtubules is essential for cell division, paclitaxel and the epothilones inhibit this process in a very specific way.

The immunosuppressant **cyclosporine** (Sect. 10.1, ◘ Fig. 10.2) blocks the activation of the immune system, the so-called helper cells. Two enzymes are involved in this process. One, cyclophilin, is a prolyl *cis–trans* isomerase. The other, calcineurin, is a Ca^{2+}/calmodulin-dependent phosphatase. Cyclosporine acts as a "putty" (or molecular glue) between these two proteins. The complex prevents the activation of helper cells and, thus, stops the stimulation of an immune response. Modern transplantation surgery would not be possible without the immunosuppressive cyclosporine and substances with an analogous mode of action.

RAS proteins play an important role in tumorigenesis. They are a family of relatively low molecular weight enzymes. RAS proteins with mutated active centers lose their ability to control cell division, and the cells divide unstoppably. They are therefore oncogenic, i.e., they cause tumors. About 50% of all lung and colon tumors have mutated *ras* genes, and about 95% of the *ras* genes in pancreatic tumors are mutated. There are other approaches to therapy. RAS proteins must move from the cytosol, the cell fluid, to the cell membrane to signal cell division. To do this, they are enzymatically tagged with a farnesyl group that anchors the protein to the cell membrane. Preventing membrane anchoring by inhibiting **farnesyl transferase** is an attractive approach for targeted cancer therapy (Sect. 26.11). It has now been shown that this principle of blocking protein farnesylation can also be used to treat parasitic infections. The farnesyl transferases of these parasites are the targets for drug development.

Tumor suppressor genes produce proteins, such as the p53 protein, that prevent cell division in the event of DNA damage. Any genetic defect in a cell that results in reduced levels of one or more of these proteins will allow cells with defective DNA to proliferate. Cell division becomes uncontrolled, and a tumor forms with additional genetic defects and uncontrolled growth.

Vascular occlusion is caused by the aggregation of platelets. Cell surface proteins called **integrins**, such as the adhesion glycoprotein $\alpha_{IIb}\beta_3$, play an important role. Two of these molecules form a complex with fibrinogen that "glues" cells together. The targeted development of small peptidomimetics (Sect. 10.6) based on an **RGD (Arg–Gly–Asp)** motif has been a major success in rational drug design (Sect. 31.2). Another system that plays an important role in cell–cell recognition between leukocytes and endothelial cells are the **selectins**. In cases of inflammation, E- and P-selectins are upregulated and presented on the endothelium, preventing leukocytes from rolling along the surfaces of the blood vessels

(Sect. 31.3). After adhesion, the leukocytes enter the vessel and migrate to the site of inflammation to fight the infection. In some diseases, excessive leukocyte infiltration leads to tissue damage. To prevent this, attempts have been made to interfere with the inflammatory cascade with compounds that block the surface expression of selectins. These receptors recognize sugar-like molecular groups on the surface of leukocytes, so the development of appropriate **carbohydrate-based antagonists** represents a suitable therapeutic concept.

For infection to occur, a surface contact must be established between the influenza virus and the host cell. The virus docks to the host cell with its capsule protein, **hemagglutinin**, to initiate endocytosis. Once inside the cell, the virus uses the protein biosynthetic machinery of the infected cell to make copies of itself. After maturation, the new virus must be expelled from the cell. To do this, the new virus buds on the cell surface and the bud is then snipped off. In the final step, viral **neuraminidase** cleaves sialic acid. This acid binds the viral hemagglutinin to the host cell. This final step can be blocked by neuraminidase inhibitors (Sect. 31.5). The inhibitors zanamivir and oseltamivir have been very successful. The **CCR5 receptor** antagonist maraviroc has been introduced for the treatment of human immunodeficiency virus (HIV) infection; the CCR5 receptor acts as an entry gate for the human immunodeficiency virus and its inhibition blocks host cell invasion.

The body's immune system has developed very effective defense mechanisms. **Antibodies** are one such defensive weapon. These proteins are able to selectively bind to foreign substances with high affinity and present them to phagocytic cells (i.e., dendritic cells and macrophages) for degradation. This sophisticated, highly specific recognition system for molecules ranging from very small low molecular weight antigens to complex macromolecular systems has been tapped for pharmaceutical therapy (Sect. 32.3). Today, about 200 artificially produced antibodies against a wide variety of targets are used in the treatment of many different diseases. There is no end in sight, as many newly developed antibodies are currently in clinical trials.

There are very few **truly "nonspecific" drugs**. Antacids, which neutralize gastric acid purely chemically, belong to this class, as do purely surface-active substances such as amphiphilic bactericides, fungicides, and hemolytics. Even the barbiturates, local anesthetics, inhalation anesthetics, and alcohol, long thought to be a nonspecific agent, have been shown to have specific mechanisms of action. Frequently, the demonstration of a specific effect has been made through the differential effects of pure enantiomers of a racemate. The β-antagonistic effect of an optically active β-blocker is associated with one enantiomer (Sect. 5.5). However, the nonspecific adverse effects on membranes are attributed equally to both enantiomers.

Is there anything new to discover? An absolute surprise was the discovery that nitric oxide(NO), a tiny gaseous molecule, is also a neurotransmitter. Substances that release NO or interfere with **NO biosynthesis** lower or raise blood pressure (Sect. 25.8). New subtypes of established receptors are constantly being discovered. The extent to which it makes sense to optimize a compound for absolute receptor specificity remains unsolved. It may well be the case that some compounds targeting multiple receptors or their subtypes are better suited for therapy than highly specific analogues. This is particularly true for compounds that bind to GPCRs. Here, the activity profile against a whole range of receptor subtypes is critical for the efficacy of a compound. Numerous GPCRs are even involved in our sense of smell, which follows this principle of multiple graded receptor responses (Sect. 29.7). This is the only way to achieve the finely tuned and nuanced diversity of perception. This is a broad area of research. To date, especially with CNS agents, only clinical research can provide the results needed to make a decision about the therapeutic utility of a compound.

22.9 Resistance and Its Origin

Pathogenic viruses, bacteria, and parasites evolve resistance to drug therapy. Inappropriate and excessive use of antibiotics in the past has led to **selection pressure for resistant strains**. Unfortunately, hospitals are the primary site for the emergence and spread of resistant strains. The spatial proximity and concentration of different pathogens is virtually unavoidable. In some cases, there are only a few effective weapons left, such as glycopeptide antibiotics. They should be used prudently and purposefully, even if this goes against the commercial interests of the manufacturer.

Most bacterial pathogens defend themselves against penicillins and cephalosporins by producing **β-lactamases** (Sect. 23.7). These are enzymes that open the four-membered lactam ring of these antibiotics to inactive cleavage products. During the long period of optimization of this class of substances, metabolically stable analogues as well as specific β-lactamase inhibitors have been developed.

The retrovirus HIV, which causes the immune deficiency disease **AIDS** (Sects. 1.3 and 24.3), transfers its genetic information from RNA back into DNA. This process is afflicted with an extremely high error rate of about one base mutation per generation. The high mutation rate leads to the rapid emergence and selection of resistant strains. In the last decade, many drugs with completely different modes of action against HIV have been introduced to the market, but resistances to many inhibitors have been observed very quickly, for example, against the HIV protease (Sect. 24.5) or reverse transcriptase inhibitors (Sect. 32.5), and even multiple resistances. The mutated viruses are even resistant to several structurally different inhibitors! Combining different drugs against the same target is of little help here. Only a combination of compounds that hit the virus at completely different instances of its life cycle offers a reprieve.

Tuberculosis is also re-emerging. Resistant pathogens require the development of new therapeutics. After the convincing success of mosquito control with DDT and therapy with synthetic antimalarials, **malaria** is making a comeback in developing countries (see Sect. 3.2).

The major problem in anticancer therapy is the development of **multidrug resistance (MDR)** during treatment. The resistance is not only to the anticancer drug, but also to several other anticancer drugs. This multidrug resistance is due to the overexpression of a transporter (Sect. 30.10), **glycoprotein 170**, which can largely eliminate structurally different xenobiotics from the cell. Although GP170 prefers cationic substances, another transporter, multidrug resistance-associated protein (MRP), eliminates amphiphilic anionic substances, i.e., compounds with polar and nonpolar character. However, amphiphilic substances are also capable of breaking the resistance of tumor cells. Quantitative structure–activity relationships show that the resistance of tumor cells to certain drugs is mainly associated with similarities in their molecular weights, i.e., the size of the inducing agent, and its lipophilicity.

22.10 Combined Administration of Drugs

Combination drugs are popular with pharmaceutical manufacturers, physicians, and patients alike. Manufacturers like them because they expand the indications for a successful drug and boost sales. Some physicians like the fact that therapy is simplified in many cases, while others oppose such **combination products**. One advantage for older patients is that they no longer have to take so many different drugs at different times of the day and in different doses, but only one or a few combination drugs. This improves **compliance**, because one of the most common reasons for treatment failure is patient behavior. Either the regular dose is forgotten, or the patient takes a break from the regimen over the weekend or while on vacation. These behaviors are particularly pronounced in elderly patients, with medications that do not have an obvious immediate effect, or with medications that have side effects that the patient subjectively perceives as unpleasant.

Clinical pharmacologists, academics, and many critically minded physicians have considerable reservations about combination products. This is understandable when one considers that a patient's attitude to a particular drug requires the observation of a dose–response relationship over a long period of time and, ultimately, individualized therapy. In a combination drug, there is always a fixed relationship between the individual components. Many combinations, such as analgesics, contain components with different modes of action. These are often misused without a strict medical indication and must, therefore, be judged critically.

There are **reasonable combinations** that even opponents to the general concept of combination therapy would accept without reservations. Among these are

- L-DOPA preparations with which the side effects can be reduced by selective combination (Sects. 9.4, 26.10, and 27.8),
- Antihypertensives and diuretics, the different mechanistic principles of which complement each other,
- Antibacterial preparations in which a dihydrofolate reductase inhibitor (Sect. 27.2) is combined with an appropriate sulfonamide,
- Hormonal contraceptives (Sect. 28.5), and
- Polyvalent vaccines, with which a single application offers protection against multiple diseases.

In the case of L-DOPA therapy, only combinations of several agents reduce side effects to a tolerable level. In the case of antihypertensives and diuretics, a single principle is often not sufficient to achieve the same effect as combinations. In the case of sulfonamide combinations and antituberculosis compounds, multiple modes of action can prevent or delay the development of resistance. An inhibitor of the P450 family of metabolic enzymes may be justified as an **adjunct to expensive drugs** or to drugs used at very high doses. In this way, the concentration of the other drug can be maintained at a higher level and for a longer period of time (Sect. 27.7). An important prerequisite for all combination drugs is an adequate therapeutic window and adapted pharmacokinetics of the components, at least those that support the actual mode of action.

22.11 Synopsis

- A relatively small portion of the druggable genome has been pharmaceutically addressed, and GPCRs overrepresent the targets for which active substances are available. Protein kinases represent a particularly promising emerging family of targets.
- Enzymes are very popular drug targets, and the natural substrates often provide the starting point for a rational drug design approach. There are three types of enzyme inhibitors: competitive inhibitors, noncompetitive inhibitors, and allosteric inhibitors. Enzyme inhibition can also be classified as reversible and irreversible. Nowadays reversible inhibition is desired, but some very important drugs are irreversible inhibitors, and some reversible inhibitors have such high affinity that they are *de facto* irreversible inhibitors.

- Receptors are also important drug targets; they can be subdivided into GPCRs, ion channels, hormone receptors, and growth factor receptors. An agonist activates the receptor, an antagonist prevents the agonist from docking at its binding site, and an inverse agonist stabilizes an inactive conformation of the receptor.
- Ion channels are extremely fast gateways for ions and can be either voltage- or ligand-gated. Ions can flow only passively with the concentration gradient through an ion channel.
- Transporters are special proteins in the membrane that can pump molecules and ions (also named ion pumps) against the concentration gradient at the expense of ATP hydrolysis. Many transporters are attractive drug targets, and others are responsible for the development of drug resistance.
- There are a large variety of known modes of action for drugs. Some of the most diverse modes of action are found in anti-infective drugs. Furthermore, tumor therapeutics exploit diverse, toxic modes of action. The goal in addressing these modes of action in terms of a therapy is to find a pathophysiological process that is unique or is as unique as possible to the disease to spare healthy tissue from damage.
- Drug resistance is an increasingly serious problem and is both an inevitable occurrence associated with using a pharmaceutical therapy, and a consequence of the misuse of anti-infectives. There are several mechanisms of resistance development in bacteria or viruses (i.e., fast genetic mutations), and in cancer therapy (i.e., aberrant transporter expression). These mechanisms are not mutually exclusive.
- The issue of combination drugs is a controversial topic. Some physicians are against them, and others are in favor of them, and both sides of the argument have good reasons. Nonetheless, some drug combinations are justifiable and help with compliance, clinical efficacy, and safety.

Bibliography and Further Reading

General Literature

G. Folkers, Ed., Lock and Key—A Hundred Years After, Emil Fischer Commemorate Symposium, Pharmaceutica Acta Helvetiae, **69**, 175–269 (1995)

A. L. Hopkins and C. R. Groom, The druggable Genome. Nature Rev. Drug Discov., **1**, 727–730 (2002)

R. Santos et al., A Comprehensive Map of Molecular Drug Targets, Nature Rev. Drug Discov., **16**, 19–33 (2017)

P. Imming, C. Sinning and A. Meyer, Drugs, their targets and the nature and number of drug targets, Nature Rev. Drug Discov., **5**, 821–834 (2006)

J. P. Overington, B. Al-Lazikani and A. L. Hopkins, How many drug targets are there? Nature Rev. Drug Discov., **5**, 993–996 (2006)

The journals *Trends in Pharmacological Sciences, Chemistry & Biology, Nature Reviews Drug Discovery* and *Pharmacon* contain in each issue a highly topical article about the mode of action of a biologically active substance

Special Literature

M. Olah, R. Curpan, L. Halip, A. Bora, N. Hădărugă, D. Hădărugă, R. Moldovan, A. Fulias, M. Mractc, T. Oprea. Chemical Informatics: WOMBAT and WOMBAT-PK: Bioactivity Databases for Lead and Drug Discovery, in Chemical Biology: From Small Molecules to Systems Biology and Drug Design, Ed. S. L. Schreiber, T. M. Kapoor, G. Wess, WILEY-VCH (2007) https://doi.org/10.1002/9783527619375.ch13b

A. Gaulton et al., "ChEMBL: a large-scale bioactivity database for drug discovery". Nucl. Acid. Res., **40** (Database issue): D1100–7 (2011)

R. B. Westkaemper, Serotonin Receptors: Molecular Genetics and Molecular Modeling, Med. Chem. Res., **3**, 269–272 (1993)

F. Saudou and R. Hen, 5-HT Receptor Subtypes: Molecular and Functional Diversity, Med. Chem. Res., **4**, 16–84 (1994)

D. J. Austin, R. Crabtree and S. L. Schreiber, Proximity versus Allostery: The Role of Regulated Protein Dimerization in Biology, Chemistry & Biology, **1**, 131–136 (1994)

J. D. Hayes and C. R. Wolf, Molecular Mechanisms of Drug Resistance, Biochem. J., **272**, 281–295 (1990)

N. D. Rawlings, F. R. Morton and A. J. Barrett, *MEROPS*: The Peptidase Database. Nucleic Acids Res., **34**, D270–D272 (2006)

M. Coll, S. H. Knof, Y. Ohga, A. Messerschmidt, R. Huber, H. Moellering, L. Rüssmann, G. Schumacher, Enzymatic Mechanism of Creatine Amidinohydrolase as Deduced from Crystal Structures, J. Mol. Biol., **214**, 597–610 (1990)

WOMBAT database: https://www.daylight.com/meetings/mug04/Oprea/Wombat.2004.1.pdf (Last accessed Nov. 20, 2024)

ChEMBL database: https://www.ebi.ac.uk/chembl/ (Last accessed Nov. 20, 2024)

Merops Database: https://www.ebi.ac.uk/merops/ (Last accessed Nov. 20, 2024)

Inhibitors of Hydrolases with an Acyl–Enzyme Intermediate

Contents

G. Klebe, *Drug Design*, https://doi.org/10.1007/978-3-662-68998-1_23

Peptidases and esterases are hydrolytic enzymes, and 2–3% of all gene products can be assigned to this group of enzymes. They, therefore, represent an important group of target proteins for the development of new drugs and are of particular importance for structure-based drug design. In recent years, many of the known human or pathogen-derived peptidases have been tested as potential targets for drug therapy.

The function of these enzymes is to cleave peptide or ester bonds, which requires a nucleophile to attack the carbonyl group of the **amide** or **ester bond** to be **cleaved**. A large number of proteins use the OH or SH groups of a serine, threonine, or cysteine for this purpose. In the following chapters, we will see other hydrolyzing enzymes that use a different mechanism. During the cleavage reaction of the hydrolases discussed in this chapter, a temporary covalent bond is formed between the substrate and the enzyme. This intermediate, called the **acyl enzyme** form, occurs with serine, threonine, and cysteine proteases, but lipases, esterases, transpeptidases, and β-lactamases also use this reaction mechanism. The design of inhibitors for these enzymes that act via an acyl enzyme intermediate will be discussed. The following two chapters will discuss peptidases that use a water molecule for the primary attack on the peptide bond to be hydrolyzed: the aspartic and the metallopeptidases. Depending on whether they cleave the amino acid chain at the *N*- or *C*-terminus or in the center, peptidases are classified as **aminopeptidases**, **carboxypeptidases**, or **endopeptidases**. Some of these proteases are relatively unspecific, while others are highly specific and cleave only very specific substrates. It is the latter enzymes that are most likely to yield a selective therapeutic inhibitor with few side effects. Bacteria and viruses have also produced their own peptidases, the inhibition of which can be exploited for chemotherapeutic treatment. Since these proteins are not endogenous to humans and, therefore, have no function in humans, their inhibition should lead to therapeutic success without the risk of severe side effects.

23.1 Serine-Dependent Hydrolases

Serine proteases are the most abundant and best studied class of peptidases. They are closely related to esterases and lipases (hydrolases), which hydrolyze ester bonds. This class of enzymes serves the human body in many ways. Some serine proteases, such as the digestive enzymes trypsin and chymotrypsin, cleave a wide range of peptides and proteins. Others, such as the coagulation enzymes thrombin and factor Xa, are highly selective and cleave only very specific substrates. Often, proteases are expressed in a nonactive precursor form, called **zymogens**. In order to convert them to their active form, sequence segments of the zymogen polypeptide chain are often cleaved, which otherwise serve as endogenous inhibitors of the activated enzyme. Release of the active form can occur either by autocatalysis (e.g., trypsin) or by other activating proteases (e.g., the coagulation cascade). A **serine side chain** in the active site plays a crucial role in the catalytic mechanism of serine proteases, esterases, and lipases. It is characterized by an exceptionally high chemical reactivity. In chymotrypsin, only this serine reacts with diisopropyl fluorophosphate (DFP, a very potent neurotoxin), while 27 other serine residues in the enzyme remain unmodified. Upon chemical conversion with DFP, the enzyme completely loses its catalytic activity.

23.2 Structure and Function of Serine Proteases

The digestive enzyme chymotrypsin was the first serine protease to have its 3D structure determined by David Blow in Cambridge, England. The numbering of the amino acids in chymotrypsin-type serine proteases is based on the sequence of chymotrypsin. The three-dimensional structures of a large number of serine proteases are now available, some of which are listed in ◘ Table 23.1. The structures show an extraordinary similarity in the active site, even for proteases that have completely different folding patterns (Sect. 14.7, comparison of trypsin/subtilisin). This so-called **catalytic triad Ser–His–Asp** is characteristic for serine proteases. In some of these enzymes, the aspartate can be replaced by a glutamate, while some transpeptidases and β-lactamases have a lysine instead of the histidine in the active site.

Because these three amino acids are far apart in the sequence, the protein must fold appropriately to bring the three side chains into close proximity. The catalytic serine, located at position 195 in the trypsin-like proteases, performs the actual attack on the amide bond to be cleaved (◘ Fig. 23.1). The oxygen atom of an unactivated hydroxyl group would not be reactive enough for this step. Its nucleophilicity, which describes its tendency to attack an electron-poor carbonyl carbon atom, is enhanced by the adjacent histidine side chain. The imidazole side chain of this histidine can accept a proton from the serine hydroxyl group, allowing the now negatively charged oxygen atom to **nucleophilically attack** the partially positively charged carbon atom of the amide carbonyl group. The neighboring aspartate group can take up a proton from the imidazole ring of the histidine and gives it back again. In this way, it compensates for the positive charge intermediately formed on the histidine residue. To stabilize the transition state formed during the attack on the carbonyl group, serine proteases have another characteristic structural motif, the so-called **oxyanion hole**. This is a small pocket next to the side chain of Ser 195, composed of two main chain NH groups (◘ Fig. 23.1). In a few cases, the terminal

Table 23.1 Serine proteases with physiological importance (X = arbitrary amino acid). The 3D structures of all listed enzymes are known

Enzyme	Cleavage site	Function or therapeutic approach
Trypsin	Arg–X, Lys–X	Digestive enzyme
Chymotrypsin	Tyr–X, Phe–X, Trp–X	Digestive enzyme
Elastase	Val–X	Tissue degradation
Thrombin	Arg–Gly	Blood coagulation
Factor Xa	Arg–Ile, Arg–Gly	Blood coagulation
Factor VIIa	Arg–Ile	Blood coagulation
Tryptase	Arg–X	Asthma
Matriptase	Arg–X	Oncology
Urokinase	Arg–X	Oncology
DPP IV	Ala–X, Pro–X	Diabetes
Furin	Arg–X	Viral infection

amide groups of asparagine or glutamine can also perform this function. The function of the oxyanion hole is to stabilize the negative charge formed on the tetrahedral transition state and to distort the geometry of the attacked carbonyl carbon atom from a **trigonal-planar to a tetrahedral** configuration. The formed transition state collapses with the release of the *C*-terminal cleavage product, which carries a free amino group at its terminus. The *N*-terminal cleavage product remains covalently bound to the protease to form an **acyl enzyme intermediate**. In a subsequent step, nucleophilic attack by a water molecule again leads to a tetrahedral transition state. This tetrahedral transition state finally collapses, releasing the *N*-terminal cleavage product. The catalytic enzyme is then ready for the next transformation.

What happens when the amino acids serine, histidine, and aspartic acid of the catalytic triad of a serine protease are individually or collectively replaced by amino acids without similar functional groups? In 1988, Paul Carter and James Wells at Genentech produced various mutants of the bacterial serine protease subtilisin (Sect. 14.8). Replacing the catalytic serine or histidine with alanine reduces catalytic activity by more than six orders of magnitude. Surprisingly, exchange of aspartic acid, whose only function is to exchange a proton with histidine, reduced catalytic activity by more than four orders of magnitude. The combined exchange of several amino acids of the catalytic triad did not result in any further reduction in catalytic activity. The triple alanine mutant, in which the catalytic triad is completely re-

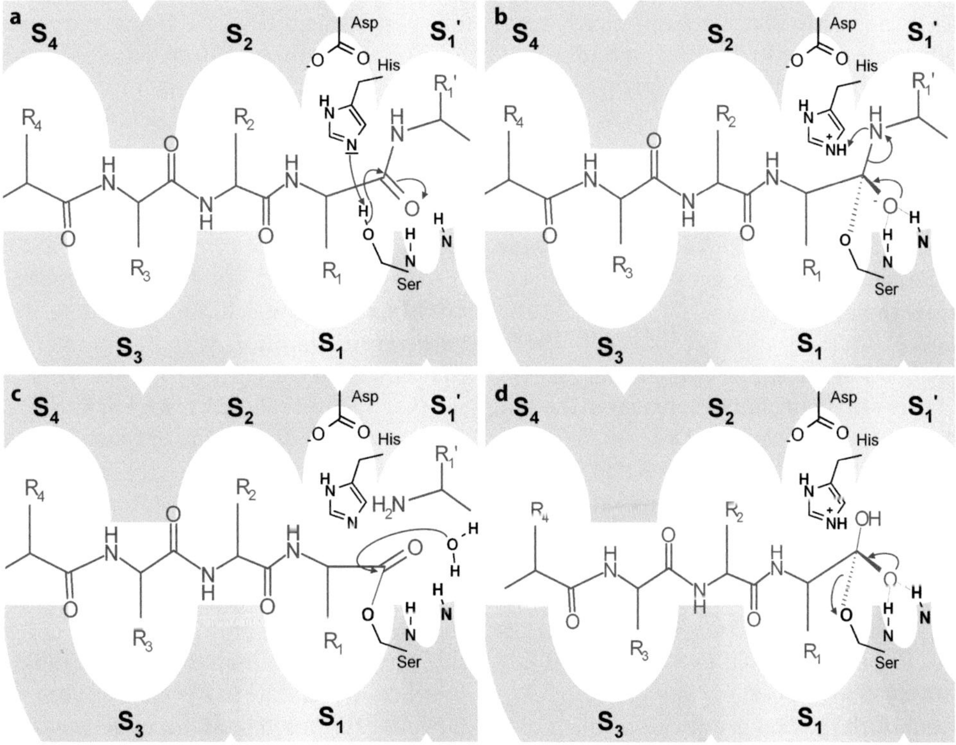

Fig. 23.1 Catalytic mechanism of serine proteases. **a** The peptide substrate binds to the enzyme in specific pockets on either side of the cleavage site. **b** The oxygen atom of the serine side chain performs a nucleophilic attack. This is facilitated by the adjacent histidine side chain, which, supported by an aspartate residue, accepts a proton from the hydroxyl group. **c** The transition state collapses to form an acyl enzyme intermediate. **d** This intermediate is hydrolyzed by the attack of a water molecule to release the *N*-terminal cleavage product

moved, still cleaves the peptide substrate more than 1000 times faster than the pure buffer solution! The remaining substrate binding sites and the oxyanion hole, whose geometry and properties stabilize the tetrahedral transition state, are responsible for this acceleration.

Now, it is certainly not difficult to destroy the binding site of an enzyme or its catalytic activity. It is more difficult to intentionally alter its specificity or function. The subtilisin mutants, in which the histidine has been replaced by an alanine, cleave substrates with the sequence –Phe–Ala–X–Phe– (X = e.g., Ala or Gln) six orders of magnitude slower than the unmodified subtilisin, with one exception: A substrate with the sequence –Phe–Ala–**His**–Phe– is cleaved only four orders of magnitude slower. The histidine of the substrate takes over to some extent the role of the histidine in the catalytic site! This process is called **substrate-supported catalysis**. The conversion is still rather slow, but the specificity of this mutant is significantly increased: The –Phe–Ala–His–Phe– sequence is cleaved 200 times faster than any of the other –Phe–Ala–X–Phe– sequences.

23.3 The S_1 Pocket of Serine Proteases Determines Specificity

Proteases recognize polypeptide chains as substrates. For this task, they use a series of more or less pronounced binding pockets on their surface, as described in Sect. 14.5. These are structurally and electronically complementary to the side chains of the corresponding residues in the substrate. As a result, the polypeptide chain of the substrate is immobilized on the surface in the vicinity of the catalytic site. Depending on the protease, the crevices on the surface look very different. Surface portions of four different serine proteases from the trypsin family are shown in ◘ Fig. 23.2. A comparison of the different serine proteases with different substrate specificities (◘ Fig. 23.3) shows that particularly the structures of the S_1 pockets of these enzymes are different. The S_1 pocket consists mainly of the sequence segments 189–195 and 214–220. There are significant differences in the side chains of the amino acid at positions 189, 216, and 226. In chymotrypsin, these are Ser 189, Gly 216, and Gly 226. They adjust the depth and shape of this pocket in a way to accommodate the aromatic side chains of the amino acids phenylalanine, tyrosine, and tryptophan. Accordingly, chymotrypsin preferentially cleaves peptide chains after one of these three amino acids. Trypsin also has a deep, spacious S_1 pocket flanked by Gly 216 and Gly 226. The negatively charged carboxylate group of Asp 189 at the bottom of the pocket is critical for the recognition of the long, positively charged side chains found in the amino acids lysine and arginine of the substrate. In elastase, the S_1 pocket is formed by the amino acids Val 216 and Thr 226. This makes the pocket much smaller. It can only accommodate amino acids with short hydrophobic side chains, such as alanine and valine. Amino acids with large groups are no longer accommodated. Amino acid 189, which is serine, is deeply buried.

The substrate specificity of the described serine proteases is primarily achieved by recognition of the **amino acid in the P_1 position**. Even very small changes can alter the substrate profile in this pocket. For example, trypsin cleaves substrates after the two basic amino acids arginine and lysine, whereas thrombin cleaves exclusively after the more basic amino acid arginine. The only difference in the S_1 pocket is that thrombin has a sodium ion next to the Asp 189 residue. Its positive charge attenuates the negative charge of the adjacent aspartate residue, thus, reducing its polarizing effect. As a result, the more basic Arg residue continues to bind in a positively charged state, while for the less basic lysine, the polarizing effect is no longer sufficient to induce protonation. This helps to discriminate between peptide chains exhibiting Lys and Arg in P_1 position (see also Sect. 4.4, ◘ Fig. 4.4). Pockets adjacent to S_1 are also important for substrate binding and selectivity. It is noteworthy that the substrate-binding pockets of serine proteases that recognize the *N*-terminal part of the substrate (unprimed side, S_1–S_4 pockets; Sect. 14.5) are more pronounced. Pockets on the unprimed side that anchor the *C*-terminal part of the substrate are much less well developed. Since the *N*-terminal cleavage product remains temporarily covalently bound to the protease as an acyl enzyme complex, this part of the substrate is bound particularly selectively.

These structural features define what a potential competitive inhibitor of a serine protease should look like: It is crucial that the S_1 pocket is filled as well as possible. The chemical composition of the parts of the inhibitor that bind in this region must be complementary to the S_1 pocket. In some cases, filling the S_1 pocket alone is sufficient to generate a selective serine protease inhibitor with respectable binding affinity. For example, in 1967, Marcos Mares-Guia and Elliott Shaw described small-molecule trypsin inhibitors with micromolar binding affinity that occupied only the S_1 pocket. It is not difficult to see that all molecules **23.1**–**23.4** in ◘ Fig. 23.4 mimic the basic amino acids arginine or lysine in the P_1 position of the substrate.

A first approach to the design of serine protease inhibitors could be based on the search for a suitable group to occupy the S_1 pocket, which could then be coupled to a chemically reactive group that binds to the catalytic serine. The various groups described in the literature for this purpose are summarized in ◘ Table 23.2. Natural products also follow this principle. The macrocyclic pentapeptide thrombin inhibitor **cyclotheonamide A** from the marine sponge *Theonella sp.* contains an α-keto function next to an amide bond. As the X-ray structure shows, this ketone group forms a tetrahedral hemiacetal structure with the OH group of the catalytic serine (◘ Fig. 23.5).

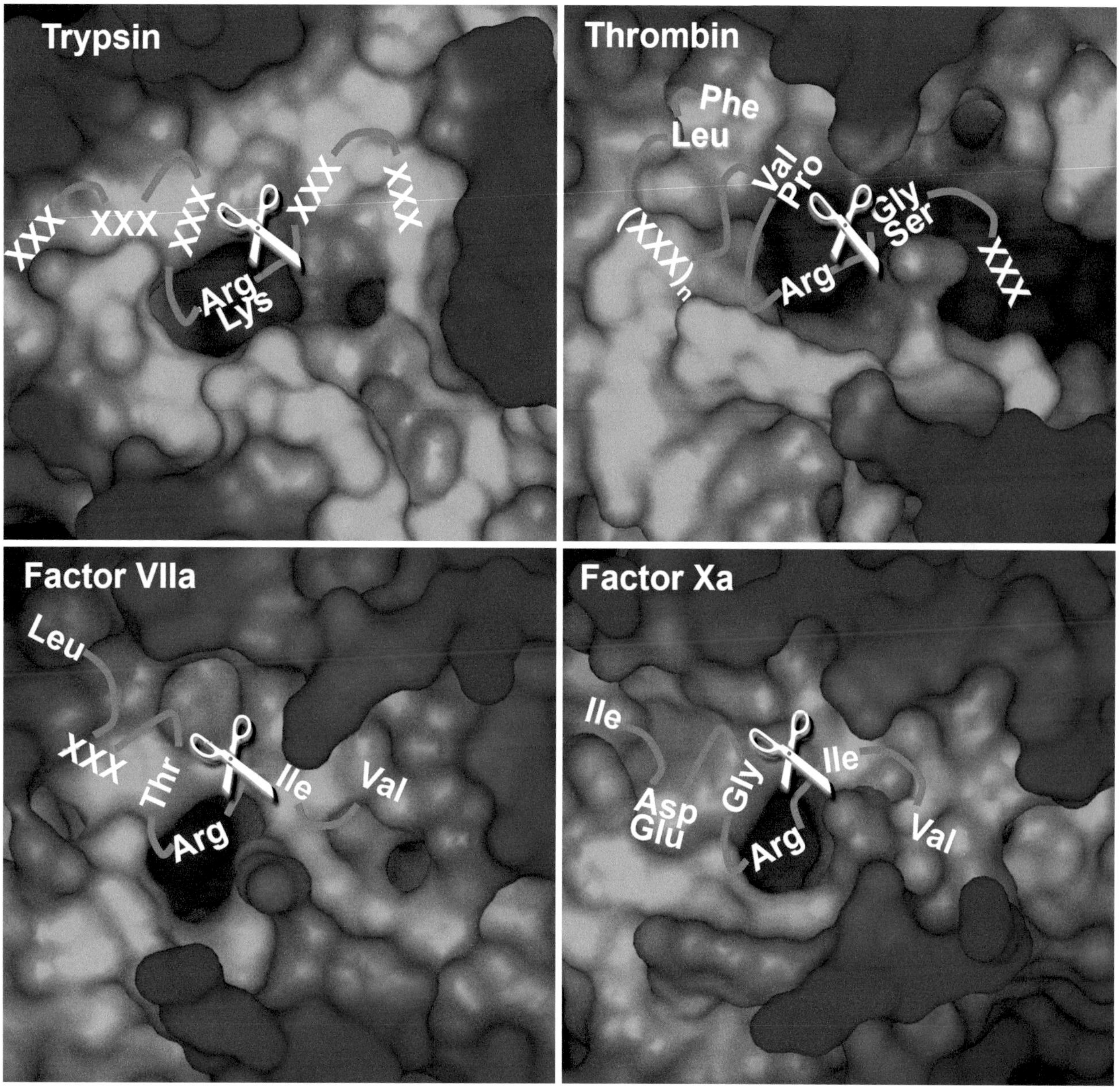

Fig. 23.2 The surfaces of the trypsin-like serine proteases trypsin, thrombin, factor VIIa, and factor Xa show a strong indentation in the area of the catalytic center. To highlight this surface pattern, the color of the surface changes from *blue* to *green* to *red* as the depth of the indentation increases. In the indentations, the exposed physicochemical properties determine the substrate selectivity of the proteases. The preferred cleavage sequences are indicated in the structures, where XXX represents any amino acid at that position. (► https://sn.pub/KPYo7E)

If the sequence of the peptide substrate of the serine protease is known, the *N*-terminal amino acid preceding the cleavage site can be coupled with one of the groups from Table 23.2 to produce a compound that is most likely to be an inhibitor. An example of this is the elastase inhibitor *N*-(methylsuccinyl)–Ala–Ala–Pro–Val–CF_3 (**23.21** in Fig. 23.14), which is derived from the substrate sequence Pro–Val. In favorable cases, the P_1 equivalent alone is sufficient, for example, in the trypsin and thrombin inhibitors **23.5** and **23.6** (Fig. 23.4). However, the usually high chemical reactivity of the functional groups in **covalently bound serine protease inhibitors**, which is necessary to interact with the catalytically active serine, can be problematic. Because of their reactivity, such groups can also react undesirably with serine residues of other enzymes, causing side effects. The design of highly potent and selective inhibitors requires at least the occupancy of the S_2, S_3, and S_4 pockets. An-

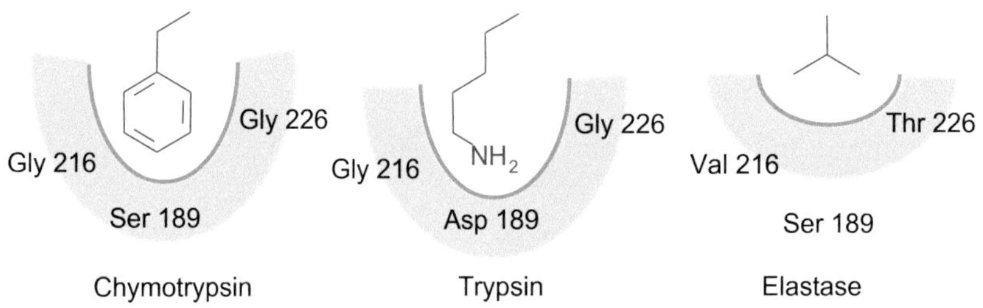

Fig. 23.3 Comparison of the S_1 pockets of chymotrypsin, trypsin, and elastase. The binding pocket of chymotrypsin is tailored for large, lipophilic side chains. The S_1 pocket of trypsin binds amino acids with positively charged side chains through its negatively charged Asp 189 residue. Because of the spatial filling of the side chains of Thr 216 and Val 226, elastase has a relatively small S_1 pocket and, therefore, binds small hydrophobic amino acids such as alanine and valine

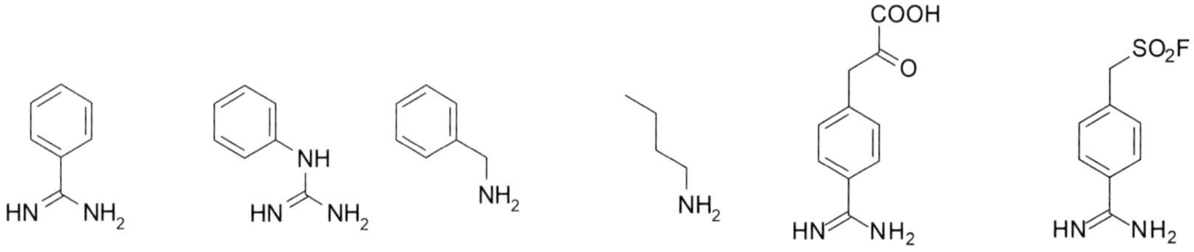

23.1 23.2 23.3 23.4 23.5 23.6

Trypsin: K_i = 18 μM 72 μM 380 μM 1500 μM Thrombin K_i = 6.5 mM
Thrombin: K_i = 220 μM

Fig. 23.4 The molecules **23.1**–**23.4** that bind in the S_1 pocket of trypsin are micromolar inhibitors. All of these molecules contain a strongly basic group that is protonated under physiological conditions; therefore, a positive charge is available to form a salt bridge to the negatively charged side chain of Asp 189. The thrombin inhibitors **23.5** and **23.6** contain an additional functional group that can form a covalent bond to the catalytically active serine

other structural feature common to all serine proteases should also be mentioned: Their substrates are bound to the peptide backbone by two antiparallel hydrogen bonds. This orientation of the two hydrogen bonds results in a pleated sheet-like geometry. In the majority of designed inhibitors, an attempt was made to mimic this pattern of hydrogen bonding (Fig. 23.6).

23.4 Seeking Small-Molecule Thrombin Inhibitors

The serine protease thrombin plays a central role in the control of **blood coagulation**. Thrombin is at the end of a complex, highly regulated cascade of serine proteases. Injury to the arterial vasculature causes membrane-bound tissue factor outside the vessel to contact the serine protease precursor factor VII in the blood. The precursor is activated to factor VIIa and initiates the coagulation cascade. Along the cascade, various factors are released, which are activated from their zymogen form by proteases from the previous step. Finally, the cascade leads to the release of von Willebrand factor, which binds to platelets and initiates blood clot formation. In addition to **extrinsic activation**, there is also an **intrinsic coagulation pathway**. It is initiated by reduced blood flow or pathologically altered vasculature. In this case, the coagulation cascade is initiated to form a platelet aggregate, which is then stabilized by a fibrin network. Factor X is found in one of the final steps where the two pathways merge. All of these steps involve serine proteases, which are potential targets for drug therapy. To date, development efforts have focused on the enzymes thrombin, factor Xa, and factor VIIa. This has already led to development candidates and marketed products for the first two.

Thrombin converts inactive **fibrinogen** into reactive **fibrin**. Together with aggregated platelets, it forms a polymer in which the different blood cells are trapped. A thrombus is formed, which is further cross-linked and stabilized by transglutaminase factor XIII (Sect. 23.8). This is an essential protective mechanism of the body to ensure wound closure. In certain diseases or situations, such as after surgery, heart attack, or to prevent stroke in patients with atrial fibrillation, it is necessary to reduce the blood's ability to clot. For this reason, there is great interest in the development of selective and, above all, orally available inhibitors of the coagulation cascade. Thrombin cleaves fibrinogen between the amino acids arginine and glycine. This sequence served as the starting point for the development of the first synthetic thrombin inhibitors, which therefore contained either an Arg or an Arg-analogous moiety. In this section, three different approaches to the development of thrombin inhibitors will be presented: substrate analogues, benzamidine analogues, and significantly modified structural analogues.

One approach for the **design of thrombin inhibitors** is provided by the $P_3 \ldots P'_3$ substrate sequence Gly–Val–Arg–Gly–Pro–Arg of fibrinogen. In the early 1970s, the Japanese group of Hamao Umezawa found that peptide aldehydes with *C*-terminal arginine residues isolated from bacteria were potent inhibitors of some trypsin-like serine proteases. The tripeptide aldehydes studied by Sándor Bajusz were derived from amino acids P_3–P_1 or P'_3–P'_1, i.e., the three amino acids "before" and "after" the cleavage site. The relative binding affinities of some peptide aldehydes are summarized in Table 23.3. Interestingly, the direct comparison of Gly–Val–Arg–H and Gly–Pro–Arg–H shows that a proline at the P_2 position inhibits thrombin about nine times more strongly. The introduction of phenylalanine instead of glycine in the P_3 position leads to an additional significant in-

Table 23.2 Reactive groups that can covalently react with the catalytically active serine

Inhibitor type	Functional group	
Irreversible	Chloromethylketone	$-COCH_2Cl$
	Sulfonylfluoride	$-SO_2F$
	Ester[a]	–COOR
	Boronic acid[a]	$-B(OR)_2$
Reversible	Aldehyde	–CHO
	Ketone	–COR (R = Alkyl, Aryl)
	Trifluoromethylketone	$-COCF_3$
	α-Ketocarboxylic acid	–COCOOH
	α-Ketoamide	–COCONHR
	α-Ketoester	–COCOOR

[a] Reversible as well as irreversible examples are known.

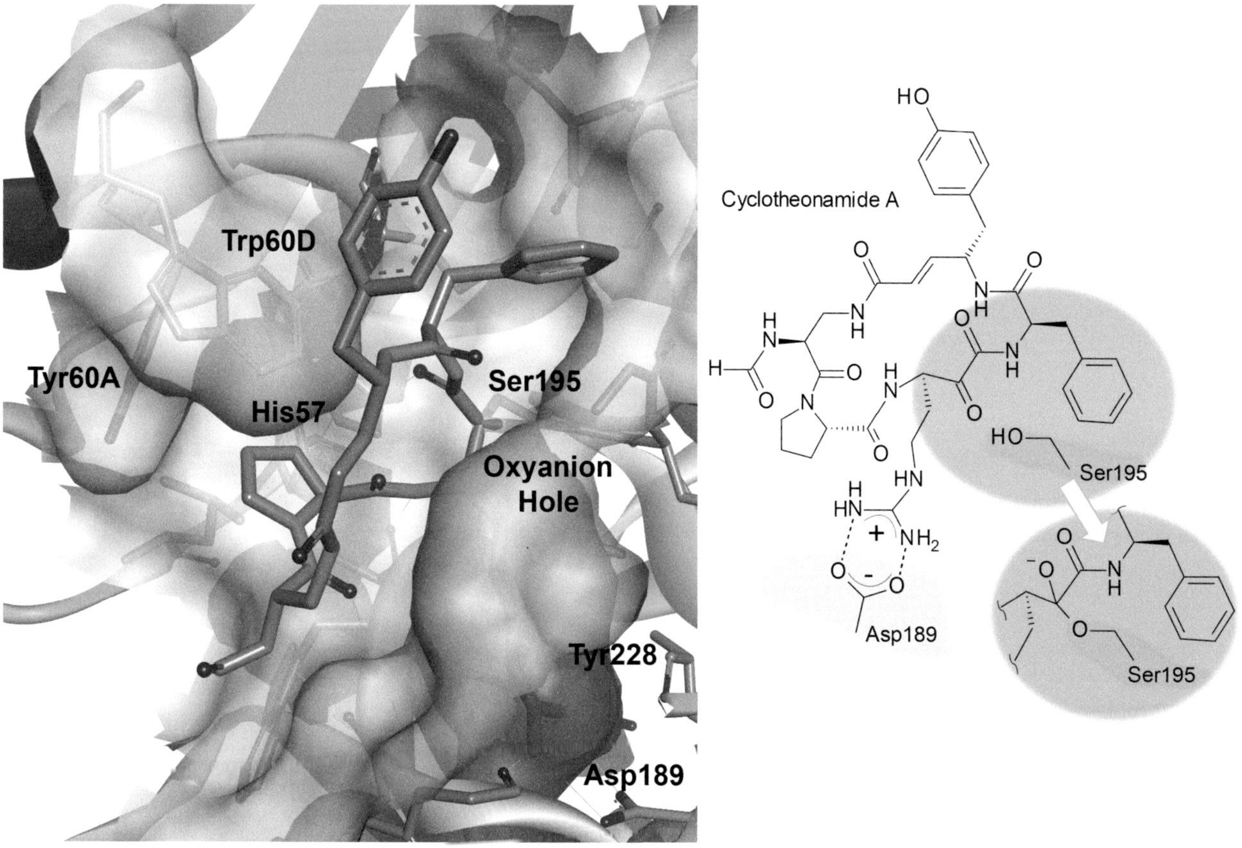

Fig. 23.5 Crystal structure of the inhibitor cyclotheonamide A with thrombin. The inhibitor forms a covalent bond to the catalytic serine with its α-keto group to form a hemiketal structure. The now negatively charged oxygen is stabilized by two hydrogen bonds in an oxyanion hole. (► https://sn.pub/zJhqBx)

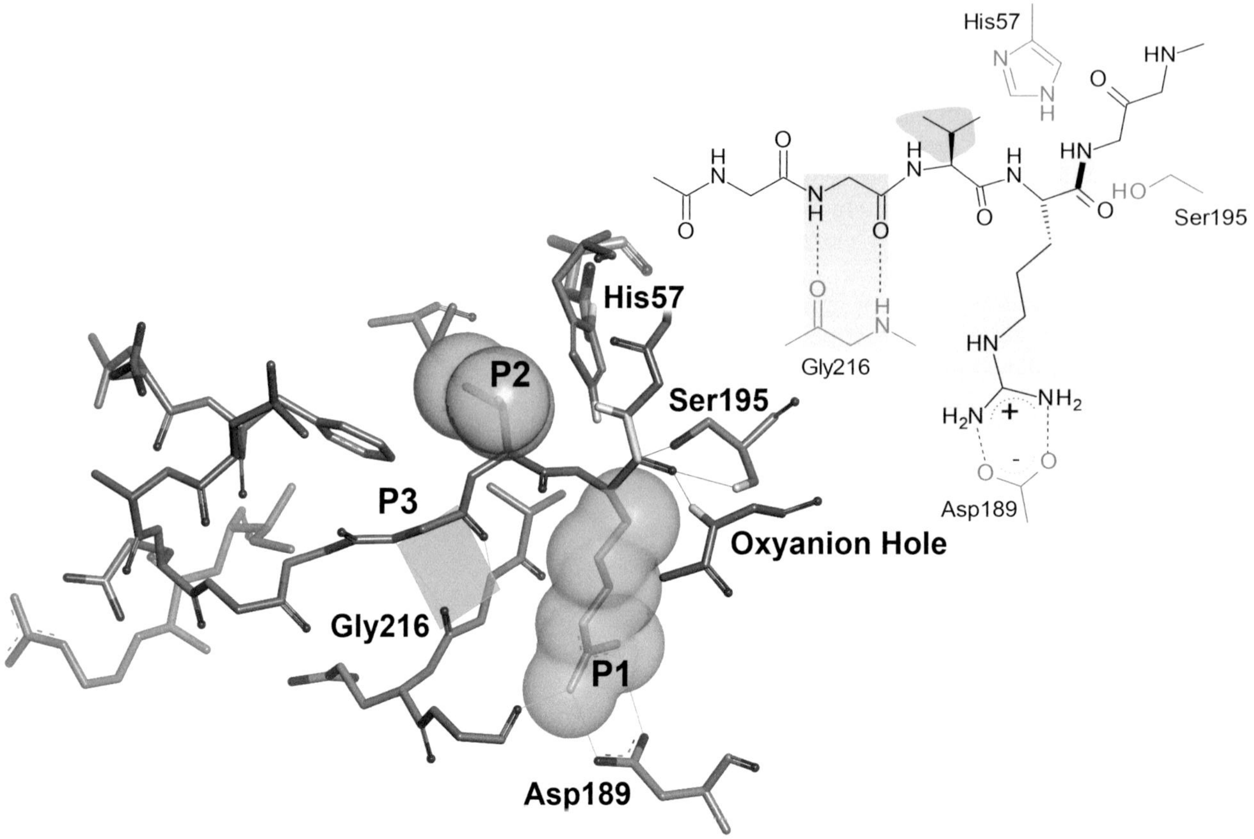

Fig. 23.6 General binding mode of a peptide chain that is to be cleaved (*gray* carbon atoms) in the catalytic site of a serine protease. The amide bond to be cleaved is highlighted in *yellow*. The substrate's P_1 (*light blue*) and P_2 groups (*green*) are shown with a surface; they bind in the S_1 and S_2 pockets of the protein. Two antiparallel-oriented hydrogen bonds (*green*) are formed to the main chain. The H-bonds to the oxyanion hole are in *purple*, and the direction of the nucleophilic attack of the Ser 195 oxygen on the carbonyl carbon is indicated in *violet blue*. (▶ https://sn.pub/yKx9nH)

crease in binding. D-Amino acids in the P_3 position were then investigated. Surprisingly, these led to a dramatic improvement in binding affinity. This result was not expected considering that the substrate sequence from P_5 to P_3 Gly–Gly–Gly–Val–Arg contains only achiral glycine residues without lipophilic side chains that can hardly form interactions corresponding to the D-Phe side chain.

Table 23.3 Relative binding affinity of tripeptide aldehydes on thrombin. Arg–H is for the aldehyde that was obtained by reducing the carboxylic acid of arginine. The larger the value of the relative inhibition, the stronger the inhibitor binds to thrombin

Peptide	Relative inhibition
Gly–Val–Arg–H	1
Gly–Pro–Arg–H	9
Phe–Pro–Arg–H	57
D-Ala–Pro–Arg–H	469
D-Val–Pro–Arg–H	1273
D-Phe–Pro–Arg–H	7370

At the time of the work described above, the spatial structure of thrombin had not yet been determined. Wolfram Bode and Milton Stubbs at the Max Planck Institute of Biochemistry in Martinsried, Germany, solved the structure of a thrombin complex with a chemically activated fibrinopeptide, Gly–Asp–Phe–Leu–Ala–Glu–Gly–Gly–Val–Arg–CH_2Cl. This peptide corresponds to the *N*-terminal portion from P_{11} to P_1 that thrombin cleaves from fibrinogen. Comparison of this structure with that of D-Phe–Pro–Arg–chloromethylketone (Fig. 23.7) provided an explanation for the surprising structure–reactivity relationship found by Sándor Bajusz. The S_3 pocket is filled by both ligands, in the case of the fibrinopeptide by the side chains of leucine and phenylalanine at positions P_8 and P_9. The peptide forms a β-turn that allows the amino acids in this sequence to be positioned in the S_3 pocket. The same pocket is ac-

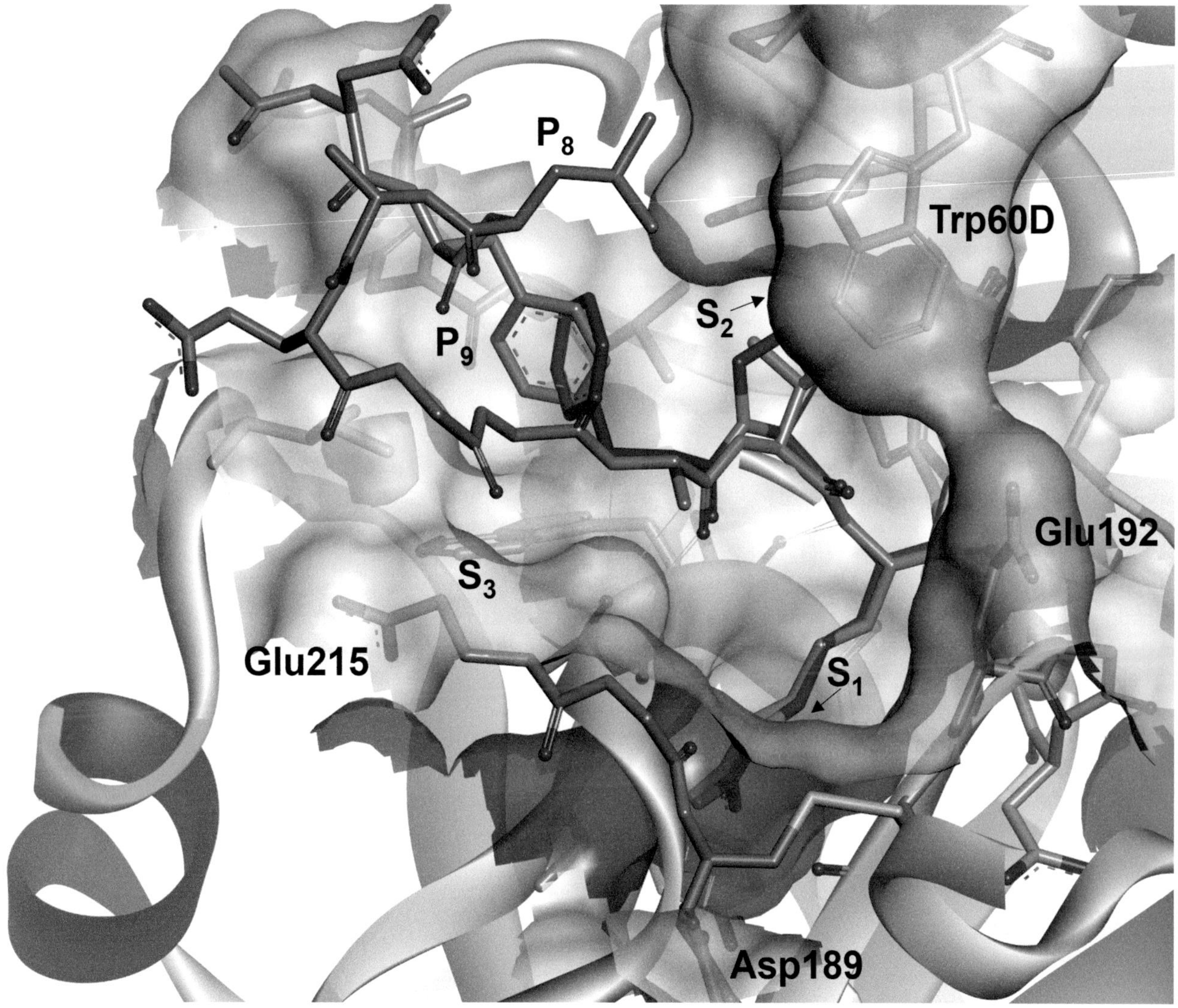

Fig. 23.7 Comparison of the binding mode of the irreversibly binding thrombin inhibitors D-Phe–Pro–Arg–CH_2Cl (*dark-red* carbon atoms) with that of the fibrinopeptide derivative (*gray* carbon atoms). Both inhibitors bind with an arginine side chain in the S_1 pocket. The S_2 pocket is occupied by a valine side chain of the fibrinopeptide. Its additional peptide chain is folded back so that the Leu and Phe side chains in positions P_8 and P_9 are oriented into the lipophilic S_3 binding pocket. In the case of D-Phe–Pro–Arg–CH_2Cl, the phenyl ring of D-Phe is also located in this pocket. (► https://sn.pub/MG65ha)

cessed by the tripeptide through the side chain of the D-amino acid at the P_3.

The compound D-Phe–Pro–Arg–H, synthesized by Bajusz, is a high-affinity thrombin inhibitor (K_i = 75 nM). However, the compound proved to be chemically unstable. This problem could be solved by *N*-methylation of the free NH_2 group. *N*-Methyl-D-Phe–Pro–Arg–H **23.7** (Gyki 14766, Efegatran, Fig. 23.8) is chemically stable.

Jörg Stürzebecher and Fritz Marquardt took a different route. They pursued the goal of managing inhibition without a covalent attachment. Their approach was based on the finding that in addition to trypsin (K_i = 18 μM), benzamidine **23.1** (Fig. 23.4, Sect. 23.3) also inhibits thrombin (K_i = 220 μM). The combination of the benzamidine group with a reactive group from Table 23.2 yielded potent thrombin inhibitors. The first low molecular weight thrombin inhibitor to be clinically tested in the 1970s was *p*-amidinophenylpyruvic acid **23.5** (Fig. 23.4, Sect. 23.3). The compound proved to be efficacious, but its selectivity was unsatisfactory. The simple benzamidine derivatives **23.8** and **23.9** (Fig. 23.8) are further typical representatives with micromolar affinity for thrombin, but without selectivity compared to trypsin.

The coupling of the benzamidine groups with a peptide structure brought significant improvement. N^{α}-(β-naphthylsulfonylglycyl)-D,L-*p*-amidinophenylalanylpiperidide **23.10** (NAPAP, Fig. 23.9) was the result

Fig. 23.8 The inhibitor **23.7** (Gyki 14766, efegatran) contains an aldehyde group that binds reversibly to Ser 195. Compounds **23.8** and **23.9** are simple derivatives of benzamidine that noncovalently inhibit the enzyme

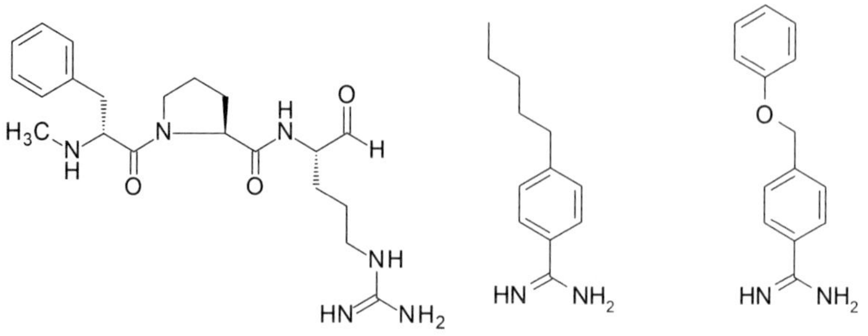

23.7 Gyki 14766, Efegatran K_i = 1.8 µM **23.8** **23.9**

of a more than 10-year-long systematic search for potent and selective thrombin inhibitors. NAPAP was the most potent representative of the class of low molecular weight thrombin inhibitors (K_i = 6 nM) for a long time, but it has only modest selectivity over trypsin.

In 1989, Wolfram Bode solved the crystal structure of thrombin with a bound inhibitor. The structure was determined first with the irreversible inhibitor D-Phe–Pro–Arg–CH_2Cl and then with NAPAP. The 3D structure of the thrombin–NAPAP complex is shown in Fig. 23.10. The racemic form was used for cocrystallization. The result that *p*-amidinophenylalanine binds to thrombin as a D-amino acid was rather surprising. The substrate consists only of L-amino acids, so it was expected that *p*-amidinophenylalanine would also bind in the L-configuration.

The groups of the ligand that form polar interactions with the protein can be deduced directly from the crystal structure. For NAPAP, these are the glycine moiety in the center of the molecule (two hydrogen bonds to the peptide backbone) and the amidinium group in the S_1 pocket. Omitting the positively charged amidinium group results in a loss of binding affinity because the salt bridge to Asp 189 cannot be formed. However, later work showed that chloro-substituted aromatic rings can also bind into the S_1 pocket and form a hydrophobic interaction with Tyr 228. At the same time, they displace a water molecule from the S_1 pocket (Fig. 4.7). Today, an arsenal of building blocks that can be used as arginine side chain mimics to fill the S_1 pocket of thrombin (Fig. 23.11) are available.

Fig. 23.9 The thrombin inhibitors NAPAP **23.10**, CRC 220 **23.11**, the latter was developed at the former Behringwerke, and **23.12** which was derived from **23.10**. The two latter compounds have distinctly better affinity to thrombin and improved selectivity relative to trypsin. The IC_{50} values for **23.10** and **23.12** are given for the racemates. Inhibitor **23.11** was measured as an enantiopure compound

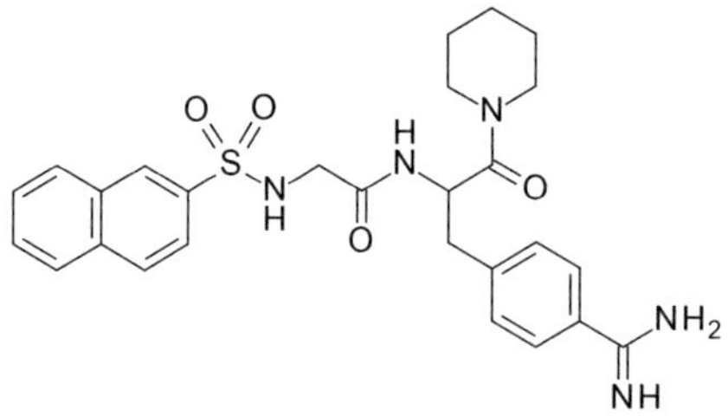

23.10

		Thrombin	Trypsin
rac-NAPAP	K_i =	0.006 µM	0.69 µM
L-Napap	K_i =	1.4 µM	25.5 µM
D-Napap	K_i =	0.0021µM	0.21 µM

Thrombin:Trypsin
1:100

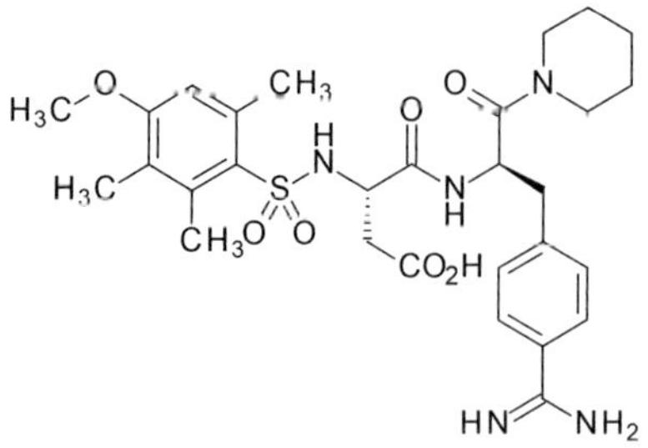

23.11 CRC220
Behringwerke
K_i = 6 nM
Thrombin:Trypsin
1:200

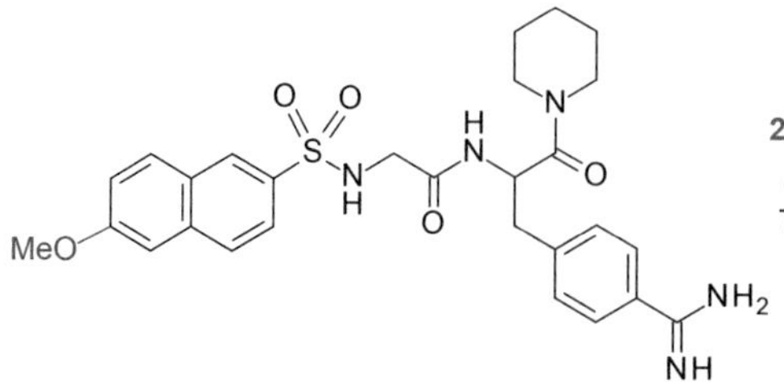

23.12 *(racemic)*
IC_{50} = 15 nM
Thrombin:Trypsin
1:600

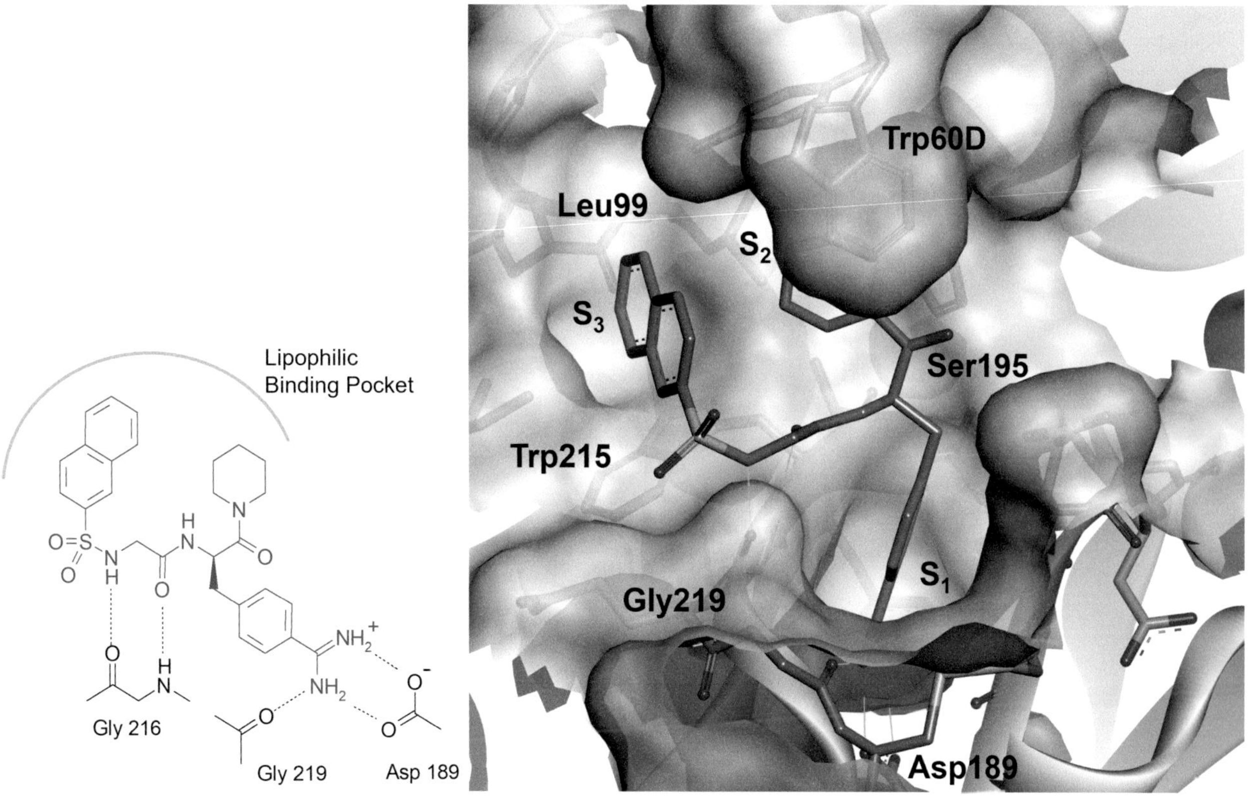

Fig. 23.10 Structure of the thrombin NAPAP complex. The most important interactions are outlined on the *left* side. The positively charged benzamidine group occupies the S_1 pocket and forms a salt bridge to the negatively charged side chain of Asp 189. Two hydrogen bonds are formed to the amino acid Gly 216. The piperidyl and naphthyl groups together occupy the two large lipophilic pockets S_2 and S_3. (▶ https://sn.pub/CPfw8M)

With its naphthyl and piperidyl side chains, NAPAP largely fills the lipophilic S_3 pocket and the spatially limited S_2 pocket (Fig. 23.10). However, it appears that even larger substituents could fit into the S_3 pocket. A weakness of NAPAP was its lack of selectivity compared to the digestive enzyme trypsin. Fortunately, the structures of NAPAP complexed with thrombin and also with trypsin are known (Fig. 23.12). A comparison of the 3D structures shows that there is a significant difference in the binding mode between the two enzymes in the S_3 pocket, resulting in a 180° flipped orientation of the naphthyl group with respect to the bond to sulfur. In thrombin, the S_3 pocket is more pronounced and surrounded by several lipophilic amino acid side chains. In trypsin, the upper end of this pocket is open and not spatially restricted. Obviously, its structuring is not necessary in the largely unspecific digestive enzyme. Therefore, the selectivity can be increased by occupying the S_3 pocket of thrombin as optimally as possible. A closer look at the thrombin–NAPAP complex reveals that an additional methoxy substituent on the naphthyl ring should be suitable for increasing selectivity. Indeed, inhibitor **23.12** binds 600 times more strongly to thrombin than to trypsin.

The compound CRC220 (**23.11**, Fig. 23.9), which fills the hydrophobic S_3 pocket much better than NAPAP, was developed at the former Behringwerke in Marburg, Germany. Because of this improved filling, CRC220 inhibits thrombin almost 200 times more effectively than trypsin.

Researchers at Hoffmann-La Roche took a different approach to the search for thrombin inhibitors. First, they focused on the optimal filling of the S_1 pocket. Benzamidine was known to be a weak thrombin inhibitor that occupies the S_1 pocket. However, it has the disadvantage of binding more strongly to trypsin (Fig. 23.4). The researchers in Basel were, therefore, looking for a small molecule that would bind more strongly to thrombin than to trypsin. More than 200 small molecules of typical fragment size were tested in this narrow search. Structures were only selected if their functional groups could interact with the negatively charged side chain of Asp 189. Guanidines, amidines, and amines were screened. *N*-Amidinopiperidine (**23.13**, Fig. 23.13) was identified as an interesting lead structure. In contrast to benzami-

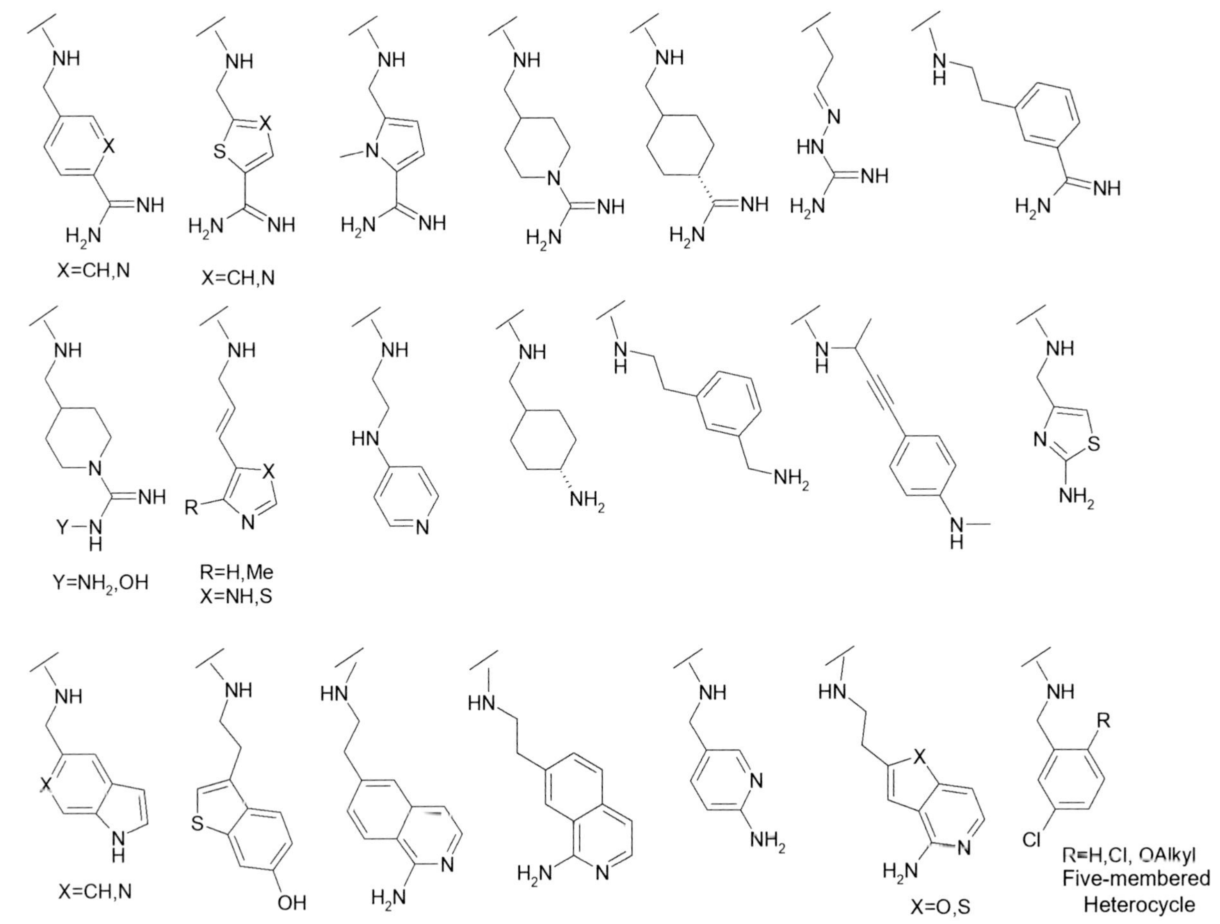

Fig. 23.11 Numerous building blocks have been developed that bind as a mimetic for arginine in the S_1 pocket of thrombin

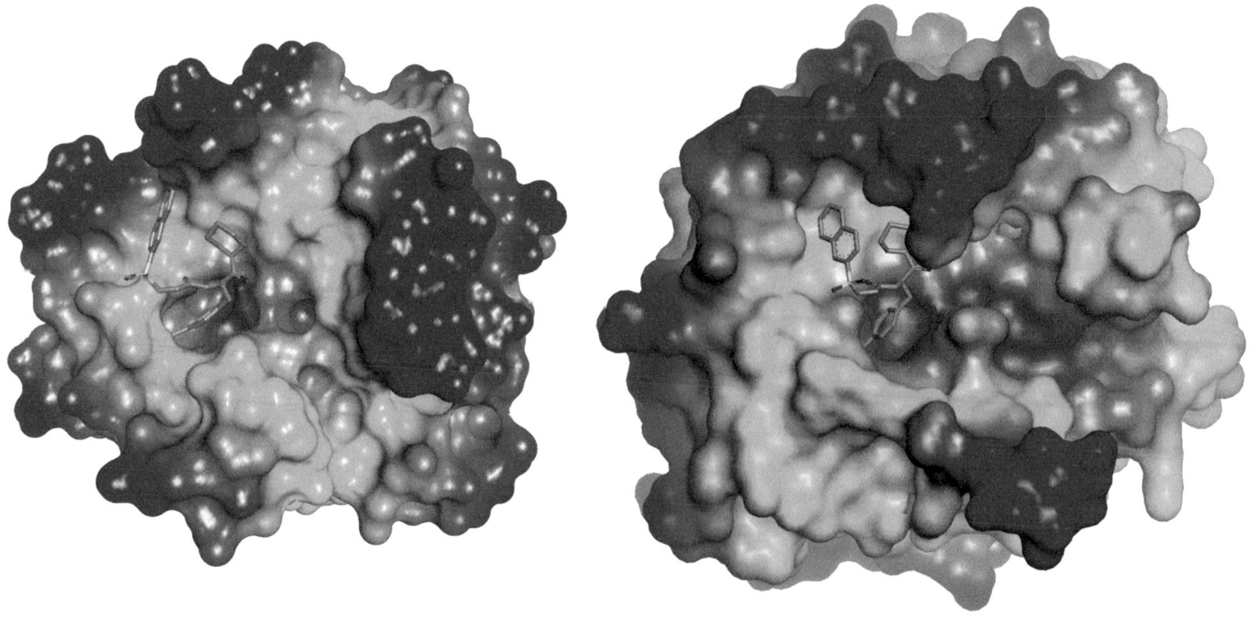

Fig. 23.12 Comparison of the 3D structures of trypsin (*left*) and thrombin (*right*), each complexed with NAPAP. The active site in thrombin is further narrowed by an additional loop from above. The depth of the pocket is, once again, color-coded (see Fig. 23.2). (► https://sn.pub/f6r1mg)

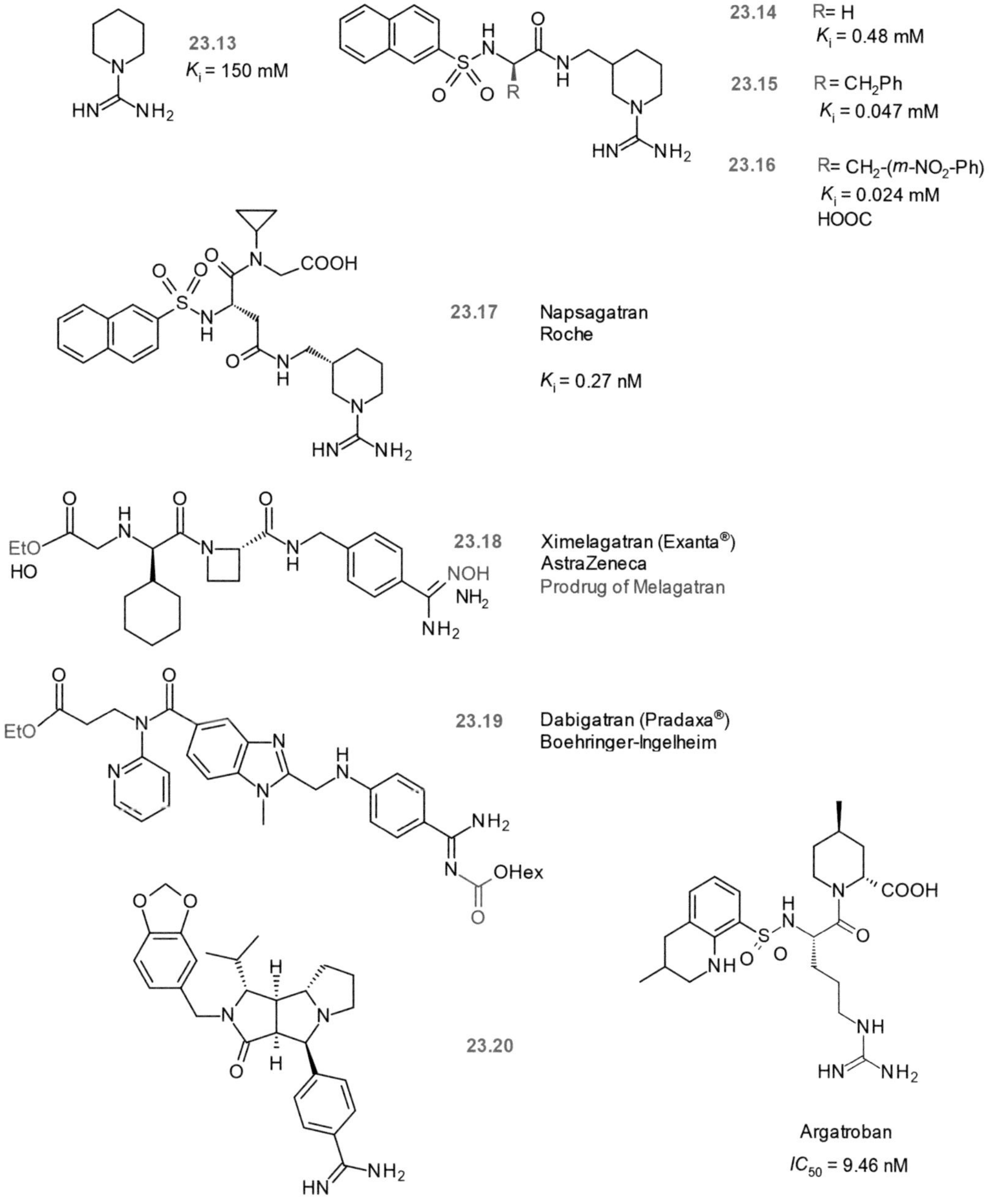

Fig. 23.13 One approach to the structure-based design of thrombin inhibitors began with **23.13** in the S_1 pocket. Compound **23.14** was derived from this lead structure. Its docking into the active site of thrombin let to the idea for the synthesis of **23.15**. Systematic variation of the side chain R yielded compounds with better binding affinity such as **23.16** and **23.17**. The compound was extensively tested in the clinic under the name napsagatran. The compound melagatran from AstraZeneca was introduced as the double prodrug ximelagatran **23.18** as the first orally available thrombin inhibitor on the market. It is derived from the tripeptide sequence D-Phe–Pro–Arg. Another orally available inhibitor, dabigatran **23.19**, was launched to market by Boehringer Ingelheim. The tricyclic inhibitor **23.20**, which was developed at the ETH in Zurich, does not possess a peptide character at all. Argatroban is approved for intravenous use

dine, amidinopiperidine binds more strongly to thrombin (K_i = 150 µM) than to trypsin (K_i = 300 µM). Systematic derivatization yielded **23.14**, a moderately active thrombin inhibitor (K_i = 0.48 µM). Based on the structural model with the protease, it seemed obvious that replacing the glycine moiety with a D-amino acid, e.g., D-Phe, should fill a lipophilic pocket and lead to a significant increase in affinity. The compound was quickly prepared and tested. Indeed, **23.15** bound ten times more strongly to thrombin. Other D-amino acids were then explored and the affinity was increased further. The high selectivity against trypsin was also encouraging; **23.16** binds 840 times more strongly to thrombin than to trypsin. The surprise came when the 3D structure of **23.14** complexed with thrombin was determined: The compound binds differently in the binding pocket than predicted! Contrary to the original assumption, the naphthylsulfonyl group swapped positions with the benzyl side chain.

The incorporation of a nonproteinogenic amino acid proved to be unfavorable from a synthetic point of view. Therefore, other central building blocks that were more readily accessible for synthesis were sought. This work ultimately led to napsagatran **23.17**, a highly potent and exceedingly selective compound. However, because it can only be administered intravenously, it never found its way into a marketed product, especially since argatroban, a marketed product for intravenous use, was discovered much earlier and was already available.

The search for small-molecule, orally available thrombin inhibitors occupied numerous large pharmaceutical companies for many years. It took a long time for AstraZeneca to launch the first orally available thrombin inhibitor, ximelagatran (**23.18**, ◘ Fig. 23.13). The compound is a double prodrug of the active substance melagatran. Its relationship to the original parent structures (e.g., the tripeptide sequence D-Phe–Pro–Arg) is still quite obvious. The head group of the arginine residue was replaced by a benzamidine, the five-membered ring of the proline was narrowed to a four-membered ring, and the terminal benzyl group was shortened to a cyclohexyl ring. The *N*-terminus was substituted with a methylenecarboxylic acid group. It proved to be extremely difficult to make the thrombin inhibitors sufficiently bioavailable and to maintain the required plasma levels for an acceptable period of time. AstraZeneca, in collaboration with Bernd Clement's group at the University of Kiel, Germany, pursued a double prodrug strategy to improve bioavailability: the terminal acid function was masked as an ester, and the benzamidine group was transformed into an *N*-hydroxyamidine. The drug, melagatran, is released in the body by ubiquitous esterases and a set of three specific reductases. AstraZenecca withdrew ximelagatran (Exanta®) after two years because of problems with liver toxicity observed in a small number of cases after weeks of use.

Many years of research into thrombin have finally led to success at Boehringer Ingelheim as well. The substance dabigatran (**23.19**, ◘ Fig. 23.13) was launched in spring 2008 for the prevention of stroke in patients with atrial fibrillation. It also has a benzamidine anchor and a pyridine group for the hydrophobic S_3 pocket. A benzimidazole moiety with an attached amide bond was chosen as a linker between these groups. As with ximelagatran, a carboxylic acid is attached to the *N*-terminus. Dabigatran has significantly less peptide character than the original lead structures. A double prodrug strategy was also used for this compound to ensure adequate bioavailability. In addition to esterification of the acid group, the amidine group was masked as a carbamoyl moiety. The prodrug is called dabigatran (Pradaxa® in the United States and Europe and Pradax® in Canada).

François Diederich's group at the ETH in Zurich has succeeded in developing a thrombin inhibitor (**23.20**, ◘ Fig. 23.13) that completely lacks the peptidomimetic character. Accurate design in the binding pocket led to an inhibitor with a central tricyclic moiety, which was readily prepared by a 1,3-dipolar addition reaction. With a benzamidine anchor for the S_1 pocket and a piperonyl moiety for the S_3 pocket, this very rigid derivative entered the realm of nanomolar inhibitors.

23.5 Design of Orally Available Low Molecular Weight Elastase Inhibitors

Human leukocyte elastase is a serine protease released in the lungs to destroy dead tissue and invading bacteria. The destructive potential of this enzyme is normally controlled by a number of endogenous inhibitors, such as α_1-protease inhibitor or leukocyte protease inhibitor. When the balance between protease and inhibitor is disturbed, for example, by genetic underexpression of an inhibitor or by airborne toxins, elastase attacks even healthy lung tissue. Cigarette smoke contains compounds that oxidize an essential methionine side chain on the endogenous α_1-protease inhibitor, thereby, deactivating the inhibitor. The chronic destruction of cells in the alveoli leads to a life-threatening disease: **emphysema**.

One possible approach for the pharmaceutical treatment of this disease is the use of elastase inhibitors. Unlike thrombin, elastase does not have a deep, pronounced S_1 pocket with an acidic amino acid through which a polar contact can be made with a potential ligand. Elastase only accepts substrates with small hydrophobic amino acids such as valine (◘ Fig. 23.3). If a large binding contribution cannot be expected from occupying the S_1 pocket, as in the case of thrombin, the catalytic serine itself can be involved in the protein–ligand interaction by forming a reversible covalent bond with the inhibitor. Such a concept was pursued at the former ICI (now part of AstraZeneca), starting with a trifluoromethyl ketone R–$COCF_3$ as a reversible covalently binding serine protease inhibitor. Starting from the substrate sequence, potent elastase inhibitors such as **23.21** and **23.22** (◘ Fig. 23.14) were found.

ICI 200880 (**23.22**) proved to be an effective elastase inhibitor in clinical trials, but it lacked oral bioavailability and had a short biological half-life. The spatial structure of the related inhibitor Ac–Ala–Pro–Val–CF_3 complexed with elastase was determined. The main interactions between elastase and the inhibitor are shown in ◘ Fig. 23.15. The inhibitor binds to elastase in a β-pleated sheet conformation with two H-bonds to Val 216 and one to Ser 214. The valine side chain fills the S_1 pocket and the carbonyl group binds covalently as a hemiketal to the side chain of Ser 195. Research has focused on nonpeptidic structures with functional groups capable of forming the same interactions as the peptidic inhibitors.

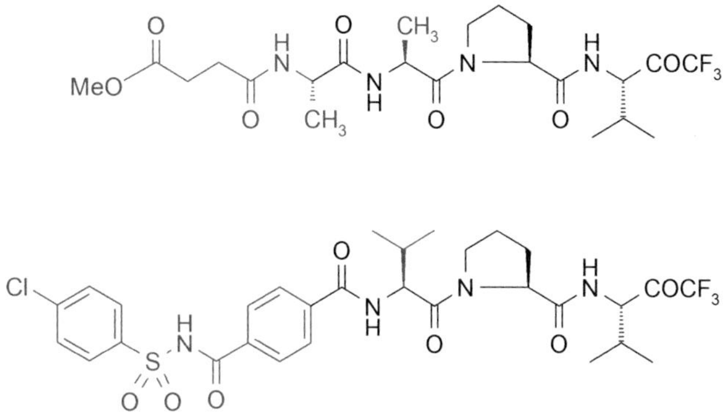

23.21

23.22
ICI 200880

Fig. 23.14 Elastase inhibitors **23.21** and **23.22** (ICI 200880) are substrate analogues. Compound **23.22** is a highly active compound, but it is not orally available

Based on the 3D structure of the protein–ligand complex, pyridones were selected as the most promising peptidomimetic replacements. The postulated binding mode of the pyridone compared to that of the peptidic inhibitors is shown in Fig. 23.15. Compounds of this chemotype were synthesized at Zeneca (now AstraZeneca) and indeed proved to be very potent elastase inhibitors. Compound **23.23** (Fig. 23.16) binds to the protein with K_i = 5.6 nM. However, this compound has several unfavorable properties. It is not orally available and inhibits chymotrypsin (K_i = 60 nM) in addition to elastase. Poor oral bioavailability was attributed to its excessive lipophilicity (log P > 4) resulting in low water solubility.

The pyrimidone class, in which a carbon atom of the heterocycle was replaced by a nitrogen atom, appeared to be synthetically simpler and, therefore, more variable. Compound **23.24** is less lipophilic (log P = 2.1) than **23.23**, ten times more water soluble, and orally available. Its binding to elastase was found to be practically unchanged (K_i = 6.6 nM), whereas chymotrypsin inhibition was much less pronounced (K_i = 1000 nM). Numerous representatives of the new class of compounds were synthesized and tested for inhibitory activity and bioavailability. It has been shown that the potency of inhibition and *in vivo* activity did not run in parallel. For example, **23.25** is a highly potent elastase inhibitor in the enzyme assay, but is not orally available. With an oral bioavailability of 60–90%, compound **23.26** (K_i = 100 nM) proved to be optimal in the animal model. The crystal structure with an analogous derivative **23.27**, which has only one additional sulfonamide group, confirmed the expected binding mode (Fig. 23.17).

The Japanese company ONO Pharmaceuticals Co. developed compound **23.28**, which is derived from **23.26**. It has a 1,3,4-oxadiazole ring instead of the trifluoromethyl group on the ketone and an unsubstituted phenyl ring on the pyrimidone. However, development of ONO-6818 was discontinued in phase II clinical trials due to abnormally elevated liver enzyme levels. However, ONO Pharmaceuticals has had success with ONO-5046, **23.29** which was developed under the name sivelestat (Elaspol®; Fig. 23.16). This inhibitor reacts specifically

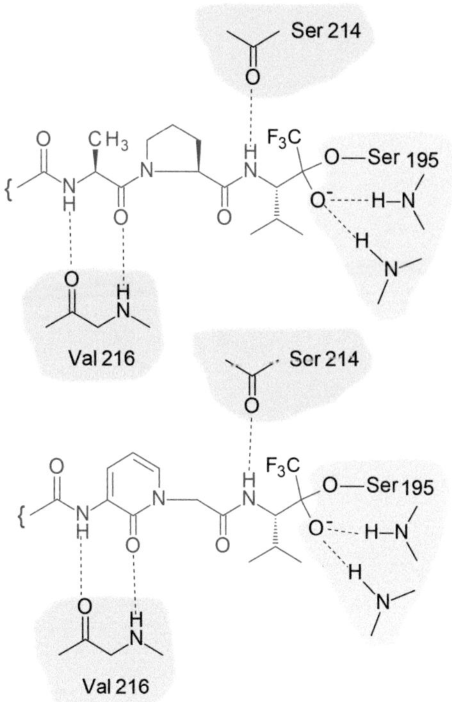

Fig. 23.15 Comparison of the binding mode of the elastase inhibitor Ac–Ala–Pro–Val–CF_3 with the postulated binding mode of the pyridone moiety (e.g., **23.23**, Fig. 23.16). Both compounds should be able to form a double H-bond to Val 216

with elastase and reversibly acylates the catalytic serine residue.

23.6 Serine Protease Inhibitors: Thrombin Was Just the Starting Point

Factor Xa and factor VIIa precede thrombin in the coagulation cascade and are being investigated as targets for **antithrombotic agents**. Both have an aspartic acid at the bottom of their deep S_1 pocket, similar to thrombin. In addition, a narrow and deep S_3 pocket flanked by aromatic amino acids (Tyr 99, Trp 215, and Phe 174) is specific to **factor Xa**. Therefore, this pocket is ideally suited

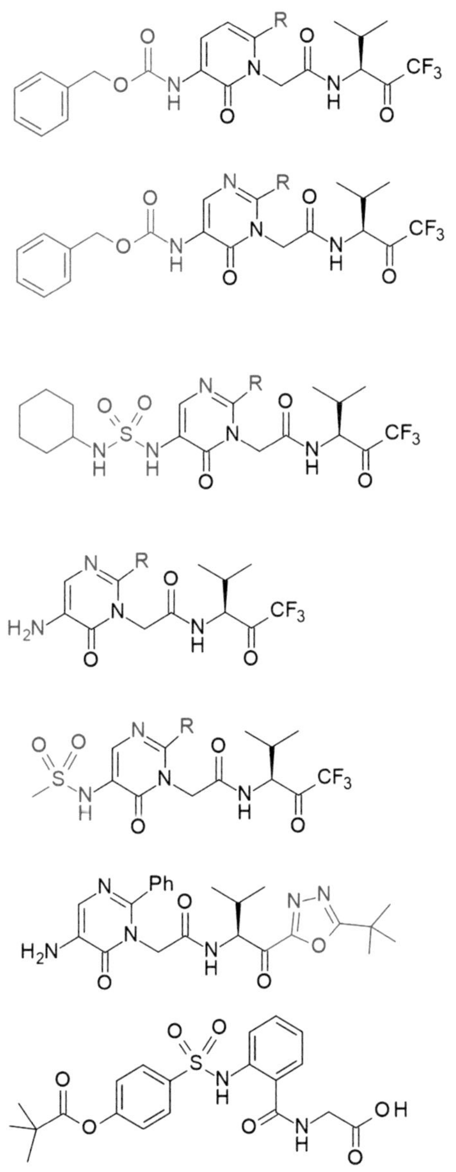

23.23 R = Phenyl
K_i = 5.6 nM

23.24 R = Phenyl
K_i = 6.6 nM

23.25 R = *p*-F-Phenyl
K_i = 1.6 nM

23.26 R = *p*-F-Phenyl
K_i = 100 nM

23.27 R = *p*-NH_2-Phenyl
K_i = 15 nM

23.28 ONO-6818

23.29 Sivelestat ONO-5046

Fig. 23.16 Development of orally available elastase inhibitors at Zeneca. The original idea of replacing the Ala–Pro moiety with a pyridone yielded **23.23**. Later, pyrimidinones were mainly investigated. An additional nitrogen atom was added to the heterocycle. Very potent compounds (e.g., **23.25**) are found in this class. Compound **23.26** has the best *in vivo* properties. The *p*-fluorophenyl group (in **23.26**) or the *p*-aminophenyl group (in **23.27**) increases the lipophilic contact with the enzyme. The compound ONO-6818 **23.28** was developed and advanced to clinical trials in Japan, where it was discontinued due to abnormally elevated liver values in treated patients. Another compound, **23.29**, was clinically tested under the name sivelestat (ONO-4056). These compounds specifically transfer an acyl group to the catalytic serine and reversibly block the enzyme

for aromatic groups on inhibitors. As mentioned above, in the mid-1990s the dogma prevailed that the S_1 pocket of trypsin-like serine proteases could only accommodate groups with a basic character. However, the binding of chloro-substituted aromatic moieties to thrombin was demonstrated at Merck & Co. in the USA. Such aromatic moieties provided a breakthrough for factor Xa inhibitors. Several research groups have been able to develop highly potent inhibitors using chlorophenyl, chloronaphthyl, or chlorothiophene substituents as binders for the S_1 pocket. By placing additional substituents in the deep aromatic S_3 pocket, inhibitors gained sufficient affinity to bind to this protease in the single-digit nanomolar range (Fig. 23.18). In addition to the development of compounds with a chlorine-substituted aromatic moiety for the S_1 pocket, inhibitors with benzamidine substituents for S_1 have also been synthesized as factor Xa inhibitors. However, it has been much more difficult to achieve sufficient selectivity with a benzamidine moiety compared to the other trypsin-like serine proteases. In addition, these derivatives encounter similar problems as the thrombin inhibitors in terms of lack of bioavailability. In Septem-

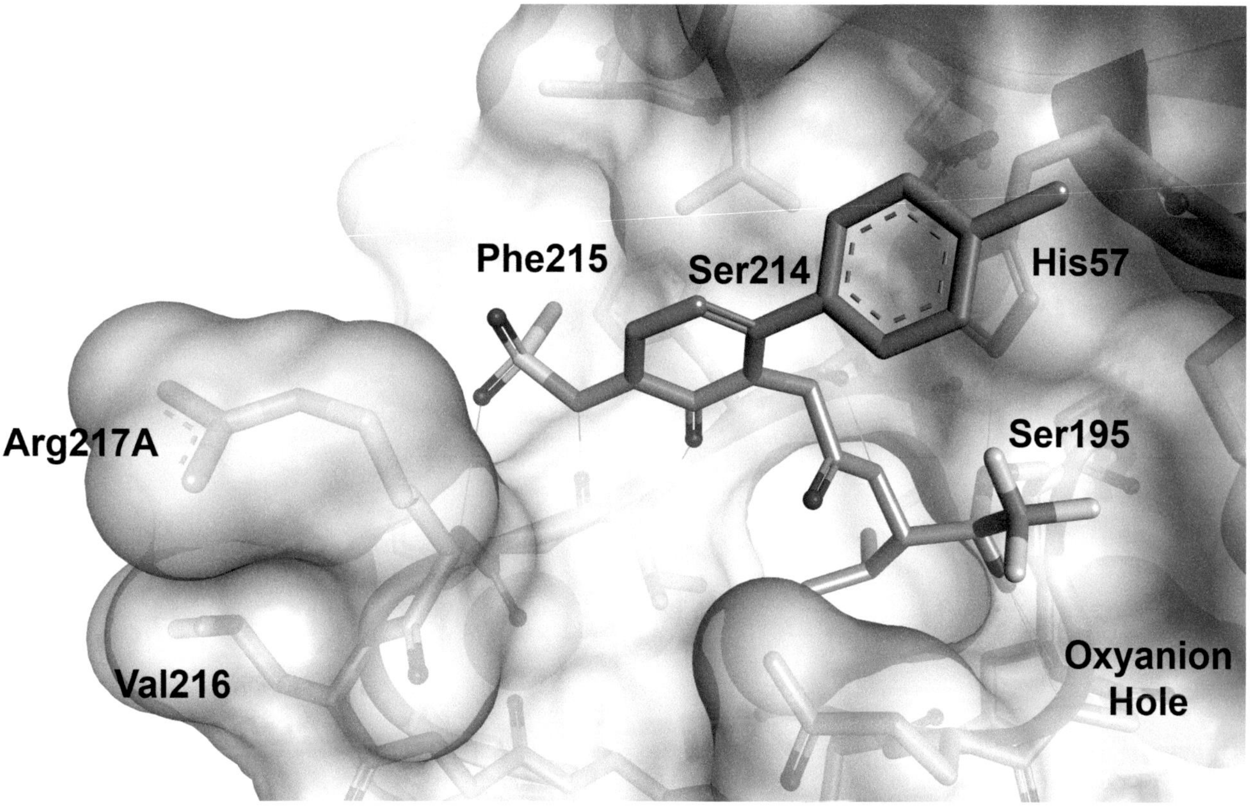

Fig. 23.17 Crystal structure of **23.27** (Fig. 23.16) complexed with elastase. The inhibitor forms two H-bonds to Val 216 and one H-bond to Ser 214. Furthermore, the oxyanion hole is occupied by an oxygen atom. (▶ https://sn.pub/Pfl2aZ)

ber 2008, Bayer launched the new factor Xa inhibitor rivaroxaban (Xarelto®, **23.30**, Figs. 23.18 and 23.19), which places a chlorothiophene substituent in the S_1 pocket.

Factor VIIa is found at the beginning of the extrinsic pathway of the coagulation cascade. This enzyme also belongs to the family of trypsin-like serine proteases, for which specific inhibitors have been sought for many years. In this case, the activation of the protease is of interest. In cases of injury, blood comes into contact with tissue. When this happens, factor VIIa and membrane-bound tissue factor can form a complex that causes a conformational change in the catalytic domain of the protease. A peptide segment adjacent to the catalytic center changes from an unfolded conformation to a helical structure. This leads to a change in the geometry of the catalytic site. Only in the complexed state does the protease have a structure that allows it to initiate the coagulation cascade. Although numerous nanomolar inhibitors are available, none of them has been able to replace the basic P_1 group at the aromatic ring.

The administration of a drug as a "direct" anticoagulant in the blood coagulation cascade always carries the risk of life-threatening and uncontrollable bleeding. For this reason, **antidotes** have recently been developed. In the field of factor Xa, Andexanet alfa, a recombinant human **factor Xa enzyme variant**, has been approved. In its catalytic center, the essential serine has been replaced by an alanine. This reduces its catalytic activity by several orders of magnitude (Sect. 23.2). It is barely able to cleave and activate prothrombin. In addition, its γ-carboxyglutamic acid (Gla) domain has been removed. However, its affinity for factor Xa inhibitors such as rivaroxaban or apixaban is virtually unchanged. It effectively removes these substances from the bloodstream. Boehringer-Ingelheim has taken a different approach in the case of thrombin. The company has developed the **antibody** idarucizumab (Sect. 32.3). In an emergency situation, it cancels out the anticoagulant effect of dabigatran **23.19** by capturing the active substance.

In addition to the serine proteases of the coagulation cascade, other proteases in this family have been selected for drug development. Drug design for these target enzymes has benefited greatly from the experience gained with thrombin inhibitors. The lessons learned from thrombin inhibitors are well transferable to the specific conditions of these proteins. **Tryptase**, **urokinase**, and **matriptase** belong to this family. Tryptase inhibitors are being investigated for the treatment of asthma, and the other two are targets for potential cancer therapeutics.

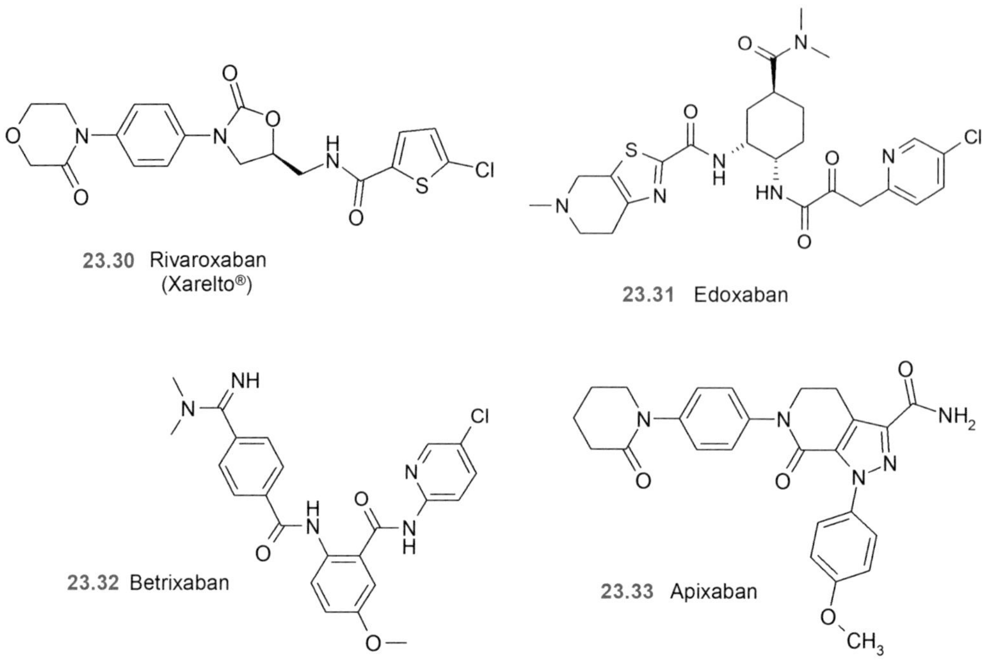

Fig. 23.18 Four potent inhibitors of factor Xa. The first three bind with a chloroaromatic group in the S_1 pocket of the enzyme. Rivaroxaban **23.30** was launched by Bayer in 2008 as the first orally available anticoagulant. Betrixaban **23.31** (Portola), edoxaban **23.32** (Daiichi Sankyo), and apixaban **23.33** (Bristol-Myers-Squibb) are three other compounds on the market. Apixaban binds with subnanomolar affinity to with a methoxy-substituted aromatic ring in the S_1 pocket of factor Xa

Tryptase is a tetramer with four trypsin-like catalytic sites. These sites are separated by several angstroms. To develop selective inhibitors, compounds were designed that carry two benzamidine-like anchor groups and a bridge long enough to connect them. In this way, two of the four sites in the tetrameric tryptase molecule are blocked simultaneously. The disadvantage of this design concept is that the developed inhibitors are very large. They are well above the molecular weight limit of 600 Da that should not be exceeded for good bioavailability.

The protease **furin** also belongs to the serine protease family, but adopts the folding of the subtilisin family (Sect. 14.8). It is involved in the maturation of proproteins. It cleaves the envelope proteins of viruses to convert them into their active form. Its involvement in the "arming" of viruses has even been reported in the tabloid press: On August 28, 2003, Germany's tabloid newspaper *BILD-Zeitung* called furin "the world's most brutal protein," which "turns epidemics into a deadly threat to humans and acts like a detonator on a bomb." Furin and other closely related subtilases cleave certain basic tetrapeptide sequences at the *C*-terminus: Arg–X–(Arg/Lys)–Arg–. Many glycoproteins of lipid-enveloped viruses are cleaved and activated at this recognition sequence. An example is the highly pathogenic avian influenza viruses, which contain such cleavage sequences in hemagglutinin, one of the surface glycoproteins. Whether the viruses can be activated depends on the availability of the omnipresent furin, and this is a prerequisite for the high pathogenic potential of the avian influenza viruses. Furthermore, other genetic combinations or requirements must be met to transform these viruses into dangerous pathogens for animals and humans. Inhibitors of furin could prevent this activation of the viruses. This concept for the development of new antiviral agents is, therefore, being intensively pursued. Furin has also been implicated in the treatment of other diseases such as cancer and cystic fibrosis. Recently, the nanomolar, highly selective and cell-permeable inhibitor BOS-318 (**23.34**, Fig. 23.20) was successfully developed to treat cystic fibrosis. Surprisingly, the crystal structure with the inhibitor bound shows a rearrangement in the catalytic center that has never been observed before. BOS-318 does not interact with the polar amino acids in the highly charged S_1 pocket. This pocket remains unoccupied and is filled with water molecules. Instead, there is a flip of Trp 254, which in thrombin and factor Xa corresponds to Trp 215 at the bottom of the S_3 pocket (Figs. 23.10 and 23.19). As a result, the S_1 pocket is largely occluded. The vacant hydrophobic position of Trp 254 is now occupied by the dichlorophenyl substituent of BOS-318. Once again, serendipity was the godfather of this most surprising design result.

In the early 1990s, an interesting observation was made that the **incretin hormones** GIP and GLP-1 (glucagon-like peptide-1), which stimulate the pancreas to release insulin after a meal, are substrates for **dipeptidylaminopeptidase IV (DPP IV)**. They are rapidly degraded by this serine aminopeptidase. Since incretins were already interesting candidates for diabetes therapy, the idea immediately arose that inhibition of DPP IV could be used as a principle for the treatment of type 2 diabetes (noninsulin-dependent diabetes). The membrane-bound protease cleaves dipeptides from its substrate when a prolyl or alanyl group is in the second position from the *N*-terminus. Vildagliptin (Galvus®) **23.35** and saxagliptin **23.36** (Onglyza®) both use a proline-related cyanopyrrolidine (Fig. 23.20) that

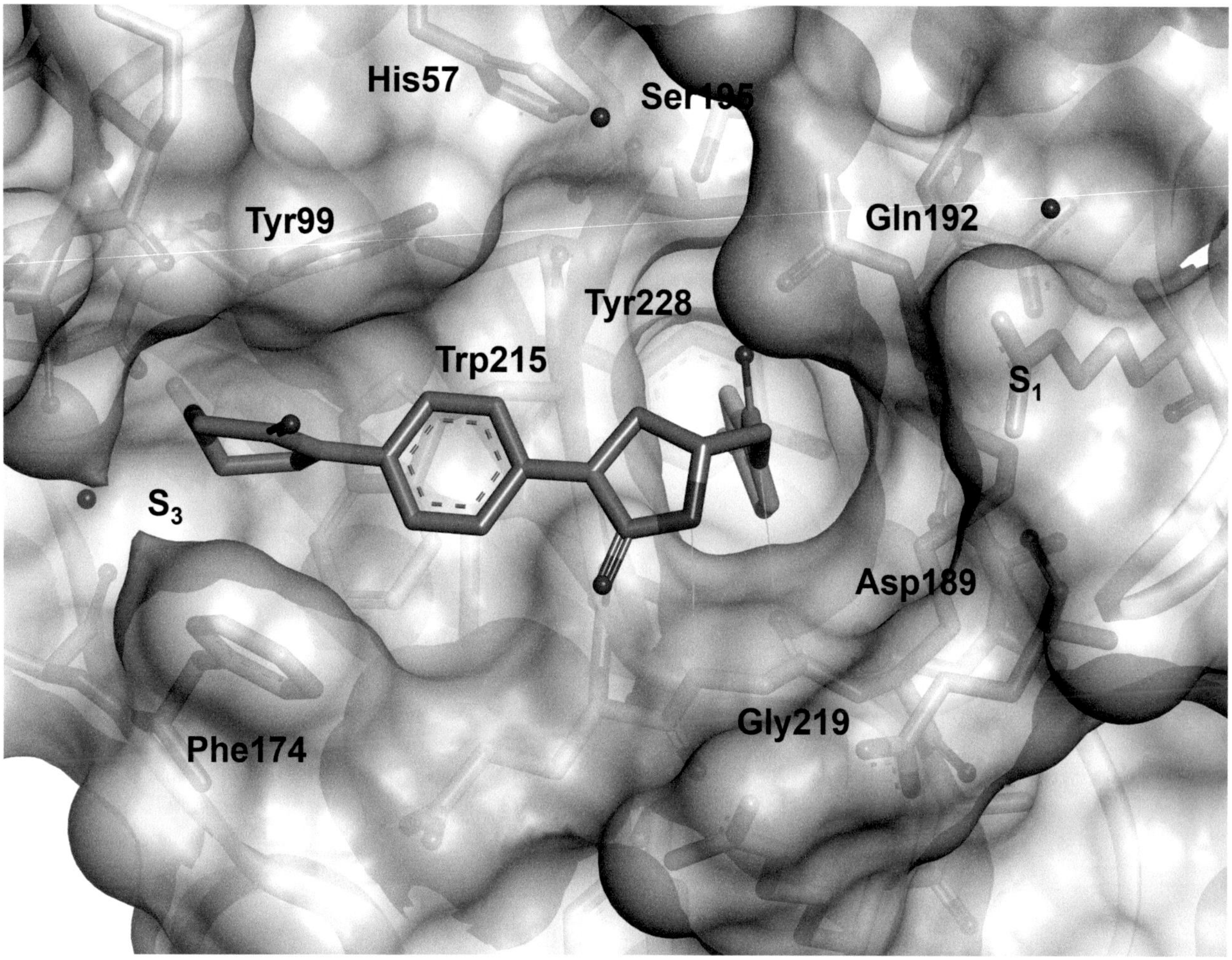

Fig. 23.19 Crystal structure of rivaroxaban **23.30** (Fig. 23.18) in factor Xa. The inhibitor's chlorothiophene group binds in the deep S_1 pocket, at the end of which Tyr 228 and Asp 189 are found. The chlorine atom forms interactions with the aromatic ring. The central phenyl ring and the terminal lactam ring of the inhibitor are found in the S_3 pocket, which is enclosed by the three aromatic groups of Tyr 99, Phe 174, and Trp 215. (► https://sn.pub/q86HYM)

can reversibly, covalently bind to the catalytic serine. Sitagliptin (Januvia®) **23.37** is another compound available for the treatment of type 2 diabetes. It blocks the protease without covalently binding to the catalytic serine. Recently, mimetics of the aforementioned incretin hormone GLP-1 have been introduced into diabetes therapy. In type 2 diabetics, GLP-1 is released at lower levels, which reduces the glucose-lowering effect. GLP-1 receptor agonists, such as semaglutide, mimic the action of GLP-1 by binding to a GPCR (Sect. 29.1). Chemically, they are slightly modified from the biologically active human oligopeptide GLP-1, which consists of 31 amino acids. The amino acid substitutions make the peptide more resistant to DPP IV protease degradation, and the addition of an 18-carbon fatty acid moiety via a short polyethylene glycol (PEG) linker results in binding to human serum albumin. This provides a human half-life of approximately 7 days and further steric protection from proteolytic degradation and rapid renal filtration. The peptides stimulate insulin secretion and reduce glucagon release, thereby delaying gastric emptying and reducing appetite. Semaglutide (Ozempic®), first introduced for the treatment of diabetes, was approved by the FDA in June 2021 as high-dose injectable semaglutide (Wegovy®) for long-term weight management in adults. The drug is currently being heralded and touted as a new way to treat obesity. Novo Nordisk, the company that developed the peptide, has become in 2024 the most valuable publicly traded company in Europe and is even generating remarkable economic growth in Denmark. The fat-away shot seems to make the dream of easy weight loss come true for many overweight people. Meanwhile, Eli Lilly and Company have launched a second drug with a similar application, tirzepatide (Mounjaro®), another modified oligopeptide.

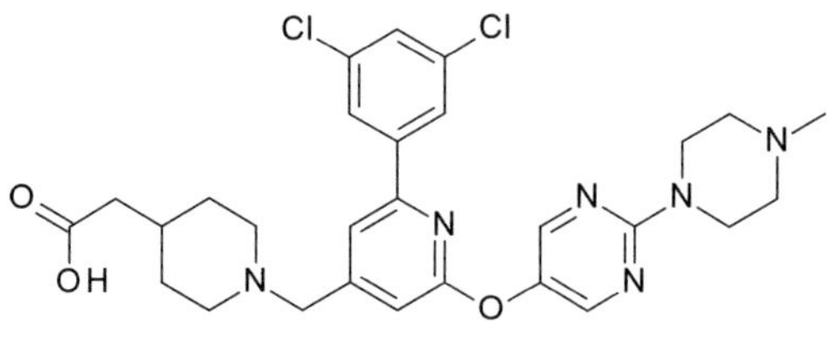

23.34 BOS-318
Furin Inhibitor IC_{50} = 1.9 nM

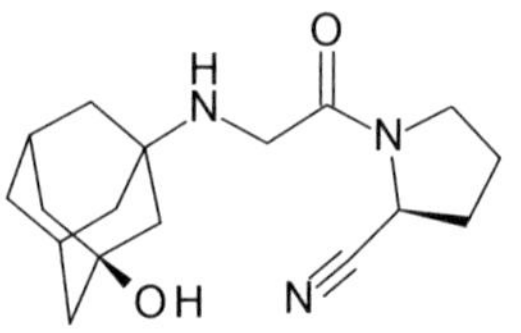

23.35 Vildagliptin (LAF 237)
Novartis

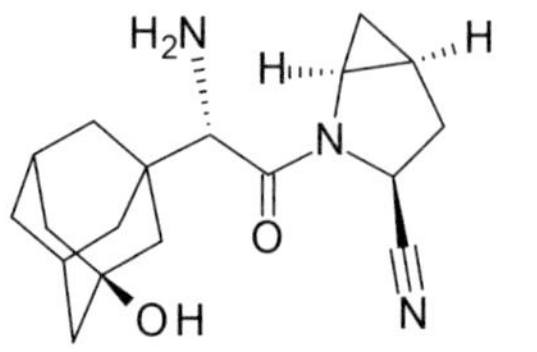

23.36 Saxagliptin (BMS-47718)

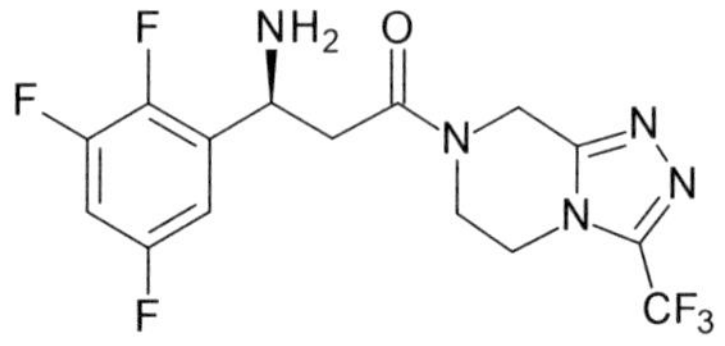

23.37 Sitagliptin (MK-0431)
Merck

Fig. 23.20 BOS-318 **23.34** is a potent and highly selective furin inhibitor with a surprising binding mode due to an unexpected protein rearrangement. Vildagliptin **23.35**, saxagliptin **23.36**, and sitagliptin **23.37** are inhibitors of the serine aminopeptidase DPP IV for the treatment of type 2 diabetes

Certainly, more serine proteases will be discovered and validated as putative drug targets in the coming years. However, the field is increasingly benefiting from the experience already gained with individual members of this protein family. When it comes to identifying new targets, this experience is helpful in quickly finding suitable lead structures as a starting point.

23.7 Serine, a Favored Nucleophile in Degrading Enzymes

Serine peptidases use the OH group of an endogenous serine as an attacking nucleophile. The adjacent histidine residue mediates the temporary proton transfer and the aspartate compensates for the intermediate charge on the imidazole ring of the histidine. A special feature is the temporary covalent bond between the *N*-terminal part of the substrate and the enzyme. Many other hydrolytically cleaving enzymes use an analogous principle. **Esterases** and **lipases** also have a catalytic triad. Occasionally, these enzymes exchange an aspartate for a glutamate. The neurotransmitter acetylcholine acts on many synapses in the vegetative nervous system and is involved in the transmission of nerve impulses. It binds to the nicotinic acetylcholine receptor and activates this ion channel (Sect. 30.4). Acetylcholine must be removed to limit the duration of the transmission process and to reset the receptor to its starting point. An imbalance in this nerve impulse transmission system leads to acute and chronic movement disorders. **Acetylcholinesterase** is responsible for the degradation of acetylcholine. Inhibitors of this enzyme are used to treat Parkinson's and Alzheimer's disease. Reversible binding inhibitors such as donepezil, rivastigmine, or galantamine exert their therapeutic effect by reducing acetylcholine degradation in the brain.

Acetylcholinesterase has a catalytic triad of serine, histidine, and glutamate. Acetylcholine (**3.46**, ◘ Fig. 3.10) is cleaved by the enzyme by transferring its acetyl group to the catalytic serine; hydrolysis slowly releases acetic acid from the esterase. The drug (*S*)-rivastigmine **23.38** (◘ Fig. 23.21) is also attacked by the catalytic serine and its carbamoyl group is transferred. Due to the increased stability of the carbamoyl enzyme complex, the esterase is subsequently deacylated very slowly and regenerated for the next transformation. This is equivalent to inhibiting the target enzyme for several hours. The suppressed degradation of acetylcholine leads to permanent excitation with muscle contraction and subsequent paralysis. Victims die from respiratory and cardiac inhibition. **Cholinesterase inhibitors** are, therefore, used as insecticides. Active ingredients such as paraoxon **23.39** (◘ Fig. 23.21), parathion (E605) **23.40**, propoxur **23.41**, or malathion **23.42** contain phosphoric acid esters or thioesters that are virtually irreversibly transferred to the catalytic serine. Because of this inhibition, acetylcholine increases to lethal concentrations in insects. In intelligence conflicts, the Novichok nerve toxins developed in Russia have gained sad notoriety. They also belong to this group of agents and carry a fluorine atom (X = F) on the phosphorus atom.

Analogous to esterases, lipases also hydrolyze ester bonds. The catalytic triad consists of a serine, histidine, and aspartate or glutamate. **Pancreatic lipase** cleaves triglycerides during the digestion of fats. Inhibitors of this intestinal enzyme are used to treat obesity. The result is a significantly reduced absorption of fats and their degradation products. Orlistat (Xenecal®, **23.43**; ◘ Fig. 23.22), a synthetic hydrogenation product of the natural product lipstatin, has a very long aliphatic side chain and a reactive β-lactone ring in its core. Serine in the catalytic site of lipase attacks the carbonyl group of the lactone ring

Fig. 23.21 (*S*)-Rivastigmine **23.38** transfers a carbamoyl group to the catalytic serine in the binding pocket of acetylcholine esterase and blocks its function because the carbamoyl–esterase complex decomposes very slowly. The acetylcholine esterase inhibitors paraoxon **23.39**, parathion **23.40**, propoxur **23.41**, or malathion **23.42** are phosphoric acids, thiophosphoric acids, or carbamic esters and are used as insecticides. They also react with the catalytic serine and form a stable covalent bond

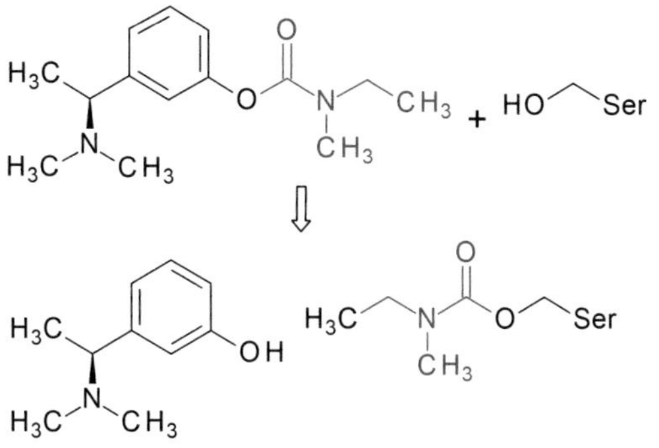

23.38 *S*-Rivastigmine

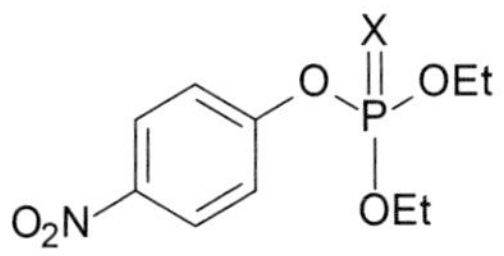

23.39 X=O
23.40 X=S Parathion, E605

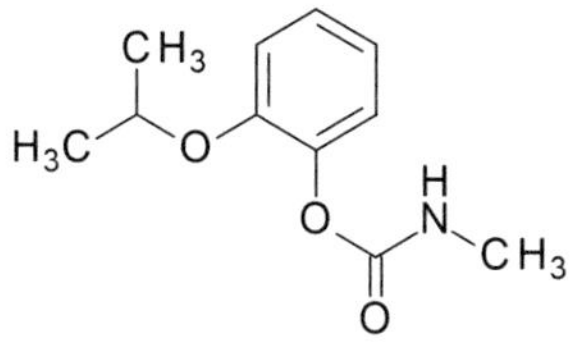

23.41 Proxopur

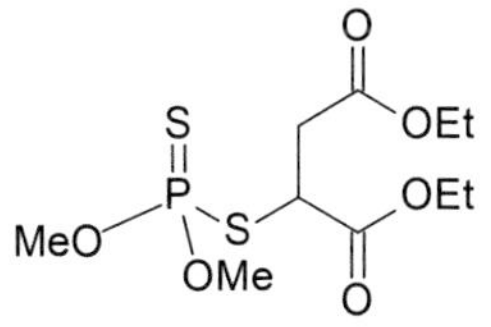

23.42 Malathion

and opens the strained ring by transformation into a stabilized acyl–enzyme complex. Once blocked, the enzyme is no longer able to break down triglycerides, resulting in a reduced ability to extract calories from food.

Lipases are often used for the **kinetic resolution** of racemates. This is usually achieved by enzymatically converting a racemic mixture of esters in which one of the two forms reacts faster than the other. An example was described in Sect. 5.4 where the lipase was used not only to hydrolyze but also to form a new amide bond. For this, the intermediate acyl–enzyme complex cannot be exposed to a water molecule as a nucleophile, but a compound with a free amino group must be available. This transformation produces a new amide bond. Bacteria use such a **transpeptidase reaction** to build their **cell wall**. This cell wall has a completely different composition than the cell wall in humans. Therefore, the enzymes used to synthesize the cell wall are bacteria-specific and particularly suited as a target for a drug therapy with few side effects.

The final step in cell wall biosynthesis is the **cross-linking** of the peptidoglycan strands. The terminal amino group of a pentaglycine chain attacks between two D-alanine residues of another peptide unit. The D-Ala–D-Ala bond is cleaved and a new peptide bond is formed between D-Ala and glycine. This cross-linking is mediated by a glycopeptide **transpeptidase**. It has a catalytic machinery very similar to that of serine proteases. In addition to a catalytic serine, the reaction center also contains a lysine and a glutamate, as well as an oxyanion hole. Penicillins **23.44** and cephalosporins **23.45** (Fig. 23.23) inhibit these transpeptidases. They have a spatial structure analogous to the D-Ala–D-Ala dipeptide and are, therefore, recognized as "false" substrates (Fig. 23.23). The β-lactam ring is opened by the attacking catalytic serine, resulting in an irreversible covalent bond to the enzyme. Cross-linking of the glycan strands is prevented and the newly synthesized cell wall does not achieve sufficient stability. It cannot withstand the osmotic pressure of the cell contents and the bacterial cell is killed.

Of the first penicillins **23.44** discovered by Alexander Fleming (Sect. 2.4), only benzyl and phenoxymethylpenicillin are still of clinical importance (Fig. 23.23). The substituents on the 6 amino group of penicilloic acid have been exchanged to improve pharmacokinetics, spectrum of activity, and acid stability. Electronegative atoms on the α carbon of the acyl function increase stability to acid-catalyzed degradation and contribute to improved oral bioavailability.

Bacteria rapidly develop resistance to penicillins. They use **lactamases**, which are enzymes structurally related to transpeptidases. Four classes of lactamases are known, three of which have a catalytic serine in the active site. Another class belongs to the zinc-dependent metalloenzymes (Chap. 25). The catalytic serine of β-lactamases is also acylated by penicillins and related cephalosporins (◻ Fig. 23.23). Up to this step, the mechanism of transpeptidases and β-lactamases is identical. However, the transpeptidases form very stable acyl enzymes, whereas the covalent intermediate of the β-lactamases is rapidly hydrolyzed. The antibiotic designed to inactivate the transpeptidase is, therefore, rendered inactive. β-Lactamases are probably descendants of the transpeptidases. They are widespread in Nature and have evolved as a result of competition between bacteria and molds. The resistance gene for β-lactamases is easily transferred between bacteria because the information is stored on an extrachromosomal plasmid. Such plasmids are transferred very rapidly.

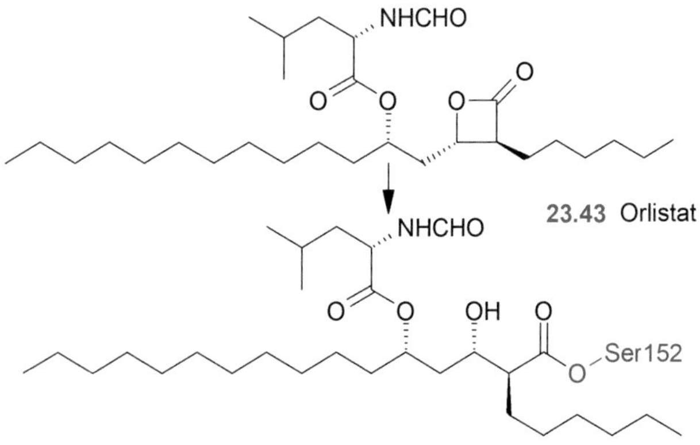

◻ **Fig. 23.22** Orlistat (Xenical®) **23.43** is a synthetic hydrogenation product of the natural product lipstatin, which has two additional double bonds. It has a reactive β-lactone ring that reacts with the catalytic serine in the catalytic site of pancreatic lipase to form an acyl–enzyme complex with ring opening. The enzyme's function is then blocked

How do β-lactamases differ from transpeptidases in that they are able to rapidly dispose of the covalently bound ring-opened penicillin? The release requires hydrolytic cleavage from the protein. This requires a well-placed water molecule in the active site that can initiate the nucleophilic attack on the acyl–enzyme species. Although the spatial architecture of transpeptidases and β-lactamases is very similar, there is little sequence identity. It has been suggested that both types of enzymes originated from a common ancestor. Thus, by targeted mutagenesis, it was possible to endow a transpeptidase with the hydrolyzing properties of a lactamase!

Only a few amino acid substitutions were necessary. It is mainly the hydrophobic amino acids such as phenylalanine and tryptophan that protect the acyl–enzyme complex from hydrolysis in the transpeptidase. They prevent the transpeptidase from accepting a water molecule at the critical position for nucleophilic attack. In contrast, polar amino acids such as glutamic acid (◻ Fig. 23.24, Glu 166) are found at the same positions in the lactamases. In contrast to the hydrophobic amino acids of transpeptidase, they anchor and activate the water molecule in the correct orientation for nucleophilic attack on the acyl–enzyme complex in lactamases. As a result, the covalent complex with the penicillin cleavage product that was

◻ **Fig. 23.23** In the last step of the bacterial cell wall synthesis, a glycopeptide transpeptidase cleaves the bond between two D-Ala–D-Ala groups and forms a new bond between D-Ala and a glycine in a peptidoglycan strand. Lactam antibiotics of the penicillin (**23.44**) or cephalosporin type (**23.45**) can block this step. The penicillin scaffold (*green*) is reminiscent of the D-Ala–D-Ala group (*orange*) and is bound analogously by the enzyme. An irreversible inhibition of the transpeptidase is achieved by a nucleophilic opening of the lactam ring with the help of the catalytic serine. (► https://sn.pub/sQuPJA)

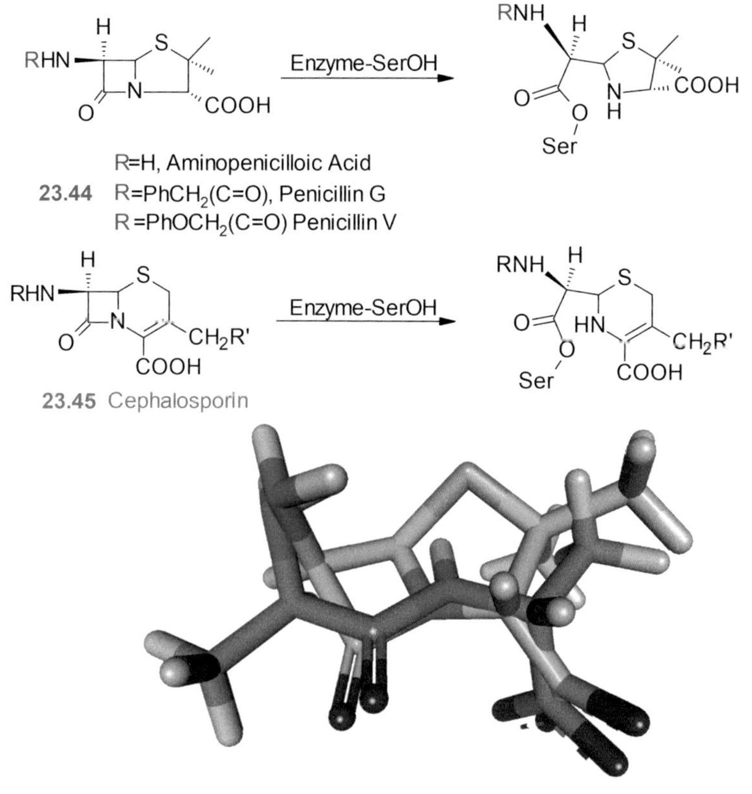

formed by ring opening in the lactamases is hydrolyzed, whereas it remains stable in the transpeptidases.

How can this lactamase-caused resistance and degradation of penicillins be stopped? Unsubstituted penicilloic acid **23.46** is rapidly cleaved by TEM-1β-lactamase (◘ Fig. 23.24). Based on structural considerations, it was suggested that a hydroxymethyl group should be added at the 6-position. This group should be in the exact position where the water molecule would start its nucleophilic attack on the acyl–enzyme form. Indeed, derivative **23.47** inactivates TEM-1β-lactamase. In the subsequently determined crystal structure, a water molecule was detected near the CH_2OH group, but it is too far away to successfully hydrolyze the acyl enzyme. The hydroxyl group, therefore, blocks the attack of a water molecule on the ester carbonyl group of the acyl enzyme.

The incorporation of such a hydroxymethyl group has been accomplished in important β-lactamase-resistant **β-lactams** such as imipenem **23.48** or meropenem **23.49** (◘ Fig. 23.25). β-Lactamases can also be irreversibly inhibited. If such an inhibitor is administered with a penicillin, the degradation of the penicillin by the lactamase is blocked, and it is available to inhibit the transpeptidase. The natural product clavulanic acid **23.50** forms an acyl–enzyme complex upon the opening of its lactam ring. By rearrangement a vinylogous urethane is formed that is resistant to hydrolysis.

With these examples, the spectrum of enzymes that use a serine as a nucleophile is far from exhausted. Viruses need cleavage enzymes. They have to cleave the polypeptide chains synthesized by the infected cell according to their own specifications into functional viral proteins. Viruses either use proteases of the infected host cell (e.g., furin) or they use their own **viral proteases**. Since the correct function of the latter enzymes is essential for the maturation of new viruses and is also virus-specific, these proteases are privileged targets for drug development. Peptidases with a catalytic serine as well as a cysteine (Sect. 23.9) are recognized. As we will see in Sect. 24.3, an aspartic protease serves other viruses. In recent years, inhibitors of the **viral NS3/4A serine protease** of the **hepatitis C virus (HCV)** have been successfully developed. An estimated 170 million people worldwide are infected with this virus. It is transmitted through contact with infected blood. If left untreated, chronic HCV infection can lead to serious liver diseases, including cirrhosis and hepatocellular carcinoma. The NS3/4A protease is responsible for the selective cleavage of the initial polypeptide chain into the individual viral proteins (NS4A, NS4B, NS5A, and NS5B). The first two inhibitors of this enzyme, boce-

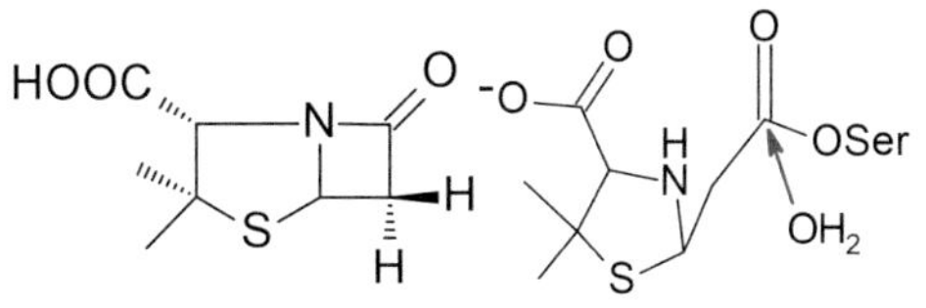

23.46 Penicillic Acid

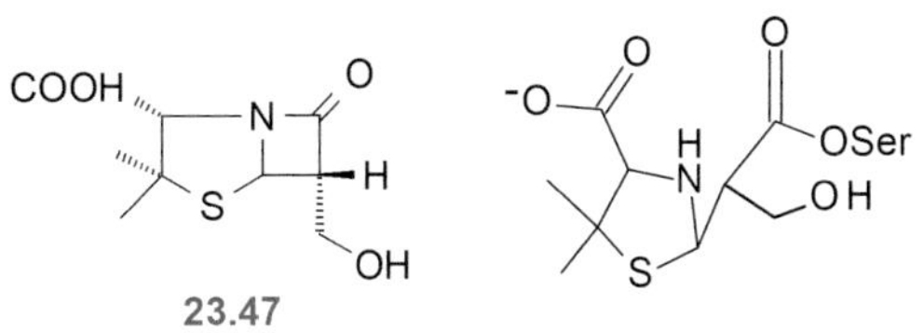

23.47

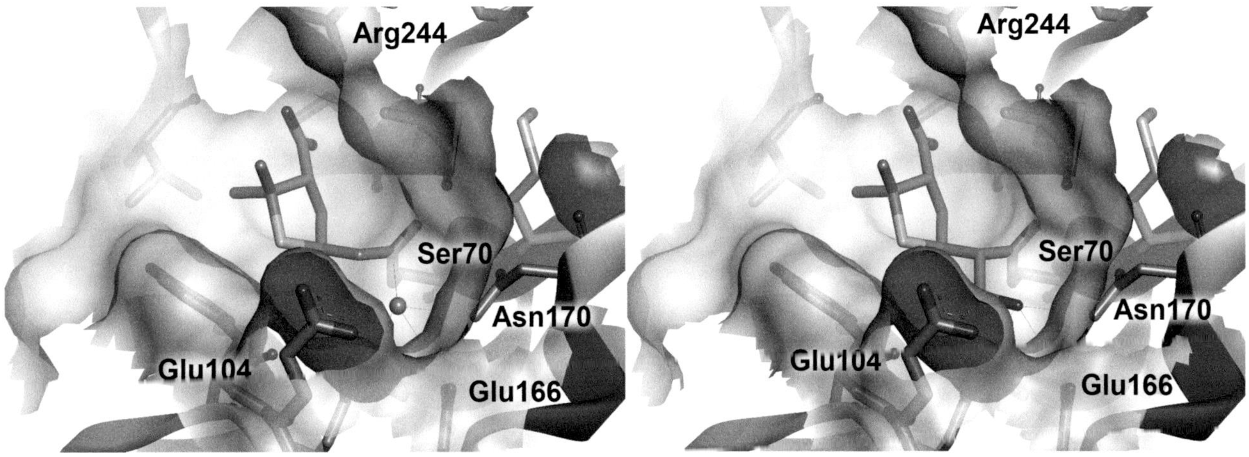

◘ **Fig. 23.24** Unsubstituted penicilloic acid **23.46** is quickly cleaved by TEM-1β-lactamase (*left*). By adding a hydroxymethyl group to the 6-position of **23.47**, a compound is obtained that forms a hydrolytically stable acyl–enzyme complex with the enzyme (*right*). A new crystal structure was determined with this compound. The hydroxyl group is found at the position where the water molecule (*orange sphere*) starts its nucleophilic attack on the acyl–enzyme intermediate (*left*, lower part of the image, modeled structure taking the coordinates from the crystal structure of the complex with **23.47**). The hydrophobic amino acids such as phenylalanine and tryptophan are found at positions 166 and 170 of the transpeptidases, which are structurally related to the β-lactamases. (► https://sn.pub/0wjVve)

previr **23.51** and telaprevir **23.52**, were approved in 2011 (◘ Fig. 23.25). They were followed by a vast number of other inhibitors (such as simeprevir, sofosbuvir, asunaprevir, danoprevir, simeprevir, paritaprevir, vaniprevir, and grazoprevir), some of which block the target enzyme covalently and others as noncovalent peptidomimetics.

The **assemblins**, another group of serine peptidases, have been found in herpes viruses. The enzyme from cytomegalovirus belongs to this group, as do those from varicella-zoster virus and herpes simplex virus. These proteases also use one serine and one histidine. An additional histidine forms the third amino acid in the triad. Despite a different folding pattern, their triad fits very well with the trypsin triad. Even the oxyanion hole is present in these viral proteases.

The carboxy-serine peptidases (**sedolisins**) are another group folded analogously to subtilisin (Sect. 14.7). They have a triad of serine, glutamate, and aspartate. A member of this family was recently discovered on the human *cnl2* gene. Mutations in this gene cause severe neurodegenerative diseases. This enzyme also contains an oxyanion hole to which, interestingly, an aspartate contributes. However, it is only in the protonated state that this residue can act as a hydrogen bond donor and negative charge stabilizer in the transition state. Since the enzymes of this family are active in a pH range of 3–5, the requirement for protonation is met.

There may be many more cleavage enzymes that use a catalytic serine to be discovered. It remains to be seen which of the discovered peptidases will be selected for pharmaceutical development. Their catalytic machinery shares the same spatial architecture in all examples. Therefore, the general principles can be transferred between the individual members of the family.

23.8 Triads in All Variations: Threonine as a Nucleophile

In addition to serine, another amino acid carries an aliphatic OH group: **threonine**. This amino acid can also be catalytically active in a protease. The proteasome is the cell's central **protein-shredding machine** and cleaves ubiquitin-labeled proteins into small oligopeptides of 3–20 amino acids. The ubiquitin tag itself is a highly conserved protein with 76 amino acids. As a cellular shredding machine, the proteasome plays a central role in protein metabolism, cell growth, and cell death. It is, therefore, an important target for the treatment of **cancer**. It is a multiprotease complex composed of more than 30 proteins and is found in both the cytoplasm and the nucleus (◘ Fig. 23.26). The proteasome is constructed like a large barrel with two lid regions that have regulatory functions; these regions control the entry of

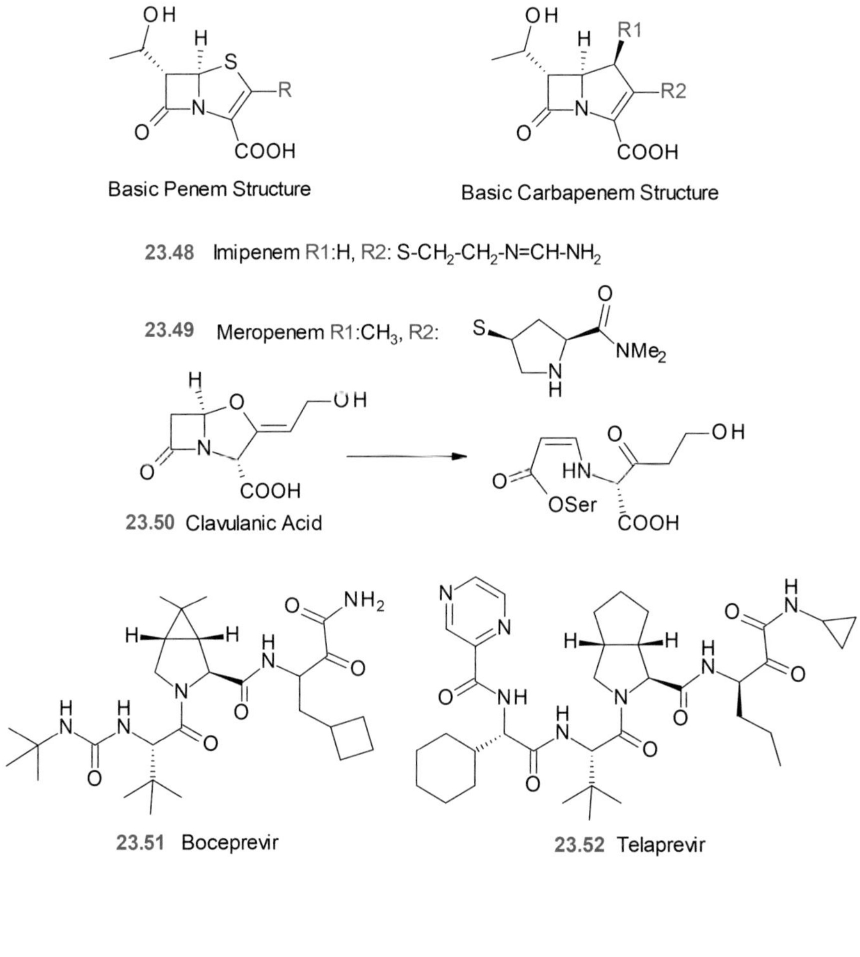

◘ **Fig. 23.25** β-Lactamase-resistant antibiotics of the penem and carbapenem type. Imipenem **23.48** and meropenem **23.49** are derived from the carbapenem type. The natural product clavulanic acid **23.50** opens its lactam ring and forms an acyl–enzyme complex with the serine residue. A hydrolysis-resistant vinyl urethane analogue is formed by a rearrangement. Boceprevir **23.51** and telaprevir **23.52** were the first two drugs approved as inhibitors of the hepatitis C virus (HCV) NS3/4A serine protease

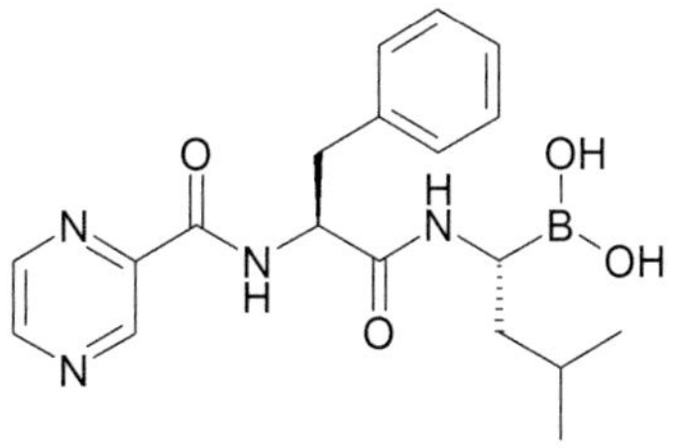

23.53 Bortezomib

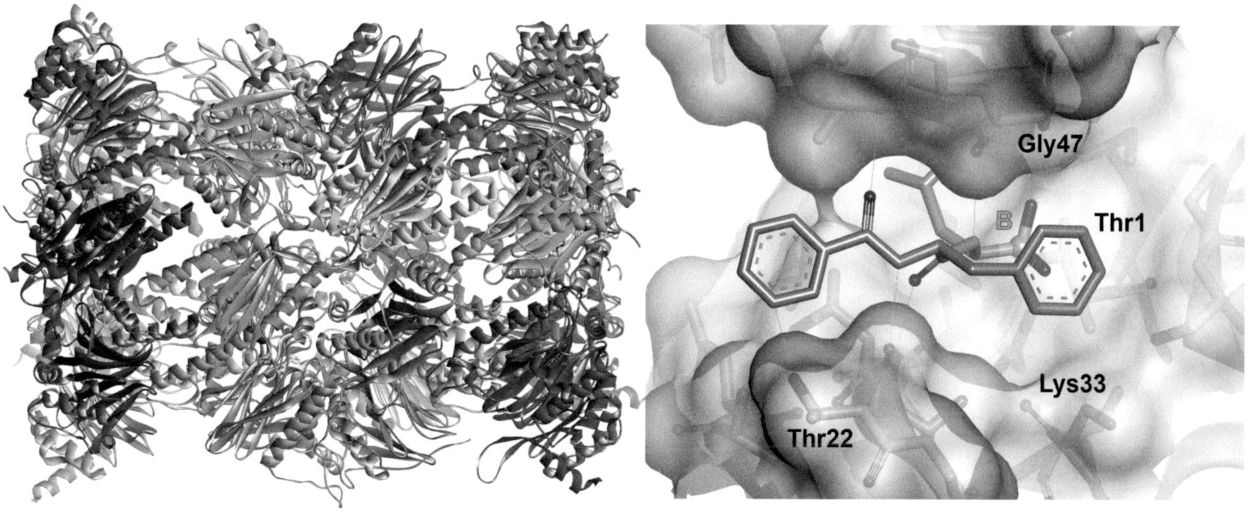

Fig. 23.26 The proteasome, a cellular shredding machine, proteolytically cleaves ubiquitinylated proteins selectively into small oligopeptides that have between 3 and 20 amino acids. The crystal structure of the 20S proteasome from yeast (subunits are shown in different colors) is shown on the *left*. Six of these units are inhibited by bortezomib (*yellow*). The boronic acid derivative bortezomib **23.53** (*right*, *gray*) reacts with the *N*-terminal Thr1 and forms a covalent boronic acid ester complex. (▶ https://sn.pub/eYXznx)

substrates into the shredder. The catalytic sites of the proteases, which have chymotrypsin-like, trypsin-like, and peptidyl-glutamyl-peptide-like substrate specificity, contain the threonine. The OH group of this threonine acts as the nucleophile. An adjacent positively charged lysine and a balancing aspartate enhance its nucleophilic strength. Since the threonine is the first amino acid at the *N*-terminus, it also carries a free amino group. This group serves as a proton acceptor in the mechanism. The nucleophilic center is complemented by two serine and one aspartate residues, which contribute to the stabilization of the transition state.

Millenium Pharmaceuticals, founded as an academic research institute, introduced bortezomib **23.53** (Velcade®) in 2006, the first agent to block the threonine protease function of the proteasome. Chemically, bortezomib is a boronic acid derivative (Fig. 23.26). The inhibitor reacts with the threonine of the catalytic triad to form a covalent bond. In addition to this reactive group, the molecule has a distinct peptide-like character. These features allow the molecule to interact with the substrate binding site in the proteasome.

Another peptide analog, carfilzomib, is in clinical trials. It carries a terminal α′β′-epoxy ketone. Upon inhibition, the threonine OH group nucleophilically attacks the keto function of the inhibitor. The adjacent *N*-terminal amino group then opens the epoxide ring. This results in an irreversible covalent bond. Although the epoxy ketone is highly reactive, carfilzomib is a highly selective proteasome inhibitor. The proximity of the nucleophilic Thr–OH of the first residue in the sequence and the *N*-terminal amino group is an unusual and exceptional combination. However, it is an essential prerequisite for the activation of this inhibitor.

The proteasome is an important target structure that appears suitable for cancer treatment. More than 20 different inhibitors are currently being developed. Bortezomib is used for the treatment of **multiple myeloma**, a type of bone marrow cancer. This cancer is based on the malignant transformation of plasma cells, the physiological function of which is to produce antibodies for immune defense. Even though bortezomib cannot heal multiple myeloma, its use can extend the life of patients for whom other therapies have failed. In multiple myeloma, the plasma cells produce massive amounts of misfolded proteins that must be digested by the proteasome. Therefore, these cells need a proteasome that functions correctly, otherwise apoptosis would be induced. However, it is desirable to block the function of the proteasome in order to utilize this mechanism for the degradation of degenerated cells. Such cells are significantly more sensitive to bortezomib therapy than normal cells. Some tumor cells also activate a transcription factor, NF-κB, which controls the proliferation and survival of the tumor cells. The proteasome is critical for the activation of NF-κB because it degrades an inhibitor of this transcription factor that acts as a kind of emergency brake on NF-κB. Therefore, the inhibition of the proteasome serves to keep NF-κB in its benign form, because its inhibiting binding partner is no longer being degraded. Bortezomib may also induce apoptosis of tumor cells by stabilizing cyclin-dependent kinase inhibitors (Sect. 26.2) and the tumor suppressor p53.

Interestingly, a protease has been discovered in bacteria that exists as a 14mer and has a spatial structure reminiscent of the proteasome. The **ClpP protein** is a serine protease involved in the degradation of cellular proteins in bacteria. Treatment with a macrolide antibiotic can cause its function to go out of control and degrade proteins in an unregulated manner. This leads to cell death in the bacteria. The company Bayer recognized this principle and used it for an **antibiotic therapy**. The goal was not to block the protease function of the ClpP protein, but rather to promote its uncontrolled effects through synthetic antibiotics.

Table 23.4 Cysteine proteases with physiological importance (X = arbitrary amino acid). The 3D structures of all of the listed enzymes are known

Enzyme	Cleavage site	Function or therapeutic use
Papain	–Val–X–X–	Model botanical enzyme from papaya
Cathepsins B, L, K, M	–Arg–X–	Inflammation
	–Gly–X–	Tumor metastasis
	–Ser–X–	Muscular dystrophy
	–Tyr–X–	Myocardial infarction
Calpains	–Lys–Ser–	Stroke
	–Arg–Thr–	Neuroprotection
	–Tyr–Ala–	Cataract
Falcipain	–Arg–Lys–	Malaria
	–Lys–X–	
Cruzipain	–Lys/Arg–	Sleeping sickness
	–Phe/Ala–	
Caspases	–Asp–X–	Rheumatoid arthritis, apoptosis, sepsis
Picornavirus 3C-proteinase	–Gln–X–	Viral infection
SARS-main proteinase	–Gln– Ser/Ala	Viral infection

23.9 Cysteine Proteases: Sulfur, the Big Brother of Oxygen as a Nucleophile in the Triad

In addition to the OH group of serine and threonine, the **thiol group of cysteine** is also capable of nucleophilic hydrolytic attack on amide bonds. Enzymes using such a cysteine possess a catalytic triad analogous to serine proteases and are called **cysteine proteases**. The first protease of this family to be structurally studied in detail was **papain**, which was isolated from the latex of papaya, the fruit of the papaya tree (*Carica papaya*). Its triad consists of a nucleophilic cysteine, a histidine, and an asparagine. The asparagine assumes the role of aspartate in serine protease. The catalytic mechanism is similar to that of serine proteases. Even the oxyanion hole (Cys 25 and Gln 19) is found in proteases of the papain family. There is evidence that the transition state is structurally similar to the acyl–enzyme intermediate. Attempts have been made to replace serine with cysteine in trypsin. The substrate binding properties (K_m) remained virtually the same, but the catalytic rate of the reaction decreased by five orders of magnitude. Although the structures are geometrically almost unchanged, the experiment shows that the difference between serine and cysteine proteases is more complicated than a simple exchange of sulfur for oxygen. The fine-tuning of structural and electronic properties is the key. Unlike the trypsin-like serine proteases, the nucleophilic cysteine exists as a preformed ion pair with its neighbor histidine.

Three families of cysteine proteases that are important targets for drug therapy have been characterized (Table 23.4). The first group is derived from papain and includes the **cathepsins**. They are proteases involved in the degradation of extracellular matrix proteins and the basal membrane. Inhibiting their function opens up a wide range of therapeutic possibilities, for example, in inflammation, tumor metastasis, bone resorption, muscle atrophy, or myocardial infarction. Another group is the **calcium-dependent calpains**, whose hydrolytic domain is folded very similar to that of papain. They are found in many cells and have different functions. Calpains occur in higher concentrations at sites of cell damage, such as after traumatic brain injury, stroke, or during the formation of cataracts in the eye. Calpains appear to be regulatory enzymes. For example, they reduce blood flow through blood vessels after injury to limit blood loss. Unfortunately, this natural protective function leads to the contrary situation during a stroke: activation of calpains reduces blood flow and parts of the brain become ischemic. Destruction of affected brain cells is the result. Specific inhibitors could counteract the over-functioning of calpains. Cysteine proteases of the papain family have also been discovered in **parasites**. Inhibition of **cruzipain** could be a concept for the treatment of sleeping sickness. **Falcipain**, which is used by the malaria parasite to digest hemoglobin, is a promising target enzyme for malaria therapy.

The second large family of cysteine proteases are the **caspases**. These are involved in the control of apoptosis, or programmed cell death. When a cell is damaged beyond the ability of natural cellular repair mechanisms to restore it to its normal state, caspases are activated to induce **apoptosis**. Dysregulation of apoptosis leads to various pathological conditions associated with cancer, immune dysregulation, or neurodegenerative disorders. Inhibitors of different caspases have potential as neuroprotective agents, as active substances for cancer treatment or for the treatment of rheumatoid arthritis.

The third family includes the **viral 3C proteases** found in picornaviruses (human rhinovirus, poliomyelitis, or hepatitis viruses) or coronaviruses (SARS). These viral proteases process the primary polypeptide chain and produce the specific viral proteins during maturation. Inhibitors of these proteases represent a concept for antiviral chemotherapy.

A special feature of **papain-type proteases** is the stereochemistry of the nucleophilic attack. In contrast to other serine and cysteine proteases, the attack occurs from the opposite side, the so-called Si face. The S_1 pocket in papain is not prominent and the P_1 group of the substrate is oriented away from the protein. In contrast, all neighboring pockets are much more prominent. Interestingly, some of the pockets on the *C*-terminal side (the primed side, S_1'–S_4') of cysteine proteases are highly structured. This can be exploited in the design of potential inhibitors. Papain prefers substrates with hydrophobic P_2 and P_3 groups. An aspartate is recognized as a P_1 group by caspases of the second folding family. For these reasons, many inhibitors developed for caspases carry a functional group with a carboxylic acid group or a corresponding mimetic at this position. The interaction with the thiol group of the catalytic cysteine is crucial for the binding of cysteine protease inhibitors to their target enzyme. Interestingly, many of the inhibitors developed attempt to involve the sulfur atom in a covalent bond. Reversible and irreversible head groups have been developed for this purpose. The inhibitor leupeptin **23.54** (◻ Fig. 23.27) is a natural product that has a terminal aldehyde function. This group reacts with the thiol group of cysteine to form a hemithioacetal. Leupeptin binds with high affinity to many members of the papain family. In addition to the aldehyde head group, many other functionalities (so-called warheads) that can be used to inhibit cysteine proteases (◻ Fig. 23.27) are known. Such irreversible inhibitors have been developed for viral proteases and have a Michael acceptor group (e.g., **23.55**). This reactive group forms an irreversible bond with cysteine and permanently shuts down the enzyme. Attempts have been made to develop inhibitors for cathepsins, calpains, and caspases that can form a reversible bond to the thiol group. Most of these structures are derived from aldehydes or ketones (**23.56**–**23.59**). From a chemical point of view, Vertex's caspase inhibitor **23.58** is interesting. It combines in a cyclic structure an aspartate-like side chain for the S_1 pocket of the enzyme and a capped aldehyde function in the form of a cyclic acetal. The aldehyde is released as the active compound from this prodrug.

An example of the successful development of a cysteine protease inhibitor in record time is the new COVID-19 inhibitor nirmatrelvir (Paxlovid®). When the coronavirus pandemic hit the world in late 2019, there was no drug or vaccine available to protect against the virus. With lightning speed, many concepts were brought to market, shattering all previously predicted drug development timelines. In less than a year, the first vector-based or mRNA vaccines (Sect. 32.4) were available. Just one year later, the first orally available **SARS-CoV-2 inhibitor**, nirmatrelvir, developed at Pfizer, was approved. Like many viruses, SARS-CoV-2 possesses a large protease (SARS-CoV-2 M^{pro}) that is required for the cleavage of mature peptide chains into virus-specific proteins. It belongs to the class of cysteine proteases. It specifically cleaves the polypeptide chain at 11 sites following a P_1-Gln residue. No human cysteine protease cuts after this amino acid. The perspective to develop a sufficiently selective inhibitor was, therefore, promising. The Pfizer team was fortunate that they did not to have to start from scratch. In 2002, when the first SARS virus infection shocked the world, Pfizer had already developed the potent but still very substrate-like inhibitor **23.60** (◻ Fig. 23.28). This inhibitor binds covalently to the substrate by forming a covalent bond to the catalytic cysteine (*green*) via an α-hydroxymethyl ketone. By the time Pfizer had the compound in hand, however, the 2002 SARS outbreak had largely subsided. The compound was not pursued further. Since the main protease of the new 2019 SARS-CoV-2 virus is virtually identical in the catalytic center, **23.60** also inhibits this protease. This set the stage for optimization into an orally available compound in March 2020. The peptidic nature of **23.60** had to be removed, as did the excessive polarity of the many H-bond donors. Pfizer initially used a nitrile and a benzothiazole-2-yl ketone in parallel as covalent anchors (◻ Fig. 23.27, groups d and h). The nitrile group eventually prevailed (circled in green). A 6,6-dimethyl-3-azabicyclohexane was used as a mimetic for the central P_2-leucine (*red*). The P_3 indole moiety was replaced by branched acyclic groups, with a methanesulfonamide and a trifluoroacetamide being the favorites. In the end, the fluorinated substituent proved superior (*blue*). Less than

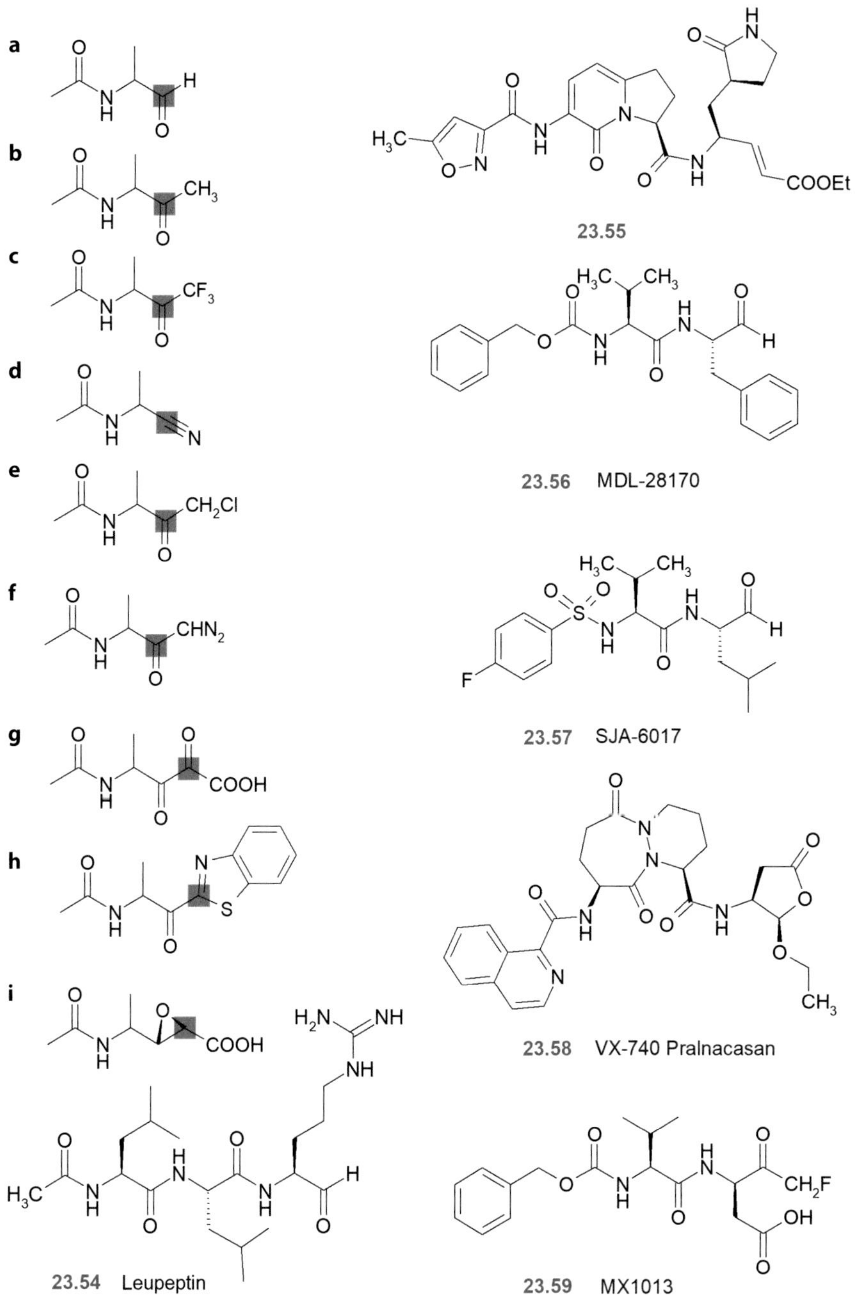

Fig. 23.27 In addition to the aldehyde head group (**a**), as found in the natural product leupeptin **23.54**, many other functionalities (**b**–**i**) have been developed. They bind reversibly or irreversibly to the catalytic cysteine (reactive site is shown in *red*), thereby, blocking cysteine proteases. Irreversible inhibitors such as **23.55** that have a Michael acceptor group are available for viral proteases. The two aldehydes **23.56** and **23.57** are development substances for the inhibition of calpains; **23.58** and **23.59** are caspase inhibitors. Compound **23.58** is a prodrug that releases an aspartate-like P_1 side chain upon ring opening and forms a hemithioacetal with the protein through its newly generated aldehyde function

six months after optimization began, **23.61** was ready for toxicological testing and in-depth clinical trials as a nanomolar inhibitor with good oral bioavailability and antiviral activity. By the end of 2021, **23.61** could be introduced into therapy under the trade name Paxlovid®. Fig. 23.28 shows the crystal structure of **23.61** with the protease.

Another group of enzymes that actually belong to the transferase family, but follow a cysteine protease-like mechanism, are the **transglutaminases**. Nine isozymes have been discovered in our genome. They are composed of four domains and contain a catalytic domain consisting of a Cys–His–Asp triad. Their function is the posttranslational modification of proteins (Chap. 26),

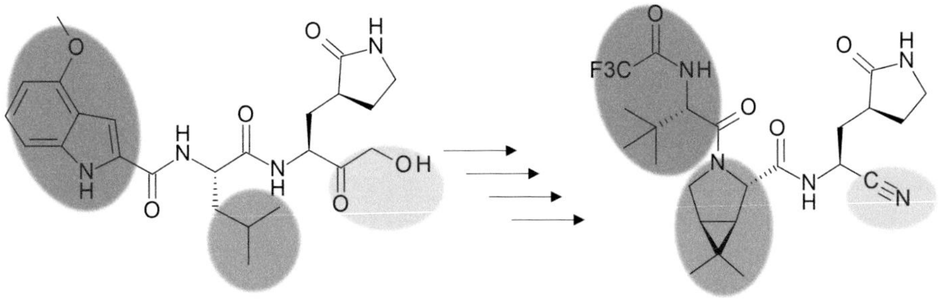

23.60 PF-00835231
K_i = 0.27 nM

23.61 PF-07321332
Nirmatrelvir
K_i = 3.11 nM

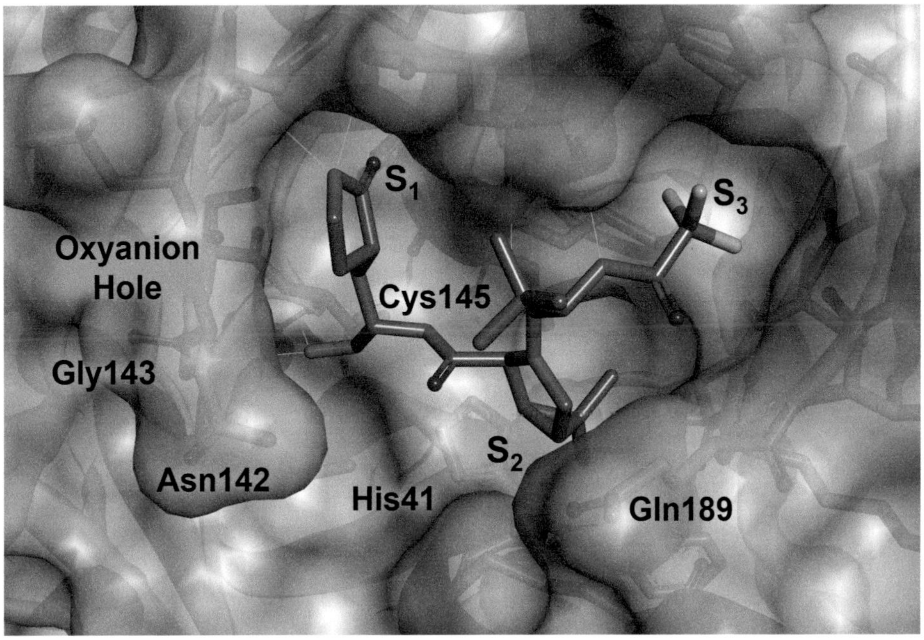

Fig. 23.28 Crystallographically determined binding mode of nirmatrelvir **23.61** to the SARS-Cov-2 M^{pro} major protease. The nanomolar inhibitor was optimized for stability and oral bioavailability from a previously discovered peptide analog **23.60** via several design steps. Nirmatrelvir binds covalently via its nitrile group to the catalytic Cys 145 and occupies the specificity pockets S_1, S_2, and S_3 with its side chains. The oxyanion hole is formed by the NH groups of Gly 143 and Cys 145. (► https://sn.pub/hNPere)

meaning they modify proteins after they have been synthesized in the ribosome. For example, they can deaminate glutamine residues to glutamate. In addition, they catalyze the **cross-linking** of chain strands on proteins by the transaminase reaction. For this purpose, the terminal amino group of a lysine is coupled to a glutamate residue to form an **isopeptide bond**. This results in a proteolytically stable cross-linking, so that transglutaminases can be compared to a "biological glue." The reaction is analogous to that of cysteine proteases. A nucleophilic cysteine first forms an acyl enzyme with the glutamine of the substrate with loss of ammonia, which is cleaved in the next step by the reactive lysine. A protein cross-link is formed. Transaminases have many functions in the body, the most important of which is to stabilize tissue proteins. In the blood coagulation cascade, the transglutaminase **factor XIII** stabilizes the initially formed clot by cross-linking (Sect. 23.4). Therefore, factor XIII inhibitors may be potent anticoagulants. Other transglutaminases are also being investigated as potential targets for drug development. **Transglutaminase-2** (TG2) plays an important role in **celiac disease**, a form of gluten intolerance. Patients with this disease are sensitive to gluten, an adhesive protein found in many grain products. They develop inflammation in the mucous membranes of the small intestine, leading to the destruction of intestinal epithelial cells and severely limiting their ability to absorb nutrients from food. TG2 inhibitors may be a therapeutic approach. Inhibitors for transglutaminases can be developed using analogous principles to those used for cysteine protease inhibitors.

23.10 Synopsis

- Serine proteases belong to the class of hydrolyzing enzymes that cleave amide or ester bonds. Depending on where they cleave a peptide chain, they are classified as amino-, carboxy-, or endopeptidases.
- Three amino acids, a serine, a histidine, and an aspartic acid that reside at quite distant positions in the sequence, are folded in characteristic proximity to one another. The hydroxyl oxygen atom of the serine nucleophilically attacks the carbonyl carbon atom of the scissile peptide bond. Its nucleophilicity is enhanced by an H-bond to an adjacent imidazole moiety of a histidine.
- The histidine accepts a proton from the nucleophilic serine OH group and is, thereby, transposed into a positively charged state. The neighboring aspartate residue compensates for the positive charge. The simultaneously created negative charge on the former carbonyl oxygen is stabilized by NH functions in the H-bond-donating oxyanion hole. Simultaneously, the carbon atom of the cleaving amide bond rearranges to a tetrahedral geometry.
- Upon release of the *N*-terminal part of the peptide substrate, the *C*-terminal part remains covalently bound as an acyl–enzyme complex. This is finally degraded via a similar mechanism that uses a water molecule as a nucleophile.
- The residues involved can be different; in particular, the nucleophilic serine can be replaced by a threonine or cysteine. The corresponding enzymes are named threonine and cysteine proteases.
- The peptide chain to be cleaved is primarily recognized in small binding pockets on the protease surface that accommodate the amino acid side chains on the *C*-terminal end adjacent to the cleavage site. Their composition determines the chemical building blocks required for inhibitor design to develop highly potent ligands for the protease.
- A number of warhead groups are known to either reversibly or irreversibly block the catalytic serine, threonine, or cysteine residue.
- The major contribution to binding affinity and ligand specificity is achieved through binding to the S_1 pocket next to the cleavage site.
- Blood coagulation is a highly regulated cascade of serine proteases. Potent inhibitors for antithrombotic therapy have been developed for thrombin and factor Xa, which participate in the last steps of the cascade.
- Whereas thrombin and factor Xa exhibit deep and well-structured S_1 pockets, elastase exhibits a flat S_1 pocket. Binding to this pocket contributes much less to the overall affinity of an inhibitor for this protease and the developed compounds all involve the catalytic serine in a reversible covalent attachment.
- To block lipases or transpeptidases, irreversible inhibition is achieved by covalent bond formation with the catalytic serine. The covalent bond is formed by ring opening of a reactive highly strained lactone or lactam ring. The latter principle is used by the penicillins and cephalosporins. A stable acyl form of the enzyme, which does not allow further conversions in the catalytic center, results. Penicillins and cephalosporins use this principle of inhibition.
- The β-lactamases, which are structurally closely related to the transpeptidases, hydrolyze the acyl–enzyme form produced by the penicillins and cephalosporins. They orient a polar glutamate residue into the catalytic center, thereby, stabilizing a water molecule in an optimal position for nucleophilic attack. In transpeptidases, this site is inaccessible to water molecules due to hydrophobic residues. Lactamase inhibitors break this resistance by blocking the water position with a polar side chain.
- Many cysteine proteases are found in bacteria, parasites, and viruses. By replacing the OH group with an SH group, the transition state in cysteine proteases appears to be closer to the acyl–enzyme form and the thiol group is presumably present in a deprotonated state. Typically, a head group that is reversibly or irreversibly covalently bound to the sulfur is used as a strategy to inhibit these enzymes.
- The transglutaminases follow a very similar enzyme mechanism as the cysteine proteases. However, instead of cleaving a peptide bond in the main chain, they form an isopeptide bond between the terminal amino group of a lysine and the carboxylate group of a glutamate. Because these bonds cause a cross-linking between different segments of the polypeptide chain, they can be compared to biological glue that make proteins more stable.

Bibliography and Further Reading

General Literature

C. Branden and J. Tooze, Introduction to Protein Structure, Garland Publ. Inc., New York (1991)

L. Polgár, The catalytic triad of serine peptidases, Cell. Mol. Life Sci. **62**, 2161–2172 (2005)

P. R. E. Mittl, M. G. Grütter, Opportunities for structure-based design of protease-directed drugs, Curr. Opin. Struct. Biol., **16**, 769–775 (2006)

L. J. Berliner, Ed., Thrombin: Structure and Function, Plenum Press, New York (1992)

S. D. Kimball, Challenges in the Development of Orally Bioavailable Thrombin Active Site Inhibitors, Blood Coagulation & Fibrinolysis **6**, 511–519 (1995)

J. A. Shafer, R. J. Gould, Eds., Design of Antithrombotic Agents, Persp. Drug Discov. Design **1**, 419–550 (1994)

R. E. Babine and S. L. Bender, Molecular Recognition of Protein–Ligand Complexes: Applications to Drug Design, Chem. Rev., **97**, 1359–1472 (1997)

T. Steinmetzer and J. Stürzebecher, Progress in the Development of Synthetic Thrombin Inhibitors as New Orally Active Anticoagulants. Curr. Med. Chem. **11**, 2297–2321 (2004)

B. Türk, Targeting Proteases: Successes, Failures and Future Prospects, Nature Reviews Drug Discov., **5**, 785–799 (2006)

G. Abbenante and D. P. Fairlie, Protease Inhibitors in the Clinic, Med. Chem., **1**, 71–104 (2005)

D. Gustafsson, R. Bylund et al., A New Oral Anticoagulant: The 50-Year Challenge, Nat. Rev. Drug Discov., **3**, 649–659 (2004)

A. Straub, S. Roehrig, A. Hillisch, Oral, direct thrombin and factor Xa inhibitors: the replacement for warfarin, leeches, and pig intestines? Angew. Chem. Int. Ed. Engl., **50**, 4574–4590 (2011)

Special Literature

B. W. Matthews, P. B. Sigler, R. Henderson, D. M. Blow, Three-dimensional structure of tosyl-alpha-chymotrypsin, Nature **214**, 652–656 (1967)

P. Carter and J. A. Wells, Dissecting the catalytic triad of a serine protease, Nature **332**, 564–568 (1988)

A. Sandner, K. Ngo, J. Schiebel, A. I. M. Pizarroso, L. Schmidt, B. Wenzel, T. Steinmetzer, A. Ostermann, A. Heine, G. Klebe, How a Fragment Draws Attention to Selectivity Discriminating Features between the Related Proteases Trypsin and Thrombin, J. Med. Chem., **64** 1611–1625 (2021)

M. Mares-Guia and E. Shaw, The Specific Inactivation of Trypsin by Ethyl p-Guanidinobenzoate, J. Biol. Chem., **242**, 5782–5788 (1967)

H. Umezawa, Enzyme inhibitors of microbial origin, Univ. of Tokyo Press (1972)

S. Bajusz, E. Barabas, P. Tolnay, E. Szell, D. Bagdy, Inhibition of thrombin and trypsin by tripeptide aldehydes. Int. J. Pept. Protein Res. **12**, 217–221 (1978)

M. T. Stubbs, H. Oschkinat, I. Mayr, R. Huber, H. Angliker, S. R. Stone, W. Bode, The interaction of thrombin with fibrinogen: A structural basis for its specificity, Europ. J. Biochem., **206**, 187–195 (1992)

J. Stürzebecher, F. Markwardt, B. Voigt, G. Wagner, P. Walsmann, Cyclic amides of *N*-alpha-arylsulfonylaminoacylated 4-amidinophenylalanine—tight binding inhibitors of thrombin, Thromb. Res. **29**, 635–642 (1983)

W. Bode, D. Turk, J. Stürzebecher, Geometry of binding of the benzamidine- and arginine-based inhibitors NAPAP and MQPA to human alpha-thrombin. X-ray crystallographic determination of the NAPAP-trypsin complex and modeling of NAPAP-thrombin and MQPA-thrombin, Eur. J. Biochem. **193**, 175–182 (1990)

K. Hilpert, J. Ackermann, D. W. Banner, A. Gast, K. Gubernator, P. Hadvary, L. Labler, K. Müller, G. Schmid, T. B. Tschopp and H. van de Waterbeemd, Design and Synthesis of Potent and Highly Selective Thrombin Inhibitors, J. Med. Chem. **37**, 3889–3901 (1994)

D. Gustafsson et al. The direct thrombin inhibitor melagatran and its oral prodrug H 376/95: intestinal absorption properties, biochemical and pharmacodynamic effects, Thromb. Res., **101**, 171–181 (2001)

D. Clement, K. Lopian, Characterization of in vitro biotransformation of new, orally active, direct thrombin inhibitor ximelagatran, an amidoxime and ester prodrug, Drug Metab. Dispos. **31**, 645–651 (2003)

U. Obst, V. Gramlich, F. Diederich, L. Weber, D. W. Banner, Design of Novel, Nonpeptidic Thrombin Inhibitors and Structure of a Thrombin–Inhibitor Complex, Angew. Chem., Int. Ed. Engl., **34**, 1739–1742 (1995)

C. A. Veale, P. R. Bernstein, C. Bryant et al., Nonpeptidic Inhibitors of Human Leukocyte Elastase. 5. Design, Synthesis, and X-Ray Crystallography of a Series of Orally Active 5-Aminopyrimidin-6-one-Containing Trifluorormethyl Ketones, J. Med. Chem. **38**, 98–108 (1995)

E. Perzborn, S. Roehrig, A. Straub, D. Kubitza, W. Mueck, V. Laux, Rivaroxaban: A New Oral Factor Xa Inhibitor. Arterioscler. Thromb. Vasc. Biol., **30**, 376–381 (2010)

L. E. J. Douglas et al., A highly selective, cell-permeable furin inhibitor BOS-318 rescues key features of cystic fibrosis airway disease, Cell Chem. Biol., **29**, 1–11 (2022)

M. S. Helfand and R. A. Bonomo, β-Lactamases: A Survey of Protein Diversity, Curr. Drug Targets—Infectious Disorders, **3**, 9–23 (2003)

M. Peimbert and L. Segovia, Evolutionary engineering of a beta-Lactamase activity on a D-Ala D-Ala transpeptidase fold, Prot. Engin., **16**, 27–35 (2003)

P. S. Langan, B. Sullivan, K. L. Weiss, L. Coates, Probing the role of the conserved residue Glu166 in a class A β-lactamase using neutron and X-ray protein crystallography, Acta Cryst. **D76**, 118–123 (2020)

I. Sánchez-Serrano, Success in translational research: lessons from the development of bortezomib, Nat. Rev. Drug Discov., **5**, 107–114 (2005)

J. Kirstein et al., The antibiotic ADEP reprogrammes ClpP, switching it from a regulated to an uncontrolled protease, EMBO Mol. Med. **1**, 37–49 (2009)

D. R. Owen, et al., An oral SARS-CoV-2 Mpro inhibitor clinical candidate for the treatment of COVID-19, Science **274**, 1586–1593 (2021)

M. Stieler, J. Weber, M. Hils, P. Kolb, A. Heine, C. Büchold, R. Pasternack, G. Klebe, Structure of Active Coagulation Factor XIII Triggered by Calcium Binding: Basis for the Design of Next-Generation Anticoagulants, Angew. Chem. Int. Ed. **52**, 11930–34 (2013)

Aspartic Protease Inhibitors

Contents

G. Klebe, *Drug Design,* https://doi.org/10.1007/978-3-662-68998-1_24

Aspartic (also aspartyl) proteases also cleave peptide bonds, but by a different mechanism. They owe their name to the presence of two aspartates, which determine the catalytic mechanism. To attack the peptide bond to be cleaved, they use a water molecule as a nucleophile, which they polarize in a suitable manner with the two aspartate residues. At the same time, these groups stabilize the transition state, balance the charges, and transfer protons. The digestive enzyme pepsin was the first member of this class of enzymes to be intensively studied. It is active at strongly acidic pH conditions between values of 1 and 5. The first 3D structure of this aspartic protease was determined in the early 1970s in the group of Alexander Fedorov. The aspartic protease family is relatively small in the human genome; it contains 15 members. Some important aspartic proteases are listed in ◘ Table 24.1.

24.1 Structure and Function of Aspartic Proteases

Pepsin preferentially cleaves peptides containing hydrophobic residues to the right and left of the cleavage site. Its spatial structure shows that **two catalytically active aspartic acid residues** are located side by side with a short distance to each other. One of these residues has an unusually low pK_a of 1.5. The other aspartic acid residue has a higher pK_a of 4.7. Thus, under the low pH conditions in the stomach, one of the side chains in the catalytic site is apparently protonated, while the other is not. This difference is crucial for the catalytic mechanism. In other aspartic proteases that function at higher pH values, a comparable difference is observed between the two groups. It is the local environment that determines the pK_a values (Sect. 4.4). On the other hand, the two aspartic acid residues are so close to each other that they can no longer be considered independent. The two aspartates behave like a coupled system, similar to a dicarboxylic acid; they are practically a diprotic acid (◘ Table 24.2). Here, the relative distance between the two acid groups determines the magnitude of the pK_a difference.

The **mechanism** of peptide cleavage by aspartic proteases is shown in ◘ Fig. 24.1. The cleavage of the amide bond occurs by **nucleophilic attack of a water molecule** on the carbonyl carbon atom. The deprotonated aspartate polarizes this water molecule. At the same time, the protonated aspartate forms an H-bond to the carbonyl group of the amide bond to be cleaved. This polarizes the C=O bond and facilitates nucleophilic attack on the carbon atom. The reaction proceeds through a tetrahedral transition state in which the oxygen atom of the nucleophilic water forms a bond to the carbonyl atom. At the same time, the proton is transferred from the water to the deprotonated aspartate. One approach to developing aspartate protease inhibitors is to mimic the **transient, unstable geminal diol transition state** with a stable molecule. Hydroxyl compounds (◘ Fig. 24.2), but also α-ketoamides and phosphinates can be used for this purpose.

To gain access to the substrate, the protease must open a mobile flap along the conversion pathway. In ◘ Fig. 24.3, the upper part of these tunnels is clipped off. The blue areas on the surface indicate the regions into which the protein directs hydrogen bond-forming groups. In the center, the two catalytic aspartate residues are hidden beneath the blue region. Adjacent are regions

◘ **Table 24.1** A few aspartic proteases and the preferred site for enzymatic cleavage

Enzyme	Cleavage site	Function
Pepsin	Phe–Phe, Leu–Phe, etc.	Digestion
Renin	Leu–Val, Leu–Leu	Blood pressure regulation
Cathepsin D	Phe–Phe, Leu–Leu, etc.	Tissue degradation
β-Secretase	Met–Asp, Leu–Asp	Proteolytic degradation of membrane proteins
Chymosin	Phe–Met	Milk curdling
HIV protease	Phe–Pro, Tyr–Pro, Phe–Tyr, Leu–Phe, Phe–Leu, Met–Met, Leu–Ala	Virus replication
Plasmepsin	Phe–Leu	Hemoglobin digestion

◘ **Table 24.2** pK_a values of a few dicarboxylic acids[a]

Dicarboxylic acid $HOOC-(CH_2)_n-COOH$	pK_a 1	pK_a 2	HOOC–COOH distance (Å)
n = 0	1.46	4.40	1.40
n = 1	2.83	5.85	2.60
n = 2	4.17	5.64	3.82
n = 3	4.33	5.52	4.95
n = 8	4.55	5.52	10.00
Z-HOOC–CH=CH–COOH	1.90	6.50	3.14
E-HOOC–CH=CH–COOH	3.00	4.50	3.80
1,2-$C_6H_4(COOH)_2$	2.96	5.40	3.14
1,3-$C_6H_4(COOH)_2$	3.62	4.60	4.93
1,4-$C_6H_4(COOH)_2$	3.54	4.46	5.71

[a] Reference values: HCOOH pK_a = 3.77; CH_3COOH pK_a = 4.76; C_6H_5COOH pK_a = 4.22

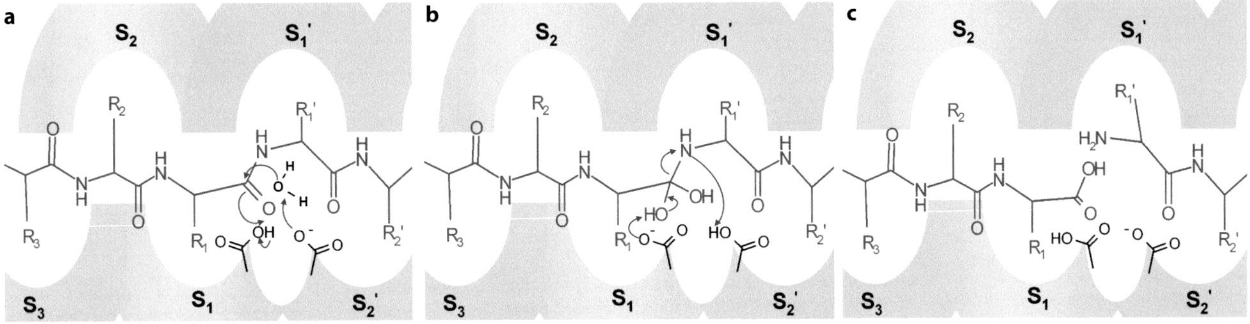

■ **Fig. 24.1** Catalytic mechanism of aspartic proteases. A water molecule, placed at the apex between the two catalytically active aspartates, is polarized by the deprotonated carboxylic acid. As a result, it acts as a nucleophile to attack the carbonyl carbon atom of the amide bond to be cleaved. The second, less acidic and protonated aspartate forms an H-bond to the carbonyl oxygen of this amide bond. This increases the electrophilicity of the carbonyl carbon (**a**). With the intermediate formation of a tetrahedral coordinated diol and the simultaneous cleavage of the former amide bond (**b**), the original peptide chain decomposes into the cleaved chain products (**c**)

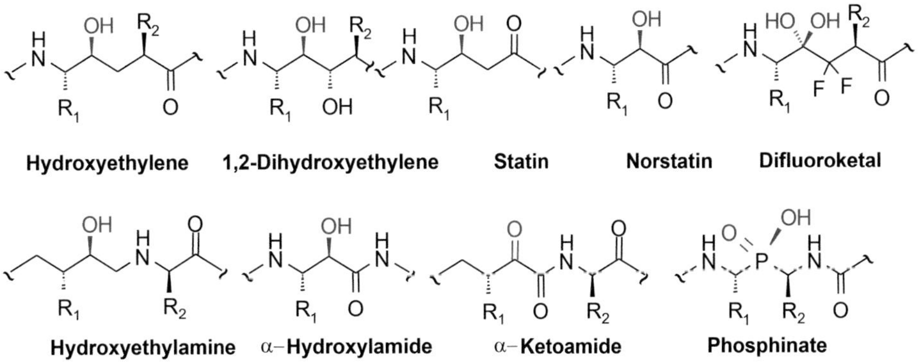

■ **Fig. 24.2** Possible transition state isosteres for the design of aspartic protease inhibitors. Hydroxyl groups are particularly well suited. Statin, a nonproteinogenic amino acid, is found in many inhibitor structures

where hydrogen bonds are formed to the peptide backbone of the substrate. A similar pattern is found on the upper half of the tunnel, which is not shown in the image because it has been clipped off. The hydrogen-bonding pattern along the tunnel is common to all aspartic proteases. Individual binding pockets to the left and right of the cleavage site are responsible for the selective recognition of substrates. They accommodate the side chains of the residues of the substrate molecules. It is striking that, in contrast to serine proteases, the pockets are well established on both sides of the cleavage site. This observation is explained by the reaction mechanism. Unlike serine proteases, no intermediate is formed that is covalently bound to the enzyme. Aspartic proteases often cleave between hydrophobic amino acids (■ Table 24.1). Since strong and directional interactions cannot form with such residues, recognition and fixation of substrate molecules over several positions to the left and right of the cleavage site is important. This explains the distinct character of the pockets on either side. Therefore, to design inhibitors, one must first find groups that mimic good interactions with the binding pockets S_3, S_2, S_1 and S_1', S_2', S_3'. One of the groups shown in ■ Fig. 24.2, which is an analogue of the **transition state**, is placed directly at the cleavage site.

Hamao Umezawa isolated one of the first potent and specific aspartic protease inhibitors, **pepstatin**, from a culture of *Streptomyces sp.* This peptide, Iva–Val–Val–Sta–Ala–Sta–OH, **24.1** (■ Fig. 24.4) is a good- to high-affinity inhibitor of many members of the aspartic protease family. It contains the nonproteinogenic amino acid **statin** with a hydroxyethyl group. The 3D structure of the pepsin–pepstatin complex shows that statin actually binds as a transition state mimicry to the catalytic aspartic acids.

24.2 Design of Renin Inhibitors

Renin is an aspartic protease that is composed of 340 amino acids. It plays a pivotal role in endogenous blood pressure regulation and in electrolyte and water homeostasis. The enzyme cleaves the peptide angiotensinogen to form the decapeptide angiotensin I (■ Fig. 24.5). This is subsequently cleaved by angiotensin-converting enzyme (ACE, Sect. 25.4), a metalloprotease, to give the octapeptide angiotensin II, which increases blood pressure. Inhibition of the enzyme renin leads to a decrease in the concentration of angiotensin I and, as a consequence, of angiotensin II. Renin inhibition, therefore, has a hypotensive effect. Because of the great therapeutic success of

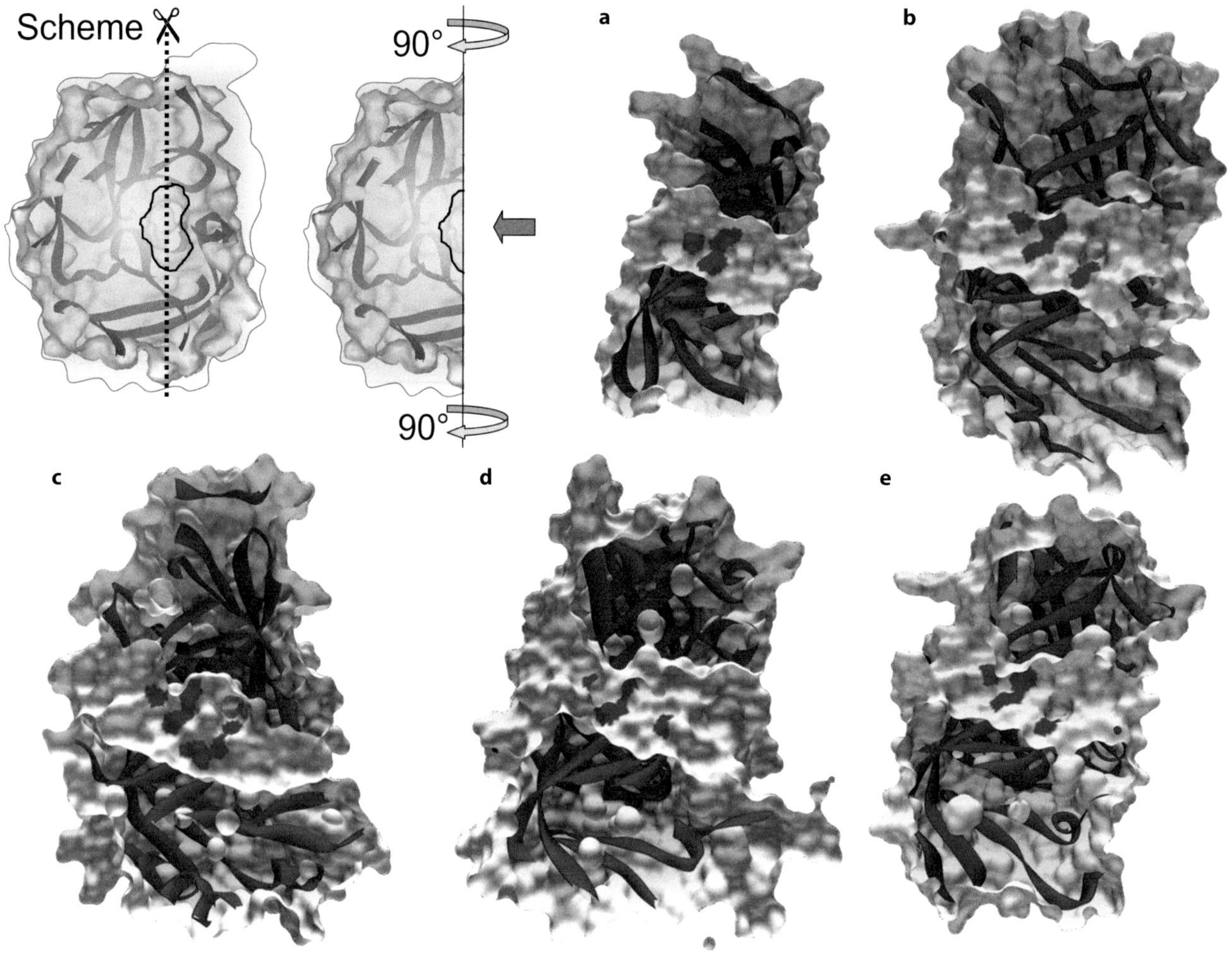

■ **Fig. 24.3** Sectional view of the binding pockets of five aspartic proteases: **a** HIV protease, **b** endothiapepsin, **c** cathepsin D, **d** plasmepsin, and **e** renin. The catalytic center extends as a tunnel through the proteases (*top left*, schematic view from the side into the tunnel). In the figures, the proteins are cut in such a way that the clipping plane passes through the center of the tunnels (*right part of the schematic drawing*, the cut protease has to be rotated by 90°). When looking from the side (in the direction of the *arrow*), only the back side of the tunnels can be seen. The protein surfaces are cut in the upper and lower part and show the course of the polypeptide chain (*red ribbons*). The *blue* areas on the backside of the tunnel refer to H-bond donor and acceptor sites beneath the protein surface to which a bound substrate is attached along its peptide backbone

ACE inhibitors, many pharmaceutical companies began searching for selective renin inhibitors in the early 1980s. Renin has an unusually high specificity. **Angiotensinogen** is the only known natural substrate of this enzyme. Therefore, it should be possible to find a highly specific renin inhibitor that does not block other enzymes and does not cause side effects, which is not the case with many other antihypertensive agents.

The starting point for the work was the peptide sequence of the substrate angiotensinogen. Renin cleaves angiotensinogen between Leu and Val. First, a suitable surrogate for the Leu–Val unit was sought that would allow the retention of the amino acids in the positions P_5 to P_3' (■ Table 24.3). The octapeptide His–Pro–Phe–His–**Leu**–**Val**–Ile–His is cleaved as a renin substrate. Replacement of the Leu–Val amide bond that is cleaved by the enzyme with the stable, isosteric groups $-CH_2NH-$

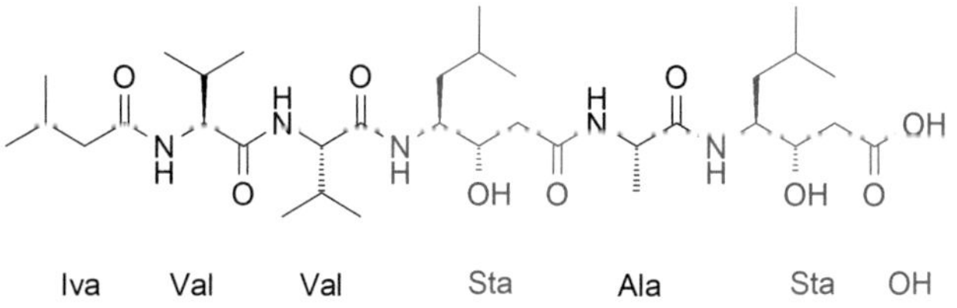

24.1 Pepstatin

■ **Fig. 24.4** Pepstatin **24.1** is an inhibitor for a large number of different aspartic proteases. (*Iva* isovaleric acid, *Sta* statin)

or $-COCH_2-$ resulted in modestly effective inhibitors. The isostere with the hydroxyethylene group, $-CH(OH)CH_2-$, was better suited as a transition state analogue, and afforded a strong inhibitor (IC_{50} = 3 nM). The incorporation of the nonproteinogenic amino acid statin

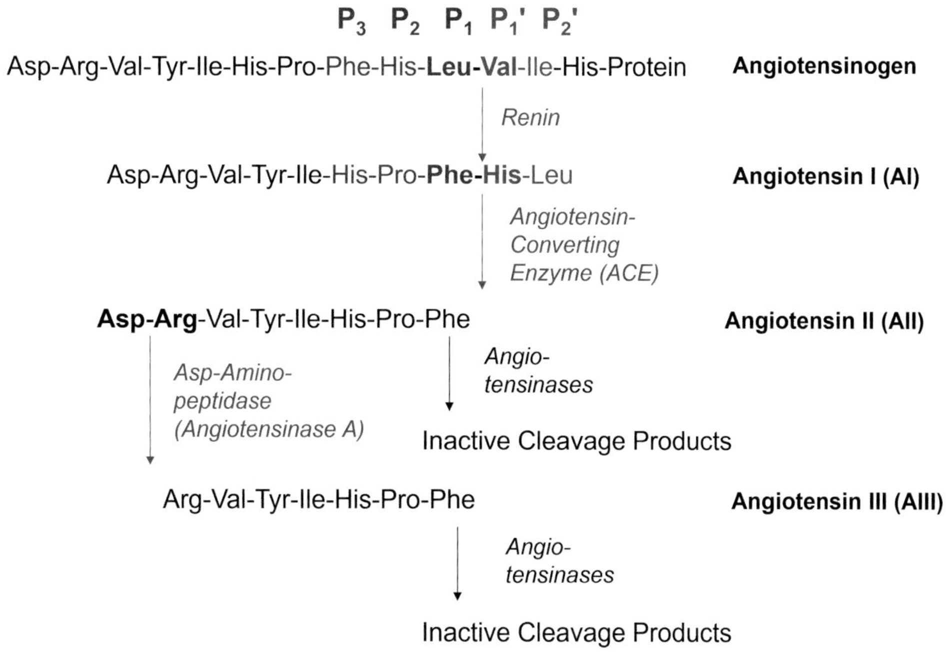

Fig. 24.5 The renin–angiotensin system. The conversion of angiotensinogen to angiotensin II (ATII), which increases blood pressure, is accomplished in two steps. Degradation by an Asp–aminopeptidase, angiotensinase A, leads to angiotensin III (ATIII), which is still biologically active. Different angiotensinases (aminopeptidases, carboxypeptidases) degrade these two peptides into inactive fragments

(see Fig. 24.4) produced a strongly binding inhibitor. As a dipeptide isostere, statin replaces the P_1–P_1' unit in the Leu–Val segment of the substrate.

The next step was the optimization of the P_1 moiety. Different groups were investigated as a replacement for the leucine side chain. The results of such a structural variation of **24.2** are listed in Table 24.4. Replacing the *iso*-butyl group with a larger cyclohexylmethylene group increased the affinity by a factor of 20. An adamantylmethylene group is obviously too large for the pocket, as the corresponding derivative only weakly inhibits the enzyme. Next, the P_2 moiety was investigated. Replacing the histidine with another group did not significantly improve the binding affinity. Nevertheless, the substitution of the basic histidine in the P_2 position was a major advance in renin research because it allowed the discovery that glycols are potent renin inhibitors. A few compounds from the **24.3** class are listed in Table 24.5. The introduction of a second hydroxyl group in the correct configuration increased the affinity by a factor of 10–200 depending on the chosen P_1 side chain. Accordingly, it was possible to find tripeptide analogues with binding constants of about 1 nM. Several companies developed renin inhibitors to the point of clinical trials. Examples are A-64662 **24.4** (Fig. 24.6) from Abbott and Ro 45-5892 **24.5** from Roche.

However, the desired goal had not yet been achieved. The compounds had short half-lives and were not orally available. It turned out that the amide bond between the P_3 residue Phe and the P_2 residue His was rapidly

Table 24.3 The replacement of the cleavable amide bond in Leu–Val (highlighted in bold) by a stable isostere leads to potent renin inhibitors. The Leu–Val group is replaced by a group in the inhibitors that the enzyme cannot cleave

Substrate/Inhibitor	IC_{50} (nM)
His–Pro–Phe–His–**Leu–Val**–Ile–His	300,000[a]
His–Pro–Phe–His–**Leu[COCH$_2$]–Val**–Ile–His	500
His–Pro–Phe–His–**Leu[CH$_2$NH]–Val**–Ile–His	200
His–Pro–Phe–His–**Statin**–Ile–His	20
His–Pro–Phe–His–**Leu[CHOHCH$_2$]–Val**–Ile–His	3

[a] Substrate, K_M value

cleaved by the digestive enzyme chymotrypsin. The high molecular weight of the compounds, which led to rapid biliary excretion, was also a problem. Further work focused on finding a suitable replacement for the P_2 and P_3 side chains.

Inhibitor stability against chymotrypsin was achieved by modification of the P_3 group, phenylalanine. The stabilities for some of these modified renin inhibitors **24.6** are summarized in Table 24.6. One compound that was no longer cleaved by chymotrypsin was obtained by using β,β-dimethylphenylalanine. This is because, in contrast to phenylalanine, the very bulky side chain no longer fits into the specificity pocket of chymotrypsin.

Table 24.4 Optimization of the P_1 side chain **R**. The binding pocket is lipophilic and obviously has just the right size for a cyclohexylmethylene group

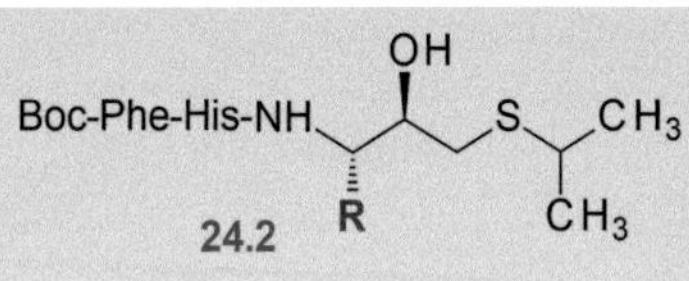

24.2

R	IC_{50} (nM)
Isobutyl	81
Cyclohexylmethylene	4
Cyclohexyl	150
Adamantylmethylene	2500
Benzyl	15

Boc = tert-butoxycarbonyl protecting group

Table 24.5 The introduction of a second hydroxyl group in the R_2 position leads to a significant increase in the binding affinity

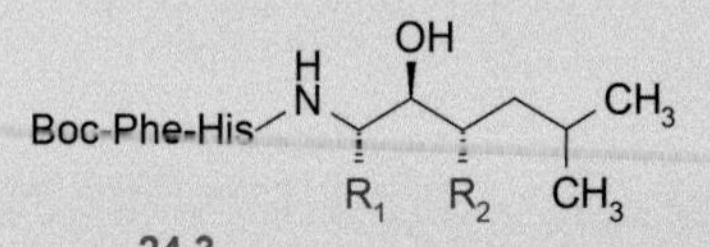

24.3

R_1	R_2	IC_{50} (nM)
Isobutyl	H	1500
	OH	11
Cyclohexylmethyl	H	10
	OH	1.5

Boc = tert-butoxycarbonyl protecting group

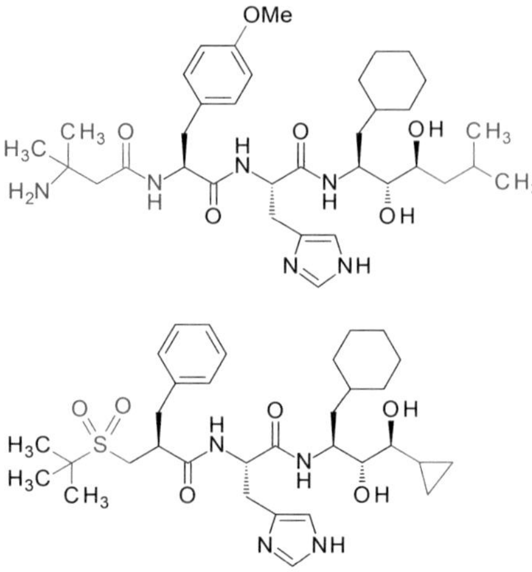

24.4 A-64662
Enalkiren
IC_{50} = 14 nM

24.5 Ro 42-5892
Remikiren
IC_{50} = 0.7 nM

Fig. 24.6 Enalkiren **24.4** and remikiren **24.5** were the first renin inhibitors to be tested in clinical trials

The extended search for possible replacements of Phe–His as the P_3–P_2 moiety led to a plethora of new non-peptidic renin inhibitors with high affinity. The introduction of a terminal basic group was highly effective, even though adequate oral bioavailability was not achieved. Typical examples are **24.8** and **24.9** (Fig. 24.7).

Despite enormous efforts, renin research stagnated worldwide because no compound could achieve the required oral bioavailability. All compounds contained at least one amide bond and their molecular weights were too high. In addition, the 3D structure of renin did not become available until the late 1980s in the laboratory of Michael James in Edmonton, Alberta, Canada. By then, it had been recognized that renin had a certain, albeit modest, sequence homology of 20–30% with aspartic proteases from fungi, for which 3D structures were known. This was the starting point for **homology modeling** in several laboratories. The first model was published by Tom Blundell's group in 1984. They used the crystal structure of **endothiapepsin** as a reference. First, the renin sequence was compared with that of other aspartic proteases to find structurally conserved regions. They then modeled the interior of the protein by replacing residues in endothiapepsin with those in renin. Deletions and insertions in the polypeptide chain had to be taken into account. The flap region was of particular importance. It opens to allow the ligand to enter and form hydrogen bonds with the protein. Its structural architecture is, therefore, important for ligand binding. Unfortunately, the renin sequence differed from that of the fungal enzymes in the flap region. A comparison of the renin model with the later determined crystal structure of renin showed good agreement, especially in the lower part of the binding pocket near the two aspartic acids. However, significant differences were found in the loops of the flap region. In the context of the overall protein architecture, these were less important. In the context of drug design, however, they were critical! Errors in the structural model had to inevitably lead to incorrect suggestions for inhibitor design.

The structure of renin complexed with the inhibitor CGP-38560 **24.10** determined by Markus Grütter and John Priestle at Ciba in Basel, Switzerland, is shown in Fig. 24.8. It was on the basis of this structure that the researchers at Ciba, now Novartis, made their breakthrough. Looking at the arrangement in the binding pocket of renin, it is apparent that the S_1 and S_3 pockets merge into one large hydrophobic cavity. The substituents in the P_1 and P_3 positions, a cyclohexylmethylene and a benzyl group, come spatially close together. Instead of spanning the molecule with its side chains on a peptidic backbone, the scientists broke the chain at the amide bond. This created a new polar moiety, a terminal charged amino group. In place of the former backbone,

Table 24.6 By modifying the P_3 substituent **R**, the stability to chymotrypsin is improved

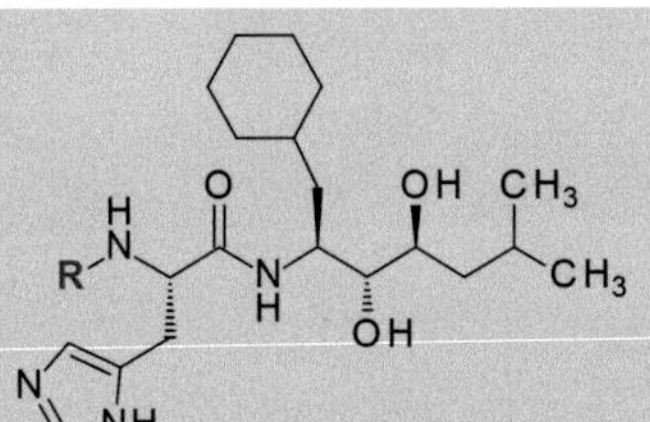

24.6

R	IC_{50} (nM)	Hydrolysis by chymotrypsin $t_{1/2}$ (min)
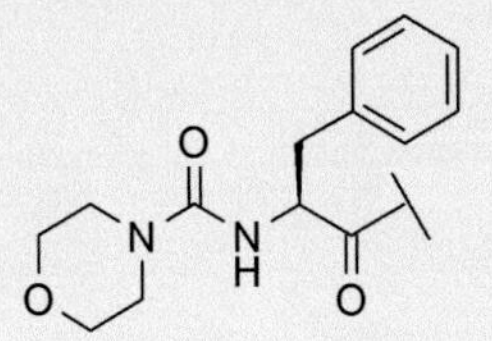	0.35	2.2
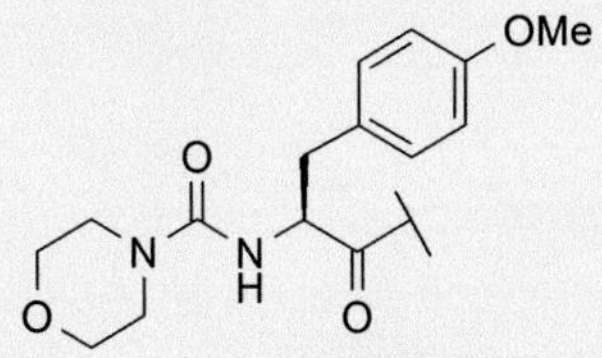	0.76	727
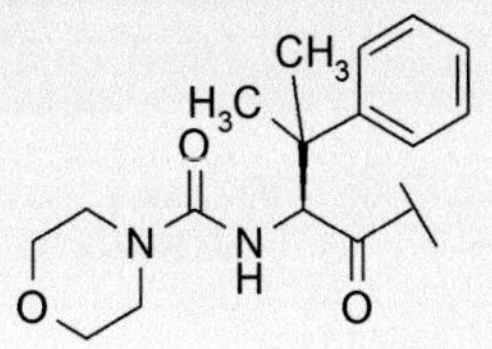	0.58	Stable

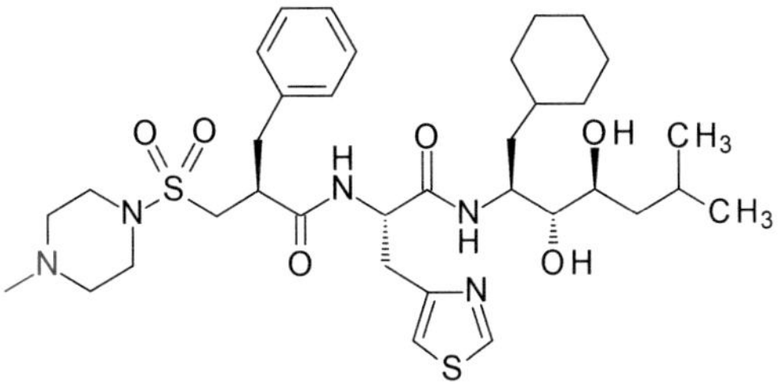

24.7 A-72517
IC_{50} = 1.1 nM

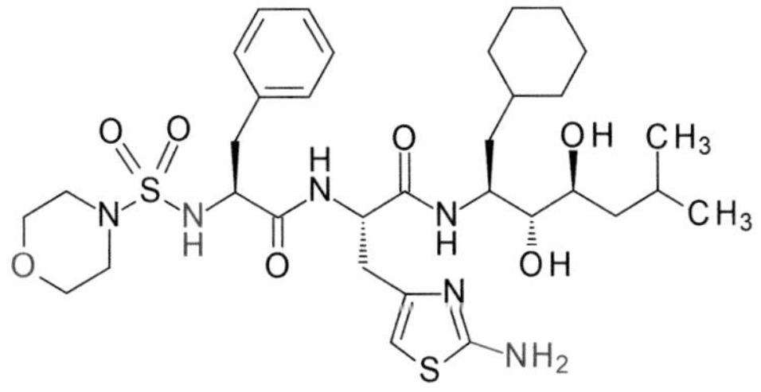

24.8 PD-134672
IC_{50} = 0.57 nM

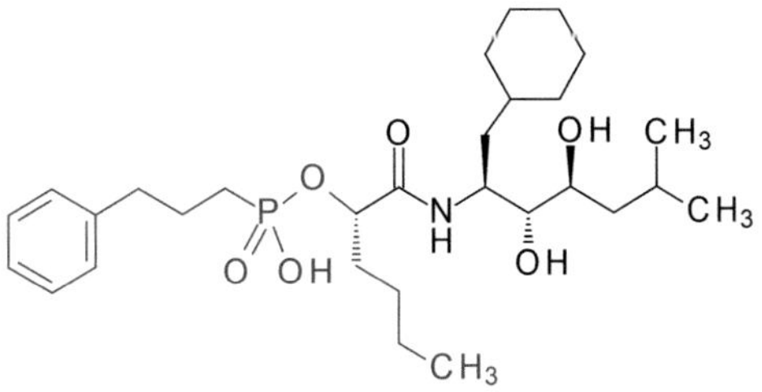

24.9 EMD-65010

Fig. 24.7 Structures **24.7**–**24.9** are a few renin inhibitors with moderate oral availability. All have in common a diol unit and a cyclohexylmethylene side chain that binds in the P_1 pocket

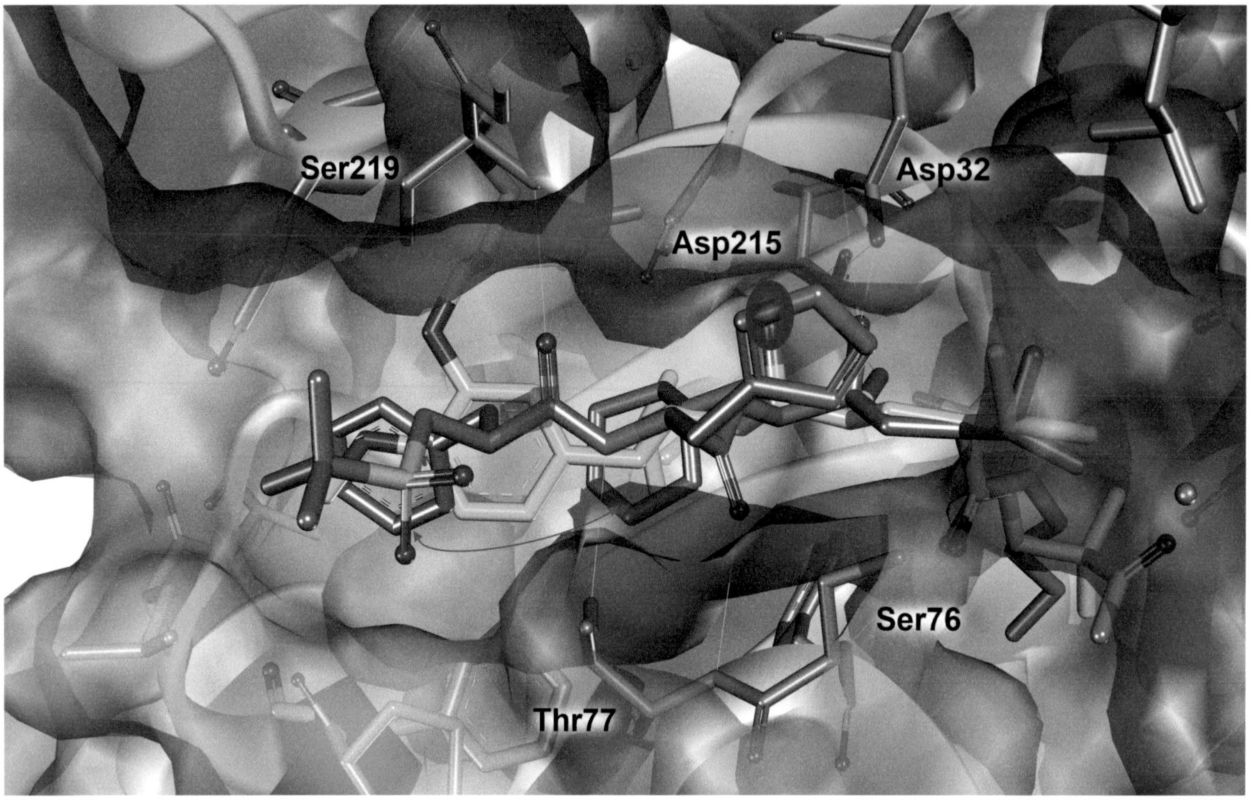

Fig. 24.8 Superposition of the crystal structures of renin complexes with the inhibitors CGP-38560 **24.10** (*gray carbon atoms*) and aliskiren **24.12** (*light-green carbon atoms*). The inhibitors bind with their peptide-like architecture in an extended, pleated-sheet-like conformation. Compound **24.10** orients its benzyl and cyclohexylmethyl groups in the broad S_3/S_1 pocket. To design aliskiren, the two hydrophobic side chains of **24.10** were chemically linked (*pink arrow*). The newly formed bond allowed the peptide chain to be cleaved and a new, polar *N*-terminus could be formed. (▸ https://sn.pub/BGi2k5)

the linkage of the molecule was redirected to the closely adjacent hydrophobic substituents in the large S_3/S_1 pocket. The result was a completely new dipeptide-like scaffold (**24.11**, Figs. 24.8 and 24.9). It had an IC_{50} of 6 nM. Finally, several steps of side chain optimization were performed on the aromatic ring and the amide linkage. The methoxypropoxy side chain occupies a slightly different pocket than the corresponding groups in CGP-38560. It results in a significant increase in binding affinity. The optimized substituent in P'_1 has little effect on *in vitro* affinity, but is critical for duration of action. A geminal substitution with two methyl groups and a terminal carboxamide group proved to be optimal for the P'_2 position. The resulting inhibitor was launched in 2006 as **aliskiren (24.12)**, the first orally available renin inhibitor. Despite being so well optimized, the compound does not have ideal bioavailability. As a result, it must be administered at relatively high doses. However, aliskiren shows virtually no binding to other aspartic proteases such as cathepsin D or pepsin.

Roche achieved another success with their work on renin, which later proved to be stimulating for the entire field of research. With remikiren **24.5**, the company had a potent inhibitor that unfortunately lacked the desired oral bioavailability. The company, therefore, initiated a new comprehensive screening program. Chlorophenylmethoxybenzyloxypiperidine **24.13** was discovered (Fig. 24.10) with an IC_{50} value of 50 µM. This structure was surprising because it did not have the typical group mimicking the transition state. The crystal structure of a very similar derivative showed that the protonated nitrogen on the piperidine ring binds between the two catalytic aspartic acids. The lipophilic chlorophenyl moiety aligns with the broad S_1/S_3 pocket normally occupied by the leucine and phenylalanine residues of the angiotensinogen substrate. Since the available space in this pocket was not yet fully occupied, the Roche researchers initially focused on structural variations in the *para*-position of the aromatic ring as a surrogate for the chlorine atom. The introduction of aromatic groups with variable chain lengths yielded derivatives with up to 100-fold improved activity. It appeared to be critical that only hydrophobic groups could be placed in this position. The best results were obtained with a propylenedioxybenzyl

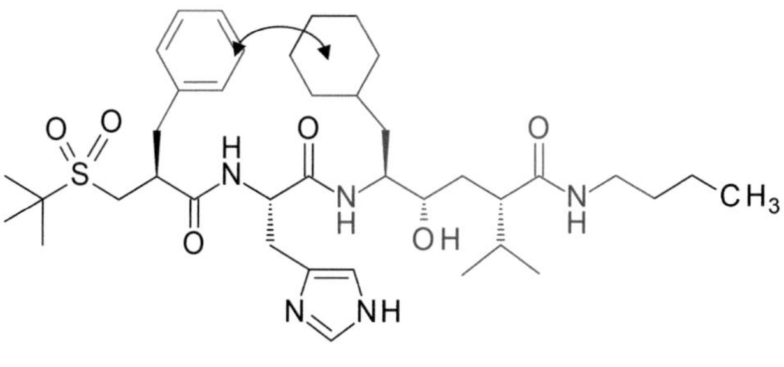

24.10 CGP-38560 IC_{50} = 2 nM

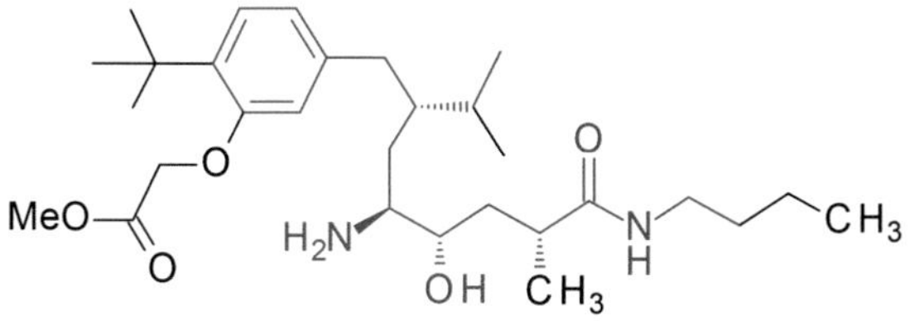

24.11 IC_{50} = 6 nM

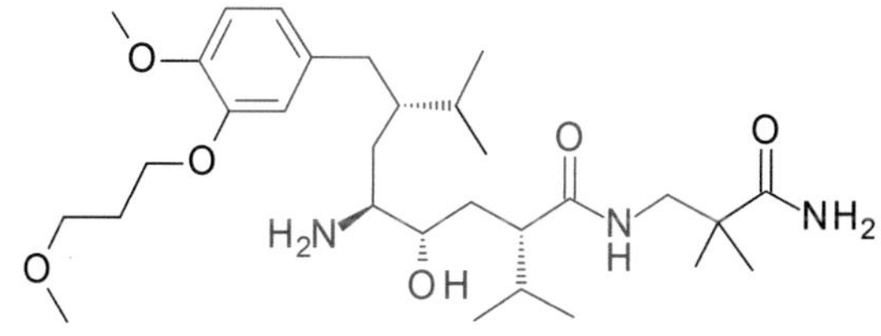

24.12 Aliskiren IC_{50} = 0.6 nM

Fig. 24.9 For the further development of **24.10** to aliskiren **24.12** as an orally available renin inhibitor the benzyl and cyclohexylmethyl side chain groups in **24.10** were tethered together to yield **24.11**. Then the peptide chain could be cleaved after the nitrogen atom, and a new polar group could be formed that binds to the catalytic center. The analogous molecular parts in both inhibitors are highlighted in *red*

side chain. This moved the compounds into the subnanomolar inhibition range. A crystal structure determination was performed on derivative **24.14**, which showed a completely unexpected binding mode (Fig. 24.11). The protonated nitrogen of the piperidine ring is still between the two aspartic acid residues, but the lipophilic naphthyl group is oriented in the broad S_1/S_3 pocket. The long hydrophobic side chain of the 4′-substituted phenyl group opens a new pocket in renin. Like all aspartic proteases, renin has a flexible flap region that collapses over the binding pocket after substrate binding. In this case, the flap is pushed outwards by the inhibitor. The enzyme adopts a geometry that is more consistent with an open-flap conformation. A hydrogen bond between Trp 39 and Tyr 75, which closes the flap, is ruptured. At the same time, the 4-phenyl group of the inhibitor occupies a region where the aromatic ring of Tyr 75 would be located if the flap were closed. This structure provided the researchers with two important pieces of information: (*i*) a nitrogen-containing heterocycle is an interesting peptidomimetic that binds to the catalytic aspartic acids, and (*ii*) inhibitors can bind to the aspartic protease

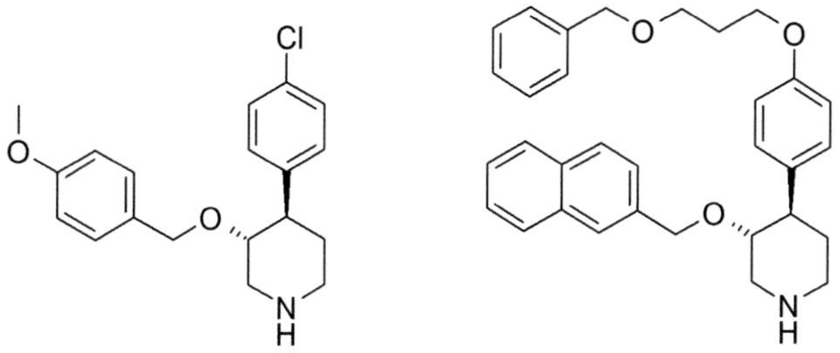

24.13 **24.14**

Fig. 24.10 A piperidine derivative **24.13** that was found in a screening campaign for renin inhibitors at Roche. A crystal structure was determined with the optimized compound **24.14**

family in more conformations than the closed-flap conformation. The open conformer can also be stabilized by an inhibitor. These exemplary studies on renin provided important information for novel work on the aspartic proteases (Sect. 24.6).

24.3 Design of Substrate Analogue HIV Protease Inhibitors

AIDS (Acquired Immune Deficiency Syndrome) is an infectious disease caused by the **human immunodeficiency virus, HIV**, which was isolated and identified in 1983 by the French virologists Luc Antoine Montagnier and Françoise Barré-Sinoussi. Many years later, in 2008, they were awarded the Nobel Prize for their work. HIV protease, which is required for viral replication, is encoded as a large proprotein in the viral genome. The function of HIV protease is to cleave the initial polypeptide chain produced in the life cycle of the virus into smaller functional proteins. Inhibitors of **HIV protease** should, therefore, be able to suppress HIV replication. The existence of HIV protease was postulated in 1985 and experimentally confirmed in 1988.

In 1989, the first 3D structure of the enzyme as well as a few enzyme–inhibitor complexes were determined. HIV protease is a homodimer made up of two identical chains. One catalytic aspartic acid comes from each chain of the homodimer. The dimeric structure of HIV protease with its twofold symmetry is shown in Fig. 24.12.

It was soon discovered that HIV protease is also inhibited by pepstatin. This was the starting point for the search for HIV protease inhibitors. Many companies already active in the renin field tested compounds generated in these programs for possible HIV protease inhibition. Starting with the nonproteinogenic amino acid statin known from the renin work, a number of active HIV protease inhibitors were discovered. As with renin, the **hydroxyethylene isostere** proved to be a particularly suitable building block. For example, H-261 **24.15** (Fig. 24.13) is a potent HIV protease inhibitor with K_i = 5 nM.

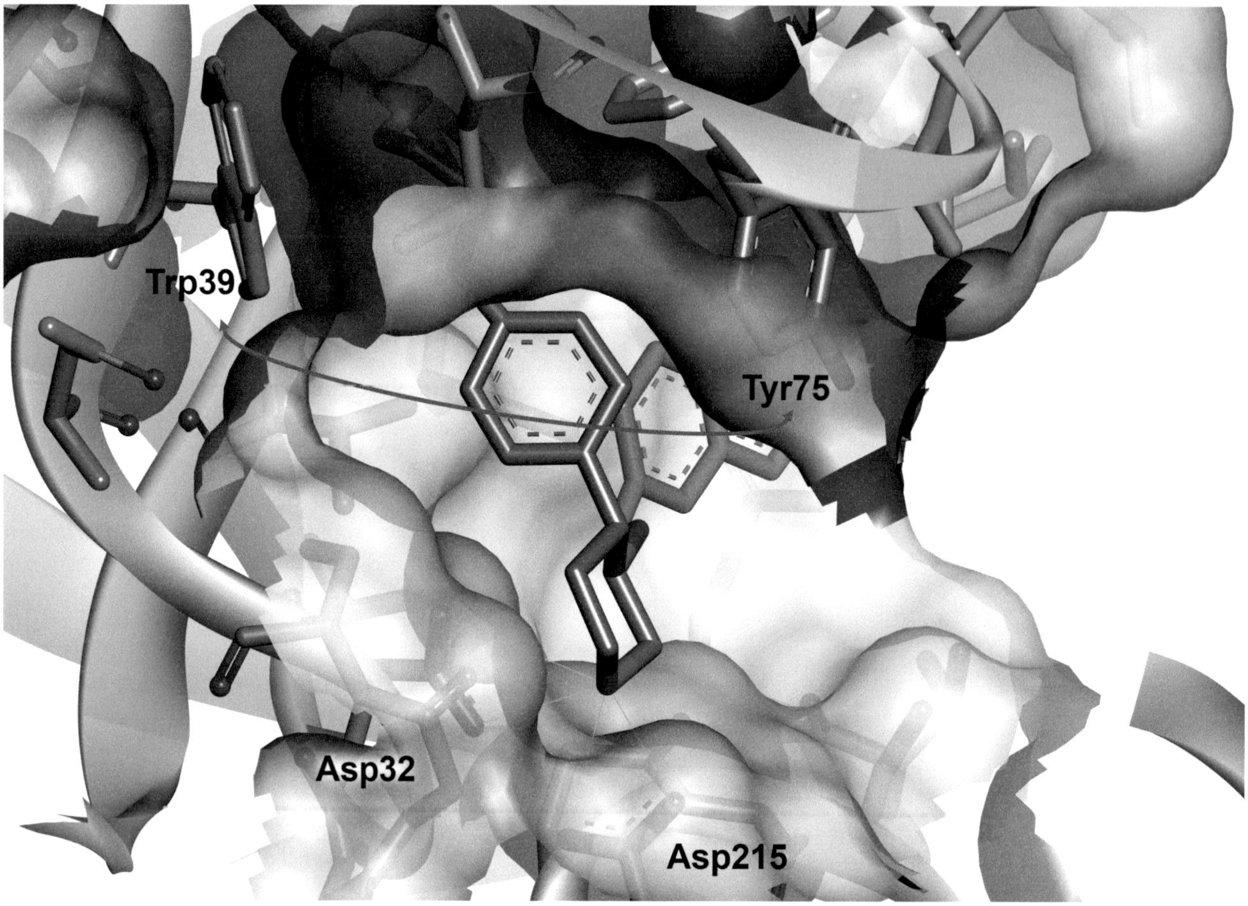

Fig. 24.11 The crystallographically determined binding mode of the piperidine lead structure **24.14** with renin. The basic nitrogen of the inhibitor binds between the two aspartic acids of the catalytic dyad. The lipophilic side chain lies in a newly opened binding pocket. It was formed by breaking a hydrogen bond that was originally present between Trp 39 and Tyr 75 in the uncomplexed protein. Both residues adopt a new position with larger distance between one another after binding of **24.14**. (► https://sn.pub/lrGrOq)

Heptapeptides have been identified as a minimal substrate for HIV protease. Ser–Leu–Asn–**Phe**–**Pro**–Ile–Val is such a substrate. Cleavage of the amide bond occurs between the amino acids Phe and Pro. Replacement of the cleavable amide bond with a hydrolytically stable hydroxyethylamino group –CHOH–CH_2–NH– led to **24.16** (JG 365, Fig. 24.13), a high-affinity HIV protease inhibitor (K_i = 0.66 nM). This compound was, however, inactive in cell culture assays. It is unable to penetrate the cell membrane to exert its antiviral effects.

Chemists at Roche have demonstrated that the design of an HIV protease substrate analogue can lead to an effective drug. Proline is often found in the P'_1 position (e.g., **24.17**, Fig. 24.14). Therefore, isosteres of analogues of the dipeptide Phe–Pro were investigated as HIV protease inhibitors. Replacement of proline by homoproline **24.18** or decahydroisoquinoline **24.19** resulted in a significant increase in potency. In addition, **24.19** showed marked selectivity towards the other aspartic proteases renin, pepsin, cathepsin D, and cathepsin E. More importantly, the compound was active in a cellular assay. It has the ability to penetrate the cell membrane. In enzyme assays, **24.19** inhibited HIV protease with K_i < 0.12 nM. Viral replication is inhibited in cell culture with EC_{50} values of 1–10 nM. The activity in cells is, therefore, on the same order of magnitude as the pure enzyme inhibition. **Saquinavir 24.19** was the first HIV protease inhibitor to complete all phases of clinical testing and receive marketing approval in November 1995. In the years since, other pharmaceutical companies have succeeded in bringing substrate-like HIV protease inhibitors to market. As a result, our drug arsenal now includes eight approved drugs (**24.19**–**24.27**) with peptide-like scaffolds (Fig. 24.15). However, nelfinavir **24.24** was withdrawn from the European market in 2007. It was noticed that tablets containing this substance had an unusual smell. Subsequent analysis revealed that the drug was contaminated with ethyl mesylate from synthesis. Because saquinavir has

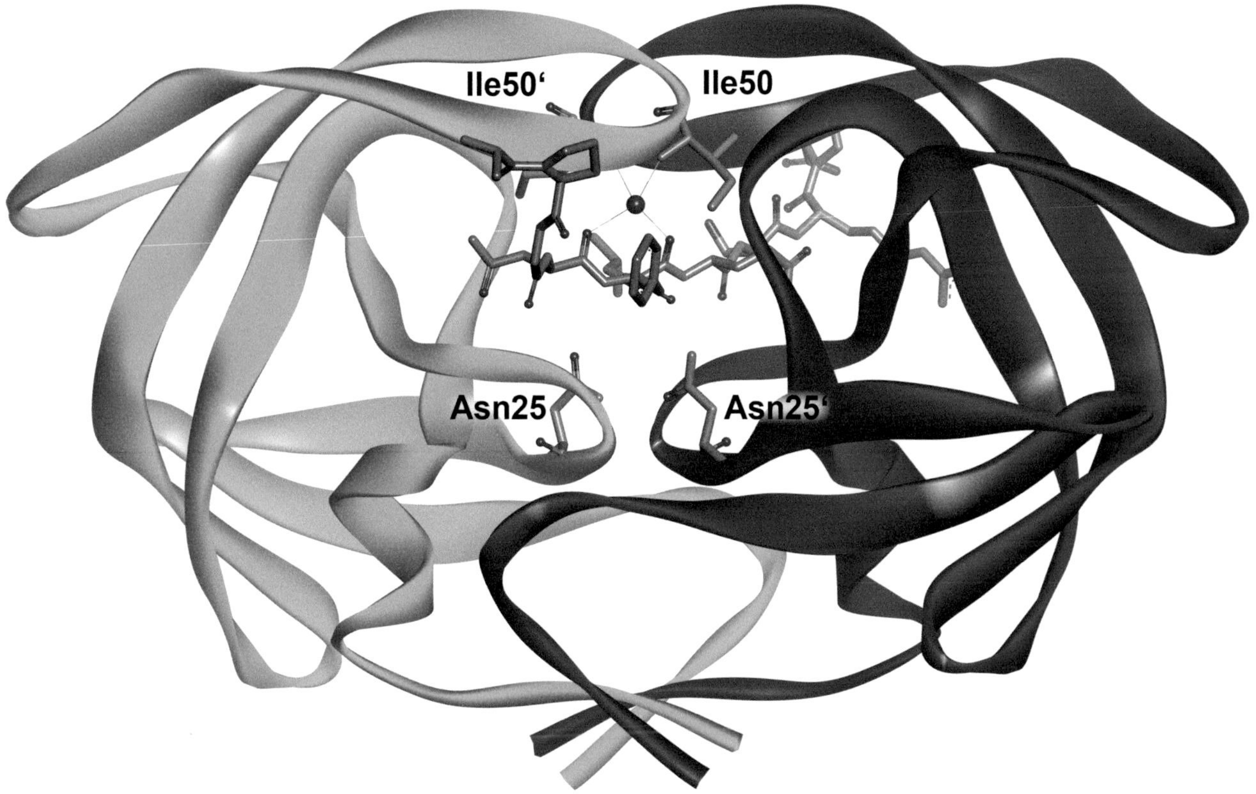

Fig. 24.12 The 3D Structure of HIV protease complexed with the peptide substrate Arg–Pro–Gly–Asn–Phe–Leu–Gln–Ser–Arg–Pro. The structure with the substrate could be obtained with a catalytically inactive enzyme variant because both acidic aspartic acids of the catalytic dyad had been mutated to asparagines. The protease exists as a C_2-symmetric homodimer. The peptide chains are shown in *green* and *red*, respectively. (▶ https://sn.pub/r7ffVB)

unsatisfactory bioavailability (3–5%), it is administered in combination with ritonavir **24.20**, a potent CYP 3A4 inhibitor (K_i = 17 nM; Sect. 27.6). This significantly minimizes the first pass effect when coadministered with saquinavir. Ritonavir has established itself as a booster for other drugs and is increasingly used in combinations with other drug molecules. For example, it is used in combination with the recently approved COVID-19 inhibitor nirmatrelvir (Paxlovid®, Sect. 23.9). Amprenavir **24.23** was withdrawn in 2004 because it was replaced by the more soluble prodrug fosamprenavir (Lexiva®).

24.4 Structure-Based Design of Nonpeptidic HIV Protease Inhibitors

The relationship to the parent substrate is clearly seen in the inhibitors introduced in the last section. The compounds are still essentially peptides. The crystal structures of the peptidic HIV protease inhibitors complexed with the enzyme all show that the inhibitors form essentially the same H-bonding pattern in the immediate vicinity of the catalytically active aspartic acid residues (Fig. 24.16). One water molecule is of particular interest because it is present in all crystal structures. This water molecule forms hydrogen bonds with both the inhibitor and the enzyme. Inhibitors designed to displace this water molecule were hoped to increase the binding affinity by the entropically favorable release of this water (Sect. 4.6). The release of water molecules is often associated with an entropic gain. Note, however, that at best this gain is not fully compensated by a comparable loss of enthalpy, leaving a net contribution to binding affinity. Moreover, it was expected that such an approach would also increase selectivity, since a water molecule with a similar function is not known to exist in the other aspartic proteases of similar pharmaceutical relevance.

At Dupont–Merck, a 3D database was searched for new scaffolds for HIV protease inhibitors. A pharmacophore pattern was derived from the crystal structure of the enzyme. The occupancy of the S_1 and S_1' pockets was considered essential for binding and interaction with the catalytic aspartates. Two lipophilic groups, separated by 8.5–12 Å and also 3.5–6.5 Å from a hydrogen bond

Fig. 24.13 The peptidic HIV protease inhibitors H-261 **24.15** and JG 365 **24.16** are potent inhibitors in the enzyme assay. They are inactive in cell culture

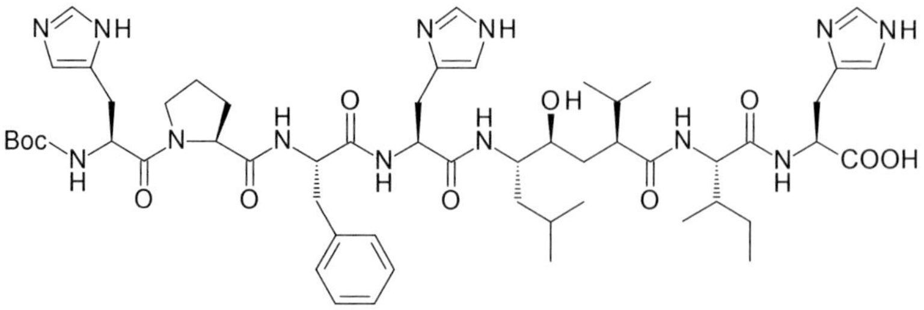

24.15 H 261

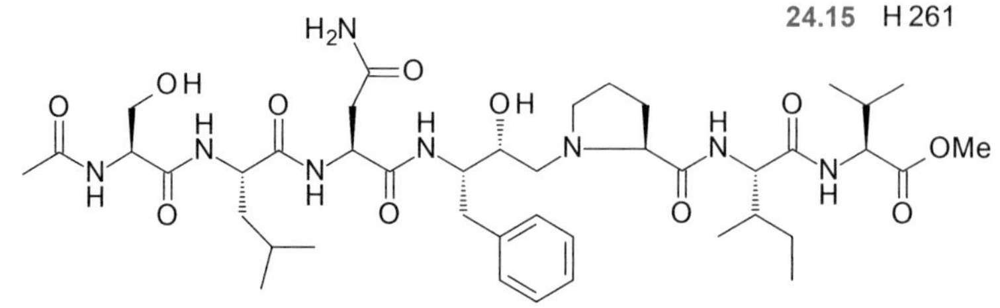

24.16 JG 365

Fig. 24.14 The stepwise optimization of the substrate analogue inhibitor **24.17** led to the highly potent HIV protease inhibitor Ro 31-8959 **24.19** via **24.18**. This compound was the first protease inhibitor to pass clinical trials and is marketed with the name saquinavir

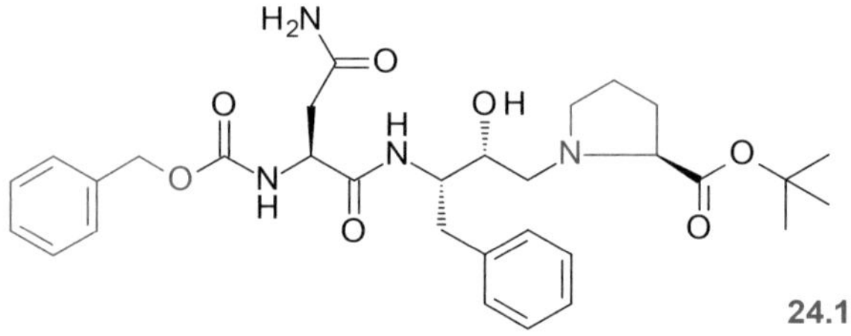

24.17 IC_{50} = 140 nM

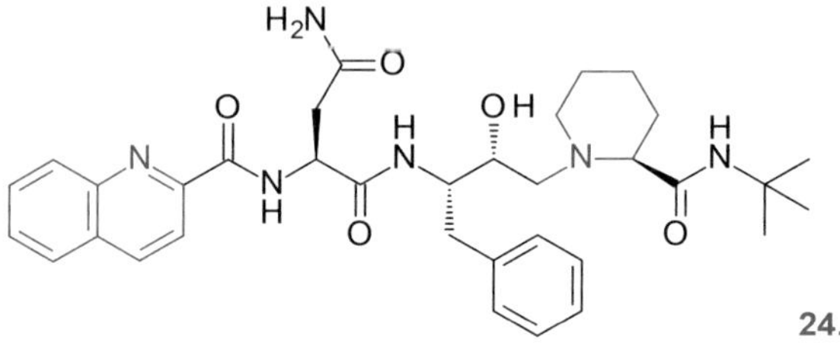

24.18 IC_{50} = 2 nM

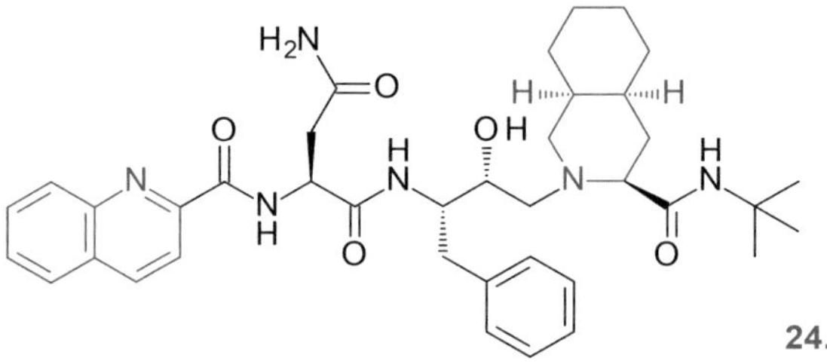

24.19 Ro 31-8959

Saquinavir
IC_{50} < 0.4 nM
K_i < 0.12 nM

acceptor or donor, were sought (Fig. 24.16, *right*). In addition, there should be a functional group between the two lipophilic groups that can displace the structurally conserved water molecule from the binding pocket. A search in the Cambridge database (Sect. 17.11) yielded a molecular scaffold derived from a substituted phenol (**24.28**). This led to the idea of using 4-hydroxycyclohexanone as the scaffold (Fig. 24.16). Modeling studies and intensive discussions with synthetic chemists eventually led to a **cyclic urea** (**24.29**) as the scaffold for the new inhibitors **24.30**–**24.33** (Fig. 24.17). The first result of this development was DMP-323 **24.32**, a small-molecule HIV protease inhibitor. The 3D structure of **24.31** complexed with the protease is shown in Fig. 24.18. It confirms the hypothesis that the carbonyl group displaces the structural water molecule and the two hydroxyl groups bind to the catalytic aspartate residues. As promising as the design of the cyclic urea as an HIV protease inhibitor seemed, no compound has yet survived all stages of clinical testing to gain regulatory approval.

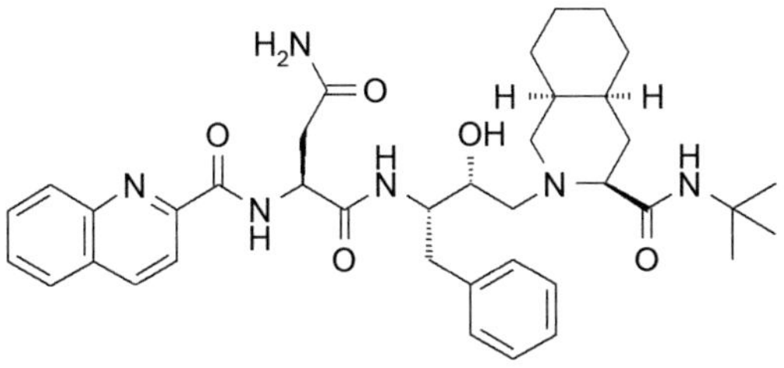

24.19 Saquinavir Invirase® (1995)

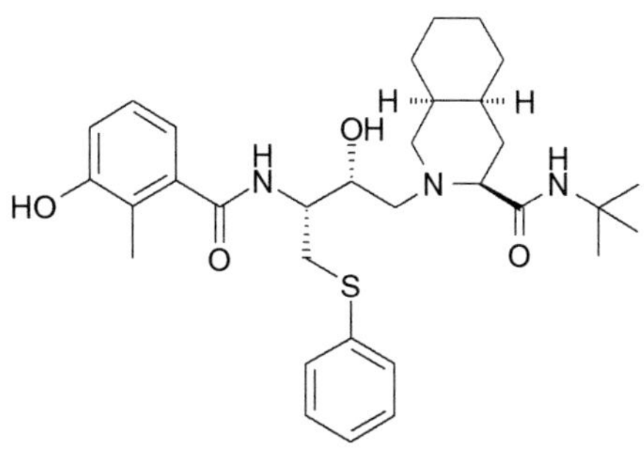

24.24 Nelfinavir Viracept® (1997)

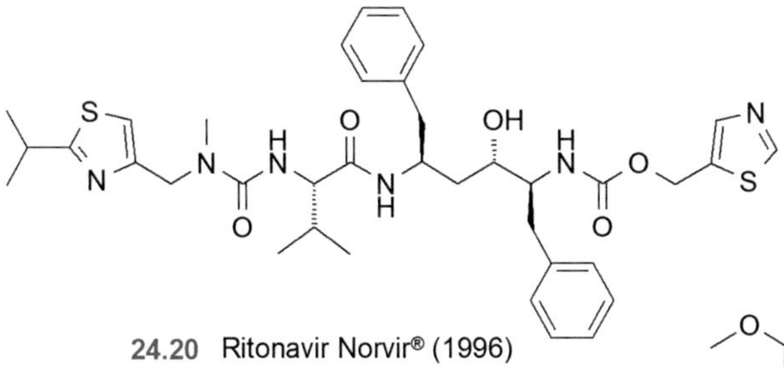

24.20 Ritonavir Norvir® (1996)

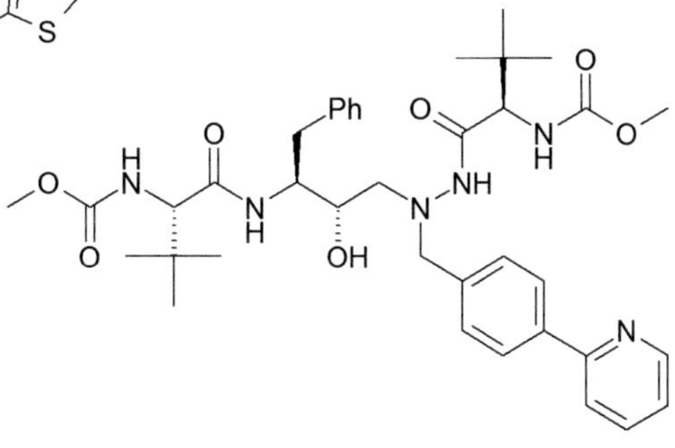

24.25 Atazanavir Reyataz®, Zrivada® (2000)

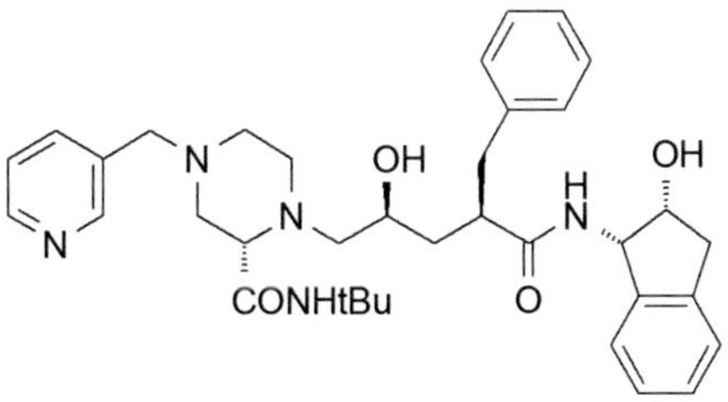

24.21 Indinavir Crixivan® (1996)

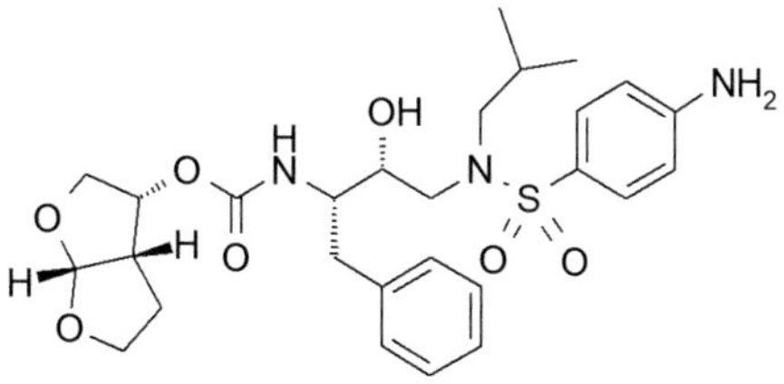

24.26 Darunavir Prezista® (2006)

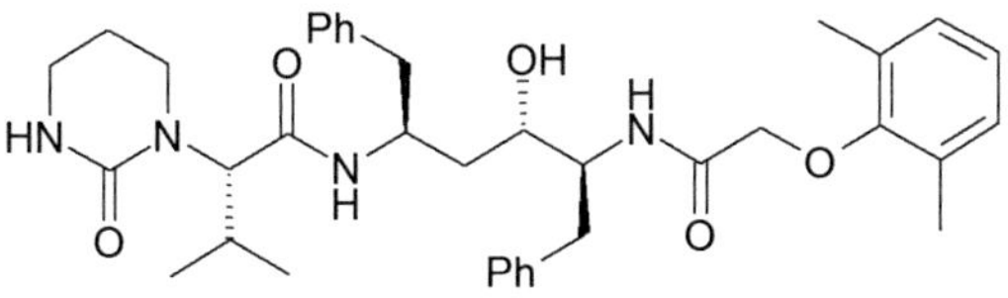

24.22 Lopinavir Kaletra®, Aluvia® (2003)

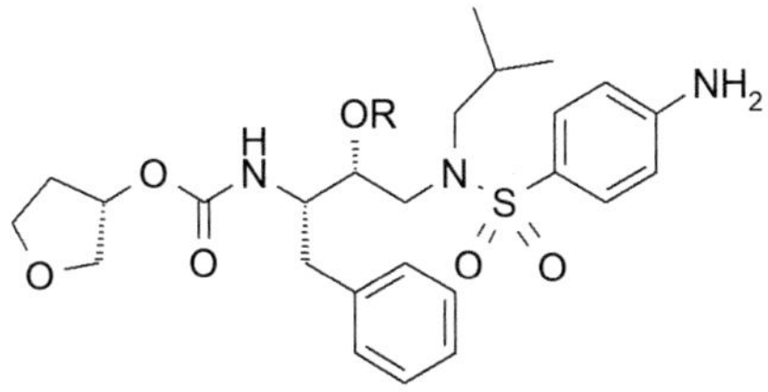

24.23 R=H Amprenavir Agenerase®, Prozei® (1999)
R=PO_3H Fosamprenavir Prodrug Lexiva® (2003)

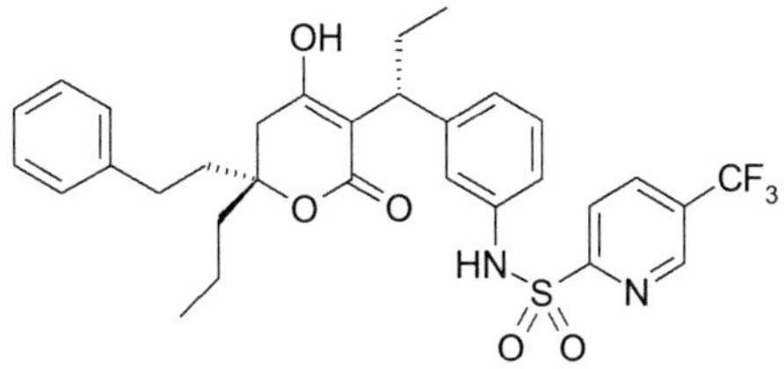

24.27 Tipranavir Aptivus® (2005)

Fig. 24.15 To date, nine new compounds for AIDS therapy have been introduced to the market, whereby nelfinavir **24.24** has been withdrawn from the market in Europe. Compounds **24.19**–**24.26** are peptide-like inhibitors, and tipranavir **24.27** alone has a completely nonpeptidic structure

A new lead structure **24.34** (K_i = 1.1 µm, Fig. 24.19) was found in a screening campaign at Parke–Davis. The spatial structure with the protease was determined with the homologous inhibitor **24.38** (Fig. 24.18). It was shown that this structure, analogous to **24.31**, displaces the water molecule in the active site and forms H-bonds to the catalytic aspartate residues as well as to the NH groups of Ile 50 and Ile 50′. The X-ray structure was used to design derivatives with improved binding properties, such as **24.36**. Modeling studies led to the idea of introducing an acidic group into the S_3 pocket to form a salt bridge with Arg 8. The corresponding compound with an OCH_2COOH group in the *para*-position of the 6-phenyl ring was synthesized and resulted in a significant increase in binding affinity. The inhibitor **24.37** (K_i = 51 nM) is achiral, has a low molecular weight, and can be prepared in three steps. The hydroxypyrone scaffold ultimately proved successful for inhibitor development. In 2005,

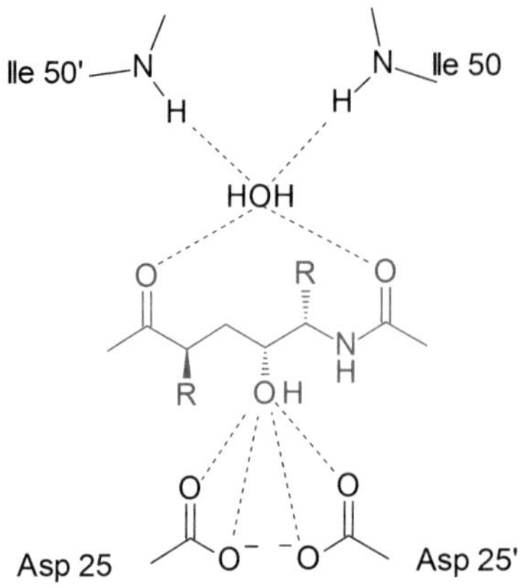

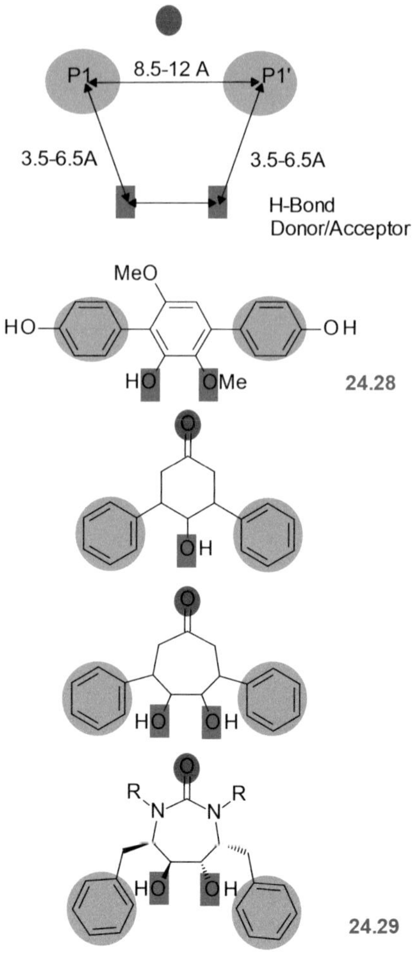

Fig. 24.16 The pattern of the hydrogen bonds between HIV protease and the peptide inhibitors in the vicinity of the catalytic aspartic acids (*left*). A water molecule is found in the binding pocket that forms two H-bonds to the inhibitor and to the protein. The hydroxyl group of the inhibitor displaces the water molecule that is involved in the catalytic process (cf. Fig. 24.1). By starting with this binding mode, the spatial pharmacophore of a potential inhibitor was defined (*right*). The search in databases of crystal structures of low molecular weight compounds was started with this pattern. It produced the substituted phenol **24.28** as a hit. From there, six- and seven-membered cyclic ketones and a cyclic urea **24.29** were developed. These derivatives could displace the structurally conserved water molecules from the binding pocket of the protease with their carbonyl groups

Boehringer Ingelheim launched tipranavir **24.27**, the first nonpeptidic HIV protease inhibitor. This compound binds to the catalytic dyad with its hydroxyl function. The carbonyl oxygen of the pyrone replaces the structural water. Side chain optimization, however, resulted in a much more complex structure than that shown in **24.34**.

24.5 The Development of Resistance Against HIV Protease Inhibitors

The first HIV protease inhibitor was developed and brought to market in less than 8 years. In the following 10 years, nine drugs were introduced as marketed products (Fig. 24.15). Compounds from completely different structural classes have been successfully developed to block the HIV protease. An arsenal of orally available, small-molecule inhibitors is now available for therapy. In addition, drug molecules have also been developed and marketed for another important viral enzyme, the reverse transcriptase (Sect. 32.5). In addition to substrate analogous inhibitors such as zidovudine (AZT **24.39**) and didanosine (DDI **24.40**), allosteric inhibitors (such as nevirapine **24.21**) are available (Fig. 24.20). HIV integrase is another enzyme that has been identified as a target for combating the HIV virus. This enzyme integrates viral DNA transcribed by reverse transcriptase into the host cell genome. Raltegravir **24.42** became the first integrase inhibitor to be approved in late 2007. Other drugs targeting this enzyme followed, including elvitegravir, dolutegravir, bictegravir, or cabotegravir. Common to these drug molecules is a hydrophobic benzyl group that occupies a hydrophobic pocket near the active site. They share a chelating triad that binds tightly to two Mg^{2+} ions, anchoring the inhibitor to the protein surface. Other drugs include enfuvirtide and maraviroc, which inhibit fusion as the virus enters infected cells (Sect. 31.5). A recent innovation with a novel mechanism of action are HIV capsid inhibitors. They interfere with the viral capsid, a protein shell that protects the genetic material and enzymes of the virus needed for replication after the virus has fused with the infected host cell. In 2022, lenacapavir became the first HIV capsid inhibitor to be approved.

The **virus has rapidly developed resistance** to reverse transcriptase inhibitors. Like other RNA viruses, HIV **replication is error-prone**. The viral reverse transcriptase makes an error approximately every 10,000 bases. In this way, the virus generates large genetic diversity, which leads

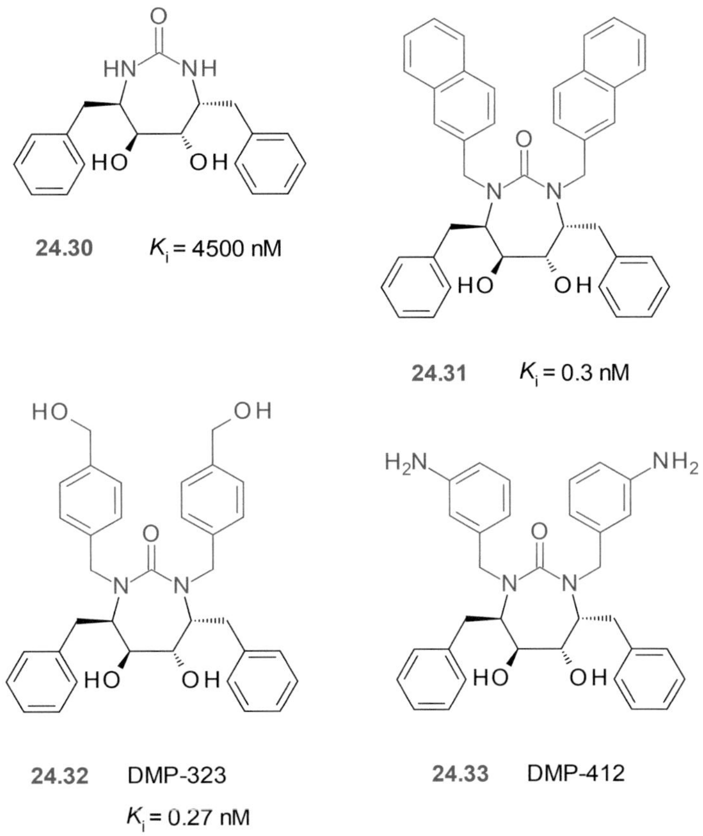

■ **Fig. 24.17** The newly discovered lead structure of the cyclic urea **24.29** was optimized stepwise via the derivatives **24.30** and **24.31** to DMP-323 **24.32** and DMP-412 **24.33** at Dupont–Merck

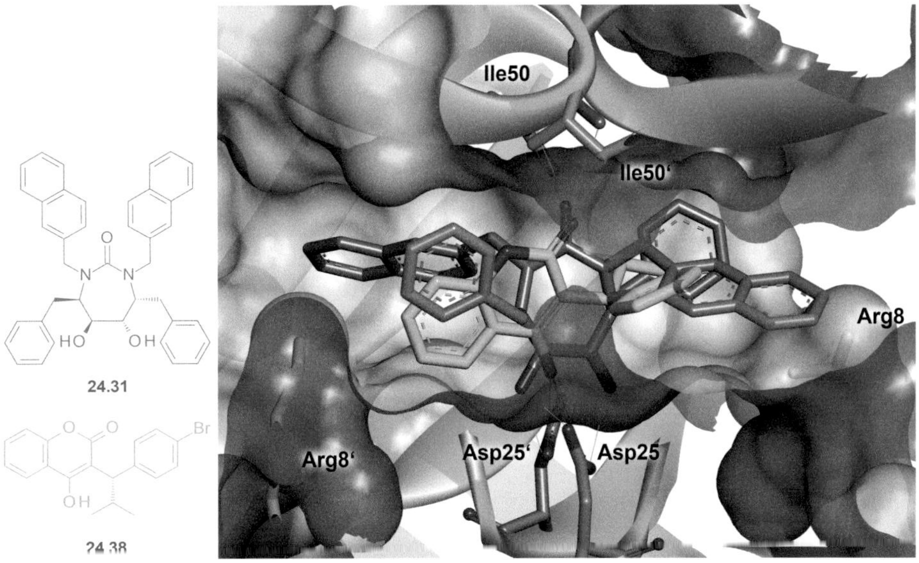

■ **Fig. 24.18** The superimposition of the crystal structures of the complexes of HIV protease with the urea-containing inhibitor **24.31** (*gray*) and the coumarin derivative **24.38** (*light green*). (► https://sn.pub/LPXIyG)

directly to resistance. With approximately 10^8–10^9 replication cycles per day, 10^5 point mutations occur in the viral protein population of an infected patient. It is, therefore, not surprising that the introduction of HIV protease inhibitors has induced a high level of resistance. Various mutations in the binding pocket further away from the active site also lead to a severe reduction in the binding affinity of HIV protease inhibitors. The positions where mutations have been observed are shown in ■ Fig. 24.21. Taking all the observed exchanges together, half of all positions in the protease are now affected. However, similar amino acid exchanges near the catalytic center are observed again and again. This is certainly due to the fact that the peptide-like inhibitors **24.19–24.26** all adopt a rather similar binding mode in the protease (see ■ Fig. 24.25).

Therefore, **combination therapy** is used to treat AIDS. The simultaneous administration of several inhibitors should lead to better suppression of viral replication.

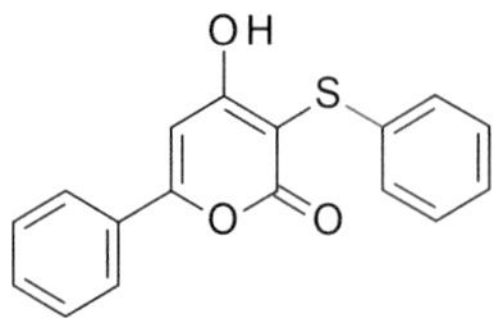

24.34
K_i = 1100 nM
IC_{50} = 3000 nM

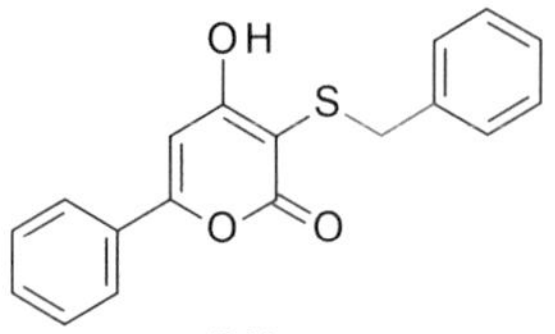

24.35
K_i = 700 nM
IC_{50} = 1670 nM

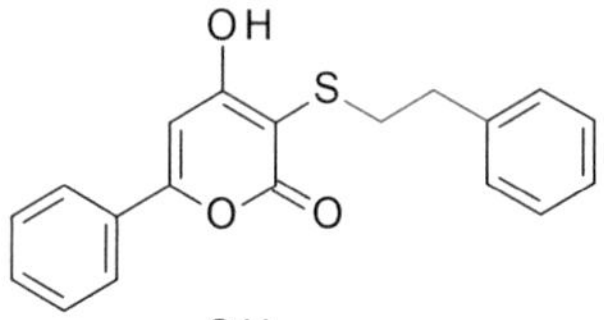

24.36
IC_{50} = 1260 nM

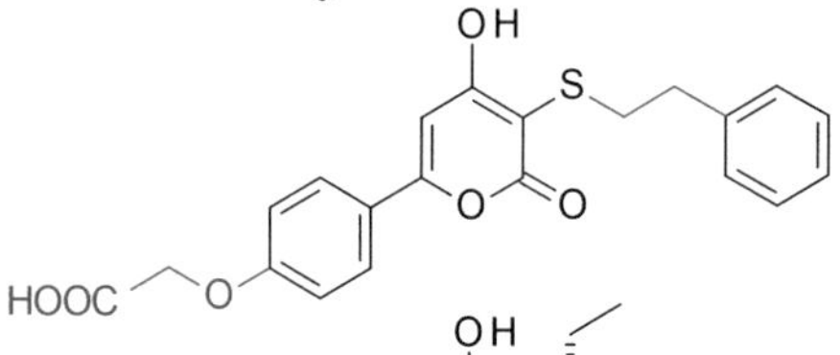

24.37
K_i = 51 nM
IC_{50} = 160 nM

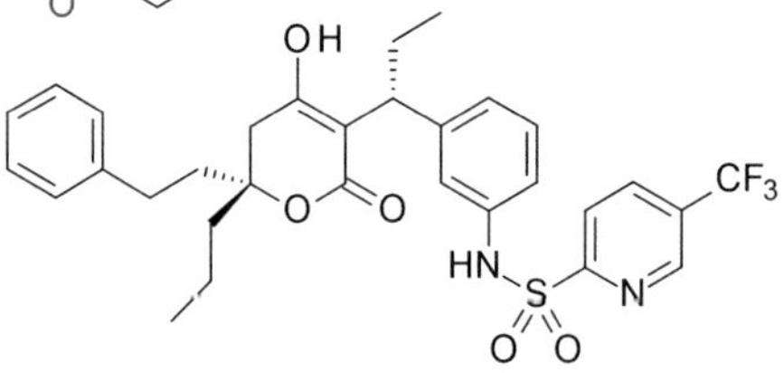

24.27 Tipranavir

Fig. 24.19 Optimization of the coumarin-like HIV protease inhibitor **24.34**, which was discovered by mass screening at Parke–Davis. The extension of the thioether side chain to **24.35** and **24.36** as well as the introduction of a carboxyl group led to **24.37**. A hydrogenated hydroxypyrone building block could be incorporated in tipranavir **24.27** at Boehringer Ingelheim. The compound represents the first nonpeptide HIV protease inhibitor for therapy

This significantly slows down and hinders the formation of resistance. The best results are achieved by combining antiviral drugs with different modes of action, such as a nucleoside and a nonnucleoside reverse transcriptase inhibitor with a protease inhibitor. Such therapies are part of the so-called **HAART** (**h**ighly **a**ctive **a**nti**r**etroviral **t**herapy) strategy, which has found application in the clinical setting.

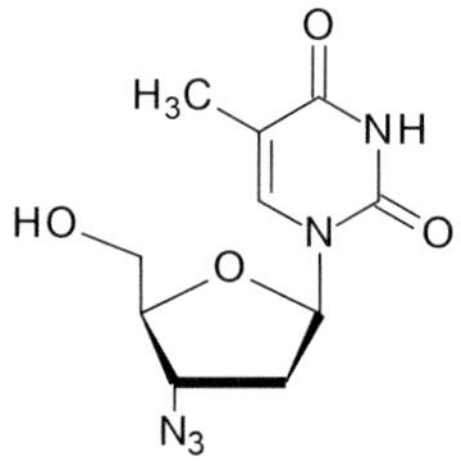

24.39 Zidovudine AZT

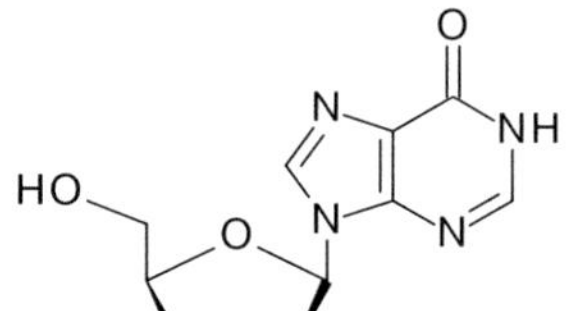

24.40 Didanosine DDI

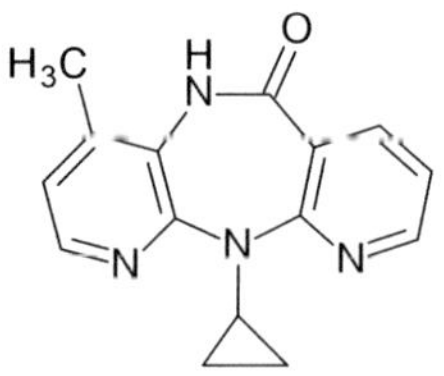

24.41 Nevirapine

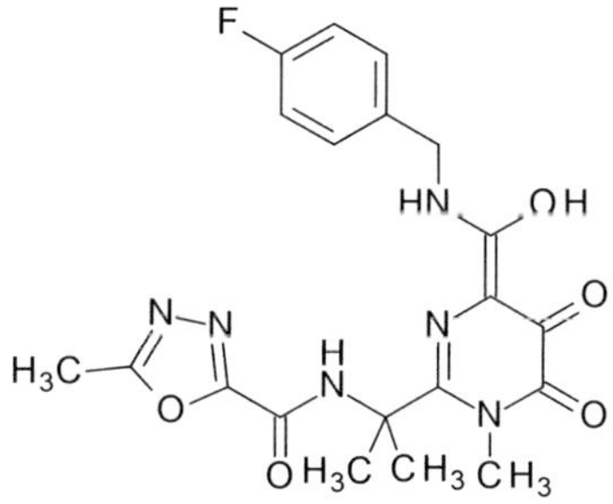

24.42 Raltegravir

Fig. 24.20 By using **HAART** therapy, which is a combination of a protease inhibitor (Fig. 24.15), a reverse transcriptase inhibitor such as **24.39**–**24.41**, or an integrase inhibitor such as **24.42**, the hope is to break through the increasingly observed resistance to the drugs in AIDS therapy

24.6 A Basic Nitrogen as a Partner for the Aspartic Acids of the Catalytic Dyad

As mentioned above, Roche discovered a piperidine derivative **24.13** (Fig. 24.10) in a comprehensive screening campaign as a renin inhibitor with micromolar activity. It was further optimized to a subnanomolar hit. Its piperidine nitrogen atom binds at the pivotal point between the two aspartic acids (Fig. 24.10). As a result of this work, secondary amines have now been intensively investigated as binding partners for the aspartic acids. A whole series of building blocks have been described (Fig. 24.22) and tested for their inhibitory potency on various aspartic proteases. In a rational design approach, a five-membered pyrrolidine ring (**24.43**) was designed instead of a six-membered piperidine. Such a ring can be placed with its nitrogen atom between the two aspartic acid residues. At the same time, the specificity pockets can be reached

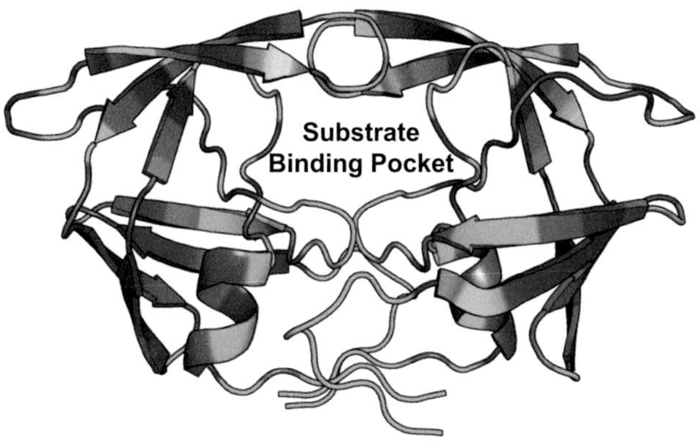

Fig. 24.21 Mutations in the amino acids in HIV protease lead to resistance to the inhibitors. The course of the polymer chain is coded in *green* or *red*. *Red* represents residues that, with high probability, have mutated, and *green* areas show little exchange. Many mutations are found in the vicinity of the active site, but some are fairly far away from the substrate-binding pocket

symmetrically with its side chains on the prime and unprime sides. As with the substrate-like inhibitors **24.19**–**24.26** (Fig. 24.15), hydrogen bond acceptor groups were included in the design on both sides of the heterocycle. As mentioned, in the special case of the HIV protease, a conserved water molecule (so-called structural water) is found in a pivotal position to mediate the interaction to the flap region. In other aspartic proteases, direct contact with the flap region is achieved. The pyrrolidine ring was, therefore, extended by aminomethylene groups on both sides and amide and sulfonamide groups were introduced as acceptor functions. At the same time, the amide nitrogen atoms served as branching points to reach the four subpockets of the protease with attached groups.

In a small series of compounds, Edgar Specker at the University of Marburg developed micro- to submicromolar lead structures for HIV protease and cathepsin D. Using the racemate of **24.45**, Jark Böttcher determined the crystal structure with HIV protease. It contained a big surprise (Figs. 24.23 and 24.24, *upper left*). The nitrogen in the pyrrolidine ring of the (*R*,*R*)-enantiomer is, as defined, at the pivotal point between the two aspartic acids. It occupies the same position as the hydroxyl groups in the transition state analogue inhibitors. However, contrary to the original concept, the inhibitor displaces the structural water from the binding pocket! The oxygen of the sulfone group forms a direct hydrogen bond to the NH group of Ile 50 in the flap. The carbonyl group of the amide bond on the opposite side is not involved in any interaction with the flap region. On the other hand, the loop of the flap adopts a distorted geometry in that the NH function of Ile 50′ results in a hydrogen bond contact with the turn of the other monomer unit. Such a geometry had never been seen before. On closer analysis, it appeared that the inhibitor did not optimally fill the S_2 to S_2' subpockets of the protease. Compared to amprenavir **24.23** (Fig. 24.15), the S_2 pocket remains virtually unoccupied. In addition, the bulky dimethylphenoxy group of the molecule appeared to protrude beyond the S_1' pocket and interfered with the loop of the flap. Despite a single-digit micromolar affinity (K_i = 1.5 μM), the inhibitor seemed "uncomfortable" in the pocket. It seemed to break almost all the "golden" rules of drug design (Sect. 4.11). Is the awkward dimethylphenoxy group responsible for the binding mode? To test this, a three-armed inhibitor **24.46** (Fig. 24.23) was synthesized. Surprisingly, the crystal structure of this inhibitor **24.46** (K_i = 52 μM) shows the same binding mode: the structural water is displaced and the loop region takes on a distorted shape, although there is obviously no large group left to interfere with the region (Fig. 24.24, *upper right*). The occupancy of the specificity pockets with this inhibitor seems to be far from optimal.

Next, an attempt was made to bring the substituents on either side of the central pyrrolidine ring closer together. Andreas Blum eliminated the two methylene linkers to use 3,4-diaminopyrrolidine **24.44** as the central scaffold (Fig. 24.22). This scaffold was symmetrically decorated on both sides with substituted sulfonamides. First, benzenesulfonic acid derivatives were optimized with respect to the substitution at the tertiary nitrogen atom (**24.47**–**24.58**; Table 24.7). In addition to the inhibitory effects on the wild-type enzyme, a rapidly induced resistant mutant variant carrying a valine instead of an isoleucine at position 84 was also studied. The enlarged pocket in the resistant mutant shows reduced binding affinity with many inhibitors due to the reduced hydrophobic contact surface. It was also shown that the wild-type enzyme does not tolerate branched groups (**24.47**–**24.50**) as well as the mutants (Table 24.7). The benzyl group proved to be the best compromise for good inhibition of both isoforms. The crystal structure of derivative **24.47** was determined (Fig. 24.24, *lower left*). The pyrrolidine nitrogen atom occupies the desired position between the two aspartic acids. The structural water is again displaced from the pocket and one of the two sulfonamide groups forms a hydrogen bond with the Ile 50 NH group in the flap region. The inhibitor sits largely symmetrically in the binding pocket. The benzyl groups on the amino group are located in the S_1 and S_1' pockets. The benzenesulfonyl group occupies the S_2 and S_2' pockets. The pockets seem to be much better filled with this inhibitor than with **24.45** or **24.46**. Nevertheless, it was obvious that the substituents in S_1 and S_1' in the *para*-position should be enlarged for optimization. An additional bromine or iodine substituent increases the affinity about sixfold compared to the wild type. The inhibition of the mutant is improved by a factor of 2. There even seemed to be enough room for larger groups in the S_2 and S_2' pockets. Indeed, a methyl group or a chlorine atom in the *ortho*-position increases the affinity by a factor of 2. This effect was not as pronounced with the mutant. Furthermore, at the end of this pocket are the acidic amino acids Asp 29 and Asp 30,

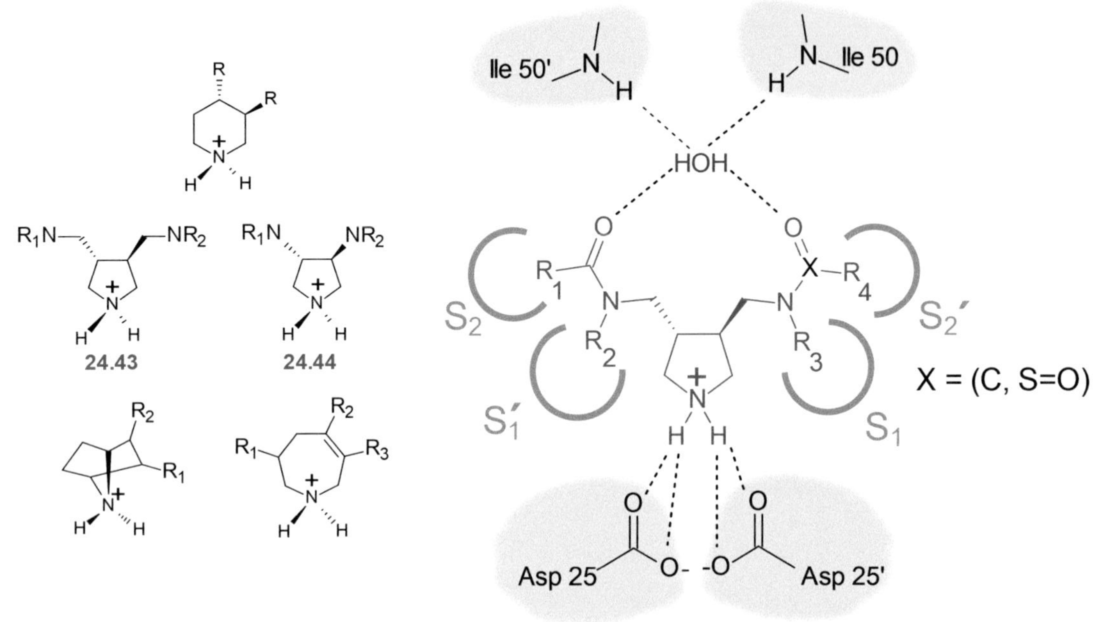

■ **Fig. 24.22** Secondary amines are promising binding partners for the aspartic acids of the catalytic dyad of aspartic proteases. In a rational design approach, the nitrogen atom of the five-membered pyrrolidine ring **24.43** was placed between the two aspartic acids. At the same time, the scaffold and its side chains symmetrically reach the specificity pockets on the primed and unprimed side of the protease. The incorporated acceptor function should form H-bonds to the structurally conserved water molecule in the flap region of HIV protease

■ **Fig. 24.23** Schematic representation of the different binding modes of the four inhibitors that are shown in ■ Fig. 24.24. The three pyrrolidine derivatives **24.45**, **24.46**, and **24.47** differ in their connecting geometry at the ring and the number of substituents. The central heterocycle was opened in **24.59**. Interestingly, the conserved structural water returns to the structure with this ligand

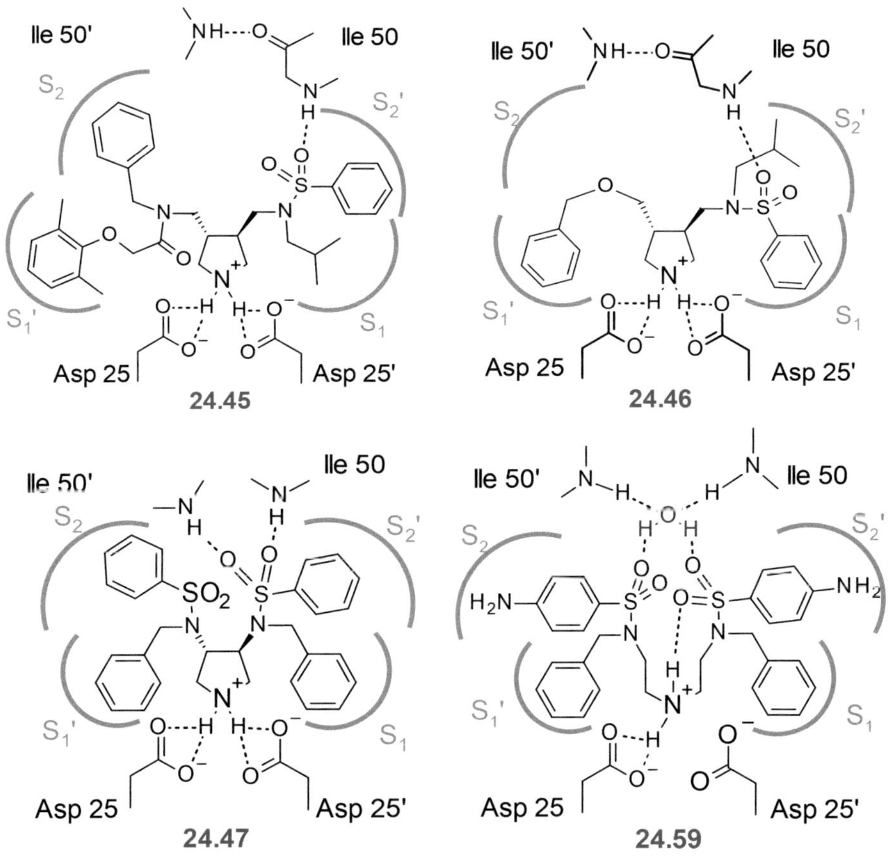

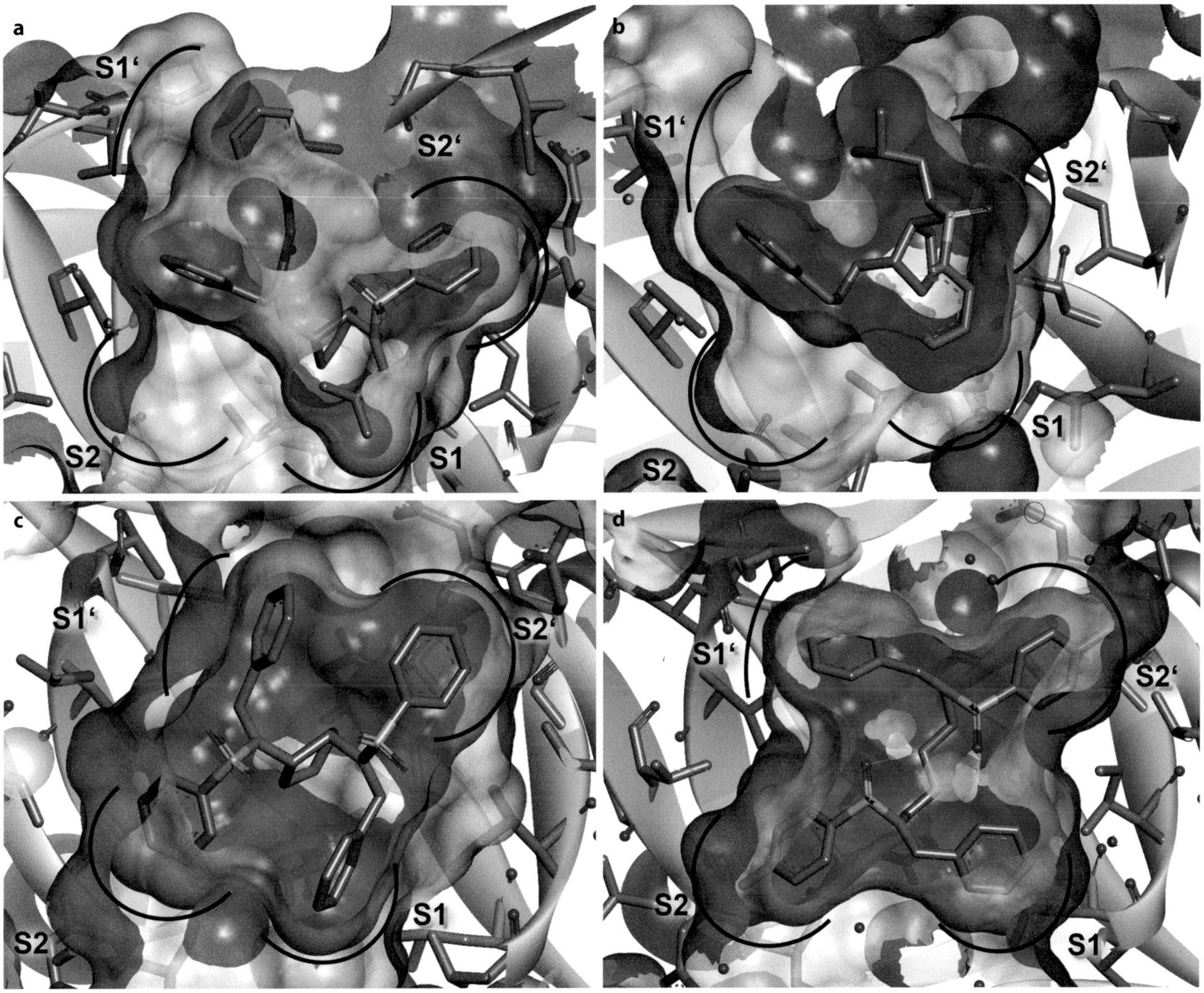

Fig. 24.24 Crystal structures of the inhibitors from Fig. 24.23 in HIV protease. **a** Compound **24.45** leaves the S_2 pocket virtually unoccupied. Its voluminous *o,o'*-dimethylphenoxy substituent only incompletely fills the S_1' pocket and seems to push against the loop in the flap region. The structural water is displaced from the binding pocket. **b** Compound **24.46** only partially fills the S_2 and S_1' pockets. Water is also displaced from this structure and the loop takes on a distorted geometry even though no unfavorable contacts are recognizable. **c** Compound **24.47** binds almost C_2 symmetrically and places its benzenesulfonyl groups in S_2 and S_2'. The *N*-benzyl groups are found in S_1 and S_1'. Here too, the structural water is displaced from the complex. **d** Compound **24.59** orients its *p*-aminobenzenesulfonyl group in S_2 and S_2'. The *N*-benzyl substituents occupy S_1 and S_1'. Both SO_2 groups form H-bonds to the structural water, which has returned in this structure. The inhibitor seems to fill the binding pocket perfectly, but it does not achieve better binding affinity than the other derivatives despite the additional NH_2 functions that form H-bonds to the protein. (► https://sn.pub/yyNuRq)

which interact with the ligand. This interaction is possible by introducing an amino or carboxamide group in the *para*-position. The binding potency to the enzyme is then increased by a factor of about 10. Further optimization led to derivative **24.58** with a CF_3 group on the P_1 benzyl group and an amide group on the P_2 substituent. It inhibits the wild-type enzyme with $K_i = 61$ nM and the mutant with 14 nM. The binding mode of these new inhibitors, based on a 3,4-diaminopyrrolidine **24.44**, differs from that of all other inhibitors currently on the market. Such a difference may provide a promising perspective for resistance breaking (Fig. 24.25).

In one of the final steps, the central heterocycle was "cut open" and replaced with open-chain secondary amines (Fig. 24.23). Two- and three-membered aliphatic chains were introduced as spacers between the central amine nitrogen atom and the two SO_2 groups designed to bind to the flap region. Inhibition constants were measured for several aspartic proteases and showed single to double digit micromolar inhibition. The crystal structure of compound **24.59** was determined

Table 24.7 By modifying the **R1** and **R2** groups on 3,4-diaminopyrrolidine **24.44**, the affinity and resistance profile is improved (*wt* wild type; *I84V* mutant)

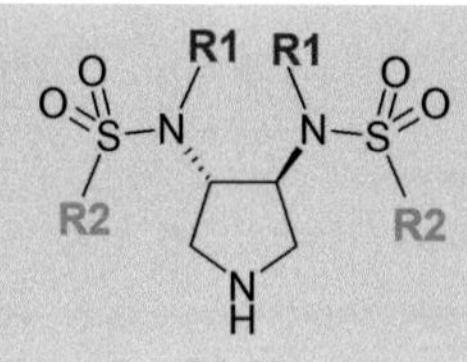

24.44

Compound	R1	R2	K_i [µM] wt	K_i [µM] I84V
24.47	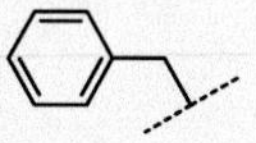	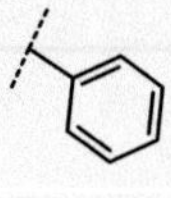	2.15	1.7
24.48	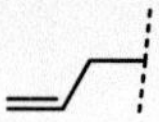	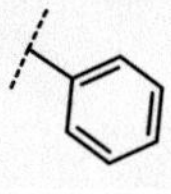	12.3	84.0
24.49	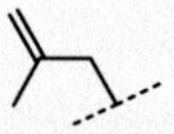	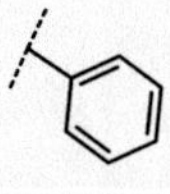	74.7	53.1
24.50	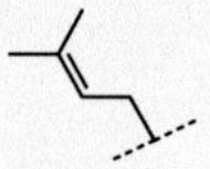	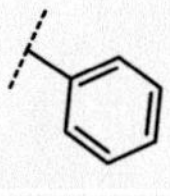	1.57	5.82
24.51	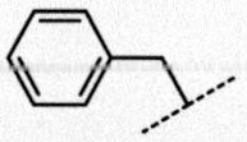	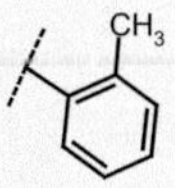	0.67	0.46
24.52	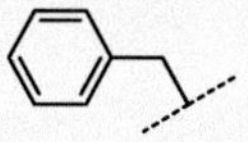	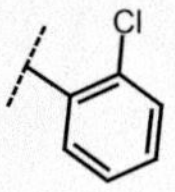	0.77	0.47
24.53	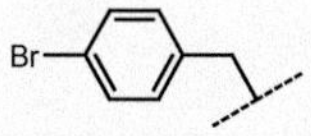	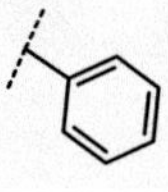	0.46	0.55
24.54	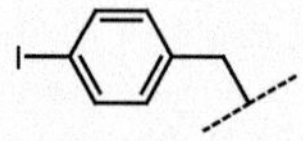	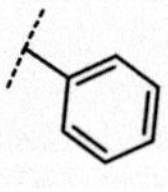	0.39	0.33
24.55	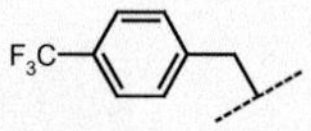	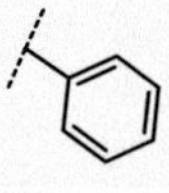	0.80	0.50
24.56	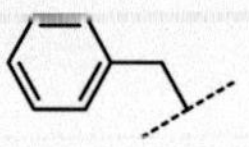	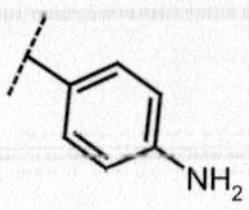	0.27	0.13
24.57	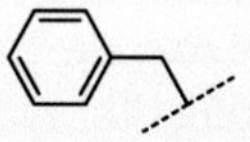	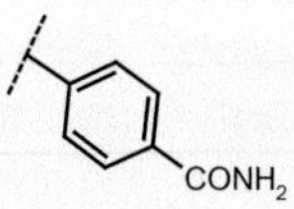	0.26	0.04
24.58	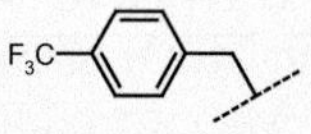	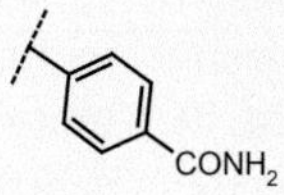	0.07	0.01

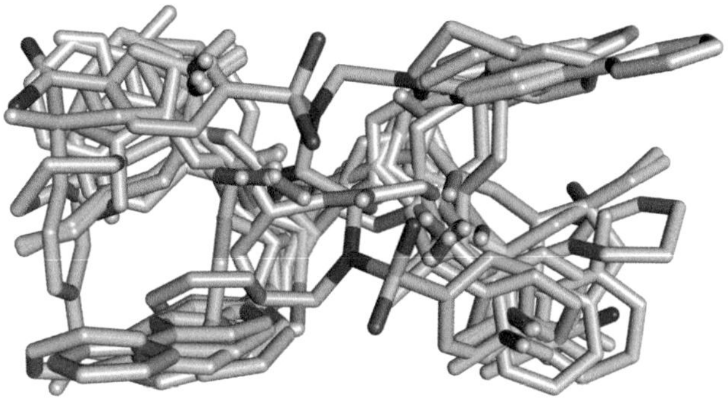

■ **Fig. 24.25** Compared to all of the currently marketed HIV inhibitors (■ Fig. 24.15, *beige*), the inhibitors based on 3,4-diaminopyrrolidine **24.44** (*light green*) adopt a deviating binding mode. As a result, a different activity profile against resistant mutants is observed

(K_i = 9.6 μM for HIV protease; ■ Fig. 24.24, *lower right*). As expected, the basic nitrogen atom binds between the two aspartic acids, but at an H-bonding distance to only one of the two aspartic acids. Interestingly, the structural water returns to the inhibitor complex and mediates a binding contact between the two sulfonyl group and the residues of the flap region.

The study of the open-chain compounds illustrates how the sterically fixed heterocycle determines the orientation of the inhibitors in the binding pocket. Its spatial demand is responsible for the displacement of the structural water from the binding pocket. As a result, the H-bond acceptor groups of the inhibitors interact directly with the flap region. The open-chain compound **24.59** gives the impression of fitting perfectly into the protease. It appears to lie completely relaxed in the binding pocket, finding partners for its polar groups in the protein and allowing the structural water to return. Although it has amino groups in the *para*-position of the benzenesulfonic acid group, which led to a tenfold increase in activity in the series with the **24.44** scaffold (■ Fig. 24.22), it binds with only micromolar affinity. Even the most beautiful binding mode will not help if the open-chain compound must first be conformationally reorganized to adopt the necessary geometry at the site of action. It loses too many degrees of freedom around rotatable bonds, and this is a high price to pay in terms of binding affinity (Sect. 4.10). The open-chain compounds suffer from this disadvantage. This emphasizes that properly pre-organized and rigidified inhibitors will have a clear advantage for entropic reasons.

24.7 Other Targets from the Family of Aspartic Proteases

In addition to the two examples of renin and HIV protease, many other members of the aspartic protease family have been validated as targets for drug development. First, **cathepsin D**, a protein involved in protein catabolism, appeared interesting and concepts for the treatment of breast cancer or muscular dystrophy were pursued. The aforementioned gastric **pepsin** was discussed as a possible therapeutic target for peptic ulcer disease. **Secretory aspartic proteases** (SAPs) from *Candida albicans* have been considered as possible target enzymes for the treatment of fungal diseases.

The field of **β-secretase** has been the focus of intense drug development as it could lead to an effective Alzheimer's therapy. The disease-causing β-amyloid protein, which leads to dangerous deposits (senile plaques) in the brain, is cleaved from a larger precursor, amyloid precursor protein (APP). In 1999, two proteases, β- and γ-secretase, were reported. As membrane-bound proteases of the aspartic protease family, they catalyze the release of **β-amyloid protein**. They are also referred to as **BACE-1** and **-2**, which is an abbreviation for **b**eta-site **A**PP-**c**leaving **e**nzymes. Drugs that inhibit these proteases could prevent the accumulation of β-amyloid and, thereby, halt the onset or progression of Alzheimer's disease. From a medicinal chemistry perspective, potent and orally available inhibitors such as verubecestat were successfully developed at Merck in the United States. However, the subsequent clinical trials were discontinued in February 2018. Unfortunately, verubecestat was not convincing in reducing cognitive decline in patients with mild to moderate Alzheimer's disease. It is unclear why the expected effect did not occur in humans in the studies.

Plasmepsins are being validated as additional aspartic protease targets for therapy. They are used by the malaria parasite in the food vacuole to degrade hemoglobin. The parasite uses the components of hemoglobin as food. Plasmepsins are used for the initial cleavage of hemoglobin and cleave the α-chain between Phe 33 and Leu 34. Several plasmepsin isoforms are involved in further cleavage to larger peptide fragments. Ten of these aspartic proteases have been found in the genome of the malaria parasite *Plasmodium falciparum*. In addition, falcipains, other cysteine proteases, and falcilysin, a zinc protease, are also involved in the degradation process. The plasmepsins show a high structural homology to cathepsin D. First lead structures have been derived using principles analogous to those of, for example, renin. Recent evidence suggests that malaria therapy based on protease inhibitors of hemoglobin degradation would need to inhibit several of the above enzymes simultaneously to achieve efficient control of the parasite. It may, therefore, be appropriate to develop **pan-inhibitors** for the simultaneous, selective silencing of several plasmepsins.

24.8 Synopsis

- Aspartic proteases possess two facing aspartate residues in their catalytic cleavage center. A water mol-

ecule, located at the apex between both aspartates, is polarized and nucleophilically attacks the carbonyl carbon atom of the amide bond to be cleaved.
- The cleavage reaction proceeds through a tetrahedral transition state with a temporarily formed geminal diol structure. Peptidomimetic inhibitors imitate this intermediate structure by using chemically stable building blocks. Hydroxyl groups embedded into hydroxyethylene or statin moieties have been especially used as transition state isosteres.
- Aspartic proteases often cleave between hydrophobic amino acids. These residues cannot form strong interactions to the specificity pockets of the protease on both sides of the cleavage site. They bind through multiple contacts, and the recognition pockets are well formed on both sides.
- Renin specifically cleaves angiotensinogen to angiotensin I. Subsequently, this product is further cleaved to the octapeptide angiotensin II, which stimulates blood pressure to increase once recognized at its receptor. Renin exhibits a large, virtually merged S_1/S_3 pocket. This gave rise to the design concept to bridge P_1/P_3 substituents in the inhibitor aliskiren and to disrupt its central peptide chain. A more polar, orally available antihypertensive agent resulted with good duration of action.
- HIV protease is a viral aspartic protease that cleaves the incipient polypeptide chain into mature proteins required for the life cycle of the virus. It is a C_2 symmetric homodimer with a structural water molecule mediating interactions between the bound substrates and the flap region closing up the catalytic site.
- Through systematic variations of the minimal substrate and introduction of transition state isosters, a variety of potent and selective drugs with peptide-like scaffolds are available for therapy. Upon administration, the virus becomes resistant through mutational modifications; meanwhile, exchanges have been reported for nearly half of all amino acid positions.
- Combination therapy, the so-called HAART strategy, is recommended for the treatment of HIV infections. This tries to achieve better suppression of viral replication through simultaneous administration of multiple inhibitors acting on different targets that are crucial for the virus.
- Multiple design attempts have been followed to depart from peptide-like inhibitors to nonpeptidic structures. To date, tipranavir is the only compound to be successfully launched to the market that has a central hydroxypyrone building block. In recent years, many scaffolds containing a basic nitrogen to address the catalytic dyad have been proposed as novel building blocks for the development of aspartic protease inhibitors.
- Aside from renin and HIV protease, cathepsin D, β- and γ-secretase, the parasitic plasmepsin proteases, and the fungal secretory aspartic protease SAP have been investigated as potential drug targets.

Bibliography and Further Reading

General Literature

W. J. Greenlee and A. E. Weber, Renin Inhibitors, Drugs, News & Perspectives, **4**, 332–339 (1991)

S. H. Rosenberg, Renin Inhibitors, Prog. Med. Chem., **32**, 37–144 (1995)

E. De Clercq, The design of drugs for HIV and HCV, Nat. Rev. Drug Discov., **7**, 1001–1018 (2007)

P. S. Anderson, G. L. Kenyon and G. R. Marshall, Eds., Therapeutic Approaches to HIV, Persp. Drug Discov. Design, Vol. 1, ESCOM (1993)

J. A. Martin, S. Redshaw and G. J. Thomas, Inhibitors of HIV Proteinase, Prog. Med. Chem., **32**, 239–288 (1995)

C. Hutchins and J. Greer, Comparative Modeling of Proteins in the Design of Novel Renin Inhibitors, Crit. Rev. Biochem & Mol. Biol. **26**, 77–127 (1991)

E. De Clercq, Toward Improved Anti-HIV Chemotherapy: Therapeutic Strategies for Intervention with HIV Infections, J. Med. Chem., **38**, 2491–2517 (1995)

E. De Clercq (Ed.) Antiviral drug strategies, 50th edn, Methods and principles in medicinal chemistry. Wiley-VCH, Weinheim (2011)

M. L. West and D. P. Fairlie, Targeting HIV-1 Protease: A Test for Drug-Design Methodologies, Trends Pharm. Sci., **16**, 67–74 (1995)

R. E. Babine and S. L. Bender, Molecular Recognition of Protein–Ligand Complexes: Applications to Drug Design, Chem. Rev., **97**, 1359–1472 (1997)

R. J. Landovitz, H. Scott and S. G. Deeks, Prevention, treatment and cure of HIV infection. Nat. Rev. Microbiol., 21, 657–670 (2023)

C. Dash, A. Kulkarni, B. Dunn and M. Rao, Aspartic Peptidase Inhibitors: Implications in Drug Development, Critical Rev. Biochem. Molec. Biol., **38**, 89–119 (2003)

J. Eder, U. Hommel, F. Cumin, B. Martoglio and B. Gerhartz, Aspartic Proteases in Drug Discovery, Curr. Pharmaceut. Design, **13**, 271–285 (2007)

A. K. Ghosh (Ed.) Aspartic acid proteases as therapeutic targets, Vol. 45, Methods and principles in medicinal chemistry. Wiley-VCH, Weinheim (2010)

Special Literature

A. R. Sielecki, A. A. Fedorov, A. Boodhoo, N. S. Andreeva, M. N. G. James, Molecular and crystal structures of monoclinic porcine pepsin refined at 1.8 Å resolution, J. Mol. Biol., **214**, 143–170 (1990)

H. Umezawa et al., Pepstatin, a new Pepsin Inhibitor produced by Actinomycetes, J. Antibiotics, **23**, 259–262 (1970)

A. R. Sielecki et al., Structure of Recombinant Human Renin, a Target for Cardiovascular-Active Drugs, at 2.5 Å Resolution, Science, **243**, 1346–1351 (1989)

J. Rahuel, J. P. Priestle, M. G. Grütter, The crystal structures of recombinant glycosylated human renin alone and in complex with a transition state analog inhibitor, J. Struct. Biol., **107**, 227–236 (1991)

B. L. Sibanda, T. Blundell, P. M. Hobart, M. Fogliano, J. S. Bindra, B. W. Dominy, J. M. Chirgwin, Computer graphics modelling of human renin. Specificity, catalytic activity and intron-exon junctions, FEBS Lett., **174**, 102–111 (1984)

C. Frazao, C. Topham, V. Dhanaraj, T. L. Blundell, Comparative modelling of human renin: A retrospective evaluation of the model with respect to the X-ray crystal structure, Pure & Appl. Chem., **66**, 43–50 (1994)

H. D. Kleinert, S. H. Rosenberg, W. R. Baker et al., Discovery of a Peptide-Based Renin Inhibitor with Oral Bioavailability and Efficacy, Science, **257**, 1940–1943 (1992)

Y. C. Li, Inhibition of Renin: An Updated Review of the Development of Renin Inhibitors, Current Opinion in Investigational Drugs, **8**, 750–757 (2007)

J. M. Wood et al., Structure-based Design of Aliskiren, a Novel Orally Effective Renin Inhibitor, Biochem. Biophys. Res. Commun., **308**, 698–705 (2003)

R. Güller et al., Piperidine-Renin Inhibitors Compounds with Improved Physicochemical Properties, Bioorg. Med. Chem. Lett., **9**, 1403–1408 (1999)

M. A. Navia et al., Three-dimensional structure of aspartyl protease from human immunodeficiency virus HIV-1, Nature, **337**, 615–620 (1989)

A. Wlodawer et al., Conserved Folding in Retroviral Proteases: Crystal Structure of a Synthetic HIV-1 Protease, Science, **245**, 616–621 (1989)

J. V. N. Vara Prasad, K. S. Para, E. A. Lunney et al., Novel Series of Achiral, Low Molecular Weight, and Potent HIV-1 Protease Inhibitors, J. Am. Chem. Soc., **116**, 6989–6990 (1994)

J. P. Vacca et al., L-735,524: An Orally Bioavailable Human Immunodeficiency Virus Type I Protease Inhibitor, Proc. Natl. Acad. Sci., **91**, 4096–4100 (1994)

J. H. Condra, W. A. Schleif, O. M. Blahy et al., *In Vivo*-Emergence of HIV-1 Variants Resistant to Multiple Protease Inhibitors, Nature, **374**, 569–571 (1995)

D. J. Kempf et al., Pharmacokinetic enhancement of inhibitors of the human immunodeficiency virus protease by coadministration with ritonavir, Antimicrob. Agents Chemother., **41**, 654–660 (1997)

P. Y. S. Lam, P. K. Jadhav, C. J. Eyermann et al., Rational Design of Potent, Bioavailable, Nonpeptide Cyclic Ureas as HIV Protease Inhibitors, Science, **263**, 380–384 (1994)

E. Specker, J. Boettcher, et al., Unexpected Novel Binding Mode of Pyrrolidine-based Aspartyl Protease Inhibitors: Design, Synthesis and Crystal Structure with HIV Protease, ChemMedChem, **1**, 106–117 (2006)

A. Blum, J. Böttcher et al., Structure-Guided Design of C_2-symmetric HIV-1 Protease Inhibitors Based on a Pyrrolidine Scaffold, J. Med. Chem., **51**, 2078–2087 (2008)

K. Ersmark, B. Samuelsson, A. Hallberg, Plasmepsins as Potential Targets for New Antimalarial Therapy, Med. Res. Rev., **26**, 626–666 (2006)

A. S. Nasamu et al., Plasmepsins IX and X are essential and druggable mediators of malaria parasite egress and invasion, Science, **358**, 518–522 (2017)

E. W. Baxter et al. 2-Amino-3,4-dihydroquinazolines as Inhibitors of BACE-1 (β-Site APP Cleaving Enzyme): Use of Structure Based Design to Convert a Micromolar Hit into a Nanomolar Lead, J. Med. Chem., **50**, 4261–4264 (2007)

M. E. Kennedy et al., The BACE-1 inhibitor verubecestat (MK-8931) reduces CNS β-amyloid in animal models and in Alzheimer's disease patients, Sci. Transl. Med. **8**, 363ra150 (2016)

Barber, J. (2018). Merck & Co. terminates Phase III study of verubecestat in prodromal Alzheimer's disease. Retrieved from https://www.merck.com/news/merck-announces-discontinuation-of-apecs-study-evaluating-verubecestat-mk-8931-for-the-treatment-of-people-with-prodromal-alzheimers-disease/ (Last accessed Nov. 21, 2024)

Inhibitors of Hydrolyzing Metalloenzymes

Contents

G. Klebe, *Drug Design*, https://doi.org/10.1007/978-3-662-68998-1_25

25

A **metal ion in the catalytic site** is required for the function of another important class of enzymes that cleave peptide and ester bonds. By coordinating the metal ion, these enzymes activate a water molecule for nucleophilic attack of the bond to be cleaved. The water molecule undergoes a drastic change in its pK_a value in this state. Zinc is by far the most commonly used metal ion in these enzymes, but iron, cadmium, cobalt or manganese are also found. The presence of a metal ion is essential for the activity of the protease or esterase. If the metal ion is removed from the enzyme by the addition of a strong complexing reagent, such as β-mercaptoethanol or ethylenediaminetetraacetic acid (EDTA), catalytic activity will no longer be observed.

Many therapeutically relevant enzymes are **metalloproteases**. The first to be mentioned are the zinc proteases, most notably angiotensin-converting enzyme (ACE). ACE inhibitors have been used for many years in the treatment of hypertension. In recent years, other metalloproteases have been identified as potential drug targets. These include endothelin-converting enzyme, neutral endopeptidases, and matrix metalloproteases (MMP; ◘ Table 25.1). Other groups of important zinc enzymes include carbonic anhydrases, zinc-containing β-lactamases, and phosphodiesterases. From the group of transferases, histone deacetylases follow the same mechanism to cleave an amide bond.

◘ **Table 25.1** Function and preferred cleavage sites of some metalloproteases

Enzyme	Cleavage site	Function
Thermolysin	X–Ala, X–Val, X–Ile	Bacterial protease
Carboxypeptidase	X–Tyr, X–Phe	Digestion
ACE	Phe–His, Phe–Leu, Pro–Phe	Transforms angiotensin I into angiotensin II, which increases blood pressure
NEP 24.11	Phe–Leu, Cys–Phe	Multifunctional (cleaves enkephalin, among others)
ECE	Trp–Val	Transforms big endothelin into endothelin, which increases blood pressure
Collagenase	Gly–Leu, Gly–Ile	Tissue remodeling
Stromelysin	Gly–Leu, Gly–Ile	Tissue remodeling

ACE angiotensin-converting enzyme, *NEP* neutral endopeptidase, *ECE* endothelin-converting enzyme

25.1 Structure of Zinc Metalloproteases

In 1967, William Lipscomb determined the 3D structure of the first zinc protease for the digestive enzyme **carbopeptidase A**. The zinc ion required for enzyme activity is complexed to two His and one Glu side chains. The fourth coordination site is occupied by a water molecule. An additional glutamate is located near the zinc ion. The same amino acids are responsible for zinc binding in many other metalloproteases. The presence of the amino acid sequence **His–Glu**–X–X–**His** (where X is any amino acid) is characteristic of most known zinc proteases. For example, it is found in collagenase, thermolysin, neutral endopeptidase 24.11, and endothelin-converting enzyme (◘ Table 25.2). The discovery of this amino acid sequence in the primary sequence of a new protein is a strong indication that it is a zinc protease. In metalloproteases or carbonic anhydrases, zinc is complexed by three histidine residues. Here, too, the fourth site is occupied by a water molecule.

In the body, zinc exists as a doubly positively charged cation, Zn^{2+}. This positive charge is used by the enzyme for amide cleavage. Ivano Bertini's group at the University of Florence, Italy, was able to determine the high-resolution structures of the uncomplexed and product-inhibited **metalloprotease MMP-12**. These structures allow the following mechanism to be deduced: In the uncomplexed metalloprotease, the zinc ion is octahedrally coordinated by three water molecules in addition to three amino acid residues (His or Glu). One of the water molecules forms an additional hydrogen bond to a neighboring glutamate. This residue, Glu 219 in MMPs, Glu 270 in carboxypeptidase, and Glu 143 in thermolysin, also polarizes the water molecule. Therefore, this water is likely to be present as an OH^- ion (◘ Fig. 25.1). The peptide substrate diffuses into the binding pocket and displaces the other two water molecules from the zinc ion. The remaining water molecule, polarized by glutamate, attacks the carbonyl group of the amide bond of the substrate to be cleaved. The substrate is held in place by hydrogen bonds to the peptide backbone at the *C*-terminal side. A geminal diol structure is formed at the reaction site, which is stabilized by the now pentacoordinated zinc ion. The actual cleavage of the amide bond is achieved and the two product molecules initially remain in the vicinity of the zinc. The glutamate residue presumably assumes the role of the proton transfer agent in this step. The cleavage product of the former *N*-terminus is coordinated to the zinc ion by an oxygen atom of the newly formed carboxylic acid function (◘ Fig. 25.2). However, it does not form any further hydrogen bonds to the protein. On the other hand, the cleavage product of the *C*-terminus forms four hydrogen bonds to the main chain of the enzyme, and its P_1' side-chain binds in the S_1' pocket. The newly formed free amino group initially remains in the vicinity of the zinc ion. It probably exists in an uncharged state next to the zinc ion. Then, the cleavage product originating from

Table 25.2 Characteristic amino acid sequences **His–Glu–X–X–His** in the active site of different metalloproteases

Enzyme	Position	Amino acids				
Thermolysin	142–146	**His**	**Glu**	Leu	Tyr	**His**
NEP 24.11	583–587	**His**	**Glu**	Ile	Thr	**His**
ECE	590–594	**His**	**Glu**	Leu	Thr	**His**
Astacin	92–96	**His**	**Glu**	Leu	Met	**His**
Collagenase	201–205	**His**	**Glu**	Phe	Gly	**His**
Stromelysin	201–205	**His**	**Glu**	Ile	Gly	**His**

NEP neutral endopeptidase, *ECE* endothelin-converting enzyme

the *N*-terminus leaves the catalytic site. It is presumably displaced by water, which takes its place next to the zinc ion. Finally, the *C*-terminal product leaves the binding pocket.

The three-dimensional structures of many zinc proteases have meanwhile been solved, including those of angiotensin-converting enzyme and many of the interesting matrix metalloproteases, such as collagenases, gelatinases, and stromelysin (Table 25.1). Four subfamilies of carbonic anhydrases are known. The therapeutically most important ones are the α-carbonic anhydrases, which fulfill important tasks in many organs and for which numerous drugs have been developed.

25.2 Key Step in the Design of Metalloprotease Inhibitors: Binding to the Zinc Ion

The zinc ion plays a key role in the catalytic mechanism. The known spatial structures of metalloprotease inhibitor complexes show that almost all highly potent inhibitors contain functional groups that bind directly to the zinc ion. If these groups are omitted, the binding affinity will decrease significantly in most cases. Therefore, the first step in the design of new inhibitors must be to search for functional groups that bind particularly well to the Zn^{2+} ion. Various groups have been described in the literature and are summarized in Fig. 25.3. Phosphonamides $-PO_2NH-$, phosphonates $-PO_2O-$, and phosphinates $-PO_2CH_2-$ can all be considered as transition state analogues of the enzyme reaction. In fact, a few potent metalloprotease inhibitors are known, such as the natural product phosphoramidon **25.1**, which contain such a group. The relative binding strength of various groups has been studied for carboxypeptidase A (Table 25.3). Similarly, various zinc-binding groups were tested for endothelin-converting enzyme. The results of these studies are shown in Table 25.4.

Remarkable variability has been observed in the binding potency of functional groups interacting with the Zn^{2+} ion. Attenuated partial charges on the zinc ion and on the anchor group are probably responsible for this effect. The zinc ion itself can be found in very different local environments (i.e., [3×His] or [2×His & 1×Glu] or [1×His & 1×Glu & 1×Cys]). Obviously, **thiol groups**, –SH, and **hydroxamic acids**, –CONHOH, are particularly well suited to contribute to a strong binding of the metalloprotease. The latter group binds as a bidentate ligand to the zinc ion. Carboxylic acids and ketones bind to the zinc ion more weakly than the above groups. Nevertheless, acids are of particular interest because acids in the form of esters are often used as orally available prodrugs (Sect. 9.2). In contrast to phosphinates and phosphonic acids, phosphonamides are chemically not very stable and are, therefore, not the first choice in the development of a new drug. On the other hand, sulfonamides are excellent zinc anchors, especially for carbonic anhydrases.

How might potential drug candidates for metalloproteases be designed? A comparison of known crystal structures (e.g., MMP-12, Fig. 25.2) shows that the

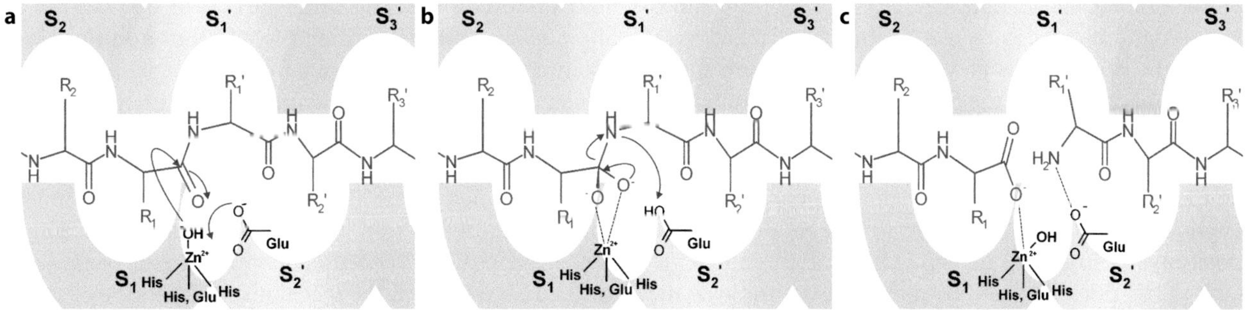

Fig. 25.1 Mechanism of peptide cleavage by a metalloprotease. The peptide substrate binds with its P_2, P_1, P'_1, and P'_2 residues in the corresponding specificity pockets of the protease. The amide group to be cleaved is found between the zinc ion and a water molecule (or OH^-), which is polarized by the acid group of the neighboring glutamate residue (**a**). This water molecule nucleophilically attacks the carbonyl carbon atom to form a tetrahedral transition state. The zinc ion is temporarily pentacoordinated and stabilizes the negative charge of the newly formed geminal diol structure (**b**). The transition state collapses with release of both cleavage products (**c**)

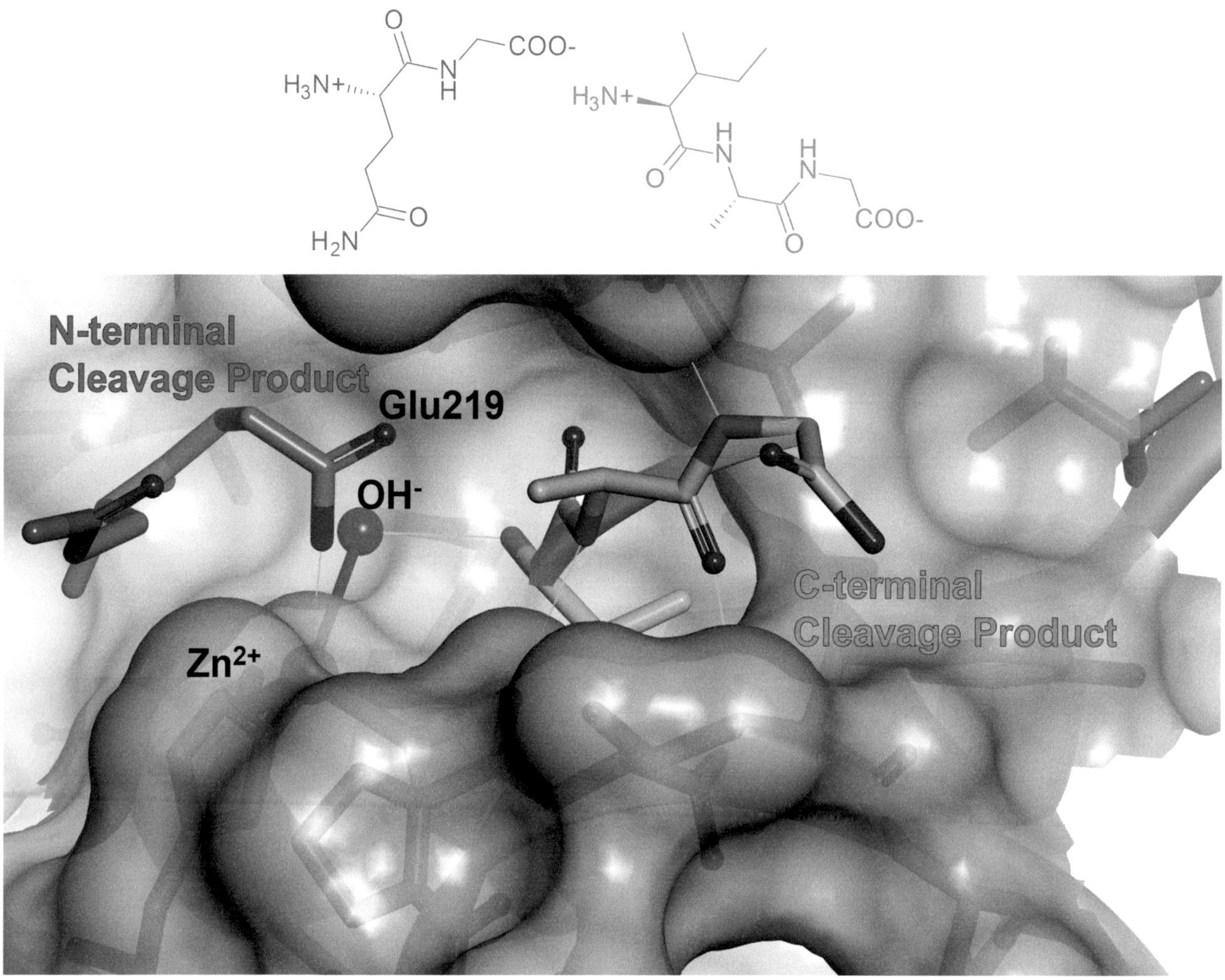

Fig. 25.2 A crystal structure of MMP-12 with both cleavage products has been determined (cf. Fig. 25.1c). The cleavage product of the former *N*-terminus (*left*, *light-red* carbon atoms) coordinates with its newly formed carboxylic acid function through an oxygen atom to the zinc ion, but it does not form hydrogen bonds to the enzyme itself. The cleavage product originating from the *C*-terminus (*right*, *light-green* carbon atoms) forms four H-bonds to the main chain of the protein and binds with its P_1' residue in the deeply formed S_1' pocket. The released amino group coordinates to Glu 219 and the water molecule that is bound simultaneously to the zinc ion. (► https://sn.pub/b8uUPw)

binding pockets in these proteins are much better defined on the primed side of the cleavage site. Therefore, inhibitor design must focus on the S_1' and its adjacent pockets on the primed side. Nevertheless, it has been shown that the occupancy of the S_1 and S_2 pockets can be very important to obtain inhibitors with adequate selectivity.

The importance of the S_1' pocket is well illustrated by the model protease thermolysin. The enzyme cleaves peptide chains before a Leu, Ile, or Phe residue at the P_1' position. By inserting the side chain of these residues into the S_1' pocket, the substrate binds to the enzyme. When analogous ligands are examined that carry instead of an aliphatic leucine side chain, the four-carbon shorter hydrogen atom of a glycine residue at P_1', the binding constant drops by a factor of 41,000! The enzyme achieves this dramatic affinity differentiation by placing the side chains in an S_1' pocket that is virtually anhydrous in the uncomplexed state. This saves the cost of desolvation of the pocket, and the binding of only four additional aliphatic carbon atoms results in this increase in affinity, which accounts for the essential part of the binding affinity (Sect. 4.6).

The choice of suitable groups to place in such a pocket is determined by the chemical composition of the pocket. In addition, the inhibitors must have a suitable head group to coordinate with the zinc ion, as described above. From the mechanism of peptide cleavage discussed in the previous section, it is easy to understand why the binding **pockets on the unprimed side** are less well established. After peptide cleavage, a peptide with a terminal acid function is formed on this side. Such a function is itself a good coordination anchor for the zinc

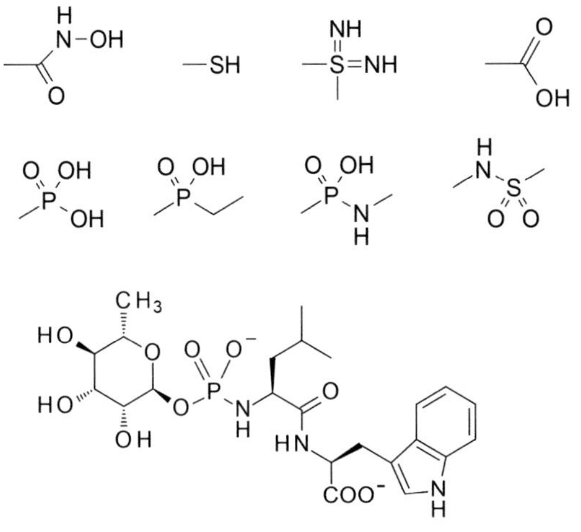

25.1 Phosphoramidon $K_i = 2.8 \times 10^{-8}$ M

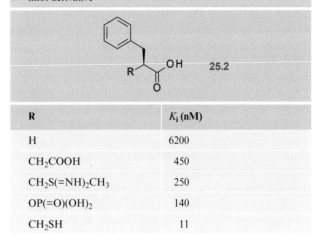

Fig. 25.3 Functional groups of metalloprotease inhibitors that are often used to bind to the zinc ion. Hydroxamic acids and thiols (*upper left*) in particular lead to highly potent inhibitors. A phosphoramide group is found in the natural product phosphoramidon **25.1** that the inhibitor uses to coordinate to the zinc ion. It inhibits thermolysin in the nanomolar range

ion. If the *N*-terminal end of the cleaved peptide were to be bound in a strongly pronounced pocket on the unprimed side, self-inhibition of the protease would result. Normally, this feature is not of interest. However, if the cleaved *N*-terminus has a low affinity for the protease, this type of inhibition will become important at high product concentrations. This may be a desirable **regulatory feedback mechanism** of Nature to temporarily slow down enzymatic turnover.

25.3 Thermolysin: Tailored Design of Enzyme Inhibitors

Thermolysin is a bacterial zinc protease of no therapeutic importance. Nevertheless, the 3D structures of thermolysin complexed with a large number of different inhibitors have been determined. The influence of many elementary factors on the strength of the protein–ligand interaction could be studied with this protease. Therefore, this enzyme is well suited for the study of 3D structure–activity relationships. Furthermore, its high stability makes it a robust object for experimental studies, and the 3D structure of thermolysin has been repeatedly used as a reference for model building of other related metalloproteases.

One of the central assumptions of structure-based drug design is the idea that the binding affinity of a ligand can be improved if the **receptor-bound conformation** can be embedded in a rigid scaffold (“**pre-organization**”). This working hypothesis was investigated in Paul Bartlett’s group using thermolysin inhibitors as an example.

Table 25.3 Binding of phenylpropionic acids **25.2** to carboxypeptidase. The strongest binding was found with the thiol derivative

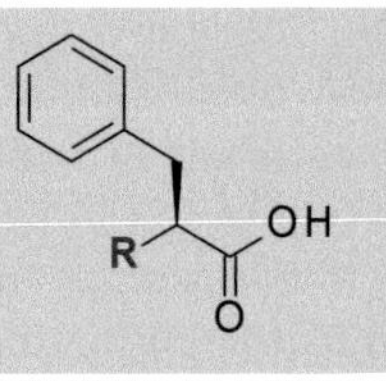

25.2

R	K_i (nM)
H	6200
CH_2COOH	450
$CH_2S(=NH)_2CH_3$	250
$OP(=O)(OH)_2$	140
CH_2SH	11

Table 25.4 Inhibition of the endothelin-converting enzyme by tryptophan derivatives **25.3**. The hydroxamic acid (R=CONHOH) as well as thiol compounds have much better affinity than the carboxylic acid derivatives

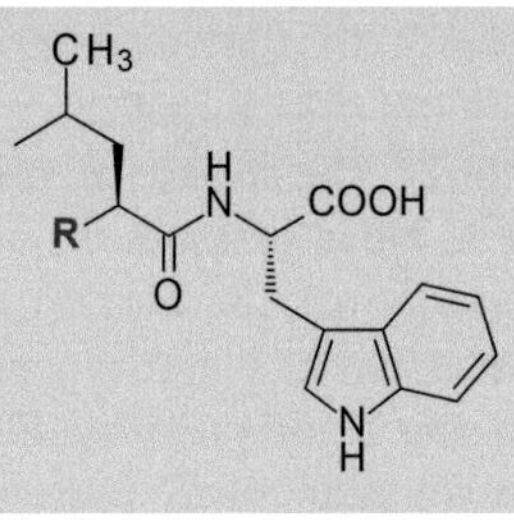

25.3

R	K_i (µM)
CONHOH	24
CH_2SH	12
COOH	> 100
CH_2COOH	> 100

The 3D structure of the complex of Cbz–Gly^P–Leu–Leu **25.4** (K_i = 9 nM, Fig. 25.4) complexed with thermolysin served as a starting point. The peptidic inhibitor binds in a conformation similar to a β-turn. Therefore, the design of a macrocyclic ligand that stabilizes this turn conformation seemed possible. The analysis of the 3D structure of this inhibitor with thermolysin revealed essential interactions. The Bartlett group then sought a rigid structural element to form a scaffold in which the conformation of the two leucine side chains remained unchanged. Chromane **25.5** (Fig. 25.4) was chosen. The additional methyl group on the ring had to be added for synthetic reasons.

A comparison of the binding constants of compounds **25.5** and **25.7** shows that the rigidification caused by the chromane group increased the binding affinity by a factor of 50. This corresponds to an energetic gain of about 10 kJ/mol. The X-ray structure analysis of the macrocyclic ligand **25.5** shows that it binds as expected. Both leucine side chains and the main chain atoms are found in the same position as in Cbz–Gly^{P}–Leu–Leu (**25.4**, ◘ Fig. 25.5). Certainly, the gain in binding energy is not only a result of the ligand becoming more rigid. The direct interaction of the chromane group with the enzyme also contributes to the affinity. The aim of the synthesis of **25.6** was to differentiate between the two effects of rigidification and the affinity gain from the chromane moiety. Compound **25.6** binds 20-times weaker to thermolysin than **25.5**. However, the 3D structure shows that the open-chain inhibitor binds to the enzyme in a different conformation. This is another example of how structures that are thought to be very similar do not necessarily bind in the same way!

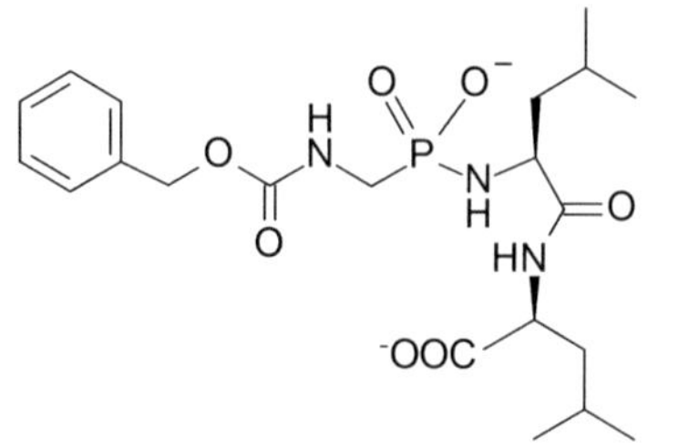

25.4 Cbz-Gly^{P}-Leu-Leu

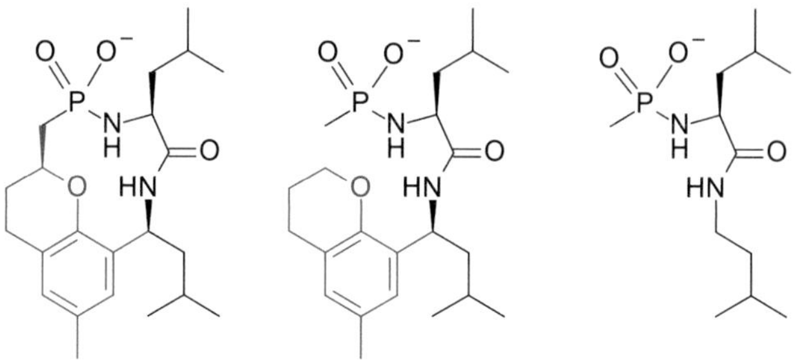

25.5 K_i = 4 nM

25.6 K_i = 80 nM

25.7 K_i = 190 nM

◘ **Fig. 25.4** The development of cyclic thermolysin inhibitors based on the open-chained inhibitor Cbz–Gly^{P}–Leu–Leu **25.4**. The cyclic inhibitor **25.5** binds 50-fold more strongly to thermolysin than the open-chain compound **25.7**. Compound **25.6** also contains the chromane scaffold, but the conformation is not enforced by a ring closure

25.4 Captopril, a Metalloprotease Inhibitor for Hypertension Therapy

Angiotensin-converting enzyme (**ACE**) converts the decapeptide angiotensin I to the octapeptide angiotensin II by cleaving off the *C*-terminal dipeptide His–Leu (◘ Fig. 24.5, Sect. 24.2). The release of this octapeptide leads to an **increase in blood pressure**. In addition, ACE catalyzes the degradation of the blood-pressure-lowering nonapeptide **bradykinin** to inactive peptides, thereby, also indirectly increasing blood pressure. This means that ACE inhibition can simultaneously prevent blood pressure increases by blocking multiple mechanisms. In 1965, Sergio Henrique Ferreira and John Robert Vane isolated a peptide mixture from the **venom of a snake**, *Bothrops jararaca* (the South American pit viper), that prolonged the blood-pressure-lowering effects of bradykinin by inhibiting a protease that degrades bradykinin in the body. This peptide (originally called **b**radykinin **p**otentiating **p**eptide, BPP) was also shown to inhibit the conversion of angiotensin I to angiotensin II. Several structurally related peptides were identified. The most active was the **teprotide** Pyr–Trp–Pro–Arg–Pro–Gln–Ile–Pro–Pro (Pyr = pyroglutamic acid). This nonapeptide was synthesized by Miguel Ondetti at Squibb. Teprotide is a potent ACE inhibitor with a binding constant of K_i = 100 nM. In clinical trials, the compound has been shown to be antihypertensive not only in animal models but also in humans. However, being a peptide, teprotide is not orally bioavailable and is, therefore, not suitable as a drug. Despite this observation, the studies demonstrated that an ACE inhibitor is an interesting compound for the **treatment of hypertension**. Further investigations showed that even dipeptides such as Val–Trp (K_i = 1.8 μM) and Ala–Pro (K_i = 230 μM) inhibit ACE, although more weakly than the nonapeptide teprotide.

The decisive breakthrough was the hypothesis of Miguel Ondetti and David Cushman that ACE was structurally similar to the metalloprotease **carboxypeptidase A**, which had been extensively studied. Lipscomb had shortly before determined the 3D structure of this enzyme. In addition, **benzylsuccinic acid** was known to be an extraordinarily potent inhibitor of carboxypeptidase A given its small molecular size (◘ Fig. 25.6). A binding mode was postulated for this molecule involving interactions with the enzyme that were also experienced by the two products of substrate hydrolysis (◘ Fig. 25.7). Ondetti and Cushman applied this concept to ACE. Whereas carboxypeptidase A cleaves off the last amino acid of a peptide, **ACE cleaves off a dipeptide**. This means that a succinic acid derivative substituted with an adequate additional amino acid should result in a potent ACE inhibitor (◘ Fig. 25.8).

Following the observation that proline as the *C*-terminal amino acid of peptidic ACE inhibitors yielded good results, carboxyalkanoylprolines were first evaluated as possible ACE inhibitors (◘ Fig. 25.9). **Succinoyl-L-proline** (**25.8**) was the first compound synthesized in this project at Squibb. As hoped, it proved to be an ACE inhibitor, but with an affinity only in the micromolar range (IC_{50} = 300 μM). Replacing the proline moiety with another amino acid did not improve binding: proline was already the optimal amino acid. Next, the length of the acid side chain was optimized. **Glutaryl-L-proline** (**25.9**) proved to be the best representative with a moderate improvement in binding (IC_{50} = 70 μM). The intro-

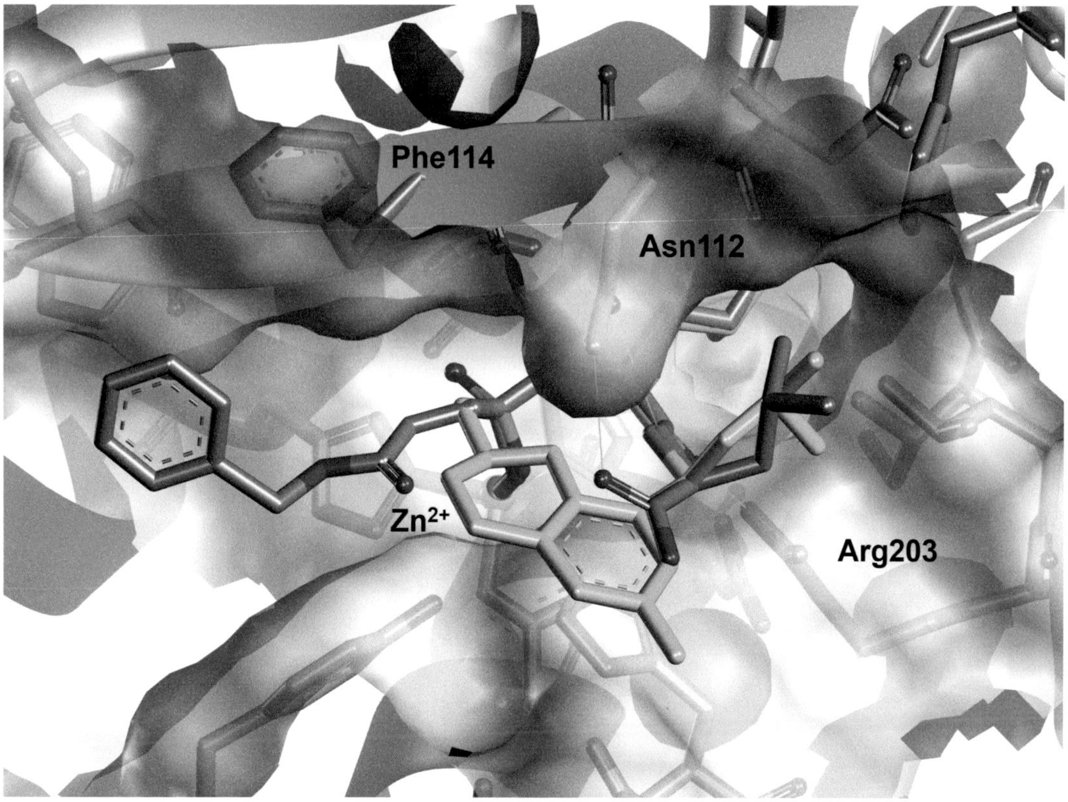

■ **Fig. 25.5** The 3D structure of a complex of thermolysin and Cbz–Gly^{P}–Leu–Leu **25.4** (*gray* carbon atoms). The leucine side chain (*right*) adjacent to a phosphate group occupies the deep S_1' pocket, which faces the interior of the protein; the second leucine residue is in the shallow S_2' pocket, which is open to the protein surface. The Cbz group is oriented in the S_1 pocket (*left*). The macrocyclic inhibitor **25.5** (*green* carbon atoms) with the chromane scaffold locks the conformation of **25.4** and analogously places the leucine side chains in S_1' and S_2'. Although this inhibitor leaves the S_1 pocket completely unoccupied, it binds to thermolysin more strongly than the open-chain compound **25.4**. (▸ https://sn.pub/fXOSRm)

duction of a methyl group in the side chain (**25.10** and **25.11**) resulted in a strong increase of the binding affinity by a factor of 15. Finally, replacing the **carboxylate** with a **thiol** group (**25.12** and **25.13**) provided the breakthrough with an order of magnitude increase in potency. The compound SQ 14225, **25.13** D-2-methyl-3-mercaptopropanoyl-L-proline binds to ACE with K_i = 1.7 nM and is orally available. SQ 14225 has been marketed for many years under the name **captopril** and has proven itself as an effective treatment for hypertension. Because lowering blood pressure significantly reduces the workload on the heart, captopril has also been used successfully in the treatment of congestive heart failure.

The compounds shown in ■ Fig. 25.10 demonstrate that both a free SH group and a free carboxylate group are necessary for the strong binding of captopril to ACE. Esterification of the carboxyl group in **25.14** or *S*-methylation in **25.15** leads to a dramatic loss of affinity, as does replacement of the amide group with a $–CH_2CH_2–$ group in **25.16** to **25.17**. Because of their susceptibility to oxidation, thiol groups are not very popular functional groups for drugs. Therefore, other anchor groups were sought.

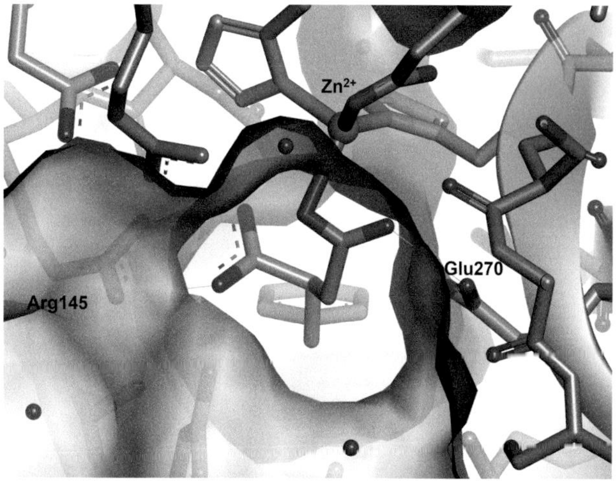

■ **Fig. 25.6** The crystal structure of the carboxypeptidase–benzylsuccinate complex. A carboxylate group binds to the zinc ion and the other forms a chelate-like salt bridge to the arginine side chain of Arg 145. The phenyl group fills a lipophilic pocket. (▸ https://sn.pub/yXJLB4)

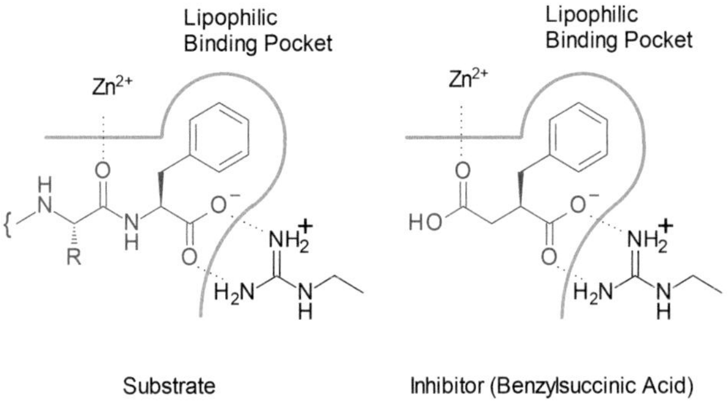

Fig. 25.7 Comparison of the binding mode of the inhibitor benzylsuccinic acid and the peptidic substrate to carboxypeptidase A. The inhibitor forms the same interaction to the enzyme as the substrate. The amide group to be cleaved was replaced by a carboxylate group

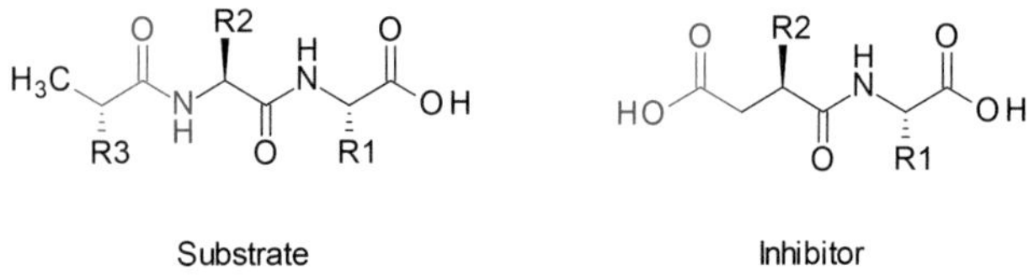

Fig. 25.8 The development of ACE inhibitors: a comparison of the substrates with the inhibitors that were investigated by Ondetti and Cushman. In the initially investigated structure, the amide bond to be cleaved is replaced by a carboxylate group

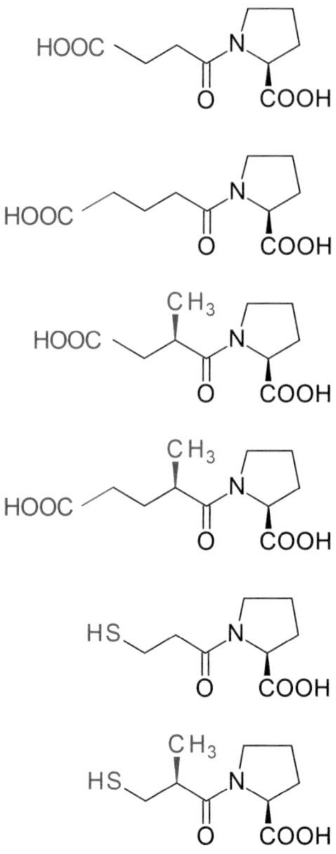

25.8 IC_{50} = 330 μM

25.9 IC_{50} = 70 μM

25.10 IC_{50} = 22 μM

25.11 IC_{50} = 4.9 μM

25.12 IC_{50} = 200 nM, K_i = 12 nM

25.13 Captopril, IC_{50} = 23 nM, K_i = 1.7 nM

Fig. 25.9 Binding of ACE inhibitors. The rationally designed lead structure **25.8** is optimized stepwise. The introduction of a methyl group in the side chain to give **25.10** as well as the replacement of the carboxylate group with a thiol are crucial for the increase in affinity. The result was captopril **25.13**

In the meantime, a whole range of effective ACE inhibitors have become available (Fig. 25.11); 17 products have found their way into clinical trials. Of particular note is enalapril **25.18** from Merck & Co. Like most other marketed products, with the exception of lisinopril, it is administered as a prodrug to increase oral availability (Sect. 9.2). As with other ethyl esters, it is rapidly converted in the body to its biologically active form, enalaprilat, the anion of the free acid. Interestingly, lisinopril **25.19** is sufficiently bioavailable even without esterification of the acid group. In addition to the acid function, the ligand also has a basic nitrogen that is protonated under physiological conditions. This makes lisinopril zwitterionic. At short distances, the two opposite charges cancel each other out so that the molecule has a more hydrophobic character with respect to the local environment (Sect. 19.4). Both enalapril and lisinopril have a much longer plasma half-life than captopril.

25.5 Finally the Crystal Structure of ACE: Does a Success Story Have to Be Rewritten?

In his seminal publication in 1977 on the design of captopril, David Cushman once again highlighted the importance of structural models for the work at Squibb:

> "The studies described above exemplify the great heuristic value of an active-site model in the design of inhibitors, even when such a model is a hypothetical one. Only when suitable information on substrate specificity and mechanism of action of an enzyme is available can one make a reasonable working hypothesis with regard to complementary functionality needed in an inhibitor."

Could he have dreamed that it would take another 25 years for this structure to be available? In 2003, Edward Sturrock's group in Cape Town, South Africa, completed the structure determination. Could it confirm the previously proposed model? Not all details of the predicted binding modes for the inhibitors were correct, but the structure provided critical insights that reinvigorated the field of ACE inhibitor research. The human enzyme is highly glycosylated. It consists of 1227 amino acids in an extracellular domain and is anchored to the cell membrane by 28 additional residues. Interestingly, it has two catalytic domains, a phenomenon that is rarely seen in enzymes and has its origin in gene duplication. The *N*-terminal domain contains 612 residues and the *C*-terminal domain 650 residues. The two domains are 60% identical. Both domains are catalytically active and

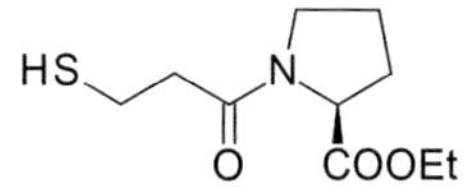

25.14 IC_{50} = 17 µM

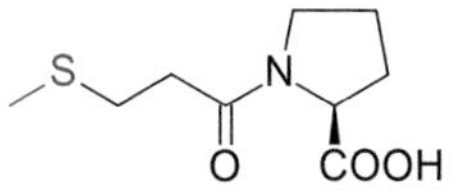

25.15 IC_{50} = 4300 µM

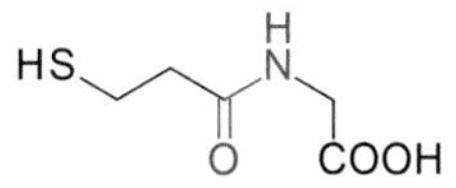

25.16 IC_{50} = 2.8 µM

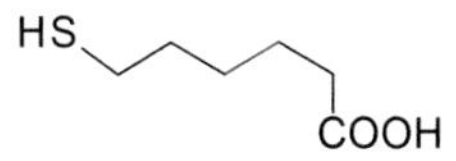

25.17 IC_{50} = 1100 µM

Fig. 25.10 A free thiol and carboxylate function are necessary for binding to ACE. Esterification of the acid group of **25.12** (Fig. 25.9) to give **25.14** reduced the binding affinity by almost two orders of magnitude. The *S*-methylation of **25.12** gives **25.15**, which has a binding affinity that is reduced by a factor of 20,000. Compound **25.17** contains merely the thiol and the carboxylate group. These two groups alone are just enough to achieve detectable binding

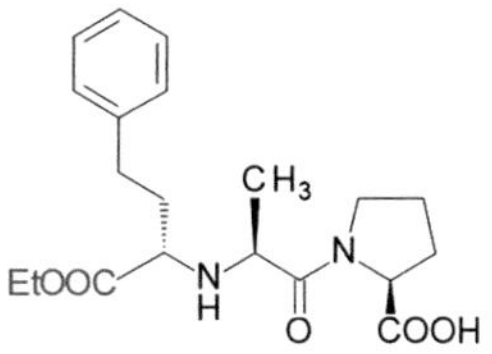

25.18 Enalapril

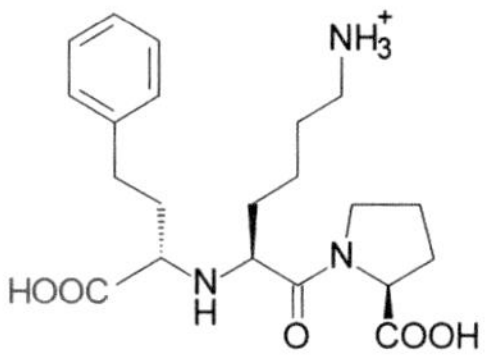

25.19 Lisinopril

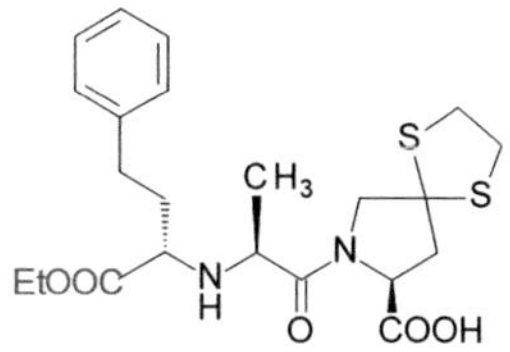

25.20 Spirapril

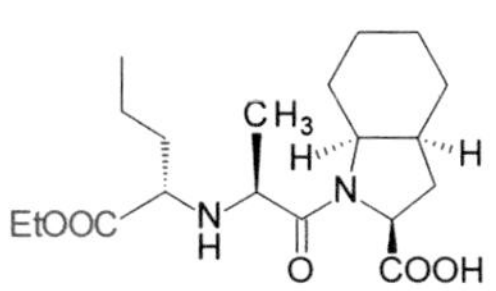

25.21 Perindopril

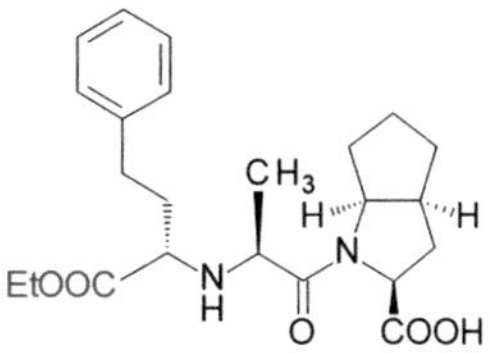

25.22 Ramipril

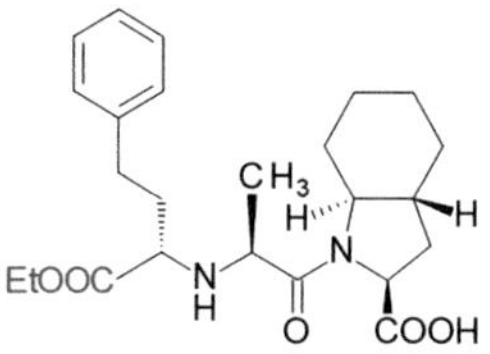

25.23 Trandrolapril

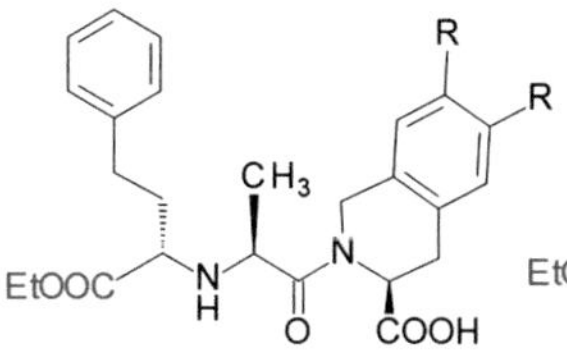

25.24 R=H Quinapril
25.25 R=OCH_3 Moexipril

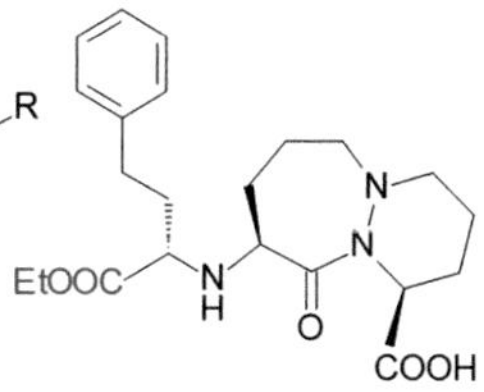

25.26 Cilazapril

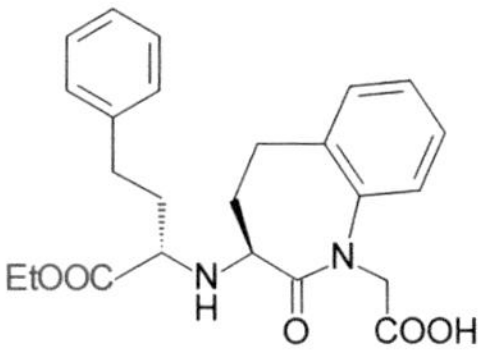

25.27 Benazepril

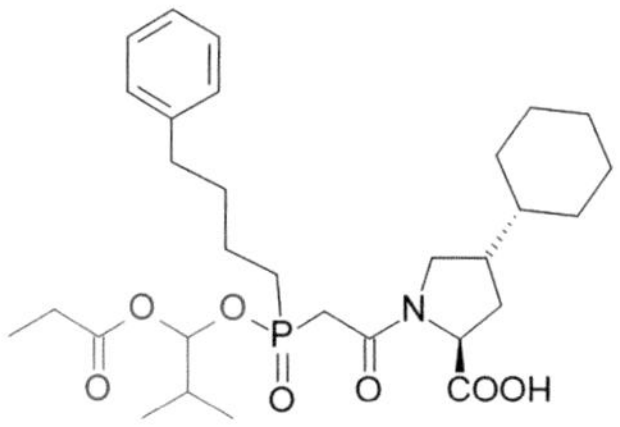

25.28 Fosinopril

Fig. 25.11 Examples of ACE inhibitors that are used in therapy

their catalytic sites differ by only a few amino acids. Nevertheless, a difference in selectivity for potential ligands is to be expected. In addition, the *C*-domain is highly dependent on the local chloride concentration, whereas the *N*-domain is much less dependent. In addition to this so-called somatic form (s-ACE), there is also a testis form (t-ACE), which is 701 amino acids long and consists of a single domain. Except for the first 36 residues, it is almost identical to the *C*-domain of the somatic form. The structure with bound lisinopril **25.19** (Fig. 25.12) was determined with this latter form. The inhibitor binds with its central acid group to the zinc ion. Its phenethyl group is located in the S_1 pocket. The lysine-like group is in the S_1' pocket and interacts with Glu 162. The proline moiety binds with its acidic group in S_2' to Lys 511 and Tyr 520.

It was then interesting to model the differences in the *N*- and *C*-domains of the s-ACE based on the t-ACE structure and to prepare the proteins by mutagenesis. Both domains bind lisinopril with very similar affinity, whereby the ligand leaves the S_2 pocket unoccupied (Table 25.5). The *N*-domain has in the S_1 and S_2 pocket Tyr 396, Asn 494, and Thr 496. In the *C*-domain, a phenylalanine, a serine, and a valine are found at these positions. In addition, there is an asparagine in this pocket in the *N* domain that limits access to the S_2 pocket by glycosylation. Therefore, it is not surprising that keto-ACE **25.29** with its bulky benzamido group interacts much better with the *C*-domain. Two other compounds are known, RXP 407 **25.30** and RXP A380 **25.31**, which bind to the two domains with a 1000-fold difference in selectivity. They are derived from phosphinic acids. Because the zinc-binding group is in the center of the molecule, these inhibitors can occupy all four pockets from S_2 to S_2' well. RXP A380 has a much larger moiety for the S_2 pocket. In addition, this molecule has an indole moiety in the P_2' position that can undergo stronger hydrophobic interactions in S_2'. At this position, the *C*-domain has an advantage over the *N*-domain: Instead of a serine, there is a hydrophobic valine at position 379, which results in stronger binding of the inhibitor to the *C*-domain.

What might be the benefit of a domain-specific inhibition of ACE? The enzyme not only converts angiotensin I to angiotensin II, but it also metabolically degrades bradykinin, which lowers blood pressure. ACE is also thought to be involved in the cleavage of other signal peptides. ACE inhibitors are generally well tolerated by patients. However, some adverse effects have been described. For example, many patients develop an

unpleasant dry cough, and occasionally life-threatening **angioedema** (acute swelling of the mucous membranes) can occur. This is thought to be related to the inhibition of the degradation of the described peptides, especially bradykinin. The catalytic activity of the *C*-domain appears to be responsible for blood pressure regulation under *in vivo* conditions, where angiotensin I is efficiently cleaved. Bradykinin, on the other hand, is cleaved equally well by both domains. By using compounds that are selective for the *C*-domain, it may be possible to lower blood pressure, while leaving a residual degradation of bradykinin intact. Excessive levels of this peptide could then be avoided. The structure determination of ACE, thus, opens up a new perspective for the development of selective inhibitors that allow efficient regulation of blood pressure according to an established principle. Hopefully, they will have fewer side effects.

25.6 Inhibitors of Matrix Metalloproteases: An Approach to Treat Cancer and Rheumatoid Arthritis?

Matrix metalloproteases (MMPs) are a family of neutral zinc endopeptidases. They play an important role in the formation and degradation of **connective tissue**, for example, after injury or during angiogenesis (the proliferation of blood vessels). In a healthy state, these proteases are kept in balance by tightly controlled mechanisms. In this way, active proteases are released from inactive precursors only when needed, or our body has sufficient endogenous inhibitors to mediate the balance between **matrix synthesis** and **matrix degradation**. In a disease state, this complex equilibrium is disrupted and various

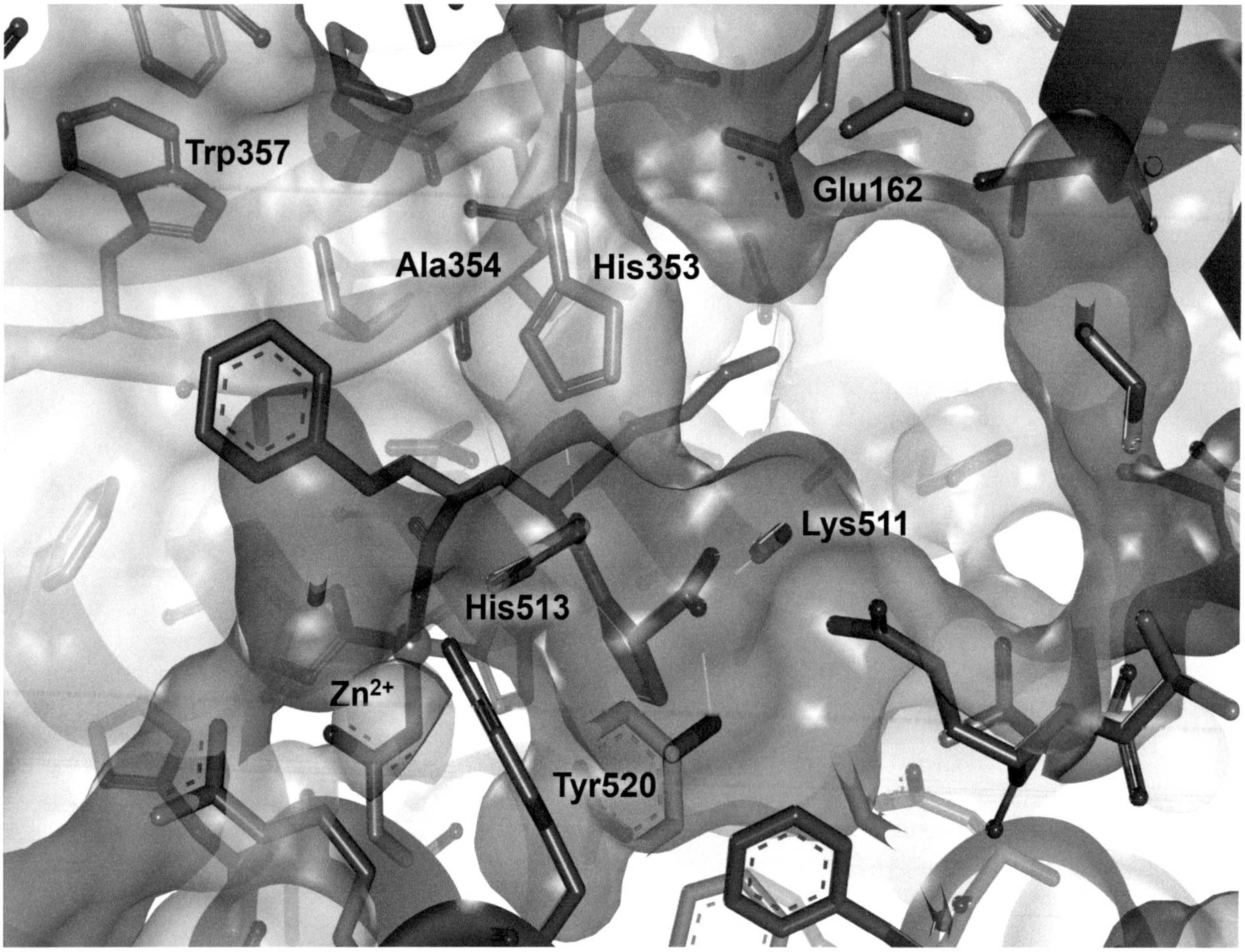

Fig. 25.12 Crystal structure of lisinopril **25.19** (Fig. 25.11) with t-ACE. The central carboxylate group of the inhibitor coordinates to the zinc ion. The NH group on the lysine residue of lisinopril forms an H-bond to the C=O group of Ala 354 in the S_1' pocket, and the carbonyl group also forms hydrogen bonds with His 353 and His 513. The terminal ammonium group of the lysine residue forms an H-bond to Glu 162. The acid group of the proline residue forms an H-bond contact with Lys 511 and Tyr 520. The phenethyl side chain is placed in the S_1 pocket. (▶ https://sn.pub/R8rq3h)

Table 25.5 Domain-specific inhibition of angiotensin-converting enzyme by structurally deviating compounds

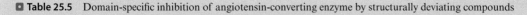

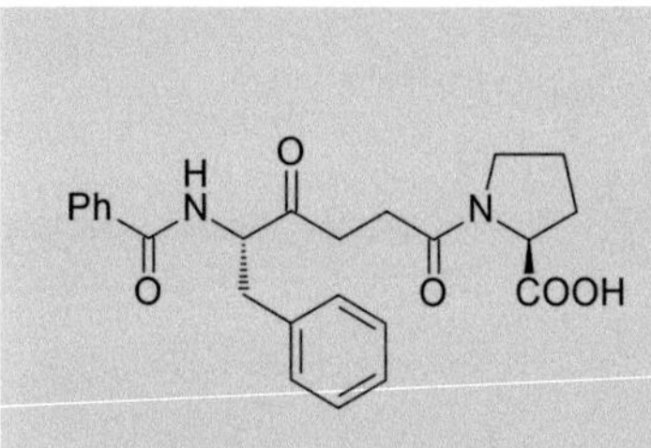

25.29 Keto-ACE

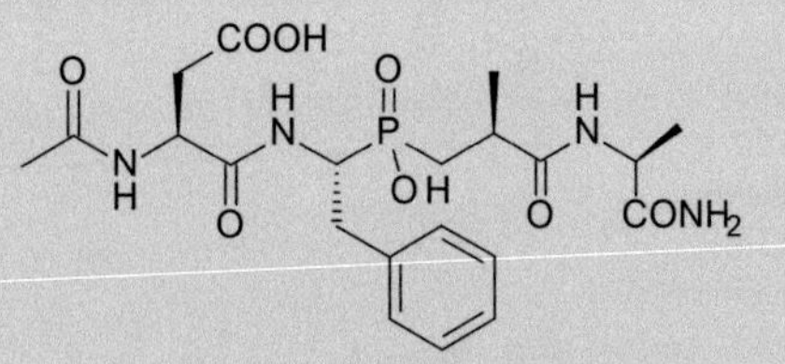

25.30 RXP407

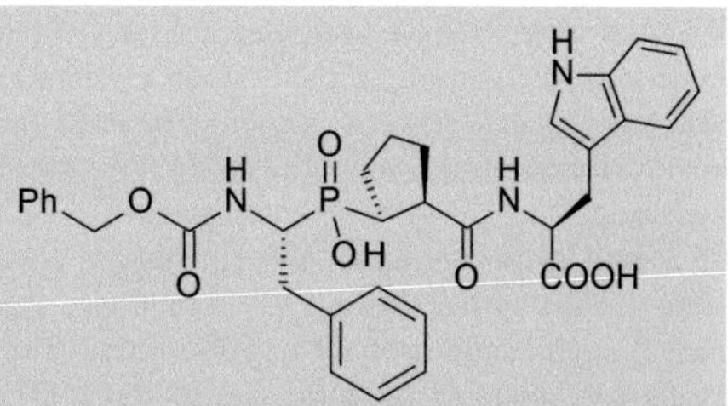

25.31 RXPA380

Compound		*N*-domain inhibition (nM)	*C*-domain inhibition (nM)
RXP A380	**25.31**	10,000	3.0
Captopril	**25.13**[a]	8.9	14.0
Enalapril	**25.18**[b]	26.0	6.3
RXP407	**25.30**	2.0	2500
Lisinopril	**25.19**[b]	44.0	2.4
Keto-ACE	**25.29**	15,000	40.0

[a] Fig. 25.9, [b] Fig. 25.11

MMPs are produced in excess. This leads to pathological situations associated with the construction and degradation of extracellular tissues.

The etiology of **rheumatoid arthritis** is based on such chronic destructive processes that lead to the loss of bone and cartilage. Cartilage tissue consists of a glycoprotein matrix that is cross-linked and reinforced by collagen. MMPs cleave these scaffold proteins. In rheumatoid arthritis, the balance between matrix synthesis and degradation appears to be lost. Excessive activity of matrix metalloproteases leads to excessive degradation of cartilage. Inhibition of these proteases may, therefore, be a promising approach to the treatment of rheumatoid arthritis. Degradation of the extracellular matrix is also critical for malignant tumor growth, tumor cell invasion, metastasis, and angiogenesis. Therefore, the inhibition of MMPs could also lead to cancer therapy.

Nearly 30 MMPs have been identified, including collagenases (MMP-1, -8, -13), gelatinases (MMP-2, -9), stromelysins (MMP-3, -10, -11), matrilysin (MMP-7), macrophage metalloelastases (MMP-12, -19), and enamelysin (MMP-20). The collagenases, gelatinases, and stromelysin recognize **collagen** as a substrate. Collagen is composed of three intertwined, left-handed α-helical chains. Each individual chain is more than 1000 amino acids long and contains the repeating sequence –(Gly–X–Y)$_n$–, where the X position is usually occupied by a proline or an alanine and the Y position by a hydroxyproline or an alanine. Collagenases cleave collagen in its native triple helical structure, gelatinases cleave collagen in a denatured form, and stromelysins are thought to cleave proteoglycans.

A number of different collagens are cleaved by collagenases between the glycine and leucine or isoleucine residues. A substrate comparison between human, bovine, mouse, and chicken showed that three amino acids to the right and left of the cleavage site are conserved. Therefore, the *N*- or *C*-terminal protected hexapeptide Ac–Pro–Leu/Gln–Gly–Leu/Ile–Leu/Ala–Gly–OEt, for example, **25.32** (Fig. 25.13), is recognized as a minimal substrate.

This established the starting point for the design of **collagenase inhibitors**. The peptide bond to be cleaved in the minimal substrate **25.32** is replaced with a non-cleavable isostere. The replacement of the amide bond between Gly and Leu with a ketomethylene group –$COCH_2$–, a hydroxymethylene group –$CH(OH)CH_2$–, or a hydroxylamine derivative led to inactive compounds in all cases. These groups are apparently unable to form a favorable interaction with the zinc ion. Finally, the use of a phosphinate group yielded a potent collagenase inhibitor **25.33**. However, if only the *N*-terminal proline is removed from this hexapeptide, the inhibitory activity will be largely lost. The search for collagenase inhibitors based on the *N*-terminal tripeptide fragment led to modestly active compounds such as **25.34**. The synthesis of potential inhibitors containing the *C*-terminal tripeptide sequence Leu–Leu–Gly–O-alkyl was much more successful. Coupling these structural elements with the potent hydroxamic acid head group to bind the zinc ion yielded collagenase inhibitors with nanomolar affinity such as Ro 31-4724, **25.35**, and Ro 31-9790, **25.36**. The X-ray structure of **25.35** complexed with human fibroblast collagenase was solved. As expected, the compound binds

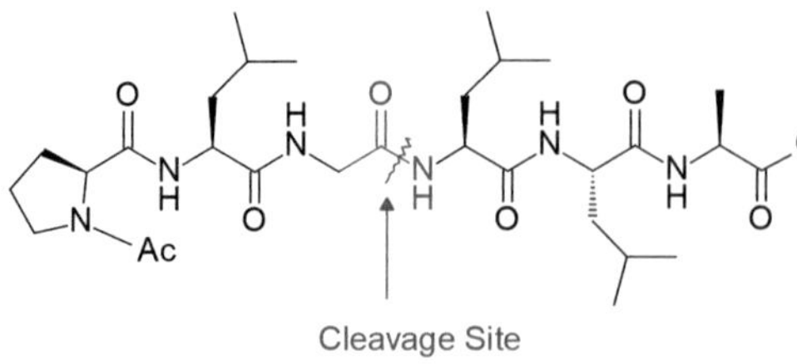

25.32 Minimal Substrate

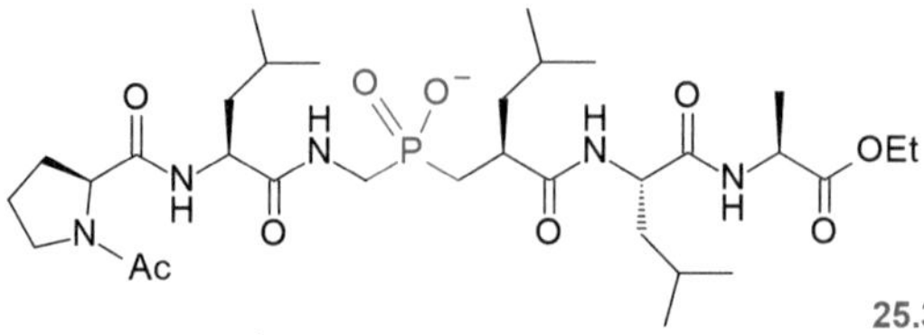

25.33 IC_{50} = 70 nM

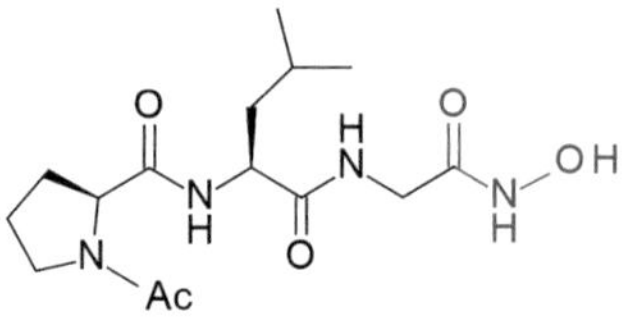

25.34 IC_{50} = 10 μM

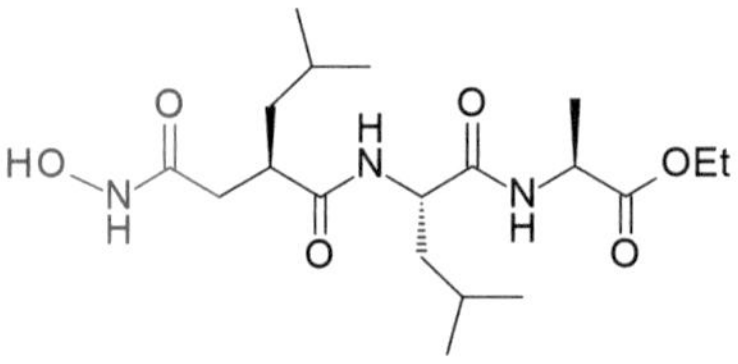

25.35 Ro 31-4724

IC_{50} = 9 nM

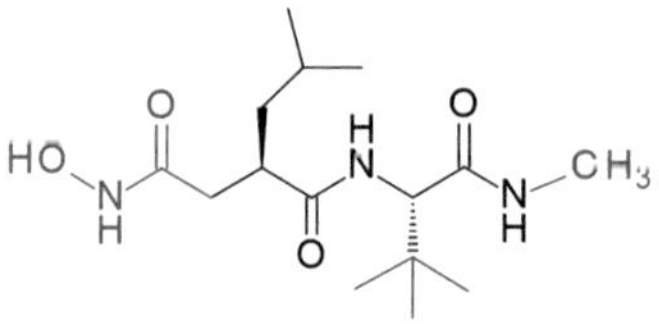

25.36 Ro 31-9790

IC_{50} = 5 nM

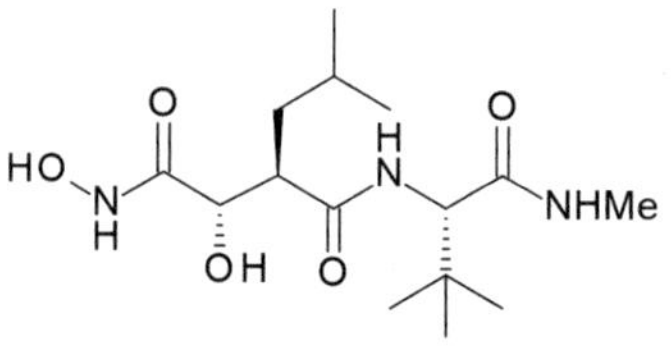

25.37 Marimastat

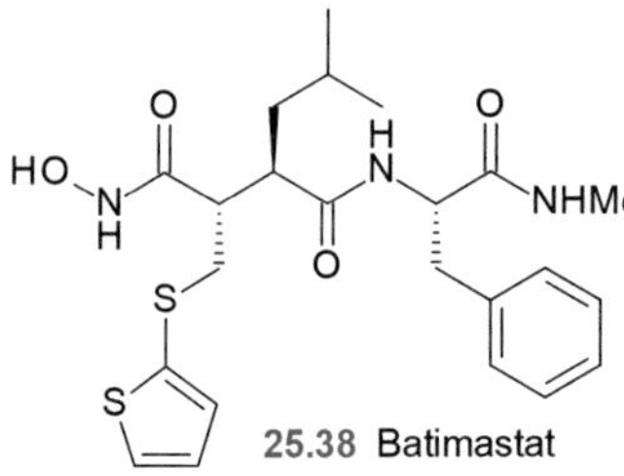

25.38 Batimastat

Fig. 25.13 Collagenase inhibitors prepared from substrate analogues. Compound **25.32** covers the substrate sequence from P_3 to P'_3. Replacement of the amide bond by a $-PO_2-$ group **25.33** leads to a potent inhibitor. Compound **25.34** contains only the three amino acids prior to the cleavage site as well as the *C*-terminal hydroxamic acid as a zinc-binding group. Compounds **25.35** and **25.36** contain the three or two amino acid side chains following the cleavage site in their structures; this time they are augmented with an *N*-terminal hydroxamic acid group. The two inhibitors marimastat **25.37** and batimastat **25.38** were in clinical trials for several years as compounds for treatment of cancer

to the zinc ion as a bidentate ligand. The leucine side chain in the P'_1 position fills the S'_1 pocket and the alanine methyl group binds in the S'_3 pocket. The leucine side chain at position P'_2, which should formally occupy the S'_2 pocket, is oriented away from the enzyme. The binding mode is shown in Fig. 25.14.

Interestingly, replacing the *iso*-butyl side chain at position P'_2 with a *tert*-butyl group in **25.36** resulted in an increase in affinity, even though the group is not in direct contact with the enzyme. This result has been attributed to conformational stabilization. The bulky *tert*-butyl group limits the mobility of the inhibitor so that the conformation adopted in the enzyme is still energetically favorable. Compound **25.36** showed some activity after oral administration in an animal model and was selected for clinical trials as a drug to treat arthritis. The structurally similar inhibitors marimastat **25.37** and batimastat **25.38** from British Biotech have been in development for many years as broad-spectrum MMP inhibitors for the treatment of cancer. Finally, in the field of matrix metalloproteases, many lead structures have been discovered and developed into potent inhibitors. Fig. 25.15 lists some of these substances (**25.39**–**25.48**), almost all of which are derived from hydroxamic acids. Hardly any peptidic character can be inferred from them. The only problem is that none of these compounds has made it through clinical trials to the market. The results of the clinical trials studies were rather sobering. Bayer's development product tanomastat **25.47** for the prevention of angiogenesis, tumor growth, and metastasis performed worse than a placebo sample. Novartis' CGS 27023A **25.48** did not fare much better.

What is the reason for the lack of success in these drug development projects so far? One of the reasons

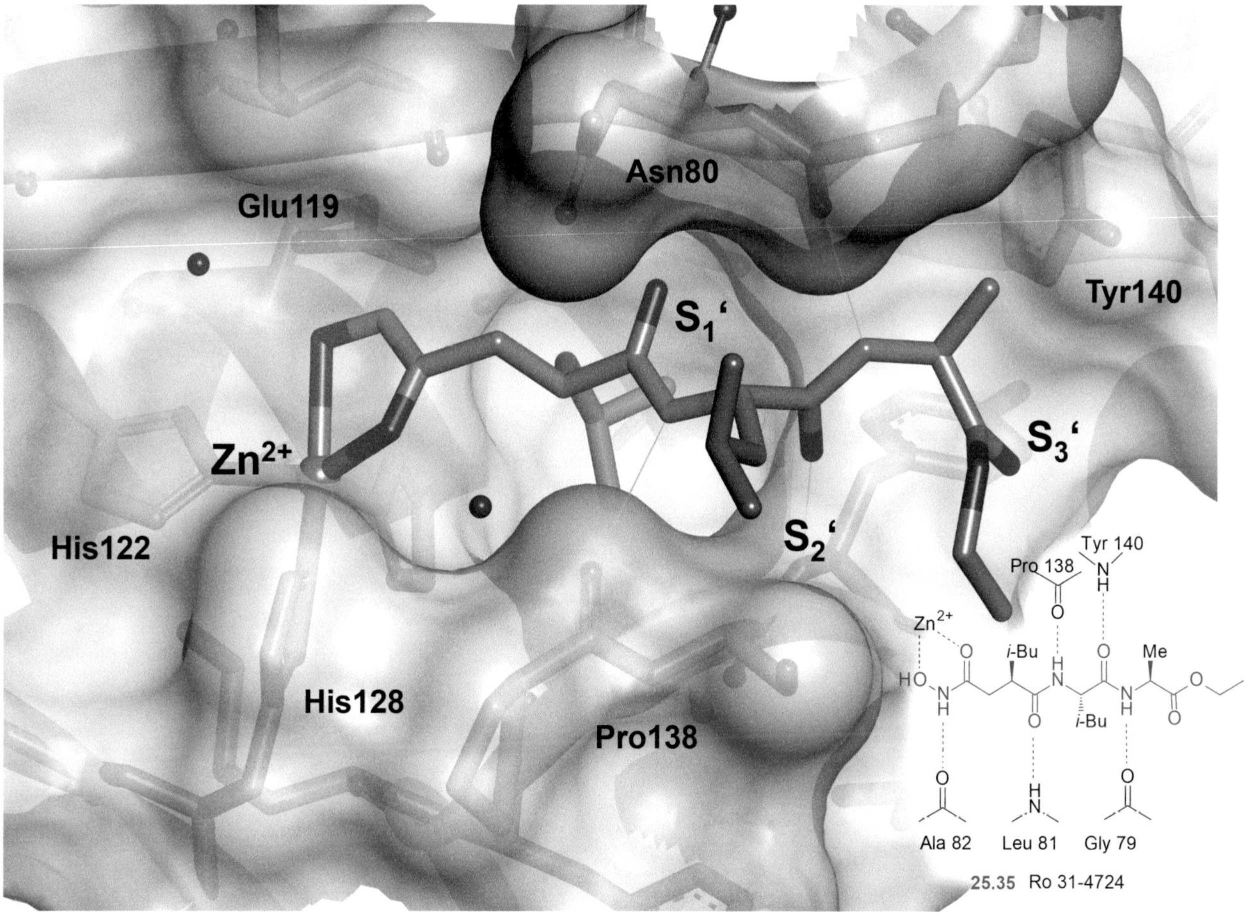

■ **Fig. 25.14** Crystal structure of collagenase with Ro 31-4724 (**25.35**, IC_{50} = 9 nM) shows the adopted binding mode. The hydroxamic acid binds in a bidentate-like manner to the zinc ion. Both amide groups form hydrogen bonds to the enzyme. The leucine side chain of the inhibitor in the P_1' position fills the S_1' pocket, which is oriented toward the protein's interior. The alanine methyl group binds in the S_3' pocket, whereas the leucine side chain in position P_2' protrudes into the solvent because the S_2' pocket is practically nonexistent. (► https://sn.pub/1LoZeU)

may be a **lack of selectivity** of the designed compounds. At the time these compounds were designed, only a few of the relevant MMPs had been identified. The different members of the MMP family are very similar to one another. **Overlapping substrate profiles** were observed. In some cases, another member of the family can **take over the task** of a protease that has been deactivated by inhibition. When comparing the proteases, it is striking that practically only the S_1' pocket is deeply buried. All other pockets S_3, S_2, S_1, S_2', S_3' are relatively shallow and easily accessible from the outside. In addition, proteins in the S_1' pocket have been shown to be highly adaptable to bound substrates and inhibitors. While this may provide an opportunity for the development of selective inhibitors, it does not generally facilitate drug development for such pockets. For the collagenase MMP-1, a conformational change at Arg 214 was shown to open a much larger S_1' pocket (■ Fig. 25.16). In the conformation determined with the first crystal structures, the available space was sufficient to accommodate a *sec*-butyl group as in **25.49**. However, after the rearrangement of the arginine, much longer biaryl ether residues (see **25.50**) can be accommodated in the S_1' pocket!

A further complication is that there is another family of zinc proteases, the **ADAM family (a disintegrin and metalloprotease**, or adamlysines), whose members share little sequence homology with MMPs but have catalytic centers that are very similar to MMPs. This family was discovered after the first MMP inhibitors were in clinical trials. **TNF-α converting enzyme (TACE)** is a member of this family. Blocking this enzyme affects the function of TNF-α, the proinflammatory cytokine that plays a central role in immune response (Sect. 29.8). The enzyme itself is being investigated as a target for drug therapy of autoimmune diseases. Cross-reactivity with MMP inhibitors is not desired. Unfortunately, it has also been shown that MMP inhibitors have no effect against advanced and late stage cancer. However, in the early days of MMP

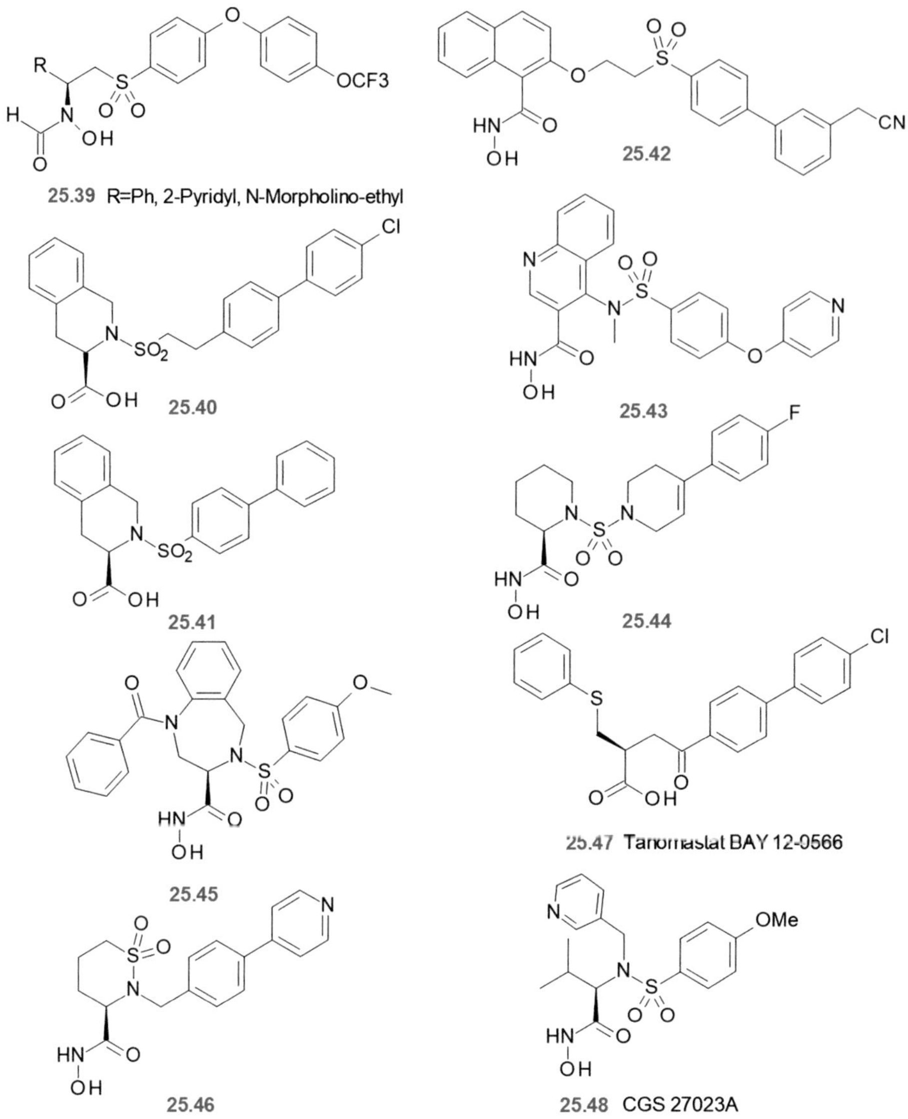

Fig. 25.15 Development candidates **25.39**–**25.48** from various companies as potent MMP isoenzyme inhibitors. Hydroxymates, inverse hydroxymates, and carboxylates were used as anchor groups for the zinc ion. Tanomastat **25.47** from Bayer and CGS 27023A **25.48** from Novartis were clinically developed for several years

research, models from an early stage of tumorigenesis were used.

Despite the availability of many potent inhibitors, drug development using members of the MMP family has not yet resulted in a successful marketed product. The problem of selectivity in this family seems to be too high an obstacle to make precise regulation of these enzymes by drugs a promising therapy.

25.7 Carbonic Anhydrases: Catalysts of a Simple but Essential Reaction

Another group of zinc-dependent enzymes that share a very similar catalytic mechanism with the zinc proteases are the carbonic anhydrases (CAs). They catalyze a very important reaction in our body, the **fixation of carbon dioxide from bicarbonate**, or the reverse reaction for the release of CO_2. In total, four different families of these enzymes are known, which are called the α-, β-, γ-, and δ-CAs. Sixteen isoforms of the α-CAs occur in mammals. Some are cytosolic and others are membrane-anchored. They are involved in many physiologically important processes such as respiration, CO_2/HCO_3^- transport between metabolizing tissues and the lungs, pH homeostasis, electrolyte secretion, biochemical reactions requiring C_1 building blocks, bone resorption and calcification, and tumor growth.

The zinc ion is located at the end of a funnel-shaped catalytic site in the **α-carbonic anhydrases**. It is held in place by three histidine residues. The fourth coordination site is occupied by a water molecule. It is strongly polarized by coordination to Zn^{2+}. Most likely, this water molecule is present as an OH^- ion. There is also a hydrogen bond acceptor group found in the OH group of Thr 199 (Fig. 25.17). The proton of this OH group at Thr 199 forms a hydrogen bond with the carboxylate group of Glu 106. The water (or OH^- ion), which has

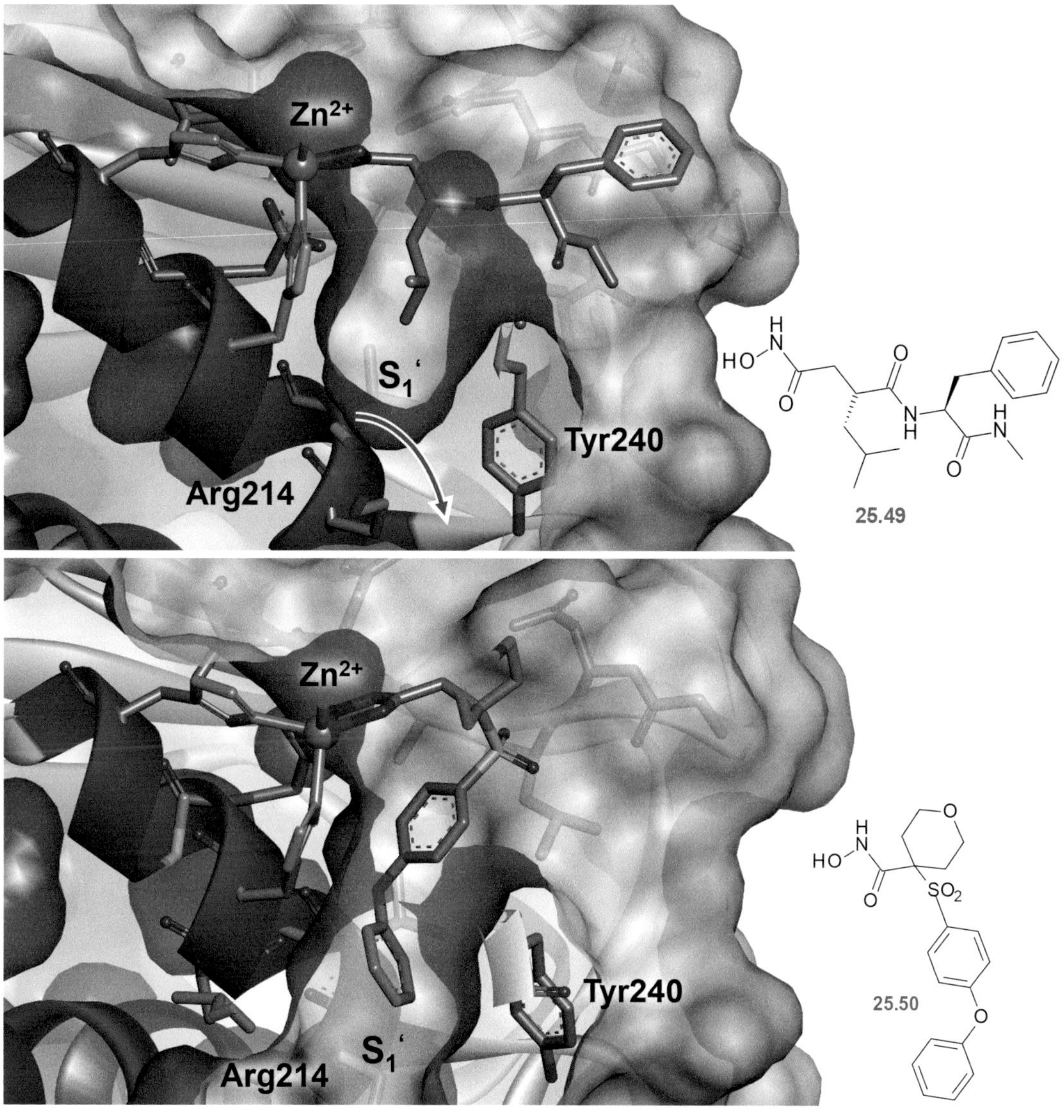

Fig. 25.16 Crystal structure of the collagenase MMP-1 with two different inhibitors **25.49** and **25.50**. Because of a conformational rearrangement of Arg 214, the S_1' pockets with their voluminous groups can be accommodated. This adaptive ability of the specificity pockets in MMPs makes the development of selective inhibitors extremely difficult. (► https://sn.pub/kLL5Ky)

greatly enhanced nucleophilicity, attacks a CO_2 molecule located in a hydrophobic niche near Val 121, Val 143, and Leu 198 at the bottom of the binding pocket. One of the oxygen atoms of the CO_2 finds a hydrogen bonding partner in the NH function of Thr 199. The newly formed bicarbonate is displaced from the temporarily pentacoordinated zinc ion, and a new water molecule takes its position at the zinc ion. A new catalytic cycle can start.

CAII is one of the fastest enzymes known. The acquisition or removal of a proton is the rate-limiting step in the reaction cycle. Carbonic anhydrases have a series of multiple histidine residues that deliver the protons from the edge of the funnel-shaped binding pocket. This arrangement also makes the funnel amphiphilic, meaning one side is hydrophobic and the other is hydrophilic. The very narrow area around the catalytic zinc ion provides only enough space for CO_2 and HCO_3^-. Putative inhibitors must be able to form equivalent interactions as the bicarbonate ion, while at the same time occupying the funnel opening. In addition to ions such as cyanide, thiocyanate, or isocyanate, especially sulfonamides, sulfamates, and sulfamides have the appropriate head group for coordination in the catalytic site. The amino group attached to these sulfur derivatives is acidic enough to easily release a proton and coordinate to the zinc ion in a charged state, analogously to the OH^- ion. The remaining proton interacts with the threonine OH group. An oxygen atom of the SO_2 function satisfies the NH function of the latter amino acid. The second S=O group expands the tetrahedral coordination state on zinc to pentavalency. An aromatic carbon that is part of a heterocyclic ring system is usually found at the fourth bond of the central sulfur atom of most known inhibitors. In some examples, there is another oxygen or nitrogen atom acting as a linker to this heterocycle.

In the case of carbonic anhydrases, the coordination of the ligands to the zinc ion in the catalytic site is essential for good binding. In this way, small ligands such as phenylsulfonamide **25.51** or its isostere thiophene-2-sulfonamide **25.52** achieve submicromolar inhibition of carbonic anhydrase II (◘ Fig. 25.18). More than 50 years ago, the replacement of these aromatic rings by other heterocycles led to the first marketed products, which were introduced into therapy as sulfonamides under the names **acetazolamide 25.53** and **methazolamide 25.54**. In 1954, acetazolamide represented the first mercury-free diuretic (Sect. 30.12). It has also been used as a systemic treatment for **glaucoma**. Glaucoma is an eye disease that causes visual field loss and, in severe cases, to blindness. It is caused by insufficient drainage of the aqueous humor from the eye. As a result, pressure builds up inside the eye, damaging the optical nerve if left untreated. Carbonic anhydrase II inhibitors reduce the production of aqueous humor and can reduce the pressure inside the eye.

Acetazolamide **25.53** and methazolamide **25.54** were used for many years to treat glaucoma. They must be administered systemically. Direct application in the form of eye drops does not work because the compounds cannot penetrate the eye from the outside. Systemic administration and low selectivity with respect to the different isoforms of carbonic anhydrase means that these enzymes are also inhibited outside of the eye. Unwanted side effects are the consequence. As a result, both of these compounds have largely disappeared from therapy today.

For a long time, it was assumed that carbonic anhydrase inhibitors could not be used as eye drops due to of their unfavorable physicochemical properties. In 1983, to general surprise, the **topically active carbonic anhydrase inhibitor 25.55** was reported for the first time. The single exchange of a methyl for a trifluoromethyl group caused this transformation! As a consequence of this discovery, the lipophilic range of a large number of carbonic anhydrase inhibitors was characterized, within which topical application is possible.

This led to the development of dorzolamide **25.57**. In fact, its design is the first example of a drug optimized by structure-based design with the support of *ab initio* calculations along with crystal structure determinations. After the X-ray structure of carbonic anhydrase II became available, structure-based design of carbonic anhydrase inhibitors began at Merck Sharp & Dohme in the

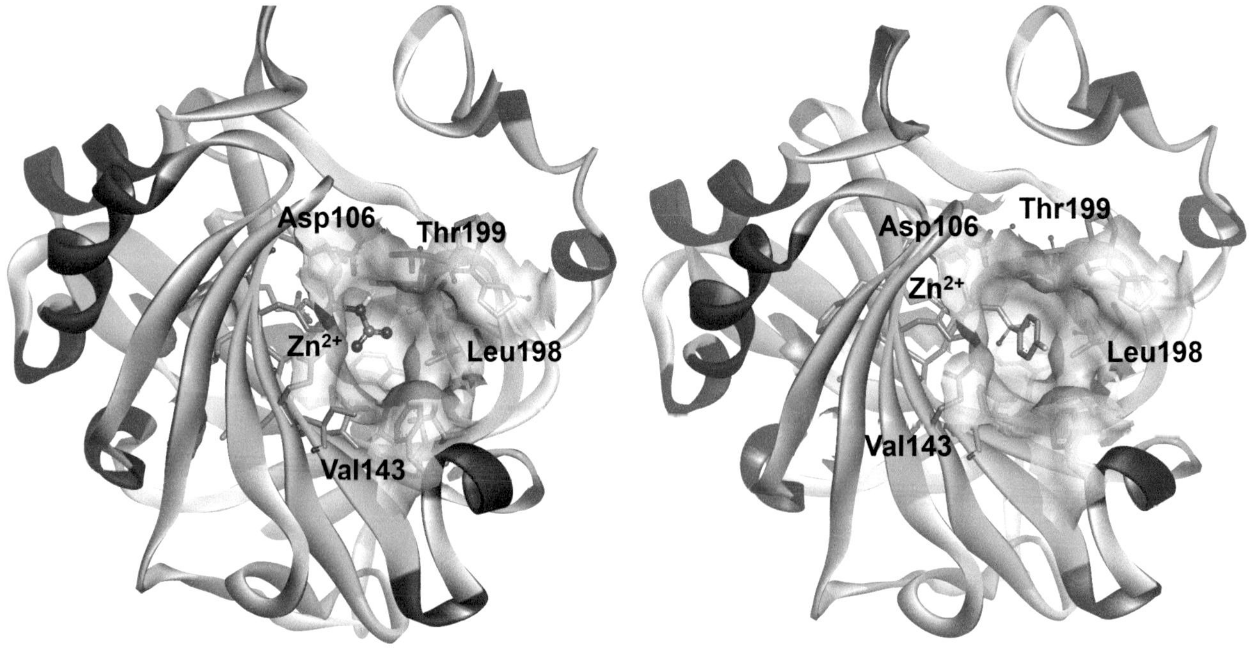

◘ **Fig. 25.17** The catalytic site in α-carbonic anhydrases is found at the end of a funnel-shaped binding pocket. There an OH^- ion which is coordinated to the Zn^{2+} ion nucleophilically attacks a CO_2 molecule. A bicarbonate ion is formed, which is held in place by Thr 199 (*left*). A sulfonamide, deprotonated at nitrogen, fits at the site of the carbonate in the very narrow binding pocket (*right*). Because of the tetravalency of the sulfur, this site can be fitted with another substituent, as is shown in the present case with a *p*-fluorophenyl group. (► https://sn.pub/mEDTzP)

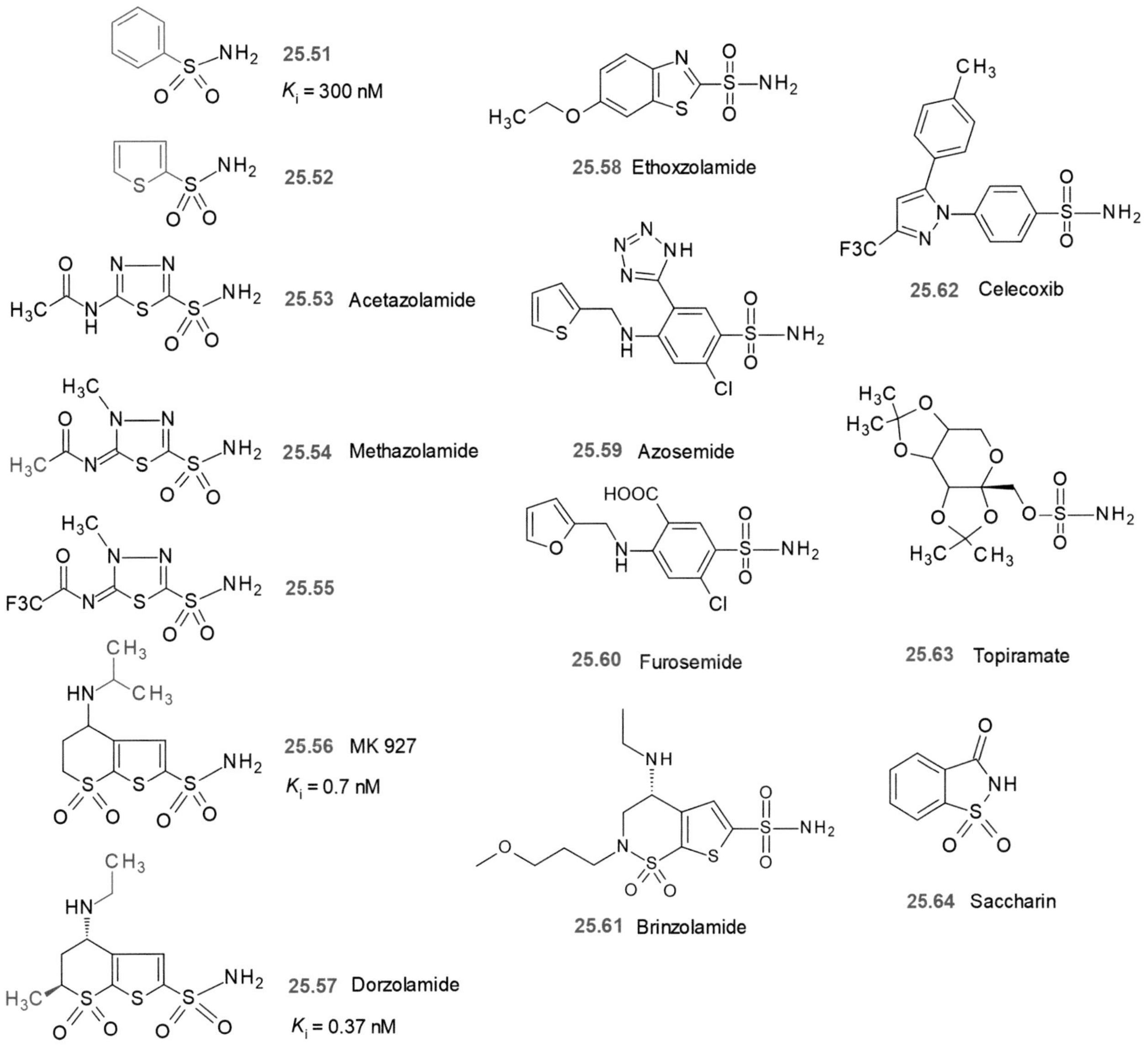

Fig. 25.18 The small aromatic sulfonamides **25.51** and **25.52** bind to carbonic anhydrase II with submicromolar affinity. By exchanging a heterocycle, acetazolamide **25.53** and methazolamide **25.54** are obtained. Both drugs were used for a long time as systemic carbonic anhydrase inhibitors for diuresis and for the treatment of glaucoma. Compound **25.55** was the first topically active CA inhibitor that is useable as eye drops. The structure-based design of new inhibitors led to the marketed product dorzolamide **25.57** by way of **25.56**. Compounds **25.58**–**25.61** are further drugs that inhibit carbonic anhydrases and are used for the treatment of glaucoma or as a diuretic. Even celecoxib **25.62**, topiramate **25.63**, and the artificial sweetener saccharin **25.64** inhibit carbonic anhydrases and this explains some of their observed side effects

mid-1980s. One of the first compounds to emerge from this effort was thienothiopyranosulfonamide **25.56** (MK 927). It binds to carbonic anhydrase with a subnanomolar inhibition constant (K_i = 0.7 nM). The crystal structure with the enzyme shows the expected coordination of the sulfonamide group to the zinc ion in the active site. In addition to hydrogen bonding, the inhibitor forms hydrophobic interactions with the protein. The observation that the *iso*-propylamino group occupies an energetically unfavorable position with an **axial orientation on the ring** was a surprise. Apparently, the compound fits better into the binding pocket in this unfavorable conformation. To enhance the affinity to the enzyme, a modification of the molecule was planned to **decrease the energetic penalty between equatorial and axial orientation** of the side chain. This was achieved by stereospecifically adding another methyl group to the six-membered ring. To compensate for the increased lipophilicity, the *iso*-propylamino group was reduced to an ethyl group. The result of this modeling was dorzolamide **25.57**. It binds to carbonic anhydrase II with K_i = 0.37 nM. Dorzolamide has successfully completed all clinical trials. It has been marketed under the name Trusopt® since 1995 and was the first topically active carbonic anhydrase inhibitor marketed for the treatment of glaucoma. Some other important drugs (**25.58–25.61**) that inhibit carbonic anhydrase are shown in ◘ Fig. 25.18. They serve as diuretics, glaucoma inhibitors, antiepileptics, and as treatments for altitude sickness, for peptic ulcer disease, or for ankylosing spondylitis (also known as Bechterew's disease, a chronic autoimmune inflammatory disease that leads to spinal fusion). Since tumors require an acidic environment, carbonic anhydrases such as CA IX and CA XII may be responsible for maintaining these conditions. Therefore, they are potential targets for cancer treatment because inhibition of CA would disrupt acid homeostasis. The 16 human α-carbonic anhydrase isoenzymes characterized to date are highly homologous. Small differences, such as the exchange of a threonine for a histidine at position 200, distinguish the CA I and CA II isoforms. Drugs must exploit these differences to achieve the desired selectivity between these isoforms (Sect. 18.14).

In the meantime, some very surprising adverse effects of known drugs can be attributed to carbonic anhydrases. To improve solubility, **terminal sulfonamide groups** have often been incorporated into drug candidates as functional groups. The analgesic celecoxib **25.62** is a cyclooxygenase II inhibitor (Sect. 27.9). It can also bind to carbonic anhydrase with nanomolar affinity through its sulfonamide group. In patients with familial adenomatous polyposis (FAP), a disease that leads to the development of polyps in the colon, a reduction in the number of tumors has been clinically observed in patients treated with celecoxib. This result may be consistent with carbonic anhydrase inhibition. One of the side effects of the antiepileptic drug topiramate **25.63** is loss of appetite. As a sulfamate, this compound is a potent mitochondrial CA V inhibitor. This isoenzyme is involved in *de novo* lipogenesis. This observation led to a thorough investigation of CA V as a possible therapeutic principle for obesity therapy. Even the very old and widely used artificial sweetener saccharin **25.64**, which contains a cyclic sulfonamide unit, can inhibit some carbonic anhydrases very strongly. Other clinically used carbonic anhydrase inhibitors, like saccharin, are known to have an unpleasant metallic aftertaste. This property is thought to be due to the inhibition of CA VI, which is produced in the oral cavity. Its inhibition affects the pH and may cause the bitter metallic taste sensation. Presumably, other drugs with terminal sulfonamide groups also have effects on carbonic anhydrases. Only time will tell whether the major problem of achieving sufficient selectivity within this class of enzymes can be solved.

25.8 A Case for Two: Zinc and Magnesium in the Catalytic Centers of Phosphodiesterases

Phosphodiesterases (PDEs) are a class of metalloenzymes with at least 12 gene families that hydrolyze the intracellular second messengers cAMP **25.65** and cGMP **25.67** (cyclic AMP and GMP) to their open-chain analogues (◘ Fig. 25.19). They are widely distributed in various tissues and organs and control important processes in the regulation of calcium channels, sense of smell, platelet aggregation, aldosterone release, cell proliferation, myocardial contractility, insulin release, inflammation modulation, smooth muscle contraction, mood, penile erectile function, or muscle metabolism. Among family members, the sequences are highly conserved.

First, the crystal structures of **PDE 4** and **PDE 5** were solved. To date, eight PDEs have been crystallographically characterized. While inhibition of PDE 4 may lead to the treatment of asthma, chronic obstructive pulmonary disease, or autoimmune diseases, PDE 5 inhibitors have been developed for the treatment of erectile dysfunction. The PDE 5 enzyme is expressed in various tissues and is specific for the hydrolysis of cGMP. In addition to the **zinc ion**, which is **essential for hydrolytic cleavage**, there is an additional magnesium ion found in the active site. The zinc ion is coordinated by two histidine and two aspartic acid residues. At the fifth position, there is a water molecule that, together with one of the two aspartic acid residues, forms a bridge to the magnesium ion (◘ Fig. 25.20). The other coordination site of the octahedrally surrounded Mg^{2+} is occupied by a water molecule. Zn^{2+} also prefers an octahedral geometry in phosphodiesterases. The sixth coordination position is occupied by a water molecule. This water molecule presumably takes over the role of the nucleophilic OH^- for the hydrolytic cleavage of the cyclic phosphodiester.

Three PDE 5 inhibitors were brought to the market as drugs to treat erectile dysfunction. Aside from sildenafil **25.69** (Viagra®), the first to be introduced by Pfizer, vardenafil **25.70** (Levitra®) and tadalafil **25.71** (Cialis®) have passed clinical trials (◘ Fig. 25.21). Interestingly, these inhibitors bind to the catalytic site of PDE 5, but do not make direct contact with the zinc ion (◘ Fig. 25.20). In fact, the binding of the basic nitrogen to the metal ion is mediated by two water molecules. The pyrazolopyrimidinone moiety in sildenafil replaces the analogous group in the natural substrate cGMP. The relationship with cGMP is even more obvious when it is considered that the 2-phenyl-substituted purines such as **25.72** served as lead structures (◘ Fig. 25.21). The pyrazolopyrimidine **25.73** or imidazotriazenone **25.74** that are contained in sildenafil and vardenafil, respectively, were developed from them. The chemically closely related vardenafil adopts a very similar binding mode as sildenafil. The structurally different tadalafil, on the other hand, adopts a distinctly different orientation.

The discovery of the effects of sildenafil was achieved once again by serendipity. The compound was in clinical trials at Pfizer for the treatment of **angina pectoris**. However, it proved to be no better than the **classic nitro compounds** (i.e., nitroglycerin or isosorbide dinitrate). These nitro derivatives release NO under reductive con-

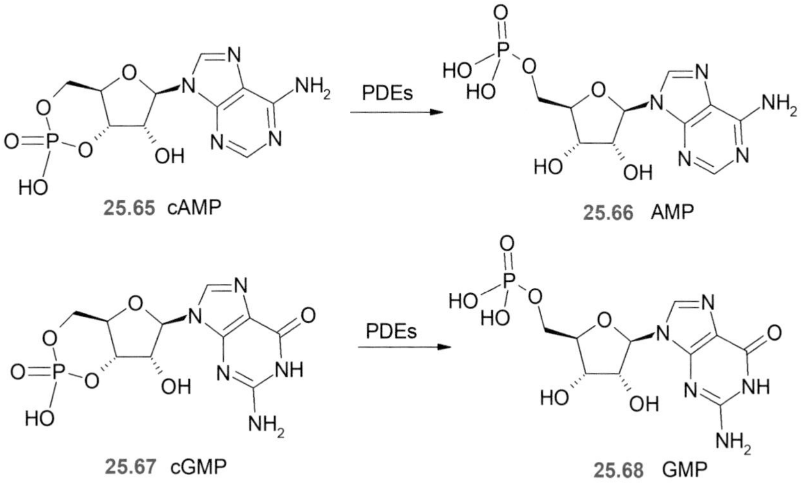

■ **Fig. 25.19** cAMP **25.65** and cGMP **25.67** are hydrolyzed into their open-chain analogues AMP **25.66** and GMP **25.68**, respectively, by phosphodiesterases

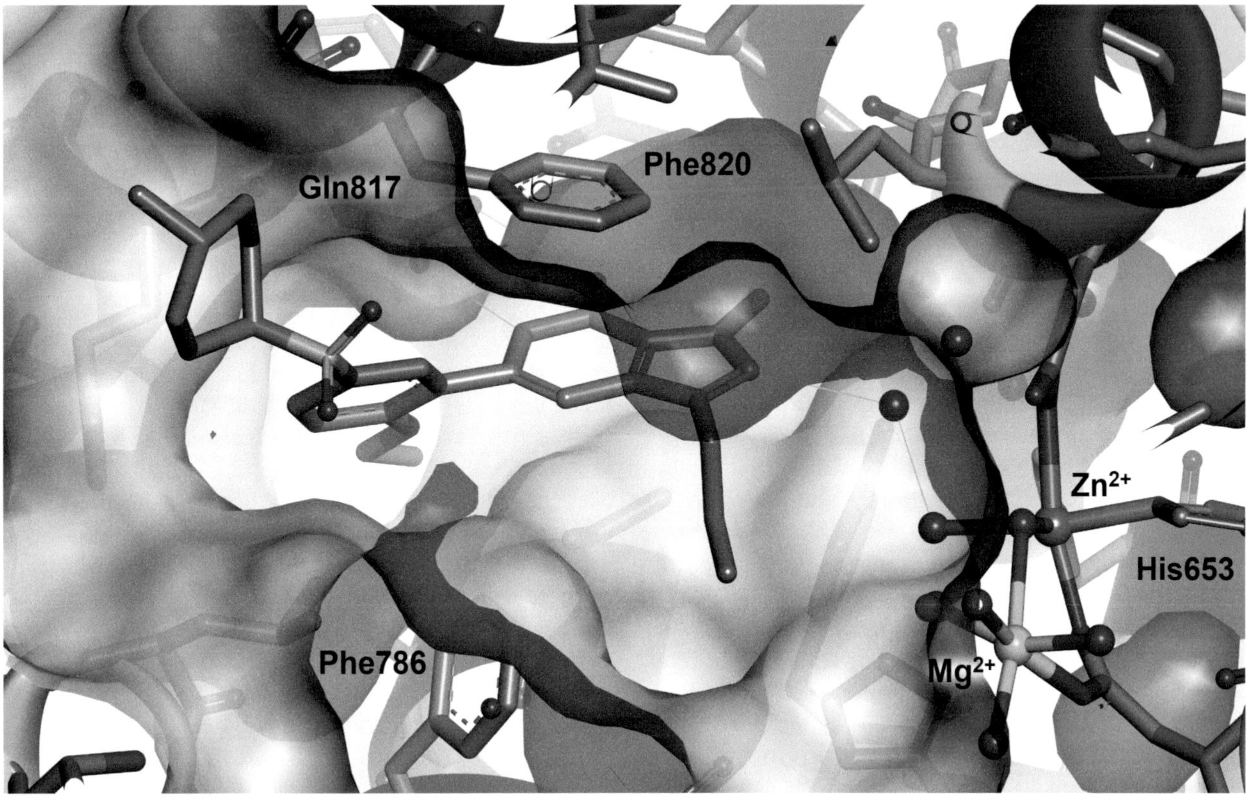

■ **Fig. 25.20** Crystal structure of sildenafil **25.69** (■ Fig. 25.21) in PDE 5. The pyrazolopyrimidinone moiety of the inhibitor is recognized by Gln 817 through two parallel hydrogen bonds and binds to the catalytic zinc ion (*blue-gray*) via a water molecule. It is found in the vicinity of a magnesium ion (*light green*), which is coordinated by five water molecules and Asp 654. A bridging water molecule is shared by Mg^{2+} and Zn^{2+}. (► https://sn.pub/mLzaFK)

ditions, which stimulates guanylate cyclase. cGMP is then formed, which in turn influences vasoconstriction. A **phosphodiesterase inhibitor** also increases cGMP levels by blocking the degradation of this second messenger. In clinical trials, however, one side effect in male probands was noteworthy: it stimulated **penile erections**. NO is released into the cavernous body of the penis and increased cGMP is produced by activation of guanylyl cyclase. This causes increased blood flow to the cavernous body and stimulates penile erection. Sildenafil enhances the effect by inhibiting the degradation of cGMP. Sildenafil was approved for the treatment of erectile dysfunction in 1998.

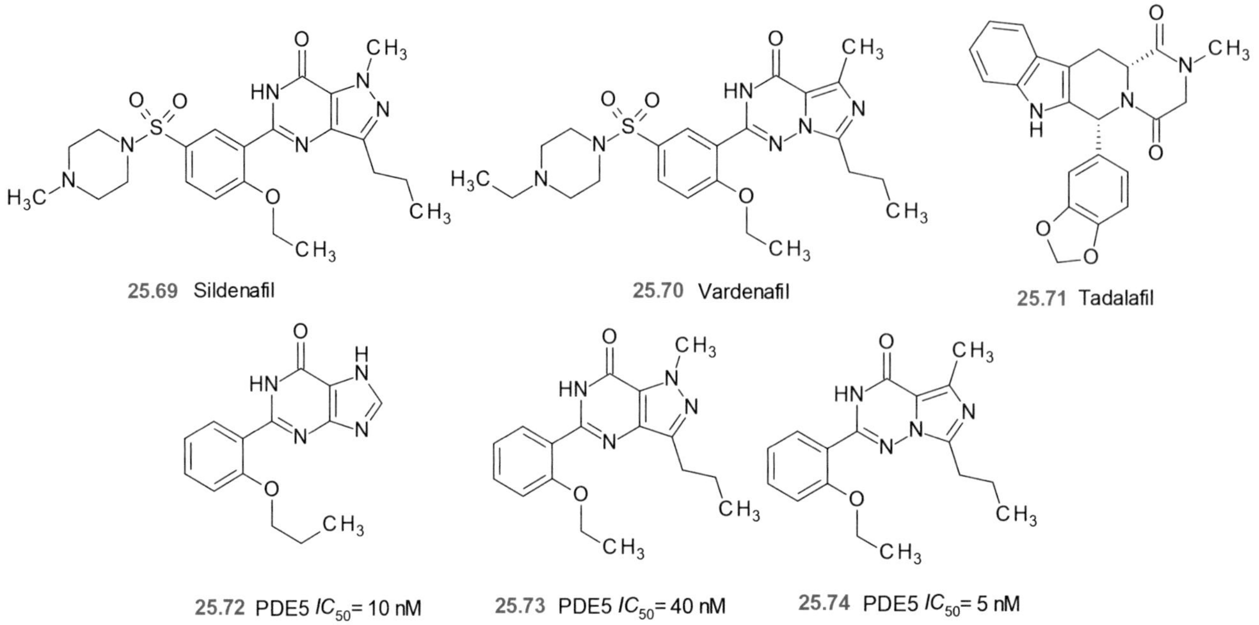

Fig. 25.21 Sildenafil **25.69**, vardenafil **25.70**, and tadalafil **25.71** represent potent PDE 5 inhibitors. The first two compounds were developed from phenyl-substituted purines such as **25.72**, and modified to pyrazolopyrimidines such as **25.73** or imidazotriazenones such as **25.74**

The market embraced Viagra® euphorically. By 2005, more than 177 million prescriptions had been registered in 120 countries around the world. In addition to PDE 5, PDE 6 is also inhibited by sildenafil, vardenafil, and tadalafil. This isoform is involved in visual processes, which explains why the use of these drugs may be associated with visual disturbances. Tadalafil has better selectivity against PDE 6, but also inhibits PDE 11 in addition to PDE 5. Sildenafil and tadalafil have another approved clinical use: They are used in intensive care units to prevent and treat pulmonary hypertension in mechanically ventilated patients.

Do PDE-5 inhibitors have another career? What helps men also seems to give cut flowers more stamina. According to experiments conducted by Heribert Warzecha at the Technical University of Darmstadt, Germany, cut gerbera flowers stay fresh longer when Viagra® is added to the water in the vase! Aspirin®, on the other hand, is less expensive and is also said to keep cut flowers fresh longer. Another study found that hamsters were able to reset their circadian rhythms faster when they had Viagra® in their blood. A higher cGMP level apparently helps the internal clock to adapt more easily to changes in external conditions. Whether Viagra® also helps to overcome jet lag after long-distance travel faster remains to be proven. These examples show that no drug is without side effects. These are often discovered only after some time in clinical trials or after practical use.

25.9 What Zinc Can Do, Iron Can Too

What makes zinc so special that it preferentially occurs in the catalytic center of so many hydrolyzing enzymes? Zinc is an ion that is commonly found in biological systems. But so is an element like iron. Zinc exists as a doubly positively charged ion. This is also the case for other ions such as Fe^{2+}, Co^{2+}, Ni^{2+}, or Cu^{2+}. Unlike the latter elements, the **zinc ion** is not redox sensitive due to its **filled d-orbitals**. When considering the reaction mechanism of an ester or amide cleavage, apart from the coordination properties, only the charge of the metal ion is critical. It serves to polarize a water molecule that initiates nucleophilic attack on the carbonyl carbon atom of the ester or amide to be cleaved. This task can also be performed by other metal ions. In fact, under **reductive conditions**, **hydrolyzing enzymes** can be found that have an **iron ion** instead of a zinc ion in the catalytic site.

New polypeptide chains synthesized in prokaryotes, mitochondria, or plastids initially carry a methionine substituted with a formyl group at the first position of the *N*-terminus. In other compartments of more complex organisms, the same proteins are formed without these formyl groups. In about one-third of all mature proteins, the methionine is cleaved by a methionine aminopeptidase. In order for the formylated chains to undergo this process, the formyl group must be removed. This is done by **peptide deformylases (PDFs)**. They carry an Fe^{2+} ion in their catalytic center and are, therefore, very sensitive to oxidation. It is only possible to exchange Ni^{2+} or Co^{2+}, but with a drastic loss of catalytic activity. On the other hand, the exchange of iron for a Zn^{2+} ion leads to a com-

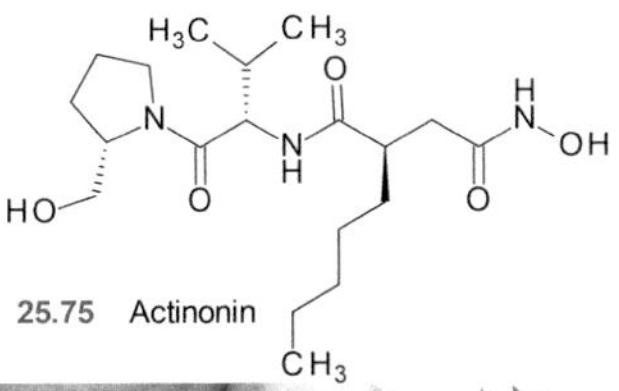

25.75 Actinonin

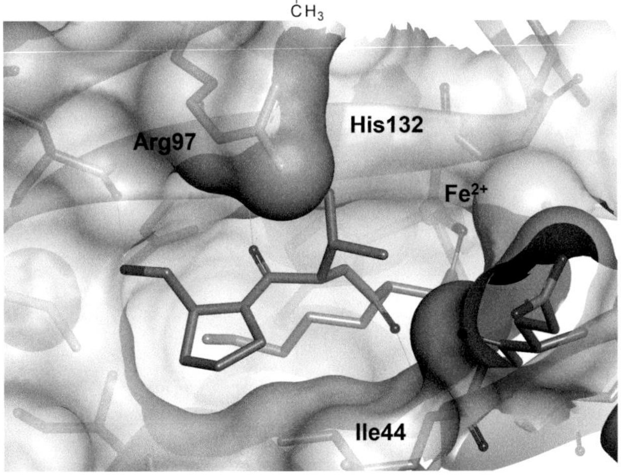

Fig. 25.22 Crystal structure of actinonin **25.75** with the peptide deformylase from *Escherichia coli*. The peptidic inhibitor binds to the Fe^{2+} ion with its hydroxamate function. Its *n*-pentyl chain replaces the methionine side chain in the natural substrate and lies in the deeply buried S_1' pocket. The iron ion is bound to a cysteine and two histidines. (▸ https://sn.pub/IzKrnZ)

plete loss of enzymatic function in almost all PDFs. Peptide deformylases are found in bacteria as well as in plant plastids and some parasites. Initially, it was thought that these enzymes were not present in humans, making them an ideal target for **antibacterial** or **antiparasitic therapy**. However, PDFs have been found **in the mitochondria** of animals and humans. This must be taken into account when developing antibiotics based on PDF inhibitors. The potent inhibitor actinonin **25.75** (Fig. 25.22) not only has antibacterial effects but also inhibits proliferation of human cells. This can lead to cytotoxic side effects, but can also be exploited for antineoplastic effects. Moreover, these inhibitors are important as herbicides.

The iron ion is tetrahedrally coordinated by two histidines and one cysteine in PDFs (Fig. 25.22). The fourth position is occupied by a water molecule. The pK_a value of this water molecule is drastically shifted by the direct coordination to the neighboring metal ion. It has an increased nucleophilicity due to the thus facilitated deprotonation. It presumably attacks the formyl peptide group to be cleaved as a hydroxide ion. The mechanism is very similar to that of proteases. The carbonyl carbon of the cleaved formyl group assumes a tetrahedral transition state. For this, the charge formed on the oxygen is stabilized by an NH of the main chain, a terminal carboxamide group of a glutamine, and the coordination to the iron. The amino group of the bond to be cleaved is bound to a glutamate by an H-bond. The polypeptide chain is cleaved with simultaneous release of the *N*-terminus. The remaining formate group leaves the metal ion coordination site and dissociates from the enzyme. Two water molecules take its place at the catalytic site. Inhibitors of this enzyme have hydroxamate groups to anchor them to the iron ion. Since the natural peptide substrate has a methionine in the P_1' position, *n*-alkyl chains with four or five carbon atoms on inhibitors are ideal in the same position. The S_1' pocket is well formed in the PDFs, but the surrounding pockets are not well characterized. This is due to the function of the proteins. A wide range of formylated substrates can be processed, meaning the amino acid sequence after the formylmethionine is arbitrarily composed. Interestingly, thiorphan also inhibits PDFs. This indicates that the thiol group can also coordinate to the iron atom. The benzyl group of the inhibitor fills the S_1' pocket of the enzyme.

25.10 Acetyl Group Cleavage Condenses Chromatin and Regulates Reading of Gene Segments: An Opportunity for Therapy?

Sect. 12.14 introduced enzymes involved in the epigenetic control of gene expression. Of particular note are the **histone deacetylases (HDACs)**, which remove an acetyl group from the terminal nitrogen of an initially acetylated lysine. This process restores the basic character of the terminal amino group of the lysine and the released amino group is then positively charged. Histones are important proteins in the cell nucleus around which the DNA is coiled as if on a spool. Because of its many phosphate groups, DNA is a highly negatively charged molecule. Histone proteins, in turn, are characterized by numerous positively charged lysine and arginine residues (Fig. 12.6). This allows them to form attractive electrostatic interactions with DNA. These interactions determine the efficiency of unwinding, so the reading process is highly dependent on the actual charge state of each binding partner. If the lysines are acetylated, the positive charges will be missing. If they are deacetylated, they will become charged and the DNA more tightly bound to the histones. As a result, the structure of chromatin in the nucleus is condensed. It is then less accessible for transcription factors to read. This addition or removal of acetyl groups **influences the process of reading specific gene segments** during gene regulation. Misregulation of the DNA reading process can be the trigger for many diseases, especially cancer. Therefore, manipulation of deacetylation has been taken up as an option in tumor therapy. First successes have already been achieved.

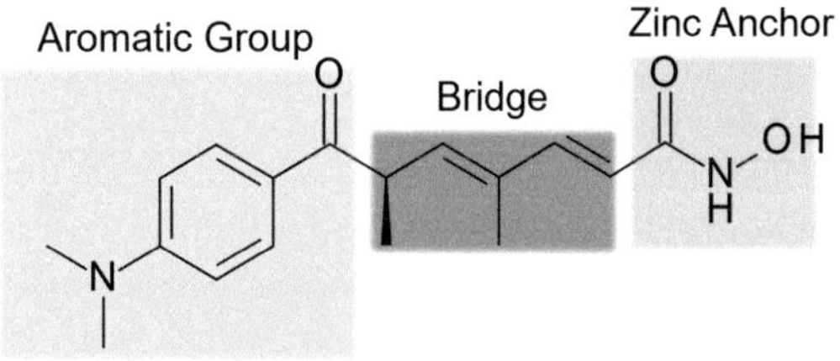

25.76 Trichostatin A

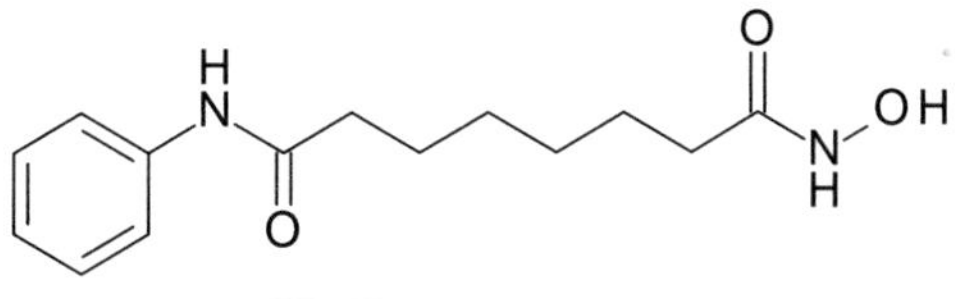

25.77 Vorinostat

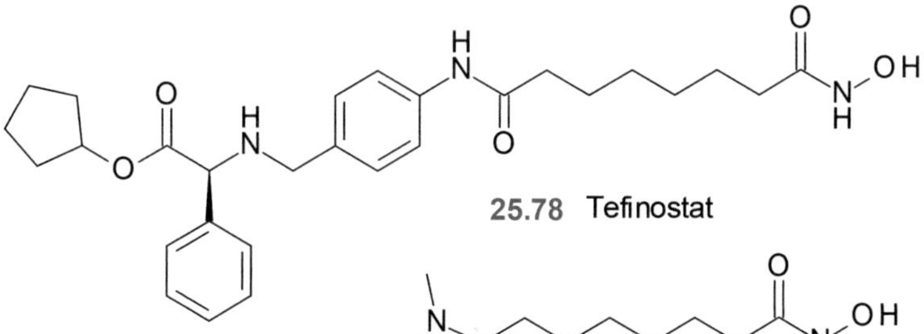

25.78 Tefinostat

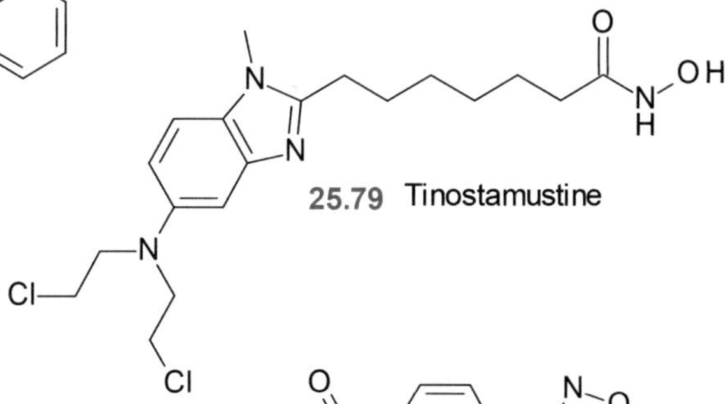

25.79 Tinostamustine

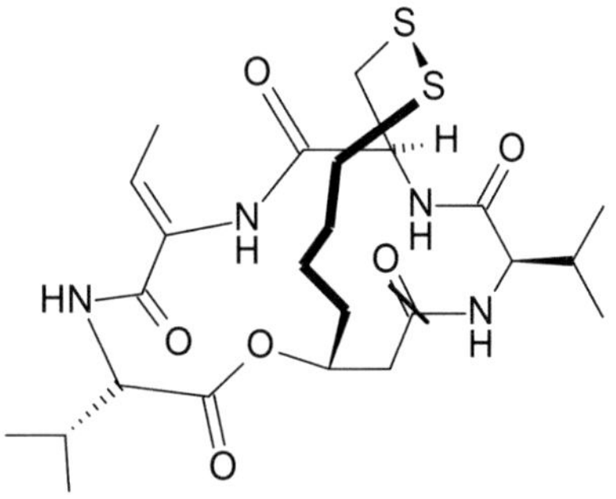

25.81 Romidepsin

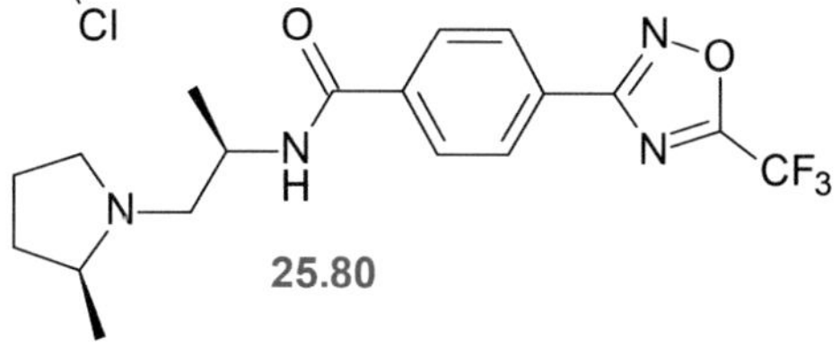

25.80

Fig. 25.23 The *Streptomyces* metabolite trichostatin A **25.76** was the first HDAC inhibitor to be discovered. As zinc-binding anchor, this molecule has a hydroxamic acid followed by a slender bridge comparable to the side chain of lysine. It bridges to an aromatic moiety that mimics the attached histone protein in the substrate. The inhibitors vorinostat **25.77**, tefinostat **25.78**, and tinostamustine **25.79** are designed according to this principle. Meanwhile, a trifluoromethyloxadiazole anchor (**25.80**) has been described as a new head group for zinc coordination. The cyclopeptidic natural product romidepsin **25.81** is a prodrug. Cleavage and reduction of its disulfide bridge results in thiol groups, of which the thiol group facing the viewer in the figure more closely binds to the zinc ion of the HDACs

For example, five compounds have been approved for the treatment of certain cancers, and many more are in clinical trials. However, because key enzymes such as HDACs regulate gene expression in many processes in our body, the side-effect profile of these approved drugs is still rather unsatisfactory.

Our organism has a total of 18 enzymes of the HDAC family. The family is subdivided into four classes. Of these, 11 are zinc-dependent metalloenzymes. They cleave the acetyl group attached to the lysine residue by an amide bond using water as a nucleophile. They thus follow a mechanism analogous to that of the zinc proteases. In the present case, one molecule of acetic acid and one lysine residue are released; this means in all cases the same substrate is converted in the cleavage step. This suggests that selectivity towards individual representatives of these enzymes is a major challenge. Seven further HDACs are called sirtuins and they use NAD^+ as a cofactor to transfer the acetyl group to the C2 position of a ribose sugar. Formally, HDACs are transferases, but due to their analogous mechanism of action, they should be mentioned here together with the zinc hydrolases.

The development of selective inhibitors for these proteins has been the subject of intense research. It seems easier to achieve selectivity against the individual classes of HDACs, since small differences in the structure of the catalytic centers can be exploited for inhibitor design. It is more difficult to find compounds that are selective against the individual members of a class. The majority of known HDAC inhibitors have a hydroxamic acid as the zinc-binding group (**25.76**–**25.79**, Fig. 25.23). Recently, a trifluoromethyl oxadiazole (**25.80**) has also been discovered for this task. In addition, the inhibitors have a more or less slender hydrophobic bridge to which an aromatic moiety is attached via a polar group (see **25.76**).

An HDAC inhibitor with a completely different structure is the cyclic peptide romidepsin **25.81**. It was isolated from a culture of *Chromobacterium violaceum* at Fujisawa Pharmaceutical Company in Tsukuba, Japan. This cyclic peptide was shown to be highly cytotoxic against several human cancer cell lines. It was characterized as a histone deacetylase inhibitor and acts as a prodrug. Inside the cell, its disulfide bond is reduced, releasing a zinc-binding thiol. This thiol then binds to the zinc ion in the binding pocket of histone deacetylase and blocks its activity.

Over the past 20 years, more than 30 HDAC inhibitors have entered clinical development, resulting in five approvals. Although the number of human HDACs is limited, these enzymes perform very different cellular functions. In addition to lysine deacetylation in various tissues, they interfere with fatty acid and polyamine deacetylation and play a role in acetyl lysine recognition. Targeted and selective inhibition of these enzymes is likely to be the key to successful therapeutic use of HDAC inhibitors.

25.11 Synopsis

- In metalloproteases, a positively charged metal ion, usually a zinc ion, activates a coordinated water molecule, which nucleophilically attacks the peptide bond to be cleaved. Through expansion of its coordination sphere, the zinc ion also polarizes the carbonyl group of the amide bond to be cleaved, and an adjacent glutamate residue helps in the transfer of protons.
- Potent inhibitors exhibit appropriate functional groups to coordinate the zinc ion efficiently; they also address the specificity pockets on the primed side that recognize the *C*-terminal part of the substrate to be cleaved.
- Angiotensin-converting enzyme (ACE) transforms angiotensin I to II by cleaving a *C*-terminal dipeptide. Rational design concepts resulted in dipeptide mimetics with a carboxylate group at a proline-like moiety and a zinc-coordinating group at the opposite end. Captopril was the first compound introduced to therapy; a large number of ACE inhibitors followed.
- The target protease ACE is composed of two slightly different catalytic domains. Aside from angiotensin I, ACE degrades other peptides such as the blood-pressure-lowering bradykinin. Undesired side effects of ACE inhibitors are related to this degradation. Because the two domains of ACE show different substrate profiles that can be translated into selective inhibition, the possibility exists to develop domain-selective active substances with efficient blood pressure regulation properties that avoid the unwanted adverse effects.
- Matrix metalloproteases (MMPs) are a large family of structurally related neutral zinc endopeptidases. They are involved in the construction and degradation of connective tissue. Several therapeutic indications have been proposed such as rheumatoid arthritis or cancer.
- The development of selective MMP inhibitors proved to be extremely difficult. The adaptive nature of the binding pockets of this protein class has proven to be challenging, and the pronounced overlapping substrate profiles are problematic because different members of the family can mutually take over the role of the protease being inhibited.
- Carbonic anhydrases are hydrolases that transform carbon dioxide to bicarbonate. They catalyze important processes from respiration to CO_2 transport, pH homeostasis, electrolyte secretion, C_1 building block delivery, bone resorption and calcification, or tumor growth.
- Due to the narrow funnel-shaped architecture of the enzyme with the catalytic zinc ion at the end, almost all α-carbonic anhydrase inhibitors feature a terminal sulfonamide group. Particularly diuretics and antiglaucoma agents, which reduce the internal eye pressure, have been brought to market.
- Phosphodiesterases (PDEs) are a small family of metalloenzymes that hydrolyze the intracellularly formed second messenger cAMP and cGMP. They are broadly distributed in different tissues and regulate many important processes.
- Inhibitors of PDE 5 such as sildenafil, originally developed for the treatment of angina pectoris, proved to be agents to stimulate penile erections via the inhibition of cGMP degradation.
- Under reductive conditions, iron (II) ions can be found instead of zinc ions in the catalytic center of hydrolyzing enzymes. In peptide deformylases, the iron ion takes a similar role as the zinc ion and helps to remove the formyl group that is found at the first position of the *N*-terminus of a newly formed polypeptide chain in prokaryotes, mitochondria, or plastids. Inhibitors of peptide deformylases are either potential antibiotics or can be exploited for antineoplastic effects.
- Histone deacetylases (HDACs) remove the acetyl group from the terminal nitrogen of an acetylated lysine, thereby, restoring the basic character of the terminal amino group and its positive charge. Histone proteins serve to efficiently coil the negatively charged DNA in the nucleus. The enhanced electrostatic interactions with the charged lysine residues strengthen DNA binding. As a result, transcription factors cannot efficiently read the gene information on the DNA. Gene expression is, thus, repressed.
- HDAC inhibitors attempt to interfere with the reading of DNA in the case of misregulation of gene expression by inhibiting lysine deacetylation and, thus, regulating gene expression. HDAC inhibitors can, therefore, be used in many diseases, but especially for cancer treatment.

25

Bibliography and Further Reading

General Literature

A. Fersht, Enzyme Structure and Mechanism, W. H. Freeman, New York, p. 416 ff. (1985)

B. W. Matthews, Structural Basis of the Action of Thermolysin and Related Zinc Peptidases, Acc. Chem. Res., **21**, 333–340 (1988)

D. H. Rich, Peptidase Inhibitors, in: Enzymes & Other Molecular Targets, P. G. Sammes, Eds., Vol. 2: Comprehensive Medicinal Chemistry, C. Hansch, P. G. Sammes and J. B. Taylor, Eds., Pergamon Press, Oxford, p. 391–441 (1990)

R. P. Becket, A. H. Davidson, A. H. Drummond, P. Huxley and M. Whittaker, Recent Advances in Matrix Metalloproteinase Inhibitor Research, Drug Discov. Today, **1**, 16–26 (1996)

B. Türk, Targeting Proteases: Sucesses, Failures and Future Prospects, Nature Reviews Drug Discov., **5**, 785–799 (2006)

T. Fischer, N. Senn, R. Riedl, Design and Structural Evolution of Matrix Metalloproteinase Inhibitors, Chem. Europ. J., **25**, 7960–7980 (2019)

Special Literature

T. A. Steitz, M. L. Ludwig, F. A. Quiocho, W. N. Lipscomb, The Structure of Carboxypeptidase A, J. Biol. Chem., **242**, 4662–4668 (1967)

I. Bertini, V. Calderone, M. Fragai, C. Luchinat, M. Maletta, and K. J. Yeo, Snapshots of the Reaction Mechanism of Matrix Metalloproteinases, Angew. Chem. Int. Ed., **45**, 7952–7955 (2006)

B. P. Morgan, D. R. Holland, B. W. Matthews and P.A. Bartlett, Structure-Based Design of an Inhibitor of the Zinc Peptidase Thermolysin, J. Am. Chem. Soc., **116**, 3251–3260 (1994)

S. H Ferreira, A bradykinin-potentiating factor (bpf) present in the venom of *Bothrops jararaca*, Br. J. Pharmacol. Chemother., **24**, 163–169 (1965)

D. W. Cushman, H. S. Cheung, E. F. Sabo and M. A. Ondetti, Design of Potent Competitive Inhibitors of Angiotensin-Converting Enzyme. Carboxyalkanoyl and Mercaptoalkanoyl Amino Acids, Biochemistry, **16**, 5484–5491 (1977)

K. R. Acharya, E. D. Sturrock, J. F. Riordan and M. R. W. Ehlers, ACE Revisited: A New Target for Structure-based Drug Design, Nat. Rev. Drug Discov., **2**, 891–902 (2003)

S. R. Bertenshaw *et al.*, Thiol and Hydroxamic Acid Containing Inhibitors of Endothelin Converting Enzyme, Bioorg. & Med. Chem. Lett., **3**, 1953–1958 (1993)

J. Hu, P. E. van den Steen, Q.-X. A. Sang and G. Opdenakker, Matrix Metalloproteinase Inhibitors as Therapy for Inflammatory and Vascular Diseases, Nat. Rev. Drug Discov. **6**, 480–498 (2007)

H. Matter and M. Schudok, Recent Advances in the Design of Matrix Metalloprotease Inhibitors, Curr. Opin. Drug Discov. Devel. **7**, 513–535, (2004)

N. Borkakoti, F. K. Winkler, D. H. Williams, A. D'Arcy, M. J. Broadhurst, P. A. Brown, W. H. Johnson and E. J. Murray, Structure of the Catalytic Domain of Human Fibroblast Collagenase Complexed with an Inhibitor, Nature Struct. Biol., **1**, 106–110 (1994)

J. R. Porter, N. R. Beeley, B. A. Boyce et al., Potent and Selective Inhibitors of Gelatinase-A, 1. Hydroxamic Acid Derivatives, Bioorg. & Med. Chem. Lett., **4**, 2741–2746 (1994)

C. T. Supuran and A. Scozzafava, Carbonic Anhydrase Inhibitors and Their Therapeutic Potential, Expert Opin. Ther. Pat., **10**, 575–600 (2000)

J. J. Baldwin, G. S. Ponticello, P. S. Anderson et al., Thienothiopyran-2-sulfonamides: Novel Topically Active Carbonic Anhydrase Inhibitors for the Treatment of Glaucoma, J. Med. Chem., **32**, 2510–2513 (1989)

J. Greer, J. W. Erickson, J. J. Baldwin, M. D. Varney, Perspective Application of the Three-Dimensional Structures of Protein Target Molecules in Structure-Based Drug Design, J. Med. Chem., **37**, 1035–1054 (1984)

A. Weber, A. Casini, A. Heine, D. Kuhn, C.T. Supuran, A. Scozzafava, G. Klebe, Unexpected Nanomolar Inhibition of Carbonic Anhydrase by COX-2-Selective Celecoxib: New Pharmacological Opportunities due to Related Binding Site Recognition, J. Med. Chem., **47**, 550–557 (2004)

K. Köhler, A. Hillebrecht, A. Innocenti, A. Heine, C. T. Supuran, G. Klebe, Saccharin, a Potent Inhibitor of Carbonic Anhydrases: An Explanation for its Metallic Aftertaste? Angew. Chem., **119**, 7841–7843 (2007)

D. P. Rotella, Phosphodiesterase 5 Inhibitors: Current Status and Potential Applications, Nat. Rev. Drug Discov., **1**, 674–682 (2002)

C. T. Supuran, A. Mastrolorenzo, G. Barbaro and A. Scozzafava, Phosphodiesterase 5 Inhibitors—Drug Design and Differentiation Based on Selectivity, Pharmacokinetic and Efficacy Profiles, Curr. Pharmaceut. Design, **12**, 3459–3465 (2006)

H. Ke, H. Wang, Crystal Structures of Phosphodiesterases and Implications on Substrate Specificity and Inhibitor Selectivity, Curr. Topics Med. Chem., 7, 391–403 (2007)

I. H. Osterloh, The discovery and development of Viagra® (sildenafil citrate). In: Dunzendorfer, U. (eds) Sildenafil. Milestones in Drug Therapy MDT. Birkhäuser, Basel (2004)

Deutsche Apotheker Zeitung, Viagra sorgt für mehr Stehvermögen bei Schnittblumen, DAZ, No. 49, p. 6, 05 December 2007

R. Jain, D. Chen, R. J. White, D. V. Patel and Z. Yuan, Bacterial Peptide Deformylase Inhibitors: A New Class of Antibacterial Agents, Curr. Med. Chem., **12**, 1607–1621 (2005)

H. Ueda, H. Nakajima, Y. Hori et al., FR901228, a novel antitumor bicyclic depsipeptide produced by *Chromobacterium violaceum* No. 968. I. Taxonomy, fermentation, isolation, physico-chemical and biological properties, and antitumor activity, J. Antibiotics, **47**, 301–310 (1994)

M. Jung, K. Hoffmann, G. Brosch, P. Loidl, Analogues of Trichostatin A and Trapoxin B as Histone Deacetylase Inhibitors, Bioorg. Med. Chem. Lett., **7**, 1655–1658 (1997)

H. Nakajima, Y. B. Kim, H. Terano, M. Yoshida, S. Horinouchi, FR901228, a potent antitumor antibiotic, is a novel histone deacetylase inhibitor, Exp. Cell Res., **241**, 126–133 (1998)

T. C. S. Ho, A. H. Y. Chan, A. Ganesan, Thirty Years of HDAC Inhibitors: 2020 Insight and Hindsight, Thirty Years of HDAC Inhibitors: 2020 Insight and Hindsight, J. Med. Chem., **63**, 12460–12484 (2020)

A. J. Stott et al., Evaluation of 5-(Trifluoromethyl)-1,2,4-oxadiazole-Based Class IIa HDAC Inhibitors for Huntington's Disease, ACS Med. Chem. Lett., **12**, 380–388 (2021)

Transferase Inhibitors

Contents

G. Klebe, *Drug Design*, https://doi.org/10.1007/978-3-662-68998-1_26

In the late 1970s, it became clear that proteins are not only translated and synthesized in ribosomes, but can also undergo postsynthesis modifications. In addition to glycosylation, phosphate groups are attached to hydroxyl functions on serine, threonine, and tyrosine residues. Later it was recognized that even histidine can be phosphorylated. It was also shown that the degree of phosphorylation of a protein can change dramatically over time in the cell. Cell proliferation was found to be strongly dependent on these changes. It became obvious to associate phosphorylation with intracellular signaling processes. ATP was identified as the source of the transferred phosphate groups. However, the bonds between the phosphate groups of ATP cannot be easily transferred to an amino acid. This reaction is kinetically too slow in aqueous solution. Nature has, therefore, developed efficient catalysts for this task: protein kinases. Similarly, the cleavage of a phosphate group from a phosphorylated amino acid is a very slow process. This process also requires efficient enzymes, the phosphatases. Thus, protein phosphorylation is a reversible process that can be "switched" in either direction by the above named enzyme classes (◘ Fig. 26.1). Although these enzymes catalyze very general reactions, their substrate recognition is highly specific. This is the only way to precisely control signal transduction processes and switch protein function on and off.

These examples do not exhaust the range of **posttranslational modifications**. Each newly synthesized protein carries an *N*-formylmethionine at its *N*-terminus. This **formyl group** is first cleaved by a deformylase (Sect. 25.9) before a methionine aminopeptidase removes the methionine residue from the peptide chain of many proteins. The attachment of sugar residues (**glycosylation**) not only improves the solubility and proteolytic stability of the protein, but also serves to label proteins with recognition characteristics that are crucial for signal transduction and intracellular transport processes. Sugar residues are particularly important for cell–cell recognition and interactions with the extracellular matrix (Sect. 31.3). The transglutaminases, which posttranslationally crosslink proteins by forming **isopeptide bonds** through glutamate and lysine side chains, were discussed in Sect. 23.9. They are also transferases, although they are mechanistically closer to cysteine proteases and were, therefore, discussed in Chap. 23. Transferases can also transfer alkyl groups. For one family of transferases, it is a **methyl group** that will be used to modify residues. Others transfer a **prenyl group**, the terpene anchor of which can be used to immobilize proteins at a membrane (Sect. 26.11). Lysine and arginine residues can be acetylated or the **acetyl group** can be removed. The latter process requires cleavage of an amide bond. This is done by histone deacetylases, which are mechanistically analogous to zinc proteases (Sect. 25.10). Finally, **ubiquitin** and **SUMO** (small ubiquitin-like modifier) should be mentioned. Ubiquitin is a polypeptide chain that marks proteins for proteolytic degradation in the proteasome (Sect. 23.8). SUMO is also a small protein that can be attached to proteins and influences processes in the cell nucleus.

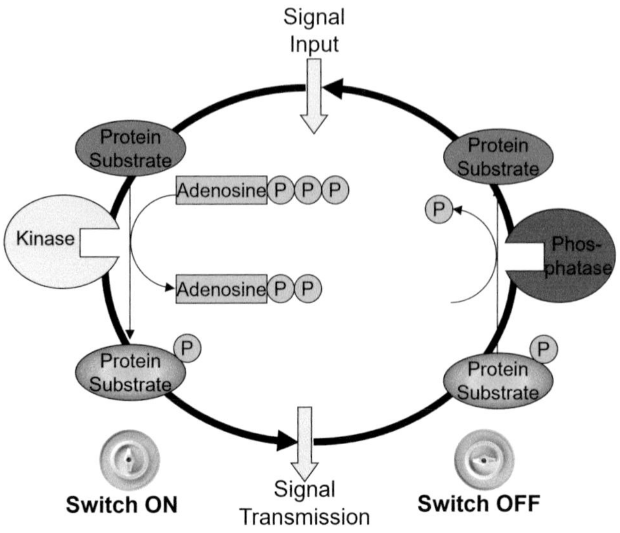

◘ **Fig. 26.1** The posttranslational phosphorylation of proteins is critical for the regulation of intracellular signaling processes, for example, cellular reproduction is highly dependent on these processes. A phosphate group (Ⓟ) is transferred from ATP (*green*) to the hydroxyl function of a serine, threonine, or tyrosine. This task of switching on protein function is performed by kinases. Conversely, phosphate groups can be removed from a phosphorylated amino acid. This step turns off the protein function. This simplified picture of switching on and off will be used in the context of this book for cellular tasks. Ultimately, however, the consequences of the transfer of a phosphate group for the function of a cell must be considered in detail in each individual case. The picture is often more complex than the simple switching on and off of a function

26.1 The Kinase "Gold Rush"

In case of disease, it sounds very attractive to target enzymes that act as switches in signaling cascades. In Sect. 12.4, kinases were identified as enzymes that are often involved in disease processes. In eukaryotes, approximately 30% of all proteins are reversibly phosphorylated. The addition of a highly charged phosphate group changes the electrostatic properties of the protein, which can lead to conformational rearrangements and, thus, the formation of new binding sites. The design of kinase inhibitors initially focused almost exclusively on the competitive displacement of ATP from its binding site. However, ATP is not the only substrate used by kinases. This molecule is the most important energy transfer system in cellular metabolism. Many cofactors use ATP as a building block to perform their cellular functions. There are approximately 2000 proteins in the human genome that use ATP as a substrate in various ways. The intracellular concentration is very high at 0.01 M. In total, the physiological turnover of ATP in an adult is 75 kg per day! Given this situation, it is reasonable to ask how specifically and selectively a binding site of a particular kinase can be blocked by an inhibitor: the same substrate, ATP, is converted by each of these enzymes, and its cellular concentration is very high. The problem is further complicated by the fact that Nature has built redundancy into many of these processes as a failsafe. If one signal transduction pathway is removed, a similar pathway can take its place by producing more of its own phosphorylated proteins. In this way, they help to correct the deficit caused by the blocked function. Is this particularly true for signaling cascades that use many different structurally similar kinases and phosphatases to transmit information? Until the early 1990s, all these problems were considered so complex and intractable that anyone who tried to develop selective kinase inhibitors as drugs was considered crazy. Since then, the tables have been completely turned. Today, a pharmaceutical company that is not working on multiple kinase projects is considered backward and not innovative! No other protein family has been investigated with so much fervor. What caused this change of heart that led to a pharmaceutical "kinase gold rush"?

26.2 Structure of Protein Kinases: More than 500 Variations with Similar Geometry

Protein kinases are one of the largest target families in the human genome. More than 530 protein kinases switch on and off the most diverse signaling pathways in our body and convert proteins from inactive to active states. They are related to one another to varying degrees by their sequence and structure and are grouped into subfamilies according to a family tree (Sect. 26.3). Kinases can also be regulated by other binding partners. **Allosteric binding sites** and **second messengers** are known to be involved in the regulation of kinase function. Inhibitory or activating proteins (e.g., cyclins) control kinase activation by complexing with the kinase domains. Autophosphorylation of kinases exerts an important influence on their conformation and the correct positioning of the catalytic residues for the transfer of the **γ-phosphate group** of ATP to the amino acids serine, threonine, tyrosine, or histidine (◘ Fig. 26.2). The conserved architecture of protein kinases is shown in ◘ Fig. 26.3. The *N*-terminal domain is composed of five β-pleated sheets. The *C*-terminal domain is predominantly α-helical and contains the substrate binding site. The two domains are connected by the hinge region. This contains the recognition motif for the adenosine moiety of ATP. The ribose moiety and the triphosphate group are bound in a cleft between the two domains and are coordinated by a magnesium ion, which is essential for the transfer mechanism. The activating loop with the **DFG (Asp–Phe–Gly)** and APE (Ala–Pro–Glu) motifs adjacent to the catalytic site is also important for the mechanism.

The structure determination of a cAMP-dependent kinase with a bound ADP and aluminum trifluoride molecule provided more detailed information about the putative reaction mechanism. The AlF_3 molecule, resembling ATP's γ-phosphate group, is located between the β-phosphate group of ADP and the serine residue of the substrate peptide chain. Both were cocrystallized as a complex with the enzyme. In addition, two magnesium ions are found in the binding pocket. Detailed information about the mechanism of phosphate transfer could be derived from these structural findings (◘ Fig. 26.4). Asp 184 of the DFG loop coordinates one of the two Mg^{2+} ions that bring the three ATP phosphate groups into the correct position for the reaction. The substrate's serine OH to be phosphorylated nucleophilically attacks the terminal γ-phosphate group. The phosphate group is then transferred forming a trigonal bipyramidal phosphorus intermediate. The adjacent Asp 166 polarizes the nucleophilic serine OH group and accepts its proton during the reaction. The positively charged residues Lys 168 of the kinase and Arg 18 of the substrate act as stabilizers. In addition, the two magnesium ions compensate for the negative charges on the phosphate groups. The aromatic rings of Phe 54 and Phe 187 shield the transition state from the aqueous environment.

Basically, there are three strategies to inhibit kinases: **blocking peptide substrate binding**, **displacing ATP** from the binding site, or **modulating an allosteric regulation site** (see below). In the first case, the formation of the protein–protein contact must be prevented because kinases recognize and bind other proteins as substrates. Inhibition of such contacts is considered to be extremely difficult due to the size of the interaction surface that is formed, especially if it is to be achieved with a small molecule (Sects. 10.6 and 29.8). In the second case, the focus is on the **competitive displacement of ATP** from the binding site. But is such a concept doomed to failure, given the many structurally similar kinases, the high concen-

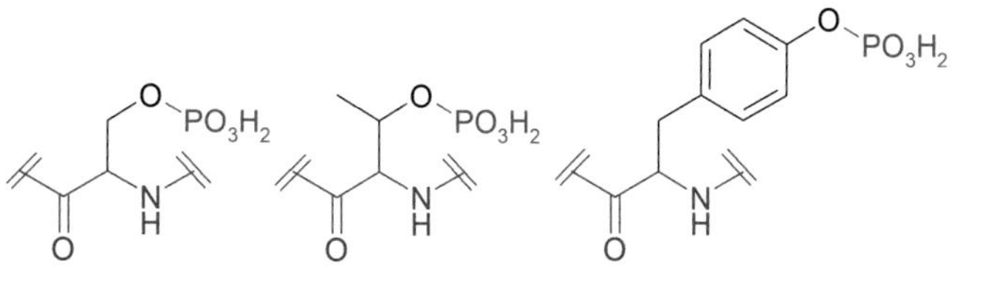

◘ **Fig. 26.2** Kinases transfer phosphate groups (*red*) to the hydroxyl function of serine, threonine, or tyrosine (*black*, peptide strand is *blue*)

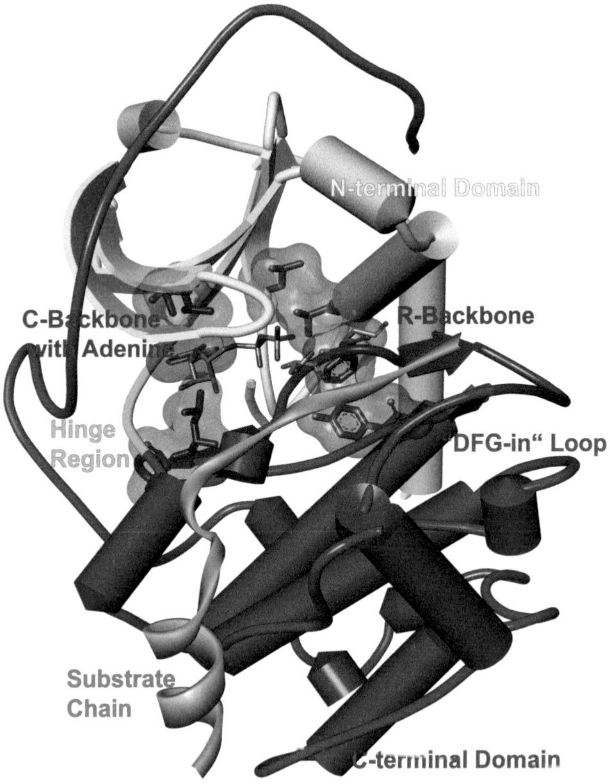

Fig. 26.3 The catalytic domains of all kinases share the same architecture. The *N*-terminal domain (*yellow*) is composed of five β-pleated sheets and helices, whereas the *C*-terminal domain (*red*) is predominantly made up of α-helical segments. Both domains are connected by the hinge region (*green*). They contain the recognition motif for the adenosine moiety of ATP (molecule in the *center*). The terminal phosphate groups are oriented near the substrate chain (*blue*), which is symbolized here by a segment of its polymer chain. It carries the Ser, Thr, or Tyr in close proximity to the phosphate groups to be transferred. Kinases are conformationally highly dynamic. A long loop containing the so-called "DFG" motif (D = Asp, F = Phe, G = Gly) rearranges upon kinase activation. It forms the hydrophobic R-spine (*brown*). The simultaneous insertion of the flat adenine building block of ATP forms the C-spine (*purple*). Both structural elements are crucial for the activation of the kinase together with the conformational change of the loop from "DFG-out" to "DFG-in" (see also Fig. 26.9) (▸ https://sn.pub/3M5Pl5)

tration of ATP in the cell, and the many other proteins that use ATP as a substrate? Some kinases are regulated allosterically. Here, the third strategy offers a possibility to interfere with the regulatory function of the kinases via the allosteric binding sites (Sect. 26.4).

26.3 Isosteric with ATP, and Selective Nonetheless?

A detailed analysis of the binding sites for ATP in a large number of kinases revealed a surprising and promising picture: there are indeed **unoccupied regions** near the ATP recognition site that differ from kinase to kinase! Two hydrophobic regions open up, one deep inside the kinase and a second on the opposite side towards the surface (Fig. 26.5). The aminopyrimidine ring of the adenine forms two adjacent hydrogen bonds to the peptide main chain in the hinge region of the kinase. A third interaction site on the polymer chain remains unused by ATP **26.1**, but may be involved in an interaction with the H-bond donor function of a ligand. The design of ATP-competitive kinase inhibitors has uncovered many interaction motifs that address the hinge region. These have been incorporated into many clinically tested kinase inhibitors (**26.2**–**26.21**, Fig. 26.6). The ubiquitous H-bonding pattern of the hinge region, which is present in all kinases, makes it difficult to endow ligands with selectivity. However, certain MAP kinases (**m**itogen-**a**ctivated **p**rotein kinases, signal transduction pathways in cell differentiation, cell growth, and cell death) offer the opportunity to design inhibitors with interesting selectivity associated with a conformational change in the hinge region. The orientation of the amide bond in this region is rotated, so that an H-bond donor function rather than an acceptor function is directed towards the bound ligands (Fig. 26.5). The **flipping of the amide bond** is possible in these kinases because of the presence of a **glycine residue** in the adjacent position. Glycine lacks a side chain at its C_α atom. Therefore, this residue has access to a much larger conformational space. Inhibitors with a dihydroquinazolinone scaffold mimicking the adenine motif of ATP can trigger this conformational flip. In the altered protein conformation, they can selectively bind to kinases that carry a glycine at this position in their sequences. If an amino acid with a side chain is present at this position, as in most other kinases, the rearrangement cannot be induced. Inhibitors that require this conformational flip to produce the specific H-bonding pattern with the hinge region will, therefore, bind only with reduced affinity to the latter kinases. In these cases, the required conformational rearrangement of the main chain is not possible for steric reasons.

The occupancy of the hydrophobic pockets on either side of the adenine-binding site (Fig. 26.5) is a generally applicable approach to design kinase inhibitors with sufficient selectivity. The pocket located deep within the protein (the so-called **back pocket**) has amino acids in its front part that can have very different properties in different kinases. These are called **gatekeeper residues**. This residue is present in all kinases, but can vary greatly in size. It can be a small amino acid such as serine, valine, or threonine, or a bulky residue such as phenylalanine,

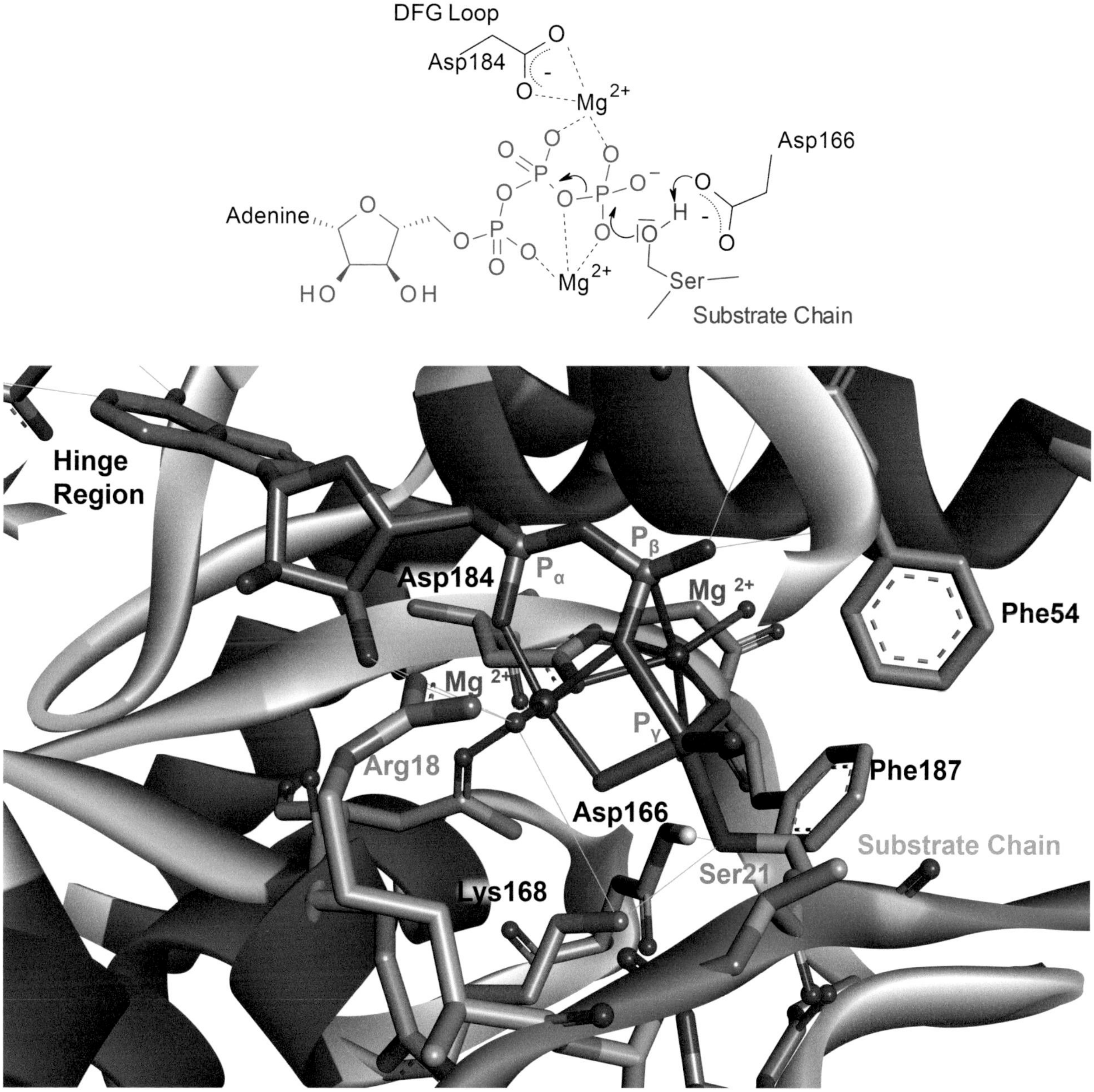

Fig. 26.4 Based on the crystal structure of a cAMP-dependent kinase with a bound ADP and aluminum trifluoride as a transition-state mimetic, the reaction steps of the phosphate group transfer from ATP (*red*) to the serine residue of the substrate (*blue*) can be modeled. Asp 184 from the DFG loop is coordinated to the β- and γ-phosphate groups of ATP via a magnesium ion. An additional Mg^{2+} helps to position the three phosphate groups correctly. Ser 21, which is to be phosphorylated, nucleophilically attacks the terminal γ-phosphate group, and a phosphorus atom is transferred with formation of a trigonal bipyramidal intermediate. The neighboring Asp 166 takes the proton from the hydroxyl group of Ser 21 during this reaction step. At the same time, the positively charged residues Lys 168 of the kinase and Arg 18 of the substrate stabilize this intermediate. (► https://sn.pub/ag9Vm2)

methionine, or tryptophan. For example, a threonine is found in the gatekeeper position in p38α and p38β kinases. A much larger methionine residue is found in the same position in the structurally similar p38γ and p38δ kinases (Fig. 26.7). Compound SB 203580 **26.3** has a *p*-fluorophenyl group at the 5-position of its indole ring. The steric demand of this group is sufficient to just allow its accommodation in the binding pocket next to the threonine. On the other hand, a methionine in this position requires such an amount of space that there is not enough room for the *p*-fluorophenyl group and the affinity of **26.3** drops significantly.

In a similar way, **26.22** benefits from a binding advantage on the p90 ribosomal S6 kinase (RSK) because

26

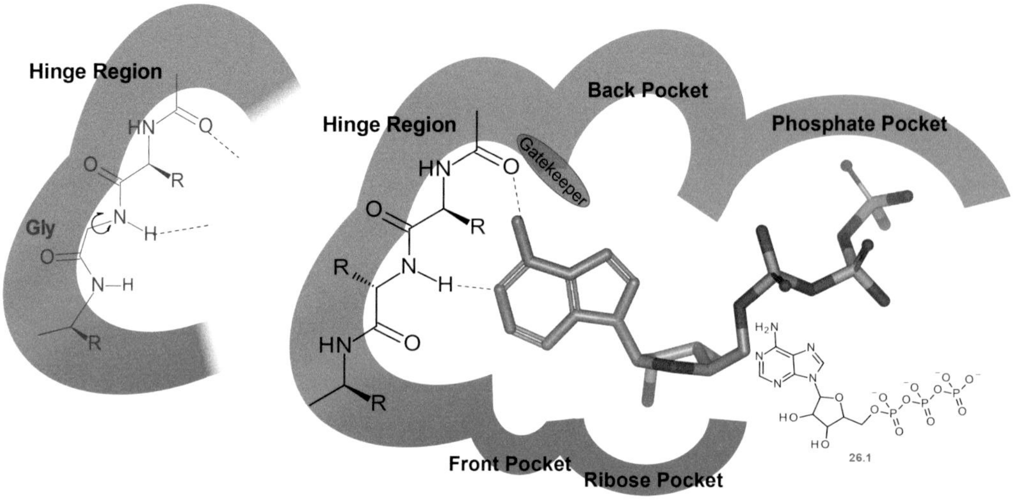

Fig. 26.5 Schematic representation of the ATP **26.1** recognition site in kinases (so-called Traxler model). The adenine moiety is recognized at the hinge region by two parallel hydrogen bonds from the peptide strand. A third carbonyl group is available for interactions but is not involved in ATP binding. Kinases with a glycine residue at this position can switch an exposed acceptor function to a donor function at this third position by flipping the adjacent amide bond (*left*). Next to the ATP-binding site, the kinases open two differently composed pockets, the so-called front and back pocket. The latter pocket is bordered by the gatekeeper residue. The residues in this pocket are not involved in ATP binding. Spatially adjacent to this pocket is the phosphate binding site

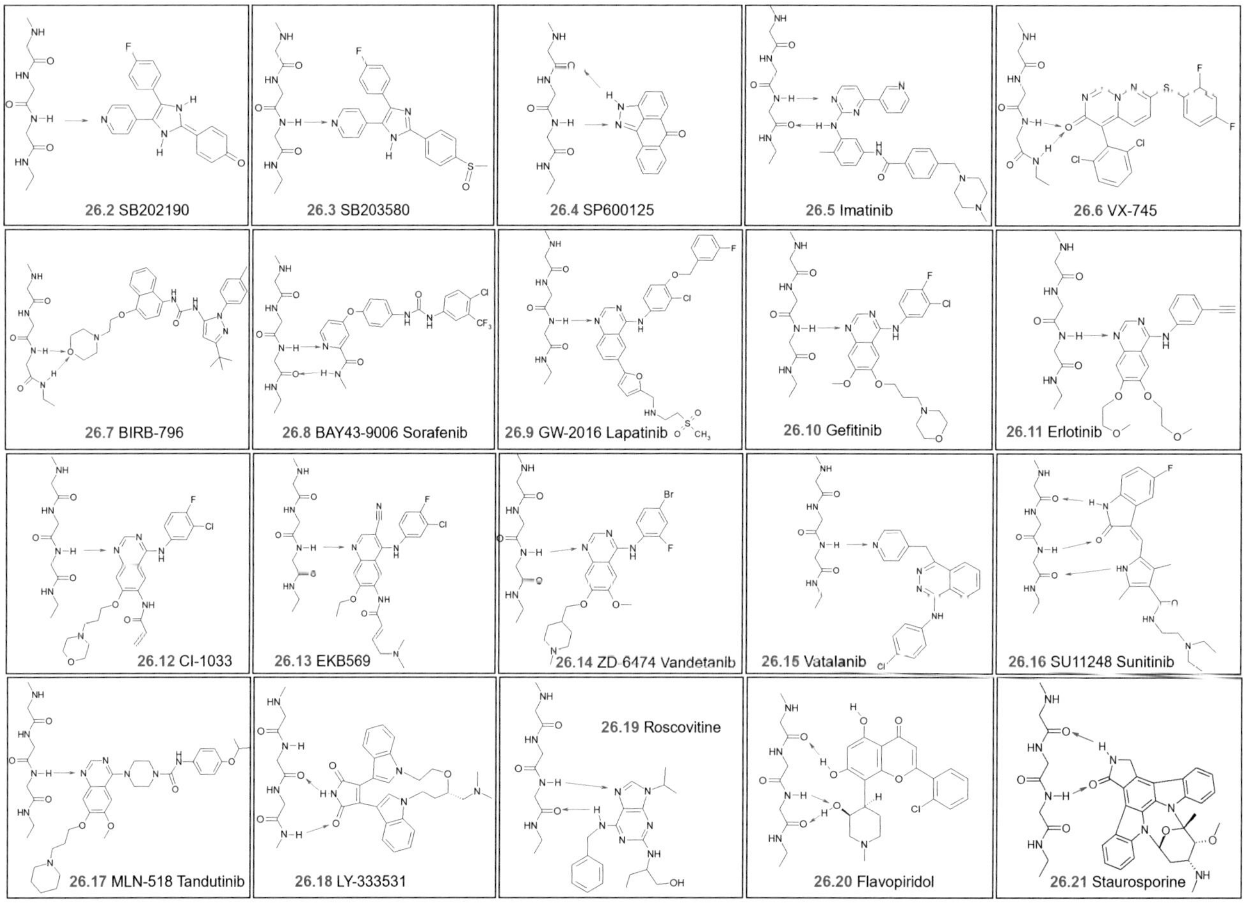

Fig. 26.6 Marketed products and development candidates of ATP-competitive kinase inhibitors **26.2**–**26.20**; staurosporine **26.21** is a natural product. All substances bind through hydrogen bonds to the peptide bonds in the hinge region of the kinases

its *p*-tolyl group has enough space in a large pocket gated by both a threonine and an adjacent cysteine (◻ Fig. 26.8). The combination of a Thr and a Cys residue at these two positions has only been discovered in three kinases in our genome. If a reactive fluoromethylene group is introduced, as in **26.22**, this group can react with the adjacent cysteine to form a stable covalent bond with the protein.

Another concept for the development of selective inhibitors exploits the conformational adaptation of kinases. During their activation, kinases undergo several steps on their way from an inactive to an active conformation (◻ Figs. 26.3 and 26.9). Interestingly, kinases show higher structural homology to one another in their active states when they have bound ATP as the uniformly identical substrate. Inhibitors that have a high affinity for the **active conformation** are, therefore, less selective than those that **stabilize an inactive conformation**. This is because the differences in the inactive conformations are much greater. Thus, the goal is to develop inhibitors that specifically bind to an inactive state of a kinase (Sect. 26.4).

Today, it is common practice to compile a so-called **inhibition** or **selectivity profile** for development candidates (◻ Fig. 26.9). Their inhibition against a large panel of kinases is measured in as many binding assays as possible. The assay results are then plotted on a **family tree** that summarizes the structural relationships between kinases from different subfamilies. The size and length of the branches reflect the degree of relationship between the kinases. The level of inhibition of each kinase is represented by circles of varying sizes, with the larger circles representing greater inhibition (◻ Fig. 26.10). It is striking that many of the compounds from ◻ Fig. 26.6 have a strong effect on individual branches of the kinase family tree. This suggests that the structural differences within a subfamily described by such a branch are often so small that no selectivity can be achieved with these compounds. As mentioned above, there is functional redundancy between kinases. If one kinase is blocked, another may take over its function by upregulating its expression. Therefore, it may be essential for a successful drug therapy that not only one member of a subfamily is blocked, but that all members are equally affected.

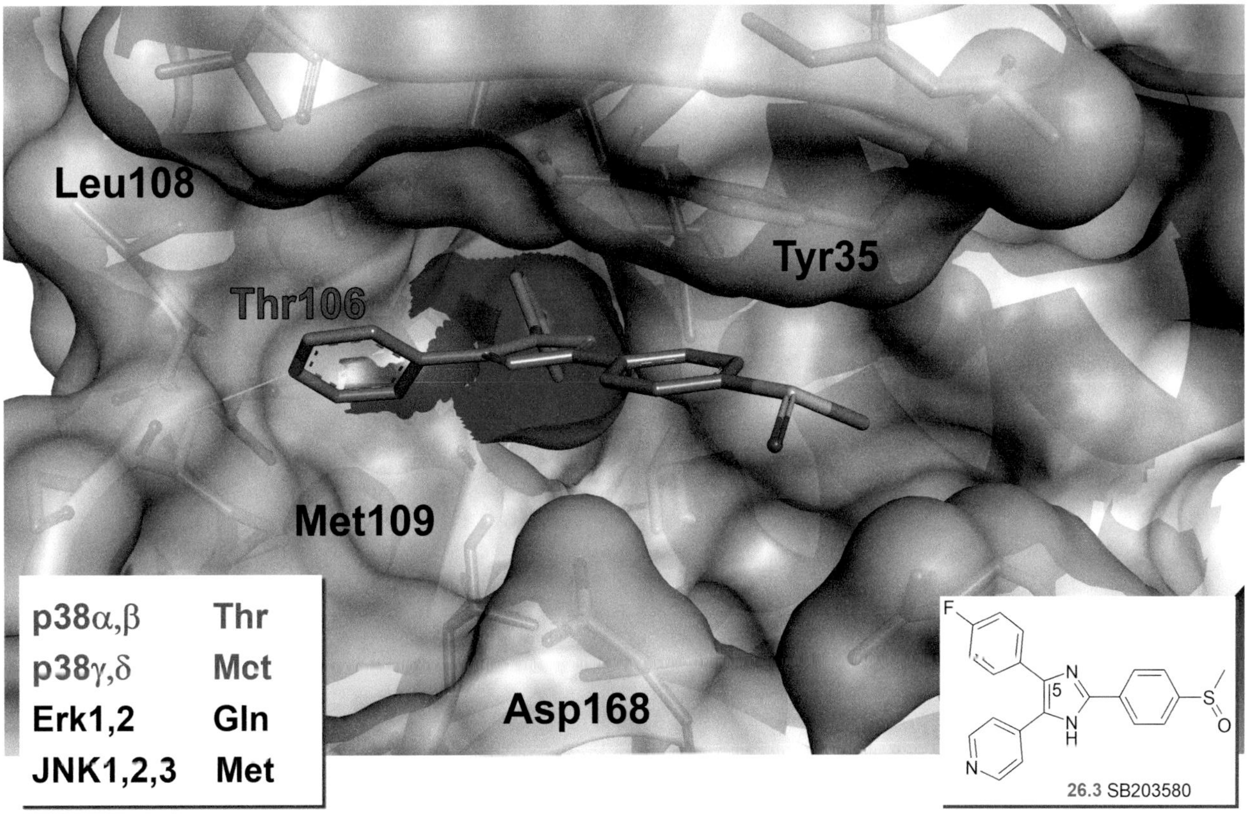

◻ **Fig. 26.7** The kinases p38α and p38β have threonine (Thr106, *violet*) as gatekeeper residues; a sterically more demanding methionine is in this position in the structurally related p38γ and p38δ kinases. SB203580 **26.3** binds with its *p*-fluorophenyl group at the central imidazole ring in a small niche next to the threonine (*green* surface, interior is *blue*). The activity is significantly reduced on other kinases with more voluminous amino acids in this position (Met, Gln) because of steric conflicts. (▸ https://sn.pub/dTeold)

The natural product **staurosporine** (**26.21**, ◘ Fig. 26.6), a highly potent alkaloid of bacterial origin, is a promiscuous inhibitor of most kinases. It binds to kinases in their active conformations. Sect. 26.6 will show how small modifications of this lead structure can nevertheless result in highly selective inhibitors.

Over the past decade, the scale of protein kinase research has exploded. More than 5000 kinase crystal structures can now be found in the public PDB database (Sect. 15.3). In addition, many thousands of inhibitor binding data have been published. More than 90 kinase inhibitors have been approved, almost all for cancer therapy. Mechanistically, our knowledge has expanded so much that it sometimes seems difficult to keep track of the different inhibition mechanisms. On the one hand, one distinguishes between the above-mentioned **"DFG-in"** and **"DFG-out" binding** of a kinase. They block the active or inactive state of a kinase. They are called **type I** and **type II** inhibitors. The transition from the inactive to the active conformation, which is accompanied by a rearrangement of the DFG loop, causes the formation of the so-called **regulatory backbone** (**R-spine**) in many kinases. Like a spine, the R-spine forms a stacked packing of hydrophobic amino acids (◘ Fig. 26.3, *brown*). This packing loses its shape in the inactive state. In the active state, however, additional binding of ATP forms a second spine of stacked amino acids called the **C-spine** (◘ Fig. 26.3, *purple*). Here, the adenine ring is included in the stack. In addition to types I and II, there are also **type III** inhibitors, which bind noncompetitively to ATP. They leave the hinge region unoccupied and are more likely to be found in the back pocket. **Type IV** inhibitors bind allosterically in a pocket far from the ATP binding site. **Type V** inhibitors are bivalent inhibitors that block two binding sites of a kinase simultaneously. **Type VI** inhibitors are covalent kinase inhibitors.

26.4 Gleevec®: Success Stories Breed Copycats!

Well into the 1980s, drug development for **cancer therapy** focused almost exclusively on processes that interfered with DNA synthesis or cell division. This led to the development of antimetabolites, alkylating compounds, microtubule disruptors, and inhibitors of DNA synthesis. These strategies attempt to attack target cells with very high division rates, such as cancer cells. The disadvantage of this type of chemotherapy is the massive side effects that severely limit the quality of life of the treated patients. In 1960, Peter Nowell and David Hungerford were the first to recognize that **chronic myeloid leukemia** is caused by a specific genetic defect. This defect causes approximately 15% of all leukemia cases. Chronic myeloid leukemia is the second most common form of chronic leukemia and is caused by a severe proliferation of white blood cells, particularly granulocytes. A reciprocal translocation between chromosomes 9 and 22 results in the shortening of chromosome 22. This is termed the **Philadelphia chromosome**. The result of this exchange is the so-called **BCR-ABL fusion gene**, which encodes a protein with constitutionally activated tyrosine kinase activity. This protein belongs to the group of receptor tyrosine kinases (Sect. 29.8) and plays an important role in the regulation of cell growth. Uncontrolled proliferation is the result of unregulated activation and the cell becomes a tumor cell. It has been shown in other leukemia models that this gene is responsible for causing this type of cancer. Therefore, it seemed that the increased kinase activity as a result of the misregulated gene was responsible for the disease. It should be possible to intervene in this overregulation with a pharmaceutical therapy. As a result, Sandoz initiated a program to develop selective inhibitors of ABL tyrosine kinases.

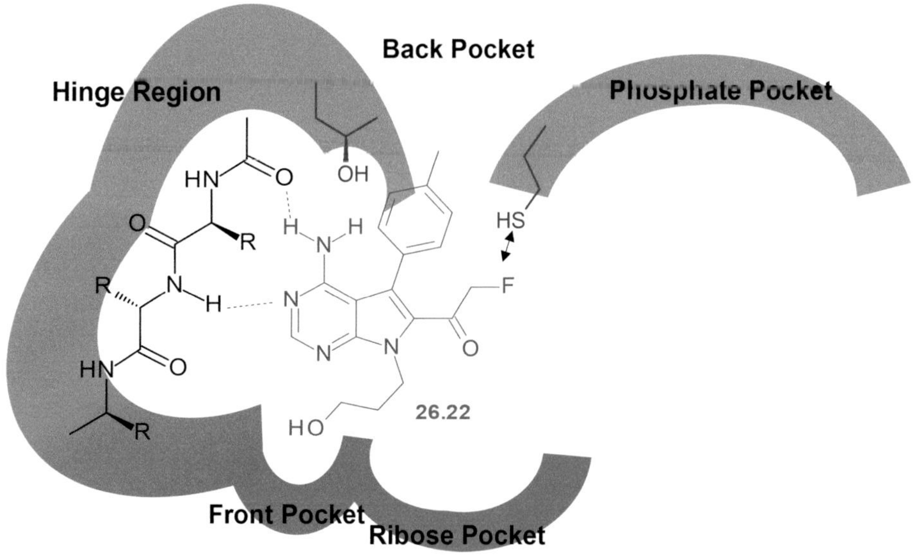

◘ **Fig. 26.8** With its *p*-tolyl group, **26.22** achieves selective binding to the p90-ribosomal S6 kinase because it finds a sufficiently large niche next to the threonine gatekeeper residue. This places the adjacent fluoromethylene group near a cysteine residue with which the inhibitor can then react. In this way, a strong covalent bond is formed with the kinase. The necessary arrangement of Thr and Cys residues has been discovered in three kinases in our genome, so that **26.22** achieves high selectivity for kinases with this amino acid composition in the back pocket

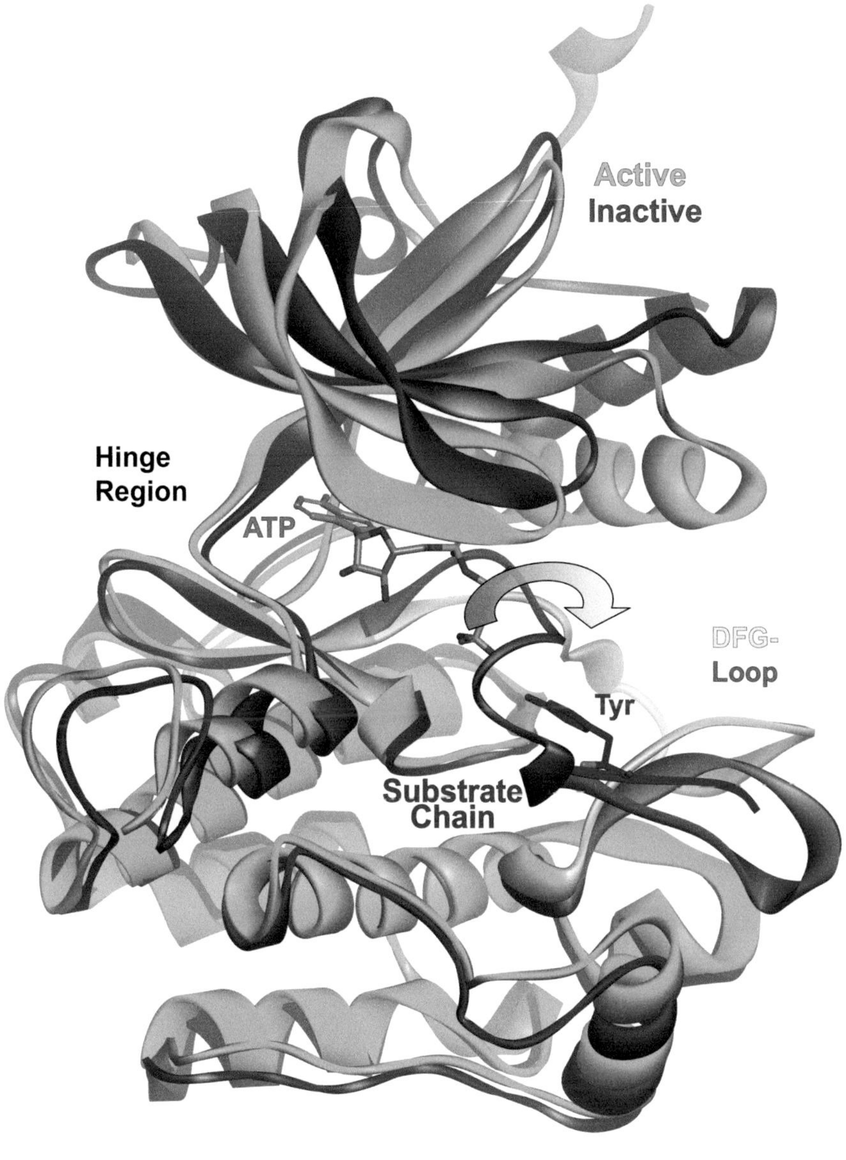

Fig. 26.9 Kinases go through multiple conformations during their activation from an inactive (*red*) to an active (*green*) state. The shown ATP molecule binds to the active conformation. For this, a complete loop, the so-called DFG loop, of the protein (inactive form, *violet*) moves from an inwards oriented geometry into an exposed orientation (*yellow*, see *arrow*). At the same time the binding site for ATP is rendered accessible and the substrate (*blue*) can bind. Interestingly, kinases possess great structural homology among themselves in this state. Therefore, inhibitors that bind with high affinity to the active conformation are less selective than inhibitors that block the inactive conformation of the kinase. (▶ https://sn.pub/XWoK8O)

The search for protein kinase C (PKC) inhibitors began in the 1980s. Phenylaminopyrimidine (**26.23**, Fig. 26.11) was identified as a good lead structure in a screening campaign. The compound was derivatized (i.e., **26.24**) and initially optimized as a PKC inhibitor. It was found that the introduction of a methyl group at position 6 (i.e., **26.26**) completely reversed the kinase inhibition. This "magic" methyl group influences the conformation between the central aromatic ring systems, which are coupled by an amino group. In the binding mode observed with ABL tyrosine kinase, the inhibitor adopts an extended conformation and the methyl group contributes to a twisted arrangement between the two ring systems.

Compound **26.26** proved to be ideal for inhibiting members of this family of tyrosine kinases. Initially, this derivative had insufficient oral bioavailability and water solubility. Therefore, an attempt was made to improve these properties by introducing polar groups such as an *N*-methylpiperazine group. Compound **26.5** proved to be optimal; it passed all phases of clinical trials and was launched in 2001 as **imatinib** (Gleevec®). The compound selectively blocks the **BCR-ABL receptor tyrosine kinase** and prevents the phosphorylation of its substrate proteins. It was later discovered that other kinases, namely the related c-Kit and PDGF receptor kinase, are also inhibited.

Why has imatinib been such a success story? First of all, the development of this inhibitor represented a **completely new approach to cancer therapy**. After all, it was treating a cancer variant with a selective therapy. The drug showed very few side effects. However, treatment with this compound is not cheap. It quickly became a blockbuster for Novartis, generating annual sales of more than a billion euros. In terms of both therapy and sales, such a success story is highly stimulating for the

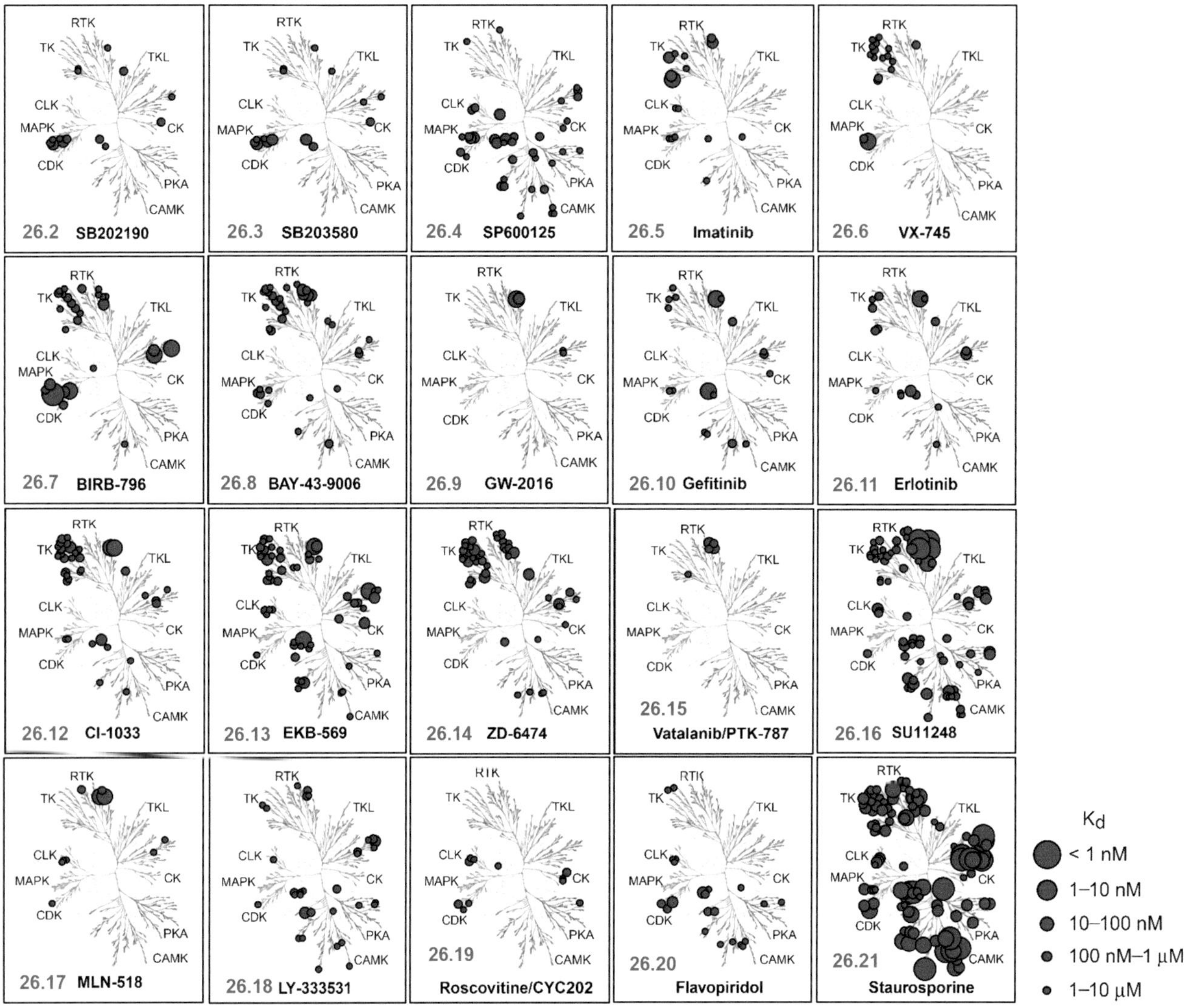

Fig. 26.10 Inhibition profile of the inhibitors **26.2**–**26.21** that were shown in Fig. 26.6 for 113 different kinases. The size of the *red circle* quantifies the strength of the inhibition. The data are shown on the kinase family tree. In this diagram, the branching and the length of the individual branches denote the degree of amino acid sequence similarity between protein kinases, grouped into families. The longer the distance in the dendrogram is, the smaller the degree of relatedness. The natural product staurosporine **26.21** is a largely unselective inhibitor, whereas **26.9** and **26.15** inhibit a few kinases very selectively. *TK* non-receptor tyrosine kinase, *RTK* receptor tyrosine kinase, *TKL* tyrosine kinase-like kinase, *CK* casein kinase family, *PKA* protein-kinase-like family, *CAMK* calcium/calmodulin-like kinase, *CDK* cyclin-dependent kinase, *MAPK* mitogen-activated kinase, *CLK* CDK-like kinase. (From M. A. Fabian et al. 2005, with kind permission from the author and publisher)

field of kinase research. Success stories breed copycats! The initial pessimism about selectivity problems and kinase redundancy seemed to have blown over. But experience has shown how difficult it is to write a similar success story. As mentioned, in the meantime, more than 90 kinase inhibitors for different indications (mostly cancer therapy) have been introduced to the market. There are also imatinib successors (see below), but no other compound has been able to achieve a similar economic and therapeutic success.

Binding of imatinib to the kinase stabilizes an inactive conformation of the enzyme. The DFG loop, which is critical for the catalytic mechanism, remains in an outwards-facing conformation (Figs. 26.9 and 26.12). The *N*-methylpiperazine group of the inhibitor, originally introduced to improve solubility, occupies a position that would be occupied by this loop in the active state. Consequently, this group is crucial for the binding mode adopted by **26.5**. A structural comparison of the kinase in complex with imatinib **26.5** and tetrahydrostaurosporine **26.27** (Fig. 26.13) is shown in Fig. 26.12. The latter inhibitor stabilizes the enzyme in its active conformation. The DFG loop takes a completely different course, resulting in the DFG sequence motif being directed inwards. The magic methyl group at the 6-position of the central phenyl ring of **26.5** forces this ring to be perpendicular to the adjacent pyrimidine ring. This geometry allows favorable hydrophobic contacts with the gatekeeper residue Thr 315, and a hydrogen bond is formed between the NH group connecting the two rings and the hydroxyl group of this threonine. The combination of optimal interaction with Thr 315 and strong binding to an inactive conformation of the protein provides the selectivity advantage of imatinib. c-Kit is the only other kinase for which imatinib has a pronounced affinity. This is explained by the high sequence homology of this kinase with BCR-ABL kinase in the DFG loop and in the ATP-binding region. In both cases, the gatekeeper residue is threonine.

In the meantime, cases of **resistance to imatinib** have developed. The observed mutations desensitize the kinase to imatinib inhibition. To date, approximately 30 mutations have been described. They are the result of single base pair exchanges in the genetic code (Sect. 12.11) and have evolved from several cell populations in which the exchanges occurred by chance or were influenced by oxidative damage to the DNA. These **variants** have become established under the **selective pressure** of imatinib blockade. The most commonly observed resistance mutation is caused by an exchange of the gatekeeper residue Thr 315 for isoleucine. Due to the larger size of the exchanged amino acid, the inhibitory effects of imatinib fail. In addition, hydrogen bonds can no longer be formed. The affinity decreases from $K_i = 85$ nM to 10 μM. In the hinge region, Phe 317 forms aromatic contacts with the pyridine ring of the inhibitor. Mutation of this residue to a leucine leads to a loss of aromatic interactions and reduces the binding affinity by a factor of three. Most of the other observed mutations are rationalized by shifting the conformation of the kinase more towards the active conformation. Consequently, the selective advantage of imatinib due to its strong binding to the inactive conformation becomes a disadvantage in terms of susceptibility to resistance mutations. Novartis has introduced a follow-up to imatinib, the structurally similar **nilotinib** (Tasigna®) **26.28** (◘ Fig. 26.13), which has an improved resistance profile. With the exception of the Thr 315 → Ile mutation, it shows good affinity for all the resistance-conferring exchanges described and stabilizes the inactive conformation of the kinase. Nilotinib, with its modified side chain containing a trifluoromethyl-substituted aromatic ring and an imidazole motif, fits better into the preformed binding pocket and achieves a higher binding affinity. The affinity advantage is thought to account for its reduced susceptibility to resistance, as small shifts from the inactive to the active conformation are better tolerated. Another compound, **dasatinib** (Sprycel®) **26.29** from Bristol-Myers Squibb, may circumvent the observed resistance to imatinib. It has a completely different mode of binding to the BCR-ABL kinase. For example, it also binds to kinases of the Scr family (a family of tyrosine kinases on the kinase phylogenetic tree that phosphorylates many cellular cytosolic, nuclear, and membrane proteins).

The native ABL kinase is posttranslationally modified with a myristic acid residue at its *N*-terminal glycine residue. The addition of this fatty acid plays an important role in the self-regulation of this kinase. The fatty acid residue occupies what is known as the **myristoyl pocket** on the catalytic domain and stabilizes the entire complex that it forms in a closed, inactive conformation. This regulatory mechanism is lost in the genetically modified BCR-ABL kinase. The *N*-terminal region in the fusion protein is replaced by a fragment of the BCR protein, leaving the BCR-ABL kinase permanently active. It has been shown that the myristoyl pocket on BCR-ABL kinase can be occupied by small molecules that weakly inhibit its activity. Subsequently, a fragment search was initiated at Novartis using the NMR method (Sect. 7.8). To further optimize the fragment hits found,

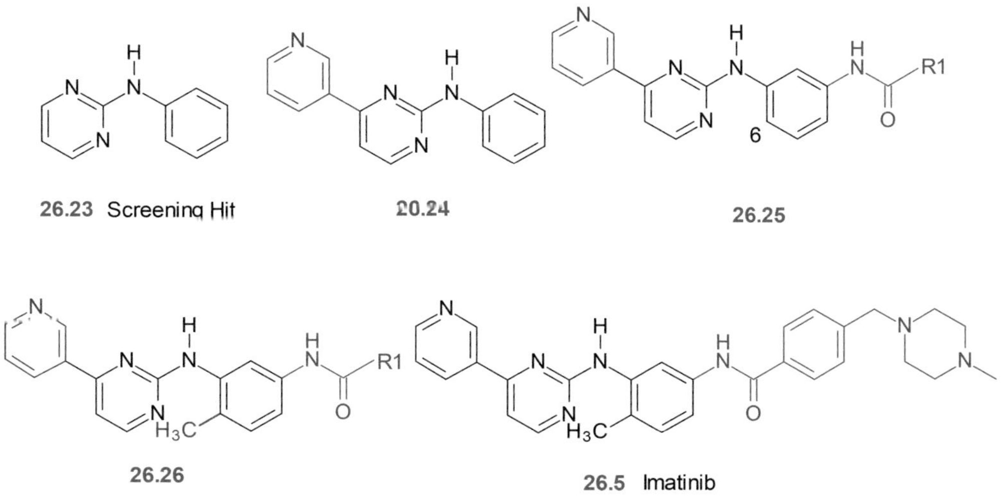

◘ **Fig. 26.11** By starting with the PKC kinase inhibition screening hit **26.23**, multiple development steps afforded imatinib **26.5**

a **conformation-sensitive assay using NMR spectroscopy** was developed. This allowed testing whether structurally enlarged and more potent binding ligands actually stabilize the closed inactive conformation of the kinase. The result of this optimization was the inhibitor **asciminib 26.30** (◘ Fig. 26.12), which highly selectively binds with subnanomolar potency. It mimics the function of the fatty acid residue in the myristoyl pocket and **allosterically stabilizes** the global inactive state of the kinase. Most importantly, its binding is unaffected by the development of resistance due to substitutions in the ATP-binding pocket, such as the gatekeeper residue. As a result, asciminib therapy is still effective in patients who have already developed massive resistance to ATP-competitive inhibitors. Using the crystal structure depicted in ◘ Fig. 26.12, it has been shown that the ATP-competitive inhibitor nilotinib **26.28** and the allosteric inhibitor asciminib **26.30** can bind simultaneously. Therefore, it is expected that an appropriate combination of both drugs with different inhibitory mechanisms will significantly impede the emergence of resistance mutations.

26.5 Tracing Selectivity: The Bump-and-Hole Method

The properties of a cell are controlled by a complex network of interwoven signaling pathways. Kinases are regulators of such information cascades. Because of the complexity of these networks, it is extremely difficult to isolate the individual signaling pathways and to tease apart the role of individual kinases. This task is further complicated by the overlapping substrate specificities of the kinases. Therefore, methods have been developed to dissect these signaling pathways using appropriate chemical probes and genetic techniques. In principle, these techniques are not limited to kinases; they can also be used to analyze the functional properties of individual members of other protein families. The structural differences between kinases that allow the design of selective inhibitors have been highlighted in detail in Sect. 26.2. The gatekeeper residue occupies a key position. The size and polarity of this residue varies from kinase to kinase. Since the gatekeeper residue is not involved in

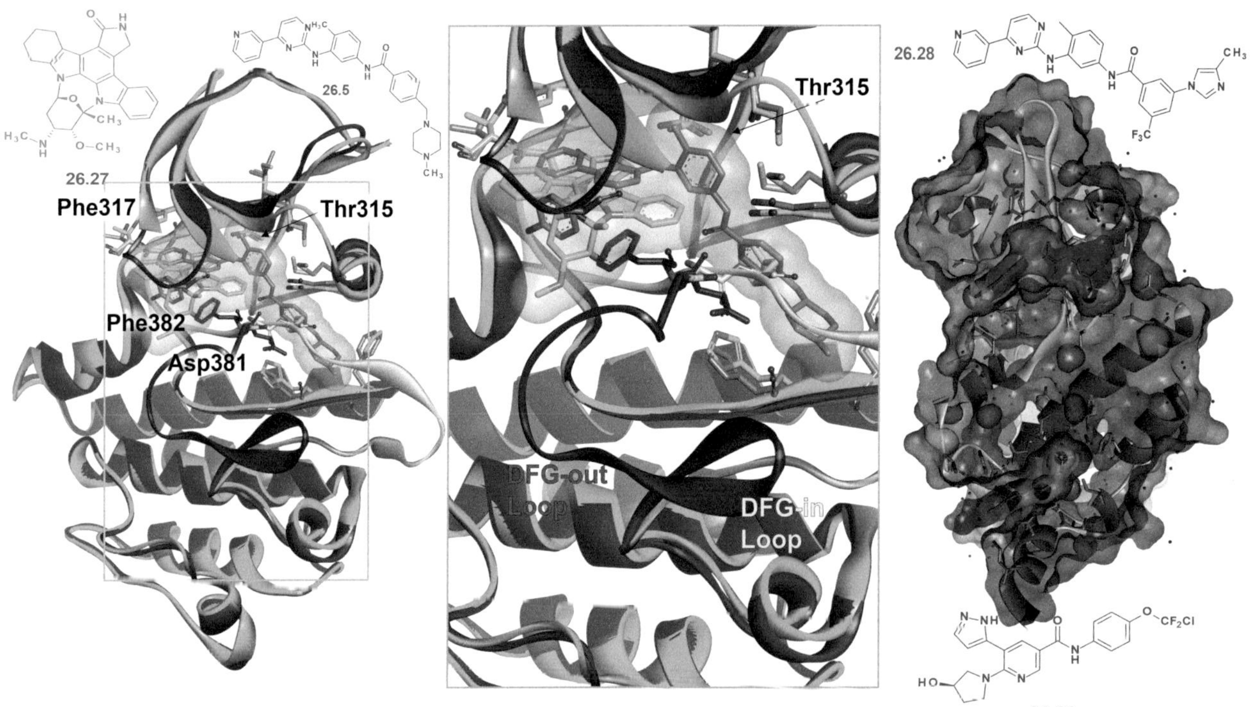

◘ **Fig. 26.12** *Left* Superimposed crystal structures of imatinib **26.5** and tetrahydrostaurosporine **26.27** (◘ Fig. 26.13) with the active (*green*) and inactive forms (*red*) of BCR-ABL receptor tyrosine kinase, respectively. *Center* A section magnification (*gray outline*) of the *left* figure is displayed. While **26.5** blocks the inactive form of the kinase (DFG-out loop, *purple*), the nonselective inhibitor **26.27** binds to the active conformation (DFG-in loop, *yellow*). With the so-called magic methyl group, imatinib faces the gatekeeper residue Thr 315, and the amino group between the two rings forms a hydrogen bond to its OH group. *Right* The crystal structure of the ternary complex of BCR-ABL kinase with the ATP-competitive inhibitor nilotinib **26.28** (*green/blue* surface) and the allosteric inhibitor asciminib **26.30** (*light blue/purple* surface) is shown. Both bind to the enzyme in its inactive form. Their binding sites are more than 20 Å apart. (► https://sn.pub/zvNzKE)

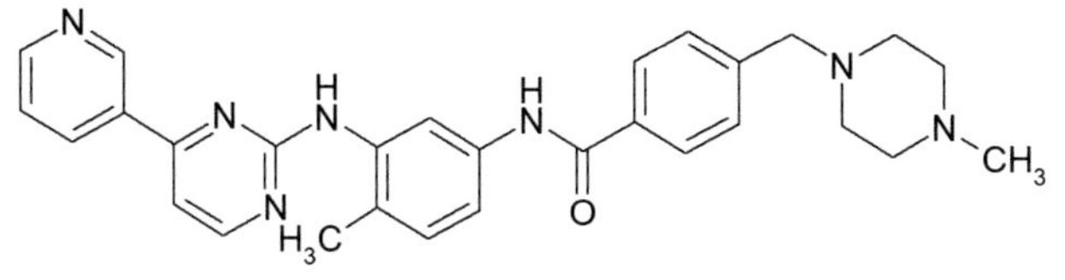

26.5 Imatinib

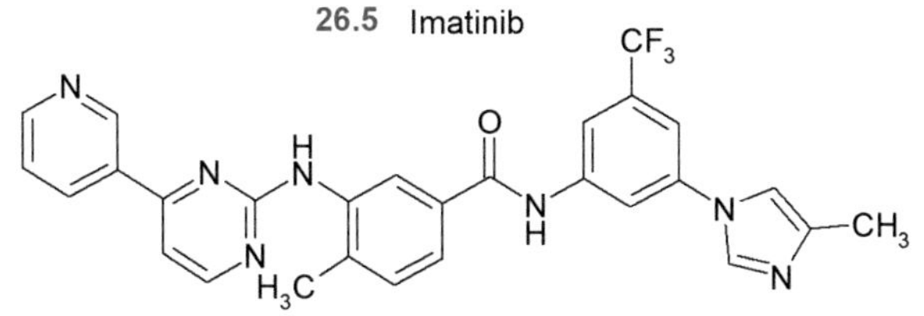

26.28 Nilotinib

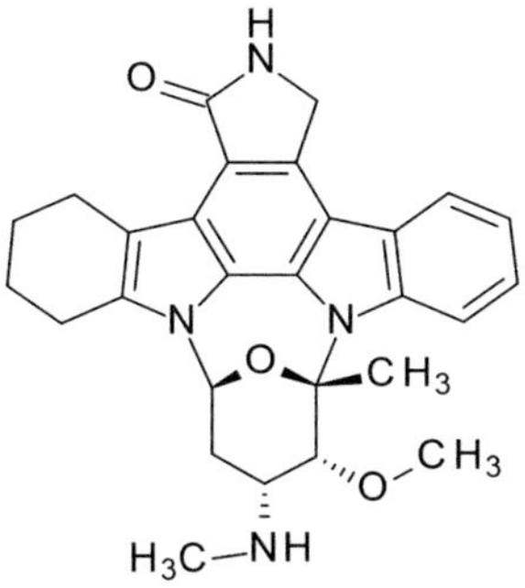

26.27 Tetrahydrostaurosporine

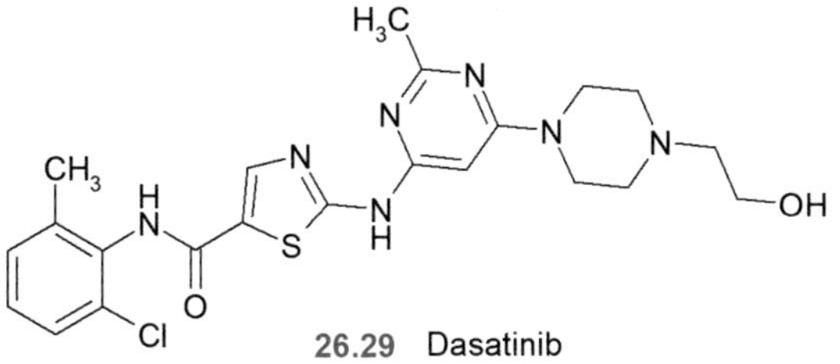

26.29 Dasatinib

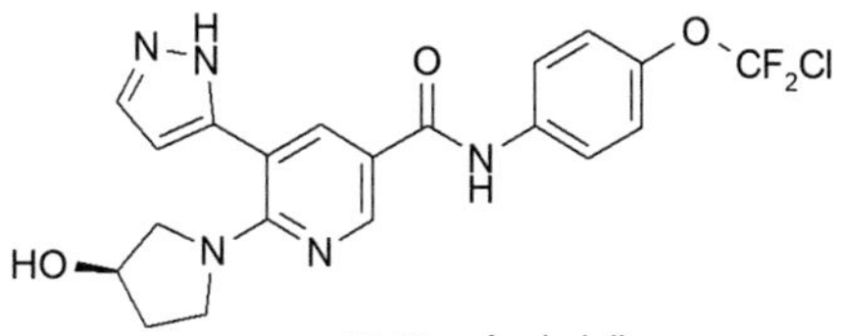

26.30 Asciminib

Fig. 26.13 Nilotinib **26.28**, which has a resistance-breaking profile, was developed as a follow-up compound for imatinib **26.5**. This compound binds with almost the same binding mode, but with stronger affinity to the BCR-ABL kinase. Dasatinib **26.29**, which was developed at Bristol-Myers Squibb, also binds to this kinase, but adopts an entirely different binding mode. Tetrahydrostaurosporine **26.27** is a nonselective inhibitor that binds to the active form of the kinase. Asciminib **26.30** was developed as the first allosteric inhibitor of BCR-ABL kinase and blocks the myristoyl pocket with subnanomolar affinity (Fig. 26.12)

ATP binding, the substrate ATP binds almost identically to all kinases with the same affinity. If the back pocket is enlarged by replacing a particular gatekeeper residue with an amino acid with a smaller side chain (e.g., Thr → Gly), the **modified kinase variant can recognize a modified ATP** with an attached side chain (i.e., **26.30**) and use this ATP surrogate as the phosphorylation reagent for the protein substrate (Fig. 26.14). This concept has been colorfully termed the "**bump-and-hole**" method. A ligand that is too large and would create a steric conflict with the protein (bump) can be converted into a well-fitting ligand if a corresponding hole is made on the side of the protein.

Of course, the technique is not limited to the phosphorylation of substrates. It can also be used to design specific inhibitors. In the research group of Kevan Shokat, formerly at Princeton and later at UCSF in San Francisco, USA, protein kinases were modified by replacing the gatekeeper residue with a glycine or alanine (Fig. 26.15). Because of this enlargement of the back pocket, the mutated kinase variant became highly sensitive to inhibition by **26.31** and **26.32**, which only weakly inhibit the wild type. This observation in an *in vitro* assay was later translated to *in vivo* conditions. The researchers used the baker's yeast *Saccharomyces cerevisiae* as a model organism. The yeast genome encodes 120 kinases, many of which are related to mammalian kinase families. One such case is the yeast cell division control protein 28 (Cdc28), member of the class of cyclin-dependent kinases (CDKs). It plays an important role in yeast reproduction and controls specific phases of the cell cycle. It shares 62% sequence identity with a comparable human enzyme, CDK2. To demonstrate the high specificity of the inhibitors **26.31** and **26.32** for the mutated kinase variant, the altered protein had to be incorporated into the yeast genome. This was done using retroviral methods established in molecular genetics (Sect. 12.15). Finally, it had to be shown that the cells of the genetically modified yeast showed normal growth. Only a 20% longer replication time was observed. Next, the inhibitor **26.32** was added to the cells of the wild-type yeast and the genetically modified yeast. The cell growth of the wild-type yeast remained unaffected, except at an inhibitor concentration above 50 µM, where a longer replication time was observed. On the other hand, the yeast with the modified *cdc28* gene showed a strong dependence on **26.32** under *in vivo* conditions. At concentrations as low as 50–100 nM, growth was reduced by 50%; at 500 nM, growth was completely arrested. Apparently, the inhibitor blocks cells at the premitotic step (cell nucleus division during cell replication), because the phenotype of these inhibited cells seemed very similar to those in which the mitotic cyclins (proteins with a key function in cell cycle control) were knocked out.

This method can be used to **study individual processes in the cell cycle** and, in particular, to determine the phase

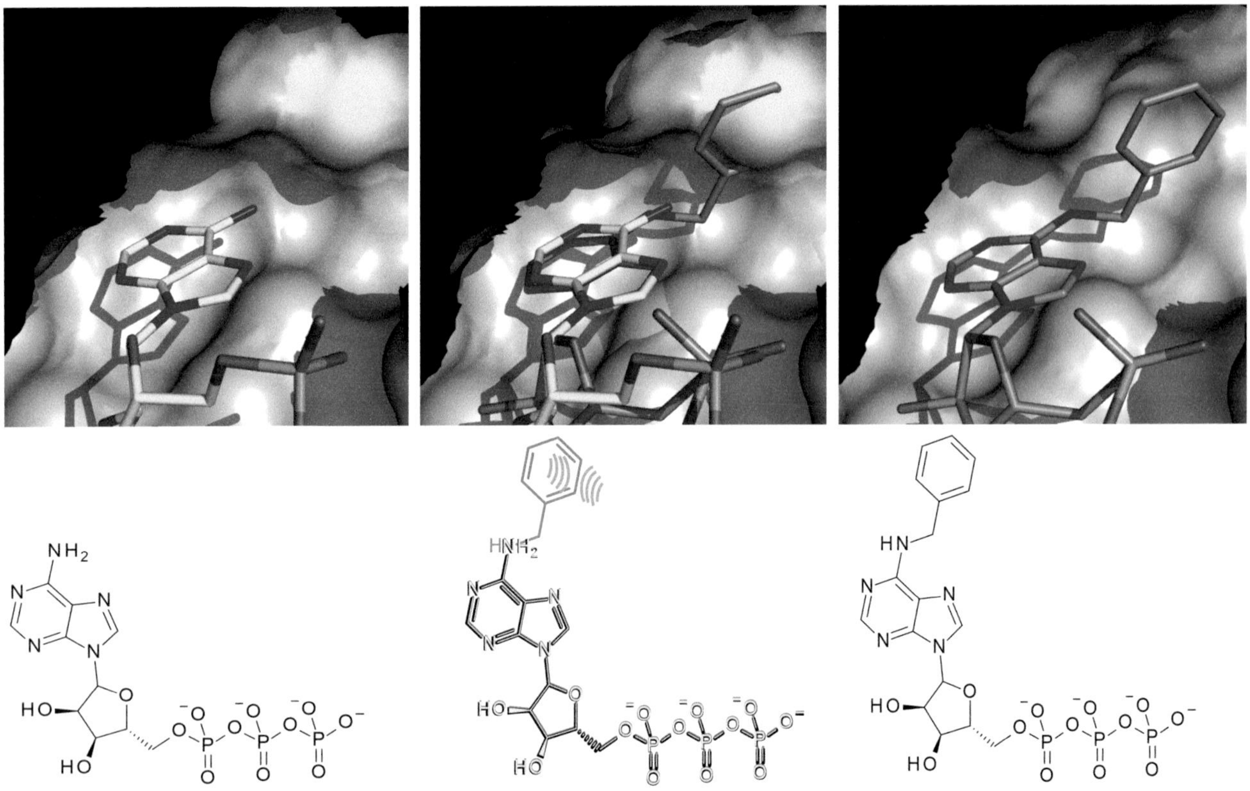

26.1 Adenosine Triphosphate ATP 26.1 26.30 26.30 Enlarged Adenosine Triphosphate

■ **Fig. 26.14** In the context of the bump-and-hole method, the *back pocket* of a kinase is enlarged by exchanging the gatekeeper residue (*yellow surface patch*) for smaller amino acids (e.g., Thr → Gly). The altered kinase (*right*) can then recognize a chemically modified ATP **26.30** with an enlarged side chain, which can subsequently be used as a phosphorylating reagent for the protein substrate

at which a specific inhibitor intervenes. This information is crucial for the development of a therapeutically effective drug. However, at the beginning of a project, there are usually no sufficiently selective inhibitors available to allow this targeted study to proceed. Furthermore, this problem is particularly acute when many proteins with high homology are found in the cell. The bump-and-hole method, a combined chemical–genetic technique, allows a **specific therapeutic validation of the biological relevance** of the target protein as well as the optimization of the inhibitor class intended for development in a model organism in an early phase of the project.

26.6 Metals Teach Kinase Inhibitors Selectivity

Metals and metal ions play an important role in biological systems, especially as catalytic centers. But can they also perform other tasks and support the design of inhibitors? In this section, we will discuss such an example. Zinc and calcium ions can contribute to the cross-linking and stabilization of proteins by acting as multidentate ligands (cf. zinc finger proteins, Sect. 28.2). Magnesium ions often serve as a kind of charge buffer to counteract the electrostatic contribution of the strongly negatively charged phosphate groups. As described in Sect. 26.2, they are involved in the phosphate transfer mechanism from ATP to the hydroxyl groups of Ser, Thr, or Tyr. In rare cases, metal ions serve as part of a ligand that binds to the biomolecule. One example is magnesium ions, which are so tightly coordinated to the β-hydroxyketo group of tetracycline **26.33** (■ Fig. 26.16) that they remain bound to the ribosome or the tet repressor during complex formation (Sect. 32.6). Another example is cisplatin **26.34**, which, through a substitution reaction on platinum, induces cross-linking in the adjacent base pairs of the DNA strands; this renders the DNA unreadable in the replication process (Sect. 14.9, ■ Fig. 14.20).

In fact, metal ions can be incorporated into drugs for very different purposes. Typically, carbon is the architectural element in drugs. However, its **coordination geometry** is rather boring. It is limited to linear, trigonal-planar, and tetrahedral geometries. A stereocenter can occur when four different substituents are on the tetrahedron (Sect. 5.2); this gives the possibility of two stereoisomers. Metals are much more exciting in this respect. By expanding their coordination sphere, they have a much

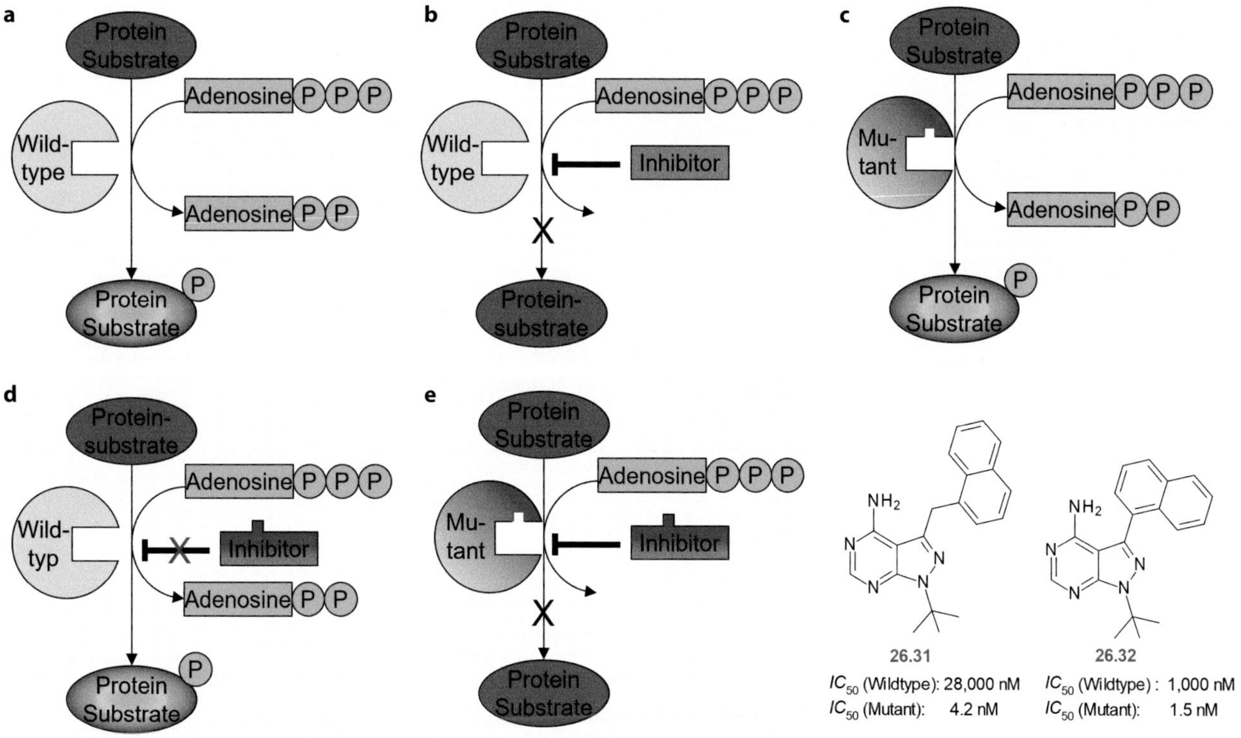

Fig. 26.15 **a** The wild type of a kinase activates a protein substrate by transferring a phosphate group. **b** When a potent inhibitor is added, phosphorylation is inhibited. **c** Exchanging a gatekeeper residue for a smaller amino acid such as glycine does not change the catalytic activity of this mutated kinase variant. **d** When an inhibitor that has an enlarged substituent to fill the pocket next to the gatekeeper residue is added to a wild-type kinase, it can barely bind to the wild-type due to steric conflicts. **e** This inhibitor could, however, block the mutated kinase variant with the enlarged pocket. The two inhibitors **26.31** and **26.32** hardly block the wild type at all, but they are able to efficiently inhibit the kinase variant with the enlarged binding pocket due to the modified gatekeeper residue

greater variety of coordination geometries at their disposal. An **octahedral center** with only six different substituents yields **30 stereoisomers**! Initially, any medicinal chemist would balk at the idea of incorporating metals as structural centers into a drug molecule. The risk that such centers might impart undesirable toxic properties to the compounds seems too great. However, if one considers metals that only form bonds with coordination partners that are inert to substitution, this argument seems less valid. Ruthenium fulfills these requirements for inert behavior very well. Why not use the advantages of a much more exciting coordination chemistry to construct a completely different molecular geometry to generate an alternative pharmacophore pattern in a very small space? The goal is to use the metal center as a scaffold and not as a partner to interact with the biomolecule. This concept, which seems unusual at first glance, has been pursued by Eric Meggers and his research group at the University of Marburg in Germany. In Sect. 26.3, staurosporine **26.21** was presented as a largely unselective inhibitor of almost all kinases. This indolocarbazole alkaloid has a molecular building block that resembles a carbohydrate and occupies a position comparable to that of the ribose ring in ATP (Fig. 26.16). On the other hand, the molecular architecture of staurosporine suggests a scaffold for a chelating ligand. If the sugar moiety is replaced by a metal center, a variety of novel and interesting scaffolds can be generated. Considering hexacoordinated metal ions, four additional coordination sites are available for further substitution.

Derivatives such as **26.35** were synthesized in the group of Eric Meggers (Fig. 26.16). They turned out to be highly potent kinase inhibitors. Interestingly, and unlike staurosporine, they have clearly graded selectivity profiles. Even complexes with cyclopentadiene groups, in which the five-membered ring covers three coordination positions, could be synthesized. Compound **26.36** was found to be selective for GSK-3 and PIM-1 kinases in an inhibition study with 57 different kinases. Compared to staurosporine (IC_{50} = 40 nM), **26.36** is ten times more potent (IC_{50} = 3 nM) at these kinases. The metal-free coordination ligand **26.37** (IC_{50} = 50 μM) and the *N*-methyl compound **26.38** (IC_{50} > 300 μM) were almost inactive. A crystal structure of the *R*-stereoisomer with PIM-1 was determined (Fig. 26.17). The structure largely agrees with the geometry of the staurosporine complex. The ruthenium complex with its carbonyl group is oriented opposite to a β-strand in the kinase fold located above the ATP binding site. The cyclopentadiene group replaces the lower part of the sugar-like moiety in staurosporine.

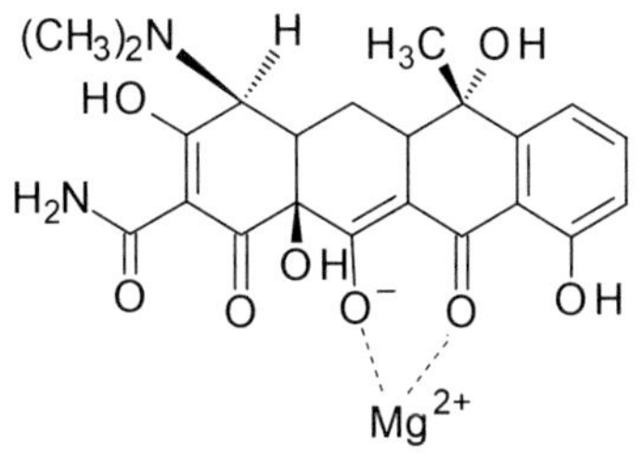

26.33 Mg^{2+} * Tetracycline

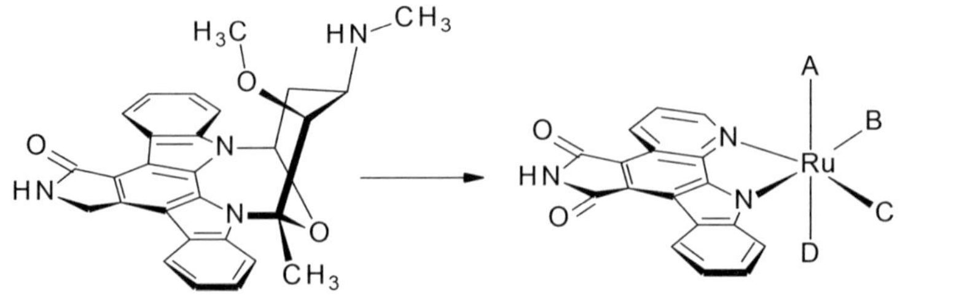

26.21 Staurosporine

26.35 Ruthenium Complex

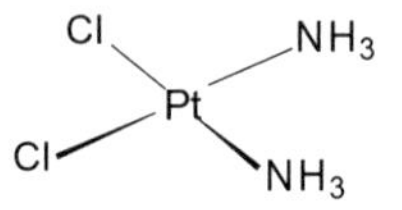

26.34 *cis*-Platin

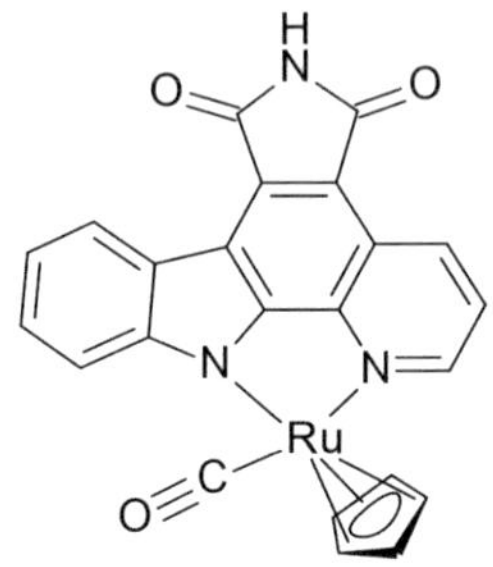

26.36 IC_{50} = 3nM

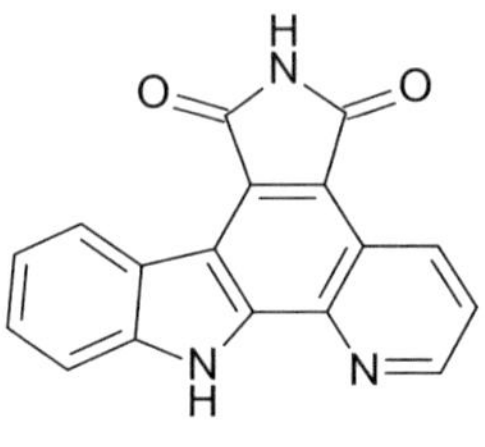

26.37 IC_{50} = 50mM

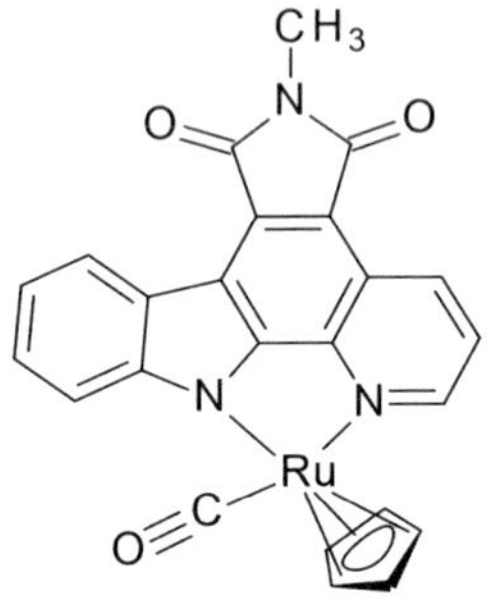

26.38 IC_{50} > 300mM

Fig. 26.16 Examples of protein ligands that bind to proteins with a tightly bound metal center. Tetracycline **26.33** chelates magnesium ions so tightly that protein binding of this ligand is achieved together with the Mg^{2+} ion. Cisplatin **26.34** binds through substitution of the chlorine atoms by the basic nitrogen atoms of the nucleotide bases of DNA. Replacement of the sugar moiety in staurosporine **26.21** led to the chelating ruthenium complex **26.35**. They proved to be potent kinase inhibitors (e.g., **26.36**). *N*-Methylation at the NH function of **26.36** leads to an almost inactive compound (**26.38**)

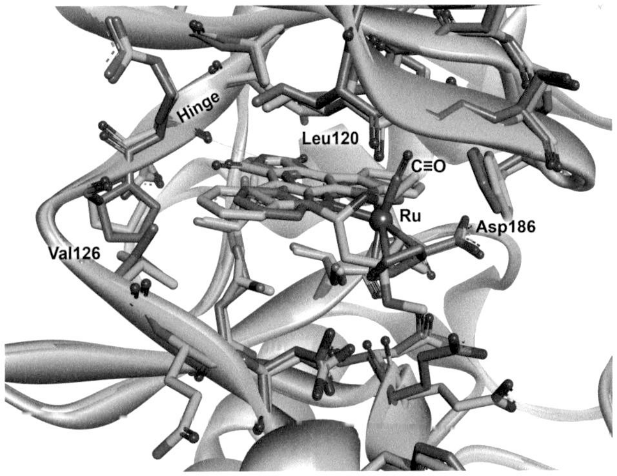

Fig. 26.17 Superposition of the crystal structures of the complex of PIM-1 kinase with the unselective inhibitor staurosporine **26.21** (*light blue*) and the selective ruthenium carbonyl complex **26.36** (*olive green*). The binding geometry is almost identical in both cases. In **26.36**, the carbonyl group is opposite to the β-strand that runs above the binding pocket. (▸ https://sn.pub/JzlPkM)

The highly similar geometries of the complexes give no obvious indication as to why the metal center converts the **promiscuous staurosporine scaffold** into **highly selective** inhibitors. The selectivity profile for other kinases can be shifted by exchanging the coordinating ligands on ruthenium and by inverting the stereochemistry. It remains unclear whether this shift is due to the strongly altered charge distribution on the scaffold or to the interactions with the polymer chain above the ATP-binding site. Interestingly, the ruthenium complexes proved to be active under *in vivo* conditions, interfering with the signaling cascade of the so-called *wnt* pathway in human cell lines and in frog and zebra fish embryos. Time will tell whether such metal complexes really open up a new perspective for drug development or whether they serve as interesting probe molecules for basic research on signaling pathways. Certainly, they will have an answer to the specific question of developing selective kinase inhibitors, but it remains to be discovered.

26.7 Phosphatases: Reversal Switch to Activate and Inactivate Proteins

Posttranslational modifications of proteins serve to regulate cellular processes. Phosphorylation by kinases usually leads to the activation of proteins; the transfer of a phosphate group switches on their biochemical function. In order to remove the phosphate group, which typically leads to the deactivation of a biochemical function, Nature has developed a counterpart to kinases: the phosphatases (■ Fig. 26.1). They can remove phosphate groups from the amino acids Ser, Thr, Tyr, and His by hydrolysis. There are **three families of phosphatases**. The **first family** removes phosphate groups from serine and threonine. It has two metal ions in its catalytic site: probably zinc and manganese or magnesium ions (■ Fig. 26.18, *left*). These are held in place by histidine and aspartic acid residues. A water molecule (or OH^-) bridges the two metal ions. It is, therefore, highly polarized and can make a nucleophilic attack on the phosphate group to be cleaved. The phosphate group also undergoes polarization and is prepared for nucleophilic attack by coordinating to the metal ions with two of its oxygen atoms. The intermediate collapses with the transient formation of a pentacoordinated phosphorus atom. The bond between the hydroxyl oxygen atom of the Ser or Thr residue and the phosphate group is cleaved. A neighboring histidine assists the cleavage by providing the necessary proton. The reaction is similar to that of phosphodiesterases (Sect. 25.8).

The **second group of phosphatases** does not use a metal ion for the cleavage reaction, but a covalent intermediate is formed during the reaction (■ Fig. 26.18, *right*). These phosphatases cleave phosphate groups from tyrosine residues. The formation of a very deep binding pocket, about 9 Å long, is characteristic of the latter phosphatases. It is fully formed only after substrate binding. A loop containing a tryptophan, proline, and aspartic acid (WPD loop) is located above the catalytic site and closes it to the outside. It contributes the catalytically important aspartic acid and is critical for substrate recognition (■ Fig. 26.18). In the closed substrate-bound state, aspartic acid forms an H-bond with the phenolic oxygen atom of the phosphotyrosine residue. This interaction polarizes the phosphate group and prepares it for nucleophilic attack. This step is accomplished by an adjacent cysteine residue located near the end of a long helix. In addition, an arginine helps to stabilize the transition state of the reaction, analogous to the oxyanion hole in serine or cysteine proteases. Similar to the acyl–enzyme complex formed in serine proteases (Sect. 23.2), the protein is transiently phosphorylated at the sulfur atom. The dephosphorylated substrate is released from the catalytic site. In the next step, a water molecule attacks and cleaves the phosphate group from the thiol group of the cysteine, which is polarized by the neighboring aspartic acid. This returns the catalyst to its initial state. The next reaction cycle can begin.

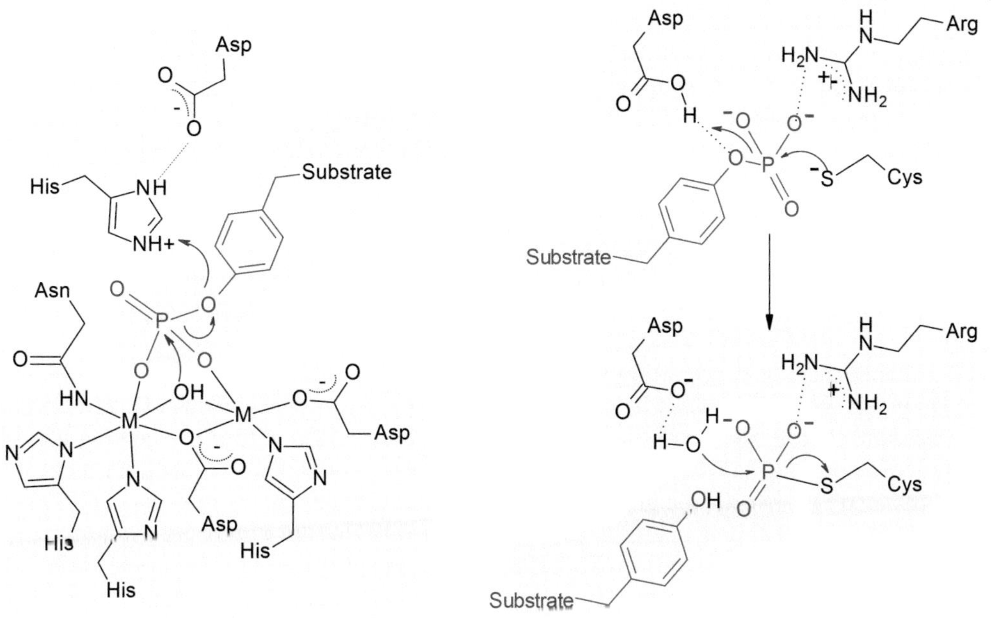

■ **Fig. 26.18** Two catalytic mechanisms have been described for the cleavage of phosphate groups from serine, threonine, and tyrosine in peptide substrates. The first group (*left*) uses two metal ions (presumably Zn^{2+} and Mn^{2+} or Mg^{2+}), which are coordinated by a histidine or aspartic acid. A water molecule (presumably in the form of an OH^- group) nucleophilically attacks the phosphate group of the substrate and initiates the cleavage. The second class of phosphatases begins the cleavage reaction with a nucleophilic attack by the thiolate group of a cysteine (*right*). The pK_a value of this cysteine is markedly shifted by the dipole moment of a helix that is pointing towards the site that accommodates the thiol group and the reaction starts from a deprotonated cysteine. Finally, a water molecule initiates the cleavage of the phosphate group from cysteine

Table 26.1 Examples for phosphatases that have been recognized as target structures for drug therapy

Family	Description	Disease, therapeutic approach
pSer, pThr	PP1, PP2A	Tumor suppression
	PP2B, PP2C	Cystic fibrosis
	(Calcineurin)	Immunosuppression
		Asthma
		Cardiovascular diseases
pTyr	PTP-1B	Diabetes, obesity
	CD45	Alzheimer's disease
	Shp2	Cancer therapy, immuno-oncology
Dual-specific phospha-tases	VHR, Cdc25	Regulation of MAP kinases Stimulation of the cell cycle Anticancer therapy

While the first and second families of phosphatases process different substrates by completely different mechanisms, there is a **third family** that functions similarly to the second group of tyrosine phosphatases. It has dual specificity and can cleave phosphate groups from serine, threonine, and tyrosine. Unlike the specific tyrosine phosphatases, it has a shorter binding pocket that allows phosphotyrosine as well as the shorter phosphoserine and phosphothreonine to reach the catalytic site.

So far, the genes for **189 phosphatases** have been discovered in our genome. In contrast to protein kinases, where folding is conserved across all catalytic domains, phosphatases show greater diversity. So far, **10 different folding patterns** have been reported for these proteins. The majority (106 examples) are phosphatases that use a thiolate group of a cysteine residue for nucleophilic attack. The second largest group with 20 examples uses the two metal ions in the catalytic center. Many phosphatases intervene in signaling cascades by **targeted dephosphorylation**. Most of them remove phosphate groups from activated proteins, thereby, deactivating the receptors involved. However, the processes can be even more complicated. Phosphorylation can also hold a protein complex in an inactivated state and release its physiological function by removing the phosphate group (see the example of Shp2 in Sect. 26.9). Often, phosphatases are active as catalytic domains in combination with larger protein assemblies of signal transduction. Since the phosphate group as well as the phosphorylated amino acids and nearby residues are involved in the interaction with the phosphatase, the selectivity problem is not as severe as with the kinases. However, the small-molecule drug to be developed competes with the recognition site of a highly polar protein substrate, which does not make its development any easier. Forty of the 189 phosphatases (21%) have been identified as targets in many different disease areas. Of these, 12 are associated with cancer. For kinases, the number is slightly higher (35%) and the proportion in cancer is also higher (20%). Some examples of drug development are summarized in Table 26.1. The example of PTB-1B, a receptor tyrosine phosphatase that has been pursued by many pharmaceutical companies as an innovative target enzyme for the treatment of diabetes and obesity, illustrates how potent inhibitors of phosphatases can be developed.

26.8 Inhibitors of PTP-1B: Treatment for Diabetes and Obesity?

Adult-onset type 2 diabetes and **obesity** are diseases that have increased alarmingly in our society in recent years. They must be considered as typical diseases of civilization. **Adult-onset diabetes** is based on increasing insulin resistance, which is observed as a reduced ability of cells in the target organ to respond to insulin. As a result, high blood insulin levels occur even when blood glucose levels are normal. Because of the resistance, the cells no longer respond as they should to the signal that insulin would send in a healthy person. Insulin causes the uptake of glucose from food into liver cells, where glucose is stored in the form of glycogen. As resistance increases, pathophysiological changes occur due to inadequate insulin control. The uptake of blood glucose into tissues and the release of glucose from the liver become imbalanced. As a result, blood glucose levels rise even higher, which can lead to complications such as coronary heart disease, retinopathy, cataracts, and vascular disease.

The other disease of civilization is much more obviously seen: obesity. The signs are a disproportionate excess of body mass. Even more alarming is the fact that obesity is by no means limited to old age. Even among young people, the number of cases of obesity is increasing dramatically. Today, about a quarter of adults worldwide are overweight. In developed countries, the numbers are much higher. In the U.S., nearly 75% are considered overweight and 40% are obese. In developing countries, too, the percentage is rising sharply. Of course, this has something to do with our changing lifestyles. An overabundance of food, often without dietary fiber, coupled with a lifestyle that requires less and less physical labor has led to this development. In addition, genetic predisposition contributes to the development of obesity.

Interestingly, the development of type 2 diabetes and obesity often occur together, increasing the health risks for the patient. The resulting symptoms are called **metabolic syndrome**. For this diagnosis, the following additional criteria apply: an abdominal girth of more than 80 cm in a woman or 90 cm in a man, and two of the following additional factors: an elevated triglyceride level (> 150 mg/dL), an elevated fasting glucose level (> 100 mg/dL), arterial hypertension (> 130/85 mmHg), and/or a reduced HDL cholesterol level (< 40–50 mg/dL; Sect. 27.3). The cost to society of this increased health risk is difficult to estimate, but it is likely to be dramatic. Therefore, great efforts have been made to find drug therapies that can counteract the metabolic syndrome and its consequences.

The correlation between insulin resistance and obesity is not yet fully understood at the molecular level. In fact, insulin is a hormone that is related to fat metabolism and influences fat deposition. For example, it influences fat storage, but insulin deficiency leads to weight loss. **Insulin** is bound to the insulin receptor, which is **autophosphorylated** by its tyrosine kinase domain in response to this signal (Sect. 29.8). This initiates a cascade of several kinases that culminates in the synthesis of the sugar-storing glycogen. The synthesis of fatty acids and proteins is also induced. **Dephosphorylation of the insulin receptor** attenuates its function. **PTB-1B tyrosine phosphatase** cleaves phosphate groups from two tyrosine residues on the receptor. This leads to deactivation of the insulin receptor and the cascade initiated by the receptor. Blocking this dephosphorylation step seems to be a rewarding concept to counteract insulin resistance. The real stimulus for the search for PTP-1B inhibitors was the observation that mice with a knocked-out *ptp-1b* gene are resistant to developing obesity despite no changes to their diet, and their insulin sensitivity is increased without any apparent negative consequences. This spectacular observation suggested that the ideal target had been found to fight the most prominent disease of civilization. This optimism was reinforced by the fact that antisense nucleotides (Sect. 32.4), which block the expression of PTP-1B, also cause an increased insulin effect. As a result, nearly every pharmaceutical company of note flocked to this enzyme to develop potent inhibitors. Within four years, more than 200 patent applications appeared in the literature!

Has PTP-1B proven to be an easy target? The mechanism of action is shown in the previous Sect. 26.7. The catalytic cysteine, which temporarily accommodates the cleaved phosphate group, aligns itself at the tip of a long helix oriented towards the catalytic site. Such a helix creates special electrostatic conditions at its terminal end (Sects. 30.2 and 30.8) and can stabilize charged species well. The catalytic center also contains an aspartic acid and an arginine. The structure with the phosphorylated tyrosine **26.39** (*green*; ◘ Fig. 26.19) is part of a substrate. The complex with this substrate could be determined because the enzyme was rendered almost catalytically inactive by replacing the catalytic Cys 215 with an analogous serine, but remained geometrically unchanged. The phosphate group is bound in a tight network of H-bonds. The phenyl ring of tyrosine is held in a hydrophobic clamp by two adjacent aromatic residues, Tyr 46 and Phe 182. These two residues also determine the depth and width of the entrance to the catalytic site of the phosphatase (◘ Fig. 26.20, *upper left*). First, an

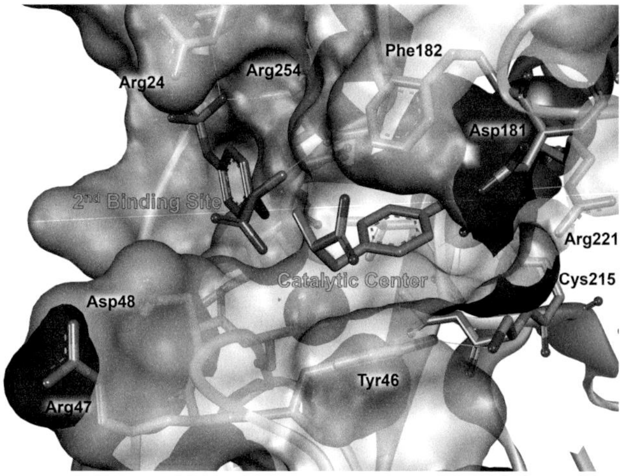

◘ **Fig. 26.19** The crystallographically determined binding mode of a phosphorylated tyrosine (**26.39**, *green*, ◘ Fig. 26.21) as a minimal mimetic for a peptide substrate in the human phosphatase PTP-1B. The phosphate group is held in place by Arg 221 and Cys 225 which is positioned for nucleophilic attack. Asp 181 is found above the Cys residue and buffers for the protonation inventory. The entrance to the binding pocket is bordered by the two aromatic residues Phe 182 and Tyr 46. The binding position of the cysteine is found at the end of a long helix. The displayed geometry is based on a crystal structure with the catalytically inactive Cys → Ser mutant. A second phosphotyrosine (*pink*) is found in the crystal structure that binds to Arg 24 and Arg 254 in a second distal pocket. Consequentially, the occupancy of this second binding pocket was important for the development of nanomolar PTP-1B inhibitors (cf. ◘ Fig. 26.21). (► https://sn.pub/QLME5k)

attempt was made to replace the phenolic oxygen atom of the tyrosine residue attaching the phosphate group of the substrate **26.39** with a nonhydrolyzable mimetic such as **26.40** (◘ Fig. 26.21). A CF_2 group was chosen to replace the oxygen atom. However, attempts have also been made to replace the fluorine on the bridging carbon with an OH group. Alternatively, dicarboxylic acids were considered as head groups. The polar properties of the compound were essentially retained, but the hydrolytic stability was significantly improved. A fragment-based screening approach using crystallography and NMR spectroscopy (Sects. 7.8 and 7.9) was used to discover oxalic anilide **26.41** and *N*-oxalylanthranilic acid **26.42** as potential phosphotyrosine mimics. The thiophene analog **26.43** proved to be a submicromolar inhibitor. Surprisingly, in the crystal structure with phosphotyrosine, a second molecule of **26.39** (*pink*) was found to be bound (◘ Figs. 26.19, 26.20, *upper left*). It binds adjacent to the first molecule (*green*) and occupies a second pocket formed by Arg 24, Arg 254, Gln 262, and Asp 48. However, the affinity for this binding site was only in the millimolar range. Nevertheless, the discovery led to the

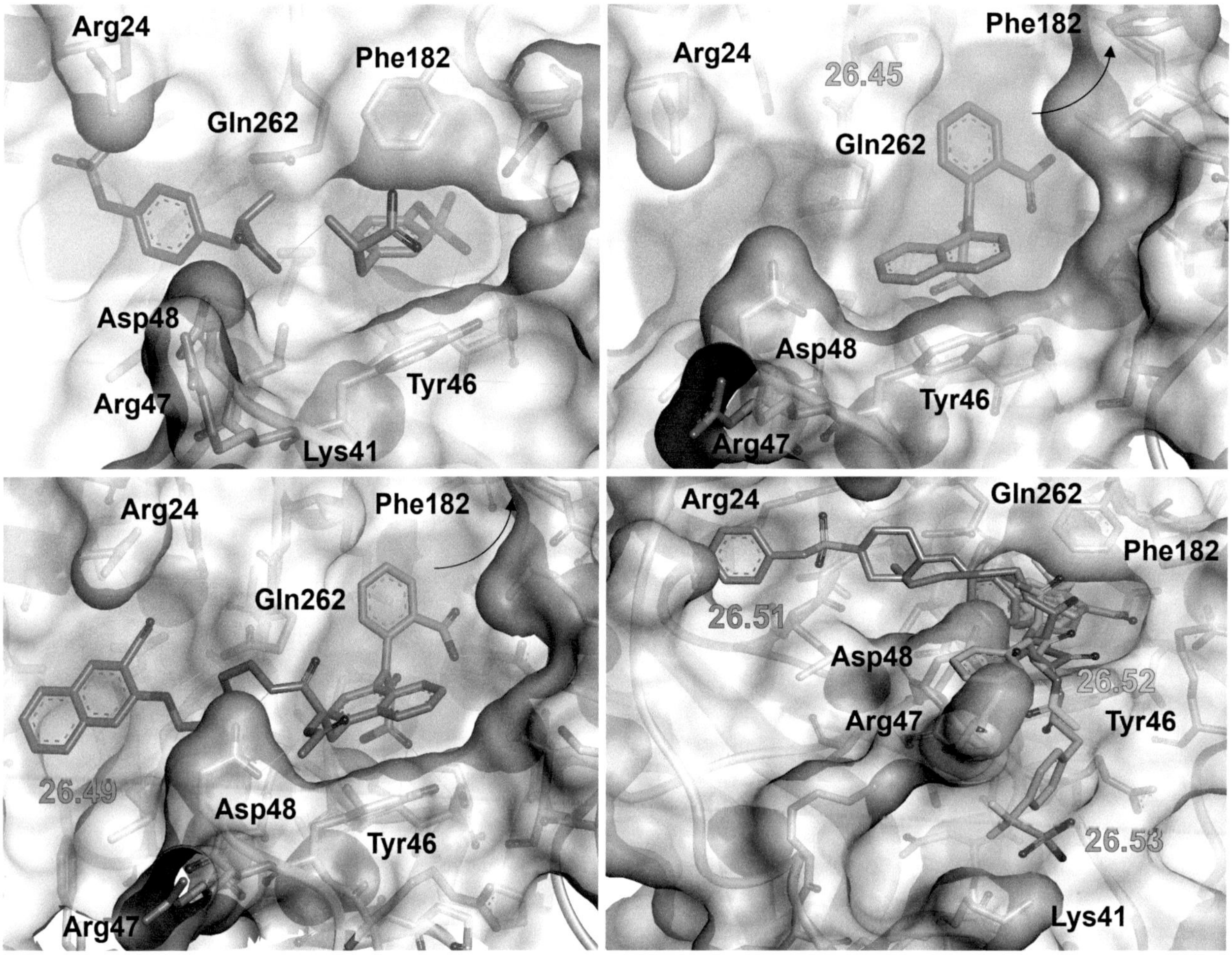

Fig. 26.20 *Upper left* Binding mode of the substrate-analogous phosphotyrosine (**26.39**, Fig. 26.21) in human PTP-1B. The phosphate group binds deeply in the catalytic center (*green*). The two hydrophobic amino acids Phe 182 and Tyr 46 form a narrow entry portal to the catalytic site. A second phosphotyrosine (*pink*) is found in the crystal structure that binds to Arg 24 and Arg 254. *Upper right* Crystal structure of an aromatic oxalic acid derivative (**26.45**) that was developed at Abbott to occupy the catalytic site (*green*). The compound induces a rearrangement of the Phe 182 side chain and opens the catalytic site to the top. *Bottom left* By chemically coupling an aromatic carboxylic acid that was discovered with the SAR-by-NMR method as a binder for the second binding site (*pink*) and a mimetic to occupy the catalytic site, a nanomolar inhibitor **26.49** (Fig. 26.21) was obtained. *Bottom right* To achieve selective binding to PTP-1B compared to the structurally very similar TCPTP, structural differences at position 41 were exploited (*light blue*). There PBP-1B has a lysine, and the related family member TCPTP has an Arg in this position. The nanomolar inhibitor **26.53** (*green*) achieves a significant selectivity advantage. **26.52** (*light blue*) binds with an uncharged head group into the catalytic center. The nanomolar inhibitor **26.51** (*purple*) also binds into the catalytic center but, like **24.49**, its oriented towards the second phosphotyrosine binding site (*pink*). (▸ https://sn.pub/D1DtFZ)

pivotal idea of coupling the active site phosphotyrosine mimetic to a molecular building block occupying this second binding site. The plan was to create inhibitors with a much higher binding affinity.

Aromatic oxalic acid derivatives such as **26.44** and **26.45** have also been worked on at Abbott as substrate mimics for binding to the catalytic site. Interestingly, the derivatives pursued by Abbott forced a conformational change at Phe 182 at the entrance, so that the top of the catalytic site is opened (Fig. 26.20, *upper right*). Abbott additionally applied their SAR-by-NMR technique (Sect. 7.8) to discover potential binders for the second binding site. Small aromatic acids such as **26.46**–**26.48** were discovered. By coupling such moieties (e.g., naphthyl carboxylic acids) and the already known mimetic **26.45** to bind to the catalytic center produced the nanomolar inhibitor **26.49** (K_i = 22 nM, Fig. 26.20, *lower left*).

This second binding site was determinant for the lead structure optimization. At Novo Nordisk, the initial oxalic acid derivatives on the thiophene ring were expanded by using Asp 48 as an additional anchor point to arrive at more potent and selective inhibitors based on scaffold **26.50**. Wyeth also focused more on the second binding site and developed a dicarboxylic acid derivative

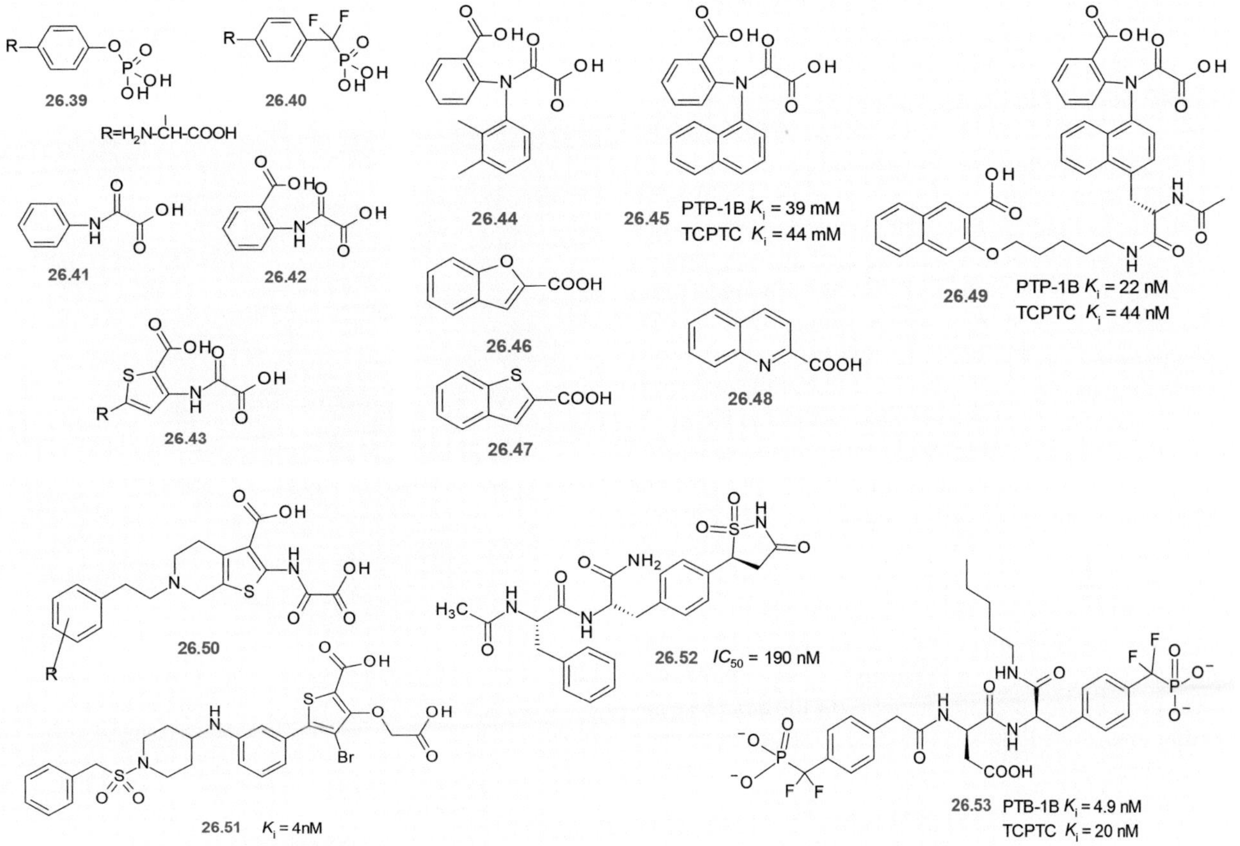

Fig. 26.21 By starting with a substrate with a terminal phosphotyrosine **26.39**, a hydrolytically stable compound **26.40** was developed. A fragment screening drew attention to the two mimetics **26.41** and **26.42**. Thiophene derivatives such as **26.43** were designed from the latter compound. At Abbott, analogous aromatic oxalic acid derivatives **26.44** and **26.45** were developed. Screening by the SAR-by-NMR method discovered aromatic carboxylic acids such as **26.46**–**26.48** as ligands for the second binding site. By chemically linking such aromatic carboxylic acids as binders for the second binding site and a mimetic for the phosphotyrosine in the catalytic site, **26.49** was obtained as a nanomolar inhibitor. Also at Novo Nordisk, the first lead structures were equipped with side chains for the second binding site (**26.50**). The thiophene derivative **26.51** orients from the catalytic center to the second binding site and achieves nanomolar affinity. Inhibitor **25.52** binds to the catalytic center with an uncharged head group. With **26.53**, a fourfold more selective inhibitor of PTP-1B than TCPTP was prepared

on a thiophene ring **26.51** as a single-digit nanomolar inhibitor. Also of note is compound **26.52** from Incyte Corporation, which binds to the catalytic center with an uncharged head group (■ Fig. 26.20, *lower right*).

The development of highly potent, PTP-1B selective, and orally available inhibitors was overshadowed by another observation. Sequence comparisons suggested that there is another phosphatase, the **T-cell protein tyrosine phosphatase TCPTP**, which is highly similar to PTP-1B. Such an observation is worrisome because the PTP-1B inhibitors in development may also inhibit this phosphatase. The crystal structure published in 2002 confirmed this suspicion: the sequence identity of the catalytic domains is 74%, and the WPD loop, which is located above the catalytic site after substrate binding, is identical. Knock-out mice lacking the *tcptp* gene are born healthy but die within 3–5 weeks of birth. More alarmingly, knocking out both the *ptp-1b* and *tcptp* genes simultaneously resulted in animals that had no chance of survival. This underscores the extreme danger that insufficiently selective PTP-1B inhibitors that also inhibit T-cell protein tyrosine kinase could lead to a life-threatening situation. The need was great. What are the structural differences between the two phosphatases that could be exploited to design sufficiently selective compounds? All of the inhibitors developed at that time showed almost equipotent affinity for both proteins. Bidentate inhibitors such as **26.53** (■ Fig. 26.21), reported in 2003, proved very interesting because they occupy the catalytic site and neglect the second binding site (■ Fig. 26.20, *lower right*). Even the sequence of this region turned out to be virtually identical to that of TCPTP. With a slightly different orientation, the new inhibitors target a lysine residue (Lys 41), which is an arginine in TCPTP. At least the nanomolar inhibitor **26.53** has a modest selectivity advantage for PTP-1B compared to TCPTP.

The Sunesis company took a completely different approach. In 2004, they reported the discovery of an **allosteric binding site 20 Å away** on the back side of the catalytic site in PTP-1B. An inhibitor that binds with micromolar affinity to the enzyme was developed for this site. It blocks its function by preventing the closure of the WPD loop. In this way, the loop cannot fold upon the substrate-binding site. The essential residues such as the catalytically active aspartic acid are not brought in the vicinity of the substrate. The most potent ligand from this series, **26.54** (IC_{50} = 8 µM), wraps itself around a phenylalanine that is found there, as proven by the crystal structure (◘ Fig. 26.22). In the structurally analogous TCPTP, a cysteine is found at this position and forms entirely different interactions with the aromatic groups of this ligand. Due to the deviating interaction pattern, this compound achieves TCPTP inhibition at only 280 µM. Perhaps blocking this allosteric binding site will open a new perspective for the selective inhibition of PTP-1B. The future must show whether the severe selectivity problem can be resolved in an appropriate way. It should be noted that all hopes for influencing this seemingly ideal target protein are currently focused on the **antisense nucleotide** mentioned above, which is currently in clinical trials (Sect. 32.4).

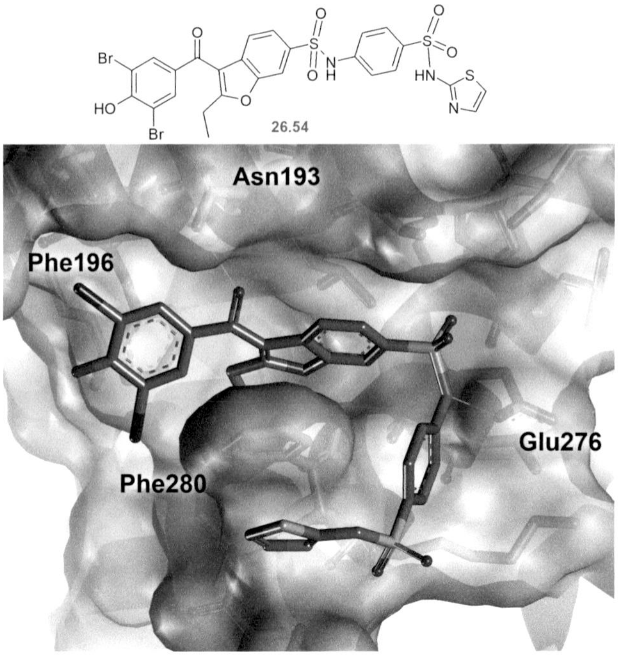

◘ **Fig. 26.22** A new allosteric binding site was discovered at Sunesis that is approximately 20 Å away from the catalytic site of the phosphatase. Compound **26.54** inhibits PTP-1B 16-fold more strongly than TCPTP. The crystal structure with PTP-1B shows that the inhibitor basically wraps itself around the exposed Phe 280. In TCPTP, a cysteine residue is found in the same position. (▶ https://sn.pub/92STnr)

26.9 Molecular Glue Inhibits the Release of Phosphatase Activity

Focusing on allosteric binding sites seems to be a much more promising approach for the development of phosphatase inhibitors than blocking the very polar catalytic center. It is inevitable that inhibitors of this site will be very polar. Therefore, they will unavoidably have bioavailability problems. As mentioned above, phosphatases are often involved as catalytic domains in larger protein assemblies for signal transduction. In terms of drug therapy, it is important whether or not such a complex is modulated in its biological function. Modulation can also be achieved by drugs that act on these complexes at a completely different site. Thus, we are not limited to blocking the active site of the phosphatase domain. A successful example of this approach is the development of the Shp2 inhibitor by Novartis. We will see that in this example, dephosphorylation removes autoinhibition and consequently releases a blocked phosphatase function. Therefore, **maintaining autoinhibition** is a promising strategy to indirectly **inhibit phosphatase activity**.

Shp2 is a phosphatase belonging to the **nonreceptor tyrosine phosphatase** subfamily. It is responsible for regulating numerous signaling pathways in normal and glioma cells. As a result, it is an anticancer target and plays an important **role in immuno-oncology**. Inhibition of Shp2 phosphatase has emerged as a promising approach for drug development against *glioblastoma multiforme* (a malignant brain tumor), a dreaded cancer with poor prognosis and low survival. As seen with PTP-1B, the high positive charge near the active site poses significant problems for inhibitor development, particularly with respect to sufficient cell permeability and bioavailability. While numerous small-molecule inhibitors of Shp2 have been described, their polar nature means that they are simply not ideal for therapeutic development.

Shp2 phosphatase consists of a catalytic phosphatase domain and two SH2 domains (◘ Fig. 26.23). In the absence of a phosphorylated tyrosine substrate, the *N*-terminal **SH2 domain** binds directly to the **phosphatase domain** and **blocks its active site**. A loop of the *N*-SH2 domain inserts itself into the catalytic center. Like a conformational switch, it either inhibits the phosphatase or binds phosphoproteins and activates the enzyme.

The Novartis researchers started with the concept of finding an allosteric inhibitor that would block the activation of the phosphatase function. They first performed a high-throughput screen using the entire Shp2 protein. Since inhibitors of the catalytic center were undesirable and should be discarded, the screening was repeated with the hits found, but now only with the truncated

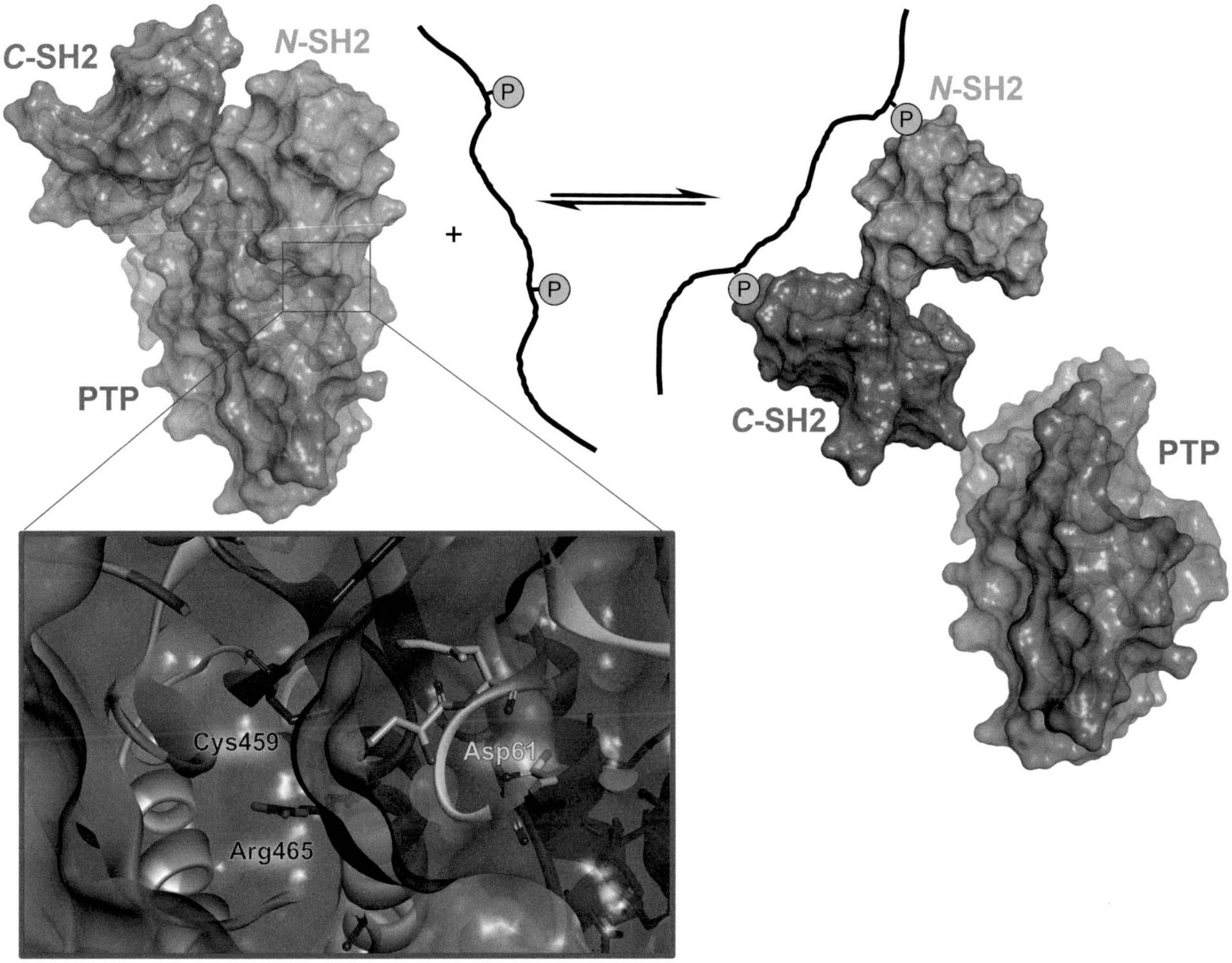

Fig. 26.23 *Left* When no phosphorylated substrate is present, the Shp2 phosphatase is self-inhibited by its *N*-SH2 domain. For this purpose, this domain binds with an exposed loop (*red box*, *yellow loop*) via Asp 61 to the catalytic center of the phosphatase domain (Cys 459, *purple*). *Right* Only in the presence of a phosphorylated substrate is the phosphatase domain exposed and can become catalytically active (based on a figure in Fortanet et al., J. Med. Chem., 59, 7773–7782 (2016)). (▶ https://sn.pub/lEIYDj)

phosphatase domain without the SH2 domains. This revealed among the nondiscarted hits from the first screen the aminopyrimidine **26.55** as a promising candidate. In the crystal structure with full-length Shp2, this hit was found to be an **allosteric binder** that interacts with all three domains of Shp2 (Fig. 26.24). Thus, it keeps Shp2 in the **autoinhibited, inactive conformation**. The aminopyrimidine **26.55** was optimized through several design cycles to the pyrazine **26.56**, a selective, well soluble, orally bioavailable, and potent Shp2 inhibitor. It has shown promising antitumor activity in animal models. The compound is now being tested in more advanced clinical trials. This success story shows that there are alternative ways to block the function of a seemingly "undruggable" phosphatase with a small molecule inhibitor. If necessary, this can also be done in an indirect way!

26.10 Inhibitors of Catechol-O-Methyltransferase

A large family of transfer enzymes are the **methyltransferases**, which add methyl groups to other biomolecules. DNA methyltransferases are an important group in this family. Their function is to chemically modify nucleobases at specific sites on DNA or RNA by transferring methyl groups. These methylations do not alter the genetic code, meaning the same amino acids are still translated into the gene product. However, they serve as a kind of label for DNA strands, e.g., to distinguish between the cell's own and foreign DNA or to distinguish between original and newly synthesized strands. Another group of methyltransferases transfer methyl groups to

26

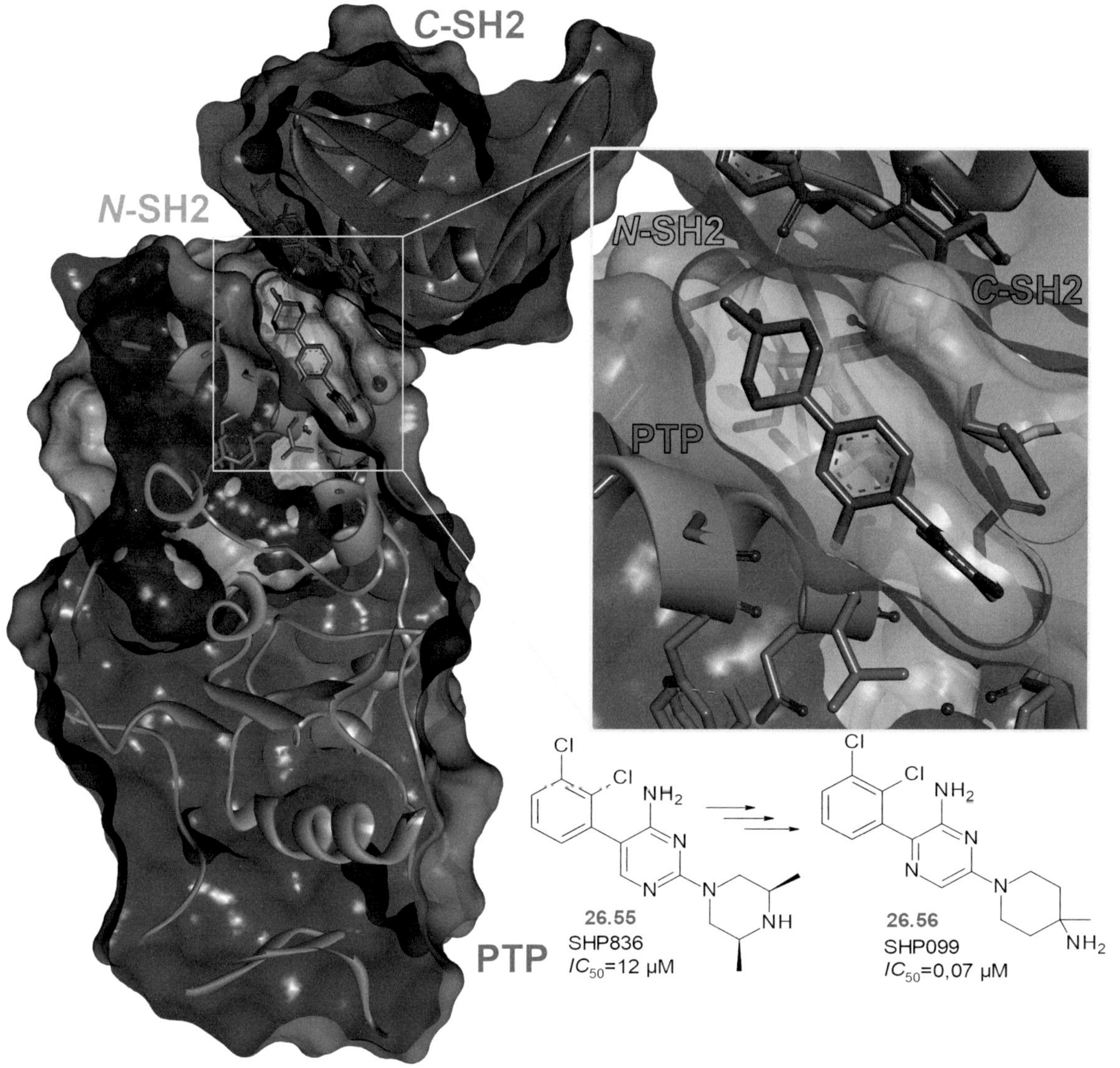

Fig. 26.24 Crystal structure of full-length Shp2 phosphatase in the inactive form autoinhibited by the *N*-SH2 domain. The compound **26.55** discovered in a screening campaign could be optimized to the potent inhibitor **26.56**, which, like a molecular glue, stabilizes the inactive form of Shp2 by simultaneously binding to the three domains (*yellow box*). (► https://sn.pub/VfL9Ak)

oxygen, nitrogen, or sulfur atoms in small biomolecules. Methyltransferases use ***S*-adenosyl-L-methionine (SAM 26.57)** as a cofactor (Fig. 26.25). In the transmethylation reaction, a highly reactive methyl group is transferred from the sulfonium group of this donor molecule to the substrate.

Inhibitors of **catechol-O-methyltransferase (COMT)** have gained importance in pharmaceutical therapy. This enzyme deactivates the endogenous function of catecholamines such as dopamine, adrenaline, or noradrenaline by transferring a methyl group to the phenolic hydroxyl group of these neurotransmitters. Polymorphisms in this enzyme have been associated with psychiatric changes that may be related to anxiety disorders and schizophrenia. Inhibitors of this enzyme are used in therapy, particularly in the treatment of Parkinson's disease. This disease, originally known as "shaking palsy," occurs primarily in older people. It is caused by a slow, progressive degeneration of dopaminergic neurons in the *substantia nigra* of the midbrain. A causal treatment of the neuronal degeneration has not yet been achieved. Therefore, attempts are being made to counteract the dopamine deficiency with exogenous replacement substances. The amino acid **L-DOPA** has already been introduced in Sect. 9.4 as a precursor of dopamine. Although it has a more polar character than dopamine, it can penetrate the blood–brain barrier because it uses an amino acid transporter to enter the brain. In practice, however, only about 1% of the administered

26.58

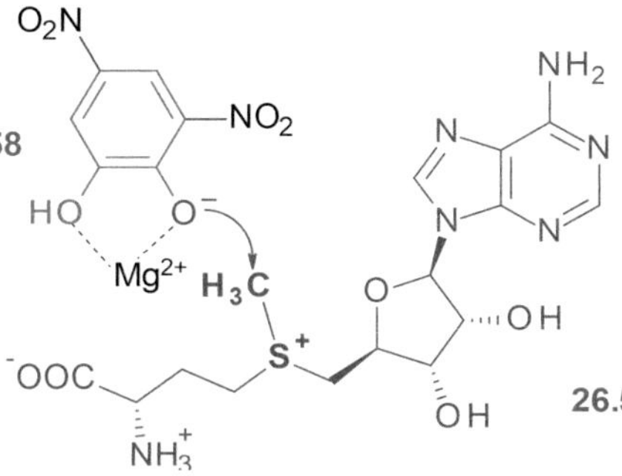

26.57 *S*-Adenosyl-L-Methionine (SAM)

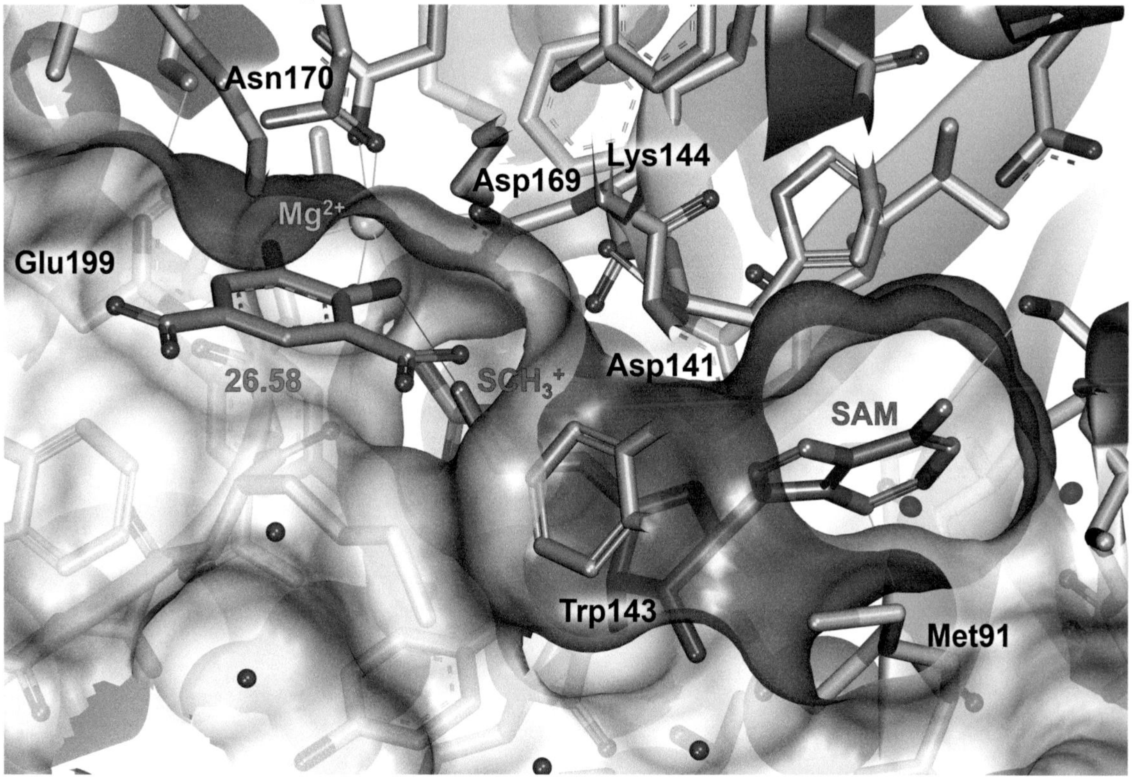

Fig. 26.25 The crystal structure of COMT with the cofactor *S*-adenosyl-L-methionine **26.57** (*magenta* carbon atoms) and the catecholamine-analogous nitro-substituted inhibitor **26.58** (*green* carbon atoms). The methyl group that is to be transferred to the phenolic oxygen atom (*red*) is within a short distance (2.63 Å, *violet line*). The phenolic oxygen, which is the nucleophile in the transfer reaction, is presumably deprotonated because of the electron-withdrawing effect of the nitro groups and the close proximity to the magnesium ion, the sulfonium group, and the ammonium group of Lys 144. The accumulated positive charges also shift the pK_a value of this hydroxyl group into the acidic range. The second phenolic OH group is probably uncharged and forms an H-bond to Glu 199. (► https://sn.pub/QTslOa)

amount reaches the brain. The vast majority is degraded in the periphery by decarboxylases. To prevent this degradation and the side effects associated with peripheral dopamine release, a decarboxylase inhibitor is administered at the same time. This inhibitor must be sufficiently polar to prevent it from crossing the blood–brain barrier (e.g., benserazide **9.39**, Fig. 9.9). This strategy significantly increases the bioavailability of L-DOPA in the brain. The drug is degraded by monoamine oxidases (Sect. 27.8) and by catechol-O-methyltransferases. COMT recognizes both L-DOPA and dopamine as substrates. They are inactivated by the transfer of a methyl group to their phenolic hydroxyl groups. Inhibition of COMT allows the bioavailability of L-DOPA to be further enhanced and a higher concentration of dopamine to be achieved in the brain.

The crystal structure of the enzyme was solved in 1994 by the group of Anders Liljas at Lund University, Sweden (Fig. 26.25). The mechanism involves a deeply buried magnesium ion that assumes an octahedral coordination geometry. The adjacent oxygen atoms of the catecholamine are chelated with the magnesium ion. This brings the phenolic oxygen atom into close proximity (2.63 Å) to

26

Fig. 26.26 Pyrogallol **26.59**, gallic acid **26.60**, or tropolone **26.61** bind to COMT with micromolar affinity. Tolcapone **26.62**, entacapone **26.63**, nitecapone **26.64**, or nebicapone **26.65** have strong electron-withdrawing groups directly on or conjugated to the aromatic ring. These compounds are nanomolar, competitive inhibitors of catecholamine. The linking of two moieties, each analogous to catecholamine or adenosine with a rigid five-membered tether (amide bond and double bond, *red*) affords the nanomolar bisubstrate-analogue inhibitor **26.66**

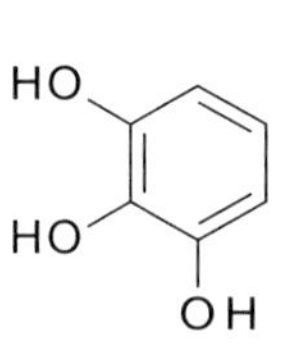

26.59 Pyrogallol

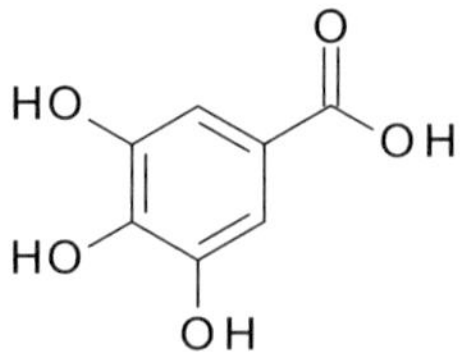

26.60 Gallic acid

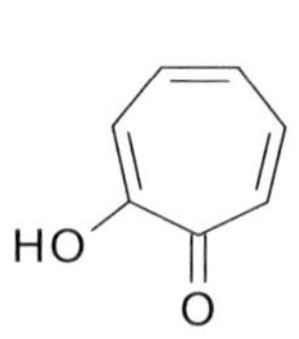

26.61 Tropolone

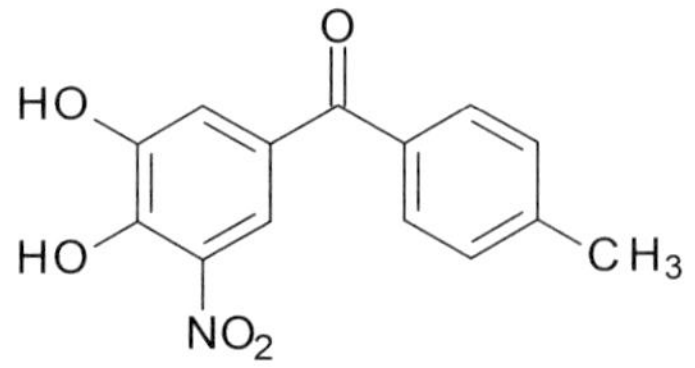

26.62 Tolcapone IC_{50}= 0.3 nM

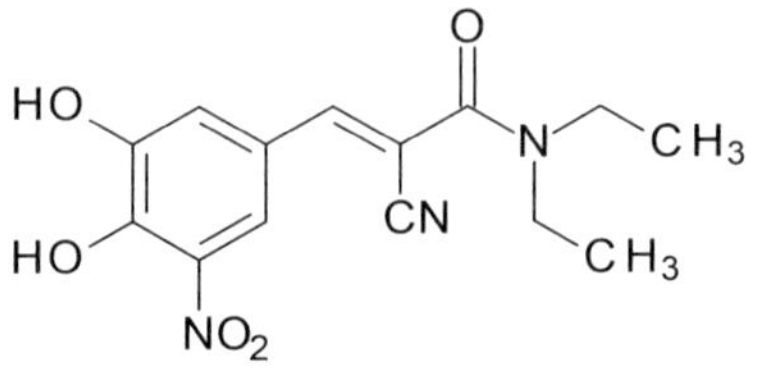

26.63 Entacapone IC_{50}=0.3 nM

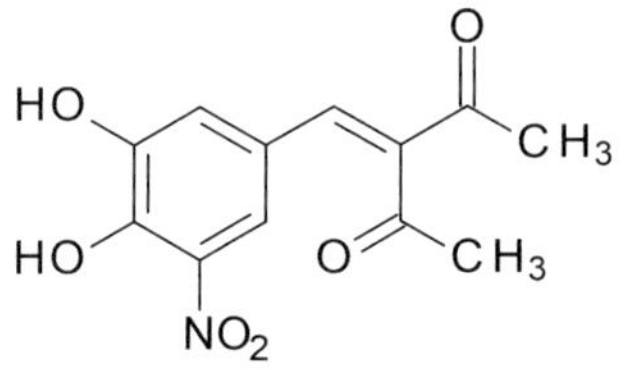

26.64 Nitecapone IC_{50}= 1 nM

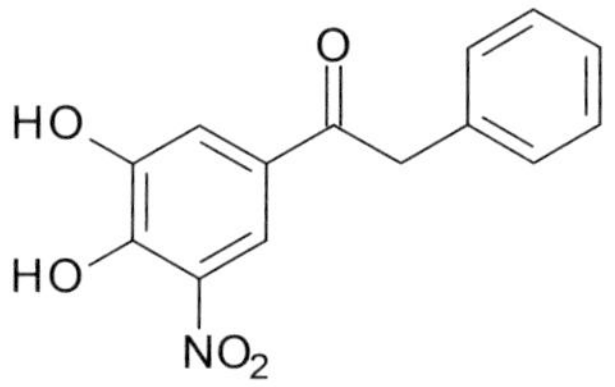

26.65 Nebicapone

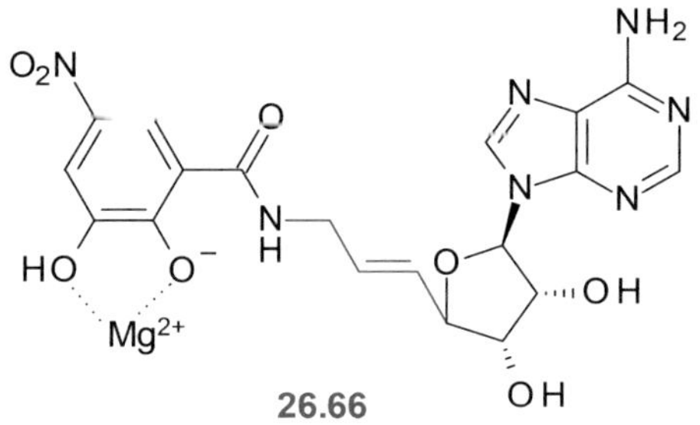

26.66

the sulfonium group. Presumably, this hydroxyl function is deprotonated due to its proximity to the magnesium ion, the sulfonium group, and the ammonium group of Lys 144, increasing its nucleophilicity for the S_N2-like transfer of the methyl group from the positively charged sulfur of SAM **26.57**. The second, probably uncharged phenolic OH group is involved in a hydrogen bond with Glu 199. The crystal structure was determined with a substrate-like inhibitor **26.58** in which the nucleophilicity of the oxygen atom is very strongly suppressed by two electron-withdrawing nitro groups (Fig. 26.25). Methyl transfer does not occur any longer.

Molecules with multiply hydroxylated aromatic rings such as pyrogallol **26.59**, gallic acid **26.60**, or tropolone **26.61** show weak, micromolar affinity for the enzyme. The introduction of strongly electron-withdrawing nitro or carbonyl groups to the aromatic rings leads to a significant increase in the affinity of these substrate-like inhibitors. The inhibitors tolcapone **26.62**, entacapone **26.63**, nitecapone **26.64**, or nebicapone **26.65** all have a substitution pattern with a nitro group in *ortho*-position to the nucleophilic hydroxyl group and a second electron-withdrawing group in *para*-position (Fig. 26.26). Crystallographic studies of these derivatives have shown that their nitro groups match that of **26.58**, which is oriented towards the SAM substrate (Fig. 26.25). The second electron-withdrawing substituent is located where the other nitro group of **26.58** is, and is oriented towards the surrounding solvent. Tolcapone **26.62** was approved in 1997 as a peripherally and centrally acting COMT inhibitor. Its therapeutic use has been severely limited due to observed liver toxicity. Concomitant administration of L-DOPA and entacapone **26.63**, which acts predominantly in the periphery, has proved more beneficial. It has been on the market since 1998 and contributes to a balanced level of L-DOPA.

All of these drugs compete with catecholamines for a magnesium ion in the binding site. In recent years, nanomolar bisubstrate inhibitors such as **26.66** have been developed. They displace both the SAM cofactor and catecholamine from the binding pocket. The original building blocks of the parent molecular model compounds can be seen in the bisubstrate inhibitors. The crystal structure of one of these inhibitors is shown in

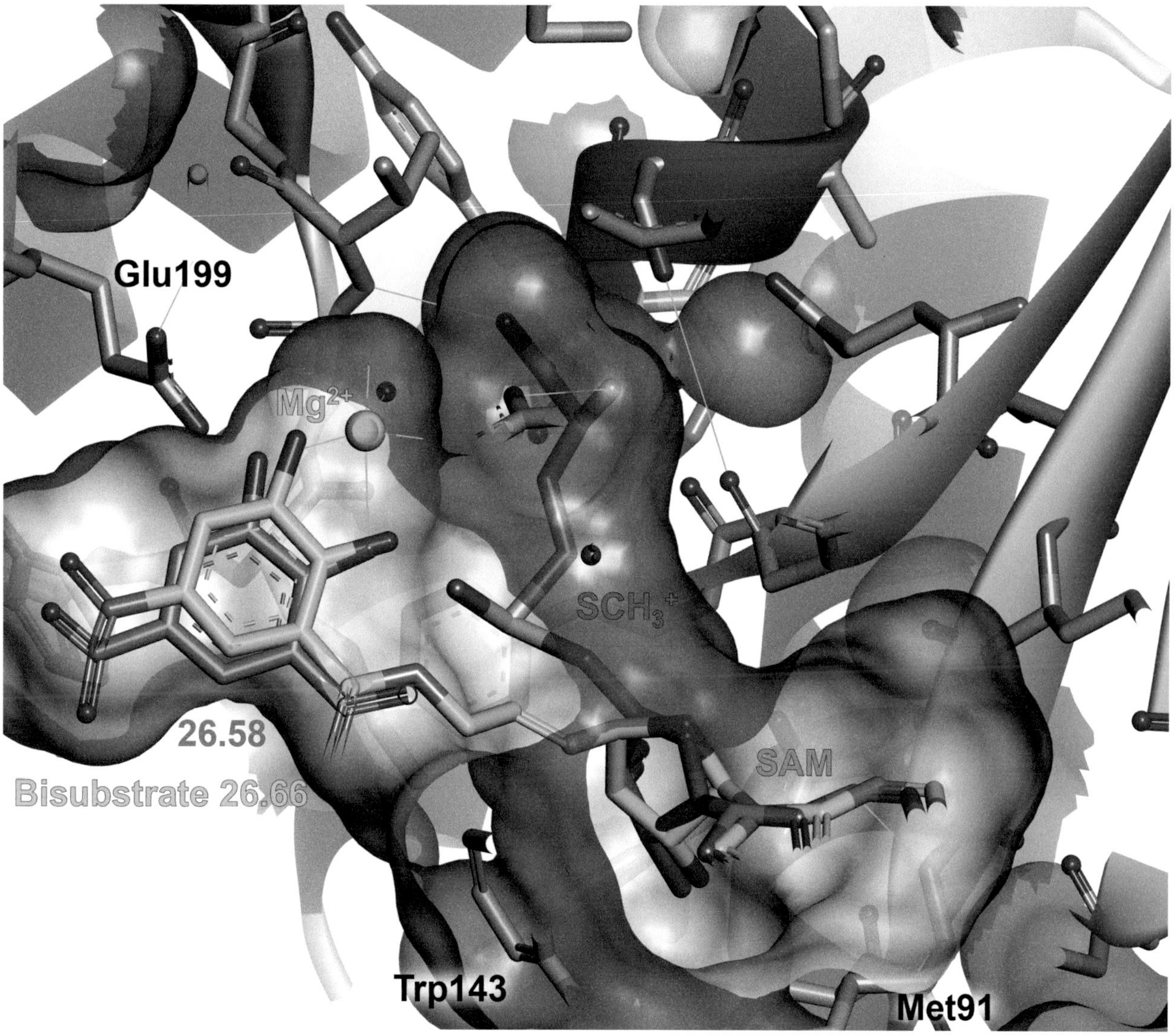

Fig. 26.27 Superposition of the crystal structures of COMT with SAM **26.57** (*magenta* carbon atoms) and the catecholamine-like inhibitor **26.58** (*green* carbon atoms) with the bisubstrate inhibitor **26.66** (*light blue* carbon atoms). (► https://sn.pub/JOQ0rm)

Fig. 26.27. Its binding geometry largely matches the adenosine part of SAM on the nucleoside side and the catecholamine side with the nitro aromatic ring. The correct choice of the connecting bridge between the two substrate-analogous parts of the molecule is crucial for the binding affinity. A rigid five-membered chain consisting of an amide group and an *E*-configured double bond represents the optimum. Transitioning to a more flexible structure by hydrogenation of the double bond reduces the binding affinity by a factor of 100. Lengthening the chain with an additional member results in a further 25-fold reduction in binding affinity. **Bisubstrate-analogue inhibitors** are expected to achieve higher selectivity for their target enzymes. In the present case, the rigid and geometrically strained linker between the two parts of the molecule is responsible for the pre-organization necessary for the binding of the pharmacophoric groups. This pre-organization of the ligand provides an advantage in binding to the receptor. Further development must show whether such bisubstrate inhibitors have a chance of entering into drug development.

26.11 Blocking the Transfer of Farnesyl and Geranyl Anchors

Kinases and phosphatases are not the only proteins that undergo posttranslational during signal transduction. The spatial location of proteins is often essential for their proper function in the cell. Some proteins need

26

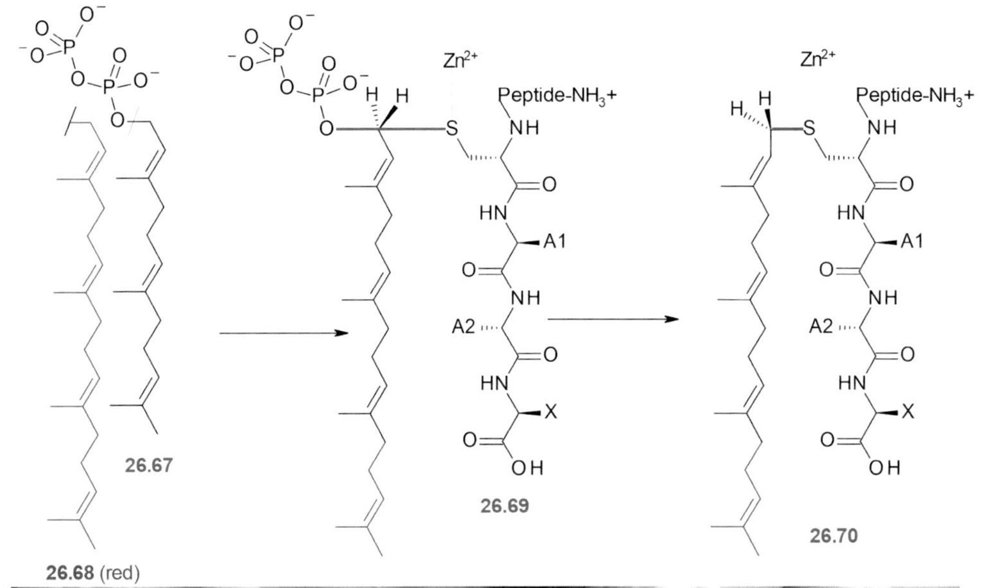

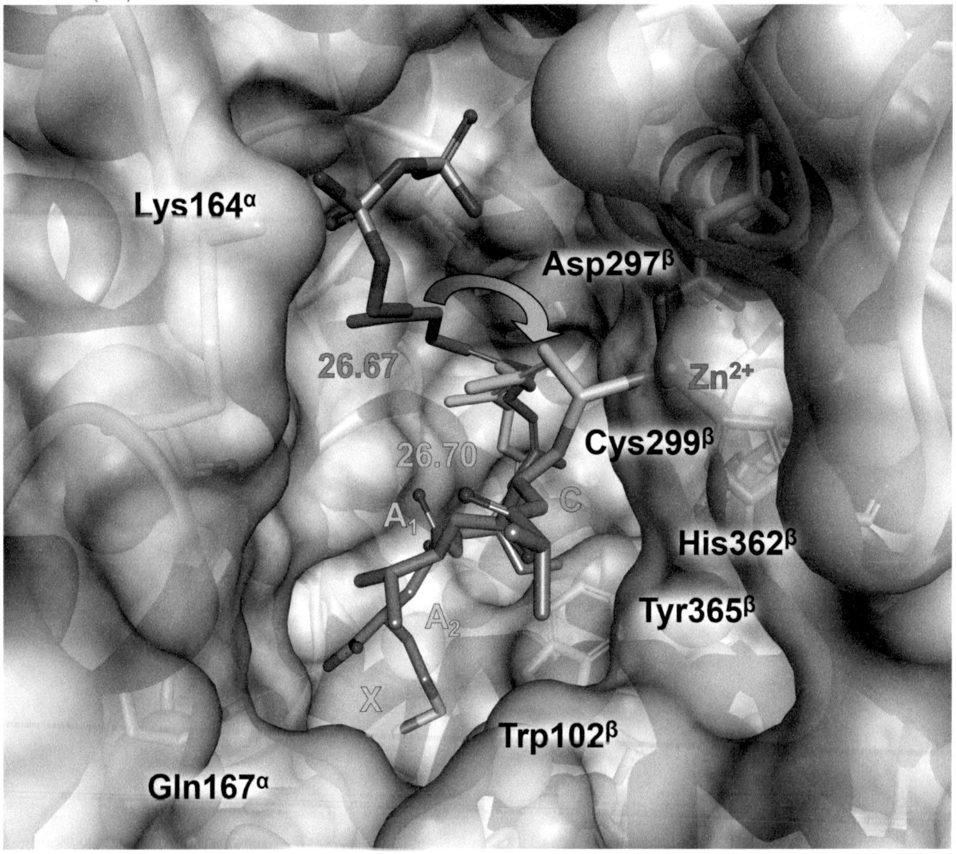

Fig. 26.28 Farnesyldiphosphate **26.67** binds to FTase and occupies a part of the large catalytic site. The crystal structure of the enzyme with this substrate was determined (colored image: farnesyldiphosphate **26.67** *dark-green*). Geranylgeranyl groups **26.68** that have an elongated isoprenyl chain (formulas: isoprenyl chain indicated in *red* instead of a *black* chain in **26.67**) are transferred by GGTase. Trp 102$^{\beta}$ and Tyr 365$^{\beta}$ border the binding pocket in FTase and determine substrate selectivity. After binding the farnesyl substrate, the peptide substrate **26.69** (*gray*) with its CAAX terminus diffuses into the binding pocket. Catalyzed by an adjacent zinc ion coordinating the cysteine residue of the substrate, the farnesyl residue is transferred to the thiol group of the cysteine. The diphosphate group is nucleophilically displaced. A crystal structure of the resulting product **26.70** (*light-green*) was also determined. It is shown in the colored image in the center, superimposed onto the binary complex. The farnesyl residue must move "forward" in the pocket (*green arrow*). The resulting product coordinates to the zinc ion. With its two aliphatic residues A_1 (here Ile) and A_2 (here Val) of the CAAX motif, the tetrapeptide moiety is recognized by the enzyme. The terminal methionine (X) forms a hydrogen bond with the carboxylate group of Glu 167$^{\alpha}$. (► https://sn.pub/9qIkq2)

Fig. 26.29 Development of compounds for the inhibition of FTase. Compound **26.71** represents a competitive inhibitor for farnesyldiphosphate **26.67**. Compounds **26.72**–**26.77** are inhibitors that bind competitively to the tetrapeptide substrate, CAAX. Only some of them (**26.72**–**26.74**, **26.76**) use their functional groups (e.g., imidazole rings) to block the zinc ion in the catalytic site. Compounds **26.75** and **26.76** inhibit FTase without direct coordination to the Zn^{2+}. Compound **26.77** blocks FTase and GGTase equipotently

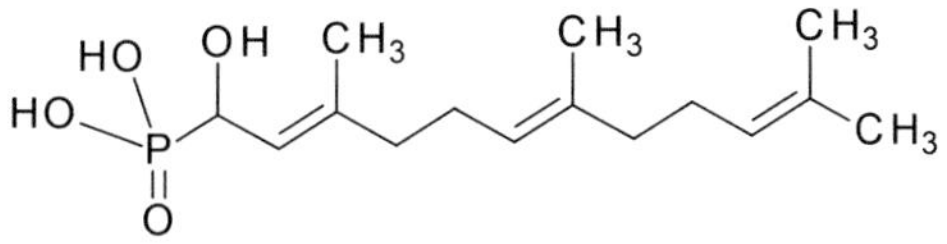

26.71 α-Hydroxyfarnesyl phosphonic Acid

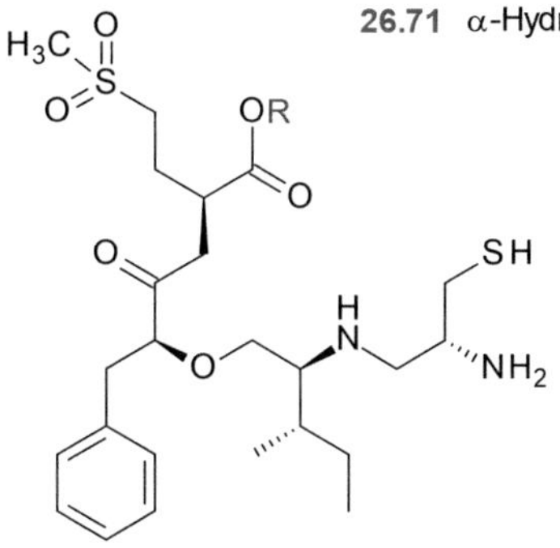

26.72 L-739750 R = H bzw. R = *i*Pr

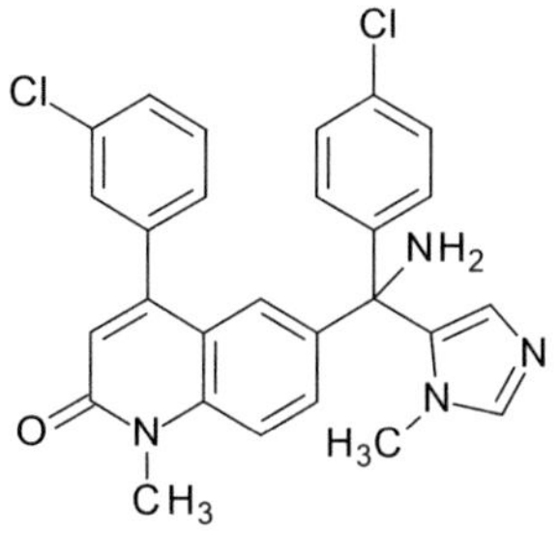

26.73 R115777 Tipifarnib

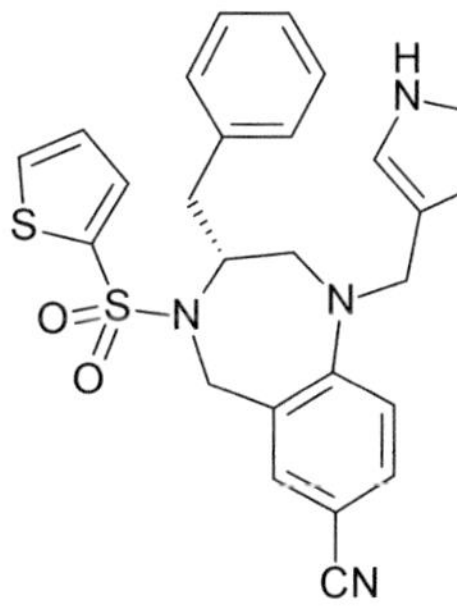

26.74 BMS-214662

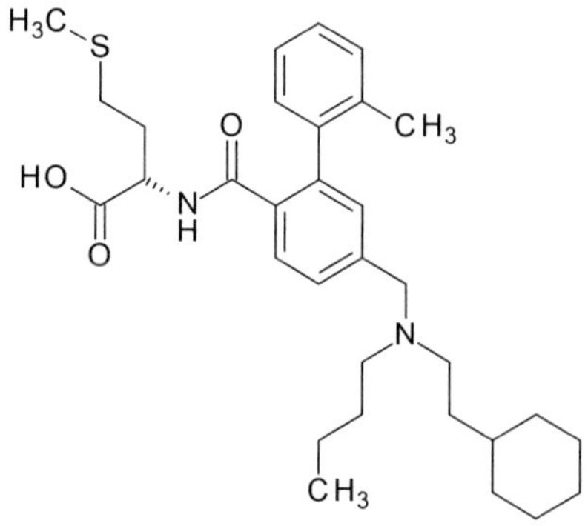

26.75 ABT-839

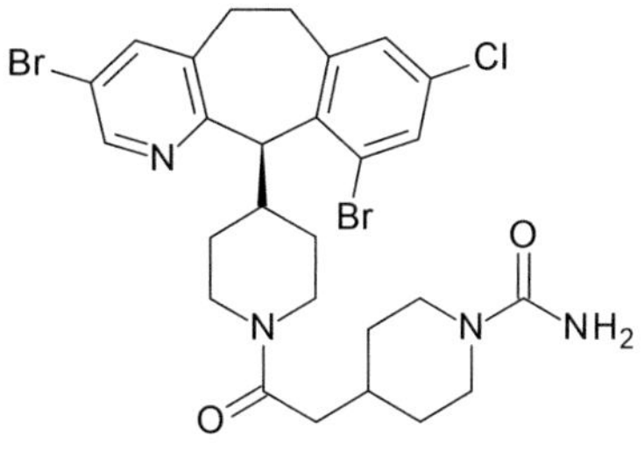

26.76 Lonafarnib

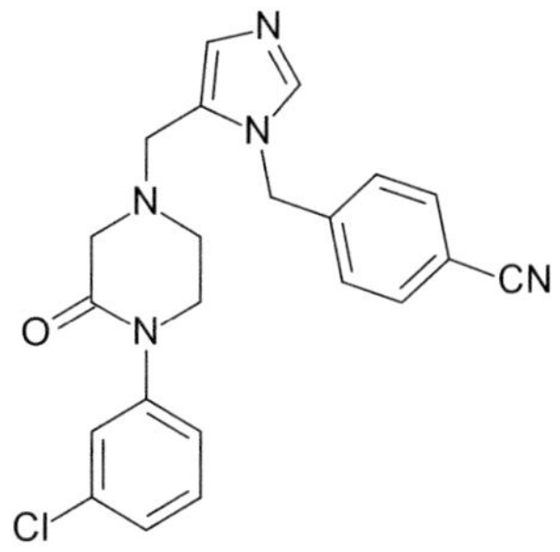

26.77 L-778123

to be **anchored to a membrane**. In addition to examples in which a portion of the polymer chain becomes immersed in the membrane, proteins are known that are anchored to the membrane by an **added farnesyl 26.67** or **geranylgeranyl 26.68** anchor. These hydrophobic anchors consist of isoprenoid units (Fig. 26.28). They attach to proteins via cysteine residues located near the *C*-terminus. Three classes of prenylating enzymes are known: the **farnesyl transferases** (FTases) and the **geranylgeranyl transferases** I and II (GGTase I and II). Substrates of these catalysts include the GTPases of the Ras, Rab, and Rho families, lamins, and the γ-subunit of G-protein heterotrimers. For FTases and GGTases to attach a prenyl anchor, the substrate proteins must have a **CAAX** sequence **26.69** at their *C*-terminus (Fig. 26.28). Here, C stands for the cysteine to which the prenyl group is transferred, and A1 and A2 are usually aliphatic amino acids. If X is serine, methionine, glutamine, or alanine, the protein will be prenylated by an FTase. A leucine at this position prefers a GGTase as a catalyst.

Meanwhile, more than 250 proteins that require the **posttranslational attachment of a prenyl tail** for their function have been discovered. Interest in these prenylating enzymes, especially FTases, began in the early 1990s. It was observed that **RAS proteins**, which in a mutated form mediate a permanent growth signal in cancer, need to be farnesylated. Only then are they active. If farnesylation is missing, RAS activity will be suppressed. After the prenyl group has been transferred in the cytoplasm to the cysteine three amino acids from the *C*-terminus, the protein enters the endoplasmic reticulum. There, the AAX tripeptide tail is proteolytically cleaved and a methyl group is transferred to the *C*-terminus through a carboxymethylation step. Finally, the prenylated protein is anchored to the cell inner membrane. FTases and GGTases contain a zinc ion in their catalytic center, which is coordinated by a cysteine, aspartate, and histidine residue. First, the farnesyl or geranylgeranyl diphosphate anchor (**26.67** or **26.68**) diffuses into the large funnel-shaped binding pocket of this enzyme. FTases and GGTases form a heterodimer with a barrel-like architecture that is almost exclusively composed of helical structural elements. FTase specifically recognizes the shorter substrate farnesyl diphosphate **26.67** because the bottom of its binding pocket is confined by Trp 102^{β} and Tyr 365^{β}. After successful binding of the prenyl substrate, the peptide chain with the **tetrapeptidic *C*-terminal CAAX** of the protein to be prenylated diffuses into the catalytic site. There, the prenyl substrate provides a large interaction surface for the incoming peptide substrate.

Next, the farnesyl chain must move towards the peptide substrate for the actual transfer reaction. The CAAX substrate occupies the fourth coordination site on the zinc ion with the thiol group of its cysteine. It also binds with its hydrophobic aliphatic side chain A_2 into the preformed binding pocket of the enzyme. The side chain A_1 protrudes into the surrounding solvent. In the structure shown in ◘ Fig. 26.28, a methionine occupies the X position and the *C*-terminal carboxylate group forms a hydrogen bond with Gln 167^{α}. The prenyl group is then transferred to the peptide chain by nucleophilic attack of the cysteine in the substrate on the carbon atom next to the diphosphate group. Finally, the prenylated product **26.70** diffuses out of the catalytic center. Interestingly, this is the rate-determining step. There is evidence that a new substrate molecule is required to displace the product from the enzyme. To do this, the product molecule takes up a new position and binds to a region of the binding pocket through which it leaves the reaction center.

According to the outlined reaction mechanism, different concepts for the development of inhibitors for this enzyme have been pursued. The first attempts were aimed at competing with the binding of isoprenoid diphosphate. For example, the isoprenoid analog α-hydroxyfarnesylphophonic acid **26.71** occupies the binding pocket similarly to farnesyl diphosphate and forms extensive interactions with the enzyme as well as with the CAAX peptide substrate. The second and most commonly used strategy is to displace the peptide substrate from the binding site. This goal can be achieved through the development of peptidomimetics. An example is L-739750 **26.72**, an ester prodrug that caused tumor regression in rats without systemic toxicity (◘ Fig. 26.29).

It has also been possible to completely abandon peptide lead structures. Examples are R115777 (tipifarnib) **26.73** from Janssen Pharma or BMS-214662 **26.74** from Bristol-Myers Squibb. Both use their imidazole groups to coordinate to the zinc ion. Compound **26.74** replaces the isopropyl group of the peptide at position A_1 with its thiophene ring. The inhibitor uses its benzyl group for the A_2 position to mimic the side chain of the isoleucine. With ABT-839 **26.75**, Abbott has found a compound that does not coordinate to the zinc ion at all. It has a methionine group at the end that is very similar to the peptide tail in position X of the natural substrate. Lonafarnib **26.76**, a tricyclic derivative, was developed at Schering-Plough; its urea group is directed to the binding site through which the processed substrate leaves the binding pocket. This inhibitor also blocks the enzyme without coordinating to the zinc ion. Compounds **26.72**–**26.76** all show a selectivity advantage for FTase. Merck has developed the nonpeptide structure **26.77**, which is a potent inhibitor of both FTase and GGTase I. Of course, as with COMT (Sect. 26.10), a strategy can be pursued that seeks to displace both substrates from the binding pocket simultaneously. The bisubstrate-analog inhibitors suffer from being very large in order to compete successfully with the two large substrates.

Clinical studies on the nonpeptidic farnesyltransferase inhibitors **26.72**–**26.77** are not advanced enough to be judged. Monotherapy with these inhibitors has been rather disappointing, although very promising results have been seen with tipifarnib **26.73** in breast cancer. It remains to be seen whether FTase inhibitors will be used as monotherapy in cancer treatment or whether they will be used more effectively in combination with other cytostatic and hormone drugs. However, in recent years, a new field of drug development has opened up for FTase inhibitors. They appear to be potential lead structures for the treatment of infectious diseases caused by pathogenic microorganisms such as *Plasmodium* (malaria), *Trypanosoma* (African sleeping sickness and Chagas disease), and *Leishmania* (leishmaniasis, kala-azar). The causative agent of fungal diseases such as *Candida albicans* can also be fought in this way. Obviously the posttranslational prenylation of their proteins is an essential step in the lifecycles of these organisms. We can hope that the sequence differences in the transferases are adequately large compared to the human enzymes to develop selective compounds.

26.12 Synopsis

- Proteins can be modified after translation in the ribosome by the attachment of groups such as phosphate, methyl, or acetyl, and larger building blocks such as prenyl or geranyl moieties or polypeptide chains such as ubiquitin or SUMO.

- Kinases transfer phosphate groups from ATP to the hydroxyl groups of Ser, Thr, or Tyr residues or the imidazole group of His. This switches on the biochemical function of the phosphorylated protein substrates; phosphatases can reverse this step by cleaving the phosphate group off again from the phosphorylated amino acid residue.
- The more than 530 human kinases act as switches in signaling cascades; thus, they seem very attractive as putative drug targets. However, their substrate ATP is present in high concentrations in cells, it is recognized by multiple proteins often with other functions, and Nature has established many processes involving kinases redundantly as a failsafe. This makes selective competitive inhibition of kinases at the ATP-binding site a difficult task.
- Kinases are rather flexible proteins that adapt to their substrates. The adenine moiety of ATP is recognized by a peptide strand in the hinge region. Pockets are found adjacent to the ATP binding site and they are called *front* and *back* pocket. They are not involved in ATP recognition but they can be exploited to endow competitive inhibitors with the required selectivity.
- Inhibitors are profiled against the kinase family (kinome) and exhibit either high selectivity against individual members or show promiscuous binding to larger groups on the phylogenetic kinase family tree. Interestingly, introduction of inert metal centers that expand the basic coordination architecture to attach pharmacophoric groups can succeed in producing highly selective compounds.
- Imatinib and its follow-up compound nilotinib bind to the inactive conformation of BCR-ABL kinase. They represent a completely new approach to cancer therapy: They cure chronic myeloid leukemia by inhibiting the product of a misregulated gene. Asciminib is a highly selective allosteric inhibitor that mimics a fatty acid residue in the myristoyl pocket and stabilizes the kinase in a globally inactive state.
- The bump-and-hole method allows a specific therapeutic validation of the biological relevance of a target protein as well as the optimization of an inhibitor class. Genetically, the target protein is modified in its substrate specificity (e.g., a kinase at its gatekeeper residue) and implemented into a model organism. Selective inhibition of this protein under *in vivo* conditions is achieved via inhibitors that are adapted to the modified binding site of the engineered protein.
- Phosphatases remove phosphate groups from Ser, Thr, Tyr, and His residues, thus, switching off the biochemical function of a substrate protein. Two catalytically different enzyme classes are known, either operating through nucleophilic attack of a water molecule, which is highly polarized by two adjacent metal ions, or through the nucleophilic attack of a cysteine residue via a pathway similar to that in cysteine proteases. In both cases, the tetrahedral phosphorous atom is nucleophilically attacked.
- PTP-1B initially appeared to be an ideal target to treat the metabolic syndrome because it involves dephosphorylation of the insulin receptor kinase. Potent inhibitors of this target with challenging druggability could be developed; however, sufficient selectivity with respect to another phosphatase, TCPTP, failed. Knock-out mice were unable to survive when the genes of both phosphatases are simultaneously turned off. A similar life-threatening situation can be anticipated with insufficiently selective inhibitors.
- The full-length phosphatase Shp2 is autoinhibited by its *N*-SH2 domain in the absence of a phosphorylated substrate. This state can be stabilized by an allosteric inhibitor that binds simultaneously to all three domains and "glues" them together. This blocks the formation of the enzymatically active form of the phosphatase.
- Catechol-O-methyl transferase is representative for the family of methyl transferases using *S*-adenosyl-L-methionine as a cofactor for methyl transfer via its sulfonium group. It transfers methyl groups to catecholamines such as dopamine, adrenaline, or noradrenaline.
- Inhibition of the methyl transferase reaction is achieved by introduction of strong electron-withdrawing groups, such as nitro groups, at the aromatic ring of the natural substrates, producing substrate-like inhibitors.
- Farnesyl and geranylgeranyl transferases transfer prenyl anchor groups onto protein substrates exhibiting a CAAX sequence on their *C*-terminus. The phosphorylated prenyl anchor is attacked by the nucleophilic cysteine thiol group, which is further polarized through the coordination to a neighboring zinc ion in the catalytic center.
- Inhibitors of farnesyl and geranylgeranyl transferases bind competitively either to the CAAX peptide substrate or the prenyldiphosphate substrate binding site. Some of them show strong peptidomimetic character and involve coordination of the zinc ion. However, completely nonpeptidic inhibitors have also been developed, some of which bind without zinc coordination.

Bibliography and Further Reading

General Literature

A. J. Bridges, Chemical Inhibitors of Protein Kinases, Chem. Rev., **101**, 2541–2571 (2001)

F. Ardito, M. Giuliani, D. Perrone, G. Troiano, L. Lo Muzio, The crucial role of protein phosphorylation in cell signaling and its use as targeted therapy, Int. J. Mol. Med., **40**, 271–280 (2017)

B. M. Klebl and G. Müller, Second-generation Kinase Inhibitors, Expert Opin. Ther. Targets **9**, 975–993 (2005)

H. Kubinyi and G. Müller, Eds., Chemogenomics in Drug Discovery. A Medicinal Chemistry Perspective, Wiley-VCH, Weinheim (2004)
R. Lorenz, J. Wu, F. W. Herberg, S. S. Taylor, R. A. Engh, Drugging the Undruggable: How Isoquinolines and PKA Initiated the Era of Designed Protein Kinase Inhibitor Therapeutics, Biochemistry, **60**, 3470–3484 (2021)
M. A. Fabian, W. H. Biggs et al., A Small Molecule-Kinase Interaction Map for Clinical Kinase Inhibitors, Nat. Biotech **23**, 329–336 (2005)
S. W. Cowan-Jacob, V. Guez, et al., Imatinib (STI571) Resistance in Chronic Myelogenous Leukemia: Molecular Basis of the Underlying Mechanisms and Potential Strategies for Treatment, Mini-Reviews in Medicinal Chemistry, **4**, 285–299 (2004)
P. J. Alaimo, M. A. Shogren-Knaak and K. M. Shokat, Chemical Genetic Approaches for the Elucidation of Signalling Pathways, Curr. Opin. Chem. Biol., **5**, 360–367 (2001)
M. J. Chen, J. E. Dixon, G. Manning, Genomics and evolution of protein phosphatases, Sci. Signal., **10**, eaag1796 (2017)
J. P. Vainonen, M. Momeny, J. Westermarck, Druggable cancer phosphatases, Sci. Transl. Med., **13**, eabe2967 (2021)
M. Köhn, Turn and Face the Strange: A New View on Phosphatases, ACS Cent. Sci., **6**, 467–477 (2020)
S. M. Stanford, N. Bottini, Targeting Tyrosine Phosphatases: Time to End the Stigma, Trends Pharmacol. Sci., **38**, 524–540 (2017)
L. Bialy and H. Waldmann, Inhibitors of Protein Tyrosine Phosphatases: Next-Generation Drugs? Angew. Chem. Int. Ed., **44**, 3814–3839 (2005)
M. J. Bonifacio, P. N. Palma, L. Almeida and P. Soares-da-Silva, Catechol-*O*-methyltransferase and Its Inhibitors in Parkinson's Disease, CNS Drug Reviews, **13**, 352–379 (2007)
C. L. Strickland and P. C. Weber, Farnesyl Protein Transferase: A Review of Structural Studies, Curr. Op. Drug Discov. Develop., **2**, 475–483 (1999)
K. T. Lane and L. S. Beese, Structural Biology of Protein Farnesyltransferase and Geranylgeranyltransferase Type I, J. Lipid Res., **47**, 681–699 (2006)
R. Roskoski Jr., Properties of FDA-approved small molecule protein kinase inhibitors: A 2024 update, Pharmacol. Res., **200**, 107059 (2024)

Special Literature

Madhusudan, P. Akamine, N.-H. Xuong and S. S. Taylor, Crystal Structure of a Transition State Mimic of the Catalytic Subunit of cAMP-dependent Protein Kinase, Nat. Struct. Biol., **9**, 273–277 (2002)
P. C. Nowell, D. A. Hungerford, Chromosome studies on normal and leukemic human leukocytes, J. Natl. Cancer Inst., **25**, 85–109 (1960)
S. W. Cowan-Jacob, G. Fendrich, et al., Structural Biology Contributions to the Discovery of Drugs to Treat Chronic Myelogenous Leukaemia, Acta Cryst., **D63**, 80–93 (2007).
A. A. Wylie et al., The allosteric inhibitor ABL001 enables dual targeting of BCR–ABL1, Nature, **543**, 733–735 (2017)
T. P. Hughes et al., Asciminib in Chronic Myeloid Leukemia after ABL Kinase Inhibitor Failure, N. Engl. J. Med., **381**, 2315–2326 (2019)
C. Bishop, J. A. Ubersax, et al. A Chemical Switch for Inhibitor Sensitive Alleles of any Protein Kinase, Nature, **407**, 395–401 (2000)
K. Islam, The Bump-and-Hole Tactic: Expanding the Scope of Chemical Genetics, Cell Chem. Biol., **25**, 1171–1184 (2018)
L. A.Witucki, X. Huang, K. Shah, Y. Liu, S. Kyin, M. J. Eck, K. M. Shokat, Mutant Tyrosine Kinases with Unnatural Nucleotide Specificity Retain the Structure and Phospho-Acceptor Specificity of the Wild-Type Enzyme, Chem. & Biol., **9**, 25–33 (2002)
E. Meggers, G. E. Atilla-Gokcumen et al. Exploring Chemical Space with Organometallics: Ruthenium Complexes as protein Kinase Inhibitors, Synlett **8**, 1177–1189 (2007)
Y. A. Puius et al., Identification of a Second Aryl Phosphate-binding Site in Protein-tyrosine Phosphatase 1B: A Paradigm for Inhibitor Design, Proc. Natl. Acad. Sci. USA, **94**, 13420–13425 (1997)
B. G. Szczepankiewicz et al. Discovery of a Potent, Selective Protein Tyrosine Phosphatase 1B Inhibitor Using a Linked-Fragment Strategy, J. Am. Chem. Soc., **125**, 4087–4096 (2003)
L. F. Iversen et al., Steric Hindrance as a Basis for Structure-Based Design of Selective Inhibitors of Protein-Tyrosine Phosphatases, Biochemistry, **40**, 14812–14820 (2001)
Andrew P. Combs et al., Structure-Based Design and Discovery of Protein Tyrosine Phosphatase Inhibitors Incorporating Novel Isothiazolidinone Heterocyclic Phosphotyrosine Mimetics, J. Med. Chem., **48**, 6544–6548 (2005)
D. P. Wilson et al., Structure-Based Optimization of Protein Tyrosine Phosphatase 1B Inhibitors: From the Active Site to the Second Phosphotyrosine Binding Site, J. Med. Chem., **50**, 4681–4698 (2007)
C. Wiesmann et al., Allosteric inhibition of protein tyrosine phosphatase 1B, Nat. Struct. Biol. & Mol. Biol., **11**, 730–737 (2004)
P. Hof et al., Crystal Structure of the Tyrosine Phosphatase SHP-2, Cell, **92**, 441–450 (1998)
J. G. Fortanet et al., Allosteric Inhibition of SHP2: Identification of a Potent, Selective, and Orally Efficacious Phosphatase Inhibitor, J. Med. Chem., **59**, 7773–7782 (2016)
R. Mitra, S. R. Ayyannan, Small-Molecule Inhibitors of Shp2 Phosphatase as Potential Chemotherapeutic Agents for Glioblastoma: A Minireview, ChemMedChem., ChemMedChem, **16**, 777–787 (2021)
J. Vidgren, L. A. Svensson und A. Liljas, Crystal Structure of Catechol-O-methyltransferase, Nature, **368**, 354–358 (1994)
C. Lerner, B. Masjost et al., Bisubstrate Inhibitors for the Enzyme Catechol-O-methyltransferase (COMT): Influence of Inhibitor Preorganization and Linker Length between the Two Substrate Moieties on Binding Affinity, Org. Biomol. Chem., **1**, 42–49 (2003)
S. B. Long, P. J. Casey, L. S. Beese, Reaction path of protein farnesyltransferase at atomic resolution, Nature, **419**, 645–650 (2002)

Oxidoreductase Inhibitors

Contents

G. Klebe, *Drug Design*, https://doi.org/10.1007/978-3-662-68998-1_27

Chemical reactions that occur through the exchange of electrons are called redox reactions. Typically, in **biochemical redox processes**, the carbon atom changes its oxidation state. In total, carbon can assume oxidation states ranging from −4 to +4. In general, oxidations convert derivatives with a significant number of directly bonded hydrogen atoms to derivatives with more contacts to nitrogen, oxygen, and sulfur. Since these bonds to the above electronegative elements are usually associated with the **introduction of polar functional groups**, redox reactions exert a decisive **influence on the physicochemical properties** of the oxidized substances. For example, water solubility is increased. This is important for the elimination of xenobiotics. Cytochrome P450 enzymes, a large group of oxidizing enzymes, are involved in the corresponding **metabolic transformations**. On the other hand, reductions are also of vital importance for the organism. In these reaction steps, reactive aldehydes or ketones are converted into alcohols, which can then be more easily conjugated and eliminated (Sect. 8.1). Transition metals, which can assume a variety of oxidation states, are predestined to serve as electron donors and acceptors in redox reactions. In biological systems, one transition metal, **iron**, is often used for this purpose. Once incorporated into a protoporphyrin scaffold, it exists in a penta- or hexavalent coordination state and can assume oxidation states between +2 and +4. It also participates in complexes with sulfur. There it forms interesting multinuclear structures called iron–sulfur clusters. In addition to iron, copper also plays a role as a mediator in biochemical redox processes.

Nature uses **cofactors** for enzyme-catalyzed redox reactions. They are embedded in the specific environment of a protein and, shielded from the surrounding solvent, carry out the electron or hydride ion transfer from the group to be oxidized to the group to be reduced. Cofactors may be tightly bound to the protein. In these cases, they are called **prosthetic groups** and do not leave the enzyme during the reaction. Other loosely bound cofactors can be taken up by the protein like the substrate, chemically modified, and finally released. These cofactors must be regenerated in another independent reaction for the next redox reaction cycle.

The **oxidoreductase** class of enzymes will be discussed in this chapter. They are involved in many electron transfer reactions and require electrons or hydrogen in the form of hydride ions. These particles are transferred by cofactors such as **NAD(P)$^+$** (nicotinamide adenine dinucleotide (phosphate)) or the flavin nucleotides **FMN** (flavin mononucleotide) and **FAD** (flavin adenine dinucleotide) and the aforementioned **iron** atom in the **heme group**. Since these enzymes are also often involved in processes that are causally related to the development of pathological disease situations, many drug therapies are aimed at inhibiting these enzyme systems.

27.1 Redox Reactions in Biological Systems Use Cofactors

As mentioned above, enzymes use cofactors to transfer electrons or hydride ions in redox reactions. **NAD$^+$/NADP$^+$ 27.1** (nicotinamide adenine dinucleotide phos-

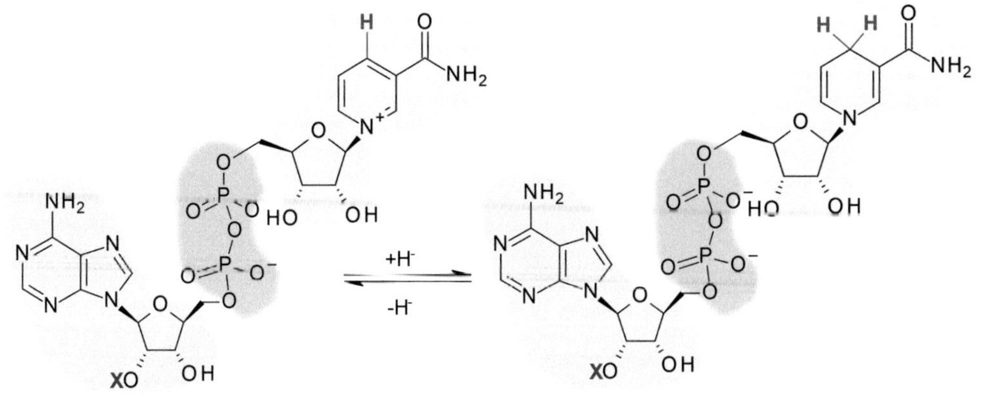

27.1 NAD$^+$ X=H
NADP$^+$ X=PO_3^{2-}

27.2 NADH X=H
NADPH X=PO_3^{2-}

Fig. 27.1 Many enzymatic redox reactions use NAD$^+$/NADP$^+$ **27.1** (nicotinamide adenine dinucleotide, P stands for a phosphate group attached to the ribose ring) and NADH/NADPH **27.2** as a cofactor for the transfer of electrons and/or hydride ions. The cofactor is made up of three components: the nicotinamide, which bears an attached ribose sugar (orange), the central diphosphate unit (violet) and the adenosine moiety (green). There are two different derivatives for **27.1** and **27.2**: one with a phosphate and one with a hydroxide group. Upon oxidation, the positively charged nicotinamide moiety takes on a hydride ion (*red*) at the 4-position; upon reduction the H$^-$ ion is released from this position

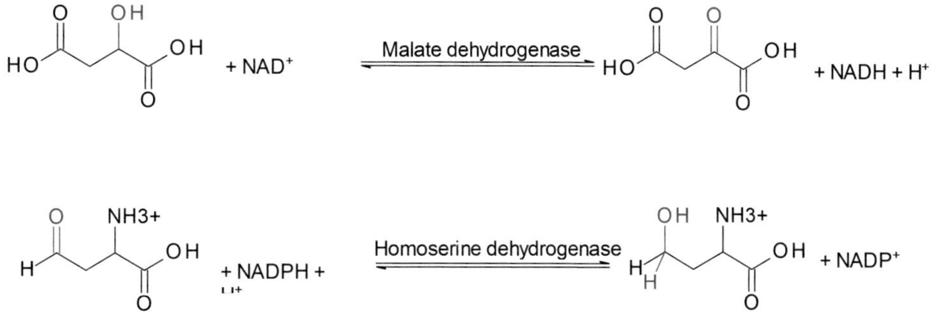

Fig. 27.2 Examples for an oxidation reaction with malate dehydrogenase (*top*) and for a reduction with homoserine dehydrogenase (*bottom*). The transformation of a hydroxyl group into a ketone function or vice versa (*red*) is carried out in both reactions

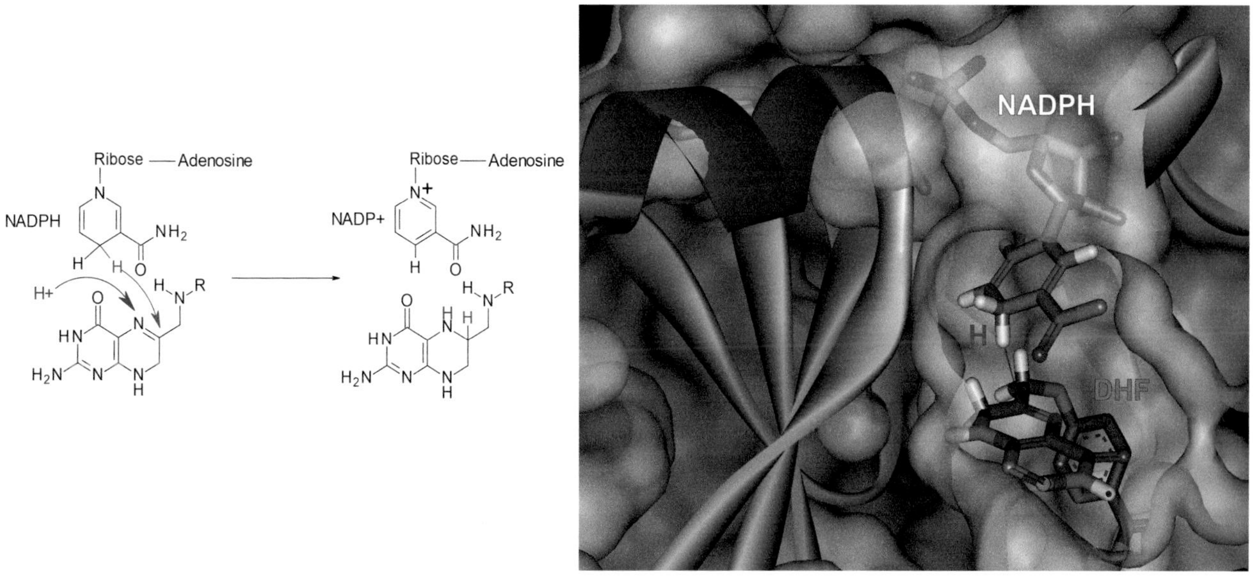

Fig. 27.3 The stereochemically unambiguous transfer of a hydride ion from the NADPH cofactor to the double bond of the substrate being reduced is accomplished deep in the protein's binding pocket. Crystal structure determination of the enzyme dihydrofolate reductase with bound dihydrofolic acid (DHF) and cofactor (NADPH) provided detailed information about the course of the reduction step. The two reaction sites come spatially very close to one another in the structure. A hydride ion is transferred from the 4-position of the reduced nicotinamide ring onto the neighboring double bond of the DHF substrate (*violet line*). (▶ https://sn.pub/ZFSx22)

phate) and **NADH/NADPH 27.2** serve as acceptors and donors of **hydride ions** (Fig. 27.1). This cofactor consists of three components: the nicotinamide with an attached ribose ring, the central diphosphate moiety, and the adenosine moiety. The latter may carry a phosphate moiety on the 2′-OH group and is then referred to as $NADP^+$/NADPH. The redox-active part is the nicotinamide moiety, a pyridine derivative. During oxidation, the positively charged $NADP^+$ accepts a hydride ion at the 4-position of the pyridine ring. In the reverse reaction, an H^- is released from the same position. A total of two electrons are transferred. $NAD(P)^+$ is loosely bound to the enzyme. It can be easily exchanged and regenerated on another protein for a subsequent reaction cycle. Typical oxidation and reduction reactions that can occur in a dehydrogenase or reductase are shown in Fig. 27.2.

In the binding pocket of such an enzyme, there is direct contact between the group to be oxidized or reduced and the nicotinamide ring. The binding site in these proteins is usually shielded from the aqueous solvent by a hydrophobic group, a loop, or an amino acid lid. On the one hand, this ensures that the stereochemistry of the hydride transfer is unambiguous. On the other hand, access to protons must be excluded, otherwise the enzyme would not be able to reduce the substrate, as elementary hydrogen would be generated. The interaction geometry for such reaction steps is shown for dihydrofolate reductase in Fig. 27.3. Although enzyme-catalyzed reactions are generally reversible, and the direction of the reaction will depend on the concentration of cofactors in the environment, **NADP/H** is, with few exceptions, involved in **reduction reactions**. Oxidation reactions are almost exclu-

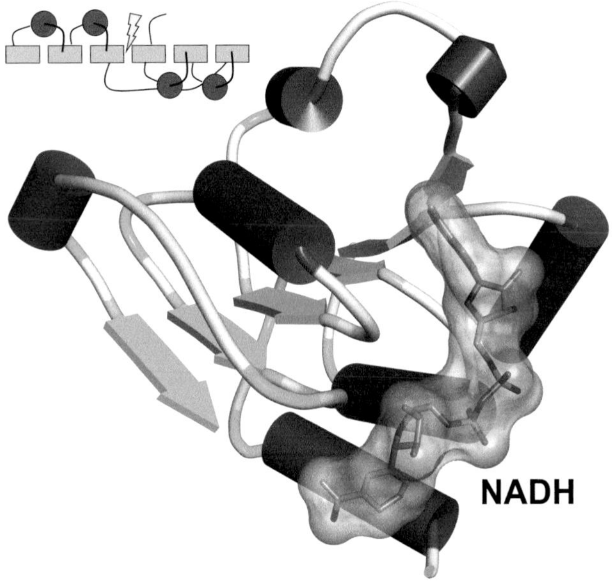

Fig. 27.4 In most NAD(P)H-dependent enzymes, the cofactor binds to the so-called Rossmann fold in a structurally conserved domain. This forms a central six-stranded pleated sheet with at least four α-helices on the top and bottom sides. The topological change of the helices from one side to the other takes place in the middle of the pleated sheet. The charged diphosphate unit of the cofactor (*yellow arrow*) binds in an extension to this position. (▶ https://sn.pub/c5pf5T)

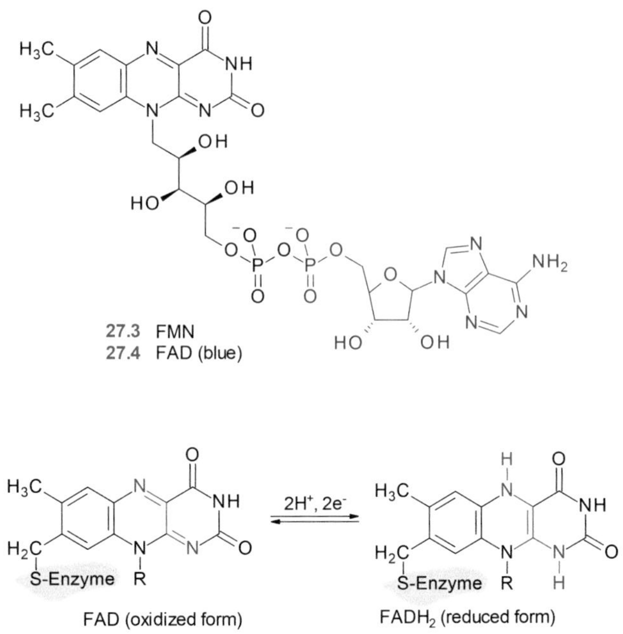

Fig. 27.5 Flavoproteins use FMN **27.3** and FAD **27.4** (extended by the *blue* part) as a cofactor. FAD is composed of an adenosine moiety, a diphosphate bridge with the carbohydrate alcohol ribitol, and the tricyclic isoalloxazine ring. This tricyclic heterocycle represents the redox-active part of the molecule and can accept or donate one or two electrons to the substrate. The cofactor is very tightly, but reversibly in some cases, anchored to the enzyme via a covalent bond

sively carried out by **NAD/H**. The majority of enzymes can distinguish unambiguously between the cofactors. This is because the additional phosphate group on the 2′-OH group of the ribose ring is specifically recognized in a binding pocket near the cofactor binding pocket as a kind of marker. The majority of NAD(P)H-dependent enzymes share a structurally similar binding domain. It consists of a total of four α-helices arranged at the top and bottom of a central six-stranded pleated sheet. A topological switch in the arrangement of the helices from one side to the other occurs in the middle of the pleated sheet (Fig. 27.4). The binding of the charged diphosphate group to the conserved nucleotide-binding moiety occurs in an extension of this position. This folding motif is called the **Rossmann fold** in honor of its discoverer, Michael Rossmann. We will revisit this nucleotide-binding domain with the enzymes dihydrofolate reductase, HMG-CoA reductase, and 11*β*-hydroxysteroid dehydrogenase in the following sections. Other folding motifs may also provide a binding site for the NADPH cofactor. A TIM barrel (Sect. 14.3) is used to bind this cofactor in aldose reductase (Sect. 27.4). Two superfamilies of proteins are known to reduce or oxidize carbonyl compounds in biological systems. The first group includes the **aldo-keto reductases**, of which aldose reductase is a representative. The second superfamily contains **short-chain dehydrogenase/reductases**, to which the 11*β*-hydroxysteroid dehydrogenase (Sect. 27.5) belongs.

Flavoproteins use FMN **27.3** and FAD **27.4** as cofactors (Fig. 27.5). They are derived from vitamin B2, riboflavin. FAD consists of an adenosine linked by a diphosphate bridge and the carbohydrate alcohol ribitol to the **tricyclic isoalloxazine ring**, which is the redox-active part of the cofactor (Fig. 27.5). This group can be reversibly reduced and oxidized by exchanging one or two electrons with the substrate. Typically, two redox equivalents are transferred and the reaction remains at the semiquinone stage, which is a stable radical. Radical intermediates are formed along the reaction pathway. To avoid damaging cells with these reactive species, flavin cofactors never exist in free solution. Instead, they are covalently anchored inside the enzymes. The flavin-dependent oxidoreductase monoamine oxidases MAO_A and MAO_B will be introduced in Sect. 27.8. Many therapeutic inhibitors act on these enzymes by binding irreversibly to the isoalloxazine ring, preventing the redox processes.

A third important cofactor is the **heme group 27.5** (Fig. 27.6), which is found primarily in proteins that use oxygen as an oxidizing agent. The heme group is found in cytochrome P450 enzymes, in cyclooxygenases, and in oxygen-transporting proteins such as hemoglobin and myoglobin. A central iron atom is embedded in a **protoporphyrin** system. It coordinates with four pyrrole

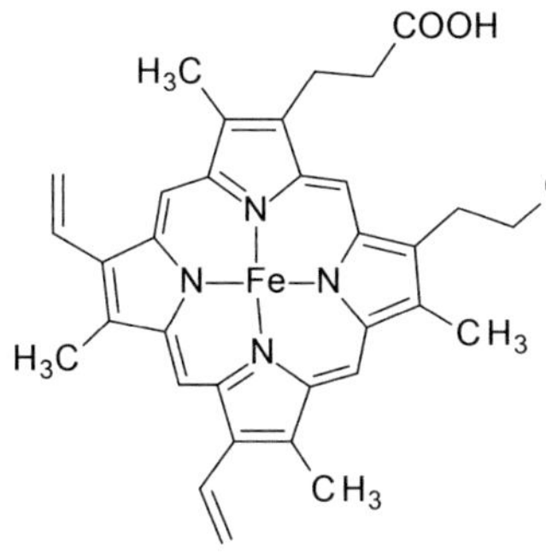

27.5 Heme

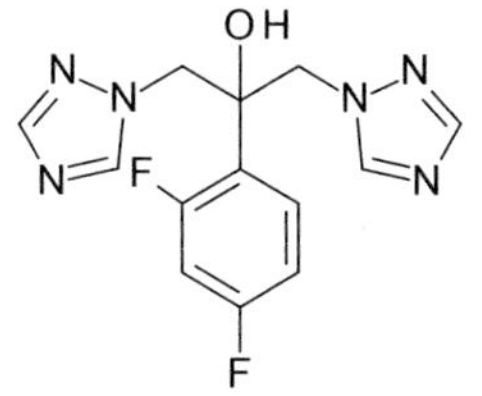

27.6 Fluconazole

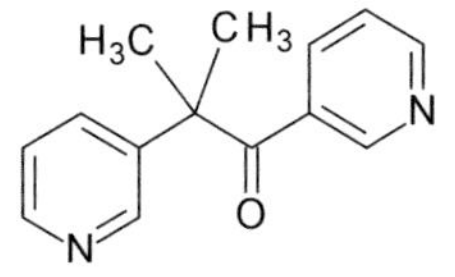

27.8 Metyrapone

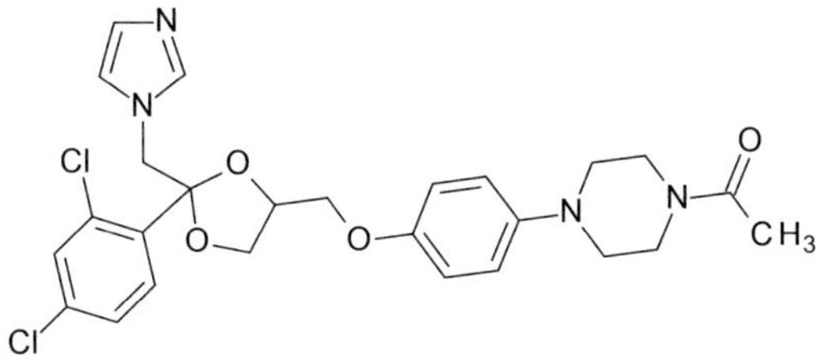

27.7 Ketoconazole CYP 3A4 K_i=15 nM

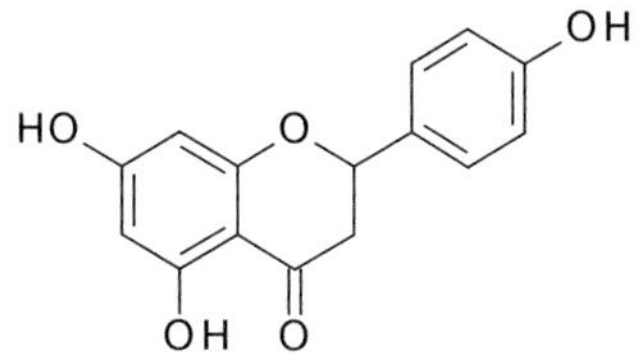

27.9 Naringenin

Fig. 27.6 The heme group **27.5** occurs as a cofactor in proteins that use oxygen as an oxidant. An iron atom is embedded in a protoporphyrin system in a quadratic–pyramidal or octahedral geometry. The four pyrrole rings form a plane. The fifth apical position is occupied by a histidine or a cysteine, and the sixth position is coordinated by a reactive oxygen species. This binding site can be blocked by a nitrogen-containing heterocycle such as a triazole or imidazole ring in fluconazole **27.6**, ketoconazole **27.7**, or by pyridine rings as in metyrapone **27.8**. Even natural products such as the flavonoid naringenin **27.9** represent examples of cytochrome inhibitors

rings in a planar geometry. The fifth apical position is occupied by a histidine or a cysteine. Electron transfer or oxidative attack of a bound oxygen molecule is accomplished through the sixth coordination site on the iron ion. The iron changes its oxidation state during the redox reaction. The heme group remains permanently bound to the protein.

The properties of the iron ion, a good partner for coordinating ligands, can be exploited to inhibit heme-containing proteins. Small molecules such as carbon monoxide or cyanide ions can attach to the sixth coordination position. This blocks the function of the protein and is responsible for the toxicity of both of these compounds. CO inhibits the binding of O_2 to hemoglobin and prevents oxygen transport in the blood, while cyanide reacts with the iron in cytochromes in the respiratory chain. Heterocycles such as imidazoles or triazoles can also coordinate to the iron ion. Potent antimycotics such as fluconazole **27.6** or ketoconazole **27.7** (Figs. 27.6 and 27.7) also follow this principle. Metyrapone **27.8**, a drug for the treatment of adrenal insufficiency, also represents a potent inhibitor of many P450 enzymes. Many natural products have been described as cytochrome inhibitors, for example, the flavonoid naringenin **27.9**, which gives grapefruit its bitter taste.

27.2 Chemotherapeutics for Cancer and Bacteria: Dihydrofolate Reductase Inhibitors

Together with thymidylate synthetase and serine transhydroxymethylase, dihydrofolate reductase forms a synthetic cycle that catalyzes the biosynthesis of thymine (Fig. 27.8). Thymine is a pyrimidine base that is a critical component of DNA (Sect. 14.9). In the first step, the nucleotide desoxyuridylate is methylated by thymidylate synthase. The methyl group comes from the enzyme cofactor, methylenetetrahydrofolate **27.10**. After successful methyl group transfer, the cofactor leaves the enzyme as dihydrofolate **27.11** and must be reduced to tetrahydrofolate **27.12**. Dihydrofolate reductase (DHFR) performs this task.

As the carrier of genetic information, DNA is produced in increased amounts when high levels of cell division are required. Cancer is an example of increased cell proliferation. Bacterial cells also reproduce at an increased rate during infections. Therefore, inhibiting this enzyme in the synthesis cycle is a target for **chemotherapy of tumor diseases**. When the target is an enzyme of a bacterial organism, a compound with **bacteriostatic activity** is obtained. Dihydrofolate reductases from different species are rather small enzymes. Depending on their origin, they consist of between 150 and 260 amino acids. The

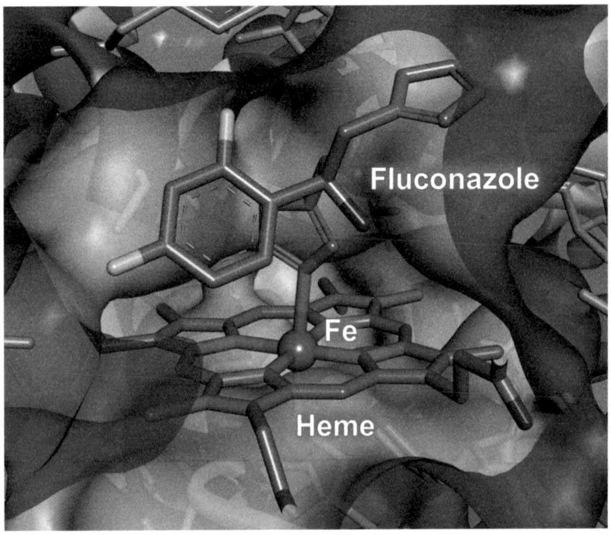

Fig. 27.7 Fluconazole **27.6** is an antimycotic and blocks the sixth coordination site on the iron ion of a cytochrome P450 enzyme. The indicated binding geometry was determined by crystallography. (▶ https://sn.pub/6eTcNi)

substrate dihydrofolate **27.11** consists of a pteridine ring, a central *para*-aminobenzoic acid, and a terminal L-glutamate moiety. Hydrogenation of the 5,6-double bond in the pteridine ring occurs stereospecifically by attack of a hydride ion with subsequent addition of a proton to the adjacent nitrogen atom. The mechanism is shown in detail in Fig. 27.3.

Even before the first crystal structure of this enzyme was determined in 1982 in the group of Joseph Kraut in San Diego, USA, methotrexate **27.13** was known to be a potent inhibitor of dihydrofolate reductase (Fig. 27.9). Aminopterin **27.14** and edatrexate **27.15** have been described as analogues. Chemically, they appear to be very similar to the natural substrate dihydrofolate **27.11**. However, there is a significant exchange of a hydrogen bond acceptor group for a donor group on the heterocycle. As explained in detail in Sect. 17.6, this results in a 90° twist in the orientation of this moiety in the binding pocket of the reductase. This prevents intimate contact with the reduced nicotinamide group of the NADPH cofactor and the double bond of the bound ligand (Fig. 27.3). Transformation is impossible; the enzyme is blocked.

Methotrexate is a potent chemotherapeutic agent used in cancer therapy to treat breast cancer, sarcomas, acute lymphoblastic leukemia, and non-Hodgkin's lymphoma. Both the natural substrate and methotrexate are highly polar compounds and must be transported into the cell via the **reduced folate carrier** (**RFC**). The ligands are then augmented with additional glutamic acid residues. A prerequisite for good and efficient inhibition of DHFR in cancer therapy is, therefore, not only strong binding to the reductase, but also highly specific uptake through the transporter. For example, derivatives **27.16** and **27.17**, which were obtained by replacing the central phenyl ring with the attached amide bond of methotrexate by a benzolactam group (Fig. 27.9), have somewhat poorer binding constants to DHFR. However, this

Fig. 27.8 Dihydrofolate reductase, thymidylate synthase, and serine transhydroxymethylase make up the synthetic cycle for the biosynthesis of thymine (TMP) from uracil (UMP). The methyl groups to be transferred (*red*) are provided by methylenetetrahydrofolate **27.10**, which is regenerated via dihydrofolate **27.11** and tetrahydrofolate **27.12**. The double bond marked in *red* is hydrogenated by dihydrofolate reductase

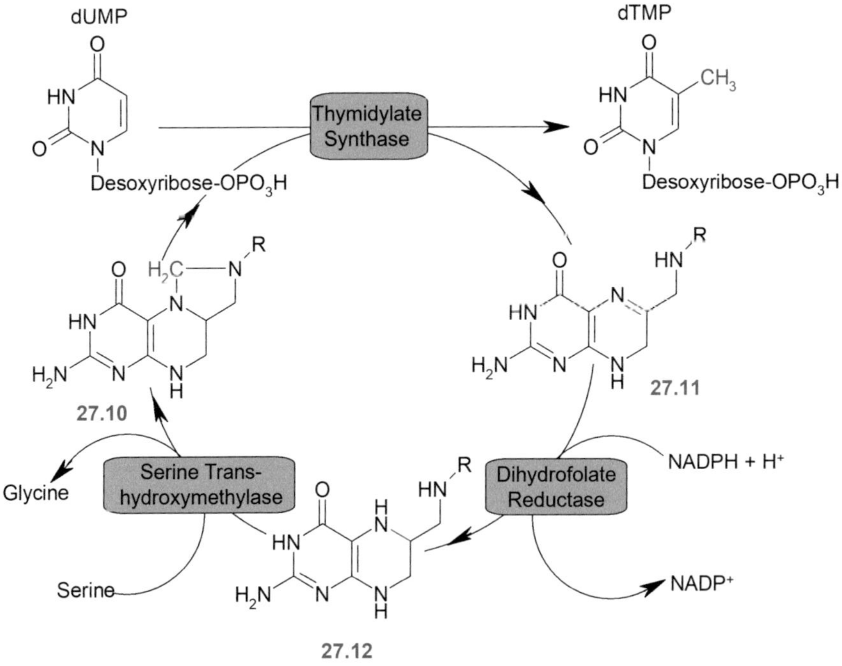

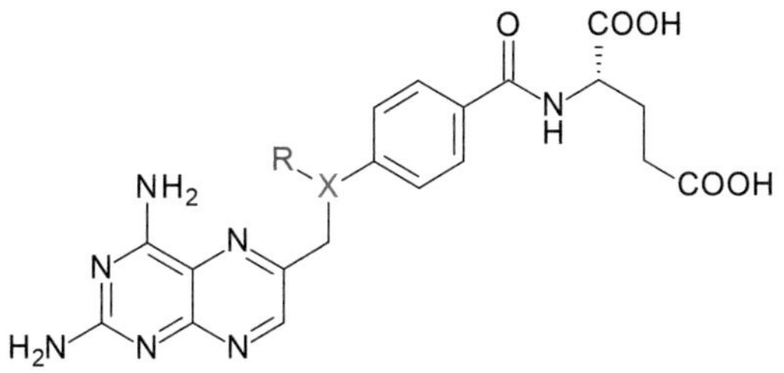

27.13 X=N, R=CH_3 Methotrexate K_i = 4.8 pM
27.14 X=N, R=H Aminopterin K_i = 3.7 pM
27.15 X=C, R=C_2H_5 Edatrexate K_i = 11 pM

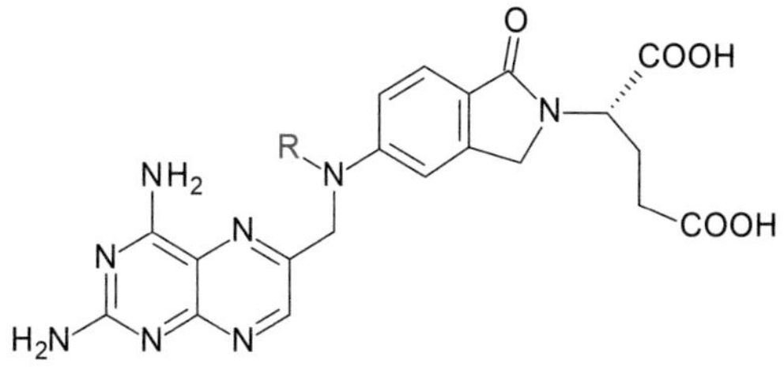

27.16 R=H K_i = 34 pM
27.17 R=CH_3 K_i = 2100 pM

Fig. 27.9 Inhibitors **27.13**–**27.17** of human DHFR that are used as chemotherapeutics in cancer therapy

is compensated for by an improved affinity to the RFC transporter, so that tumor growth can be suppressed equipotently by these compounds. The RFC transporter and the highly potent binding of folic acid analogs to this receptor offers another perspective for cancer treatment. This transporter is expressed on malignant cells to ensure that their increased need for folic acid is met. Because of the tight binding of folic acid derivatives to this receptor and the subsequent internalization of its substrates, there is a possibility that folic acid derivatives could carry an additional molecular cargo that would be piggybacked into the cell. Upon arrival, this cargo could be chemically unloaded and, if a potent anticancer drug, could unleash its destructive effects inside the tumor cell (Table 27.1).

Table 27.1 Binding constants of a few dihydrofolate reductase (DHFR) inhibitors for the human enzyme and the reduced folate carrier (RFC) and cell growth inhibition in tumor tissue

Compound	DHFR K_i (pM)	RFC K_i (μM)	Cell growth IC_{50} (nM, 72 h)
27.13	4.8 ± 0.45	4.7 ± 1.3	14 ± 2.6
27.14	3.7 ± 0.35	5.4 ± 0.09	4.4 ± 0.10
27.16	34 ± 3.0	0.28 ± 0.10	5.1 ± 0.25
27.17	2100 ± 200	1.1 ± 0.11	140 ± 5.0

In addition to chemotherapeutic agents for cancer treatment, **bacteriostatic inhibitors** such as trimethoprim **27.19** are directed against the corresponding bacterial enzymes. Some of these nonclassical antifolate inhibitors (**27.18**–**27.23**) are listed in Fig. 27.10. Structurally, the relationship to the natural substrate is obvious. The first heterocycle is the same as in methotrexate, so an identical binding mode is observed for this moiety. For all DHFRs in different species, an aspartate or glutamate is conserved, using an interaction with the positively charged nitrogen atom in the ring and the exocyclic 3-amino group. The amino group in the 1-position finds interaction partners in two carbonyl groups in the protein backbone (Figs. 17.7 and 17.12). In contrast to methotrexate, trimethoprim-like antibiotics have a more hydrophobic group as the second ring moiety. This grouping is critical for the selective inhibition of DHFRs in bacteria. At therapeutic doses, trimethoprim inhibits bacterial but not human dihydrofolate reductase (Table 27.2). For bacteria, the inhibitory concentrations range from a factor of 60 (*Neisseria gonorrhoeae*, the causative agent of gonorrhea) to 50,000 (the intestinal bacterium *Escherichia coli*) times lower than for human DHFR. This enormous specificity was puzzling at first, because trimethoprim binds to all these enzymes in a very similar way. Even the amino acids directly involved in ligand binding have very similar physicochemical properties.

An explanation for this observation comes from evidence with mutants of *Escherichia coli* in which the amino acids that bind the inhibitor are unchanged, but which nevertheless bind trimethoprim less strongly than wild-type DHFR. In fact, the inhibitor binds with an unchanged geometry. Trimethoprim is positively charged at physiological pH. Charges in the environment of the binding site should, therefore, decisively influence the affinity of these ligands. In one of the DHFR mutants, the negatively charged Glu 118 is replaced by a neutral glutamine. Despite a distance of about 15 Å between the oppositely charged groups, the absence of a negative charge in the extended environment causes a loss of affinity by a factor of 4–5. An even more pronounced effect was observed in the double mutant, in which a leucine about 8 Å away is additionally exchanged for a positively charged arginine. Because of this additional unfavorable charge change, the inhibition constant is about 200-fold lower than in the wild type (Table 27.2).

A comparison of chicken DHFR (from liver) with *Escherichia coli* DHFR shows that seven amino acid side chains within 10–16 Å of the positively charged nitrogen atom on the ligand change their charge: two from negative to neutral and another five from neutral to positive. The vicinity of the binding site in chicken DHFR has, therefore, become unfavorable for the addition of a positively charged molecule with respect to seven charge units. Accordingly, it is not only the direct contacts that are responsible for the strength of the protein–ligand interaction, but rather the **electrostatic interactions in the remote environment**.

Fig. 27.10 Bacteriostatic inhibitors **27.18–27.23** of bacterial dihydrofolate reductase

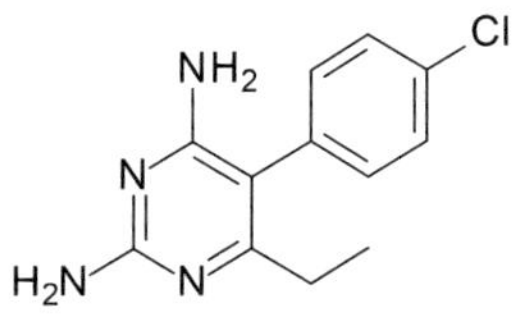

27.18 Pyrimethamine

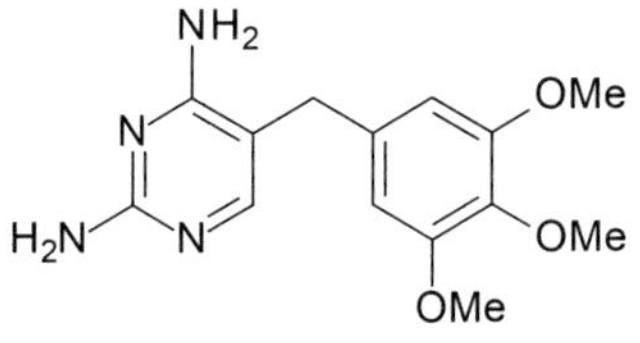

27.19 Trimethoprim

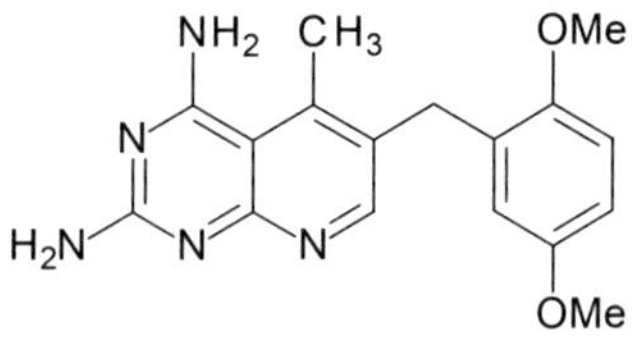

27.20 Piritrexim

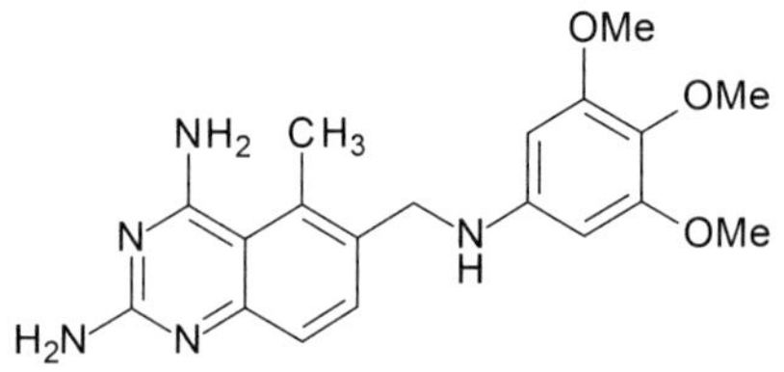

27.21 Trimetrexate

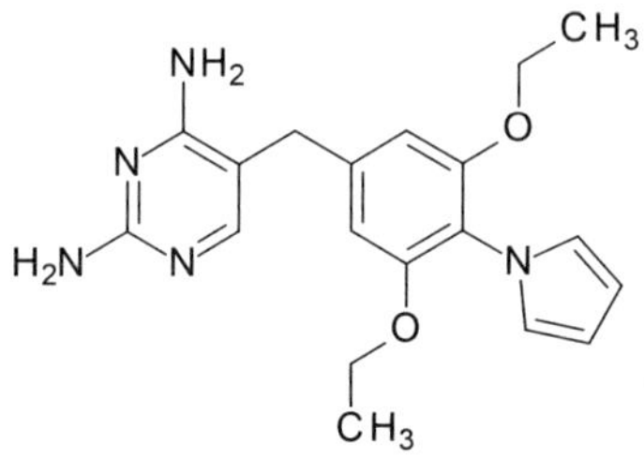

27.22 Epiroprim

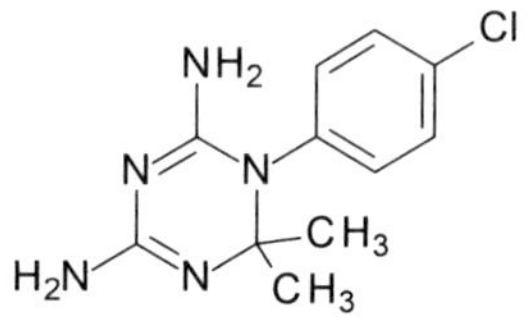

27.23 Cycloguanil

Table 27.2 Dissociation constants K_d of trimethoprim **27.19** for dihydrofolate reductase from different species

Species	K_d (nM)
Escherichia coli	0.02
Escherichia coli, Gln118 mutant	0.09
Escherichia coli Arg28/Gln118 double mutant	3.8
Lactobacillus casei	0.4
Neisseria gonorrhoeae	15
Chicken	3500
Mouse	3500
Cattle	330
Human	1000

In another model study, the 3-methoxy group in trimethoprim was replaced by an unsaturated acidic side chain in **27.24** (Fig. 27.11). The modified derivative shows significantly improved affinity (by a factor of 5000) and, thus, selectivity for the bacterial enzyme from *Pneumocystis jirovecii* compared with the enzyme from rodents. Crystal structures of inhibitor **27.24** were obtained with the bacterial and vertebrate enzymes. Asn 64 in the vertebrate enzyme is replaced by Phe 369 in the bacterial enzyme. This change results in a more hydrophobic and less charged environment in the bacterial reductase and, thus, causes increased selectivity for the modified trimethoprim derivative.

In the bacterial enzyme, there is a tighter and more favorable contact between the unsaturated triple bond of the ligands and the aromatic ring of phenylalanine. A comparable contact to Asn 64 in the rodent enzyme cannot achieve this contribution. In the pioneering era of structure-based drug design in the early 1980s, DHFR was the model protein par excellence. As a result, much of the knowledge that shapes our current understanding of selectivity phenomena has been accumulated on this enzyme.

27.3 HMG-CoA Reductase Inhibitors: The Changing Fate of Drug Development

Among the leading causes of death in most European countries and the United States are coronary heart disease (CHD), atherosclerosis, the myocardial infarction (MI) and stroke associated with them. CHD has multifactorial genetic causes and is also a common disease in the developed world. Risk factors include obesity, smoking, high blood pressure, and elevated levels of fibrinogen and cholesterol. **High levels of cholesterol** are found in the plaques that constrict and occlude the blood vessels. The preponderance of academic opinion is that lowering cholesterol levels is a reasonable treatment strategy, so cholesterol-lowering drugs are often prescribed. Cholesterol has various functions in the construction of cell membranes (Sect. 4.2) and serves as a starting material for the synthesis of steroid hormones and bile acids (Sect. 28.3). The brain, adrenal glands, skeletal muscle, skin, blood, and liver have an increased need for cholesterol. Between 0.9 and 2 g of this substance are needed daily. About one-third comes from food, and the rest is synthesized in the liver.

Statins are a group of drugs that inhibit **cholesterol biosynthesis**. Hardly any other class of compounds illustrates the success and failure of drug development in pharmaceutical research as well as statins. There is a fine line between astronomical financial success through gigantic sales figures and catastrophic crashes that can bring a company to the brink of financial ruin. The de-

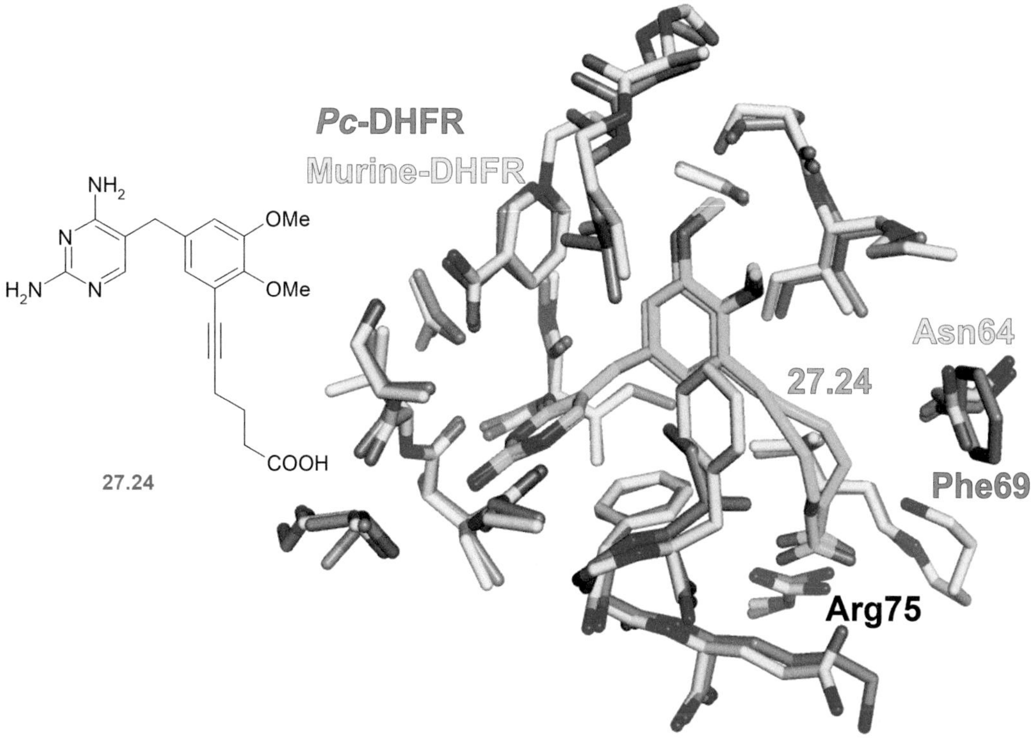

Fig. 27.11 The exchange of the 3-methoxy group in trimethoprim **27.19** for an unsaturated aliphatic side chain in **27.24** gives an affinity that is improved by a factor of 5000 and selectivity for the bacterial enzyme from *Pneumocystis jirovecii* compared to the mouse enzyme. An Asn 64 residue is found in the crystal structure of the vertebrate enzyme in the place where Phe 69 is found in the bacterial enzyme. This exchange to a more hydrophobic and less-charged environment in the bacterial enzyme results in the selectivity advantage for **27.24**

velopment of statins began in the 1950s. The American company Merck & Co. began intensive work on the biochemistry of lipid metabolism. In 1956, Karl Folkers and Carl Hoffman, both from Merck, discovered **mevalonic acid 27.25**, an intermediate in the **biosynthesis of cholesterol 27.26** (Fig. 27.12). However, the importance of the substance and the enzyme 3-hydroxy-3-methylglutaryl coenzyme A reductase (**HMG-CoA reductase**), which transforms HMG-CoA to mevalonic acid, was not recognized at the time. The enzyme reduces the substrate, which is composed of two acetate units, using two equivalents of NADPH in the rate-determining step of the biosynthetic pathway.

As a therapeutic approach to lowering cholesterol levels, Merck initially pursued a **basic ion exchange resin** (cholestyramine) that has a high affinity for bile acids. Since bile acids are synthesized from cholesterol, the removal of bile acids from the intestine results in more dietary cholesterol being used to replace these substrates, thus, reducing the overall blood cholesterol levels. The success of clofibrate **27.27** (Fig. 27.12, Sect. 28.6) began in the 1960s. This drug reduces elevated triglyceride levels and, to a lesser extent, cholesterol levels. However, long-term observations showed that the number of deaths was higher in the group of patients treated with clofibrate than in the control group. Cases of liver cancer have also been observed in animal studies.

During this time, interest focused on the body's own systems that serve as transport vehicles for hydrophobic substances in blood plasma. They consist of lipids and proteins and are classified according to their density. In particular, the importance of **low-density lipoproteins (LDL)**, which consist mainly of apolipoprotein B-100, has been recognized. They serve to transport water-insoluble cholesterol. They carry most of this freely circulating substance. LDLs are easily oxidized. In this form, it is taken up by macrophages in the arterial wall and stored. This overloading of the macrophages leads to the formation of foam cells, which can rupture and, together with the coagulation process, lead to the **deposition of plaques** and even to the complete occlusion of an artery. The result is hardening of the arteries (atherosclerosis). If such a plaque ruptures, it can block an artery at another location, resulting in a heart attack, stroke, renal failure, or angina pectoris. In short, high LDL levels are associated with an increased risk of atherosclerosis. Interestingly, a high HDL (high-density lipoprotein) level is beneficial and may even influence plaque disaggregation. Conversely, high LDL levels are particularly dangerous in patients with genetic hypercholesterolemia and apolipoprotein B deficiency. They have an extremely high risk of developing atherosclerosis. One therapeutic approach to reducing this risk is to lower blood cholesterol levels.

Beginning in 1973, Merck & Co. and other companies began to investigate the influence of hydroxylated steroids on cholesterol biosynthesis. At Merck, *in vitro* cell assays were developed to test compounds that inhibited cholesterol biosynthesis, specifically HMG-CoA reductase, the enzyme at the rate-determining step of the synthetic pathway. Although these compounds showed activity *in vitro*, they did not show efficacy in animal studies. At the same time, Akiro Endo and colleagues at Sankyo Japan began screening extracts from 8000 microorganisms. The most active compound, simultaneously isolated at Beecham in England, was compactin **27.28** (mevastatin, ◘ Fig. 27.13). In early 1979, Endo filed a Japanese patent for another microbial HMG-CoA reductase inhibitor, **monacolin K**, without knowing its structure. In the fall of 1978, Merck was also investigating microbial extracts. In the second week of the experiment, they found what they were looking for. The compound was isolated in February 1979, and a patent for lovastatin **27.29** (◘ Fig. 27.13), complete with structural details, was filed in June 1979. The compound was identical to monacolin K. The Merck patent was granted in the United States in late 1980 and later in other countries. In a few countries, the patent was granted to Sankyo instead. The reason for this difference in recognition was a different interpretation of **time priorities**. Sankyo filed the patent (**first-to-file**) 4-months earlier. Merck was awarded the patent in the US and many other countries because they could prove a 3-month earlier date of invention (**first-to-invent**).

Merck began clinical trials with lovastatin in April 1980, but they were discontinued in September of that year. The reason for this was rumors that lovastatin had caused tumors in dogs. Toxicity studies with lovastatin showed no evidence of this, and the rumors could not be confirmed. Nevertheless, the project was initially halted. In July 1982, Merck negotiated an agreement with the FDA to allow lovastatin to be used clinically by selected investigators. The use of lovastatin would be limited to treatment-resistant cases with severely elevated cholesterol levels because these patients were at particularly high risk of heart attack and stroke. The therapeutic effects on LDL cholesterol and total cholesterol levels were convincing and the side effects were minimal. Chronic toxicology and clinical studies were resumed. An application for approval was submitted in November 1986. A total of 160 volumes of preclinical and clinical data were submitted to the FDA. Only 9 months later, the drug was approved and the compound became a blockbuster with sales in the billions.

Years later, the crystal structure of the target enzyme, HMG-CoA reductase, was determined. In its active form, the enzyme is a tetramer. Each monomer consists of three subunits. The *N*-terminal domain has an anchor that attaches the enzyme to the membrane of the endoplasmic reticulum. The smaller *S*-domain, which contains the binding site for reduced NADP(H), is nested within the larger *L*-domain. The *S*-domain adopts the geometry of a Rossmann fold. The extended HMG-CoA molecule binds to the *L*-domain. The pantothenic acid moiety protrudes deep into the protein, while the ADP portion is located in a pocket with positively charged residues on the surface of the protein. The actual binding site for hydroxymethylglutaric acid (HMG) is located between the *L*- and *S*-domains. The product of the first reduction step, mevaloyl-CoA, has a negatively charged oxygen atom that is stabilized by a neighboring Lys 691 in the enzyme (◘ Fig. 27.14). The thiolate temporarily released from the CoA group is stabilized by His 752, which is presumably protonated. The activity of HMG-CoA reductase can be regulated by phosphorylation. A serine residue in the vicinity of the bound $NADP^+$ cofactor is phosphorylated. This is thought to reduce the affinity for NADP(H). The energetically demanding cholesterol biosynthesis in the cell can be curtailed by this step.

Like the discovered natural products, the statins were designed as structural analogues of the carboxylic acid

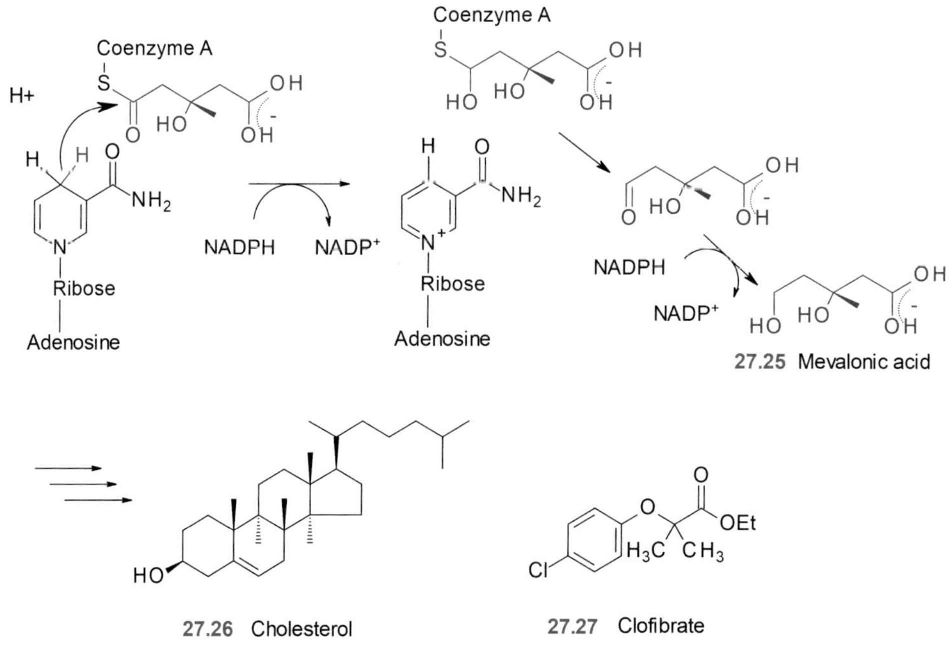

◘ **Fig. 27.12** 3-Hydroxy-3-methylglutaryl-coenzyme A reductase (HMG-CoA reductase) transforms HMG-coenzyme A into mevalonic acid **27.25** (*blue*) by consuming two equivalents of NADPH. The reaction takes place in two steps. In the first step, the thioester is reduced to a thioacetal, which is hydrolyzed to mevaldehyde. In the next step another equivalent of NADPH is used to reduce the newly formed aldehyde function to an alcohol. Finally, mevalonic acid is transformed via multiple steps to cholesterol **27.26**. Clofibrate **27.27** is a PPARα agonist (Sect. 28.6)

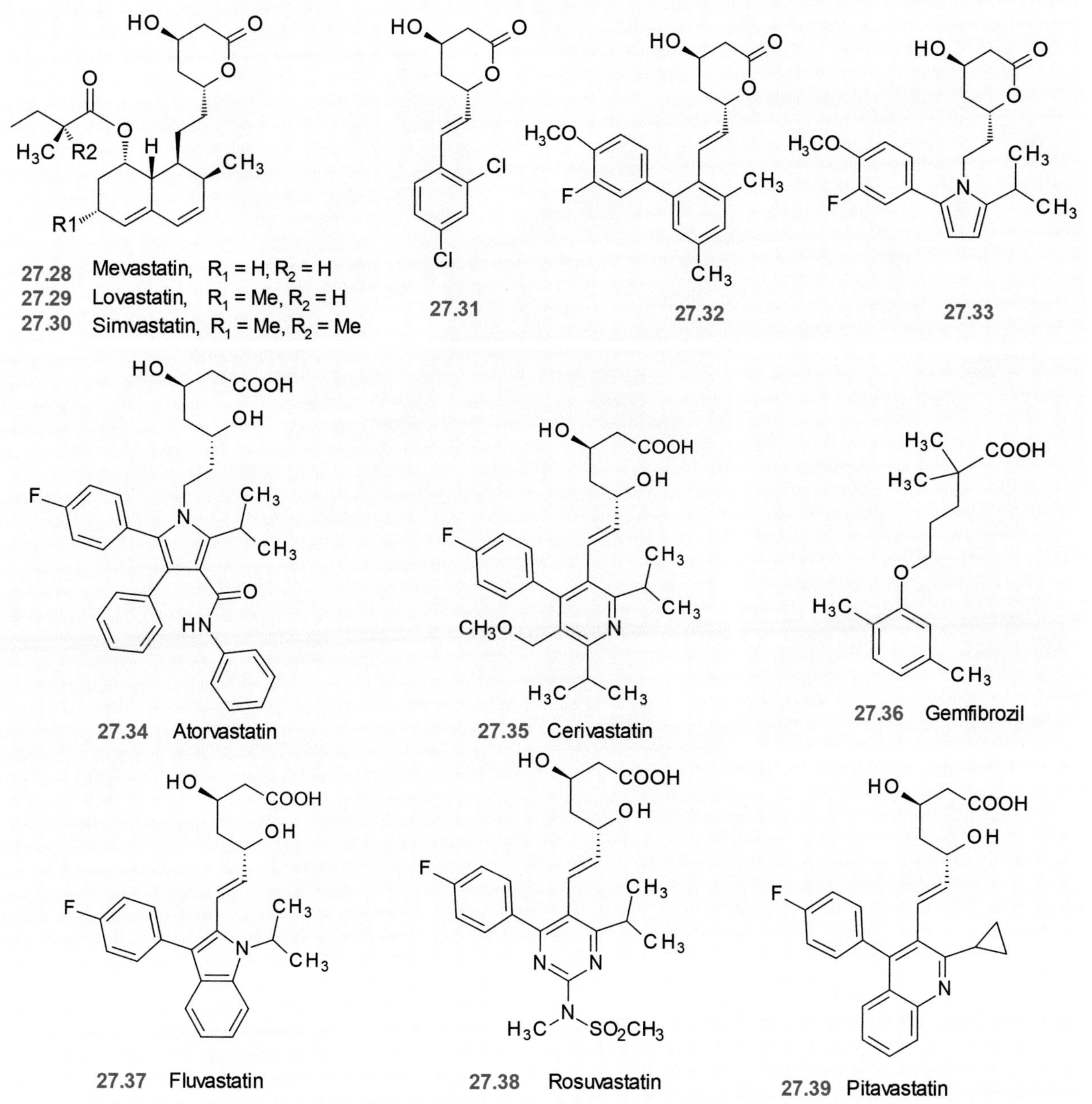

Fig. 27.13 The natural products mevastatin (compactin) **27.28** and lovastatin **27.29** inhibit cholesterol biosynthesis at the HMG-CoA reductase step. Simvastatin **27.30** a partial-synthetic analogues was developed later. The development of fully synthetic statins starting from **27.31** and **27.32** at Merck, USA led to **27.33**, which also attracted attention at Parke-Davis. This led to the fully synthetic HMG CoA reductase inhibitors atorvastatin **27.34**, cerivastatin **27.35**, fluvastatin **27.37**, rosuvastatin **27.38**, and pitavastatin **27.39**. Compared to lovastatin, they use an open-ring form to achieve lower lipophilicity with reduced central nervous system side effects. The opened lactone ring is the actual active form of lovastatin and its analogues (Sect. 9.2). Co-administration of cerivastatin **27.35** with the fibrate gemfibrozil **27.36** results in a fivefold increase in the plasma level of cerivastatin due to reciprocal blockade of its degradation by cytochrome CYP 3A4

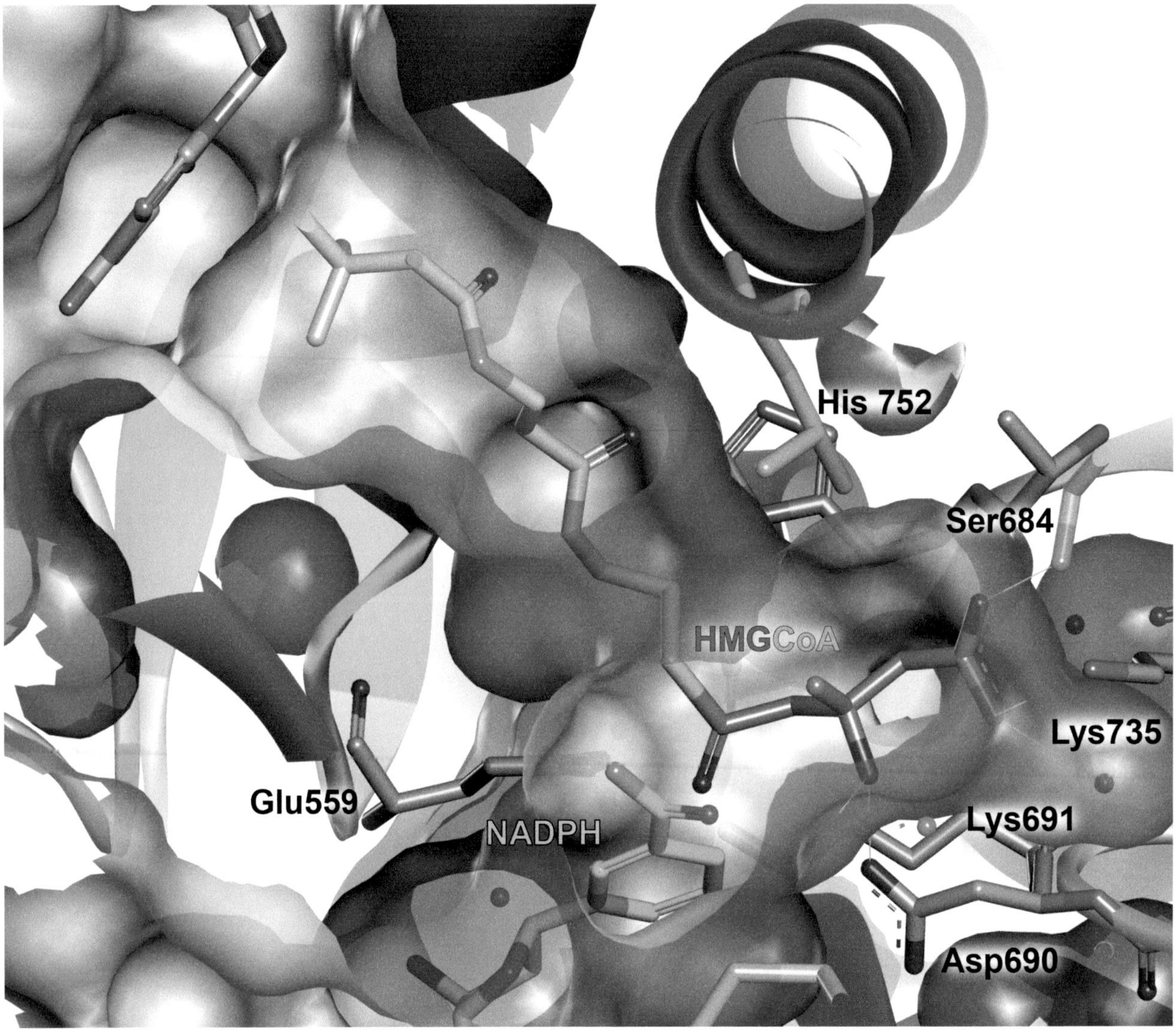

■ **Fig. 27.14** The crystal structure determination of HMG-CoA reductase was accomplished with the bound NADPH cofactor (*green*) and HMG-coenzyme A (*pink*). The nicotinamide ring of the cofactor lies underneath the thioester bond of HMG-coenzyme A. The hydride ion is transferred from there in the first reduction step (cf. ■ Fig. 27.12). (▸ https://sn.pub/x7EbxX)

chain of 3-hydroxy-3-methylglutaryl-CoA. They inhibit the reductase competitively, but their affinity is a thousand times higher than that of the natural substrate. The statins mevastatin **27.28**, lovastatin **27.29**, and the later developed simvastatin **27.30** (■ Fig. 27.13) are prodrugs with a lactone structure, which is opened in the mucous membranes of the gastrointestinal tract or in the liver to the actual active substances.

A comparison of the structure of the reductase with the substrate and the inhibitor simvastatin shows that the extended 3,5-dihydroxycarboxylate moiety is located at the position of the HMG portion (■ Fig. 27.15). All later fully synthetic statins have a carbocyclic or heteroaromatic ring system at the C5 atom of the dihydroxycarboxylate moiety, separated by a two-membered linker. This moiety binds in a region where the thiol side chain of the pantothenic acid moiety comes to rest. The high structural variation of this moiety in the more recently developed fully synthetic statins (**27.34**, **27.35**, **27.37**–**27.39**, ■ Fig. 27.13) emphasizes that although this part of the molecule contributes to the affinity of the inhibitors to the enzyme, no specific interactions are formed in the binding pocket that opens to the outer surface. The heteroaromatic moieties differ significantly from the prototypes originally obtained from microorganisms (**27.28**–**27.30**).

But how did these heteroaromatic, fully synthetic inhibitors come about? In the mid-1980s, Merck experi-

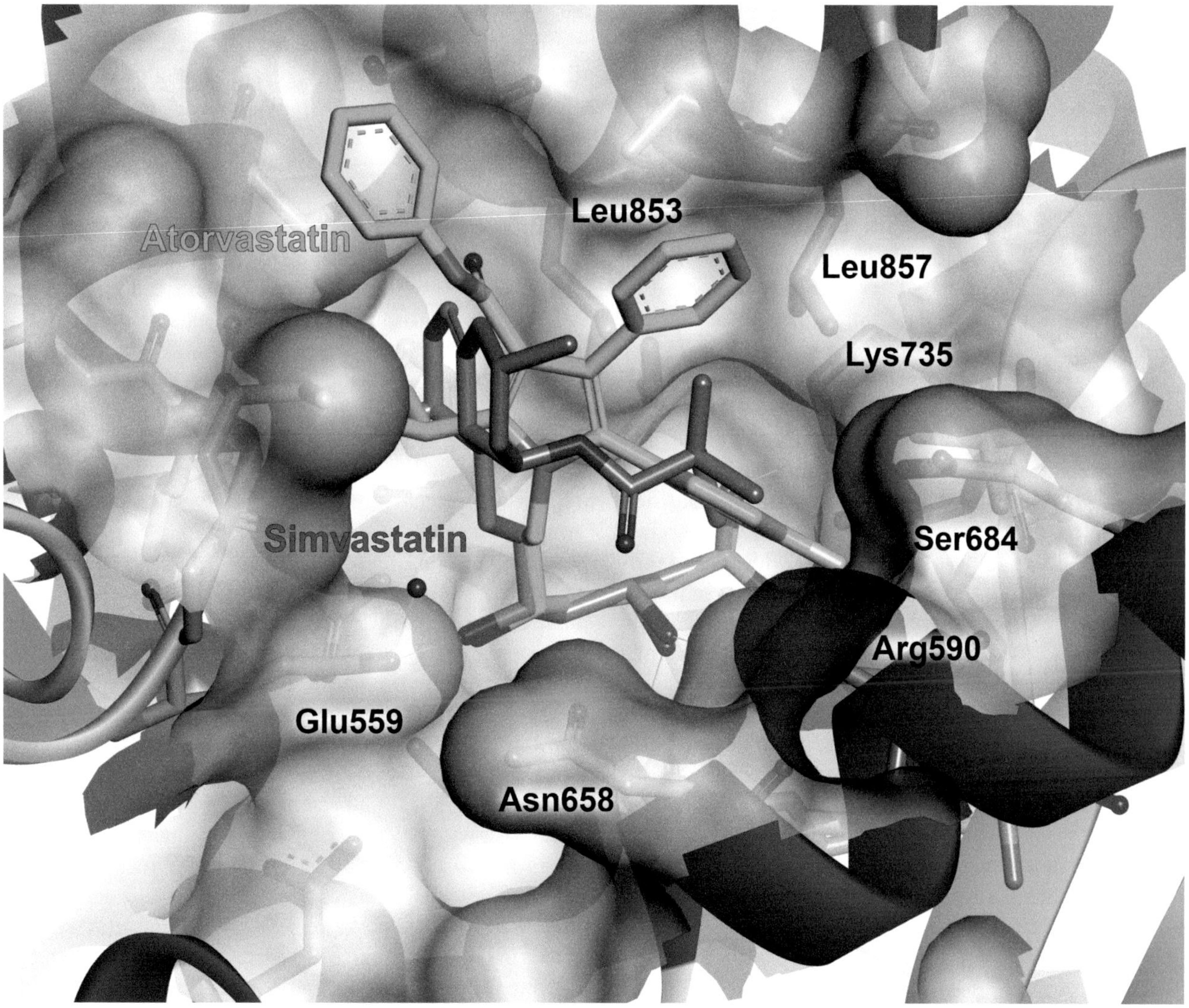

Fig. 27.15 Superposition of the structures of HMG-CoA reductase in complex with simvastatin **27.30** (*gray*) and atorvastatin **27.34** (*green*). Both inhibitors bind Lys 745, Ser 684, and Arg 590 with their mevalonic acid analogue moiety, just as the natural substrate does. The remainder of the molecule, which is very different in the natural product-like simvastatin and the fully synthetic atorvastatin, binds to the region occupied by the CoA residue in the substrate complex. The NADPH pocket remains unoccupied in the structures. (► https://sn.pub/67z25R)

mented with synthetic analogues of the first generation of statins. Derivatives were discovered that still carried the lactone building block on the ethylene linker chain. However, they were provided with substituted aromatic groups (**27.31**, **27.32**). The extension to the biphenyl scaffold was guided by the idea that mevastatin also carried a substituent in this spatial direction. The competitors at Parke-Davis adopted this concept and replaced the central aromatic moiety with a pyrrole ring (**27.33**). Further decoration of this five-membered ring in the 3- and 4-positions improved the effect. The optimum proved to be **27.34**, which was launched in 1997 as **atorvastatin**.

Developed at Parke-Davis, at the time part of the Warner-Lambert Group, the drug was transferred to Pfizer through a company acquisition. There, atorvastatin became a success story par excellence. It became the best-selling drug of all time (Sortis® and Lipitor®). In 2004, it accounted for half of the statin market share. In both 2006 and 2007, Pfizer earned US$ 14 billion from the drug. Sales in Germany have been low due to a healthcare reform that introduced a fixed copayment for statins. Cerivastatin **27.35** (Lipobay®) became a similar cash cow for Bayer Corporation. This product was launched in Germany in 1997 and in other European countries and the United States. In late 1998, the German Federal Institute for Drugs and Medical Devices (BfArM) reported deaths associated with cerivastatin therapy. After further deaths were reported in the United States and Germany,

Bayer withdrew the drug from the market in mid-2001. What happened? The deaths occurred as a result of rhabdomyolysis, an acute disintegration of skeletal muscle and concomitant renal failure due to toxic muscle metabolites. This adverse event has been associated with overdosing, particularly with the combination of cerivastatin and gemfibrozil **27.36**, a fibrate. Gemfibrozil increases the plasma level of cerivastatin by a factor of five and may itself cause myopathy. The cause of death was considered to be an overdose of cerivastatin due to simultaneous inhibition of the degradation mechanism of both compounds by the metabolizing cytochrome CYP 3A4 (Sect. 27.6). Patients were informed of the risk by the enclosed package leaflet describing the correct use of the medication, and pharmacists distributing the drug in the US were also informed. Cerivastatin was considered a growth product for Bayer's pharmaceutical business. Shortly after its approval, it achieved sales of 2.5 billion euros. About six million people worldwide were taking the drug. Its withdrawal had far-reaching consequences for the Bayer Corporation, which had to struggle with it for several years. The recall itself caused resentment because the press and shareholders were informed before physicians and pharmacists. This approach was certainly suboptimal, as it easily contributed to a loss of public confidence in the pharmaceutical industry and suggested purely commercial intentions. Thus, the two drugs Sortis® and Lipobay® illustrate the fine line between success and failure in the pharmaceutical business and the risks associated with launching a new drug despite a known and established principle.

27.4 Hitting a Moving Target: Aldose Reductase Inhibitors

The alarming increase in cases of **type 2 diabetes mellitus** has already been mentioned in Sect. 26.8. In Germany alone, more than 7 million people have been diagnosed with diabetes. With approximately 500,000 new cases each year, this number could rise to 12 million by 2040. Statistically, the risk of premature death is up to 2.6 times higher and life expectancy is reduced by about 5 to 6 years compared to people without this metabolic disease. Treatment of diabetes and its complications costs billions of dollars and represents a massive economic and healthcare burden. Acquired diabetes, which manifests itself as an increasing resistance of cells to insulin, will lead to serious long-term consequences if not controlled by substitution therapy. These manifest themselves in secondary complications, such as increased atherosclerosis (Sect. 27.3), which raises the risk of heart attack and stroke. The long-term consequences of poorly controlled blood glucose levels preferentially affect cells in tissues that do not control their glucose uptake with insulin. This is particularly true of cells in the vascular system, nerves, eyes, and kidneys. Exogenous insulin administration does not directly help these cells because they are unable to downregulate their glucose uptake. Premature blindness, kidney damage, and peripheral vascular disease can result, and may require limb amputation to treat.

One way to intervene in blood glucose regulation, in addition to significant changes in diet and lifestyle, is to administer exogenous insulin (Sect. 32.2). However, even the most rigorous insulin replacement therapy cannot match the effectiveness of endogenous insulin. Repeated episodes of injurious hyperglycemia may occur, particularly affecting insulin-independent cells. Despite therapy, people with diabetes can expect long-term complications. These consequences affect the quality of life of elderly patients in particular. Therapeutic approaches that can reduce long-term complications are, therefore, being sought.

One approach is to intervene in the so-called **polyol pathway**. In this pathway, glucose **27.40** is reduced to sorbitol and then oxidized to fructose (■ Fig. 27.16). The first step is catalyzed by aldose reductase and the second by sorbitol dehydrogenase. The conversion by **aldose reductase** is the rate-determining step. It proceeds by consuming NADPH, which is oxidized to $NADP^+$. NADH/NAD^+ is needed as a cofactor in the next step of dehydrogenase. It has long been debated whether overloading the polyol pathway with high levels of glucose would lead to an increased concentration of polar reaction products in the cell. As a result, the cell would experience increased osmotic pressure, which would be alleviated by increased water uptake. However, this would result in cell swelling and increased **osmotic stress** on the membrane. However, the **oxidative stress** experienced by the cell due to the overloaded polyol pathway appears to be more serious. The increased flux of glucose along this pathway requires increasing amounts of NADPH and NAD^+, thus, putting a severe strain on the homeostasis of these redox-active substances. The body has to protect itself against reactive oxygen species, which are formed as a byproduct of the approx. 400–800 L of oxygen we take in daily. These species have cell-damaging potential. When the production of these aggressive oxygen derivatives exceeds the detoxification capacity of the cell's endogenous antioxidant systems, we speak of oxidative stress. The main defense system is glutathione, which is oxidized to glutathione disulfide species under oxidative conditions. To be available for defense, it must be regenerated by glutathione reductase, which consumes NADPH. If only a small amount of NADPH is available, the capacity of the glutathione defense system will be quickly exhausted and the cell will show signs of oxidative stress. Today, this is mainly associated with inadequate regulation of glucose homeostasis.

Inhibition of aldose reductase is one way to avoid overloading this pathway. A **genetic polymorphism** that

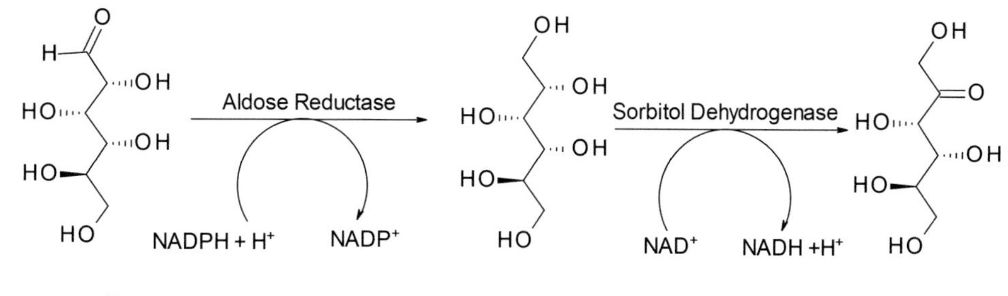

Fig. 27.16 D-Glucose **27.40** is transformed by aldose reductase to D-sorbitol and further by sorbitol dehydrogenase to D-fructose along the polyol pathway

has been found to be associated with the risk of diabetic complications has provided evidence for the therapeutic relevance of such a strategy. Variations in the genes that carry the information for aldose reductase lead to increased or decreased expression of aldose reductase in affected individuals. An increased supply of the enzyme leads to increased NADPH consumption. Carriers of the allele for increased expression appear to have an increased susceptibility to diabetic complications. On the other hand, a reduced supply of the enzyme corresponds to a reduced prevalence. This is a clear indication that reducing aldose reductase activity is a useful strategy to avoid the long-term complications of inadequately controlled blood glucose levels.

Interestingly, aldose reductase appears to have another role in the cell. It is capable of reducing a wide range of aldehyde substrates. This is an important task in the detoxification mechanism. Aldehyde reductase, an enzyme related to aldose reductase, performs a similar function. Blockade of both enzymes leads to serious problems because **the removal of toxic and highly reactive aldehydes** from the cell is suppressed. Therefore, a potent aldose reductase inhibitor must be designed to be as selective as possible. However, the very broad substrate promiscuity of aldose reductase makes this goal nontrivial. This reductase has an amazing ability to adapt to substrates of widely varying sizes. The enzyme's binding pocket opens as much as possible to accommodate substrates, apparently without significant energetic cost. But the same properties that allow the enzyme to flexibly adapt to substrates of different sizes can also be used to inhibit it. Thus, attempting to block the function of aldose reductase with various inhibitors can be compared to shooting at a moving and evasive target.

First, the **architecture and mode of action of aldose reductase** should be considered. It uses NADPH as a cofactor and, unlike the majority of reductases, does not have an NADPH-binding domain with a Rossmann fold, but rather a TIM barrel geometry. The active site is located at the lid of this barrel structure. It is divided into a relatively rigid catalytic center, known as the anion-binding pocket, and the specificity pocket, which is adaptable to the substrate (Fig. 27.17). In the anion-binding pocket, the reduction step takes place with the transfer of a hydride ion from NADPH to the aldehyde group of the substrate. The product is an alcohol. A number of head groups have been discovered that can mimic the geometry of the reduction step. Carboxylic acids derived from acetic acid are commonly incorporated into many lead structures (Fig. 27.18). However, the pK_a of these compounds is unfavorable for their bioavailability. Therefore, other groups have been sought that can also take on an anionic character, but have more favorable pK_a qualities for transport and distribution (Sect. 19.4). Hydantoins have proven to be very useful in this regard. Other variations are shown in Fig. 27.18.

Another interesting property of aldose reductase is how one enzyme can adapt to so many different substrates. Protein–ligand complexes of four different inhibitors are shown in Fig. 27.19 (cf. Fig. 27.18, **27.50**, **27.44**, **27.48**, and **27.45**), each binding to a different protein conformer. In Sect. 4.7, it was pointed out that two extreme cases of transient pocket opening are discussed. One is **conformational selection**, in which the binding ligand selects the appropriate pocket from a dynamic equilibrium of protein conformers to occupy and stabilize. The other is **induced fit**, in which the ligand first docks at the protein surface and then unlocks the required pocket for immediate binding. Presumably, processes of solvation and desolvation of the ligand and pocket have a crucial influence on the energy balance of this occupancy of transiently opening pockets. Recent evidence suggests that in the case of aldose reductase, the mechanism follows an induced-fit adaptation. It is mainly hydrophobic portions of the ligands that occupy the transient pocket in aldose reductase.

To better understand the conformational diversity of possible geometries of an enzyme, MD simulations can be performed. In Sect. 15.8, the example of aldose reductase was used to show how such a study can be performed. Typically, there are side chains of a small number of amino acids that allow these conformational adaptations and, thus, induce the opening and closing of entire regions of the binding pocket. The simulations also give an idea of the energy balance associated with pocket opening. They also provide some clues as to the probability of the process occurring, and whether it might only be noticeable at elevated temperatures.

Crystal structures are very important here. They give a picture of what the endpoints of the binding will look

27

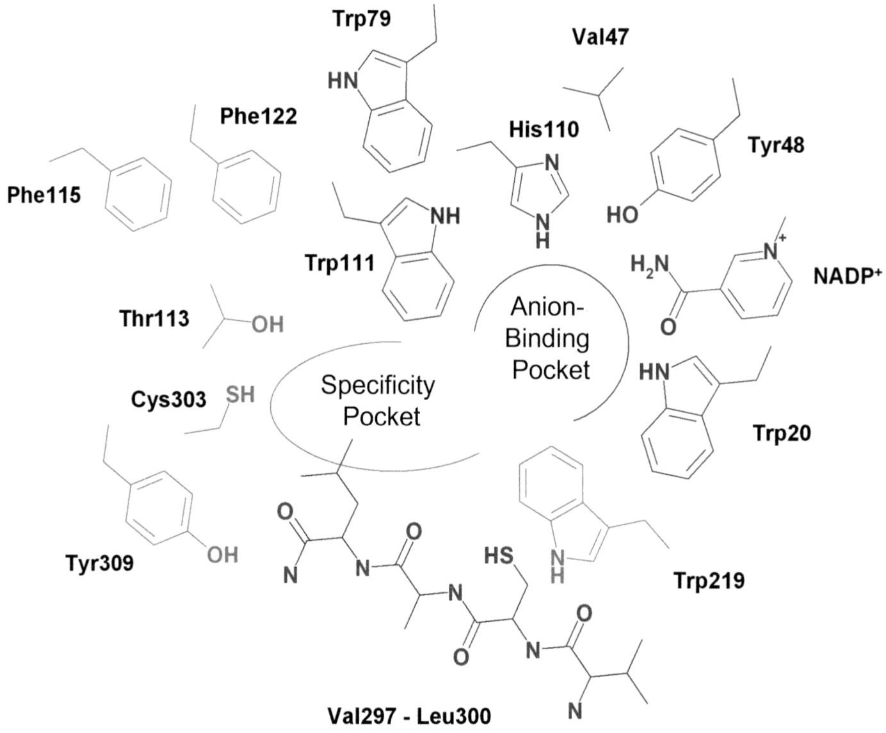

Fig. 27.17 The binding pocket of aldose reductase is divided into a catalytic (*blue*) and a specificity (*orange*) pocket. The cofactor NADPH/$NADP^+$ binds below the catalytic pocket. Phe 122, Trp 219, and Leu 300 are primarily responsible for the structural adaptability of the specificity pocket. Trp 20 can also undergo conformational changes in the catalytic pocket. The segment Val 297–Leu 300 (*red*) belongs to a loop that exhibits particularly high adaptability

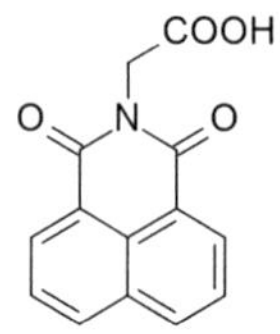
27.41 Alrestatin

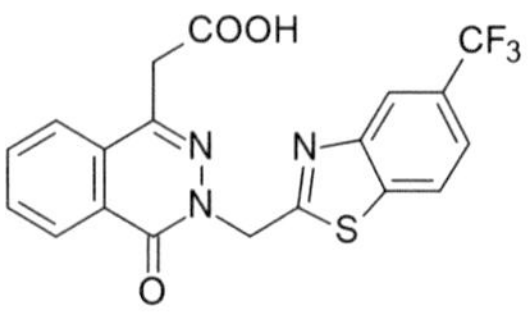
27.42 Zopolrestat

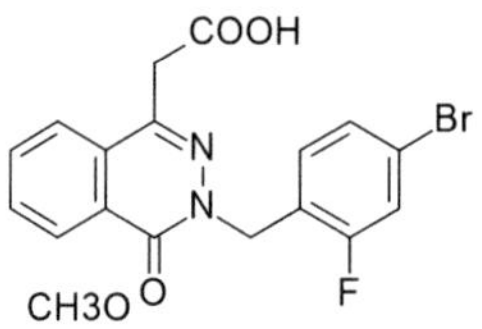
27.43 Ponalrestat

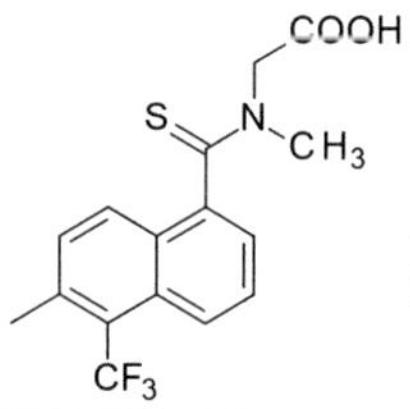
27.44 Tolrestat

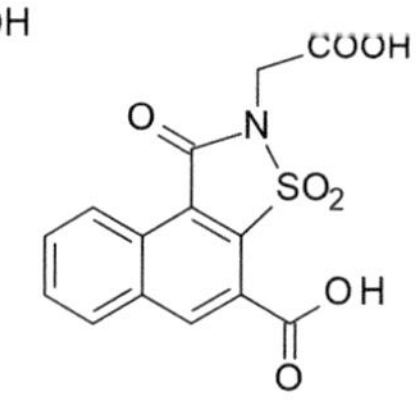
27.45

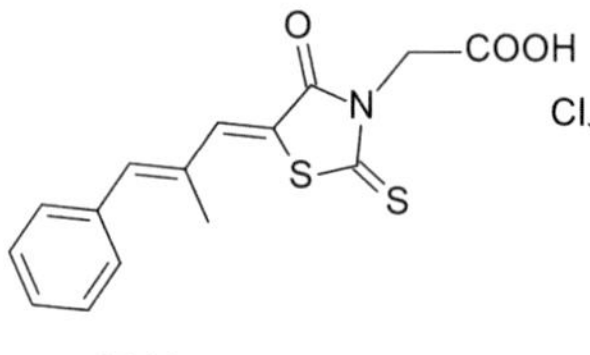
27.46 Epalrestat

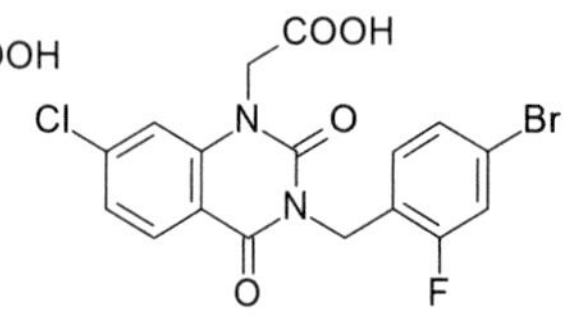
27.47 Zenarestat

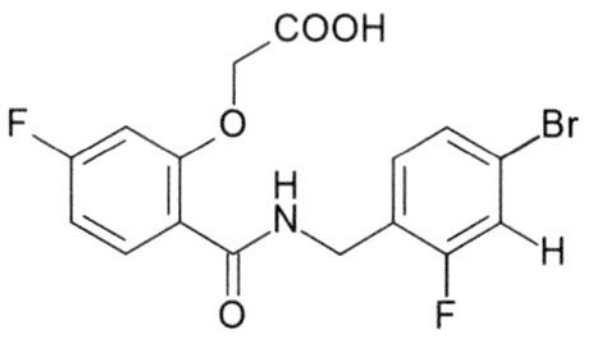
27.48 IDD594

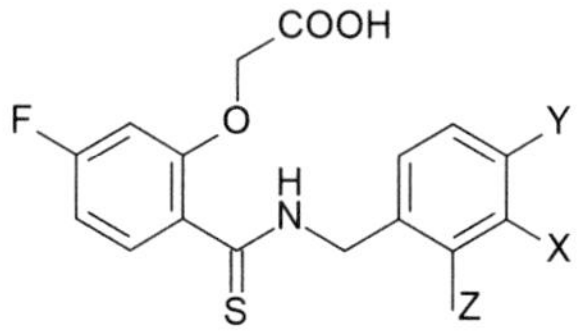
27.49 IDD series
X: - NO_2
X: - COOH
X: - H
X: - $SOCH_3$

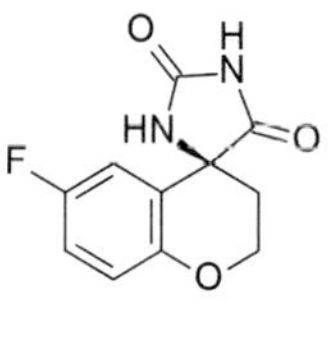
27.50 Sorbinil

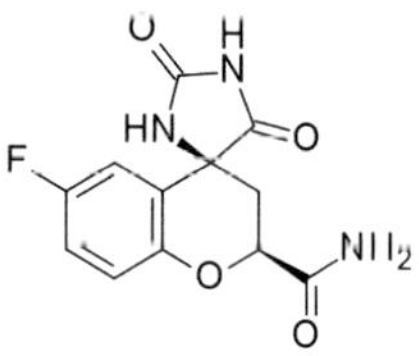
27.51 Fidarestat

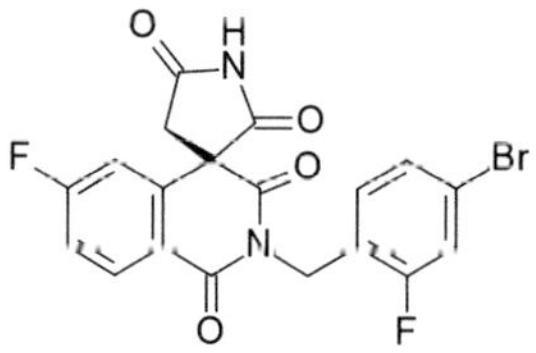
27.52 Minalrestat

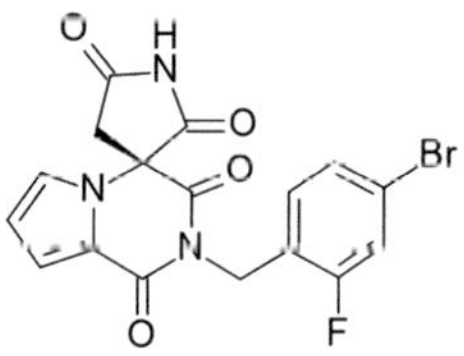
27.53 Ranirestat

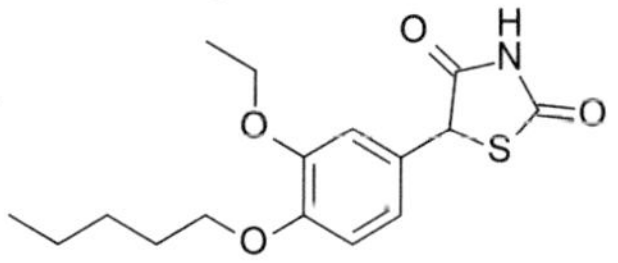
27.54 Risarestat

Fig. 27.18 Synthetic inhibitors **27.41**–**27.54** of aldose reductase. Only epalrestat **27.46** has become a marketed product. The crystal structures of **27.50**, **27.44**, **27.48**, and **27.45** are shown in Fig. 27.19. The series **27.49** of analogous IDD derivatives partly bind like **27.48** in the opened transient specificity pocket, while some remain outside and the pocket does not open. However, both binding geometries are in some cases found side by side in the same crystal structure of aldose reductase

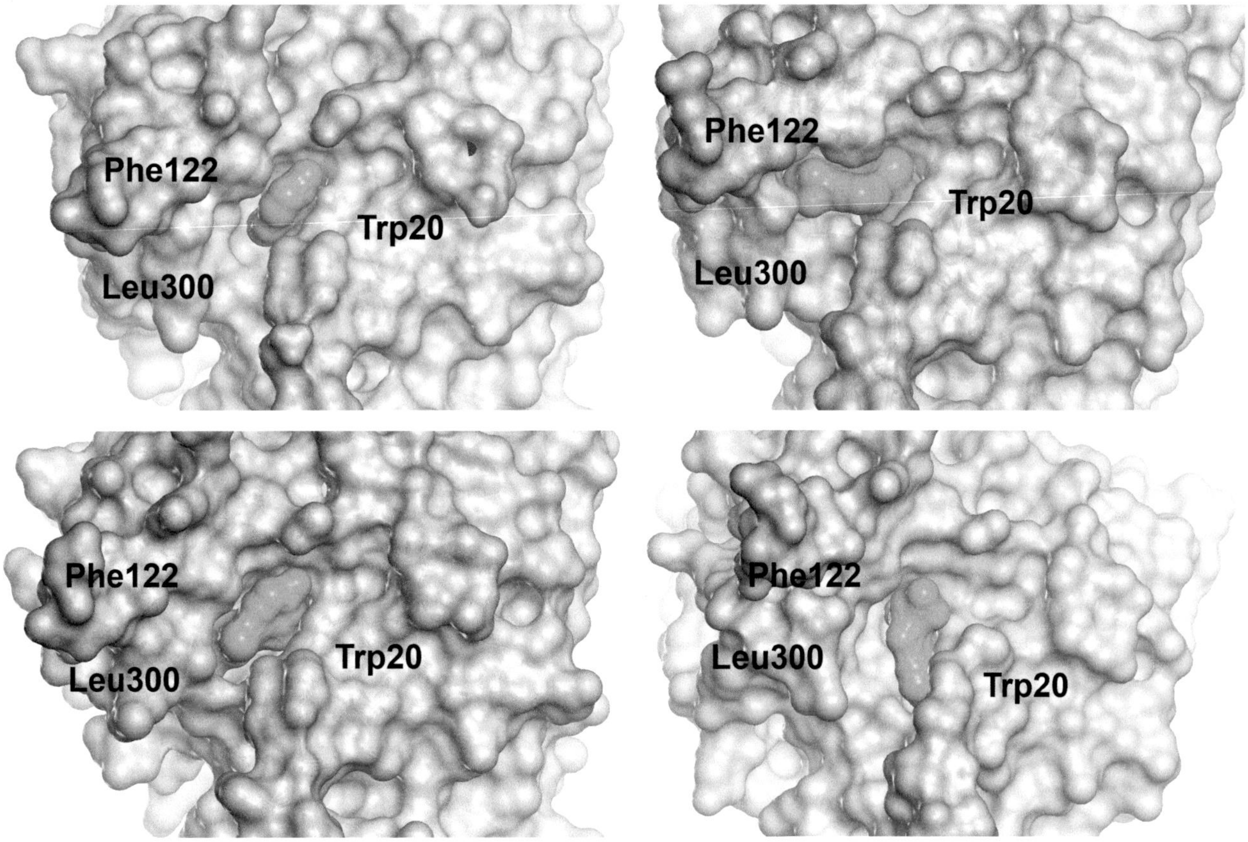

Fig. 27.19 Crystal structures of aldose reductase with sorbinil **27.50** (*upper left*), tolrestat **27.44** (*upper right*), IDD594 **27.48** (*lower left*), and **27.45** (*lower right*, fomulas see Fig. 27.18). All inhibitors bind to a different conformer of the protein. Above all, the residues Trp 20, Phe 122, and Leu 300 undergo significant spatial rearrangements and open up structurally altered subpockets in the enzyme. (▸ https://sn.pub/vNLRrq)

like, at least for a given ligand. Fig. 27.20 shows the binding geometry of sorbinil **27.50** (Fig. 27.19, *upper left*). The inhibitor blocks the catalytic center with its hydantoin group and is located above the nicotinamide ring in the cofactor. Interestingly, it leaves the specificity pocket closed. Phe 122 and Leu 300 face each other like two wings of a door and seal the region of the specificity pocket behind them. The series of IDD inhibitors with the parent scaffold **27.49** has been best studied. The derivative **27.48** with an *ortho*-fluoro and a *para*-bromo substituent on the terminal phenyl ring opens the pocket and binds with a nanomolar binding affinity (Fig. 27.19, *lower left*). The same is true for the highly potent derivative with a *meta*-nitro group (**27.49**). In contrast, the isosteric derivative with a *meta*-carboxylic acid group in the same position as the nitro group remains outside the pocket and this ligand binds with a 1000 times lower potency. On the other hand, the micromolar derivative with an unsubstituted phenyl ring is able to open the pocket. However, in its crystal structure, only 80% of all pockets are occupied, the remaining 20% of pockets are closed and left unoccupied. When the phenyl ring is substituted with a *meta*-methylsulfoxy group, this submicromolar ligand is found side by side in the crystal structure in both arrangements with open and closed pocket. These examples highlight the complexity of substituent effects on transient pocket opening. Perhaps this is how the enzyme controls which substrates it converts efficiently and which it does not.

The development of potent aldose reductase inhibitors has been ongoing for many years. Numerous candidates have successfully entered clinical trials (Fig. 27.18). Unfortunately, most of them were discontinued at this stage. Often, adverse effects or lack of efficacy led to this decision. In 1992, ONO Pharmaceutical Co. in Japan succeeded in bringing epalrestat (Kinedak®) **27.46** to market for the treatment of diabetic neuropathy. Many other derivatives, such as fidarestat **27.51**, ranirestat **27.53**, zopolrestat **27.42**, ponalrestat **27.43** and zenarestat **27.47**, made it into phase II clinical trials. In some cases, development was discontinued at this stage or the studies have not yet been completed.

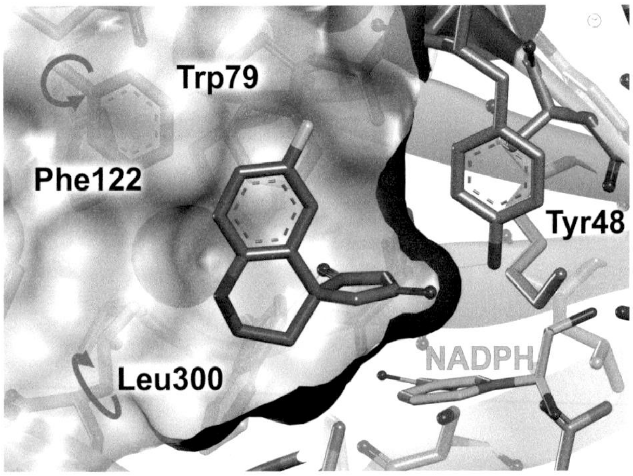

■ **Fig. 27.20** Crystallographically determined binding geometry of sorbinil **27.50** to aldose reductase. The inhibitor's hydantoin group binds above the nicotinamide ring of the cofactor to Tyr 48, Trp 79, and His 110. The specificity pocket remains closed during this binding. This pocket can be opened by twisting the side chain of Phe 122 and Leu 300 out of space. (▶ https://sn.pub/VuQGSY)

It is surprising that such a modest therapeutic benefit has been achieved despite intensive research and what appears to be a valid therapeutic principle. Aldose reductase, however, holds another record. It is probably the best-characterized protein in terms of structural and physicochemical properties. With the inhibitor IDD594 **27.48**, a crystal structure with a resolution of 0.66 Å was determined, showing almost every water molecule and H-atom (see ■ Fig. 13.9), and a well-resolved neutron structure is also available. Hardly any other enzyme has been characterized by so many quantum-chemical calculations and MD simulations. Thermodynamic and mutational studies have provided insight into the energetics of the protein's adaptive behavior. But the extensive knowledge of its properties has not yet helped to find a reliable and broadly applicable therapy for late-onset diabetic complications by using an appropriate inhibitor.

27.5 11*β*-Hydroxysteroid Dehydrogenase

Isoforms of 11*β*-hydroxysteroid dehydrogenase (11*β*-HSD) are jointly responsible for the conversion of the biologically active glucocorticoid **cortisol 27.56** to the biologically inactive 11-keto form, **cortisone 27.55** (■ Fig. 27.21). Two isoenzymes, 11*β*-HSD1 and 11*β*-HSD2, belonging to the **short-chain dehydrogenase/reductase** superfamily, were found. Their sequence identity is only 15%. Chemically, they are opponents. 11*β*-HSD1 is widely distributed, with increased expression in the liver and adipose tissue. The enzyme acts as a reductase, consuming NADPH to form active cortisol from inactive cortisone, which binds to and activates the glucocorticoid receptor (Sect. 28.5). On the other hand, as a dehydrogenase, 11*β*-HSD2 oxidizes cortisol to inactive cortisone by consuming NAD^+. In doing so, it protects the mineral corticoid receptor from overexposure to this active hormone. This is especially important in the colon and kidneys. Overactivation of this receptor by cortisol, in addition to aldosterone, leads to increased renal absorption of sodium and chloride ions. The result is water retention and an increase in blood pressure. A congenital genetic defect that causes mutations in 11*β*-HSD2 can lead to an inherited form of **hypertension**. The mutated enzyme is less efficient. The result is an excess of cortisol. The receptor becomes overloaded and causes elevated blood pressure. Interestingly, **glycyrrhizin 27.59** (■ Fig. 27.22), one of the constituents of **licorice**, is a potent 11*β*-HSD2 inhibitor. Excessive consumption of this candy, which is made from the root of *Glycyrrhiza glabra*, can, in the worst cases, lead to temporary symptoms similar to those of the congenital genetic defect.

The short-chain dehydrogenases/reductases adopt a Rossmann folding pattern. The presence of a **Tyr–Lys–Ser triad**, present in almost all members of this family, is critical for the catalytic mechanism. The sequence of the reduction reaction is shown in ■ Fig. 27.23. A hydride ion is transferred from the nicotinamide ring of NADPH to the carbonyl group being reduced. The carbonyl function is involved in a network of hydrogen bonds that is responsible for its polarization for nucleophilic H^- ion attack. The hydroxyl group of a tyrosine residue acts as a proton donor. In addition, the ammonium group of an adjacent lysine immobilizes the OH groups of the sugar moiety and facilitates proton transfer by lowering the pK_a of the tyrosine residue (■ Figs. 27.23 and 27.24). The different isoforms of 11*β*-HSD are suitable for both oxidation (dehydrogenases) and reduction (reductases) steps, which are catalyzed by very similar mechanisms.

Endocrinologists have long recognized a phenotypic similarity between the relatively rare Cushing's syndrome and the **metabolic syndrome**, which is common in industrialized countries. Cushing's syndrome occurs as a result of excessive cortisol production and results in a "full moon face" and adrenocortical (central fat distribution) obesity. The alarming increase in obesity in the industrialized world and the concomitant increase in type 2 diabetes have already been discussed in the previous two sections and in Sect. 26.8. Interestingly, elevated cortisol levels have been found in the adipose tissue of obese individuals compared to lean individuals.

Apparently, there is a tendency for increased 11*β*-HSD1 activity in the adipose tissue of people who are predisposed to obesity. Resistance to diet-induced obesity was observed in genetically modified mice lacking 11*β*-HSD1 activity. The mice showed improved lipid and

Fig. 27.21 The two isoforms of HSD1 and HSD2 of the 11β-hydroxysteroid dehydrogenase transform inactive cortisone **27.55** into active cortisol **27.56** and vice versa. In rodents, the same enzyme pair transforms 11-dehydrocorticosterone **27.57** into corticosterone **27.58**

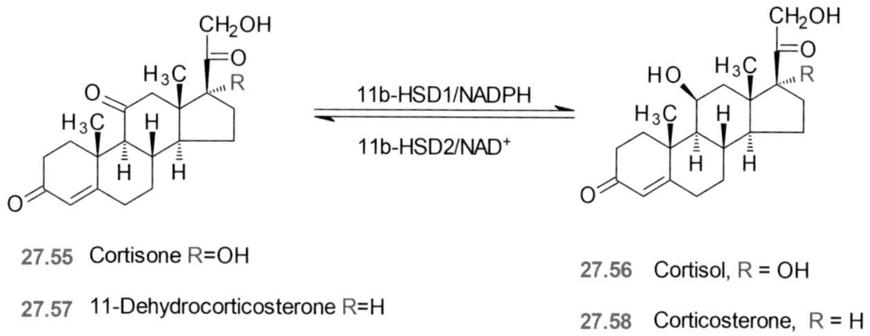

lipoprotein levels and increased insulin sensitivity in the liver. On the other hand, transgenic mice with induced overexpression of 11β-HSD1 in adipose tissue showed increased insulin resistance. These results suggest that lowering 11β-HSD1 activity may be a promising therapeutic principle for the treatment of metabolic syndrome. This strategy seems particularly attractive because elevated 11β-HSD1 levels are found only in certain tissues, particularly adipose tissue. In addition, the unselective 11β-HSD inhibitor carbenoxolone **27.60** (Fig. 27.22) increases insulin sensitivity in the liver in both healthy and diabetic volunteers without increasing glucose metabolism in the periphery.

As a result, programs to develop selective 11β-HSD1 inhibitors were established in numerous pharmaceutical companies. The arylsulfonamidothiazolenes were the first class of compounds. BVT-2733 **27.61** (Fig. 27.22) was derived from this series as a promising development candidate. These inhibitors have an aminothiazole substructure that binds to the serine and tyrosine residues of the catalytic triad, mimicking the geometry of the 11β-ketone function of the natural steroid. The crystal structure of a representative of this class, **27.62**, is shown in Fig. 27.25. Other classes of compounds use an amide group for the keto function, a urea moiety, or a heterocycle as a mimetic for the keto function in the substrate. Sulfones derived from adamantane (e.g., **27.63**) and sulfonamides have been developed at Abbott to mimic the hydrophobic steroid backbone. The hydrophobic adamantane moiety replaces the C- and D-rings

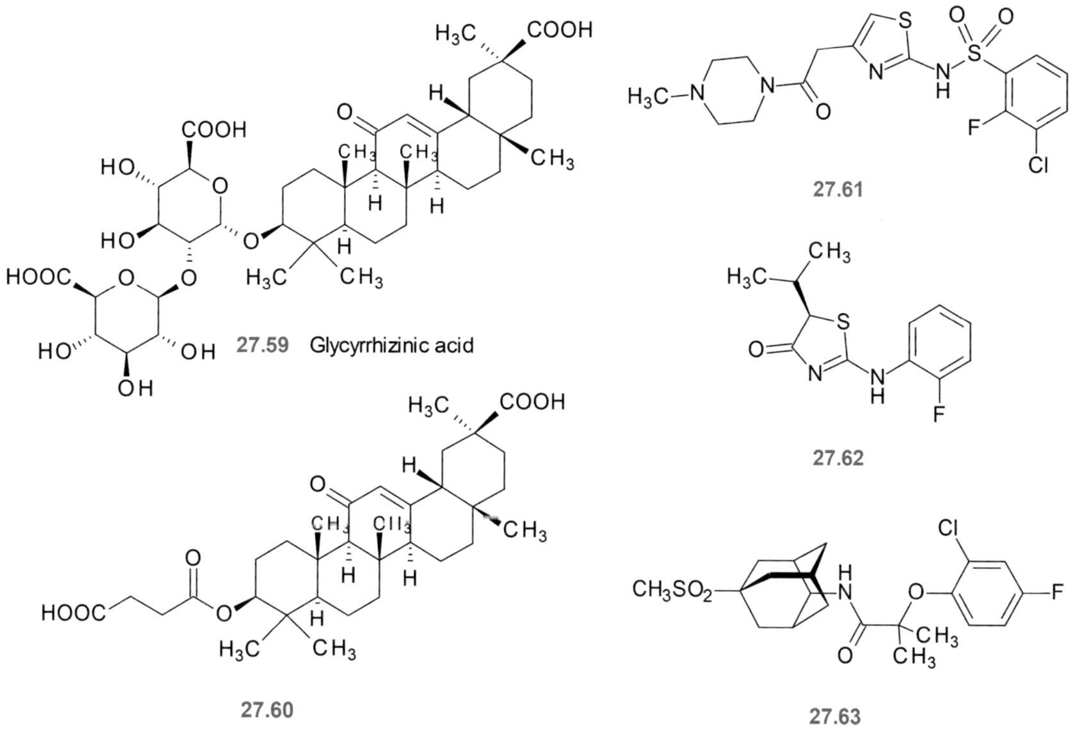

Fig. 27.22 The contents of licorice, glycyrrhizin acid **27.59**, represents a potent inhibitor of both 11β-HSD isoforms. Its derivative, carbenoxolone **27.60**, is also able to block both isoforms of 11β-HSD. Arylsulfonamidothiazole BVT-2733 **27.61** was developed as an inhibitor of 11β-HSD1. The crystal structure shown in Fig. 27.25 was obtained with an analogous compound, **27.62**. The adamantyl sulfone **27.63** with a central amide bond was developed at Abbott

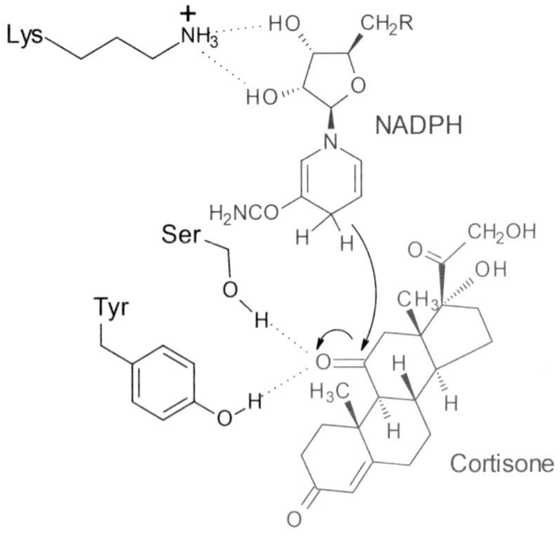

■ **Fig. 27.23** 11*β*-HSD1 uses NADPH as a cofactor for the reduction of cortisone to cortisol. The substrate's carbonyl group is involved in a hydrogen bond with a neighboring Ser and Tyr residue. In this way, it is prepared for the nucleophilic attack by the hydride ion. In addition, the positive charge of a neighboring Lys residue polarizes the carbonyl group (cf. ■ Fig. 27.25) and reduces the pK_a value of tyrosine, which serves as a proton donor

of the steroid backbone (■ Fig. 27.25). The terminal sulfone group mimics the 17-$COCH_2OH$ substituent, and **27.63** binds to Ser 190 and Tyr 183 of the catalytic triad with the carbonyl function of its central amide group. The example of this enzyme also underscores the point that structurally very different molecular scaffolds can mimic the geometry and properties of a steroid to successfully block the binding pocket and, thus, the catalytic mechanism.

27.6 The Cytochrome P450 Enzyme Family

The family of cytochrome P450 enzymes plays a central role in drug metabolism. The fundamentals of drug distribution, transport, and degradation were discussed in Sect. 9.2. The architecture and mode of action of these monooxygenases, especially their interaction with small-molecule drugs, will be presented here. The **cytochrome P450s (CYPs)** are a superfamily of heme proteins that carry out biochemical transformations as monooxygenases, usually by introducing oxygen to the substrate to be oxidized. They have an iron-containing protoporphyrin system as a prosthetic group in their center. A cysteine residue occupies the fifth, apical position of the iron. An oxygen is intermediately bound to the sixth co-

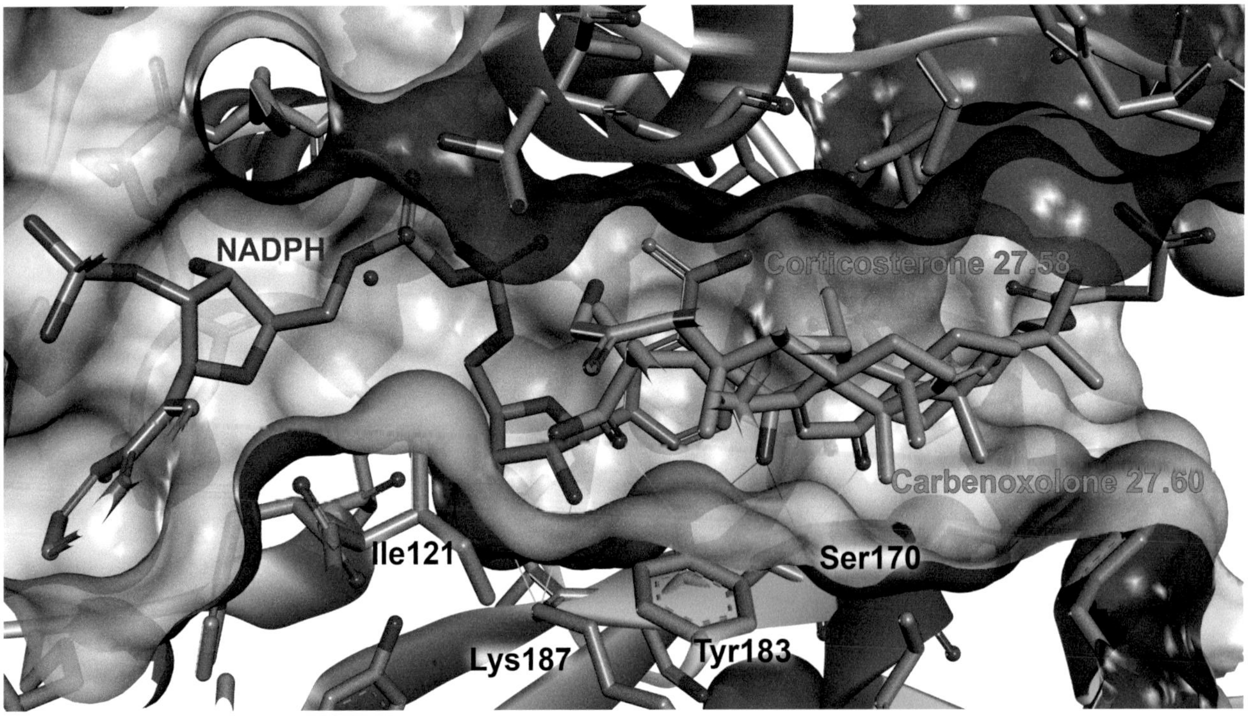

■ **Fig. 27.24** Crystal structure of human 11*β*-HSD1 with carbenoxolone **27.60** (*violet*). The inhibitor binds competitively to cortisol, the natural ligand. The binding geometry of corticosterone **27.58** (*green*) was extracted from the crystal structure with the murine enzyme and superimposed on human 11*β*-HSD1. This indicates the binding geometry of the natural substrate. (▸ https://sn.pub/ciMeJI)

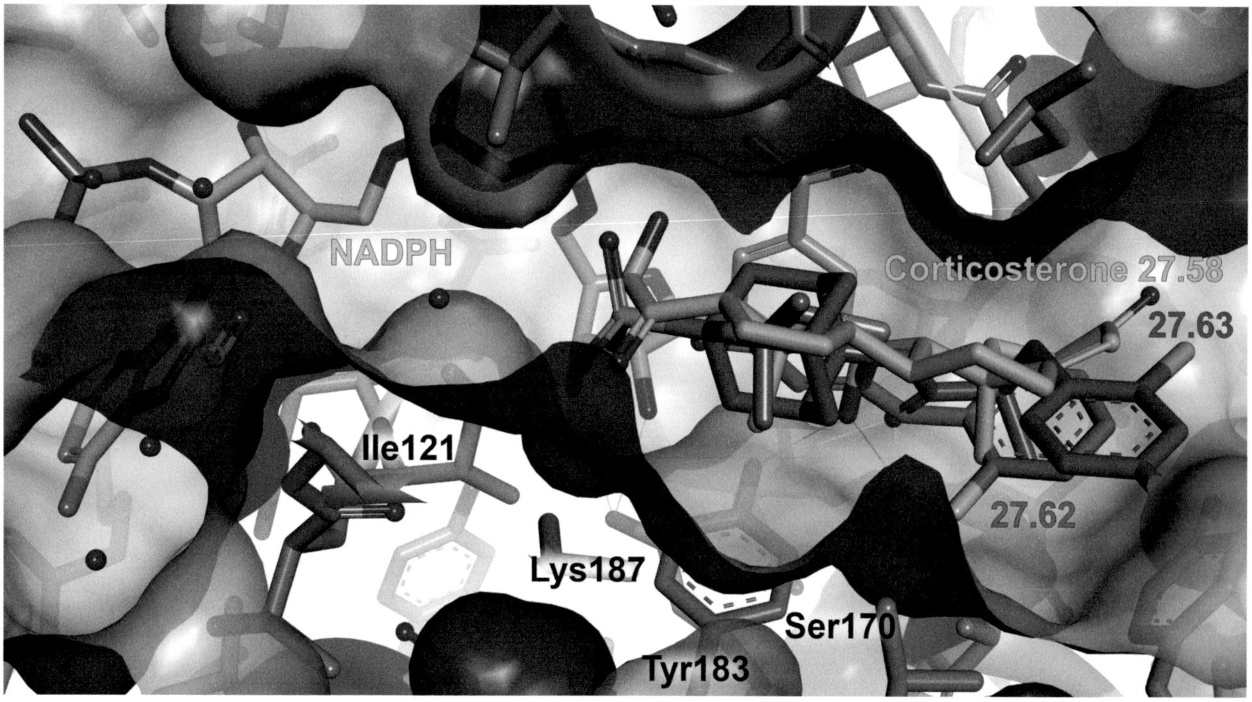

Fig. 27.25 Crystallographically determined binding geometries of the inhibitors **27.62** (*beige*) and **27.63** (*gray*) together with the binding mode of corticosterone **27.58** (*green*) taken from the crystal structure of the murine enzyme. Despite entirely different molecular scaffolds, the inhibitors largely occupy the steroid's position. They bind with their amide (**27.63**) or amide-like (**27.62**) groups to Ser 170 and Tyr 183 of the catalytic triad. Lys 187 holds the ribose moiety in position and polarizes the oxygen functionality of the neighboring Tyr 183. (► https://sn.pub/QplnCF)

ordination position and is introduced into the substrate from there. The name comes from a typical absorption band at 450 nm observed when the complex is blocked with carbon monoxide. The proteins consist of about 500 amino acids. To date, more than 6000 genes for CYPs have been described in Nature. In humans, 17 families have been characterized, subdivided into 57 isoenzymes. A combination of numbers and letters is used to name the proteins, with the first number indicating the family, the letter the subfamily, and the second number the isoform. In the body, they are found primarily in the liver, lungs, and gastrointestinal tract. This provides clues to their function: primarily to intervene in the **metabolism of xenobiotics**. Some CYPs perform important transformations on endogenous substrates, such as CYP 2R1 in vitamin D metabolism, CYP 19A1 (aromatase) in steroid metabolism, or CYP 2J2 and CYP 5A1 (thromboxane synthase) in eicosanoid metabolism. Xenobiotic compounds are transformed into more water-soluble and, thus, more easily excreted substances in so-called **phase I reactions**. Usually, these transformations serve the detoxification of compounds, but in a few cases they can also lead to the toxification of the substrate (Sect. 9.1). Some typical reactions catalyzed by CYPs are listed in Fig. 27.26.

The catalytic cycle of P450 enzymes is NADPH dependent. Initially, the **iron ion in the heme center** is in the +3 oxidation state. The substrate diffuses into a reaction cavity that is virtually completely shielded from the outside (Fig. 27.27). A mobile helical sequence segment provides access to the catalytic site and also acts as a lid over the catalytic site. An NADPH reductase transfers the first electron to the cytochrome, where it reduces the iron atom. Molecular oxygen then coordinates with the iron. In the next step, the NADPH reductase releases a second electron. A proton is then accepted and a Fe^{2+} species is formed, which homolytically cleaves the C–H bond of the substrate being oxidized, with concomitant release of the water molecule. An OH group is stereoselectively transferred from the iron to the carbon atom being oxidized. The iron returns to its original +3 state and the oxidized product can leave the binding pocket. Even today, the details of the reaction are not yet fully understood. However, it has been shown that some of the P450 enzymes are capable of **extreme adaptations** to their substrates. Not only can one substrate molecule be incorporated into the binding pocket. It is even possible to accommodate two different molecules. As we will see, this has far-reaching consequences for drug metabolism. Most CYPs are located in the **liver**. In mammals,

27

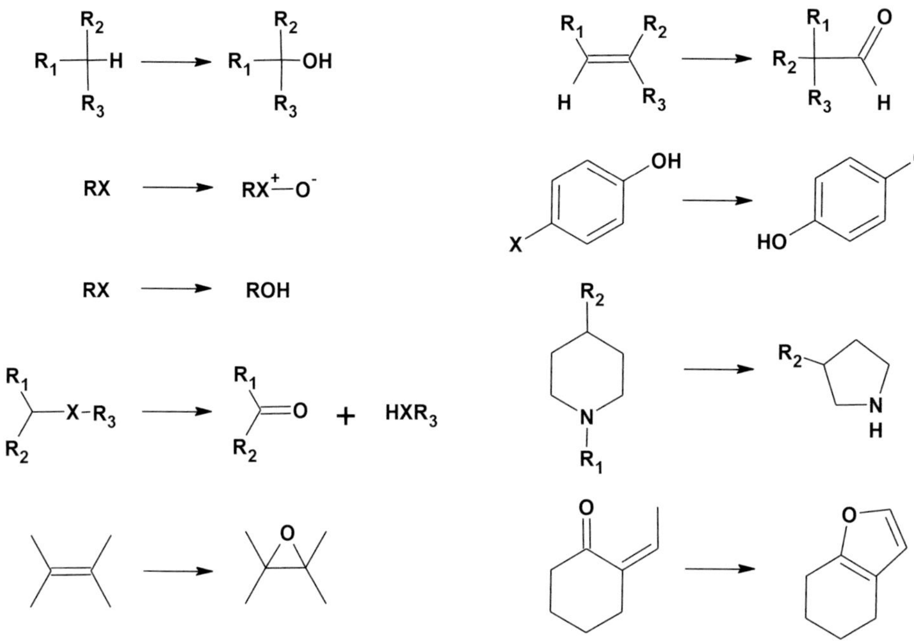

Fig. 27.26 Examples of typical oxidation reactions catalyzed by cytochrome P450 enzymes; X stands for heteroatoms such as nitrogen or sulfur

they are anchored to the membrane of the endoplasmic reticulum. The distribution of CYPs in different families is shown in Fig. 27.28. Considering their role in drug metabolism, CYP 3A4, CYP 2D6, and CYP 2C9 are the most important ones (Table 27.3). CYP 3A4 in particular has a highly adaptive structure. Its binding pocket expands from 900 Å^3 in the uncomplexed state to 2000 Å^3 in the complexed state upon binding of erythromycin (Fig. 27.27). Moreover, erythromycin must completely rearrange itself in the binding pocket, since in the experimentally determined crystal structure the oxidized group is still 17 Å away from the heme center.

P450 enzymes can be blocked by a variety of compounds. Compounds containing heteroaromatic rings such as **imidazole or triazole** tend to inhibit them. Fluconazole **27.6** and ketoconazole **27.7** are potent CYP 3A4 inhibitors (Fig. 27.6). Other examples are the flavonoids such as **naringenin 27.9** found in grapefruit juice. They are metabolized by CYPs to active inhibitors and ultimately bind irreversibly to various CYPs, especially CYP 3A4. Other examples are listed in Table 27.3. These inhibitory properties must be considered when a CYP inhibitor is administered with a drug that is metabolized by this enzyme. Because of limited metabolism, the plasma concentration of the coadministered drug may increase, with serious consequences in terms of the fraction of the dose that is in the body (see cerivastatin **27.33**, Sect. 27.3). On the other hand, such a factor can be used to reduce the dose of an expensive drug such as cyclosporine (Sect. 10.1). Coadministration of ketoconazole **27.7** allows lower doses of this immunosuppressant because cyclosporine is metabolized by CYP 3A4.

The organism must be flexible in its response to xenobiotic exposure and adapt its degradation mechanisms. Therefore, **CYPs** can also be **induced**, i.e., the body upregulates the availability of a particular CYP isoform when needed. This induction is thought to have several mechanisms. One possibility is the stimulation of the **transcription factor PXR**, which belongs to the group of nuclear receptors. Such a case is presented in Sect. 28.7. For example, one of the components of St. John's wort, hyperforin, induces increased expression of CYP 3A proteins. As a result, the metabolism of drugs that are degraded by this CYP family is increased. This can lead to a fall below therapeutically adequate doses. Especially when treatment with St. John's wort is discontinued, a dangerous situation may arise for the patient. Other examples of inducers of different CYP isoforms are listed in Table 27.3. In addition to metabolism by alcohol dehydrogenase, **alcohol** is also **degraded** by CYP 2E1. This enzyme is upregulated with excessive alcohol consumption, especially in **chronic alcoholics**, and is available in increased levels for alcohol metabolism. This explains the tolerance that gives heavier drinkers the ability to "hold a drink" better. However, if these drinkers want to treat their hangover with **paracetamol** (or in the US acetaminophen) the next morning, problems can occur. Paracetamol **27.64** is partially metabolized by CYP 2E1, and this enzyme forms the toxic intermediate **27.65** (Fig. 27.29). If the intermediate is present in low concentrations, it can be converted with available glutathione and detoxified. If paracetamol is extensively metabolized by this pathway, the amount of glutathione available will be insufficient and toxicity symptoms may occur. This risk is greatest in heavy drinkers, in whom CYP 2E1 levels are permanently elevated by continuous induction and in whom paracetamol is predominantly metabolized by this pathway.

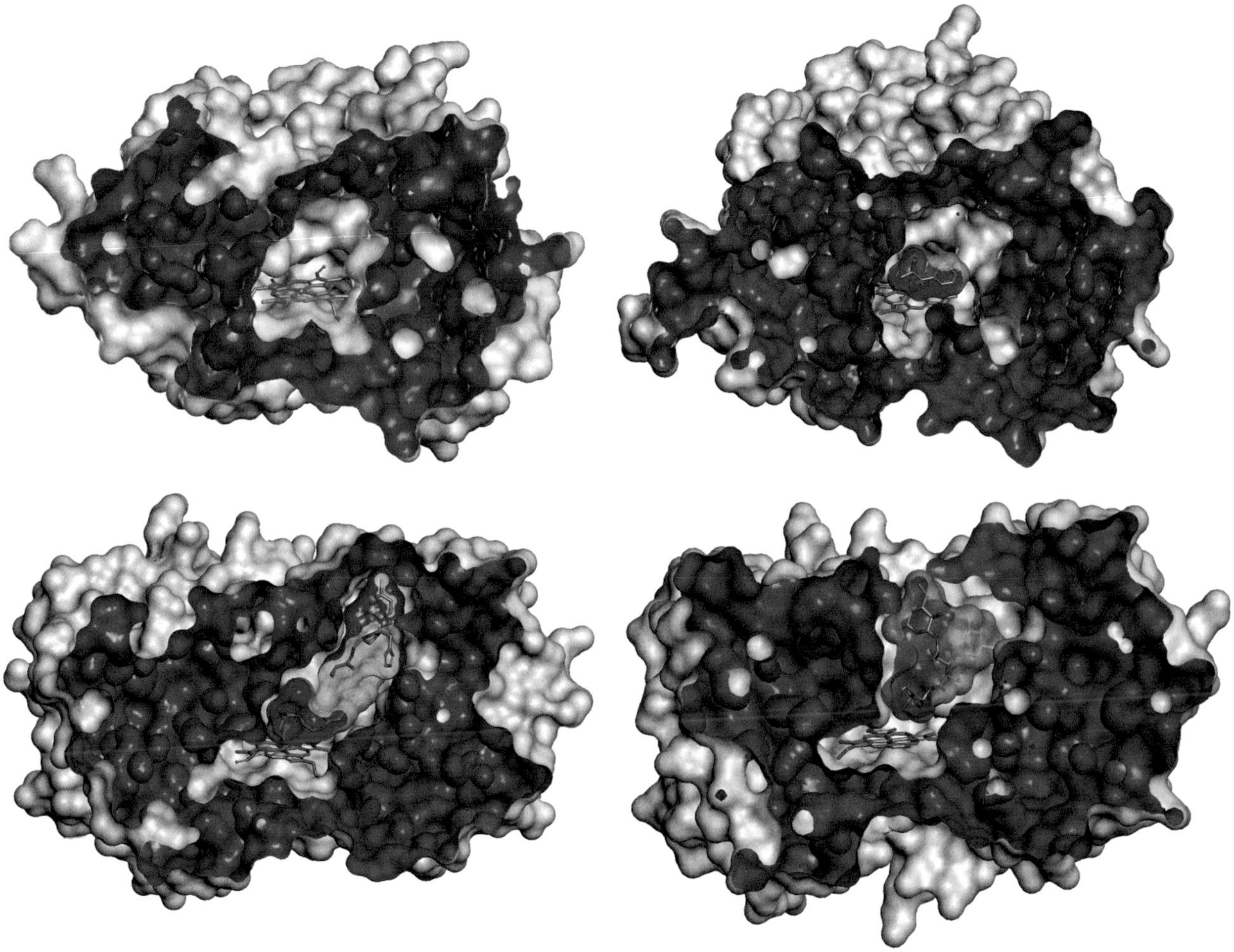

Fig. 27.27 Crystal structures of human CYP 3A4 in an uncomplexed state (*upper left*), with bound metyrapone **27.8** (*upper right*), with ketoconazole **27.7** (*bottom left*), and with erythromycin **32.36** (*bottom right*). The protein is shown with a *white* surface that is *red* colored on the interior. For clarity, the polypeptide chain has been omitted. The ligands are shown with their own surfaces (outside *green*, inside *blue*). In the case of ketoconazole, two ligands bind to the protein (the second molecule is shown with a *violet* surface and a *cyan* interior). CYP 3A4's binding pocket, which is nearly fully closed to the exterior (cf. *upper row*), has proven itself to be extremely adaptive. It is only because of this that the enzyme can take on ligands with entirely different sizes and shapes. (► https://sn.pub/zC3Z5n)

Fig. 27.28 Percentage of CYP P450 enzymes involved in drug metabolism and their relative distribution. *Left* A study from 2002 compiled the data for the relative portion of the different CYP enzymes that take part in the metabolism of the 200 best-selling drugs. *Center* Proportion of the different CYP enzymes in the small intestines. *Right* Relative distribution of CYP enzymes over the different P450 families in humans

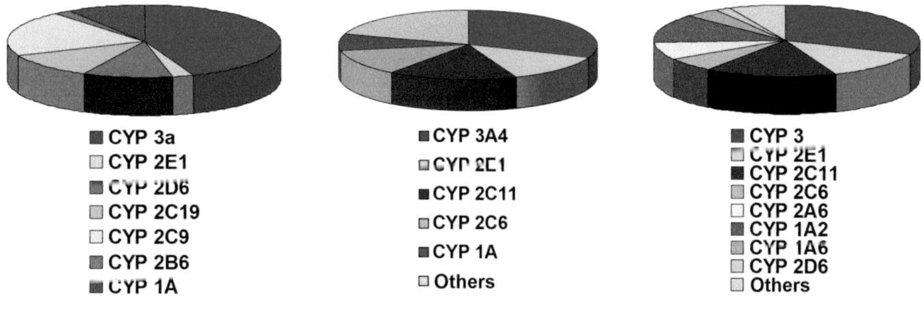

27

Table 27.3 Examples of drugs that act as substrates, inhibitors, or inducers of CYP 3A4, CYP 1A2, and CYP 2D6

	Substrate	Inhibitor	Inducer
CYP 3A4	Amitriptyline Clarithromycin Cyclosporine Dexamethasone Carbamazepine Terfenadine Ethinylestradiol	Ketoconazole Cimetidine Ciprofloxacin Erythromycin Fluconazole Ritonavir Grapefruit juice	Barbiturate Carbamazepine Glucocorticoids Phenobarbital Rifampicin St. John's wort
CYP 1A2	Caffeine Amitriptyline Paracetamol Theophylline Verapamil	Cimetidine Ciprofloxacin Grapefruit juice	Insulin Omeprazole Aromatic hydrocarbons Smoking
CYP 2D6	Amitriptyline Captopril Chlorpromazine Codeine Imipramine Metoprolol Propafenone Debrisoquine	Cimetidine Haloperidol Clotrimazole Quinidine Ritonavir	Dexamethasone

Saturation and upregulation of cytochromes by **induction** or **inhibition of cytochromes** by **drug–drug interactions** represent a serious potential hazard in drug metabolism. Therefore, efforts are made in drug design to estimate the metabolic profile of a development candidate. It would be desirable to know where a compound is metabolized and whether cytochrome inhibition is to be expected, especially of the most important enzymes. Crystal structure determination of the major human CYPs was aggressively pursued. The information obtained was rather sobering. The proteins have such extremely adaptable properties that it seems virtually impossible to predict plausible binding modes in order to estimate inhibition data. Even predicting which parts of a molecular scaffold are preferentially metabolized and which metabolites are to be expected has not become any easier, despite the large number of crystal structures. At present, routine structural determination of each development candidate with these proteins seems rather utopian. Moreover, it has been shown that not only binary but also ternary complexes can be formed with one or two different ligands. Only time will tell how the methodology will develop in this area. However, the current state of the art allows the estimation of metabolic properties with empirical QSAR models and 3D comparisons (Chaps. 17 and 18). The **program MetaSite**, developed by Gabriele Criciani at the University of Perugia in Italy, attempts to find the best-fitting patterns by considering possible complementary interaction patterns on the surface of the ligand and in the CYP binding pocket. Several ligand conformations are considered for this approach. Next, concepts about possible binding modes in the binding pocket of the metabolizing cytochromes are developed that estimate the spatial accessibility for oxidative attack by the iron atom at the different sites in the studied ligand. In addition, the technique accesses a system of rules for assessing the reactivity of organic molecules similar to those introduced to construct the Hammett equation (Sect. 18.2). Both concepts rank the individual sites in a molecule in terms of the probability of metabolic transformation. The combination makes it possible to estimate the metabolic properties of drugs surprisingly well.

27.7 What Makes Slow and Fast Metabolizers Different?

A standardized prediction of drug metabolism is virtually impossible for the simple reason that we are all different. The **equipment with cytochrome P450 enzymes varies** from person to person. This is due in part to varying enzyme concentrations in our bodies and in part to **polymorphisms** (Sect. 12.11) that cause different metabolic behaviors in different individuals. This has been extensively studied for the enzymes CYP 2D6, CYP 2C9, and CYP 2C19. For example, CYP 2C9 is absent in 1–3% of Caucasians. These individuals have difficulty metabolizing *S*-warfarin, and prodrugs such as codeine, tramadol, and losartan are not activated. Variations in the polymorphism of CYP 1A2 are responsible for the fact that different individuals react differently to caffeine. Several polymorphisms have been described for CYP 3A4. The best-studied example of a correlation between genetic variability and mode of action is the degradation of debrisoquine **27.66** to 4-hydroxydebrisoquine **27.67**, an antihypertensive drug (◘ Fig. 27.30). This drug is metabolized by CYP 2D6. Caucasians can be divided into **slow**, **extensive**, and **ultrafast metabolizers** based on their ability to metabolize this drug. A study conducted in Sweden found the distribution shown in ◘ Fig. 27.30. This has implications for prescribing this drug. If a standard dose is given to all patients, extensive metabolizers will do well. In slow metabolizers, the plasma level will be too high, which can lead to adverse effects. In ultrarapid metabolizers, the level required for therapy will hardly be reached and the desired effect of the drug will not be achieved. At this point, it would be ideal if the physician or pharmacist could read directly from the **patient's gene chip** to which group of metabolizers the person prescribed the drug belongs. If an alternative antihypertensive were available that is metabolized by a different CYP, the patient might benefit from switching to a different drug. There are already chips on the market that can

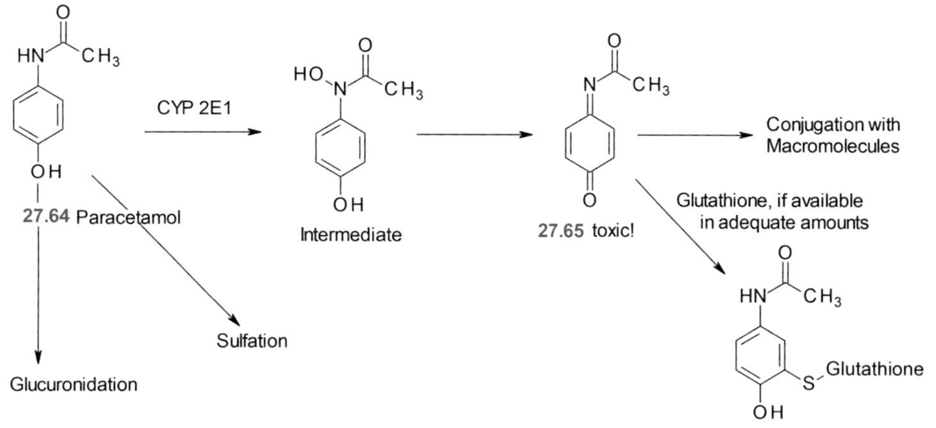

Fig. 27.29 In addition to alcohol dehydrogenase, alcohol is metabolized by CYP 2E1. This enzyme is overexpressed because of induction in chronic alcoholics. The analgesic paracetamol **27.64** is partly metabolized by CYP 2E1. Compound **27.65** is formed in the process as a toxic intermediate. At low concentrations, it can be detoxified by glutathione. If, however, elevated levels of paracetamol enter this pathway because of CYP 2E1 upregulation, the supply of glutathione will be insufficient, and poisoning may occur

record a patient's genetic CYP profile. It must be said, however, that the information on the genome encoding a particular enzyme is not sufficient to assign an individual to a metabolic group. The **genotype** is not important for metabolic efficiency, i.e., the individual genetic complement of coding proteins, but rather the actual amount of a protein expressed. This determines the **phenotype**. In addition, the phenotype can vary according to a person's lifestyle and state of health. Just think of the induction of CYP 2E1 in heavy drinkers. This patient profile becomes important when the **therapeutic window** for the use of a drug is very narrow. The therapeutic window is the concentration difference required to achieve the desired effect while avoiding a toxic dose (Sect. 19.7).

It should also be mentioned that genetic differences in the **cytochrome complement** are not the only factors leading to variable metabolic behavior. Transferases (Chap. 26), which transfer acetyl groups, sugar moieties, or methyl groups, also play an important role. These enzymes are likewise differentially expressed in the general population, which is divided, for example, into **fast and slow acetylators**. More attention needs to be paid to metabolism and genetic and phenotypic variability. The distribution of metabolic characteristics among subject groups should also be better controlled in clinical trials. This is the only way to obtain reliable data on the therapeutic breadth of a drug before it is widely used in therapy. Another serious problem is that, especially

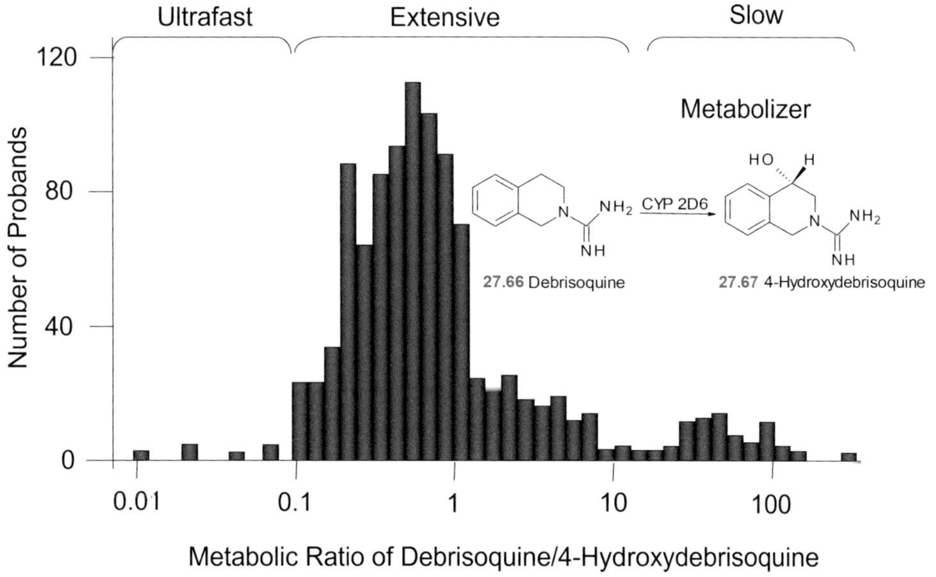

Fig. 27.30 Correlation between genetic variability and the metabolism of the antihypertensive debrisoquine **27.66** to hydroxydebrisoquine **27.67**. The Caucasian population metabolizes this drug with CYP 2D6 and is divided into slow, extensive, and ultrafast metabolizers. If a standard dose of the drug is prescribed, the extensive metabolizers would respond well. On the other hand, the same dose will lead to a plasma level that is too high for the slow metabolizers, which can lead to side effects. The ultrafast metabolizers will barely reach a plasma level that is adequate for therapy, and the desired effect of the drug will not be achieved

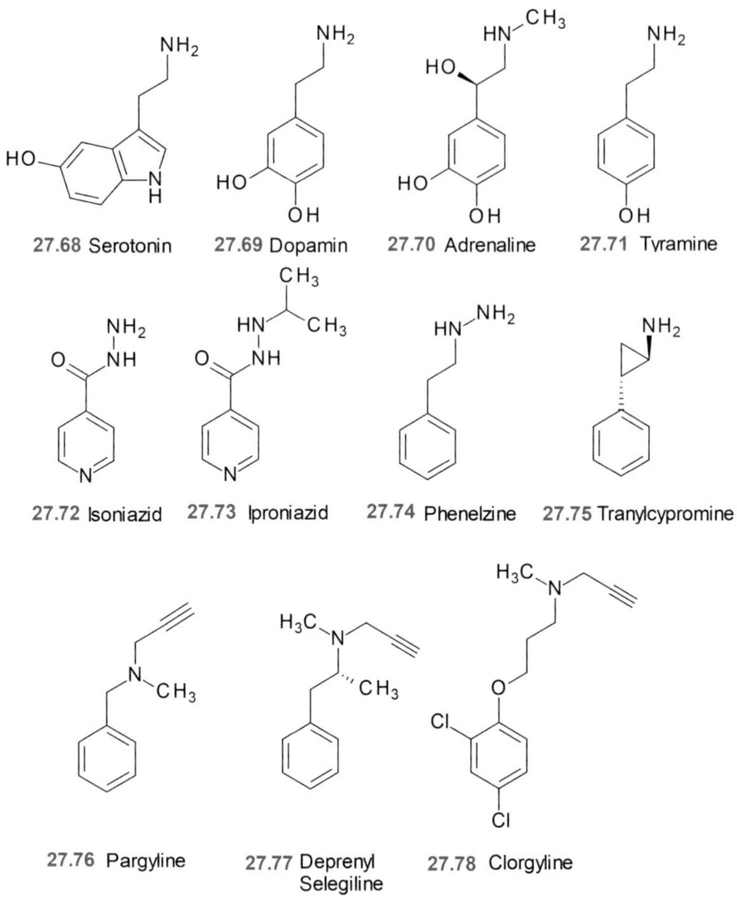

Fig. 27.31 Serotonin **27.68**, dopamine **27.69**, adrenaline **27.70**, and tyramine **27.71** are metabolized by MAOs. At first, the hydrazide derivatives such as isoniazid **27.72**, iproniazid **27.73**, or the hydrazine phenelzine **27.74** were discovered as inhibitors. Follow-up drugs such as tranylcypromine **27.75** react with the FAD system of the flavoenzyme by a ring-opening reaction. Others such as pargyline **27.76**, L-deprenyl (or selegiline) **27.77**, and clorgyline **27.78** react through their propargyl groups

in older patients, a single drug is rarely administered; it is not uncommon for more than 10 drugs to be used simultaneously. However, how these drugs interact with each other in terms of the metabolism of a particular individual is generally unknown. They must be determined step by step during therapy.

27.8 Blocking the Degradation of Neurotransmitters: Monoamine Oxidase Inhibitors

Monoamine oxidases are examples of oxidoreductases that have a FAD molecule in the active site as a cofactor. Two isoenzymes, MAO_A and MAO_B, have been characterized. There is a sequence identity of 70% between the two forms. These enzymes are located in the mitochondrial membrane. They were first described in 1928 as tyramine oxidases. In addition to tryptamines such as serotonin **27.68**, which are preferentially metabolized by MAO_A, both isoforms can metabolize dopamine **27.69**, adrenaline **27.70**, and tyramine **27.71** (Fig. 27.31). Their function is to degrade these **neurotransmitters**, often colloquially referred to as "happiness hormones," in the synaptic cleft by **oxidative deamination**. The neurotransmitters are released into the synaptic cleft and bind to a postsynaptic G-protein-coupled receptor (Fig. 22.7). To terminate the stimulation, the neurotransmitters are removed from the synaptic cleft by a transporter and shuffled back into the presynaptic cell. There they are either stored in a vesicle or chemically degraded by monoamine oxidase. Inhibition of MAO_A or MAO_B reduces the oxidative deamination of the transmitters. As a result, they remain available longer for nerve transmission. This therapeutic principle can be exploited in diseases in which brain metabolism or neurotransmission are disturbed. Examples include depression, Alzheimer's disease, and Parkinson's disease.

Inhibiting MAO_A increases serotonin levels. Drugs that inhibit this isoform are used to treat major depression. Blocking MAO_B increases dopamine levels, which may be an approach to treating dementia and Parkinson's disease. The catalytic reaction produces H_2O_2, an aldehyde, and free ammonia. The peroxide can play an important role as a source of hydroxyl radicals in metabolic processes, which, depending on the amount, can have either a protective or destructive effect. The other decomposition products also have biological functions. However, there may be a risk of cytotoxic effects if their concentrations are too high.

The first MAO inhibitors were discovered by chance. Isoniazid **27.72** was synthesized by Hans Meyer and Josef

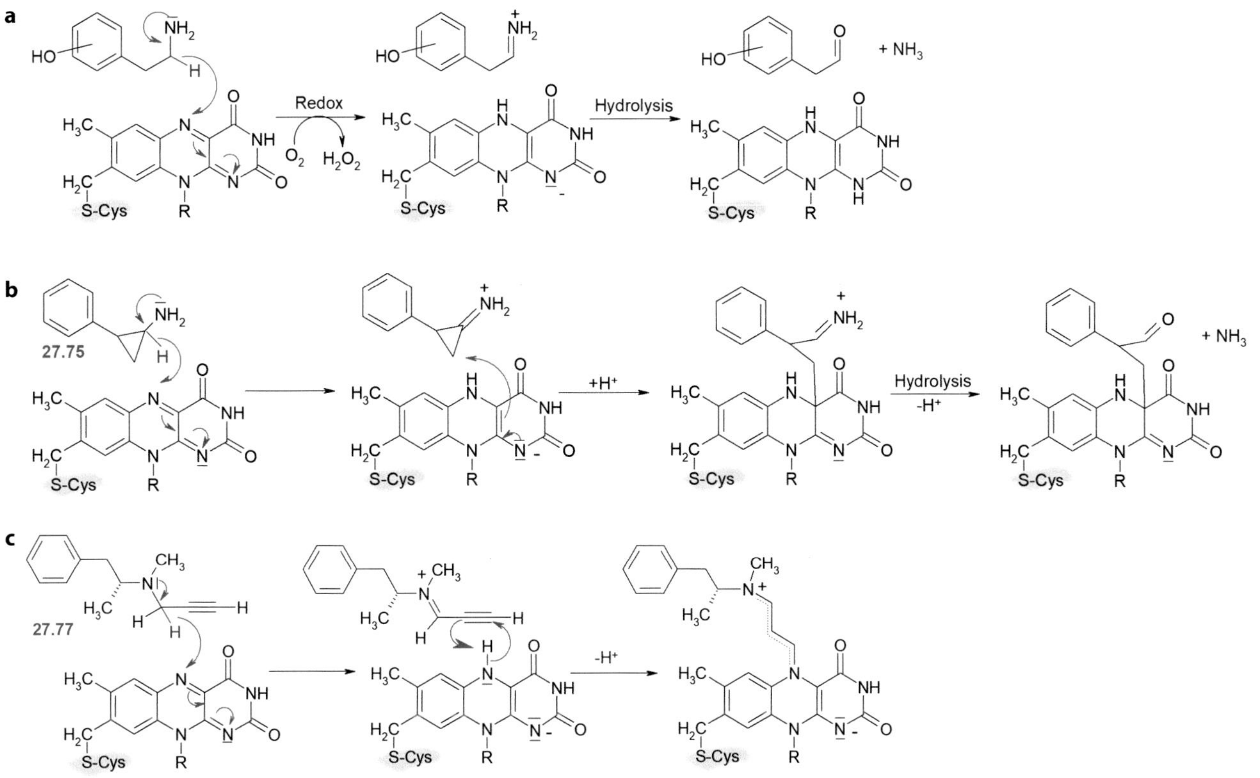

■ **Fig. 27.32** Possible mechanism for the deamination and inhibition of MAO enzymes. **a** Biogenic amines are transformed into iminium compounds in a redox reaction by hydrogen abstraction next to the amino group. Formally, a hydride ion is transferred to the oxidized form of the FAD system. Hydrolysis of the iminium ion produces ammonia and an aldehyde. The prosthetic group is reoxidized with molecular oxygen, whereby H_2O_2 is formed. **b** Tranylcypromine **27.75** reacts with the oxidized form of the FAD system upon ring opening and forms a covalent bond to the C4a carbon atom on the ring. **c** Derivatives such as L-deprenyl **27.77** transfer one of their propargylic hydrogens onto the oxidized form of the FAD scaffold. A covalent bond is formed with the N5 nitrogen atom. A delocalized electron system between the FAD molecule and the inhibitor is formed

Mally at the University of Prague in 1912 (■ Fig. 27.31). Its antibiotic activity was recognized during World War II. It is still a component of tuberculosis therapy today. A hydrazide-substituted derivative, iproniazid **27.73**, was developed at Roche. It was launched under the name Marsilid®. Shortly after its introduction, a mood-lightening side effect was observed in tuberculosis patients. On this basis, it was prescribed to patients suffering from depression. As the only available treatment for these patients was electroconvulsive therapy, iproniazid was quickly hailed as "Drug of the Year." However, there were deaths due to liver toxicity, and it was withdrawn from the market in 1960. The success of this compound stimulated the search for other inhibitors that lacked the side effects. For example, phenelzine **27.74**, tranylcypromine **27.75**, and pargyline **27.76** (■ Fig. 27.31) resulted from this work. They react with the FAD system in the enzyme, rendering it useless for the electron transfer mechanism (■ Fig. 27.32b, c).

In the meantime, both isoforms have been characterized by crystallography. Interestingly, MAO_B is a dimer and MAO_A is a monomer. The complex of MAO_B with tranylcypromine **27.75** has been structurally investigated. How the compound forms an irreversible covalent bond with the flavin scaffold is shown in ■ Fig. 27.33. Many other MAO inhibitors have a propargylamine group in their scaffold. This moiety reacts with the nitrogen of the central FAD ring to form a covalent bond. A delocalized electron system across multiple bonds is formed (■ Figs. 27.32 and 27.33).

L-Deprenyl **27.77** selectively blocks MAO_B, while clorgyline **27.78** selectively inhibits the MAO_A isoform. Both isoforms are very similar near the FAD binding site. Differences occur only in the region of the pocket where the biogenic amine binds as substrate. Of the 20 amino acids that make up this region, seven are structurally different. In particular, the two residues Ile 199 and Tyr 326 in MAO_B are replaced by Phe 208 and Ile 335 in MAO_A. They give the binding pocket a different shape. In MAO_A, the pocket is shorter, but wider and shallower. It is bordered from below by Phe 208 and can easily accommodate the 2,4-dichlorophenoxy group of clorgyline. Conformationally, the substituent of **27.78** must align with the continuation of the aliphatic chain of the inhibitor so that its phenyl ring and the attached chain are all in one plane. This is accomplished by the conformational

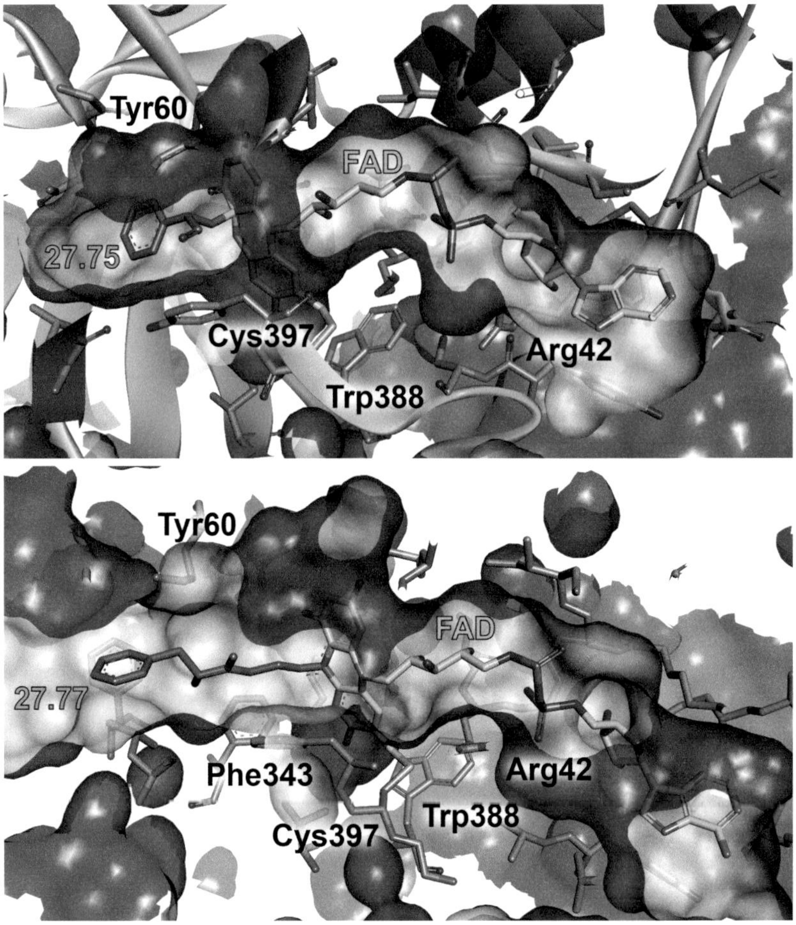

Fig. 27.33 *Upper row* Crystal structure of MAO_B complexed with covalently bound tranylcypromine **27.75**. The inhibitor is attached to the FAD system through the C4a carbon atom. *Bottom row* Crystal structure of MAO_B complexed with covalently bound L-deprenyl **27.77**, which is coupled to the cofactor through the N5 nitrogen atom. A delocalized electron system between the FAD molecule and the inhibitor is formed. (► https://sn.pub/pmLiSt)

properties of the phenoxymethyl group, which favors planar attachment to the chain despite the presence of an *ortho*-substitution (Fig. 27.34, *upper row*). In MAO_B, the pocket takes the form of a deep crevice into which a phenyl ring can be inserted along its edge. The laterally limited volume of the pocket is due to the large residue Tyr 326. Instead of a phenyl ring at the bottom of the pocket as found in MAO_A (cf. Phe 208 in MAO_A), the pocket wall is limited here by the residue Ile 199. This residue is thought to have the properties of a flexible entry port. The phenethyl group of **27.77**, on which both *ortho*-positions must remain unsubstituted, allows the ligand deprenyl to adopt the required conformation with its terminal aromatic ring perpendicular to the aliphatic chain (Fig. 27.34, *bottom row*).

In addition to the irreversible, covalently binding inhibitors, reversibly binding inhibitors such as moclobemide **27.79**, befloxatone **27.80**, or toloxatone **27.81** (Fig. 27.35) are also known. They also occupy the part of the binding pocket that accommodates the biogenic amine substrate. However, they do not form a covalent bond with the FAD scaffold.

MAO inhibitors are used primarily as antidepressants and in the treatment of Parkinson's disease. The antidepressant effect is mainly achieved by specific inhibition of MAO_A in the central nervous system. In the brain, dopamine, noradrenaline, and serotonin levels are increased. Parkinson's disease therapy, which is usually combined with an L-DOPA strategy (Sect. 26.10), focuses on inhibiting MAO_B because this isoform is overexpressed in the brains of patients with Parkinson's disease. Since both isoforms metabolize dopamine equally well, selective MAO_B inhibitors are used to target the Parkinson's etiology.

In addition to the above-mentioned liver toxicity observed with first-generation hydrazide-type antidepressant MAO inhibitors, **hypertensive crises** were also observed as a result of acute blood pressure dysregulation. This led to their withdrawal from the market. Liver toxicity was largely avoided with compounds such as tranylcypromine **27.75** or pargyline **27.76** (Fig. 27.31), but hypertensive crises continued to occur. These could be provoked by an increased concentration of tyramine in the body, especially when certain **tyramine-rich foods** (e.g., cheese, which causes the so-called **cheese effect**, or wine) were consumed and the metabolizing enzymes were irreversibly blocked by an MAO inhibitor. The result is an elevated concentration of noradrenaline, which activates the vascular system and can lead to arrhythmias

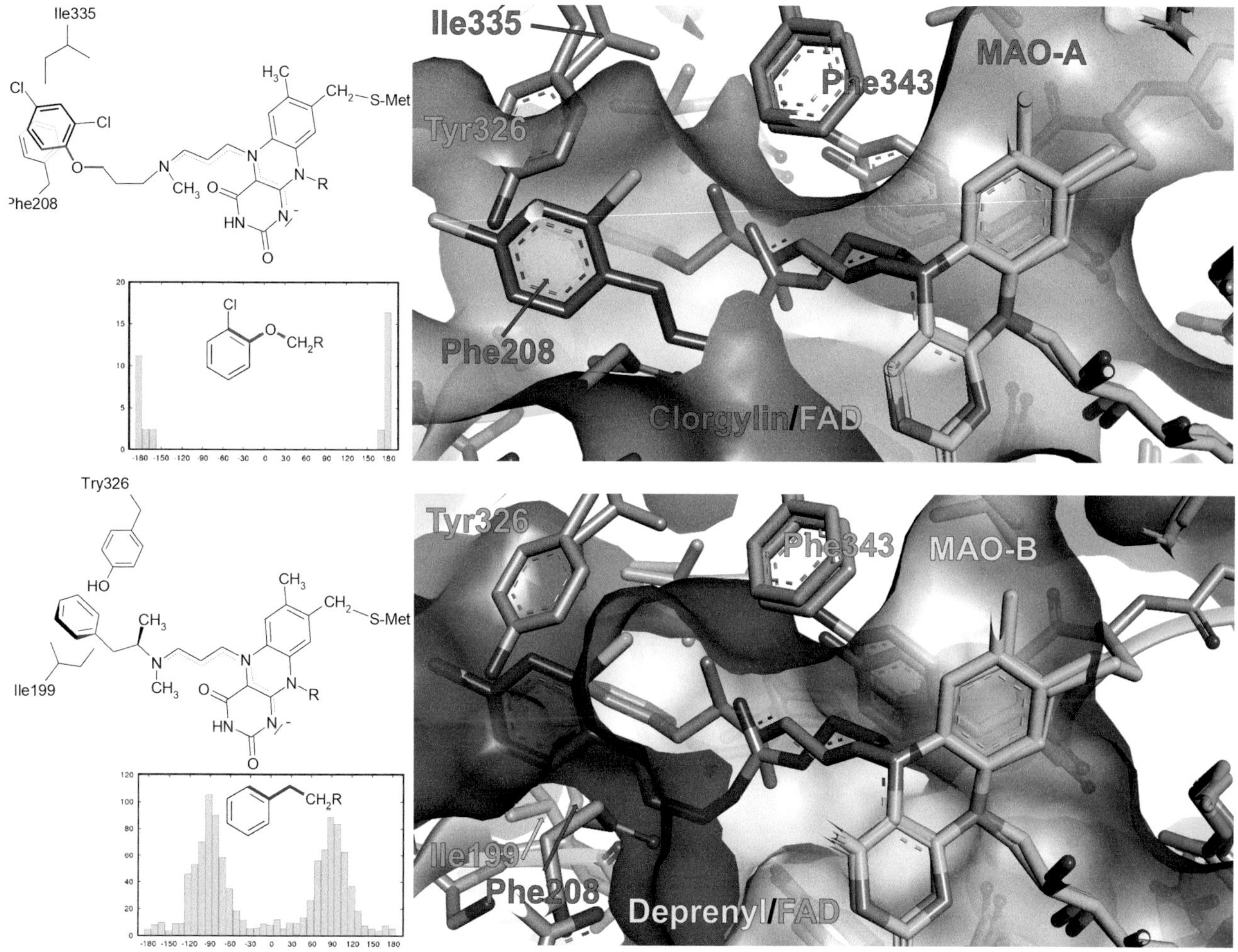

◘ Fig. 27.34 MAO_A (*upper row*) and MAO_B (*bottom row*) differ in the binding region of the biogenic amine. The shape of the binding pocket is mainly determined by the exchange of **Phe 208**$_{\mathbf{MAOA}}$ → **Ile 199**$_{\mathbf{MAOB}}$ and **Ile 335**$_{\mathbf{MAOA}}$ → **Tyr 326**$_{\mathbf{MAOB}}$. The 2,4-dichlorophenoxymethyl moiety of the selective inhibitor clorgyline **27.78** (*violet*) binds in the broad and shallow binding pocket, which is bordered by Phe 208, in the complex with MAO_A (residues are *violet*). The inhibitor's aliphatic chain lies in the same plane as the aromatic ring. This chain conformation relative to the ring is the preferred geometry for this group. A statistical evaluation of the geometry of the *ortho*-chlorophenoxymethyl group in small-molecule crystal structures indicates torsion angles (*red*) of preferentially ±180°. The binding pocket in the complex of MAO_B (*orange* residues) with the selective inhibitor L-deprenyl **27.77** is severely limited by Tyr 326 and opens only as a narrow crevice. The phenyl group of the inhibitor (*gray*) submerges into this crevice. For this, the aromatic ring must adopt an orientation that is 90° perpendicular to the attached aliphatic sidechain. A geometry similar to the one of the dichlorophenoxy group in clorgyline is not possible for steric reasons (cf. superimposed geometry of **27.78**). On the other hand, the deprenyl's phenyl ring cannot bind to MAO_A with the same "submerged" edge-on geometry because a steric conflict with Phe 208 would occur. Also here, a statistical analysis of the torsion angles (*blue*) shows a clear preference for values ±90°, which corresponds exactly to the desired perpendicular orientation of the plane of the phenyl ring to the chain. (▸ https://sn.pub/zjY9aD)

or heart attacks. Reversible MAO_A inhibitors can avoid this problem to some extent. They block the enzyme sufficiently in the central nervous system to achieve the desired antidepressant effect. In the periphery, tyramine displaces the reversible inhibitor, allowing tyramine to be degraded.

Oxazolidinones are a new group of antibacterial agents that are thought to inhibit the peptidyl transferase center in the bacterial ribosome (Sect. 32.7). A representative of this group is linezolid **27.82** (◘ Fig. 27.35). Because of its structural similarity to the reversible MAO inhibitors, toloxatone **27.81** is also an MAO_A inhibitor. Therefore, administration of this compound may induce hypertensive crises as described above. An attempt is, therefore, being made to develop oxazolidinones with adequate selectivity for the bacterial target without such side effects.

In the previous two sections, the importance of cytochrome P450 enzymes in drug metabolism was introduced. MAO enzymes also play a certain part in this

Fig. 27.35 Examples of reversible MAO enzyme inhibitors **27.79**–**27.81**. The antibiotic linezolid **27.82** bears structural similarity to the oxazolidinones **27.80** and **27.81**. It also blocks MAO_A. MAO enzymes also play a role in drug metabolism. Citalopram **27.83**, sertraline **27.84**, and triptans such as **27.85** are metabolized by these enzymes

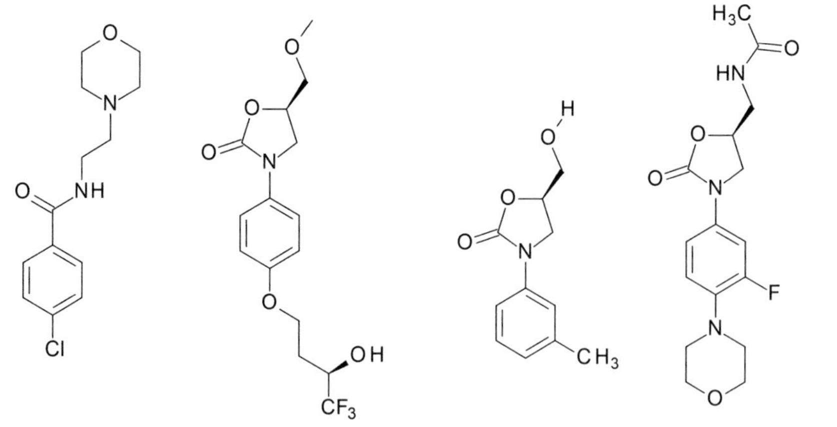

27.79 Moclobemide 27.80 Befloxatone 27.81 Toloxatone 27.82 Linezolid

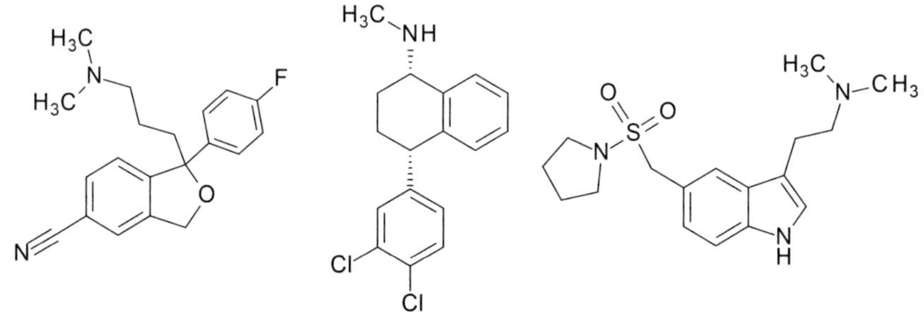

27.83 Citalopram 27.84 Sertraline 27.85 Almotriptan

metabolism. Compounds such as citalopram **27.83**, sertraline **27.84**, or triptans such as **27.85** are MAO substrates and are, therefore, metabolized by these enzymes.

27.9 Cyclooxygenase: A Key Enzyme in Pain Sensation

The organism synthesizes many important signaling molecules from components of the lipid membrane. Phospholipids are the starting materials from which arachidonic acid **27.86** (Fig. 27.36) is formed. This chain-like molecule of 20 carbon atoms has a carboxylate group as its only polar function. It is characterized by four isolated *cis* double bonds. In order to produce paracrine hormones such as the **prostaglandins 28.87–27.93** with sufficient water solubility, arachidonic acid must be oxidized. Oxygen-containing functional groups must be transferred. This task is performed by **cyclooxygenases** (COX). These are bifunctional enzymes that catalyze the conversion to prostaglandins in a two steps mechanism. First a cyclooxidation takes place, then a peroxidase reaction (Fig. 27.36). Due to its poor solubility in water, arachidonic acid diffuses directly from the membrane into the reaction center of cyclooxygenase. The enzyme submerges into the membrane. Three helices are used to allow it to "swim" on the membrane. These helices anchor the protein to the membrane, but do not cross the membrane as is often observed with membrane-anchored proteins. There are two isoforms, COX-1 and COX-2, whose amino acid sequences are 65% identical (Fig. 27.37). Their catalytic sites are almost identical. They are active as dimers. Access to the catalytic center is through a long channel that opens directly into the membrane environment. The natural substrate arachidonic acid **27.86** is taken up in this way. The channel is slightly narrower in COX-1 than in COX-2 because a central isoleucine is replaced by a valine. The arachidonic acid that has diffused into the channel is converted to the endoperoxide PGG_2 (**27.87**) by the addition of oxygen at C11 and C15 (Fig. 27.36).

The **heme cofactor** near the reaction channel is essential for the transformation. Its fifth coordination site is occupied by histidine. The oxidative oxygen species is bound to the sixth position. The dioxygen species is transferred as a hydroperoxide in a two-electron reaction. Tyr 385 acts as an intermediate **tyrosyl radical for the electron transfer**, abstracting a hydrogen atom from C13 (Fig. 27.38). The temporarily present, unsaturated radical adds the peroxide group to the allyl position on C11. A cyclic peroxide is then formed with C9; C8 reacts with nearby C12 to form a 5-membered carbocycle. Another hydrogen abstraction at C13 initiates a peroxide transfer to C15, which is also in an allylic position. The peroxide is then converted to a hydroxyl function in a subsequent reduction step catalyzed by peroxidase

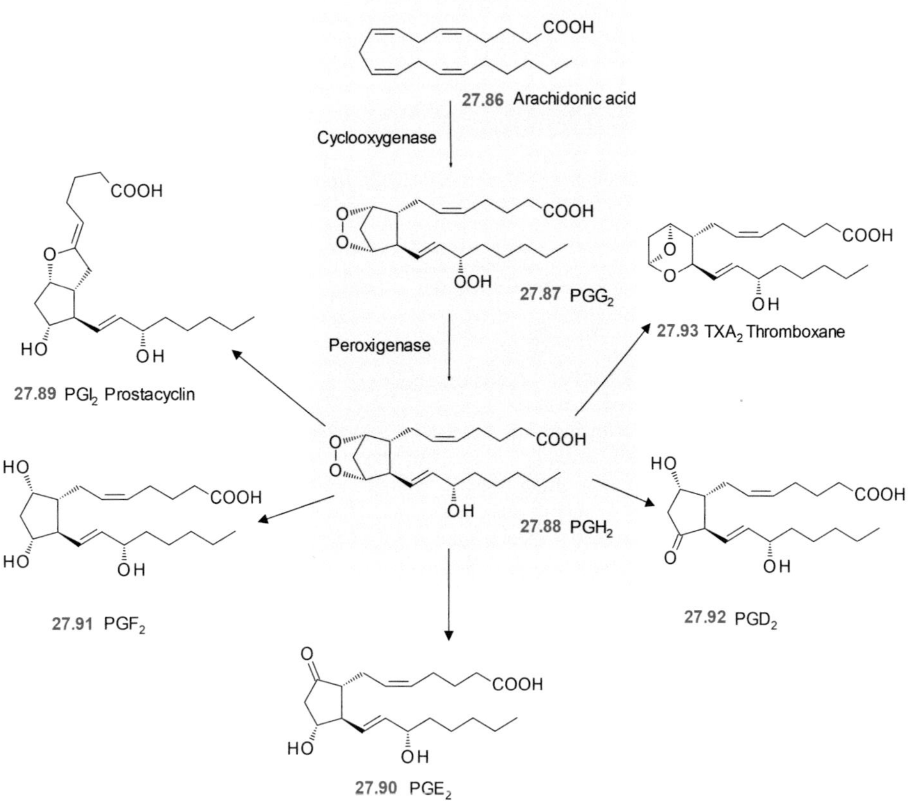

■ Fig. 27.36 Arachidonic acid **27.86** is transformed into the prostaglandin PGH_2 **27.88** by the bifunctional enzyme cyclooxygenase using a cyclooxidation and a peroxidase step. PGH_2 is the starting material for the synthesis of a variety of prostaglandins **27.89**–**27.93**, which are formed by specific synthases

activity. It is assumed that the peroxidase reaction site of the enzyme, located on the opposite side, is accessed from the outside of the protein in the vicinity of the endoplasmic reticulum for this reaction step. The oxidized substrate, which is now much more polar, has to diffuse from the arachidonic acid channel to the peroxidase reaction site.

Tyr 385, located deep in the protein near the heme center, is critical for the overall reaction. It catalyzes the oxidation and reduction steps according to the changing oxidation states of the iron. At the same time, the oxygen species to be transferred are supplied from this center. In the COX enzymes, two enzymatic processes take place that are tightly linked. One dioxygen species is needed as a reagent for cyclooxygenase activity. Tyrosine has a special role because it is coupled to both activities as an intermediate radical. The radical state of this residue is formed during the peroxidase reaction and initiates the cyclooxygenase reaction via the homolytic abstraction of hydrogen atoms. The crystal structures of COX-1 with the superimposed arachidonic acid substrate **27.86** (violet) and the product PGH2 **27.88** (gray) are shown in ■ Fig. 27.39. For clarity, important parts of the protein environment have been omitted from the figure.

PGH_2 **27.88** is the central starting material for the synthesis of a number of products derived from arachidonic acid (■ Fig. 27.36). A variety of synthases are involved in the transformations that yield the vari-

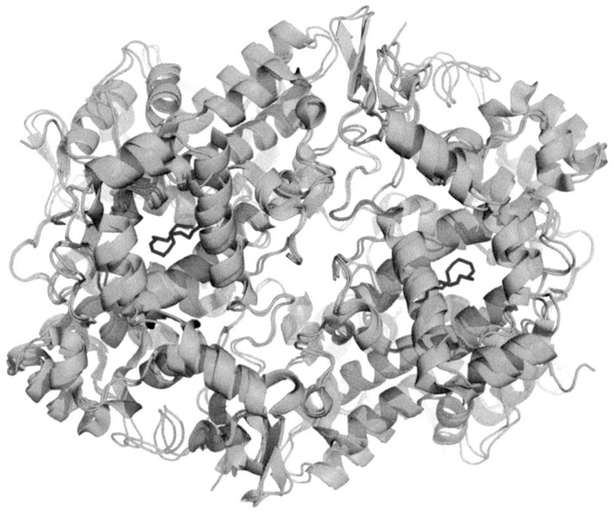

■ Fig. 27.37 Two isoenzymes COX-1 (*green*) and COX-2 (*blue*), which have 65% sequence homology, are known. They are catalytically active as dimers and submerge into the membrane with a ring of the hydrophobic helices (coming out of the page and facing the reader). This ring represents an opening to the channel through which arachidonic acid **27.86** (*dark blue*) can diffuse from the membrane into the catalytic site. The superposition of the crystal structures of both isoforms is shown from the direction of the membrane

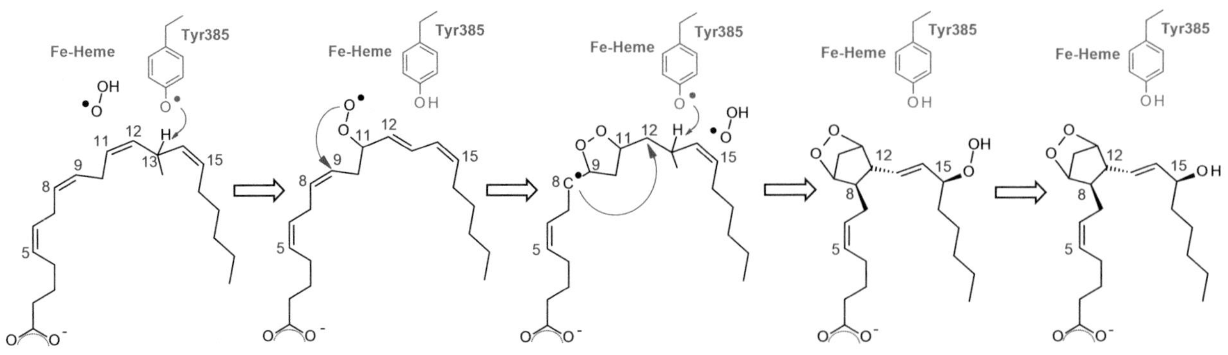

27.86 Arachidonic acid

27.87 PGG_2

27.88 PGH_2

Fig. 27.38 The chemical transformation of arachidonic acid **27.86** to PGG_2 **27.87** and PGH_2 **27.88** occurs by an attack of the tyrosyl radical 385 on the C13 carbon atom, from which a hydrogen atom is abstracted. The intermediately formed, unsaturated radical adds a peroxide group to C11. A cyclic peroxide is formed with C9 by a ring-closing reaction. The tyrosyl radical abstracts another hydrogen atom from C13, and C8 forms a carbocycle upon reaction with C12 to produce PGG_2 **27.87**. The product leaves the binding pocket and is further chemically transformed to PGH_2 **27.88** in a peroxidase reaction

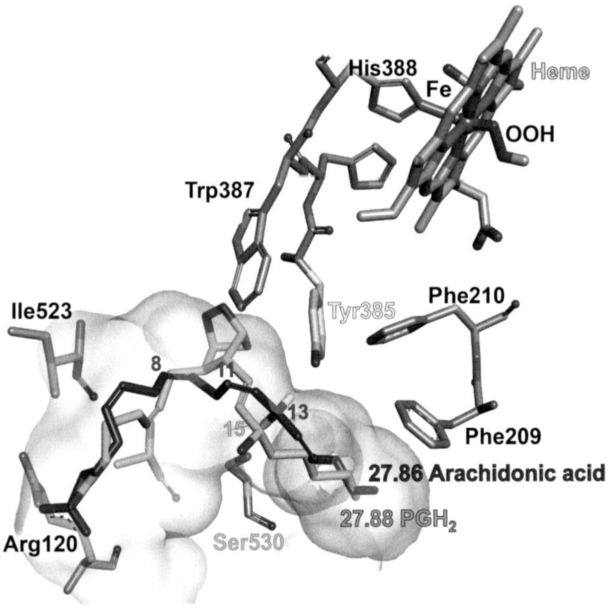

Fig. 27.39 Superposition of arachidonic acid **27.86** (*violet*) and PGH_2 **27.88** (*light blue*) in the reaction channel of COX. The heme center to which oxygen is bound is at the *top, right* side. Tyr 385 (*yellow*) is responsible for the hydrogen atom abstraction from C13 of arachidonic acid. The atoms of the protein are largely removed for clarity, and the reaction channel is indicated with a *transparent* surface. The displayed geometry is based on the crystallographically determined complexes of COX with arachidonic acid and PGH_2. (▸ https://sn.pub/nVS1d3)

ous **prostaglandins**. COX catalyzes the rate-determining step, which explains its central role in the regulation of inflammatory processes. Prostaglandins are called **inflammatory mediators**. Prostaglandins PGI_2 **27.89** and PGE_2 **27.90** increase vascular permeability. This leads to tissue swelling and redness as a result of increased blood flow. Nociceptive nerve endings are sensitized and the perception of pain is increased. In the stomach, PGI_2 and PGE_2 are involved in the regulation of mucous membranes and gastric acid production. PGE_2 is also associated with the occurrence of fever in inflammatory processes. The prostaglandin PGF_2 **27.91** is associated with reproductive processes. At the onset of labor, COX-2 is expressed at elevated levels in the placenta. The produced PGE_2 is involved in stimulating the uterus to contract. PGD_2 **27.92** takes on the task of regulating contractions in the bronchial airways. PGH_2 is a starting material for the synthesis of thromboxane TXA_2 **27.93**. It is formed by COX-1, which is present in platelets. TXA_2 binds to the thromboxane receptor, a GPCR (Sect. 29.1), and activates platelet aggregation. This last step initiates cellular coagulation and serves to close injured blood vessels (Sects. 23.4 and 31.2). Like TGD_2, thromboxane also causes smooth muscle contraction in the pulmonary vasculature. In addition, the prostaglandins are associated with important regulatory processes such as renal perfusion, body temperature regulation, modulation of the immune response, and regulatory processes in the ovarian cycle.

Because of their central role in regulating such diverse processes in tissues, the enzymes of the prostaglandin synthesis cycle are ideal candidates for drug therapy. As mentioned above, there are two isoforms of cyclooxygenases. **COX-1** is ubiquitously expressed in all tissues. It is **constitutive**, that is, its production is largely independent

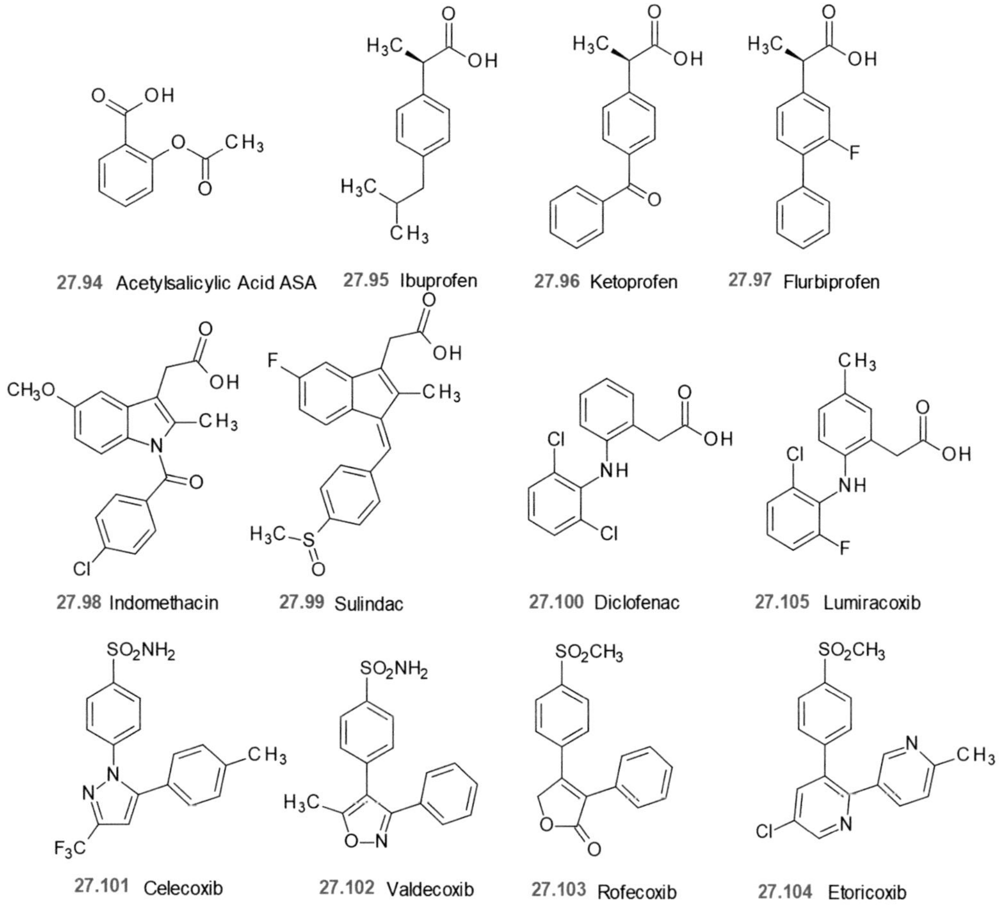

Fig. 27.40 Inhibitors of COX isoenzymes. Acetylsalicylic acid **27.94** and the arylacetic acids or propionic acids **27.95–27.100** are unspecific inhibitors of both isoforms. After the discovery of the induced COX-2, the coxibs **27.101–27.104** were developed as selective inhibitors of this isoform. Rofecoxib was withdrawn from the market due to an increased risk of cardiovascular diseases. Lumiracoxib **27.105**, which is structurally identical to diclofenac with the exception of a Cl/F exchange and an additional methyl group, was introduced to the market as a COX-2 selective inhibitor

of cell type, cell stage, or other external influences. This isoform is found exclusively in platelets. COX-1 is found in the endothelial cells of normal blood vessels, whereas the COX-2 isoform is found in the endothelial cells of proliferating blood vessels, in inflamed tissue, and at sites of atherosclerotic damage. In addition, COX-2 is highly expressed in some tumor cells, where it may play a role in tumor growth. It is involved in the production of prostacyclin PGI_2 in the kidney, which then activates renin production (Sect. 24.2). COX-1 is located in the renal cortex and produces PGE_2 and PHI_2, which increase renal perfusion and glomerular filtration rate. Overdoses of COX-1 inhibitors can, therefore, have harmful side effects on renal function. The expression of the second isoform, **COX-2**, can be **induced** in many different ways, and the amount present in the cell depends strongly on the state of the cell and its environment. Presumably, COX-2 diverged from COX-1 by gene duplication long before the development of vertebrates, and both isoforms evolved in parallel. It was first speculated in 1972 that there might be two isoforms. Twenty years later, the new form was found and sequenced. Its structure determination enabled the development of specific inhibitors for both forms. The first selective COX-2 inhibitors were introduced into therapy in 1999.

Cyclooxygenase inhibitors are very old drugs that have been used in therapy for a very long time (Fig. 27.40). **Acetylsalicylic acid (ASA) 27.94** was the first to find widespread use (Sect. 3.1). It has an interesting mechanism of action because it inhibits both isoforms equally well by **irreversible acetylation** of Ser 530 (Fig. 27.41). ASA diffuses into the COX reaction channel. It most likely forms a salt bridge with Arg 120 and transfers its acetyl group to the nearby OH group of Ser 530, similar to the reaction observed in a serine hydrolase. The channel is then irreversibly blocked and the enzyme is permanently inactivated. The function of COX in the inhibited cells can only be restored by resynthesis of the enzyme. This means, for example, that thromboxane production is permanently blocked in platelets, which lack a nucleus and are, therefore, unable to synthesize proteins. The platelets are no longer able to biosynthesize the COX enzymes. For the 8–12 day life cycle of platelets, their ability to provide **thromboxane A_2** (TXA_2) to initiate aggregation is, therefore, significantly reduced. This effect is responsible for the blood-thinning effect of Aspirin®. Patients are, therefore, asked prior to surgery whether they have taken Aspirin® in the previous week. Salicylic acid, which lacks the acetyl group, is a weak but reversible inhibitor of COX that competes with arachidonic acid. When Ser 530 is mutated to Ala, the enzyme is catalytically fully

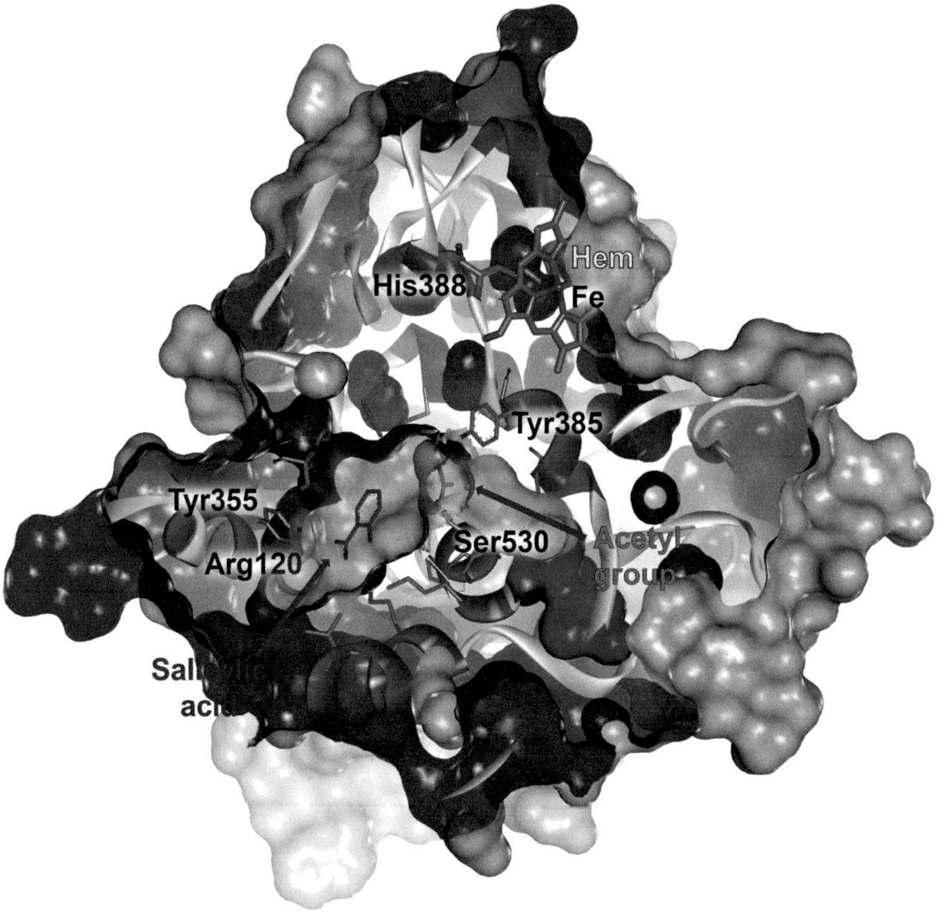

Fig. 27.41 The most probable binding mode of acetylsalicylic acid (ASA) **27.94** with COX-1. ASA binds in the middle of the reaction channel (*gray* surface) that is normally occupied by the natural substrate, arachidonic acid **27.86**. The channel spans through the protein with a bent shape from the lower left. ASA forms a salt bridge with Arg 120 and reacts with the OH group of Ser 530 by transferring its acetyl group. This blocks the channel irreversibly. The additional volume that the acetyl group blocks is indicated with a *violet* surface (interior is *yellow*). The displayed geometry is based on a crystal structure that was determined with a bromine derivative of ASA. (▸ https://sn.pub/nVS1d3)

active. However, the mutant is only weakly inhibited by ASA.

In addition to ASA, arylacetic and propionic acids are another group of slightly selective and reversible COX inhibitors worth mentioning. Members of this class include ibuprofen **27.95**, ketoprofen **27.96**, flurbiprofen **27.97**, indometacin **27.98**, sulindac **27.99**, and diclofenac **27.100** (Fig. 27.40). Ibuprofen also binds in the arachidonic acid channel and forms a salt bridge with its terminal carboxylic acid function to Arg 120. Other important COX inhibitors include oxicam, anthranilic acid, and pyrazole derivatives. They are called **NSAIDs** (**n**on**s**teroidal **a**nti-**i**nflammatory **d**rugs). The mechanism of action of paracetamol (acetaminophen), a very old and widely used analgesic, has long been associated with COX enzymes. Now, however, it appears that this drug may act by conjugating to arachidonic acid through amidation with its metabolite, *p*-aminophenol and, thus, intervening in the pain cascade. The newly formed *N*-arachidonoyl-*p*-aminophenol is a nanomolar vanilloid and CB1 receptor antagonist, both examples of GPCRs, and the cellular uptake of the analgesically active anandamide (arachidonoylethanolamide) is inhibited.

Because COX-1 is constitutively expressed in all tissues, unselective COX inhibitors also act where prostaglandins are needed for other nonpain-related functions. One example is the production of **prostacyclin 27.89**, which is responsible for regulating mucus production in the stomach. COX inhibitors block its synthesis, and as an undesirable side effect, the protection of the gastric epithelial cells against the highly acidic milieu is lost. The result is gastric irritation, which can lead to serious complications.

When it was discovered in the early 1990s that the expression of COX-2 is upregulated at the site of pain, hopes were high that a **pain therapy free of side effects**

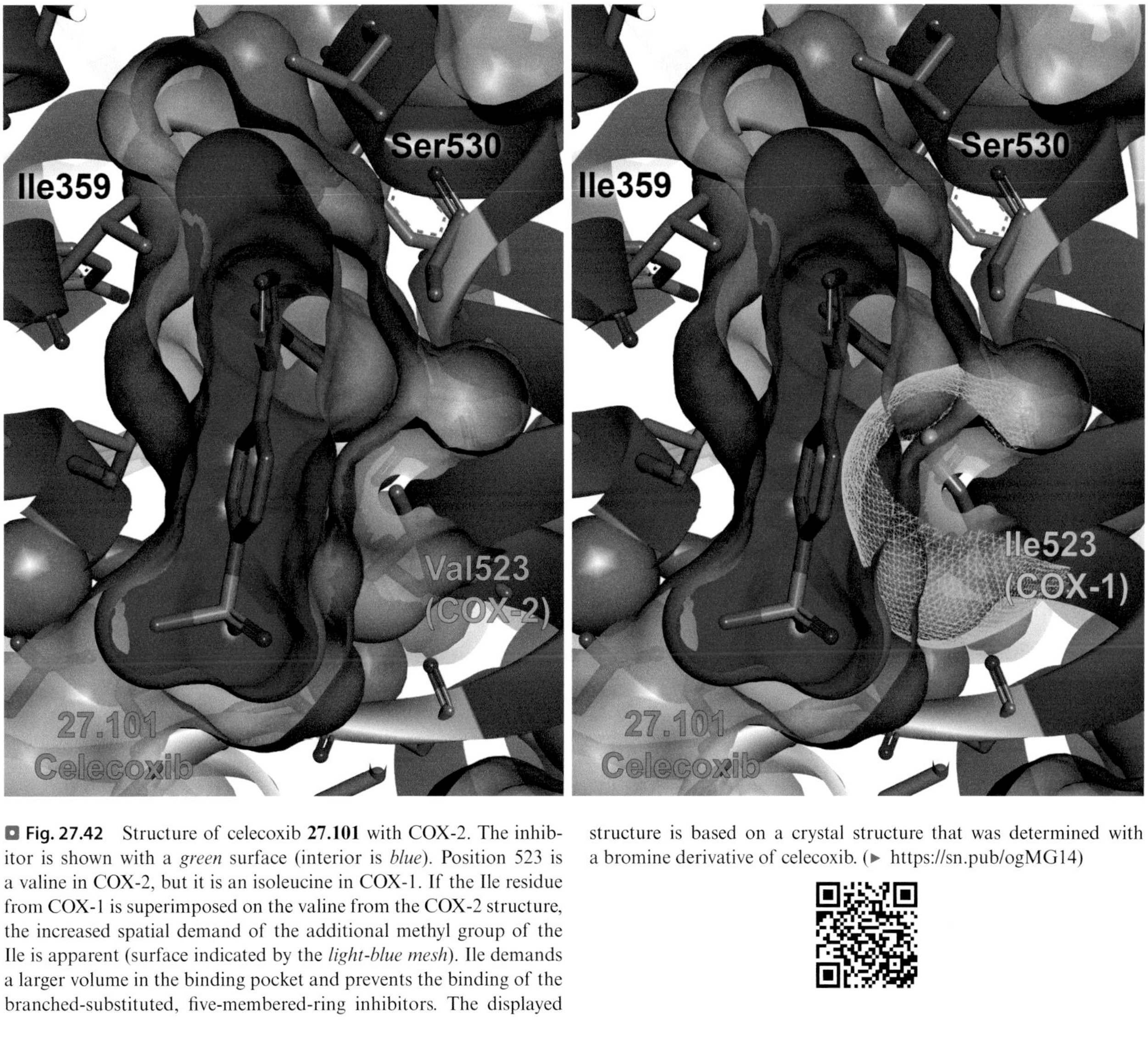

Fig. 27.42 Structure of celecoxib **27.101** with COX-2. The inhibitor is shown with a *green* surface (interior is *blue*). Position 523 is a valine in COX-2, but it is an isoleucine in COX-1. If the Ile residue from COX-1 is superimposed on the valine from the COX-2 structure, the increased spatial demand of the additional methyl group of the Ile is apparent (surface indicated by the *light-blue mesh*). Ile demands a larger volume in the binding pocket and prevents the binding of the branched-substituted, five-membered-ring inhibitors. The displayed structure is based on a crystal structure that was determined with a bromine derivative of celecoxib. (▶ https://sn.pub/ogMG14)

could be achieved by selectively inhibiting this enzyme. Careful analysis of the two enzymes revealed small but significant differences: at position 523, COX-1 has an Ile residue, whereas COX-2 has a Val residue. In addition, at position 503, a Phe residue in COX-1 is exchanged for a Leu residue in COX-2. What can be expected in terms of selectivity from such **small differences as the methyl group exchange** at position 523? At the very least, the binding pocket of COX-2 is 17% larger, and there is a new subpocket in the arachidonic acid channel (◘ Fig. 27.42). It stands to reason that structurally larger inhibitors could be developed that take advantage of the additional subpocket. Such inhibitors can no longer inhibit COX-1 due to the steric conflict caused by the isoleucine residue at position 523. The first generation of successfully developed COX-2 inhibitors **27.101**–**27.104** all have a similar structure (◘ Fig. 27.40). In the center is either a five- or six-membered ring, usually functionalized with aromatic substituents. This results in a branched structure that better mirrors the larger binding pocket of COX-2 than COX-1.

In practice, it was shown that the selective COX-2 inhibitors left COX-1 uninhibited, and side effects such as gastric bleeding or renal dysfunction were almost completely eliminated. The first substances to reach the market were celecoxib **27.101**, valdecoxib **27.102**, and rofecoxib **27.103** (◘ Fig. 27.40). Their indications ranged from rheumatism to osteoarthritis, chronic polyarthritis, and ankylosing spondylitis (Bechterew's disease). All of these diseases are associated with severe pain. Rofecoxib **27.103** (Vioxx®) quickly achieved billion-dollar sales. However, the drug was withdrawn from the market in 2004 because significant side effects were observed in patients on long-term therapy. In particular, there was an increased risk of cardiovascular disease, especially heart attack, unstable angina pectoris, and stroke. As a result, Merck & Co. experienced a 29% drop in profits in 2004. By March 2006, there were already 10,000 lawsuits. How-

27

ever, shortly after the withdrawal of rofecoxib **27.103**, Merck launched a new COX-2 inhibitor, etoricoxib **27.104**.

Overall, this raises the question of whether the side effects seen with rofecoxib **27.103** are typical of all COX-2 inhibitors. The cardiovascular risk must be weighed against the risk of gastric bleeding that can occur with acetylsalicylic acid, diclofenac, ibuprofen, or indometacin. Rofecoxib belongs to the first generation of COX-2 inhibitors, all of which have a five-membered ring in the center. In 2006, lumiracoxib **27.105** (Prexige®), a COX-2 inhibitor, was launched. This COX-2 inhibitor is structurally similar to the less selective diclofenac **27.100** except for an additional methyl group and a Cl/F exchange. Apparently, the additional methyl group exploits the second structural difference of Phe 503 in COX-1 versus the more mobile Leu in COX-2 to achieve a selectivity advantage. However, due to an unacceptable side effect profile, lumiracoxib has since been withdrawn from the market. Celecoxib **27.101**, etoricoxib **27.104**, and parecoxib, a prodrug of valdecoxib **27.102** (masked with a propionyl group at the sulfonamide nitrogen), are currently in therapeutic use. The example of the coxibs has impressively demonstrated how even small differences between COX-1 and COX-2 can be successfully translated into selective agents of a new class of compounds through systematic design.

27.10 Synopsis

- Enzyme-catalyzed redox reactions use different cofactors to accomplish the electron or hydride transfer from the group being oxidized to the group being reduced. The most important cofactors in oxidoreductases are dinucleotides such as the nicotinamides $NAD(P)^+$ or the flavin derivatives FMN and FAD and the iron-containing protoporphyrin ring system in heme enzymes.
- The nicotinamide moiety in $NAD(P)^+$ is an *N*-substituted pyridine derivative that either accepts or releases a hydride ion in the 4-position. The cofactor binds in many oxidoreductases to a conserved fold motif, the nucleotide-binding Rossmann fold.
- Dihydrofolate reductase is involved in the biosynthesis of thymine. Inhibitors competitive with the binding site of the natural substrate dihydrofolate have been developed as potent chemotherapeutics in cancer therapy, or as bacteriostatics to fight bacterial infections.
- Reduction of the cholesterol blood level is a strategy to fight coronary heart disease and atherosclerosis as high excess of cholesterol is found in plaques constricting and, thus, occluding blood vessels.
- HMG-CoA reductase is involved in the biosynthesis of precursors of cholesterol. The substrate, composed of two acetate units, is reduced by using two equivalents of NADPH. Inhibitors, the statins that occupy the cofactor binding site, were first derived from natural compounds discovered by screening substances from microorganisms. Later fully synthetic derivatives were developed that evolved into the best-selling drugs ever.
- Aldose reductase, an NADPH-dependent reductase lacking a Rossmann-folded nucleotide-binding domain, is involved in the polyol pathway, along which glucose is metabolized to sorbitol and subsequently to fructose. Overloading this pathway results in increased production of polar compounds, which creates osmotic stress and oxidative stress as a result of high reductase activity.
- Long-term consequences of a poorly controlled blood glucose level in the case of type 2 diabetes preferentially affect cells that do not control their glucose uptake by insulin. Inhibition of aldose reductase is a viable principle to reduce long-term complications.
- Aldose reductase is able to reduce a broad scope of different aldehyde substrates. This is achieved by a highly adaptive binding pocket, which also allows the development of inhibitors showing largely deviating scaffolds and binding modes.
- Cortisol is transformed to cortisone and vice versa via two isoforms of 11*β*-hydroxysteroid dehydrogenase, which is a NADPH-dependent reductase that takes on a Rossmann fold. 11*β*-HSD1 inhibition has been suggested as a promising therapy concept to treat metabolic syndrome.
- The cytochrome P450 enzymes are a superfamily of heme proteins that carry out biochemical transformations as monooxygenases by introducing oxygen onto a substrate being oxidized. They are particularly involved in the metabolism of xenobiotics and a large production of the administered drug molecules are metabolized in CYP 3A4, CYP 2D6, and CYP 2C9.
- The CYP enzymes are highly adaptive and accommodate substrates of significantly different sizes. They can be inhibited by drug molecules, particularly those containing heteroaromatic rings that coordinate the catalytic iron ion in the heme center. Their expression can be induced and, thus, upregulated by xenobiotics activating, for instance, the PXR transcription factor.
- Because the equipment with cytochrome P450 enzymes varies with geno- and phenotype, this polymorphism causes varying metabolic behavior between different individuals. Differentiation into slow, extensive, and fast metabolizers has consequences for the prescription and required dose level of a given drug metabolized by the involved CYPs.
- Because the activity of a metabolizing CYP enzyme can be further modulated either by inhibition or induction by coadministered drugs or by xenobiotics ingested with food, there may be serious conse-

quences with respect to the dose level present in the body, which may result in undesirable and dangerous side effects or unexpected failure of drug action.

- Monoamine oxidases MAO_A and MAO_B are FAD-dependent oxidases and metabolize important neurotransmitters such as dopamine, adrenaline, or serotonin. Inhibition of these enzymes can help in the therapy of depression, Alzheimer's disease, or Parkinson's disease.
- Most of the current MAO inhibitors are activated by an initial redox step and a covalent attachment is formed to the FAD cofactor via a highly reactive intermediate; this leads to an irreversible chemical modification of its redox properties.
- The membrane-associated cyclooxygenases COX-1 and COX-2 synthesize the endoperoxide PGG_2, which is a precursor to a large variety of prostaglandins, from arachidonic acid. Prostaglandins are an important class of paracrine hormones and are also referred to as inflammatory mediators.
- COX contains a heme center, and PGG_2 is synthesized through a cyclooxidation step involving radical intermediates. In a subsequent peroxidation step involving release and diffusion of the substrate to another reaction site, PGG_2 is further modified to PGH_2.
- COX is inhibited by nonsteroidal anti-inflammatory drugs such as acetylsalicylic acid, ibuprofen, indometacin, or diclofenac. They bind to the reaction channel and block access of the natural substrate arachidonic acid.
- Acetylsalicylic acid transfers its acetyl group irreversibly to a channel-exposed hydroxyl group of Ser 530 in a reaction similar to that in serine hydrolases. As a consequence, in cells lacking a nucleus such as thrombocytes, prostaglandin synthesis and its products such as thromboxane are permanently blocked for the lifetime of the cell.
- Two isoforms of COX exist. COX-1 is ubiquitously expressed in all tissues and constitutively present. Due to its multiple involvement in many physiological processes overdosing of COX-1 inhibitors can exert severe side effects. COX-2 is induced and found in endothelial cells of proliferating blood vessels, inflamed tissue, sites of atherosclerotic damage, and in some tumor cells. This makes selective COX-2 inhibition a prospective therapeutic principle.
- COX-1 and COX-2 differ in the reaction channel by the crucial exchange of an isoleucine for a valine residue. The additional volume created in COX-2 by the absent methyl group gives rise to the development of size-extended furcated inhibitors, the coxibs. Their indications range from rheumatism, osteoarthritis, chronic polyarthritis, to ankylosing spondylitis; all of these diseases are associated with severe pain.

Bibliography and Further Reading

General Literature

D. C. N. Chan and A. C. Anderson, Towards Species-specific Antifolates, Curr. Med. Chem., **13**, 377–398 (2006)

A. Gangjee and H. D. Jain, Antifolates—Past, Present and Future, Curr. Med. Chem. Anti-Cancer Agents, **4**, 405–410 (2004)

P. R. Vagelos, Are Prescription Drug Prices High? Science, **252**, 1080–1084, (1991)

J. A. Tobert, Lovastatin and Beyond: The History of HMG-CoA Reductase Inhibitors, Nat. Rev. Drug Discov., **2**, 517–526 (2003)

P. Oates, Aldose Reductase, Still a Compelling Target for Diabetic Neuropathy, Curr. Drug Targets, **9**, 14–36 (2008)

S. P. Webster and T. D. Pallin, 11*β*-Hydroxysteroid dehydrogenase type 1 inhibitors as Therapeutic Agents, Expert Opin. Ther. Patents, **17**, 1407–1422 (2007)

F. Hoffmann and E. Maser, Carbonyl reductases and Pluripotent Hydroxysteroid Dehydrogenases of the Short-chain Dehydrogenase/Reductase Superfamily. Drug Metab. Rev., **39**, 87–144 (2007)

D. C. Lamb, M. R. Waterman, S. L. Kelly and F. P. Guengerich, Cytochromes P450 and Drug Discovery, Curr. Opin. Biotech., **18**, 504–512 (2007)

R. Weinshilboum and L. Wang, Pharmacogenomics: Bench to Bedside, Nat. Rev. Drug Discov., **3**, 739–748 (2004)

L. C. Wienkers and T. G. Heath, Predicting *in vivo* Drug Interactions From *in vitro* Drug Discovery Data, Nat. Rev. Drug Discov., **4**, 825–833 (2005)

M. B. H. Youdim, D. Edmondson and K. F. Tipton, The Therapeutic Potential of Monoamine Oxidase Inhibitors, Nat. Rev. Neurosci., **7**, 295–309 (2006)

R. J. Flower, The Development of COX-2 Inhibitors, Nat. Rev. Drug Discov., **2**, 179–191 (2003)

C. Michaux and C. Charlier, Structural Approaches for COX-2 Inhibition, Mini-Rev. Med. Chem., **4**, 603–615 (2004)

J. A. Mitchell and T. D. Warner, COX Isoforms in the Cardiovascular System: Understanding the Activities of Non-steroidal Anti-inflammatory Drugs, Nat. Rev. Drug Discov., **5**, 75–86 (2006)

Special Literature

V. Cody, J. Pace, K. Chisum and A. Rosowsky, New Insights into DHFR Interactions: Analysis of *Pneumocystis carinii* and Mouse DHFR Complexes with NADPH and Two Highly Potent 5-[ω-Carboxy(alkyloxy)] Trimethoprim Derivatives Reveals Conformational Correlations with Activity and Novel Parallel Ring Stacking Interactions, Proteins, **65**, 959–969 (2006)

J. T. Bolin, D. J. Filman, D. A. Matthews, R. C. Hamlin, J. Kraut, Crystal structures of *Escherichia coli* and *Lactobacillus casei* dihydrofolate reductase refined at 1.7 A resolution. I. General features and binding of methotrexate. J. Biol. Chem., **257**, 13650–13662 (1982)

A. Rosowsky, R. A. Forsch and J. E. Wright, Synthesis and *in vivo* Antifolate Activity of Rotationally Restricted Aminopterin and Methotrexate Analogues, J. Med. Chem., **47**, 6958–6963 (2004)

D. E. Wolf, C. H. Hoffman, P. E. Aldrich, H. R. Skeggs, L. D. Wright, K. Folkers, beta-Hydroxy-beta-methyl-gamma-valerolactone (mevalonic acid) a new biological. J. Am. Chem. Soc., **78**, 4498–4499 (1956)

A. Endo, The discovery and development of HMG-CoA reductase inhibitors, J. Lipid Res., **33**, 1569–1582 (1992)

E. S. Istvan, M. Palnitkar, S. K. Buchanan, J. Deisenhofer, Crystal Structure of the Catalytic Portion of Human HMG-CoA Reductase: Insights into Regulation of Activity and Catalysis, EMBO J., **19**, 819–830 (2000)

B. D. Roth, The Discovery and Development of Atorvastatin, a Potent Novel Hypolipidemic Agent, Prog. Med. Chem., **40**, 1–22 (2002)

G. E. Stokker et al., 3-Hydroxy-3-methylglutaryl-coenzyme A Reductase Inhibitors. 1. Structural Modification of 5-Substituted 3,5-Dihydroxypentanoic Acids and Their Lactone Derivatives, J. Med. Chem., **28**, 347–358 (1985)

B. D. Roth et al., Inhibitors of Cholesterol Biosynthesis. 1. trans-6-(2-Pyrrol-l-ylethyl)-4-hydroxypyran-2-onesa, Novel Series of HMG-CoA Reductase Inhibitors. 1. Effects of Structural Modifications at the 2- and 5-Positions of the Pyrrole Nucleus, J. Med. Chem., **33**, 21–31 (1990)

M. Ekroos and T. Sjögren, Structural Basis for Ligand Promiscuity in Cytochrome P450 3A4, Proc. Natl. Acad. Sci. USA, **103**, 13682–13687 (2006)

A. K. Daly, Pharmacogenetics of the Cytochromes P450, Curr. Top. Med. Chem., **4**, 1733–1744 (2004)

G. Cruciani, E. Carosati, et al., MetaSite: Understanding Metabolism in Human Cytochromes from the Perspective of the Chemist, J. Med. Chem., **48**, 6970–6979 (2005)

L. Bertilsson L, Y. Q. Lou, et al. Pronounced Differences between Native Chinese and Swedish Populations in the Polymorphic Hydroxylations of Debrisoquin and S-Mephenytoin, Clin. Pharmacol. Ther., **51**, 388–397 (1992)

L. De Colibus, M. Li, et al., Three-dimensional Structure of Human Monoamine Oxidase (MAO A): Relation to the Structure of rat MAO A and human MAO B, Proc. Natl. Acad. Sci. U.S.A., **102**, 12684–12689 (2005)

G. A. FitzGerald, COX-2 and Beyond: Approaches to Prostaglandin Inhibition in Human Disease, Nat. Rev. Drug Discov., **2**, 879–890 (2003)

M. A. Windsor, P. L. Valk, S. Xu, S. Banerjee, L. J. Marnett, Exploring the molecular determinants of substrate-selective inhibition of cyclooxygenase-2 by lumiracoxib, Bioorg. Med. Chem. Lett., **23**, 5860–5864 (2013)

Agonists and Antagonists of Nuclear Receptors

Contents

G. Klebe, *Drug Design*, https://doi.org/10.1007/978-3-662-68998-1_28

With all the excitement surrounding the structure-based design of enzyme inhibitors, it is important to remember that less than half of the prescription drugs available today act on enzymes. Many other drugs target receptors, transporters, pores, or ion channels. Most receptors mediate the transfer of information from the outside to the inside of the cell. Activating or blocking them changes the state of the cell. In this way, the cells can take on modulatory functions. Transporters, pores, and ion channels serve to transport selected substances across the membrane, especially substances that cannot cross the membrane by passive diffusion due to their polar character. Like many of the receptors, these proteins are embedded in the cell membrane. Before turning to these classes of membrane-bound targets, another class of receptors found in the interior of the cell should be considered. Nuclear receptors are controlled by specific ligands. An endogenous hormone must first enter the cell to achieve activation. This usually occurs by passive diffusion across the membrane. The ligands must, therefore, have adequate lipophilic or amphiphilic properties or be substrates of transporters.

28.1 Nuclear Receptors Are Transcription Factors

Nuclear receptors are soluble receptors found in the cytosol. As **transcription factors**, they regulate the expression of specific genes in the nucleus and are, therefore, responsible for the production of proteins. They bind directly to DNA and play an important role in **gene regulation** during embryonic development, cell growth, differentiation, and specialization. Malfunction of these receptors leads to diseases with uncontrolled cell growth (e.g., cancer), metabolic disorders (diabetes or obesity), or reproductive disorders (infertility). They are activated by hormones. These natural ligands, which include steroid hormones as well as lipophilic ligands such as retinoic acid, various fatty acids, triiodothyronine, vitamin D, prostaglandins, bile acids, and phospholipids, must passively cross the cell membrane barrier (◘ Fig. 28.1). Once at the site of action, they bind to the ligand-binding domains of nuclear receptors. From a drug design perspective, these receptors are interesting targets because the natural ligands are the **typical size of a drug molecule**. As a result, 13% of all drugs approved through 2006 act at these receptors. This number is exceeded only by the G-protein-coupled receptors, which will be discussed in

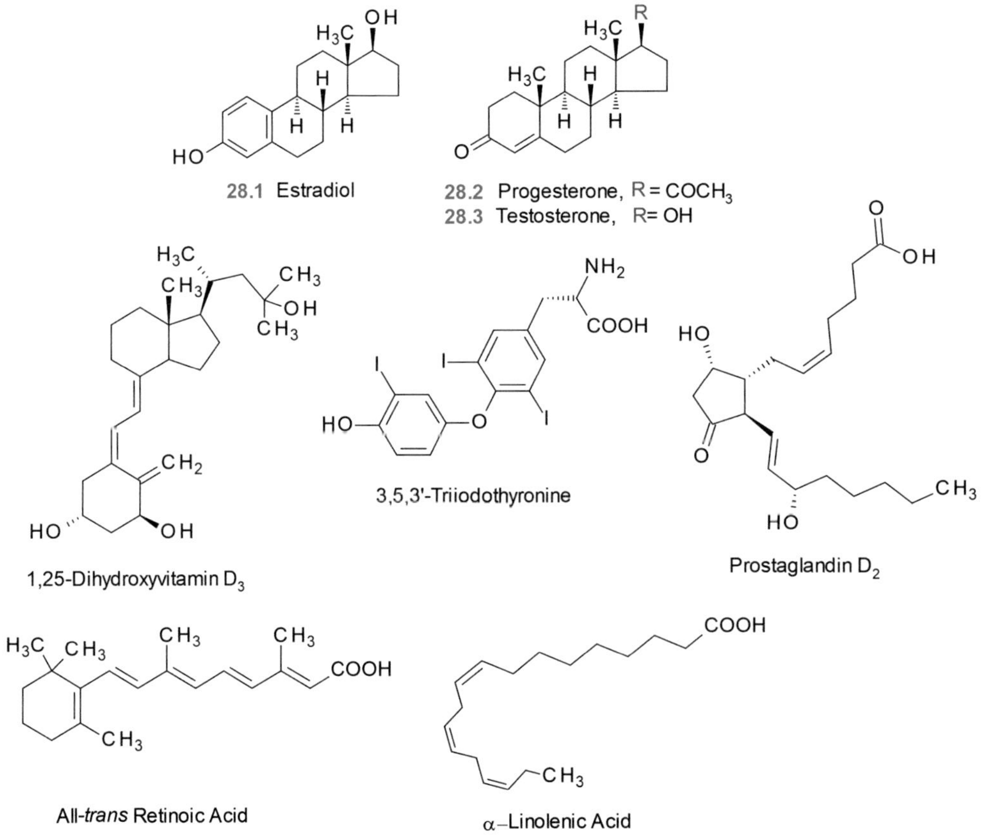

◘ **Fig. 28.1** Some examples of natural ligands of nuclear receptors, for example, steroids such as estradiol **28.1**, progesterone **28.2**, and testosterone **28.3** as well as molecules such as retinoic acid, fatty acids, triiodothyronine, vitamin D, or prostaglandins

the next chapter. At first glance, these target structures seem ideal, but the biological control of gene expression is very complex. The receptors not only have ligand-dependent domains, but also ligand-independent domains that activate transcription. Once the receptors translocate to the nucleus, coactivators, corepressors, and transcription factors contribute to the regulation of gene expression. Both upregulation and downregulation can be achieved. They also appear to interact with other signal transduction pathways, such as those controlled by, for instance, NF-κB or the activator protein AP-1. On the other hand, the nuclear receptor protein family appears to be simple in terms of molecular diversity. There are 48 genes in our genome that encode the different receptors. However, drugs have so far only been developed for 14 of these receptors.

28.2 The Structure of Nuclear Receptors

Nuclear receptors are all constructed according to the same blueprint. They contain three domains. The *N*-terminal A/B region is the most variable in the family. It contains the transactivation domain and is involved in the ligand-independent recognition of cofactors and further transcription factors. This is followed by the **DNA-binding domain**, which contains about 70 amino acids and two **zinc finger motifs**. This domain is the most conserved in the entire gene family. The *C*-terminal domain ends with the **ligand-binding region**, which contains about 250 amino acids. It contains the binding site for small ligands and provides an additional regulatory element for the recognition of coactivators and other transcription factors.

Nuclear receptors can be divided into two groups. The first includes steroid receptors that form a **homodimer** to be activated. The second large group includes receptors that form a **heterodimer** with the **promiscuous retinoid X receptor** (**RXR**) to function. Dimerization may occur in response to ligand binding, or the bound ligand may accelerate and stabilize dimer formation. Some nuclear receptors are present in the cytosol as inactive complexes with heat shock proteins. Ligand binding stimulates the disassembly of these initially inactive complexes and triggers the signal for translocation to the nucleus. There, the dimeric receptor binds with its DNA-binding domains to a **DNA response element**, which is deposited as a promoter or repressor region on the target gene. In the unbound state, some of the receptors can bind **corepressors** containing, for example, one of the histone deacylases (HDACs) described in Sect. 25.10. By deacetylating the basic amino acids on the histone proteins, they make the DNA less accessible and reading by transcription factors is blocked. When an agonist is bound as a ligand, the corepressor is replaced by a **coactivator**, such as a histone acetyltransferase. As an opponent of HDACs, it weakens the binding of DNA to the then more acetylated and less charged histone proteins. Transcription becomes possible. The complex formed with the coactivator, thus, serves as a trigger signal for the onset of transcription and gene expression; dimerization is achieved in response to the binding of an agonist, or the dimer formation is stabilized by the bound agonist.

Each DNA-binding domain recognizes a specific pattern of six bases in the major groove of DNA by using a two-helix motif (Sect. 14.9). This pattern is located mirror-symmetrically on both complementary strand segments (◘ Fig. 28.2) in opposite directions. Two **zinc fingers** stabilize the two-helix motif. For this, the zinc ion coordinates tetrahedrally to four neighboring cysteine residues, allowing cross-linking within the protein strand.

The **ligand-binding domains** of the nuclear receptors also follow a common construction principle. They consist of 12 helices. The sequence at the end of the 12th helix has a particular task. It opens and closes access to the ligand-binding pocket like a door. In doing so, it undergoes a spatial rearrangement that provides an activation signal for the receptor (Sect. 28.4).

The ligand-binding pockets in the nuclear receptors encompasses about 400–600 $Å^3$. They have polar amino acids at both ends and a belt of hydrophobic residues in the center. In receptors that form heterodimers with the retinoic acid receptor RXR (◘ Fig. 28.2), the ligand-binding pockets are significantly larger. For the peroxisomal proliferator-activating receptors PPAR, it can be up to 1300 $Å^3$. It is Y-shaped and, like its endogenous ligands such as eicosapentaenoic acid, linolenic acid (◘ Fig. 28.1), or arachidonic acid, is highly hydrophobic. Despite their common architecture and the rather large variation in volume for ligand uptake, many of the ligand-binding domains have a surprising selectivity for recognizing their ligands. This selectivity will be discussed in more detail in the next section.

28.3 Steroid Hormones: How Small Differences Translate to the Receptor

Male and female sex hormones and corticosteroids are substances with strikingly similar structures. All are derived from an identical scaffold. On a grand scale, Nature manages to produce a wide range of biological effects with minimal structural variation. One mistake can have fatal consequences. The difference between estradiol **28.1**, progesterone **28.2**, and testosterone **28.3** will be examined in more detail. On the aromatic ring of estradiol, there is a hydroxyl group in the first ring (A-ring) of the steroid scaffold, which is changed to a carbonyl group in the partially hydrogenated ring of progesterone and testosterone. The aromatic A-ring of the female hormone estradiol adopts a planar structure, but the ring in the male hormone testosterone forms a half-chair (◘ Fig. 28.3).

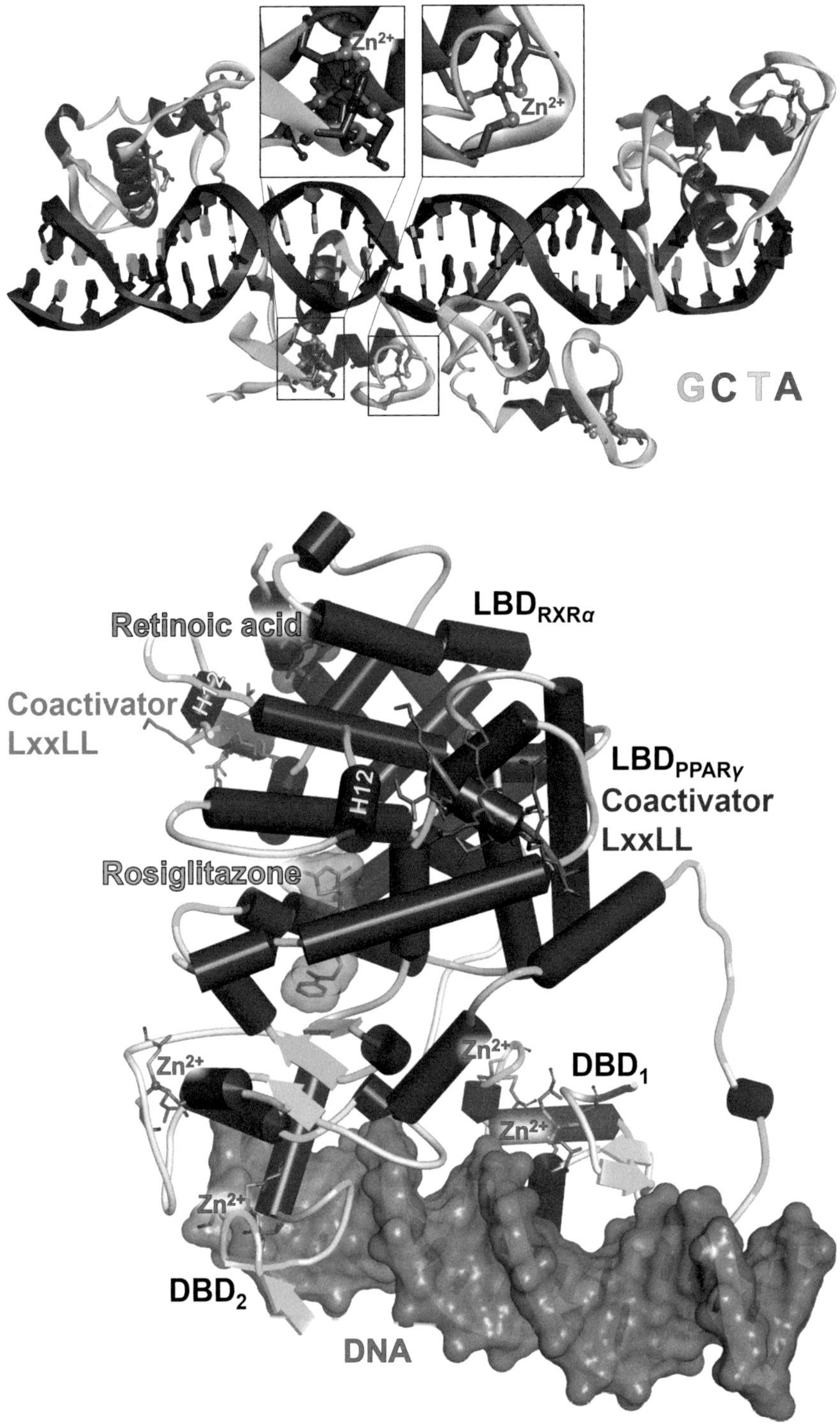

Fig. 28.2 *Upper row* Detail of the crystal structure of the DNA-binding domain of the estrogen receptor. With a motif of two helices (*red*), the receptor uses a zinc finger (highlighted in *yellow*, shown in total four times with opposite direction along the DNA) in the major groove of the DNA (backbone *purple*, bases color-coded, cf. Fig. 14.18) to recognize a specific pattern of six bases. The motif of the two helices is cross-linked via zinc ions (*blue-gray*, see insets) coordinated tetrahedrally by spatially adjacent cysteine residues. *Lower row* For the heterodimeric PPARγ–RXRα receptor, a crystal structure with a bound DNA fragment (*blue-gray* surface) was successfully determined. The receptor binds with its zinc fingers (DBD1, DBD2) in the major groove of the DNA and reads the response sequence. The agonists rosiglitazone (*green*, **28.32**, Fig. 28.14) and retinoic acid (*brown*) are bound to the ligand-binding domains (LBD–PPARγ, LBD–RXRα). They activate a conformation of helix 12 (H12) that releases the coactivator binding site for the recognition motif LxxLL. The structure could be crystallized with short peptide sequences containing this motif (*blue*, *purple*). (▸ https://sn.pub/Ydg6jE)

There is also a methyl group on carbon atom 10 in the male hormones and in progesterone. The 19-methyl group is missing at this position due to the aromatic character of the first ring in estradiol. The 19-methyl group shields this first ring from above and takes up a relatively large amount of space in progesterone and testosterone.

How does the receptor detect this small difference? As shown in the crystal structure of the estrogen receptor with bound estradiol, the hydroxyl group on the aromatic A-ring is involved in a hydrogen-bonding network with Glu 353 and, via a water molecule, with Arg 394 (◘ Fig. 28.4). Glu 353 is most likely deprotonated and recognizes the hormone by the donor functionality of its hydroxyl group. In the structure of the progesterone receptor, there is a glutamine at the same position (◘ Fig. 28.5). It forms a hydrogen bond with the carbonyl group in the A-ring of progesterone through the amino group of its terminal carboxamide. The water-mediated H-bond to an arginine is also found in this receptor. Therefore, a glutamate to glutamine exchange results in a change from a hydroxyl group to a carbonyl function in the hormone. **H-bond donors and acceptors are exchanged in pairs**! How is the additional volume requirement of the 19-methyl group recognized by the receptor? Hydrophobic amino acids in the central region form the contact surface for the bound hormone. In the estrogen receptor, two bulky, terminally branched leucine residues shield the space above the A-ring. Their rigid geometry efficiently limits the volume of the binding pocket (◘ Fig. 28.4). Two methionine residues are found at the same position in the progesterone receptor (◘ Fig. 28.5). They are also bulky and hydrophobic. However, their linear structure allows them to conform to the shape of the bound ligand, leaving a small volume for the 19-methyl group. The structure of the androgen receptor with testosterone is also available. The same amino acids as in the progesterone receptor are present to recognize the carbonyl function on the A-ring and are in close proximity to the 19-methyl group. Each receptor achieves the required selectivity for its specific ligand through these small but distinct changes. For example, the difference in the side chain at C17 helps distinguish progesterone from testosterone.

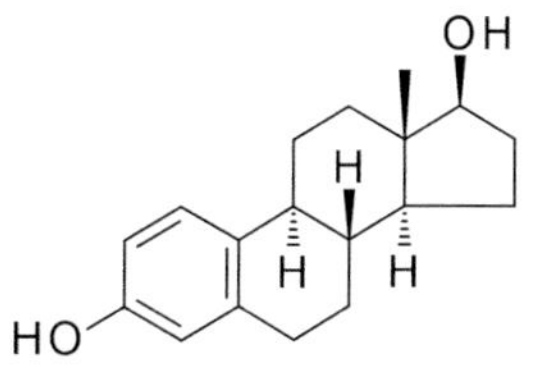

28.1 Estradiol

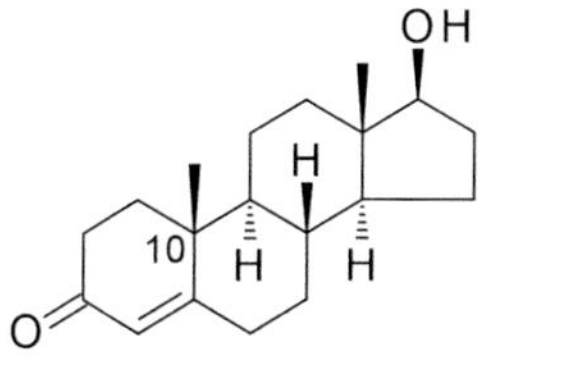

28.3 Testosterone

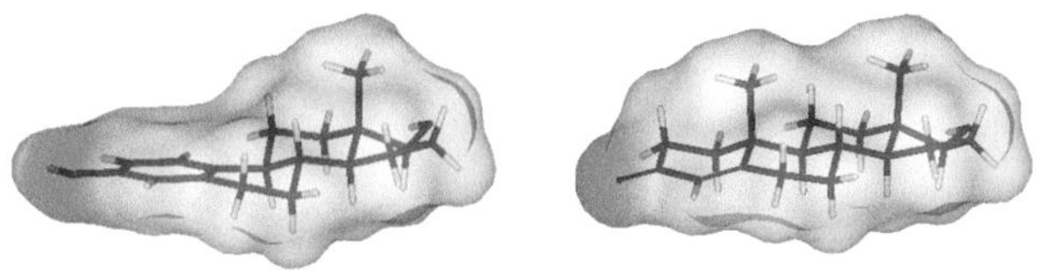

◘ **Fig. 28.3** The difference between the female hormone estradiol **28.1** and the male hormone testosterone **28.3** consists of a change from a hydroxyl group on the aromatic ring of estradiol to a carbonyl group in a partially hydrogenated ring of testosterone. The aromatic A-ring of estradiol takes on a planar structure, whereas the A-ring in testosterone forms a half-chair. A methyl group occurs on carbon C10 in the male hormone that gives the molecule additional volume

28.4 Helix Open, Helix Closed: How Agonists and Antagonists Are Differentiated

The **ligand-binding domains** of nuclear receptors consist of 12 helices (◘ Fig. 28.6). The 12th and last helix in the sequence, also called the AF-2 helix (activation function 2), closes the entrance to the ligand-binding pocket like a terminal gate. For a ligand to gain access, helix 12 must undergo a spatial rearrangement. When an agonist is bound, the gate closes again. At the same time, this rearrangement opens the recognition site for the coactivator. This coactivator interacts with the now available surface segment through a Leu–x–x–Leu–Leu (or short **LxxLL** motif, where x stands for any amino acid) binding motif, which is a segment of an amphiphilic helix (◘ Fig. 28.8). Binding of an antagonist suppresses the rearrangement of helix 12, which can no longer close the entrance area and block the recognition site for the LxxLL motif of the coactivator. Signal transduction does not occur, the receptor does not translocate to the nucleus, and the response element does not bind to DNA. At the molecular level, the difference between **agonist and antagonist** has been most thoroughly studied for the estrogen receptor. When the natural agonist estradiol **28.1** or a synthetic substitute such as diethylstilbestrol **28.4** bind (◘ Fig. 28.7), helix 12 is in its active position, allowing the coactivator to bind to the peptide recognition motif. Asp 351 plays an important role in stabilizing this helix position. It is located in the middle of the rather long helix 3 and exactly opposite to the *N*-terminal end of helix 12. The three NH groups protruding over the edge of the helix end are 3–4 Å away from the carboxylate groups of this acidic residue. Positions that lie opposite such a helix end are predestined for the stabilization of negative charges. This is due to a strong dipole moment that is formed along the axis of the helix. Antagonists such as 4-hydroxy-tamoxifen **28.5** or raloxifene **28.6** have side chains that remain in the entrance channel after ligand binding and prevent closure by helix 12. The antagonists carry a basic group at the end of this side chain. It is most likely in a positively charged state and forms a salt bridge with Asp 351. In this way, the antagonists are able to compensate for the negative charge of the acidic amino acid.

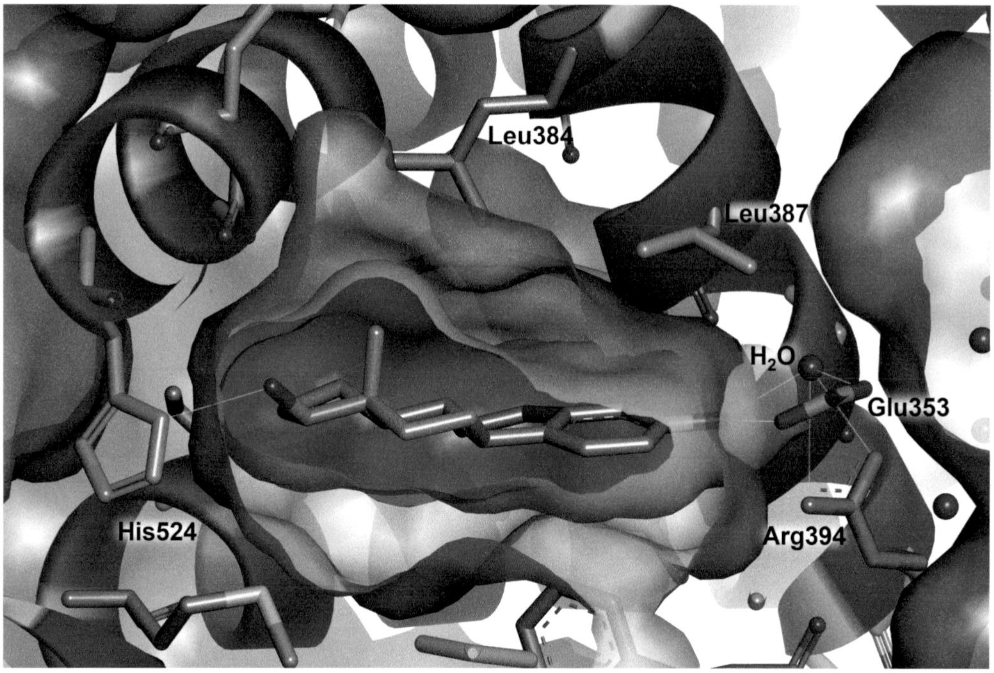

Fig. 28.4 Section of the crystal structure of the estrogen receptor with bound estradiol (surface is *green*, interior is *blue*). The hydroxyl group of the aromatic A-ring forms an H-bond to Glu 353 and a water-mediated H bond to Arg 294. The volume above the planar, aromatic A-ring is limited by Leu 384 and Leu 387. (► https://sn.pub/0tENuz)

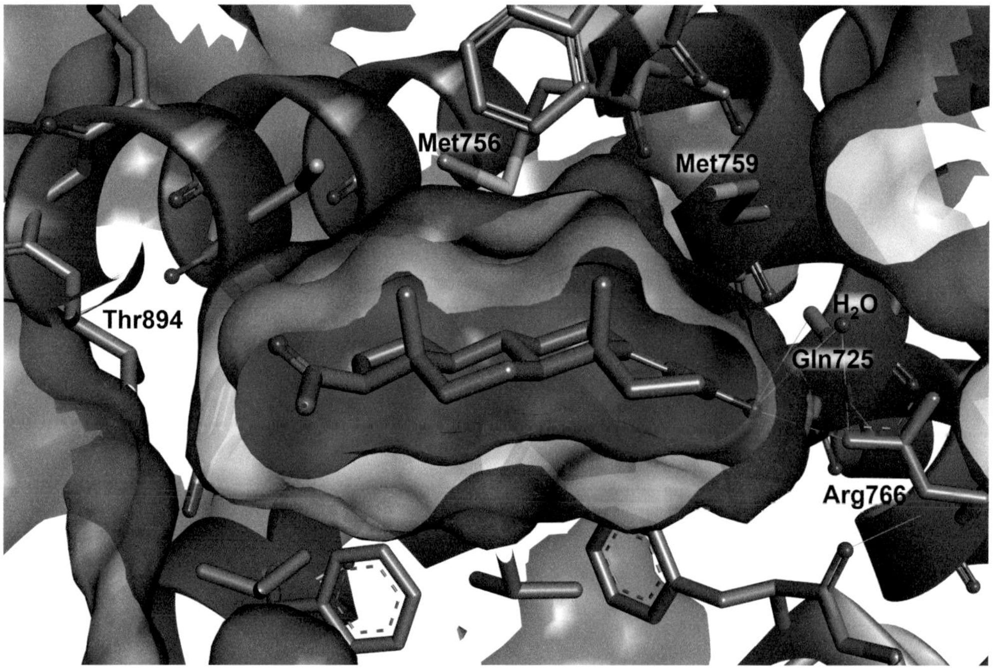

Fig. 28.5 Section of the crystal structure of the progesterone receptor with bound progesterone (surface is *green*, interior is *blue*). The carbonyl group on the partially hydrogenated A-ring accepts an H-bond from Gln 725 and binds to Arg 766 through a water molecule. Because of the 19-methyl group above the A-ring, the steroid occupies a larger volume that is limited by Met 756 and Met 759, which have more flexible and therefore better adaptive side chains. (► https://sn.pub/TS469i)

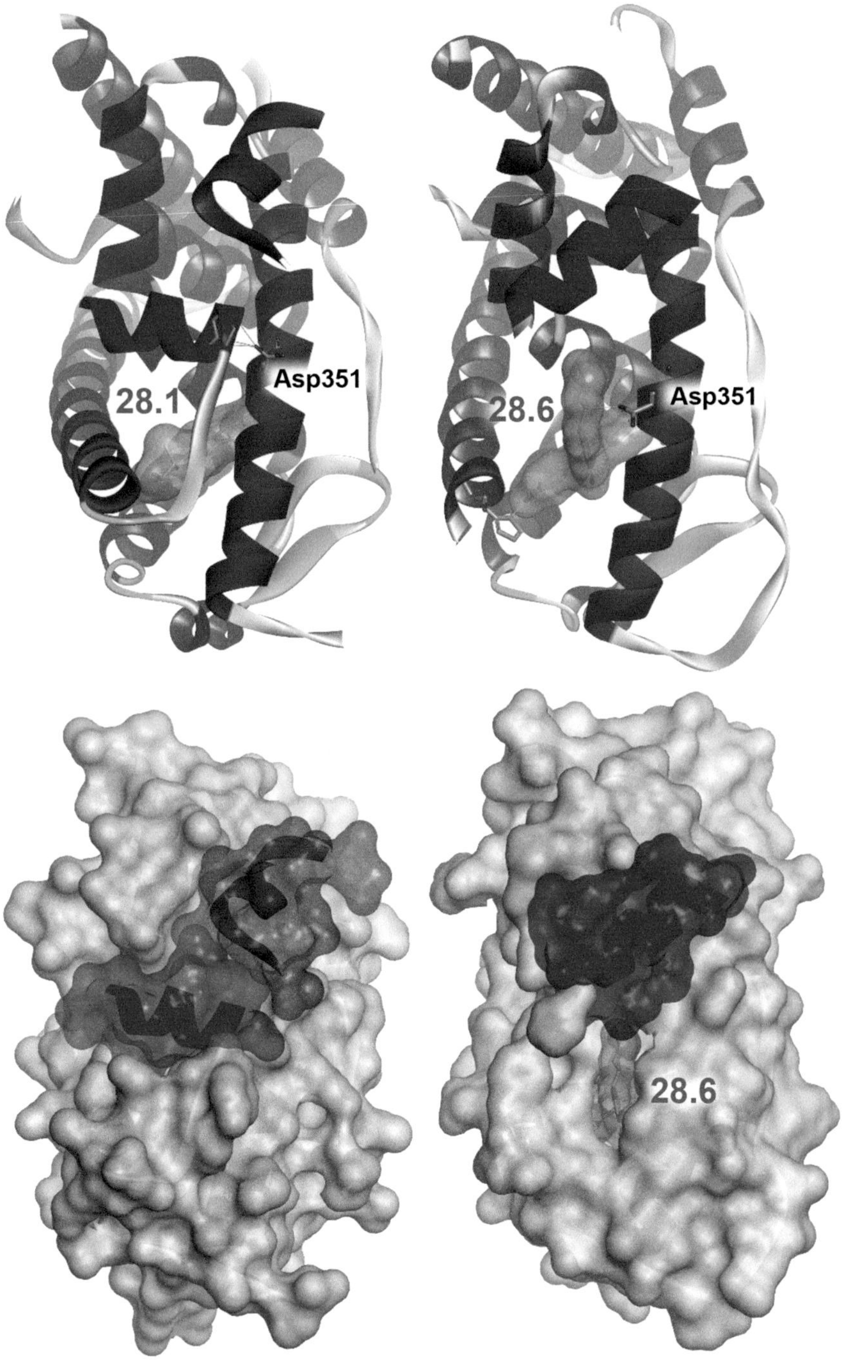

Fig. 28.6 The ligand-binding domain of the nuclear receptors is constructed from 12 helices. Upon binding an agonist such as estradiol **28.1**, the 12th and last helix (*blue*) closes like a gate over the entrance to the ligand-binding pocket (*upper* and *lower row*, *left*). Asp 351 orients on the tip of the helix and stabilizes it in the active position. At the same time, the recognition site is opened for the coactivator with the helical LxxLL motif (*violet*) to bind to the receptor. Upon binding an antagonist such as raloxifene **28.6**, helix 12 cannot close up the entrance channel (*upper* and *lower row*, *right*). The terminal basic group of the antagonists forms a hydrogen bond to Asp 351. (► https://sn.pub/HFjfyW)

The orientation of helix 12 in the active position during agonist binding is an important prerequisite for the availability of the recognition site for the LxxLL motif on the surface of the coactivator. Cocrystallization of the 11-membered peptide with the estradiol-bound receptor has been achieved (Fig. 28.8). The peptide adopts a helical geometry and aligns the critical leucine residues in a hydrophobic groove on the receptor surface. Three amino acids from helix 12 help to guide part of this surface. Once again, a negatively charged carboxylate group, in this case Glu 448 on helix 12, is positioned on the opposite side of the *N*-terminal end of the helical segment of the LxxLL peptide. Again, the electrostatic interaction stabilizes the intermolecular contact.

28.5 Agonists and Antagonists of Steroid Hormone Receptors

Steroid hormones are produced in endocrine adenocytes, such as in the adrenal glands, testes, or ovaries, and released into the bloodstream. There they circulate freely, often by binding to a transport protein. Far from the site of production, they reach the target cells for which the signal is intended. Their lipophilic nature allows them to passively cross membranes. Once in the cytosol, they bind to the appropriate steroid receptor. There are five classes of steroid receptors: glucocorticoid, mineralocorticoid, androgen, estrogen, and progesterone receptors. Two subtypes of estrogen receptor (α-ER and β-ER) have been discovered, which differ in the exchange of a leucine for a methionine and a methionine for an isoleucine in the vicinity of the binding site of the C- and D-ring of the steroid scaffold. The binding affinity to their receptors is extremely high, typically 0.05–50 nM. As a result of binding, the gene expression described in the previous sections is initiated. The cellular response to these processes occurs within hours to days. In addition to this direct control of gene expression, steroid hormones can also initiate rapid regulatory processes in cells. This is done by binding to receptors on the outside of the cell. These receptors, which belong to the class of G-protein-coupled receptors or dimerizing receptors with a tyrosine kinase domain, are discussed in Chap. 29.

As an example, the function of the **estrogen receptor** will be examined in more detail. Estrogen controls the menstrual cycle in women of childbearing age. In addition to this function, estrogen reduces the risk of coronary heart disease and helps maintain bone density. After menopause, at an age of about 50 years, the ovaries stop producing estrogen so that women at this age have an increased risk of coronary heart disease and osteoporosis. Overall, the body's hormonal homeostasis must find a new balance. This is often accompanied by unpleasant physical and psychological symptoms of menopause. Hormone replacement therapy was proposed as a solution in the 1960s. The body is supplied with estradiol **28.1** or an analogous receptor agonist. For example, diethyl-

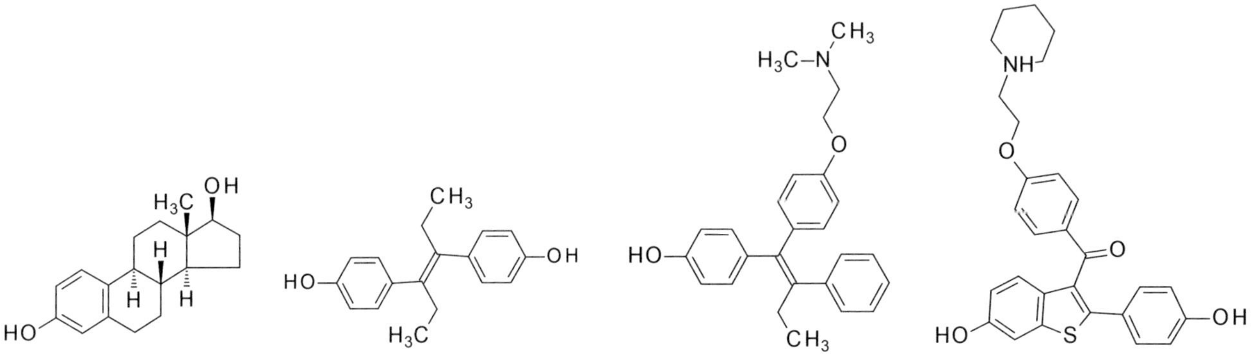

28.1 Estradiol 28.4 Diethylstilbestrol 4-OH 28.5 4-Hydroxy-Tamoxifen 28.6 Raloxifene

Fig. 28.7 Estradiol **28.1** and diethylstilbestrol **28.4** are estrogen receptor agonists, while the 4-hydroxy form of tamoxifen **28.5** and raloxifene **28.6** are antagonists

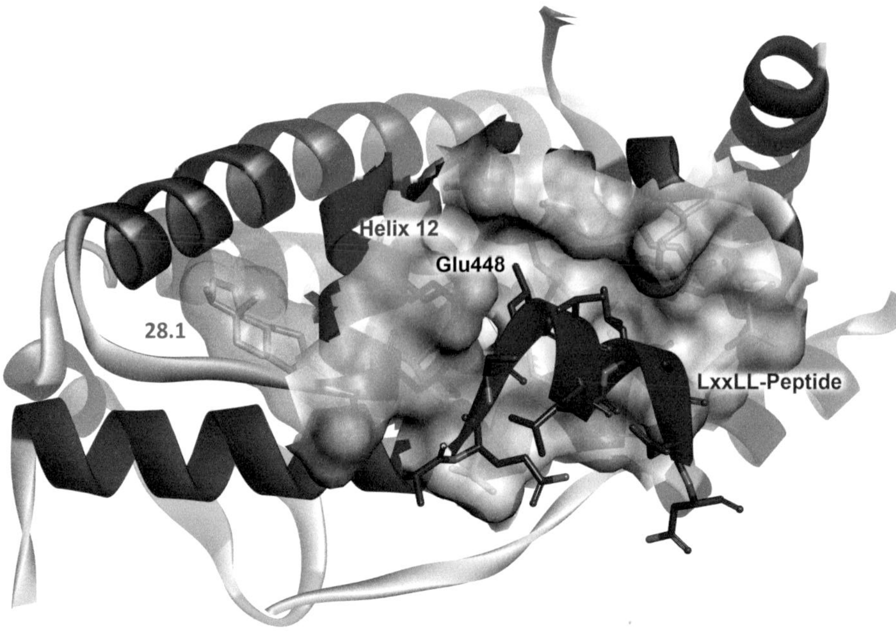

Fig. 28.8 The recognition site of the LxxLL motif on the surface of the coactivator in this crystal structure is reflected in the 11-membered peptide with the estradiol-bound receptor. The peptide takes on a helical geometry and orients its three leucine residues in the hydrophobic groove on the surface. Three amino acids of helix 12 (*blue*) form a part of this surface. Glu 488 on helix 12 binds to the LxxLL motif at the tip of the *N*-terminal end of the helix. (▸ https://sn.pub/5P5ZWZ)

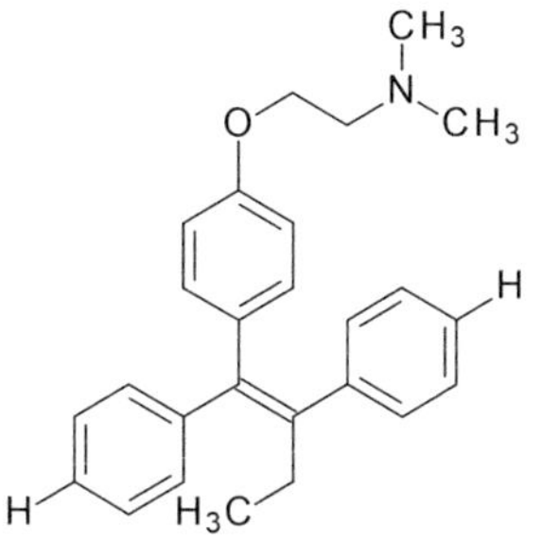

28.5 Tamoxifen

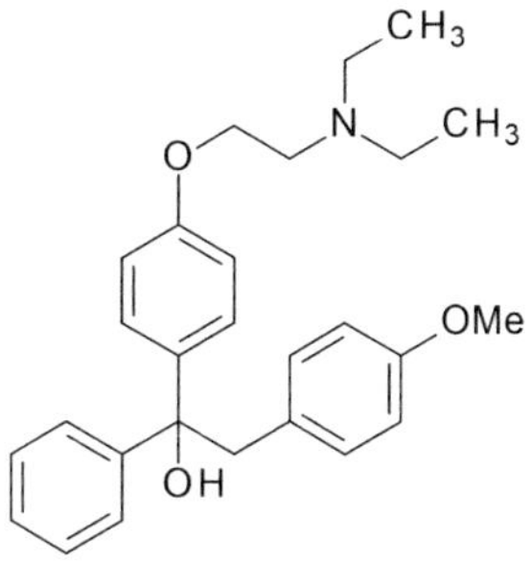

28.7

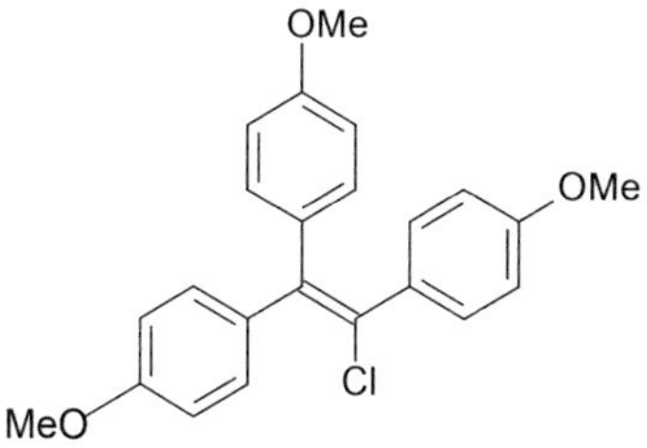

28.8

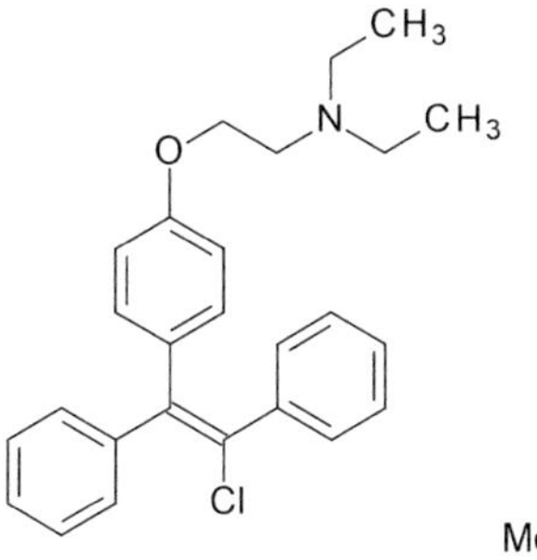

28.9 Clomiphen

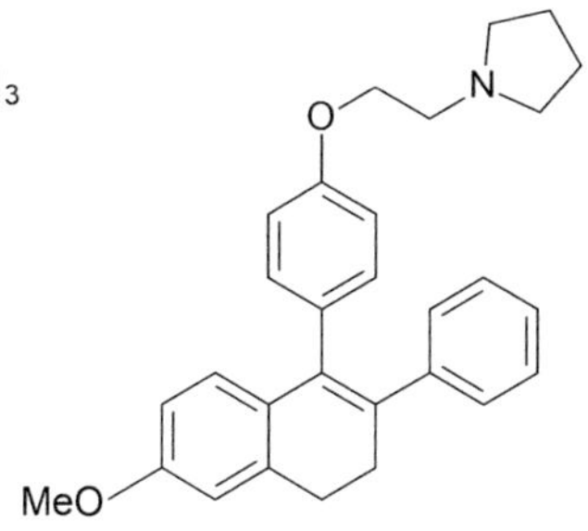

28.10 Nafoxidine

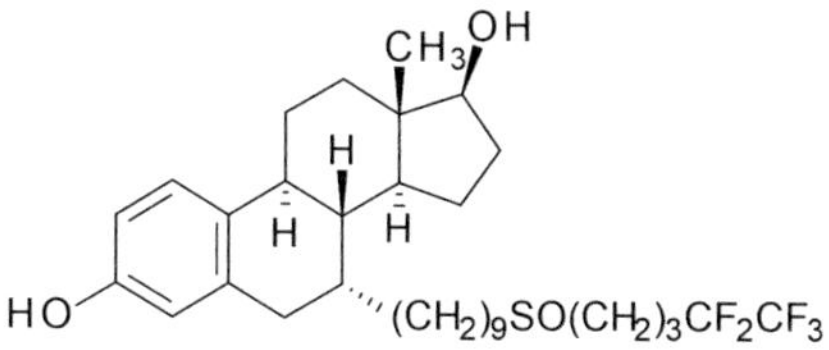

28.11 Fulvestrant

Fig. 28.9 Tamoxifen **28.5** was developed from compound **28.7**, which originated in cardiovascular research. The marketed product has a hydrogen atom in the 4-position, but the actual active substance is the metabolic oxidation product, the 4-hydroxy derivative of **28.5**. Fulvestrant **28.11** does not show the same resistance that has been observed with tamoxifen

stilbestrol **28.4**, which is related except that it lacks a steroid scaffold, was once used but is no longer prescribed because of an increased risk of cancer.

Long-term use of hormone replacement therapy significantly increases the risk of breast cancer. This devastating finding was demonstrated in a study involving one million nurses in the United States. The relationship between ovarian function and the development of breast cancer was described over a hundred years ago. In 1936, Antonie Lacassagne speculated that the action of estrogen antagonists could lead to the prevention of breast cancer. Once again, the discovery of the first antagonists was purely serendipitous. Compound **28.7** was synthesized in the late 1950s in Merrel, USA, as part of a cardiovascular research program (Fig. 28.9). Because of its chemical similarity to **28.8**, a synthetic estrogen surrogate known at the time, it was also tested in an estrogen activity test. This effect was not seen, but rather the opposite: antiestrogen activity. Clomiphene **28.9** was obtained by slight structural modification. This compound was introduced in the 1960s to induce ovulation for the treatment of female infertility. It was not initially intended to be used to prevent breast cancer. Development of nafoxidine **28.10** was also abandoned because of severe side effects. In England, ICI had been pursuing a program since 1940 to develop nonsteroidal estrogen substitutes for breast cancer therapy. Given the focus on contraceptives in the 1970s, it was a stroke of luck that tamoxifen **28.5** emerged from this program in 1973 and was approved for the treatment of breast cancer. The compound quickly proved to be a breakthrough in the treatment of breast cancer. Today, it is estimated that the use of tamoxifen has saved the lives of one million women each year in the industrialized world.

The fact that tamoxifen is a prodrug was only discovered later. The actual active ingredient is obtained by hydroxylation at the 4-position with a CYP enzyme (Sect. 27.6). It stood ripe for further development, which resulted in **raloxifene 28.6** (Fig. 28.7). All derivatives with an antagonistic profile carry a side chain with a basic group. As explained in Sect. 28.4, this side chain blocks the refolding of helix 12 into the active position. The example of raloxifene also shows how complex the effects on the whole organism can be. Raloxifene was originally developed for the treatment of breast cancer. However, this goal was abandoned in the late 1980s because the compound showed no advantages over tamoxifen. However, it proved to be an effective drug in the treatment and prevention of osteoporosis. It also reduced the risk of breast cancer. Raloxifene is classified as a selective estrogen receptor modulator (SERM). Compounds with such a profile are believed to have great potential as hormone replacement therapy without increasing the risk of osteoporosis, coronary heart disease, or breast cancer.

Often, the full profile of a compound becomes apparent only after long-term use. Tamoxifen had the disturbing result that 50% of breast tumors started to grow again after long-term therapy. The development of resistance is explained by the fact that the estrogen receptor is phosphorylated by protein kinase A. This does not pre-

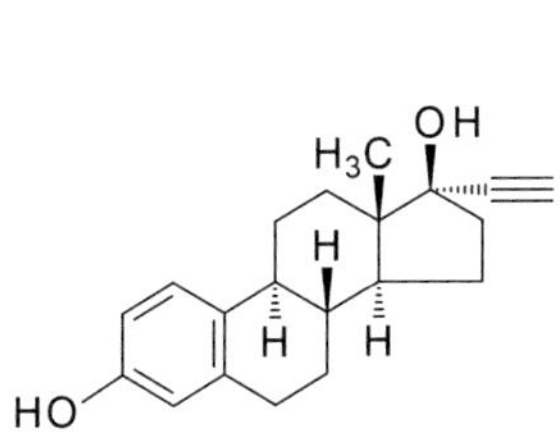

28.12 Ethinylestradiol

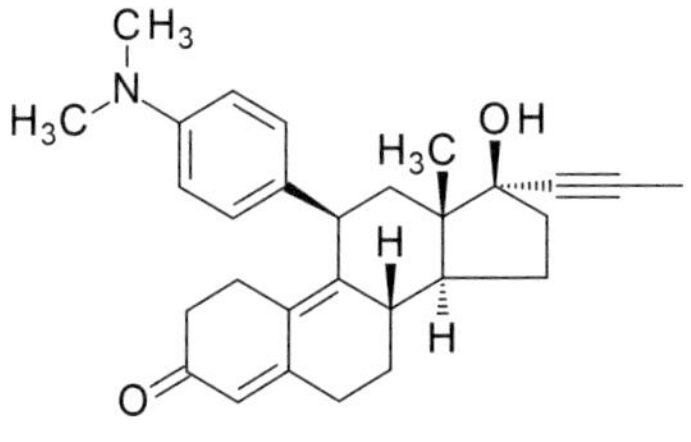

28.13 Mifepriston RU486

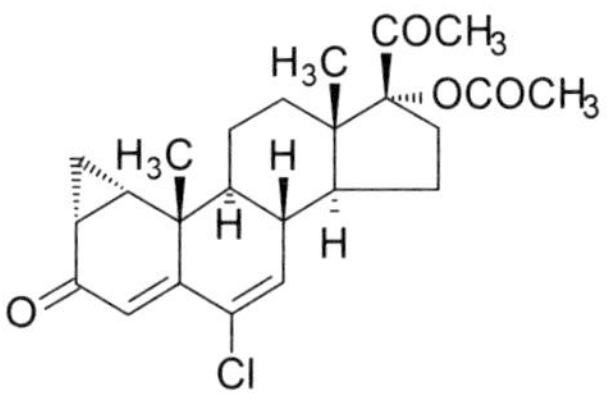

28.14 Cyproterone Acetate

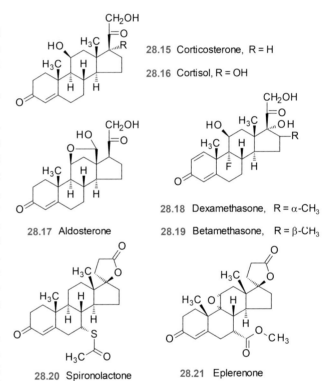

◼ **Fig. 28.10** The introduction of a 17β-ethinyl group leads to orally active steroids, for example, ethinylestradiol **28.12**. The progesterone receptor antagonist mifepristone **28.13** acts as an antigestagen: a "morning after pill." The antiandrogen cyproterone acetate **28.14** has gained importance in the specific therapy of prostate cancer

vent tamoxifen from binding, but the antagonistic effect is reversed. Fulvestrant **28.11** appears to offer a solution to this problem, as resistance has not yet been observed (◼ Fig. 28.9).

The biological functions of the estrogen and **progesterone receptors** are closely linked. While estrogen **28.1** (follicle hormone) promotes and controls egg maturation in the proliferation phase and indirectly induces ovulation, progesterone **28.2** (*corpus luteum* hormone) is formed in the secretory phase of the menstrual cycle. It controls cyclic changes in the uterus and endometrium, reduces fertility, and maintains an already intact pregnancy. Gestagens, progesterone receptor agonists, and estrogen derivatives were introduced as contraceptive hormones (◼ Fig. 28.10). In the 1950s, Carl Djerassi and Gregory Pincus laid the foundation for **oral contraception**. It is based on the timed administration of a combination of an estrogen and a gestagen, which suppresses ovulation, the release of a mature egg, at mid-cycle. A progesterone antagonist, **mifepristone 28.13** (RU486), which, like estrogen antagonists, has a nitrogen function on its side chain, was discovered at Roussel Uclaf during the search for glucocorticoid receptor antagonists. Its use as the "**morning after pill**" is highly controversial in many countries due to its antigestagenic effects. To terminate an intact pregnancy, a single dose of 600 mg of mifepristone is administered, followed 36–48 h later by a prostaglandin to induce uterine contractions. This combination leads in 96% of cases to an abortion by the 7th week of pregnancy. Side effects may include prolonged bleeding and, in rare cases, cardiac dysfunction. Opponents of this substance can take comfort in the fact that, for these reasons alone, it is not suitable for widespread use.

The male hormone testosterone **28.3** acts as an agonist on the androgen receptor. It is responsible for the development of secondary male characteristics, intervenes in the process of spermatogenesis, and regulates protein synthesis. The ability of androgens to increase the size of skeletal muscle cells has led to their use as an anabolic hormone to improve performance in competitive sports, bodybuilding, and animal breeding. Antiandrogens such as cyproterone acetate **28.14** are used in the treatment of prostate cancer.

In addition to sex hormones, there are other active substances in the steroid class. In addition to the cardiac

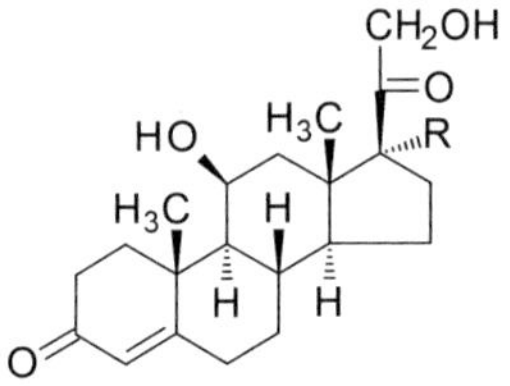

28.15 Corticosterone, R = H

28.16 Cortisol, R = OH

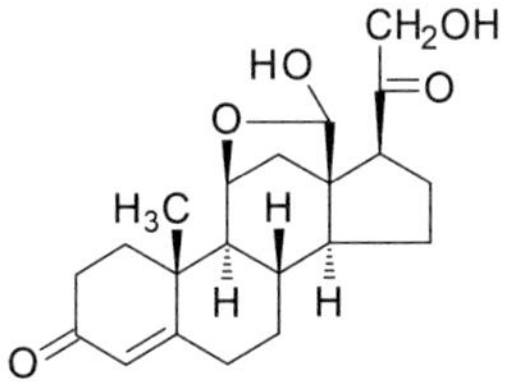

28.17 Aldosterone

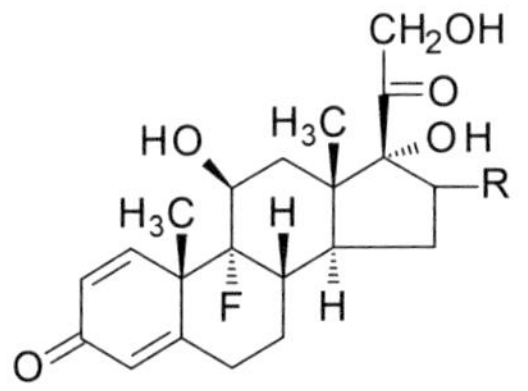

28.18 Dexamethasone, R = α-CH_3

28.19 Betamethasone, R = β-CH_3

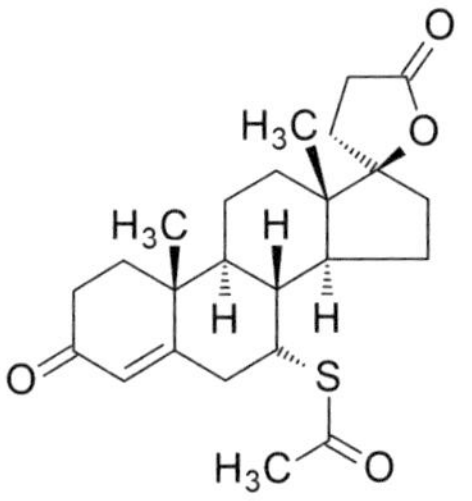

28.20 Spironolactone

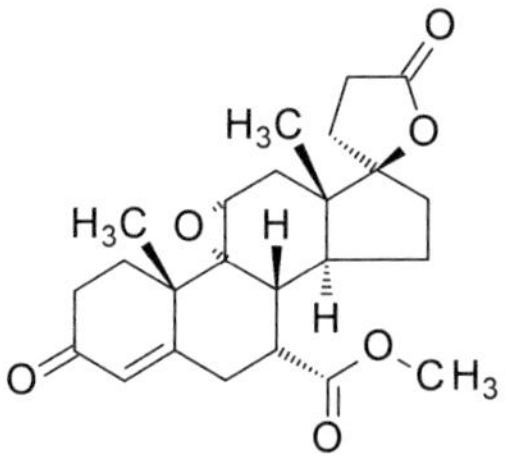

28.21 Eplerenone

◼ **Fig. 28.11** Corticosterone **28.15** and cortisol (hydrocortisone) **28.16** are glucocorticosteroids. They regulate the release of glucose, both by stimulating gluconeogenesis and by inhibiting its metabolic degradation. A stress-induced release of cortisol leads to rapid release of glucose as an energy source. The mineralocorticoid aldosterone **28.17** is responsible for the regulation of the water and electrolyte homeostasis. The naturally occurring glucocorticoids act in an anti-inflammatory manner, but they have mineralocorticoid side effects. Dexamethasone **28.18** and betamethasone **28.19** are "pure" glucocorticoids. They have 30-times stronger anti-inflammatory activity and the mineralocorticoid side effects of cortisol are absent. The diuretic spironolactone **28.20** achieves its effect by a competitive displacement of aldosterone from its receptor. Eplerenone **28.21** is a mineralocorticoid receptor antagonist and is used for the treatment of hypertension and congestive heart failure glycosides found in plants, the adrenal corticosteroids or corticoids are of great importance. If the adrenal glands fail, the absence of these substances can lead to death, or in the case of under- or overfunctioning of the adrenal glands, to severe illness. They are classified into glucocorticoids and mineralocorticoids by their binding to the respective nuclear receptors. The basic scaffold is very closely related to progesterone **28.2**, although they carry more functional groups (**28.15**–**28.17**; ◘ Fig. 28.11). The natural agonists of both receptors are cortisol **28.16** and aldosterone **28.17**. The therapeutic importance of glucocorticoids was initially underestimated. It was not until specific drugs without mineralocorticoid side effects such as dexamethasone **28.18** and betamethasone **28.19** became available that broad therapeutic application became possible. Glucocorticoids affect metabolism, interfere with water and electrolyte homeostasis, and influence the cardiovascular and nervous systems. They are anti-inflammatory, immunosuppressive and anti-allergic. Highly active forms are used in emergency cases of anaphylactic shock or sepsis. They also have serious side effects. Their use requires strict attention to indication and dosage.

Mineralocorticoids affect water and electrolyte homeostasis. They increase the absorption of sodium ions in the kidney and increase the excretion of potassium. Ligands for the mineralocorticoid receptor can be used as diuretics. Potassium-sparing diuresis can be achieved with the structurally related spironolactone **28.20** which competitively displaces aldosterone from its receptor. The selective antagonist eplerenone **28.21** is used as a selective agent in the treatment of hypertension and congestive heart failure.

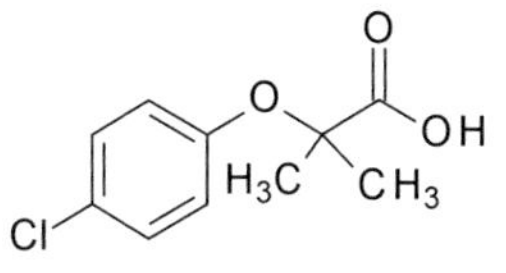
28.22 Clofibric Acid

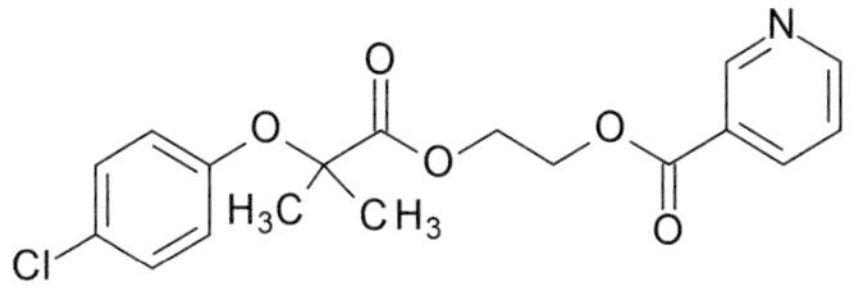
28.23 Etofibrate

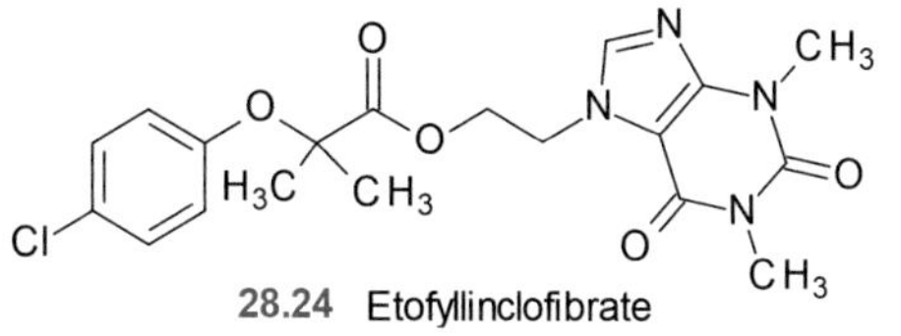
28.24 Etofyllinclofibrate

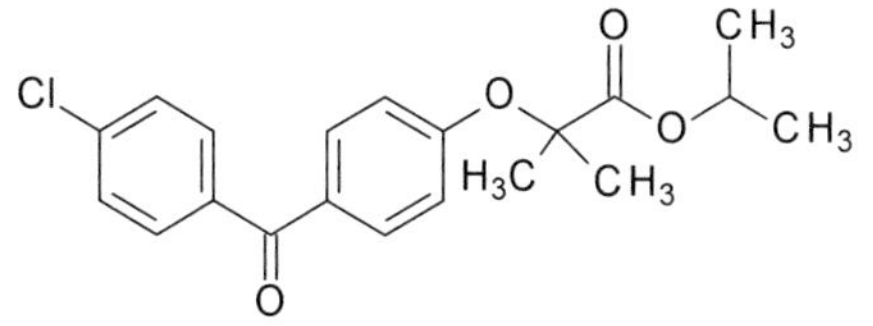
28.25 Fenofibrate

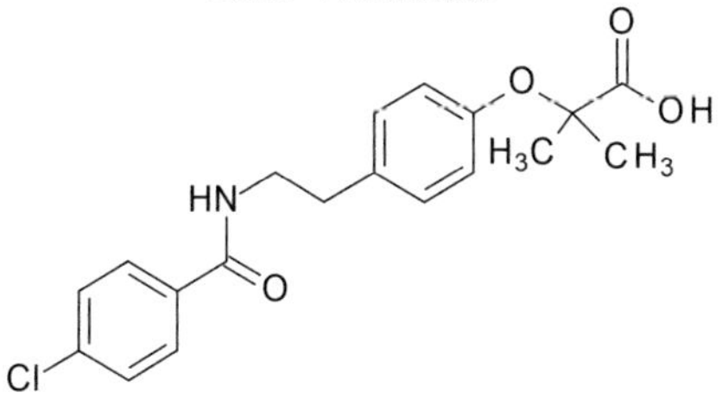
28.26 Bezafibrate

◘ **Fig. 28.12** Activation of the peroxisomal proliferator-activated receptor PPARα increases fat metabolism. Ligands such as the fibrates **28.22**–**28.26** activate this receptor and act as lipid-lowering agents

28.6 Ligands of PPAR Receptors

Of the RXR heterodimer receptors, **peroxisomal proliferator-activated receptors (PPARs)** have become important drug targets. Three subtypes are distinguished, PPARα, PPARβ/δ, and PPARγ, of which three isoforms are known for the γ-type. Originally, PPARβ was described as a subtype. The δ-form was thought to be a fourth subtype. However, it turned out that there are only three types in the human genome. The term PPARβ/δ has, therefore, become established. Their natural ligands are metabolites of fatty acids, prostaglandins, leukotrienes, cholesterol, and bile acids. These receptors serve as sensors to control biosynthesis and metabolism in lipid homeostasis. They are also involved in the release of cytokines such as TNF-α and other mediators from adipocytes.

PPARα is mainly found in the liver. Its activation increases the **degradation of fatty acids** in this organ. PPARα is found primarily in brown adipose tissue and in organs such as the liver, kidney, heart, and skeletal muscle. Artificial ligands for this receptor type are lipid-lowering compounds from the group of **fibrates 28.22–28.26** (◘ Fig. 28.12). A crystal structure with the bound agonist **28.27** and antagonist **28.28** was determined for the PPARα receptor (◘ Fig. 28.13). As in the case of the estrogen receptor, it is again helix 12 that orients itself over the entrance gate of the ligand upon agonist binding. The terminal acid group of the agonist forms a hydrogen bond to Tyr 464 and stabilizes helix 12 in the active position. Antagonist **28.28** is extended by a propionamide group. It blocks the refolding of helix 12 to the active position. In the unfolded geometry, it fits into a different region of the receptor surface. It also controls the release of the binding site for corepressors and coactivators. Helix 12 is capable of forming different conformations in these receptors, depending on the ligand bound. This suggests that the simple classification of ligands as agonists or antagonists is too simplistic. Rather, we are dealing with **ligand-specific transcriptional control**. Furthermore, helix 12 has a much more complex

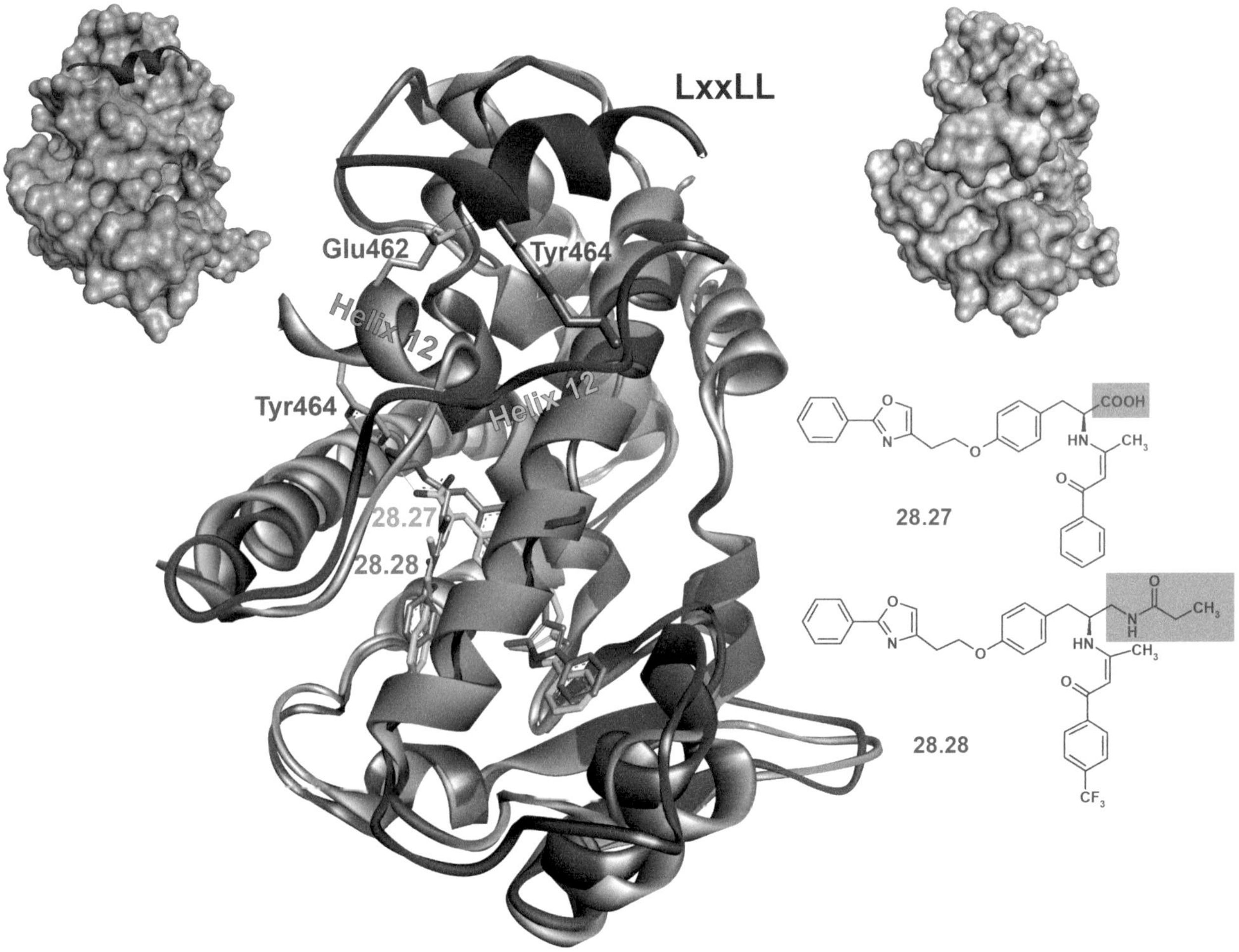

Fig. 28.13 Superimposed crystal structures of the PPARα receptor with bound agonists (*green*) and antagonists (*brown*). The terminal acid group of agonist **28.27** (*light-green*) forms a hydrogen bond to Tyr 464 and stabilizes helix 12 (*oriented to the left*) in the active position. This provides access to the recognition site for the LxxLL peptide motif of the coactivator (*violet*). Glu 462 stabilizes the helical segment of this peptide strand. The antagonist **28.28** is extended by a propionic acid amide group. The binding pose of this ligand (*light-brown*) blocks the placement of helix 12 into the active position. Helix 12 remains unfolded and adopts now an *orientation to the right*. (► https://sn.pub/iqtOSx)

role than merely acting as a switch between active and inactive receptors.

In the case of the **PPARγ** receptor, substances with an **agonist profile** can **increase glucose metabolism** by decreasing the free concentration of fatty acids and the release of mediators that inhibit insulin secretion. As a result, they **counteract insulin resistance**. This is believed to be the main cause of the massive increase in type 2 diabetes. The γ-subtype is present in all adipose tissues. A number of thiazolidinedione derivatives have been developed as PPARγ agonists. The starting point was the lipid-lowering agent clofibrate **28.22** which is a PPARα agonist (◘ Fig. 28.14). In 1979, ciglitazone **28.31**, the first insulin sensitizer, was developed at Takeda via compounds **28.29** and **28.30**. GlaxoSmithKline followed with rosiglitazone **28.32** and Takeda introduced another compound, pioglitazone **28.33**, shortly thereafter. Both are highly selective for PPARγ. They are administered as racemates because the stereocenter is isomerized in the organism. However, the crystal structure with the bound ligand showed that the *S*-enantiomer is bound by the receptor. It was also possible to cocrystallize a peptide with the LxxLL recognition motif with this structure. Again, helix 12 makes the binding pocket available in the active position and stabilizes the helical segment of the LxxLL recognition peptide by positioning Glu 471. The action of glitazones is often associated with problematic side effects. For this reason, their use in therapy is nowadays limited. Rosiglitazone was withdrawn from the market due to an increased risk of myocardial infarction. Pioglitazone is suspected of causing bladder cancer.

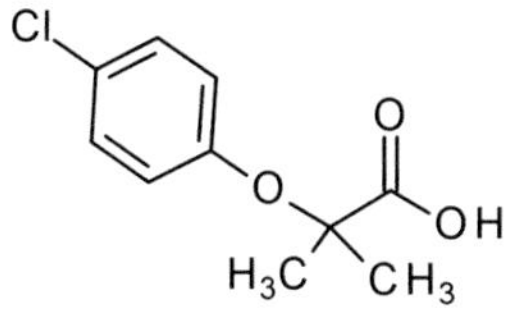

28.22 Clofibric Acid

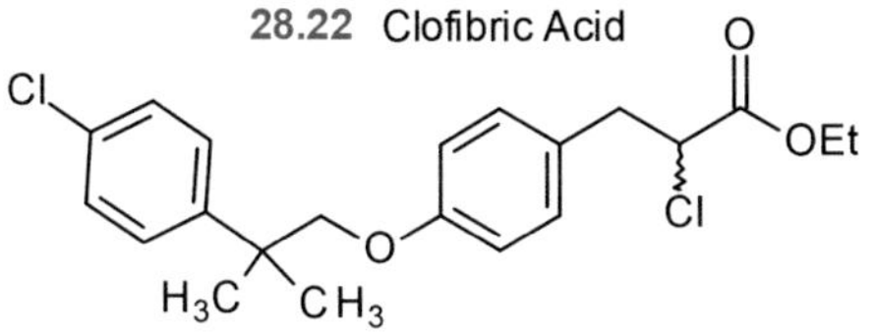

28.29

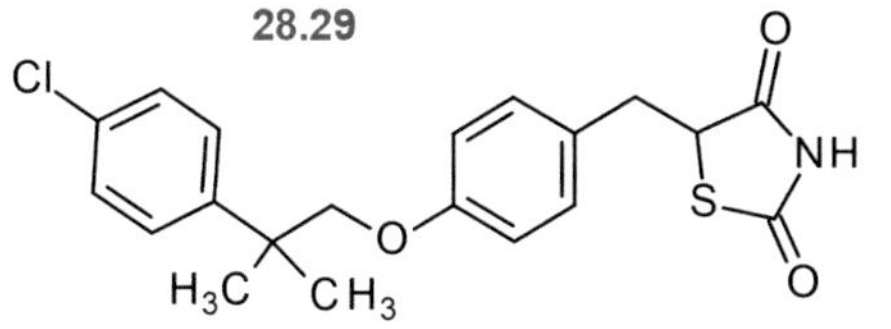

28.30

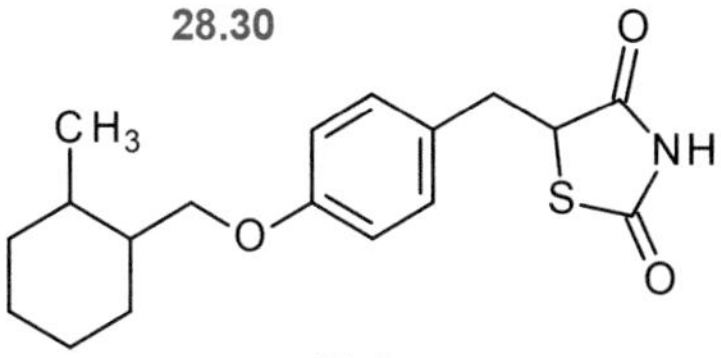

28.31 Ciglitazone

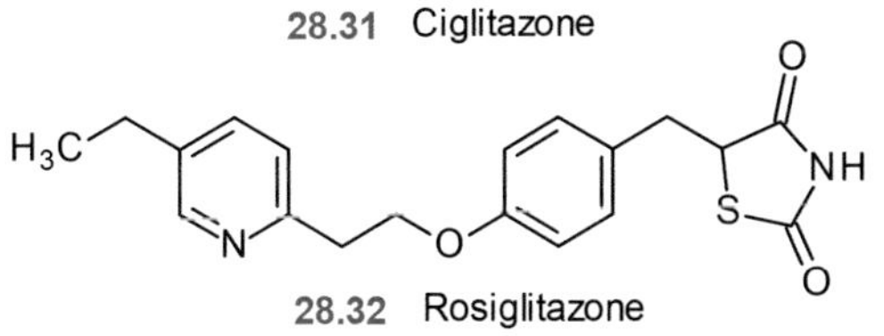

28.32 Rosiglitazone

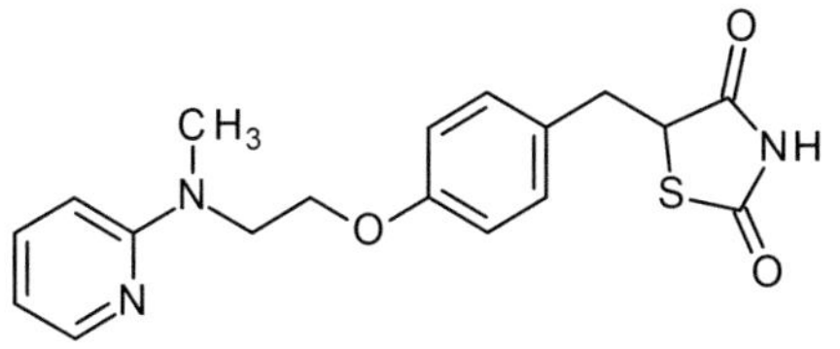

28.33 Pioglitazone

Fig. 28.14 The insulin sensitizer ciglitazone **28.31** was developed at Takeda as a PPARγ ligand through compounds **28.29** and **28.30**, which themselves were developed from the lipid-lowering drug clofibrate **28.22**, a PPARγ agonist. Rosiglitazone **28.32** and pioglitazone **28.33** act on the same receptor

PPARs are also a potential target for cancer therapy. In addition to unsaturated fatty acids such as eicosapentaenoic acid, linolenic acid, or arachidonic acid, PPARβ/δ is also controlled by prostacyclin (Sect. 27.9) as a natural ligand. This receptor is present in almost all tissues. Its expression is regulated by several oncogenic signaling pathways. It is often overexpressed in tumor cells. It promotes the proliferation of tumor cells, in part by inhibiting apoptosis. Antagonists of this receptor could, therefore, represent a new concept for the development of antitumor agents.

28.7 Ligands of Nuclear Receptors Stimulate Metabolism

Cytochrome P450s were discussed as metabolic enzymes in Sect. 27.6. They perform the initial oxidative attack on exogenous xenobiotics. They attach polar groups to lipophilic compounds to prepare them for renal elimination. Drugs can also induce their own **metabolism** or that of other xenobiotics. This property is based on the increased expression of cytochrome P450 enzymes in liver and gastrointestinal cells. This process is mediated by the nuclear receptors **PXR (pregnane X receptor)** and **CAR (constitutive androstane receptor)**. When activated by the binding of an inducing drug, the nuclear receptors bind to a xenobiotic response element in the promoters of specific cytochromes and induce their transcription and expression. The increased biosynthesis of cytochromes results in elevated metabolic activity. This property of certain drugs must be taken into account when prescribing them, especially with regard to possible interactions with other drugs that are simultaneously administered (Sect. 27.7).

The pregnane X receptor can be activated by ligands of very different sizes (Fig. 28.15). For example, phenobarbital **28.34** or the cholesterol-lowering compound SR12813 **28.35** can activate the receptor. Paclitaxel **28.36**, the macrolide rifampicin **28.37**, and the natural compound hyperforin **28.38** from St. John's wort, which are all much larger, fit into the same binding pocket. Molecules with different volumes are apparently able to activate PXR. The steroid receptors were introduced as highly selective proteins that can discriminate between the sole exchange of an OH function for a carbonyl group or between the absence or presence of a methyl group. The architecture of the corresponding receptors allows this high selectivity. The pregnane X receptor belongs to the same family and has an analogous folding. However, Nature has introduced small modifications in the spatial geometry and secondary structural elements that allow this folding architecture to transition **from a highly selective to a promiscuous** type of receptor that now barely distinguishes between "large" and "small" (Fig. 28.16). Forty-five amino acids have been added between helix 1 and 3 of the ligand-binding domain, and helix 2 has been replaced by a multistranded pleated sheet. This structural element is significantly enlarged compared to the estrogen receptor. In addition, helix 6 is unfolded and exists as a long loop. These changes in the architecture of the general folding pattern of nuclear receptors have the consequence that the ligand-binding pocket, located in the center of the protein, is endowed with **pronounced adaptive properties**. This allows PXR

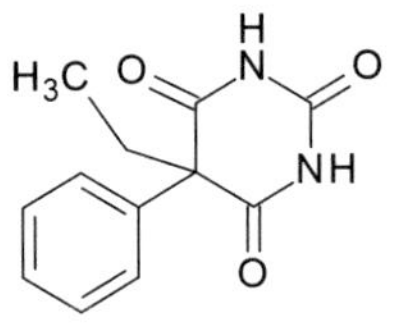

28.34 Phenobarbital

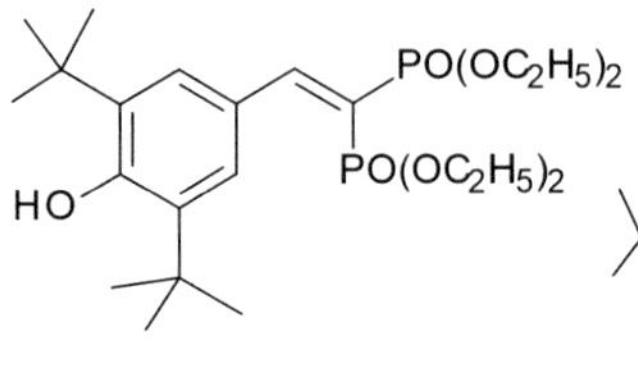

28.35 SR12813

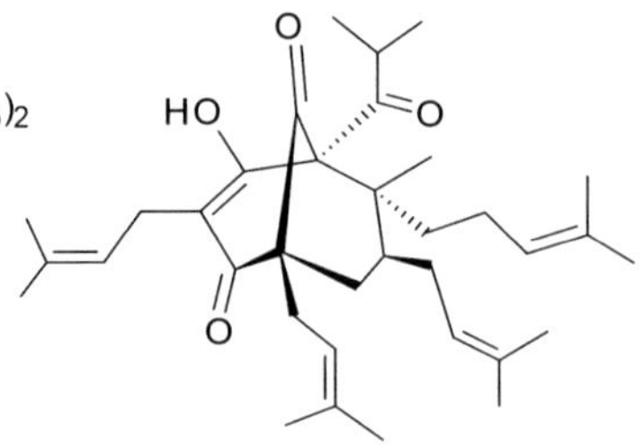

28.38 Hyperforin

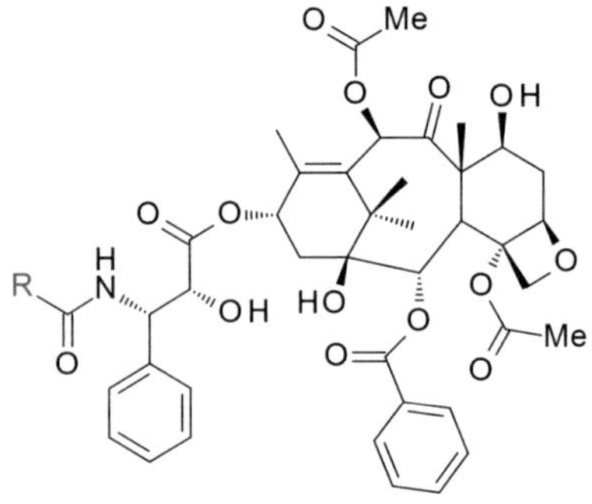

28.36 Paclitaxel R=Phenyl
28.40 Docetaxel R=t-Butyl

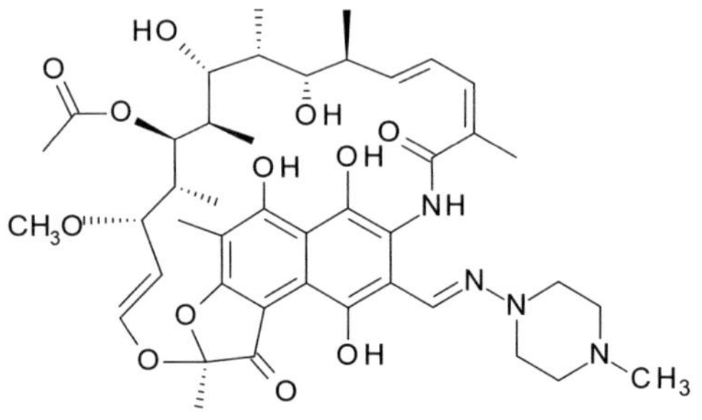

28.37 Rifampicin

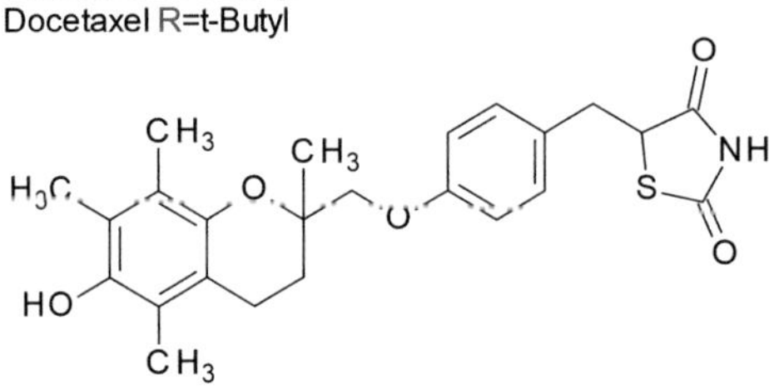

28.39 Troglitazone

Fig. 28.15 Upon binding an activator, the pregnane X receptor induces the expression of cytochrome P450s from the CYP 3A family, which metabolize numerous drugs. Small ligands such as phenobarbital **28.34** and the cholesterol-lowering SR12813 **28.35** as well as large natural products such as paclitaxel **28.36**, hyperforin **28.38**, or the macrolide rifampicin **28.37** activate PXR. The insulin sensitizer troglitazone **28.39** was withdrawn from the market because of its activity on the PXR receptor. Small changes such as the exchange of a phenyl group for a *tert*-butyl group on paclitaxel can be enough to suppress this activating property

to bind to structurally diverse xenobiotics as agonists. The CAR is related to the PXR and is also activated by a large number of chemically diverse ligands such as barbiturates, chlorpromazine, paracetamol, or the heme metabolite bilirubin. Activation may also occur without direct ligand binding by promoting translocation from the cytosol to the nucleus. Crystal structures of CAR suggest that the central binding pocket is about half the size of that of PXR, so activating ligands also appear to be smaller for this receptor (about 600 Å^3 for CAR compared to 1200–1600 Å^3 for PXR).

What are the consequences of these observations that PXR is an activator to induce cytochrome P450 expression? First of all, CYP 3A proteins, which preferentially control the metabolism of drugs, are produced at an accelerated rate. In the last section, the glitazones were discussed as insulin sensitizers. Troglitazone **28.39**, another potent diabetes drug, was withdrawn from the market because it activates PXR. As a result, this compound increased the production of CYP 3A4, which metabolizes troglitazone to a potentially toxic quinone that can lead to liver damage. The same behavior was not observed with rosiglitazone **28.32** or pioglitazone **28.33**. Even paclitaxel activates PXR, so that this chemotherapeutic agent is eliminated from the body at an accelerated rate by the additional CYP 3A4 produced. Replacing the terminal phenyl group with a *tert*-butoxy group leads to docetaxel **28.40**, which can no longer activate PXR. Today, efforts are being made to prevent potential PXR activation that could lead to enhanced CYP 3A4 metabolism early in drug development. The use of a natural product such as hyperforin **28.38** from St. John's wort, which is commonly used to treat mild depression, must be considered because it is a potent PXR activator. It leads to increased metabolism of other drugs such as hormonal contraceptives, HIV protease inhibitors, statins, or coumarin-like anticoagulants. This can greatly reduce the therapeutic success of these other compounds.

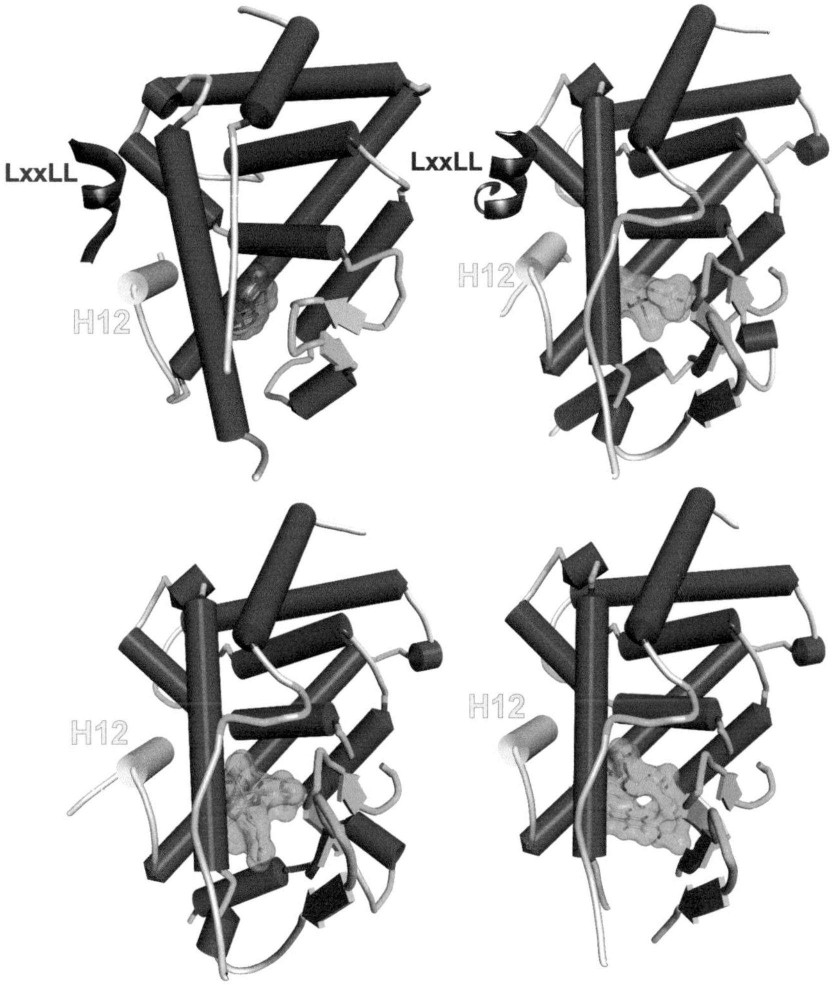

Fig. 28.16 Schematic representation of the polypeptide chain in the highly selective estrogen receptor (*upper row*, *left*) and the promiscuous pregnane X receptor. An insertion of 45 amino acids occurs in PXR that renders the structural portion in the lower right part of the protein extremely adaptive. Because of this, the receptor can bind ligands of very different size. *Upper row*, *right* Crystal structure with bound SR12813 **28.35**. *Lower row*, *left* Crystal structure with bound hyperforin **28.38**. *Lower row*, *right* Crystal structure with bound rifampicin **28.37**. For comparison, the estrogen receptor bound to estradiol is shown (*upper row*, *left*). Here, as in the case of the PXR complex with **28.35**, crystallization was achieved in the presence of a peptide (*purple*) containing the LxxLL sequence. (► https://sn.pub/WGy6OY)

28.8 Synopsis

- The nuclear receptors are a family of 48 members in humans that are present as soluble proteins in the cytosol. They are transcription factors and play an important role in gene regulation. They form either homo- or heterodimers and are activated by small molecules such as steroid hormones, retinoic acid, fatty acids, triiodothyronine, vitamin D, prostaglandins, bile acids, or phospholipids.
- Nuclear receptors exhibit a ligand and a DNA-binding domain; however ligand-independent domains are also involved in the activation of transcription. The activated receptor, stimulated through agonist binding, migrates to the cell nucleus and recruits coactivators, corepressors, and additional transcription factors to regulate gene expression.
- The ligand-binding domains can exhibit impressive selectivity in the recognition of their ligands. Steroid receptors can distinguish the tiny structural differences between male and female hormones in terms of H-bond donor/acceptor functional group exchanges and presence or absence of the 19-methyl group.
- Agonist and antagonist binding induce a different orientation of helix 12, which closes the entrance to the ligand-binding site. Antagonist binding hampers reorientation of this helix across the entrance gate and simultaneously blocks the recognition site for the binding of a helical LxxLL motif found on the surface of the coactivator.
- Agonists and antagonists of the steroid receptors are important drugs which interfere with the menstrual cycle as contraceptives, act in anticancer therapy, show anti-inflammatory, immunosuppressive, or antiallergic activity on the glucocorticoid receptor, or act as diuretics or hypertensive agents on the mineralocorticoid receptor.
- PPAR receptors occur as several subtypes and form heterodimers with the retinoic receptor upon activation. Agonists of the PPARγ receptor can induce an increase in glucose metabolism. They are used as insulin sensitizers in diabetes therapy.
- The transcription and expression of cytochrome P450 enzymes involved in the metabolism of xenobiotics can be regulated by the nuclear receptors PXR and CAR. These receptors can be activated by a variety of

structurally rather diverse xenobiotics, which leads to increased biosynthesis of cytochromes and, therefore, elevated metabolic activity.
- In contrast to the stereochemically highly specific nuclear steroid receptors, the pregnane X receptor shows pronounced promiscuous binding of structurally highly diverse activators. Binding of small to large ligands is accomplished by an additional highly adaptive structural element in this receptor.

Bibliography and Further Reading

General Literature

H. Gronemeyer, J.-Å. Gustafsson and V. Laudet, Principles for Modulation of the Nuclear Receptor Superfamily, Nat. Rev. Drug Discov., **3**, 950–964 (2003)

J. T. Moore, J. L. Collins and K. H. Pearce, The Nuclear Receptor Superfamily and Drug Discovery, ChemMedChem, **1**, 504–523 (2006)

E. Ottow and H. Weinmann, Eds., Nuclear Receptors as Drug Targets (Vol. 39 in Methods and Principles in Medicinal Chemistry, R. Mannhold, H. Kubinyi and G. Folkers, Eds.), Wiley-VCH, Weinheim (2008)

L. Michalik et al. International Union of Pharmacology. LXI. Peroxisome Proliferator-Activated Receptors. Pharmacol. Rev., **58**, 726–741 (2006)

Special Literature

L. F. Fieser, M. Fieser, Steroids, Reinhold, New York (1959)

R. Hirschmann, Medicinal chemistry in the golden age of biology: lessons from steroid and peptide research. Angew. Chem. Int. Ed. Engl., **30**, 1278–1301 (1991)

Y. E. Timsit, M. Negishi, CAR and PXR: The xenobiotic-sensing receptors. Steroids, **72**, 231–246 (2007)

V. Chandra, P. Huang, Y. Hamuro, S. Raghuram, Y. Wang, T. P. Burris, F. Rastinejad, Structure of the Intact PPAR-γ–RXR-α Nuclear Receptor Complex on DNA, Nature, **456** 350–357 (2008)

A. Lacassagne, Hormonal pathogenesis of adenocarcinoma of the breast, Am. J. Cancer, **27**, 217–225 (1936)

V. C. Jordan, Tamoxifen: A Most Unlikely Pioneering Medicine, Nat. Rev. Drug Discov., **2**, 205–213 (2003)

J. Owens, Growing Concern for Tamoxifen, Nat. Rev. Drug Discov., **3**, 647 (2004)

C. Djerassi, The Pill, Pygmy Chimps, and Degas' Horse: The Autobiography of Carl Djerassi. BasicBooks, New York (1992)

T. M. Willson and S. A. Kliewer, PXR, CAR and Drug Metabolism, Nat. Rev. Drug Discov., **1**, 259–266 (2002)

V. Zoete, A. Grosdidier, O. Michelin, Peroxisome proliferator-activated receptor structures: Ligand specificity, molecular switch and interactions with regulators. BBA-Mol Cell. Biol., **L 1771**, 915–925 (2007)

C. D. Buchman, S. C. Chai, T. Chen, A Current Structural Perspective on PXR and CAR in Drug Metabolism, Exp. Opin. Drug Metab. Toxicol., **14**, 635–647 (2018)

Agonists and Antagonists of Membrane-Bound Receptors

Contents

G. Klebe, *Drug Design,* https://doi.org/10.1007/978-3-662-68998-1_29

Messenger molecules are responsible for **carrying and transmitting information** between cells. These messengers can be as small as single ions or as large as signal peptides or even proteins. Even light quanta can serve this purpose. To transmit their signals to the cell, they bind extracellularly as ligands to a membrane-bound receptor. There is hardly any other way for these messengers to transmit their signals, because messenger molecules such as dopamine, histidine, or adrenaline, but also peptides and proteins such as insulin, interleukins, angiotensin, endothelin, or neurokinin cannot cross the cell membrane. Inside the cell, the receptors transmit the signal of ligand binding by changing their conformational state. Upon activation, the bound ligand stabilizes an active conformation of the receptor. However, ligand binding can also leave the conformational equilibrium unchanged or stabilize the receptor in an inactive conformation. This may result in a lack of signal transduction or a reduction in the basal activity of a receptor. Either of these possibilities may be a worthwhile therapeutic approach. In the former case, we speak of **agonists** (activation), in the latter of **antagonists** (no change) or **inverse agonists** (reduction). A large group of membrane-bound receptors are the **G-protein-coupled receptors** (**GPCRs**), which cross the membrane with seven helices. Agonists stimulate GPCRs to activate the coupled cellular **G-protein**, which triggers subsequent processes within the cell. These subsequent processes can be either activating or inhibitory in nature. A second group consists of receptors that also penetrate the cell membrane. In many cases, their activation requires **dimerization** of the receptor. However, oligomerization to form larger assemblies has also been observed. Many of these receptors carry **intracellular tyrosine kinase domains** that phosphorylate each other for signal transduction. This puts them in a state where a phosphorylation cascade can also turn on the function of downstream proteins. Again, these downstream processes can be either activating or inhibitory in nature. **Cytokines** and **interleukins** bind to another set of oligomeric membrane receptors. They also trigger kinase-dependent signaling pathways intracellularly as a result of ligand binding. Members of the GPCR family are the targets of about one-third of our current drug repertoire. Among these, those stimulated by small natural ligands are particularly targeted. For the other groups of membrane-bound receptors, the number of currently available drugs is much more limited. These receptors are all targeted by large endogenous ligands, making the development of a competing small xenobiotic drug molecule extremely difficult (Sects. 7.10 and 29.8).

29.1 The Family of G-Protein-Coupled Receptors

The **G-protein-coupled receptor** (**GPCR**) family is the largest group of integral membrane proteins in the human genome. Approximately 800 members of this family have been identified, about half of which are receptors that control our olfactory sensation (Sect. 29.7). GPCRs mediate the flow of information across the membrane and respond to a wide variety of extracellular signals. These can be triggered by light, protons or single ions, but also by small biogenic amines (neurotransmitters), hormones, prostaglandins, signaling peptides, and even proteins. Once activated, GPCRs can trigger various cascades of intracellular processes. The transition from an inactive to an active state involves **conformational changes** of the receptor. These states are in thermodynamic equilibrium. The individual conformers that result are different and determine which signaling pathway is activated. Bound ligands, in turn, can select certain conformers and, thus, regulate the activity of the receptors. In the simplest picture, ligands are classified as **agonists** if they stabilize an active state of the receptor. If an agonist does not achieve full activation of the receptor, it is called a **partial agonist**. Most receptors have a basal activity even without a bound agonist. This basal activity of a receptor function can be reduced or even completely suppressed by **inverse agonists**. An **antagonist** can bind to an active or inactive state, but it cannot turn off the **basal activity**. Once bound, it blocks the uptake of an agonist and, thus, prevents its action. In addition to the actual binding site for these endogenous signaling molecules (so-called orthosteric sites), several other binding sites have been identified to which **allosteric modulators** can bind. These are known as negative allosteric modulators (NAM) and positive allosteric modulators (PAM). Molecules interacting at these sites may be of therapeutic interest, as they may have a positive or negative effect on the action of signaling molecules. In any case, the binding of the endogenous signaling molecule is transmitted to the cell by setting a specific receptor conformation.

How is the conformational change of the receptor translated into a cellular function? Intracellularly, a **heterotrimeric G-protein** binds to the receptor, to which the conformational change of the receptor is transmitted. Depending on whether the trimer is an **activating, stimulating or inhibiting G-protein**, very different signaling pathways are triggered in the cell. Four main families of G-proteins are known, classified as G_s, $G_{i/o}$, $G_{q/11}$, or $G_{12/13}$. As long as a guanosine **di**phosphate (GDP) molecule is bound to the α-subunit of the G-protein, Gα forms an inactive heterotrimer with the Gβγ dimer. As a result of receptor activation, the α-subunit exchanges GDP for GTP (guanosine **tri**phosphate). Subsequently, the α-subunit undergoes conformational changes that lead to dissociation of the α- and βγ-subunits. The dissociated subunits can then modulate the activity of several downstream effector proteins. Gα subunits target effectors such as adenylyl cyclases, cGMP phosphodiesterase, phospholipase C, and RhoGEFs (proteins with guanine nucleotide exchange factor activity that regulate small GTPases of the Rho family). In contrast, Gβγ subunits recruit G-protein-coupled receptor kinases to the membrane. They control inwardly rectifying potassium channels, voltage-gated calcium channels, adenylyl cyclases, phospholipase C, phosphoinositol-3-kinase, and G-protein-coupled receptor kinases (GRKs). As long as GTP is bound to the α-subunit, the G-protein is in an active state. Slow hydrolysis of the bound GTP to GDP returns the G-protein to its inactive state. The separated Gα and Gβγ subunits then reassemble into the original inactive trimer. The fact that several subtype families of each subunit are known results in a high degree of molecular diversity. These building blocks can be combined in many different ways to form different trimers.

The receptor remains activated as long as the externally bound ligand is present in the receptor-binding pocket and continues to trigger new G-proteins inside the cell to perform their function. In this way, signal amplification is achieved. On the other hand, the rates of the terminating hydrolysis of GTP to GDP regulate the amount of active G-protein and, thus, contribute to the intensity and duration of the receptor signal. The activated effector proteins can release a so-called **second messenger** into the cell. This second messenger is then responsible for the actual effect, such as the activation of other proteins, especially kinases, or the control of an ion channel. The best-known effector protein, **adenylyl cyclase**, generates adenosine 3′,5′-cyclophosphate (**cAMP**) from adenosine triphosphate (ATP). The resulting cAMP can then activate kinases such as protein kinase A or MAP kinases. It can also have a stimulatory effect on channels. Other second messengers are guanosine 3′,5′-cyclophosphate (**cGMP**), inositol 1,4,5-triphosphate (**IP_3**), diacylglycerol (**DAG**), **arachidonic acid**, or simply **Ca^{2+}** ions. Their formation can be partially triggered by the aforementioned $G_{q/11}$ proteins, which activate phospholipase Cβ. Subsequently, the second messengers DAG and IP_3 are formed in several steps. In addition to the activating G_s proteins, inhibitory $G_{i/o}$ proteins are also known. They inhibit adenylyl cyclase and, thus, reduce the supply of the second messenger cAMP. Another family of G-proteins ($G_{12/13}$) activates Rho proteins that regulate the actin–myosin cytoskeleton. The contraction of muscle cells, for example, is controlled by this signaling pathway.

Recent research has increasingly shown that a particular receptor preferentially triggers one signaling pathway. In many cases, however, the receptor is capable of communicating with different G-proteins. Drugs that are structurally somewhat different from the endogenous ligand can induce modest changes in the conformation and dynamics of the receptor upon binding. This allows them to transmit signaling information into the cell in a slightly different way. The result is a small difference compared to the binding of the endogenous ligand. Such "biased ligands" are designed to induce or suppress a specific function of the targeted GPCR in a controlled and selective manner. In principle, they could potentially avoid unwanted side effects. They are, therefore, currently under intensive investigation. It is hoped that such **biased GPCR ligands** will preferentially activate or deactivate one of several possible signaling pathways after binding to the receptor. In addition, attempts are being made to discover **allosteric binding sites** on the receptors (see above) and to occupy them with ligands that also allow such "biasing" towards a slightly modulated receptor response. It is hoped that this will enrich the range of available pharmaceuticals with substances that have ideally tailored profiles for therapy.

The misregulation of **GPCRs** is associated with **many disease patterns** because of their central role in mediating information during changes in the cell's state and function. Therefore, GPCRs represent common targets for drugs. There are five classes of GPCRs: rhodopsin-like (class A), adhesins (class B1), secretins (class B2), glutamates (class C), and Frizzled/Taste2 (class F). Among these, class A is by far the largest and most important. Numerous subtypes are known for the individual receptors. They differ in their tissue distribution, ligand specificity, and also in the subsequent signaling pathways initiated by the different G-proteins. This fact explains why drugs targeting these receptors have been successful in completely different indications. Adrenaline and noradrenaline (Sect. 1.4) act on the so-called adrenergic receptors. In 1948, Raymond Ahlquist demonstrated that the different effects of adrenaline on diverse organs are attributed to two different types of these receptors, α- and β-receptors. Later, they were subdivided into α_1- and α_2-, and β_1- and β_2-receptors and even further **subtypes**. For example, depending on the tissue, the β_2-adrenergic receptor is associated with asthma, hypertension, or heart attack. These differences helped very much in the development of specific drugs, for example, β-agonists or β-antagonists (β-blockers, Sects. 8.5 and 29.3). The **serotonin receptor** displays an exceedingly complex spectrum of different subtypes; these are also called 5-HT receptors according to the chemical structure of serotonin (5-hydroxytryptamine; ◘ Table 29.1). Their dysregulation is associated with disease patterns such as migraine, pulmonary hypertension, depression, schizophrenia, eating disorders, and nausea and vomiting.

Initially, GPCRs were thought to function as monomers. However, this picture has evolved. Today, it is known that the formation of homo- and heterodimers represents an additional regulatory and controlling signal for the differentiation of the pharmacological cell response. In addition, phosphorylation of the receptor on its inner cell surface and subsequent binding to an **arrestin** can desensitize its function. This step is necessary for receptors to be transported from the cell surface into intracellular vesicles, e.g., for temporary storage. From there, the receptors can either be proteolytically degraded or delivered back to the cell surface when needed. Can these steps also be targeted by extracellularly delivered drugs? This is an exciting question that future drug discovery will have to answer.

Approximately 30% of all drugs on the market today exert their effects on GPCRs. Therefore, it is all the more desirable that more information about the spatial geometry of these structures becomes available. In 2020, 27 small molecules acting on GPCRs were among the 100 best-selling drugs. There are now 475 approved drugs for 108 members of the 398 GPCRs that are not olfactory receptors. These are mainly compounds targeting aminergic and opioid receptors. In 2017, 66 compounds were in clinical trials that mainly target peptide receptors. Given the immense interest in this structural class, there is a particular desire to obtain reliable information on the spatial structure and activation mechanism of GPCRs.

29.2 Rhodopsins Provide the First Models of G-Protein-Coupled Receptors

Crystallographic structure determination of G-protein-coupled receptors has proven to be extremely difficult over the years. As membrane-bound proteins, with the *N*-terminus on the extracellular side and the *C*-terminus in the cytosolic side of the cell, they are not easy to remove from their natural membrane environment and transform into a crystal suitable for structure determination. Special lipid phases are used to stabilize the protein and promote crystal formation. They also have distinct loop regions on either side of the membrane that bridge the individual segments that cross the membrane. These loops are critical for function and also for intact spatial architecture. Furthermore, the production of sufficient quantities of these receptors for structure determination has been a major problem.

Table 29.1 There are 14 different subtypes of serotonin receptors, including one ion channel (**5-HT$_3$**[a]). The therapeutic potential for treating hypertension, migraine, schizophrenia, depression, anxiety, vomiting, and gastrointestinal dysmotility has only been partially exploited so far (according to the IUPHAR Receptor Database. International Union for Basic and Clinical Pharmacology)

Receptor	Gene	Type, function	Possible therapeutic indication
5-HT$_{1A}$	*5-ht$_{1A}$*	G_i/G_o-coupled, inhibitory, reduces cAMP level, inhibits adenylyl cyclase	Treatment of CNS diseases such as anxiety and depression
5-HT$_{1B}$	*5-ht$_{1B}$*	G_i/G_o-coupled, inhibitory, reduces cAMP level, inhibits adenylyl cyclase	Neuronal inflammatory processes, migraine
5-HT$_{1D}$	*5-ht$_{1D\alpha}$*(h), *5-ht$_{1D\beta}$* ≙ *5-ht$_{1B}$* (R)	G_i/G_o-coupled, inhibitory, reduces cAMP level, inhibits adenylyl cyclase	Neuronal inflammatory processes, migraine
5-HT$_{1E}$	*5-ht$_{1E}$*	G_i/G_o-coupled, inhibitory, reduces cAMP level, inhibits of adenylyl cyclase	Neuronal inflammatory processes, migraine
5-HT$_{1F}$	*5-ht$_{1F}$*	G_i/G_o-coupled, inhibitory, reduces cAMP level, inhibits adenylyl cyclase	Neuronal inflammatory processes, migraine
5-HT$_{2A}$	*5-ht$_{2A}$*	$G_{q/11}$-coupled, excitatory, increases IP_3 and DAG level	CNS disease, atypical antipsychotic, wound healing, arterial hypertension
5-HT$_{2B}$	*5-ht$_{2B}$*	$G_{q/11}$-coupled, excitatory, increases IP_3 and DAG level	Influence on smooth muscle of the cardiovascular system
5-HT$_{2C}$	*5-ht$_{2C}$*	$G_{q/11}$-coupled, excitatory, increases IP_3 and DAG level	CNS disease, atypical antipsychotic, wound healing, arterial hypertension
5-HT$_3$	***5-ht$_3$***	**Ligand-gated *ion channel*, depolarization of cell membrane**	**Suppression of cytostatic-induced emesis**
5-HT$_4$	*5-ht$_4$*	G_s-coupled, excitatory, increases cAMP level	Gastrointestinal tract, irritable bowel syndrome
5-HT$_{5A}$, 5-HT$_{5B}$	*5-ht$_{5A}$*, *5-ht$_{5B}$*	G_i/G_o-coupled, inhibitory, reduces cAMP level, inhibits adenylyl cyclase	Circadian rhythm
5-HT$_6$	*5-ht$_6$*	G_s-coupled, excitatory, increases cAMP level	Involved in memory and learning
5-HT$_7$	*5-ht$_7$*	G_s-coupled, excitatory, increases cAMP level	Regulation of the day/night rhythm

HT, *ht* 5-hydroxytryptamine (= serotonin), *R* rat, *h* human, *CNS* central nervous system, *IP3* inositol 1,4,5-triphosphate, *DAG* diacylglycerol

[a] Serotonin receptor subtype that is a ligand-gated ion channel in bold

In 1990, the publication of the structure of **bacteriorhodopsin** provided the first clues to the principle structure of GPCRs, which belong to the group of seven-transmembrane receptors. Using high-resolution electron microscopy (Sect. 13.6), Richard Henderson determined the structure using two-dimensional crystals. Bacteriorhodopsin is not a GPCR *per se*, but rather a proton pump that establishes a pH gradient across the membrane. Like all GPCRs, however, it is made up of seven transmembrane helices. Bacteriorhodopsin shares negligible sequence homology with human GPCRs. Nevertheless, the structure was used extensively in the early 1990s to model a large number of pharmacologically relevant GPCRs. This difficult work stimulated the development of many technical advances at the time to reliably model multiple sequence alignments, model helix properties, and incorporate a large number of mutation data and binding profiles from ligand series.

It was another decade before the first high-resolution structure of a true GPCR was determined. **Bovine rhodopsin** is a particularly favorable case because it occurs in high concentrations in bovine eyes. By diligently collecting bovine eyes from slaughterhouses, sufficient protein material could be obtained. Rhodopsin (also known as visual purple) is stabilized by covalently bound 11-*cis*-retinal. It is a light-regulated receptor. A high-resolution structure determination was performed in the inactive state, where the receptor is found in the dark. No signal would be transmitted to the cell. Years later, the structure of the photoactivated receptor was determined. This structure provided information about the activation process and the conformational change of the receptor. Rhodopsin was also used to determine the structure of a phosphorylated receptor bound to arrestin for the first time (Sect. 29.4).

29.3 Structure of the Human β_2-Adrenergic Receptor

In 2007, the first structure determination of a **human GPCR** was achieved. Two structures of the human **β_2-adrenergic GPCR** were described by Brian Kobilka at Stanford University in collaboration with Raymond Stevens at the Scripps Research Institute in La Jolla, USA, and colleagues at the MRC in Cambridge, England. As mentioned above, producing sufficient quantities of these receptors for structure determination has been a major problem for many years. GPCRs are now expressed from human HEK or insect cells. In addition, the proteins can be specifically modified by mutations or the introduction of fusion proteins (see below) to optimize their expression, stability, and aggregation behavior for biochemical and structural biology studies. Another obstacle was crystallization: the receptors have a high degree of flexibility and proteolytic instability. The third intracellular loop proved to be particularly critical. The scientists had to use a trick. A specific antibody was found that

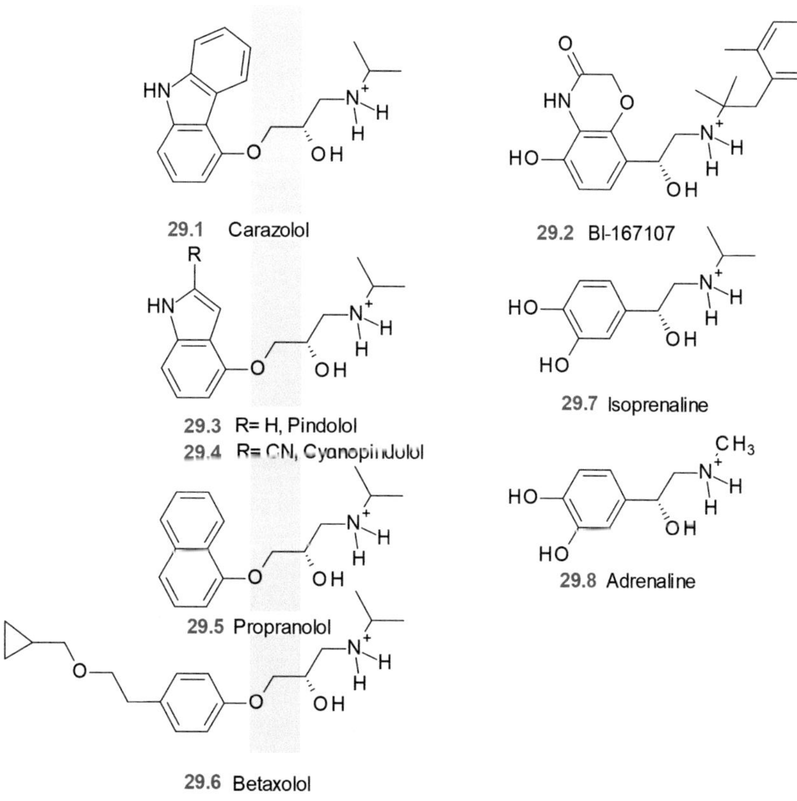

Fig. 29.1 Ligands of the human β_2-adrenergic receptor: carazolol **29.1**, BI-167107 **29.2**, pindolol **29.3**, cyanopindolol **29.4**, propranolol **29.5**, betaxolol **29.6**, isoprenaline **29.7**, and adrenaline **29.8**. The substances on the *left* are β-blockers, whereas agonists of the receptor are listed on the *right*. The side chains of the agonists are two chain members shorter than those of the antagonists/inverse agonists (*yellow*)

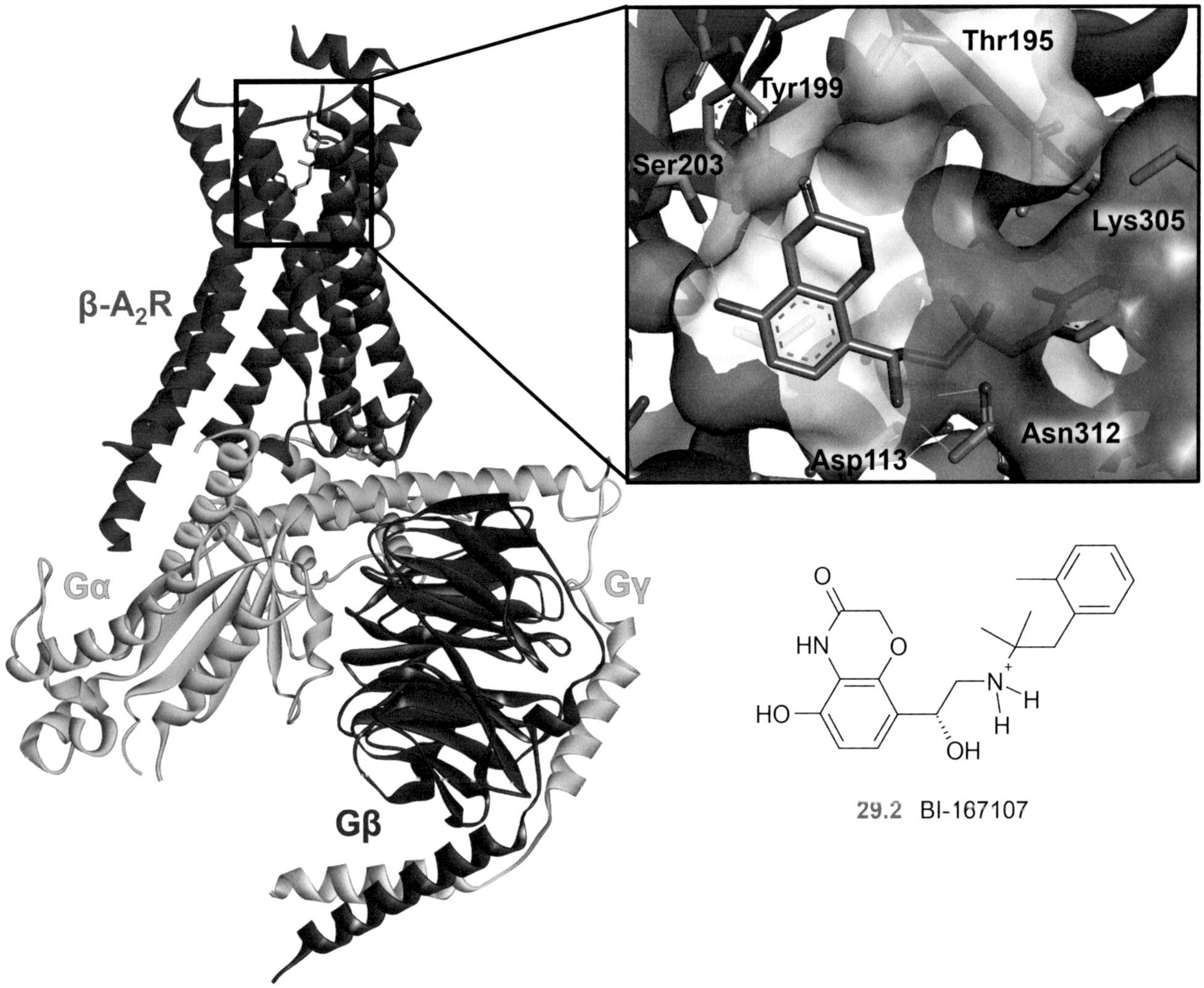

29.2 BI-167107

Fig. 29.2 Crystal structure of the human β_2-adrenergic receptor with the bound agonist BI-167107 **29.2**. The receptor (*red*) could be crystallized with a bound G_s-protein consisting of the α- (*light blue*), β- (*purple*), and γ-subunit (*yellow*). An antibody fragment was added to stabilize the structure; it is omitted from the figure. The binding pocket of the receptor containing the agonist BI-167107 **29.2** is shown on the *right*. The ligand binds to Ser 203 on helix H3 via its lactam nitrogen and phenolic OH group. The OH group in the side chain interacts with Asn 312 on H7. The protonated nitrogen forms a salt bridge with Asp 113 on H3. (► https://sn.pub/psPaBH)

binds to this loop and stabilizes the receptor in its native functional structure. In another strategy, the critical third loop was cut out of the receptor and replaced with a protein that has long been known to crystallize well, T4 lysozyme. The resulting fusion protein with the bridging T4 lysozyme in place of the critical third loop shows largely unchanged pharmacological properties. Both structures were crystallized together with the partially inverse agonist carazolol **29.1** (Fig. 29.1). The techniques for stabilizing the receptors have been continuously refined. In addition to antibody stabilization and the replacement of flexible loops with building blocks from stable protein structures, targeted point mutations were used to make the receptors sufficiently stable for crystallization. As a major achievement, Brian Kobilka, in collaboration with many other colleagues, was able to present the complete structure of the β_2-adrenergic receptor with the bound agonist BI-167107 (**29.2**) and a G_s protein in 2011 (Fig. 29.2). Brian Kobilka and Robert Lefkowitz were awarded the 2012 Nobel Prize in Chemistry for their structural, functional, and regulatory studies to understand the molecular mechanisms of GPCRs.

BI-167107 **29.2** binds with its ammonium group to Asp $113^{3.32}$ and with its OH group to Asn $312^{7.39}$ (the nomenclature with the superscript numbers is intended to make clear on which helix and in which position the amino acid is located). It was known from mutation studies on Asp 113 that the exchange of asparagine reduces the activation of the G-protein by four orders of magni-

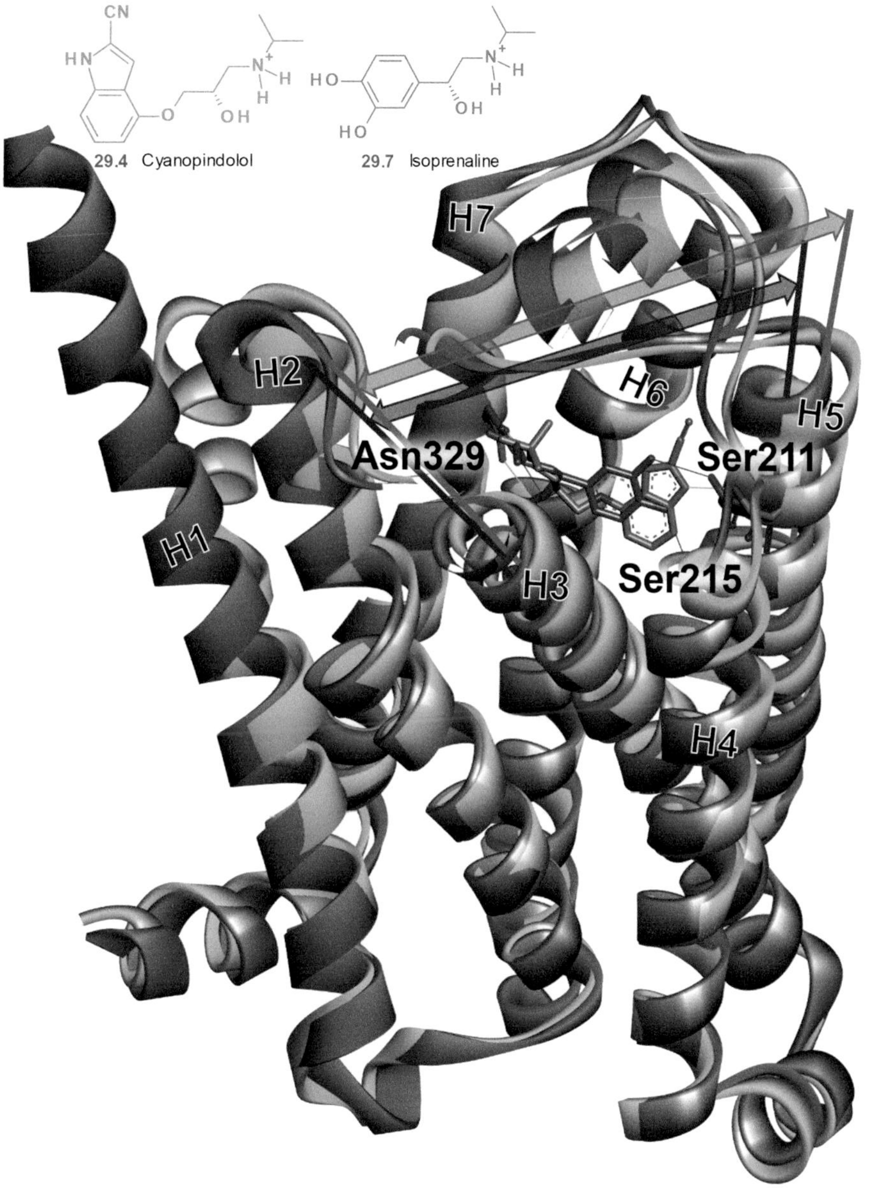

Fig. 29.3 Superposition of the crystal structures of the β_1-adrenergic receptor with the agonist isoprenaline **29.7** (*reddish brown*) and the antagonist cyanopindolol **29.4** (*green*). The agonist forms H-bonds to Ser 211 and Ser 215 using its phenolic OH groups. The binding pocket is contracted by tilting helices H5 and H7 by about 1 Å. This movement of the helices (indicated by the helix axes) is conveyed to the binding site of the G-protein. As a result, the receptor is activated and binding of the G-protein is improved. (▸ https://sn.pub/Qo0Q0d)

tude. The mutation of Asn 312 to a nonpolar amino acid such as alanine or phenylaniline leads to the collapse of the receptor function, whereas an amino acid with a polar side chain (threonine or glutamine) partially preserves the function of the receptor. The heteroaromatic moiety of **29.2** forms two hydrogen bonds to Ser $203^{5.42}$ via its phenolic OH group and the lactam nitrogen. This residue has also been identified in mutagenesis studies as critical for catecholamine agonist binding. For aryloxyaminopropanol-type β-blockers with a nitrogen heterocycle, such as pindolol **29.3**, it was known that their affinity for the β-adrenergic receptor decreases significantly when this serine is replaced by another residue. In the central part, β-blockers are surrounded by numerous contacts with hydrophobic amino acids (Val $114^{3.32}$, Phe $290^{6.52}$, Phe $193^{5.32}$). This explains why the ligands in this region usually have an aromatic building block (Fig. 29.1).

Many β-blockers display poor selectivity for the subtypes of the β-adrenergic receptors (β-AR). Nonetheless, such selectivity is highly desirable because, for example, the β_1-receptor is found in the cardiac vasculature and the β_2-subtype is found in the bronchi. Efficient β_1-receptor inhibition reduces the contractility and frequency of the heart. At the same time, though, bronchoconstriction by blocking the β_2-receptors is undesirable. Interestingly, all amino acids surrounding carazolol in the binding site of the β_2-receptor are conserved in the β_1-receptor. The 94 observed exchanges between the β_1- and β_2-receptors are all in the loop regions. Therefore, it is assumed that the pharmacological differences that are exploited in selective ligands such as betaxolol **29.6** are located in the entrance region to the binding site and cause small changes in the helix packing.

The next interesting question concerned the activation mechanism and the involved structural changes of the receptors. Comparing the inverse agonist (■ Fig. 29.1, *left*) carazolol **29.1** with the agonist (*right*) isoprenaline **29.7**, it is obvious that the two hydroxyl groups of the catechol form hydrogen bonds with Ser $204^{5.43}$ and Ser $207^{5.46}$. In addition, Asn $293^{6.55}$ and Tyr $308^{7.35}$ have been described as critical for agonist binding. However, in the carazolol-bound receptor structure, these residues are too far apart to interact efficiently with an agonist such as isoprenaline. The same structural expansion in the side chain is also observed with the other antagonists or inverse agonists **29.3**–**29.6** (■ Fig. 29.1, *left*, marked in *yellow*). Instead, all agonists **29.2**, **29.7**, and **29.8** (*right*) have a shorter side chain. This suggests that the receptor must undergo a conformational transformation upon activation to successfully discriminate between agonist binding on the one hand and antagonist or inverse agonist binding on the other hand.

29

In the group of Gebhard Schertler and Chris Tate in Cambridge, England, the structure of the β_1-receptor was elucidated using the antagonist cyanopindolol **29.4** (■ Fig. 29.3). The researchers used a thermostable receptor variant from turkey with six amino acid substitutions. Its overall structure does not differ from that of the β_2-receptor, but the stability of the inactive form has been improved. Overall, the β_1-receptor has a lower intrinsic activity than the β_2-analogue. This increased intrinsic activity of the β_2-receptor is physiologically important. The Thr264Ile mutant of the β_2-receptor occurs as a polymorphism in humans. It has lower intrinsic activity compared to the β_1-receptor and is associated with a certain heart disease.

In addition to the cyanopindolol antagonist, a crystal structure of a complex with the agonist isoprenaline **29.7** was determined. As expected, the binding of this agonist with the shorter side chain leads to a contraction of the binding pocket. As a result, helices H5 and H7 move towards each other (■ Fig. 29.3). Either the agonist or the antagonist forms two hydrogen bonds with its aminopropanol group to Asn 329 on H7. The expanded antagonist, cyanopindolol **29.4**, uses its indole NH function to interact with the side chain of Ser 211 on H7. The catecholamine isoprenaline **29.7**, on the other hand, uses its two aromatic OH groups to hydrogen bond with Ser 211 and Ser 215 to pull H5 into the agonist-bound conformation. The relative displacement of these two helices contributes to the transition of the receptor from the inactive to the active state and is propagated to the binding site of the G-protein. This improves the interaction with the cellular macromolecular binding partner. The polymorphism mentioned above, which reduces the intrinsic activity of the β_2-receptor, also affects the contact between H5 and H7 in this receptor.

A critical look back at modeling before the structure of the β_2-adrenergic receptor was known is warranted. As described above, the rhodopsin structure was used as a first reference. Therefore, it is possible to compare the previously constructed homology models of the β_2-adrenergic receptor with the later determined crystal structures. It is important to note that the models correctly reflect the overall topology of the receptor. Many observations from the mutation experiments could be interpreted correctly. However, for modeling the exact binding mode of a ligand, the models could not provide the necessary level of detail. It turns out that the models all show a greater structural similarity to the template structure of rhodopsin than to the actual structure of the β_2-receptor. This result is of course thought-provoking. However, it also illustrates the difficulty of modeling complex and, above all, highly flexible proteins. This makes the question of how communication between a GPCR and the coupled G-protein works all the more interesting.

29.4 How Does a GPCR Communicate with Its Macromolecular Protein Partners in the Cell?

How are the **structural shifts of the helices transmitted** to the binding site of the G-protein? As mentioned above, one of the best studied receptors is **rhodopsin**. Biophysical studies have shown that the activation of the β-adrenergic receptor is analogous to that of rhodopsin. In the case of the light-dependent receptor, activation is triggered by a *cis*–*trans* isomerization of covalently bound retinal (■ Fig. 29.4). The retinal binding site is located in the same region as the ligand-binding site in other GPCRs. In the inactive state, a network of charge-assisted hydrogen bonds is formed between two glutamate residues (Glu $134^{3.49}$, Glu $247^{6.30}$) and one arginine residue (Arg $135^{3.50}$). This network is located several angstroms away from the retinal binding site in the lower part of the receptor near the cytosolic end. The interaction is known as an "ionic lock" and connects transmembrane helices 3 and 6. It is connected to the retinal site by a chain of hydrogen bonds mediated by a network of several water molecules. Upon light activation, the ionic lock ruptures, resulting in the translocation of helices 3 and 6 to the cytosolic side. At the same time, the local water structure is altered. These translocations trigger the activation of the heterotrimeric G-protein on the intracellular side.

Detailed glimpses into the activation mechanism have allowed a comparison of the structures of a stabilized Glu113/Gln-mutated variant of the **active and inactive rhodopsin**. After photoactivation, the receptor binds one retinal molecule in the all-*trans* configuration. At the same time, the activated receptor complexed with an 11-residue peptide corresponding to the interacting epitope of the α-subunit of the G-protein was characterized. The β-ionone ring of retinal is shifted by 4.3 Å in the di-

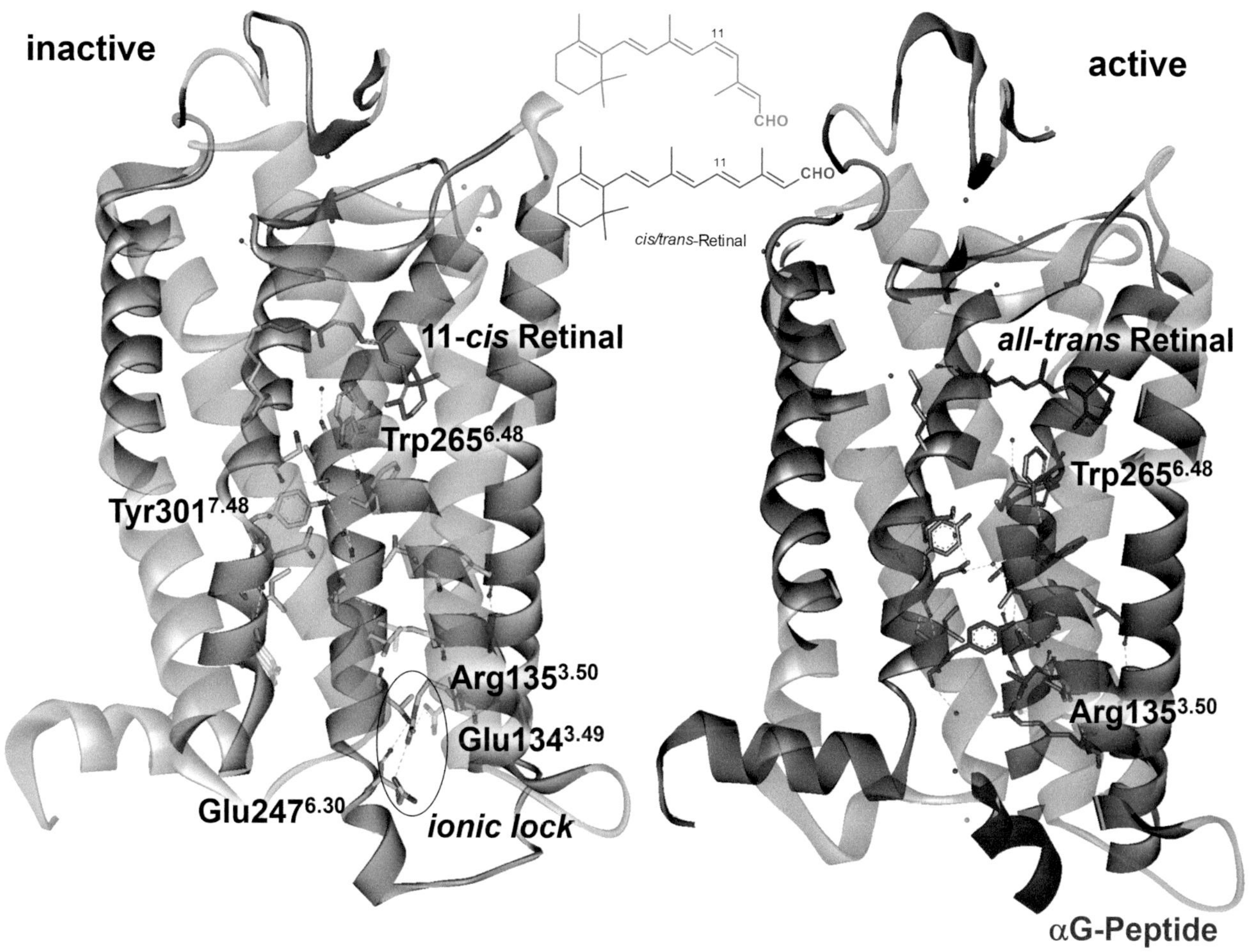

Fig. 29.4 Comparison of the crystal structures of inactive (*green*) and active (*brown*) rhodopsin. The photoactivation is triggered by retinal, when its cyclohexene ring shifts because of an isomerization of the 11-double bond from a *cis* to a *trans* configuration. This movement is translated to Trp 265 and from there through a cascade of water-mediated H-bonds all the way to the "ionic lock," which is made up of Glu 134, Arg 135, and Glu 247. The salt bridges are dissolved, and the binding site for the peptidic binding epitope that mimics the α-domain of the G-protein is established. (► https://sn.pub/wnwTaC)

rection of a gap between helices H5 and H6 in activated rhodopsin. At the same time, Trp265$^{6.48}$ is also moved from its initial position in the ground state. This transition requires a global restructuring of the orientation of the helices, and the water-mediated interaction network between H6 and H7 is disrupted and rearranged. The cytosolic end of helix 6 moves by twisting away from the center of the helix bundle H1- H4 and H7. The side chains from Tyr223$^{5.58}$ and Tyr306$^{7.53}$ orient themselves towards the interior of the receptor. They form new contacts to the water network, and undergo interactions with the highly conserved E(DRY) motif at the cytosolic end of helix 3. The salt bridges between the side chains of Glu134$^{3.49}$, Arg135$^{3.50}$, and Glu247$^{6.30}$, which form the ionic lock, open and allow access to the binding site for the peptide epitope that mimics the α-domain of the G-protein.

After ligand binding on the extracellular side, the pharmacologically relevant GPCRs presumably undergo very similar spatial shifts of their helix ends at the binding site in the contact region with the G-protein. Most likely, the activation process is a multistep cascade in which individual conformational states are passed through. Therefore, in the case of GPCRs, drug binding and the resulting **therapeutic profile** of a given drug must be viewed as a process in which, in addition to the spatial fit of a ligand, the influence and **modulation of the conformational dynamics** of the receptor are important. The modulation of the dynamic properties provides an additional quality of importance to these receptors and the kind of signal to be transmitted. For the enzymes considered so far, modulation of the dynamics by a bound drug molecule may be of lesser importance for the therapeutic response, since the enzymes are made to tightly bind the transition state of the catalyzed reaction and the drug should block this step.

29

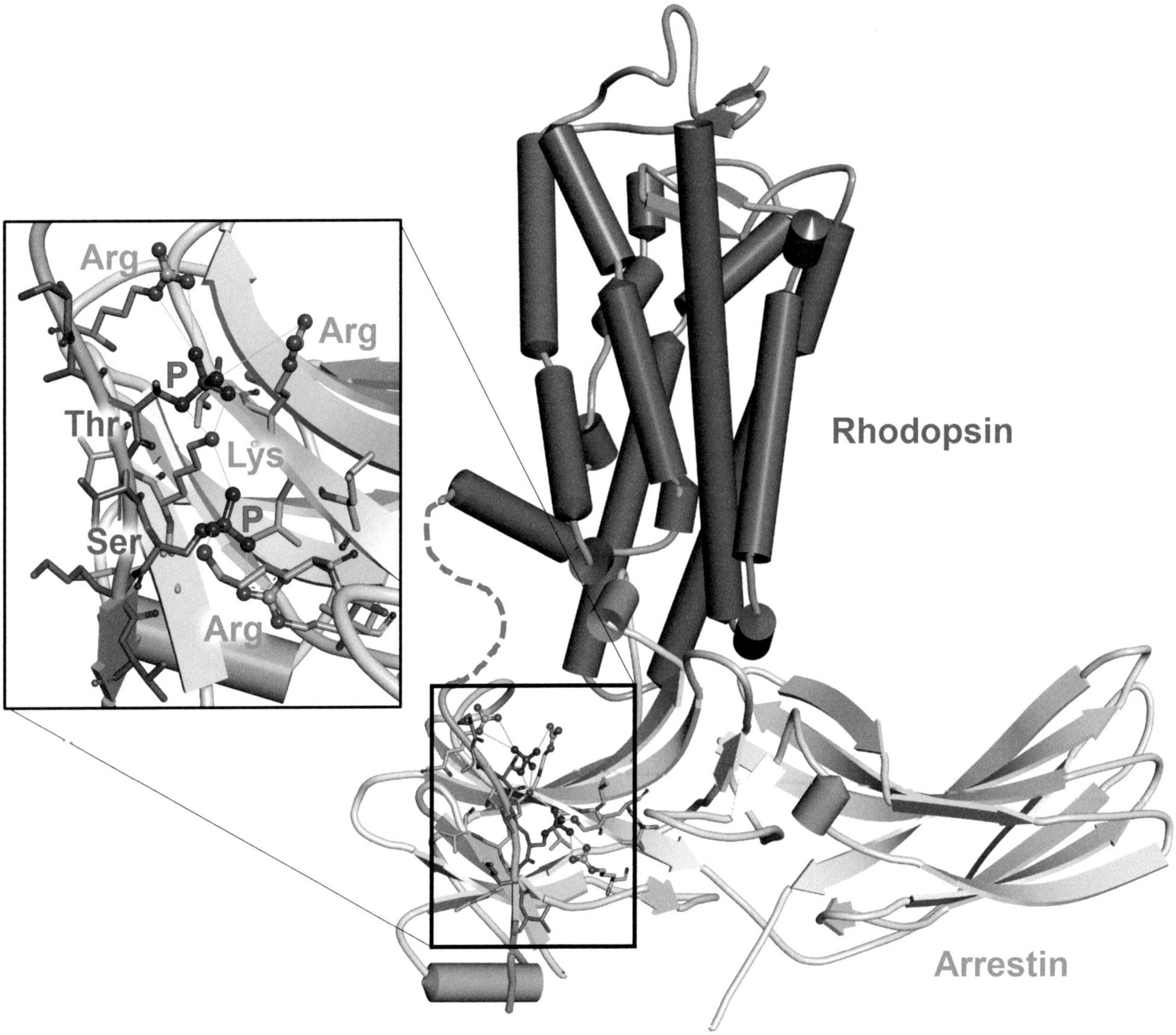

Fig. 29.5 Crystal structure of rhodopsin (*purple*) with bound arrestin (*turquoise*). Rhodopsin has been phosphorylated three times at its *C*-terminus, with the phosphate groups being attached to a threonine and a serine residue in the structure. Arrestin, which is composed of β-sheets, contacts the *C*-terminus of the receptor via electrostatic interactions using the phosphate groups. For this purpose, arrestin uses several arginine and lysine residues (see detailed view on the *left*). (► https://sn.pub/z0vMH0)

As described above, in response to the binding of an agonist, a GPCR triggers intracellular processes that lead to the formation of secondary messengers and result in very different signaling pathways. These processes can also be turned off by activating **GPCR-specific kinases** that phosphorylate a GPCR at its intracellular *C*-terminus. This triggers binding to an **arrestin** protein. As a result of this complex formation, signal transduction is prevented and the receptors can be internalized in the cell, bound in vesicles. Once again, rhodopsin was used to experimentally determine the first structure of a phosphorylated receptor bound to arrestin (Fig. 29.5). The structure was determined on a large number of nanocrystals using serial diffraction experiments with the X-ray free-electron laser (XFEL).

The receptor, phosphorylated at threonine and serine residues, forms electrostatic interactions with charged arginine and lysine residues of arrestin. The *C*-terminus of the receptor chain complements the β-sheet geometry of arrestin with an additional strand. Five amino acids of the receptor chain could not be resolved due to intrinsic disorder (*dashed line*). Arrestin occupies the binding pocket of the receptor on the cytosolic side with a loop that is filled by a loop of the Gα-subunit when complexed with a G-protein.

29.5 Peptide-Binding Receptors: Development of Angiotensin II Antagonists

More than in almost any other area of drug discovery, drug design approaches have been used to find and optimize potent ligands for G-protein-coupled receptors. In particular, thousands of examples of ligand-based optimizations (see Chap. 17) could be presented here. For example, many selective agonists and antagonists have been developed for receptors that recognize a small neurotransmitter as their natural ligand, such as the dopamine, histamine, or serotonin receptors. Bioisosteric substitutions by groups of the endogenous ligand have been the subject of ever-new variations. The resulting molecules have been systematically rigidified to reveal the protein-bound conformation of a drug molecule. However, it is important to remember that the actual receptor structures were unknown for all of these examples. Therefore, modeling had to focus exclusively on the ligand-based approaches. Over time, especially after the structure of bacteriorhodopsin became available, additional information about the receptor environment was incorporated. At the time, however, this information was based only on models of the receptors. With advances in crystallography and, more recently, cryo-electron microscopy, a wealth of more than 400 structure determinations of GPCRs has become available over the last 15 years. Many of these structures include a bound ligand. Most of these studies use the geometric and proteolytic stabilization techniques mentioned above, including antibody stabilization, replacement of fragile loops with robust protein surrogates such as T4 lysozyme, or introduction of stabilizing point mutations.

What is the picture that emerges? Most of the structures belong to the **rhodopsin family** (class A). They have a ligand-binding pocket near the extracellular end of the transmembrane helix bundle. The second of the connecting extracellular loops shows the greatest structural variance in the family. It is located close to the entrance of the ligand-binding pocket. The seven helical bundles within a given GPCR family are spatially very similar. Thus, ligand selectivity between members of a GPCR family is determined by differences in side chains rather than the conformation of the backbone of the transmembrane helix bundle and the folding of the second loop. Functionally different ligands induce **small spatial shifts** or **modulations of the dynamics** within the helix bundle. These are transmitted as a signal to the binding site of the partner proteins on the cytosolic side. They control which G-protein or whether arrestin is bound after phosphorylation. Interestingly, further cavities and pockets are found between the helix bundles and towards the adjacent membrane region. In principle, they are large enough to accommodate low molecular weight ligands. Future research must elucidate whether these pockets are functionally important. Ligands occupying these pockets may also influence the regulatory function of GPCRs. All of these variations present additional challenges for the design of specific ligands that interfere with receptor function. Increasingly better resolved structures indicate that embedded water molecules also play an important role in GPCRs. It is not only the fit of a ligand into a given binding pocket that must be optimal. The design also requires correct modulation of the dynamics and spatial arrangement of the helices, including the water molecules involved. It should not be forgotten that some of these receptors already have a distinct basal activity. A ligand is then expected to be able to regulate this activity to the pharmacologically intended extent.

As a result of the growing knowledge base of GPCR structures, structure-based design techniques are increasingly applied today (Chap. 20). Examples that follow a ligand-based approach are still in practice, but are becoming less and less important. For this reason, we will not present another example from this area of ligand-based drug design on an aminergic or opioid receptor. Instead, we will focus on a receptor that is targeted by small peptide ligands. Here, too, small-molecule drugs could be found that play a crucial role in today's therapy.

The importance of the **renin–angiotensin–aldosterone system** for the treatment of hypertension has already been discussed in Sects. 24.2 and 25.4. The vasoactive octapeptide **angiotensin II**, Asp–Arg–Val–Tyr–Ile–His–Pro–Phe is formed from angiotensinogen by the action of renin and angiotensin-converting enzymes (ACE). Blocking this system at any stage of the cascade results in a decrease in blood pressure. First, renin secretion from the kidney can be suppressed by inhibiting the β-adrenergic receptors. Then, renin and ACE can be blocked by inhibitors, and of course the use of an **angiotensin II antagonist** thwarts the binding of angiotensin II to the **AT_1 receptor**. ACE, a relatively unspecific protease, cleaves other peptides such as bradykinin, enkephalin, and substance P in addition to angiotensin I. These reactions are suppressed by ACE inhibitors (Sect. 25.5), which have long been used therapeutically. As a result, about 5–10% of all ACE-treated patients develop a dry cough as a troublesome side effect. This is due to the inhibition of the bradykinin metabolism. An angiotensin II antagonist leaves bradykinin levels unchanged. An intervention is made at the terminal end of the cascade with an AT_1 receptor antagonist, also turning off the effects of angiotensin synthesized in the body by other proteases independent of renin–ACE.

In 1971, the octapeptide **saralasin**, **Sar**–Arg–Tyr–**Val**–His–Pro–**Ala** (Sar = sarcosine, or *N*-methylglycine) was identified as the first specific **angiotensin II antagonist**. This peptide acts to lower the blood pressure in patients with high renin levels, but it is not orally available. Moreover, it has a short half-life and other undesirable properties. Therefore, saralasin was not suitable as

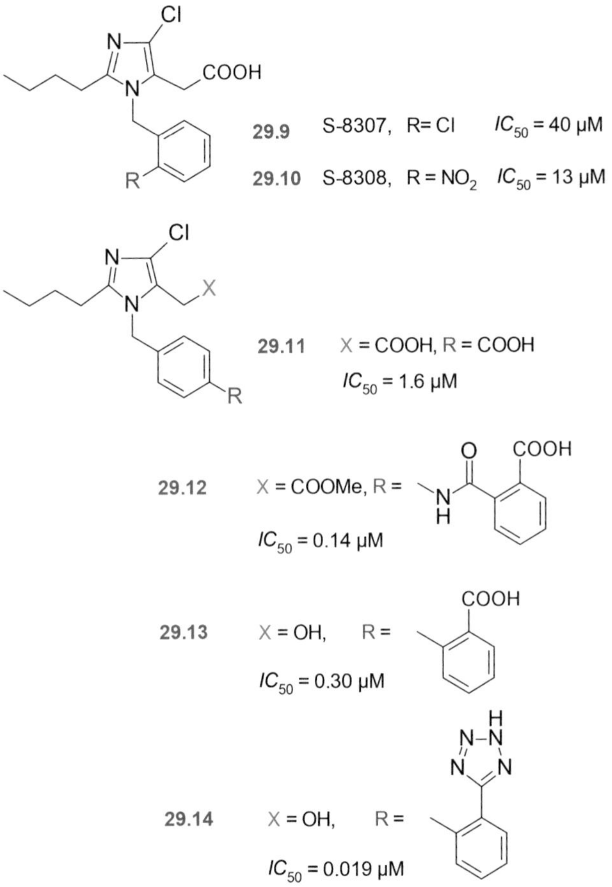

Fig. 29.6 The most important intermediates in the development of the angiotensin II receptor antagonist losartan. The basic structure of the angiotensin II antagonists **29.9** and **29.10**, published in a Takeda patent, was retained. Variations in the R substituents were based on a superposition of the Takeda structure with a model of the receptor-bound conformation of angiotensin II (Fig. 29.9, *bottom row, left*). Compounds **29.13** and **29.14** are orally available angiotensin II receptor antagonists. Losartan **29.14** successfully completed clinical trials and has been available for clinical use since 1994

a drug. Attempts to find nonpeptidic antagonists based on saralasin and other peptides were futile.

Little progress was made in this area of research until the early 1980s. At that time, Takeda's angiotensin II antagonist efforts were abandoned in favor of an ACE project. In 1982, however, the publication of two patents provided the impetus for further research. The nonpeptidic antagonists S 8307 **29.9** and S 8308 **29.10** (Fig. 29.6) covered by these patents were only weakly active, but the first nonpeptidic antagonists caused a sensation.

Numerous companies investigated these new lead structures, Dupont included. Because of the extensive basic research on the peptidic structures, a broad knowledge about the conformation of angiotensin II and many analogues was available. The Takeda structure was compared with the assumed receptor-bound conformation of angiotensin II. Structural superposition led to the conclusion that a modification of the Takeda structure at the *para*-position of the benzyl group would be the most promising strategy to increase the affinity. The result of these considerations was the synthesis of **29.11**. The compound is 10 times more potent than S-8307 and S-8308.

Further systematic variations at this position led to **29.12**, which binds another 10 times more potently to the AT_1 receptor (IC_{50} = 140 nM). The first compounds in this substance class produced a dose-dependent reduction in blood pressure in rats, but were not orally available. The biphenyl derivative **29.13** made the breakthrough to oral availability. The slightly poorer binding to the receptor is unimportant in view of this essential property. Replacing the carboxyl group on the aromatic ring with a more lipophilic tetrazole isostere finally led to DuP 753, **29.14** (losartan), which binds to the receptor with 19 nM, is orally available, and has a very long half-life. Losartan successfully passed all of the clinical trials and has been marketed as Lozaar® since 1994. Losartan was the first angiotensin II receptor antagonist to be approved for the treatment of hypertension. Only one year later, Novartis followed their colleagues at Dupont with valsartan **29.15**. In the meantime, an entire class of drugs, known as **sartans**, have received approval for therapy (Fig. 29.7).

However, after several years of clinical use, not all sartans have proven to be equally effective, for example, in the treatment of congestive heart failure. A comparative study of more than 5000 patients in Sweden showed that candesartan produced better therapeutic results than losartan. Ninety percent of patients with this disease treated with candesartan survived one year and 61% survived five years. In contrast, 83% of patients treated with losartan survived one year and 44% survived five years. The higher affinity of candesartan for the AT_1 receptor compared to losartan is striking, as is its longer persistence at the site of action, which is 10–30 times longer. Perhaps these parameters reflect the increased efficacy of candesartan (cf. Sect. 8.8).

29.6 Do Peptidic Agonists and Small-Molecule Antagonists Bind at the Same Position of the AT_1 Receptor?

The blood-pressure-increasing effect of angiotensin II is based on its binding to the AT_1 receptor, of which two isoforms have been described. As a result, arterial vasoconstriction is provoked. Aldosterone is released to control the electrolyte homeostasis, the cardiac contractility is increased, and the glomerular filtration rate of blood through the kidney is regulated. The AT_2 receptor, which belongs to the same family of GPCRs, is associated with other regulatory processes.

The octapeptide angiotensin II **29.20**, which binds as an agonist, served as a reference compound for the development of low molecular weight antagonists. As de-

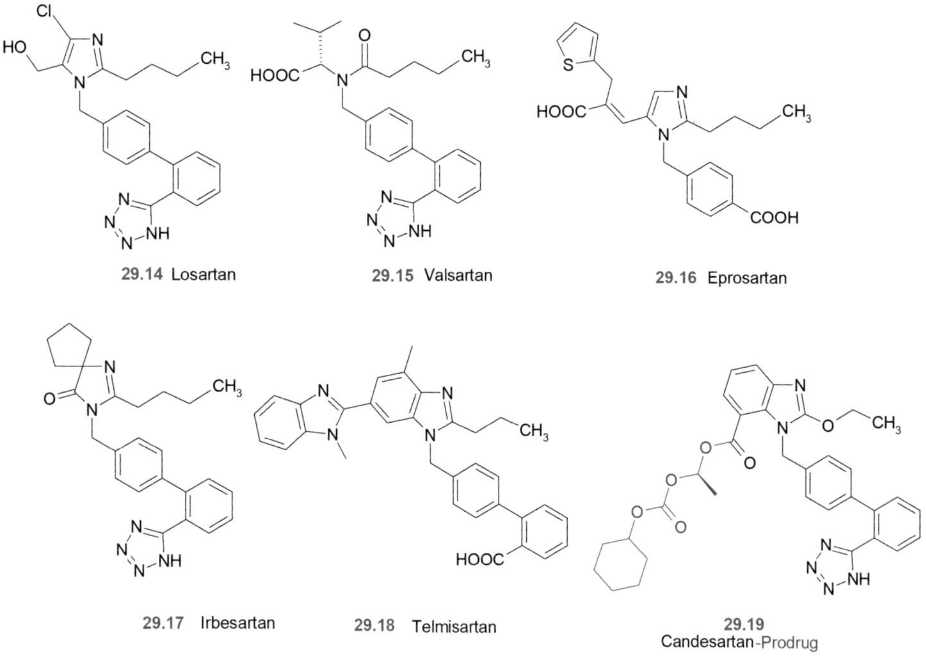

Fig. 29.7 Losartan **29.14** was the first angiotensin II receptor antagonist; shortly thereafter valsartan **29.15** followed. Eprosartan **29.16**, irbesartan **29.17**, telmisartan **29.18**, and candesartan **29.19** are further representatives of the sartan class. Candesartan is a prodrug; the *red-colored* portion is cleaved to release the actual active substance

scribed above, the design hypothesis assumed an overlap of the *C*-terminal amino acids, Ile–His–Pro–Phe, with the sartan scaffold (Figs. 29.8 and 29.9 *lower left*). Although this comparison appeared to describe a successful working hypothesis, it was later challenged by a **mutagenesis study**. Point mutations in the AT_1 receptor revealed that the amino acids affecting angiotensin II binding were located in the three extracellular loops and the *N*-terminal sequence segment. In contrast, mutations that altered losartan binding were located within the transmembrane portion of the receptor. Overlapping binding regions for the peptide agonist and the small-molecule antagonists, therefore, seemed implausible. This notion of divergent binding regions was supported by a 1995 study in the frog *Xenopus laevis*. Assay data indicated that while the frog AT_1 receptor recognizes the octapeptide with nanomolar affinity, losartan binds only in the double-digit micromolar range. The antagonist, which is highly potent at the human receptor, therefore, fails at the frog receptor. This observation prompted the scientists to incorporate the putative antagonist binding site of the human receptor into the initially unresponsive amphibian receptor. By targeted mutation of 13 amino acids in the transmembrane region, believed to be important for losartan binding in the human receptor, the frog receptor suddenly developed a high affinity for the antagonist. This experiment also suggested a spatial separation of the binding regions for the peptidic agonist and the small-molecule antagonists.

Twenty years later, structural research on GPCRs has made further progress. X-ray structure determinations are available for the AT_1 and AT_2 receptors. They show that the sartan binding site is located deep inside the transmembrane part of the receptor. But what about the binding of octapeptide **29.20**? In 2020, a structural determination of the peptide with the AT_1 receptor was successful. With all possible caution, a comparison of the binding proportions should be made to evaluate the original design hypothesis.

Superimposition of the structures of the AT_1 receptor with the bound octapeptide **29.20** and with a low molecular weight ligand of the sartan type (**29.21**) is shown in Fig. 29.8. Accordingly, the peptide places its *C*-terminus into the binding site in the transmembrane region and overlapping binding positions of the sartan and the peptide are observed. Comparing the initial design hypothesis with the experimental binding modes finally observed in the AT_1 receptor, the following picture emerges (Fig. 29.9, *bottom left* and *right*). The *C*-terminal acid group of peptide **29.20** and the tetrazole moiety of **29.21** do not directly overlap. Nevertheless, they form H-bonding contacts with the amino acids Lys 199 and Arg 167, respectively. The first ring of the biphenyl moiety of **29.21** fills a similar volume segment as the aromatic ring of the terminal Phe 8 of **29.20**. The five-membered ring of Pro 7 does not overlap with the second aromatic ring of the biphenyl unit, but rather with the heterocycle of the partially hydrogenated naphthyridinone in **29.21**. There is no obvious overlap in the region of His 6 and Ile 5. The present example must be viewed historically. On the one hand, it should be emphasized that the peptide is an agonist and the sartans are antagonists. As described above, current experience with numerous GPCRs has shown that agonists and antagonists occupy overlapping binding regions. However, there are geometrical deviations that can lead to relevant pharmacological differences and these are

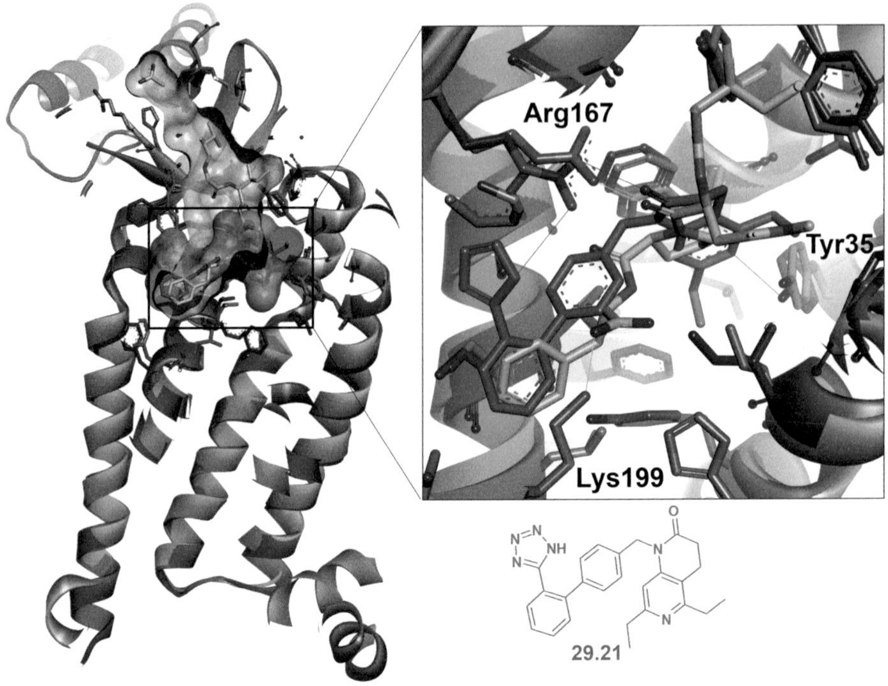

Fig. 29.8 Overlay of the crystal structures of the AT_1 receptor with the bound octapeptide angiotensin II **29.20** (*yellow*; inner surface *yellow*, outside *dark red*) and the low molecular weight sartan-like ligand **29.21** (*green*; inner surface *blue*, outside *green*). For crystallization, the receptor was stabilized with an antibody (not shown in the figure). The *C*-terminal acid function of **29.20** forms a hydrogen bond to Lys 199, and the tetrazole of **29.21** interacts with Arg 167. Structure determination suggests that the binding domains of angiotensin II and the sartans overlap spatially. (► https://sn.pub/SHuM3N)

29

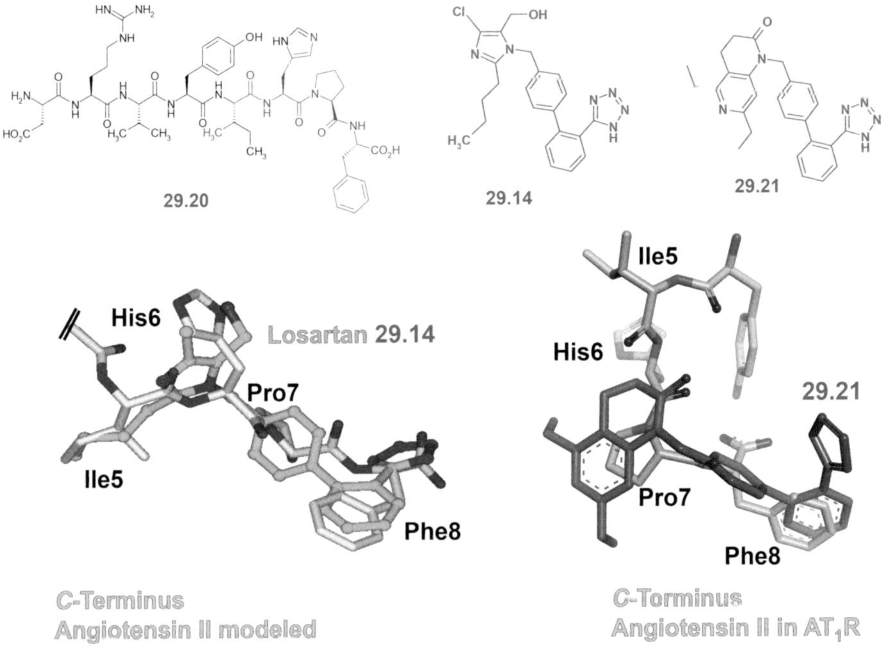

Fig. 29.9 As a working hypothesis for the original design of the sartans, the *C*-terminal part (*red*) of the octapeptide angiotensin II **29.20** (*yellow*; *bottom left*) was compared with the structure of losartan (**29.14**, *green*). The *n*-butyl side chain describes the isoleucine residue, and the imidazole ring with the attached CH_2OH group falls on the histidine. The proline and the phenyl ring of phenylalanine are described by the biphenyl substituent. The tetrazole represents an isoster of the terminal acid function. The structures determined many years later at the AT_1 receptor confirm a comparable binding region of **29.20** and the sartan-like ligand **29.21** in the receptor, but the exact spatial position of the interacting groups differs significantly (*bottom right*)

associated with conformational changes of the receptor. In addition, the example highlights the importance of solid experimental structural information, especially for the correct interpretation of mutagenesis data. For the pragmatist, though, it remains that sometimes incorrect or inaccurate design hypotheses have their value and can guide development in the right direction.

29.7 Lessons Taught by the Nose: We Smell with GPCRs

The wealth of nuances that our **sense of smell** can perceive is impressive. It is almost poetic to try to describe the gradations of scent with words. Our sense of smell is probably the easiest biological system to illustrate when it comes to biological activity and its dependence on the spatial chemical structure of molecules. With each breath, volatile molecules are drawn into our noses and brush against olfactory receptors. There they leave a nuance-rich signal that is translated into a multifaceted sense of smell in the brain. It has long been known that the shape of molecules is associated with a particular odor. Elliptical molecules, for example, have a camphor-like odor. Long stretched molecules are described as having an ether-like smell, and a floral character requires a structure that resembles the shape of a violin case. However, even small structural changes can have impressive effects on our sense of smell (Sect. 5.7).

The understanding of our sense of smell is based on the work of Linda Buck and Richard Alex, who were awarded the Nobel Prize in Medicine in 2004. Olfactory molecules are perceived by olfactory cells in the mucous membrane of the nasal cavity. Different olfactory cells are depolarized and activated by different odors, which means that the receptor proteins on the cells can distinguish between structurally different odors by their affinity for them. The GTP-dependent activity of an adenylate cyclase increases as a signal in the cell. This is interpreted as a clear indication of the involvement of intracellular G-proteins in the olfactory process.

Linda Buck and Richard Alex sought after a family of G-protein-coupled receptors expressed in the olfactory mucosa of rats. They were quickly successful. It is now known that there are about 1000 **GPCR-type olfactory receptors** in mice and about 350–400 different receptors in humans. They make up about 1–5% of the mammalian genome. Despite their similar function, their amino acid sequences vary greatly.

This hypervariability is consistent with the nuanced recognition and binding of odorants with very different structures. But are 350–400 different receptor variants really enough for such diversity? Only one type of olfactory receptor is expressed per neuron, meaning that each neuron has a single olfactory receptor gene at its disposal. By studying the response profile of different neurons generated by recognizing structurally modified odorants, the following result was obtained: each olfactory receptor can recognize several odorants. On the other hand, a given odorant is recognized by multiple receptors, but the induced receptor response varies in intensity. This means that different odors are registered by different combinations of receptors, and an attenuated signal distribution is generated. This corresponds to a **combinatorial coding** of the sense of smell. This trick of encoding olfactory signals in **composite receptor profiles** makes it possible to discriminate an almost unlimited number of odors. This is the secret of the versatility of our sense of smell. The magical world accessible to mice is almost unimaginable. They have almost three times as many olfactory receptors as humans. Perhaps even Jean-Baptiste Grenouille from Patrick Süskind's novel *Perfume: The Story of a Murderer* would pale in comparison.

It is well known that the olfactory sense can vary between people. Some scents are inaccessible to some people. Other individuals experience a scent as disgusting and offensive, while others describe it as pleasant and welcoming. This, for example, can be seen in case of the steroid **androstenone**. Androstenone is an important component of typical male body perspiration. It is a metabolite of the male sex hormone testosterone and acts as an attractant in various mammals. In wild boar, it stimulates the female boar and makes her eager to make love to her boar. According to a study from China with a test group of almost 400 people described in the literature, the sense of smell is divided into three population groups of approximately equal size. While one third of the subjects could not smell the compound at all, another third found the substance repellent, and the remaining 30% found it to smell pleasantly of vanilla. The author of this book was able to conduct a much larger study over several years in lectures and talks with 1921 people, mostly from Central Europe (64.4% female, 35.6% male). Almost half (45.0%) of those tested did not perceive the odor at all (40.5% of the women, 53.2% of the men). Another 43.5% found it quite unpleasant and repulsive (47.7% of female subjects, 35.8% of male subjects). On the other hand, 11.5% found androstenone to have a pleasant vanilla scent (there was virtually no gender difference here).

As an explanation for this discrepancy, genetic differences in the olfactory receptors activated by androstenone were found. The receptor OR7D4 showed the strongest sensitivity to the steroid. Variations in single base pairs in the genome of individuals (so-called **s**ingle **n**ucleotide **p**olymorphisms, SNPs, Sect. 12.11) were found in the search for **genetic polymorphisms**. Two coupled exchanges occur most often that cause the exchange of an arginine for a tryptophan, and a threonine for a methionine. Both expression forms (alleles) of the mutated receptor show a reduced response to the odorants under *in vitro* conditions. Interestingly, the probands that carry a mutated receptor seem not to perceive the scent, or they perceive it only weakly. People with the unchanged gene variation are overly sensitive to the odorant. Therefore, it could be in the genes that some men cannot smell each other! The differences between the studies from China and Central Europe may be indicative of genetic differences that exist between populations from different parts of the world. Such differences were discussed in the chapter on CYP enzyme metabolism (Sect. 27.7).

The exchange of a serine for an asparagine at another position in the OR7D4 receptor increases sensitivity to the steroid. Interestingly, this exchange, along with four other mutations in the receptor, is what distinguishes humans from chimpanzees. Recognizing potential rivals in the wild for a male chimpanzee or a sexual partner for a female chimpanzee may be more important than for us humans. Perhaps Nature has equipped chimpanzees with a more sensitive olfactory receptor for androstenone for this reason.

Two aspects from the study of the sense of smell can be translated to the effects of drugs on GPCRs. Synthetic agonists and antagonists compete for the same binding site in cases of GPCRs that are regulated by small biogenic amines in particular. Typically, the synthetics are larger and interact with a greater number of amino acid residues than the endogenous competitor. **Polymorphisms** based on single base substitutions have also been described for these receptors. Therefore, a reduced sensitivity to the effects of these drugs on the mutated receptors must be expected. As a result, this is noticeable when, within one group of patients, variations in the therapeutic window are found. As already discussed for drug metabolism, this could lead to an individually adjusted dosage of a drug.

The other aspect that was illuminated by the research on olfactory receptors is the **combinatorial composition of a binding profile** made up of the individual interaction signals from the different receptors. Multiple subtypes of pharmacologically relevant GPCRs are known, for example, the serotonin receptor, which are expressed on the cells. Efforts have indeed been made to develop highly selective ligands for these subtypes, but this is not an easy task when the **receptor subtypes** are particularly closely related. There is always attenuated binding to all related receptors. Therefore, the signal that reaches the cell is a composite of information from all individual binding profiles. These profiles are distinct for different ligands and provide a **divergent spectrum of pharmacological activity**. This makes it extremely difficult to estimate the therapeutic value of a development candidate in this area prior to clinical trials. It may be that these ligands have just the right balance of multiple subtypes to achieve their therapeutic value. By analogy, one fragrance may develop its sublime potential by stimulating an optimally graded shotgun of multiple olfactory receptors, while another may not exceed the modest level of a cheap perfume!

29.8 Receptor Tyrosine Kinases and Cytokine Receptors: Where Insulin, EPO, and Cytokines Display Their Activity

GPCRs are not the only receptors that transduce extracellular signals. Another large group of membrane-bound receptors are the classes of **dimerizing or oligomerizing receptors** that **bind growth factors, cytokines, or interleukins**. These receptors carry a tyrosine kinase domain on the cytosolic side. Therefore, the **intracellular tyrosine kinases** of these receptors can also be considered as allosterically regulated enzymes with regulatory domains located outside the cell. The ligands for these receptors, the growth factors, are themselves proteins with approximately 50–400 amino acids. By binding, presumably initially to a monomeric building block of the receptor, oligomerization is achieved. Inside the cell, conformational changes are induced in the two assembled tyrosine kinase domains, leading to autophosphorylation of the receptor at multiple sites. This triggers the recruitment of additional adapter proteins that are also activated by phosphorylation. These processes lead to a subsequent kinase-dependent signaling cascade and activate processes in the nucleus that regulate gene expression. Approximately 20 classes have been characterized in the family of dimerizing tyrosine kinase receptors.

A large group of these are the **insulin-like growth factor receptors** (IGFRs). Insulin, a protein of 51 residues consisting of two chains, binds to such a receptor. A peculiarity of IGFRs is that they are permanently dimeric. Disulfide bridges link the two halves of the receptor. For another group, the epidermal growth factor receptors, it was discovered that heterodimers can form between receptor subtypes. Not all tyrosine kinases of these receptors are functional, so that additional receptor regulation of the subsequent signaling cascade is possible by dimerization of different units.

The therapeutic goal to be achieved by activating or inhibiting these receptors can be very different. Stimulation of the insulin receptor is a concept for the treatment of diabetes mellitus because this receptor regulates the uptake of glucose into the cell. On the one hand, the focus can be on modified insulin derivatives (Sect. 32.2), which stimulate the receptor analogously to natural insulin. Alternatively, the tyrosine kinase activity of the receptor can be activated. The nonpeptidic fungal metabolite L-783281 **29.22** (◘ Fig. 29.10) was discovered in a cell-based screening assay at Merck & Co. When administered orally, this compound reduces blood glucose levels in a diabetic mouse model. The binding of specific small-molecule ligands may offer a promising perspective for the development of oral insulin replacement therapy.

Insulin-like receptors, such as the **epidermal growth factor receptor**, are interesting targets for tumor therapy. Their expression is upregulated in tumors. They stimulate cell growth and prevent programmed cell death (apoptosis). In these cases, blocking the function of these receptors is of interest. To date, all attempts to develop small-molecule inhibitors that block the recognition of growth factors on the surface segment of the dimerizing receptor have failed. The interaction between the large interaction surface and the protein to be recognized is a nontrivial problem for drug design (Sect. 7.10, 10.6 and below). Nevertheless, highly **specific antibodies** have been found that compete with the natural ligand and prevent receptor binding (Sect. 32.3). An alternative concept is the inhibition of intracellular tyrosine kinases. In this case, some success has been achieved. Gefitinib **29.23**, a tyrosine kinase inhibitor (Sect. 26.3) for the epidermal growth factor receptor, has found clinical application (◘ Fig. 29.10). **Antisense nucleotides** (◘ Fig. 32.4) are alternative therapeutic concepts, as can gene silencing with siRNA (Sect. 12.7) to reduce the expression rate of the target receptor.

In addition to the signaling cascades used by tyrosine kinase receptors, living organisms also use **cytokines** for signal transduction. These are also protein-like signaling molecules that often adopt a folding pattern consisting of a bundle of four helices (◘ Fig. 29.11). Cytokines are secreted by many cells. Their main function is to transmit signals in the immune system. They are, therefore, involved in immune and inflammatory responses, and infectious diseases. They also signal leukocytes and macrophages to migrate to sites of inflammation. Their role in cell differentiation and proliferation makes them important in cancer therapy.

Cytokines are recognized by receptors on the cell surface that are also coupled to a protein kinase inside the cell. Through these kinases, they are able to initiate cellular processes. This can lead to the up- or downregulation of gene expression. Cytokines are also known as **interferons**, **interleukins**, and **chemokines**. Interferons stimulate cells involved in immune defense, particularly during viral infections. Interleukins were originally thought to be involved only in communication between leukocytes, but because they are also involved in modulating cell growth and death, they are also used in the treatment of cancer. Chemokines are signaling molecules that attract immune cells to sites of inflammation.

From a therapeutic point of view, cytokines themselves or functional surrogates are of interest. Either stimulation or inhibition of their receptors can be a therapeutic concept. Since we are dealing with receptors that are regulated by proteins, it is difficult to intervene with small molecules. Therefore, native cytokines are used as

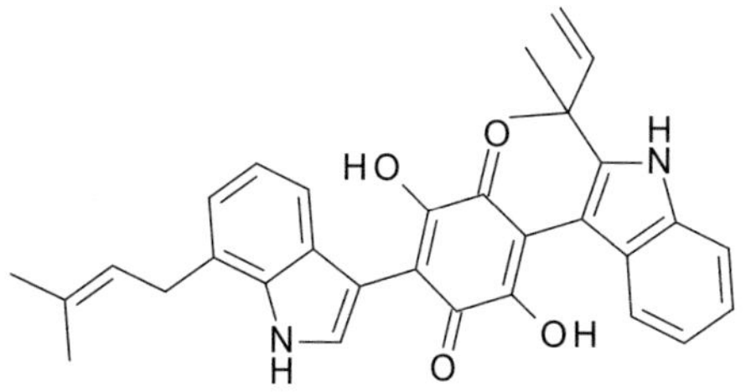

29.22 L-783 281

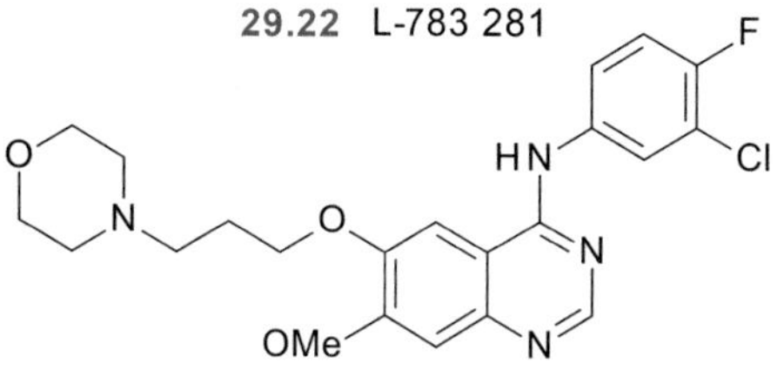

29.23 Gefitinib

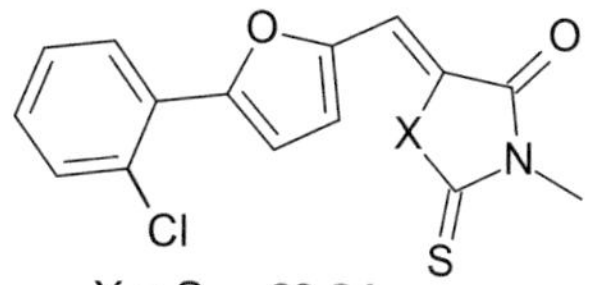

X = S **29.24**
X = NH **29.25**

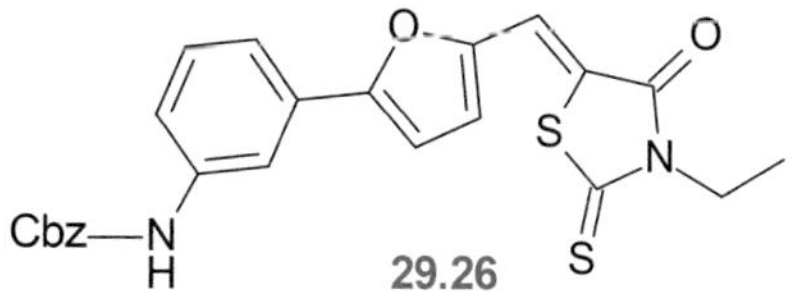

29.26

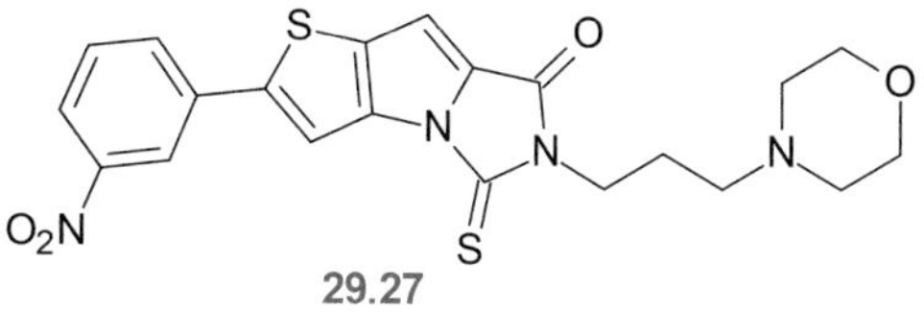

29.27

◘ **Fig. 29.10** The fungal metabolite L-783281 **29.22** was discovered as an insulin mimetic. It stimulates the tyrosine receptor kinase of the insulin receptor and has antidiabetic properties. Gefitinib **29.23** inhibits the tyrosine kinase domain of epidermal growth factor. The TNF inhibitors **29.24**–**29.26** with a rhodanine heterocycle were discovered in the screening and optimized to **29.27**. The binding mode of the latter compound is shown in ◘ Fig. 29.12

therapeutic drugs. Erythropoietin (EPO) stimulates the production of red blood cells and is used to treat anemia in dialysis patients or after aggressive chemotherapy. It has also been abused by athletes for doping purposes. Interferons INF-α and INF-β are used to treat multiple sclerosis and chronic viral hepatitis. An artificial TNF-α receptor to intercept TNF-α can be used to treat rheumatism and chronic arthritis. For this reason, these substances are top sellers among the so-called "biologicals" (Chap. 32). Anakinra, a genetically engineered interleukin-1 receptor antagonist, was approved for rheumatoid

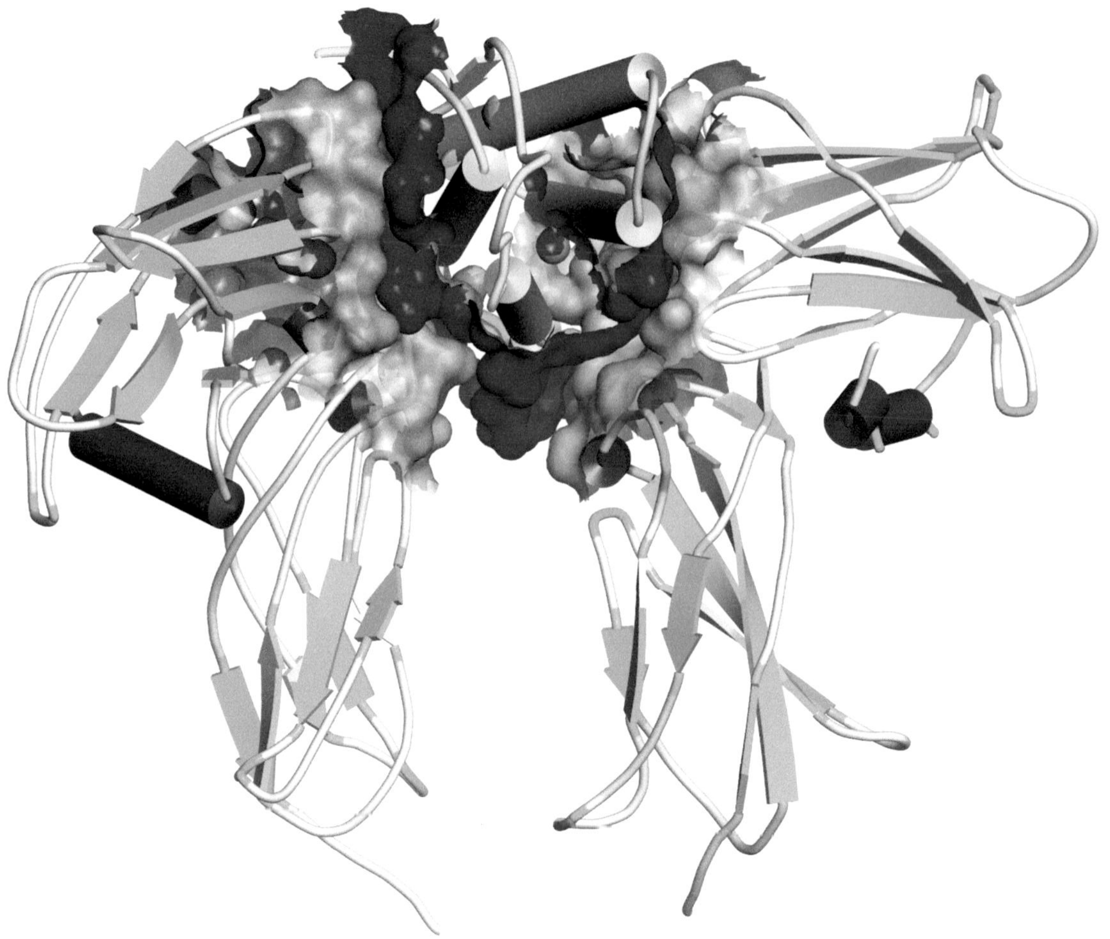

Fig. 29.11 Crystal structure of erythropoietin (EPO) with the ligand-binding domain of the erythropoietin receptor (EPOR). The structural elements are represented schematically. EPO adopts a tetrahelical bundle folding pattern (*yellow*). The dimeric receptor is basically constructed from β-pleated sheets. The contact surface between receptor and ligand are shown in *white* and *blue* (interior is *yellow*). (► https://sn.pub/4K9ABp)

arthritis in 2002. As an antagonist, it blocks the inflammatory effects of interleukin IL-1.

Finally, two examples will be presented that provide a perspective on how small molecules could also be developed as agents targeting these receptor systems. The first is **TNF-α**, a **proinflammatory cytokine** with a trimeric structure. It binds to two TNF receptors, the constitutively expressed TNF-R1 and the inductively regulated TNF-R2. In this way, TNF-α can activate signaling cascades that can induce apoptosis, inflammation, cell division, or immune responses. TNF-α has been implicated in the pathogenesis of many diseases, including sepsis, diabetes, cancer, osteoporosis, multiple sclerosis, rheumatoid arthritis, and inflammatory bowel disease. Increased TNF-α activity can lead to an excessive immune response with extensive damage to surrounding tissue. At Dupont, a screening assay was established to discover small-molecule compounds that block the complex formation of TNF-α with TNF-R1. Micromolar inhibitors were found that shared a rhodanine heterocycle as a structural building block (**29.24**–**29.26**, Fig. 29.10). A brief optimization led to, among others, IV703 (**29.27**), which blocks the receptor at 270 nM. Interestingly, the binding of **29.27** appeared to be irreversible. When the assays were repeated in the absence of light, only 50 μM inhibition was observed, but this was now reversible. Thus, photochemically enhanced inhibition was detected, resulting in micromolar reversible inhibition without light, but nanomolar irreversible inhibition with light.

A crystal structure has been determined with IV703 (**29.27**). As a trimer, TNF forms a complex with three receptor subunits (Fig. 29.12). When crystals of the TNF receptor are exposed to **29.27**, the ligand reacts by forming a covalent bond with the nitrogen of the main chain of Ala 62, thus, sterically preventing the binding of TNF-α to its receptor. It can be assumed that **29.27**

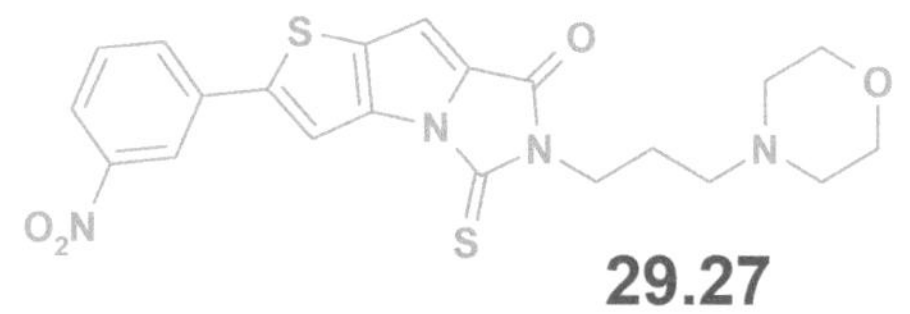

29.27

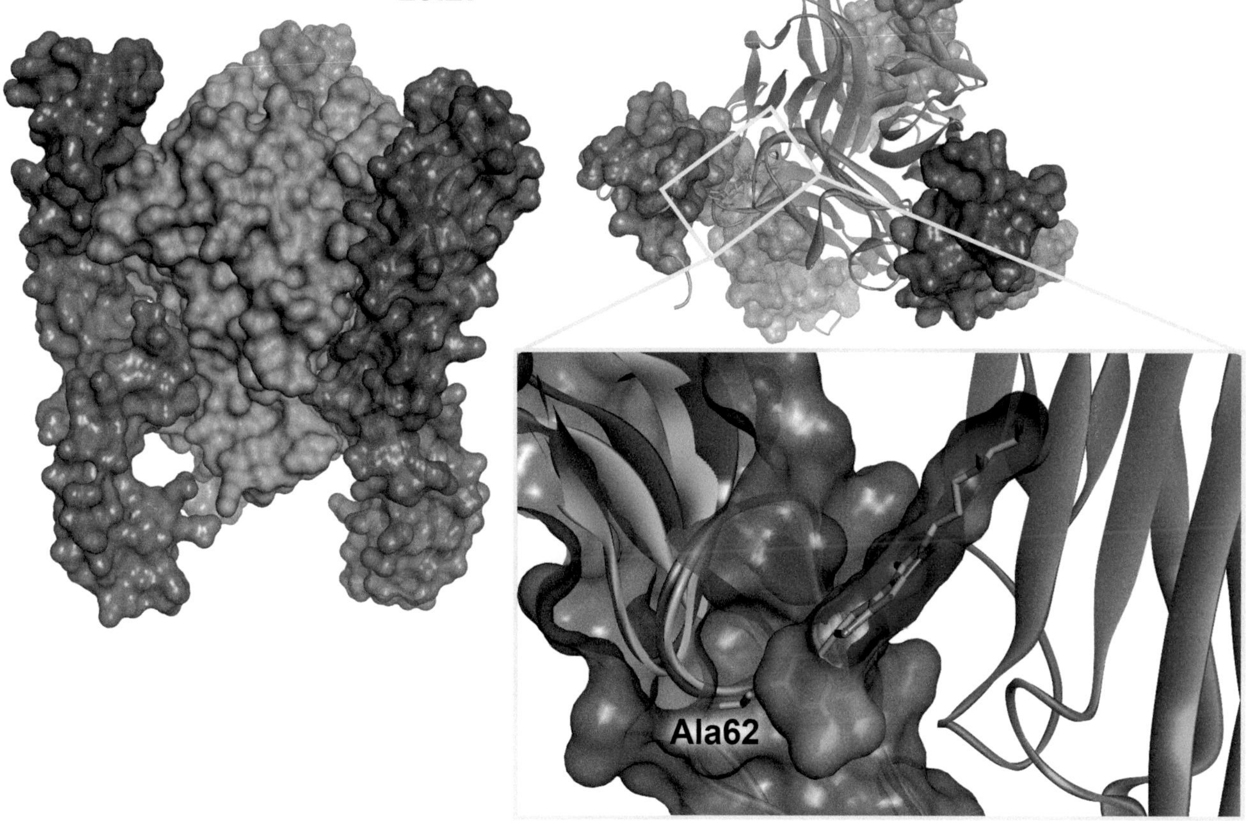

Fig. 29.12 *Left* In the crystal structure, the trimeric TNF-α (*green*) binds to three identical subunits of the TNF receptor (*brown*). In a screening campaign and after some design cycles of medicinal chemistry optimization, the ligand **29.27** was obtained. It binds with micromolar affinity to the receptor in the dark and blocks the interaction of TNF-α with its receptor (*right*). In the figure, superposition of the TNF–receptor complex and the inhibited complex with **29.27** is shown. Upon exposure to light, **29.27** reacts with the amide nitrogen of Ala 62 (*right*; *yellow* insert) and forms an irreversible covalent linkage. Thus, nanomolar inhibition results. The low molecular weight ligand prevents formation of the TNF–receptor complex for steric reasons. (► https://sn.pub/NQCXAP)

first finds a noncovalent binding site near Ala 62 with micromolar affinity. Subsequently, the irreversible covalent blockade of the receptor is probably formed by a photochemical radical reaction. However, the complex with **29.27** proves that macromolar complex formation can be thwarted with a small molecule. If noncovalent inhibition is desired, then the micromolar binding in the dark will need to be further enhanced. The assays suggest that such noncovalent inhibition should be possible.

However, the formation of the TNF–receptor contact can also be disrupted by modifying the TNF itself. Thus, a so-called phage display method was used (Sect. 11.2). For this purpose, a library of modified surface proteins of a bacteriophage is prepared to which different peptide tails are added. This is achieved by inserting appropriate gene fragments into the bacteriophage gene. The expressed surface proteins with the added peptide sequence were further semisynthetically modified at the cysteine residues and then tested for biological activity. The peptidic ligand **29.28** was isolated from a library as a nanomolar inhibitor. The crystal structure shows that **29.28** disrupts the TNF trimer to form a dimer. The dimer no longer binds to the TNF receptor.

Similarly, the small-molecule inhibitor **29.29** dissociates the TNF trimer into a dimer. This compound was discovered in a screening campaign. Like **29.28**, **29.29** binds to the center between two monomeric units of TNF-α (Fig. 29.13). The "wonder drug" suramin **6.18** (Fig. 6.4) was introduced in Chap. 6. It originally came from dye research. It has since been shown to also inhibit

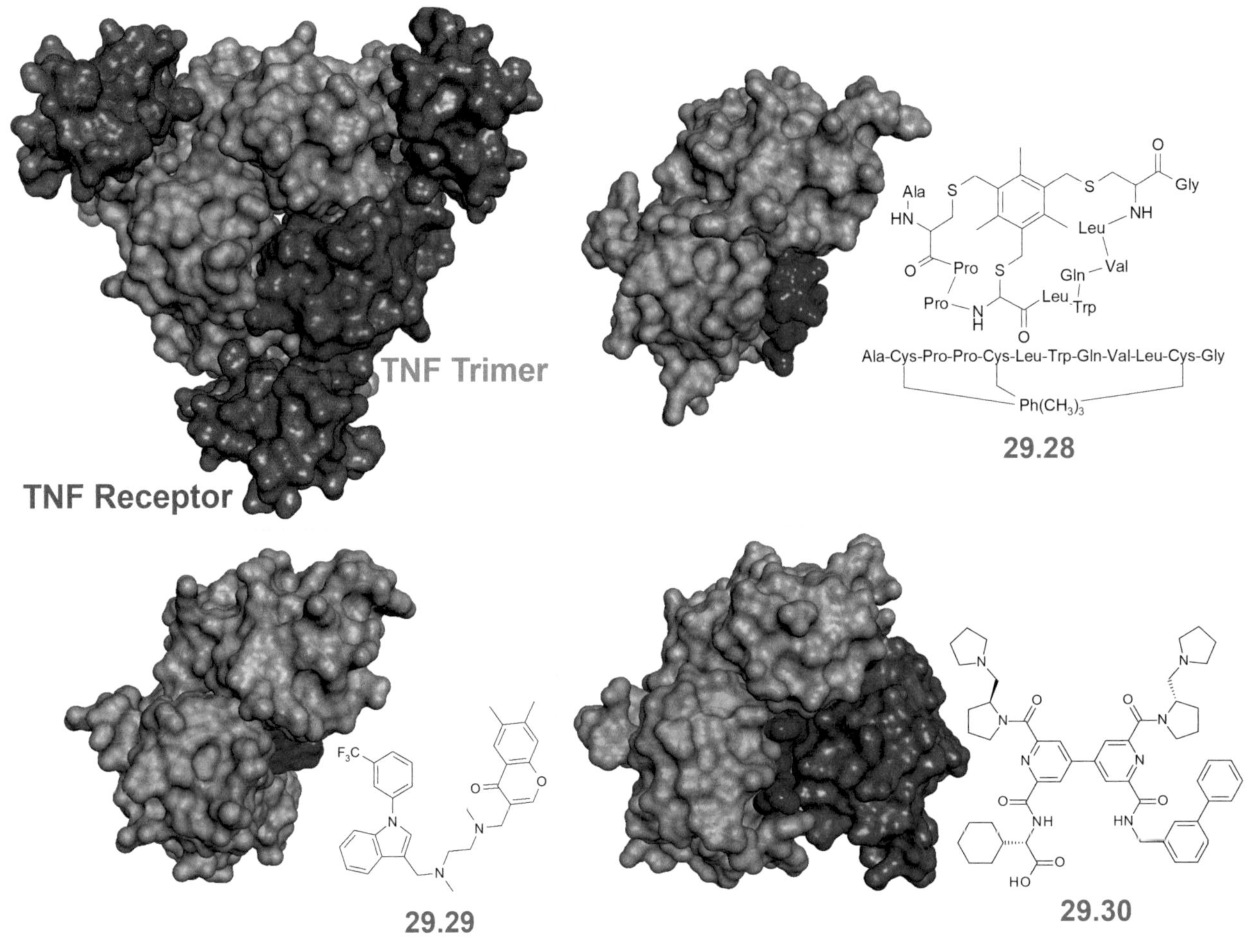

Fig. 29.13 Crystal structures of the trimeric TNF receptor complex (*green* and *brown surfaces*, *upper left*) and two modified TNF cytokines (*upper right* and *lower left*). Compounds **29.28** and **29.29** (*blue surfaces*) are two inhibitors that break up the TNF trimer into a dimer. The dimer can no longer bind to the receptor. With **29.30**, an inhibitor was discovered that expands the trimer of a structurally TNF-analogous cytokine to such an extent that binding to the receptor is prevented (*lower right*)

the TNF/TNF–receptor complex. Therefore, libraries of large "multiarmed" ligands were synthesized and tested against cytokines of the TNF family. The TNF-structurally analogous cytokines CD40 and **29.30** were selected from the library. Interestingly, compound **29.30** binds while maintaining its trimeric structure. It also positions itself in the center between the monomer units, expanding the mutual distance of the monomer units to such an extent that recognition of the trimer at the cytokine receptor is no longer possible.

A second example is the inhibition of complex formation between **interleukin-2** (**IL-2**) and its **heterotrimeric receptor**. T-helper cells of the immune system release IL-2 after an antigen binds to the MHC II receptor on an antigen-presenting cell (Sect. 31.7). This triggers an intracellular signaling cascade that leads to T-cell activation and clonal expansion. Inhibitors of the complex formation of IL-2 with its receptor can be used to suppress the immune response. In a medicinal chemistry initiative at Roche, compound Ro26-4550 **29.31** was discovered as a single-digit micromolar inhibitor. It was not pursued further, presumably due to a lack of cellular activity. Sunesis Pharmaceuticals was able to elucidate the exact binding mode of **29.31** to understand the inhibition mechanism. The compound binds to a conformationally flexible structural motif of IL-2, thereby, altering the contact surface of this cytokine. As a result, the cytokine is no longer recognized at its receptor. Fig. 29.14 shows the receptor complex with bound IL-2 on the left. The small-molecule ligand **29.31** binds with its biarylalkyne moiety to IL-2 in the left part of the binding groove. The interleukin then undergoes a conformational change in this hydrophobic part of the binding site. Conversely, IL-2 remains spatially unchanged in the part of the pocket that accommodates the guanidine moiety of the ligand. Sunesis wanted to explore whether the disulfide tethering method (Sect. 7.10) would reveal suitable molecular building blocks leading to potent IL-2 inhibitors. Several amino acids on the IL-2 surface were replaced by cysteines.

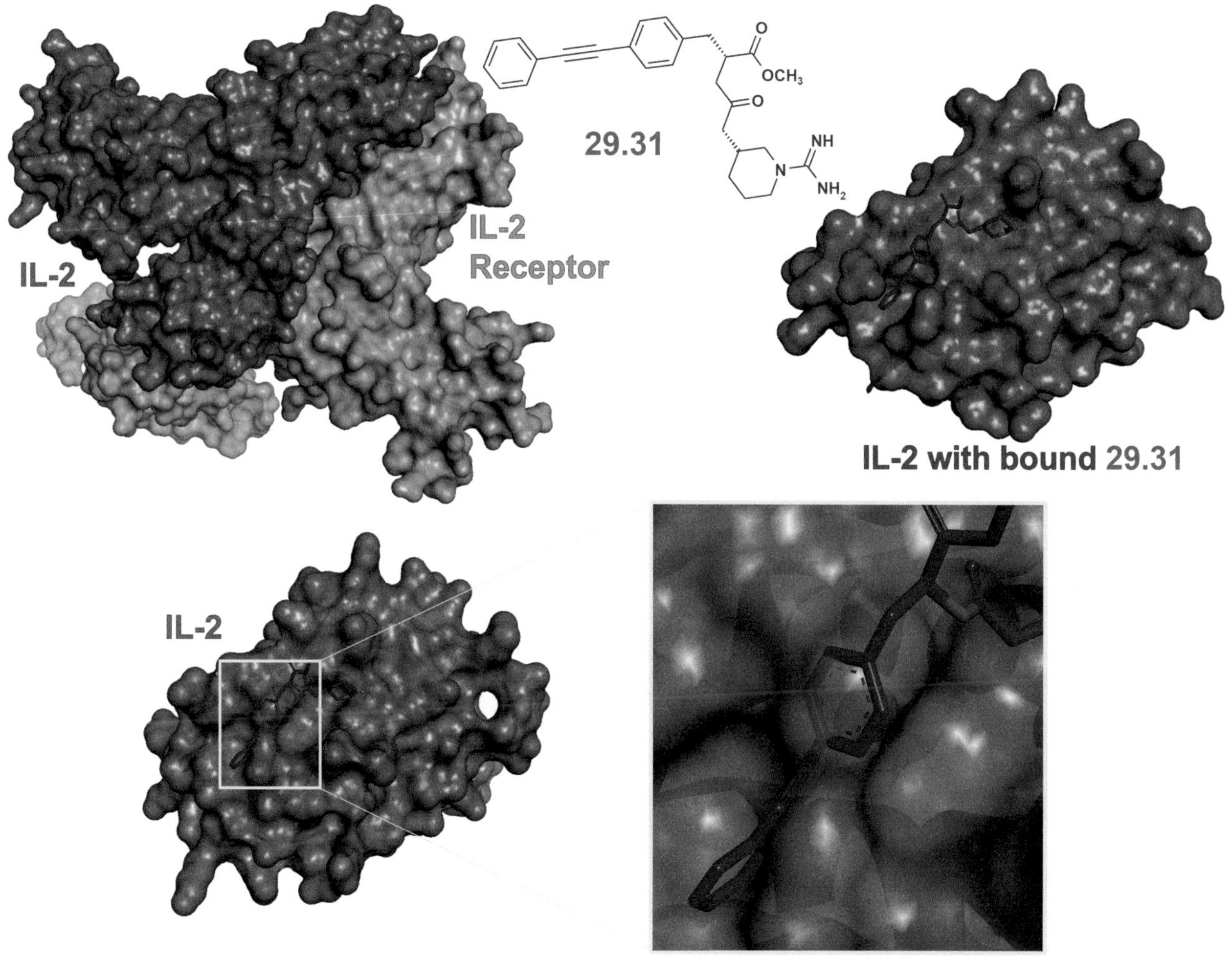

Fig. 29.14 Crystal structure of the complex of IL-2 (*purple*) with its heterotrimeric receptor (*turquoise*, *upper left*). The small-molecule ligand **29.31** was discovered by screening at Roche and binds to IL-2 with micromolar affinity. To achieve this binding, the cytokine undergoes a conformational change in the surface region that accommodates the biarylalkyne moiety (*upper right*, IL2 with *light-brown surface*). Without this conformational change, the ligand would experience steric repulsion on the surface of IL-2 (*lower left*, superposition, IL-2 with *purple surface*, steric clash is shown in the detailed insert). (► https://sn.pub/lIA5h7)

Libraries containing several thousand disulfide derivatives were allowed to react with the modified cytokine. They found what they were looking for at the exchange sites Tyr 31/Cys and Lys 44/Cys. Fig. 29.15 shows the structures with these fragments coupled via disulfide bonds. The fragments bind similarly to the terminal groups of **29.31** in the two anchor regions described. Including suggestions for additional fragments that had attracted attention on other single mutant cysteines, Sunesis was able to develop a 60 nM inhibitor of cytokine binding to its receptor with **29.32**.

These two examples demonstrate that blockade of signal transduction can be achieved by binding a small-molecule ligand to the cytokine itself or to the surface of a corresponding receptor.

29.9 Synopsis

- Membrane-bound receptors of the family of G-protein-coupled receptors (GPCRs) and oligomeric receptors with attached tyrosine kinase domains transmit information from outside of the cell into its interior, thus, enabling signal transduction cascades.
- GPCRs represent a large family of proteins in the human genome; the family has about 800 members and are targeted by 30% of the marketed drugs. GPCR activation is initiated by the binding of an extracellular ligand and transmitted through conformational transformations to the cytosolic side. It leads to structural changes at the binding site where the G-protein interacts with the receptor.

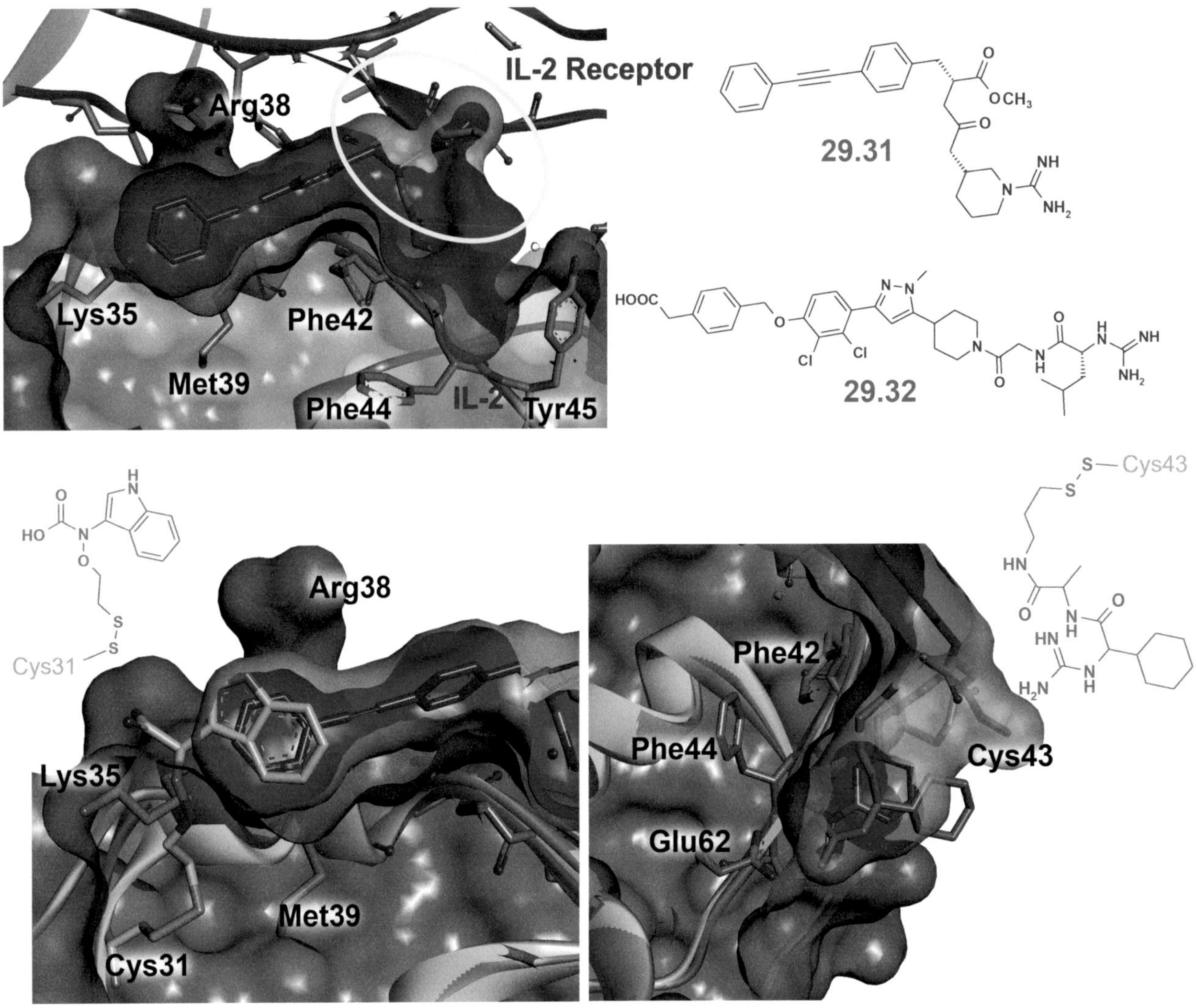

■ **Fig. 29.15** *Upper row* Superimposition of the native IL-2/IL-2 receptor structure with the IL-2 receptor conformationally altered by binding of **29.31** (*green surface, blue inside*). The altered surface (*light brown*) prevents the cytokine from interacting with its receptor. Steric conflicts occur (see inside *yellow ellipse*). *Lower row* Using the disulfide tethering method, two fragments (*green*) were found at the mutated Cys 31 (*lower left*) and Cys 43 (*lower right*), which spatially fill the binding region of the terminal ends of **29.31** (*purple*). Based on additional fragments found by tethering and medicinal chemistry optimization, **29.32** was designed to bind with nanomolar potency to IL-2, thus, efficiently blocking receptor binding. (► https://sn.pub/P5NiRR)

- Extracellular ligands can feature the properties of an agonist by activating the receptor and stabilizing an active conformation. An antagonist prevents the binding of an agonist; however, it does not turn off the basal activity of the receptor. An inverse agonist is able to reduce or fully suppress the receptor function and stabilizes an inactive conformation of the receptor.
- GPCRs exist in many subtypes and their dysregulation is associated with numerous diseases. How they affect the organism depends strongly on the tissue in which they are expressed.
- The first structural insights into GPCRs were obtained from bacteriorhodopsin, a seven-transmembrane receptor that is a proton pump and not a real GPCR.
- Rhodopsin, the light-regulated receptor in the eye, is currently the best studied GPCR. Structures of its inactive and active states are known. They show how activation is transmitted via conformational rearrangements of amino acid residues along with restructuring of a water network leading to movements of transmembrane helices and breaking of salt bridges ("ionic lock"). This changes the structural properties of the recognition site of the G-protein.

Furthermore, the structure of rhodopsin when phosphorylated at the *C*-terminus was elucidated with an arrestin bound to the cytosolic side. GPCRs are desensitized by arrestin binding and can be stored intracellularly in vesicles.

- Crystal structures of the β-adrenergic receptors explain binding features of classical β-blockers and suggest tiny structural transformations from the inactive to active state.
- Antagonists of the peptide-binding AT_1 receptor were developed based on the working hypothesis that the antagonists bind in a structurally analogous manner as the four *C*-terminal residues of the angiotensin II peptide. The resulting sartans are potent antihypertensives. Mutation experiments carried out later suggested that the peptide agonist and the small-molecule antagonists bind to the receptor in distinct regions. Recent structural studies of the AT_1 receptor reveal a common binding site. However, the binding modes deviate from a direct overlap with the *C*-terminus. Thus, an incorrect working hypothesis can lead to a successful drug design.
- The wealth of nuances of our sense of smell is achieved by a simultaneous recognition of odorants at multiple GPCRs with composite and attenuated receptor profiles.
- Genetic polymorphism of the odorant receptors results in an attenuated sensitivity of individuals for different scents.
- Composite receptor profiles and attenuated sensitivity due to genetic polymorphism can be expected for GPCRs targeted by marketed drugs, too.
- Dimerizing or oligomerizing receptors bind growth factors and cytokines and carry a tyrosine kinase domain on the cytosolic side. Upon activation, the kinase domain starts autophosphorylation, which initiates kinase-dependent signaling cascades.
- Activation or suppression of oligomerizing receptors needs ligands that interfere with the binding of macromolecular endogenous ligands. Antibodies have successfully been raised to compete with the natural ligands. Furthermore, small-molecule kinase inhibitors have been developed to block the function of the attached cytosolic tyrosine kinase domain.
- Low-molecular weight ligands can also be found that modify either the receptor or the macromolecular endogenous ligand in such a way that complex formation is blocked. Disulfide tethering and fragment-based approaches have achieved initial successes in the rational design of such compounds.

Bibliography and Further Reading

General Literature

D. M. Rosenbaum, S. G. F. Rasmussen, B. K. Kobilka, The structure and function of G-protein-coupled receptors, Nature, **459**, 356–363 (2009)

D. Hilger, M. Masureel, B. K. Kobilka, Structure and dynamics of GPCR signaling complexes, Nature Struct. Mol. Biol., **4**, 4–12 (2018)

M. Persechinoa, J. B. Hedderich, P. Kolb, D. Hilger, Allosteric modulation of GPCRs: From Structural Insights to in silico Drug Discovery, Pharmaco. Therapeut., **237**, 108242 (2022)

A. S. Hauser, M. M. Attwood, M. Rask-Andersen, H. B. Schiöth, D. E. Gloriam, Trends in GPCR drug discovery: New agents, targets and indications, Nat. Rev. Drug Discov., **16**, 829–842 (2017)

S. B. Gacasan, D. L. Baker, A. L. Parrill, G protein-coupled receptors: the evolution of structural insight, AIMS Biophys., **4**, 491–527 (2017)

J. B. Hedderich, M. Persechino, K. Becker, F. M. Heydenreich, T. Gutermuth, M. Bouvier, M. Bünemann, P. Kolb, The pocketome of G-protein-coupled receptors reveals previously untargeted allosteric sites, Nat. Comm., https://doi.org/10.1038/s41467-022-29609-6 (2022)

R. R. Rexler *et al.*, Nonpeptide Angiotensin II Receptor Antagonists: The Next Generation in Antihypertensive Therapy, J. Med. Chem., **39**, 625–656 (1996)

P. B. M. W. M. Timmermans, P. C. Wong, A. T. Chiu and W. F. Herblin, Nonpeptide Angiotensin II Receptor Antagonists, Trends Pharm. Sci., **12**, 55–61 (1991)

L. B. Buck, Unraveling the sense of smell (Nobel lecture). Angew. Chem. Int. Ed. Engl., **44**, 6128–6140 (2005)

Special Literature

R. Henderson, P. N. T. Unwin, Three-dimensional model of purple membrane obtained by electron microscopy. Nature, **257**, 28–32 (1975)

D. M. Rosenbaum, V. Cherezov et al., GPCR Engineering Yields High-Resolution Structural Insights into β_2-adrenergic Receptor Function, Science, **318**, 1266–1273 (2007)

V. Cherezov et al., High-resolution Crystal Structure of an Engineered Human β_2-adrenergic G Protein-coupled Receptor, Science, **318**, 1258–1265 (2007)

S. G. Rasmussen et al., Crystal Structure of the Human β_2-Adrenergic G-protein coupled Receptor, Nature, **450**, 383–387 (2007)

S. G. F. Rasmussen et al., Crystal structure of the β_2-adrenergic receptor-Gs protein complex, Nature, **477**, 549–556 (2011)

T. Warne, et al., Structure of a β_1-adrenergic G-protein-coupled Receptor, Nature, **454**, 486–491 (2008)

T. Warne et al., The structural basis for agonist and partial agonist action on a β_1-adrenergic receptor. Nature, **469**, 241–244 (2011)

J. Standfuss J, P. C. Edwards et al. The structural basis of agonist-induced activation in constitutively active rhodopsin. Nature, **471**, 656–661 (2011)

X. E. Zhou et al., Identification of Phosphorylation Codes for Arrestin Recruitment by G Protein-Coupled Receptors, Cell, **170**, 457–469 (2017)

P. B. M. W. M. Timermanns et al. Angiotensin II Receptors and Angiotensin II Receptor Antagonists, Pharmaco. Rev., **45**, 205–242 (1993)

H. Ji, W. Zheng, Y. Zhang, K.J. Catt, K. Sandberg, Genetic Transfer of a Nonpeptidic Antagonist Binding Site to a Previously Unresponsive Angiotensin Receptor, Proc. Nat. Acad. Sci. USA, **92**, 9240–9244 (1995)

A. Keller, H. Zhuang, Q. Chi, L.B. Vosshall und H. Matsunami, Genetic Variations in a Human Odorant Receptor Alters Odour Preception, Nature, **449**, 468–472 (2007)

H. Asada, A. Inoue, F. M. Ngako Kadji, C. Suno, J. Aoki, S. Iwata, The Crystal Structure of Angiotensin II Type 2 Receptor with Endogenous Peptide Hormone, Structure, **28**, 418–425 (2020)

R. Bianco et al. Rational Bases for the Development of EGFR inhibitors or Cancer Treatment, Int. J. Biochem. & Cell Biol., **39**, 1416–1431 (2007)

P. De Meyts and J. Whittaker, Structural Biology of Insulin and IGF1 Receptors: Implications for Drug Design, Nat. Rev. Drug Discov., **1**, 769–783 (2002)

N. Wilkie, P. B. Wingrove, J. G. Bilsland, L. Young, S. J. Harper, F. Hefti, S. Ellis, S. J. Pollack, The non-peptidyl fungal metabolite L-783,281 activates TRK neurotrophin receptors, J. Neurochem., **78**, 1135–1145 (2001)

D. W. Banner, A. D'Arcy, W. Janes, R. Gentz, H. J. Schoenfeld, C. Broger, H. Loetscher, W. Lesslauer, Crystal structure of the soluble human 55 kd TNF receptor-human TNF beta complex: implications for TNF receptor activation, Cell, **73**, 431–445 (1993)

P. H. Carter et al., Photochemically enhanced binding of small molecules to the tumor necrosis factor receptor-1 inhibits the binding of TNF-α, Proc. Nat. Acad. Sci. USA, **98**, 11879–11884 (2001)

J. W. Tilley et al., Identification of a Small Molecule Inhibitor of the IL-2/IL-2Rα Receptor Interaction Which Binds to IL-2, J. Am. Chem. Soc., **119**, 7589–7590 (1997)

M. R. Arkin et al., Binding of small molecules to an adaptive protein–protein interface, Proc. Nat. Acad. Sci. USA, **100**, 1603–1608 (2003)

A. C. Braisted, J. D. Oslob, W. L. Delano, J. Hyde, R. S. McDowell, N. Waal, C. Yu, M. R. Arkin, B. C. Raimundo, Discovery of a potent small molecule IL-2 inhibitor through fragment assembly. J. Am. Chem. Soc., **125**, 3714–3715 (2003)

M. R. Arkin, J. A. Wells, Small-Molecule Inhibitors of Protein–Protein Interactions: Progressing towards a Dream, Nat. Rev. Drug Discov., **3**, 301–317 (2004)

D. J. Stauber et al., Crystal structure of the IL-2 signaling complex: Paradigm for a heterotrimeric cytokine receptor, Proc. Nat. Acad. Sci. USA, **103**, 2788–2793 (2006)

Ligands for Channels, Pores, and Transporters

Contents

G. Klebe, *Drug Design,* https://doi.org/10.1007/978-3-662-68998-1_30

The cell is the smallest structural and functional unit of all living organisms. Single-cell organisms contain only one such unit. In complex organisms, such as humans, there are 10^{13}–10^{14} cells. Because of their constitution, **cells** are capable of **metabolism**. They have a complex architecture that is directly related to their function. Due to the **high degree of differentiation** in higher organisms, it is not possible to speak of a typical, representative cell. Every cell is surrounded by a cell membrane. This membrane ensures that the cell is an individual, self-contained unit. Signals have to be transmitted across this membrane. Systems that accomplish this task were discussed in Chaps. 28 and 29. However, the exchange of substances must also be possible so that the cell can be supplied with the substances necessary for its function. The **selective permeability of the membrane** is of particular importance. Amphiphilic compounds can passively diffuse through the membrane. For example, the steroid hormones discussed in Chap. 28 have this property. Polar compounds such as amino acids, peptides, or sugars do not passively cross the membrane, but are essential for the maintenance of the cell. For this reason, the cell is equipped with special **transporters**, some of which are highly selective and some of which are surprisingly promiscuous. Since the transport of polar compounds in most cases occurs against a concentration gradient, it is usually only possible by consuming energy. Therefore, Nature couples the task of such a transport with an energetically favorable reaction. In biological systems, the **hydrolysis of the triphosphate** unit of **ATP** serves this purpose (Sect. 4.3).

Another group of charged particles, **ions**, are of fundamental importance for the regulatory function of cells. Without special protein systems, however, they could not permeate the membrane. If there are different concentrations of certain ions inside and outside the cell, there will be a difference in the **electrochemical potential**. Changes in membrane permeability to ions play a crucial role in cell stimulation and signal transduction. Nerve and muscle cells in particular respond to such stimuli with specific changes in their state. For example, the contractions of muscle cells determine the heartbeat. Nerve cells transmit stimuli over short or long distances and serve to distribute information in the central nervous system.

The establishment and maintenance of such **concentration gradients** across the membrane requires the transport of the relevant ions across the membrane barrier. It is primarily the **ion pumps** that establish a concentration gradient across the membrane barrier. They are relatively slow and energy consuming. Therefore, their function is coupled to an energetically favorable reaction, usually the hydrolysis of ATP. Ion pumps achieve a transport rate of 10^{2}–10^{4} particles per second. Although their local density of 10^{3}–10^{5} molecules per μm^2 is relatively high, ion pumps are much too slow for fast switching of cellular processes. Therefore, there are **specific ion channels** responsible for the selective and passive passage of ions along a concentration gradient. They achieve a flow rate of 10^{6}–10^{8} ions/s, which is only slightly below the diffusion rate. Their occupancy density in the membrane is much lower at 1–10 molecules/μm^2. Ion channels are either **voltage- or ligand-gated** and allow the membrane potential to change within milliseconds. Once the pumps have established an electrochemical gradient across the membrane, the opening of a particular ion channel leads to a flow of ions across the membrane in a concentration gradient for purely entropic reasons.

The cell must also regulate its **water homeostasis**. Some individual water molecules can diffuse directly across the membrane. However, to transport larger amounts of water across the membrane, specific pores called **aquaporins** are required to regulate the entry and exit of water according to the given osmotic pressure gradient.

These systems of specific particle transport across a membrane will be considered in detail in this chapter. They are all **integral membrane proteins**. Examples of these membrane proteins that have been characterized by structural biology will be presented. Ligands that provide an approach to therapeutically relevant regulation of these proteins will be discussed. Furthermore, bacterial transport systems that alter the permeability of membranes can act as antibiotics to combat other microorganisms.

30.1 Electric Potential and Ion Gradients Stimulate Cells

Surely everyone is familiar with the construction of **electrochemical redox cells** from electrochemistry. If a U-shaped vessel with an ion-permeable glass membrane is filled on both sides of the membrane with solutions of different concentrations of, for example, copper sulfate, and copper metal plates are placed in both solutions, a voltage difference can be measured. The voltage can be calculated using the Nernst equation. Since identical half-cells are used on both sides of the membrane, only the logarithm of the concentration difference on both sides and the number of migrating charges determines the potential. As a thought experiment, imagine that the different sides of the vessel are filled with potassium and sodium chloride solutions. The separating membrane should be permeable only to potassium ions and impermeable to sodium ions. The system will try to equalize the **concentration gradient** of Na^+ and K^+ ions for entropic reasons. However, the membrane only allows this for potassium ions. Therefore, an excess of positive ions accumulates on one side of the membrane and a deficit on the other. As in the first case, the potential difference can be calculated from the concentration difference at the interface using the Nernst equation. After a short time,

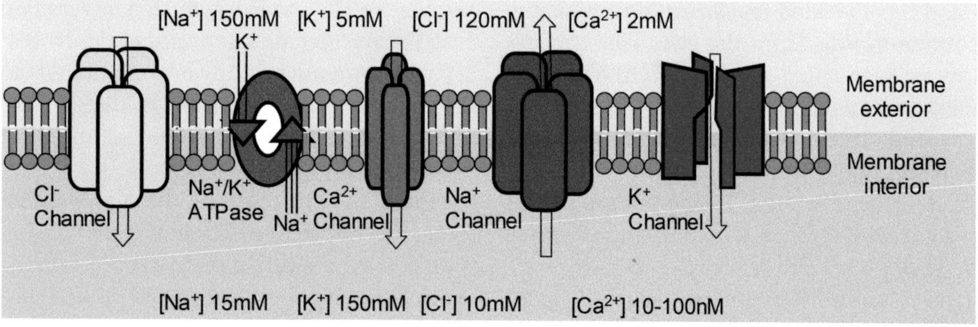

Fig. 30.1 Various pumps and ion channels provide the calibration of ion gradients across the cell membrane, creating a potential difference across the membrane. They can be either ligand-gated or voltage-gated. Potassium channels (*blue*) are highly selective for potassium ions and are largely responsible for the calibration of the resting potential. An action potential is triggered by the opening of fast sodium (*red*) and calcium (*green*) channels, leading to depolarization. A pump (*violet*) that exchanges three Na^+ ions for two K^+ ions restores the resting Na^+/K^+ ion concentration. Chloride channels (*yellow*) allow the influx of Cl^- ions, which hyperpolarizes the cell and prevents depolarization. The concentrations in mol/L (M) given on both sides of the membrane correspond to the approximate values in the resting state

the net migration of potassium ions stops because the effort to equalize the concentration difference is counteracted by the **repulsive electrical potential difference** that builds up and repels further potassium ions from migrating. Only a few potassium ions have actually migrated until the system reaches equilibrium. A dynamic equilibrium is established, which means that the number of ions migrating from one side to the other or back is exactly the same.

In living Nature, it is mainly sodium, potassium, calcium, and chloride ions that create such potential differences between the inside and outside of the cell. First of all, ion pumps are responsible for creating a concentration gradient across the membrane. If the cell membrane were only permeable to potassium ions, a 30-fold concentration gradient of K^+ ions between the inside and the outside would result in a voltage of −90 mV (Figs. 30.1 and 30.2). This is the situation in the resting state of a cell. As described in the thought experiment, the outflow of K^+ ions through a highly specific potassium channel creates this voltage difference. The −90 mV mentioned above is not measured, but a resting membrane potential of about −70 mV is observed (Fig. 30.2). Since other ions also have a certain permeability, the actual measured membrane potential at any given time reflects a complex mixture of different contributions from individual ions and their conductivities (Fig. 30.1). To stabilize the cell in a particular phase, for example, in the resting state, the cellular ion distribution is maintained by **Na^+/K^+ ATPases**. They pump ions against the electrochemical concentration gradient by consuming ATP. For each transport process, three sodium ions are pumped out of the cell and two potassium ions are pumped in. There is another such pump that determines the concentration of calcium ions.

When the cell is stimulated by a so-called **action potential**, the membrane permeability for the individual ions changes. First, the permeability for sodium ions changes dramatically. In the resting state, the concentration of sodium ions in the extracellular space is about ten times higher than in the cell interior. At a threshold potential of about −60 mV, the sodium channels open and the membrane potential is temporarily shifted into the positive range to about +40 mV by **depolarization**. Even before the Na^+ equilibrium potential of about +60 mV is quickly reached, the rapid influx of sodium ions stops because the sodium channel closes again. Finally, the potential changes in the direction of the rest-

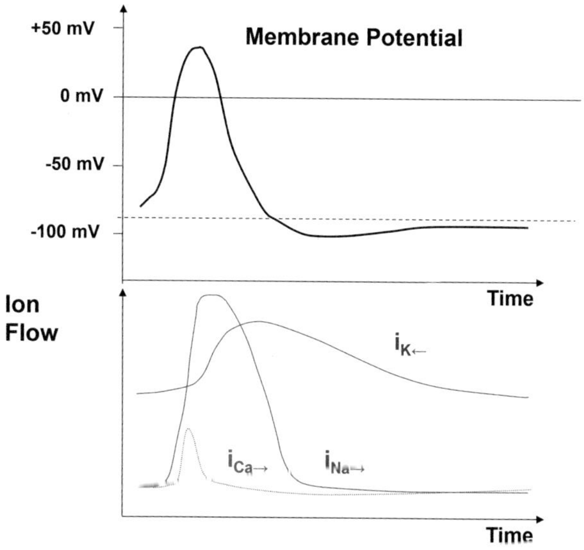

Fig. 30.2 The membrane potential in the resting state is about −70 mV and is stabilized by the efflux of potassium ions (i_{K^+}←, *blue*). Upon excitation, a fast sodium channel opens. The influx of sodium ions (i_{Na^+}→, *red*) shifts the membrane potential by about 100 mV in the positive region. When this value is reached, the sodium channel closes. The efflux of potassium ions repolarizes the cell and shifts the potential below the threshold of the resting membrane potential (hyperpolarization). In cells that have calcium channels, their opening can also contribute to depolarization and, therefore, to an action potential ($i_{Ca^{2+}}$→, *green*)

ing potential and the so-called **repolarization** causes the outflow of potassium ions from the cell. This process is controlled by voltage-gated K_V potassium channels. The membrane potential drops slightly below the resting membrane potential, and the potassium channels close (◘ Fig. 30.2). When the membrane potential has fallen to more negative values than the resting potential, this is called **hyperpolarization**. The excitability of the cell is reduced in this state.

Calcium ions can contribute to the stimulation and action potential of a cell. They play a crucial role in heart muscle cells, for example. The extracellular calcium concentration is significantly higher than the intracellular concentration. Therefore, an additional influx of Ca^{2+} ions into the cell can enhance depolarization (◘ Fig. 30.1). Ca^{2+} enters the cell through highly selective **calcium channels**. A slowing of the depolarization can be achieved by blocking the opening of these sodium or calcium channels. Some local anesthetics work by inhibiting sodium channels in nerve cells. Calcium channel blockers reduce the influx of Ca^{2+} ions. This slows down the upstroke of the action potential in the heart muscle cells, and the heart muscle works more economically. For this reason, drugs such as nifedipine, diltiazem, and verapamil are used to treat hypertension or arrhythmias (Sect. 30.4).

The electrophysiological processes described in this section are greatly simplified. Depending on the function and tissue-specific location of the cell in question, several ion-specific channels are at work to achieve a finely tuned setting of the required membrane potential.

30.2 Molecular Function of a Potassium Channel at the Atomic Level

The finely attenuated setting of ion gradients across the membrane, which determines the overall membrane potential, shows that channels must have **high selectivity** for individual ions. The difference between the ions that must pass is very small. Sodium and potassium ions have the same charge and differ in size by only a little more than 0.35 Å. Their **hydration enthalpies** are slightly different, but the geometry of their hydration shells is different. The larger potassium ion is surrounded by eight water molecules, while the sodium ion prefers six nearest neighbors. How can a protein efficiently exploit this small difference to create a selective ion filter? The discrimination achieved is impressive: only one sodium ion is smuggled in for every 10,000 potassium ions!

Ion channels are huge molecular constructions. They are embedded in the membrane. It is extremely difficult to remove them from the membrane and embed them in a crystal lattice with auxiliary material without destroying them. Once this is done, a crystal structure can be determined. Roderick MacKinnon achieved this masterpiece in 1998 at Rockefeller University in New York. Just 5 years later, he was awarded the Nobel Prize. Today, cryo-electron microscopy is available as another method for structure determination. In recent years, it has provided exciting insights into the fascinating world of these membrane channels.

First, the structure of the **KcsA potassium channel** from the bacterium *Streptomyces lividans* was determined. The channel consists of a homotetramer and crosses the membrane with two long helices per monomer (◘ Fig. 30.3). The *C*-terminal end of another shorter helix is oriented in a cavity in the middle of the channel. Such a helix forms a dipole moment due to the periodic orientation of well-aligned amide bonds along the protein backbone (Sect. 14.2). The preferred binding site for a positive charge is formed at the end of such a helix (◘ Fig. 30.4). Four of these helices are oriented towards the cavity inside the channel. The potassium ions, surrounded by a shell of eight water molecules, are essentially pulled out of the cytosol. This allows the potassium ions to enter the hydrophobic membrane environment. However, this does not discriminate against sodium ions. After the acceleration process into the channel, a selec-

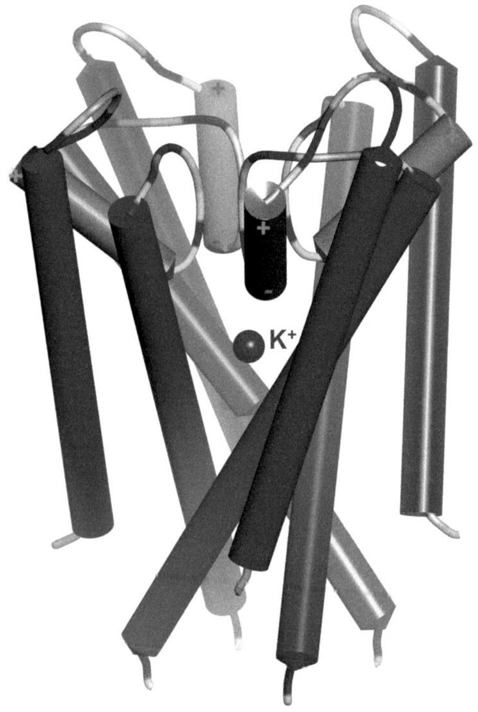

◘ **Fig. 30.3** The four shorter helices (*bright red*) of the tetrameric potassium channel orient their negatively polarized *C*-terminus towards the binding site where the potassium ion (*violet sphere*) sheds its water shell. They draw the positively charged ions into the ion channel and stabilize them in the interior. (► https://sn.pub/bu2aBz)

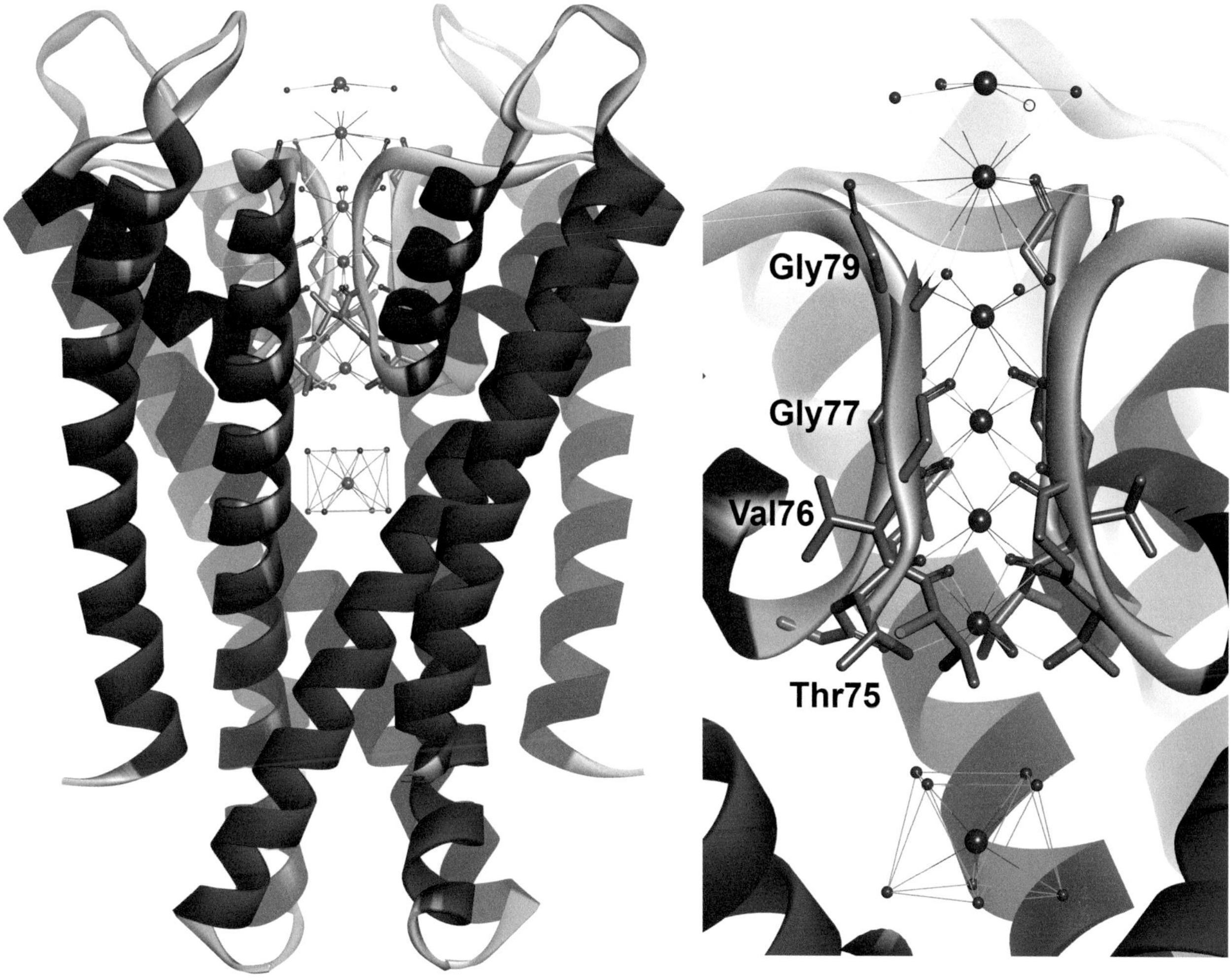

Fig. 30.4 Crystal structure of the bacterial potassium channel KcsA in the open state. The channel forms a tetramer (*left*), each monomer is constructed from three helices. Two of these helices (*red*) traverse the entire membrane whereas the third, shorter helix (*blue-violet*) is oriented towards a cavern in the interior of the channel. There, the potassium ions (*violet spheres*), which are surrounded by eight water molecules, shed their water shell and enter the selectivity filter (*right*). A potassium ion with its square-antiprismatic coordination is shown before entering into the filter. The carbonyl groups from the main chain adopt the octavalent coordination sphere wrapping around the potassium ion with similar geometry and transfer the ion across the membrane. (▶ https://sn.pub/hAr03O)

tivity filter is applied. This requires the potassium ions to shed their hydration shells. During the structure determination, it was possible to capture some potassium ions in the channel. A potassium ion has a **square-antiprismatic water shell** just before entering the selectivity filter (Fig. 30.4). The latter is constructed so that four oxygen atoms from four threonine residues (Thr 75) of the tetramer occupy four coordination sites of the eight fold coordinated potassium ion. Four adjacent carbonyl oxygen atoms of the main chain (Thr 75, Val 76, Gly 77, and Gly 79) act as additional coordinating ligands. This motif of four carbonyl groups forming a ring is arranged as a **selectivity filter** and is repeated three times. These ring-oriented carbonyl groups adopt a relative arrangement that replaces the perfectly square-antiprismatic coordination of the water molecules in the potassium ion coordination sphere. Nature has ingeniously reconstructed the coordination geometry of the potassium ion. With the architecture constructed from the TVGYG motif, it has achieved impressive selectivity. The orientation of the coordinating groups simply does not fit around lithium or sodium ions. Nature preferred to use main-chain carbonyl groups for this task. Side-chain oxygen functionalities would be too spatially flexible for this purpose. Only side-chain OH groups from four threonine residues, which have a certain degree of mobility, open the entrance.

Meanwhile, the structure of a channel from *Bacillus cereus*, which has the same general architecture but lacks the selectivity for potassium ions, has been determined.

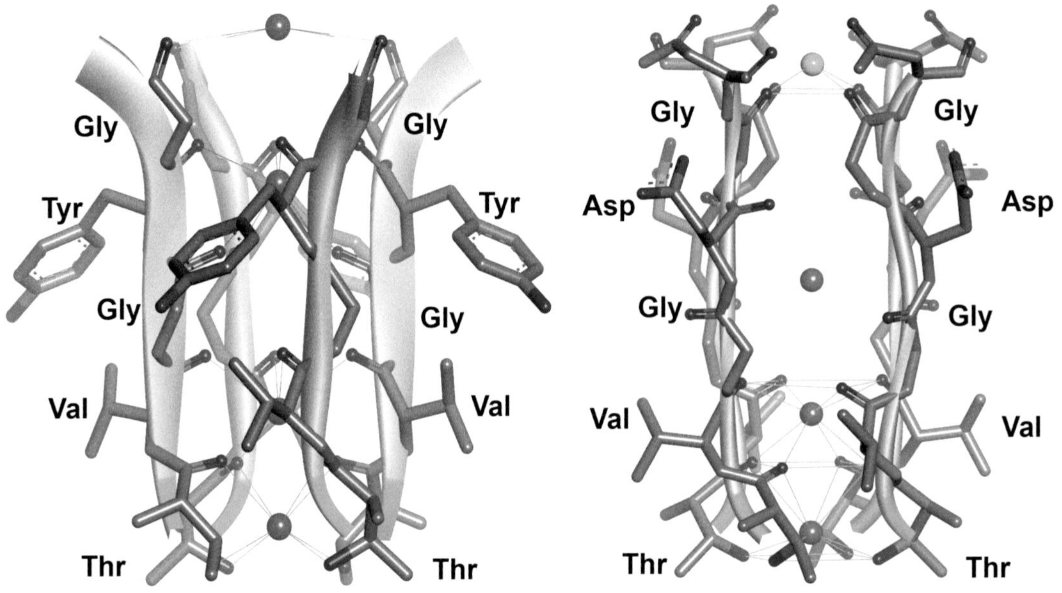

30

Fig. 30.5 Comparison of the tetrameric ion filter of the highly selective potassium channel KcsA from *Streptomyces lividans* (*left*) and the sodium and potassium-permeable channel from *Bacillus cereus* (*right*). The selective channel forms a tetramer from a TVGYG motif; a TVGDG is found at the same place in the Na^+/K^+ channel. Both channels have the same geometry in the lower part formed by threonine and valine residues. The backbone carbonyl groups of the amino acids Gly–Tyr are rotated towards the interior in the potassium-selective channel and contribute to the filter, whereas the C=O groups from the four Gly–Asp motifs are rotated away in the unselective channel. It opens to a chamber that can accommodate an ion, but does not achieve selectivity filtering. (▶ https://sn.pub/JJdImS)

It can hardly distinguish between K^+ and Na^+ ions. The structure of the selectivity filter in this protein is different. The tyrosine residue in the TVGYG motif of the potassium channel has been replaced by an aspartic acid to form TVGDG. This has far-reaching consequences. In the lower part, formed by the threonine and valine residues, the geometry of the filter remains largely unchanged in both channels. In the potassium-selective channels, the backbone carbonyl groups of the following amino acids are turned inwards and contribute to the filter (Fig. 30.5). In nonselective channels, these carbonyl groups are turned away from this area and open a chamber that can accommodate an ion, but cannot achieve selectivity filtering.

The mechanism of opening and closing of the potassium channel has been further elucidated by structural studies of channels from other organisms. A change in the membrane potential of about 50 mV causes the channel to open. Since this voltage difference occurs over a distance of about 50 Å, it causes an enormous effect of about 100,000 V/cm. Obviously, parts of the channel that sense the voltage difference become highly positively charged and float like paddles on the outside of the membrane. A change in the voltage across the membrane causes these paddles to move, initiating the opening or closing of the channel. A kink in one of the extended transmembrane helices enables this process. A combination of a kink and turn movement of about 30° of the helical end in each subunit of the tetramer causes closure or opening of the channel. A highly conserved glycine residue is found at the kink position. The absence of a side chain gives this amino acid greater conformational flexibility. Therefore, glycine is predominantly involved in conformational switches.

One group of potassium channels is ATP-gated. Their structural architecture is much more complex than the described bacterial channel. Two genes, *Kir6.1* and *Kir6.2*, are known to encode the pore-forming part of the ATP-dependent channel. The channels are hetero-octamers, each consisting of four Kir channel proteins and four regulatory units. The latter are called **sulfonylurea receptors** because they can be blocked by sulfonylureas. ATP binds to the Kir subunit and the channel closes. In a multistep process, ATP is hydrolyzed to ADP and the ATP induced closure is reversed. Dissociation and rebinding of Mg–ADP keeps the channel open. The state of the channel, therefore, depends on the ATP/ADP ratio in the cell. Drugs such as pinacidil **30.1**, diazoxide **30.2**, or levcromakalim **30.3** are known to stabilize the channel in the open state (Fig. 30.6). Pinacidil is used to treat high blood pressure, and diazoxide is used in the therapy of Langerhans islet cell tumors. In contrast, the large group of sulfonylureas (Fig. 30.6) blocks the regulatory subunit and leads

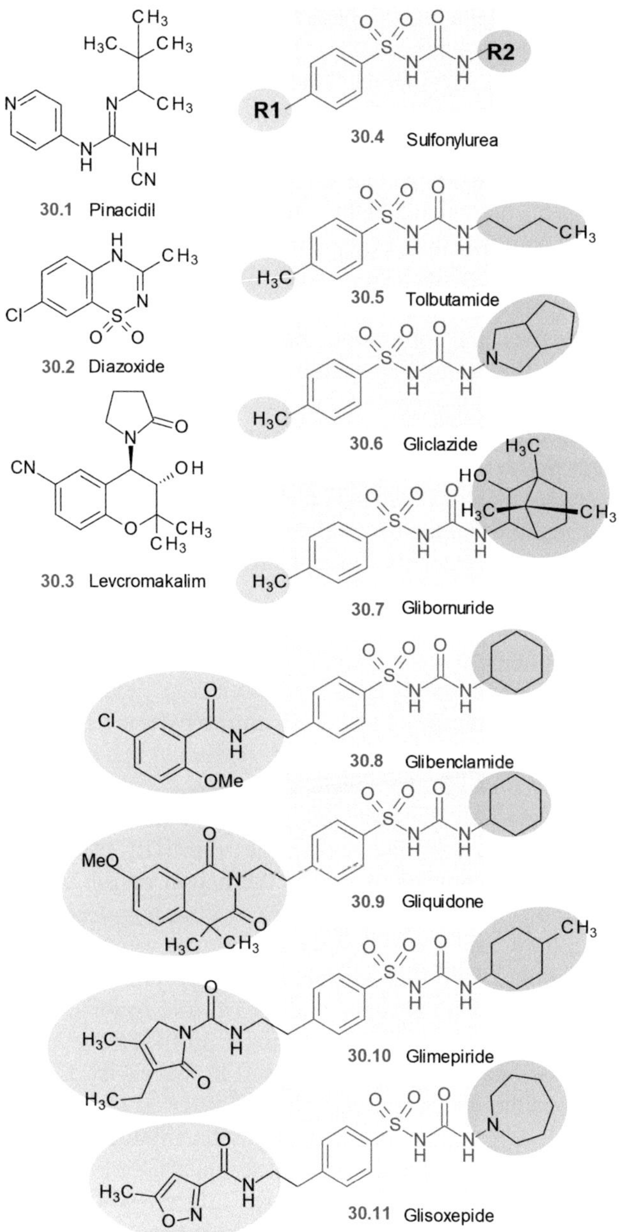

Fig. 30.6 Pinacidil **30.1**, diazoxide **30.2**, and levcromakalim **30.3** are potassium channel openers. In contrast, sulfonylureas such as **30.4** block the regulatory subunit of the ATP-dependent potassium channel in the insulin-producing cells of the pancreas. The basic scaffold of the sulfonylureas can be broadly varied on both termini (**30.5**–**30.11**) with aliphatic (R1, *green*), aromatic, or other cyclic groups (R2, *blue*)

to closure of the attached potassium channel in the insulin-producing cells of the pancreas. Elevated glucose concentrations stimulate insulin secretion from pancreatic β-cells. This release occurs in response to a number of intracellular metabolic and electrophysiological processes. Glucose enters the β-cells via the GLUT-2 transporter. There it is phosphorylated and extensively metabolized. This leads to an increase in the intracellular ATP/ADP ratio and is associated with the closure of ATP-dependent potassium channels. The membrane potential is depolarized. When the threshold potential of about −50 mV is reached, the voltage-gated calcium channels open. The influx of calcium triggers an action potential. Finally, at the end of this complex cascade, insulin is released by the fusion of insulin-containing granules (membrane-bound lysosomal vesicles) with the cell wall. Insulin secretion occurs in two phases. The first, immediate release occurs by fusion and opening of the vesicles located in the direct vicinity of the membrane. In the second phase, insulin granules must first be reconstituted from cellular storage. This is caused by a signaling cascade initiated by the signaling of some GPCRs (e.g., GLP1-R, GPR119). These receptors are stimulated by incretin hormones such as GLP-1 (Sect. 23.6). Activation of the downstream adenylate cyclase (Sect. 29.1) finally leads to an increase in intracellular calcium and, thus, to increased insulin release. Sulfonylureas block the regulatory subunit of ATP-dependent K^+ channels and mediate an increase in intracellular calcium concentration. The resulting effect is analogous to an increase in the ATP/ADP ratio and, as a result, stimulates insulin secretion from pancreatic β-cells. This is used as a therapeutic principle in the treatment of **type 2 diabetes mellitus**. A side effect of sulfonylureas is the risk of insulin secretion despite low glucose levels. This can lead to life-threatening hypoglycemia. Therefore, agonists of the GLP1-R and GPR119 receptors have been developed that do not release insulin at low glucose levels.

30.3 Binding Undesirable: The hERG Potassium Channel as an Antitarget

In September 2007, the drug clobutinol **30.12** was withdrawn from the market after more than 45 years of use in therapy (Fig. 30.7). This drug was used to treat dry cough. It is estimated that approximately 200 million patients used it over the years. It was even converted to an over-the-counter drug. Later clinical studies in healthy adults raised suspicions that the drug could cause **cardiac arrhythmias**, which in the worst cases could be fatal. Several other well-known drugs have also been withdrawn from the market. Terfenadine **30.13**, astemizole **30.14**, sertindole **30.15**, thioridazine **30.16**, grepafloxacin **30.17**, and cisapride **30.18** have been withdrawn or their use severely restricted (Fig. 30.7). In all of these cases, the risk of a rare but life-threatening cardiac arrhythmia was the reason for withdrawal. This is all the more disturbing because these are drugs that are not usually used to treat life-threatening conditions, but rather to treat conditions such as chronic cough, allergies, infections, or gastrointestinal disorders.

What happens that suddenly causes arrhythmias, which in the worst case can lead to death, especially during physical activity? The depolarization and repo-

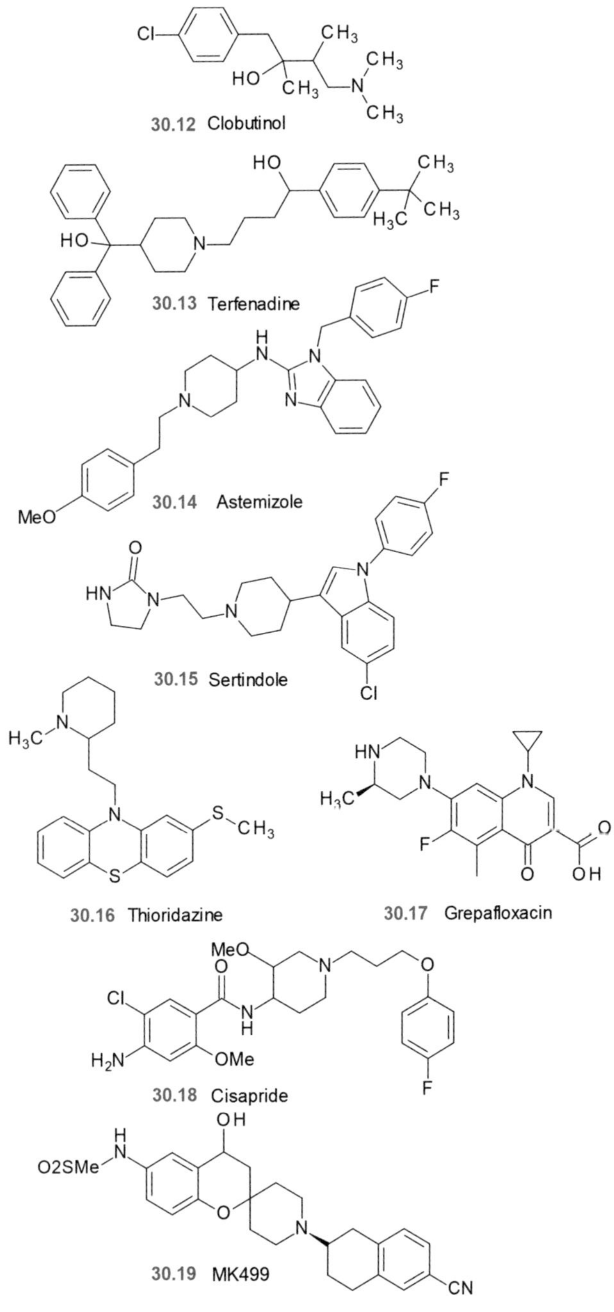

Fig. 30.7 Clobutinol **30.12** was withdrawn from the market after 45 years of clinical use because of the risk of provoking an arrhythmia. Terfenadine **30.13**, astemizole **30.14**, sertindole **30.15**, thioridazine **30.16**, grepafloxacin **30.17**, and cisapride **30.18** met the same fate and were either withdrawn, or their indications for use were severely limited. MK499 **30.19** is a potent class II antiarrhythmic agent, and it binds to the hERG potassium channel

larization of heart muscle cells, as described above, is regulated by the influx and efflux of sodium and potassium ions through ion channels. Therefore, some drugs are known to act as antiarrhythmics by blocking the sodium channel. Other drugs inhibit the potassium channel and prolong the action potential of cells. A **prolongation of the QT interval**, the time between the beginning and end of the ejection phase of the heartbeat, can result in a dangerous arrhythmia. This can lead to sudden racing of the heart (*torsades de pointes tachycardia*, Fig. 30.8), ventricular fibrillation, and cardiac arrest. The prolongation of the QT interval is caused by blockade of a potassium channel, the **hERG** (*human ether-à-go-go-related* gene) channel. The channel was discovered as a result of detailed genetic studies of patients with inherited long QT syndrome. An unwanted side effect of a drug can cause the same condition if the hERG channel is inhibited by the drug being administered. Although this side effect is rare, it is extremely dangerous in acute cases. It has been estimated that approximately 3000 deaths per year in the U.S. can be attributed to such adverse events. In order to avoid such side effects, attempts are now being made to eliminate binding to the hERG channel as early as the drug development stage. The hERG channel is related to the bacterial KcsA channel discussed in the previous section.

An alanine scan was performed to determine which amino acids are critical for inhibition. The altered binding of the potent class II antiarrhythmic MK499 **30.19** was tested. Two aromatic residues in this channel, Tyr 652 and Phe 656, were found to be critical. They are located on the four subunits in the interior of the broad cavity before entering the selectivity filter (see Fig. 30.3, approximately at the height of the potassium ion position). Furthermore, the binding decreased even more when four additional residues were replaced by alanine. With this information, models of the hERG channel were constructed and meanwhile confirmed by a cryo-EM structure with astemizole **30.14**. The residues identified as critical are all aligned in this cavity. Drugs responsible for prolonging the QT interval fit into the model, suggesting a sticky hydrophobic interaction with the aromatic residues. It allows the construction of a superposition model of the known inhibitors. It shows that inhibitors bind with an extended geometry and have a central charged basic nitrogen atom. This atom is in the center of a pyramidal arrangement formed by three to four hydrophobic aromatic moieties. This spatial pattern has been further refined using structure–activity relationships. It serves as a kind of reference to check whether newly designed compounds could possibly bind to the hERG channel. The goal of this design is not to optimize, but to prevent binding. The hERG channel is, therefore, considered an **antitarget**. In addition to these design considerations, the actual hERG channel inhibition of synthesized compounds is measured today. In this way, an attempt is made at an early stage of drug discovery to avoid the bitter and very expensive surprise of finding serious side effects later on.

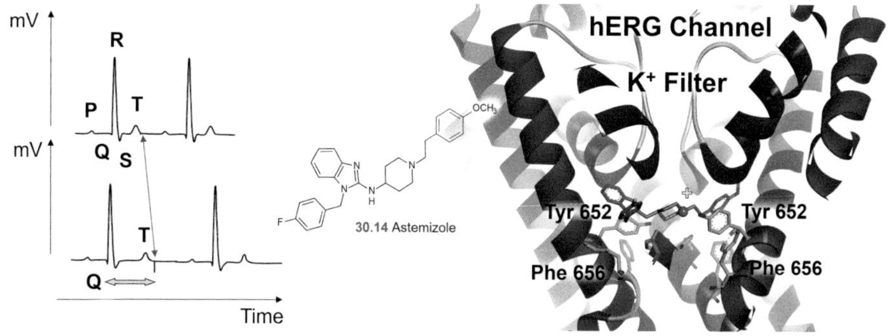

Fig. 30.8 *Left* The prolongation of the QT interval between the beginning (Q) and end (T) of the heart's ejection phase can lead to fatal arrhythmias, including sudden tachycardia (heart racing), ventricular fibrillations, and cardiac arrest. *Right* The cryo-EM structure of the hERG channel shows two aromatic residues, Tyr 652 and Phe 656, on each of the four long helices, oriented towards the interior of the broad cavity before the entrance gate of the potassium selectivity filter. Astemizole **30.14** is also bound in this cavity. It places its charged basic nitrogen beyond this entrance and the attached hydrophobic aromatic moieties interact with the aromatic residues in the channel. (► https://sn.pub/xamjf9)

30.4 Electromechanical Control of Voltage-Dependent Ion Channels: How Small Ligands Tighten of Loosen a Hydrophobic Belt in Ion Channels

In recent years, **voltage-gated sodium and calcium channels** have been structurally characterized by X-ray structure analysis and cryo-electron microscopy (Sect. 13.6). These evolutionarily related transmembrane-signaling proteins play important roles in action potential initiation, neurotransmission, excitation–contraction coupling, and many other physiological processes. Voltage-gated sodium (Na_V) channels **initiate action potentials** in nerve, muscle, and other electrically excitable cells. As described in Sect. 30.1, sodium channels are rapidly activated during depolarization of cells such as nerve or muscle fibers, shifting the membrane potential into the positive range. After a few milliseconds, the sodium channels are rapidly inactivated. The sodium conductance returns to nearly its initial value. Prolonged or repeated depolarizations put the sodium channels in a closed, inactivated state. Overall, three functional states are distinguished: the "**closed-resting**" state, the "**open-activated**" state, and the "**closed-inactivated**" state.

Sodium conductance is mediated by an ion selectivity filter that controls the entry of sodium ions and restricts the passage of other ions. Na_V channels are regulated by intracellular signaling pathways. Because of the importance of their function, they are interesting targets for drug therapy of neurological, psychiatric, and cardiovascular diseases. Drugs such as local anesthetics, neurotoxins, alkaloids, or antiarrhythmics can be used to treat diseases such as epileptic seizures, nerve pain, cardiac arrhythmias, or muscle diseases. In many types of excitable cells, voltage-gated calcium channels (Ca_V channels) are activated during an action potential. As a result, they conduct calcium ions into the cells. They increase depolarization and, thus, contribute to the control of physiological processes such as contraction, neurotransmission, secretion, and gene transcription. They also have an ion selectivity filter, which has a slightly different amino acid composition than in Na_V channels. Classical calcium channel blockers are used to treat cardiovascular diseases such as arrhythmia, hypertension, and angina pectoris.

The channels embedded in the cell membrane consist of four homologous domains. In bacteria these are identical; in more advanced organisms they are related but not identical. Each domain is composed of six helical segments. These include the voltage-sensing domains, which are formed by a bundle of four transmembrane α-helices. They sense the potential difference across the membrane. The voltage-dependent activation is explained as an outwards movement of the Arg-residue-containing structural components of the voltage sensors in response to a change in the electric field. This leads to the activated state and a specific conformation of the channel. In addition, the four domains contribute to the pore, along which ions pass through the membrane. These sections are connected by loop regions that play a critical role in mechanical control and conformational adjustment. In simplified terms, there are three functional states of these channels: The closed-resting state, the open-activated state, and the closed-inactivated state. However, there are probably many more states that such a channel can adopt. Interestingly, different drugs respond to the different conformational states. This results in different profiles for the flow of ions across the membrane, which may be important for therapy.

The topology of the folded sodium and calcium channels is very similar (Fig. 30.9). In detail, the selectivity is achieved by the respective amino acid composition. Similar to the potassium channels discussed earlier (Sect. 30.3), the Na_V and Ca_V channels have a fourfold architecture. The sodium and calcium ions travel long dis-

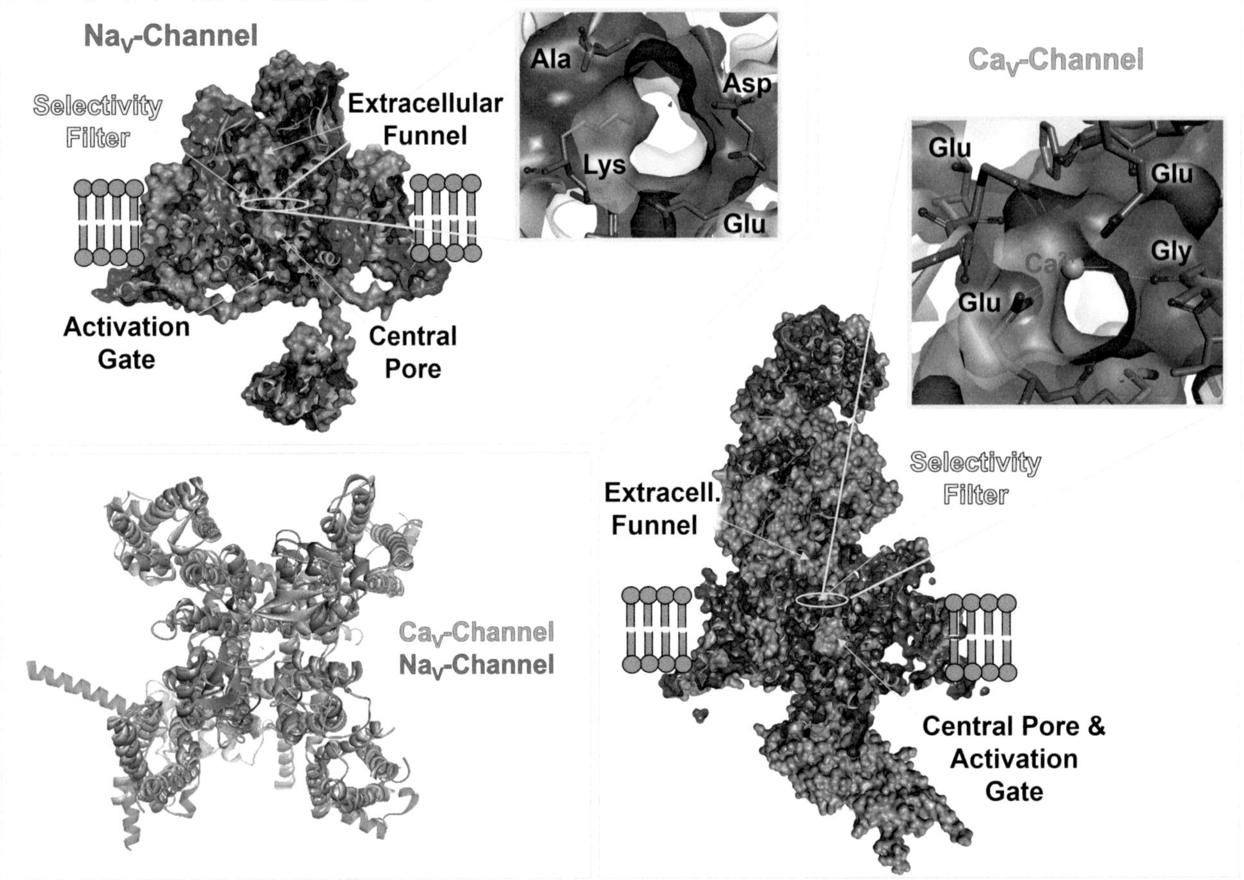

Fig. 30.9 *Upper left box* Cryo-EM structure of the human Na$_V$1.7 channel. The central part of the channel is embedded in the membrane. The membrane is spanned by an opening that starts extracellularly with a funnel-shaped entrance and leads via the selectivity filter (*yellow ellipse*) to a central pore. In mammalian channels, the selectivity filter is formed by the amino acids Asp–Glu–Lys–Ala (DEKA) (*inset, top center*). This is followed by a central pore and the activation gate. The mechanical locking of the channel is located in this region (see Fig. 30.10). *Bottom left box* The structure in the central part, which is embedded in the membrane, is largely conserved from protozoa to mammals. Ca$_V$ channels (*ochre*) adopt the same folding topology as the human Na$_V$ (*purple*), compare superimposition of the Na$_V$ and Ca$_V$ channels from rabbit. *Right box* The Ca$_V$ channel from rabbit was also studied by cryo-EM. Comparable cavities open along the membrane passage. The selectivity filter shows an increased abundance of glutamic acid residues (*inset top right*). (▶ https://sn.pub/Wj5heV)

tances through the channel together with their hydration shell. The ions interact with the protein environment via their water shells, that means the coordinated water molecules mediate contact with the surrounding amino acids. Presumably, the water shell is only partially removed in the selectivity filter, where either carbonyl groups of the protein backbone or carboxylate groups of Asp and Glu residues coordinate the ions. The resolution of the structures obtained so far is not sufficient to determine these geometries precisely. Cryo-EM data make it difficult to determine water positions, and negatively charged residues can only be determined with reduced accuracy.

Both sodium and calcium ions prefer an octahedral coordination sphere of water molecules. However, calcium can extend its coordination number up to eight surrounding water molecules, and the variety of coordination polyhedra adopted is significantly larger. This includes the square antiprismatic geometry preferred by potassium ions. Again, the fourfold geometry is determined by the coordination chemistry of the passing ions and is, therefore, reflected in the structure of the channels.

In addition to the size of the hydration shells, there is another important criterion to distinguish between sodium and calcium ions: Both ions differ by one unit in their charge state. In mammalian sodium channels, there is an Asp–Glu–Lys–Ala motif in the **selectivity filter** (DEKA, Fig. 30.9). The two negatively charged acidic residues of Asp and Glu generate a strong negatively charged electrostatic field. The Lys residue partially compensates for the excess negative charge and, thus, exerts a repulsive effect on cations trying to pass through the channel. This repulsion is obviously still tolerated for sodium ions, which have only a single positive charge. For the doubly positively charged Ca^{2+} ions, the repulsion is probably too great to allow efficient passage. They are rejected. Comparison of the sequence data of the selectivity filters of sodium- and calcium-specific channels shows that in the case of calcium channels, there is always a higher number of up to four Asp or Glu residues in the selectivity filter. This increases the electrostatic attraction.

The detailed control of voltage-gated channels occurs through a complex interplay of multiple conformational changes, resulting in an electromechanically coupled system. The voltage-sensing segments detect the applied membrane potential. They transmit this information to the mechanical locking of the loops, which ultimately determines the diameter of the pore for the passage of ions into the interior of the cell. ◘ Fig. 30.10 shows the open, closed, and resting inactivated states of a bacterial Na_V channel. A bundle of helices opens with a motion reminiscent of the opening of an iris aperture in a camera. This occurs in response to voltage-dependent conformational changes in the voltage sensors. The isoleucine residues at position 217 contribute to the closure. The transition from the closed to the open conformation implies an opening of less than 1 Å to 10.5 Å (◘ Fig. 30.10). This large opening is sufficient for the passage of hydrated sodium ions without a significant energy barrier.

Such a complicated, highly controlled protein system offers a variety of potential interaction sites for drug molecules. For example, its electrostatic properties can be modulated by bound ligands. The mobility of loop regions can be locked into specific states. In particular, neurotoxic peptides from spiders, scorpions, and sea anemones attack the voltage-sensing segments. They block the channel in a particular state, so that the pore is locked in either an open state or a closed state. The channel pore itself can also be blocked by bound ligands. This prevents ions from passing through. For example, the highly toxic tetrodotoxin from the fugu fish **30.20** introduced in Sect. 6.3 binds in the DEKA selectivity filter of the human sodium channel (◘ Fig. 30.11). Its blockade means that action potentials are no longer activated. Nerve and muscle excitation is prevented and respiratory paralysis occurs. **Antiarrhythmic drugs** such as propafenone **30.21** or flecainide **30.22** block the open Na_V channel in the central pore between the selectivity filter and the activation gate. The influx of sodium ions into the myocardial cells is prevented, thereby, calming or normalizing muscle excitation. The **local anesthetic** lidocaine **30.23** binds to both the open and closed forms and stabilizes the channel in the inactivated closed state. It is located in the central pore directly at the intracellular exit of the narrow ion selectivity filter. The protonated amino group of lidocaine faces upwards into the selectivity filter where it interacts with the backbone carbonyl groups of Thr 175. This amino

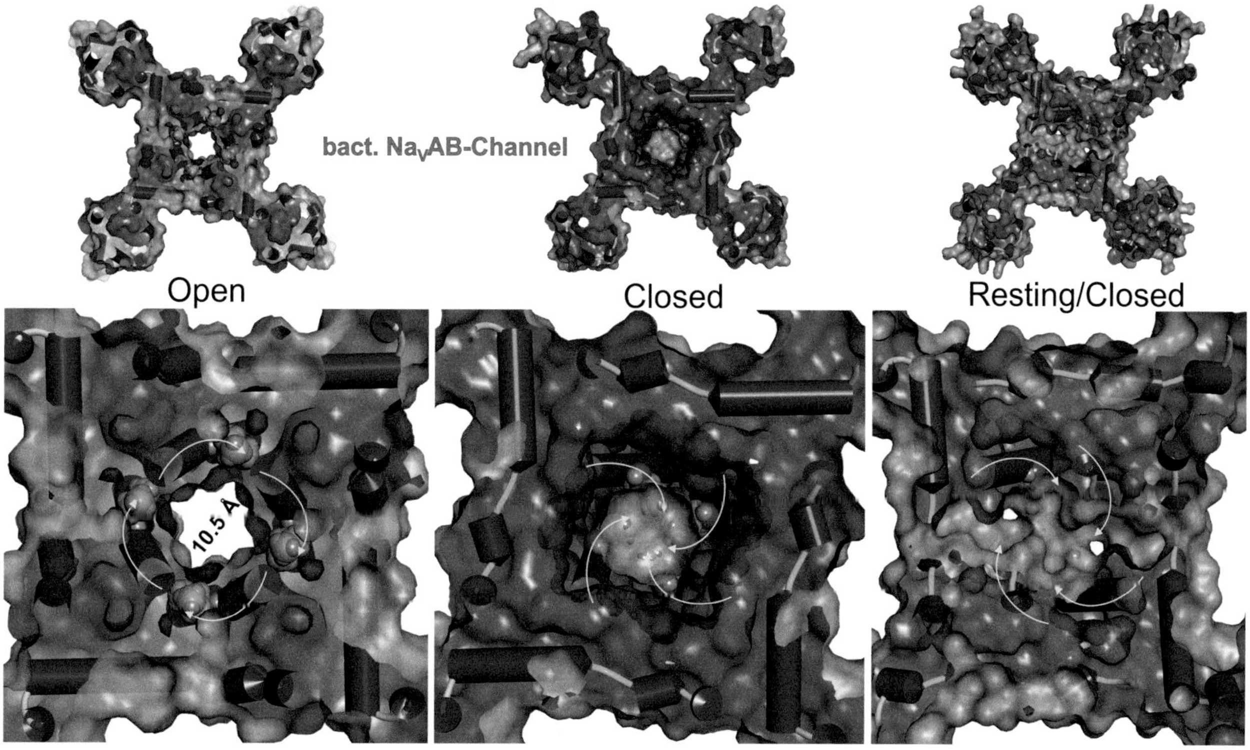

◘ **Fig. 30.10** Crystal structures of the central membrane unit of a bacterial Na_VAB channel in the open, closed, and resting/closed states. The channel is shown in the top view from the extracellular side. The closure takes place in the lower part of the channel. The locking mechanism is triggered by a rotational movement of several helices. They perform a movement reminiscent of the opening of an iris aperture in a camera. At each of the four domains, the side chain of the Ile 217 residue contributes to the closure (shown in van der Waals representation, C-atoms *green*). The channel is opened to more than 10 Å, allowing sodium ions with their hydration shells to pass freely

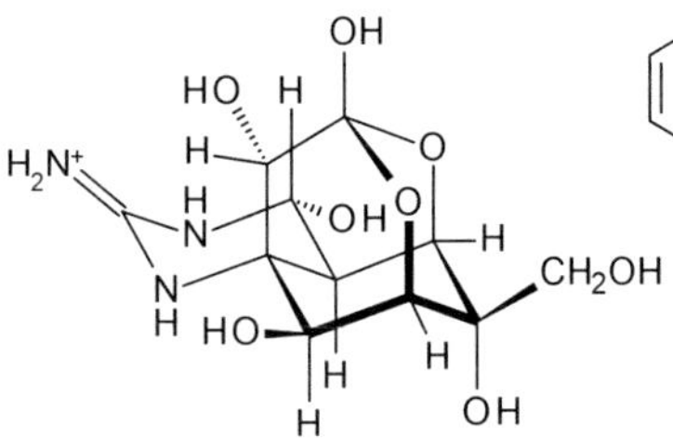

30.20 Tetrodotoxin

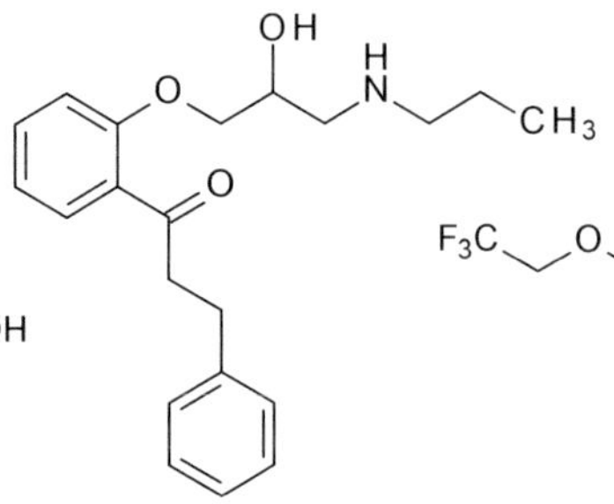

30.21 Propafenone

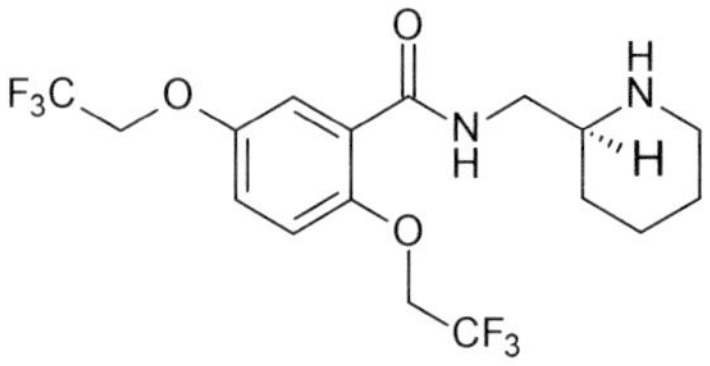

30.22 Flecainide

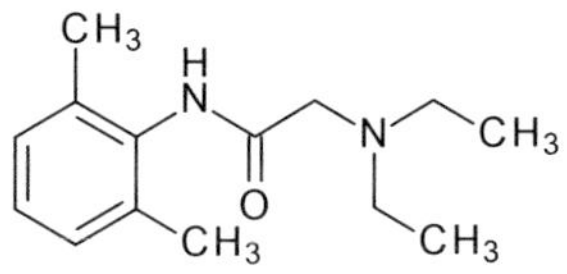

30.23 Lidocaine

Fig. 30.11 Modulators of the Na_V channel that occupy the central pore and block the channel. Tetrodotoxin **30.20**, toxin from the fugu fish, propafenone **30.21**, flecainide **30.22**, and lidocaine **30.23** bind below the selectivity filter in the central pore

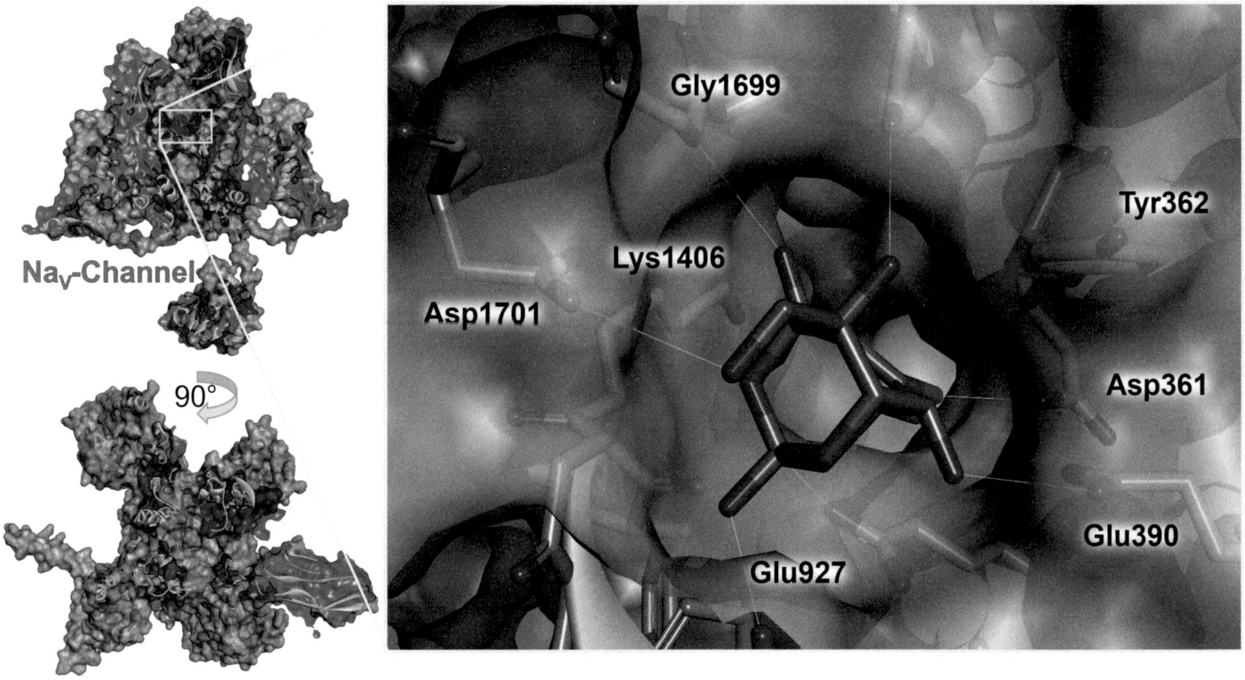

Fig. 30.12 Cryo-EM structure of the human $Na_V1.7$ channel with the highly toxic tetrodotoxin **30.20** from the fugu fish. The alkaloid has an adamantane-like parent scaffold and is decorated with a large number of functional groups to form hydrogen bonds. Ideally placed counter groups of the toxin are found at the lower end of the selectivity filter, where it blocks the channel for the passage of sodium ions. (► https://sn.pub/zcJqmJ)

acid represents the final coordination site for sodium ions entering the selectivity filter. Presumably, lidocaine is delivered into the channel through the activation gate, which opens over 10 Å to the membrane (Fig. 30.12).

For many years, three chemically distinct classes of compounds have been identified as so-called **calcium blockers** for **L-type calcium channels**: dihydropyridines, benzothiazepines, and phenylalkylamines. The most prominent representatives are nifedipine **30.24**, diltiazem **30.25**, and verapamil **30.26**. It has long been debated whether there are overlapping binding modes for the three compound classes and whether they should be described as channel blockers or calcium antagonists. Another surprising observation was that stereoisomers of the dihydropyridine Bay-K8644 **30.27** (Sect. 5.5) exert opposite biological effects: One isomer inactivates the channel; the other prolongs its open duration. A structural determination of the Ca_V channel from the rabbit has now shed light on these questions (Figs. 30.13 and 30.14).

Verapamil 30.26 and **diltiazem 30.25** bind to the channel in overlapping regions located in the central pore on the intracellular side of the ion selectivity filter (Fig. 30.14). Both span the central pore and, thus, directly block ion flow. They can, therefore, be described as **channel blockers**. In contrast, the binding site of the **dihydropyridines** is located at a distance of about 10 Å on the

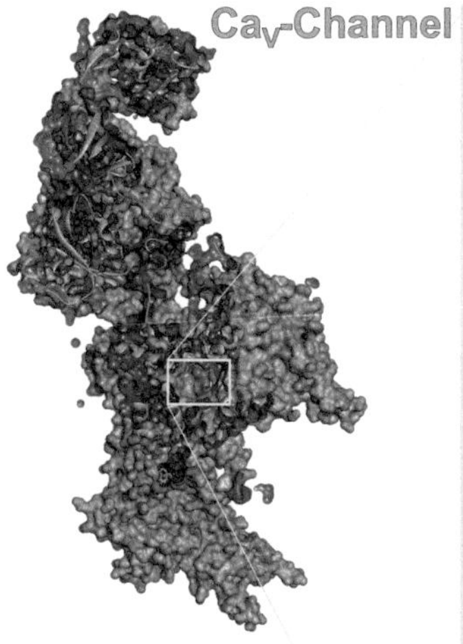

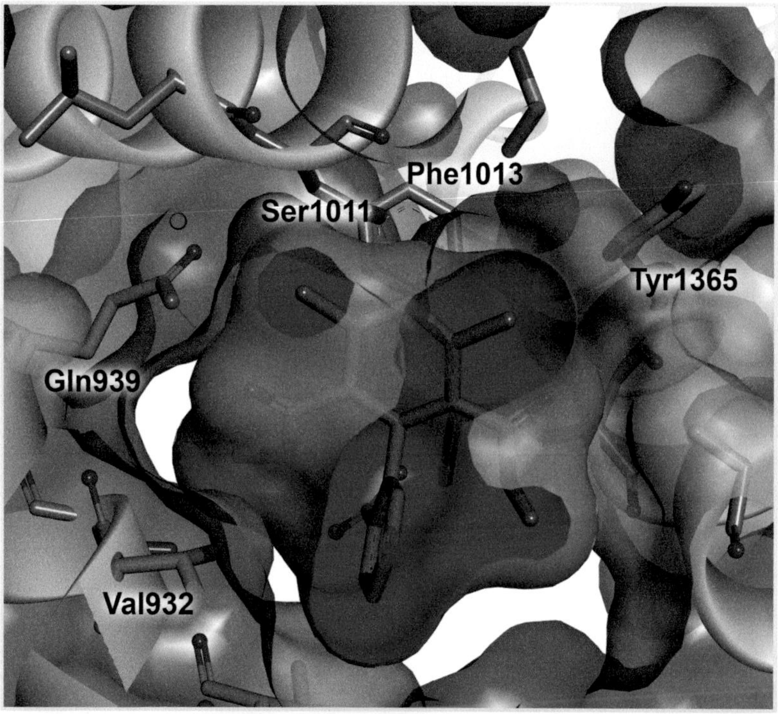

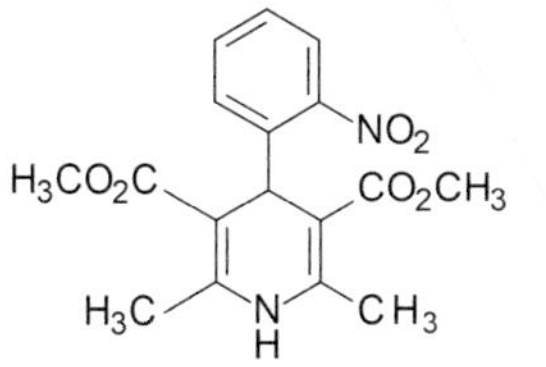

30.24 Nifedipine

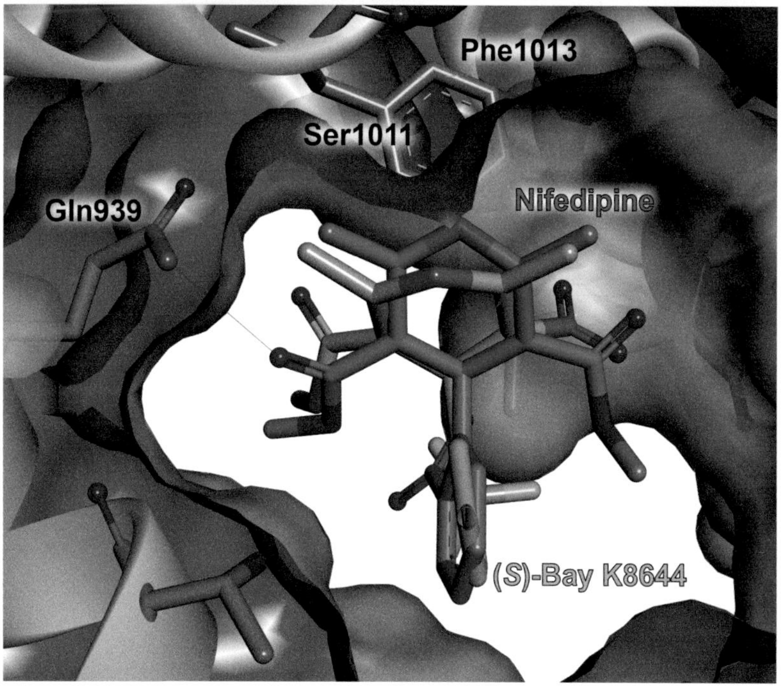

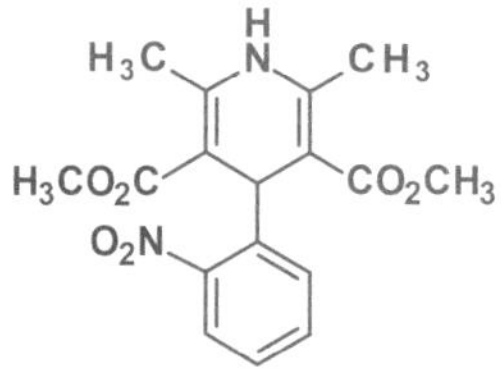

30.24 Nifedipine

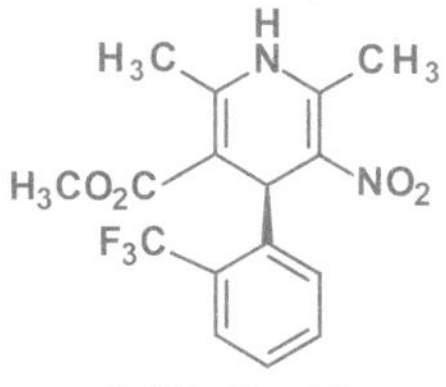

30.27 Bay K8644

Fig. 30.13 *Upper row* Cryo-EM structure of the rabbit Ca_V channel with the bound calcium antagonist nifedipine **30.24**. The dihydropyridine binds to the lipid-facing surface of the central pore between the domains of two adjacent voltage sensors. This highlights the indirect allosteric inhibition of the pore by nifedipine, and the compound should, therefore, be classified as an antagonist. *Bottom row* The stereoisomers of Bay-K8644 **30.27** (Sect. 5.5) show opposite effects. The *R*-enantiomer inactivates the channel, whereas the *S*-analogue prolongs its opening. The superposition of the binding modes of nifedipine **30.24** and (*S*)-Bay-K8644 **30.27** shows an almost perfect overlap despite the divergent ring conformations. The other stereoisomer (*R*)-Bay-K8644 should bind identically to nifedipine. Since the nitro and ester groups exchange positions, this would result in a binding mode analogous to that of nifedipine. A comparable action profile with stabilization of the inactivated channel can be expected. However, if the electron-withdrawing nitro group is oriented into the hydrophobic protein environment, as in the *S*-isomer, this results in agonistic activation of the channel. (► https://sn.pub/d9ptVi)

lipid-facing surface of the pore between the domains of two neighboring voltage sensors. This result confirms the indirect **allosteric mechanism** of pore blockade by dihydropyridines. They can, therefore, be described as **antagonists** that achieve their function through allosteric effects.

A superimposition of the binding modes of nifedipine **30.24** and (*S*)-Bay-K8644 **30.27** (◘ Fig. 30.13, *bottom row*) shows an almost perfect alignment of the dihydropyridine rings and the perpendicularly oriented phenyl rings, despite the different ring conformations. Nifedipine is achiral due to its symmetrical substitution. Nevertheless, from the superposition it can be estimated that the *R*-stereoisomer of Bay-K8644 binds almost identically to nifedipine. Only the nitro and ester groups exchange positions. This means that the ester group in this stereoisomer has the same ori-
30 entation as in nifedipine. As a result, it can stabilize the inactivated channel in the same way as nifedipine. However, if the electron-withdrawing nitro group is oriented towards the hydrophobic protein environment, as observed for the *S*-isomer, this will lead to agonistic activation of the channel. It can be seen that even small effects uncovered by structure determination of these giant channels can explain a design hypothesis and may guide future work.

30.5 Tiny Ligands Gate Giant Ion Channels

There are several classes of **ligand-gated ion channels**. One of the most important classes, the Cys-loop superfamily, includes the **nicotinic acetylcholine receptor**, the **5-HT$_3$ receptor**, and the **inhibitory glycine** and **GABA$_A$ receptors** (Sect. 30.7). The first two are excitatory receptors that respond to acetylcholine and serotonin. They are essential for the transmission of fast nerve impulses

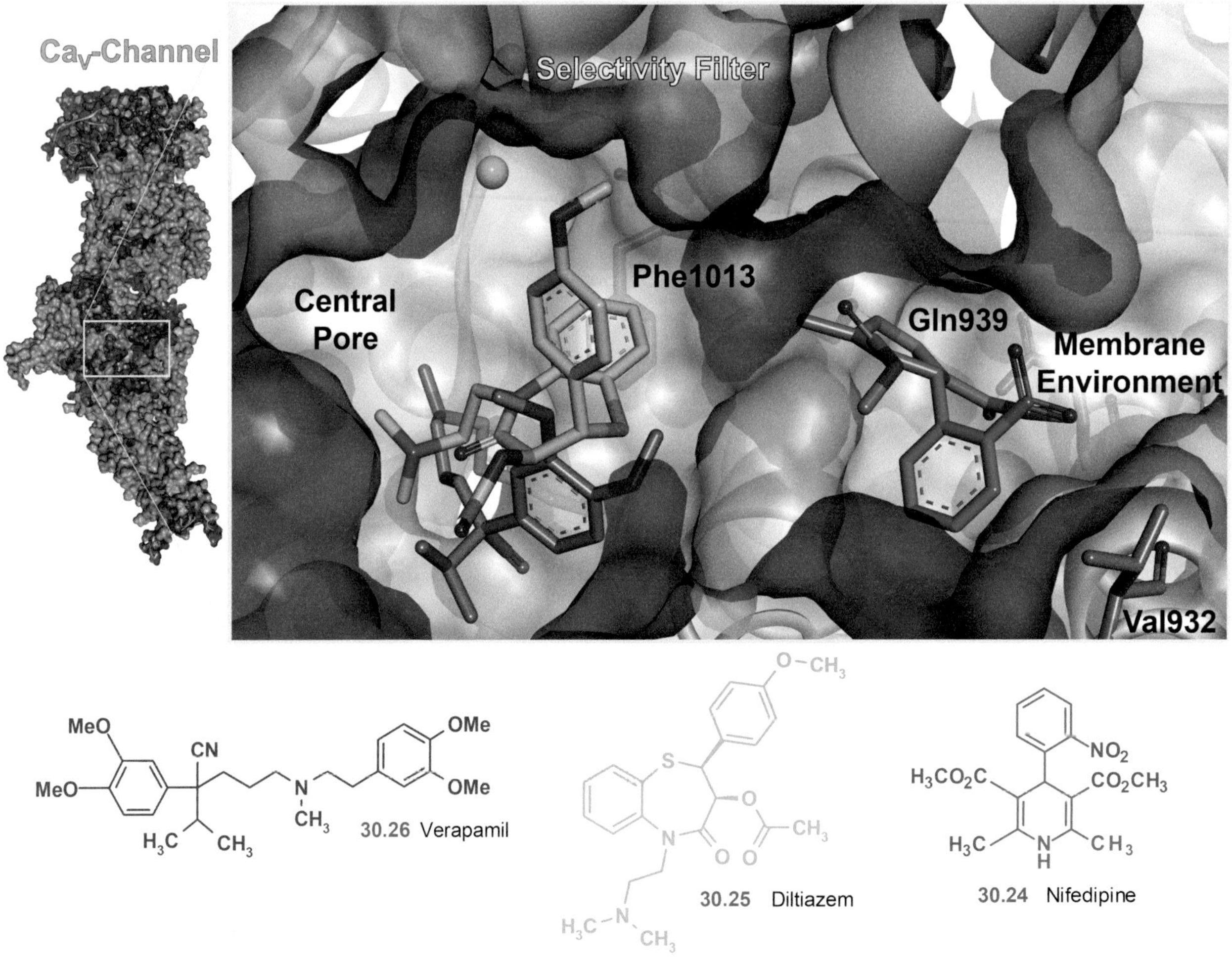

◘ **Fig. 30.14** Superposition of the binding modes of verapamil **30.26**, diltiazem **30.25**, and nifedipine **30.24** in the rabbit Ca$_V$ channel determined by the cryo-EM method. While verapamil and diltiazem bind below the selectivity filter in the central pore and block the passage of calcium ions, nifedipine finds a binding site several angstroms away near the exit port to the membrane. This puts nifedipine in contact with the domains of two adjacent voltage sensors. This leads to an indirect allosteric mechanism of pore blockade by the dihydropyridines. They are, therefore, classified as calcium antagonists. (► https://sn.pub/LRI7dO)

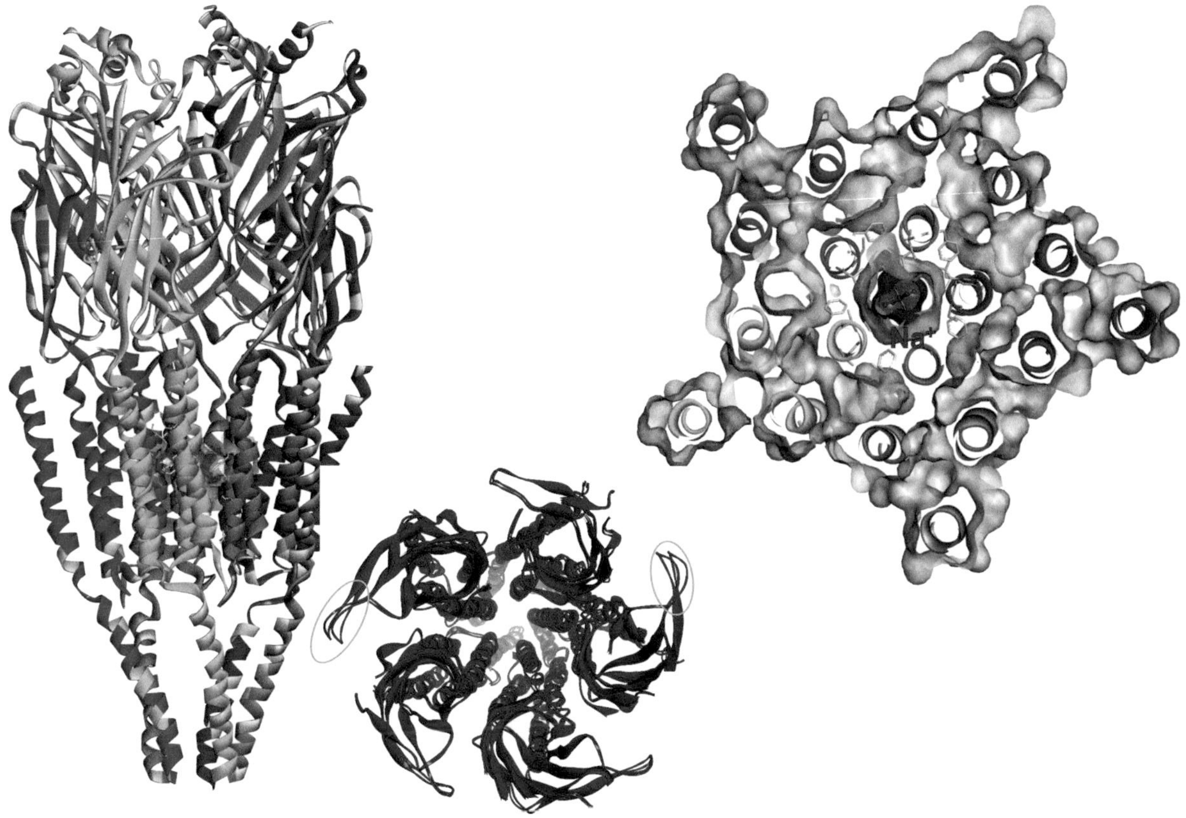

■ **Fig. 30.15** In the cryo-EM structure, the nicotinic acetylcholine receptor has a diameter of 80 Å and is 125 Å long (*left*). It is a pentamer of five subunits. The central region, consisting of four helices per monomer, spans the membrane. The interior of the channel is extracellularly polar and has many acidic amino acids (shown in *yellow*). Two of the extracellular domains bind the neurotransmitter. *Center* Comparison of the structures with (*blue*) and without (*red*) bound acetylcholine shows the inwards movement of two *C*-loops (*green ellipse*). This movement is transferred to the cytosolic part formed by helices. At the narrowest point inside the channel (*right*), it tapers to 6 Å when closed (center, marked by *white surface*). There, a belt of hydrophobic residues restricts the passage of cations. When opened, the helices rearrange themselves in a concerted rotational motion and widen the channel passage by about 3 Å, enough to allow monovalent cations with their hydrate shell to pass. At the narrowest region a sodium ion with its octahedral hydration shell has been modeled into the structure. (▸ https://sn.pub/E5XcmS)

at synapses. The inhibitory glycine and $GABA_A$ receptors are controlled by glycine and γ-aminobutyric acid, respectively. All ion channels of the **Cys-loop family** have a common architecture. They form a pore in the membrane, and open in response to the binding of an agonist, allowing the passive flow of ions. They have a pentameric structure, although the composition of this heteropentamer can vary. In the case of nicotinic acetylcholine receptors (nAChR), which preferentially form a channel for monovalent cations, a large number of different subtypes with different expression patterns are formed from a set of 17 homologous subunits (10 α, 4 β, 1 γ, 1 δ, and 1 ε unit). They have a wide range of functional properties with different pharmacological characteristics. They are expressed in numerous tissues, including the brain, skeletal muscle, white blood cells, and cochlear hair cells. Parallel to this widespread distribution, nAChRs mediate diverse physiological functions, including sensory perception, muscle contraction, immunomodulation, and sound discrimination. In the 1970s, smokers were found to have an increased density of nAChRs in the brain due to receptor stabilization by the agonist nicotine. This observation has been linked to the development of smoking addiction.

Structurally, four transmembrane helices come together to form one of the five transmembrane domains. They encircle the ion channel in their interior. Each **pentamer** has two extracellular ligand-binding domains. In the center of the five transmembrane domains, the innermost of the five helices, called M2 helices, form the channel. They have hydrophobic amino acids such as valine, phenylalanine, and leucine in their center, which are responsible for opening and closing the channel.

Thanks to the pioneering work of Nigel Unwin, we have detailed insights into the construction of a channel of this family, the nicotinic acetylcholine receptor. Using an electron microscope on two-dimensional crystals, he was able to obtain a picture of this ligand-gated ion channel in the closed state (■ Fig. 30.15, *left*) from the electric organ of the electric ray species *Torpedo* with a resolution of 4 Å. It crosses the membrane with five subunits (α2βγδ) that make up the ion channel, each composed of four transmembrane helices. It protrudes about 60 Å across the membrane into the synaptic cleft. This is where the ligand-binding domain, made up of β-pleated sheets, is located; it carries the binding pocket for the neurotransmitter acetylcholine. After loading the two-dimensional crystal with acetylcholine, Unwin was able to observe the resulting conformational changes. He registered **spatial rearrangements of the *C*-loops** near the acetylcholine binding region (◘ Fig. 30.15, *center*). They are transferred into the transmembrane part of the ion channel. In this way, the receptor "senses" ligand binding and transmits it to the ion permeability pore, which is 30 Å away. The pore remains open after ligand binding. The M2 transmembrane helices, located directly inside the channel, carry bulky, hydrophobic amino acids. In the closed state, they surround the channel center like a hydrophobic belt (◘ Fig. 30.15, *right*). The remaining opening of about 6 Å is too narrow to allow Na^+ or K^+ ions with their hydration shells to pass through. Since the channel does not provide a polar environment, similar to the potassium channel described in Sect. 30.3, the ions cannot simply shed their hydration shells for passage. The membrane permeability remains locked for them.

The binding of the ligand acetylcholine, together with the movement of the *C*-loops, initiates a cascade of conformational changes that are transmitted to the M2 helices. This is accomplished by a concerted rotation of all five M2 helices by approximately 15°. As a result, the pore expands by about 3 Å, allowing ions with their hydration shells to pass through in the open state. To illustrate this, a sodium ion with its octahedral water shell has been inserted into the pore in ◘ Fig. 30.15 (*right*). Unwin's work provided the first fascinating insight into the function and dynamics of a ligand-gated ion channel. Huge protein structures respond to the binding of a comparatively tiny agonist. Information is transmitted over long distances. We will see that other channels of the Cys-loop family also work according to this principle.

30.6 Ligands Gate as Agonists and Antagonists: The Function of an Ion Channel

In the meantime, the extracellular domains of a number of acetylcholine receptors have been successfully isolated and crystallized with bound ligands. Analogous to the receptor described in Sect. 30.5, these acetylcholine-binding proteins form a pentamer. The binding protein of the Californian sea slug (*Aplysia californica*) exists as a homopentamer. Its structure was determined using agonists and antagonists. This provided insight into two very interesting aspects.

First, the structures illustrate the molecular origins of the conformational rearrangement that is translated from the ligand-binding domain to the isthmus of the ion channel, which then opens or closes. In the example above, agonists and antagonists differ greatly in size. The agonist nicotine **30.28** is the main alkaloid in tobacco (◘ Fig. 30.16). At low doses, it stimulates neurotransmission, while at high doses, it causes permanent depolarization and blocks neurotransmission. Epibatidine **30.29** occurs naturally in the skin of the Ecuadorian poison dart frog and has potent analgesic properties. *Lobelia inflata*, a flowering plant, contains α-lobeline **30.30**. The leaves of this plant were smoked by Native Americans to treat asthma. In higher doses, the compound is extremely toxic. As shown in the structural comparison of the unbound and bound acetylcholine receptor, binding of a small agonist causes the long *C*-loop to cover the binding site on the receptor (◘ Fig. 30.17). This gives the extracellular domain of the receptor a more compact structure, especially at the contact points between the extracellular domains. When an inhibitor (also known

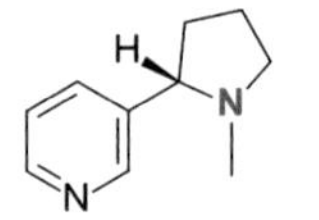

30.28 Nicotine

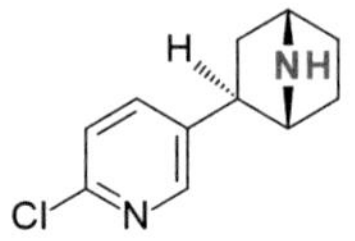

30.29 Epibatidine

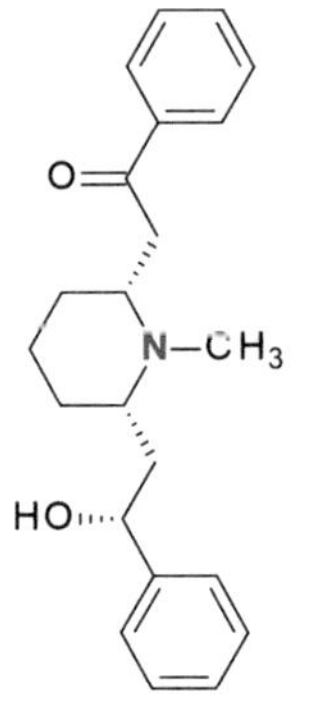

30.30 α-Lobeline

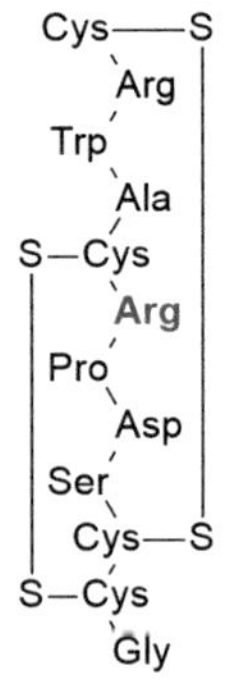

30.31 α-Conotoxin

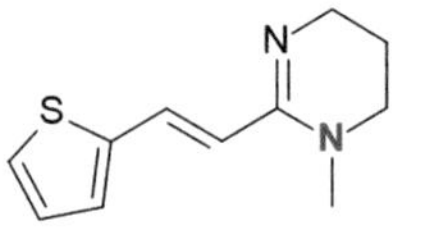

30.33 Pyrantel

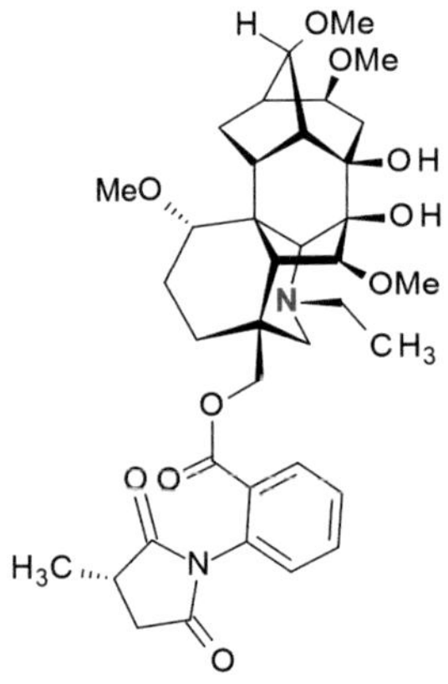

30.32 Methyllycaconitine

◘ **Fig. 30.16** As agonists, nicotine **30.28**, epibatidine **30.29**, and α-lobeline **30.30** open the nicotinic acetylcholine receptor. The dodecapeptide α-conotoxin **30.31** and the diterpene alkaloid methyllycaconitine **30.32** block the receptor as antagonists. All bind to the ligand-binding domain of the pentameric receptor. They have functional groups (*red*) that can exist in a positively charged state. Pyrantel **30.33** is also an agonist and is used as a vermifuge to eradicate ascarids and oxyurans in the intestine

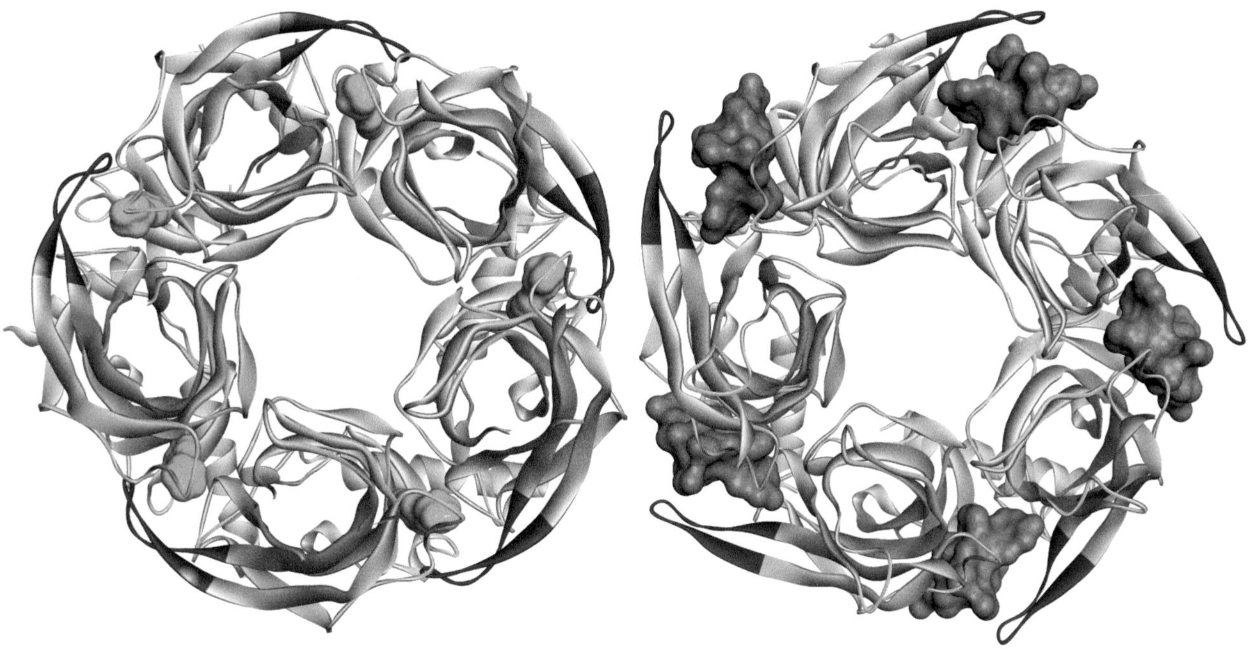

■ **Fig. 30.17** Binding of an agonist (here epibatidine, **30.29**) or antagonist (here α-conotoxin **30.31**) to the ligand-binding domain of the nicotinic acetylcholine receptor either places a loop (*red*) directly on the binding site (*left*) or leaves it spread out by about 10 Å (*right*). This enhances the interactions between the extracellular domains of the pentamer during agonist binding. This conformational signal is propagated to the channel's constriction site about 30 Å away, where it causes the channel to remain closed in the case of antagonist binding or to open in the case of agonist binding. (▶ https://sn.pub/ydoma3)

as an antagonist) such as the peptide α-conotoxin **30.31** binds, this loop remains spread apart for steric reasons and the domain contacts do not solidify. The peptide α-conotoxin is used as a venom by a carnivorous sea snail that lives in tropical oceans. Since the snail cannot bite, it shoots its venom, which is packaged in small chitin-coated arrows that even have barbed hooks, through a kind of blowpipe. The diterpene alkaloid methyllycaconitine **30.32** from the seeds of the medicinal plant larkspur has the same effect. When this ligand binds to the receptor protein, a movement of more than 10 Å is registered (■ Fig. 30.17). This difference is cascaded to the narrowest part of the channel and regulates sodium ion permeability. Considering the analogously constructed chloride channels (Sect. 30.7), we will learn how this signal is transmitted to close the channel.

As a second aspect, these structures provide insight into how structures that are chemically completely different can induce the same effect on a receptor. The antagonist α-conotoxin **30.31**, a dodecapeptide with two intramolecular disulfide bridges, binds with the geometry shown in ■ Fig. 30.18. The herbal alkaloid methyllycaconitine **30.32** binds in the same pocket, but the binding areas of the peptide and alkaloid overlap only in the central region. The agonists epibatidine **30.29**, nicotine **30.28**, and α-lobeline **30.30** also bind only in this region, but they occupy a much smaller area. The diterpene and the above-mentioned agonists all have a secondary or tertiary basic nitrogen atom that is most likely protonated in the binding pocket. In all of these structures, this nitrogen atom is located near a tryptophan residue. There, an H-bond can be formed to the carbonyl group of the amino acid, and a cation–π interaction with the neighboring aromatic ring can also play an important role. The peptide α-conotoxin does not have a chemically comparable nitrogen atom. It places, however, a positively charged arginine residue in the vicinity of the tryptophan, providing a comparable binding relationship. These structures are certainly an extreme example of bioisosteres. They illustrate, however, the diversity that living nature creates to arrive at the same goal through entirely different molecular scaffolds. Medicinal chemists can only learn from these creative solutions!

Pyrantel **30.33** (■ Fig. 30.16) is an antihelmintic used to treat ascariasis and enterobiasis (intestinal roundworm and pinworm infections, respectively) that is related in size and structure to the above-mentioned agonists. It binds to the nicotinic acetylcholine receptor of the worms. As a result, the ion channel opens, leading to depolarization. As a result, neuromuscular blockade is initiated and the worms are paralyzed. They can then be flushed out of the infected intestines. Because the drug is poorly absorbed from the gastrointestinal tract, it is safe and well tolerated in humans.

30

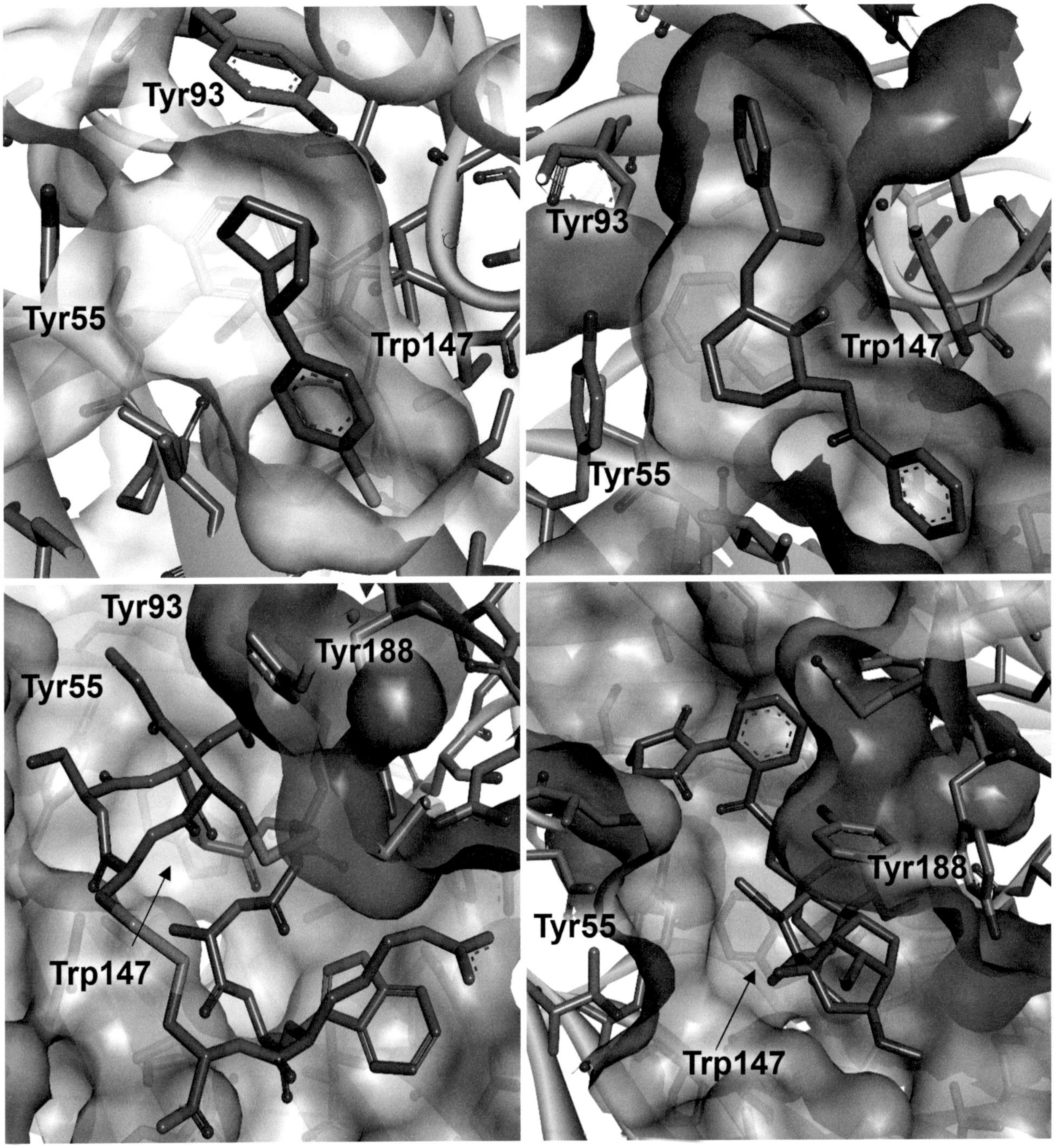

Fig. 30.18 Crystal structure of the ligand-binding domain of the nicotinic acetylcholine receptor with the bound agonists epibatidine **30.29** (*upper left*) and α-lobeline **30.30** (*upper right*), the peptidic antagonist α-conotoxin **30.31** (*bottom left*), and the diterpene alkaloid methyllycaconitine **30.32** (*bottom right*). Despite their very different sizes, they all occupy the same binding site. In the case of the agonists, a loop (Fig. 30.17) lies across the binding site, which spreads apart in the case of the antagonists. All of these ligands have a positive charge that allows them to form a cation–π interaction with the aromatic ring of a nearby tryptophan. (► https://sn.pub/WLqMxI)

30.7 Power Brake Boosters for GABA-Gated Chloride Channels

Glycine and **GABA$_A$ receptors** are **inhibitory neuroreceptors** because they regulate the influx of chloride ions. This leads to **hyperpolarization** and lowers the voltage-dependent excitability, which makes depolarization of the cell more difficult. Both of these receptors are regulated by the low molecular weight ligands glycine **30.34** or γ-aminobutyric acid (GABA) **30.35** (◘ Fig. 30.19).

In vertebrates, GABA$_A$ receptors mediate neuronal inhibition in the central nervous system. Dysfunction results in abnormal ionic currents associated with epilepsy, insomnia, anxiety, and chronic pain. Like the other members of the Cys-loop superfamily, the GABA$_A$ receptor is a **heteropentamer**, with subunits from seven different classes available for assembly. In the majority of cases, an α-, β-, and γ-subunit is incorporated simultaneously (◘ Fig. 30.20). The chloride ions are "pulled in" by the increased number of positively charged amino acids in the extracellular entrance region of the receptor. As in the nAChR, the channel is closed by a belt of hydrophobic residues that control the opening diameter. The central helices receive the conformational signal to open by binding of the endogenous ligand GABA, which slows down the excitability of the cells by opening the channel. For the channels described here (the structural data shown below are from the largely similar α1β3γ2 and α1β2γ2 types), the two GABA binding sites are located in the extracellular region on the contact surfaces between the β- and α-subunits (labeling follows counterclockwise rotation). As with nAChR, it is again the *C*-loop that moves conformationally closer to the domains during agonist binding, contributing to their spatial compaction (◘ Fig. 30.20). In total, more than 180 Å^2 of additional contact area is buried between the domains. This signal is propagated to the belt of hydrophobic amino acids in the same way as in nAChR. In the GABA$_A$ receptor, this belt consists of five leucine residues. They are located on the long M2 helices that line the inside of the channel. When GABA binding is blocked by the competitive and spatially more demanding antagonist bicuculline **30.36** (◘ Fig. 30.21), the *C*-loop will not move. At the position of the Leu belt, the channel diameter is then significantly smaller compared to the situation with bound GABA: the channel remains closed.

GABA$_A$ receptors are among the most important drug targets. They have numerous binding sites to which compounds with anticonvulsant, anxiolytic, analgesic, sedative, and anesthetic properties bind. Anesthetics modulate the activity of these receptors and lead to stabilization of the channel in an open state. Cholesterol and other steroids can also have this effect. For example, the synthetic pregnane steroid alfaxalone **30.37** opens the GABA$_A$ receptor for a prolonged period of time. The channel can also be inhibited by mechanically "blocking" it. This is equivalent to antagonizing the receptor. For example, the natural product picrotoxin, a plant alkaloid mixture of the more potent component picrotoxinin **30.43** (see Sect. 17.1) and picrotin, follows this principle of action. The channel blocker **30.43** is located in the channel below the locking Leu belt (◘ Fig. 30.20).

The pharmacological action of drugs such as **barbiturates** or **benzodiazepines** (◘ Fig. 30.19) is also based on this regulation. Like the endogenous ligand GABA, they affect the dynamic properties of the receptor. The benzodiazepines also bind to the extracellular domain of the receptor and enhance the effect of GABA by stabilizing the channel in the open state. They are, thus, positive **allosteric regulators** and are also called "brake boosters." Barbiturates, although structurally similar, are probably better described as receptor agonists. They

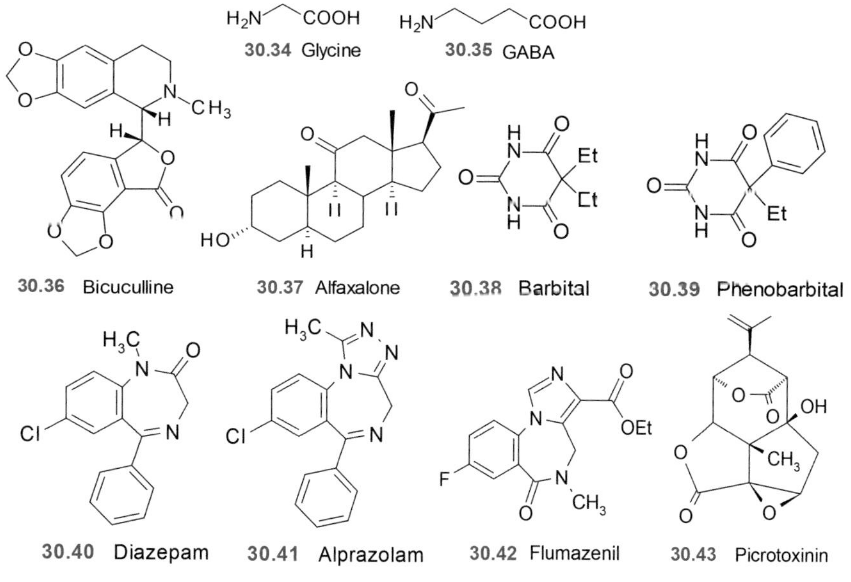

◘ **Fig. 30.19** Glycine **30.34** and γ-aminobutyric acid (GABA) **30.35** regulate ligand-dependent chloride channels. Bicuculline **30.36** displaces GABA from its binding site and abolishes its effect. Alfaxalone **30.37** and barbiturates such as barbital **30.38** and phenobarbital **30.39** open the GABA$_A$ receptor for a longer period of time. Benzodiazepines such as diazepam **30.40** and alprazolam **30.41** enhance the GABA effect and activate the channel through allosteric regulation. Flumazenil **30.42** displaces benzodiazepines from their binding site and neutralizes their effect. It, therefore, serves as an antidote. Picrotoxinin **30.43** is a plant alkaloid that blocks the chloride channel

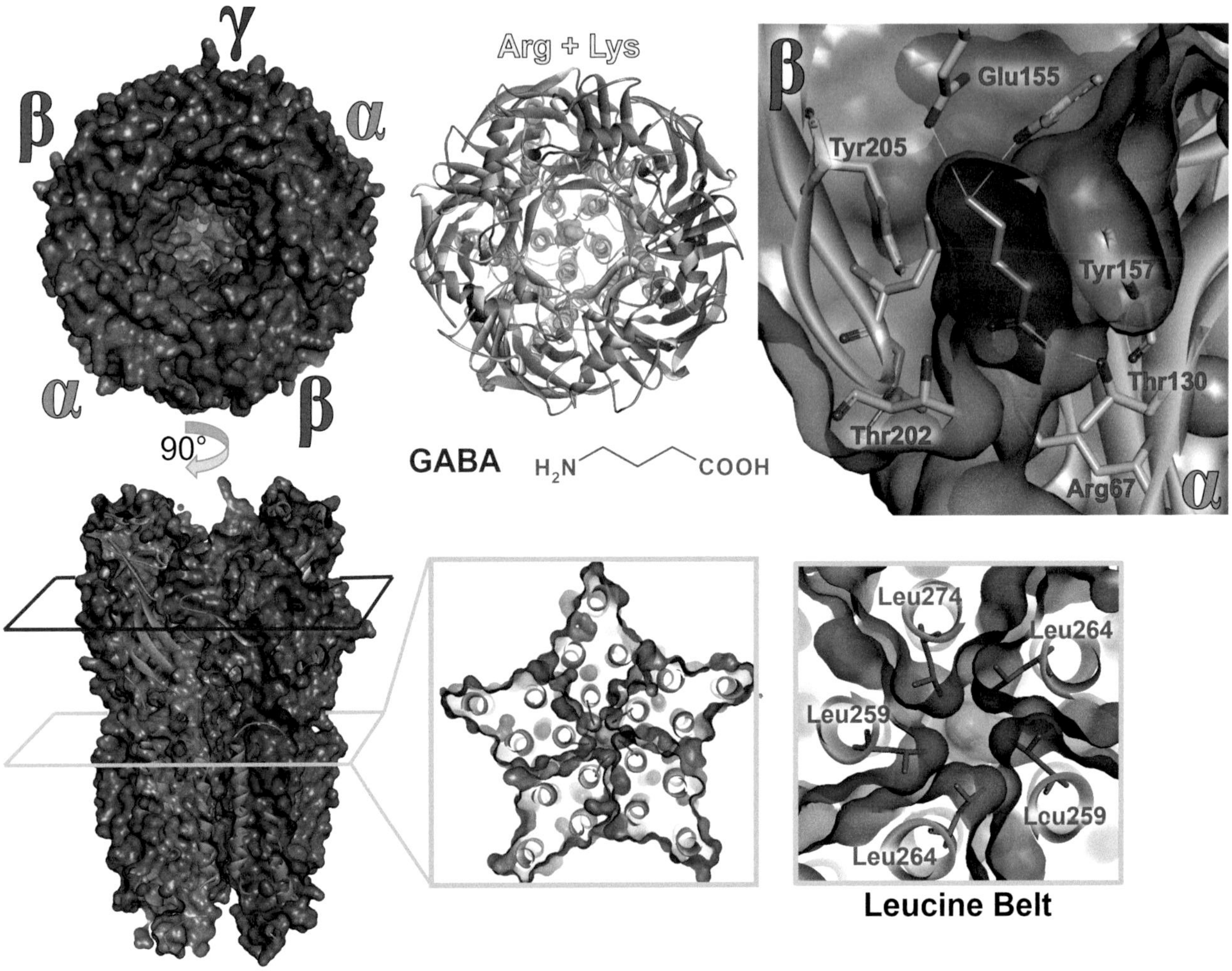

Fig. 30.20 Cryo-EM structure of the $GABA_A$ receptor showing two bound GABA molecules (*purple surface*) and the channel blocker picrotoxinin **31.43** (*green surface*). The subtype shown is composed of two α1-, two β3-, and one γ2-subunit. An increased number of positively charged residues are visible in the entrance funnel on the extracellular side (*yellow*). They attract the chloride ions electrostatically. In the extracellular region (*violet "section plane"*), the two binding sites of the allosteric regulator γ-aminobutyric acid (GABA, *violet*) **30.35** are located in each of the contact areas between the β- and α-subunits (counterclockwise labeling). GABA forms two salt bridges with the surrounding amino acids (Arg 67 and Glu 155, *upper right*). The other contacts are hydrophobic in nature due to aromatic amino acids. The gating leucine belt is located deeper in the channel between the long M2 helices (*yellow "section plane"*). Depending on the diameter of the opening, chloride ions with their water shells are either allowed to pass or they are retained. In the structure shown, the channel is blocked by the inhibitor picrotoxinin **31.43** (*green*). (▶ https://sn.pub/A5QrAn)

keep the channel open longer. The binding sites of all these ligands are located on the contact surfaces between different subunits. However, some of them are located at different heights in the receptor and on contact surfaces between different subunits (see Fig. 30.23). This also influences their specific profile of action.

Fig. 30.22 illustrates the amplification effect of benzodiazepines. The structure with the competitive ligand bicuculline **30.36** is compared with that of bound GABA together with the benzodiazepine alprazolam **30.41**. The binding of GABA triggers the inwards movement of the *C*-loops. The contact between the β- and α-subunits is increased, resulting in a slight rotational movement of the extracellular domains. The benzodiazepine binding site is also located at this level of the receptor. However, it is located between the α- and γ-subunits. This contact is weakest in the ligand-free pentamer. Benzodiazepine binding, therefore, strengthens the subunit contacts and enhances the rotational movement. Thus, it supports the effect of channel opening in terms of a positive allosteric regulation. How is the rotation in the extracellular part transmitted to the channel part in the membrane? The rotation is transmitted to the attached long M2 helices, which carry the critical leucine residues of the pore's closure belt. The leucine side chains are concertedly rotated outwards as the pore opens, and the belt widens at the narrowest point of the channel between the helices. Chloride ions can pass through with

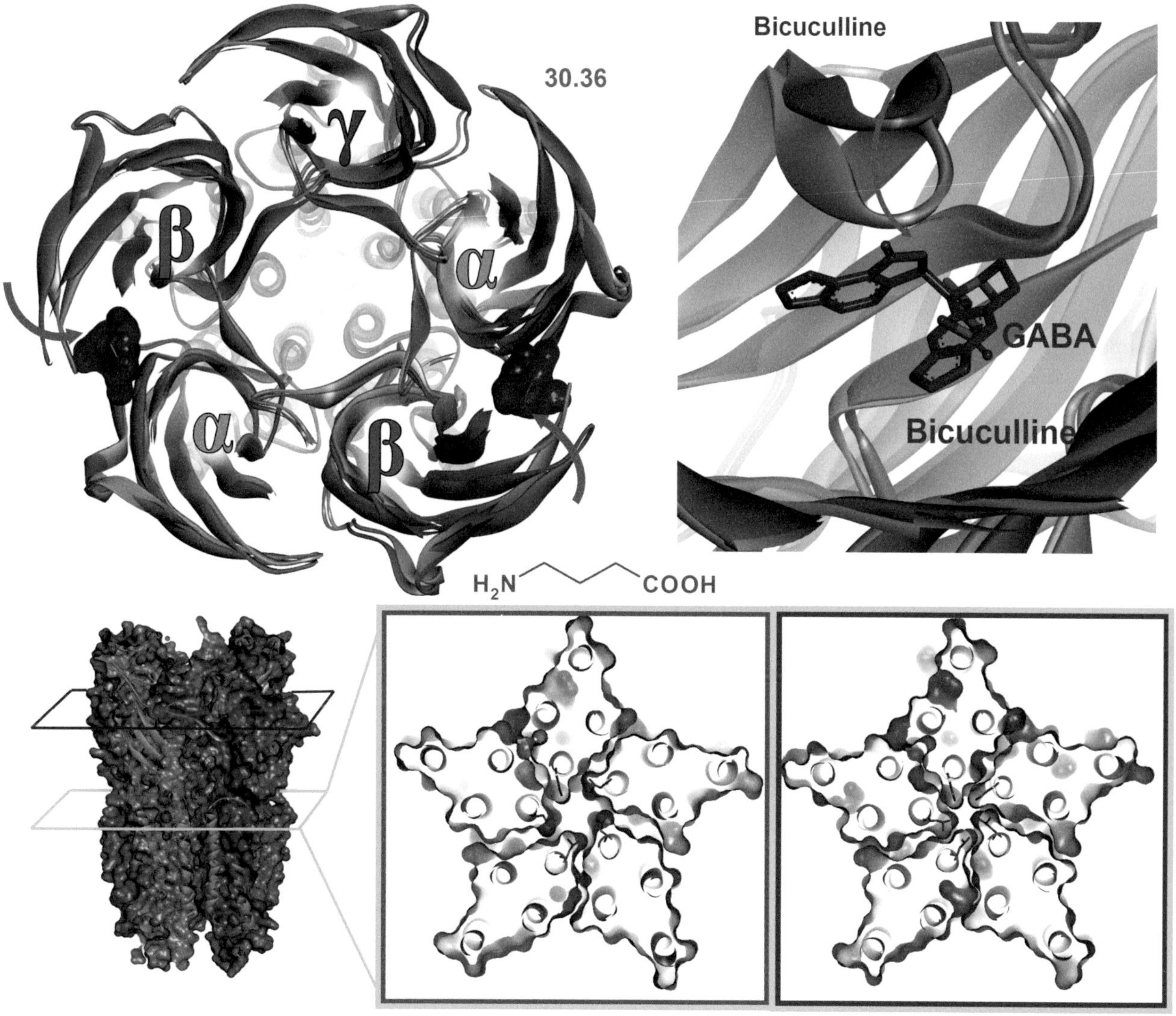

Fig. 30.21 Comparison of the cryo-EM structures of the $GABA_A$ receptor with bound GABA (*violet*, and picrotoxinin, cf. Fig. 30.20) and the competitive antagonist bicuculline **30.36** (*purple*), respectively. The binding sites of the two ligands overlap. The much larger bicuculline prevents the inwards movement of the *C*-loop (*violet arrows*). When this conformational signal is triggered, it is transmitted to the leucine belt located deeper in the channel (*yellow "section plane"*). In the case of bicuculline, this signal is not released and the channel remains closed. This can be seen in the direct comparison of pore diameters observed for bound GABA (*bottom center*) and bicuculline (*bottom right*). (▶ https://sn.pub/iU9Fdy)

their hydrate shells and flow into the cell. The hyperpolarization of the cell increases.

What is the difference between benzodiazepines and barbiturates in their effect on the channel? Barbiturates do not bind to the extracellular domain at the contact surface between the α- and γ-subunits (*"purple section plane"*) like the benzodiazepines. Instead, their binding site is located deeper in the channel in the upper region of the transmembrane helices. There, phenobarbital **30.39** is localized at the contact surface between α- and β- as well as γ- and β-subunits (Fig. 30.23, middle column, *green "section plane"*). Interestingly, in addition to the usual occupancy in the extracellular region (*purple "section plane"*), there are even three binding sites for diazepam in this part. However, the only binding site between the α- and β-subunits is exclusively occupied by phenobarbital. As a consequence, the channel at the site of the leucine belt (*yellow "section plane"*) in the structure of phenobarbital is obviously opened further and longer than in the structure of diazepam **30.40**. The structure determinations shown can only capture the structural relationships. Nothing can be said about the duration of channel opening. The concentration of the added ligand may also be important for the multiplicity of occupied binding sites when preparing the samples for structural determination. For diazepam, however, it is known that the quality of its effect profile also depends on the dose used. This may be reflected in the number of occupied binding sites.

For benzodiazepines, the compound flumazenil **30.42** is known to have an antagonistic effect profile. It disrupts the amplifying effect of benzodiazepines and neutralizes their sedative action. It is, therefore, used as an antidote for benzodiazepine overdose. It lacks the phenyl substituent at the 5-position. It also binds to the benzodiazepine binding site in the extracellular region (◘ Fig. 30.24, *right*). It can, therefore, displace the other benzodiazepines from this site. Due to its structure and binding mode, it interacts more strongly with the α-subunit, so that the stabilizing contact to the γ-subunit is weaker here. Comparing the structure in the region of the leucine belt, there are only small differences in the channel opening compared to the diazepam binding. It is possible that the difference is due to the kinetics and to a temporally more labile stabilization of the channel opening.

Barbiturates and benzodiazepines have sedative, hypnotic, anxiolytic, anticonvulsant, muscle relaxant, and anterograde amnestic (suppressing memory of the time after gradual recovery of consciousness) effects. Barbiturates have lost their importance as sleeping pills and tranquilizers, mainly because of their increased risk of addiction and suicidal abuse. Today, they have been replaced by the better-tolerated benzodiazepines, which are associated with a markedly lower risk of suicide when taken alone.

Indeed, hardly any other substance class has illustrated the concept of **bioisosteric replacement** as thoroughly as the **benzodiazepines**. As a result, a wide range of derivatives are available that, due to their different activity profiles and kinetics, open up a variety of therapeutic approaches for the treatment of insomnia, sedation, anxiety, and spasticity, as hypnotics or muscle relaxants. All compounds have a seven-membered 1,4-diazepine ring with a benzene nucleus fused to it. The benzene nucleus can also be replaced by a thiophene ring as a bioisoster. Most benzodiazepines have a lactam moiety in the central seven-membered ring. This can be replaced by an amidine group or a fused heterocyclic five-membered ring. In many derivatives, the lactam nitrogen has an aliphatic substituent. Another unsaturated structural element is

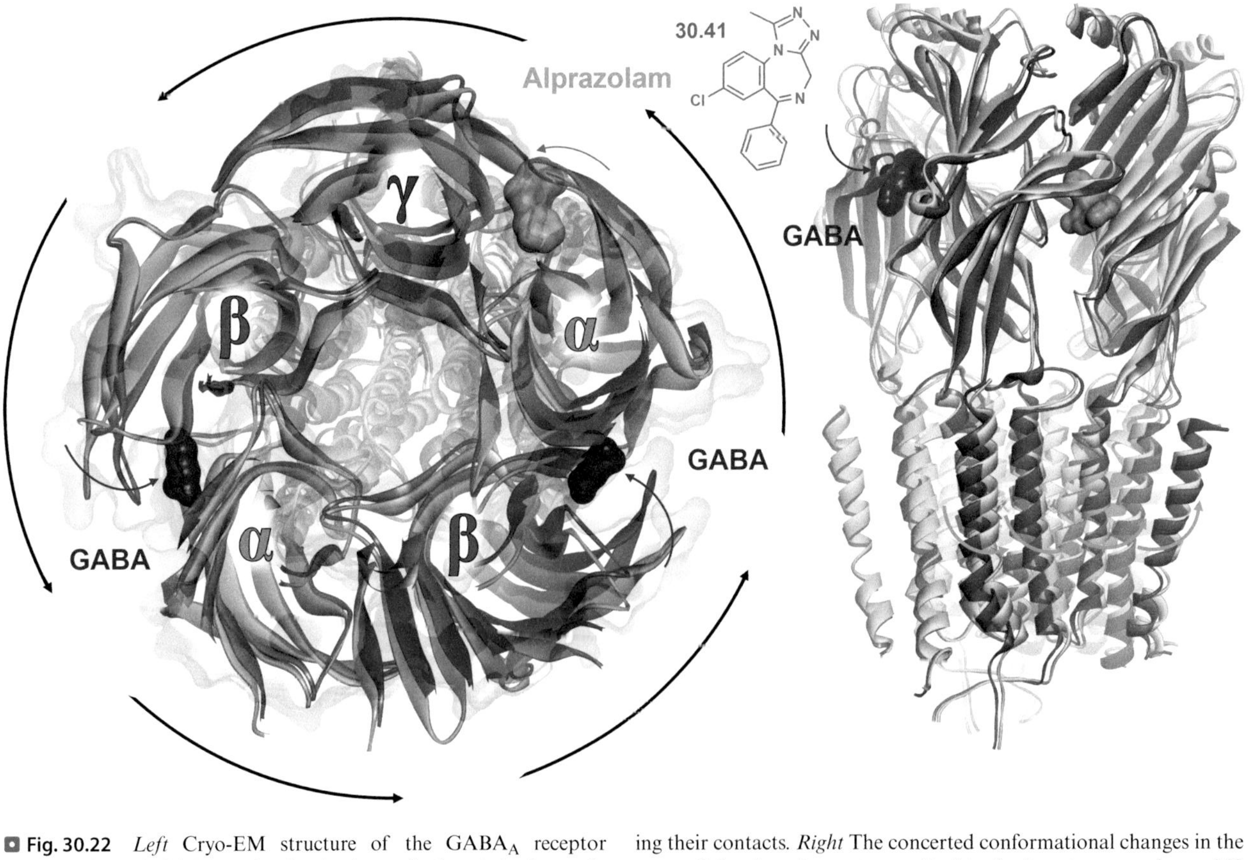

◘ **Fig. 30.22** *Left* Cryo-EM structure of the $GABA_A$ receptor with two bound GABA molecules (*violet surface*) and the benzodiazepine alprazolam **30.41** (*green surface*). The structure is shown in comparison with that complexed by the competitive bicuculline (cf. ◘ Fig. 30.21, the superimposed bicuculline structure is shown with *lighter colors*). The benzodiazepine binding site is at the same level in the receptor as the GABA ligand-binding site. GABA binding triggers the inwards movement of the *C*-loops on the β-subunits. This is transmitted as a slight rotation to the extracellular domains, which move "inwards" to the adjacent interfaces with the α-subunits, strengthening their contacts. *Right* The concerted conformational changes in the extracellular domain are transmitted to the long transmembrane M2 helices, inducing a counterclockwise rotation (*cyan arrows*, see also second part of the video to ◘ Fig. 30.21). The figure shows an overlap of the structure with that of bound bicuculline (highlighted in *yellow* for distinction). The spatial divergence of the long helices is evident. The benzodiazepines bind between the interfaces of the α- and γ-subunits. Thereby, they increase the contact between these two domains, thus, supporting the described rotational movement. This leads to the well-known positive allosteric regulatory effect

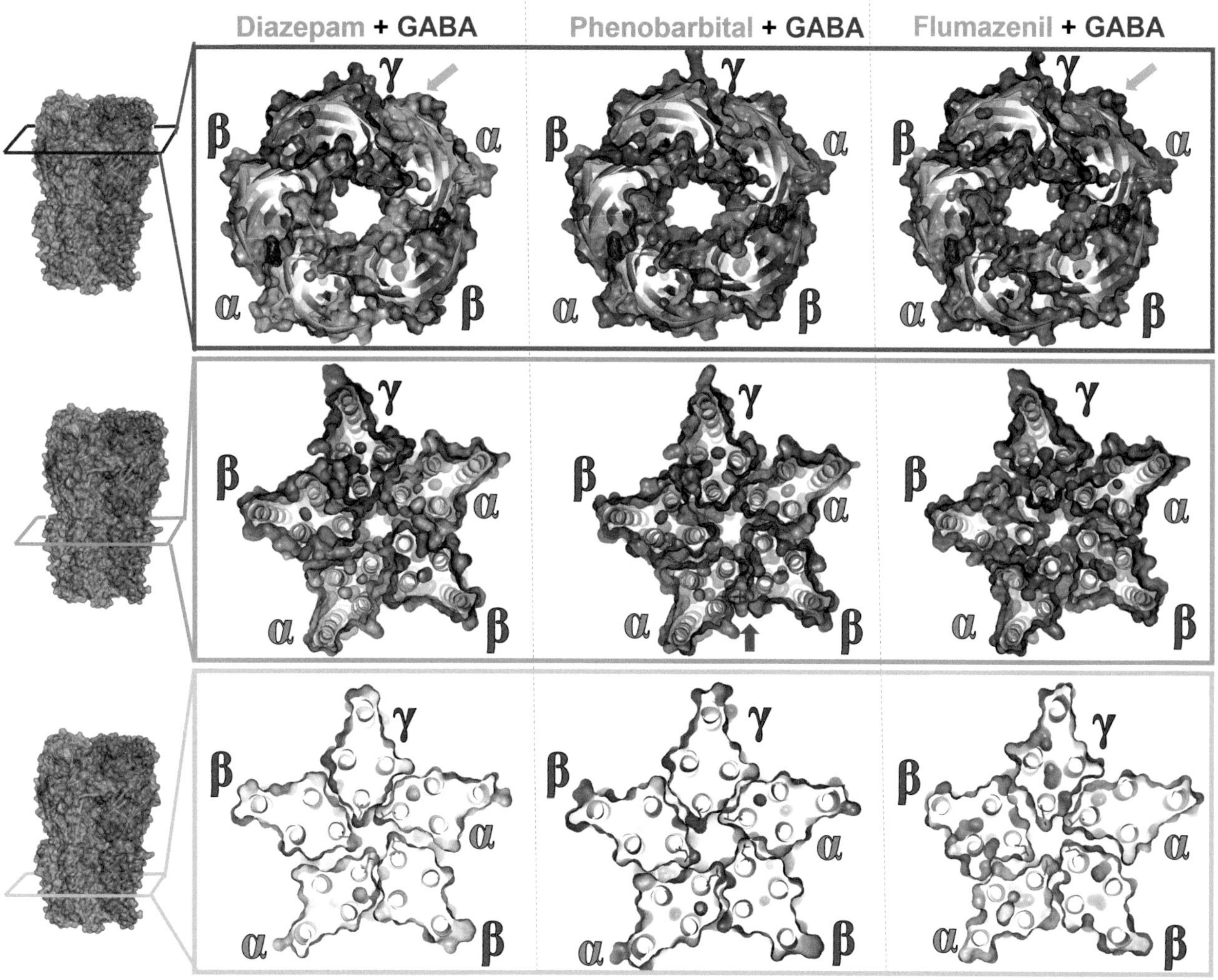

◘ Fig. 30.23 Cryo-EM structures of the $GABA_A$ receptor with bound GABA **30.35** and diazepam **30.40** (*left*), phenobarbital **30.39** (*center*), and flumazenil **30.42** (*right*). Sections through the receptor perpendicular to the channel axis are shown. In the extracellular domain region (*purple "section plane"*), the binding sites of GABA (*purple surface*) are located between the β- and α-subunits (counterclockwise labeling). The benzodiazepines diazepam **30.40** and the antidote flumazenil **30.42** occupy a binding site between the α- and γ-subunits (*green surface* and *green arrow*). Phenobarbital **30.39** is not found at this level. In the upper region of the transmembrane helices (*green "section plane"*), we find three binding sites for diazepam and two for phenobarbital. For the barbiturate, there is an exclusive binding site between the α- and β-subunits (*red arrow*). At the level of the leucine belt (*yellow "section plane"*), the channel is most open for phenobarbital binding, more so than for diazepam or flumazenil binding. (▸ https://sn.pub/719r4Y)

a C=N bond in the seven-membered ring, which can also be formed as an *N*-oxide function. In the fused benzene nucleus, the 7-position in the *para*-position to the lactam nitrogen is often blocked by chlorine, bromine, or nitro substituents. Such a group can help to adjust lipophilicity, but it also blocks the activated position against metabolism and reduces the electron density in the benzene nucleus. In the protein environment (◘ Fig. 30.24, *right*), this chlorine atom is located between the polar groups of His 102 and Asn 60, forming stabilizing contacts. In addition, a phenyl group is often found as a substituent at the 5-position of the central seven-membered ring. On this attached phenyl ring, a 2′-substituent also serves to increase lipophilicity, but has a conformational effect as well. It fixes a twisted arrangement of the adjacent rings. This orientation of the attached phenyl ring is also adopted by diazepam in the protein without *ortho*-substituents. The 3-position in the seven-membered ring is also interesting. As an enantiotopic position, a chemical change at this position, for example, by oxidation, leads to the introduction of a stereogenic center. Its chirality is irrelevant for the administration of the compound (e.g., oxazepam), as the compound racemizes easily in aqueous solution. The 3-hydroxylation leads to more hydrophilic derivatives,

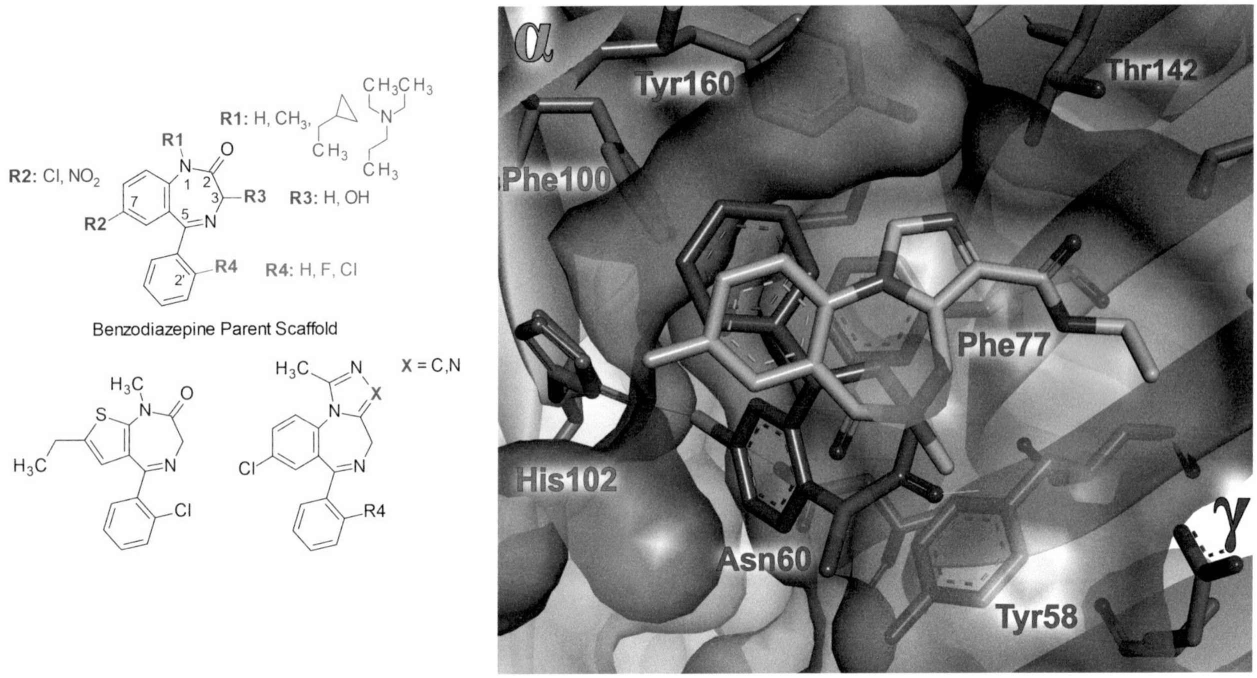

30

Fig. 30.24 *Left* Systematic bioisosteric replacement with exchange of substituents R1–R4 yields numerous benzodiazepines with divergent efficacy profiles and kinetics for therapy. *Right* Binding site of benzodiazepines between α and γ-subunits. Diazepam **30.40** (*gray*) and flumazenil **30.42** (*green*, formulas see Fig. 30.19) have significantly different binding modes despite a similar chemical structure. Flumazenil interacts mainly with the α-subunit, diazepam with both the α- and γ-subunits. Diazepam places its chloro substituent between His 102 and Asn 60 (*purple line*) and forms polar interactions. The carbonyl oxygen of the amide bond finds a hydrogen-bonding partner in the OH group of Ser 205 (not shown in the image). These are the only polar interactions in an otherwise very hydrophobic binding site characterized by many aromatic amino acids. (▶ https://sn.pub/5AFikG)

which are absorbed more slowly. Benzodiazepines with increased lipophilicity (alkylation at N1, chloro substituents in the 7- and 2′-position) lead to faster achievement of the active concentration in the central nervous system. This increases the sedative-hypnotic component. In contrast, increased hydrophilicity (unsubstituted N1 atom, 3-hydroxylation, no 2′-halogenation) is required for the tranquilizing profile.

Cryo EM structures can contribute greatly to the structural and mechanistic understanding of Cys-loop ion channels. Although it has been possible to develop a wealth of pharmacologically valuable drugs for the $GABA_A$ receptors without knowledge of the structures, the binding geometries allow many aspects of their action to be understood retrospectively. Although the α-, β-, and γ-subunit receptors presented here represent the majority of GABA channels found, the toolbox for their assembly contains a large number of other subunits. It is known that some of these combinations are not sensitive to benzodiazepines. Structural biology will have to show in the future whether they lack appropriate binding sites or whether they are structurally too different to allow binding. It is hoped that further structural data will allow the design of selective compounds. These should then have specific therapeutic properties, e.g., only an anxiolytic or sedative component.

The **glycine receptors (GlyRs)** mentioned above have a very similar pentameric structure. Binding of the neurotransmitter glycine opens the central pore and allows chloride ions to enter the cell. In terms of domain composition, the GlyRs appear simpler. Homogeneous pentamers consisting of five identical α-subunits (α1 to α4 are known) and heteropentamers with an inserted β-subunit are known. The receptor is activated like the $GABA_A$ and nACh receptors by rearrangement of the *C*-loop after glycine binding. Agonists, antagonists, and synthetic modulators have been described. For example, the highly toxic alkaloid strychnine displaces the agonist glycine and, thus, prevents GlyR activation. The inhibitory effect of glycine is lost. This leads to overexcitation of spinal nerves and muscle rigidity.

Dysregulation of GlyRs is a major cause of chronic pain. Pharmacological interventions that restore the inhibitory function of GlyRs could provide a therapeutic solution. However, targeted regulation of the system controlled by GlyRs requires precise identification of the GlyR subtypes involved in chronic pain conditions of different origins. Unfortunately, there is currently a lack of subtype-specific ligands to gain this insight into the role of specific GlyR subtypes in disease states. This opens up a wide field for future targeted drug discovery.

30.8 The Mode of Action of a Voltage-Gated Chloride Channel

More detailed insights into the structure of a class of **voltage-gated chloride channels** have also been obtained. Nine isoforms of these **ClC channels** are present in our genome. They take on numerous physiological functions, for example, the control of the resting potential in skeletal muscle and nonexcitable cells. Moreover, they exert an influence on the absorption of sodium chloride from the kidney into the blood stream or they are involved in processes that are necessary for the establishment of an acidic milieu. Malfunction and genetically caused mutations in these channels are associated with diseases such as myotonia, which is associated with pathological muscle stiffness, or particular forms of epilepsy, neuropathy, and osteopetrosis (a bone disease).

In 2003, Roderick MacKinnon's research group solved the crystal structure of a bacterial ClC channel. The structure of the human channel has since been determined, and it is very similar to the bacterial version. The channel is composed of two identical subunits that are linked by a twofold symmetry. Interestingly, this membrane protein does not have long helices oriented perpendicular to the membrane. Rather, the 18 helices of this channel are tightly packed and tilted up to 45° to the membrane axis. The channel pore is reminiscent of an hourglass. The pore widens to an atrium on the intracellular and extracellular sides, where positively charged arginine residues are found in the vicinity (◘ Fig. 30.25). The channel narrows in the center over a distance of about 15 Å. A selectivity filter together with a conserved glutamate residue is found at the apex. This residue acts as a gatekeeper. In addition, the ends of two **antiparallel-oriented helices** terminate exactly there. They form a preferred binding site for a negative charge. For this to occur, the helices must have an opposite orientation compared to that in the potassium channel. Here, they have their *N*-terminal ends oriented to the narrowest point in the channel. As in the potassium channel, the dipole moments generated along the helices create a special binding site for negatively charged ions. The carboxylate group of Glu 148 is located exactly at this point in the crystal structure. If this residue is exchanged for a neutral glutamine, the position will be freed and the glutamine will assume a different position. Instead, a bound chloride ion is found at this position. Mutation to glutamine leaves the channel in a permanently open state. The two structures are thought to describe the open and closed states of the ClC channel. The fact that the Gln 148 mutant shows a chloride ion in this position underlines the importance of the special position between the two oppositely oriented helix ends for the stabilization of a negative charge.

In addition to this chloride ion, two other chloride ions were found in both the open and closed channels. One is located deep inside the pore and has completely shed its solvation shell. It is stabilized by two NH groups from the main chain and the OH groups from Ser 107

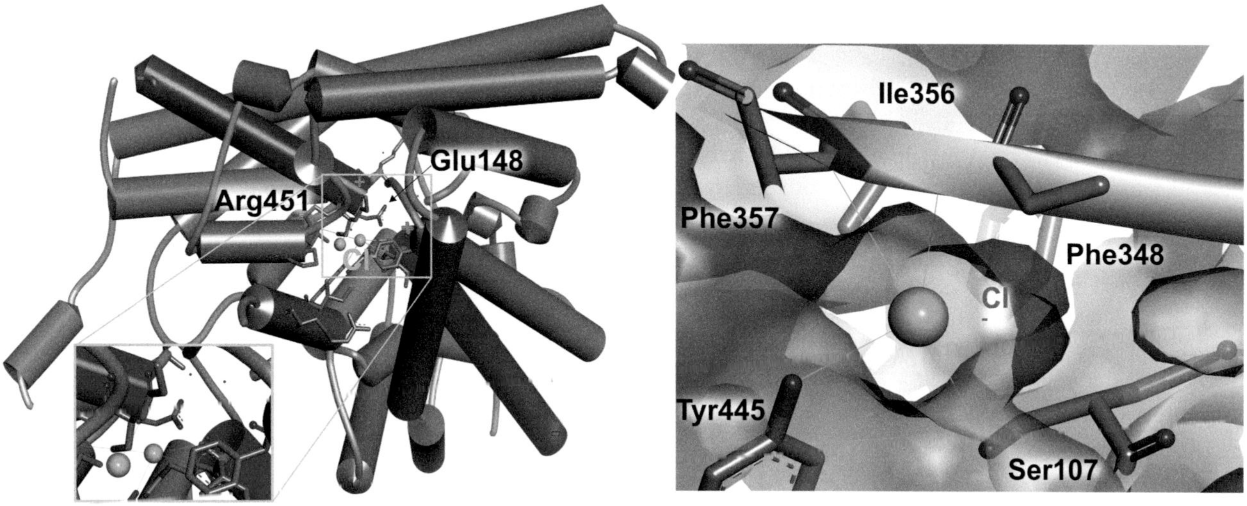

◘ **Fig. 30.25** *Left* Two long helices orient their positively polarized, *N*-terminal ends towards the narrowest position in the channel in the crystal structure of the voltage-gated ClC channel. Glu 148 is found at this position (*see inset*), which acts as a gatekeeper and opens and closes the channel. A conformational rearrangement of the negatively charged residue opens passage for chloride ions. *Right* Upon passage, the chloride ion sheds its water shell. This is replaced by coordination to the hydroxyl groups of Ser 107 and Tyr 445 and two contacts to NH groups from the main chain. (► https://sn.pub/B4a24f)

and Tyr 445 (Fig. 30.25). The other chloride ion is located at the entrance and is still partially solvated by water molecules.

Regulation by glutamate as a placeholder allows the channel to open and close in response to external signals. The structurally related human ClC-0 channel is voltage-gated when the potential inside the cell shifts to the positive range. An adjacent negative potential closes the channel. When the extracellular chloride ion concentration increases, the channel opens. The same can be observed when the pH of the environment drops. It is possible that the glutamate residue changes its protonation state as it swings out of the pore apex to make way for the chloride ion. This would explain its regulatory function during pH adjustment and the stoichiometric exchange of Cl^- for H^+. ClC channels are specific for monovalent anions. In addition to chloride, Br^-, I^-, NO_3^-, and SCN^- can also pass through, although with reduced permeability. Since the latter ions play a minor role in biological systems, a pronounced selectivity is not necessary. However, divalent ions such as sulfate and hydrogen phosphate are denied passage. There are also substances that are known to block the channel. The exact experimental characterization of their binding mode is still lacking. They may be the starting point for the development of selective modulators of ClC channels.

30.9 ATP Hydrolysis Fuels Ion Flux Against Concentration Gradients

The ion channels introduced so far open in response to a voltage- or ligand-gated signal. Ions then flow across the membrane in the applied **electrochemical concentration gradient** for the duration of the opening. As a result, the gradient slowly decreases and the flow would come to a stop after some time. In order to keep the cells excitable for their function, the gradient has to be reestablished again and again. How can this be achieved? This requires energy. In biology, this comes mainly from the hydrolysis of the energy-rich molecule adenosine triphosphate (ATP). Its degradation to diphosphate releases a free enthalpy of −30.5 kJ/mol. Ion channels capable of restoring an electrochemical concentration gradient are, therefore, coupled to an **ATPase function**. They are, thus, also called **ion pumps**. We will see that the transporters presented in the next section also use ATP hydrolysis for their pumping process.

The structures of some members of the class of so-called **P-type ATPases** have been elucidated. These include the **sodium–potassium pump**, which restores the concentration gradient of sodium and potassium ions across the cell membrane. Digitoxin exerts its effect by binding to this pump (Sect. 6.1). **Calcium ATPases** also belong to this family. Another member, the **proton pump** in the parietal cells of the stomach, plays an important role in drug research. This is where the important drug molecules of proton pump inhibitors act. All P-type ATPases share a common structure (Fig. 30.26). They consist of two subunits. The much larger α-subunit crosses the membrane with 10 transmembrane helices. It has one extracellular domain and three on the cytosolic side. These contain the ATP-binding site, a phosphorylation site, and the so-called actuator domain. The β-subunit consists of a long helix and is responsible for translocation to the membrane. The channels pass through four different states in one pumping cycle, which differ in their affinity for the ions to be pumped.

The **H^+/K^+-ATPase**, the proton pump, will be discussed in more detail. It performs the electroneutral exchange of protons and potassium ions with ATP consumption in the parietal cells of the stomach. Successful **proton pump inhibitors** such as omeprazole **30.44** or pantoprazole **30.45** (Fig. 30.26, *right*) act on this pump (Sect. 9.5). For this purpose, they undergo an acid-catalyzed prodrug rearrangement that is triggered only in the acidic environment of the parietal cells. This specifically generates the reactive species that forms a **covalent disulfide bridge** to Cys 813. In this way, the pump is uniformly and irreversibly inhibited over an extended period of time. Despite its very successful use in therapy, its pH-dependent activation and short plasma half-life have been repeatedly questioned. As an alternative, the search has focused on noncovalently binding drug molecules that do not require activation. In recent years, the first compounds, the so-called **p**otassium **c**ompetitive **a**cid **b**lockers (P-CABs), have been introduced to the market. They bind at the same site as omeprazole or pantoprazole near Cys 813 and appear to be characterized by a faster onset of action and improved blocking of acid secretion.

The starting point was the compound SCH28080 **30.46**, which had already been described in the 1980s. The compound was found to cause liver toxicity and, therefore, not further pursued. However, a crystal structure with this compound could be determined. These results allowed the development of **30.46** further into improved imidazopyridines. Tegoprazan **30.47** and soraprazan **30.48** are now approved for therapy. They were used for cryo-EM structure determinations. Fig. 30.27 shows the binding mode with **30.47**. The drug molecule binds to the proton pump in the outwardly open E2P state near the upper end of the ion channel. This is also where Cys 813 is located, to which omeprazole **30.44** binds irreversibly by forming a disulfide bridge. The binding pocket is largely hydrophobic. Overall, the three imidazopyridine derivatives show a conserved binding mode. The binding of SCH28080 **30.46**, which binds with an IC_{50} = 2.1 μM, could be improved to 0.30 μM for soraprazan **30.48**. Presumably, the potency of this compound is enhanced by correct fixation of the termi-

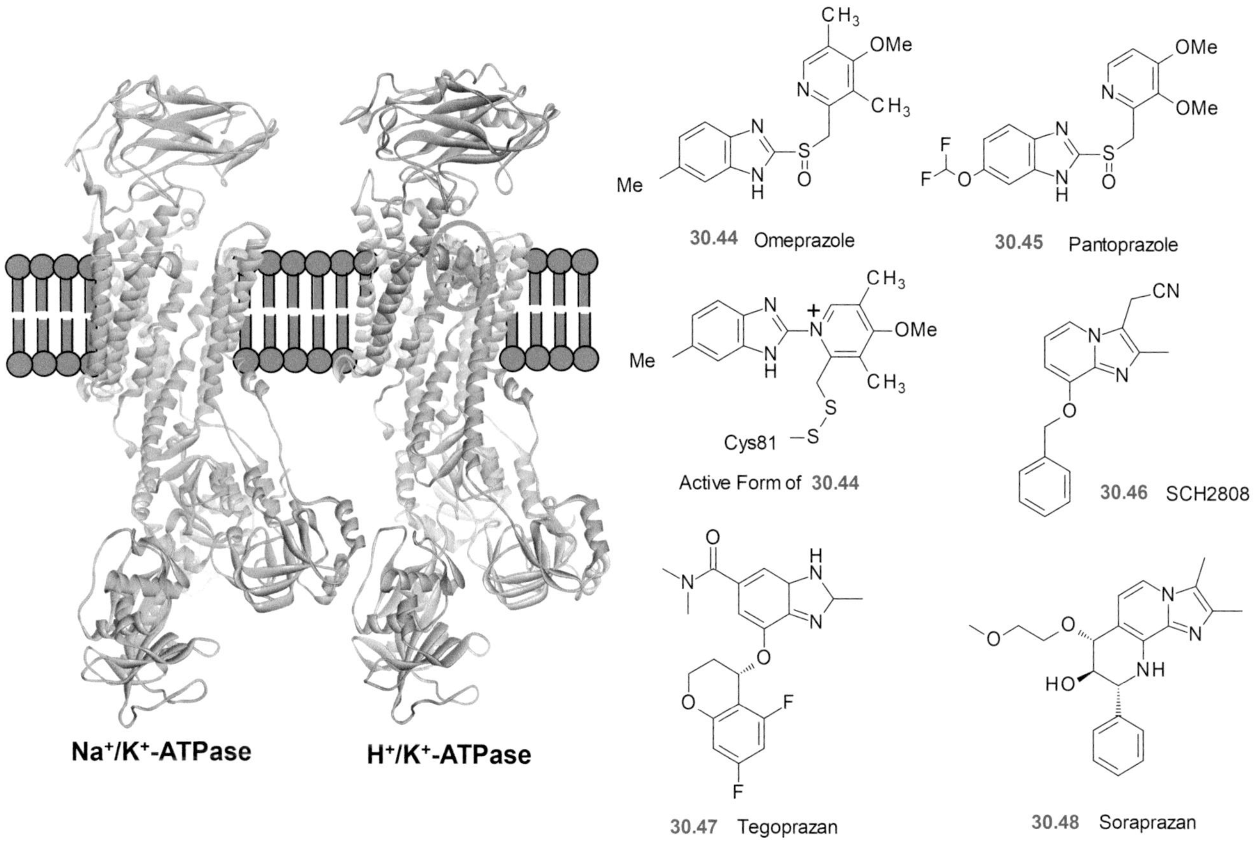

Fig. 30.26 *Left* Crystal structure and cryo-EM structure of Na^+/K^+-ATPase (*green*) and H^+/K^+-ATPase (*beige*). Both pumps have a very similar folding pattern. The binding site of the proton pump inhibitors is located near the upper rim of the membrane (*green ellipse*). *Right* Omeprazole **30.44** and pantoprazole **30.45** are prodrugs that form a covalent disulfide bridge to the ATPase after rearrangement in an acidic environment. In the meantime, noncovalently binding inhibitors have also been found. Based on the lead structure SCH28080 **30.46**, the two approved inhibitors tegoprazan **30.47** and soraprazan **30.48** were developed

nal phenyl ring in the bound conformation. The conformationally flexible ethylene linker is spatially fixed by the fused six-membered ring. The elucidation of the mechanism of action of the P-CABs at the molecular level will contribute to the rational design of improved and more potent compounds. They should help in the treatment of acid-related gastrointestinal diseases and in the eradication of the *Helicobacter pylori* bacterium in affected patients.

30.10 Transporters: The Gatekeepers to the Cell

All cells need to be able to selectively transport endogenous and exogenous compounds across the cell membrane. A large class of proteins that perform this task are the **membrane transporters**. They carry, for example, hormones, amino acids, bile acids, uric acid, or lipids across the membrane barrier. Mutations in these transporters are associated with severe genetic diseases such as adrenoleukodystrophy (which causes neurological degeneration) or retinal degeneration. An important group of transporters is responsible for the efficient reuptake of released neurotransmitters from the synaptic cleft (Sect. 22.7, Fig. 22.7) into the presynaptic nerve cell. This reuptake can be blocked by drugs. Serotonin and noradrenaline transporter reuptake inhibitors, for example, have been the subject of intensive and successful research in the pharmaceutical industry. Often these inhibitors have an additional mode of action as antagonists against the corresponding receptor on the postsynaptic side. These receptors belong to the GPCR family and are divided into a wide range of subtypes (see Table 29.1). Because they bind at completely different sites with structurally apparently related binding sites, these inhibitors have different pharmacological profiles and side effect spectra.

In addition to importing compounds into cells, transporters are also responsible for removing exogenous compounds from the cell. The majority of drugs belong to the group of exogenous or xenobiotic compounds. Drug resistance often develops during therapy, for example, against drugs used to treat infections. The development of **multiple drug resistance (MDR)** also involves transporters that are presumably up-regulated in order to expel

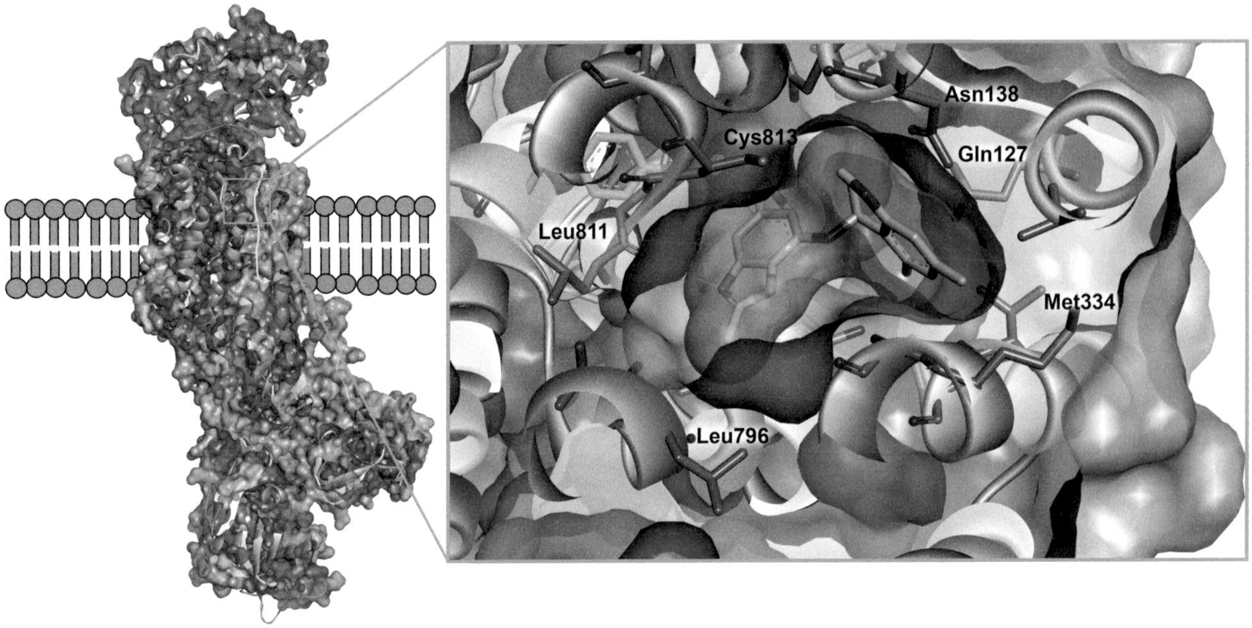

Fig. 30.27 Cryo-EM structure of the H^+/K^+-ATPase with the inhibitor tegoprazan **30.47**. The drug binds near the opening of the channel (*yellow line*) to the upper rim of the membrane (*left*, *green box*). The binding site is located below Cys 813, the residue where the thiol group of omeprazole **30.44** forms a covalent bond to the proton pump. Overall, the binding pocket is largely formed by hydrophobic amino acids. (► https://sn.pub/GUAtP4)

drugs from the cell. They either use a proton gradient to transport the drug, or their transport is coupled to the energetically favorable hydrolysis of ATP (in the case of ATP synthase-binding cassette [ABC] transporters). The latter group of **ABC transporters** represents a large family of proteins that import and export a broad range of substances, including amino acids, ions, sugars, lipids, and drugs. To date, 46 of these ABC transporters have been identified in humans. They consist of at least two nucleotide-binding domains (NBD) and two transmembrane domains (TMD). Several structures of these ABC transporters have been determined. They are very similar in structure. Again, the binding of ATP is essential for the function of the transporters. However, the TMDs are crucial for the actual membrane passage. There are transporters for both hydrophobic and hydrophilic compounds. In the case of hydrophilic compounds, they provide a shield against the hydrophobic membrane environment.

The best studied transporter is the human MDR-ABC transporter **P-glycoprotein GP170** (MDR1/ABCB1). Like a hydrophobic vacuum cleaner, it removes lipids as well as a wide range of drug molecules from the cell. Recently, structures determined by cryo-electron microscopy (Sect. 13.6) have provided crucial information about the mechanism of action. The transporter spans the membrane with two arms, each consisting of six helices (Fig. 30.28). Their **nucleotide-binding domains** (NBDs) are located on the cytosolic side. During the transport cycle, the TMDs undergo dramatic conformational changes. At the beginning of a transport cycle, before ATP is incorporated into the two NBPs, the two transmembrane domains spread apart. The shape of the molecule resembles an inverted letter "V". It is 136 Å long and its top spans the membrane. The arms spread 70 Å apart and the two ATP-binding sites are separated by 30 Å. At the apex of the "V", the transporter opens a chamber. It can accommodate molecules to be transported directly from the cell or from the rim region of the inner leaflet of the membrane. The chamber appears to be highly adaptive, which explains the wide substrate diversity of the transporter for the uptake of very different molecules. In this state, there is no access to this binding chamber from the extracellular side. Substrate binding stimulates the uptake of ATP by the NBDs. This is thought to be the signal for dimerization of the two NBDs. On the way to dimerization, a structure has been identified in which the central chamber is closed (Fig. 30.28, *center*). In this structure, the arms are still far apart. After rearrangement and formation of the contact between the two NBDs, which is propagated to the transmembrane helices by a twist, the previously closed chamber opens to the extracellular side. This extracellular conformation has been captured by the structurally analogous bacterial multidrug exporter Sav1866 (Fig. 30.28, *right*). The substrate can leave the chamber, allowing hydrophobic substrates to migrate into the

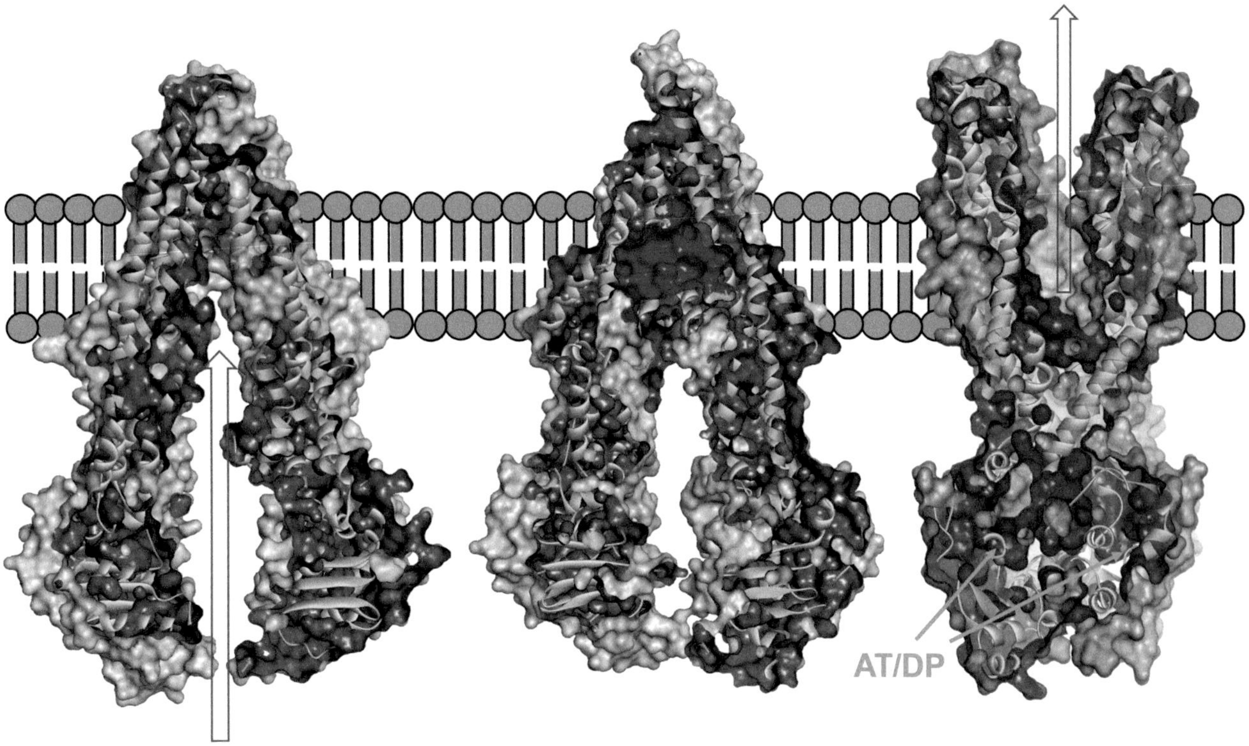

Fig. 30.28 Crystal or cryo-EM structures of different situations along the transport cycle of ABC transporters. They are based on structures determined of the human ABCB1, a chimeric ABCB1 and the bacterial Sav1866 transporter. The cycle begins with the transporter open inwards (*left*). At the apex of the V-shaped protein, a chamber opens to accommodate the ligand to be transported. After binding of ATP to the nucleotide-binding domains (NBDs) located at the cytosolic end, the chamber closes (*center*) and the NBDs domains dimerize. This motion is transmitted to the long helices. Subsequent opening of the chamber to the extracellular side allows the ligand to diffuse out (*right*). After hydrolysis of ATP (*green*) to ADP, the transporter opens again to accept a new ligand for the next transport cycle. (► https://sn.pub/oCgMHa)

outer leaflet of the membrane. The release of substrates requires a reduced binding affinity. This is probably due to changes in the contacts between protein and substrate during the transition from the inward-facing to the outward-facing conformation. This step is probably also supported by ATP hydrolysis. After ATP hydrolysis, NBD dimerization is disrupted. This causes the transporter to revert to the inwards conformation and the next transport cycle can begin.

The development of **resistance due to transporters** is a serious problem in drug therapy. It is, therefore, all the more important to study the molecular criteria that make molecules good substrates for these transporters. Consequently, it may be possible to understand how to modify molecules so that they are no longer good substrates. This task is not trivial because the binding pockets in these transporters appear to be highly adaptive, so that the typically small changes in a drug molecule that are tolerable in terms of its mode of action will not affect its binding behavior to the transporters. On the other hand, potent inhibitors of these transporters can be sought. Some compounds, such as *R*-verapamil (Sect. 2.6), have been discovered for this purpose. However, their clinical use to break resistance has proven problematic because inhibiting the transporters also prevents their natural function. On the other hand, it should not be forgotten that the inducible and heterologous expression of these transporters represents a crucial defense mechanism of cells against xenobiotics. It is not without reason that Nature has developed such a highly efficient and flexible protective mechanism. It is, therefore, possible that these transporters are not an ideal drug target in humans. However, when it comes to fighting bacteria and parasites, the picture is quite different. Bacteria and parasites also use such transporters to attack drug molecules (see Sect. 3.2). The weapons currently used against bacteria and parasites will eventually become ineffective. Recently, attempts have been made to inhibit parasite and bacterial transporters in order to break resistance. If these goals are achieved, it will be a double success. On the one hand, it would break resistance to older and well-proven therapeutic drugs. On the other hand, the undesirable pathogens would be additionally damaged because the transporter would no longer be available as a defense mechanism against unwanted foreign substances that are potentially harmful to them. It remains to be seen whether this concept, which is currently in the research phase, will have the desired success.

30.11 Membrane Passage in Bacteria: Pores, Carriers, and Channel Formers

Gram-negative bacteria are surrounded by two membranes: an inner plasma membrane and an outer cell membrane. These are separated by the periplasmic space. Although most proteins penetrate the inner membrane with a helical sequence segment, interesting pores are found in the outer membrane that have a pleated sheet structure. They are among the most abundant proteins in bacteria. Each of these openings, called **porins**, represents a water-filled channel that allows passive diffusion of nutrient building blocks and waste products out of the cell. Their diameter is limited, preventing the selection of potentially toxic compounds. The porin structure of the bacterium *Rhodobacter capsulatus* was first elucidated by the research group of Georg Schulz and Wolfram Welte in Freiburg, Germany (◘ Fig. 30.29). The pore exists as a trimer in which the monomers are packed together in a triangular shape. Each pore is formed by a 16-stranded up-and-down β-barrel (Sect. 14.3), and the individual β-strands adopt an antiparallel orientation. The β-barrel is a common folding pattern in enzymes. Usually, however, only up to eight pleated sheets come together to form the tightly packed core of the barrel-like structure. Due to the large number of strands, there is enough space in the porin to open a passage to the interior. However, it is partially closed by a long loop that limits the remaining eyelets to a maximum diameter of 8 Å. The eyelet region is almost entirely composed of positively and negatively charged amino acids that are oriented to opposite sides of the pore. This orientation of charged groups also contributes to the selection of molecules that can pass through the pore.

Bacteria also synthesize small peptide-like systems that penetrate the membranes of other organisms and in doing so also offer a possibility for the passage of, for example, ions. These systems are termed **transport antibiotics**. They render the membrane permeable in different ways. The antibiotic **gramicidin A** is an oligopeptide made of 15 amino acids that have alternating L- and D-configurations. The peptide forms a tube-shaped helical structure and traverses the membrane as a dimer (◘ Fig. 30.30). This creates a channel with a diameter of 4 Å in the interior. It is highly permeable for monovalent cations such as Na^+ and K^+. On the other hand, multivalent cations and anions are prevented from entering. Up to 10^7 cations per second can pass through this channel, a transport rate that is only a factor of 10 below the diffusion rate in water. The cations must shed their hydration shells. Then they apparently slide through the opening along the amide bonds, which are oriented parallel to the channel axis. The side chains of the hydrophobic amino acids orient themselves in the surrounding lipid membrane.

The depsipeptide **valinomycin** follows a completely different mode of action. It is made up of valine, lactate, and hydroxyisovalerate residues. It encapsulates the potassium ion with its polar groups, which are oriented towards the interior. It presents its hydrophobic groups to the outside. When wrapped in such a chelate-ligand complex, charged ions can pass through the membrane barrier inside the encapsulating hydrophobic particle. In addition to valinomycin, other such carriers are known, such as **nonactin** (◘ Fig. 30.31). These **transport antibiotics** alter the ion permeability of bacterial cell membranes and intracellular compartments. As a result, they can cause bacterial cells to die. Valinomycin, for example, accumulates in mitochondrial membranes, increases potassium influx, and thereby disrupts mitochondrial energy homeostasis and ATP synthesis. The transport antibiotics are important as combination drugs for external use, e.g., for the treatment of oropharyngeal infections.

The **lipopeptide daptomycin** has been introduced into therapy to fight Gram-positive bacteria. The cyclic peptide, composed of 11 amino acids, penetrates the bacterial cell membrane with its hydrophobic side chain. It

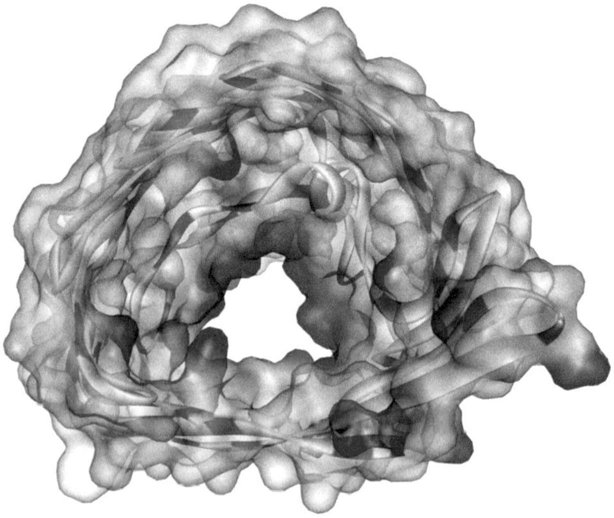

◘ **Fig. 30.29** Crystal structure of the porin from the bacteria *Rhodobacter capsulatus*. Each pore of the trimeric proteins (only one monomer is shown) is made up of a 16-stranded "up-and-down" β-barrel. The pore traverses the membrane along the view axis and broadens to about 8 Å. It is flanked by positively (*blue*) and negatively (*red*) charged amino acids that establish an electrical field gradient across the membrane. (► https://sn.pub/lr54fp)

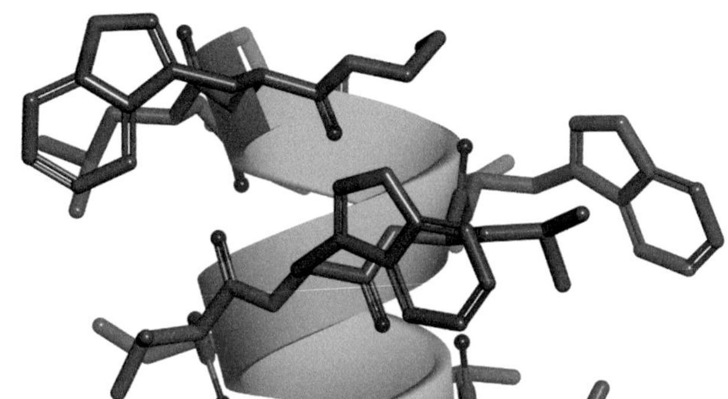

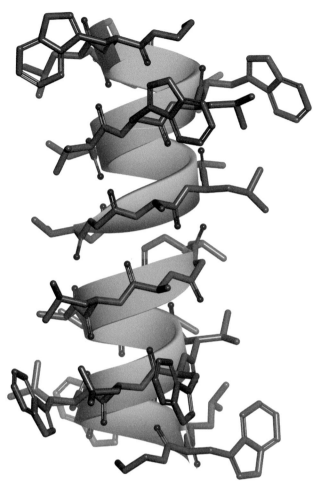

■ **Fig. 30.30** Gramicidin A (Val–Gly–Ala–Leu–Ala–Val_3–(Trp–Leu)$_3$–Trp–ethanolamine) is made up of 15 alternating L- and D-configured amino acids and forms a narrow channel of about 4 Å through the membrane along which monovalent cations such as Na^+ and K^+ can migrate. (► https://sn.pub/HdjHkS)

forms channels for ions by oligomerization. This makes the cell membrane permeable for potassium ions. Their efflux leads to depolarization and ultimately to bacterial cell death. Peptides with 20–25 amino acids such as magainin (Locilex®) use an analogous mechanism to form amphipathic helices in the membrane.

30.12 Aquaporins Regulate the Cellular Water Inventory

The cellular lipid bilayer is a barrier to water molecules. Despite an **osmotic gradient** across the membrane, simple diffusion is limited to a few individual water molecules or they cross in association with other particles. Large amounts of water molecules cannot actively cross the membrane. In 1992, Peter Agre's group in Baltimore, MD, USA, discovered a 28-kDa protein in the erythrocyte membrane that turned out to be a **water pore**. It is used only for water transport, neither ions nor other small molecules such as glycerol or urea can pass through it. The direction of water flow is determined by osmotic pressure alone. The first **aquaporin** discovered in erythrocytes was named AQP1.

Since then, more than 100 aquaporins have been discovered in all kinds of organisms. Humans alone have more than ten isoforms, seven of which are used at different sites in the kidney. Some aquaporins are exclusively specialized for water, while others, despite their similar architecture, also allow the transfer of small molecules such as glycerol and urea. The discovery of aquaporins has revolutionized our understanding of the regulation of **water homeostasis**. For this achievement, Peter Agre was awarded in 2003 the Nobel Prize in Chemistry.

Sequence analysis of aquaporins reveals an architecture composed of two nearly identical segments. Each half contains a highly conserved Asn–Pro–Ala– (NPA) motif. The functional aquaporin unit is a **tetramer** in which each monomeric unit encloses a pore. The crystal structure shows that each pore consists of six transmembrane helices. The channel extends like a hose through the protein, widening on the extracellular and cytosolic sides to a 15 Å funnel-shaped vestibule (■ Fig. 30.32, *left*). In the middle, it narrows to a diameter of 2.8 Å. The vestibules contain many polar but mostly uncharged amino acids. A chain of accessible backbone carbonyl oxygen atoms and NH groups of Asn residues lines one side of the pore. The opposite wall consists of hydrophobic residues. This pattern of H-bonding groups along the pore wall is thought to be involved in the passage of migrating water molecules. They move as if on a rope ladder, temporarily forming H-bonds along the way (■ Fig. 30.32, *right*).

The chain of exposed C=O groups, together with the amphipathic nature of the tubular channel, also forms the **selectivity filter**. The geometry of the inner-facing carbonyl groups is reminiscent of the selectivity filter in the potassium channel. Since the carbonyl groups are only on one side of the channel, they cannot completely replace the hydration shell around a cation. A cation that enters the pore is, therefore, too large to pass through the pore. Complete shedding of the water shell is not possible. The amphiphilic channel lacks circularly arranged polar anchor points to coordinate an ion.

A histidine and an arginine are found at the smallest isthmus. A phenylalanine is found on opposite side. These three amino acids are highly conserved among the porins that specialize in water permeability. Because of the charge on **His** and **Arg**, they cause a further sieving for positively charged ions, even H_3O^+. Negatively charged ions are so strongly repelled by the many negatively po-

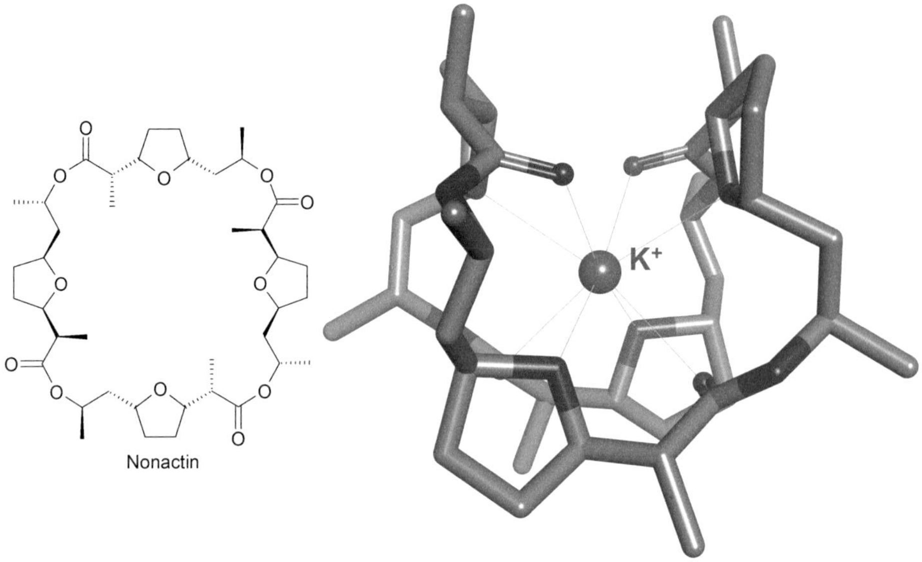

Fig. 30.31 Nonactin represents a chelating ligand to coordinate potassium ions. It wraps optimally around the ion and can penetrate the membrane as a chelated complex. Thus, this antibiotic penetrates the membrane with an exterior surface that appears hydrophobic on the surface. (▶ https://sn.pub/t2eiTD)

larized carbonyl groups that their passage is energetically too unfavorable. The channels that allow glycerol to pass in addition to water have an additional diameter of 1 Å at their narrowest point. At the same time, the histidine conserved in the exclusively water-permeable channels is replaced by a glycine. Overall, the glycerol-permeable channel has a slightly more hydrophobic character.

Aquaporins occur virtually ubiquitously in our bodies, but in greater numbers and diversity in the kidney. To achieve rapid control over their function, they are partially **stored in vesicles**. When needed, the vesicles fuse with the cell membrane. This increases the number of active aquaporins. The water channels represent an excellent target structure for therapeutic intervention. In addition to the development of diuretics, their use in the treatment of glaucoma, obesity, or to combat angiogenesis in tumors has been discussed. They have also become the focus of research as a target for the development of drugs to treat parasitic infections. Interestingly, **mercury salts** have long been used as **diuretics**. The thiol group of an accessible cysteine residue is located in the upper pore region of AQP1 (■ Fig. 30.32, *left*). Presumably, the mercury ion blocks the pore by coordinating to this cysteine. Because of their toxicity, mercury salts are certainly not drugs of choice. However, recent research has shown that the discovery of small molecule modulators is not an easy task. On the one hand, the narrow channel limits the molecules that can be used. On the other hand, the hydrophilic atrium is difficult to address. In addition, the development of functional assays does not seem to be trivial. It remains to be seen whether research will find potent and selective alternatives to the mercury salts. Such alternatives should specifically interfere with the regulation of aquaporins in order to treat a disease associated with their dysfunction.

30.13 Synopsis

- Cells require material exchange across the membrane. Amphiphilic compounds can diffuse through the membrane of their own accord. For the transfer of polar compounds, cells are equipped with special transporters that sometimes exhibit remarkable selectivity, but sometimes also have broad promiscuity.
- For the biologically relevant ions (Na^+, K^+, Ca^{2+}, Cl^-) special ion channels exist that allow ions to flow along a concentration gradient building up an electrochemical potential across the membrane.
- Cells are stimulated by action potentials. In the resting state, a potential of −70 mV is maintained. The extracellular sodium ion concentration is tenfold higher than in the cell interior. At a potential of about −60 mV, fast sodium ion channels open allowing Na^+ influx and gradually shift the potential to +40 mV (depolarization). They close at this value.
- Repolarization of the cell results from an efflux of potassium ions through slow and highly selective potassium channels. The potential returns to negative values. When the membrane potential falls to more negative values than the resting state, it is referred to as hyperpolarization. In some cells, this state can also be induced by Cl^- influx through chloride channels. In other cells, an influx of Ca^{2+} ions through specific channels can intensify depolarization across the membrane.
- KcsA potassium channels cross the membrane as tetramers with long helices. Four helices that are oriented with their *N*-terminal ends into a central cavern drag the positively charged ions across the membrane. The potassium ions, which are coordinated with eight water molecules in a square-antiprismatic geometry, pass through a selectivity filter by shedding

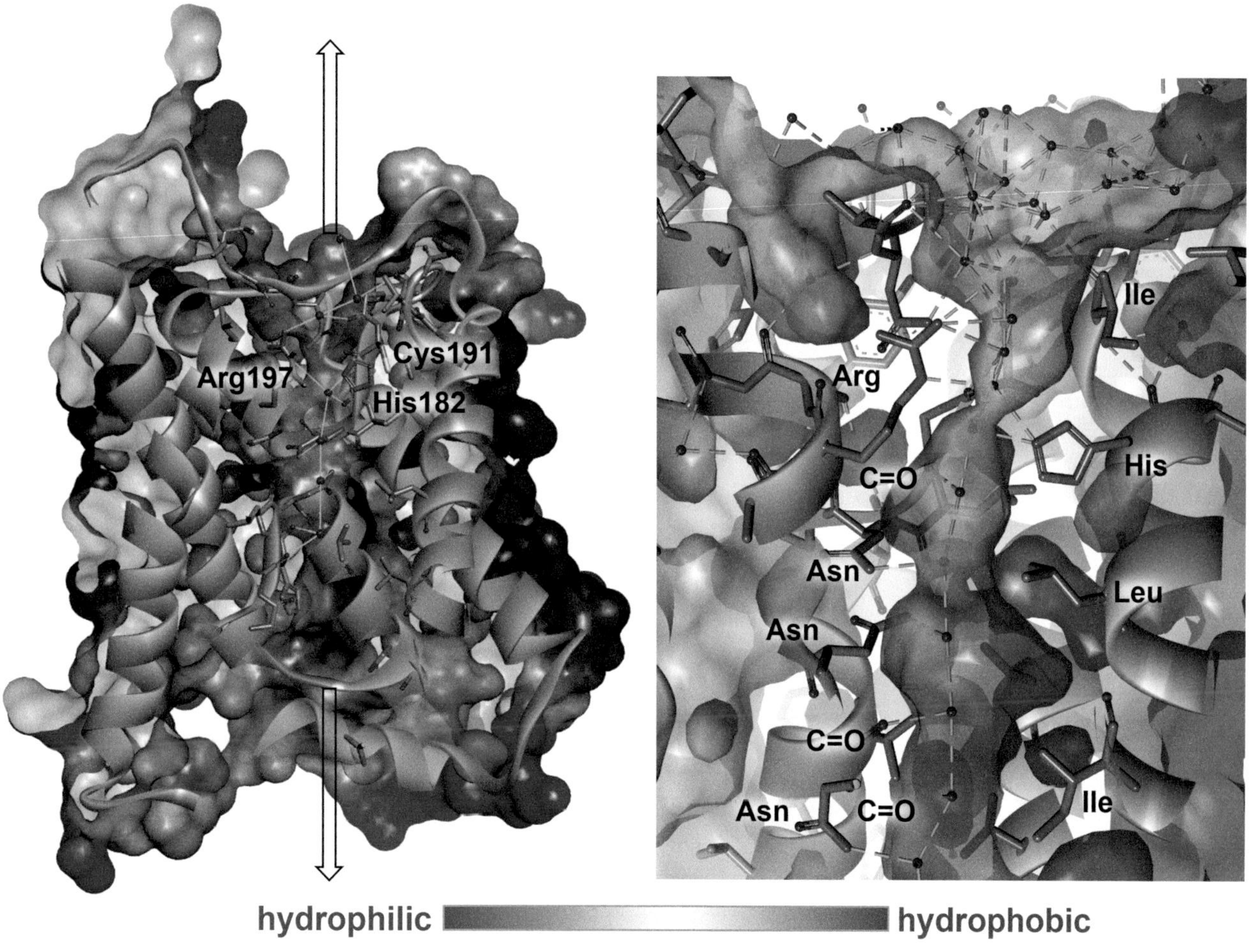

Fig. 30.32 *Left* Crystal structure of bovine aquaporin-1. Only one pore of the fourfold protein is shown. It widens in a funnel-like manner towards the extracellular (top) and cytosolic sides. At the narrowest point, the pore tapers to about 2.8 Å. Opposite to each other at this position are a positively charged His and Arg residue. They prevent the passage of positively charged ions, including H_3O^+. Near the narrowest position is a cysteine residue that can be complexed by mercury ions. This binding blocks the channel and explains the diuretic effect of mercury salts used in therapy many years ago. *Right* The structure of a yeast aquaporin was determined with a resolution of 0.88 Å. It reveals details of the water structure. As if lined up on a string, the water molecules can move through the channel through hydrogen bonds to carbonyl oxygen atoms of the main chain and NH_2 groups of Asn side chains. The H-bonding partners protrude into the channel from one side, and the opposite wall is formed by hydrophobic amino acids. The surface is colored according to local hydrophobicity (*brown*) or hydrophilicity (*blue*). (► https://sn.pub/dZfYsV)

their solvation shell. The protein provides a fourfold arrangement of backbone carbonyl groups that perfectly replaces the water coordination sphere around potassium ions, thus, achieving impressive selectivity over other cations.

- Some potassium channels are ATP-dependent with several domains and regulatory units. Sulfonylureas can block the regulatory unit on the pancreatic β-cells that is responsible for insulin secretion. This is exploited as a therapeutic principle for the treatment of type 2 diabetes mellitus because blocking the regulatory unit results in enhanced insulin secretion from the β-cells.
- Depolarization and repolarization of the heart muscle cells are important for correct control of the heart beat frequency. Drug molecules with a particular pattern of aromatic moieties and a central basic nitrogen can block the hERG channel, a potassium channel involved in the regulation of heart beat. Because fatal arrhythmias can occur, potential binding to the hERG channel as an antitarget is avoided in the early phase of drug discovery.
- Voltage-gated sodium and calcium channels contribute to the depolarization of cells. They consist of four transmembrane domains. They pass through different states by changing their conformation and open or close like an iris aperture by the movement

of helices with hydrophobic residues. At its center, the ion channel has a selectivity filter and a central pore. Inhibitors block this pore below the selectivity filter. Local anesthetics such as lidocaine bind to Na_V channels, while calcium blockers such as verapamil or diltiazem bind to Ca_V channels. The dihydropyridines provide indirect allosteric blockade and bind between two adjacent domains of the Ca_V channels.

- Ligand-gated ion channels of the Cys-loop superfamily are giant pentamers consisting of 20 transmembrane helices. The extracellular ligand-binding domains, which also have a pentameric geometry, contain binding sites for agonists and antagonists. When an agonist is bound to the contact surface between two domains, the so-called *C*-loop is attracted to the ligand-binding domain and strengthens the contact between the domains. This imparts a slight rotational motion to the transmembrane helices, which carry hydrophobic residues such as valine or leucine as a kind of belt at their narrowest point. The channel pore expands by several angstroms due to concerted rotation of the five innermost helices. This allows sodium, potassium, or chloride ions to pass through the channel.
- The ligand-binding domains of the pentameric nicotinic acetylcholine receptor are opened by agonists such as acetylcholine, nicotine, or epibatidine, whereas antagonists (or blockers) such as α-conotoxin keep it closed. $GABA_A$-chloride channels are opened by the agonist GABA. Allosteric regulators such as benzodiazepines strengthen an additional contact within the pentameric domains. They, thus, enhance the effect of the endogenous ligand GABA. The excitability of the cells is blocked by the opening of the chloride channel.
- Depending on the substitution pattern, benzodiazepines act as sedatives, hypnotics, anxiolytics, anticonvulsants, muscle relaxants, or anterograde amnestics. Their effects are neutralized by the competitive binding antagonist flumazenil. Barbiturates also strengthen the domain contacts, but in the transmembrane region and achieve a stronger and longer lasting opening of the chloride channel.
- The voltage-gated ClC chloride channels orient two extended helices with their *N*-terminal ends towards the center of the channel. Together with a conserved glutamate residue at the apex, they achieve the required selectivity, possibly via an intermediate change of protonation state of a glutamate residue taking the role of a gatekeeper.
- Ion pumps adjust ion equilibria against a concentration gradient. They couple this process to the energy-providing hydrolysis of ATP. The Na^+/K^+ and K^+/H^+ ion pumps belong to the class of P-type ATPases. They are the target structure of cardioactive digitoxins and proton pump inhibitors such as omeprazole. The latter binds on the extracellular side near Cys 813, resulting in covalent irreversible inhibition.
- Transporters shuffle endo- and exogenous compounds across the cell membrane. Transport is usually coupled with the energetically favorable hydrolysis of ATP to allow membrane passage against a concentration gradient. Particularly the human MDR-ABC transporter P-glycoprotein GP170 is upregulated in drug resistance and removes a broad palette of drug molecules from the cell.
- Bacteria have developed special transporter systems to either allow access to cells or to penetrate the membrane of other organisms. One class of pores is formed by large β-barrels of parallel-oriented strands that open a passage to the interior. Other systems either wrap around cations to form hydrophobic carriers on their exteriors, or penetrate into membranes with helix-forming elements to build-up channels that make them permeable for ions.
- Despite an osmotic gradient, water molecules cannot passively diffuse across the membrane. Water homeostasis is regulated by aquaporins, which are channels that extend like a hose through the membrane-bound protein. They have 15Å-wide funnel-shaped vestibules on either side and narrow to a diameter of 2.8 Å in the middle. There, a His and an Arg residue prevent the passage of H_3O^+ ions.
- On both sides, the aquaporin opens like a funnel to the cytosol or extracellular space. A chain of exposed C=O and NH groups lines one side of the pore, allowing water molecules to move along by H-bonds like on a rope ladder. The opposite channel wall is formed by hydrophobic residues. For rapid regulation of water balance, aquaporins are partially stored in vesicles and released into the cell membrane as needed.

Bibliography and Further Reading

General Literature

R. MacKinnon, Nobel Lecture, Potassium Channels and the Atomic Basis of Selective Ion Conduction, Angew. Chem. Int. Ed. Engl., **43**, 4265–4277 (2004)

M. Mark, Sulfonylharnstoffe und Glinide, Pharm. u. Zeit, **31**, 252–262 (2002)

M. C. Sanguinetti and J. S. Mitcheson, Predicting Drug-hERG Channel Interactions That Cause Acquired Long QT Syndrome, Trends in Pharmacol. Sci., **26**, 119–124 (2005)

W. A. Catterall, M. J. Lenaeus, T. M. G. El-Din, Structure and Pharmacology of Voltage-Gated Sodium and Calcium Channels, Annu. Rev. Pharmacol. Toxicol., **60**, 133–154 (2020)

Y. Zhao, G. Huang, J. Wu, Q. Wu, S. Gao, Z. Yan, J. Lei, N. Yan, Molecular Basis for Ligand Modulation of a Mammalian Voltage-Gated Ca^{2+} Channel, Cell, **177**, 1495–1506 (2019)

G. Wisedchaisri, T. M. G. El-Din, Druggability of Voltage-Gated Sodium Channels—Exploring Old and New Drug Receptor Sites, Front. Pharmacol., **13**, No. 858348 (2022)

N. Unwin, Acetylcholine Receptor Channel Imaged in the Open State, Nature, **373**, 37–43 (1995)

N. Unwin, Nicotinic Acetylcholine Receptor at 9 Å Resolution, J. Mol. Biol., **229**, 1101–1124 (1993)

N. Unwin, Structure and Action of the Nicotinic Acetylcholine Receptor Explored by Electron Microscopy, FEBS Lett., **555**, 91–95 (2003)

M. Cascio, Modulating Inhibitory Ligand-gated Ion Channels, The AAPS Journal, **8**, E353–361 (2006)

W. A. Sather and E. W. McCleskey, Permeation and Selectivity in Calcium Channels, Ann. Rev. Physiol., **65**, 133–159 (2003)

C. Higgins, Multiple Molecular Mechanisms for Multidrug Resistance Transporters, Nature, **446**, 749–757 (2007)

D. J. Triggle, M. Gopalakrishnan, D. Rampe and W. Zheng, Voltage-gated Ion Channels as Drug Targets, (Vol. 29 in Methods and Principles in Medicinal Chemistry, R. Mannhold, H. Kubinyi and G. Folkers, Eds.), Wiley-VCH, Weinheim (2006)

R. J. Vaz and T. Klabunde, Eds., Antitargets. Prediction and Prevention of Drug Side Effects (Vol. 38 in Methods and Principles in Medicinal Chemistry, R. Mannhold, H. Kubinyi and G. Folkers, Eds.), Wiley-VCH, Weinheim (2008)

Special Literature

D. A. Doyle, J. M. Cabral, R. A. Pfuetzner, A. Kuo, J. M. Gulbis, S. L. Cohen, B. T. Chait and R. MacKinnon, The Structure of the Potassium Channel: Molecular Basis of K^+ Conduction and Selectivity, Science, **280**, 69–77 (1998)

N. Shi, S. Ye et al., Atomic Structure of a Na^+- and K^+-conducting Channel, Nature, **440**, 570–574 (2006)

T. Asai et al., Cryo-EM Structure of K^+-Bound hERG Channel Complexed with the Blocker Astemizole, Structure, **29**, 203–212 (2021) and S0969-2126(24)00368-X (2024)

G. Huang et al., High-resolution structures of human Nav1.7 reveal gating modulation through α-π helical transition of S6IV, Cell Reports, **39**, No. 110735 (2022)

L. Tang et al., Structural basis for Ca^{2+} selectivity of a voltage-gated calcium channel, Nature, **505**, 56–61 (2014)

J. Wu et al., Structure of the voltage-gated calcium channel Cav1.1 complex, Science, **350**, 1491–1499 (2015)

M. J. Lenaeus et al., Structures of closed and open states of a voltage-gated sodium channel, Proc. Natl. Acad. Sci. USA, **114**, E3051–E3060 (2017)

V. P. San Martín, A. Sazo, E. Utreras, G. Moraga, G. E. Yévenes, Glycine Receptor Subtypes and Their Roles in Nociception and Chronic Pain, Frontiers Mol. Neurosci., **15**, No. 848642 (2022)

S. G. Aller et al., Structure of P-Glycoprotein Reveals a Molecular Basis for Poly-Specific Drug Binding, Science, **323**, 1718–1722 (2009)

T. Shinoda, H. Ogawa, F. Cornelius, C. Toyoshima, Crystal structure of the sodium–potassium pump at 2.4 Å resolution. Nature, **459**, 446–450 (2009)

S. Tanaka et al., Structural Basis for Binding of Potassium-Competitive Acid Blockers to the Gastric Proton Pump, J. Med. Chem., **65**, 7843–7853 (2022)

S. Masiulis et al., $GABA_A$ receptor signalling mechanisms revealed by structural pharmacology, Nature, **565**, 454–459 (2019)

J. J. Kim et al., Shared structural mechanisms of general anaesthetics and benzodiazepines, Nature, **585**, 303–308 (2020)

K. Wang et al., Structure of the human ClC-1 chloride channel, PLoS Biol., **17**, (4): e3000218 (2019)

C. Altamura et al., Skeletal muscle ClC-1 chloride channels in health and diseases, Pflügers Archiv – Europ. J. Physiology, **472**, 961–975 (2020)

R. J. Dawson, K. P. Locher, Structure of a bacterial multidrug ABC transporter, Nature, **443**, 180–185 (2006)

J. A. Olsen et al., Structure of the human lipid exporter ABCB4 in a lipid environment, Nature Struct. Mol. Biol., **27**, 62–70 (2020)

A. Alam, J. Kowal, E. Broude, I. Robinson, K. P. Locher, Structural insights into substrate and inhibitor discrimination by human P-glycoprotein. Science, **363**, 753–756 (2019)

R. Dutzler, Structural Basis for Ion Conduction and Gating in ClC Chloride Channels, FEBS Lett., **564**, 229–233 (2004)

M. S. Weiss, G. E. Schulz, Structure of porin refined at 1.8 Å resolution, J. Mol. Biol., **227**, 493–509 (1992)

A. L. Lomize, V. I. Orekhov, A. S. Arsenev, Refinement of the spatial structure of the gramicidin A ion channel, Bioorg. Khim., **18**, 182–200 (1992)

M. Dobler, The Crystal Structure of Nonactin, Helv. Chim. Acta, **55**, 1371–1384 (1972)

H. Sui, B. G. Han, J. K. Lee, P. Walian, B. K. Jap, Structural basis of water-specific transport through the AQP1 water channel, Nature, **414**, 872–878 (2001)

U. Kosinska Eriksson, G. Fischer, R. Friemann, G. Enkavi, E. Tajkhorshid, R. Neutze, Subangstrom Resolution X-Ray Structure Details Aquaporin-Water Interactions, Science, **340**, 1346–1349 (2013)

D. F. Savage, R. M. Stroud, Structural Basis of Aquaporin Inhibition by Mercury, J. Mol. Biol., **368**, 607–617 (2007)

Ligands for Surface Receptors

Contents

G. Klebe, *Drug Design*, https://doi.org/10.1007/978-3-662-68998-1_31

In Chap. 29, receptors that allow signal transduction from the outside to the inside of the cell were discussed. These systems initiate numerous processes within the cell that alter its state. In addition to this type of information exchange, a cell must have other ways to stay in constant contact with its environment. To accomplish this task, they have many other **surface receptors**. For example, the cell's **integrin receptors** not only receive signals from outside the cell, they can also send signals into the environment. When a cell moves, for example, in a blood vessel or in tissue, it must remain in constant communication with the environment as it translocates. This is how leukocytes find their way to sites of infection as part of the **immune response to pathogens**. To do this, they receive signals from the environment through their surfaces using special surface receptors. In viral diseases, a virus attempts to attach to a host cell and eventually enter the cell. Recognition of endogenous cell surface receptors or specific adhesion molecules occurs first before the target cell can be reprogrammed for invasion. After viral maturation and replication, the new virus must be budded and released from the infected host cell (exocytosis). Surface-exposed proteins also regulate this process. Drugs can be used to intervene in both processes: the **attack** and the **release of viruses**. Our immune system uses specific surface proteins to distinguish between healthy and diseased cells. Manipulating these processes leads to **immune stimulation**. The structure and function of these surface receptors will be discussed in this chapter. How specific ligands can suppress or reprogram the actual tasks of these surface receptors to lead to a successful therapeutic concept will be explained.

31.1 The Family of Integrin Receptors

Integrin receptors are responsible for bidirectional communication between cells. By regulating cell–cell and cell–matrix contacts, they are involved in a variety of biological processes such as growth, tissue repair, angiogenesis, inflammation, and hemostasis. From a therapeutic point of view, integrins are, therefore, of great interest as cell adhesion receptors. As surface-exposed receptors, they penetrate the membrane and possess an architecture with intra- and extracellular domains. With their extracellular part, which is easily accessible to possible potential drugs, they interact with the **extracellular matrix** and mediate **cell adhesion**. This property could already be used to reconstruct the contact between bones or bone implants and the surrounding tissue. Improved remodeling of the tissue around bones can be achieved by the adhesion of the extracellular domains of integrin receptors or by fixation of ligands that stimulate these receptors.

Integrins are found on almost all cell types in mammals. The **integrin family** is divided into numerous subtypes, and several subtypes can be expressed simultaneously on the same cell. They respond rapidly, in less than a second, to external signals. They have the complex structure of a heterodimeric membrane protein, with α- and β-subunits, each consisting of multiple domains. Some subtypes have an additional insertion domain. Several divalent calcium and magnesium ions, forming the so-called **m**etal-**i**on-**d**ependent **a**dhesion **s**ite (**MIDAS**), are essential for the function of integrin receptors. To date, 18 α- and 8 β-subunits have been characterized in humans. They can be combined to form heterodimers with different compositions. To date, 24 different combinations of these subunits have been identified in integrin receptors. The nomenclature for these receptors follows the general convention: they are referred to as $\alpha_x\beta_y$ receptors, where x is a Roman numeral and y is an Arabic number.

Signal processing occurs through a complex scheme of multiple sequential **conformational transformations**. The completed transformation is reminiscent of the opening of a pocket knife (◻ Fig. 31.1). The folded receptor geometry first changes to a twisted geometry as the knife blade and handle spread apart, and then to an open horseshoe-like shape. This geometry is presented when the receptor is in an active state. The extracellular domain of the activated receptor is available for interactions with other proteins. Binding occurs via a so-called β-propeller-like domain and an insertion domain (I-like domain, ◻ Fig. 31.1), which is brought into the active state by the outlined conformational changes. At

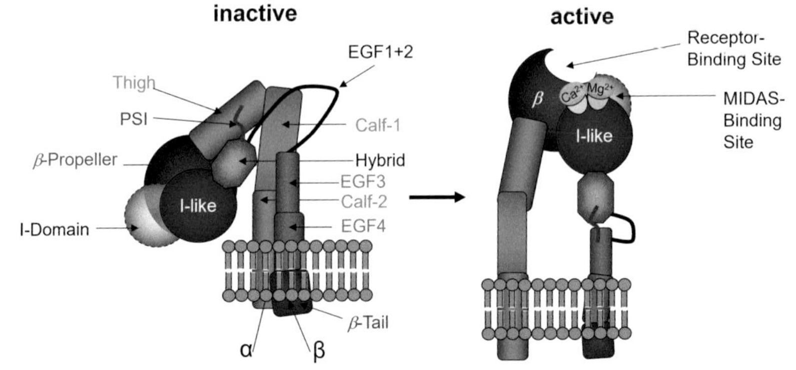

◻ **Fig. 31.1** Integrin receptors have a complex structure consisting of a membrane-bound heterodimer of α- and β-subunits. Each subunit is composed of several domains, and some subtypes have an additional insertion domain (I-domain). Signal processing occurs through a complex scheme of sequential conformational changes that progress from an inactive folded structure to an active horseshoe shape. Several divalent calcium and magnesium ions forming a metal-ion-dependent adhesion site (MIDAS) are essential for function. The receptor-ligand-binding region is located on the β-propeller domain and an I-like domain the same time, the active conformation makes the MIDAS-binding site available. The structural considerations described are based on crystal structure determinations of the individual domains of the receptor. The assembly of these individual building blocks provides an overview of the composition of the overall structure. Nonetheless, more accurate ideas about the individual conformations of intermediates that the receptor goes through during its activation are eluded by this approach.

The structure and function of the $\alpha_{IIb}\beta_3$ **receptor** will be considered in detail. For this receptor, fibrinogen receptor antagonists have been successfully developed and introduced into therapy. The $\alpha_{IIb}\beta_3$ receptor plays an important role in the coagulation cascade. It is found on the surface of blood platelets (thrombocytes). In the resting state, there are approximately 50,000–70,000 inactive copies of this receptor. When an injury occurs that stimulates clotting, an additional 50,000 receptors are transferred from the interior to the surface and conformationally activated. The receptor can now bind ligands that contain a specific motif: an **Arg–Gly–Asp (RGD)** sequence. Fibrinogen, a dimeric soluble plasma protein, contains such a motif and reacts with the $\alpha_{IIb}\beta_3$ integrin receptor on the surface of activated platelets. This cross-linking leads to platelet aggregation and initiates the formation of a **thrombus** for wound closure, known as the primary or cellular hemostasis. Through a second docking site on the platelet, the forming blood clot binds to von Willebrand factor, which is produced by endothelial cells. This contact creates a permanent connection between the aggregating platelet and the vessel wall.

Blocking the surface receptors on the platelet terminates the clotting process. Since this process is necessary throughout the body, internal bleeding could be the con-

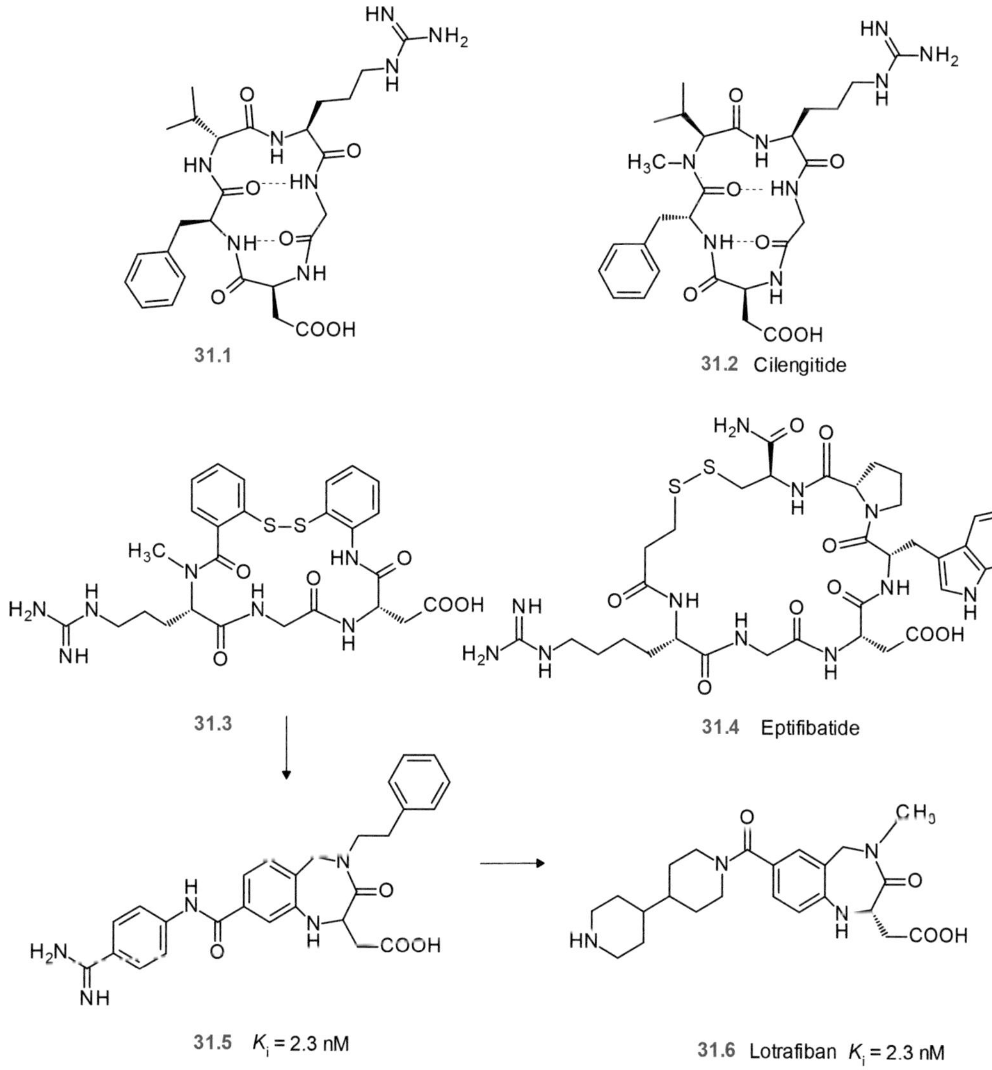

Fig. 31.2 The bound conformation of the RGD motif of the natural ligand fibrinogen to the $\alpha_{IIb}\beta_3$ integrin receptor subunits could be determined by using structurally rigid cyclopeptides. They served as the first lead structures for the development of nonpeptidic receptor antagonists such as the benzodiazepines **31.5** and **31.6**. Cilengitide **31.2** began clinical trial phase III several years ago. The cyclopeptide eptifibatide **31.4** was introduced to therapy as a drug

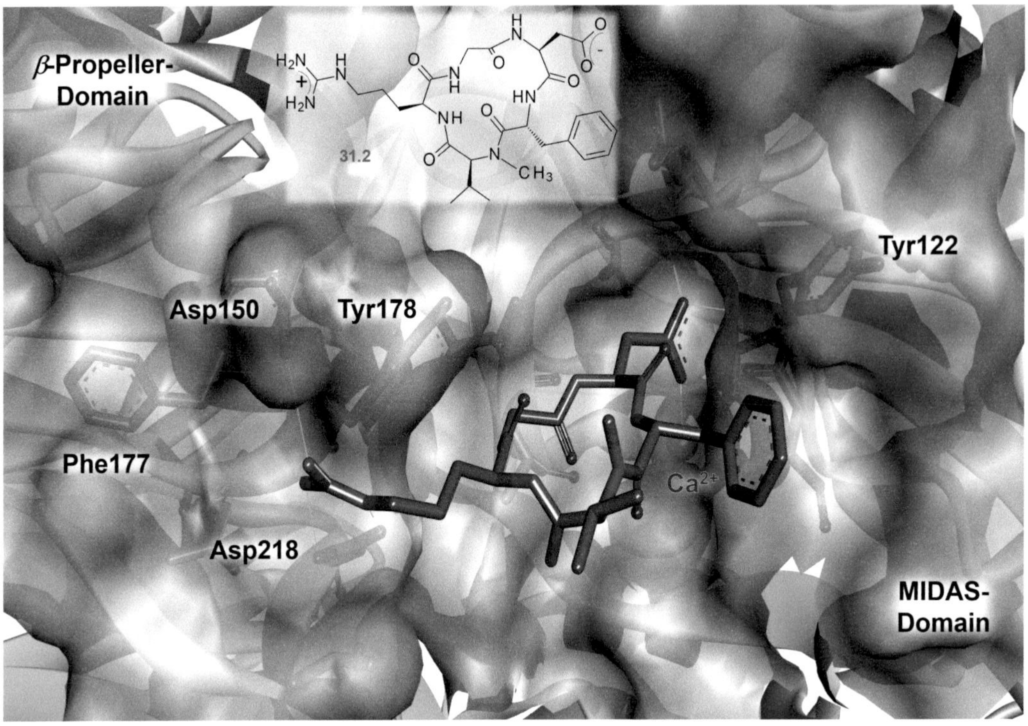

Fig. 31.3 Crystal structure of the $\alpha_{IIb}\beta_3$ integrin receptor with the cyclopeptide **31.2** cilengitide. The structure confirms that the peptide is in a β-turn conformation at the receptor. The RGD motif of the peptide binds in an extended geometry with its arginine residue between two aspartic acids in the propeller domain and with the aspartic acid residue to the metal ions in the MIDAS-binding site. (► https://sn.pub/pHxCti)

sequence. A snake, the common saw-scaled viper or carpet viper (*Echis carinatis*), uses this active principle in its venom to subdue its prey. Because they are often found near human settlements in Africa and Asia, their bite has been fatal to some members of our species. Their venom is a 49-residue peptide with an RGD sequence in the center. A drug based on this inhibitory principle would be desirable to achieve a local anticoagulant effect. This is of interest in the context of angina pectoris, myocardial infarction, stroke, atherosclerosis, or in emergency medicine to prevent ischemic complications.

31.2 Successful Design of Peptidomimetic Fibrinogen Receptor Antagonists

As described in the previous section, antagonists of the $\alpha_{IIb}\beta_3$ integrin receptor, found on the surface of thrombocytes, represent a rewarding point of attack for the development of anticoagulants. Fibrinogen initiates the coagulation cascade by interacting with the $\alpha_{IIb}\beta_3$ receptor with a sequence containing the tripeptide motif **Arg–Gly–Asp (RGD motif)**. The first step was to determine the conformation in which this tripeptide binds to the receptor. For this purpose, **cyclic pentapeptides** containing an Arg–Gly–Asp sequence were synthesized in the group of Horst Kessler at the Technical University of Munich, Germany. The peptide cyclo-(Arg–Gly–Asp–Phe–D-Val) **31.1** (◻ Fig. 31.2) was found to be a high-affinity ligand of the $\alpha_{IIb}\beta_3$ receptor with an inhibition constant of $IC_{50} = 2$ nM. NMR spectroscopic studies suggested that this cyclic pentapeptide adopts a β-loop conformation. If the configuration of the neighboring amino acids Val and Phe is swapped, this preferred conformation is retained. As it turned out, Val and Phe do not interact with the receptor protein (◻ Fig. 31.3). Additional *N*-methylation of valine improved the properties of the cyclopeptide. A few years later, a crystal structure of the αIIbβ3 receptor with the pentapeptide cyclo-(Arg–Gly–Asp–D-Phe–MeVal) **31.2** was determined (◻ Fig. 31.3), confirming the bound geometry. The compound is currently in phase III clinical trials under the name **cilengitide** and has been tested for many years for the treatment of malignant brain tumors (*glioblastoma multiforme*).

Other highly potent peptidic structures were discovered, including the cyclic peptide **31.3** with a disulfide bridge at SmithKline Beecham. Another cyclopeptide,

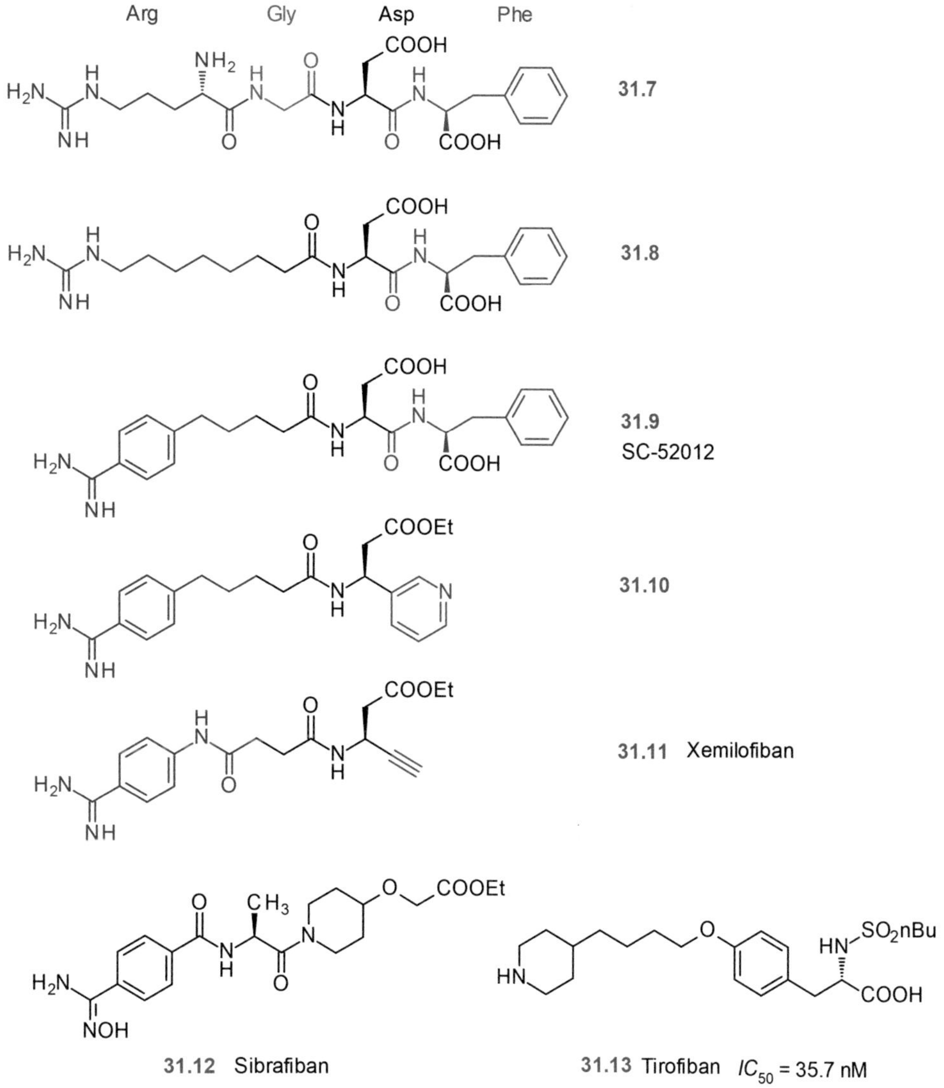

Fig. 31.4 By starting with the linear peptide Arg–Gly–Asp–Phe **31.7**, xemilofiban **31.11** was obtained by stepwise modification. The ethinyl group instead of the pyridine ring does not change the binding affinity but it significantly increases the bioavailability. A similar development candidate, sibrafiban **31.12**, was already tested but was not pursued to a marketed product because of bleeding problems. Tirofiban **31.13** was introduced to the market for the emergency prevention of ischemic complications due to a thrombus in the course of a stroke or heart attack

eptifibatide **31.4**, also stabilized by a disulfide bridge, was introduced into therapy in 1999 by COR Therapeutics under the name Integrilin®. However, the original goal of developing nonpeptidic small-molecule structures had not yet been achieved. Therefore, small organic molecules with functional groups that could mimic the orientation of the side chains of arginine and aspartic acid in **31.1**–**31.4** were sought.

SmithKline Beecham researchers focused on benzodiazepine derivatives. This structural class had two favorable properties. First, benzodiazepines have been extensively studied in synthetic chemistry and many derivatives are readily available. On the other hand, benzodiazepines are rigid and, therefore, well suited for conformational stabilization. In addition, they have been extensively studied as β-turn mimetics (Sect. 10.5). A comparison of several benzodiazepine derivatives with the lead peptidic structure suggested that derivative **31.5** should be able to position the Arg and Asp side chains exactly as in **31.3**. Indeed, **31.5** proved to be a potent fibrinogen receptor antagonist (K_i = 2.3 nM). Further modifications led to lotrafiban **31.6** as a candidate for clinical trials (Fig. 31.2). The compound later failed in the clinical setting due to inadequate efficacy and isolated fatalities.

The Searle group went in a somewhat different direction (Fig. 31.4). The starting point was the peptide Arg–Gly–Asp–Phe (**31.7**, IC_{50} = 29 μM). In a first step, the dipeptide fragment Arg–Gly was replaced by an 8-guanidinooctanoyl group (**31.8**, IC_{50} = 3 μM). Inspired by the result with thrombin inhibitors that an alkylguanidine group can be replaced by a benzamidine, such a moiety was introduced. This resulted in a dramatic increase in binding affinity (**31.9**, IC_{50} = 0.072 μM). Although this compound was not orally available, SC-52012 was Searle's first fibrinogen receptor antagonist to enter clinical trials for intravenous administration. The goal of the work was not to further increase the binding affinity, but rather to im-

31

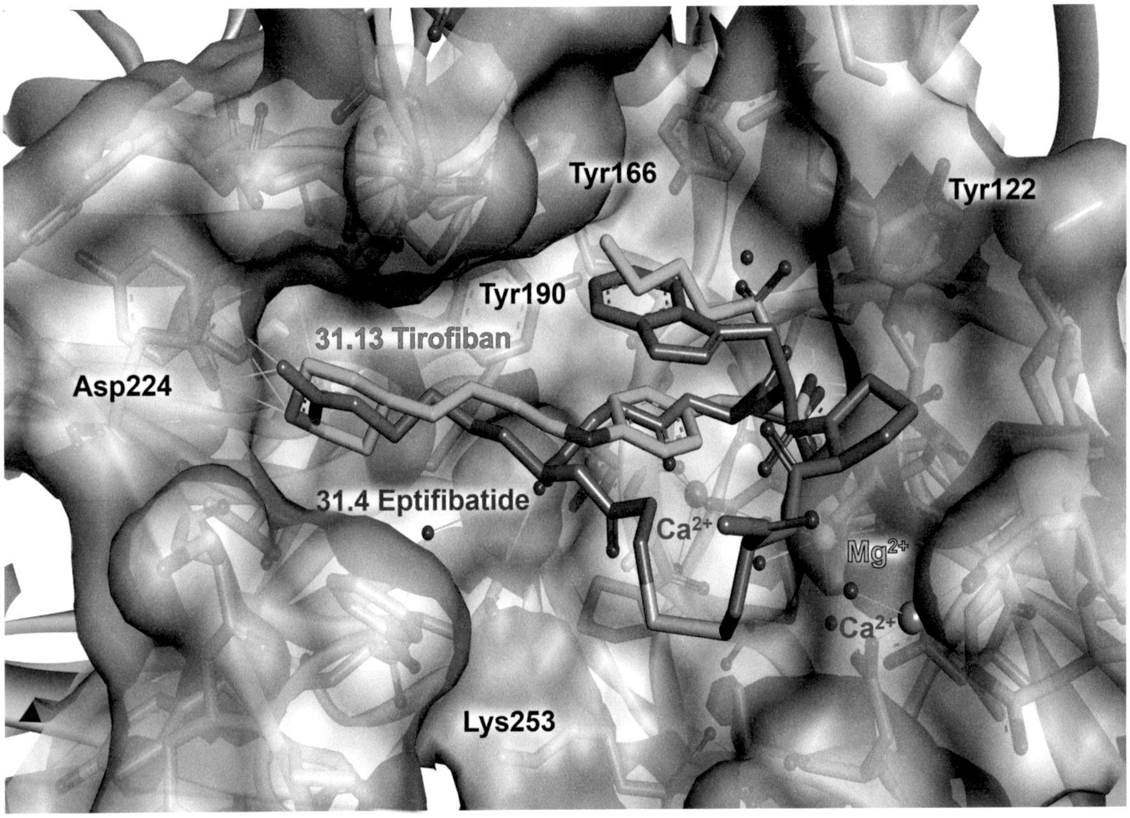

Fig. 31.5 Superposition of the crystal structures of eptifibatide **31.4** with tirofiban **31.13** and the $\alpha_{IIb}\beta_3$ integrin receptor. Peptidic as well as nonpeptidic marketed products bind on one side to the aspartic acid residues of the propeller domain and on the side opposite to the metal ions of the MIDAS-binding site. The example demonstrates how amino acid residues can be replaced by other nonpeptidic groups. (▶ https://sn.pub/6QrCn9)

prove the bioavailability. For this purpose, derivatives with lower molecular weight were investigated. It was shown that the *C*-terminal amino acid phenylalanine could be replaced by a simple pyridine ring without a massive loss of affinity. By additional esterification of the carboxylate group, the Searle group obtained a compound with weak oral activity. Compound **31.10** is a prodrug that is rapidly converted in the body by esterases to the free carboxylate, which is the actual active ingredient (IC_{50} = 0.15 μM for the free carboxylate). Finally, aminobenzamidino succinates were investigated. The idea here was to increase affinity by forming an additional H-bond to the receptor by reintroducing an amide group. Indeed, **31.11** is a highly potent fibrinogen receptor antagonist (IC_{50} = 0.067 μM for the free acid). The compound is well absorbed after oral administration. Searle took xemilofiban **31.11**, as the compound was later named, into clinical trials, which were discontinued in phase III.

The work at Roche had led to the comparable development candidate sibrafiban **31.12**. A double prodrug entered clinical trials. The company conducted a broad study with this compound in 9000 high-risk patients. At low doses, the effect was comparable to that of ASA (Aspirin®). At higher doses, bleeding problems increased significantly. Development of this compound was therefore discontinued. Despite numerous clinical trials with a large number of development candidates, only Merck has introduced a **nonpeptidic receptor antagonist**, tirofiban **31.13** (Aggrastat®), for use in **emergency medicine** to prevent ischemic complications associated with a thrombus formed as a result of a stroke or myocardial infarction. According to the established RGD pharmacophore pattern of a basic group, a bridge, and an acidic group, **31.13** was designed as an inhibitor with an IC_{50} = 375 nM (Fig. 31.5) by replacing the benzamidino group with a piperidine ring and abandoning the amide group in the linker between the basic group and the acidic function. Because of the insufficient oral availability, it is administered intravenously. Time will tell whether fibrinogen receptor antagonists will gain importance in the therapy of thrombotic diseases beyond their use in emergency medicine.

31.3 Selectins: Surface Receptors Recognizing Carbohydrates

Leukocytes, or white blood cells, are transported throughout the body with the bloodstream. Their main function is to defend against pathogens during inflammatory processes. To accomplish this task, they must first be slowed down in the normal blood flow in the vessels adjacent to the site of inflammation (◘ Fig. 31.6). This deceleration is manifested in a slower rolling behavior of the leukocytes. Surface receptors on the rolling leukocytes are involved in this stopping process. On the other hand, in cases of inflammation and in the vicinity of the actual site, **selectins** are increasingly expressed on the cell surface of the endothelium. Temporary contacts, which are weak but highly selective **sugar–protein interactions**, are responsible for the deceleration. Finally, the leukocytes are stopped completely. Integrins on the leukocytes interact with intercellular adhesion molecules (ICAMs) on the endothelium. In the final step, the leukocytes leave the blood vessel (extravasation). After migrating to the site of inflammation, they fight the infection by releasing cytokines and degradation substances. The latter attack the site of inflammation both oxidatively and proteolytically.

Some inflammatory processes lead to vascular damage by excessive leukocyte infiltration, for example, in the setting of myocardial infarction (reperfusion), chronic inflammation such as rheumatoid arthritis, atherosclerosis, diabetic angiopathy, or cancer metastasis. In such situations, a therapeutic concept that intervenes in the **inflammatory cascade** is well suited to reduce excessive leukocyte infiltration. This can be achieved by binding an antagonist to the selectins.

Selectins belong to the large group of lectins, a family of complex glycoproteins. They form interactions with carbohydrate structures and are capable of anchoring between cells and/or cell membranes. Selectins are a subset of these glycoproteins. They are classified as **E-, L-, and P-selectins**. They are structurally related and differ in the number of certain repeat sequences (short consensus repeats). In addition to a *C*-terminal cytoplasmic portion, they have a transmembrane domain. The binding site for carbohydrate molecules is located on a lectin domain at the *N*-terminus. The structure of such selectin domains is shown in ◘ Fig. 31.7.

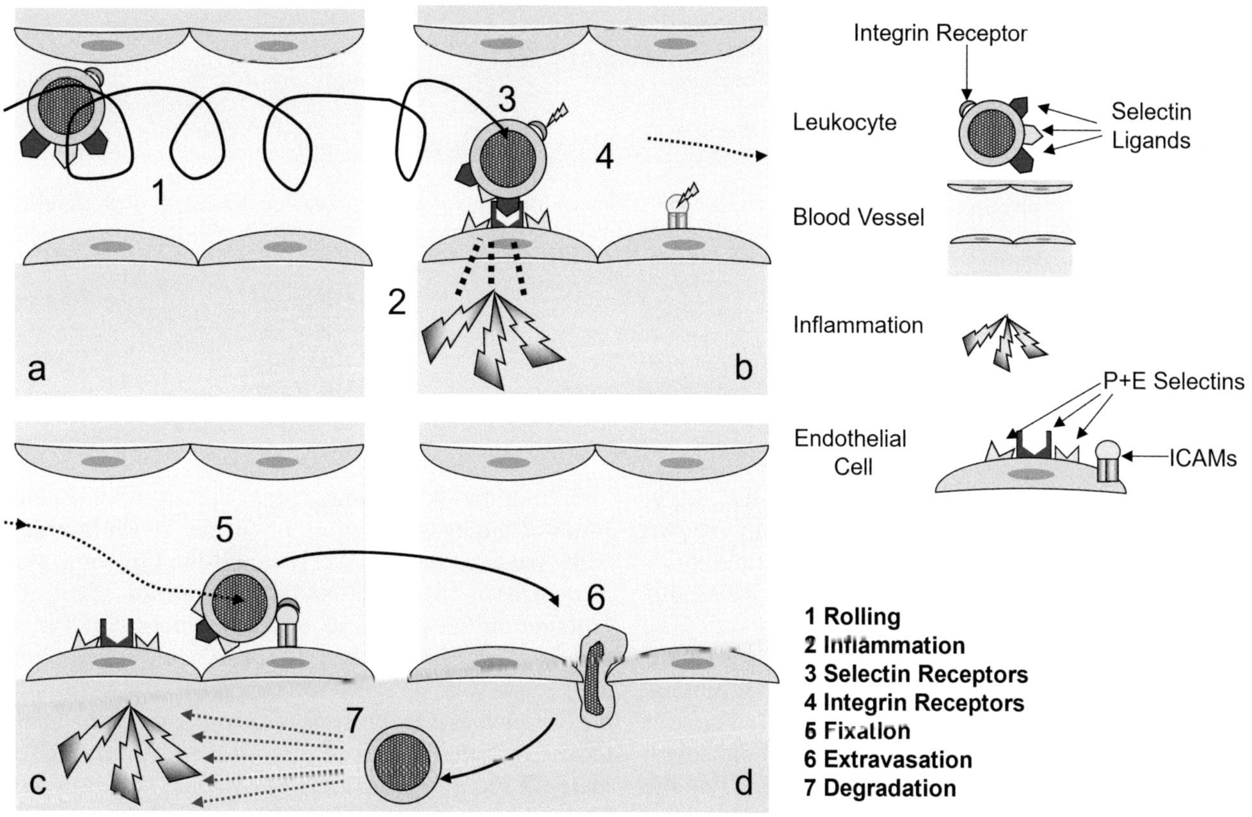

◘ **Fig. 31.6** *Upper left* Leukocytes are transported through the body in blood vessels with the bloodstream (*1*). When the vessel passes a site of inflammation (*2*), the leukocytes are stopped from the normal blood flow. *Upper right* They change their rolling behavior by interacting with selectin receptors (*3*), which are increasingly expressed on the endothelium near the site of inflammation. Integrin receptors on the surface of the leukocytes are activated (*4*). *Lower left* The leukocytes are fully fixed by binding of integrin receptors to intracellular adhesion molecules (ICAM; *5*). *Lower right* Leukocytes leave the vasculature (*6*) and migrate into adjacent tissues to the site of inflammation, which they fight by releasing cytokines and degrading substances such as oxidants and proteases (*7*)

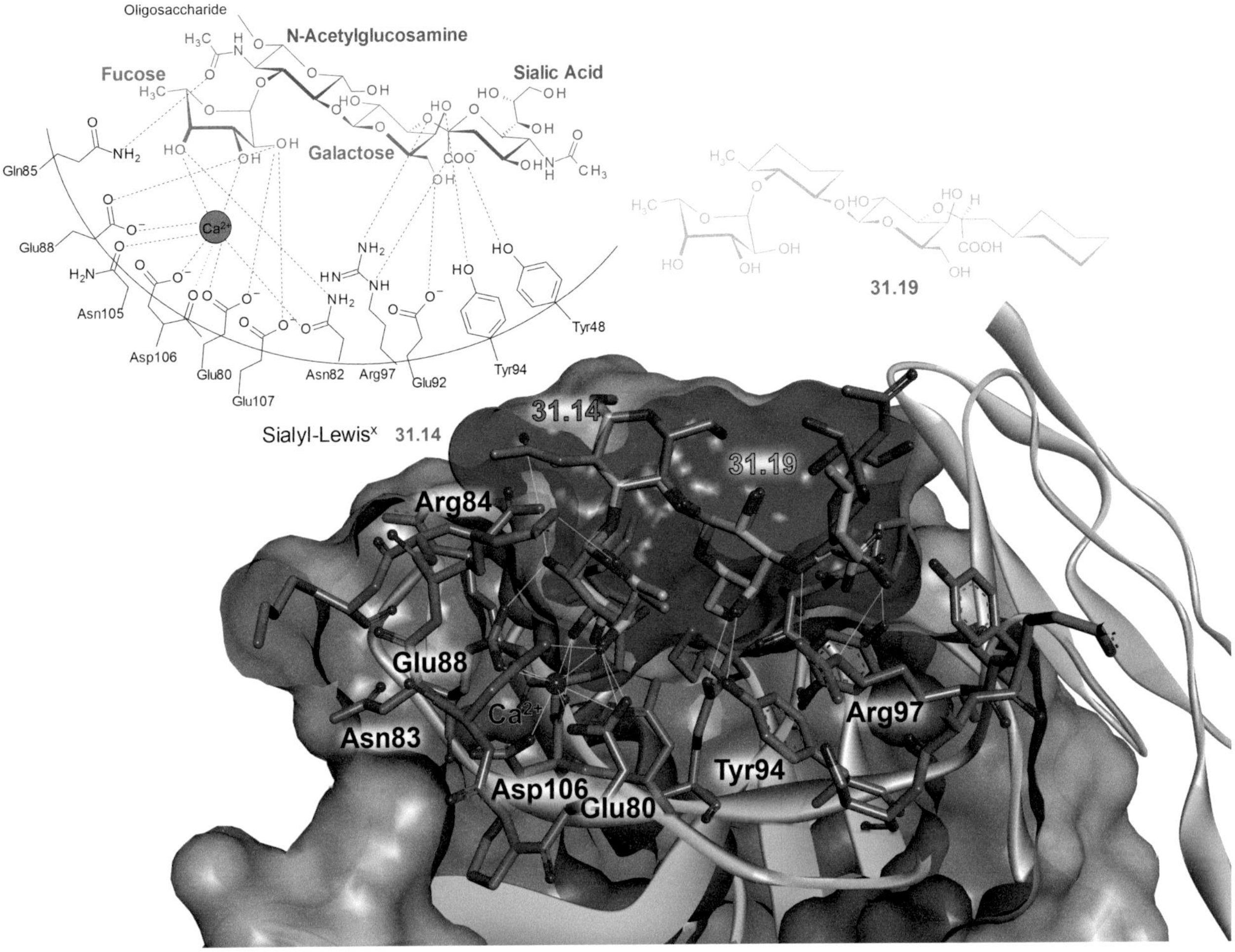

Fig. 31.7 The crystallographically determined binding mode of sialyl-Lewis^X **31.14** exposed binding epitope of the PSGL-1 protein to the selectin surface domain (*bottom*). The four carbohydrate moieties: *N*-acetylglucosamine (*violet*), fucose (*green*), galactose (*blue*), and sialic acid (*red*) form numerous hydrogen bonds with their oxygen atoms to the protein in a shallow, bowl-shaped binding pocket (*upper left*). A calcium ion (*purple sphere*) is involved in the binding and interacts with multiple protein residues as well as with the ligand's fucose moiety. Rational design allowed the development of the glycomimetic **31.19** (*yellow*), which binds to the protein with 19 µM (*upper right*). (► https://sn.pub/99w1ZI)

The endogenous ligand of selectins is PSGL-1, a glycoprotein on the surface of leukocytes. As an exposed binding epitope, the PSGL-1 protein has multiple copies of a motif consisting of four sugar molecules. This motif is termed **sialyl-Lewis^X 31.14** and abbreviated sLe^X. The four sugar moieties consist of an *N*-acetylglucosamine, a fucose, a galactose, and a sialic acid. Their binding mode is shown schematically in Fig. 31.7. The four sugar molecules form numerous hydrogen bonds with their hydroxyl groups to a flat, bowl-shaped binding site on the protein. A calcium ion, interacting with several exposed residues in the binding pocket, binds directly next to the sLe^X-binding epitope. It also contacts the fucose. The sLe^X-binding epitope is unsuitable as a drug because it is easily degraded by glycosidases and has a relatively low binding affinity (IC_{50} = 4 mM). Therefore, compounds that mimic sugar binding were sought. Initially, the hydroxyl group of fucose, which interacts with Asn 82, Glu 80, Asp 106, and the Ca^{2+} ion, was recognized as critical for binding. In addition, the acidic function on the sialic acid, which interacts with Tyr 48 and Ser 9, was focused on. These two polar ligand-binding regions were designed to be coupled to a hydrophobic biphenyl moiety. Instead of fucose, the synthetically more accessible mannose was used and the inhibitor **31.15** (Fig. 31.8) with an IC_{50} = 500 µM was obtained. The affinity was further improved by a factor of 5 by adding a second structurally similar group to give **31.16**.

A different approach was taken at Revotar Biopharmaceuticals. Compound **31.15** was used as a reference. Smaller, multiply hydroxylated aromatic rings were sought to replace the mannose moiety. A pyrogallol sub-

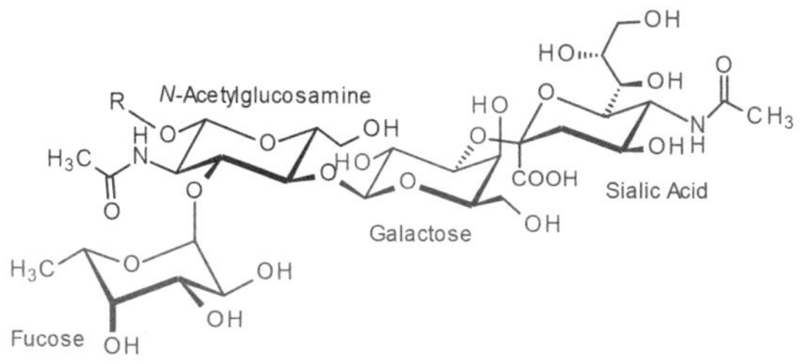

31.14 Sialyl-Lewis^x IC_{50} = 4 mM

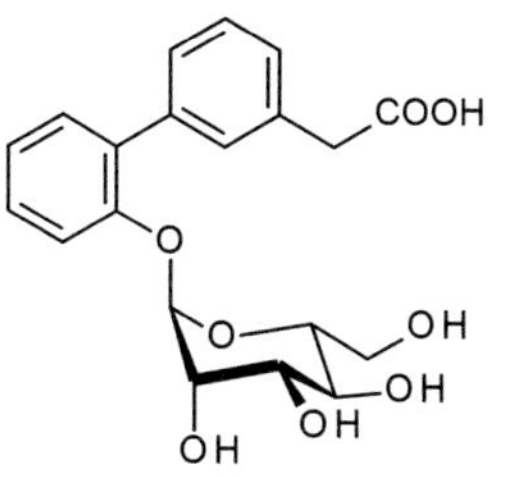

31.15 TBC265 IC_{50} = 500μM

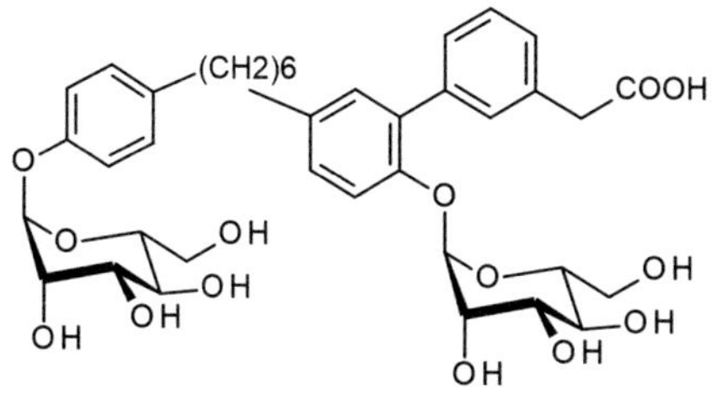

31.16 Bimosiamose IC_{50} = 95μM

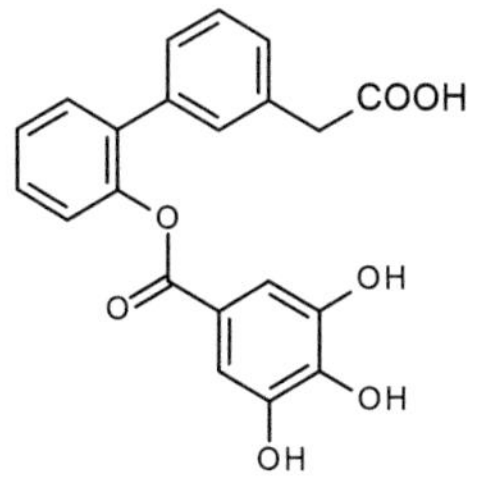

31.17 IC_{50} = 1μM

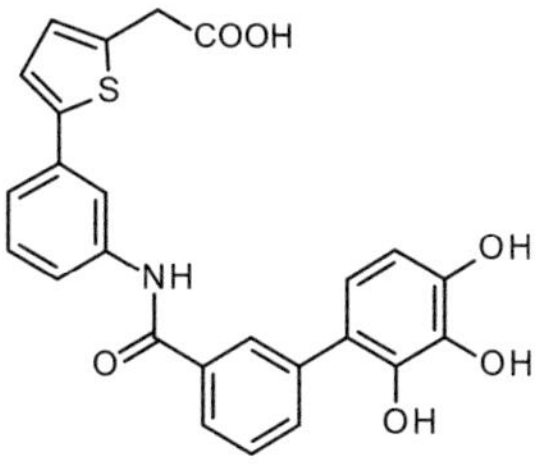

31.18 IC_{50} = 0.75μM

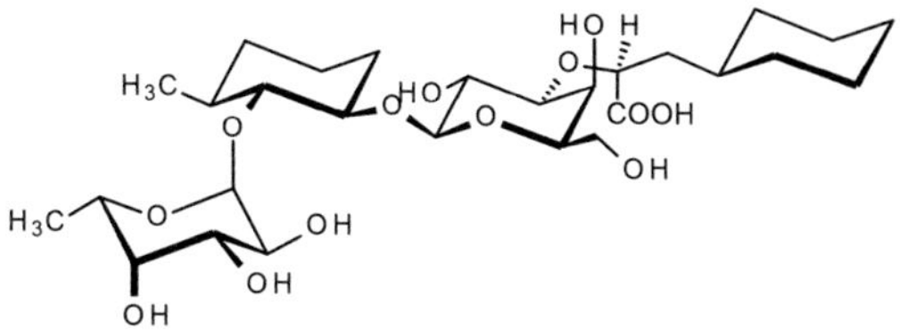

31.19 K_d = 19μM

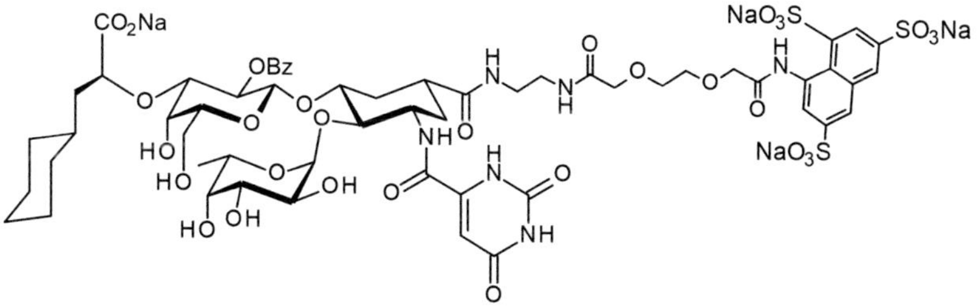

31.20 GMI-1070 Rivipansel IC_{50} (E-Selectine) = 4.3 μM

Fig. 31.8 A micromolar lead structure **31.15** was developed from sialyl-Lewis^X **31.14** by replacing fucose with mannose and adding a hydrophobic bridge with a terminal acid group. Its affinity can be improved by adding a second structurally analogous building block to bimosiamose **31.16**. Starting from a pyrogallol scaffold, the completely nonsugar-like structures **31.17** and **31.18** with submicromolar affinity were developed. The micromolar glycomimetic **31.19** attempts to mimic the sialyl-Lewis^X scaffold. Rivipansel **31.20** was derived from this lead structure. The compound serves as a therapeutic agent for the treatment of acute sickle cell crisis

stituent was found to be the best mimetic. Linked to the biphenyl residue, **31.17** shows IC_{50} values for L-selectin in the low micromolar range and for P-selectin in the nanomolar range. The scaffold was further optimized. The introduction of an enlarged linker between the two terminal anchor groups and replacement of a phenyl ring with thiophene led to **31.18**. This compound shows an *in vitro* affinity in the upper nanomolar range with a molar mass below 500 Da. Considering the very shallow and wide open binding pocket, this is a remarkable binding affinity towards the target protein for such a small antagonist. Beat Ernst's group at the University of Basel, Switzerland, designed the micromolar mimetic **31.19** from the sialyl-Lewis^X structure. It inspired the development of the *pan*-selectin inhibitor GMI-1070 **31.20**. The compound was commercialized by GlycoMimetics with Pfizer as rivipansel. It inhibits E-selectin-mediated adhesion to leukocytes, thereby, improving blood flow. Rivipansel represents a new therapeutic intervention for acute painful sickle cell crises (Sect. 12.13). Patients with sickle cell anemia have a mutated form of hemoglobin that tends to aggregate. As a result, their leukocytes take on a characteristic sickle shape, which significantly impairs their ability to roll.

31.4 Fusion Inhibitors Impede Viral Invasion

Because viruses lack their own metabolic and reproductive machinery, they are forced to hijack a host cell for these tasks. They do, however, contain the program and information for their reproduction in the form of their own DNA or RNA. To gain access to a host cell, they must dock onto the cell and their envelope must **fuse with the host cell membrane**. We will discuss an example. The human immunodeficiency virus fuses with T-lymphocytes, initiating an AIDS infection (Fig. 31.9).

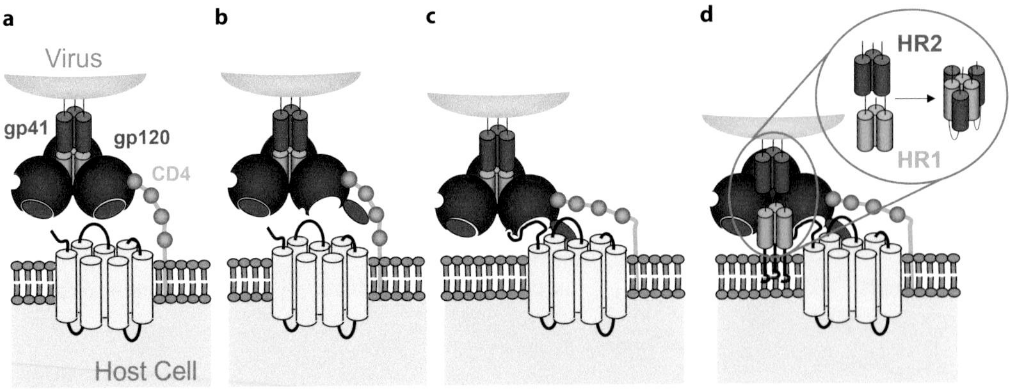

Fig. 31.9 An AIDS infection is initiated by an attack of the human immunodeficiency virus (*orange*) on T-lymphocytes (*gray*; **a**). It uses a trimer of its surface proteins containing gp120 (*violet*) and gp41 subunits (*red/green*) for this purpose. The gp120 protein binds to the endogenous CD4 receptor (*blue*). A conformational change in the gp120 protein takes place (**b**). For this, an interaction with the CCR5 or CXCR4 coreceptors (*yellow*), which are in the vicinity of the CD4 receptor, is formed (**c**). Both receptors belong to the GPCR class. By binding to these chemokine receptors, the sewing-pin-like "warhead" gp41, which consists of three segments in a helix bundle (*red/green*), undergoes a conformational change. The virus penetrates the membrane of the host cell with this helix bundle, and the fusion process is initiated (**d**). Finally, the initially extended peptide chains assemble and compress themselves into a tight bundle of six helices. This brings the virus and the host cell even closer together (**d**, inset)

31

The virus has a diameter of approximately 120 nm (1200 Å). More than 70 glycoproteins are embedded in its membrane envelope. Each of these surface proteins consists of gp120 and gp41 subunits, which are arranged as trimers. The gp41 subunit protrudes from the membrane envelope like a pin used in sewing, whereas the gp120 subunit is a nearly spherical outer head of the pin. Both subunits have been characterized structurally and biologically. The gp120 protein, which is composed of folded sheets and helices, acts as a mooring anchor for the virus. It binds to the CD4 receptor on the surface of T-lymphocytes. A conformational change then occurs in the gp120 protein. This initiates the subsequent interaction with the nearby CCR5 or CXCR4 coreceptor. Binding to these chemokine receptors triggers another conformational change in the sewing-pin-like "warhead" on the viral envelope. The monomers that make up the trimeric helix bundle and form the gp41 subunit are each composed of three segments, the HR1, HR2, and FP domains. The virus penetrates the host cell membrane with its fusion domains. The bundle of three HR1 domains provides three grooves on its surface that are optimally suited to accommodate the HR2 domains (Figs. 31.9 and 31.10). For this purpose, they must adopt a helical geometry. The three initially extended and parallel oriented HR1 and HR2 peptide chains "zip" together to form a **compact bundle of six helices**. This zipping together causes the membranes of the virus and host cell to be pulled together. This initiates the **fusion process** of the envelopes.

Can the fusion process be blocked to stop the onset of infection? The tightly packed bundle of HR1 helices provides a groove on the surface to accommodate the helical HR2 peptide. At Duke University in Durham, North Carolina, USA, peptides were synthesized that mimicked the sequence of the HR2 domain. In 1996, one of these peptides was discovered by the company Trimeris. DP178, a 36 residue peptide, like the HR2 peptide, is able to dock into the available groove on the HR1 peptide and block gp41 from zipping. The lead structure was further developed in collaboration with Roche into the drug **enfuvirtide**, a 36 amino acid peptide (Ac–Tyr–Thr–Ser–Leu–Ile–His–Ser–Leu–Ile–Glu–Glu–Ser–Gln–Asn–Gln–Gln–Glu–Lys–Asn–Glu–Gln–Glu–Leu–Leu–Glu–Leu–Asp–Lys–Trp–Ala–Ser–Leu–Trp–Asn–Trp–Phe-NH_2) with a molecular weight of 4492 Da. It was launched under the name Fuzeon® as the first **fusion inhibitor** for viral diseases. It must be injected subcutaneously and is currently used as a substitute therapy when resistance to HAART therapy (Sect. 24.5) has developed. The interactions of the helical structure for the HR2 peptide strand with the bundle of HR1 domains have stimulated the search for **small-molecule fusion inhibitors**. In particular, the three hydrophobic amino acids Trp 628, Trp 631, and Ile 635 are responsible for the contact between the helical strands. So far, only a few relatively highly charged structures, such as **31.21** and **31.22** that can form contacts and block zipping (Fig. 31.11) have been found by screening. Since other projects have successfully found small-molecule inhibitors, such as the helix mimetic for the BCL-XL protein described in Sect. 10.6, that compete with the contact between a helix and an extended groove, there is hope that small-molecule lead structures will also be found here. Time will tell if resistance develops to these compounds as well.

At this point, it is worth mentioning that there is another important coreceptor for cell entry processes that

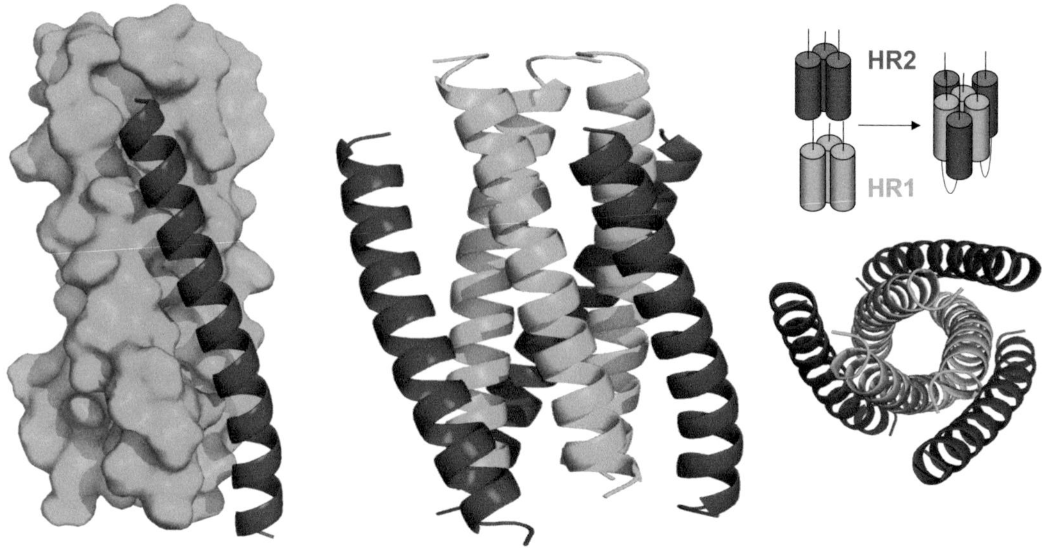

Fig. 31.10 The bundle of three HR1 domains (*green*) makes three grooves on its surface available that are optimally suited to accept the HR2 domains (*red*) once these have transformed to a helical geometry. Three initially extended and parallel-oriented peptide chains fold together and form a tight bundle of six helices. This tying together pulls the membranes of the virus and the host cell together

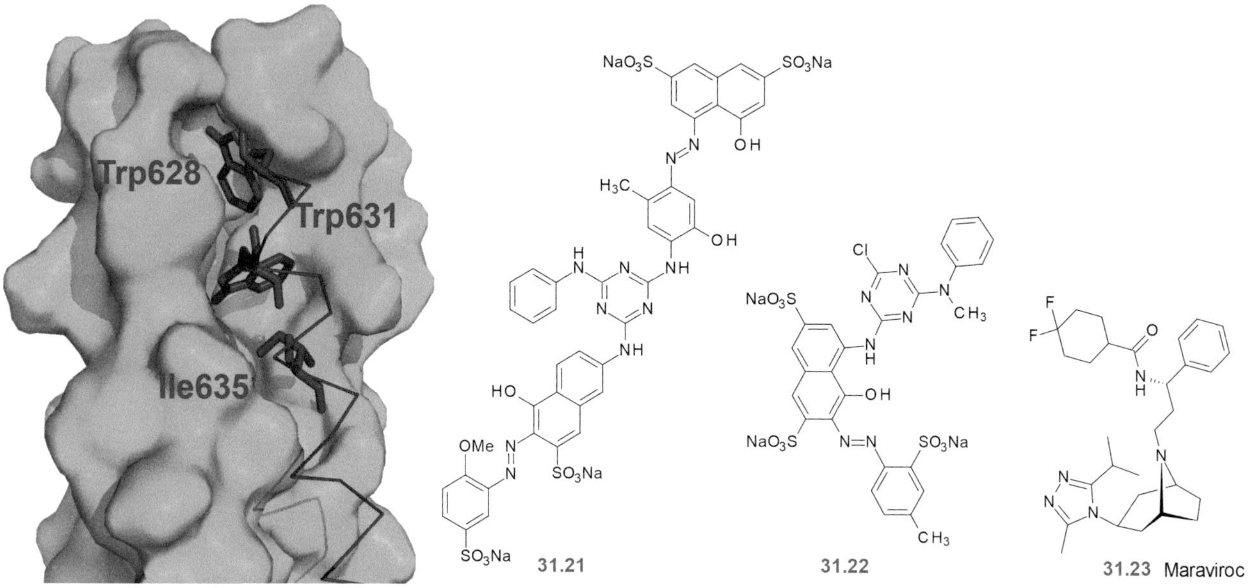

Fig. 31.11 The HR2 peptide strand (*red*) interacts with the three hydrophobic amino acids Trp 628, Trp 631, and Ile 625 with the bundle structure of the HR1 domain (*green*). In screening, the multiply charged structures **31.21** and **32.22** were discovered as mimics that can block the bundle-type packing of the helices. Maraviroc **31.23** antagonizes a cytokine receptor that is involved in initiating the fusion process between the human immunodeficiency virus and T-lymphocytes (Fig. 31.9)

can also be antagonized with small-molecule ligands: the **chemokine receptor CCR5** (Fig. 31.9). Chemokine receptors belong to the class of GPCRs (Sect. 29.1). The function of CCR5 can be suppressed with ligands such as maraviroc **31.23**, which was introduced for therapy by Pfizer in 2007. Viral fusion processes can also be suppressed with this approach.

31.5 Neuraminidase Inhibitors Prevent Budding of Mature Viruses

As described in the previous section, viruses are incapable of living an autonomous life. Therefore, they are forced to find a host cell that they can reprogram for their own reproduction and exploit for their own metabolism. Viruses store their genome and, thus, their blueprints in single-stranded or double-stranded DNA or RNA, de-

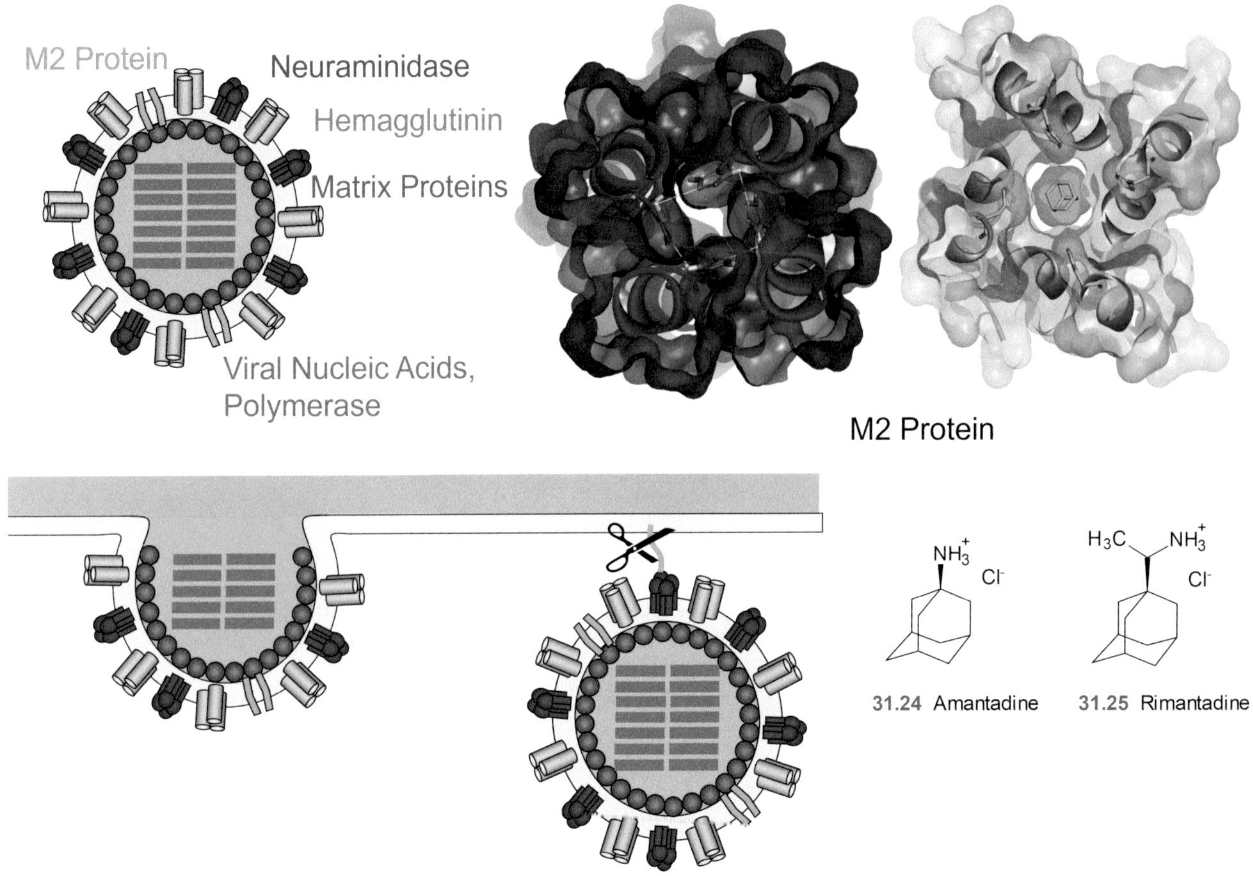

■ **Fig. 31.12** *Upper left* In addition to the docking protein hemagglutinin (*blue*), of which 18 subtypes are known, the envelope of the influenza virus contains a neuraminidase (*red*), which has eleven variations (N1–N11), and the M2 proton channel protein (*green*). *Upper right* This latter pore is formed by a bundle of four helices. It is closed by four adjacent histidine residues (*red*), which form a ring of hydrogen bonds between them (*yellow lines*). When the His residues are protonated, the H-bonding network ruptures and the channel opens (*green*). However, the channel can be blocked by the drug molecules amantadine **31.24** and rimantadine **31.25**. *Bottom left* Upon maturation and budding of a newly formed virus, the glycolytic activity of neuraminidase is needed to detach from the host cell (*gray*) in the last step by cleaving a sugar chain (*green*). (▶ https://sn.pub/K1Zr14)

pending on the virus type. These nucleic acids are located inside the virus and are surrounded by a protein coat, the so-called capsid, which, depending on the virus type, may also consist of lipid building blocks. Glycoproteins are embedded in the capsid. Antibody-mediated defense mechanisms are directed specifically against the proteins in the capsid. Viruses encode the information for numerous enzymes in their genome that are specifically required for their replication. They can also have channel proteins that allow the transfer of substances between the inside of the virus and its environment.

One of the most common viral diseases, influenza, is caused by the **influenza virus**. This virus belongs to the family of enveloped viruses, and three subtypes, A, B, and C, are known. Influenza viruses are airborne and are usually spread by sneezing. In addition to infecting humans, influenza viruses can be ingested and transmitted by animals. Initially, there is no transmission from one species to another. However, such transmission routes from animals to humans and vice versa are observed in regions where these species live in close contact with one another. Once in the respiratory tract of a new victim, influenza viruses adhere to mucous membranes using **hemagglutinin** proteins found on their surfaces. In addition to the hemagglutinin docking protein, the viral envelope also contains **neuraminidase** and the **M2 proton channel protein** (■ Fig. 31.12). Proteins can sometimes vary significantly in their sequential amino acid constitution and still be able to perform the same functions. These variations are called subtypes. Eighteen subtypes are known for hemagglutinin (**H1–H18**), and eleven variants of the surface enzyme neuraminidinase (**N1–N11**) have been characterized. New variants are constantly being created from new combinations, which then work their way through the population. In recent years, the H5N1 (avian flu) and H1N1 (swine flu) variants have kept us on

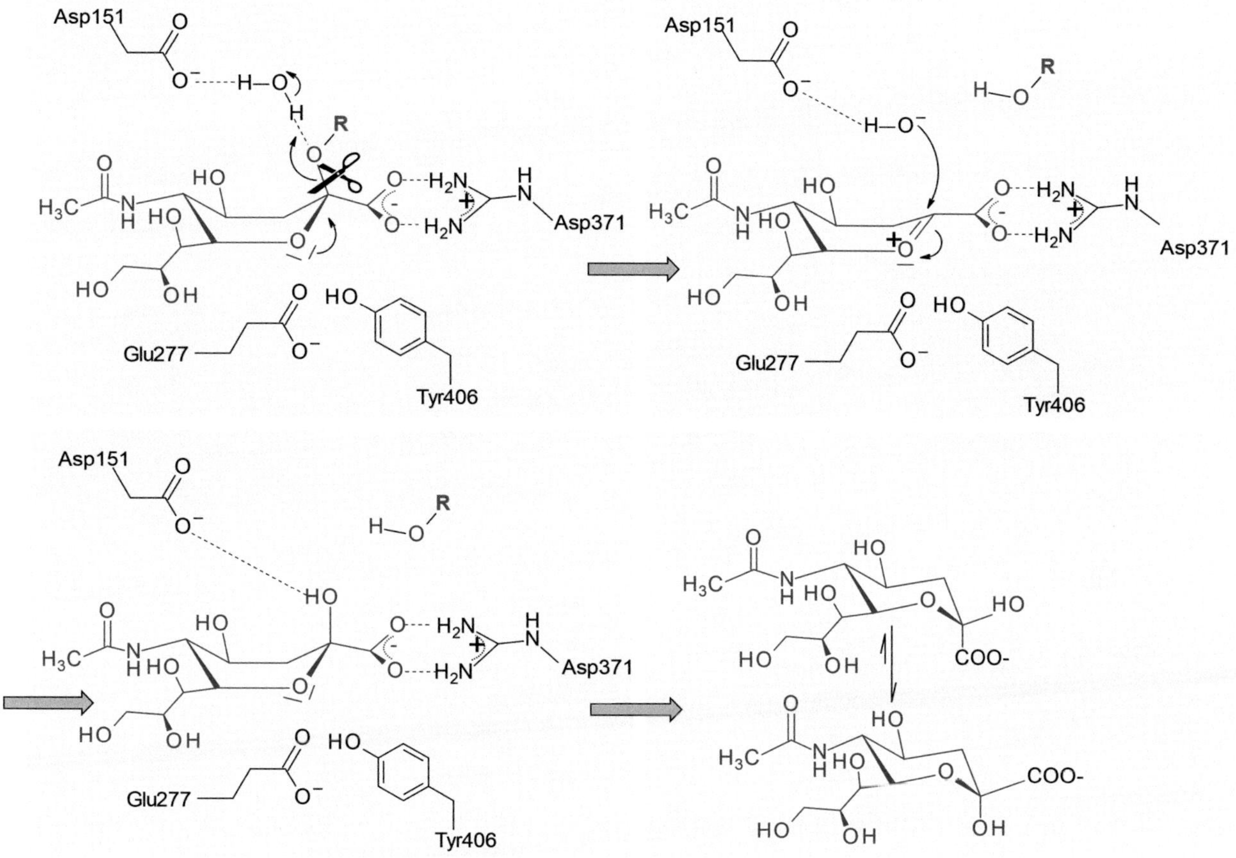

Fig. 31.13 Reaction mechanism of the glycolytic cleavage of a sialic acid residue. The residue is at the end of a carbohydrate chain that couples the virus to the host cell. The sialic acids binds to viral neuraminidase. The glycosidic bond is cleaved from the remaining sugar chain with assistance from the two neighboring acidic amino acids, Glu 277 and Asp 151. A sialosyl cation is formed that is temporarily stabilized by Tyr 406. The sugar is released after transfer of an OH group to the trigonal center. The stable stereoisomer is formed by ring opening and reformation of the cyclic sugar

our toes. Their development is also attributed to a jump from the corresponding animal species to humans. The observed H5N1 variant proved to be particularly pathogenic, but the mechanism of infection in humans and **interspecies crossing** was less efficient. The H1N1 swine flu of autumn 2009 was particularly infectious, but the clinical course was less severe. Unfortunately, this picture can be quickly altered by small changes in the viral proteins or when new interspecies crossings may occur.

A distinction is made between **antigen drift** and **antigen shift**. In drift, genetic changes occur, usually as a result of copying errors due to an error-prone transcription process of the viral genome. The virus slowly and randomly alters its surface proteins. Because the infected host organism develops its own antibodies, or the immune system is stimulated to produce such antibodies by a **flu vaccine**, even small changes in the capsid proteins can be adequately recognized and rendered harmless. Antigen shift is much more dangerous; it is caused by the exchange of genetic information between virus species or subtypes. It can occur especially between species, that is during transmission from animals to humans. Because this route results in new combinations of surface proteins, it is difficult for the immune system to generate sufficiently high antibody titers quickly enough to render such a modified virus harmless. Such antigenic shifts can lead to **pandemics**. As mentioned above, they usually originate in regions where many species such as ducks, chickens, pigs, cats, dogs, and humans live together in close proximity. Because of the lifestyle, high population density, and traditional animal husbandry practices where animals and humans live under the same roof, the East Asian regions of southern China or on the American continent in Mexico have repeatedly proven to be incubators for such genetically diverse virus forms. There have been many pandemics in the past. The most severe worldwide was certainly the so-called Spanish flu of 1918, which claimed at least 25 million lives. The influenza viruses of this pandemic had a particularly virulent subtype, H1N1. In 1957, there was a pandemic with the H2N2 subtype, and in 1968, it was the H3N2 combination that was particularly dangerous. The last influenza pandemic warning from the WHO was issued in the autumn of 2009, after a new H1N1 variant (the

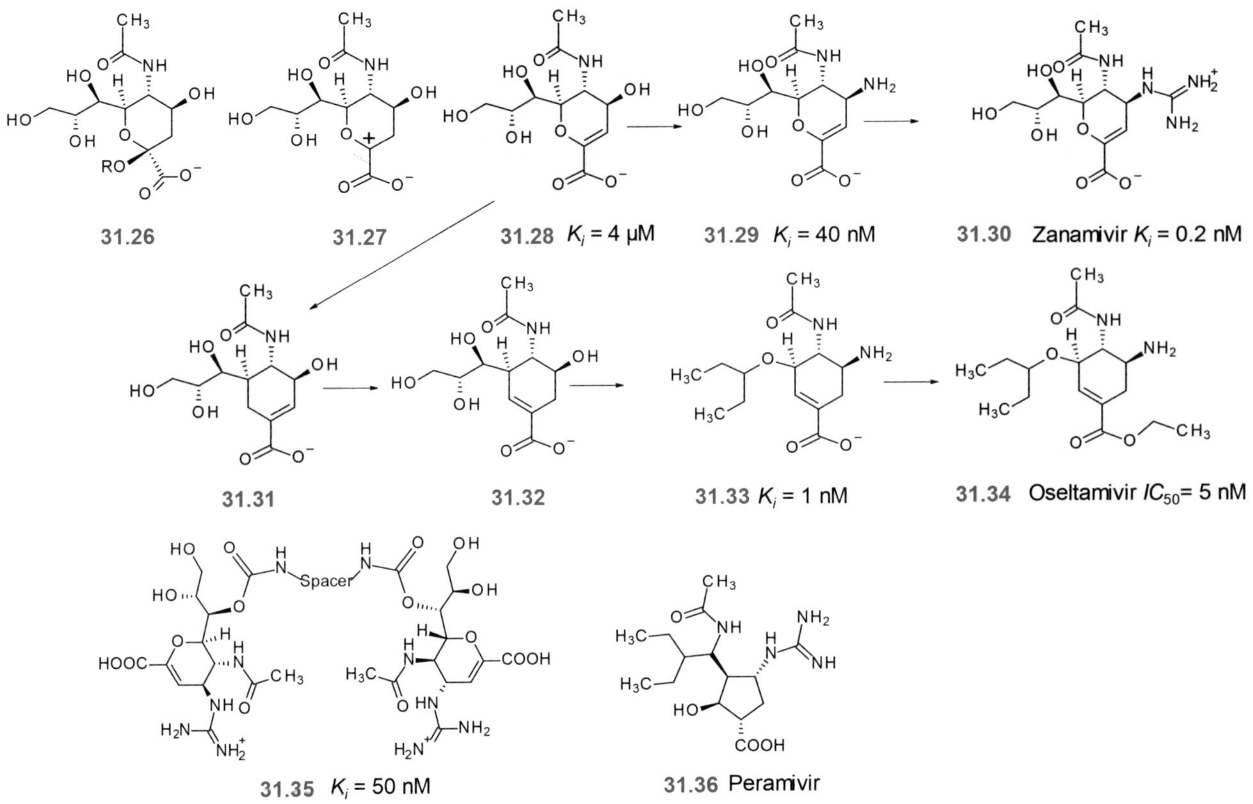

Fig. 31.14 The development of the neuraminidase inhibitors zanamivir **31.30** and oseltamivir **31.34**. Compound **31.28** was developed as a stable structural analogue to the sialosyl cation **31.27**. By exchanging the OH group for an NH_2 group, **31.29** is formed with K_i = 40 nM. The introduction of a guanidinium group to form **31.30** brought a further improvement in the activity. Carbocyclic analogues **31.31** and **31.32** were synthesized at Gilead Sciences and further optimized to **31.33** by exchanging an OH for an NH_2 group. To improve the bioavailability, an ester prodrug, **31.34** was introduced as oseltamivir into therapy. A depot form was developed with **31.35**. Peramivir **31.36** is an intravenously administered neuraminidase inhibitor

same subtype as the 1918 Spanish flu) emerged in Mexico in the form of the so-called swine flu. A year later, we knew better. This variant, in its observed form, turned out to be not nearly as dangerous as initially expected. The H5N1 avian influenza virus is currently spreading among cattle, particularly in the United States. Experts are concerned that the virus could be transmitted to humans and, in a worst-case scenario, reach high levels of infection and pathogenicity.

The current preventive therapy for influenza is **vaccination**. The vaccine contains parts of the surface proteins hemagglutinin and neuraminidase or matrix proteins as antigens and stimulates the immune system to produce antibodies. The production of a new vaccine takes some time and is very expensive. Therefore, an attempt is made to estimate which viral subtypes might be involved before a seasonal flu wave strikes. Viral envelope proteins are isolated from these subtypes and used to develop a vaccine for the next vaccination campaign. This is exactly what was done in the summer of 2009 to prepare a vaccine against swine flu type H1N1 for the more densely populated northern hemisphere in time for the winter. At the same time, the population was asked not to neglect the vaccines against virus strains from previous years and to obtain adequate protection through such vaccination. Inspired by the success of RNA vaccines against coronavirus, scientists are now also working on RNA vaccines against influenza viruses that encode parts of the envelope proteins of these viruses (Sect. 32.4).

Three surface proteins can be targeted with small molecules. The drugs amantadine **31.24** and rimantadine **31.25** (Fig. 31.12), which block the **proton channel protein M2**, are already quite old. The target protein is a pore that is permeable to protons. It is opened and regulated in a pH-dependent manner by a ring of four histidine residues. When the four histidine residues are deprotonated, the pore is closed by a contiguous network of H-bonds formed between the histidines. When the histidines are protonated and exist in a charged state, their spatial orientation changes and the H-bond network is disrupted. As a result, the channel of the M2 protein is opened. The two ligands **31.24** and **31.25** are not very specific and do not allow an efficient therapy. In addition, several resistance mutations have been observed that abolish the action of these drugs.

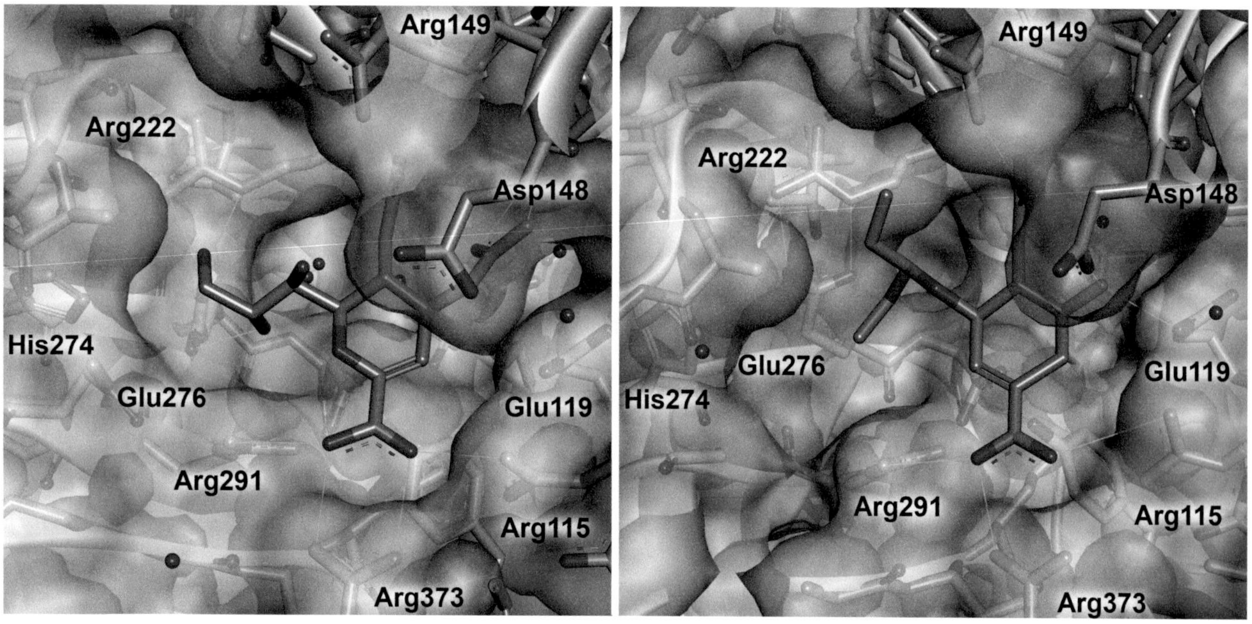

Fig. 31.15 The crystallographically determined binding geometries of zanamivir **31.30** (*left*) and oseltamivir **31.34** (*right*) in neuraminidase. The acidic function of the inhibitors is anchored by Arg 115, Arg 291, and Arg 373. The opposite *N*-acetyl group interacts with Arg 149. Asp 148 forms a layered geometry with the guanidinium group from zanamivir, whereas the amino group found in the same position in oseltamivir forms a hydrogen bond to Asp 148. Compound **31.30** forms a hydrogen bond to Glu 276 by using its glycerol groups, whereas the more hydrophobic *iso*-pentylether group in oseltamivir induces a rearrangement of Glu 276 to form a salt bridge to Arg 222. (► https://sn.pub/SbhgCP)

The docking protein **hemagglutinin** initially appeared to be an ideal target structure for binding a small-molecule ligand. If this protein could be rendered nonfunctional, viral infection would be stopped at the stage of host cell penetration. Unfortunately, this protein changes so much through constant mutations that it is difficult to develop ligands for long-term use.

Therefore, **neuraminidase** remains as another target structure. It does not play a role in viral entry into the host cell. On the other hand, it regulates the budding of a newly formed virus and, in particular, its detachment from the host cell. In the **final step of detachment**, the newly formed virus is still attached to the host cell by a sugar chain (Fig. 31.12). The last two sugar moieties of this anchor are a galactose and a sialic acid **31.26** (or *N*-acetylneuraminic acid) coupled by a glycosidic oxygen bridge (Figs. 31.13, 31.14). The virus uses its neuraminidase to permanently release itself from the host cell. Since this protein must specifically recognize sialic acid, the virus cannot change its structure too much without losing the efficient recognition of sialic acid and the catalytic glycosidic cleavage process of its neuraminidase. It would risk sacrificing its own ability to survive.

Neuraminidase has an **enzymatic glycosidase function**. Such enzymes have a dyad of two aspartate or glutamic acid residues. We have already seen the mechanism of such an enzyme in Sect. 21.3. Neuraminidase attacks the glycosidic bond between the terminal sialic acid and the galactose with its Glu 277 residue. A cation **31.27** (Fig. 31.13) is formed, which is temporarily stabilized by the adjacent residue Tyr 406. The crystal structure of influenza neuraminidase with a sialic acid analog **31.28** was determined in 1983 (Fig. 31.15). It mimics the transition state of the enzymatic reaction. Compound **31.28** blocks the protein with $K_i = 4\ \mu M$. In order to find other key positions for additional functional groups in the binding pocket, the GRID program of Peter Goodford (Sect. 17.10) was applied. This method suggested that there is a favorable position for a large positively charged group in the vicinity of the 4-OH group and the adjacent residues Glu 119 and Glu 227. Replacing this OH group with an aliphatic amino group led to **31.29**, which results in a stronger hydrogen bond to the protein. The binding affinity improved into the nanomolar range. If the amino group is modified to a guanidino group, as in **31.30**, the two adjacent glutamate residues can be involved in an interaction with the ligand. Compound **31.30** binds to the protein with $K_i = 0.2$ nM. The compound was clinically developed by GlaxoSmithKline (GSK) and **zanamivir 31.30** was marketed under the name Relenza® in 1999. Due to its high polarity, the drug has poor oral bioavailability. It can only be administered by inhalation. A special inhalation device had to be developed to deliver the drug.

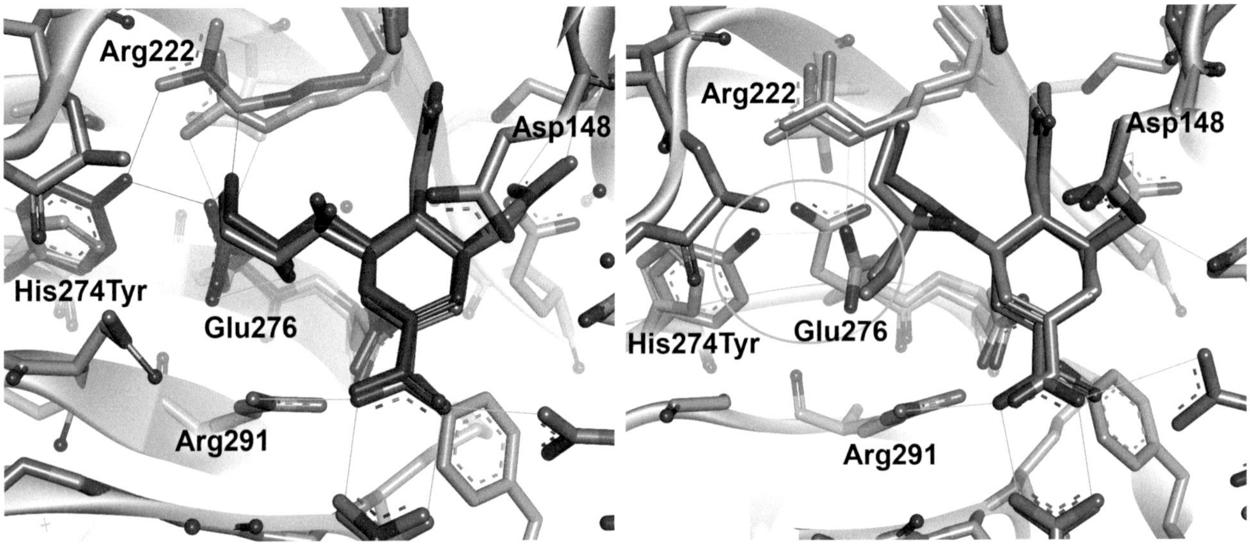

31

Fig. 31.16 Crystallographically determined binding geometry of zanamivir **31.30** (*left*) and oseltamivir **31.34** (*right*) in the His274Tyr resistance mutant (*left*: *purple*, *right*: *blue-violet*, *wildtype orange*) of neuraminidase. The exchange of His for Tyr has little effect on the binding mode of zanamivir, which carries the same side chain as the natural substrate. In contrast, Glu 276 in the complex with oseltamivir, which carries an altered side chain, can no longer adopt the same binding geometry in the His274Tyr mutant (*green cycle*). This shifts the H-bonding network and the affinity of oseltamivir for the mutant enzyme drops from 0.3 to 85 nM. The closer the structure of the inhibitor stays to the natural substrate, the lower is the risk of a loss of affinity due to a resistance mutation (see also the video to Fig. 31.15)

Nevertheless, the goal was to bring an orally available drug to market. A different approach was taken at Gilead Sciences to reduce the high polarity of zanamivir (Fig. 31.14). First, the central pyran ring of **31.28** was replaced by a carbocycle (**31.31**) and the double bond was shifted one position. This allows **31.32** to better mimic the transition state of the reaction. Next, the glycerol function was exchanged for a more hydrophobic *iso*-pentyl ether group in **31.33**. To improve the bioavailability of **31.33**, a prodrug strategy was chosen. Esterification of the free acid function resulted in a new orally available inhibitor, **oseltamivir 31.34**. Roche licensed this compound in 1999 and launched it as Tamiflu® (Fig. 31.14).

After almost 10 years of clinical use, the first cases of **resistance to oseltamivir** and **zanamivir** have been described. Even cross-resistance to both drugs has occurred. The effects of such mutations can be illustrated by a His→Tyr exchange at position 274 that creates oseltamivir resistance (Fig. 31.16). The reorientation of Glu 276 required for oseltamivir binding is blocked in the viral-mutated variant. This results in a significant reduction in binding affinity. Significantly less resistance has been described for zanamivir. Perhaps this is because it is structurally more similar to the sialic acid substrate, making it more difficult for the virus to evolve a mutation without compromising its ability to bind its own substrate. Such a concept is certainly the silver bullet to prevent rapid resistance to a new promising drug. Perhaps it is also the case that zanamivir has been largely protected from resistance development because its inhalation route of administration is less convenient and, therefore, it has simply been administered less frequently.

Follow-up drugs are already being sought. A bivalent zanamivir **31.35** has been described. It must be administered less often, analogous to a depot form. For special circumstances, peramivir **31.36**, which is administered intravenously, was developed from a furanose derivative and, like zanamivir, has insufficient oral availability. During the last H1N1 pandemic ("swine flu"), peramivir was in the final stages of clinical trials and received emergency approval for parenteral treatment of severe cases. Its binding also requires a rearrangement of the Glu 276 side chain, as does oseltamivir. Therefore, cross-resistance between the two drugs has already been described.

31.6 Stopping the Common Cold: Inhibitors for the Capsid Protein of Rhinovirus

The common cold or mundane "sniffle" is caused by **rhinoviruses**, which belong to the family of **picornaviruses** (RNA viruses, *pico* = small). They are nonenveloped viruses that do not have a lipid coat. Their genome is contained on a single positive-strand RNA packaged in an icosahedral capsid. Their diameter is approximately 200–300 Å. The rhinoviruses prefer temperatures between 3 and 33 °C, which means that their growth is inhibited at higher temperatures, such as body temperature. They thrive in cool weather, and infections are most common on cold, wet days. They infect our noses in particular

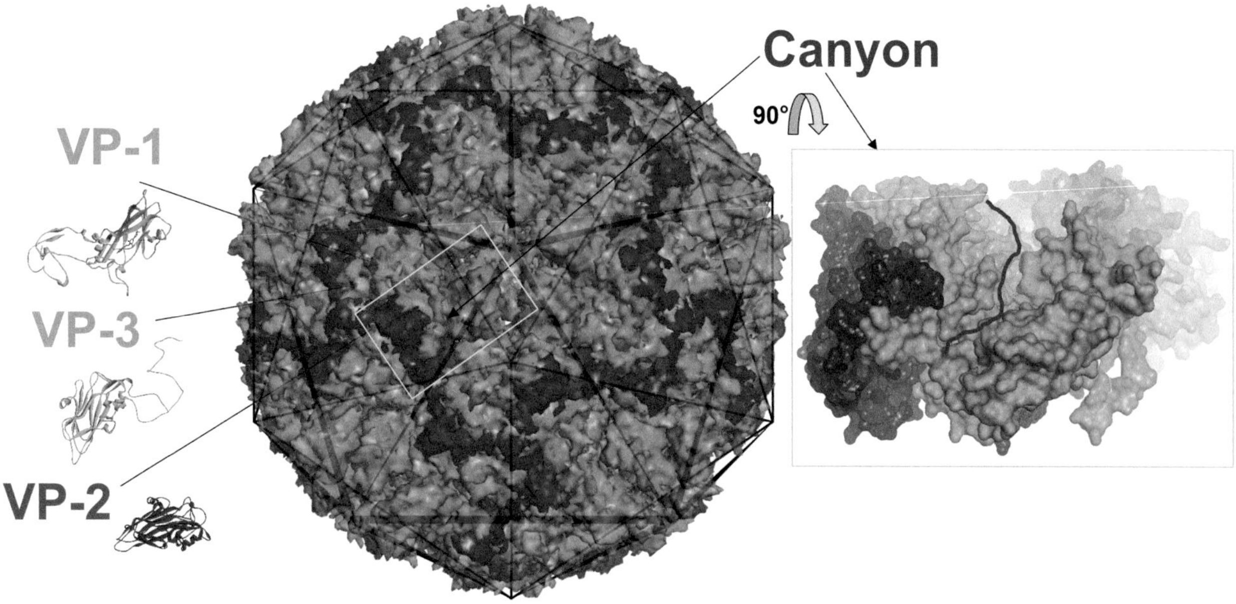

Fig. 31.17 The picornavirus capsid has an icosahedral geometry. The triangular faces of the icosahedron are highly symmetrically formed by three viral surface proteins, VP-1 (*green*), VP-2 (*red*), and VP-3 (*light blue*). A fourth chain, VP-4 (*purple*), joins VP-2 and is directed towards the interior of the viral capsid. A deep groove forms on the surface of the viral capsid proteins. This groove, referred to as "canyon," is shown in the *inset on the right* with an orientation rotated by 90°. The *blue line* indicates the course of the canyon. It is important for the recognition and binding of adhesion proteins (ICAM-1, Fig. 31.18) of the infected host cell. (▶ https://sn.pub/uCKSSz)

because there the local cool temperature provides them with an ideal breeding ground. Rhinoviruses are usually transmitted by direct contact through contaminated hands (known as a smear infection). People whose immune systems are suppressed by a temporary constitutional weakness, or children whose immune systems are not yet fully developed, are particularly susceptible to infection. The incubation period for the common cold is only 12 h, after which the first newly formed viruses leave the infected host cells. Rhinoviruses are strictly localized to the nose and throat. Humans respond to the viral attack with an inflammatory response of the mucous membranes. The nose becomes red and swollen, and its temperature rises. There is a general feeling of being unwell with headache and fatigue. Often, the primary viral infection is accompanied by a secondary bacterial infection or infection with a much more pathogenic virus, which can pose a real health risk.

In general, our immune system can handle rhinoviruses without intervention and within a week the body has defeated the viral invaders. There are more than 100 viral serotypes. This term refers to variations in the surface proteins of these viruses that repeatedly force the immune system to produce new antibodies to defend against them. While the common cold rarely poses a real long-term health threat to us, it is important to remember that it is a major economic burden. It is estimated that over 40 million work days are lost each year due to cold-related absenteeism!

Picornaviruses belong to one of the largest families of viruses, which includes the rather harmless rhinoviruses as well as very dangerous viruses such as **poliovirus**, hepatitis A and B viruses, or viruses that can cause meningitis, myocarditis, or encephalitis. **Foot-and-mouth virus disease** also belongs to this family. Although this virus is not dangerous to humans, it can quickly spread to epidemic proportions among cloven-hoofed animals such as cattle, pigs, or sheep. It has often threatened the livestock of entire regions, especially when highly virulent strains with malignant progression and high mortality occur. Surviving animals are often left with permanent heart muscle damage. The poliovirus, which posed a massive threat to humanity until the 1960s, was largely defeated with the introduction of a successful oral vaccine. In a poliomyelitis infection, also known as infantile paralysis, the poliovirus attacks the nerve cells in the spinal cord that control muscles, causing permanent paralysis or even death. The victory lap against this virus will only continue if future generations show a high degree of discipline in prophylactic vaccination. Complacency easily creeps in when the acute threat of such a disease fades from the public's view because of fewer cases. Effective protection through vaccination can only be achieved with a sufficiently high level of immunization of the population. As we have seen recently, measles is re-emerging in children

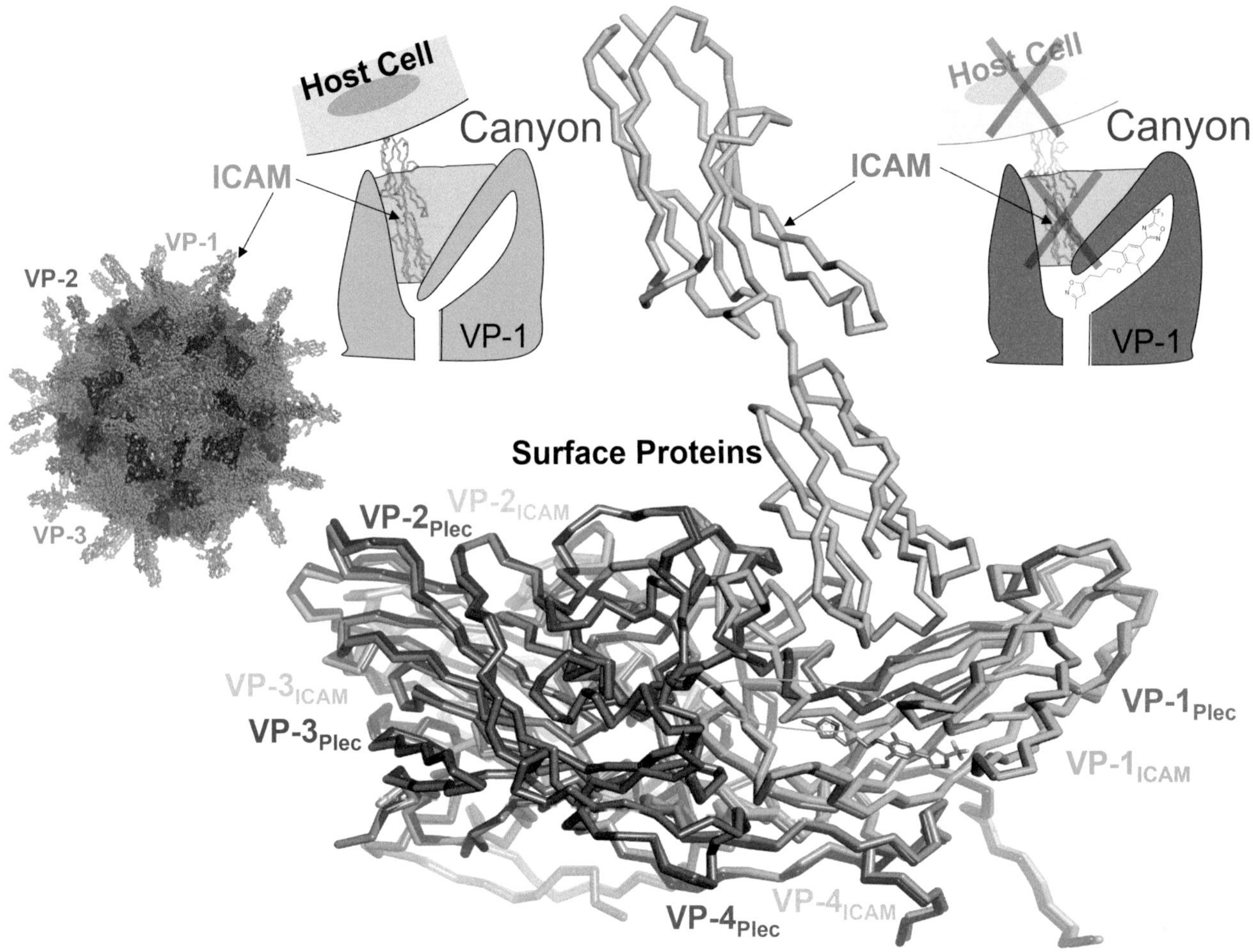

Fig. 31.18 Superposition of the crystal structure of the capsid proteins VP-1 (*green*), VP-2 (*red*), VP-3 (*light blue*), and VP-4 (*violet*) with bound pleconaril (*darker colors, index Plec*) or with bound adhesion protein ICAM of the infected host cell (*lighter colors, index ICAM*). The course of the Cα chain of the protein is shown. Binding of an antiviral compound such as pleconaril (**31.42**, Fig. 31.19) below the canyon causes the chain of VP-1 to deviate near the binding site of pleconaril (*orange oval*). This changes the conformation in the canyon to such an extent that binding of the host cell adhesion protein can no longer occur (*cf. schematic drawings*). (▶ https://sn.pub/hvezQa)

at an alarming rate, largely due to a decline in immunization discipline, fueled in part by unscientific conspiracy theories in countries where vaccination is not mandatory.

The 30-Å thick capsids of picornaviruses are composed of 180 polypeptide chains, of which 60 copies are identical. This is due to the icosahedral architecture of the virus (Fig. 31.17). The three polypeptide chains VP-1, VP-2, and VP-3 are arranged on a **20-faced icosahedron**. Between the faces are grooves about 25 Å deep, called canyons. A fourth short peptide chain, VP-4, is attached to VP-2. It is oriented inside the virus and has no contact with the outside. The viral genome contains 7500 bases that encode the capsid proteins, two proteases, a polymerase, an ATPase, and four other proteins. The **surface-exposed canyons** are particularly important in the region formed by VP-1. On the one hand, the canyon forms binding sites for adhesion molecules (ICAM-1, see Sect. 31.3) found on the surface of infected host cells. Since the virus must not change its surface composition too much there, antibodies from vaccination sera (Sect. 32.1) are specifically directed against this part of the canyon. Such a strategy can be very successful with viruses that have a less broad serotype distribution than rhinoviruses. On the other hand, the canyon near VP-1 has an opening to the interior of the viral capsid, which is important for the release of the viral genome. Michael Rossmann's research group at Purdue University, Lafayette, USA, studied the proteins of the viral capsid in detail. Using cryo-electron microscopy (Sect. 13.6), they were able to determine a structure of the capsid protein with an adhesion molecule. This complex was determined at reduced resolution, but it shows that the binding of the adhesion molecule ICAM-1 occurs in the deep cleft of the canyon (Fig. 31.18).

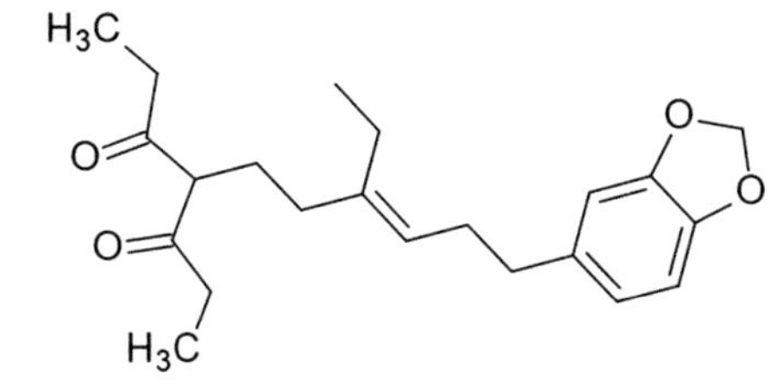

31.37 β-Diketone Mimetic of Juvenile Hormone

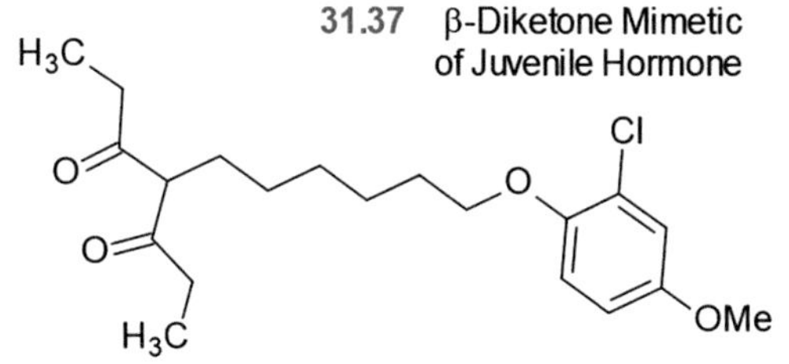

31.38 Arildon

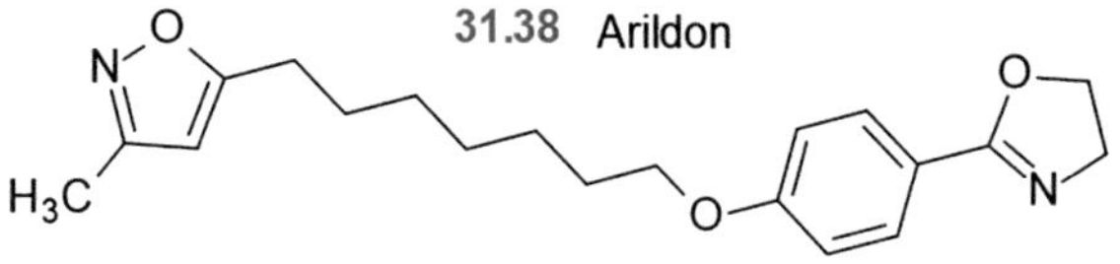

31.39 Disoxaril

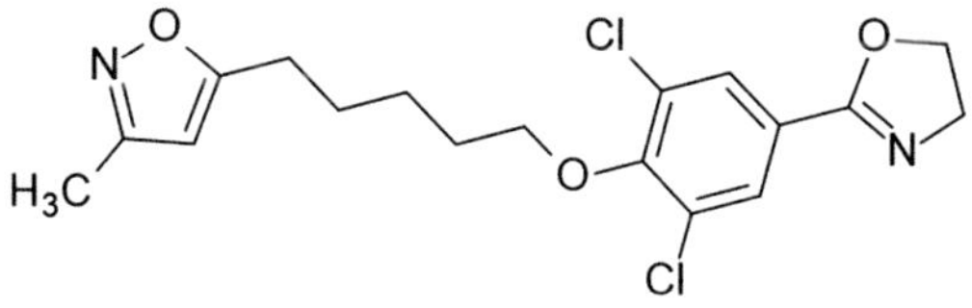

31.40 WIN 54954

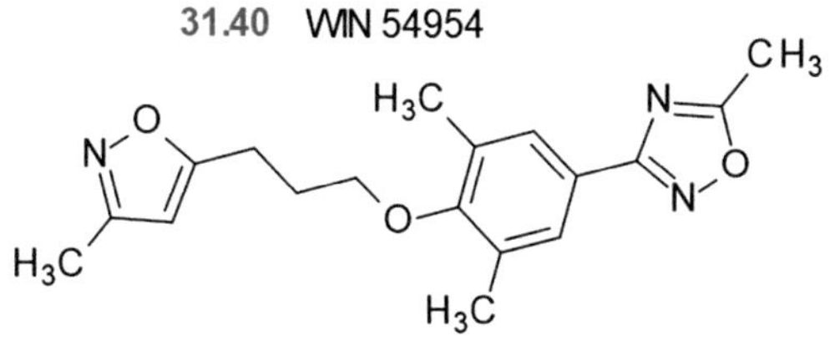

31.41 WIN 61893

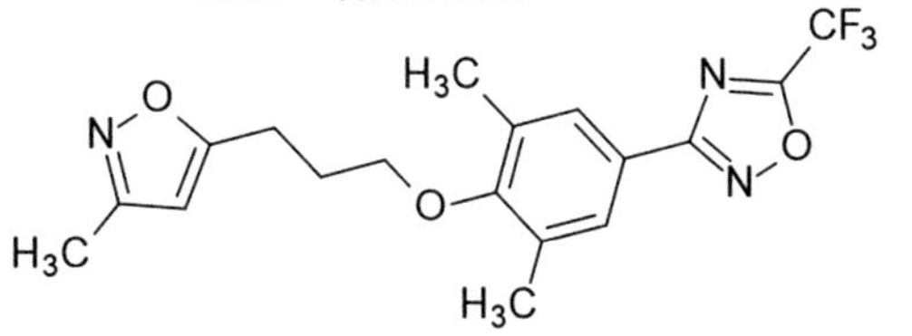

31.42 WIN 63843 Pleconaril

Fig. 31.19 β-Diketones (**31.37**, **31.38**) that showed antiviral activity against picornaviruses were prepared in the course of a synthetic program towards developing juvenile hormone mimetics. By introducing a terminal heterocycle, varying the chain length, and blocking positions on the heterocycles to improve the metabolic stability (**31.39**–**31.41**), pleconaril **31.42** could be developed as an inhibitor of the viral attack on host cells

Because of the common architecture of picornaviruses, **antiviral therapy** against these viruses can be developed using the same concepts. One strategy initially developed and pursued at Sterling-Winthrop was to target the elongated pocket found below the VP-1 canyon (Fig. 31.18). Accommodation of antiviral compounds in this pocket causes a conformational change at the bottom of the canyon. This change in geometry alters the interactions with the adhesion molecule on the surface of the infected host cell. Stable contact with the host cell is no longer possible, and the virus cannot transfer its viral RNA to the infected cell. The viral infection is terminated.

The first lead structures at Sterling-Winthrop were β-diketones (such as **31.37**) synthesized as intermediates in a research project for the development of juvenile hormone (Fig. 31.19). Arildone **31.38** was derived from this lead compound and successfully blocked the replication of the poliovirus. Because the β-diketone moiety had unsatisfactory chemical and metabolic stability, it was replaced by an oxazole ring. Further optimization led to disoxaril **31.39**, which blocked viral infection by several picornaviruses in animal models.

By the end of the 1980s, Michael Rossmann had already succeeded in elucidating the binding geometry of the Winthrop compounds complexed with viral capsid proteins. The structures showed the occupancy of an extended pocket below the canyon. Disoxaril **31.39** had no activity against rhinovirus, and its bioavailability of less than 15% seemed unsatisfactory. Further development led to derivatives with a di-*ortho*-substitution on the central phenyl ring. WIN54954 **31.40** is significantly more potent and has better bioavailability. However, it does not have the desired broad potency against all viral strains and its metabolic stability still leaves something to be desired. The methyl derivative **31.41** with the terminal oxadiazole ring also lacks the desired stability. Only the replacement with a CF_3 group led to success. The compound pleconaril **31.42** entered clinical trials. Fig. 31.20 shows how it binds to the capsid protein. Its administration to 2100 patients with rhinovirus infection showed that the duration and severity of the disease were shorter and milder, respectively, than in a placebo group. However, the FDA denied marketing approval for pleconaril in 2002. Concerns were raised about the drug's safety. There was evidence of complications in women using oral contraceptives.

For a disease such as the common cold, from which our bodies can recover without drugs, it is certainly appropriate to look very closely at the **benefits and risks** of a drug. Undoubtedly, a compound like pleconaril can reduce the inappropriate use of antibiotics or prevent a serious secondary bacterial infection. Such a compound may also help asthma and COPD (chronic obstructive pulmonary disease) patients with an infection. Schering-Plough has conducted additional clinical studies with the compound, which is used as a nasal spray. Knowledge of the compound's mechanism of action, which may be transferable to other picornaviral diseases, could be very valuable. It could help in the development of drugs for other infectious diseases with much higher health risks. However, there is unlikely to be a similarly lucrative mar-

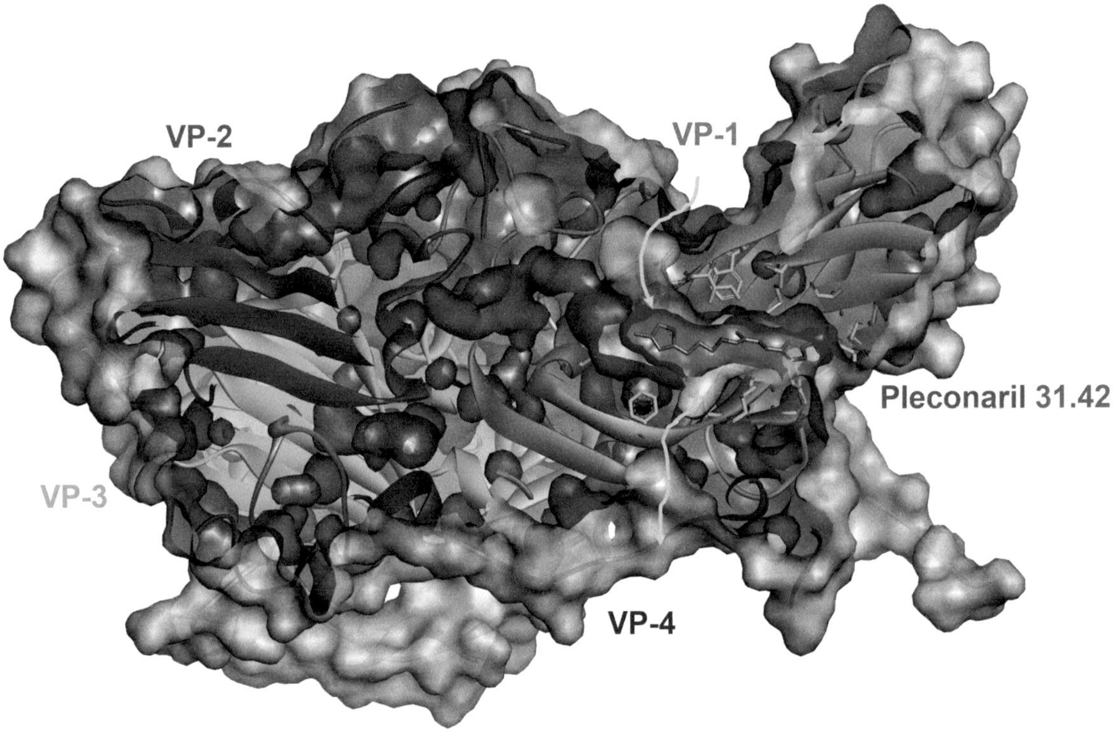

Fig. 31.20 Crystal structure of the viral capsid proteins of rhinovirus HRV-14 with the inhibitor pleconaril **31.42** (*blue surface*). The antiviral agent binds to VP-1 (*green*) below the canyon into a narrow, elongated pocket formed by numerous hydrophobic amino acids. It blocks an entry channel through the viral capsid into the interior of the virus (*pathway indicated as orange line*). Furthermore, the drug induces a conformational change of VP-1 by its binding and, thereby, interferes with the recognition of the adhesion protein of the host cell to be infected (Fig. 31.18). (▶ https://sn.pub/3EqPLZ)

ket for these. ViroPharma is considering filing for FDA approval of pleconaril based on successful clinical trials for the **treatment of meningitis** caused by enteroviruses.

31.7 MHC Molecules: Where the Immune System Presents Peptide Fragments

Our immune system defends us against the harmful invasion of antigens and eliminates cells that are either infected or transformed into a potentially pathological state. It is divided into **unspecific** and **specific defense mechanisms**, which are served by **cellular** and **humoral** (in body fluids) components. The nonspecific defense mechanisms attempt to deactivate pathogens and foreign substances upon initial contact. We have various glycoproteins and interferons in the circulating blood and tissues that carry out the initial attack as the so-called **humoral complement system**. Their defensive efforts are not targeted and serve to degrade the foreign substances (cf. Sect. 31.3). For example, they adhere to bacteria and create an opening in their membrane that allows fluid and salts to flow in. This causes the bacterial cell to swell and eventually burst. Lysozyme is another factor that enzymatically hydrolyzes the cell walls of certain bacteria. In addition, interferons are released, which have an immune-stimulating effect on neighboring cells. In these cells, proteins are produced that initiate entirely different mechanisms to fight foreign substances.

In fact, the body has an additional, very effective and specific protective barrier that must first be developed and "trained." This immunological defense mechanism is activated when a harmful material is recognized as such. The resulting immune response consists mainly of three types of cells: **macrophages**, **B-lymphocytes**, and **T-lymphocytes**. These defense mechanisms are highly specific and usually result in immunity. As a result, the body becomes insensitive to foreign materials with which it has previously come into contact. In the humoral defense, this task is performed by antibodies (Sect. 32.3). They are formed 5–7 days after an immunocompetent B-lymphocyte comes into contact with an antigen. After the initial contact with the invader, effector cells are formed to produce antibodies, as well as memory cells that continue to circulate in the blood. Upon renewed exposure, the immune system will be ready to attack, even if the antigen was recognized years ago.

Vertebrates have developed an adaptive system for cellular defense that distinguishes between healthy and infected cells. T-lymphocytes, also known as T-cells, play a critical role in the **cellular immune response**. They belong to the group of white blood cells. Produced in the stem cells of the bone marrow, they mature in the thymus into the actual T-cells. They carry T-cell receptors on their surface that are responsible for recognizing antigens. By scanning cells for the characteristic "diseased" or "healthy," they find these antigens in the form of peptide sequences that are presented on the surface of the cells by **MHC molecules (major histocompatibility complex)**. There are two different types of MHC molecules, called class I and class II. MHC-I molecules present peptides of 8–10 residues, derived primarily from the cytosol of a nucleated cell. MHC-II molecules use longer peptides that are mainly produced during endosomal protein degradation. They are found on specialized antigen-presenting cells such as macrophages or B-cells. Both classes of MHC molecules "present" their antigens to T-helper cells, which then regulate the immune response to foreign antigens. The name MHC stands for "tissue compatibility complex." It was first recognized that MHC molecules, by presenting foreign proteins, initiate the rejection reaction in organ transplantation. For this reason, compatibility typing of donor and recipient antigen patterns is performed prior to a planned organ transplantation. It is now known that the cellular immune system uses MHC molecules to distinguish between healthy and infected cells. Similar to the learning process of B-lymphocytes, T-lymphocytes generate daughter cells after initial contact with an antigen. These serve as long-lived memory cells and can quickly initiate a defense when re-exposed to an antigen.

The developmental pathway for class I MHC will be discussed in more detail (◘ Fig. 31.21). MHC-I molecules are normally loaded with peptides derived from cytosolic proteins. They are cleaved into **peptide fragments of 8–10 amino acids** in the proteasome, a kind of cellular shredder (Sect. 23.8). In healthy cells, peptides are formed exclusively from endogenous cellular proteins. If the cell is infected with a virus or has undergone a transformation that has resulted in mutated proteins, foreign or altered peptide fragments will be produced. These are also bound to MHC molecules and presented on the cell surface. The T-cells can immediately recognize which cell has been infected with a virus or has degenerated into a diseased state. The loading process of MHC molecules with peptides produced by the proteasome takes place in the endoplasmic reticulum (ER). The peptides are channeled into the ER by a specific transporter (TAP). The loaded, membrane-anchored MHC molecules are then transported to the cell surface by vesicles that fuse with the ER and/or the cell. If the foreign or virus-infected cell or tumor cell carries such an antigenic peptide fragment in complex with an MHC-I molecule, it is recognized by

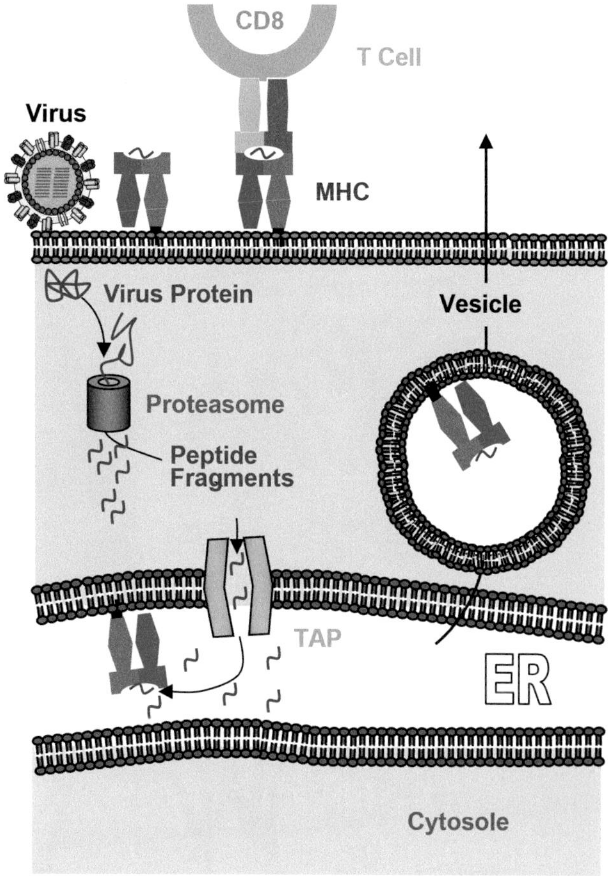

◘ **Fig. 31.21** When a cell is infected with a virus, viral proteins are found in the cytosol (*gray*). They are degraded in the proteasome along with endogenous cellular proteins and cut into peptide fragments. These fragments are transported to the endoplasmic reticulum (ER) by the TAP transporter (*light-green*). There, the membrane-bound MHC class I molecules (*blue-violet*) are loaded with peptides of 8–10 residues in length. Enclosed in vesicles, the peptide-presenting MHC-I molecules are transferred to the cell surface and anchored to the membrane. T-cells (*green*) scan through the presenting MHC molecules by forming a complex with their T-cell receptors and recognize whether the presented fragments are from endogenous or foreign proteins. If the protein is of foreign origin, or an endogenous protein that has been overexpressed (e.g., in tumor cells), an immune response will be initiated

so-called $CD8^+$ lymphocytes. **Cytotoxic T-killer cells** belong to this cell type and release cytokines, pore-forming perforins, and proteases after binding. As a result, they lyse the recognized diseased cell and induce apoptosis.

It is at this point that interest in the molecules of cellular immune defense becomes attractive for drug therapy. Cancer is a common cause of death, especially in advanced age. Surgical resection, chemotherapy, and radiation therapy with tissue-destroying effects are the mainstay of cancer treatment. Increasing knowledge about the role of the immune system in controlling malignant degeneration, the molecular interaction of antigen-presenting tumor cells with immune cells, and the view of antigen processing as presentation have

31

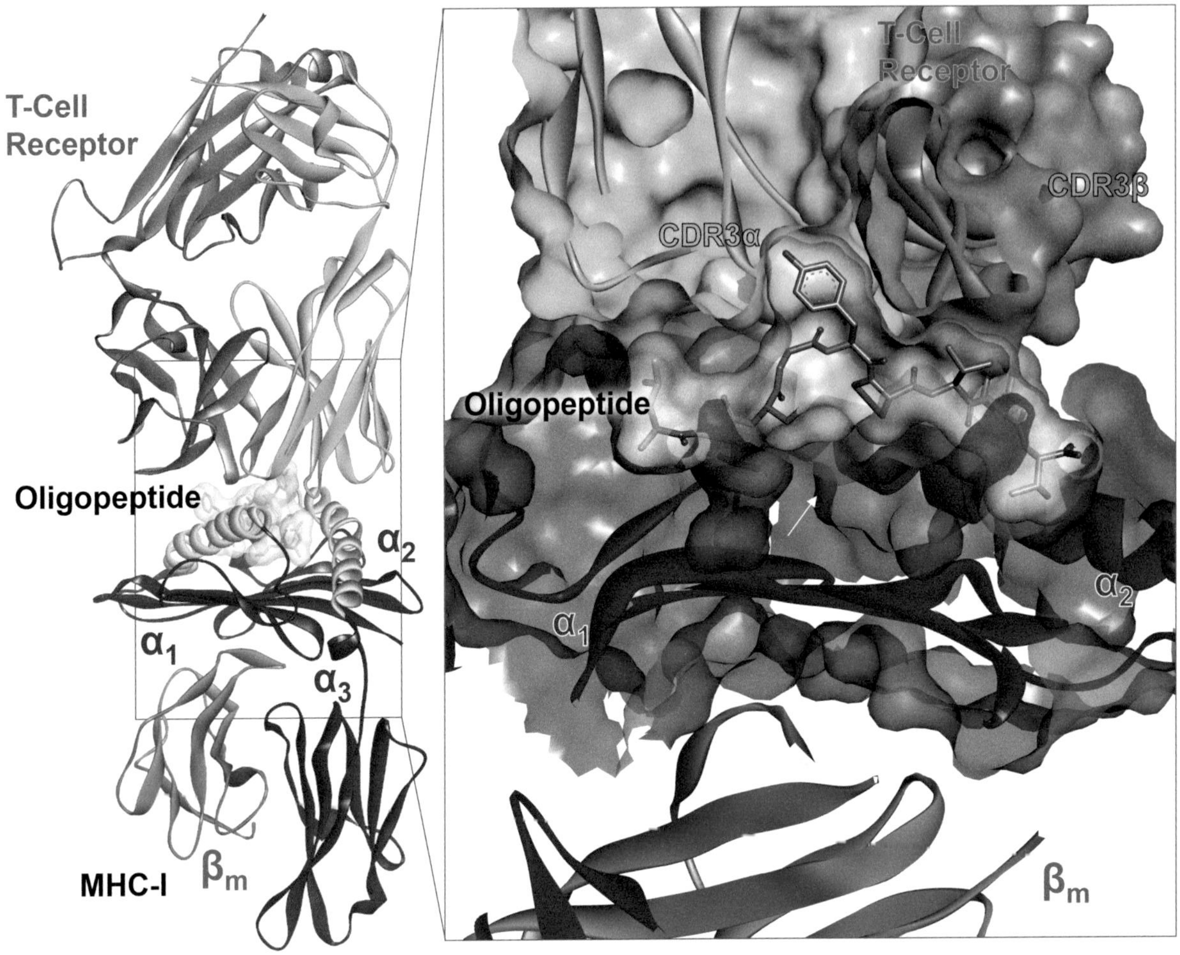

Fig. 31.22 Crystal structure of the complex of an MHC-I molecule with a bound nonapeptide Leu–Leu–Phe–Gly–Tyr–Pro–Val–Tyr–Val (*gray*) and the T-cell receptor: total structure (*left*), peptide-binding site (*right*). The MHC-I molecule is composed of a heavy chain with the domains α_1, α_2, and α_3 (*violet*) and a light β_m-chain (*blue*). The pleated-sheet structure formed by the α_1/α_2-domains forms a bowl that is open above and is bordered by two long, parallel-oriented α-helices (*yellow*). It accommodates the antigen peptide fragment and presents its upper face to the T-cell receptor. This heterodimeric receptor, made up of one α- (*light-blue*) and a β-chain (*gray*), also has a pleated-sheet-like geometry. It recognizes the amino acid tyrosine in position 5 of the antigen peptide with its hypervariable loops CDR3α and CDR3β. (► https://sn.pub/TX08AY)

opened up entirely new perspectives in **tumor therapy**. The immune system recognizes and destroys many cells that have degenerated into tumor cells. However, tumor cells use a variety of strategies to evade the immune response. One approach to developing a drug therapy is to stimulate the immune response using specific tumor antigens. **Peptide-like vaccines** are being developed that can stimulate an immune response against the tumor cells. The first successes of such a therapy have been described in patients with melanoma (malignant degeneration of pigment cells). Peptidic vaccines target the antigen-presenting MHC-I molecules complexed with the $CD8^+$ T-cell receptors. After being stimulated by dendritic cells to become T-effector cells, T-cells are qualitatively and quantitatively capable of capturing peptides presented on somatic cells. On the one hand, foreign proteins are recognized; on the other hand, the cytotoxic killer cells are able to filter out cells that overexpress endogenous proteins due to the high density of presented peptides.

These properties can be exploited to design peptide vaccines. By offering a large amount of endogenous peptides, immune tolerance to native proteins should be overcome. Killer cells stimulate a specific and enhanced immune response against the degenerated tumor cells. The goal of the development of such specific vaccine serums is to **replace endogenous peptides** with analogues that can induce the same or an **exaggerated immune stimulation**, but that also have a much better stability and bioavailability due to the incorporation of nonproteinogenic amino acids or peptidomimetic groups.

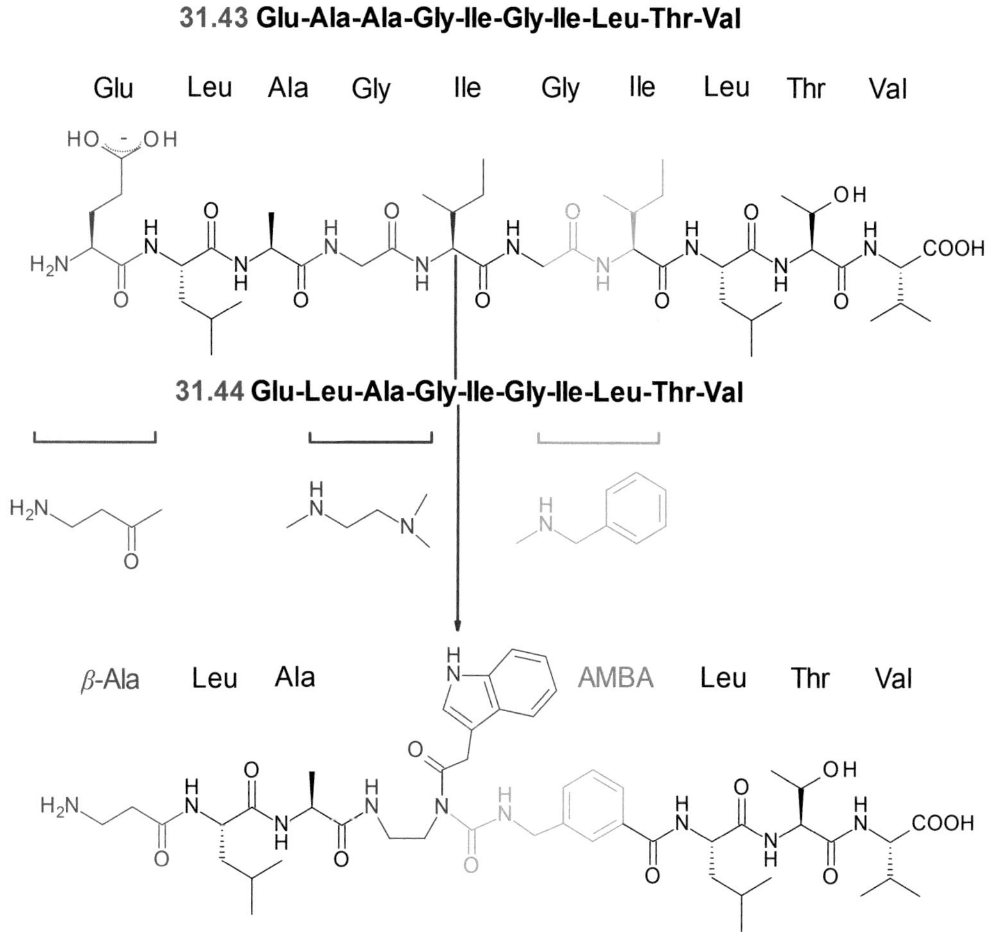

31.45

Fig. 31.23 The decapeptide **31.43**, isolated from patients, was optimized to **31.44** by substituting an alanine for a leucine at position 2 to develop peptidomimetics as candidates for an immunostimulatory vaccine against melanoma. It was used in clinical vaccine trials. By stepwise replacement of the building blocks in **31.44**, a peptidic lead structure was modified into a stabilized peptidomimetic that binds identically to the MHC molecule, but elicits an enhanced immune response by binding to the T-cell receptor. Four modifications were made in the stepwise development of **31.45**. The *N*-terminal glutamic acid was replaced by a β-alanine (*red*) to improve stability. Replacement of the first Gly–Ile moiety by a 2-aminoethylene (*blue*) group together with a change to a CO–CH_2 indoyl group (*violet*) increased the immune response at the T-cell receptor. Replacing the second Gly–Ile unit with the peptidomimetic moiety 3-aminomethylbenzoic acid (*green*, AMBA) allowed the peptide backbone to follow the same course in space

The architecture of the complex of an MHC-I molecule with a presented peptide and the T-cell receptor will be considered in more detail (Fig. 31.22). The MHC-I molecule is composed of a heavy chain (approximately 360 amino acids) and a light chain (90 amino acids). The heavy chain is membrane anchored and consists of three domains, α_1, α_2, and α_3. The α_1- and α_2-domains form a kind of bowl with a bottom consisting of a six-stranded antiparallel pleated sheet. The bowl is bordered by two long helices oriented parallel to each other. Between the helices there is a crevice to accommodate the antigen peptide fragment. Peptides with a length of 8–10 Å fit into this area. Peptide binding is largely due to hydrogen bonds at the *N*- and *C*-termini. MHC-I molecules are highly polymorphic in their amino acid composition, even having the same architecture, so that interactions with the backbone are primarily responsible for their binding in the crevice, and these interactions can be formed by peptides in general. In the middle sequence segment, the antigen peptide protrudes slightly from the binding pockets of the α_1/α_2-domains. Residues at the beginning and end of the oligopeptides orient themselves in the small pockets formed by the MHC-I molecule. They determine the binding affinity of each peptide to the protein. The bulging residues in the middle do not contribute much to binding to the MHC-I molecule, but they are crucial for the recognition and interaction with the T-cell receptor. The sequence of the β_m-chain is virtually invariant, and most genetic modifications occur in the α-chain. In addition, polymorphisms (Sect. 12.11) have been discovered in this chain, which are found to vary from individual to individual. They determine the

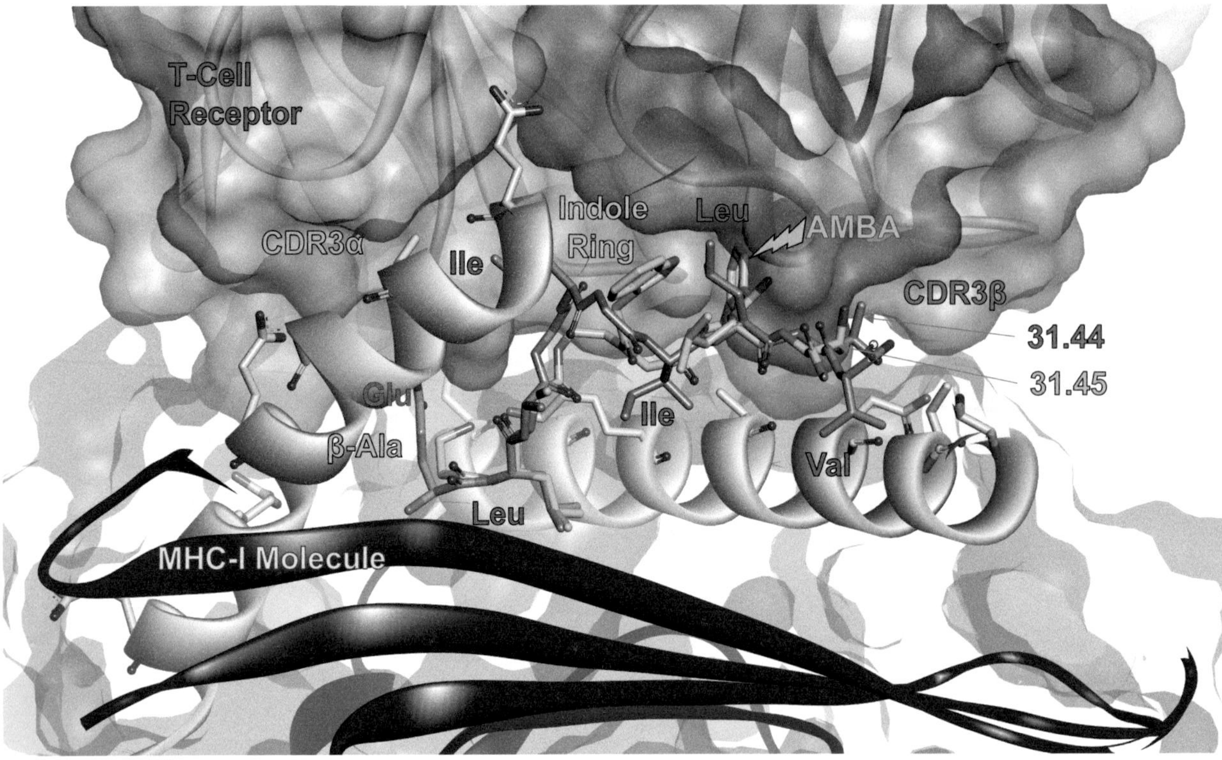

31

Fig. 31.24 Crystallographically determined binding geometry of the reference peptide **31.44** (*ochre*) superimposed with the designed peptidomimetic **31.45** (*light-green*) in the binding pocket of the ternary complex with the MHC-I molecule (*yellow*) and the T-cell receptor (α-*chain light-blue*, β-*chain gray*). With its hypervariable loops CDR3α and CDR3β, the receptor recognizes the phenyl ring of the AMBA moiety and the CO–CH_2–indoyl group of the peptidomimetic. (▶ https://sn.pub/bJ4tf5)

tissue compatibility between donor and recipient in organ transplantation. Susceptibility to infections and autoimmune diseases may also be explained by these variations.

MHC-I molecules bind the antigenic peptide based on its sequence. They force the peptide into an extended conformation, exposing the peptide's central amino acid residues for molecular recognition by a T-cell receptor. The T-cell receptor is a heterodimeric transmembrane glycoprotein found exclusively on T-cells. It is composed of an α- and a β-chain. The folding pattern of the two chains is similar to that of the light chains in antibodies (Sect. 32.3). The antigen binding site is located in the loop region between the individual folded sheets of the domains. These loops are hypervariable and determine the recognition properties of each receptor. In a complex with the MHC-I molecule, the receptor lies diagonally above the peptide-binding site. With its variable loops CDR3α and CDR3β, it mainly covers the amino acid residues of the antigen peptide that point away from the MHC-I molecule. At the same time, the T-cell receptor is in contact with the surface portions of the flanking helices of the MHC-I molecule.

The design of **peptidomimetics** as candidates for a **vaccine therapy** to stimulate the immune response will be illustrated on the case of the melan-A/MART-1 antigens. These antigens are presented on the surface of melanoma tumor cells by an MHC-I complex. The nonapeptide Ala–Ala–Gly–Ile–Gly–Ile–Leu–Thr–Val and the decapeptide Glu–Ala–Ala–Gly–Ile–Gly–Ile–Leu–Thr–Val **31.43** were isolated from melan-A of patients with this disease (Fig. 31.23). Both oligopeptides bind with low affinity to the MHC-I molecule. An exchange of alanine for leucine at position 2 significantly increases the binding affinity. The leucine-carrying peptide **31.44** is significantly more immunogenic than **31.43**. It was, therefore, selected for a clinical vaccine trial on melanoma patients. As a peptide, however, it has a low stability in the organism and is rapidly degraded.

The group of Francine Jotereau and Stéphane Quideau in Bordeaux, France, therefore, set out to develop a peptidomimetic. It should have the same binding affinity to the MHC molecule, the same or better affinity to the T-cell receptor, and be significantly more stable. Since only a crystal structure of the binary complex of **31.44** with the MHC-I molecule was available, a model without the T-cell receptor was developed using a ter-

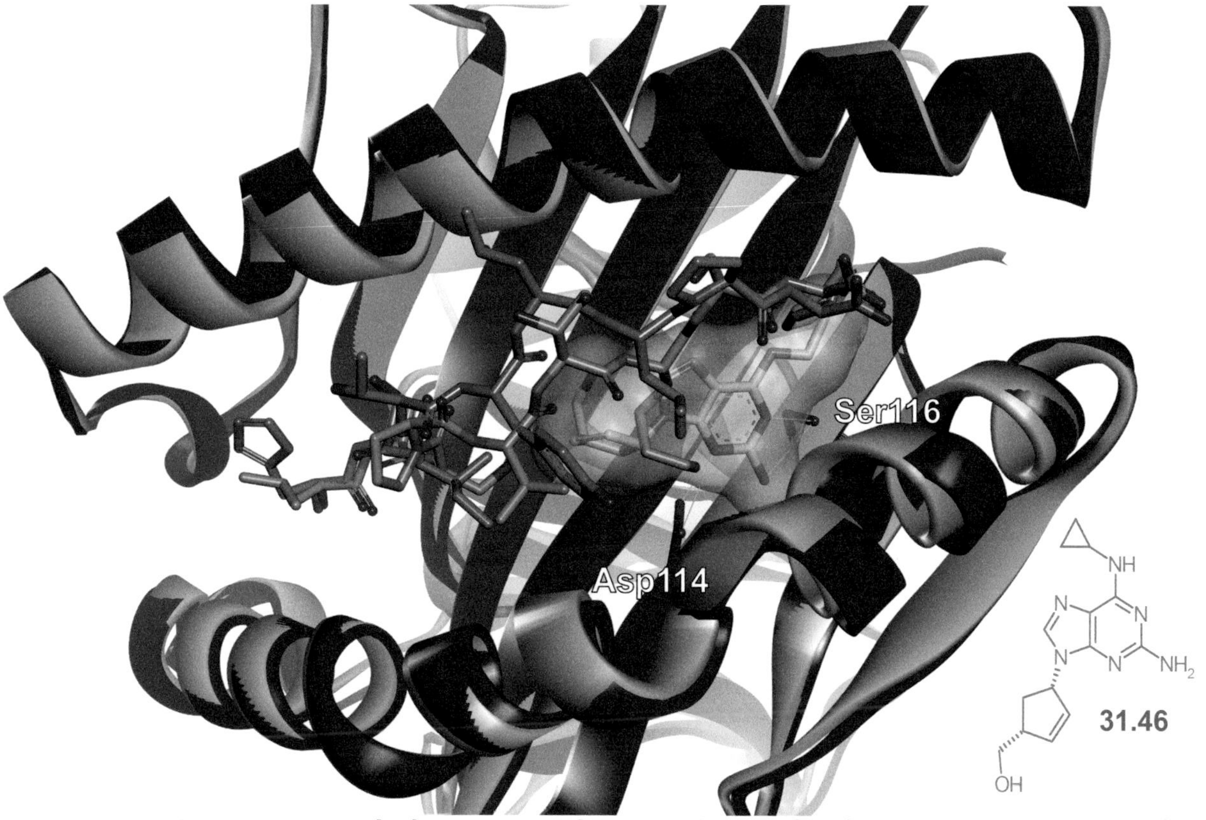

Fig. 31.25 Crystallographically determined binding geometry of the MHC-I molecule (HLA-B*57:01 allele) loaded with the self peptide Leu–Ser–Ser–Pro–Val–Thr–Lys–Ser–Phe (*green*). Shown is the superimposed structure of the ternary complex of the MHC-I molecule with the bound nucleoside Abacavir **31.46** (*gray with yellow surface*) and the modified peptide His–Ser–Ile–Thr–Tyr–Leu–Leu–Pro–Val. The original self-peptide can no longer bind because of the sterically demanding Lys and Phe residues. (► https://sn.pub/7HW3x2)

tiary complex with a structurally similar peptide. A glutamate residue in the first position was to be replaced by a peptidase-stable moiety. The choice was β-alanine, which is hardly proteolytically cleavable. The leucine at position 2 and the valine at position 10 should be retained as they are crucial for anchoring in the MHC molecule. A spatially conserved orientation of the backbone scaffold was expected. The group proceeded step by step. The amino acid at position 5, which is an isoleucine in the reference peptide **31.44**, seemed to be critical for the interaction with the CDR3 loop of the T-cell receptor. The *sec*-butyl group of the isoleucine was replaced by an aromatic moiety, with an indole group proving to be optimal. Next, the central peptide bonds of the Gly–Ile–Gly–Ile motif were modified by reduction. In the end, an *N*-(2-aminoethyl) bridge with the attached indoyl moiety was chosen for the first Gly–Ile segment. The second Gly–Ile unit could be replaced by the well-known peptidomimetic group 3-aminomethylbenzoic acid (AMBA). As a result of this optimization, the peptidomimetic **31.45** was obtained, which shows almost the same binding affinity to the MHC-I molecule as the reference peptide **31.44**. Its crystallographically determined binding mode to the MHC-I molecule is shown in ◘ Fig. 31.24. In an immune response stimulation assay, this compound induced the strongest release of γ-interferon of all the compounds tested. This is likely due to a more intense interaction with the T-cell receptor. Further development will show whether **31.45** is a promising lead structure for the development of peptidomimetic vaccines for immunotherapy against melanoma tumors.

At the end of this chapter, two examples are given to illustrate the opportunities and risks associated with therapeutic intervention in the patient's own immune system. The first example concerns the molecular basis of **drug hypersensitivity**. It is associated with the misdirected elimination of the body's own cells by the T-cells. The MHC-I molecules on these T-cells present peptide fragments that are mistakenly recognized as "diseased" or "foreign." How does this false loading of

MHC-I molecules occur? The antiviral drug **abacavir 31.46** (Fig. 31.25) is a nucleoside analog that suppresses HIV replication (Sect. 32.5). The use of abacavir leads to severe immune-mediated hypersensitivity with nonspecific gastrointestinal and respiratory symptoms in about 8% of treated patients. These are associated with fever and rash. Repeated treatment with abacavir may lead to life-threatening reactions. Hypersensitivity is associated with the presence of a specific variant, the HLA-B*57:01 allele, carried by approximately 8% of all people (alleles are the different gene variants at a specific location on a chromosome). The amino acid composition of the MHC-I molecules in our **h**uman **l**eukocyte **a**ntigen (HLA) system varies according to genetic makeup. Carriers of the HLA-B*57:01 allele produce MHC-I molecules that have a specific binding site for abacavir in the bowl-shaped peptide fragment presenting pocket. This changes the repertoire of self peptides that are bound and presented in the peptide-binding crevice, as they bind simultaneously with abacavir. As a result, the T-cell receptor is misdirected. Healthy cells that presented the "right" peptides before treatment with abacavir are recognized as "foreign" after administration of **31.46**, as they now present different peptides.

31

The second example concerns tumor therapy. It is important that the immune system responds neither too weakly nor too strongly to pathogenic or cancerous cells. Otherwise, some of these cells could survive the attack, or if the defense is too strong, healthy tissue could be attacked (autoimmune reaction). Fine-tuned control is therefore essential. To prevent overreactions, the immune system relies on immune checkpoints on immune cells: After successfully fighting a pathogen, the immune system sends signaling molecules that attach to the checkpoints and stop the activity of the immune cells. Some tumor cells know how to suppress this mechanism by binding to the checkpoints themselves. Antibodies have been developed to inhibit these checkpoints. This allows the immune response to be unleashed and attack the now unmasked cancer cells. In recent time, these checkpoint inhibitors have shown impressive results and are seen as one ray of hope in the fight against cancer using the body's own immune system.

Another approach is the **CAR-T cell therapy**. It is a modern and innovative cancer treatment that can currently be used for certain blood and lymphatic cancers. The term CAR-T cell stands for **c**himeric **a**ntigen **r**eceptor **T cells**. As described above, T-cell receptors read the peptides presented on the cells to be screened. This is how they detect infection with foreign proteins. Cancer cells also present peptide fragments. However, these are derived from the body's own protein material. Therefore, solely the amount of peptides presented is indicative for cell malignancy. This makes detection difficult and cancer cells can easily hide from being eliminated by the immune system. The goal of CAR-T cell therapy is to ensure that T-cells reliably find tumor cells and render them to cell death.

First, white blood cells are collected from a cancer patient. The T-cells are isolated outside the body and genetically engineered with chimeric antigen receptors. Using viral gene transfer vectors, the genetic information for the chimeric receptor is transferred to the T-cells and incorporated into their genome (Sect. 12.13). In this way, the genetic information for the chimeric receptor is also passed on to the daughter cells during their activation and division. The introduced chimeric receptors contain the antigen-binding domain of an antibody recognition site (Sect. 32.3) that specifically identifies, for example, CD19, a receptor marker on tumor cells. Finally, the genetically modified CAR-T cells are infused back into the patient. The modified T-cells can now recognize the cancer cells via their antigen-specific receptors and drive them to cell death. From a regulatory perspective, the use of genetically modified cells is complex and requires good clinical monitoring of patients. The price of treatment with currently approved products is in the six-figure euro range. Nevertheless, the CAR-T approach is considered to have great potential. Checkpoint inhibitors and CAR-T cell therapy will change therapy in oncology. And yet the CAR-T procedure is not limited to cancer therapy. Initial successes have already been achieved in the fight against autoimmune diseases, in which antibodies are directed against the cell nucleus components of the body's own cells.

31.8 Synopsis

- Integrin receptors are responsible for the bidirectional communication between cells. They are cell-surface exposed and possess intra- and extracellular domains of complex architecture. Upon activation, these receptors undergo a series of sequential conformational transformations.
- Integrins are formed as heterodimers of 18α- and 8β-subunits. Transition to the active conformation makes the MIDAS-binding site available, which is comprised of divalent calcium and magnesium ions. The activated receptor is accessible for interactions with other proteins.
- In the case of the $\alpha_{IIb}\beta_3$ receptor, which is present on the surface of platelets, recognition of an Arg–Gly–Asp (RGD) motif leads to activation and subsequently to platelet aggregation, an initial step in thrombus formation.
- Blockage of the surface receptors on blood platelets leads to an arrest in the coagulation process. Compounds either of cyclic peptide or open-chain peptidomimetic structure, comprising a mimic of the RGD motif, have been successfully designed as po-

tent fibrinogen-receptor antagonists for the therapy of thrombotic events.
- To defend against pathogens during inflammatory processes, leukocytes, which are transported through vessels with the blood stream, have to be stopped and fixed through sugar–protein interactions with selectins.
- Intervention in the inflammatory cascade can help in disease situations resulting from damage by excessive leukocyte infiltration. Inhibitors of the selectins on the surface of leukocytes have been developed as low molecular weight surrogates of the corresponding sugar moieties on the endogenous PSGL-1 proteins.
- Viruses gain entry into host cells by docking and subsequently merging their envelope with the host cell membrane. First contact with the human immunodeficiency virus occurs via the CD4 receptor and is assisted by contacts to cytokine receptors. A trimeric helical bundle with a sewing-pin-like structure accesses the host cell as a kind of warhead and a helical peptide stretch refolds and zips together to initiate the fusion process. Structurally similar peptides such as enfuvirtide can block the zipping and, thus, work as fusion inhibitors.
- Influenza is a viral disease caused by influenza viruses. These enveloped viruses exhibit the docking protein hemagglutinin, the glycosidase neuraminidase, and the M2 proton channel on their surface. For hemagglutinin and neuraminidase, various subtypes (H1–H18, N1–N11) are known, and they are constantly changing, thus, giving rise to antigen drift and shift. They force the immune system to constantly adapt and produce new antibodies.
- The influenza virus conquers the host cell, exploits its machinery for reproduction, and, after reassembling, buds from the host cell. Final release from the host cell is catalyzed by its neuraminidase, which cleaves the sugar bond between galactose and terminal sialic acid.
- Two potent inhibitors, zanamivir and oseltamivir, have been developed as antiviral drugs to block the catalytic cleavage site of the enzyme. Oseltamivir uses an aliphatic side chain to improve its bioavailability. This induces a rearrangement of amino acids near the introduced side chain. A resistance mutation prevents this rearrangement and reduces the affinity for oseltamivir.
- The common cold is caused by rhinoviruses belonging to the class of nonenveloped single-stranded RNA picornaviruses. Their capsid, formed as a regular 20-faced icosahedron encompassing the RNA, is constructed from four surface proteins. They show a structured surface with deep canyons that bind to cellular adhesion proteins. A stretched-out pocket is found underneath the canyon to which small molecules can be bound. They induce a small shift in the canyon, and this prevents efficient binding to the cell-adhesion proteins.
- Even though pleconaril showed efficacy fighting the common cold, FDA approval was not granted due to the risk of interference with oral contraceptives.
- The cellular immune system can scan cells for the characteristic of being "healthy" or "diseased" via interactions of surface-exposed MHC molecules and hypervariable recognition loops of the $CD8^+$ T-cell receptor on T-lymphocytes.
- Antigen-presenting MHC-I molecules are loaded by 8- to 10-residue-long peptide sequences originating from protein degradation in the proteasome. In this manner, peptides fragments from foreign (viral attack) or altered proteins are also exposed.
- To use cellular immune defense mechanisms in drug therapy (e.g., for cancer treatment), strategies have to be followed that break the evasion of immune response. This can be achieved by applying peptidomimetic surrogates as vaccines of the MHC-exposed peptide stretches that stimulate the immune response on tumor cells. In CAR-T therapy, T-cells are genetically engineered outside of the patient's body so that they carry additional receptors that recognize specific markers on the tumor cells. The cells are then readministered to the patient.
- Drug hypersensitivity can occur because drugs also bind to the peptide-binding site of MHC molecules, altering the scope of peptide fragments presented. Cells that were previously recognized as the body's own are then considered foreign or degenerate, and an immune response is initiated.

Bibliography and Further Reading

General Literature

M. Shimaoka and T. A. Springer, Therapeutic Antagonists and Conformational Regulation of Integrin Function, Nat. Rev. Drug Discov., **2**, 703–716 (2003)

S. A. Andronati, T. L. Karaseva and A. A. Krysko, Peptidomimetics—Antagonists of the Fibrinogen Receptors: Molecular Design, Structures, Properties and Therapeutic Applications, Current Medicinal Chemistry, **11**, 1183–1211 (2004)

R. J. D. Hatley, S. J. F. Macdonald, R. J. Slack, J. Le, S. B. Ludbrook, P. T. Lukey, An av-RGD Integrin Inhibitor Toolbox: Drug Discovery Insight, Challenges and Opportunities, Angew. Chem. Int. Ed., **57**, 3298–3321 (2018)

W. S. Somers, J. Tang, G. D. Shaw and R. T. Camphausen, Insights into the Molecular Basis of Leukocyte Tethering and Rolling Revealed by Structures of P- and E-Selectin Bound to SLeX and PSGL-1, Cell, **103**, 467–479 (2000)

S. R. Chhabra, A. S. Abdul Rahim and B. Kellam, Recent Progress in the Design of Selectin Inhibitors, Mini Reviews in Medicinal Chemistry, **3**, 679–687 (2003)

T. Matthews, M. Salgo et al., Enfuvirtide: The First Therapy to Inhibit the Entry of HIV-1 into Host CD4 Lymphocytes, Nat. Rev. Drug Discov., **3**, 215–225 (2004)

B. J. Doranz, S. W. Baik, R. W. Doms, Use of a gp120 Binding Assay to Dissect the Requirements and Kinetics of Human Immunodeficiency Virus Fusion Events, J. Virology, **12**, 10346–10358 (1999)

M. von Itzstein, The War Against Influenza: Discovery and Development of Sialidase Inhibitors, Nat. Rev. Drug Discov., **6**, 967–974 (2007)

G. Kolata, The story of the great influenza pandemic of 1918 and the search for the virus that caused it. Touchstone, New York (2001)

A. M. De Palma, I. Vliegen, E. De Clercq and J. Neyts, Selective Inhibitors of Picornavirus Replication, Med. Research Reviews **28**, 823–884 (2008)

E. Lazoura and V. Apostolopoulos, Rational Peptide-based Vaccine Design for Cancer Immunotherapeutic Applications, Curr. Med. Chem. **12**, 629–639 (2005)

Special Literature

J. A. Zablocki, J. G. Rico, R. B. Garland et al., Potent *in vitro* and *in vivo* Inhibitors of Platelet Aggregation Based Upon the Arg-Gly-Asp Sequence of Fibrinogen. (Aminobenzamidino)succinyl (ABAS) Series of Orally Active Fibrinogen Receptor Antagonists, J. Med. Chem., **38**, 2378–2394 (1995)

T. W. Ku, F. E. Ali, L. S. Barton et al., Direct Design of a Potent Non-peptide Fibrinogen Receptor Antagonist Based on the Structure and Conformation of a Highly Constrained Cyclic RGD Peptide, J. Am. Chem. Soc., **115**, 8861–8862 (1993)

C. Mas-Moruno, F. Rechenmacher, H. Kessler, Cilengitide: The First Anti-Angiogenic Small Molecule Drug Candidate. Design, Synthesis and Clinical Evaluation, Anti-Canc. Agents Med. Chem., **10**, 753–768 (2010)

R. Kranich, A. S. Busemann et al., Rational Design of Novel, Potent Small Molecule Pan-Selectin Antagonists, J. Med. Chem., **50**, 1101–1115 (2007)

R. C. Preston, R. P. Jakob, F. P.C. Binder, C. P. Sager, B. Ernst, T. Maier, E-selectin ligand complexes adopt an extended high-affinity conformation, J. Mol. Cell Biol., **8**, 62–72 (2016)

J. Chang, J. T. Patton, A. Sarkar, B. Ernst, J. L. Magnani, P. S. Frenette, GMI-1070, a novel pan-selectin antagonist, reverses acute vascular occlusions in sickle cell mice, Blood, **116**, 1779–1786 (2010)

S. Jiang, Q. Zhao and A. K. Debnath, Peptide and Non-peptide HIV Fusion Inhibitors, Curr. Pharmaceut. Design, **8**, 563–580 (2002)

P. W. Smith, S. L. Sollis, et al., Novel Inhibitors of Influenza Silaidases Related To GGI67, Bioorg. Med. Chem. Lett., **6**, 2931–2936 (1996)

J. N. Varghese, W. G. Laver, P. M. Colman, Structure of the influenza virus glycoprotein antigen neuraminidase at 2.9 Å resolution, Nature, **303**, 35–40 (1983)

M. A. Williams, W. Lew, et al., Structure-Activity Relationships of Carbocyclic Influenza Neuraminidase Inhibitors, Bioorg. Med. Chem. Lett., **7**, 1837–1842, (1997)

C. U. Kim, W. Lew, et al., Influenza Neuraminidase Inhibitors Possessing a Novel Hydrophobic Interaction in the Enzyme Active Site: Design, Synthesis and Structural Analysis of Carbocyclic Sialic Acid Analogues with Potent Anti-influenza Activity, J. Am. Chem. Soc., **119**, 681–690 (1997)

C. U. Kim, W. Lew, et al., Structure-Activity Relationship Studies of Novel Carbocyclic Influenza Neuraminidase Inhibitors, J. Med. Chem., **41**, 2451–2460 (1998)

P. R. Kolatkar et al. Structural Studies of Two Rhinovirus Serotypes Complexed with Fragments of their Cellular Receptor, EMBO J., **18**, 6249–6259 (1999)

Y. Zhang et al., Structural and Virological Studies of the Stages of Virus Replication that are Affected by Antirhinovirus Compounds, J. Virol., **78**, 11061–11069 (2004)

D. N. Garboczi et al., Structure of the Complex between Human T-cell Receptor, Viral Peptide and HLA-A2, Nature, **384**, 134–141 (1996)

C. Douat-Casassus, N. Marchand-Geneste, E. Diez, N. Gervois, F. Jotereau und S. Quideau, Synthetic Anticancer Vaccine Candidates: Rational Design of Antigenic Peptide Mimetics That Activate Tumor-Specific T-Cells, J. Med. Chem., **50**, 1598–1609 (2007)

C. Douat-Casassus, O. Borbulevych, M. Tarbe, N. Gervois, F. Jotereau, B. M. Baker, S. Quideau, Crystal Structures of HLA-A*0201 Complexed with Melan-A/MART-126(27L)-35 Peptidomimetics Reveal Conformational Heterogeneity and Highlight Degeneracy of T Cell Recognition, J. Med. Chem., **53**, 7061–7066 (2010)

D. A. Ostrova et al., Drug hypersensitivity caused by alteration of the MHC-presented self-peptide repertoire, Proc. Natl. Acad. Sci. U.S.A., **109**, 9959–9964 (2012)

Types of Influenza Viruses: https://www.cdc.gov/flu/about/?CDC_AAref_Val=https://www.cdc.gov/flu/about/viruses/ (Last accessed Nov. 25, 2024)

31

Biologicals: Peptides, Proteins, Nucleotides, and Macrolides as Drugs

Contents

G. Klebe, *Drug Design*, https://doi.org/10.1007/978-3-662-68998-1_32

The importance of peptides, proteins, sugars, and nucleotides for functional processes in our bodies has been discussed in many chapters of this book. Exogenous small molecules can be used to try to regulate or interfere with the processes in which these endogenous substances are involved. On the other hand, the question can be raised as to whether the administration of the endogenous biomolecules themselves may be a promising therapeutic concept in the case of some diseases. This is especially true for diseases in which a particular endogenous substance is insufficiently produced by the organism or is produced but is not functional, e.g., due to an amino acid mutation. The perspective of selectively producing polypeptides and proteins with specific properties in sufficient quantities was only opened up by genetic engineering methods (Chap. 12).

As part of a strategy to use endogenous proteins and peptides as drugs, it may be useful to slightly modify the native substances to give them additional properties, such as a longer half-life, better stability, or higher bioavailability. A serious problem that often arises is that peptides and proteins have far too poor stability and bioavailability for oral administration. Nevertheless, there are many promising areas of application, such as the treatment of digestive disorders with the administration of lipases. The issue of bioavailability is also different for skin diseases than for oral and systemic administration. However, the skin also has a protective enzymatic barrier that sensitive biomolecules cannot easily cross. For hospital use, the treating physician can easily choose intravenous administration where this issue is less critical. This problem will be discussed in more detail using the example of insulin, the daily exogenous administration of which is essential for patients with diabetes.

Another pharmaceutical concept for the administration of exogenously administered biomolecules exploits the principle of the body's own immune defense. The body uses macromolecular structures, the antibodies, to recognize and specifically deactivate pathogenic substances. A drug therapy can copy this principle to fight pathogens or malignant cells using the same concept. The antibody proteins of the humoral defense system are not orally bioavailable due to their size and must be administered intravenously. They cannot cross the cell membrane except by special endocytosis. Their use is, therefore, mostly limited to mechanisms that influence biological processes from outside the cells.

Recognition of an oligonucleotide or segment of DNA or RNA is critical for many biological processes. Transcription factors (Sect. 28.2) are involved, as are enzymes that translate DNA into RNA or RNA into DNA. DNA is a huge molecule that can only be stored in the cell if it is efficiently packaged. In Sect. 12.14, we learned that DNA can be wrapped around histone proteins. In bacterial cells, DNA is brought into a compact form by overspiraling, which is the addition of extra turns in the DNA molecule. This requires the breaking of a strand, which is catalyzed by the enzyme topoisomerase with the consumption of ATP. The function of such proteins, which require precise recognition of DNA or RNA, can be blocked by drugs that act as molecular wedges in the complex structure.

Other drug concepts have been developed in the area of nucleotides that interfere with the translation of RNA or the transcription of DNA. Oligonucleotide or nucleoside analogues have been developed for this purpose. The goal of transcription into mRNA is to ultimately translate it into the amino acid sequence of a protein in the ribosome. Microorganisms have developed diverse strategies to outcompete rivals, which are often other microorganisms or parasites. They have a multienzyme complex that allows them to build complex, often macrocyclic compounds according to combinatorial principles. They achieve their cyclic construction in the form of larger, multimembered rings through lactone formation. Many of these compounds, called macrolides, block the ribosomes of hostile organisms. Other members of this family inhibit cell-cycle processes. Again, these natural products, sometimes with minor chemical modifications, can be used as biologicals in therapy. A few prime examples of such biologicals are discussed in this chapter.

32.1 Gene-Technological Production of Proteins

Endogenous proteins have long been used in substitution therapy. In the past, animal pancreatic material was used as an **insulin source** for the therapy of diabetes mellitus; this insulin differed from human insulin by one amino acid (porcine insulin) or three amino acids (bovine insulin). Although these insulins are suitable for therapy and techniques exist to replace the structurally different amino acid in porcine insulin with that in human insulin, all the slaughterhouses in the world would not be sufficient to supply all diabetics with the insulin they need. Factor VIII deficiency in people with hemophilia used to be compensated by blood transfusions. Today, only recombinant proteins are used, because the risk of viral contamination is too high with products derived from human blood. It was often recognized too late that the factor VIII batches were contaminated with hepatitis viruses and human immunodeficiency virus (HIV), the virus that causes acquired immunodeficiency syndrome (AIDS). For this reason, efforts were made very early on to produce human proteins using gene technology. The first protein produced in this way was human insulin from the bacterium *Escherichia coli*, which was introduced into therapy by Eli Lilly in 1982. Although Hoechst in Frankfurt, Germany, had also developed a promising method for industrial production, it was not until 1994 that production could begin. It took that long in Germany for all the objections to the manufacturing license to be resolved. Since then, the following have been manufactured using gene technology: insulin, human growth hormone, a hepatitis B vaccine, tissue plasminogen activator, and many other proteins. ◻ Table 32.1 provides an overview of some of the most important proteins produced by gene technology in the pharmaceutical industry.

Production in bacteria and cell cultures is only one way to produce human proteins. For several years now, efforts have been made to introduce genetic information into animals. In the Netherlands, Herman the bull became famous. The animal, which died in 2004, carried in its genome the information for human lactoferrin, a component of human milk that protects young children from gastrointestinal infections. Patients with weakened immune systems, such as those suffering from AIDS or undergoing chemotherapy, can also benefit from lactoferrin. Herman's female offspring produce milk containing lactoferrin. The company Genzyme Transgenics has bred transgenic sheep. They produce tissue plasminogen activator (tPA) in their milk, which is used to dissolve blood clots, at levels of up to 3 grams per liter of milk. This reduces production costs from several hundred dollars per gram in cell culture to a few dollars. The next step could be to transfer genetic information into agricultural crops.

◻ **Table 32.1** Important proteins manufactured by the pharmaceutical industry using gene technology, their indications, and the manufacturers (at the time the drug was launched)

Drug	Indication	Manufacturer
Insulin	Diabetes	Novo-Nordisk, Eli Lilly, Hoechst, and others
Growth hormone	Growth disorders	Pharmacia, Novo-Nordisk, Eli Lilly, and others
Hepatitis B vaccine	Vaccine	SmithKline Beecham, Merck & Co.
Tissue plasminogen activator	Thrombolysis	Genentech, Boehringer Ingelheim
α-Interferon	Viral hepatitis, leukemia, diverse tumors, AIDS	Sumitomo, Schering-Plough, Roche, and others
Erythropoietin (EPO)	Anemia in renal failure	Amgen, Johnson & Johnson, Chugai, and others
Factor VIII	Hemophilia	Baxter, Cutter/Miles (Bayer)
Granulocyte colony-stimulating factor (G-CSF)	Chemotherapy	Amgen, Chugai, Sankyo, Immunex, and others
Glucocerebrosidase	Gaucher's disease	Genzyme

Imagine a field of sugar beets that can produce insulin or another human protein in large quantities!

In addition to producing proteins to treat diseases where a specific protein needs to be replaced and the drug discovery process mentioned earlier (Chap. 12), gene technology is playing an important role in:

- Antibodies and vaccines,
- Enzymes for medical diagnostics,
- Proteins for biosensors, and
- Proteins for biotechnological processes, for instance, the enzymatic production of optically active intermediates (Chap. 5).

32.2 Tailored Modifications to Insulin

Diabetes is caused by a deficiency of the hormone insulin produced by the pancreas. This polypeptide consists of two chains of 21 (A-chain) and 30 (B-chain) amino acids cross-linked by three disulfide bridges (two inter-chain and one intra-chain linkage). It acts as an agonist at the insulin receptor, which is structurally related to the group of growth hormone receptors (Sect. 29.8). Frederick Banting and Charles Best first isolated insulin in pure form in 1921. Two years later, it was successfully used therapeutically to save a 13-year-old boy from certain death. Since then, insulin isolated from the pancreas of pigs and cattle has been used therapeutically. Since 1982, genetically engineered insulin has been available. Companies that use gene technology to produce insulin have developed concepts for improving the properties of human insulin through targeted modifications. Of particular interest has been longer-acting insulin. Such a depot effect can be achieved, for example, by reducing the solubility of the protein. Due to its predominantly acidic amino acids, the isoelectric point of insulin is at pH = 5.5. Since the solubility of a peptide or protein is lowest at its isoelectric point, shifting this value to the neutral point at pH = 7 should result in a decrease in the solubility of insulin. In practice, this concept can indeed be successfully used. The introduction of an additional arginine in the B-chain (Arg B31) does not change the biological properties, but leads to an increased depot effect. If another arginine is added at B32, this modified insulin is no longer active. It crystallizes after injection and is, thus, not available in sufficient quantities. X-ray structure analysis of the double-mutated insulin shows that the crystal packing is more stable than in normal insulin because of additional contacts. An additional amino acid exchange that weakens these contacts has resulted in an insulin analogue with optimal depot properties. It only needs to be administered once a day. The *C*-terminal end of the B-chain of insulin was also modified at Eli Lilly. No amino acids were added, but the penultimate amino acids Pro 28 and Lys 29 were swapped. This double mutant has the same hypoglycemic properties as native human insulin, but surprisingly acts much faster. The result is an insulin that is rapidly absorbed, short-acting and, therefore, well controlled. This is a huge advantage for patients because the time between injection and meal is significantly shorter.

32.3 Monoclonal Antibodies as Vaccines, Chemotherapeutics, and Receptor Antagonists

The structure and function of our **immune system**, which defends us against foreign substances, termed antigens, was discussed in Sect. 31.7. It is divided into an unspecific and a specific defense. The specific immune response is divided into the **humoral and cell-specific systems**. The role of MHC molecules in complex with the T-cell receptor as a control system to recognize and eliminate diseased and healthy cells was discussed in detail (Sect. 31.7). **Antibodies** play a similar role in the humoral system by recognizing foreign substances, which are then delivered to phagocytic cells such as macrophages for degradation. Analogous to the cell-specific system, the humoral defense is a "learning" system. Once a foreign substance has been recognized, this information is remembered by the immune system in the form of memory cells that can initiate an immediate response when confronted with the same invader, even if the second contact occurs years later. Antibody-producing cells can only produce one specific antibody at a time. There are about 10^{12} different antibodies in the human body. When an antigen appears, only those immune cells are selected for proliferation that produce antibodies to intercept the antigen. Once recognized, invaders are bound to their surface and delivered by the immune system for disposal.

Antibodies share a common structural architecture. Their geometry is roughly comparable to the shape of the letter Y. The branches of the Y are composed of two identical copies, a light and a heavy chain, linked in the center by multiple disulfide bridges (◘ Fig. 32.1). The light chain folds into two domains (V_L and C_L) with a distinct pleated-sheet-like architecture. The heavy chains V_H and C_{H1} adopt a very similar spatial structure. These regions are called **F_{ab} domains** (ab for antigen binding). Then there is the stem of the Y, which also has a pleated sheet-like structure consisting of two chains (C_{H2} and C_{H3}). This is called the **F_c domain** (c for constant). This last domain contains the recognition regions where the antibody is detected by phagocytic cells.

At both ends of the bifurcated Y are loop regions, three of which have been proven to be highly variable in length and sequence among different antibodies. These **hypervariable loops**, or **complementarity-determining regions (CDRs)**, allow antibodies to provide binding sites for very different antigens. Like the fingers of a hand, these variable loops grasp and wrap around the antigen. Eight CDR loops are shown in different colors in ◘ Fig. 32.2. Two antibody structures are shown that bind to two very different antigens, even though they are folded almost identically. One structure captures phosphocholine **32.1**, a small antigen, while the other recognizes and binds the protein lysozyme (129 amino acids) via a large surface patch (◘ Fig. 32.3). Phosphocholine orients its charged quaternary ammonium group to interact with two glutamic acid residues and one asparagine. The terminal phosphate group forms H-bonds to a tyrosine and an arginine residue. The interface between the antibody and lysozyme is approximately 20 × 30 Å in size. Seventeen residues of the antibody are in direct contact with 16 residues of lysozyme. Only a few antigen residues penetrate deeper into the antibody surface and form hydrogen bonds at their ends. Their ability to bind highly efficiently to chemical structures of widely varying size and composition appears to make antibodies ideal for detecting and eliminating disease-causing foreign substances and malignant or degenerate cells. To use them in diagnostics or therapy, they must be purposefully designed against specific antigen surface structures and produced in sufficient quantities.

The development of suitable antibodies can be accomplished in a donor organism. Antibody-producing cells can be isolated from the serum of an immunized mammal and purified. To obtain larger quantities of antibody-producing cells, an attempt can be made to culture the cells. Under these conditions, however, the cells grow for only a few generations and then die. In 1975, Georges Köhler went to the laboratory of César Milstein in Cambridge, England, to improve the production of antibodies in cell culture. There, the idea emerged to hybridize normal antibody-producing cells with easily reproducing tumor cells to create **hybridoma cells**, thus, combining the properties of both cell types. Once again, serendipity helped. Köhler decided to use murine cells. It was later discovered that these cells fuse with tumor cells 100 times better than other cells. The hybridoma cells produced the desired antibodies and continued to divide for unlimited generations. They became immortal antibody-producing cells. This method of producing **monoclonal antibodies** has since become a billion-dollar business. Georges Köhler and César Milstein were awarded the Nobel Prize. But that was all they got. They did not patent their method, nor did they try to start a company to profit from their invention.

Antibodies produced in this way are useful for **medical diagnostics**. They can also be used to treat cancer or septic shock, for example. In general, they are used to fight diseases in which a protein needs to be neutralized in the body. A problem arises when antibodies are isolated from an animal organism. They can act as antigens themselves, triggering an immune response. **Chimeric proteins**, which are combinations of mouse antibodies with parts of human antibodies, can help. Even more

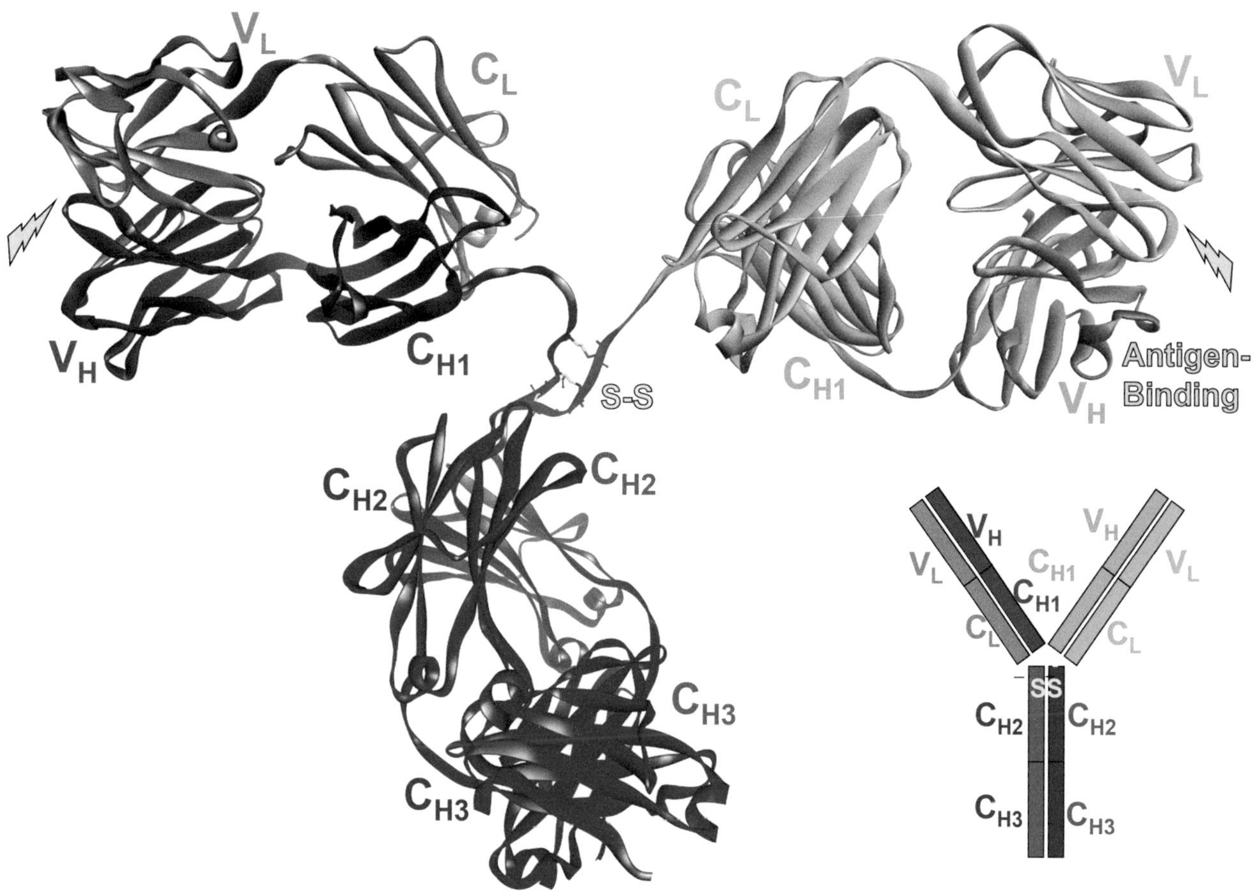

Fig. 32.1 Crystal structure of a complete immunoglobulin G (IgG) antibody. The two F_{ab} regions form the left and right branches (*red*, *green*) of the Y-shaped molecule. They are made up of a light (*lighter color*) and a heavy chain (*darker color*). The antigen-binding site (*light-blue arrow*) is found at the end of both branches. It is formed by eight loop regions. The F_c domain (*blue*) is connected through a hinge region with multiple disulfide bridges. It forms the trunk of the Y-shaped molecule. Two chain strands with a pleated-sheet architecture are positioned against one another here too. The schematic construction of the antibody with the same color codes is shown below on the right. (▸ https://sn.pub/F3VQjA)

elegant is the so-called **humanization**, in which only the variable antigen-binding site of the mouse antibody is coupled with a human antibody. *In vitro* production of completely human antibodies with certain viruses is another method. As with the production of other important proteins, human antibodies can be produced in sheep's milk. Companies such as Genzyme Transgenics have bred transgenic sheep that produce human monoclonal antibodies in their milk.

A major application for antibodies is the prevention and treatment of diseases with **vaccines**. The development of gene technology has made a major contribution to the production of vaccines. For example, **vaccines for hepatitis B** used to be isolated from the blood of chronically infected patients, a very laborious and dangerous technique. However, a vaccine does not require the entire virus. For the recognition by an antibody, it is sufficient to reproduce a typical surface portion of the viral envelope. The genetic information for this region is extracted from the virus and inserted into plasmids (Sect. 12.1). The envelope protein portion is then produced just like any other protein in bacteria or other suitable cells. Gene technologically produced vaccines against AIDS and other viral and bacterial diseases are being intensively investigated. Even diseases caused by parasites, such as malaria, are being fought with this concept. There was great hope for the development of a vaccine against the HI-virus. Unfortunately, after more than 40 years of intensive research, no successful vaccine has yet been developed. The virus exists in many variants, continually masks itself from the body's immune system, and carries sugar structures on its surface. This makes it very difficult for antibodies to recognize and eliminate the virus.

In recent years, antibodies have become increasingly important as drugs. Countless examples are in clinical development, and many have been successfully used in therapy. More and more recombinantly produced antibodies are coming onto the market, usually with tongue-twisting

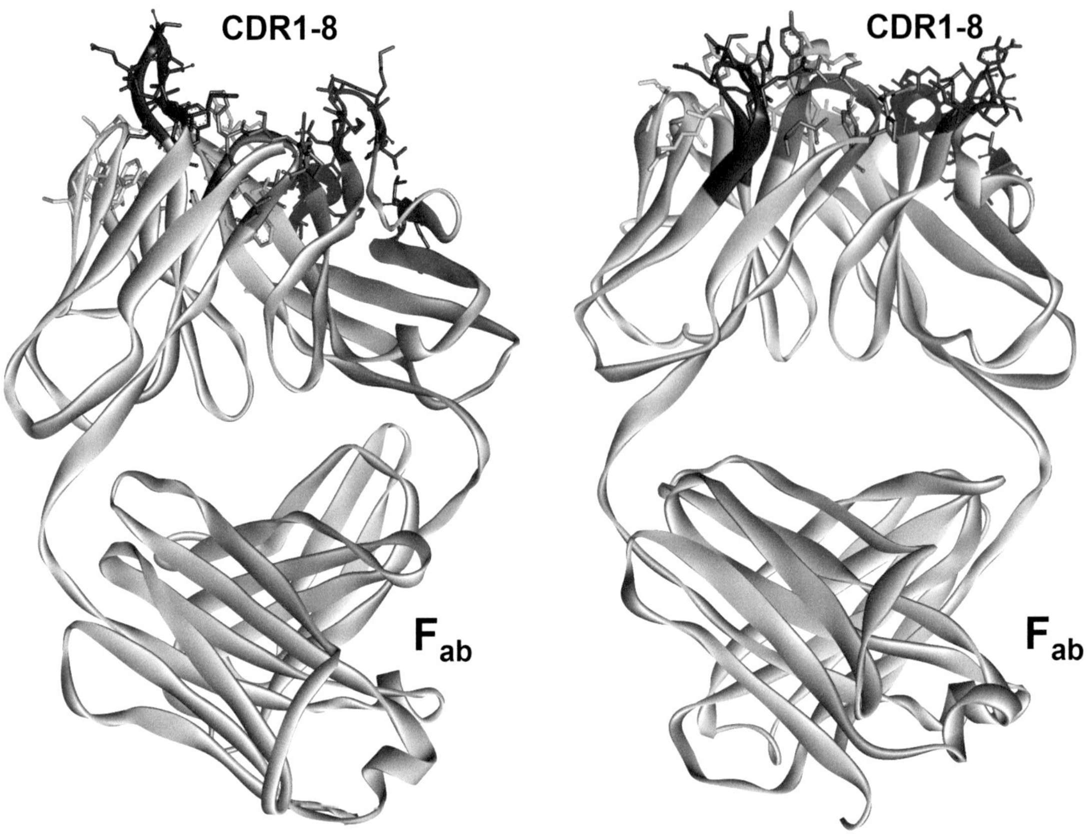

Fig. 32.2 Comparison of the crystal structures of two F_{ab} domains that were released by proteolytic cleavage with papain. The eight variable loop regions that form the antigen-binding site are represented with different colors. The structure on the *left* binds a small molecule as an antigen, whereas the structure on the *right* recognizes the surface of a protein as a foreign substance. (► https://sn.pub/ycLmWE)

names ending in "mab" (for **mo**noclonal **anti**body). As mentioned above, antibodies have the great advantage that they can be raised specifically against virtually any surface structure. They then fish the corresponding antigen out of the organism in a highly selective manner and, once bound, deliver it via the phagocytic cells to the normal degradation pathway of the immune system. Along the way, not only are unwanted invaders neutralized, but cancer cells or unwanted signaling and regulatory proteins can also be removed from the organism. On the other hand, like a drug, antibodies can block a cell-specific receptor as a proteinogenic signaling molecule. They can be exploited as a sort of sniffer hound and, in combination with a sophisticated molecular ferry, can transport a drug molecule to the site of action. Once there, the transported molecule is released in a very high local concentration to unfold its therapeutic effect.

Increasing efforts are being made to specifically **design antibody recognition loops**. Computer-aided design methods are now sufficiently advanced, and the database of structurally determined proteins with F_{ab} regions has grown immensely, to enable the rational design of CDR regions for antibodies. However, there is great competition for the experimental design of antibodies using the immune systems of laboratory animals. The spectacular successes of structural biology on large and fragile membrane proteins would never have been possible without the use of such antibodies to stabilize labile loop regions (Sect. 29.3). Animals of the llama family are preferred for this purpose, as these organisms produce structurally simpler antibodies (so-called nanobodies).

One limitation to the use of antibodies should not go unmentioned. The cell membrane is generally an insurmountable barrier for antibodies. Antibodies are, therefore, limited to recognizing structures on the cell surface or can only be directed against substances dissolved extracellularly. If they are used to influence intracellular processes, this can be achieved, for example, by interacting with surface receptors that are at the beginning of a signaling cascade. Of course, it must be ensured that the antibodies bind specifically to the correct receptors on the desired cells in the correct compartment of our body. Antibodies can only gain access to cells through **receptor-mediated endocytosis**. In this process, noncellu-

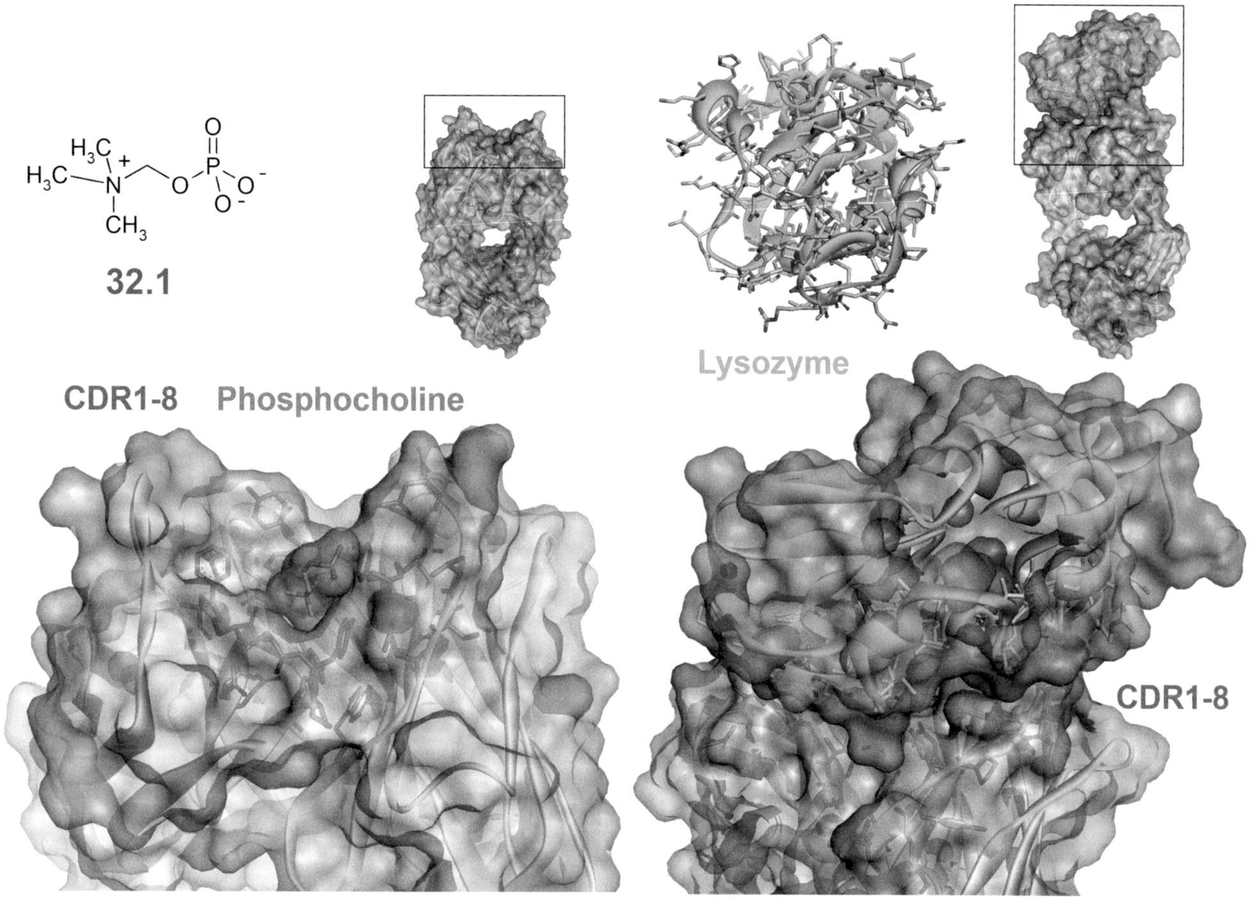

Fig. 32.3 *Left* The contact surfaces with the bound antigens are shown for both of the F_{ab} domains displayed in Fig. 32.2. The small-molecule phosphocholine (*green surface*) is bound in a deep pocket in the antibody. It deeply penetrates the *violet-colored* surface of the antibody. *Right* In the case of the antigen lysozyme, a 20 × 30 Å contact surface is formed and 16 or 17 residues of both binding partners, respectively, are involved in the interaction. A large contact area of the antigen (*green*) is in close proximity with the surface of the antibody (*purple*). (► https://sn.pub/syq1kO)

lar material is taken up into the cell by invagination and constriction of parts of the cell membrane, forming vesicles or vacuoles. We will see that the new antibody–drug conjugates take advantage of this mechanism.

A recently published study points to another potentially promising application of antibodies. In the search for new recombinant antibodies that can be used as monomeric, dimeric, and secretory antibodies to capture the receptor-binding domain of the SARS-CoV-2 spike protein, it has been shown that treatment and prevention of coronavirus infections can be achieved by nasal administration of the antibodies. The approach does not involve immunization of the host, as done through active vaccination, but rather the administration of "ready-to-use" antibodies that can trap the virus in the respiratory tract. The concept is effective in a mouse model, but it remains to be seen whether it can be used in humans.

A few selected examples will illustrate the wide range of applications for antibodies. **Cetuximab** is a genetically engineered antibody. Its antigen-binding site is directed against the epidermal growth factor receptor, to which it then binds and blocks. This receptor is overexpressed on solid tumor cells. When the antibody reaches its site of action, it binds specifically to the extracellular domain of the receptor (Fig. 32.4). There, it completely blocks the binding of the endogenous ligand, the epidermal growth factor. The subsequent steps of the signaling cascade that stimulate cell division are then turned off. In addition to this therapeutic inhibition of signal transduction, the antibody-labeled cells are delivered to phagocytic cells for disposal.

The strategy described for cetuximab is used by many antibodies. In addition to cancer therapy, antibodies are also used to suppress excessive or unwanted immune responses. **Basiliximab** and **daclizumab** target the α-subunit of the IL-2 receptor on T-cells. Only after their activation do T-cells form the α-subunit and the binding affinity for IL-2 increases significantly. After organ transplantation,

T-cells are activated by the presentation of donor antigens specifically directed against the new organ. This initiates processes that ultimately lead to rejection of the transplanted organ. By using specific antibodies against the α-subunit of the IL-2 receptor, it is possible to selectively eliminate from the bloodstream only those T-cells that are directed against the transplanted organ for the immune response.

The **tumor necrosis factor**, TNF-α, is an important indicator in the pathogenesis of infectious and neoplastic diseases, including autoimmune diseases (Sect. 29.8). Elimination or suppression of this factor represents a beneficial therapeutic option for the treatment of these diseases. TNF-α binds to a receptor composed of three identical domains. To inactivate TNF-α, the extracellular, soluble ligand binding site of this receptor is detached and then fused to the F_c portion of an antibody. The result is a soluble hybrid protein that specifically recognizes TNF-α via the original receptor domain and, like an antibody, delivers it to phagocytic cells via the F_c domain. These hybrid molecules have been introduced into therapy by Amgen under the name etanercept (Enbrel®) for the treatment of autoimmune rheumatic diseases and severe forms of psoriasis.

Another interesting concept to improve the efficacy of antibodies has been exploited in cancer therapy. The **CD20 antigen** is overexpressed on the surface of cancer cells. A specific antibody could be developed to find the CD20 cells in the body. This therapy can be combined with local **radiotherapy**. The CD20 antibodies are labeled with a radioactive metal ion. Depending on the half-life of the isotope, the decay of the unstable nuclide leads to a local release of ionizing radiation. Either α- or β-particles are emitted, according to the nuclear reaction. This radiation can exert its tissue-destroying effect in the immediate vicinity of the tumor, causing the tumor to die (◘ Fig. 32.5). Attaching a radioactively labeled antibody to a cancer cell with the CD20 antigen ensures that high concentrations of ionizing radiation are emitted only there. Typically, radioactive ions such as copper, yttrium, rhenium, lutetium, bismuth, or astatine are used. The stable coupling between the metal ion and the antibody is achieved by a suitable chelating ligand. In addition, the chelating ligand backbone is equipped with a reactive group that can be covalently coupled to the surface of the F_c domain of the antibody, for example, via an exposed lysine residue. **Tositumomab** and **ibritumomab** are two antibodies raised against CD20. They carry ^{131}I or ^{90}Y as a radioactive source. This therapeutic approach combines radiation therapy with the body's own immune system.

As described above, antibodies can detect cancer cells with a high degree of accuracy via proteins specifically expressed on their surface. Could this also be used for targeted chemotherapy? The disadvantage of conventional chemotherapy is that the highly effective cytostatic drugs are not tumor-specific. They flood the entire organism and affect all cells. It is true that the rapidly dividing cancer cells are affected much more severely. However, the side effects of killing other rapidly dividing but healthy cells are severe and limit the dosage of the chemotherapy drug. Is there a way to limit chemotherapy to the tumor cells only? To do this, an antibody directed against a marker protein on the surface of a cancer cell must be chemically linked (or "conjugated") to a highly potent anticancer agent. This strategy combines the advantages of the highly specific targeting ability of antibodies with the potent killing effect of a cytotoxin. Once the antibody portion of such an **antibody–drug conjugate (ADC)** has bound to the target antigens on the cancer cell, a signal is triggered to internalize the ADC via receptor-mediated endocytosis. Once inside the

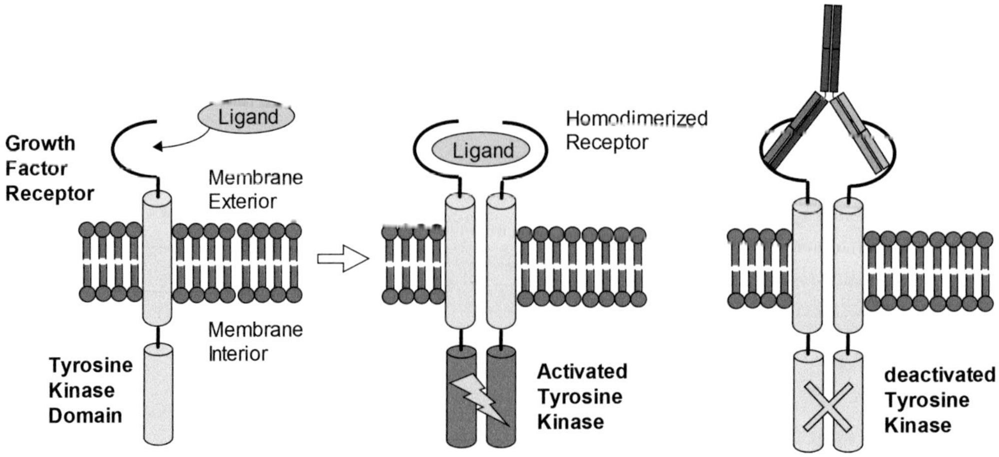

◘ **Fig. 32.4** *Left* The epidermal growth factor receptor is stimulated by binding a macromolecular ligand. *Center* Autophosphorylation initiates the intracellular tyrosine kinase cascade and the signal is transmitted into the cell. *Right* A specific antibody that was raised against the surface structure of the receptor can bind so tightly to the receptor that it blocks the uptake of the natural ligand. The signal cascade is antagonized, and signal transmission does not occur lysosome, a cell organelle with an acidic pH and many digestive enzymes, the conjugate is degraded. As linkers, specific peptide sequences, acid-labile hydrazones or disulfide bridges are used. Once released, the anticancer drug can then undergo a disulfide exchange with glutathione, which is present in high concentrations in cancer cells. The chemotherapeutic agent can subsequently develop its cytotoxic effects and cause the cancer cell to enter into cell death (apoptosis). Since the composition of these antibody–drug conjugates follows a modular principle, many combinations can be realized according to the outlined concept. The development of many new targeted chemotherapeutic approaches can, therefore, be expected in the future. The first ADCs have already been approved and more than 100 candidates are in clinical trials.

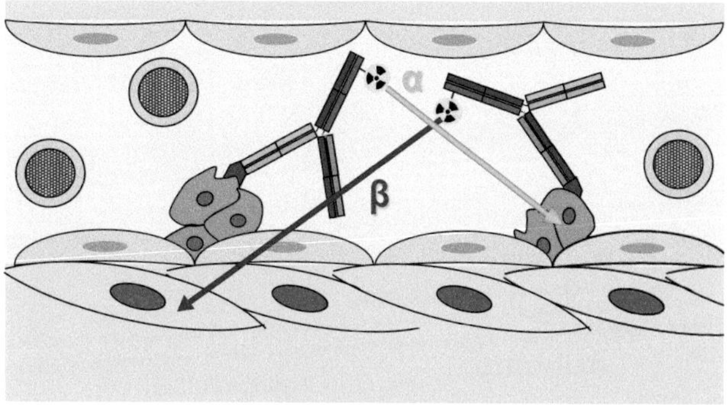

Fig. 32.5 An antibody raised against the surface protein (*red*) from tumor cells (*orange*) finds such cells in the organism. If a metal ion chelator carrying a radioactive isotope is covalently coupled to the antibody, a radiation source can be specifically brought into the direct vicinity of the malignant cancer cell. Ionizing radiation from nuclear decay is released where it exerts its tissue-destroying effects locally. The tumor tissue is treated with radiation therapy directly at the site of the tumor

32.4 Antisense Oligonucleotides and mRNA as Drugs?

It is desirable to suppress the unwanted effects of certain proteins in cancer, excessive immune responses, and septic shock, but also in other diseases such as hypertension, emphysema, or pancreatitis. This can be achieved in many different ways. At the protein level, enzymes can be blocked by inhibitors, and receptors can be blocked by antagonists or inverse agonists. These modes of action have been discussed in detail in previous chapters. It is also possible to intervene at the DNA level by inhibiting protein biosynthesis. Soluble (cytosolic) receptors, such as steroid receptors, act directly on DNA by regulating specific gene segments (Sect. 28.1). Agonists and antagonists of these receptors, thus, indirectly control the biosynthesis of proteins via the translation of the corresponding genes. In this way, it is not necessary to prevent the function of a protein, but rather to **prevent its biosynthesis**.

As discussed in Sect. 12.7, there is another way to block the production of a particular protein: by intervening at the level of the **messenger RNA (mRNA)**. When a protein is expressed, the double-stranded DNA is first transcribed into mRNA. Only one of the two DNA strands, namely the so-called "sense" strand that carries the genetic information, is utilized. The resulting mRNA is single-stranded. After docking to the ribosome (Sect. 32.7), the final step is the translation of the base sequence into the amino acid sequence of the encoded protein. This step can be prevented by the addition of a complementary mRNA so-called "antisense" oligonucleotide strand. When a length of 12–28 bases is reached, the mRNA forms a double strand with the complementary sequence. The hybrid resulting from base pairing is either digested by RNase H (see below) or this sequence segment cannot be read during protein biosynthesis. As a result, the cell does not produce the protein. Nature uses an analogous principle in RNA interference, where it suppresses gene expression by using short RNA sequences of 20–23 bases: RNA silencing (Sect. 12.7).

The mRNA can be complexed with an antisense DNA or RNA segment. This leads to the enzymatic digestion of the mRNA. Another possibility is the preparation of an antisense mRNA segment that competes with native mRNA in the ribosomal protein synthesis. In viral diseases, it is possible to synthesize a complementary oligonucleotide sequence that is targeted directly against individual viral genes.

As simple as the principle of complementary complexation of a nucleic acid sounds, it is difficult to put into practice. **Oligonucleotides** (Fig. 32.6) are very polar and highly negatively charged due to their sugar–phosphate backbone; they **cannot penetrate the cell membrane** without assistance. Their backbone must be chemically modified, for example, by replacing the oxygen atoms on the phosphate groups with sulfur. Although this simple modification results in greater stability against nucleases, it also leads to poorer complexation of the complementary mRNA.

In addition to many other modifications of this type, e.g., to carbonates, carbamates, acetals, imines, or oximes, the sugar building blocks have also been chemically modified. Methylation or methoxyethylation of the 2′-OH group of the ribose ring results in reduced toxicity and increased stability against **RNase H**. This enzyme is responsible for degrading RNA that is required for gene expression and is no longer used after reading. By cleaving the bonds in the sugar–phosphate backbone, the nuclease disassembles the mRNA into its monomeric components. The desired improved stability can also be achieved with so-called locked nucleic acids (LNA) by forming a cyclic ether between the 2′-OH group and C4 of the ribose ring.

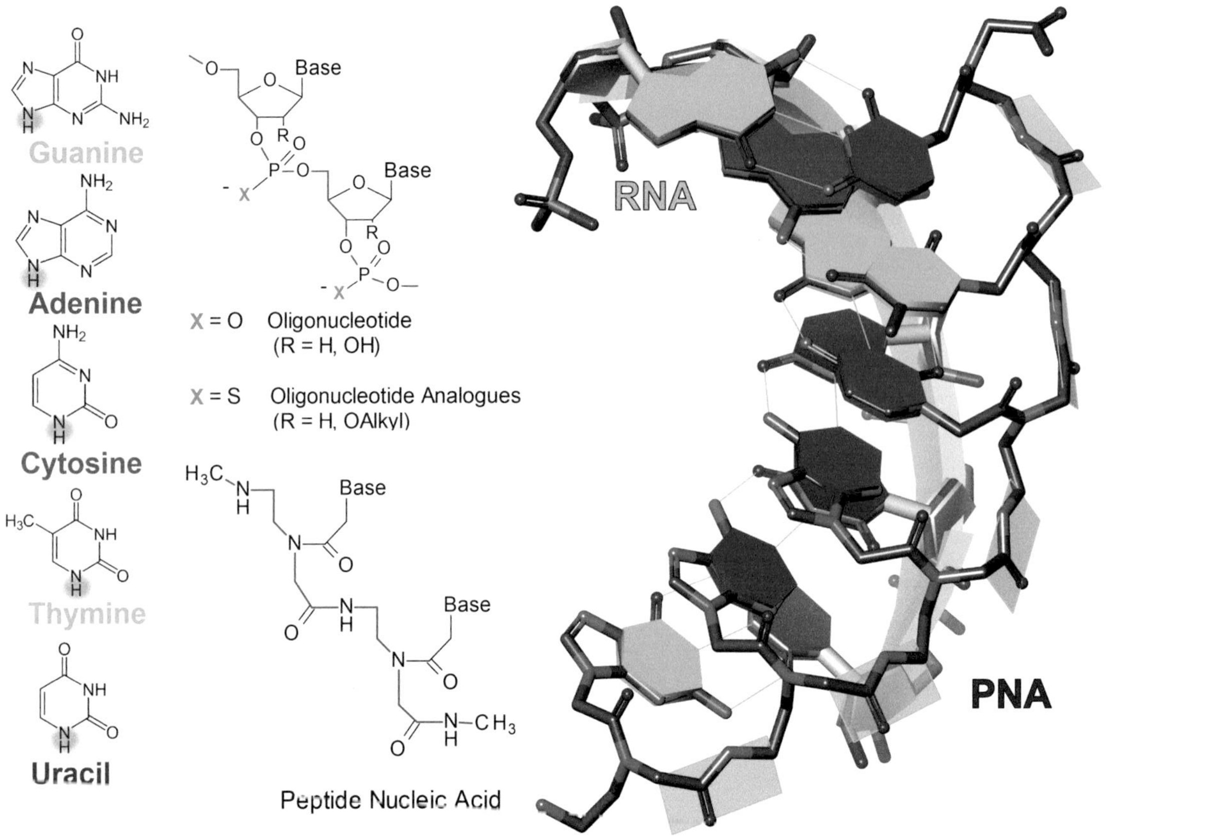

Fig. 32.6 Modifications are performed on the backbone of the oligonucleotide strand to reduce its polarity and increase its metabolic stability. A complete exchange of the ribose phosphate chain is accomplished by using an oligogylcine peptide strand. Such a PNA shows a high degree of geometric analogy to the RNA strand. As the crystal structure (*right*) of an RNA (*light-gray arrow* represents the phosphate sugar strand) and PNA (*dark-gray colored* carbon atoms with amide bonds highlighted by *green planes*) double strand shows, both scaffolds can successfully hybridize with each other. On the *left*, the formulas of the nucleobases are listed, attachment point of the sugar–phosphate building block is marked in *red*. (▶ https://sn.pub/D29uVy)

A more far-reaching exchange is the replacement of the sugar–phosphate groups with an **oligoglycine strand**. The resulting peptide nucleic acids (PNA) can be readily complexed with an oligonucleotide strand. Fig. 32.6 shows the crystal structure of a double-stranded hybrid of a RNA and a PNA strand. The PNA strands are characterized by high biological stability and low toxicity, but due to their poor solubility, there are problems with their uptake into cells. Chimeric structures of LNA/PNA with DNA/RNA oligomers have also been considered as alternatives.

The important criteria that an antisense drug must fulfill are as follows:

- Simple chemical synthesis,
- Adequate *in vivo* stability,
- Good membrane permeability and distribution in the organism,
- Adequate intracellular half-life,
- Strong and sequence-specific binding to the target mRNA,
- Good nuclease stability, and
- No unspecific binding to other biological macromolecules.

Antisense therapy can be administered both locally and systemically. Local administration allows for a high concentration of the antisense nucleotide at the site of action. In 1998, Novartis Ophthalmics introduced fomivirsen (Vitravene®) as the first antisense nucleotide for the treatment of cytomegalovirus retinitis. This disease occurs as an opportunistic infection in immunocompromised AIDS patients. The compound is administered directly into the vitreous humor and prevents the production of viral proteins by binding to viral mRNA. The company discontinued marketing in 2002 for financial reasons.

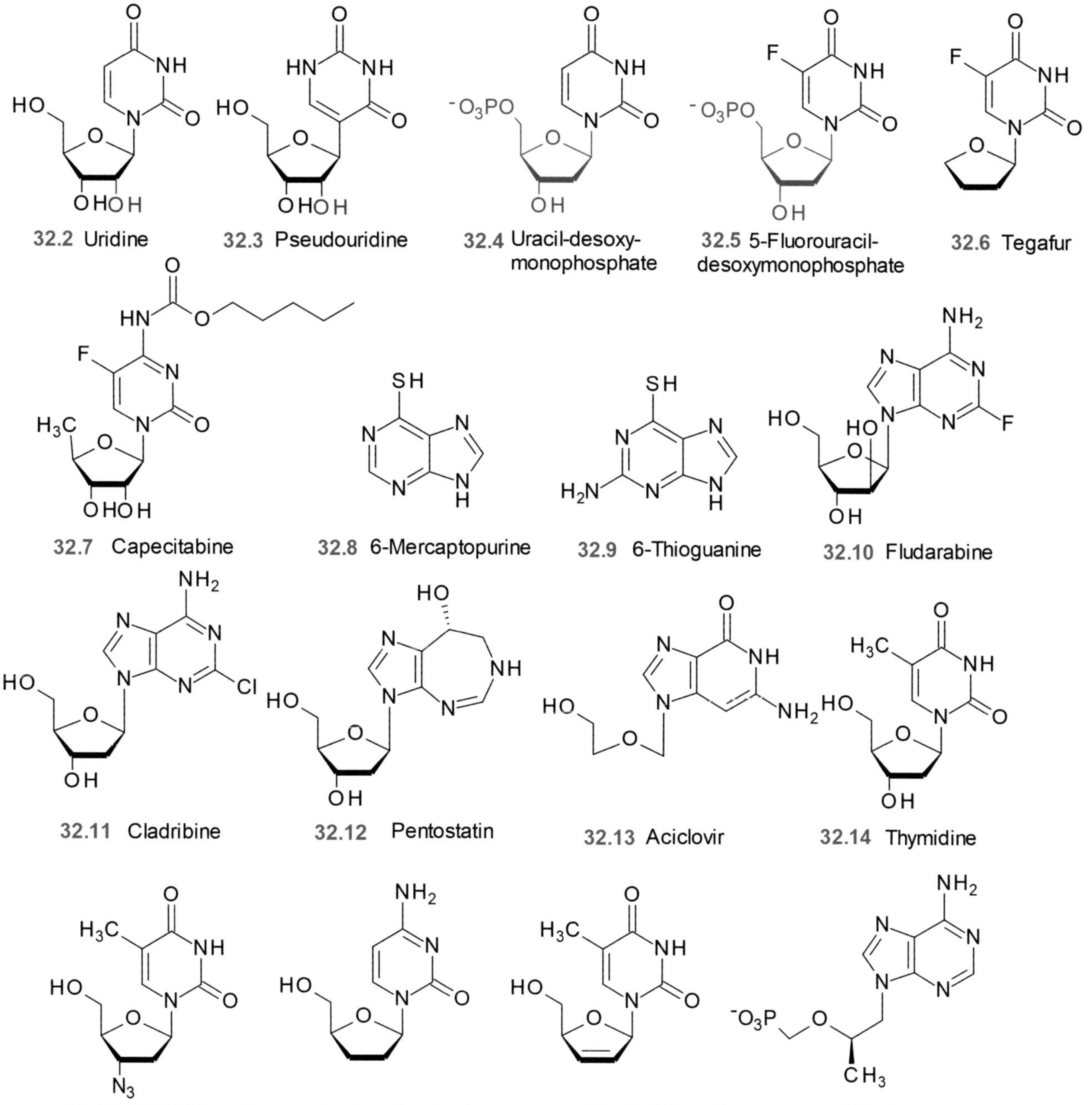

Fig. 32.7 Nucleoside analogue inhibitors of thymidylate synthase, diverse deaminases, and reverse transcriptase

Other local therapies target skin diseases such as psoriasis. Systemic use is usually directed at the treatment of various cancers. Antisense nucleotides have been developed against the mRNA of the BCL-2 protein, which is expressed in many malignant diseases. Other approaches target TGF-β2 (transforming growth factor β2) because this protein is not only responsible for tumor growth and metastasis but also protects tumor cells from attack by the body's immune cells (Sect. 31.7). Antisense nucleotides are also used to combat inflammatory diseases (Crohn's disease, ulcerative colitis, and asthma) and metabolic syndrome. In Sect. 26.8, we discussed the great promise of an antisense strategy to block the expression of the phosphatase PTP-1B.

It is worth noting that **antisense DNA technology** is already well established in plants and is an important tool for elucidating specific metabolic pathways. Rather than using an mRNA nucleotide, antisense DNA is loaded onto small gold particles and "shot" into the cell. Transcription of the antisense DNA leads to the production of antisense mRNA, which then forms a complex with the "sense" mRNA, preventing the biosynthesis of the corresponding protein. The first genetically modified food products to be developed in this way were the long-lasting Flavr-Savr tomatoes.

As described above, mRNA is the direct carrier of genetic information. It, therefore, provides the blueprints for the amino acid sequences of the proteins to be synthesized by the ribosome (Sect. 32.7). This raises the question of whether **mRNA itself** can be used **as a drug** to produce desired proteins. It would be possible to produce all kinds of proteins for therapy. For a long time, mRNA was considered far too unstable a molecule to be successfully used as a drug. It is produced in the body and cleared from noncoding inserts. Shortly after it has migrated to the ribosome and been read, it is rapidly degraded. Synthetic methods have been successfully developed to produce this polymer. But how do you get this fragile and extremely polar molecule into a cell? Experiments in which human cells were treated with mRNA embedded in fat vesicles showed that the cells actually produced the encoded proteins. Similar results were obtained in experiments in which such a preparation was injected into mice. Katalin Karikó, one of the first to realize the dream of using mRNA for therapy, together with Drew Weissman, made a decisive step towards stabilizing mRNA. Human mRNA contains slightly modified building blocks within the mRNA as variants. The nucleoside with uridine **32.2** can be replaced by one with pseudouridine **32.3** (◘ Fig. 32.7). If the mRNA contains only uridine, this would indicate that it comes from a viral organism and must be attacked by the immune system. If pseudouridine is used instead of uridine, the mRNA will no longer be considered foreign. It is also stabilized with the appropriate nucleoside. In 2023, Katalin Karikó and Drew Weissman were jointly awarded the Nobel Prize in Physiology and Medicine for their discoveries concerning nucleoside base modifications, which enabled the development of effective mRNA vaccines, e.g., against COVID-19 (see below).

Several biotech companies have embraced the concept of using mRNA as a blueprint for desired proteins in therapy. Their primary goal was to fight cancer. When the first coronavirus infections were described in China in late 2019, the companies BioNTech, CureVac, and Moderna decided to use their technology platform, which had been developed primarily for cancer, to develop an **mRNA vaccine** against COVID-19. Their strategy was to focus on the spike protein on the surface of the virus. Similar to what we have seen in Sect. 31.5, the primary docking contact of the virus with the cells to be infected is via such a protein. The mRNA vaccine should, therefore, contain the blueprint for this spike protein. Once the vaccine has been injected into a person and has migrated into the cells, the ribosome produces the spike protein or the encoded surface parts of it. The humoral and cellular immune systems then respond by producing antibodies and stimulating T-cells. This builds up protection in case a person is actually infected with the virus. The body's own defense mechanisms can then immediately intervene and fight the virus. To achieve long-term protection, specific antibody-secreting cells must be developed in long-lived plasma cells in the bone marrow. These cells then produce the antibodies against the surface proteins of the invading virus. Apparently, there are differences in how fast these protective antibodies wane in response to different viruses, and this determines the duration of protection that can be achieved. There are two important aspects to the development of RNA vaccines.

First, a stabilized protein construct had to be designed for the spike protein against which our immune system would build a defense. Second, the right lipid envelope for the mRNA had to be found to ensure sufficient stabilization and successful endocytosis of the mRNA. The involved companies were able to draw on their many years of experience in the development of mRNA cancer therapeutics. In less than a year, both BioNTech and Moderna were able to obtain approval for a vaccine that provided excellent protection. The vaccine developed at CureVac did not achieve the same efficacy. This may be due to the exclusive use of uridine in the vaccine. There is no doubt that these vaccine developments have saved countless lives worldwide. It illustrates the potential of using mRNA for therapy. All the more reason to hope that this commercial success will help the companies in the search for cancer vaccines in the future.

Vaccines based on mRNA are not limited to antiviral therapy. As mentioned, the companies originally focused on the development of cancer therapy using the body's own immune system. In contrast to antiviral therapy, cancer vaccines are not used to prevent an infection, but to fight an existing cancer by arming the immune system against the tumor tissue. This requires the extraction of genetic material from a patient's tumor cells. This is then used to develop a personalized mRNA-based vaccine. The vaccine contains the blueprint for the patient's individual antigen. As with COVID-19 therapy, this stimulates the immune system to recognize and attack the tumor. Most progress has been made in the development of mRNA vaccines against malignant melanoma. Promising approaches against glioblastoma have been reported. The therapeutic potential of mRNA vaccines in oncology is enormous, but much research remains to be done in this field.

32.5 Nucleosides and Nucleotides as False Substrates

As monomeric DNA and RNA building blocks, nucleosides play an analogous role in the construction of oligonucleotides and genes as amino acids do in the construction of proteins. As carriers of genetic information and the coding instructions for protein biosynthesis, DNA and RNA are essential biomolecules for a variety of processes in our bodies. Interventions in the synthesis of these biomolecules, especially in processes that are necessary for the production of large quantities of these molecules, can provide important principles for drug therapy. Inhibition of these processes is of particular interest. This is possible with molecules that are very similar to nucleosides, but modified at key positions. As **false substrates**, they are recognized by the enzymes as starting material for DNA and RNA biosynthesis, but lead to the termination of synthesis in subsequent steps. Increased synthesis capacity is particularly necessary in reproducing cancer cells and proliferating viruses. Limiting the rate of synthesis of these molecules can lead to an effective strategy in the fight against cancer and viral infections.

Nucleosides are constructed from a **purine** (adenine and guanine) or a **pyrimidine base** (cytosine and thymine in DNA or cytosine and uracil in RNA) and a **pentose**. If the OH group is missing from the 2-position of the cyclic five-membered-ring sugar, the nucleoside is used as a building block for DNA. The hydroxylated form serves RNA as a monomeric building block. By transforming the exocyclic hydroxymethylene group into a phosphate ester, a nucleo**s**ide becomes a nucleo**t**ide.

The biosynthesis of thymine was introduced in Sect. 27.2. The enzyme thymidylate synthase adds a methyl group to the pyrimidine base uracil to convert it to thymine (◘ Fig. 27.8). When a slightly modified substrate is presented to thymidylate synthase, this molecule is recognized and bound by the enzyme. However, further biosynthesis is stopped. For this reason, such pyrimidine analogues are used as **chemotherapeutic agents** in tumor therapy. The replacement of the hydrogen by fluorine atom in the 5-position of the uracil scaffold **32.4** to form the 5-fluorouracil derivative **32.5** is initially not recognized because of the very similar size of H and F (◘ Fig. 32.7). Thus, the modified base is metabolized via the mono- and diphosphate to the 5-fluoro-2′-deoxyuridine diphosphate. After cleavage of one phosphate group, it is taken up by thymidylate synthase as a false substrate. There it reacts with Cys 146 to form a covalent bond and irreversibly blocks the enzyme. Tegafur **32.6** is a prodrug of 5-fluorouracil that is activated in the liver by CYP 3A4. As an advantage over 5-fluorouracil, tegafur can be administered orally as a chemotherapeutic agent and used as an ambulatory palliative chemotherapy. Capecitabine **32.7** is another prodrug for the treatment of colorectal cancer. It must be activated in several steps in tumor tissue. After cleavage of the carbamate group and exchange of the NH_2 function for a carbonyl group by cytidine deaminase, fluorouracil is released, which can be further biotransformed.

Several purine base analogues have also been described such as 6-mercaptopurine **32.8** or 6-thioguanine **32.9**. After biotransformation and phosphorylation, they competitively inhibit purine biosynthesis. Accordingly, nucleosides such as fludarabine **32.10**, cladribine **32.11**, and pentostatin **32.12** inhibit adenosine deaminase and are used as chemotherapeutics for leukemia.

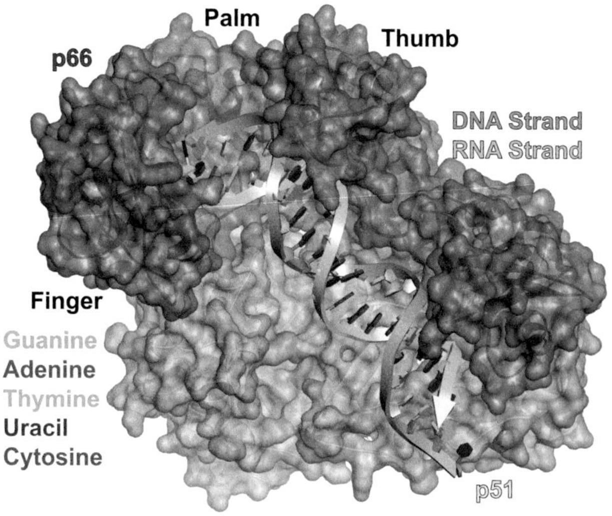

◘ **Fig. 32.8** Crystal structure of the HIV reverse transcriptase. The protein is made up of a p66 (*purple*) and a p51 (*yellow*) subunit. A hybrid double strand of DNA (*pink*) and RNA (*bright-green*) is positioned in the protein structure. The palm area, where the polymerase activity of the transcriptase is carried out, lies between the finger and thumb area. (▶ https://sn.pub/KGpvBc)

Antivirals have a completely different mechanism of action. Because viruses store a program for reproduction and proliferation, but lack their own metabolism, they must exploit the infected host cell for their own purposes. They do this by reprogramming the infected host cell to produce the necessary viral components. This requires the introduction of viral genetic information into the genome of the infected cell. Depending on the type of virus, this task is performed by a **reverse transcriptase (RT)** or a **DNA polymerase**. These enzymes require RNA/DNA nucleosides as starting material for synthesis or translation. If a **false substrate** is offered as a nucleotide building block, it can cause the host cell's synthesis machinery to stop replicating the viral genome. This is a powerful principle for treating viral infections. The recent coronavirus pandemic has reminded us of the threat of suddenly emerging new viruses. To be better prepared for such a threat, it would be desirable to have **broad-spectrum antivirals** available. If there is a chance to develop such substances, they will probably resemble such false substrates of nucleosides.

The group of herpes viruses stores their genes on double-stranded DNA that is synthesized by a viral DNA polymerase. If this viral polymerase is offered a false substrate that is very similar to the natural nucleosides, but unsuitable to continue the nascent chain construction, DNA synthesis will be terminated. It is important that such a drug molecule has sufficient selectivity for the vi-

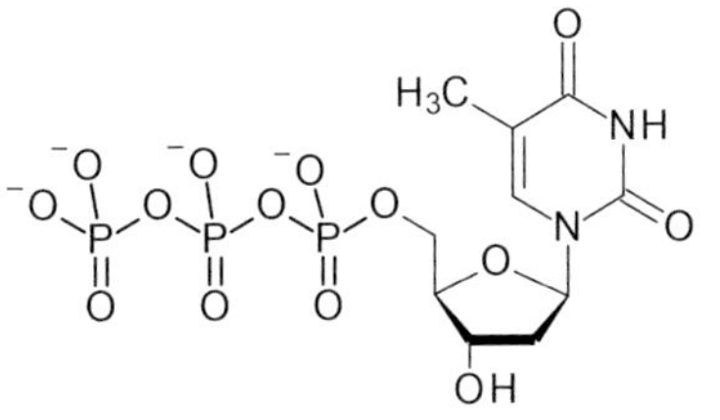

Thymidine-5'-triphosphate (TTP)

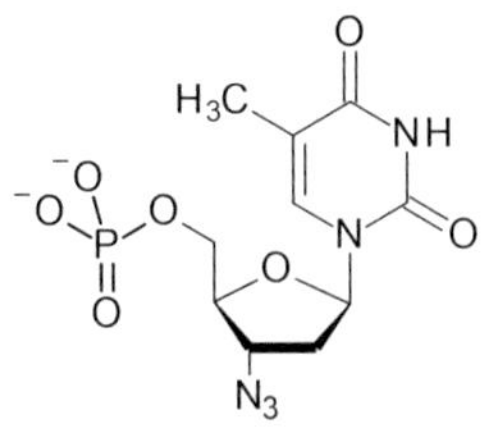

AZT-5'-monophosphate

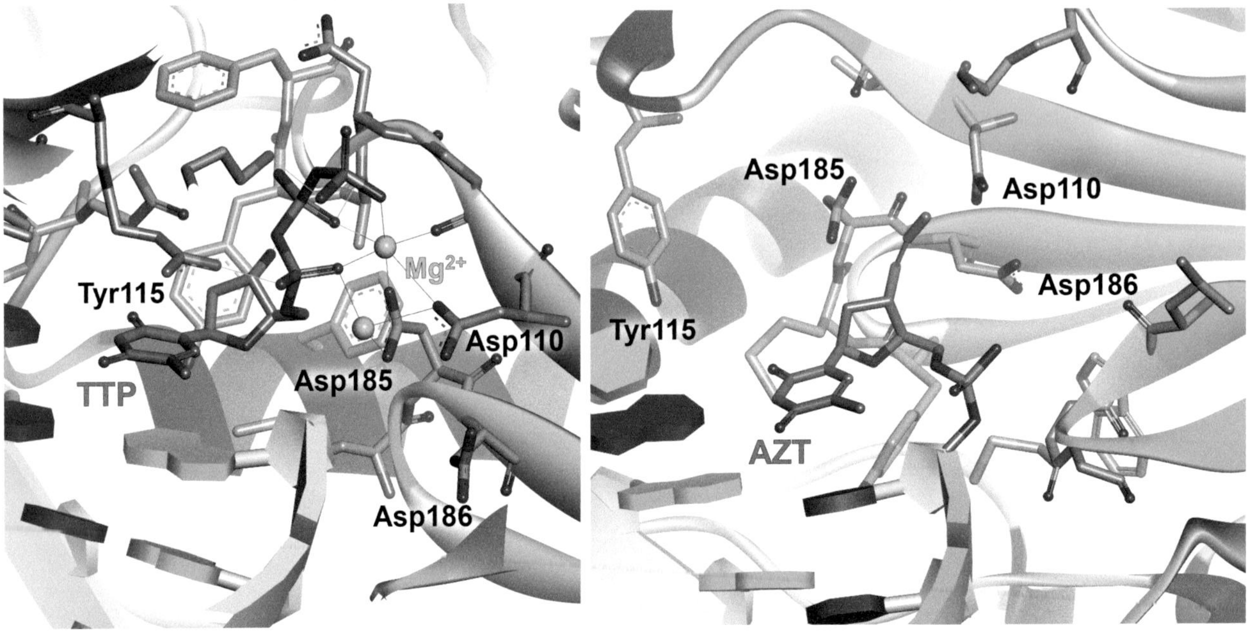

Fig. 32.9 *Left* Crystal structure of reverse transcriptase with a covalently attached DNA strand. Thymidine-5′-triphosphate (TTP) together with two magnesium ions are found at the polymerase site. As the reaction proceeds, this TTP substrate is added to the backbone of the phosphate sugar chain of the newly synthesized DNA. *Right* Binding mode of AZT–monophosphate in the binary complex with reverse transcriptase and the DNA strand. The AZT substrate is added to the nascent DNA strand. In the subsequent step, the chain elongation stops because the azide functional group is unsuitable for the addition of the next phosphate group. (▶ https://sn.pub/GrkfUy)

32

ral polymerase so that the cell's own DNA polymerase is not excessively inhibited at the same time. Looking at the backbone of a DNA strand, the OH groups at the 5′ and 3′ positions are critical for building up the chain structure. Drugs designed to induce **chain termination** during DNA replication are preferentially modified at the pentose ring in the 3′ position. In Sect. 9.5, acyclovir **32.13** was introduced as a prodrug for the treatment of viral infections (Fig. 32.7). Formally, the five-membered ring sugar in this nucleoside is opened up and reduced in structure. As a result, the OH group at the 3′ position is missing. However, this guanoside analogue is still initially phosphorylated by a viral thymidine kinase. Since this kinase is only present in virus-infected cells, the guanosine analogue is exclusively phosphorylated to the 5′-monophosphate in these cells. The further transformation to the triphosphate is carried out by endogenous kinases. Once activated, the triphosphate is incorporated into the nascent DNA strand by the viral DNA polymerase with hydrolysis of the triphosphate. In the subsequent step, however, further attachment of a nucleoside building block is impossible, and the chain is terminated because the necessary 3′-OH group is missing.

In so-called **retroviruses**, a large group of enveloped viruses, the genetic information is stored in the form of a single RNA strand. These viruses are the cause of a few widespread infectious diseases. They infect animals and humans, but most are specialized on a particular host. In humans, it is above all the human immunodeficiency virus that represents a deadly threat.

In order to replicate, retroviruses must transcribe their RNA into DNA and integrate it into the genome of the host cell. For this purpose, they possess the following enzymes: a **reverse transcriptase** (**RT**) and an **integrase**. The principle of reverse transcriptase was first described independently by Howard Temin and David Baltimore in 1970, for which they received the Nobel Prize in 1975. The discovery overturned the previously accepted dogma that information in biology must always flow from DNA to RNA to protein. The RT first synthesizes a hybrid RNA–DNA strand. The enzyme uses its **DNA polymerase function** to do this. However, it reads the synthesis protocol from its own single-stranded RNA. The hybrid must then be converted into pure double-stranded DNA. To do this, the RT uses a second domain that has an RNase H function. Proteins with this activity are used to degrade RNA after it has been read during protein biosynthesis and is no longer needed. The remaining single-stranded DNA is finally completed to a double-stranded DNA by the DNA polymerase activity of the RT. The newly formed DNA with the viral blueprint is then integrated into the chromosome of the host cell by the integrase enzyme.

Since its discovery and structural characterization, HIV reverse transcriptase has been a preferred target enzyme for drug design and will be discussed in more detail in the following. The enzyme is a heterodimer composed of a p66 and a p51 subunit (◘ Fig. 32.8). Both subunits are encoded by the *gag-pol* gene and are cleaved from the primary gene product by HIV protease. The p66 subunit carries the residues for polymerase and RNase activity. The p51 domain is important for the structural architecture of the protein and completes the binding site for double-stranded DNA and the DNA–RNA hybrid strand. The architecture of the p66 subunit can be compared to the shape of a hand. It can be divided into finger, thumb, and palm regions. To perform its function, the RT must undergo significant conformational changes. In particular, the thumb and finger regions must rearrange to grasp the DNA strand and accommodate the next nucleotide triphosphate to be incorporated into the DNA sequence. The crystal structure of HIV-RT together with the RNA–DNA hybrid strand is shown in ◘ Fig. 32.8. By artificially anchoring the DNA strand covalently with the enzyme, it was possible to determine the crystal structure of a tertiary complex of protein, DNA, and the newly incorporated nucleoside triphosphate (◘ Fig. 32.9, *left*). The nucleotide to be incorporated is coordinated by two magnesium ions through its phosphate group and brought into position at the end of the nascent DNA strand. The two magnesium ions that mediate the binding are held in place by two aspartic acid residues in position 110 and 185.

RT can be inhibited by structurally modified nucleoside analogues. Azidothymidine or zidovudine **32.15** (AZT) was approved in 1987 as the first HIV-RT inhibitor (◘ Fig. 32.7). This thymidine analogue has an azide functional group in place of the 3′-hydroxyl group. First, the modified analogue is phosphorylated and incorporated into the DNA strand, just as the natural substrate thymidine **32.14** would be (◘ Fig. 32.9, *right*). The azide function orients itself into the binding region of both aspartic acids and induces a rearrangement of these residues. When the next nucleotide is incorporated into the nascent DNA strand, the chain is terminated. In the absence of the 3′-OH group, the next required backbone phosphoester bond cannot be formed.

In addition to AZT **32.15**, a whole series of other nucleoside analogues **32.16**–**32.18** have been developed, all of which either have a chemically modified 3′-OH group or lack a substituent in the 3′-position (◘ Fig. 32.7). The open-chain inhibitor tenofovir **32.18** has already a terminal phosphate group. To exert its effect, it is also converted to a triphosphate and then incorporated as a false substrate. As with HIV protease, the high mutation rate of the virus poses a challenge to the development of RT drugs. Resistant strains emerge very quickly, rendering inhibition by a potent nucleoside analogue useless. First, residues in the immediate vicinity of the catalytic site mutate, leading to better discrimination between the natural and false substrates. At the site of action, the two substrates compete with each other. Their local concentration and binding affinity determine whether the true or false substrate is incorporated. Here, a small shift in binding affinity relative to the false substrate can lead to significant effects. An additional resistance-breaking mechanism is achieved when the growth-terminated DNA strand with the false substrate is phosphorolytically degraded at an accelerated rate. Mutations to enhance this degradation step have also been observed. It is a constant struggle to modify and adapt drugs to evade emerging resistance mechanisms. This is the only way to demonstrate their therapeutic advantage over the constantly changing protein of the pathogen.

32.6 Molecular Wedges Destroy Protein–Nucleotide Recognition

In many biological processes, the molecular recognition of an oligonucleotide or a piece of DNA or RNA on a protein is of crucial importance. In particular, transcription factors bind to DNA and, thus, initiate its translation (Sect. 28.2). As seen in the last section, there are also enzymes that synthesize DNA and RNA or transcribe RNA and DNA into each other. DNA is a huge molecule that can only be accommodated in a cell if it is efficiently packaged. In Sect. 12.14, we saw that DNA can be coiled onto histone proteins, for example, in an extremely space-saving manner. Bacterial cells use a different principle. There, the DNA is brought into a more compact form by overcoiling, that is, introducing additional turns into the molecule. However, this requires a break in the strand. The enzyme topoisomerase catalyzes this process by consuming ATP. A number of proteins that require **precise recognition of DNA or RNA** for their function can be blocked by drugs that insert themselves into the complex structure **like a molecular wedge**. In the following, three examples of such cases will be presented.

In the first example, we return to the human immunodeficiency virus **reverse transcriptase** from the last section.

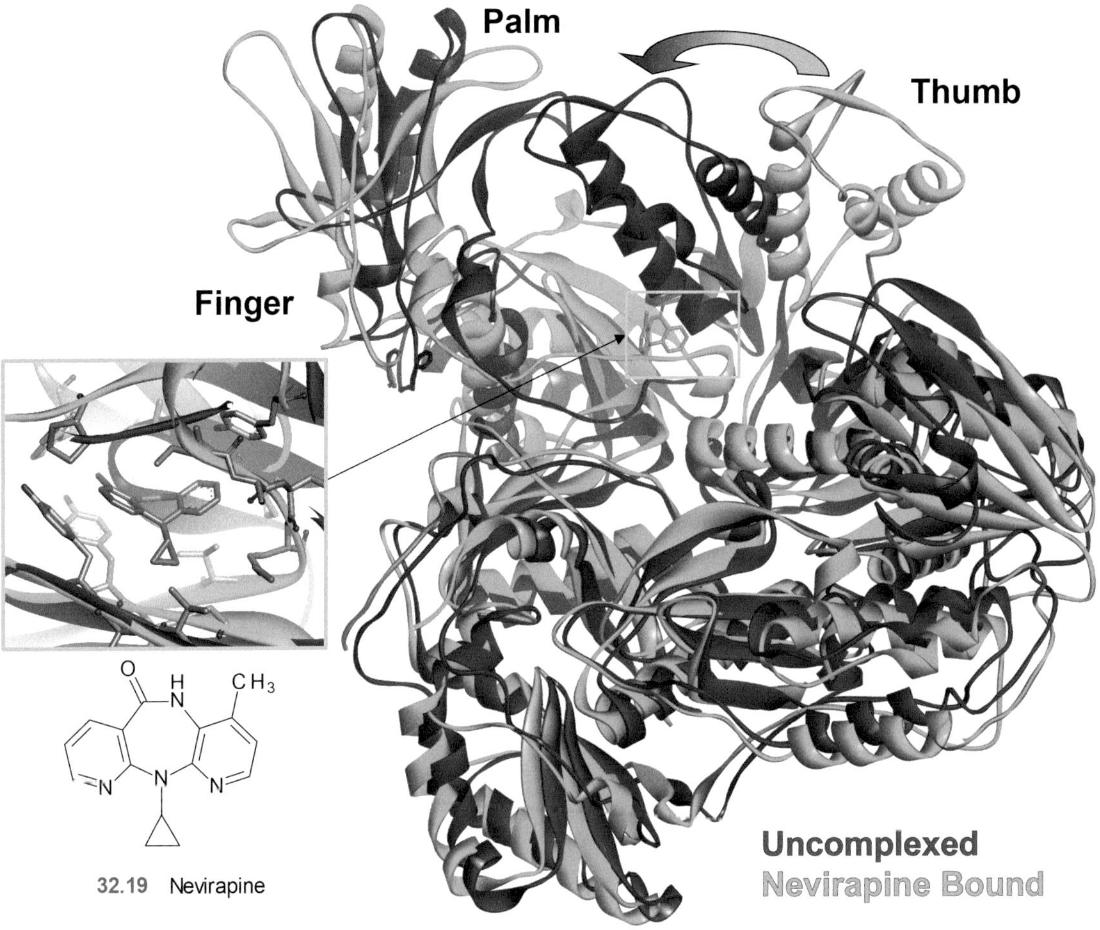

Fig. 32.10 Nevirapine **32.19** was discovered as an allosteric inhibitor of reverse transcriptase in screening. The rigid molecule binds to the protein in a small, hydrophobic pocket adopting a butterfly-like conformation (detailed view, *left*). Like a wedge, the occupancy of this pocket leads to the fixation of the open conformation of the enzyme (*green*). The thumb and finger regions remain far from one another. Upon binding the RNA–DNA hybrid double strand, both of these regions must move towards each other (*arrow*: *green* → *red*) to grasp the double helix. The allosteric inhibitor prevents this movement and does not allow the protein to rearrange into its active conformation. (▸ https://sn.pub/0L3TAR)

In addition to the binding of an invalid substrate to the catalytic center, which leads to the termination of the polymerization of the nucleotide chain, another inhibition mechanism was discovered by chance during a screening campaign. It causes an **allosteric** enzyme **blockade** that is **not competitive** with the natural nucleosides. A hydrophobic pocket in the palm region of the protein can open and accommodate small organic molecules. Like a wedge, it locks the enzyme in a broadly **open conformation** that prevents the protein from accepting the RNA–DNA hybrid strand (Fig. 32.10). Thus, these allosteric inhibitors do not prevent the uptake of nucleoside triphosphate substrates. However, the reaction step that results in the incorporation of the nucleotide into the growing DNA strand is prevented. The small, allosteric binding site is formed by aromatic and hydrophobic residues that almost exclusively come from the p66 subunit. Interestingly, the binding pocket accepts ligands that are chemically very different (Fig. 32.11). The first-discovered inhibitors nevirapine **32.19**, TIBO **32.20**, and loviride **32.21** adopt a butterfly-like geometry in the binding pocket.

Resistance mutations have also been rapidly observed in this allosteric binding site. They change the shape and aromatic character of the binding pocket and rapidly lead to a decrease in the binding affinity of the allosteric inhibitors. At Janssen Pharmaceuticals in Beerse, Belgium, under the direction of Paul Janssen and in close collaboration with the research group of Edward Arnold at Rutgers University in New Jersey, USA, a triazine or pyrimidine moiety was incorporated as a structural element into **32.25**–**32.26**, starting with loviride **32.21** and indolylthioureas (ITU) **32.24** (Fig. 32.11). The new derivatives were systematically analyzed by crystallography. To the great surprise of the scientists, different binding

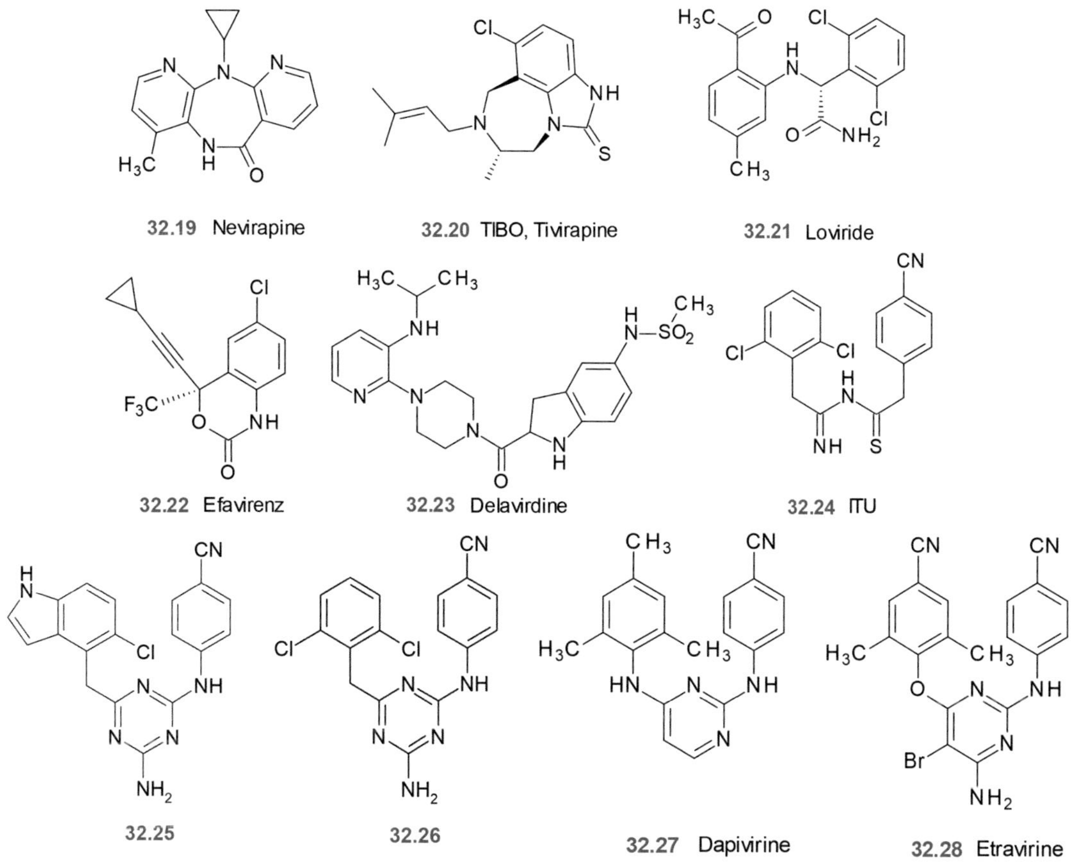

Fig. 32.11 Nonnucleosidic, allosterically acting reverse transcriptase inhibitors

modes were demonstrated for the structurally very similar derivatives **32.25** and **32.26** (Fig. 32.12). In the context of evading resistant mutants, this result is ideal. It is much more difficult for viruses to effectively evolve resistant mutants against compounds that undergo adaptive, chameleon-like binding modes. Consequently, the researchers took advantage of this behavior. Compounds were developed that had the ability to jiggle into alternative binding modes. On the other hand, they had enough conformational degrees of freedom to adapt to small changes in the enzyme (so-called wiggling), for example, when a small amino acid is replaced by a larger one in a mutation. This led to the development of dapivirine **32.27** and etravirine **32.28**, which have an impressively invariable resistance profile compared to their predecessors. Etravirine (Intelence®) was approved as an active ingredient in 2008. Vaginal rings containing dapivirine were launched as an effective way for women in Africa to protect themselves against HIV infection. This example shows that adaptive inhibitors in particular have a clear advantage. This is especially true when developing compounds that have a high tolerance profile against a broad range of mutant variants of a viral protein.

Another class of molecular wedges that disrupt protein–nucleotide recognition are quinolone carboxylic acids, or **quinolones** for short. Quinolones are an important class of antibiotics used to treat infections caused by Gram-negative bacteria. They attack **gyrase**, an enzyme belonging to the group of topoisomerases that catalyzes the **overspiralization of bacterial DNA**. This overspiralization of DNA is caused by the addition of extra turns and is necessary to pack the molecule into the bacterial cell as efficiently as possible. Gyrase must twist the cyclic bacterial chromosome around itself so that the DNA forms a noose around the enzyme. To introduce an additional turn, the enzyme must make a temporary break in the DNA double strand. The topologically lower end of the chopped strand must be moved to the upper end and reconnected. The double-stranded DNA is cleaved in such a way that there is an offset of four base pairs. The 5′-end of the freed phosphate group is temporarily coupled to a tyrosine residue (Tyr 118, Fig. 32.13) by a covalent phosphoester bond. The 3′-end with its OH group remains noncovalently bound in the vicinity of one of the acidic residues of the magnesium binding site formed by Glu 433, Asp 508, and Asp 510.

The first representative of the quinolones was nalidixic acid **32.29**, which was introduced into therapy in 1962 for the treatment of urinary tract infections (Fig. 32.14). The 1-alkyl-4-pyridone-3-carboxylic acid

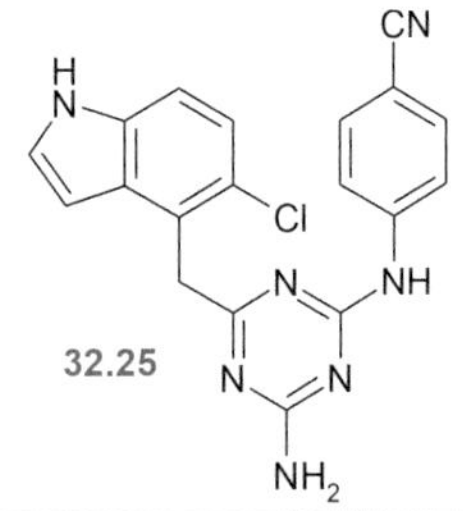

32.25

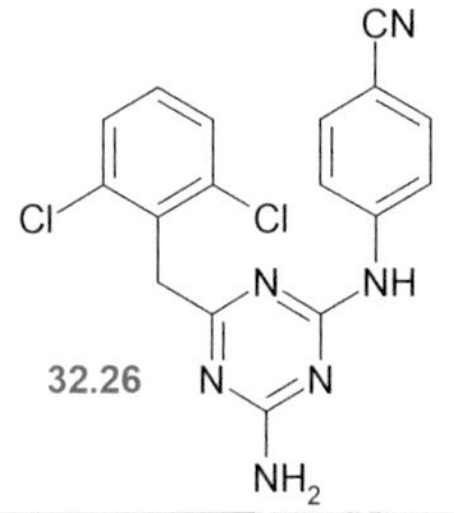

32.26

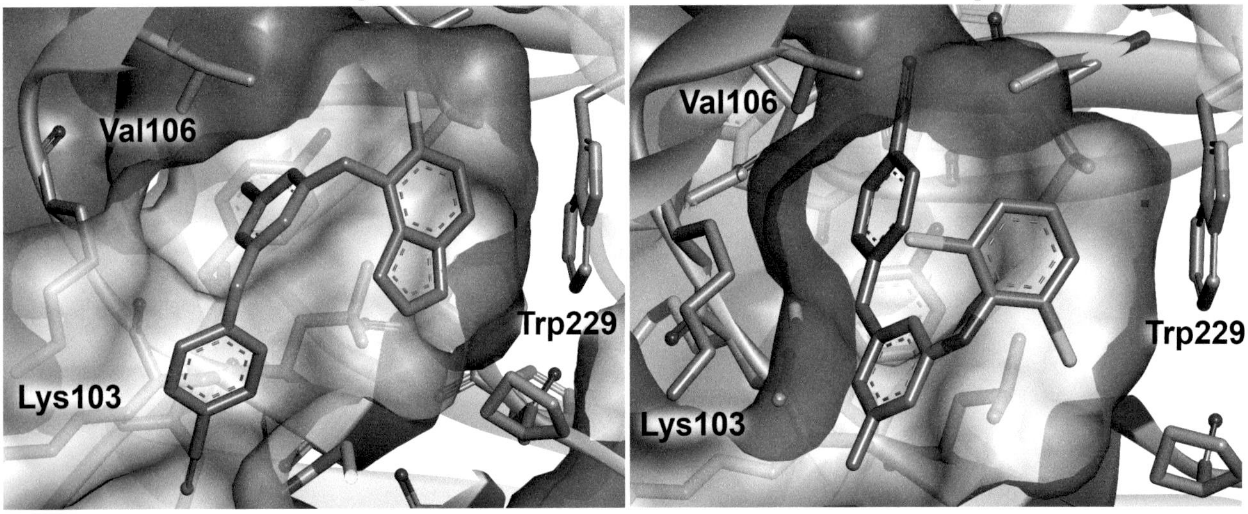

Fig. 32.12 The two triazines **32.25** and **32.26** block the allosteric binding site of HIV reverse transcriptase. Surprisingly, the ligands, which have very similar chemical structures, adopt entirely different binding modes. Clinical candidates were developed from this compound series that have a remarkable resistance-breaking profile. This is attributed to the multiple binding modes of the adaptive ligands able to adjust to a binding pocket that has been altered by mutagenesis. (► https://sn.pub/zHdrDR)

scaffold **32.30** was varied in further drug development, especially around the 1-alkyl group and in the 7-position with basic piperazine-like groups. The addition of a fluorine at the 6-position led to a significant improvement in the activity. Important representatives of this drug class are ciprofloxacin **32.31** and moxifloxacin **32.32**.

The structure of the protein–DNA complex with moxifloxacin was determined in 2009. Two antibiotic molecules intercalate between the two cleaved ends of the DNA (Fig. 32.13). Like a wedge, they prevent the cleaved ends of the double strand from rejoining. Their planar heteroaromatic scaffold is sandwiched on either side by a guanine from one strand and an adenine from the other. The cyclopropyl group is located in a pocket formed by Ser 79 and Asp 83. The development of resistance has been observed as a result of mutations in these residues. In particular, the replacement of Ser 79 by larger residues such as Phe or Tyr led to a reduction in activity due to steric reasons. The basic ring substituent at position 7 is oriented between the base pairs four positions further in the sequence from the cleavage site and is located in a region accessible to solvent. This explains why this group could be widely varied in the context of quinolone development. The 6-fluoro group is oriented away from the protein and DNA; presumably its electron-withdrawing properties are needed to optimally adjust the electron density of the central aromatic moiety for stacking with the neighboring bases. Interestingly, the 3-carboxyl and the 4-keto groups, which are common to all quinolones, are oriented away from the magnesium binding site mentioned above, so that involvement of these groups in chelation of the metal ion seems unlikely.

The development of **resistance to tetracyclines (32.33,** Fig. 32.15) will be considered as a third example of such a molecular wedge that interferes with protein–DNA recognition. Tetracyclines inhibit ribosomal function, which will be introduced in the next section. Interestingly, tetracyclines bind to a **transcription factor**, the **Tet repressor**, which regulates the supply of the transport protein TetA. It is responsible for expelling foreign substances from bacterial cells, including tetracyclines. As long as the Tet repressor is bound to the gene segment that codes for the transport protein, its expression is suppressed. When tetracycline binds to the repressor, it loses its affinity for the regulatory DNA segment. Like a switch, it falls off the DNA and gene expression is initiated. The transport protein is produced and the antibiotic is expelled from the cell. Resistance occurs because

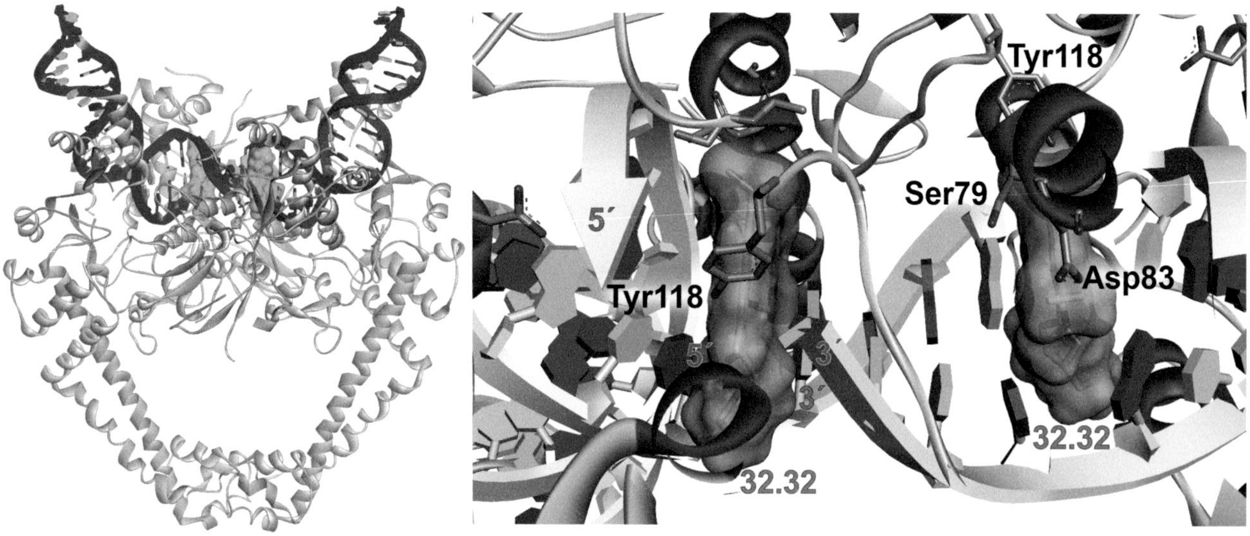

Fig. 32.13 *Left* Crystal structure of the topoisomerase (topo IV from *Streptococcus pneumoniae*, *gray ribbon model*) with two oligomeric DNA sequences (*blue* and *violet*) and two bound moxifloxacin molecules (*green surface*). The protein must wrap the ring-shaped bacterial DNA around itself like a noose for overspiralization. The two DNA segments in the crystal structure emulate this orientation. To achieve an extra turn in the DNA, the double strand must be broken. *Right* This cut occurs with an offset of four base pairs. In doing so, the 5′-end of the free phosphate group is temporarily covalently attached to Tyr 118. The 3′-end remains noncovalently bound in the vicinity of the magnesium ion (near Asp 508). The 1-cyclopropyl group stays in the direction of Ser 79. Exchanging this residue for a Phe or Tyr leads to resistance to this antibiotic. The large basic group at position 7 orients itself towards the outside of the complex into the surrounding solvent. (► https://sn.pub/DL4SgZ)

the concentration of tetracycline required to block the ribosome can no longer be reached in the bacterial cells.

Interestingly, tetracycline, together with a bound magnesium ion, positions itself like a wedge between the helices of the repressor and causes a conformational change (Fig. 32.15). The repressor functions in a similar way to the zinc fingers discussed in Sect. 28.2. As a dimer, the protein reads from the two palindromic DNA sequences, which are two helix–turn–helix motifs arranged nearly symmetrically with a spacing of 36 Å. The wedging by tetracycline causes a widening of the separation of the helix–turn–helix motif to 40 Å. The DNA base sequence can no longer be read correctly. The repressor loses its affinity for the gene segment and the production of the transport protein is initiated. Tetracyclines act practically like a switch and can specifically regulate gene expression. This property is used in molecular biology to selectively turn on gene expression.

32.7 Macrolides: Microbial Warheads as Potential Cytostatics, Antimycotics, Immunosuppressants, or Antibiotics

Biomolecules not only control and regulate the function of organisms, they can also be used as **chemical weapons** in the fight for survival. In particular, microorganisms such as bacteria and fungi produce a variety of unusual substances that they use against their competitors. These enemies, often other bacteria and fungi, should be destroyed to win the ongoing battle for limited resources. Microorganisms also pose a threat to human health. In the days before modern drug research, **infectious diseases** were the leading cause of death (Sect. 1.3). This makes it all the more obvious that the structure and mode of action of these microbial weapons should be studied in detail in order to explore their potential for drug therapy, for example, against bacterial pathogens.

Microorganisms have a unique multienzyme complex, which does not exist in humans, for the synthesis of these complex, often **macrocyclic substances** with peptidic character; their synthesis is independent of the peptide and protein synthesis in the ribosome described below. The size of the compounds produced ranges from a few hundred to a thousand daltons. This violates Lipinski's "Rule of 5", but the macrolides are still well bioavailable. In addition to the 20 proteinogenic amino acids, the multienzyme complex (the so-called **nonribosomal peptide synthesis machinery**) uses many other amino acids and small synthetic building blocks, often with unusual stereochemistry, as starting materials. In addition, peptide construction and ring closure are achieved not only through the formation of amide bonds, but also through the closure of ester bonds. The multienzyme complex for these syntheses is **modular** and composed of several **function-specific domains**. Depending on the product to

Fig. 32.14 Nalidixic acid **32.29** was the first quinolone to be approved; all quinolones have the 1-alkyl-4-pyridone-3-carboxylic acid scaffold **32.30**. Variations were made in the 1-, 6-, and 7-positions. A fluorine in the 6-position proved to be beneficial for activity. Aliphatic, basic heterocycles were substituted in the 7-position. Two important representatives of this drug class are ciprofloxacin **32.31** and moxifloxacin **32.32**

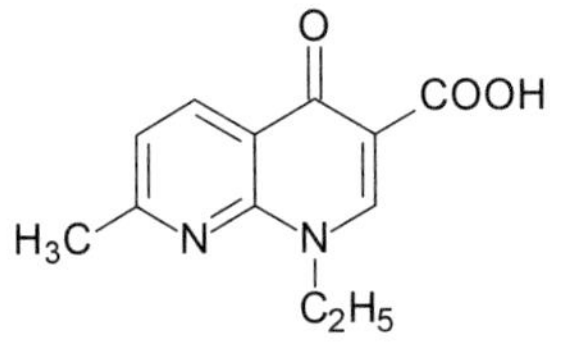

32.29 Nalidixic Acid

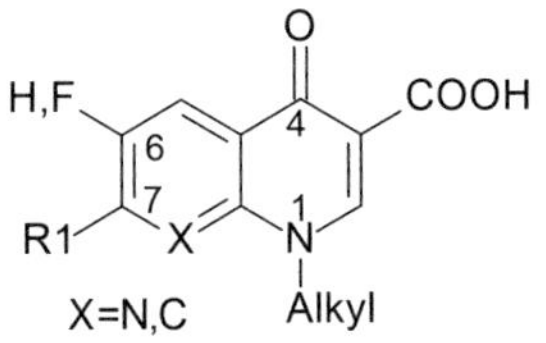

32.30 R1= Aliphatic Heterocycle

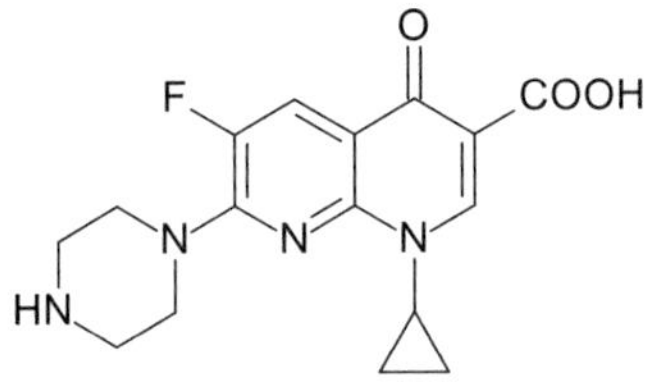

32.31 Ciprofloxacin

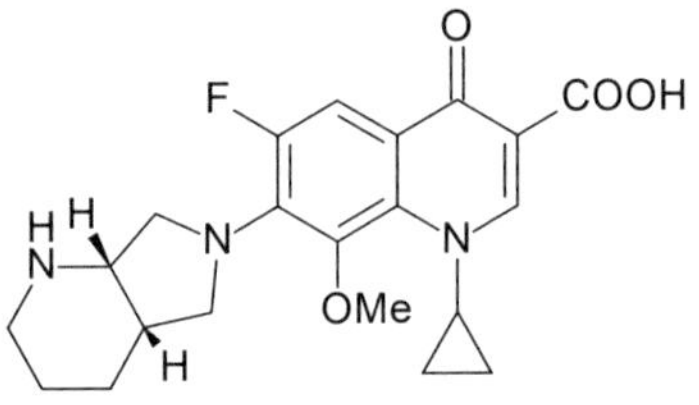

32.32 Moxifloxacin

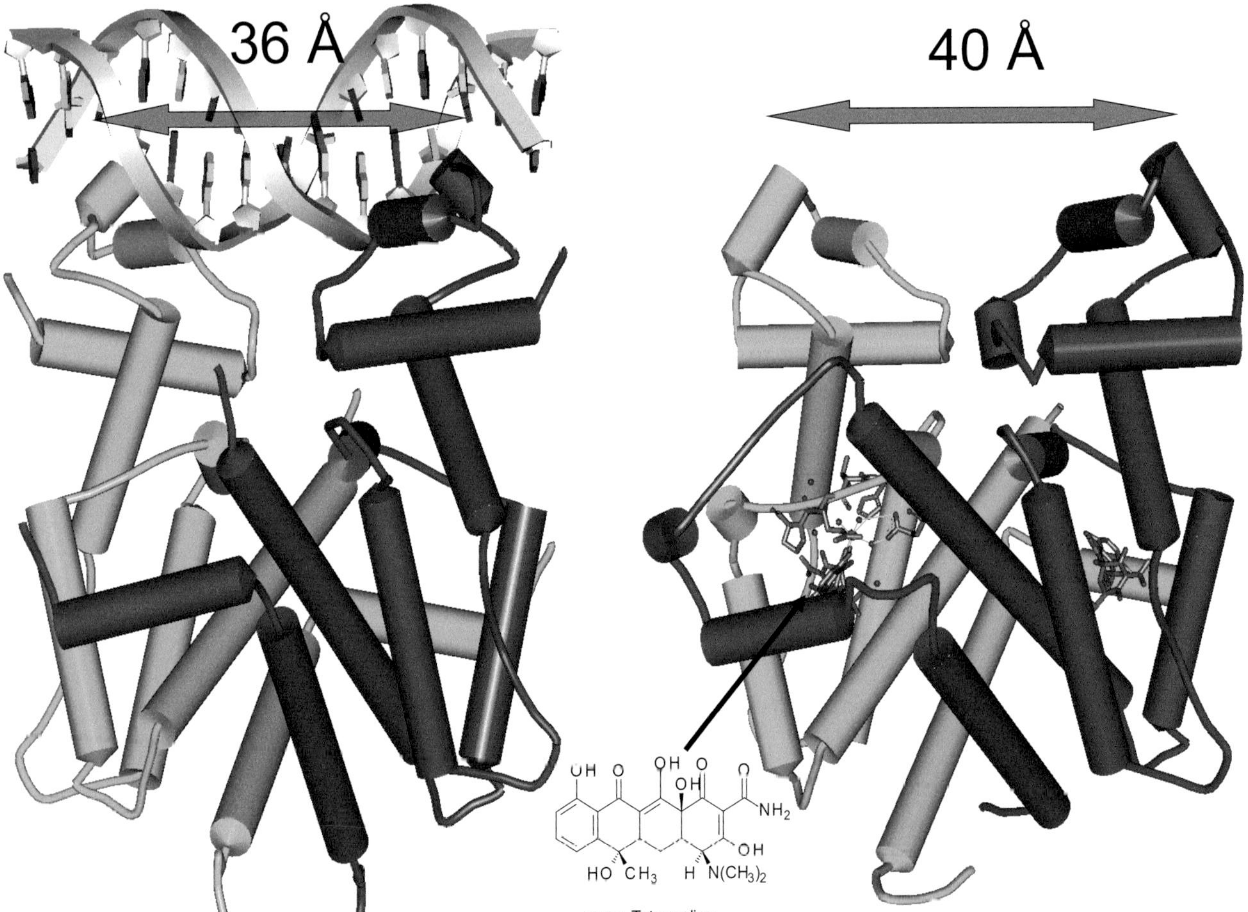

Fig. 32.15 Crystal structures of the Tet repressor with a bound DNA sequence segment (*left*). On the *right*, the tetracycline **32.33** has inserted itself into the repressor. The protein is a dimer composed exclusively of helices (*red* and *green cylinders*). With two helix–turn–helix motifs arranged *C2*-symmetrically to each other, the repressor grabs the sequence segments on the DNA that occur palindromically to each other. When a tetracycline molecule is inserted like a wedge into each of the two monomers of the repressor, a conformational expansion of the helical protein occurs. This increases the relative distance between the two reading motifs to 40 Å. This distance becomes too large to recognize sequence segments in successive turns of the DNA, which has a pitch of 36 Å. As a result the repressor does not bind any longer to the DNA. (► https://sn.pub/Kqdxdr)

be formed, these domains are assembled in the complex with the required multiplicity. A single module consists of domains for recognition, activation, and incorporation of specific substrate components into the desired product. They provide the basic function of extending the nascent peptide. In addition, new synthetase domains are constantly being discovered that allow deviation from a simple linear synthetic sequence. Synthetic products resulting from the use of such multienzyme complexes often exhibit variations in the peptide backbone that allow branching and ultimately **macrocyclization**. Another synthetic route that produces similarly complex and pharmacologically interesting natural products is the **polyketide synthesis pathway**. It does not use amino acids, but is a modification of fatty acid biosynthesis. The C2 units of decarboxylated malonyl-CoA are used as starting materials.

Many of the compounds synthesized in this way are macrocycles of variable ring size. Relatively small 9-membered rings up to 30- or 40-atom rings have been discovered. **Macrolides** with 14- to 16-membered rings are mainly used as **antibiotics** for the treatment of bacterial infections.

However, macrocycles can also intervene in completely different mechanisms, for example influencing the **cell cycle**, the **integrity of cell membranes**, or stimulating the **immune system**. The macrocyclic undecapeptide cyclosporine (Sect. 10.1) has made organ transplantation possible. Its administration prevents the donor organ from being rejected as foreign tissue by the recipient. Cyclosporine acts as an immunosuppressant by inhibiting both the humoral and cellular immune response and suppressing the release of interleukin-2 (IL-2) from T-cells. The lack of IL-2 release prevents the maturation of T-cells into cytotoxic killer cells (Sect. 31.7). After penetration, cyclosporine binds to the cytosolic protein cyclophilin. The resulting binary complex inhibits the calcium-dependent phosphatase activity of the calcineurin–calmodulin complex, which is responsible for dephosphorylating a activating nuclear factor. As a result, this transcription factor is not translocated to the nucleus and IL-2 synthesis is blocked.

Macrolides such as nystatin, natamycin or amphotericin B associate with ergosterol in the cell membrane of fungi. Through this antifungal principle, they affect membrane integrity and make the cell membrane permeable to potassium ions. This can lead to cell death in the corresponding fungus. Rhizopodin, sphinxolide B, kabiramide C, and jaspisamide A interact with actin polymerization. They disrupt the development of the cytoskeleton and have cytostatic effects. Zearalenone was discovered in the group of fungal toxins and shows an effect comparable to that of an estrogen.

The largest group of macrolides acts against **ribosomal function**. In this synthetic machinery, genetic information is translated into the production of new proteins.

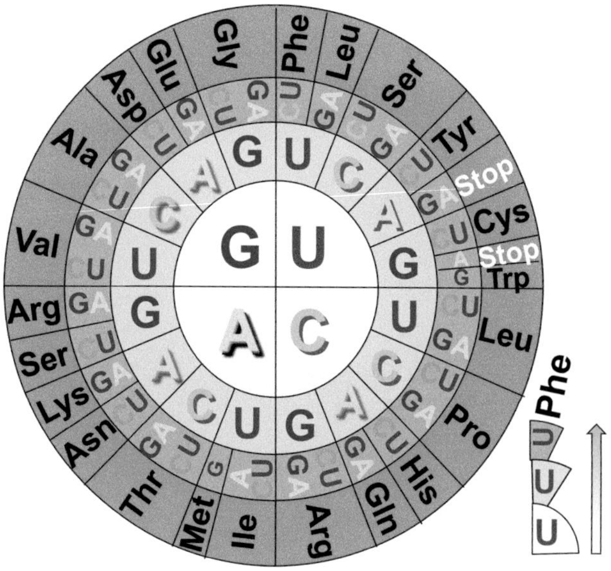

Fig. 32.16 In principle, 64 triplets can be formed from four bases: guanine (G), uracil (U), adenine (A), and cytosine (C). In the diagram, these are oriented from the inside to the outside. To decode an amino acid, start with the central quadrant, for example U, and then take a base from the first ring (*light gray*), for example, a U again. The third base is taken from the second, *dark-gray* ring. If it is also a U, the code will be UUU for phenylalanine. Three triplets are interpreted as a stop codon (UAG, UAA, UGA). Since there are 20 proteinogenic amino acids, up to six codons can encode a single amino acid (e.g., Arg or Leu). Tryptophan (UGG) and methionine (AUG) are encoded by a single triplet. A few enzymes, such as glutathione peroxidase, have a selenocysteine in the active site. This 21st proteinogenic amino acid is encoded by the UGA codon in certain contexts; UGA usually serves as a stop codon

Because of its central importance for the maintenance of all life, the ribosome has been the focus of intensive research for many years. This large and complex natural complex was discovered in the 1950s, and more than 25 years ago, the group of Ada Yonath at the Weizmann Institute in Israel began to crystallize and determine its structure. In small steps, more and more information about the spatial structure of this ribonucleoprotein complex was deciphered from the diffraction data. The real breakthrough came in 2000, when the crystal structure of the large subunit was determined at a resolution of 2.4 Å by Tom Steitz's group at Yale University in New Haven, USA. In the same year, Ada Yonath's group and, independently, Venkatraman Ramakrishnan's group at the MRC in Cambridge, England, solved the small subunit. The three researchers were awarded the 2009 Nobel Prize in Chemistry for this magnificent *tour de force*. The first high-resolution structure analysis was performed on the ribosome from the very robust bacteria *Thermus thermophilus* and *Haloarcula marismortui*. More recently, the ribosome from the eubacterium *Deinococcus radiodurans* has proven to be a versatile and easy-to-crystallized workhorse. Many structure determinations of complexes with antibiotic macrolides have been determined with

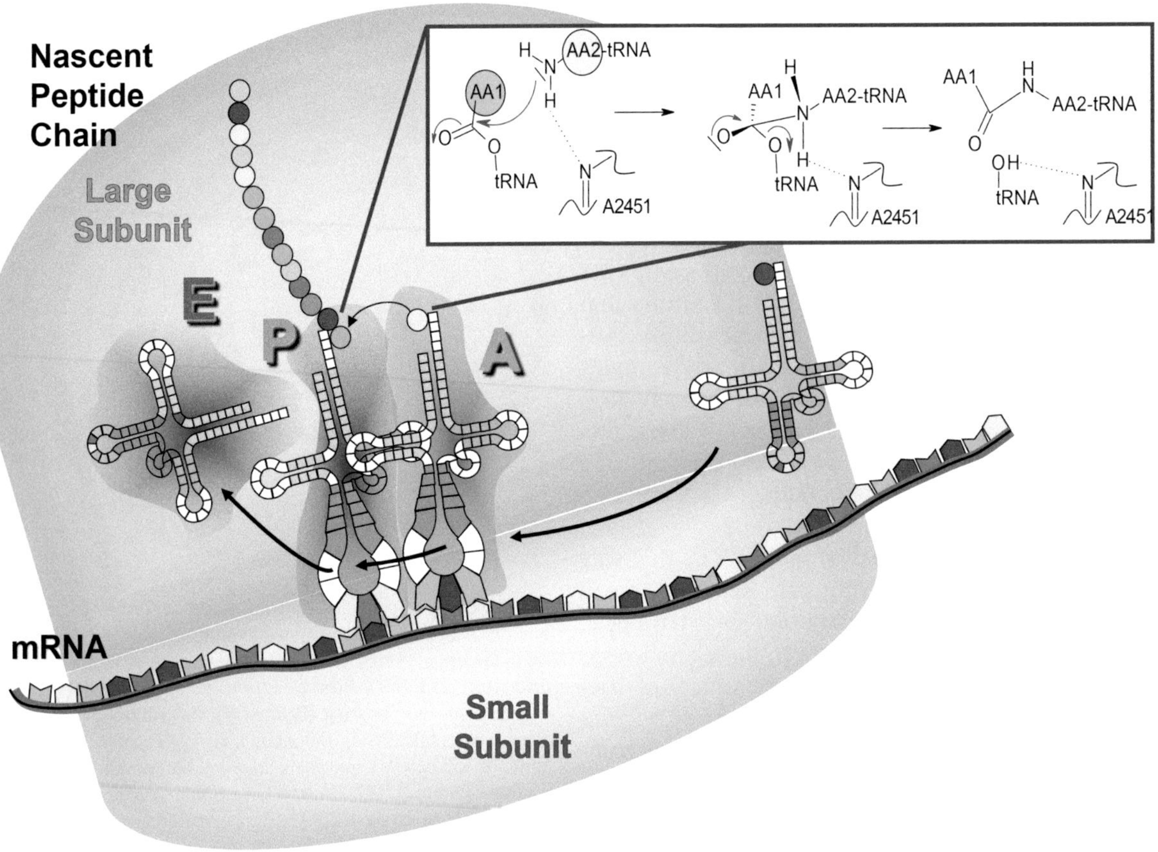

Fig. 32.17 The mRNA carries the genetic translation for the synthesis of new proteins in the ribosome on a single strand. The tRNAs are loaded with one of the 20 proteinogenic amino acids according to the codon in the anticodon loop. The ribosome has three tRNA binding sites, the A-, P-, and E-sites. The A-site accepts the aminoacylated RNA, the P-site binds the peptidyl tRNA, and the tRNA leaves the ribosome through the E-site. The energy required for the formation of the polypeptide chain is provided by the coupled GTPase activity. To be properly recognized, the tRNA in the A- or P-site must have a complementary base triplet in its anticodon loop. In the peptidyl transferase center of the ribosome, a new amide bond is formed between the amino acid in the P- and A-site. The amino group of the AA2 amino acid of the aminoacylated tRNA performs the nucleophilic attack on the carbonyl group of the AA1 amino acids of the peptidyl tRNA. A trigonal geometry is formed at the carbonyl carbon atom via an intermediate tetrahedral transition state. The surrounding nucleosides, e.g., A2451, are responsible for polarization and stabilization of the temporarily charged transition state

this system (see below). It shows a high sequence homology to the ribosomes of important pathogenic organisms.

The surprise was great after the first high-resolution structure determination. The ribosome is indeed a molecular complex of proteins and nucleic acids, but because of its catalytic function it should not be called an "enzyme" but rather a "**ribozyme.**" Proteins do not catalyze the crucial steps of synthesis. It is the RNA molecules that perform this function. This fact proves that the ribosome is evolutionarily one of the oldest catalysts in living Nature. Despite its stately size of over two million daltons, it is highly conserved and occurs with great similarity in archaebacteria, prokaryotes, and highly developed eukaryotes. Organisms from these three domains of life share a common origin that dates back more than 3.5 billion years! Because of its central importance for protein production, it is not surprising that the ribosome has become a prominent target structure for the chemical weapons of microorganisms. They bind to a few vulnerable sites on the ribosome, thereby, disabling its function. These binding sites are located near the mechanistically important active sites.

To understand the importance of these sites, we need to look at how the ribosome works. The blueprints for our proteins are stored as the genome on DNA (Sect. 12.3). In order to translate this information into proteins, eukaryotes must first transcribe it onto a strand of RNA, from which the noncoding regions are then cut out. The resulting transcript, called mRNA, travels out of the nucleus to be **translated into protein in the ribosome**. In prokaryotes, protein synthesis can begin immediately. At first glance, the genetic code on DNA is a pure sequence of four nucleic acids: guanine, adenine, thymine, and cytosine. Uracil takes the place of thymine in RNA. Each of these three bases codes for an amino acid, and several **codons** can code for the same amino acid (Fig. 32.16).

The translation of **base triplets** on the mRNA and the synthesis of a protein take place in the ribosome

(◘ Fig. 32.17). When a particular triplet reaches the catalytic site of the ribosome, its **tRNA** counterpart is recruited. The tRNA carries the so-called **anticodon loop** in an exposed loop that binds complementarily to the nucleobases of the mRNA triplet (◘ Fig. 32.18). Each triplet in the anticodon loop uniquely encodes one of the 20 proteinogenic amino acids that are loaded onto the 3′ end of the tRNA. Each new protein starts with a methionine. The starting point on the mRNA is, therefore, the base sequence AUG. As a result, the P-site of the ribosome accommodates a tRNA with the pattern UAC in its anticodon loop. This tRNA carries the amino acid methionine. The next triplet code on the mRNA is CGC in our example. This leads to a tRNA with the sequence GCG to be bound at the A-site, next to the P-site. Such a tRNA is loaded with the amino acid arginine. The two amino acids at the end of the loaded tRNAs align in the catalytic **peptidyl transferase center** (◘ Fig. 32.19). There, peptide bond formation between the two amino acids is catalyzed, and the first amide bond in the backbone of the new protein is formed. The individual steps of the reaction mechanism are reminiscent of the reaction sequence in proteases. However, it occurs in the reverse order, and substrate recognition in the catalytic center is exclusively performed by the nucleic acids (◘ Fig. 32.17). After the methionine transfer, the discharged tRNA leaves the P-site through the neighboring E-site. The tRNA from the A-site migrates to the neighboring P-site. This corresponds to a progression of the sequential information on the mRNA. The empty A-site is now loaded with a novel tRNA whose base triplet in the anticodon loop is complementary to the next triplet sequence of the mRNA. The new protein grows according to this synthesis sequence and exits the ribosome through the ribosomal tunnel (◘ Fig. 32.19). If the ribosome encounters a triplet sequence corresponding to a stop codon, protein synthesis will be terminated.

Biosynthesis is performed with breathtaking speed. No more than 50 ms are required for one synthesis cycle. As mentioned above, the ribosome is a mixed complex consisting of two thirds RNA and one third protein and it is organized in two subunits. The **small subunit (30S** in prokaryotes) is responsible for interpreting the genetic code. The **large subunit (50S** in prokaryotes) adds the individual amino acids to the nascent peptide chain according to the blueprints on the mRNA.

As mentioned above, the giant ribosome is blocked by **antibiotics** at a few vulnerable sites. Although antibiotics have distinct structural differences, they bind to overlapping regions composed of ribosomal RNA molecules. In addition to the large group of **macrolides**, other ligands with a completely different chemical structure have been found to block this region of the 50S subunit. These include chloramphenicol **32.34** and clindamycin **32.35** (◘ Fig. 32.20). Both bind near the **peptidyl transferase center** and compete with tRNA for the A- and

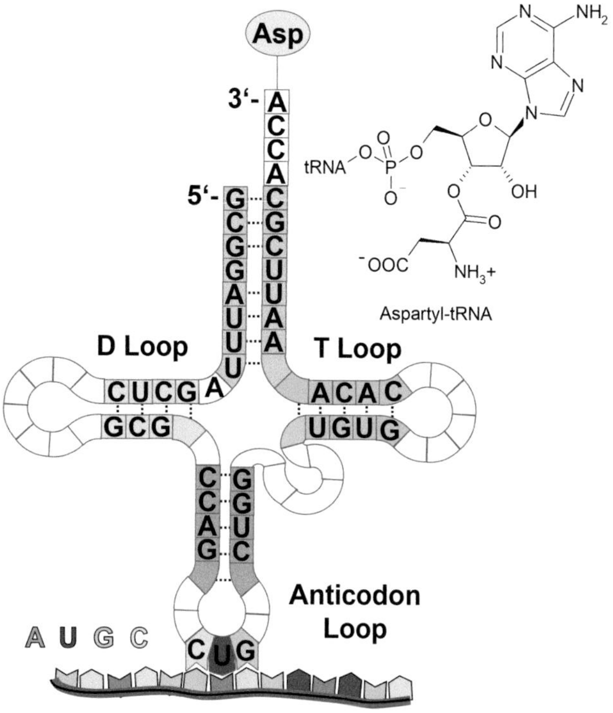

◘ **Fig. 32.18** The schematically depicted clover-leaf-shaped tRNA consists of 80 nucleotides that are paired into double strands over several sequence segments. In its actual spatial structure, the folded tRNA more closely resembles the shape of the letter L, with the bases on certain loops oriented outwards. Of particular importance is the anticodon loop, which contains the coding base triplet. In this example, the anticodon sequence is CUG, which corresponds to the GAC codon and encodes aspartic acid. The base at the 3′-end is always an adenosine. The 2′-OH group of its ribose moiety carries the amino acid to be transferred, attached through an ester bond

P-sites. Tetracyclines (e.g., **32.33**) and aminoglycosides (e.g., **6.14**, Sect. 6.4, ◘ Fig. 6.3) also attack the ribosome, but inhibit the function of the 30S subunit. Macrocyclic substances **32.36**–**32.41** bind to the entrance of the **ribosomal tunnel**, which is not far from the peptidyl transferase center. Their inhibitory effect is exerted by blocking the growth of the nascent polypeptide. Depending on their size, they allow the synthesis of protein fragments of up to 3–7 amino acids before the synthesis succumbs.

The most important of these compounds is erythromycin **32.36**, a macrolactone with a 14-membered ring. The scientist Abelardo Aguilar from the Philippines sent soil samples from the province of Iloilo to the Eli Lilly and Company in 1949. A metabolic product was isolated that showed antibiotic activity. The natural product was commercialized in 1952 under the name Iloson®. Its total synthesis was a challenge for synthetic chemists. The first total synthesis of erythromycin from simple starting materials was achieved in 1981 by Robert Woodward's research group. The compound is well tolerated but has inadequate acid stability. The free OH group at the 7-position reacts with the 10-carbonyl group by intramolecular ketalization. This step initiates the degradation of

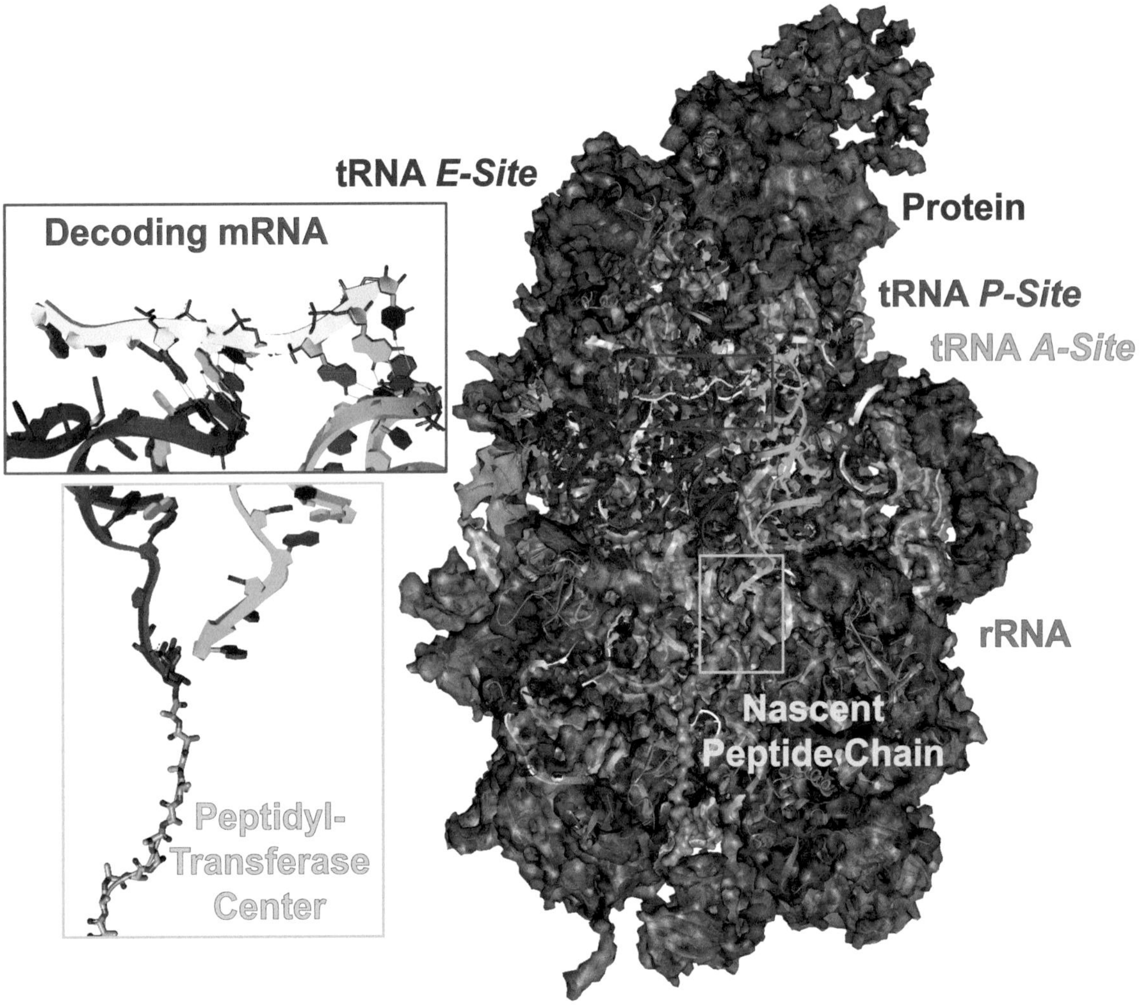

Fig. 32.19 View of the human mitochondrial ribosome based on a cryo-EM structure. RNA parts are shown in *white/gray*, protein parts in *orange/brown*. The three tRNAs at the A- (*green*), P- (*red*), and E-sites (*purple*) and a short segment of single-stranded mRNA (*yellow*) are bound in the structure. The nascent peptide chain (*yellow*) generated in the peptidyl transferase center winds its way through the ribosomal tunnel to the surface of the ribosome. The detailed view in the *red box* indicates the region where the mRNA is translated. The anticodon loops of the tRNAs at the A- and P-sites form H-bonds to the complementary nucleotide bases on the mRNA. In the peptidyl transferase center (detailed view in the *orange box*), the amide bond of the newly formed polypeptide chain is synthesized. New amino acids to be added are loaded into the P-site at the 3′ end of the tRNA. In bacterial ribosomes, macrolides bind to the anterior part of the peptide tunnel and stop chain synthesis after a few synthesis steps (see Fig. 32.22). (► https://sn.pub/WM6p5l)

the compound to products that are inactive as antibiotics. Therefore, erythromycin must be administered in the form of gastric acid-resistant tablets. Clarithromycin **32.37** is derived from erythromycin by ether formation at the 7-OH group. This suppresses instability under acidic conditions. Similarly, roxithromycin **32.38** achieves comparable stability by replacing the 10-carbonyl group with an oxime. In azithromycin **32.39**, the lactone ring is extended to 15 members and the carbonyl group is replaced by a methylamino group, which is not susceptible to attack by the OH group.

The spectrum of susceptibility of Gram-positive pathogens to these macrolides is somewhat different, in part because of differences in bioavailability. Erythromycin is well tolerated for topical use. Therefore, it is often used for skin diseases. Clarithromycin, roxithromycin, and azithromycin are acid stable and have better tissue penetration. They are often used to treat respiratory infections and infections of the ears, nose, and throat. Erythromycin and clarithromycin are potent cytochrome P450 CYP 3A4 inhibitors (Sect. 27.6). Therefore, the metabolism of many other drugs that are metabolized by

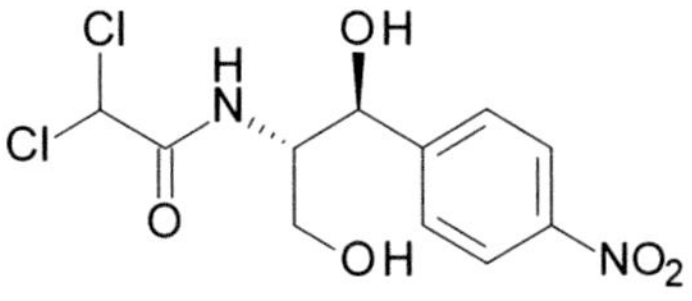

32.34 Chloramphenicol

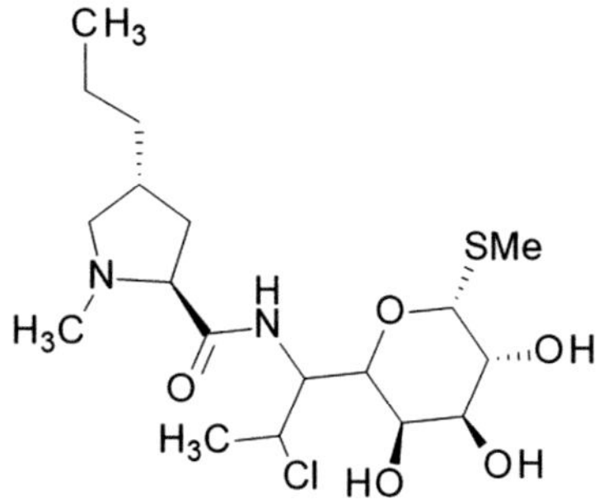

32.35 Clindamycin

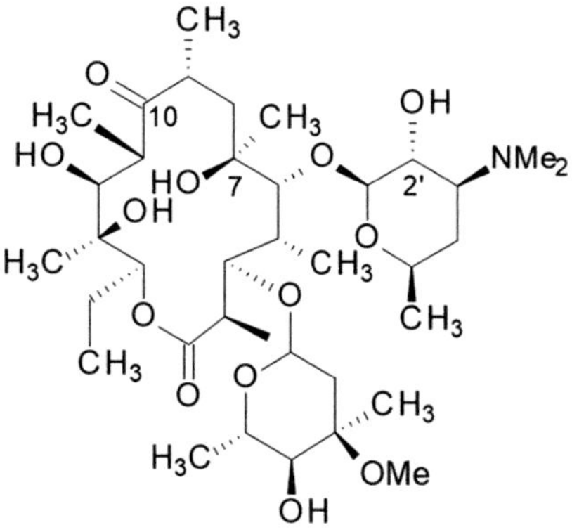

32.36 Erythromycin

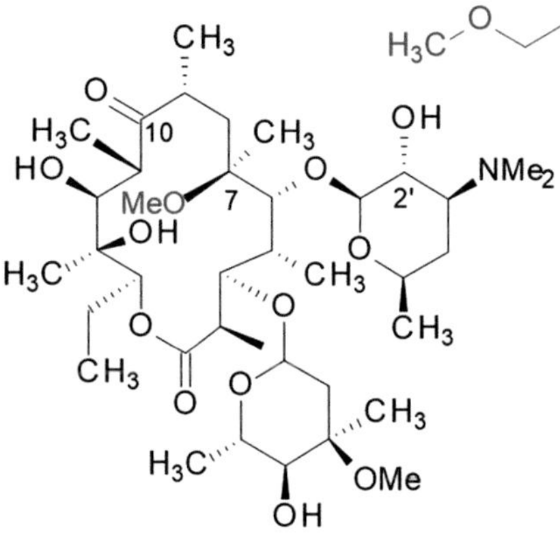

32.37 Clarithromycin

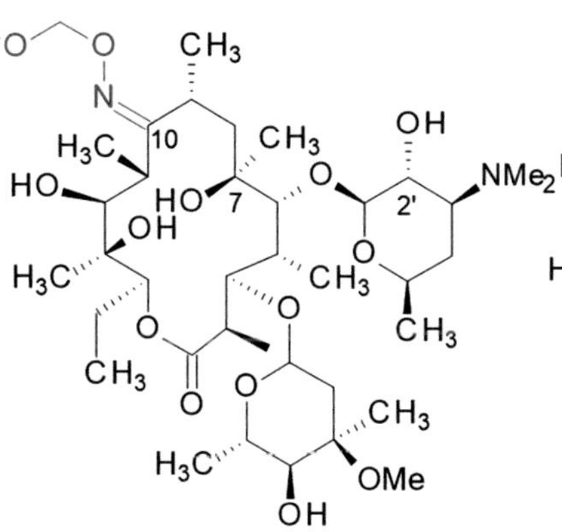

32.38 Roxithromycin

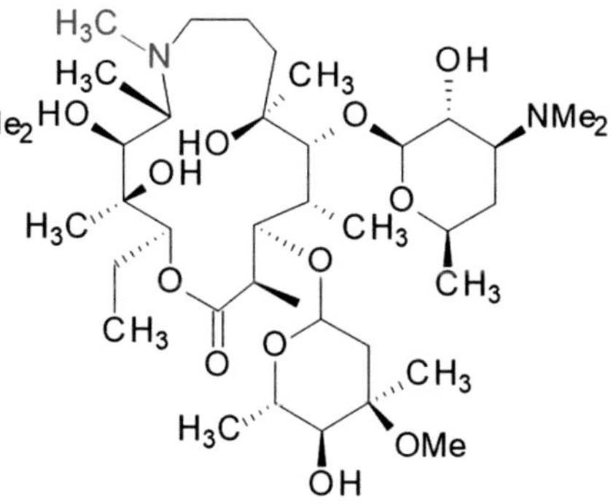

32.39 Azithromycin

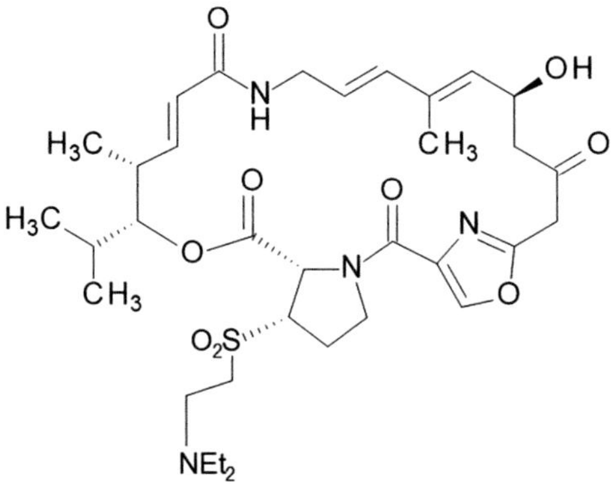

32.40 Dalfopristin

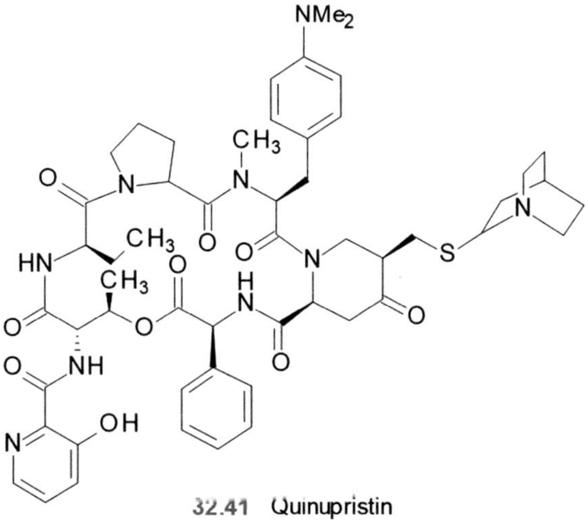

32.41 Quinupristin

Fig. 32.20 Chemical structures of a few antibiotics that bind to the 50S subunit of the ribosome. The substances **32.36–32.41** represent macrolides

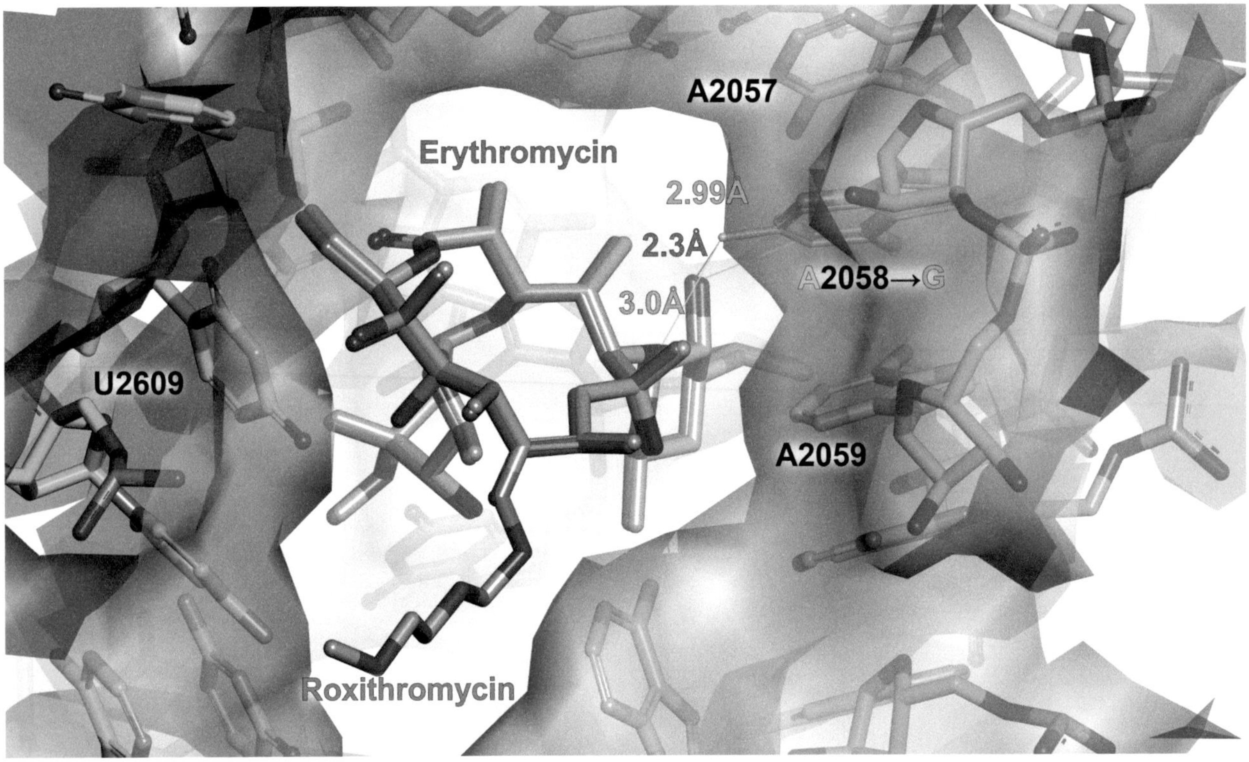

Fig. 32.21 Crystallographically determined binding geometry of erythromycin **32.36** (*blue-gray*) or roxithromycin **32.38** (*brown*) at the beginning of the peptide tunnel near the peptidyl transferase center (compare Fig. 32.22). An essential hydrogen bond is formed from the 2′-OH group of the amino sugar moiety to adenosine 2058 (*green*, 2.99 Å). A resistance mutation of A2058 to guanosine (*orange*) brings an amino group into close proximity to the macrolide. The repulsive distance of 2.30 Å (*purple*) indicates unfavorable interactions. With a distance of 3.02 Å, the distance between the amino group and the ether oxygen is hardly favorable. As a result, the binding of macrolides to the A → G resistance mutant decreases by five orders of magnitude. (► https://sn.pub/QyRcDw)

this enzyme may be blocked. Failure to take this into account may result in a dangerous increase in the concentration of concomitant drugs.

The binding modes of erythromycin **32.36** and roxithromycin **32.38** are shown in Fig. 32.21. As mentioned, they block the exit tunnel of the nascent peptide chain near the peptidyl transferase center in the ribosome. This region is made up exclusively of RNA building blocks, and binding occurs largely through strong van der Waals contacts with the tunnel wall. A crucial interaction is observed between the nucleoside adenosine 2058 and the 2′-OH group of the amino sugar group of the macrolides in the form of an H-bond. The development of **resistance** also plays an important role in the use of these antibiotics. Replacing the adenine base by a guanine reduces the inhibitory potential of erythromycin by five orders of magnitude. For steric reasons, a guanine at position 2058 leads to a repulsive interaction with the ribosome (Fig. 32.21). This exchange is observed in resistant mutants of clinical pathogens. Interestingly, eukaryotes also have a guanine at this position. This fact explains why the 14-membered macrolides have good **selectivity** for inhibiting bacterial ribosomes, because they possess an adenine at this position.

Many examples in this book have shown how small molecules find their intended site of action among many macromolecular targets, based on their appropriate steric construction and also their correct placement of interacting functional groups. Some readers may have wondered whether it is possible for two ligands to exert their influence on a target structure by **synergistic binding**. In fact, such cases do exist. Many of them have probably not been recognized, especially those where the affinity of the two components is very different. The mode of action of such potentiating effects has been characterized only in very few cases. One such example will be discussed as a final case. The macrocyclic streptogramins A and B, dalfopristin **32.40** and quinupristin **32.41**, bind to the ribosome in close proximity (Fig. 32.22). Quinupristin **32.41**, like erythromycin, is located at the entrance of the ribosomal tunnel. This allows very short peptides to still be synthesized by the ribosome. Dalfopristin **32.40** also prevents these synthesis steps by binding to the peptidyl transferase center, and the tRNA molecule is no longer accom-

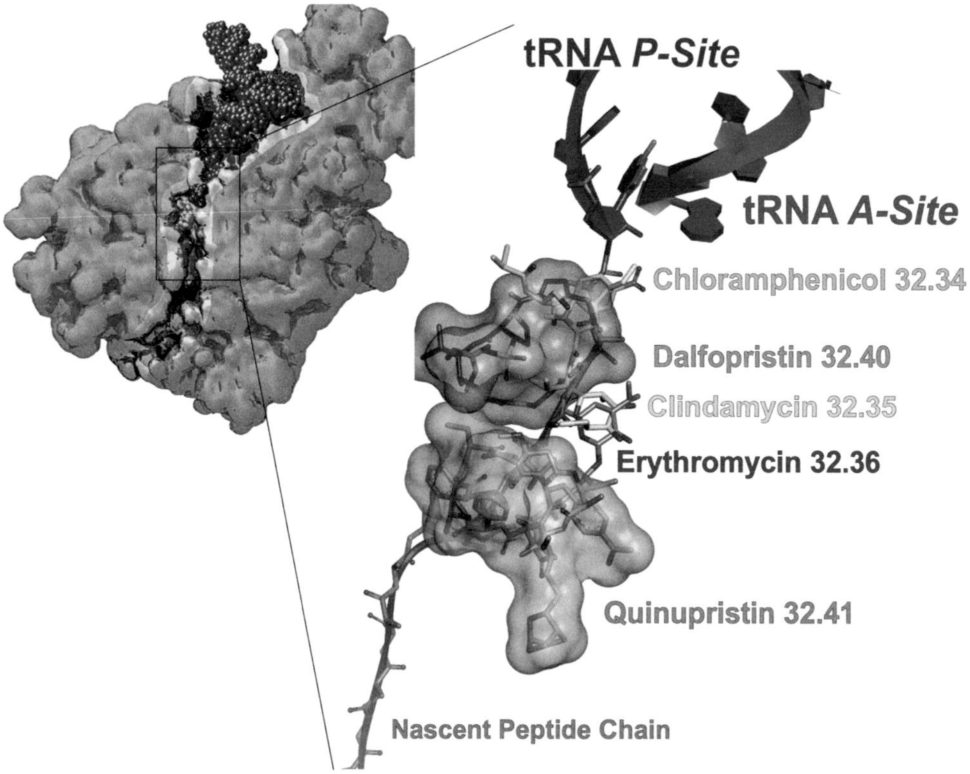

Fig. 32.22 The binding sites of dalfopristin **32.40** and quinupristin **32.41** are in close proximity to the peptidyl transferase center (*top left, black box*). They block the passage through the ribosomal tunnel. Both macrolides are in contact via a shared hydrophobic surface region. Their binding geometries have been determined crystallographically and are shown on the *right* as a blow-up with transparent surfaces around dalfopristin **32.40** (*brown*) and quinupristin **32.41** (*green*). For clarity, the RNA portions of the surrounding ribosome have been omitted. For comparison, the binding positions of erythromycin **32.36** (*blue-gray*), clindamycin **32.35** (*yellow*), and chloramphenicol **32.34** (*light blue*) are superimposed. For orientation, the 3′-termini of the tRNAs in the P- and A-sites (*blue*) and the nascent peptide chain (*violet*) are also superimposed. However, their binding positions will be blocked once the antibiotics are present. Quinupristin binds similarly to erythromycin in the ribosomal tunnel. Dalfopristin binds similarly to chloramphenicol to the peptidyl transferase center and blocks the incorporation of tRNAs into the A- and P-sites, respectively (*left image* taken from J. M. Harms et al., BMC Biology 2, 4 (2004), reprinted with the kind permission of the authors and publisher). (► https://sn.pub/ZBEeLP)

modated. If the binding position of dalfopristin is compared with that of chloramphenicol **32.34** (Fig. 32.22), a very similar volume area is occupied. The mutually enhanced binding of the two macrolides is explained by a pronounced hydrophobic contact surface that reduces the solvent-accessible surface. Furthermore, an altered conformation is observed for the highly conserved, catalytically important residue U2585. This results in a stable distortion of the peptidyl transferase center. This additional effect contributes to the synergistic inhibition of the ribosome when both macrolides are bound simultaneously. Both compounds were launched in 2000 as a 70:30 dalfopristin/quinupristin mixture under the brand name Synercid®. The drug is a potent antibiotic against highly resistant strains of bacteria. It is noteworthy that quinupristin and dalfopristin, when administered together, are up to sixteen times more effective at inhibiting susceptible bacteria than either of the two agents when used alone.

32.8 Synopsis

- Recombinantly produced proteins are used in substitution therapy, particularly when the endogenous protein is insufficiently or nonfunctionally produced by the organism.
- Diabetes is caused by a deficiency in the hormone insulin. Nowadays gene-technologically produced insulin has also been improved in its properties by mutational changes for either longer action or as a quickly absorbed form for shorter action.
- Antibodies specifically recognize foreign substances via their surface properties, bind them efficiently, and deliver them to phagocytic cells such as macrophages for degradation. They share a common architecture roughly comparable to the form of the letter Y. The branches and trunk are made of barrel-like pleated-sheet geometries. The antigen-recognizing regions

are found at the tips of the branches and are composed of several hypervariable loops that form the complementarity-determining regions to bind the antigen.
- The ability of antibodies to bind efficiently to nearly any chemical structure makes them ideal for the detection and culling of disease-causing foreign substances and malignant or degenerated cells. Recombinantly manufactured monoclonal antibodies specifically tailored for the recognition of protein surface determinants are used to selectively capture antigens out of the organism and to deliver them to the usual degradation pathway of the immune system.
- Antibodies can be raised against surface proteins on tumor cells or to compete with endogenous macromolecular ligands for cell surface receptors; once generated, these antibodies block and interfere with subsequent steps of signaling cascades. Their specific recognition properties can be exploited for targeting because the antibody scaffold can be chemically linked to other therapeutic principles, for example, the local exposure of instable nuclides for tissue-destroying radiation therapy.
- Protein biosynthesis requires reading single-stranded mRNA. Hybridization with short sequences of antisense oligonucleotides results in base pairing and leads either to digestion of the resulting double strand by RNase H, or the double strand simply cannot be read during protein biosynthesis.
- Due to their polar character, antisense oligonucleotides are insufficiently bioavailable and suffer from low chemical stability. Chemical modifications either of the phosphate backbone, conformational locking of the ribose moiety, or chemical functional group replacements improve their properties for successful drug applications.
- Nucleosides and nucleotides with crucial chemical modifications can still be recognized by enzymes as false substrates. Once bound to the catalyst, they can be covalently attached to the active site to irreversibly block the protein, or they can be incorporated as a building block in a polymer chain reaction. Due to sophisticated changes in their scaffold, for instance, by placing an azide group in the 3′-position of the ribose ring, the chain reaction is terminated and reproduction of the viral genome is stopped.
- Delivered in the form of a vaccine, mRNA can provide the blueprint for therapeutically important proteins in the body. For example, viruses produce surface proteins that stimulate the production of antibodies. In the event of a subsequent infection with the virus, they are available as a defense mechanism. During the coronavirus pandemic, vaccines against the spike protein of this pathogen were developed very quickly.
- Aside of chemically modified nucleotides as false substrates of the polymerase reaction, the HIV reverse transcriptase can be blocked allosterically by inhibitors that fix the enzyme in a broadly opened conformation; this prevents recognition of the nascent RNA–DNA hybrid strand.
- The enzyme gyrase catalyzes the overspiralization of bacterial DNA. The DNA has to wrap around the enzyme and through temporary cutting and reconnection of the backbone additional turns are introduced. The quinolones intercalate into the cut DNA like a wedge and prevent the cleaved ends of the double strands from rejoining.
- Tetracyclines inhibit ribosomal function; however, by high-affinity binding to the Tet repressor, they can initiate gene expression of a transport protein that expels foreign substances including tetracyclines themselves from the bacterial cell. As a consequence, resistance occurs because the tetracycline concentration in the bacterial cell falls below the level needed to block ribosomal function.
- Macrolides have been developed by microorganisms to fight other bacteria and fungi by blocking ribosomal function. The ribosome, a large ribonucleoprotein, operates as a ribozyme and, according to the base triplets read from the single-stranded mRNA, assembles the nascent polymer chain of the protein to be synthesized in its peptidyl transferase center.
- Several classes of antibiotics are known that block the ribosome at a few vulnerable points, such as in the peptidyl transferase center or the ribosomal peptide tunnel.
- Resistance to potent ribosomal inhibitors results in many cases from the exchange of nucleosides at positions where the ribosome performs crucial interactions with the bound inhibitors. A single adenine to guanine exchange can cause the inhibitory potential to drop by several orders of magnitude. For similar reasons, high species selectivity can be achieved with respect to the inhibition of bacterial or human ribosomes.

Bibliography and Further Reading

General Literature

Zündorf and T. Dingermann, Vom Rinder-, Schweine-, Pferde-Insulin zum Humaninsulin: Die biotechnische und gentechnische Insulin-Herstellung, Pharm. u. Zeit, **30**, 27–32 (2001)

Alyas, J. et al., Human Insulin: History, Recent Advances, and Expression Systems for Mass Production. Biomedical Research and Therapy, **8**, 4540–4561 (2021)

D. E. Milenic, E. D. Brady, M. W. Brechbiel, Antibody-Targeted Radiation Therapy, Nat. Rev. Drug Discov., **3**, 488–498 (2004)

O. H. Brekke, I. Sandlie, Therapeutic Antibodies for Human Diseases at the Dawn of the Twenty-First Century, Nat. Rev. Drug Discov., **2**, 52–62 (2003)

J. Kurreck, Antisense Technologies: Improvement through Novel Chemical Modifications, Eur. J. Biochem., **270**, 1628–1644 (2003)

T. Aboul-Fadl, Antisense Oligonucleotides: The State of the Art, Curr. Med. Chem., **12**, 2193–2214 (2005)

K. Das, P. J. Lewi, S. H. Hughes, E. Arnold, Crystallography and the Design of Anti-AIDS Drugs: Conformational Flexibility and Positional Adaptability are Important in the Design of Non-nucleoside HIV-1 Reverse Transcriptase Inhibitors, Progress in Biophysics & Mol. Biol., **88**, 209–231 (2005)

E. Grabar, U. Bahnsen, Das Ende aller Leiden: Wie RNA-Therapien die Behandlung von Krebs, Herzerkrankungen und Infektionen revolutionieren, Quadriga-Verlag (2022), ISBN 978-3-86995-116-4.

Y. K. Kim, RNA therapy: rich history, various applications and unlimited future prospects. Exp. Mol. Med., **54**, 455–465 (2022)

V. Vivet-Boudou, J. Didierjean, C. Isel, and R. Marquet, Nucleoside and Nucleotide Inhibitors of HIV-1 Replication, Cell. Mol. Life Sci., **63**, 163–186 (2006)

S. A. Sieber, M. A. Marahiel. Molecular mechanisms underlying nonribosomal peptide synthesis: approaches to new antibiotics. Chem Rev., **105**, 715–38 (2005)

A.Yonath, A. Bashan, Ribosomal Crystallography: Initiation, Peptide Bond Formation, and Amino Acid Polymerization are Hampered by Antibiotics, Annu. Rev. Microbiol., **58**, 233–251 (2004)

J. Lin et al., Ribosome-targeting antibiotics: modes of action, mechanisms of resistance, and implications for drug design, Annu. Rev. Biochem. **87**, 451–478b (2018)

D. N. Wilson, Ribosome-targeting antibiotics and mechanisms of bacterial resistance, Nat. Rev. Microbiol., **12**, 35–48, (2014)

Special Literature

T. Forst, Schnell wirkende Insulinanaloga, Pharm. u. Zeit, **30**, 118–123 (2001)

M. Schubert-Zsilavecz and M. Wurglics, Insulin glargin – ein langwirksames Insulinanalogon, Pharm. u. Zeit, **30**, 125–130 (2001)

G. Köhler, C. Milstein, Continuous cultures of fused cells secreting antibody of predefined specificity, Nature, **256**, 495–497 (1975)

J. Graham, M. Muhsin and P. Kirkpatrick, Cetuximab, Nat. Rev. Drug Discov., **3**, 549–550 (2004)

I. Zündorf and T. Dingermann, Kineret®, Enbrel®, Remicade® und Co., Rekombinante Wirkstoffe bei Rheumatoider Arthritis, Pharm. u. Zeit **5**, 376–383 (2003)

H. Marcotte et al., Conversion of monoclonal IgG to dimeric and secretory IgA restores neutralizing ability and prevents infection of Omicron lineages, Proc. Natl. Acad. Sci. USA, 121(3), e2315354120 (2024)

D. C. Nguyen, I.T. Hentenaar et al., SARS-CoV-2-specific plasma cells are not durably established in the bone marrow long-lived compartment after mRNA vaccination, Nat. Med., https://doi.org/10.1038/s41591-024-03278-y (2024)

K. Dhuri, C. Bechtold, E. Quijano, H. Pham, A. Gupta, A. Vikram, R. Bahal, Antisense Oligonucleotides: An Emerging Area in Drug Discovery and Development. J. Clin. Med., **9(6)**, 2004 (2020)

M. M. Swarbrick, P. J. Havel et al., Inhibition of Protein Tyrosine Phosphatase-1B with Antisense Oligonucleotides Improves Insulin Sensitivity and Increases Adiponectin Concentrations in Monkeys, Endocrinology, **150**, 1670–1679 (2009)

R. G. Nanni, J. Ding, A. Jacobo-Molina, S. H. Hughes, E. Arnold, Review of HIV-1 reverse transcriptase three-dimensional structure: Implications for drug design, Persp. Drug Discov. Design, **1**, 129–150 (1993)

T. D. M.Pham, Z. M. Ziora, M. A. T. Blaskovich, Quinolone antibiotics. Medchemcomm., **10**,1719–1739 (2019)

W. Saenger et al., The tetracycline repressor—a paradigm for a biological switch, Angew. Chem. Int. Ed., **39**, 2042–2052 (2000)

F. Schlünzen, R. Zarivach et al., Structural Basis for the Interaction of Antibiotics with the Peptidyl Transferase Centre in Eubacteria, Nature, **413**, 814–821 (2001)

R. B. Woodward, et al. Asymmetric Total Synthesis of Erythromycin. Synthesis of an Erythronolide A Seco Acid Derivative via Asymmetric Induction. J. Am. Chem. Soc., **103**, 3210, 3213, 3215 (1981)

J. L. Hansen, J. A. Ippolito et al., The Structures of Four Macrolide Antibiotics Bound to the Large Ribosomal Subunit, Molecular Cell, **10**, 117–128 (2002)

J. M. Harms, F. Schlünzen, P. Fucini, H. Bartels, A. Yonath, Alterations at the Peptidyl Transferase Centre of the Ribosome Induced by the Synergistic Action of the Streptogramins Dalfopristin and Quinupristin, BMC Biology, **2**, 4 (2004)

I. Laponogov, M. K. Sohi et al., Structural insights into the quinolone–DNA cleavage complex of type II topoisomerase, Nat. Struct. Mol. Biol., **16**, 667–669 (2009)

F. Schlünzen, R. Zarivach et al. Structural basis for the interaction of antibiotics with the peptidyl transferase centre in eubacteria. Nature, **413**, 814–821 (2001)

Service Part

G. Klebe, *Drug Design*, https://doi.org/10.1007/978-3-662-68998-1

Illustration Source References

The crystal structures used for the various images can be downloaded from: https://doi.org/10.2210/pdb**XXXX**/pdb, where **XXXX** refers to the four-letter PDB codes listed below with the figure numbers, e.g.: https://doi.org/10.2210/pdb3OOG/pdb

Cover image: Crystal structures of human protein kinase A with ligands derived from fragment discovery and optimization (PDB codes: 3OOG, 3OVV, 3OXT, 3P0M) summarized in the Ph.D. thesis of Dr. Helene Köster, Univ. Marburg (2012), (▶ https://d-nb.info/102718376X/34).

Figure explaining the images: Various representations of the crystal structure of carbonic anhydrase II with p-fluorophenylsulfonamide (PDB code: 1IF4), created with the computer graphics program Discovery Studio Visualizer V20.1.0.19295 from Biovia, Copyright 2019, Dassault Systemes Biovia Corp, analogous to the other molecular representations in this book.

Fig. Introduction: Crystal structure of thrombin with the ligand NAPAP (PDB code: 1DWD), background starry sky in July 2015, taken by Mr. Hagen Glötter at the Höfingen Observatory, Germany (▶ https://www.sternwarte-hoefingen.de/2015/07/sternenhimmel-im-juli/). We thank Mr. Glötter and the working group of the Höfingen Observatory for the permission to use the image.

Fig. 1.4 From Noe CR, Bader A (1993) Chem Britain 29:126–128.

Fig. 4.1 Segment of the crystal structure of the complex of the retinol-binding protein with retinol (PDB code: 1RBP).

Fig. 4.4 Crystal structure of ligand **4.1** with different protonation state in thrombin and trypsin (PDB codes: 6TDT, 6SYB) from J. Med. Chem. 63, 3274–3289 (2020).

Fig. 4.8 Crystal structures of a series of ligands of the basic structure 4.5 on thermolysin (PDB codes: 4N66, 4MZN, 5JS3) from J. Med. Chem. 59, 10530–10548 (2016). The figure was taken from the article "Rational Design of Thermodynamic and Kinetic Binding Profiles by Optimizing Surface Water Networks Coating Protein-Bound Ligands" by Stefan G. Krimmer, Jonathan Cramer, Michael Betz, Veronica Fridh, Robert Karlsson, Andreas Heine, and Gerhard Klebe with kind permission of the American Chemical Society (▶ https://pubs.acs.org/doi/10.1021/acs.jmedchem.6b00998).

Fig. 4.9 after P. R. Andrews et al., J. Med. Chem. 27, 1648–1657 (1984).

Fig. 4.10 Crystal structures with the water arrangement in the S1 binding pocket of trypsin before ligand binding (PDB code: 5MNZ/5MNF), after binding of 2-aminopyridine (PDB code: 5MNX/5MNB), aniline (PDB code: 5MNY/5MNA), benzylamine (PDB code: 5MO1/5MNL), *N*-amidinopiperidine (PDB code: 5MO2/5MNO) and benzamidine (PDB code: 5MO0/5MNH).

Fig. 4.13 Crystal structures of ligands with and without salt bridge in thrombin (PDB code: 2ZFF, 2ZDA) and tRNA guanine transglycosylase (PBB code: 4Q8T, 4Q8W).

Fig. 5.10 Crystal structure of *Candida antarctica* lipase with two enantiomers of the transition-state-analogue inhibitor from Bocola et al. (2003) Protein Eng 16:319–322.

Fig. 5.14 From Caner H et al. (2004) Drug Discov Today 9:105–110.

Fig. 5.17 Segment from the crystal structure of a complex of trypsin with BX5633 (PDB codes: 1MTS and 1MTU).

Fig. 5.18 Segment from the crystal structure of an inhibitor complex of carboanhydrase II (PDB code: 1CIL and Greer J et al. (1994) J Med Chem 37:1035–1054).

Fig. 5.19 Segment from the crystal structure of the complex of the retinoic acid receptor hRARγ with BMS270394/5 (PDB codes: 1EXX and 1EXA).

Figure before Chap. 6: Announcement poster from the research group of the author on the occasion of a conference in 2003 in Rauischholzhausen, Marburg, Germany.

Fig. 7.6 Functionality of the micro-thermophoresis process, illustration redrawn similarly to J. Mol. Struct, 1077, 101–113 (2014).

Fig. 7.7 Functionality of the switchSENSE process, illustration redrawn similarly to Laborwelt, 11, 21–22 (2010).

Fig. 7.9 Segments from the NMR structure of stromelysin and two fragments **7.1** and **7.2**, and the common product **7.3** (Hajduk PJ et al. (1997) J Am Chem Soc 119:5818–5827, the coordinates were kindly provided by P. Hajduk at Abbott).

Fig. 7.10 Segment from the crystal structures of thermolysin with different bound molecular probes (PDB codes: 1FJQ (acetone), 1FJU (acetonitrile), 8TLI (isopropanol), 1FJW (phenol)), and with bound benzyl succinic acid (PDB code: 1HYT).

Fig. 7.14 Superposition of the crystal structures of thymidylate synthase with *N*-tosyl-D-proline derivates (PDB codes: 1F4C, 1F4D, 1F4E).

Fig. 10.10 Segment of the NMR structure of the BCL-X_L complex with a 16-residue peptide from the BAK protein (PDB code: 1BXl).

Fig. 10.14 Illustration redrawn similarly to Bartlett PA (1992) Caveat user manual, San Francisco.

Figure Before Chap. 11: © Dr. Dirk Bossemeyer, German Cancer Research Center, Heidelberg, Germany.

Fig. 11.3 Illustration redrawn similarly to Christen HR, Vögtle F (1992) Organische Chemie, 2nd edn, vol II, Fig. 24.5, p 131. Otto Salle & Sauerländer.

Fig. 11.4 Illustration redrawn similarly to Gallop MA et al. (1994) Fig. 2, Applications of combinatorial libraries. J Med Chem 37:1233–1251.

Fig. 11.9 Superposition of the crystal structures of acetylcholinesterase with a *syn* and *anti* click chemistry reaction product (PDB codes: 1Q83, 1Q84).

Fig. 11.10 Illustration redrawn similarly to Ramström O, Lehn J-M (2002) Fig. 1, Nat Rev Drug Discov 1:27–36.

Fig. 12.4 Crystal structure of a complex of a VHL-E3 ligase, a PROTAC molecule and a BRD4 bromodomain (PDB code: 5T35).

Fig. 12.5 Fig. 6 from Lottspeich F (1999) Angew Chem 111:2630–2647; reprinted with kind permission of the author and publisher.

Fig. 12.6 Redrawn similarly to an illustration of Fonds der Chemischen Industrie im Verband der Chemischen Industrie e. V., Mainzer Landstraße 55, 60329 Frankfurt am Main, Biotechnologie – kleinste Helfer – große Chancen.

Fig. 12.7 Crystal structure of a histone protein with DNA segment (PDB code: 1EQZ).

Fig. 13.2 Crystal packing of the structure with the reference code FUXBIJ (Cambridge crystallographic database).

Fig. 13.3 Taken from Hargittai I, Hargittai M (1995) Symmetry through the eyes of a chemist, 2nd edn, Figs. 8–23. Springer, New York, p 363; reprinted with kind permission of the author and publisher.

Fig. 13.4 Taken from Pohl RW (1983) Einführung in die Physik, 18th edn, vol 1, Mechanik, Akustik and Wärmelehre, Fig. 380, p 198; reprinted with kind permission of the author and publisher.

Fig. 13.5 Taken from Glusker JP, Trueblood KN (1972) Crystal structure analysis, a primer, Fig. 5. Oxford University Press, New York, p 19.

Figs. 13.6 and **13.7** Illustration redrawn similarly to Keller E (1982) Chem unserer Zeit 16:71–88, Figs. 7 and 25; reprinted with kind permission of the author and publisher.

Fig. 13.9 Electron density of the crystal structure of aldose reductase (PDB code: 1US0).

Fig. 13.10 Diffraction pattern from a crystal of a small molecule and a protein (**a**), reprinted with kind permission of Bruker AXS GmbH (**b**), reprint with permission of the author and publisher, taken from Boese R (1989) Chem unserer Zeit 23:77–85, Fig. 11, (**c**), Reprint from HBZ, Berlin with kind permission of the synchrotron staff (**d**), Section through the electron density observed in a thermolysin complex (PDB code: 5M9W) (**e**), Section through the electron density in a thrombin complex (PDB code: 2ZC9) (**f**), Folding pattern of aldose reductase (PDB code: 1US0) (**g**), Thermal ellipsoids for crystal structure of formamide, reference see Fig. 14.1 (**h**), Protein folding pattern with superimposed color-coding by B-factors (**i**), Crystal packing of thrombin (PDB code: 2ZDA) (**k**).

Fig. 13.11 Crystal packing of 2,4,6-tris(4-pyridyl)-1,3,5-triazine – $ZnCl_2$ and with the bound glucuronidated metabolite of gemfibrozil (entries deposited in the CSD: CCDC 809387 and CCDC 1983849).

Fig. 13.15 NMR structure of a domain of the guanine nucleotide exchange factor (PDB code: 1B64).

Fig. 14.1 Stevens ED (1978) Acta Crystallogr B34:544–551, Fig. 1; reprinted with kind permission of the publisher (CSD entry: CCDC 1159372).

Fig. 14.2 Redrawn after Zubay G (1988) Biochemistry, 2nd edn, Fig. 2.7, p 66 and Fig. 2.10, p 68. MacMillan, New York.

Fig. 14.4 Redrawn after Zubay G (1988) Biochemistry, 2nd edn, Fig. 2.12, p 70 and Fig. 2.15, p 73. MacMillan, New York.

Fig. 14.5 Taken from Lesk A (1991) Protein architecture, Fig. 4.1, part **b** and **c**, Oxford University Press, Oxford; reprinted with kind permission of the publisher.

Fig. 14.7 Kindly provided by Prof. R. Zimmer, LMU Munich (prepared with the MolScript program; protein structures with den PDB codes: 1TIM, 4FXN, 1I1B, 3MBA, 2RHE, 2STV, 1UBQ, 1APS, 256B).

Fig. 14.8 Redrawn after Branden C, Tooze J (1991) Introduction to protein structure, Fig. 5.2, p 60, Figs. 5.14, 5.15, p 69, Fig. 5.17, p 71, Fig. 5.19, p 72. Garland, New York, and Zubay G (1988) Biochemistry, 2nd edn, Fig. 2.26, p 82. MacMillan, New York.

Fig. 14.9 Crystal structures of triosephosphate isomerase (PDB code: 1TIM) and flavodoxin (PDB code: 3FXN).

Fig. 14.10 Redrawn after Zubay G (1988) Biochemistry, 2nd edn, Fig. 2.12. MacMillan, New York, p 70, and Illustration of a crystal structure of a Fab fragment with phosphocholine (PDB code: 2MCP).

Fig. 14.13 Taken from Vyas K, Monahar H, Venkatesan K (1990) J Phys Chem 94:6069–6073, Fig. 1; reprinted with kind permission of the publisher.

Fig. 14.14 From a template, the source is unknown, redrawn by the author.

Fig. 14.15 Taken from Bürgi HB, Dunitz JD (1994) Structure correlation, vol 2, Fig. 13.24. Wiley, p 585; reprinted with kind permission of the publisher.

Fig. 14.16 Distribution of H-bond donor and acceptor groups around an imidazole moiety; entry from the IsoStar database. Cambridge Crystallographic Data Centre. ► https://www.ccdc.cam.ac.uk/solutions/software/isostar/.

Fig. 14.17 Crystal structures of trypsin (PDB Code: 3PTB) and subtilisin (PDB code: 1SBC).

Fig. 14.20 Crystal structures von DNA oligonucleotide strands with cisplatin and daunorubicin (PDB code: 1A2E and 1AL9).

Fig. 15.1 (1994) Discover manual, Part 1, Fig. 3.5, San Diego.

Fig. 16.1 Redrawn after from Christen HR, Vögtle F (1992) Organische Chemie, vol I, 2nd edn, Fig. 2.3, p 71. Otto Salle & Sauerländer.

Figure before Chap. 17: Announcement poster from the research group of the author on the occasion of a conference in 2005 in Rauischholzhausen, Marburg, Germany.

Fig. 17.1 Taken from Mackay MF, Sadek M (1983) from Aust. J. Chem. 36:2111–2117, Fig. 1; reprinted with kind permission of the publisher (CSD entry: CCDC 1124665).

Fig. 17.3 Superposition with cryo-EM-Structure of $GABA_A$-Receptor with bound Picrotoxinin (PDB code: 6HUJ).

Fig. 17.7 Superimposition of the crystal structure of dihydrofolate reductase with dihydrofolate and methotrexate (PDB codes: 1DHF, 3DFR).

Fig. 17.9 Taken from Seidel W, Meyer H, Kazda S, Dompert W (1984) Fig. 6. In: Seydel J (ed) QSAR and strategies in the design of bioactive compounds. Wiley, pp 366–369; reprinted with kind permission of the publisher.

Fig. 17.10 Segment of the crystal structures of thermolysin with different, bound molecular probes (cf. Fig. 7.8) superimposed with "hot spots" from a calculation from DrugScore.

Fig. 17.11 Distribution of hydrogen-bond donor groups around a carboxylic acid, ester, keto, and ether grouping from the Isostar database ▶ https://www.ccdc.cam.ac.uk/solutions/software/isostar/ Cambridge Crystallographic Data Centre.

Fig. 17.12 Superposition of the crystal structures of DHFR with MTX (PDB code: 3DFR) with the distribution of hydrogen-bonding geometries from IsoStar ▶ https://www.ccdc.cam.ac.uk/solutions/software/superstar/ Cambridge Crystallographic Data Centre.

Fig. 18.4 Redrawn after Cramer RD, Patterson DE, Brunce JD (1988) J Am Chem Soc 110:5959–5967, Fig. 1.

Figs. 18.9 and **18.10** From Weber A et al. (2006) J Chem Inf Model 46:2737–2760, Figs. 7 and 8.

Fig. 20.5 Superposition of the crystal structures of three cytochrome *c* enzymes (PDB code: 3C2C, 5CYT, 155C).

Fig. 20.6 Redrawn after Verlinde CLMJ, Hol WGJ (1994) Structure 2:577–587.

Fig. 20.7 Redrawn after Böhm HJ (1993). In: Wermuth CG (ed) Trends in QSAR and molecular modelling, vol 92. ESCOM Science, Leiden, Fig. 3, p 30.

Fig. 21.2 Crystal structure of TGT with bound tRNA (PDB code: 1Q2S).

Fig. 21.6 Crystal structure of TGT with bound **21.3** (PDB code: 1ENU).

Fig. 21.7 Crystal structure of TGT with bound **21.9** (PDB code: 1N2V).

Fig. 21.10 Crystal structure of TGT with bound **21.14** (PDB code: 1Y5V).

Fig. 21.11 Crystal structure of TGT with bound **21.13** (PDB code: 2BBF).

Fig. 21.16 Crystal structure of TGT with bound **21.23** and **21.24** (PDB codes: 3GE7, 3EOS).

Fig. 21.17 Crystal structures of TGT with bound **21.22, 21.24** and **21.25** (PDB codes: 4GKT, 4PUJ, 5I02).

Fig. 21.20 Crystal structures of TGT with bound **21.26** (left) or **21.14**, **21.27** and **21.28** (right) in two packing dimers (PDB codes: 4FSA, 5I07, 6YGZ, 5LPP, 5LPQ, 5LPS, 5LPT).

Fig. 21.26 Crystal structure of TGT wild type and Y330C mutant (PDB codes: 1PUD, 4JBR).

Fig. 21.28 Crystal structures of TGT with the sulfolane **21.34** and the fluorine derivative **21.35** (PDB codes: 7A3X, 7A6D).

Figs. 21.29 and **21.30** Crystal structures of TGT with bound **21.21**, **21.38** and **21.39** (PDB codes: 4PUK, 5UTI, 6RKQ).

Figure before Chap. 22: Announcement poster from the research group of the author on the occasion of a conference in 2007 in Rauischholzhausen, Marburg, Germany.

Fig. 22.1 After data from R. Santos et al., A Comprehensive Map of Molecular Drug Targets, Nature Rev. Drug Discov., 16, 19–33 (2017).

Fig. 22.4 Segment from the crystal structure of a complex of creatinase with carbamoyl sarcosine (PDB code: 1CHM).

Fig. 23.2 Binding pocket from the crystal structures of trypsin (PDB code: 1PPC), thrombin (PDB code: 1DWD), factor VIIa (PDB code: 1W7X) and factor Xa (PDB code: 2P93).

Fig. 23.5 Segment of the crystal structure of the complex of thrombin with cyclotheonamide A, an inhibitor from the marine sponge *Theonella* sp. (PDB code: 1TMB).

Fig. 23.6 Modeled geometry from a crystal structure of thrombin with fibrinopeptide (PDB code: 1FPH).

Fig. 23.7 Superposition of the crystal structures of thrombin with fibrinopeptide (PDB code: 1FPH) and PPACK (PDB code: 1PPB).

Fig. 23.10 Segment from the crystal structure of the complex of thrombin with NAPAP (PDB code: 1DWD).

Fig. 23.12 Comparison of the crystal structures of NAPAP with trypsin and thrombin (PDB code: 1PPC and 1DWD).

Fig. 23.17 Segment from the crystal structure of the complex of elastase with a pyridone-like inhibitor (PDB code: 1EAT).

Fig. 23.19 Segment from the crystal structure of the complex of factor Xa with rivaroxaban (PDB code: 2W26).

Fig. 23.24 Segment from the crystal structure of the complex of 1β-lactamase (PDB code: 1TEM).

Fig. 23.26 Crystal structure of the yeast proteasome with bortezomib (PDB code: 2F16).

Fig. 23.27 Segment from the crystal structure of nirmatrelvir and the SARS-Cov-2 M^{pro} main protease (PDB-Code: 7RFW).

Fig. 24.3 Crystal structures of the aspartic protease cathepsin D (PDB code: 1LYB), endothiapepsin (PDB code: 4ER1), HIV protease (PDB code: 5HPV), plasmepsin (PDB code: 1SME), and renin (PDB code: 4APR).

Fig. 24.8 Superposition of the crystal structure of renin with CGP-38560 (PDB code: 1RNE) and aliskiren (PDB code: 2V0Z).

Fig. 24.11 Segment from the crystal structure of renin with a piperidine-like inhibitor (PDB code: 1UTH).

Fig. 24.12 Crystal structure of HIV protease with a peptide substrate (PDB code: 1MT9).

Fig. 24.18 Superposition of the crystal structures of HIV protease with a urea-like (PDB code: 1HVR) and coumarin-like inhibitor (PDB code: 1UPJ).

Fig. 24.24 Crystal structures of HIV protease with inhibitors with a secondary amine nitrogen atom (PDB codes: 1XL2, 3BHE, 2PQZ, 3BGB).

Fig. 24.25 Superposition of the ligands in the crystal structure with HIV protease ritonavir (PDB code: 1HXW), atazanavir (PDB code: 2AQU), darunavir (PDB code: 1T3R), amprenavir (PDB code: 1HPV), indinavir (PDB code: 1HSG), nelfinavir (PDB code: 1OHR), saquinavir (PDB code: 1HXB), lopinavir (PDB code: 2O4S), tipranavir (PDB code: 2O4P), and **24.44** (PDB code: 2QQN).

Fig. 25.2 Segment from the crystal structure of the complex of matrix metalloproteinase MMP-12 with the cleavage product of the protease reaction (PDB code: 2OXZ).

Fig. 25.5 Segment from the superposed crystal structures of the complexes of thermolysin with the inhibitor Cbz-Gly^{P}-Leu-Leu (PDB code: 5TMN) and a cyclized inhibitor (PDB code: 1PE5) derived from it.

Fig. 25.6 Segment from the crystal structure of the complex of carboxypeptidase with benzylsuccinate (PDB code: 1CBX).

Fig. 25.12 Segment from the crystal structure of the complex of lisinopril with t-ACE (PDB code: 1O86).

Fig. 25.14 Segment from the crystal structure of the complex of fibroblast collagenase with Ro 31–4724 (PDB code: 2TCL).

Fig. 25.16 Segment from the crystal structures of the complexes of fibroblast collagenase with a peptidic (**25.49**) and a nonpeptidic inhibitor (**25.50**, PDB codes: 1HFC and 966C).

Fig. 25.17 Segment from the crystal structure of the complex of carboanhydrase II with *p*-fluorophenylsulfonamide and modeled geometries of a carbonylation in CA II (PDB code: 1IF4).

Fig. 25.20 Segment from the crystal structure of the complex of phosphodiesterases 5 and sildenafil (PDB code: 1UDT).

Fig. 25.22 Segment from the crystal structure of the complex of peptide deformylase from *Escherichia coli* with actinonin **25.75** (PDB code: 1G2A).

Fig. 26.3 Crystal structure of the cAMP-dependent protein kinase (PDB code: 1L3R).

Fig. 26.4 Modeled geometries of the transition state on the coordinates of the crystal structure of the cAMP-dependent protein kinase (PDB code: 1L3R).

Fig. 26.7 Crystal structure des complex of MAP kinase p38 with SB203580 (PDB code: 1A9U).

Fig. 26.9 Superposition of the inactive and active form of the tyrosine kinase domains of the human insulin receptor (PDB codes: 1IRK and 1IR3).

Fig. 26.10 From a figure out of Fabian MA et al. (2005) Nat Biotechnol 23:329–336; reprinted with kind permission of the author and publisher.

Fig. 26.12 Superposition of the crystal structures of BCR-ABL protein kinase with bound imatinib (Gleevec®) **26.5** and tetrahydrostaurosporine **26.27** (PDB codes: 2HYY and 2HZ4) and with nilotinib **26.28**/asciminib **26.30** (PBD Code: 5MO4).

Fig. 26.14 Segments from the crystal structures of Src kinase with ANP and the mutated Src-kinase with N6-benzyl-ADP (PDB codes: 1KSW and 2SRC).

Fig. 26.17 Superposition of the crystal structures of Ser/Thr-Kinase PIM-1 with staurosporine and a ruthenium complex (PDB codes: 1YHS and 2BZH).

Fig. 26.19 Segment from the crystal structure of human tyrosine phosphatase PTP-1B (PDB code: 1PTY).

Fig. 26.20 Segments from the crystal structures of human tyrosine phosphatase PTP-1B with different inhibitors (PDB codes: 1PTY, 1NO6, 1NNY, 1N6W, 2QBP, 2CM7).

Fig. 26.22 Segment from the crystal structure of human tyrosine phosphatase PTP-1B with an allosteric inhibitor (PDB code: 1T4J).

Fig. 26.23 Crystal structure of the full-length Shp2 phosphatase with C-SH2 and N-SH2 domains (PDB code: 2SHP).

Fig. 26.24 Crystal structure of full-length Shp2 phosphatase with C-SH2 and N-SH2 domains and SHP099 (PDB code: 2EHR).

Fig. 26.25 Crystal structure of COMT with a substrate-analogue inhibitor and *S*-adenosyl-L-methionine (PDB code: 1VID).

Fig. 26.27 Superposition of the crystal structures of COMT (PDB code: 1VID and 1JR4).

Fig. 26.28 Superposition of the crystal structures of FTase with farnesyl diphosphate and the farnesylated tetrapeptide CAAX (PDB codes: 1FT2 and 1D8D).

Fig. 27.3 Segment from the crystal structure of dihydrofolate reductase from *Lactobacillus casei* with methotrexate (PDB code: 3DFR).

Fig. 27.4 Segment from the crystal structure of horse liver alcohol dehydrogenase with bound NADPH (PDB code: 1HET).

Fig. 27.7 Segment from the crystal structure of cytochrome P450 14-α-steroldemethylase (CYP51) from *Mycobacterium tuberculosis* in complex with fluconazole **27.6** (PDB code: 1EA1).

Fig. 27.11 Superposition of the binding pocket of *Pneumocystis jiroveci* and murine DHFR (PDB codes: 2FZI and 2FZJ).

Fig. 27.14 Segment from the crystal structure of HMG-CoA reductase with bound HMG-CoA and mevalonic acid (PDB codes: 1DQA, 1DQ9).

Fig. 27.15 Segment from the crystal structure of HMG-CoA reductase with bound inhibitors simvastatin and atorvastatin (PDB codes: 1HW9, 1HWK).

Fig. 27.19 Segment from four crystal structures from aldose reductase with sorbinil, tolrestat, IDD594, and **27.46** (PDB codes: 1AH0, 2FZD, 1US0, 2NVD).

Fig. 27.20 Binding pocket from the crystal structure of aldose reductase with sorbinil (PDB codes: 1AH0).

Fig. 27.24 Crystal structure of human 11β-HSD1 with carbenoxolone superimposed with the complex of murine 11β-HSD1 with bound corticosterone (PDB codes: 2BEL, 1Y5R).

Fig. 27.25 Segment from the crystal structures of human 11β-HSD1 in complex with two inhibitors and the complex of murine 11β-HSD1 with bound corticosterone (PDB codes: 2ILT, 2RBE, 1Y5R)

Fig. 27.27 Crystal structures of human CYP 3A4 uncomplexed and in complex with metyrapone, erythromycin, and ketoconazole (PDB codes: 1W0E, 1W0G, 2J0D, 2V0M).

Fig. 27.30 Redrawn after Fig. 3 in Weinshilboum and Wang (2004) Nat Rev Drug Discov 3:739–748.

Fig. 27.33 Crystal structures of human MAO_B in complex with tranylcypromine and L-deprenyl (PDB codes: 1OJB and 2BYB).

Fig. 27.34 Crystal structures of human MAO_A in complex with clorgyline (PDB code: 2BXR) and MAO_B with L-deprenyl (PDB code: 2BYB).

Fig. 27.37 Superposition of the crystal structures of cyclooxygenase-1 and -2 in complex with arachidonic acid (PDB codes: 1PRH and 1CVU).

Fig. 27.39 Segment from the crystal structures of cyclooxygenase with arachidonic acid (PDB code: 1DIY) and prostaglandin PGH2 (PDB code: 1DDX).

Fig. 27.41 Segment from the crystal structures of cyclooxygenase-1 with a bromine analogue of acetylsalicylic acid (PDB code: 1PTH).

Fig. 27.42 Segment from the crystal structures of cyclooxygenase-2 with a bromine analogue of celecoxib (PDB code: 6COX).

Fig. 28.2 Crystal structure of the DNA-binding domain of the estrogen receptor with a bound oligonucleotide strand (PDB code: 1BY4) and crystal structure of the heterodimeric PPARγ-RXRα receptor with bound DNA oligomer, rosiglitazone, and retinoic acid (PDB code: 3DZY).

Fig. 28.4 Segment from the crystal structures of the ligand-binding domain of the estrogen receptor with bound estradiol (PDB code: 1ERE).

Fig. 28.5 Segment from the crystal structure of the ligand-binding domain of the progesterone receptor with bound progesterone (PDB code: 1A28).

Fig. 28.6 Comparison of the crystal structures of the estrogen receptor with bound estradiol and raloxifene (PDB codes: 2J7X and 1ERR).

Fig. 28.8 Segment from the crystal structure of the ligand-binding domain of the estrogen receptor with bound estradiol and the LxxLL binding motif (PDB code: 2J7X).

Fig. 28.13 Superposition of the crystal structures of the ligand-binding domain of the PPARγ receptors with a bound agonist (PDB code: 1K7L) and antagonist (PDB code: 1KKQ).

Fig. 28.16 Schematic course of the secondary structural elements in the crystal structures of the estrogen receptors (PDB code: 2J7X) and three examples for the PXR receptor (PDB codes: 1NRL, 1M13, 1SKX).

Fig. 29.2 Crystal structure of the human β2-adrenergic receptor with attached trimeric G protein (PDB code: 3SN6).

Fig. 29.3 Superposition of the crystal structures of the human β1-adrenergic receptor with the bound antagonist cyanopindolol **29.4** (PDB code: 2VT4) and the agonist isoprenaline **29.7** (PDB code: 2Y03).

Fig. 29.4 Crystal structures of a mutant of the inactive (PDB code: 1GZM) and active (PDB code: 2X72) rhodopsin.

Fig. 29.5 Crystal structure of the complex of *C*-terminally phosphorylated rhodopsin and bound arrestin (PDB code: 5W0P).

Fig. 29.8 Superposition of the crystal structures of the AT_1 receptor with the bound octapeptide angiotensin II **29.20** (PDB code: 6OS0) and the low molecular weight sartan-like ligand ZD7155 **29.21** (PDB code: 4YAY).

Fig. 29.13 Crystal structure of the erythropoietin receptor with bound erythropoietin (EPO; PDB code: 1CN4).

Fig. 29.12 Superimposition of the crystal structures of TNF with its trimeric receptor (PDB code: 1TNR) with a structure of the TNF receptor to which the small molecule inhibitor **29.27** has covalently bound (PDB code: 1FT4).

Fig. 29.13 Crystal structures of TNF with its trimeric receptor (PDB code: 1TNR) and the cytokine split into a dimer with the substances **29.28** and **29.29** (PDB codes: 4TWT and 2AZ5). With **29.30**, TNF-analogous cytokines form a highly expanded structure (PDB code: 3LKJ).

Fig. 29.14 Crystal structure of IL-2 with its heterotrimeric receptor (PDB code: 2ERJ). The binding of **29.31** alters the surface of IL-2 in such a way that a binding region is formed and the structurally altered IL-2 no longer binds to the receptor (PDB code: 1M48).

Fig. 29.15 The receptor complex (PDB code: 2ERJ) can no longer form due to the binding of **29.31** to IL-2 (PDB code: 1M48). By introducing Cys residues in positions 31 and 43, tether fragments (PDB codes: 1M4A and 1M4B) were discovered and crystallographically elucidated.

Fig. 30.3 Schematic representation of the crystal structure of the bacterial potassium channel KcsA (PDB code: 1K4C).

Fig. 30.4 Segment of the crystal structure of the bacterial potassium channel KcsA (PDB code: 1K4C).

Fig. 30.5 Segment of the crystal structures of a selective and unselective bacterial potassium channel (PDB codes: 1K4C, 2AHY).

Fig. 30.8 Cryo-EM structures of the hERG channel with bound Astemizole (PDB code: 8ZYO).

Fig. 30.9 Cryo-EM structures of the human Na_V-1.7 channel and the Ca_V channel from the rabbit (PDB codes: 7W9M and 6JP5).

Fig. 30.10 Crystal structures of the bacterial Na_VAB channel in the open, closed and resting/closed state (PDB codes: 5VB2, 5VB8 and 4EKW).

Fig. 30.12 Cryo-EM structure of the human Na_V1.7 channel with bound tetrodotoxin **30.20** (PDB code: 7W9M).

Fig. 30.13 Cryo-EM structures of the Ca_V channel from the rabbit with bound nifedipine **30.24** and Bay K8644 **30.27** (PDB codes: 6JP5 and 6JP8).

Fig. 30.14 CryoEM structures of the CaV channel from the rabbit with bound nifedipine, verapamil and diltiazem (PDB codes: 6JP5, 6JPA and 6JPB).

Fig. 30.15 Crystal structure (electron diffraction) of the nicotinic acetylcholine receptor in the closed state from the electric organ of an electric ray (PDB code: 2BG9). The structures of the acetylcholine-free (PDB code: 4AQ5) and acetylcholine-bound (PDB code: 4AQ9) channels are shown superimposed.

Fig. 30.17 Crystal structures of the ligand-binding domain of the nicotinic acetylcholine receptors from the California sea slug (*Aplysia californica*) with bound α-conotoxin (PDB code: 2BYP) and epibatidine (PDB code: 2BYQ).

Fig. 30.18 Segment from the crystal structures of complexes of the ligand-binding domains of the nicotinic acetylcholine receptor from the California sea slug (*Aplysia californica*) with bound epibatidine (PDB code: 2BYQ), α-lobeline (PDB code: 2BYS), α-conotoxin (PDB code: 2BYP), and methyllycaconitine (PDB code: 2BYR).

Fig. 30.20 Cryo-EM structure of the $GABA_A$ receptor with bound GABA/picrotoxinin (PDB code: 6HUJ).

Fig. 30.21 Cryo-EM structures of the $GABA_A$ receptor with bound GABA/picrotoxinin or bicuculline (PDB codes: 6HUJ and 6HUK).

Fig. 30.22 Cryo-EM structures of the $GABA_A$ receptor with bound bicuculline or GABA/alprazolam (PDB codes: 6HUK and 6HUO).

Fig. 30.23 Cryo-EM structures of the $GABA_A$ receptor with bound GABA/diazepam, GABA/phenobarbital, GABA/flumazenil (PDB codes: 6X3X, 6X3U and 6X3W).

Fig. 30.24 Cryo-EM structures of the $GABA_A$ receptor with bound GABA/diazepam and GABA/flumazenil (PDB codes: 6X3X and 6X3U).

Fig. 30.25 Crystal structure of the bacterial ClC channel from *Escherichia coli* (PDB code: 1OTS).

Fig. 30.26 Crystal structure or cryo-EM structure of the Na^+/K^+-ATPase (PDB code: 2ZXE) and H^+/K^+-ATPase with the inhibitor tegoprazan (PDB code: 7W47).

Fig. 30.27 Cryo-EM structure of the H^+/K^+-ATPase with the inhibitor tegoprazan (PDB code: 7W47).

Fig. 30.28 Crystal and cryo-EM structures of the human, chimeric ABCB1 transporter, and the bacterial ABC transporter Sav1866 (PDB codes: 4M1M, 6QEE and 2HYD).

Fig. 30.29 Crystal structure of the pore from the bacteria *Rhodobacter capsulatus* (PDB code: 2POR).

Fig. 30.30 Crystal structure of gramicidin A (PDB code: 1GRM).

Fig. 30.31 Model of nonactins based on a crystal structure (CSD Refcode: NONKSC, deposited as CCDC 1222314).

Fig. 30.32 Crystal structure of bovine aquaporin 1 (PBD code: 1J4N) and the high-resolution structure of a yeast aquaporin (PDB code: 3ZOJ).

Fig. 31.3 Segment from the crystal structure of the $\alpha_{II}\beta_3$ integrin receptor with the cyclopeptide **31.2** cilengitide (PBD code: 1L5G).

Fig. 31.5 Superposition of the segment from the crystal structure of the $\alpha_{II}\beta_3$ integrin receptor with eptifibatide (PBD code: 1TY6/2VDN) and tirofiban (PBD code: 1TY5/2VDM).

Fig. 31.6 Redrawn after an image of the research group of Prof. B. Ernst, University of Basel (▶ http://www.pharma.unibas.ch/molpharm/index.html).

Fig. 31.7 Segment from the crystal structure of sialyl-Lewis^X and the mimetic **31.19** with an E-selectin (PDB code: 4CSY and 4C16).

Fig. 31.9 Schematic drawing analogous to Doranz et al. (1999) J Virol 12:10346–10358.

Figs. 31.10 and **31.11** Segment from the crystal structure of the gp41 protein (PBD code: 1AIK).

Fig. 31.12 Crystal and NMR structure of the M2 channel in open and closed state (PDB codes: 3C9J and 2RLF).

Fig. 31.15 Segments from the crystal structures of neuraminidase with zanamivir **31.30** (PDB code: 1A4G) and oseltamivir **31.34** (PDB code: 2HT8).

Fig. 31.16 Segments from the crystal structures of a resistance variant of neuraminidase with zanamivir

31.30 (PDB code: 3CKZ) and oseltamivir **31.34** (PDB code: 3CL0).

Fig. 31.17 Crystal structure of the HRV-14 virus (PDB code: 1NCQ).

Fig. 31.18 Superposition of the crystal structures of the capsid proteins of HRV-14 in complex with pleconaril (PDB code: 1NA1) and the cryoelectron-microscopically determined complex with domains of the adhesion protein (PDB code: 1D3I).

Fig. 31.20 Crystal structure HRV-14 capsid protein in complex with pleconaril (PDB code: 1NA1).

Fig. 31.22 Crystal structure of the tertiary complex of a MHC I molecule with a nonapeptide and the T-cell receptor (PDB code: 1BD2).

Fig. 31.24 Superimposed crystal structures of the ternary complex of 31.44 (ochre) and the binary complex of **31.45** (green) with the MHC-I molecule (PDB codes: 1AO7 and 3O3A).

Fig. 32.25 Superposition of the crystal structures of the MHC-I molecules with peptides and abacavir (PDB codes: 3UPR, 2RFX).

Fig. 32.1 Crystal structure of a complete IgG antibody (PBD code: 1IGT).

Figs. 32.2 and **32.3** Comparison of the crystal structures of the F_{ab} domain of an antibody with phosphocholine (PBD code. 2MCP) and lysozyme (PBD code: 1FBI).

Fig. 32.5 Redrawn after Fig. 3 in Milenic D.E. et al. Nat Rev Drug Discov 3:488–498 (2004).

Fig. 32.6 NMR structure of an oligomeric double strand of RNA and PNA (PBD code: 176D).

Fig. 32.8 Crystal structure of the HIV reverse transcriptase with a bound RNA–DNA hybrid double strand (PBD code: 1HYS).

Fig. 32.9 Structure comparison of the crystal structures of HIV reverse transcriptase with a bound DNA double strand and bound thymidine-5′-triphosphate (PBD code: 1RTD) and AZT (PBD code: 1N5Y).

Fig. 32.10 Superposition of the crystal structures of HIV reverse transcriptase in an unbound and a nevirapine-bound state (PBD code: 1DLO and 1VRT).

Fig. 32.12 Crystal structures of HIV reverse transcriptase with two allosterically acting triazines (PBD code: 1S9E and 1S9G).

Fig. 32.13 Crystal structure of topoisomerase IV from *Streptococcus pneumoniae* with bound moxifloxacin (PDB-Code: 3FOF).

Fig. 32.15 Crystal structure of the Tet-repressor with bound DNA–oligonucleotide (PDB-Code: 1QPI) and tetracycline (PDB-Code: 1BJY).

Fig. 32.19 Cryo-EM structure of the human mitochondrial ribosome (PDB code 7A5I).

Fig. 32.21 Crystallographically determined binding mode of erythromycin and roxithromycin in the ribosome (PBD code: 1JZY and 1JZZ).

Fig. 32.22 Figure on the left taken from J.M. Harms et al., BMC Biology 2:4 (2004), Fig. 3, with kind permission of the publisher and the authors. Superposition of the crystallographically determined binding modes of dalfopristin, quinupristin (PDB code: 1SM1), erythromycin (PDB code: 1JZY), chloramphenicol (PDB code: 1K01), clindamycin (PDB code: 1JZX) and the nascent peptide chain (PDB code: 7A5I).

The molecular representations and videos were generated with the software program Discovery Studio Vizualizer v20.1.0.19295, Copyright 2019, Dassault Systèmes Biovia Corp., San Diego, CA, USA (▶ https://3ds.com/products-services/biovia/products). For some representations, the program PyMol (Schrödinger, L., & DeLano, W. (2020). PyMOL. Retrieved from ▶ http://www.pymol.org/pymol) was used. For the schematic drawings in the figures, either Powerpoint from Microsoft, Redmond, WA, USA, or the program MDL ISIS Draw 2.5, ACDLabs Software, Toronto, Ontario, Canada, or its successor program DRAW v22.1 from Dassault Systèmes Biovia Corp., San Diego, CA, USA (especially chemical formulas, ▶ https://www.3ds.com/products-services/biovia/products/scientific-informatics/biovia-draw/) were used. The videos were edited with the program Stages, Version 11.8.13, Copyright AquaSoft (▶ www.aquasoft.de) 1999–2022, Hegelallee 19, 14467 Potsdam, Germany and combined with additional images and texts. The spoken text was created with the help of the voice of "William" by the AI program Narakeet™ from Video Puppet Limited, Copyright Video Puppet Limited 2018–2023, version 2.33.19 (▶ https://www.narakeet.com/).

Name Index

F

G

H

I

J

K

L

M

N

O

P

Q

R

S

T

U

V

W

Y

Z

Subject Index

B

C

D

E

F

G

H

J

K

L

N

O

P

T

U

V

W

X

Z